Lecture Notes in Computer Science 3314

Commenced Publication in 1973
Founding and Former Series Editors:
Gerhard Goos, Juris Hartmanis, and Jan van Leeuwen

Editorial Board

David Hutchison
  Lancaster University, UK
Takeo Kanade
  Carnegie Mellon University, Pittsburgh, PA, USA
Josef Kittler
  University of Surrey, Guildford, UK
Jon M. Kleinberg
  Cornell University, Ithaca, NY, USA
Friedemann Mattern
  ETH Zurich, Switzerland
John C. Mitchell
  Stanford University, CA, USA
Moni Naor
  Weizmann Institute of Science, Rehovot, Israel
Oscar Nierstrasz
  University of Bern, Switzerland
C. Pandu Rangan
  Indian Institute of Technology, Madras, India
Bernhard Steffen
  University of Dortmund, Germany
Madhu Sudan
  Massachusetts Institute of Technology, MA, USA
Demetri Terzopoulos
  New York University, NY, USA
Doug Tygar
  University of California, Berkeley, CA, USA
Moshe Y. Vardi
  Rice University, Houston, TX, USA
Gerhard Weikum
  Max-Planck Institute of Computer Science, Saarbruecken, Germany

Jun Zhang Ji-Huan He Yuxi Fu (Eds.)

Computational and Information Science

First International Symposium, CIS 2004
Shanghai, China, December 16-18, 2004
Proceedings

 Springer

Volume Editors

Jun Zhang
University of Kentucky, Department of Computer Science
773 Anderson Hall, Lexington, KY 40506-0046, USA
E-mail: jzhang@cs.uky.edu

Ji-Huan He
Donghua University, College of Science
1882 Yan-an Xilu Road, Shanghai 200051, China
E-mail: jhhe@dhu.edu.cn

Yuxi Fu
Shanghai Jiaotong University, Department of Computer Science
1954 Hua Shan Road, Shanghai 200030, China
E-mail: fu-yx@cs.sjtu.edu.cn

Library of Congress Control Number: 2004116721

CR Subject Classification (1998): D, F, G, H, I

ISSN 0302-9743
ISBN 3-540-24127-2 Springer Berlin Heidelberg New York

This work is subject to copyright. All rights are reserved, whether the whole or part of the material is concerned, specifically the rights of translation, reprinting, re-use of illustrations, recitation, broadcasting, reproduction on microfilms or in any other way, and storage in data banks. Duplication of this publication or parts thereof is permitted only under the provisions of the German Copyright Law of September 9, 1965, in its current version, and permission for use must always be obtained from Springer. Violations are liable to prosecution under the German Copyright Law.

Springer is a part of Springer Science+Business Media

springeronline.com

© Springer-Verlag Berlin Heidelberg 2004
Printed in Germany

Typesetting: Camera-ready by author, data conversion by Scientific Publishing Services, Chennai, India
Printed on acid-free paper   SPIN: 11368984   06/3142   5 4 3 2 1 0

Preface

The 2004 International Symposium on Computational and Information Sciences (CIS 2004) aimed at bringing researchers in the area of computational and information sciences together to exchange new ideas and to explore new ground. The goal of the conference was to push the application of modern computing technologies to science, engineering, and information technologies to a new level of sophistication and understanding.

The initial idea to organize such a conference with a focus on computation and applications was originated by Dr. Jun Zhang, during his visit to China in August 2003, in consultation with a few friends, including Dr. Jing Liu at the Chinese Academy of Sciences, Dr. Jun-Hai Yong at Tsinghua University, Dr. Geng Yang at Nanjing University of Posts and Communications, and a few others. After several discussions with Dr. Ji-Huan He, it was decided that Donghua University would host CIS 2004.

CIS 2004 attempted to distinguish itself from other conferences in its emphasis on *participation* rather than *publication*. A submitted paper was only reviewed with the explicit understanding that, if accepted, at least one of the authors would attend and present the paper at the conference. It is our belief that attending conferences is an important part of one's academic career, through which academic networks can be built that may benefit one's academic life in the long run.

We also made every effort to support graduate students in attending CIS 2004. In addition to set reduced registration fees for full-time graduate students, we awarded up to three prizes for the *Best Student Papers* at CIS 2004. Students whose papers were selected for awards were given cash prizes, plus a waiver of registration fees.

We received approximately 450 papers. All papers were reviewed by anonymous referees, members of the Scientific Committee, and the Co-chairs. Eventually 190 papers were selected for publication in the CIS 2004 proceedings. Papers were submitted by authors from 21 different countries and areas, symbolizing the true international nature of this symposium.

Many people did a lot of work to make CIS 2004 possible. We are unable to recount their names one by one. Most of them helped CIS 2004 in the form of reviewing some submitted papers. Their time and efforts spent on making CIS 2004 successful is greatly appreciated. Special thanks are due to Laurence T. Yang for help in the proceedings publication negotiation with Springer, and to Dr. Jeonghwa Lee for categorizing the accepted papers.

The CIS 2004 Scientific Committee was co-chaired by Drs. Jun Zhang, Ji-Huan He, and Yuxi Fu. Dr. Zhang was responsible for the overall organization of the conference, including forming the scientific committee, inviting the keynote speakers, calling for papers, handling most of the submitted papers, contacting the publishers, and preparing the final publications. Dr. He was responsible for organizing the local committee, applying for initial funding, arranging the conference site, handling some of the submitted papers, and collecting registration fees. Dr. Fu was mainly responsible for external funding and industrial sponsorship.

CIS 2004 was jointly sponsored by Donghua University, Shanghai Jiaotong University, and the Laboratory for High Performance Scientific Computing and Computer Simulation at the University of Kentucky. We would like to thank the institutions for their generous support.

September 2004   Jun Zhang
                                                                         CIS 2004 Co-chair

Organizing Committee

International Scientific Committee

Michael Berry, University of Tennessee, USA
Xue-Bin Chi, Chinese Academy of Sciences, China
Mehdi Dehghan, Amirkabir University of Technology, Iran
Tony Drummond, Lawrence Berkeley National Laboratory, USA
Yuxi Fu, Shanghai Jiaotong University, China (Co-chair)
George Gravvanis, Hellenic Open University, Greece
Qingping Guo, Wuhan University of Technology, China
Murli M. Gupta, George Washington University, USA
Ji-Huan He, Donghua University, China (Co-chair)
Katica (Stevanovic) Hedrih, University of Nis, Yugoslavia
Zhongxiao Jia, Tsinghua University, China
Hai Jin, Huazhong University of Science and Technology, China
Sangbae Kim, Hannam University, South Korea
Wai Lam, Chinese University of Hong Kong, China
Ming-Lu Li, Shanghai Jiaotong University, China
Ming-Chih Lai, National Chiao Tung University, Taiwan
Zhongze Li, Chinese Academy of Sciences, China
Jing Liu, Chinese Academy of Sciences, China
Guang Meng, Shanghai Jiaotong University, China
Zeyao Mo, IAPCM, China
Kengo Nakajima, University of Tokyo, Japan
Jun Ni, University of Iowa, USA
Mohamed Othman, University Putra Malaysia, Malaysia
Yi Pan, Georgia State University, USA
Haesun Park, University of Minnesota, USA
Padma Raghavan, Pennsylvania State University, USA
Dinggang Shen, University of Pennsylvania, USA
Pengcheng Shi, University of Science and Technology, Hong Kong, China
Jie Wang, Nanjing University of Technology, China
Wei Wang, University of North Carolina-Chapel Hill, USA
Dexuan Xie, University of Wisconsin-Milwaukee, USA
Geng Yang, Nanjing University of Posts and Communications, China
Laurence Tianruo Yang, St. Francis Xavier University, Canada
Jun-Hai Yong, Tsinghua University, China
Jae Heon Yun, Chungbuk National University, South Korea
Xiaodong Zhang, National Science Foundation, USA
Jennifer J. Zhao, University of Michigan-Dearborn, USA
Hong Zhu, Fudan University, China
Jianping Zhu, University of Akron, USA
Jun Zhang, University of Kentucky, USA (Co-chair)
Albert Zomaya, University of Sydney, Australia

Local Organizing Committee

Guang Meng, Shanghai Jiaotong University, China (Chair)
Juan Zhang, Donghua University, China (Secretary-General)
Yu-Qin Wan, Donghua University, China (Secretary)
Hong-Mei Liu, Donghua University, China (Secretary)

Referees

Many people spent their valuable time on reviewing the submitted papers. We would like to thank them for their help. The following is an incomplete list of CIS 2004 referees:

Gulsah Altun, Woo Jeong Bae, Deng Cai, Jiaheng Cao, Ke Chen, Kefei Chen, Wufan Chen, Yan Qiu Chen, Fuhua Cheng, Kwang-Hyun Cho, Bong Kyun Choi, Soo-Mi Choi, Se-Hak Chun, Larry Davis, Chris Ding, Yiming Ding, Yongsheng Ding, Yi Dong, Donglei Du, Hassan Ebrahimirad, Pingzhi Fan, Minrui Fei, Zongming Fei, Xiaobing Feng, Tongxiang Gu, Klaus Guerlebeck, Karimi Hamidreza, Young S. Han, Jianmin He, Yoshiaki Hoshino, Lei Hu, Qiangsheng Hua, Haining Huang, Maolin Huang, Xiaodi Huang, Ryu Ishikawa, Christopher Jaynes, N. Jeyanthi, Hao Ji, Yi Jiang, Hai Jin, Tao Jin, Yong-keun Jin, Han Jing, Jiwu Jing, Michael A. Jones, Jan Kaminsky, Oya Kalipsiz, Jiten Chandra Kalita, Ning Kang, Sung Ha Kang, Yun-Jeong Kang, Samir Karaa, Cheol-Ki Kim, Heechern Kim, Hyun Sook Kim, Jaekwon Kim, Kyungsoo Kim, Min Hyung Kim, Sangbae Kim, Yongdeok Kim, Wonha Kim, Andrew Klapper, Myeong-Cheol Ko, Oh-Woog Kwon, Sungho Kwon, Young Ha Kwon, Wai Lam, Zhiling Lan, Dong Hoon Lee, Eun-Joo Lee, Hong Joo Lee, Hyung-Woo Lee, Jeonghwa Lee, Kun Lee, Guido Lemos, Beibei Li, C.C. Li, Guojun Li, Jiguo Li, Minglu Li, Shuyu Li, Rui Liao, Chunxu Liu, Haifeng Liu, Huafeng Liu, Jundong Liu, Caicheng Lu, Liuming Lu, Linzhang Lu, RongXing Lu, Aarao Lyra, Kaveh Madani, D. Manivannan, Timo Mantere, R.K. Mohanty, Mohammad Reza Mostavi, Juggapong Natwichai, Michael K. Ng, Jun Ni, DaeHun Nyang, Mohamed Othman, Yi Pan, Hyungjun Park, Soon Young Park, Bingnan Pei, Dehu Qi, Ilkyeun Ra, Moh'd A. Radaideh, Chotirat Ann Ratanamahatana, John A. Rose, Hossein Rouhani, Chi Shen, Dinggang Shen, Wensheng Shen, Dongil Shin, Taeksoo Shin, Yeong Gil Shin, Bo Sun, Dalin Tang, Jason Teo, R. Thandeeswaran, Haluk Topcuoglu, Bruno Torresani, Changhe Tu, Jie Wang, Morgan Wang, Yong Wang, Xin Wang, Yu-Ping Wang, Zheng Wang, Ziqiang Wang, Yimin Wei, Yimin Wen, M. Victor Wickerhauser, Yilei Wu, Nong Xiao, Shuting Xu, Yinlong Xu, Yun Xu, Geng Yang, Huaiping Yang, Ruigang Yang, Yun Yang, Leslie Ying, Jun-Hai Yong, Kyung Hyun Yoon, Yijiao Yu, Yao Yuan, Yu-Feng Zang, Yiqiang Zhan, Naixiao Zhang, Yanning Zhang, Yufang Zhang, Yuqing Zhang, Jennifer Jing Zhao, Hongjun Zheng, Kun Zhou, Hong Zhu, Jianping Zhu, Qiaoming Zhu, Albert Zomaya

Table of Contents

High Performance Computing and Algorithms

High Order Finite Difference Schemes for the Solution of Elliptic PDEs
   Pierluigi Amodio, Ivonne Sgura 1

An Algorithm for Optimal Tuning of Fuzzy PID Controllers on
Precision Measuring Device
   Jia Lu, Yunxia Hu 7

A Grid Portal Model Based on Security and Storage Resource Proxy
   Quan Zhou, Geng Yang 13

Optimal Designs of Directed Double-Loop Networks
   Bao-Xing Chen, Wen-Jun Xiao 19

A QoS-Based Access and Scheduling Algorithm for Wireless Multimedia
Communications
   Bin Wang 25

Feedforward Wavelet Neural Network and Multi-variable Functional
Approximation
   Jing Zhao, Wang Chen, Jianhua Luo 32

The Distributed Wavelet-Based Fusion Algorithm
   Rajchawit Sarochawikasit, Thitirat Wiyarat, Tiranee Achalakul 38

Alternating Direction Finite Element Method for a Class of Moving
Boundary Problems
   Xu-Zheng Liu, Xia Cui, Jun-Hai Yong, Jia-Guang Sun 44

Binomial-Tree Fault Tolerant Routing in Dual-Cubes with Large
Number of Faulty Nodes
   Yaming Li, Shietung Peng, Wanming Chu 51

The Half-Sweep Iterative Alternating Decomposition Explicit
(HSIADE) Method for Diffusion Equation
   J. Sulaiman, M.K. Hasan, M. Othman 57

An Effective Compressed Sparse Preconditioner for Large Scale
Biomolecular Simulations
   Dexuan Xie 64

A Study on Lower Bound of Direct Proportional Length-Based DNA Computing for Shortest Path Problem
  Zuwairie Ibrahim, Yusei Tsuboi, Osamu Ono, Marzuki Khalid 71

Key Management for Secure Multicast Using the RingNet Hierarchy
  Guojun Wang, Lin Liao, Jiannong Cao, Keith C.C. Chan 77

Open Middleware-Based Infrastructure for Context-Aware in Pervasive Computing
  Xianggang Zhang, Jun Liao, Jinde Liu 85

Boundary Integral Simulation of the Motion of Highly Deformable Drops in a Viscous Flow with Spontaneous Marangoni Effect
  Wei Gu, Olga Lavrenteva, Avinoam Nir 93

Solving Separable Nonlinear Equations with Jacobians of Rank Deficiency One
  Yun-Qiu Shen, Tjalling J. Ypma................................. 99

Optimal Capacity Expansion Arc Algorithm on Networks
  Yuhua Liu, Shengsheng Yu, Jingzhong Mao, Peng Yang............. 105

Solving Non-linear Finite Difference Systems by Normalized Approximate Inverses
  George A. Gravvanis, Konstantinos M. Giannoutakis 111

An Adaptive Two-Dimensional Mesh Refinement Method for the Problems in Fluid Engineering
  Zhenquan Li 118

High Order Locally One-Dimensional Method for Parabolic Problems
  Samir Karaa 124

Networked Control System Design Accounting for Delay Information
  Byung In Park, Oh Kyu Kwon 130

Eidon: Real-time Performance Evaluation Approach for Distributed Programs Based on Capacity of Communication Links
  Yunfa Li, Hai Jin, Zongfen Han, Chao Xie, Minna Wu.............. 136

Approximate Waiting Time Analysis of Burst Queue at an Edge in Optical Burst-Switched Networks
  SuKyoung Lee 142

A Balanced Model Reduction for T-S Fuzzy Systems with Uncertain
Time Varying Parameters
    Seog-Hwan Yoo, Byung-Jae Choi 148

Genetic Algorithms with Stochastic Ranking for Optimal Channel
Assignment in Mobile Communications
    Lipo Wang, Wen Gu 154

A MPLS-Based Micro-mobility Supporting Scheme in Wireless Internet
    SuKyoung Lee 160

A Novel RBF Neural Network with Fast Training and Accurate
Generalization
    Lipo Wang, Bing Liu, Chunru Wan 166

Basic Mathematical Properties of Multiparty Joint Authentication in
Grids
    Hui Liu, Minglu Li 172

GA Based Adaptive Load Balancing Approach for a Distributed System
    SeongHoon Lee, DongWoo Lee 182

A Novel Approach to Load Balancing Problem
    Chuleui Hong, Wonil Kim, Yeongjoon Kim 188

Asynchronous Distributed Genetic Algorithm for Optimal Channel
Routing
    Wonil Kim, Chuleui Hong, Yeongjoon Kim 194

High-Level Language and Compiler for Reconfigurable Computing
    Fu San Hiew, Kah Hoe Koay 200

A Parallel Algorithm for the Biorthogonal Wavelet Transform Without
Multiplication
    HyungJun Kim 207

Algorithms for Loosely Constrained Multiple Sequence Alignment
    Bin Song, Feng-feng Zhou, Guo-liang Chen 213

Application of the Hamiltonian Circuit Latin Square to the Parallel
Routing Algorithm on 2-Circulant Networks
    Yongeun Bae, Chunkyun Youn, Ilyong Chung 219

A Distributed Locking Protocol
    Jaechun No, Sung Soon Park 225

A Study on the Efficient Parallel Block Lanczos Method
Sun Kyung Kim, Tae Hee Kim 231

Performance Evaluation of Numerical Integration Methods in the Physics Engine
Jong-Hwa Choi, Dongkyoo Shin, Won Heo, Dongil Shin 238

A Design and Analysis of Circulant Preconditioners
Ran Baik, Sung Wook Baik 245

An Approximation Algorithm for a Queuing Model with Bursty Heterogeneous Input Processes
Sugwon Hong, Tae-Sun Chung, Yeonseung Ryu, Hyuk Soo Jang, Chung Ki Lee 252

Improved Adaptive Modulation and Coding of MIMO with Selection Transmit Diversity Systems
Young-hwan You, Min-goo Kang, Ou-seb Lee, Seung-il Sonh, Tae-won Jang, Hyoung-kyu Song, Dong-oh Kim and Hwa-seop Lim .. 258

Design of a Cycle-Accurate User-Retargetable Instruction-Set Simulator Using Process-Based Scheduling Scheme
Hoonmo Yang, Moonkey Lee 266

An Authentication Scheme Based Upon Face Recognition for the Mobile Environment
Yong-Guk Kim, Taekyoung Kwon 274

A Survey of Load Balancing in Grid Computing
Yawei Li, Zhiling Lan 280

Fractal Tiling with the Extended Modular Group
Rui-song Ye, Yu-ru Zou, Jian Lu 286

Shelling Algorithm in Solid Modeling
Dong-Ming Yan, Hui Zhang, Jun-Hai Yong, Yu Peng, Jia-Guang Sun 292

Load and Performance Balancing Scheme for Heterogeneous Parallel Processing
Tae-Hyung Kim 298

A Nonlinear Finite Difference Scheme for Solving the Nonlinear Parabolic Two-Step Model
Weizhong Dai, Teng Zhu 304

Analysis on Network-Induced Delays in Networked Learning Based
Control Systems
  Lixiong Li, Minrui Fei, Xiaobing Zhou 310

A New Boundary Preserval and Noise Removal Method Combining
Gibbs Random Field with Anisotropic-Diffusion
  Guang Tian, Fei-hu Qi 316

The Geometric Constraint Solving Based on Mutative Scale Chaos
Genetic Algorithm
  Cao Chunhong, Li Wenhui 324

Genetic Algorithm Based Neuro-fuzzy Network Adaptive PID Control
and Its Applications
  Dongqing Feng, Lingjiao Dong, Minrui Fei, Tiejun Chen 330

Formalizing the Environment View of Process Equivalence
  Yuxi Fu, Xiaoju Dong 336

A Scalable and Reliable Mobile Agent Computation Model
  Yong Liu, Congfu Xu, Zhaohui Wu, Yunhe Pan 346

Building Grid Monitoring System Based on Globus Toolkit:
Architecture and Implementation
  Kejing He, Shoubin Dong, Ling Zhang, Binglin Song 353

History Information Based Optimization of Additively Decomposed
Function with Constraints
  Qingsheng Ren, Jin Zeng, Feihu Qi 359

An Efficient Multiple-Constraints QoS Routing Algorithm Based on
Nonlinear Path Distance
  Xiaolong Yang, Min Zhang, Keping Long 365

The Early and Late Congruences for Asymmetric $\chi^{\neq}$-Calculus
  Farong Zhong 371

Improvement of the Resolution Ratio of the Seismic Record by
Balanced Biorthogonal Multi-wavelet Transform
  Wenzhang He, Aidi Wu, Guoxiang Song 379

Computer Modeling and Simulations

Formally Specifying T Cell Cytokine Networks with B Method
  Shengrong Zou 385

Three-Dimensional Motion Analysis of the Right Ventricle Using an
Electromechanical Biventricular Model
  Ling Xia, Meimei Huo 391

Growing RBF Networks for Function Approximation by A DE-Based
Method
  Junhong Liu, Saku Kukkonen, Jouni Lampinen 399

Dual-Source Backoff for Enhancing Language Models
  Sehyeong Cho 407

Use of Simulation Technology for Prediction of Radiation Dose in
Nuclear Power Plant
  Yoon Hyuk Kim, Won Man Park 413

A Numerical Model for Estimating Pedestrian Delays at Signalized
Intersections in Developing Cities
  Qingfeng Li, Zhaoan Wang, Jianguo Yang 419

Feature Selection with Particle Swarms
  Yu Liu, Zheng Qin, Zenglin Xu, Xingshi He 425

Influence of Moment Arms on Lumbar Spine Subjected to Follower Loads
  Kyungsoo Kim, Yoon Hyuk Kim 431

Monte Carlo Simulation of the Effects of Large Blood Vessels During
Hyperthermia
  Zhong-Shan Deng, Jing Liu 437

A Delimitative and Combinatorial Algorithm for Discrete Optimum
Design with Different Discrete Sets
  Lianshuan Shi, Heng Fu 443

A New Algebraic-Based Geometric Constraint Solving Approach: Path
Tracking Homotopy Iteration Method
  Wenhui Li, Chunhong Cao, Wan Yi 449

A BioAmbients Based Framework for Chain-Structured Biomolecules
Modelling
  Cheng Fu, Zhengwei Qi, Jinyuan You 455

Stability of Non-autonomous Delayed Cellular Neural Networks
  Qiang Zhang, Dongsheng Zhou, Xiaopeng Wei 460

Allometric Scaling Law for Static Friction of Fibrous Materials
  Yue Wu, Yu-Mei Zhao, Jian-Yong Yu, Ji-Huan He 465

Flexible Web Service Composition Based on Interface Matching
  Shoujian Yu, Ruiqiang Guo, Jiajin Le 471

Representation of the Signal Transduction with Aberrance Using Ipi Calculus
  Min Zhang, Guoqiang Li, Yuxi Fu, Zhizhou Zhang, Lin He 477

The Application of Nonaffine Network Structural Model in Sine Pulsating Flow Field
  Juan Zhang 486

Biological and Medical Informatics

Microcalcifications Detection in Digital Mammogram Using Morphological Bandpass Filters
  Ju Cheng Yang, Jin Wook Shin, Gab Seok Yang, Dong Sun Park 492

Peptidomic Pattern Analysis and Taxonomy of Amphibian Species
  Huiru Zheng, Piyush C. Ojha, Stephen McClean, Norman D. Black, John G. Hughes, Chris Shaw 498

Global and Local Shape Analysis of the Hippocampus Based on Level-of-Detail Representations
  Jeong-Sik Kim, Soo-Mi Choi, Yoo-Joo Choi, Myoung-Hee Kim 504

Vascular Segmentation Using Level Set Method
  Yongqiang Zhao, Lei Zhang, Minglu Li 510

Brain Region Extraction and Direct Volume Rendering of MRI Head Data
  Yong-Guk Kim, Ou-Bong Gwun, Ju-Whan Song 516

Text Retrieval Using Sparsified Concept Decomposition Matrix
  Jing Gao, Jun Zhang 523

Knowledge-Based Search Engine for Specific 3D Models
  Dezhi Liu, Anshuman Razdan 530

Robust TSK Fuzzy Modeling Approach Using Noise Clustering Concept for Function Approximation
  Kyoungjung Kim, Kyu Min Kyung, Chang-Woo Park, Euntai Kim, Mignon Park 538

Helical CT Angiography of Aortic Stent Grafting: Comparison of
Three-Dimensional Rendering Techniques
  Zhonghua Sun, Huiru Zheng 544

A New Fuzzy Penalized Likelihood Method for PET Image
Reconstruction
  Jian Zhou, Huazhong Shu, Limin Luo, Hongqing Zhu 550

Interactive GSOM-Based Approaches for Improving Biomedical Pattern
Discovery and Visualization
  Haiying Wang, Francisco Azuaje, Norman Black 556

Discontinuity-Preserving Moving Least Squares Method
  Huafeng Liu, Pengcheng Shi 562

Multiscale Centerline Extraction of Angiogram Vessels Using Gabor
Filters
  Nong Sang, Qiling Tang, Xiaoxiao Liu, Wenjie Weng 570

Improved Adaptive Neighborhood Pre-processing for Medical Image
Enhancement
  Du-Yih Tsai, Yongbum Lee 576

On the Implementation of a Biologizing Intelligent System
  Byung-Jae Choi, Paul P. Wang, Seog Hwan Yoo 582

Computerized Detection of Liver Cirrhosis Using Wave Pattern of
Spleen in Abdominal CT Images
  Won Seong, June-Sik Cho, Seung-Moo Noh, Jong-Won Park 589

Automatic Segmentation Technique Without User Modification for 3D
Visualization in Medical Images
  Won Seong, Eui-Jeong Kim, Jong-Won Park 595

Adaptive Stereo Brain Images Segmentation Based on the Weak
Membrane Model
  Yonghong Shi, Feihu Qi 601

PASL: Prediction of the Alpha-Helix Transmembrane by Pruning the
Subcellular Location
  Young Joo Seol, Hyun Suk Park, Seong-Joon Yoo 607

Information Processing in Cognitive Science
  Sung-Kwan Je, Jae-Hyun Cho, Kwang-Baek Kim 613

Reconstruction of Human Anatomical Models from Segmented Contour
Lines
   Byeong-Seok Shin 619

Efficient Perspective Volume Visualization Method Using Progressive
Depth Refinement
   Byeong-Seok Shin 625

Proteomic Pattern Classification Using Bio-markers for Prostate
Cancer Diagnosis
   Jung-Ja Kim, Young-Ho Kim, Yonggwan Won 631

Deterministic Annealing EM and Its Application in Natural Image
Segmentation
   Jonghyun Park, Wanhyun Cho, Soonyoung Park 639

The Structural Classes of Proteins Predicted by Multi-resolution
Analysis
   Jing Zhao, Peiming Song, Linsen Xie, Jianhua Luo 645

A Brief Review on Allometric Scaling in Biology
   Ji-Huan He 652

On He Map (River Map) and the Oldest Scientific Management Method
   Ji-Huan He 659

A Novel Feature Selection Approach and Its Application
   Gexiang Zhang, Weidong Jin, Laizhao Hu 665

Applying Fuzzy Growing Snake to Segment Cell Nuclei in Color Biopsy
Images
   Min Hu, Xijian Ping, Yihong Ding 672

Evaluation of Morphological Reconstruction, Fast Marching and a
Novel Hybrid Segmentation Method
   Jianfeng Xu, Lixu Gu 678

Data and Information Sciences

Utilizing Staging Tables in Data Integration to Load Data into
Materialized Views
   Ahmed Ejaz, Revett Kenneth 685

HMMs for Anomaly Intrusion Detection
   Ye Du, Huiqiang Wang, Yonggang Pang 692

String Matching with Swaps in a Weighted Sequence
Hui Zhang, Qing Guo, Costas S. Iliopoulos 698

Knowledge Maintenance on Data Streams with Concept Drifting
Juggapong Natwichai, Xue Li 705

A Correlation Analysis on LSA and HAL Semantic Space Models
Xin Yan, Xue Li, Dawei Song 711

Discretization of Multidimensional Web Data for Informative Dense Regions Discovery
Edmond H. Wu, Michael K. Ng, Andy M. Yip, Tony F. Chan 718

A Simple Group Diffie-Hellman Key Agreement Protocol Without Member Serialization
Xukai Zou and Byrav Ramamurthy 725

Increasing the Efficiency of Support Vector Machine by Simplifying the Shape of Separation Hypersurface
Yiqiang Zhan, Dinggang Shen 732

Implementation of the Security System for Instant Messengers
Sangkyun Kim, Choon Seong Leem 739

Communication in Awareness Reaching Consensus Without Acyclic Condition
Ken Horie, Takashi Matsuhisa 745

A High-Availability Webserver Cluster Using Multiple Front-Ends
Jongbae Moon, Yongyoon Cho 752

An Intelligent System for Passport Recognition Using Enhanced RBF Network
Kwang-Baek Kim, Young-Ju Kim, Am-Suk Oh 762

A Distributed Knowledge Extraction Data Mining Algorithm
Jiang B. Liu, Umadevi Thanneru, Daizhan Cheng 768

Image Retrieval Using Dimensionality Reduction
Ke Lu, Xiaofei He, Jiazhi Zeng 775

Three Integration Methods for a Component-Based NetPay Vendor System
Xiaoling Dai, John Grundy 782

A Case Study on the Real-Time Click Stream Analysis System
  Sangkyun Kim, Choon Seong Leem 788

Mining Medline for New Possible Relations of Concepts
  Wei Huang, Yoshiteru Nakamori, Shouyang Wang, Tieju Ma 794

Two Phase Approach for Spam-Mail Filtering
  Sin-Jae Kang, Sae-Bom Lee, Jong-Wan Kim, In-Gil Nam 800

Dynamic Mining for Web Navigation Patterns Based on Markov Model
  Jiu Jun Chen, Ji Gao, Jun Hu, Bei Shui Liao 806

Component-Based Recommendation Agent System for Efficient Email
Inbox Management
  Ok-Ran Jeong, Dong-Sub Cho 812

Information Security Based on Fourier Plane Random Phase Coding
and Optical Scanning
  Kyu B. Doh, Kyeongwha Kim, Jungho Ohn, Ting-C Poon 819

Simulation on the Interruptible Load Contract
  Jianxue Wang, Xifan Wang, Tao Du 825

Consistency Conditions of the Expert Rule Set in the Probabilistic
Pattern Recognition
  Marek W. Kurzynski 831

An Agent Based Supply Chain System with Neural Network Controlled
Processes
  Murat Ermis, Ozgur Koray Sahingoz, Fusun Ulengin 837

Retrieval Based on Combining Language Models with Clustering
  Hua Huo, Boqin Feng 847

Lightweight Mobile Agent Authentication Scheme for Home Network
Environments
  Jae-gon Kim, Gu Su Kim, Young Ik Eom 853

Dimensional Reduction Effects of Feature Vectors by Coefficients of
Determination
  *Jong-Wan Kim, Byung-Kon Hwang, Sin-Jae Kang, Hee-Jae Kim,
  Young-Cheol Oh* 860

A Modular k-Nearest Neighbor Classification Method for Massively
Parallel Text Categorization
  Hai Zhao, Bao-Liang Lu 867

Avatar Behavior Representation and Control Technique: A Hierarchical
Scripts Approach
    Jae-Kyung Kim, Won-Sung Sohn, Soon-Bum Lim, Yoon-Chul Choy .. 873

Analyzing iKP Security in Applied Pi Calculus
    Yonggen Gu, Guoqiang Li, Yuxi Fu 879

General Public Key m-out-of-n Oblivious Transfer
    Zhide Chen, Hong Zhu 888

Determining Optimal Decision Model for Support Vector Machine by
Genetic Algorithm
    Syng-Yup Ohn, Ha-Nam Nguyen, Dong Seong Kim, Jong Sou Park .. 895

A Mobile Application of Client-Side Personalization Based on WIPI
Platform
    SangJun Lee 903

An Agent Based Privacy Preserving Mining for Distributed Databases
    Sung Wook Baik, Jerzy Bala, Daewoong Rhee 910

Geometrical Analysis for Assistive Medical Device Design
    Taeseung D. Yoo, Eunyoung Kim, Daniel K. Bogen, JungHyun Han .. 916

Hybrid Genetic Algorithms and Case-Based Reasoning Systems
    Hyunchul Ahn, Kyoung-jae Kim, Ingoo Han 922

Papílio Cryptography Algorithm
    *Frederiko Stenio de Araújo, Karla Darlene Nempomuceno Ramos,
    Benjamín René Callejas Bedregal, Ivan Saraiva Silva* 928

A Parallel Optical Computer Architecture for Large Dadatbase and
Knowledge Based Systems
    Jong Whoa Na 934

Transaction Processing in Partially Replicated Databases
    Misook Bae, Buhyun Hwang 940

Giving Temporal Order to News Corpus
    Hiroshi Uejima, Takao Miura, Isamu Shioya 947

Semantic Role Labeling Using Maximum Entropy
    *Kwok Cheung Lan, Kei Shiu Ho, Robert Wing Pong Luk,
    Hong Va Leong* 954

An Instance Learning Approach for Automatic Semantic Annotation
Wang Shu, Chen Enhong 962

Interpretable Query Projection Learning
Yiqiu Han, Wai Lam 969

Improvements to Collaborative Filtering Systems
Fu Lee Wang 975

Looking Up Files in Peer-to-Peer Using Hierarchical Bloom Filters
Kohei Mitsuhashi, Takao Miura, Isamu Shioya 982

Application of Web Service in Web Mining
Beibei Li, Jiajin Le 989

A Collaborative Work Framework for Joined-Up E-Government Web Services
Liuming Lu, Guojin Zhu, Jiaxun Chen 995

A Novel Method for Eye Features Extraction
Zhonglong Zheng, Jie Yang, Meng Wang, Yonggang Wang 1002

A Q-Based Framework for Demand Bus Simulation
Zhiqiang Liu, Cheng Zhu, Huanye Sheng, Peng Ding 1008

A Revision for Gaussian Mixture Density Decomposition Algorithm
Xiaobing Yang, Fansheng Kong, Bihong Liu 1014

Discretization of Continuous Attributes in Rough Set Theory and Its Application
Gexiang Zhang, Laizhao Hu, Weidong Jin 1020

Fast Query Over Encrypted Character Data in Database
Zheng-Fei Wang, Jing Dai, Wei Wang, Bai-Le Shi 1027

Factoring-Based Proxy Signature Schemes with Forward-Security
Zhenchuan Chai, Zhenfu Cao 1034

A Method of Acquiring Ontology Information from Web Documents
Lixin Han, Guihai Chen, Li Xie 1041

Adopting Ontologies and Rules in Web Searching Services
He Hu, Xiaoyong Du 1047

An Arbitrated Quantum Message Signature Scheme
Xin Lü, Deng-Guo Feng 1054

Fair Tracing Without Trustees for Multiple Banks
Chen Lin, Xiaoqin Huang, Jinyuan You 1061

SVM Model Selection with the VC Bound
Huaqing Li, Shaoyu Wang, Feihu Qi 1067

Computational Graphics and Visualization

Unbalanced Hermite Interpolation with Tschirnhausen Cubics
Jun-Hai Yong, Hua Su 1072

An Efficient Iterative Optimization Algorithm for Imaging Thresholding
Liju Dong, Ge Yu 1079

Computing the Sign of a Dot Product Sum
Yong-Kang Zhu, Jun-Hai Yong, Guo-Qin Zheng 1086

Bilateral Filter for Meshes Using New Predictor
*Yu-Shen Liu, Pi-Qiang Yu, Jun-Hai Yong, Hui Zhang,
Jia-Guang Sun* 1093

Scientific Computing on Commodity Graphics Hardware
Ruigang Yang 1100

FIR Filtering Based Image Stabilization Mechanism for Mobile Video Appliances
Pyung Soo Kim 1106

p-Belief Communication Leading to a Nash Equilibrium
Takashi Matsuhisa 1114

Color Image Vector Quantization Using an Enhanced Self-Organizing Neural Network
Kwang Baek-Kim, Abhijit S. Pandya 1121

Alternate Pattern Fill
*Xiao-Xin Zhang, Jun-Hai Yong, Lie-Hang Gong, Guo-Qin Zheng,
Jia-Guang Sun* 1127

A Boundary Surface Based Ray Casting Using 6-Depth Buffers
Ju-Whan Song, Ou-Bong Gwun, Seung-Wan Kim, Yong-Guk Kim ... 1134

Adaptive Quantization of DWT-Based Stereo Residual Image Coding
Han-Suh Koo, Chang-Sung Jeong 1141

Finding the Natural Problem in the Bayer Dispersed Dot Method with Genetic Algorithm
Timo Mantere 1148

Real-Time Texture Synthesis with Patch Jump Maps
Bin Wang, Jun-Hai Yong, Jia-Guang Sun 1155

Alternation of Levels-of-Detail Construction and Occlusion Culling for Terrain Rendering
Hyung Sik Yoon, Moon-Ju Jung, JungHyun Han 1161

New Algorithms for Feature Description, Analysis and Recognition of Binary Image Contours
Donggang Yu, Wei Lai 1168

A Brushlet-Based Feature Set Applied to Texture Classification
Tan Shan, Xiangrong Zhang, Licheng Jiao 1175

An Image Analysis System for Tongue Diagnosis in Traditional Chinese Medicine
Yonggang Wang, Yue Zhou, Jie Yang, Qing Xu 1181

3D Mesh Fairing Based on Lighting and Geometric Conditions for Interactive Smooth Rendering
Seung-Man Kim, Kwan H. Lee 1187

Up to Face Extrusion Algorithm for Generating B-Rep Solid
Yu Peng, Hui Zhang, Jun-Hai Yong, Jia-Guang Sun 1195

Adaptive Model-Based Multi-person Tracking
Kyoung-Mi Lee 1201

A Novel Noise Modeling for Object Detection Using Uncalibrated Difference Image
Joungwook Park, Kwan H. Lee 1208

Fast and Accurate Half Pixel Motion Estimation Using the Property of Motion Vector
MiGyoung Jung, GueeSang Lee 1216

An Efficient Half Pixel Motion Estimation Algorithm Based on Spatial Correlations
HyoSun Yoon, GueeSang Lee, YoonJeong Shin 1224

Multi-step Subdivision Algorithm for Chaikin Curves
Ling Wu, Jun-Hai Yong, You-Wei Zhang, Li Zhang 1232

Imaging Electromagnetic Field Using SMP Image
  Wei Guo, Jianyun Chai, Zesheng Tang 1239

Support Vector Machine Approach for Partner Selection of Virtual Enterprises
  Jie Wang, Weijun Zhong, Jun Zhang 1247

Author Index 1255

High Order Finite Difference Schemes for the Solution of Elliptic PDEs[*]

Pierluigi Amodio[1] and Ivonne Sgura[2]

[1] Dipartimento di Matematica, Università degli Studi di Bari,
70125 Bari, Italy
amodio@dm.uniba.it
[2] Dipartimento di Matematica, Università degli Studi di Lecce,
73100 Lecce, Italy
ivonne.sgura@unile.it

Abstract. We solve nonlinear elliptic PDEs by stable finite difference schemes of high order on a uniform meshgrid. These schemes have been introduced in [1] in the class of Boundary Value Methods (BVMs) to solve two-point Boundary Value Problems (BVPs) for second order ODEs and are high order generalizations of classical finite difference schemes for the first and second derivatives. Numerical results for a minimal surface problem and for the Gent model in nonlinear elasticity are presented.

Keywords: Finite differences, Boundary Value Methods, Elliptic PDEs.

1 Introduction

Let us consider a two-dimensional *nonlinear elliptic PDE* formulated as:

$$\begin{cases} -div(M(|\nabla u|)\nabla u) = \lambda, & (x,y) \in \Omega, \\ u(x,y) = g(x,y) & \text{on } \partial\Omega \end{cases} \quad (1)$$

where $\Omega = [x_0, x_f] \times [y_0, y_f]$, $\lambda \in \mathbb{R}$, $|\cdot|$ stands for the Euclidean norm in $\mathbb{R}^2$, div is the divergence operator and $M(\cdot)$ is sufficiently regular in its argument. Problems of this kind describe, for example, mathematical models for elastomers and soft tissues (see e.g. [2]) and are usually solved by finite elements and mixed-finite element techniques (see e.g. [3]). In this paper, we propose to solve (1) by stable finite difference schemes of high order on a regular domain with uniform meshgrid. The main idea is the application in more spatial dimensions of the new classes of Boundary Value Methods (BVMs) introduced in [1] to solve two-point BVPs for second order ODEs. For this reason, in the first part of the paper we report on these formulae and their main properties. In the second part, we show how these schemes can be applied along each space dimension and then combined to solve an elliptic PDE. We show the performance of these techniques on a minimal surface problem and on the Gent model in nonlinear elasticity.

---

[*] Work supported by GNCS and MIUR (60% project). The work of I. S. was partially supported by the Progetto Giovani Ricercatori Università di Lecce-MIUR 2001/2002.

2 Numerical Approximation for the ODE Case

Let us consider the nonlinear two-point BVP

$$f(x, y, y', y'') = 0, \quad x \in [x_0, x_f], \quad y(x_0) = y_0, \quad y(x_f) = y_f, \quad (2)$$

and the discretization of the interval $[x_0, x_f]$ by means of a constant stepsize $h = (x_f - x_0)/(n+1)$ such that $x_i = x_0 + ih$, $i = 0, \ldots, n+1$, $x_{n+1} \equiv x_f$.

To approximate the derivatives in (2), we consider the following $k-1$ finite difference schemes on a non-compact (long) stencil of $k+1$ points

$$y''(x_i) \approx \frac{1}{h^2} \sum_{j=-s}^{k-s} \alpha_{j+s}^{(s)} y_{i+j}, \quad y'(x_i) \approx \frac{1}{h} \sum_{j=-s}^{k-s} \beta_{j+s}^{(s)} y_{i+j}, \quad s = 1, \ldots, k-1, \quad (3)$$

such that each formula requires s initial values and $k-s$ final values. For all s, the coefficients $\alpha_{j+s}^{(s)}$ and $\beta_{j+s}^{(s)}$ have to be calculated in order to attain the maximum possible order k.

We note that, if $k = 2$, then $s = 1$ and the obtained formulae correspond to the traditional central differences to approximate the second and the first derivatives [4]. For order $k > 2$, we can define higher order approximation schemes to solve (2) such that all the formulae in (3) need to be used. In fact, by following the approach of the BVMs, first introduced in [5], for each derivative we select one scheme among the $k-1$ in (3) by fixing the number $s = \bar{s}$ of initial conditions to be required. This scheme is called *main method* and it is used to approximate each derivative at the points x_i, for $i = \bar{s}, \ldots, n + \bar{s} - k + 1$. The formulae for $s = 1, \ldots, \bar{s} - 1$ are used to approximate the derivatives in the first $\bar{s} - 1$ points and the formulae for $s = \bar{s} + 1, \ldots, k - 1$ are used for the approximation in the last $k - \bar{s} - 1$ points of the interval.

We remark that high order finite difference schemes on non-compact stencils typically present complicated numerical formulations near boundaries (see e.g. [6]) that can reduce the global order of approximation. For this reason, compact schemes aimed at achieving higher accuracy without extra points have been preferred in many applications (see e.g. [7]).

In [1] new classes of BVMs of even order have been defined to discretize the derivatives in (2). For the second derivative it has been proved that the coefficients in (3) satisfy the symmetry property $\alpha_j^{(i)} = \alpha_{k-j}^{(k-i)}$, for $i = 1, \ldots, k-1$, $j = 0, \ldots, k$. Therefore, by imposing $\bar{s} = k/2$ a generalization of the central differences formulae is obtained. For example, if $k = 4$ the following approximations are obtained:

$$y''(x_1) \approx \frac{1}{h^2}\left(\frac{11}{12}y_0 - \frac{5}{3}y_1 + \frac{1}{2}y_2 + \frac{1}{3}y_3 - \frac{1}{12}y_4\right),$$

$$y''(x_i) \approx \frac{1}{h^2}\left(-\frac{1}{12}y_{i-2} + \frac{4}{3}y_{i-1} - \frac{5}{2}y_i + \frac{4}{3}y_{i+1} - \frac{1}{12}y_{i+2}\right), \quad i = 2, \ldots, n-1,$$

$$y''(x_n) \approx \frac{1}{h^2}\left(-\frac{1}{12}y_{n-3} + \frac{1}{3}y_{n-2} + \frac{1}{2}y_{n-1} - \frac{5}{3}y_n + \frac{11}{12}y_{n+1}\right).$$

Similarly, the coefficients for the approximation of the first derivative satisfy $\beta_j^{(i)} = -\beta_{k-j}^{(k-i)}$, for $i = 1, \ldots, k-1$, $j = 0, \ldots, k$. In [1] three different schemes of even order have been considered. They correspond to high order extensions of the classical *backward, forward and central difference schemes*. For this reason, they have been called Generalized Backward Differentiation Formulae (GBDFs) (see also [5]), Generalized Forward Differentiation Formulae (GFDFs) and Extended Central Difference Formulae (ECDFs), respectively. For example, if $k = 4$ we have the formulae

$$y'(x_i) \approx \tfrac{1}{h}\left(-\tfrac{1}{4}y_{i-1} - \tfrac{5}{6}y_i + \tfrac{3}{2}y_{i+1} - \tfrac{1}{2}y_{i+2} + \tfrac{1}{12}y_{i+3}\right), \quad i = 1, \ldots, \rho-1,$$

$$y'(x_i) \approx \tfrac{1}{h}\left(\tfrac{1}{12}y_{i-2} - \tfrac{2}{3}y_{i-1} + \tfrac{2}{3}y_{i+1} - \tfrac{1}{12}y_{i+2}\right), \quad i = \rho, \ldots, \sigma-1,$$

$$y'(x_i) \approx \tfrac{1}{h}\left(-\tfrac{1}{12}y_{i-3} + \tfrac{1}{2}y_{i-2} - \tfrac{3}{2}y_{i-1} + \tfrac{5}{6}y_i + \tfrac{1}{4}y_{i+1}\right), \quad i = \sigma, \ldots, n.$$

where the following values of (ρ, σ) characterize the method

GBDF: $(n-1, n)$; ECDF: $(2, n)$; GFDF: $(2, 3)$.

In the rest of this paper the combinations of the generalized central differences for the second derivative with the above three classes of methods for the first derivative will be called D2GBDFs, D2ECDFs and D2GFDFs.

If $Y = [y_1, y_2, \ldots, y_n]^T$ is the unknown vector and $Y(x) = [y(x_1), \ldots, y(x_n)]^T$ is the vector of the exact solution, we have the following approximations

$$Y''(x) \approx \frac{1}{h^2}\widetilde{A}\widetilde{Y}, \qquad Y'(x) \approx \frac{1}{h}\widetilde{B}\widetilde{Y},$$

where $\widetilde{Y} = [y_0, Y^T, y_f]^T$, and $\widetilde{A}$ and $\widetilde{B}$ are the $n \times (n+2)$ matrices containing the coefficients of all the schemes used for the approximation. A conditioning analysis on the linear test problem $y'' - 2\gamma y' + \mu y = 0$ is reported in [1]. The main result is that, for $\mu < 0$, the D2ECDFs yield well conditioned matrices for all $\gamma \in \mathbb{R}$ and for all h, whereas the D2GBDFs and D2GFDFs require the stepsize restrictions described in Table 1. Note that the well-conditioning regions become wider for increasing orders.

Table 1. Stepsize restrictions required to obtain well-conditioning

method	$\gamma < 0$	$\gamma > 0$
D2GBDF	$h \le -q/\gamma$	no restr.
D2GFDF	no restr.	$h \le q/\gamma$

order	4	6	8
q	1	17/12	16/9

In [1] many numerical examples on second order ODE-BVPs show that these schemes have good convergence properties also for nonlinear problems.

3 Application to Elliptic PDEs

Since elliptic problems in two or more dimensions usually present both second order, first order and mixed partial derivatives, it seems to be a good idea to apply the above schemes along each space variable and to combine them to have an overall finite difference method of high order. In the following two-dimensional examples we discretize the rectangular domain $[x_0, x_f] \times [y_0, y_f]$ by means of different stepsizes $h_x = (x_f - x_0)/(n+1)$ and $h_y = (y_f - y_0)/(m+1)$. The discrete solution in the internal points is set in the unknown vector $U = (U_1, U_2, \ldots, U_m)^T$, $U_j = (u_{1j}, \ldots u_{nj})$ where $u_{ij} \approx u(x_i, y_j)$. Therefore, the following approximations are considered: $U_{xy}(x,y) \approx \frac{1}{h_x h_y}(\tilde{B}_m \otimes \tilde{B}_n)\tilde{U}$,

$$U_{xx}(x,y) \approx \frac{1}{h_x^2}(\tilde{I}_m \otimes \tilde{A}_n)\tilde{U}, \qquad U_{yy}(x,y) \approx \frac{1}{h_y^2}(\tilde{A}_m \otimes \tilde{I}_n)\tilde{U},$$

$$U_x(x,y) \approx \frac{1}{h_x}(\tilde{I}_m \otimes \tilde{B}_n)\tilde{U}, \qquad U_y(x,y) \approx \frac{1}{h_y}(\tilde{B}_m \otimes \tilde{I}_n)\tilde{U}$$

where $\tilde{I}_r = [0_r, I_r, 0_r]$, $0_r \in \mathbb{R}^r$, $\tilde{A}_r$ and $\tilde{B}_r$ are the same matrices of dimension $r \times (r+2)$ defined in Section 2, and $\tilde{U}$ contains also the boundary values

$$\tilde{U} = (\tilde{U}_0, \tilde{U}_1, \ldots, \tilde{U}_{m+1})^T, \qquad \tilde{U}_i = (u_{0j}, \ldots u_{n+1,j}).$$

Both the examples proposed in this section are nonlinear and their discretization yields a nonlinear discrete problem that we solve by the Picard's iteration:

$$\tilde{\Delta} U^{(p+1)} = f(U^{(p)}), \qquad p = 0, 1, \ldots, \qquad U^{(0)} \text{ given}, \qquad (4)$$

where $f(U^{(p)})$ contains the nonlinear terms, A_r is a square quasi-Toeplitz band matrix with bandwidth $k+1$ (k order of the method), obtained by eliminating the first and last column of $\tilde{A}_r$, and $\tilde{\Delta} = \frac{1}{h_x^2}(I_m \otimes A_n) + \frac{1}{h_y^2}(A_m \otimes I_n) \in \mathbb{R}^{m \cdot n} \times \mathbb{R}^{m \cdot n}$ approximates the Laplace operator. Numerical experiments show that the considered approach could be effective when the solution is smooth. Otherwise a continuation strategy with respect to the parameter in the PDE is needed. The linear system at each step is solved by means of a direct solver and the stopping criteria used are $\|U^{(p+1)} - U^{(p)}\| \leq tol$ and $\|\tilde{\Delta} U^{(p+1)} - f(U^{(p)})\| \leq tol$, for a fixed tolerance tol. The numerical results are reported only for the D2GBDFs and D2ECDFs, since D2GFDFs exhibit results very similar to the D2GBDFs.

3.1 A Minimal Surface Problem

This problem is obtained by setting

$$M(|\nabla u|) = (\sqrt{1 + |\nabla u|^2})^{-1} \quad \text{and} \quad \lambda = 0,$$

and can be formulated as

$$(1 + u_y^2)u_{xx} - 2u_x u_y u_{xy} + (1 + u_x^2)u_{yy} = 0.$$

This is a well known elliptic problem and it is usually solved by means of finite element methods.

We consider an example with the exact solution $u(x,y) = \log\left(\frac{\cos(y-0.5)}{\cos(x-0.5)}\right)$ and Dirichlet boundary conditions on the square domain $\Omega = [0,1] \times [0,1]$. The problem is solved starting in (4) from $U^{(0)} \equiv 0$ and using $h_x = h_y$, that is $m = n$. The convergence behaviour and the order estimates are given in Table 2.

Table 2. Minimal surface problem - Convergence behavior ($tol =$ 1e-13)

method	n	order 4 error	rate	iterat.	order 6 error	rate	iterat.	order 8 error	rate	iterat.
D2GBDF	9	4.65e-06		21	4.59e-07		20	6.83e-08		19
	19	2.63e-07	4.15	21	9.73e-09	5.56	20	6.35e-10	6.75	19
	39	1.09e-08	4.60	21	1.26e-10	6.27	20	2.67e-12	7.89	19
D2ECDF	9	4.63e-06		17	4.58e-07		17	6.83e-08		17
	19	2.69e-07	4.11	17	9.66e-09	5.57	17	6.34e-10	6.75	17
	39	1.14e-08	4.56	17	1.24e-10	6.28	17	2.67e-12	7.89	17

3.2 The Gent Model in Nonlinear Elasticity

The Gent model is used in nonlinear elasticity to describe the anti-plane shear deformations for a class of isotropic and incompressible hyperelastic rubber-like materials [2]. To investigate the mechanical response of the material the PDE (1) has to be solved with

$$M(|\nabla u|) = \mu(1 - J_m^{-1}|\nabla u|^2)^{-1}, \quad \mu > 0, \quad J_m > 0,$$

for $(x,y) \in \Omega = [0,L] \times [0,1]$ and $u(x,y) = 0$ for $(x,y) \in \delta\Omega$. If $\alpha = J_m^{-1}$, the equation (1) is equivalent to the following expression:

$$[1 + \alpha(u_x^2 - u_y^2)]u_{xx} + [1 - \alpha(u_x^2 - u_y^2)]u_{yy} + 4\alpha u_x u_y u_{xy} = \frac{\lambda}{\mu}[1 - \alpha(u_x^2 + u_y^2)]^2.$$

For increasing values of the parameter $\lambda > 0$ an interior localization occurs that represents the zone of fracture of the material. We consider two applications with $L = 1$ and $L = 4$. In the first case the localization is at the point $(0.5, 0.5)$ and the solution is a downward cone that, for increasing values of λ, tends to an up-side down pyramid. If the rectangular domain is considered, the localization is along the line $(x, 0.5)$ and, in addition, boundary layers appear in the y-direction.

We set $J_m = 97.2$, $\mu = 1$ and we solve both cases for $\lambda = 50$ by using $h_x = h_y$. To achieve convergence a continuation technique starting from $\lambda = 10$ has been used. In Tables 3–4 we report the meshsizes, the number of Picard iterations occurred (with $tol = 10^{-10}$) along the continuation and an error estimate calculated with respect to the solution obtained by the method of order 8 in the same class and on the same mesh. We emphasize that the two methods behave

Table 3. Gent problem with $\lambda = 50$ on a square domain

		D2GBDF6		D2ECDF6	
n	m	error	iterates	error	iterates
19	19	5.37e-3	(10-50-30)	5.19e-3	(10-47-27)
39	39	7.99e-5	(10-47-20)	7.85e-5	(10-48-20)
79	79	1.34e-5	(10-48-14)	1.33e-5	(10-48-14)

Table 4. Gent problem with $\lambda = 50$ on a rectangular domain

		D2GBDF6		D2ECDF6	
n	m	error	iterates	error	iterates
9	39	1.87e-1	(14-56-257-134)	1.87e-1	(14-49-179-124)
19	79	5.87e-3	(14-53-146-103)	5.64e-3	(15-52-142-105)
39	159	8.29e-5	(15-52-143-85)	8.13e-5	(15-52-142-85)

similarly, but the continuation strategy is required to obtain the convergence of the D2GBDF.

As a general comment, we can conclude that the new schemes introduced in the ODE framework seem to be very promising also to solve elliptic PDEs accurately. Future developments of this research concern the approximation with variable stepsize, a suitable strategy of mesh variation and numerical comparisons with compact finite difference schemes (see e.g. [7]).

References

1. Amodio, P., Sgura, I.: High order finite difference schemes for the solution of second order BVPs. J. Comput. Appl. Math. (2004) in press.
2. Horgan, C., Saccomandi, G., Sgura, I.: A two-point boundary-value problem for the axial shear of hardening isotropic incompressible nonlinearly elastic materials. SIAM J. Appl. Math. **62** (2002) 1712–1727
3. Quarteroni, A., Valli, A.: Numerical Approximation of Partial Differential Equations. Springer Series in Comput. Math. 23, Springer-Verlag, Berlin (1994)
4. Ascher, U., Mattheij, R., Russell, R.: Numerical Solution of Boundary Value Problems for ODEs. Classics in Applied Mathematics 13, SIAM, Philadelphia (1995)
5. Brugnano, L., Trigiante, D.: Solving ODEs by Linear Multistep Initial and Boundary Value Methods. Gordon & Breach, Amsterdam (1998)
6. Fornberg, B., Ghrist, M.: Spatial finite difference approximations for wave–type equations. SIAM J. Numer. Anal. **37** (1999) 105–130
7. Ge, L., Zhang, J.: Symbolic computation of high order compact difference schemes for three dimensional linear elliptic partial differential equations with variable coefficients. J. Comput. Appl. Math. **143** (2002) 9–27

An Algorithm for the Optimal Tuning of Fuzzy PID Controllers on Precision Measuring Device

Jia Lu and Yunxia Hu

Department of Computer Science and Information System,
University of Phoenix, 5050 NW 125 Avenue,
Coral Springs,
FL 33076,
U.S.A.
clujia@email.uophx.edu

Abstract. A new computability methodology was proposed for the fuzzy proportional integral derivative (PID) controllers based on the theoretical fuzzy analysis and the downhill simplex optimization. The paper analyzes the algorithm of downhill simplex searching of the optimization objective functions. The input and objective function of downhill factors were selected for constructing optimal decision rules for the fuzzy logic controller. An optimizer was built around the simplex algorithm that it minimized a simplex within an N-dimensional. The sampling rate is 0.1 and the controllers are implemented under a 0.5 second time delay. The simulation confirmed the viability of the algorithm in its effectiveness of the adaptive fuzzy logic controller.

1 Introduction

There have been a number of proposed methods in tuning fuzzy logic control [1] [2]. Genetic algorithms are searching algorithm to optimize the rule table of conventional fuzzy logic controller [3]. Genetic algorithm tuning is to design coding scheme, which refers to select tuning parameters and encode them into a bit-string representation [4]. However, the genetic algorithm is attempting to describe some system an engineer faces the fact that all processes and events in the system cannot be fully described and identified. This is why approximate algorithms are always needed to determine for industrial and medical fields. In addition to the controller structure design, another important issue is parameter tuning. There exist some problems to obtain optimal parameters due to lack of the information between a model and its real system. Downhill simplex searching algorithm was used for multi-dimension optimization problems such as quantitative analysis of convergence beam electron diffraction. Downhill is a local optimization algorithm that can be used to find the nearest local minimum as soon as possible [5]. The downhill simplex algorithm was chosen in this research for optimizing the design of fuzzy logic controllers. It determined the control decision for the process and played a key role in the fuzzy logic controller.

2 Algorithm Searching

2.1 Searching Space

An easy way we used was to set up a simplex for starting with an initial starting point P_0, and the other N initial points.

The downhill simplex method now takes a series of steps, but the most steps just moving the point of the simplex where the function is largest through the opposite face of the simplex to a lower point. When combined with other operations, the simplex was used to change direction on encountering a valley at an angle and contract in the neighborhood of a minimum. The three vertices of the polytope are denoted by P_l, P_a, P_h, and their function are f_l, f_a and f_h respectively. The indices are selected as $f_l < f_a < f_h$. The object was to search a point at which the function value is minimum. The best point can be used as the conjugate point of the worst point. A line is connected between point P_l and P_a, and let P_h be the middle point bisecting the line segment. A line is connected between P_h and P_a, such as the point P_r satisfies $P_h P_a = P_a P_r$. The point P_r lies on the segment $P_h P_a$ is the conjugate point of P_r. Therefore, f_r is the objective value of point P_r. If $f_r \geq f_a$, the new point P_r goes far from the worst point P_h. Therefore, the size of the search step needs to be decreased slightly. A point P_s is selected between P_h and P_r. If $f_r < f_a$, the new point P_r is really much better than the worst point P_h. The new point P_e can be chosen in the extension line of the P_h and P_r. If $f_e \leq f_r$, then the new point P_e can be chosen as P_s, otherwise, P_r will be P_s.

2.2 Searching for Fuzzy PID Controller

Combining the downhill simplex searching algorithm with the µ-law surface tuning of fuzzy lookup table, we setup tuning parameters the fuzzy PID controller. The number of dimensions in the downhill simplex algorithm is determined by the number of tuning parameters. A set of parameters was constructed a point in the downhill simplex algorithm for the fuzzy PID controller. In order to start a downhill simplex searching, *N+1* points were used. Let all the points be represented by P_i, where $i = 1... N+1$ and the worst point is P_{N+1}.

If $0<\beta<1$, then P_s is a constant point. If $\beta = 1$, P_s is the conjugate point of the worst point. If $\beta > 1$, P_s is an expanse point of the worst point. The contraction along all dimensions towards the low point.

The idea of simplex search is to replace the point having the highest function value with another point repeatedly until some terminal condition is satisfied. For the process of simplex search, we used y_i as the function value at point P_i, l and h represent the minimum and maximum value in current round of computation respectively; that is, $y_i = \min(y_l)$, $y_h = \max(y_h)$. Let P be the average of the 3 points, and each cycle of this method starts with a reflection point P* of P_h that the point with the largest function value. Based on the function value of P*, there were four possible operations to change the current simplex to explore the scale of the function efficiently in the multidimensional space.

The positive constant α is the reflection coefficient set to 1 in the research. P* is on the line connecting P_h and $\overline{P}$. From the research, we measured the values of every point in the set of parameters under tuning. We used the sampling of errors to cover the whole trip of the process. Then we took the average values for the tracking system with the corresponding parameters.

1) If y* value is in interval 1, then the process will go to expansion operation.
2) If y* value is in interval 2, then the process will replace P_h with P*. It ends of the operation process.
3) If y* value is in interval 3, then the process will replace Ph with P*. It will go to contraction operation.
4) If y* is in interval 4, then the process will directly go to contraction.

Second, we need to define the expansion point P** and its values using the following formula. The reflection and expansion are away from P_h. P** = $\overline{P}$ + γ (P* - $\overline{P}$), and y** = $f(p^{**})$.

γ was the reflection coefficient, which was set to 2.0. If y** is in interval 1, then we need to replace P_h with P**, and the process will be terminated. Otherwise, the process will replace P_h with the original reflection point p* and finish the operation process. Third, we need to define the contraction point P** using the following formula that it is the contraction along the direction connecting P_h and ⁻P. P** = $\overline{P}$ + β (P_h - P), y** = $f(p^{**})$. β is the value of reflection coefficient. It is set to 0.5. If y** is in interval 1, 2, or 3, then we need to replace P_h with P** to finish the process. Otherwise, the process will go to the operation of shrinkage. Finally, we need to define the shrinkage. Shrinkage is toward P_l along all directions that it replaces P_i with P_i = (P_i + P_l)/2. The process will be repeated until a given stopping criterion is reached, and then it will terminate the process. In this research, there are three constants: α=1, β=0.5, γ=2 that will be used for the coefficients of reflection, contraction, and expansion. The optimal values will be selected for these coefficients. The other way of the process is used to change the coefficient α, β, and γ using random numbers that makes a wider area of the inputs. Any optimizer can be used for maintain consistent in standard downhill simplex routine.

We selected the type of inputs and the objective function for tuning the fuzzy logic controller. In tuning of the controller, the tracking input is selected according to the typical steps: the target starts from stationary to move, speeds up steadily until reaches its highest value. The tracking inputs of speed and position were used for simulation. The functions for tracking input, we considered an important factor to minimize the cost factor for maintaining the tracking output within the dimension tolerance of target. To improve the performance of the PID controller, instead of a linear function and tuning two constants, the fuzzy PID controller can also be realized in the form of the input ranges and the dimensional lookup table.

In the precision measurement system, the program running time and target moving speed can be used as one of the research strategy. The selection of sampling rate can be used to determine the tracking performance and the reliability of measurement. If the system samples too slowly, it may not be able to respond to moving tracking error signal promptly. In this case, it will lead to the failure in following the moving target. On the other hand, sampling too fast also causes problem because the control program needs a certain time period to complete the computation loop that contains decision table manipulation, output signal defuzzification, and research data recording. If the sampling rate is even faster than a single computation loop, it will mess up the software operation and lower the system reliability and data validity. Therefore, it is good way to make use of PC build-in interrupts turns out, interrupt rate of intervals with the same length, and synchronize the sampling rate with the software speed that can be had enough time to catch up the moving target. In the research, we used membership functions and the control rules indirectly, and we used decision table tuning to tune the scaling gain and the surface shape. A P controller is the simplest close-loop controller that can be written as the following: u (k) = K_p x e (k) where e (k) is the tracking error and u (k) is the output at the discrete time k. u(k) is linearly related to e(k), and the relationship between them can be graphically represented by a straight line. The law of u (k) is the control function, and e (k) is the tracking error and error change of the laser tracking system. According to the control law, we can draw the control surface using Fuzzy Logic Toolbox 2.0. Similar to the one dimension, we can get a better performance by changing the K_p and K_D with the functions in the control system. If the control plane is used as two dimension decision table for the membership functions and control rules in the fuzzy PD controller, the modification will be equal to the tuning of fuzzy PD controller.

3 Results

We selected the type of inputs and the objective function for tuning the fuzzy logic controller. In tuning the controller, the tracking input was selected according to the typical steps: the target started from stationary to move, and speeded up steadily until it reached its highest value. The tracking inputs of speed and position were used for simulation. The selection of sampling rate can be used to determine the tracking performance and the reliability of measurement. If the system samples too slowly, it may not be able to respond to the moving tracking error signal promptly. In this case, it would lead to the failure in following the moving target. If the sampling rates were even faster than that of a single computation loop, it would mess up the software operation and reduce the system reliability on data validity. Therefore, it is a good way to make use of PC built-in interrupt turns out, interrupt rate of intervals with the same length, and synchronize the sampling rate with the software speed that can be given enough time to catch up with the moving target. We had more flexibility in tuning different K_{ps} in the fuzzy P controller that it changed the shape of the control line than the constant K_p.

A third order linear model was chosen in the simulation. The sampling rate was 0.1 and the controllers were implemented under a 0.5 second time delay. The input was applied with a linear combination of raise, peak, and settle time based on the objective function. The function shows the initial and final surface of the fuzzy PID controller lookup table. Combined with these two tuning parameter inputs, we had the different downhill simplex tuning of fuzzy PID.

Downhill tuning was applied to tune the fuzzy PID controller. All of the control gains reached their maximum when the error was small. The starting point was to identify the fuzzy logic controller input and output variables with those of the controllers. The basic scheme of an incremental fuzzy PID controller with the error and change in error were the input fuzzy logic controller, and the increment of the control was the fuzzy logic controller output. The parameters chosen to tune the fuzzy logic controller were the scale factors that weighted the input and output variables, respectively. A set of optimal parameters of fuzzy PID was achieved in the simulation. The maximum tracking speed when the target was 1 meter away can reach 4 meter/minute. It is important to be aware whether the input signals were saturated in their universes.

A suitable value of the gain on error was always possible to calculate all combinations of inputs before putting the controller into operation. The relation between all input combinations and their corresponding outputs were arranged in a look-up table. We proved this tuning strategy using the third linear mode. The parameters were obtained by the mode. A fuzzy PID controller with parameters $K_p = 0.3$, and $K_d = 1$ or the equivalent FLC is auto-tuned for the system response.

A third order linear model is chosen in the simulation. The sampling rate is 0.1 and the controllers are implemented under a 0.5 second time delay. The input is applied with a linear combination of raise, peak, and settle time based on the objective function.

For the research, the precision measuring device in a tracking mode with feedback errors that are sampled in steady state. The sine waveform graphics based on the mathematical model and the structure for the simulation system showed the control results of the initial fuzzy controller. The error was equal to 0.01 and the change in error was equal to −0.04. The percentage of this error proved that the membership functions according to the real time error analysis did improve the system's control performance in terms of reducing the system errors tremendously. They were constructed due to the fixed sampling interval. The change in error always fell into several fixed values, which made the result of change in error in a reduced resolution. The different reduction of average errors we got were 36.3%, 27%, 20.3%, 14.5%, and 9.8%. It was observed that the procedure could be repeated to get the optimal membership function relocation until the membership function modification stopped changing the error and error. Each time, the reduction of average error would be reduced within error and change in error until the reduction of average error got closer to zero percentage. Therefore, we could see the reduction values come from the modification of membership functions. During the tuning, the membership functions for both input error and change in error were adjusted conveniently for satisfying the static system requirements.

The experiment showed that it was possible to control in a certain extent of the variation of the control surface. With the input and output, the flexible fuzzy controller can just be looked up in an appropriate function of the tuning and stability in the closed loop system. The results were also obviously decreasing the system's tracking errors with the membership functions. We map the corresponding membership functions using different input of data, which the results showed us that the algorithm of membership functions could be used to improve the control performance of a real control system. Membership functions were evenly distributed before the auto-tuning started. We should modify the membership functions in precision measuring device using the algorithm, although error signals stretch out to the value of 0.2, but the main portion lied between [-0.01 0.01], and the majority of change in error in the interval of [-0.01 0.01] as the entire data range spreads out to ± 0.02. This is also case of using the algorithm to tune of membership functions accordingly. After the change, we tracked the target again starting from the ending point of the controller, and also tuned the control parameters with average error 0.035101 and the reduction in average error was 9.8%. The reduction came only from the modification of membership functions, which it convinced us to use the algorithm of downhill simplex can improve the control performance of the system. It was also found the optimal membership functions relocation being reached with the approach.

4 Conclusion

The fuzzy logic controller with a linear rule base can be considered as a fuzzy linear controller. The fuzzy transfer function is defined to describe the influence of input scaling gains on the output. The initial performance is satisfactory and very close to the well-tuned linear equivalent. After further slight tuning, the performance could be further improved. Through the fuzzy logic controller has similar characteristics to its linear rule base, it appears more robust on the overall performance, especially for increasing the precision measuring device of the system processes. Downhill simplex algorithm can be applied to various forms of fuzzy PID controller, but it requires good initial conditions.

References

[1] Nedler, J. A., Mead, R.: Simplex method for function minimization, Computer Journal. 7 (1965) 308-410
[2] Zhi, H. D.: Massively parallel simulated annealing embedded with downhill. IEEE Computer Society. 3 (1999) 297-302
[3] Viljamaa, Pauli., Koivo, N.: Fuzzy logic in PID gain scheduling, Third Congress on Fuzzy and Intelligent Technologies EUFIT'95, Aachen, Germany. 3 (1995) 25-31
[4] Han, X. L.: A comparative design and tuning for conventional fuzzy control. IEEE Transactions on Systems, MAN, and CYBERNETICS. 27 (1997) 884-887.
[5] Prahlad, V.: DNA coded GA for the rule base optimization of a fuzzy logic controller. Proceeding of the 2001 IEEE Congress on Evolutionary Computation Seoul, Korea (2001) 321-329

A Grid Portal Model Based on Security and Storage Resource Proxy[1]

Quan Zhou and Geng Yang

Department of Computer Science and Technology,
Nanjing University of Posts and Telecommunications,
Nanjing 210003, China
{b994621, yang}@njupt.edu.cn

Abstract. Grid computing is recently developed concept in high performance computing. Security and resource management is the two key technologies in a grid. In this paper, we describe a grid portal model based on security proxy and storage resource broker. We set up a testing environment and make some improvements to the proxy model. We also point out several disadvantages of current mechanism, and make some corresponding improvements.

Keywords: Grid Computing, Grid Portal, Security Proxy, Storage and Data Management.

1 Introduction

Recently, grid computing[1] becomes an important research domain for high performance computing and Internet applications. The "Grid" is a term used to describe the software infrastructure that links multiple computational resources such as people, computers, sensors and data. Grid Portals, based on standard Web technologies, are increasingly used to provide user convenience interfaces for Computational and Data Grids. This paper describes a grid portal based on security and resource proxy. First, we introduce the chief concepts of grid portal and its basic architecture. Then we emphasize on its key technology, including online credential repository and storage resource broker. Finally, we discuss our implementation of a testing environment based on the grid portal we described.

2 Grid Portal Architecture

Grid Portals are a common approach to providing user interfaces to varied grid applications. Based on the standard Web technology, a grid portal allows the use of a standard Web browser as a simple graphical client for grid applications. Presently,

---

[1] This work is supported by The Natural Science Foundation of Jiangsu Province(BK2004218, BK2003106) and The Foundation of QingLan Project of Jiangsu Province. Corresponding author: Quan ZHOU, Email: b994621@njupt.edu.cn

most grid portal toolkits use Globus Toolkit as their basic grid operations system. There are a wide variety of ways to design and build grid portals. A basic architecture is shown in Figure 1. In this architecture, all scripts are implemented in Perl/CGI. The portal scripts are responsible for processing the incoming CGI data and constructing the correct data formats for the specific GridPort functions.

As shown in Figure 1, Grid Portal usually use a set of grid services that includes the PKI based Grid Security Infrastructure (GSI)[2], Globus Resource Allocation Management, GSI-FTP, and the Storage Resource Broker (SRB).

GSI software is a set of libraries and tools that allows users and applications to access resources securely. GSI focuses primarily on authentication and message protection, defining single sign-on algorithms and protocols, cross-domain authentication protocols, and delegation mechanisms for creating temporary credentials for users and for processes executing on a user's behalf. GRAM, a resource management component of the Globus Toolkit, is a module for job submission in grid environment. For data transmission, not only GSI-FTP (a basic component of Globus Toolkit) is given, but also the Storage Resource Broker (SRB) can be used. The grid portal is based on two basic proxy: security proxy (MyProxy) and resource proxy (SRB). The two key technologies will be discussed in detail in the following sections.

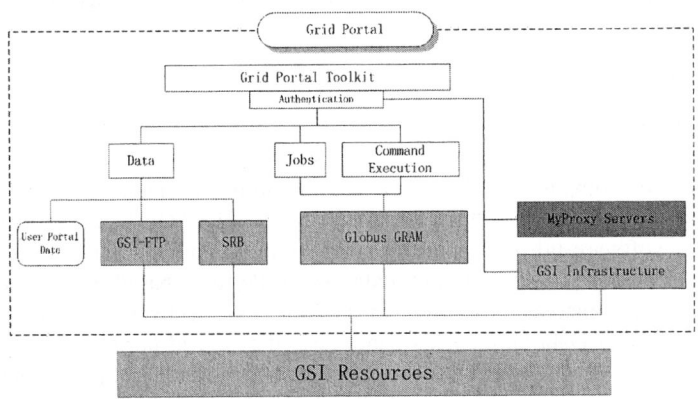

Fig. 1. Grid Portal Architecture

3 Security Proxy

3.1 Existing Mechanism Constraints

Traditional grid security infrastructure based on PKI has two constraints when it comes to meeting grid portal requirements. First, since grid credentials are typically stored as files and the private key must be kept private, a user must have secure access to the file system to use their credentials. This results in a user being unable to access their credentials when away from their private system. Second, not all applications that are desired for Grid use are GSI-enabled, and hence lack the ability to do proxy credential delegation. Due to the above constraints, SDSC (San Diego Supercomputer Center) has developed an online credential repository, called MyProxy, designed to bridge incompatibility between Web and grid security, thus enabling grid portals to use GSI-protected resource in a secure, scalable manner.

3.2 The Credential Repository System

The MyProxy credential repository system consists of a repository server and a set of client tools that can be used to delegate to and retrieve credentials from the repository [3]. The implementation of MyProxy includes delegation credential to the repository and retrieval of credential from repository.

1) Delegation to the Repository

A user starts by using the myproxy-init client program along with his permanent credentials to contact the repository and delegates a set of proxy credentials to the server along with authentication information and retrieval restrictions (See Figure 2).

The user identity is also typically different from the user's Distinguished Name (DN) in their Grid credentials, as it is actually hand-typed by the user at later times. To do so, the user identity becomes more memorable and concise than a typical DN.

2) Retrieval of Credential from Repository

The user uses the myproxy-get-delegation program to contact the server and request a delegation of the user's credentials. After verifying this information and checking any restrictions that the user presented with the delegation, the repository will in turn delegate a proxy credential back to the user (See Figure 2).

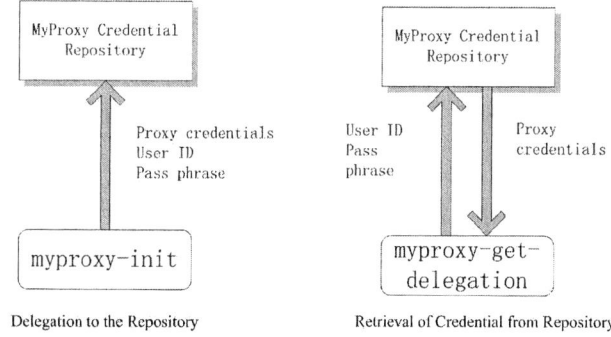

Fig. 2. Delegation to the repository & retrieval of credential from repository

The GridPort project at SDSC uses Perl CGI scripts that simply wrap the myproxy-get-delegation program with the appropriate user identity and pass phrase passed on from the HTTPS request header.

Currently, the restriction that each credentials in the repository has a lifetime is not uncommon for grid jobs to run for a period of time that exceed the credential lifetime. We plan to investigate mechanisms to enable MyProxy to securely support long-running applications.

4 Storage Resource Broker

We manage data for portal user by using 3 mechanisms: the SDSC Storage Resource Broker (SRB) for collection management; GSI-FTP for secure file transfers; and simple file upload, download, and transfers via GridPort software.

The SRB is a client-server middleware that provides a facility for collection-building, managing, querying, accessing, and preserving data in a distributed data grid framework. The SRB provides a means to organize information stored on multiple heterogeneous systems into logical collections for ease of use. The SRB, in conjunction with the Meta data Catalog (MCAT), supports location transparency by accessing data sets and resources based on their attributes rather than their names or physical locations [4].

SRB is a federated server system, with each SRB server managing/brokering a set of storage resources. This implementation provides several advantages: location transparency, improved reliability and availability, logistical and administrative reasons, fault tolerance, integrated data access, persistence and so on. Figure 3 presents the mechanism how SRB mediates between portals and storage resources. Each client has his own local file space, distributed files are managed by SRB/MCAT. First, the client posts HTTP request to Portal Web Service, where exists correspondent webID. Then, the Portal Web Service interacts with SRB/MCAT servers to find where grid data resource is. Grid users use their user proxy id to contact SRB/MCAT to map into SRB-ID, then interact with remote resource. All of the steps are under the control of GSI mechanism.

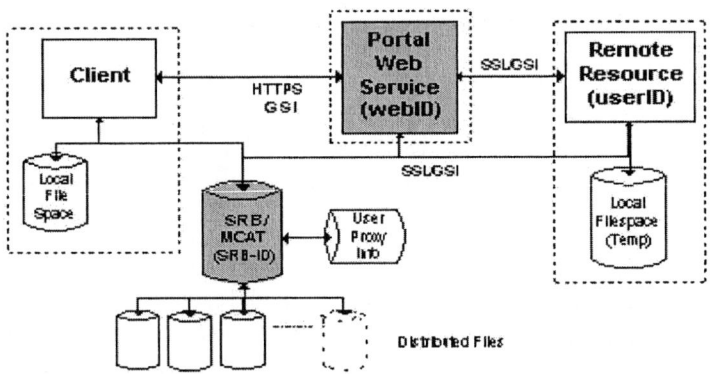

Fig. 3. SRB between portals and storage resources

5 Implementation

In our testing environment, a data grid is set based on Globus Toolkit 2.4, Myproxy 1.3, SRB, Gridport v2.0, Iptables firewall and Red Hat Linux 9.0 operation system. Considering the grid security, we build Globus CA, Myproxy Server, and one of the SRB servers on the same node. Iptables strict firewall is used to protect this node from foreign attack. To set the myproxy server, we append a environment variable $MYPROXY_SERVER=full.grid.njupt.edu.cn (the myproxy server node's FQDN) on all the nodes in our grid. The federal SRB server consists of full.grid.njupt.edu.cn and bluespan.grid.njupt.edu.cn,. We build high level firewall to protect the key node (full.grid.njupt.edu.cn) with iptables software based on linux kernel 2.4.x. Figure 4 shows the basic architecture of our testing environment.

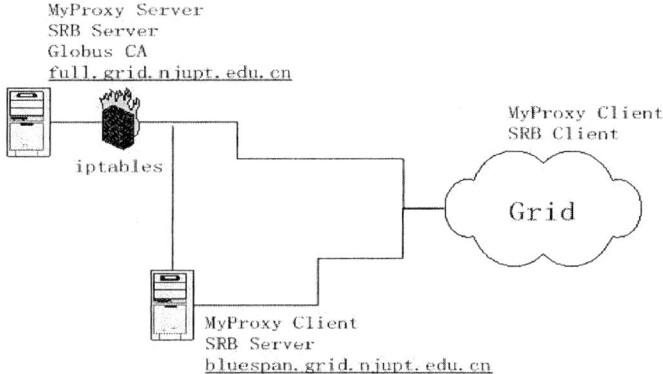

Fig. 4. Testing environment

To allow all any client, including anonymous client, with a valid MyProxy pass phrase to retrieve credentials gives users the flexibility to set their own policies on their credentials. We modify the MyProxy configuration as follows:

```
# MyProxy configuration file
#
# Grid Security Project
# Last modified March 30, 2004
#
# Accepted Credentials
#
# Which credentials is the server willing to accept and
store?
#
accepted_credentials "/O=Grid/O=MyGrid/*"
#
# Authorized Retrievers
#
# Who is authorized to retrieve credentials from the
repository?
#
authorized_retrievers "*"
```

We also do some research and development on Perl/CGI to develop the Web site ourselves to make user convenient to access to our grid everywhere they are. Figure 5 presents the home page of the portal. It provides user to log in the grid we set up via X.509 certificates or MyProxy delegations.

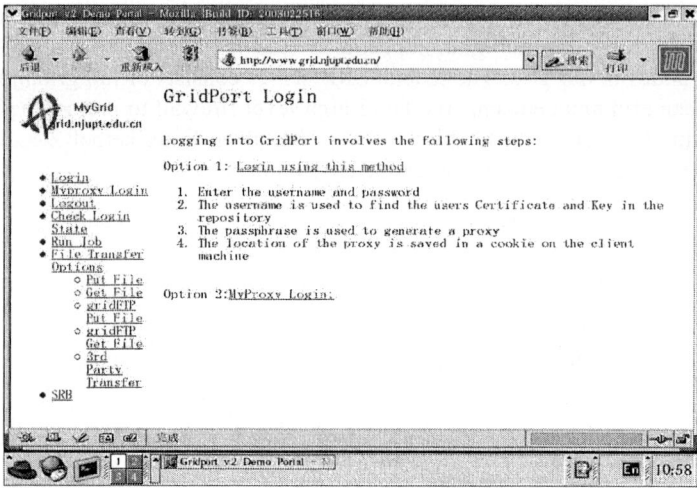

Fig. 5. Home Page of the Portal

6 Conclusions

Grid portal is proved to be the standard interface between user and grid. The grid portal based on security proxy and storage resource broker can strongly support scalable, dynamic and flexible grid environment. Currently, there are still some limitations: the grid portal software is based on Perl/CGI, which is known to have low performance on webservers than technologies such as Java and Servlets. And the security mechanism does not support long-time-running jobs. We will continue with ongoing upgrades planned for the grid portal based on security and resource proxy system, including developing some practicable solutions for security proxy mechanisms and full integration of the SRB data collection.

References

[1] Foster, I. and C. Kesselman, eds. *The Grid: Blueprint for a New Computing Infrastructure.* 1999, Morgan Kaufmann.
[2] [R. Butler, D. Engert, I. Foster, C. Kesselman, S. Tuecke, J. Volmer, and V. Welch. *Design and deployment of a national-scale authentication infrastructure.* IEEE Computer, 33(12):60-66,2000
[3] Lorch, M., Basney J., and Kafura D., *A Hardware-secured Credential Repository for Grid PKIs.* 4th IEEE/ACM International Symposium on Cluster Computing and the Grid, Chicago, Illinois, April 19-22, 2004
[4] Wan M., Arcot Rajasekar, Reagan Moore, Phil Andrews, *A Simple Mass Storage System for the SRB Data Grid.* Proceedings of the 20th IEEE Symposium on Mass Storage Systems and Eleventh Goddard Conference on Mass Storage Systems and Technologies, San Diego, April 2003.

Optimal Designs of Directed Double-Loop Networks*

Bao-Xing Chen[1] and Wen-Jun Xiao[2]

[1] Dept. of Computer Science, Zhangzhou Teacher's College,
Zhangzhou, Fujian, 363000 P. R. China
cbaoxing@hotmail.com
[2] Dept. of Computer Science, South China University of Technology,
Guangzhou, 510641 P. R. China
wjxiao@scut.edu.cn

Abstract. Let n, s be positive integers such that $2 \leq s < \frac{n}{2}$. A directed double-loop network $G(n; 1, s)$ is a directed graph (V, E), where $V = Z_n = \{0, 1, 2, \ldots, n-1\}$, $E = \{i \to i+1 \pmod{n}, i \to i+s \pmod{n} | i = 0, 1, 2, \ldots, n-1\}$. An attention deserving problem is: for a given positive integer n, how to choose an integer s such that the diameter of $G(n; 1, s)$ is minimal. An $O(k^{2.5} n^{0.25} \log n)$ algorithm, where $O(k)$ is upper-bounded by $O(n^{0.25})$, is given in this paper to determine s such that the diameter of $G(n; 1, s)$ is minimal.

Keywords: directed double-loop networks, diameter, optimal design, k-tight optimal.

1 Introduction

Double-loop networks have been widely used in designs of local area networks and communication computer networks. Many researchers are interested in the double-loop networks [1-15]. Their interests mainly focus on routing [9,10], diameters [11,12,15] and designs of optimal double-loop networks [1-8,13,14].

Let $d(i, j)$ be the length of a shortest path from node i to node j. Let $d(n; 1, h)$ denote the diameter of $G(n; 1, s)$ and $d(n) = \min\{d(n; 1, s) | 2 \leq s < n/2\}$. Let $\lceil x \rceil$ denote the minimum integer which is bigger than or equal to x. As $G(n; 1, s)$ is vertex symmetric, $d(n; 1, s) = \max\{d(i, j) | 0 \leq i, j < n\} = \max\{d(0, i) | 0 \leq i < n\}$. By using L-shape tiles, Wong, C. K. and Coppersmith, D. [11] gave a lower bound of $\lceil \sqrt{3n} \rceil - 2$ for $d(n)$ in 1974. From then on, people use L-shape tiles to construct k-tight optimal infinite families of double-loop networks (DLN), to compute the diameter of DLN, and to find a shortest path for any two nodes of DLN.

---

* This work was supported by the Natural Science Foundation of Fujian Province(No. F0110012) and the Scientific Research Foundation of Fujian Provincial Education Department(No. JA03144).

Aguilo, F. and Fiol, M. A. [7] gave an algorithm to find $s_1, s_2 \in Z_n$ such that the digraph $G(n; s_1, s_2)$ had the minimum diameter. The running time complexity of the algorithm was $O(k^3 \log n)$, where $O(k)$ was upper-bounded by $O(n^{0.25})$.

Liu, H. P., Yang, Y. X., and Hu, M. Z. [4] gave an $O(n \log n)$ algorithm to find $s \in Z_n$ such that the digraph $G(n; 1, s)$ had the minimum diameter.

An $O(k^{2.5} n^{0.25} \log n)$ algorithm, where $O(k)$ is upper-bounded by $O(n^{0.25})$, is given in this paper to determine s such that the diameter of $G(n; 1, s)$ is minimal. As $O(k)$ is upper-bounded by $O(n^{0.25})$, the algorithm given in this paper is better than the one given in [4].

2 Definitions and Some Lemmas

We use Z to denote the set of integers, Z^+ to denote the set of nonnegative integers.

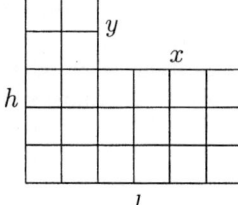

Fig. 1. $L(n; l, h, x, y)$

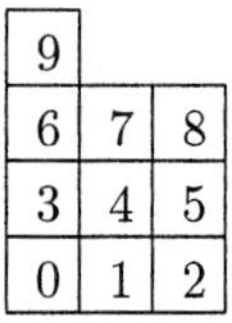

Fig. 2. The L-shape tile of $G(10; 1, 3)$

In order to compute the diameter of $G(n; 1, s)$, we only need to calculate the distances from node 0 to all other nodes. Wong, C. K. and Coppersmith, D. [11] gave the following construction of a pattern for those distances.

"In the first quadrant ($x \geq 0, y \geq 0$) of Euclidean plane, we proceed to fill the lattice points (i. e., x, y being integers) with the points reachable from 0 in 0 step, 1 step, 2 steps, and so on. Note that to reach a point only the total number of 1's and the total number of s's are material, not their ordering. At the lattice point (x, y) we fill the value k determined by the equation $x1 + ys \equiv k \pmod{n}$, meaning that in a total of $x + y$ steps the point k can be reached. In general, we proceed in the following manner: Start from the origin $(0, 0)$, then the line $(1, 0), (0, 1)$, and then the line $(2, 0), (1, 1), (0, 2)$, and so on. At each point (x, y) if the value k has not appeared so far, we write it down, otherwise we just leave a blank. We stop when all values of k, where $0 \leq k \leq n-1$, have been accounted for."

It was proved in [11] that the pattern determined by $G(n; 1, s)$ is an L-shape (see Fig. 1).

Definition 1. *The pattern shown in Fig. 1 is called an L-shape tile with four parameters l, h, x, y, where $l, h, x, y \in Z^+$, and $l, h \geq 2$, $0 \leq x < l$, $0 \leq y < h$.*

This L-shape tile is denoted by $L(n;l,h,x,y)$. We use $d(L(n;l,h,x,y))$ to denote $\max\{l+h-x-2, l+h-y-2\}$. If $d(L(n;l,h,x,y)) = lb(n)+k$, then the L-shape tile $L(n;l,h,x,y)$ is called k-tight. An L-shape tile $L(n;l,h,x,y)$ determined by $G(n;1,s)$ is written as $L(G(n;1,s))$.

The L-shape tile $L(10;3,4,2,1)$ shown in Fig. 2 is determined by $G(10;1,3)$.

Definition 2. *An L-shape tile $L(n;l,h,x,y)$ is called realizable, if there is a directed double-loop network $G(n;1,s)$ such that $L(G(n;1,s)) = L(n;l,h,x,y)$.*

Definition 3. *Suppose that n is a positive integer and $n \geq 5$. If $d(n) = lb(n)+k$, then n is called k-tight optimal.*

Lemma 1. [2] *For a given positive integer T, the interval $[4, 3T^2+6T+3]$ can be written as:*

$$\bigcup_{t=1}^{T} \bigcup_{i=1}^{3} I_i(t),$$

where $I_1(t) = [3t^2+1, 3t^2+2t]$, $I_2(t) = [3t^2+2t+1, 3t^2+4t+1]$, $I_3(t) = [3t^2+4t+2, 3t^2+6t+3]$. Moreover $n \in I_i(t)$ if and only if $lb(n) = 3t-2+i$.

Lemma 2. [13] *Let $h' = h-y$, $l' = l-x$. $L(n;l,h,x,y)$ is realizable if and only if $\gcd(y,h') = 1$, $l > y, h \geq x$. Furthermore if the sufficient and necessary conditions are satisfied, then $L(n;l,h,x,y)$ is realized by a unique directed double-loop network $G(n;1,s)$, where $s \equiv \alpha l - \beta l' \pmod{n}$, and $\alpha y + \beta h' = 1$ for some integers α, β.*

Lemma 3. [2] *Suppose that $L(n;l,h,x,y)$ is an L-shape tile. Let $z_0 = |x-y|$. Then $d(L(n;l,h,x,y)) \geq \sqrt{3n - 0.75z_0^2} + 0.5z_0 - 2$.*

3 Main Results

For any given positive integer n, we can get $d(n)$ by using the following algorithm:

```
For s = 2 to n/2 - 1, calculate the diameter of G(n;1,s) by using the al-
gorithm given in [12].
```

By using the above algorithm, we can find $s_0 \in \{2, 3, \cdots, \lceil \frac{n}{2} \rceil - 1\}$ such that the digraph $G(n;1,s_0)$ has the minimum diameter. As $d(n;1,n+1-s)=d(n;1,s)$ (Theorem 7 [1]), we have $d(n) = d(n;1,s_0)$.

With the aid of a computer and by using the above algorithm, we have the following Lemma 4.

Lemma 4. *When $n \leq 300000$, we have $d(n) \leq lb(n) + 4$ and only 27 integers such that $d(n) = lb(n) + 4$. These 27 4-tight optimal double-loop networks are $G(53749; 1, 985)$, $G(64729; 1, 394)$, $G(69283; 1, 1764)$, $G(94921; 1, 515)$, $G(101467; 1, 2438)$, $G(103657; 1, 528)$, $G(142441; 1, 635)$, $G(147649; 1, 617)$, $G(159076; 1, 676)$, $G(196435; 1, 746)$, $G(200507; 1, 3532)$, $G(210488; 1, 6696)$,*

$G(225109;\ 1,\ 5650)$, $G(226131;\ 1,\ 7632)$, $G(235171;\ 1,\ 791)$, $G(236671;\ 1,\ 6374)$, $G(244909;\ 1,\ 13376)$, $G(253882;\ 1,\ 964)$, $G(256507;\ 1,\ 963)$, $G(260941;\ 1,\ 8969)$, $G(266281;\ 1,\ 875)$, $G(267107;\ 1,\ 858)$, $G(271681;\ 1,\ 3063)$, $G(279827;\ 1,\ 5363)$, $G(281939;\ 1,\ 851)$, $G(284431;\ 1,\ 869)$, $G(291883;\ 1,\ 881)$. *The former 4 4-tight optimal double-loop networks were reported in* [14].

By Lemma 4 we can prove the following Theorem 1.

Theorem 1. *Suppose that n is k-tight optimal. If an L-shape tile $L(n; l, h, x, y)$ is k-tight, then $|x - y| \leq 2k + 6$.*

By Lemma 2 and Theorem 1 we have the following Lemma 5.

Lemma 5. *Suppose that $n = 3t^2 + At + B \in I_i(t)$ and n is k-tight optimal. Let $l = 2t + a$, $h = 2t + b$, $x = t + a + b - i - k$, $y = t + a + b - i - k + z$, where $0 \leq z \leq 2k + 6$. Then a k-tight L-shape tile $L(n; l, h, x, y)$ (or $L(n; l, h, y, x)$) is realizable if and only if the indeterminate equation (1) has an integer solution (a_0, b_0) and $2t + a_0 > t + a_0 + b_0 - i - k + z \geq 0$, $2t + b_0 \geq t + a_0 + b_0 - i - k \geq 0$, $\gcd(2t + b_0, t + a_0 + b_0 - i - k + z) = 1$ (or $2t + a_0 > t + a_0 + b_0 - i - k \geq 0$, $2t + b_0 \geq t + a_0 + b_0 - i - k + z \geq 0$, $\gcd(2t + b_0, t + a_0 + b_0 - i - k) = 1$).*

$$(a + b - i - k)(a + b - i - k + z) - ab + (A + z - 2i - 2k)t + B = 0 \quad (1)$$

Theorem 2. *Let $H(k, z, t) = (i + k)^2 - (i + k)z + z^2 + 3[(2i + 2k - A - z)t - B]$. The indeterminate equation (1) has an integer solution if and only if there exist two nonnegative integers s and m such that $4H(k, z, t) = s^2 + 3m^2$.*

If $4H(k, z, t)$ can be represented as $s^2 + 3m^2$, where $s, m \in Z^+$, then the following four pairs are solutions of the indeterminate equation (1) and at least two of them are integer solutions:

(1) $a = \frac{j+s+3m}{6}$, $b = \frac{j+s-3m}{6}$; (2) $a = \frac{j+s-3m}{6}$, $b = \frac{j+s+3m}{6}$;
(3) $a = \frac{j-s+3m}{6}$, $b = \frac{j-s-3m}{6}$; (4) $a = \frac{j-s-3m}{6}$, $b = \frac{j-s+3m}{6}$;
where $j = 4i + 4k - 2z$.

From Theorem 2 we know that the following Algorithm 1 can find out all integer solutions of the indeterminate equation (1).

```
Algorithm 1: Suppose that t,A,B,i,k,z are given integers. In the foll-
owing we will find out all integer solutions of the equation (1).
  Step 1: H := 4(i+k)^2 - 4(i+k)z + 4z^2 + 12[(2i+2k - A - z)t - B];
  Step 2: if H ≥ 0 then
    s := √H ;
    do while s ≥ 0
       m := √(H - s^2)/3 ;
       if H == s^2 + 3m^2 then
         Find out integer pairs in the four pairs of Theorem 2;
       endif
       s := s - 1;
    enddo
  endif
```

Note that the running time complexity of the algorithm 1 is determined by the value of $\sqrt{H}$ and the value of H is mainly determined by $12*(2i+2k-A-z)t$. Thus the running time complexity of the algorithm 1 is $O(\sqrt{kt})$, i.e., $O(\sqrt{k}n^{0.25})$. So from Theorem 2 we get the following Theorem 3.

Theorem 3. *For any given integers t, A, B, i, k, z, Algorithm 1 can find out all integer solutions of the indeterminate equation (1). The running time complexity of the algorithm 1 is $O(\sqrt{k}n^{0.25})$.*

For any given positive integer n ($n \geq 5$), the following Algorithm 2 will find out a k-tight optimal double-loop network $G(n;1,s)$.

```
Algorithm 2: For any given positive integer n (n ≥ 5), we will find out
a k-tight optimal double-loop network G(n;1,s).
  Step 1: Given n, calculate t,A,B,i such that n = 3t² + At + B ∈ Iᵢ(t). Let
k := 0;
  Step 2: Do while .t.
    z := 0;
    Do while z <= 2k + 6
    /* For any given integers t,A,B,i,k,z, Algorithm 1 can find out all
integer solutions of the indeterminate equation (1). */
      For an integer solutions (a₀,b₀) of the indeterminate equation (1),
let l := 2t + a₀;  h := 2t + b₀;  x := t + a₀ + b₀ - i - k;
      y := t + a₀ + b₀ - i - k + z;  h' := h - y;
      if (x ≥ 0, h > y ≥ 0, l > y and gcd(y,h') = 1) or (l > y ≥ 0, h > x ≥ 0,
h ≥ y and gcd(x,h - x) = 1) then
        L(n;l,h,x,y) or L(n;l,h,y,x) is realizable; goto step 3;
      endif
      z := z + 1;
    enddo
    k := k + 1;
  enddo
  Step 3: For a realizable L-shape tile L(n;l₀,h₀,x₀,y₀) we just found, we
can find out two integers α and β such that αy₀ + β(h₀ - y₀) = 1. Let s ≡
αl₀ - β(l₀ - x₀) (mod n). Then we get G(n;1,s), which is a k-tight optimal
double-loop network.
```

Theorem 4. *For any given positive integer n ($n \geq 5$), Algorithm 2 can find out a k-tight optimal double-loop network $G(n;1,s)$. The running time complexity of Algorithm 2 is $O(k^{2.5}n^{0.25}\log n)$.*

With the aid of a computer and by using Algorithm 2, we find out that there are 153 5-tight optimal integers when $n \leq 10000000$. They are 417289, 526429, 858157, 1302637, 1368379, 1498333, 1507507, 1562149, 1572187, 1585333, 1620149, 1624085, 1648837, 1820047, 1832749, 1913203, 1940597, 2063833, 2143921, 2205637, 2226769, 2366981, 2511661, 2551397, 2773321, 2828101, 2886739, 2926097, 3130549, 3147077, 3225553, 3349743, 3504857, 3509749, 3545867, 3586741, 3630413, 3649189, 3863273, 4029797, 4032493, 4226743, 4269343, 4282909, 4333223, 4335347, 4343606, 4379069, 4457293, 4472131, 4475545, 4494557, 4550419, 4564933, 4565153, 4800031,

4803541, 4916373, 4943467, 5080441, 5100967, 5134069, 5284307, 5458921, 5560873,
5605427, 5637901, 5670869, 5683259, 5702509, 5759077, 5809849, 5892629, 5899027,
5921329, 5989249, 6005741, 6062029, 6110827, 6350429, 6390793, 6473537, 6505839,
6533653, 6575167, 6605357, 6666961, 6704059, 6835893, 6920707, 6921077, 6921565,
7117013, 7272697, 7276777, 7295053, 7302173, 7315269, 7425289, 7541929, 7593397,
7617121, 7631221, 7667467, 7675253, 7705783, 7733041, 7823495, 7824149, 8011667,
8102293, 8104009, 8124297, 8305807, 8351836, 8375197, 8408707, 8467493, 8550321,
8568124, 8600936, 8639023, 8659213, 8679027, 8686149, 8727733, 8823511, 8849477,
8871373, 9029681, 9055117, 9097969, 9155057, 9197197, 9197749, 9204749, 9231617,
9247505, 9279149, 9279413, 9309497, 9344659, 9393953, 9394393, 9470143, 9472057,
9478977, 9607633, 9697471, 9739907, 9859849, 9870097, 9881161.

There are only 3 6-tight optimal integers when $n \leq 10000000$. They are 7243747, 8486867, 9892013.

References

[1] Hwang, F. K., Xu, Y. H.: Double loop networks with minimum delay. Discrete Mathematics, **66**(1987) 109-118
[2] Li, Q., Xu, J. M., Zhang, Z. L.: Infinite families of optimal double loop networks. Science in China (Series A), **23**(1993) 9:979-992 (in Chinese)
[3] Xu, J. M.: Designing of optimal double loop networks. Science in China (Series E), **29**(1999) 3: 272-278 (in Chinese)
[4] Liu, H. P., Yang, Y. X. , Hu, M. Z.: On the construction of tight double loop networks. Practising & Theory in System Engineering, **12** (2001) 72-75 (in Chinese)
[5] Shen Jian, Li Qiao: Two theorems on double loop networks. Journal of China University of Science and Technology, **25** (1995) 2:127-132
[6] Esque, P., Aguilo, F., Fiol, M. A.: Double commutative-step digraphs with minimum diameters. Discrete Mathematics, **114** (1993) 147-157
[7] Aguilo, F., Fiol, M. A.: An efficient algorithm to find optimal double loop networks. Discrete Mathematics, **138** (1995) 15-29
[8] Erdos, P., Hsu, D. F.: Distributed loop networks with minimum transmission delay. Theoretical Computer Science, **100** (1992) 223-241
[9] Mukhopadhyaya, K., Sinha, B. P.: Fault-tolerant routing in distributed loop networks. IEEE Trans. on Comput., **44** (1995)12:1452-1456
[10] Chen, Z. X., Jin, P.: On the [+1]-link-prior shortest path and optimal routing for double-loop networks. Journal of computer research & development, **38** (2001)7: 788-792 (in Chinese)
[11] Wong, C. K., Coppersmith, D.: A combinatorial problem related to multimodule memory organizations. J. ACM **21** (1974) 392-402
[12] Ying, C., Hwang, F. K.: Diameters of weighted double loop networks. Journal of Algorithm, **9** (1988) 401-410
[13] Shen Jian, Li Qiao: Two theorems on double loop networks. Journal of China University of Science and Technology, **25** (1995) 2:127-132
[14] Xu, J. M., Liu, Q.: One infinite family of 4-tight optimal double loop networks. Science in China (Series A), **33** (2003)1:71-74
[15] Rodseth, O. J.: Weighted multi-connected loop networks. Disc. Math., **148** (1996) 161-173

A QoS-Based Access and Scheduling Algorithm for Wireless Multimedia Communications

Bin Wang

Depart. of Radio Engineering, Southeast University, Nanjing, P. R. of China
wellbeing_wb@hotmail.com

Abstract. In this paper, an overview of existing MAC scheduling algorithms is briefly presented. Then considerations concerning QoS of wireless communication are discussed. In order to improve the QoS of wireless communication, we proposed a new QoS-based algorithm for wireless access and scheduling in centralized network where all traffics are controlled by the base station. The algorithm uses TDMA in the radio air interface, but it can also be adapted to other access methods, such as CDMA. Finally, the simulation is performed to show the QoS guarantee with the new algorithm.

1 Introduction

The media access algorithms can be classified into two categories: centralized and distributed[1]. Distributed algorithm is the contention-based protocol, where the access to media depends on the contention of each station, while the centralized algorithm assigns the access opportunity to the station according to the access request.

To meet QoS requirements in wireless applications, the distributed algorithms such as EDCF adopt the idea of traffic categories[3]. The traffic with higher priority is assigned with more media share by reducing the probability of the lower priority traffic access. Although the EDCF can meet the QoS requirements statistically, it cannot strictly provide QoS guarantee. What's more, the QoS of the real-time traffic keeps decreasing as more traffics join the contention.

It is found that distributed algorithms are not suitable to strictly guarantee QoS requirements, so the traditional time-stamp based algorithms such as EDD[5] and WFQ[6] cannot be used. Comparatively, centralized protocols perform better than distributed ones from the QoS point of view.

In the centralized network, BS(Base Station) selects from all applicants according to their queue length and traffic priority indicated in their access requests, PCF [2] and HCF [4] are such kind of algorithms which can meet QoS requirements in some degree, but still cannot guarantee strictly time-bounded traffic.

To solve the problems mentioned above, we proposed a QoS-based access and scheduling algorithm to enhance the QoS performance in wireless multimedia network. The algorithm absorbs the merits of both centralized and distributed algorithms to serve real-time and non-real-time traffics with different QoS priorities [7].

This paper is organized as following. Section 2 introduces the main idea and principle of our algorithm. Section 3 describes our access and scheduling algorithm in detail. Section 4 presents some simulation results. Section 5 gives some conclusions.

2   Radio Resource Management for Multimedia Communications

The criteria for classification of different QoS includes: throughput, BER, acceptable delay, acceptable delay jitter. However, from users' point of view, the main concern is the sensitiveness of the traffic delay[8]. Thus, the QoS can also be categorized into classes: background, interactive, streaming and conversational.

In this algorithm, we categorize the streaming and conversational classes into real-time services, which are provided with higher guarantee compared to the non-real-time traffics such as the web browsing of interactive class and email of background classes. As depicted in the figure 1, traffics are arranged in periodic super frames, and each super frame can be divided into three parts, access period, real-time service transmit period and best-effort service transmit period.

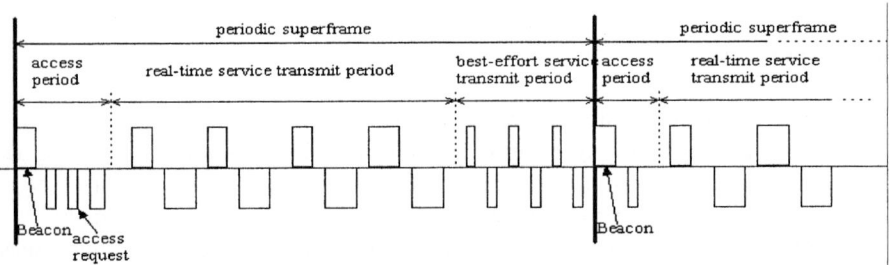

Fig. 1. The structure of the resource arrangement in time

All traffics are managed by the BS. The downlink traffics(from BS to MS) are directly sent to MS in compound frames(data and polling parts), while the uplink traffics(from the MS to BS) have to wait until the BS polls them.

Because real-time services have demanding requirements on time delay, we have to assign them fixed time slots and scan them periodically. The difficulties of the scheduling algorithm are how to poll the MS in time to guarantee the QoS of real-time traffics while at the same time to reduce the overhead of polling as much as possible, in other words, to reduce the chance of polling idle MSs. This problem will be solved by the algorithm proposed in the next section.

We adopt the best-effort policy to serve non-real-time traffics in the round-robin manner.

3 QoS-Based Access and Scheduling Algorithm

3.1 Access Algorithm

If only the best-effort traffics are supported by the system, the module of access control is not necessary. But in order to guarantee specific QoS requirement under limited system capacity, the module controlling the new arriving access requests is required. We all have the suffering of unbearable delay when we watch online TV, the fact is that the required output of the port overweighs the output the system can afford, so no one can get desired service quality from it. The problem comes from the lackness of protection mechanism for the QoS of the ongoing traffic. In other words, we need the mechanism allowing the system to reject new access request before the QoS requirements of the ongoing traffics and the newly arrived traffics cannot be longer satisfied. In our algorithm whether the access can be accepted is determined by two factors: service QoS requirement and current traffic load of the system (Figure 2).

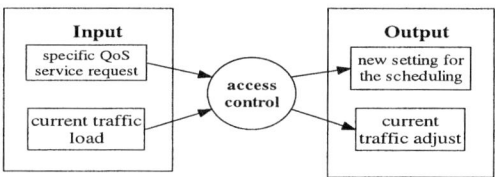

Fig. 2. Access control system

The duty of the beacon in our algorithm is of no difference to the regular one, it maintains the synchronization of local timers of the MSs and delivers protocol related parameters. Upon receiving the beacon frame, the MSs that want to initiate new traffics start the access requests. The collision of the access requests can be avoided to some extent by the slotted ALOHA.

Access requests can be classified accordingly into real-time and non-real-time ones. After successfully receiving the access request frames, all the non-real-time access requests are accepted. The requests of real-time traffic, however, have to be checked whether the current system capacity can accommodate them. If it is big enough, the requested traffic will be activated by the subsequently polling. Otherwise, the BS access control module will deny the request.

3.2 Scheduling Algorithm

With the protection mechanism of the access control, we can schedule the accepted traffics efficiently. To each newly accepted traffic, resources are assigned. The traffic in higher QoS class has the priority to be served first. In the same QoS class, the traffic accepted earlier has higher priority than later ones. The interactions of different modules are presented in Figure 3.

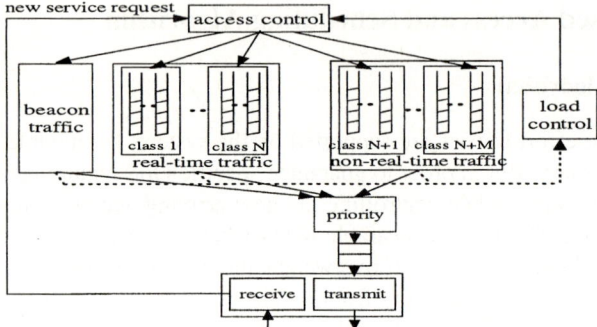

Fig. 3. QoS-based MAC access and scheduling algorithm

To real-time or say time-bounded traffics, traffic delay must be strictly controlled. The bearable delay of certain QoS class can be calculated by $T_i = L_i / R_i$, where L_i is the length of the packet transmitted, and R_i is the average bit rate of the traffic, i.e., the length of the frame transmitted each time is 128Bytes, while the real-time traffic rate is 64Kbps, so the period of this QoS class traffic is $\frac{128 \times 8}{128 \times 1000} = 8ms$. Each class of real-time traffic has its own T_i and our algorithm picks the greatest common divisor of all the T_i as the period of the super frame.

In summary, the whole access and scheduling algorithm can be described as:

1. Once new service requests arrive, all the best-effort service requests are accepted. To real-time service requests, the access control module has to check the current traffic load of the system to see whether the left channel capacity can accommodate the new traffic with required QoS, if the QoS requirement of the new access request cannot be met, the access is denied.
2. If the real-time service access request has been accepted, the system will assign the appropriate type of channel(time slot in fact) to the new connection. In this algorithm, the system creates a new queue in the requested QoS class, the scheduling and load control functions are adjusted accordingly.
3. Real-time traffics with the highest QoS class are served first. As to the traffic MS_i (ith mobile station), the algorithm detects whether there are downlink frames for MS_i or the period of the traffic T_i has expired. If T_i has expired, BS will poll the MS_i. If meantime there are downlink frame for the MS_i, the polling frame is made into a compound frame. Otherwise, the traffic with lower priority is served.
4. After finishing real-time traffics, the time slots left in the super frame are allocated among the non-real-time traffics.
5. The assignment of left time slots among non-real-time traffic depends on the QoS classes of the non-real-time traffic, which is similar to the operation of the real-time traffics, the traffics with higher level of QoS are served first. In the same class, traffics are served by the mechanism of first-in-first-serve, which is implemented by a cyclic array. So the traffic is not served this super frame will have a relatively higher priority in the subsequent ones.

The beacon frame is used to trigger access requests from MS, MSs compete in the subsequent uplink slots for access using the slotted ALOHA.

4 Simulation

To evaluate the performance of the access and scheduling algorithm we proposed, we set up a simulation model depicted in Figure 4 based on OPNET simulator.

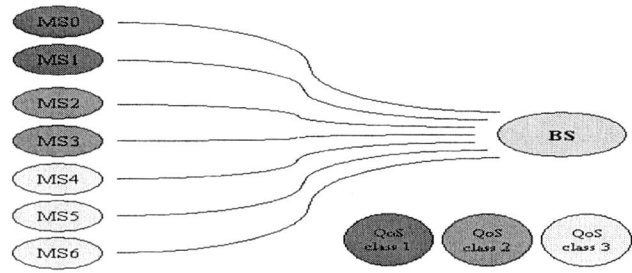

Fig. 4. The scenario for simulation

This scenario consists of 8 entities--one BS and 7 MSs, and MSs can be classified into 3 classes as shown in Figure 4, where the QoS class 1 represents 64kbps duplex voice traffic and the QoS class 2 represents 128kbps downlink video stream, and both classes belong to real-time traffics. The QoS class 3 represents the non-real-time traffic. In each class, the MS with the lower sequence number has higher priority.

Table 1. Simulation parameters for whole system

Propagation delay	6us
Processing delay	0us
Length of simulation running	6hours
Data rate	1Mbps
Duty cycle	16ms

Table 2. Simulation parameters for each MS

	QoS class	L_i (bytes)	R_i (kbps)	T_i (ms)	Starting time(ms)
MS0	1	128	64	16	6
MS1	1	128	64	16	17
MS2	2	512	128	32	33
MS3	2	512	128	32	49
MS4	3	128	~	~	65
MS5	3	128	~	~	71
MS6	3	128	~	~	87

The parameters for system simulation can be seen in Table 1 and the simulation parameters for each MS are listed in Table 2. By simulation, the average delays, maximum delays and delay variances of real-time traffics are shown in Table 3.

Table 3. Simulation results for real-time traffic

	MS0	MS1	MS2	MS3
Average delays(ms)	5.1	7.3	18.1	19.3
Maximum delays(ms)	8.9	11.7	24.3	26.5
Delay variances(ms^2)	3.3	3.8	7.7	10.4
Packets transmitted in time	100%	100%	100%	100%

The results of simulations in Table 3 show that the both average delays and maximum delays of the real-time traffics are all within the limit of service requirements(the delays are all lower than the required traffic period T_i with the comparison of Table 2 and Table 3), and the percentages of packets transmitted in time also convince the efficiency of our algorithm.

Table 4. Simulation results for non-real-time traffic

	MS4	MS5	MS6
Average delays(ms)	63.6	63.9	63.8
Average throughput(kbps)	16.1	16.0	16.0

Table 4 shows that the real-time traffics grab most of the bandwidth from non-real-time traffics to guarantee its QoS, and the share of the remaining time slots among non-real-time traffics is fair, which can be seen from average delays and throughput.

5 Conclusions

In this paper, we proposed a QoS-based access and scheduling algorithm for wireless multimedia communications in packet-switched network. The main focus of the paper is to suppress the delays of the real-time traffics under the required QoS limit. The algorithm triggers the real-time traffic by events of either downlink traffic arrival or traffic bearable delay expiration. The non-real-time traffics are transmitted at the rest bandwidth(time slots), the scheduling of non-real-time traffics adopts the round-robbin algorithm. Moreover, we adopted the access control mechanism to secure the QoS requirements of real-time traffics by denying the new traffic request which cannot be satisfied and could undermine the quality of existing traffics.

References

1. Ramjee, P.,Marina, R: Technology Trends in Wireless Communications. Artech House
2. Wireless LAN Medium Access Control and Physical Layer Specifications. IEEE 802.11 standard, 1999.
3. Stefan, M., Sunghyun: IEEE 802.11e Wireless LAN for Quality of Service
4. H.L. Truong, i: The IEEE 802.11e MAC for Quality of Service in Wireless LANs
5. Alan, D: Analysis and Simulation of a Fair Queueing Algorithm. Proc. ACM SigComm'89
6. Duan-Shin, L: Weighted Fair Queueing and Compensation Techniques for Wireless Packet Switched Networks. *IEEE Trans. on Vehicular Tech.* on 10/29/03.
7. Habetha, J: 802.11a versus HiperLAN/2 –A Comparison of Decentralized and Centralized MAC Protocols for Multihop Ad Hoc Radio Networks.
8. P. Coverdale: Itu-T Study Group 12: Multimedia QoS requirements from a user perspective. Workshop on QoS and user perceived transmission quality in evolving networks, Oct 2001.

Feedforward Wavelet Neural Network and Multi-variable Functional Approximation[1]

Jing Zhao[1,2], Wang Chen[3], and Jianhua Luo[1]

[1] Biomedical Engineering Department, School of Life Science and Technology,
Shanghai Jiaotong University, Shanghai 200240, P.R. China
zjane_cn@yahoo.com.cn
[2] Basic Department,
[3] Department of Automation,
Logistical Engineering University, Chongqing 400016, P.R. China

Abstract. In this paper, a novel WNN, multi-input and multi-output feedforward wavelet neural network is constructed. In the hidden layer, wavelet basis functions are used as activate function instead of the sigmoid function of feedforward network. The training formulas based on BP algorithm are mathematically derived and training algorithm is presented. A numerical experiment is given to validate the application of this wavelet neural network in multi-variable functional approximation.

1 Introduction

Artificial neural network is a kind of mathematical model simulating the structural and functional features of brain neuron network of mankind and having the ability of self-modification. Neural network has the ability of approximating any function at the fuzzy, noisy and non-linear conditions [1]. But it has its own deficiency such as slow convergence speed and getting local minimum value. On the other hand, just as Fourier series, wavelet series has the ability of functional approximation. Furthermore, wavelet transformation is a good tool for time-frequency localization. It could characterize non-stationary signal locally and self-adaptively. In 1992, Zhang Qinghua and Benveniste presented the concept of wavelet neural network (WNN) by integrating wavelet analysis with neural network [2]. Combining the localization feature of wavelet transform and the self-modification function of neural network, WNN self-adaptively modifies the parameters of basis wavelets through training. So it could overcome the backwards of neural network and could approximate functions more rapidly and efficiently. The WNN proposed by Zhang Qinghua is a mono-output network and the activate function of the output layer is a linear function. In this paper, we present a novel WNN, a multi-input and multi-output feedforward wavelet neural

---

[1] This work is supported by 863 Project of China (No.2002AA234021) and 973 Project of China (No. 2002CB512800).

network. In addition, a nonlinear function, sigmoid function is chosen as activate function of the output layer. This makes WNN has stronger approximation functions.

2 The Structure of Wavelet Neural Network

For a BP neural network with only one hidden layer of neurons, using basis wavelets as its activate functions of hidden layer, we get a multi-input and multi-output wavelet neural networks in Fig. 1.

This WNN has m, p, n nodes in the input layer, hidden layer and output layer respectively. Its structural parameters are following:

Connection matrix between input layer and hidden layer: $W^{(1)} = (w_{jk}^{(1)})_{p \times m}$

Connection matrix between hidden layer and output layer: $W^{(2)} = (w_{ij}^{(2)})_{n \times p}$

Bias vector of the hidden layer neurons: $\Theta^{(1)} = (\theta_1^{(1)}, \theta_2^{(1)}, ... \theta_p^{(1)})$

Bias vector of the output layer neurons: $\Theta^{(2)} = (\theta_1^{(2)}, \theta_2^{(2)}, ... \theta_n^{(2)})$

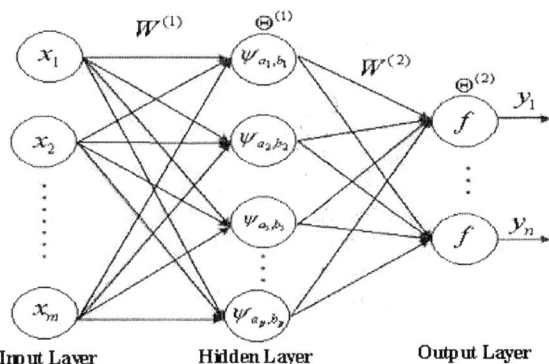

Fig. 1. The Structure of Wavelet Neural Network

Suppose the activate functions of the hidden layer neurons are a group of basis wavelets. This means the activate function of the jth node in the hidden layer is:

$$\psi_{a_j,b_j}(t) = \frac{1}{\sqrt{a_j}} \psi(\frac{t-b_j}{a_j}) \quad j=1,2,...,p \qquad (1)$$

in which the mother wavelet $\psi(t)$ is localized both in time and frequency and could be chosen as different function according to the feature of the problem.

According to the desire of the problem, the activate function of the output layer neurons could be chose as linear function or sigmoid function. In this paper, $f(t)$ is chosen as sigmoid function.

So the parameters of the wavelet neural network in Fig. 1 are:

$$(W^{(1)}, W^{(2)}, \Theta^{(1)}, \Theta^{(2)}, a_1, \cdots, a_p, b_1, \cdots, b_p)$$

These parameters could be adjusted through training.

3 Training Formulas Derivation for WNN

Backpropagation method is the most frequently used technique for training a feed-forward network. It involves two passes through the network, a forward pass and a backward pass. The forward pass generates the network's output activities and the backward pass involves propagating the error initially found in the output nodes back through the network to assign errors to each node that contributed to the initial error. Once all the errors are assigned, the weights are changed so as to minimize these errors. Since the WNN in Fig. 1 is derived from a feedforward neural network, we use backpropagation method to train this network.

For the WNN in Fig. 1, when the input vector is $X = (x_1, x_2, \cdots, x_m)$, we get the output of the jth node in hidden layer:

$$\psi_{a_j,b_j}(\sum_{k=1}^{m} w_{jk}^{(1)} x_k - \theta_j^{(1)}) = \psi_{a_j,b_j}(net_j^{(1)}) = \frac{1}{\sqrt{a_j}} \psi(\frac{net_j^{(1)} - b_j}{a_j}) \qquad (2)$$

in which $net_j^{(1)} = \sum_{k=1}^{m} w_{jk}^{(1)} x_k - \theta_j^{(1)}$ \qquad (3)

The output of the ith node of output layer is:

$$y_i = f(\sum_{j=1}^{p} w_{ij}^{(2)} \psi_{a_j,b_j}(net_j^{(1)}) - \theta_i^{(2)}) = f(net_i^{(2)}), \qquad (4)$$

where $net_i^{(2)} = \sum_{j=1}^{p} w_{ij}^{(2)} \psi_{a_j,b_j}(net_j^{(1)}) - \theta_i^{(2)}$ \qquad (5)

From equation (4) we get the output vector of the WNN: $Y = (y_1, y_2, \cdots, y_n)$.

Suppose we have Q training samples. For each sample q, the desired output vector is $D_q = (d_{q1}, d_{q2}, \cdots d_{qn})$, the output vector of the WNN is $Y_q = (y_{q1}, y_{q2}, \cdots, y_{qn})$. With these Q training samples, we train the WNN through batch learning process. Then the main goal of the network is to minimize the total error E of each output node i over all training samples q:

$$E = \frac{1}{2} \sum_{q=1}^{Q} \sum_{i=1}^{n} (d_{qi} - y_{qi})^2 \qquad (6)$$

From equations (2)–(6), we could deduce the partial derivatives of the error E to each parameter as following.

$$\frac{\partial E}{\partial w_{ij}^{(2)}} = -\sum_{q=1}^{Q}(d_{qi}-y_{qi})\cdot y_{qi}\cdot(1-y_{qi})\cdot \psi_{a_j,b_j}(net_{qj}^{(1)}) \tag{7}$$

$$\frac{\partial E}{\partial w_{jk}^{(1)}} = -a_j^{-1}\sum_{q=1}^{Q}[\psi'_{a_j,b_j}(net_{qj}^{(1)})\cdot x_{qk}\cdot \sum_{i=1}^{n}(d_{qi}-y_{qi})\cdot y_{qi}\cdot(1-y_{qi})\cdot w_{ij}^{(2)}] \tag{8}$$

$$\frac{\partial E}{\partial \theta_i^{(2)}} = \sum_{q=1}^{Q}(d_{qi}-y_{qi})\cdot y_{qi}\cdot(1-y_{qi}) \tag{9}$$

$$\frac{\partial E}{\partial \theta_j^{(1)}} = \sum_{q=1}^{Q}[\psi'_{a_j,b_j}(net_{qj}^{(1)})\cdot \sum_{i=1}^{n}(d_{qi}-y_{qi})\cdot y_{qi}\cdot(1-y_{qi})\cdot w_{ij}^{(2)}] \tag{10}$$

$$\frac{\partial E}{\partial a_j} = \sum_{q=1}^{Q}\{[\frac{a_j^{-1}}{2}\psi_{a_j,b_j}(net_{qj}^{(1)})+\frac{net_{qj}^{(1)}-b_j}{a_j^2}\psi'_{a_j,b_j}(net_{qj}^{(1)})] \\
\cdot \sum_{i=1}^{n}(d_{qi}-y_{qi})\cdot y_{qi}\cdot(1-y_{qi})\cdot w_{ij}^{(2)}\} \tag{11}$$

$$\frac{\partial E}{\partial b_j} = a_j^{-1}\sum_{q=1}^{Q}[\psi'_{a_j,b_j}(net_{qj}^{(1)})\sum_{i=1}^{n}(d_{qi}-y_{qi})\cdot y_{qi}\cdot(1-y_{qi})\cdot w_{ij}^{(2)}] \tag{12}$$

where $\psi'_{a_j,b_j}(net_j^{(1)}) = \frac{1}{\sqrt{a_j}}\psi'(\frac{net_j^{(1)}-b_j}{a_j})$.

In order to speed convergence and avoid vibration, we use backpropagation algorithm with momentum. Suppose learning rate is η and momentum is α ($0<\alpha<1$), we get adjusting formulas for network parameters as following:

$$w_{ij}^{(2)}(t+1) = w_{ij}^{(2)}(t) - \eta\frac{\partial E}{\partial w_{ij}^{(2)}} + \alpha[w_{ij}^{(2)}(t)-w_{ij}^{(2)}(t-1)] \tag{13}$$

$$w_{jk}^{(1)}(t+1) = w_{jk}^{(1)}(t) - \eta\frac{\partial E}{\partial w_{jk}^{(1)}} + \alpha[w_{jk}^{(1)}(t)-w_{jk}^{(1)}(t-1)] \tag{14}$$

$$\theta_i^{(2)}(t+1) = \theta_i^{(2)}(t) - \eta\frac{\partial E}{\partial \theta_i^{(2)}} + \alpha[\theta_i^{(2)}(t)-\theta_i^{(2)}(t-1)] \tag{15}$$

$$\theta_j^{(1)}(t+1) = \theta_j^{(1)}(t) - \eta\frac{\partial E}{\partial \theta_j^{(1)}} + \alpha[\theta_j^{(1)}(t)-\theta_j^{(1)}(t-1)] \tag{16}$$

$$a_j(t+1) = a_j(t) - \eta\frac{\partial E}{\partial a_j} + \alpha[a_j(t)-a_j(t-1)] \tag{17}$$

$$b_j(t+1) = b_j(t) - \eta\frac{\partial E}{\partial b_j} + \alpha[b_j(t)-b_j(t-1)] \tag{18}$$

4 Algorithm for WNN Training

From the discussion above, we get algorithm for training the WNN in Fig. 1 as following:

(1) Initializing network parameters: For each parameter, give a little random number as its initial value;
(2) Supplying training sample set: Give input vectors $X_q = (x_{q1}, x_{q2}, \cdots, x_{qm})$ and desired output vectors $D_q = (d_{q1}, d_{q2}, \cdots d_{qn})$, $q=1,2,\ldots Q$;
(3) Self-training of the network: Compute network outputs according to current network parameters and input vectors in training sample set:

$$y_{qi} = f(\sum_{j=1}^{p} w_{ij}^{(2)} \psi_{a_j,b_j}(net_{qj}^{(1)}) - \theta_i^{(2)}) = f(net_{qi}^{(2)}), \quad q=1,2,\ldots Q;$$

(4) Computing the error: Compute the error between network outputs and desired outputs over all training samples q:

$$E = \frac{1}{2}\sum_{q=1}^{Q}\sum_{i=1}^{n}(d_{qi} - y_{qi})^2$$

If E is less than ε, the little positive number given in advance, stop network training. Otherwise, go to (5).
(5) Computing gradient vectors: Compute the partial derivatives of the error E to each parameter according to equations (7) ~ (12).
(6) Modifying network parameters: Modify network parameters according to equations (13)~(18). Go to (2).

5 Numerical Example

Approximate the function $f(x, y) = (x+y)e^{0.5x}$, $x \in [0,1]$, $y \in [0,2.5]$.

The sample data set is composed of 400 input-output pairs. A WNN of 2 nodes in input layer, 30 nodes in hidden layer and 1 node in output layer is constructed. The activate function of the jth node in the hidden layer is:

$$\psi_{a_j,b_j}(t) = \frac{1}{\sqrt{a_j}}\psi(\frac{t-b_j}{a_j}) \quad j=1,2,\ldots,30$$

in which the mother wavelet $\psi(t)$ is chosen as Mexican Hat wavelet:

$$\psi(t) = (1-t^2)e^{-t^2/2}$$

The activate function of the output layer neurons is chosen as sigmoid function:

$$f(t) = \frac{1}{1+e^{-t}}$$

After 300 times of iteration, the final mean square error is 0.0557.

To compare with the performance of this WNN, we construct another standard neural network (NN) with the same structure. The only difference is the activate function of the hidden layer of the standard neural network is sigmoid function. After 300 times of iteration, the final mean square error of this standard neural network is 0.2996. In Fig 2, we give the original function with 400 input-out pairs and the approximated function with 1600 input-output pairs respectively by WNN and NN. The proposed WNN has better performance from Fig. 2.

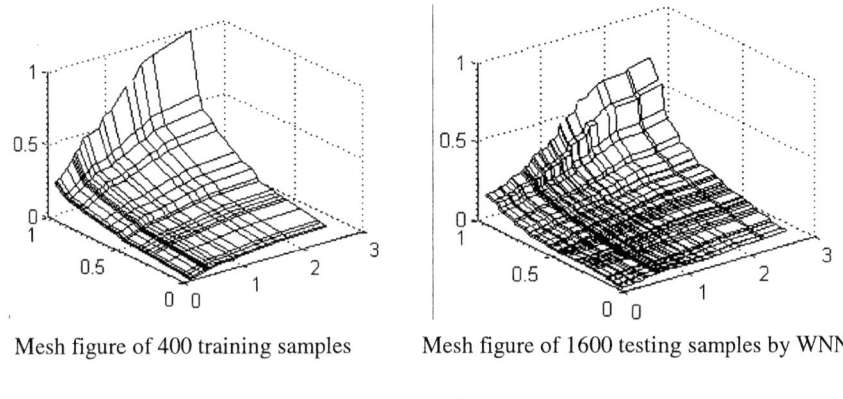

Mesh figure of 400 training samples   Mesh figure of 1600 testing samples by WNN

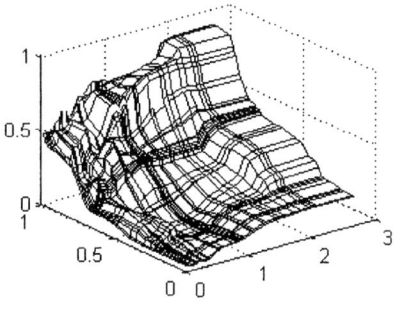

Mesh figure of 1600 testing samples by NN

Fig. 2. Mesh Figures of the Original Function and the Approximated Function

References

1. Cybenko, G.: Approximation by Superpositions of a Sigmoid Function. Math of Control. Signals and Systems. 2 (1989) 303-314
2. Zhang, Q., Benveniste, A.: Wavelet Networks. IEEE Trans. on NN. 3 (1992) 889-898
3. Zhang, Q.: Using Wavelet Network in Nonparametric Estimation. IEEE Trans Neural Networks 2 (1997) 227-236
4. Daubechies, I.: Ten Lectures on Wavelets. CBMS-NSF Regional Series in Applied Mathematics. Philadelphia PA: SIAM (1992)

The Distributed Wavelet-Based Fusion Algorithm

Rajchawit Sarochawikasit, Thitirat Wiyarat, and Tiranee Achalakul

Department of Computer Engineering, King Mongkut's University of Technology,
Thonburi, Bangkok 10140, Thailand
{raj, tiranee}@cpe.kmutt.ac.th, wthitirat@hotmail.com

Abstract. This paper describes a distributed algorithm for use in the analysis of the multi-spectral satellite images. The algorithm combines the spatial-frequency wavelet transform and the maximum selection fusion algorithm. It fuses a multi-spectral image set into a single high quality achromatic image that suits the human visual system. The algorithm operates on a distributed collection of computers that are connected through LAN. A scene of Chiangmai, Thailand, taken from the Landsat ETM+ sensor is used to assess the algorithm image quality, performance and scalability.

1 Introduction

Multi-spectral image fusion is the process of combining images from different wavelengths to produce a unified image, removing the need for frame by frame evaluation to extract important information. Image fusion can be accomplished using a wide variety of techniques that include pixel, feature, and decision level algorithms [3]. At the pixel level, raw pixels can be fused using image arithmetic, band-ratio methods [8], wavelet transforms [6], maximum contrast selection techniques [7], and/or the principal/independent component transforms [2], [5]. At the feature level, raw images can be transformed into a representation of objects, such as image segments, shapes, or object orientations [4]. Finally, at the decision level, images can be processed individually and an identity declaration is used to fuse the results [4]. Many image fusion techniques in recent literatures often utilize the multiscale-decomposition-based methods. The methods generally involve transformation of each source image from the spatial domain to other domains, such as frequency or spatial-frequency domains. The composite representation is then constructed using a wide variety of fusion rules, and the final fused image can be obtained by taking an inverse transformation. Several multi-scale transforms provide both spatial and frequency domain localization. However, we chose to study the wavelet transform as it provides a more compact representation, separates spatial orientation in different bands, and efficiently de-correlates interesting attributes in the original image.

Our fusion algorithm is based on a variation of the Daubechies discrete wavelet transform [9] and the implementation of the maximum coefficient fusion system. From the experiments, we found that the DWT introduced a relatively complete set of embedded information with little noise and also relatively efficient in computing. The wavelet theory is used as a mathematical foundation to produce coefficient components for each source image. Then, a composite representation is constructed, based on the maximum absolute coefficient selection. The model is capable of joining the composite discrete wavelet, supporting the sharpness and brightness changes, edges and lines boundaries or even feature in the image set. In our approach, the coefficients are associated with one another in the same scale, which can be called a single-scale grouping scheme. The fusion algorithm can be described in Fig. 1.

The 2-level decomposition of Mallat's Algorithm for DWT is utilized in our work. The DWT is applied recursively over the spatial data. Each input image will be decomposed and down-sampled using the low pass and high pass digital FIR filters. First, the row of the input image is convolved, and the column is then down-sampled to obtain two sub-images whose horizontal resolutions are reduced by a factor of 2. Both sub-images are then filtered and down-sampled into four output images producing high-high, high-low, low-high, and low-low bands.

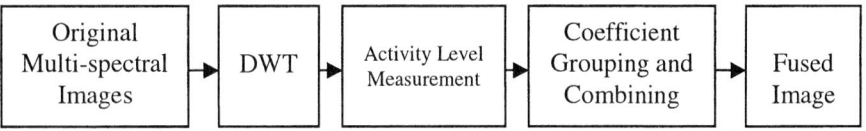

Fig. 1. The fusion algorithm

The activity level is then measured. The level reflects the energy in the space between the coefficients. Our work employed the coefficient-based activity (CBA). The technique considers each coefficient separately. After the decomposition process, we obtain a set of coefficients in several decomposition levels. We, then, experimented with the single-scale grouping method [10], which joining the coefficients from the same decomposition scale.

After achieving the approximate coefficients and the group of detail coefficients, we create the resulting image by fusing multiple components using the maximum selection rule. Let I_n be the coefficient matrix that represents the original image of frame n. I^h_n represents the high frequency components, and I^l_n represents the low frequency components. G_n denotes the gradient of the high frequency component I^h_n, $G_n = gradient(I^h_n)$. The fused coefficient of all high frequency components, F_h, can be calculated as shown in the following code fragment.

The fused coefficient of all low frequency components, F_l, can be calculated as follows: $F_l = \max(I^l_n)$. Using the fused components of the low and the high frequency coefficients, the resulting image can then be generated by taking the inverse transform of both fused components.

```
max = 0
for ( i = 0; i < n ; i++)
  if (G_max < G_i)
    max = i ;
F_h  =  I^h_max;
```

To demonstrate the algorithm, it was applied to an 8-band multi-spectral image collected with the Landsat Satellite. These images correspond to the mixture of urban and foliated area taken at wavelengths between 0.45 and 12.5 micron. Fig. 2

shows a single multi-spectral image via a representative sample of frames picked from the 8 spectral bands. Notice that the band on the left shows significant contrast on the forestry and road systems. However, since this image is hidden in a data set, an automated method is required to extract the information without frame-by-frame inspection.

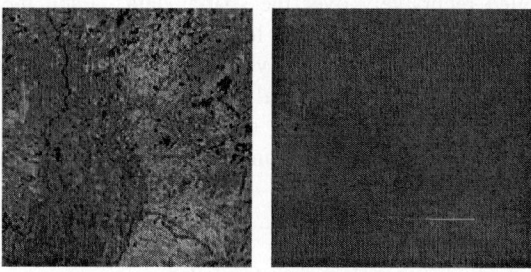

Fig. 2. The Landsat ETM+ of Chiangmai, Thailand

Fig. 3 shows the resulting fused image obtained through our algorithm. Notice that the image quality, especially the contrast level and the image details are visibly enhanced compared to the original images. The result suggests that the fusion system is capable of summarizing most of the important information and put it into the resulting image. The calculated SNR values are all above 40 db, which states the minimal information loss in the resulting image.

The wavelet transform has high computational costs because it involves pixel convolution. The performance requirement discourages the use of our algorithm in real-time applications. To improve performance, we explored a concurrent algorithm employing low-cost, commercial-off-the-shelf computers connected using a standard LAN.

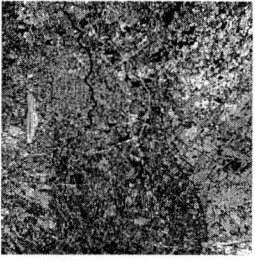

Fig. 3. The resulting image

2 Concurrent Algorithms

The concurrent algorithm decomposes each image frame into sub-images, which can be operated on relatively independently. Each sub-image consists of a set of pixels.

The allocation of sub-images to processors is managed through a variant of the manager/worker technique. This strategy employs a manager thread that performs the above decomposition, and distributes sub-cubes to a set of worker threads. Each worker performs relatively independent components of the overall image transformation. A manager thread coordinates the actions of the workers, gathers partial results from them, and assembles the final resulting image.

The main abstract code of the algorithm is shown in Program 1 and is executed at every processor on the network. The manager and workers are executed as independent threads with a single thread per processor. The manager abstract code is shown below in Program 2. The manager thread serves primarily to synchronize and accumulate partial results from the workers. The manager loads original image frames and then distributes them to a set of workers (line 4&5). It also synchronizes the calculation by making sure that the partial results are received from all workers before moving on to the next stage in the algorithm. When the partial results are returned (line 6), the manager applies the maximum selection rule (stated in the previous section) to form a fused coefficient set (line 7&8). The fused coefficient set is then divided into subsets (line 9) and the subsets are distributed once again to perform the inverse transformation (line 10&11). The final result is assembled after all the workers send back the partial results (line 12) and then is displayed (line 13).

Each worker thread waits for the manager to send its part of image (line 1). Once the sub-image arrives, the convolution is performed to filter and to down sample the sub-image using Mallat's Algorithm (line 2). The activity level measurement and the coefficient single-scale grouping are then performed (line 3&4). The resulting coefficient matrix is sent back to the manager to be fused (line 5). The worker then waits for its next set of data (line 6). Once received, it applies the inverse wavelet transform to convert the coefficient set back to the spatial domain (line 7). The results are sent to the manager for displaying (line 8).

```
main() {
   p = getMyProcessorId()
   if(p == 0) {
     numSubImages = getNumSubImages(numWorker, numPixel)
     manager(numSubimages, numWorkers)
   }
   foreach remaining available processor
   worker();
}

manager(numSubImages, numWorkers) {
1   coeffCube = [][]
2   coeffFused = []
3   finalImage = []
4   foreach worker i {
5      send (i, aSubImage)
6      coefficientMatrix [i] = recv(i)
   }
7   coeffCube = build (coeffMatrix [])
8   coeffFused = maxSelection(coeffCube)
9   subCoefficient = sizeof(coeffFused) / numWorkers
10  foreach worker i {
11     send(i, aSubCoefficient)
```

```
12        finalImage = merge (finalImage, recv (i))
   }
13   display(finalImage)
}
worker(numSubImages, numWorkers) {
1    aSubImage = recv (manager)
2    coeffMatrix = convolve(aSubImage)
3    coeffMatrix = activityMeasure(coeffMatrix)
4    coeffMatrix = coeffGrouping(coeffMatrix)
5    send (manager, coeffMatrix)
6    subCoefficient = recv(manager)
7    subImage = inverseTransform(subCoefficient)
8    send (manager, subImage)
}
```

3 Performance Evaluations

In this section we study the algorithms scaling properties. The performance of the algorithm when generating the results presented in section 1 was measured on a networked workstations of eight nodes. Each node is a Silicon Graphics O2 running at 300 MHz with a RAM of 128 MB. Fig. 4 shows the speed up gained as a function of the number of processors, plotted against the ideal speedup.

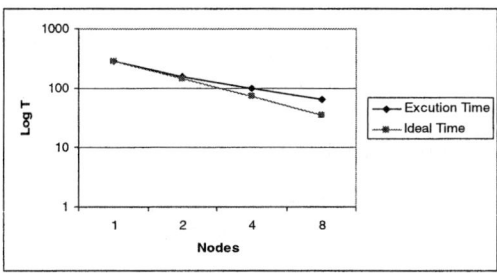

Fig. 4. Performance

To reduce the communication overheads, the next version of the concurrent code will add granularity as another variable. We suspect that by adjusting the granularity, we will be able to find the optimum grain size for the data set. Communication and computation overlapping concept can be utilized and thus reduce the effect of the overheads. Applying the dynamic load balancing should also help as the load can be transferred to faster processors introducing the best system utilization.

4 Conclusion

This paper has described an image fusion model and its concurrent algorithm based on the wavelet transform. The algorithm has been applied to the 8-band Landsat

ETM+ data set. The experiments demonstrated acceptable results in terms of performance and scalability. A more complete experimental and analytical study is in progress. In addition, we are currently exploring the concept of dynamic load balancing to speedup the concurrent algorithm. We expect that by fine-tuning the concurrent code, the algorithm will enable real-time satellite image fusion for land use classification application.

References

1. Chandy L. M., Taylor S., An Introduction to Parallel Programming, Jones and Bartlett publishers, Boston, 1992.
2. Gonzales R. and R. Woods, "Digital Image Processing", Addison-Wesley Publishing Company, pp 81 – 125, 1993.
3. Hall D.L., Mathematical Techniques in multisensor Data Fusion, Artech House Inc., Boston, 1992.
4. Hall D.L, "An Introduction to Multisensor Data Fusion," Proceedings of the IEEE, Vol. 85, No. 1, January 1997, pp. 6-23.
5. Lee T., Independent Component Analysis: Theory and Applications, kluwer Academic Publishers, Boston, 1998.Li H., Manjunath B. A., Mitra S. K., "Multisensor Image Fusion Using the Wavelet Transform," Graphical Models and Image Processing, Vol. 57, No. 3, May 1995, pp. 235-245.
6. Li H., B.S. Manjunath, S.K. Mitra, "Multisensor Image Fusion Using the Wavelet Transform", Graphical Models and Image Processing, vol. 57, No. 3, pp. 235-245, 1995.
7. Peli T., K. Ellis, and R. Stahl, "Multi-Spectral Image Fusion for Visual Display", SPIE vol. 3719 Sensor Fusion: Architectures, Algorithms, and Applications III, Orlando, FL, pp. 359-368, April 1999.
8. Richards J. A., and Jai X., Remote Sensing Digital Image Analysis: An Introduction, New York, NY: Springer, 1998.
9. W.H. Press, S.A. Teukolsky, W.T. Vetterling, and B.P. Flannery, Numerical Recipes in C 2nd edn., (Cambridge University Press, Cambridge, 1995).
10. Zhang Z. and R.S. Blum,"A Categorization of Multiscale-decomposition-based Image Fusion Schemes with a Performance Study for a Digital Camera Application", Proceedings of the IEEE, pp. 1315-1328, August 1999.

Alternating Direction Finite Element Method for a Class of Moving Boundary Problems

Xu-Zheng Liu[1], Xia Cui[2], Jun-Hai Yong[3], and Jia-Guang Sun[3]

[1] Department of Computer Science and Technology,
Tsinghua University, Beijing, 100084, P.R. China
[2] Laboratory of Computational Physics, Institute of Applied Physics
Computational Mathematics, Beijing, 100088, P.R. China
[3] School of Software, Tsinghua University, Beijing, 100084, P.R. China

Abstract. An alternating direction finite element scheme for a class of moving boundary problems is studied. Using coordinate transformations of the spatial variants, a new domain independent of the time is obtained and an ADFE scheme on the new domain is proposed. Then the unique solvability of the approximation scheme is proved, and optimal H^1 and L^2 norm space estimates and $O((\Delta t)^2)$ estimate for the temporal variant are obtained.

1 Introduction

In many problems in science and engineering, the locations of one or more of the domain boundaries are not fixed. Solutions of these problems require the tracking of moving boundaries. The moving boundaries that are not only irregular but also changing during the computation present a challenge to many numerical methods. In spite of this difficulty, there are still many research works on these problems [1] [2]. Usually most of the numerical procedures need fairly heavy computation for the multi-dimensional moving boundary problems. Alternating direction (AD) method is an efficient way to deal with multi-dimensional problems, because it can reduce such complex problems to a series of simple one-dimensional problems, hence greatly eliminate the calculation. However, to our best knowledge, in spite of the obvious merit of the AD method, there is still no related study on applying it to the numerical simulation for multi-dimensional moving boundary problems. A probable reason for this phenomenon may come from the fact that the irregularity and persistent variation of the boundary bring much difficulty to the practical realization of the AD algorithm. In fact, for a variety class of moving boundary problems, AD procedure is feasible. In this paper, we will consider the AD method for a class of moving boundary problems. For those problems, firstly coordinate transformations are used to change the practical domain to a fixed one which is called as computational domain. The boundaries of this computational domain are independent of time. Then AD calculations are done on the computational domain. During the calculation procedure, the corresponding evaluations can be turned back to the original domain whenever and wherever needed. To obtain numerical solutions with high precision, herein, we use ADFE (Alternating Direction Finite Element) method for the AD simulation, because it keeps both the the advantages of AD method (efficiency) and FE method (high accuracy). ADFE method is first developed by Douglas et al [3] in 1971, and many related publications can be found since then [4]. The idea of coordinate transformations extends from [5], where Lesiant P. and Touzani R. study one-dimensional Stefan problem. Moreover, this strategy of the paper may be extended to more general numerical schemes, such as ADFD (Alternating Direction Finite Difference), etc., as long as they are efficient to solve the derived problems after the coordinate transformation procedure.

Consider the moving boundary problem written as:

$$\begin{aligned} &u_t(X,t) - \nabla \cdot (a(X,t,u)\nabla u(X,t)) \\ &+ \sum_{i=1}^{2} b_i(X,t,u)\frac{\partial u(X,t)}{\partial x_i} = f(X,t,u), \quad && X \in \Omega(t) \subset R^2, t \in J, \\ &u(X,0) = u_0(X), \quad && X \in \Omega(0) \subset R^2, t = 0, \\ &u(X,t) = 0, \quad && X \in \partial\Omega(t), t \in J, \end{aligned} \quad (1.1)$$

where $\Omega(t) = \{X = (x_1, x_2), x_i \in (s_{i,1}(t), s_{i,2}(t)), i = 1, 2)\}$, $J = [0, T]$. The given functions f, u_0 are assumed to be properly smooth functions. The boundary functions $s_{i,j}(t)(i,j = 1,2)$ and their first derivatives $\dot{s}_{i,j}(t)(i,j = 1,2)$ are continuous. Additionally the two boundary functions $s_{i,1}(t), s_{i,2}(t)$ ($i = 1,2$) of $\Omega(t)$ do not intersect with each other.

The remaining part of the paper is organized as follows. An ADEF scheme for (1.1) is proposed in Section 2, corresponding approximation and stability properties are derived in Section 3, hence the convergence of the scheme is obtained. Further discussion is presented in the last section.

Here and below, K will be a generic positive constant and may be different each time it is used. ϵ is an arbitrary small constant. Denote $(\phi, \psi)_t = \int_{s_{2,1}(t)}^{s_{2,2}(t)} \int_{s_{1,1}(t)}^{s_{1,2}(t)} \phi\psi \, dx_1 dx_2$, $(\phi, \psi) = \int_0^1 \int_0^1 \phi\psi dy_1 dy_2$, $\dot{\phi}(t) = \frac{d\phi(t)}{dt}$, and $D\phi = \frac{\partial^2 \phi}{\partial y_1 \partial y_2}$. Divide $[0,T]$ into L small equal intervals, define the timestep $\Delta t = \frac{T}{L}$, and $t_n = n\Delta t$. Let $\phi_t^n = \phi_t|_{t=t_n}$, $\phi^n(\psi) = \phi(Y, t_n, \psi)$, $\partial_t \phi^n = \frac{\phi^{n+1} - \phi^{n-1}}{2\Delta t}$, $\partial_{tt}\phi^n = \frac{\phi^{n+1} - 2\phi^n + \phi^{n-1}}{(\Delta t)^2}$ and $d_t\phi^n = \frac{\phi^{n+1} - \phi^n}{\Delta t}$. Hence $\partial_t \phi^n = \frac{\Delta t}{2}\partial_{tt}\phi^n + d_t\phi^{n-1}$, this relation is useful in the future estimate.

2 ADFE Schemes

Firstly, the variational formulation of (1.1) can be described as:

Finding $u \in L^2(J, H_0^1(\Omega(t)))$ with $u_t \in L^2(J, H^{-1}(\Omega(t)))$ such that

$$\begin{aligned} &(u_t, v)_t - (a(X,t,u)\nabla u(X,t), \nabla v)_t \\ &+ \sum_{i=1}^{2}(b_i(X,t,u)\tfrac{\partial u(X,t)}{\partial x_i}, v)_t = (f(X,t,u), v)_t, \quad && \forall v \in H_0^1(\Omega(t)), t \in J, \\ &(u(X,0), v)_t = (u_0, v)_0, \quad && \forall v \in H_0^1(\Omega(0)). \end{aligned} \quad (2.1)$$

Secondly, we introduce the coordinate transformations: $y_i = \frac{x_i - s_{i,1}(t)}{s_{i,2}(t) - s_{i,1}(t)}$ ($i = 1, 2$), namely, $x_i = s_{i,1}(t) + (s_{i,2}(t) - s_{i,1}(t))y_i$ ($i = 1, 2$), then $y_i \in (0, 1)$, and the domain $\Omega(t)$ is transformed into $\hat{\Omega} = (0, 1) \times (0, 1)$. Denote $Y = (y_1, y_2)$, $\hat{u} = \hat{u}(Y, t) = u(X, t)$, $\hat{a}(\hat{u}) = \hat{a}(Y, t, \hat{u}) = a(X, t, u)$, $\hat{b}_i(\hat{u}) = \hat{b}_i(Y, t, \hat{u}) = b_i(X, t, u)$ ($i = 1, 2$), $\hat{f}(\hat{u}) = \hat{f}(Y, t, \hat{u}) = f(X, t, u)$. Set $p_i(\hat{u}) = p_i(Y, t, \hat{u}) = \frac{\hat{a}(\hat{u})}{(s_{i,2}(t) - s_{i,1}(t))^2}$, $q_i(\hat{u}) = q_i(Y, t, \hat{u}) = \frac{\hat{b}_i(\hat{u})}{s_{i,2}(t) - s_{i,1}(t)} - \frac{\dot{s}_{i,1}(t)}{s_{i,2}(t) - s_{i,1}(t)} - \frac{\dot{s}_{i,2}(t) - \dot{s}_{i,1}(t)}{s_{i,2}(t) - s_{i,1}(t)} y_i$ ($i = 1, 2$). The variational formulation (2.1) is equivalent to:

Finding $\hat{u} \in L^2(J, H_0^1(\hat{\Omega}))$, $\hat{u}_t \in L^2(J, H^{-1}(\hat{\Omega}))$ such that

$$\begin{aligned}(\hat{u}_t, \hat{v}) - \sum_{i=1}^{2}((p_i(Y, t, \hat{u})\tfrac{\partial \hat{u}(Y,t)}{\partial y_i}, \tfrac{\partial \hat{v}}{\partial y_i}) \\ + \sum_{i=1}^{2} q_i(Y, t, \hat{u})\tfrac{\partial \hat{u}(Y,t)}{\partial y_i}, \hat{v}) = (\hat{f}(Y, t, \hat{u}), \hat{v}), & \quad \forall \hat{v} \in H_0^1(\hat{\Omega}), \\ (\hat{u}(Y, 0), \hat{v}) = (\hat{u}_0, \hat{v})_0, & \quad \forall \hat{v} \in H_0^1(\hat{\Omega}). \end{aligned} \quad (2.2)$$

Divide $\hat{\Omega}$ into $M_1 \times M_2$ small equal intervals and denote $h_1 = 1/M_1$, $h_2 = 1/M_2$, $h = max\{h_1, h_2\}$. Let $\mu_i = span\{\gamma_1^i(y_i), \gamma_2^i(y_i), ..., \gamma_{M_i}^i(y_i)\} \subset H_0^1([0, 1])$ ($i = 1, 2$) be the one dimensional k degree finite element space along the direction y_i. Define the tensor finite element space $\mu = \mu_1 \otimes \mu_2$. Let $\hat{U}^n = \sum_{i,j} \Gamma_{i,j}^n \gamma_i^1(y_1)\gamma_j^2(y_2)$, $p_i^n(\hat{U}) = p_i(Y, t_n, \hat{U}^n)$, $q_i^n(\hat{U}) = q_i(Y, t_n, \hat{U}^n)$ ($i = 1, 2$), $f^n(\hat{U}) = f(Y, t_n, \hat{U}^n)$. Denote p^* as an upper bound for the functions p_1 and p_2, let λ be a properly selected positive constant and satisfy the inequality $\lambda > \frac{1}{4}p^*$, then we propose an alternating direction discrete scheme of (2.2):

Finding $\hat{U}^n \in \mu$ satisfies

$$\begin{aligned}(\partial_t \hat{U}^n, \hat{v}) + \sum_{i=1}^{2}(p_i^n(\hat{U})\tfrac{\partial \hat{U}^n}{\partial y_i}, \tfrac{\partial \hat{v}}{\partial y_i}) + \sum_{i=1}^{2}(q_i^n(\hat{U})\tfrac{\partial \hat{U}^n}{\partial y_i}, \hat{v}) \\ + \lambda(\nabla(\hat{U}^{n+1} - 2\hat{U}^n + \hat{U}^{n-1}), \nabla \hat{v}) \\ + 2\lambda^2(\Delta t)^3(D\partial_{tt}\hat{U}^n, D\hat{v}) = (f^n(\hat{U}), \hat{v}), & \quad \forall \hat{v} \in \mu, \\ (\hat{U}^0, \hat{v}) = (\hat{u}_0, \hat{v}), & \quad \forall \hat{v} \in \mu.\end{aligned} \quad (2.3)$$

Set $C_{irs} = (\gamma_r^i(y_i), \gamma_s^i(y_i))$, $A_{irs} = ((\gamma_r^i(y_i))', (\gamma_s^i(y_i))')$, for $r, s = 1, ..., M_i$; let $C_i = (C_{irs})_{r,s}$, $A_i = (A_{irs})_{r,s}$ be $M_i \times M_i$ matrices ($i = 1, 2$). Equivalently, (2.3) can be rewritten into a vector form which will give a clear hint of alternating direction:

$$(C_1 + 2\lambda \Delta t A_1) \otimes (C_2 + 2\lambda \Delta t A_2)(\Gamma^{n+1} - 2\Gamma^n + \Gamma^{n-1}) = 2\Delta t \Psi^n, \quad (2.4)$$

where, $\Psi^n = -\sum_{i=1}^{2}(p_i^n(\hat{U})\tfrac{\partial \hat{U}^n}{\partial y_i}, \tfrac{\partial \hat{v}}{\partial y_i}) + \sum_{i=1}^{2}(q_i^n(\hat{U})\tfrac{\partial \hat{U}^n}{\partial y_i}, \hat{v}) + (f^n(\hat{U}), \hat{v}) - (d_t \hat{U}^{n-1}, \hat{v})$, Γ^{n+1} is the unknown vector. Noticing that C_i and A_i are independent of time, thus (2.4) only needs to be decomposed once at each time step. Therefore, the calculation is highly economical.

3 Approximation and Stability Properties

We define the Ritz Projector of the solution $\hat{u}$ in space μ as $\tilde{u}$, where $\tilde{u}$ satisfies

$$\sum_{i=1}^{2}(p_i(\hat{u})\frac{\partial(\hat{u}-\tilde{u})}{\partial y_i},\frac{\partial \hat{v}}{\partial y_i}) + \sum_{i=1}^{2}(q_i(\hat{u})\frac{\partial(\hat{u}-\tilde{u})}{\partial y_i},\hat{v}) + \kappa(\hat{u}-\tilde{u},\hat{v}) = 0, \quad \forall \hat{v} \in \mu, \quad (3.1)$$

and κ is a proper positive constant. Set $\hat{u} - \tilde{u} = \eta$, then similar as [6], we can get the following approximation property:

$$\|\eta_t\|_{L^2(L^2)} + \|\eta\|_{L^\infty(L^2)} + h\|\eta\|_{L^\infty(H^1)} = O(h^{k+1}). \quad (3.2)$$

Denote $\xi^n = \tilde{u}^n - \hat{U}^n$, then $\hat{u}^n - \hat{U}^n = \eta^n + \xi^n$. Subtracting (2.3) from (2.2), and noting the equality (3.1), we derive the error equation:

$$\sum_{i=1}^{3} L_i^n =: (\partial_t \xi^n, \hat{v}) + [\sum_{i=1}^{2}(p_i^n(\hat{U})\frac{\partial \xi^n}{\partial y_i}, \frac{\partial \hat{v}}{\partial y_i}) + \lambda(\nabla(\xi^{n+1} - 2\xi^n + \xi^{n-1}), \nabla \hat{v})]$$

$$+ 2\lambda^2 (\Delta t)^3 (D\partial_{tt}\xi^n, D\hat{v})$$

$$= (\partial_t \hat{u}^n - \hat{u}_t^n - \sum_{i=1}^{2}[q_i^n(\hat{u}) - q_i^n(\hat{U})]\frac{\partial \tilde{u}^n}{\partial y_i} - \sum_{i=1}^{2} q_i^n(\hat{U})\frac{\partial \xi^n}{\partial y_i}$$

$$+ [f^n(\hat{u}) - f^n(\hat{U})] - \partial_t \eta^n + \kappa \eta^n, \hat{v}) - \sum_{i=1}^{2}([p_i^n(\hat{u}) - p_i^n(\hat{U})]\frac{\partial \tilde{u}^n}{\partial y_i}, \frac{\partial \hat{v}}{\partial y_i})$$

$$+ [\lambda(\Delta t)^2 (\nabla \partial_{tt} \hat{u}^n, \nabla \hat{v}) + 2\lambda^2 (\Delta t)^3 (D\partial_{tt} \tilde{u}^n, D\hat{v})] =: \sum_{i=1}^{3} R_i^n. \quad (3.3)$$

Taking $\hat{v} = \partial_t \xi^n$ as a test function, multiplying (3.3) by $2\Delta t$ and summing for $n = 1, 2, ..., N-1$, and estimating the derived terms one by one, we show for the left hand, there are

$$2\Delta t \sum_{n=1}^{N-1} L_1^n = 2\Delta t \sum_{n=1}^{N-1} \|\partial_t \xi^n\|^2, \quad (3.4)$$

$$2\Delta t \sum_{n=1}^{N-1} L_2^n = 2\Delta t \sum_{n=1}^{N-1} \sum_{i=1}^{2}([p_i^n(\hat{U}) - 2\lambda]\frac{\partial \xi^n}{\partial y_i}, \frac{\partial(\partial_t \xi^n)}{\partial y_i})$$

$$+ \sum_{n=1}^{N-1} \lambda(\|\nabla \xi^{n+1}\|^2 - \|\nabla \xi^{n-1}\|^2) =: L_{2,1}^N + L_{2,2}^N. \quad (3.5)$$

Obviously, $L_{2,2}^N = \lambda(\|\nabla \xi^N\|^2 - \|\nabla \xi^{N-1}\|^2) - \lambda(\|\nabla \xi^1\|^2 - \|\nabla \xi^0\|^2)$. Using summation by parts, the following estimate holds:

$$L_{2,1}^N \leq \tfrac{1}{2}\|p^{N-1}(\hat{U}) - 2\lambda\|_{L^\infty}(\|\nabla \xi^{N-1}\|^2 + \|\nabla \xi^N\|^2)$$

$$+ \tfrac{1}{2}\|p^0(\hat{U}) - 2\lambda\|_{L^\infty}(\|\nabla \xi^0\|^2 + \|\nabla \xi^1\|^2) + K\Delta t \sum_{n=1}^{N-1} \|\nabla \xi^n\|^2, \quad (3.6)$$

where $\|p^l(\hat{U}) - 2\lambda\|_{L^\infty} = max\{\|p_1^l(\hat{U}) - 2\lambda\|_{L^\infty}, \|p_2^l(\hat{U}) - 2\lambda\|_{L^\infty}\} (l = N-1, 0)$. Next, noticing that $\partial_{tt}\xi^n = (d_t\xi^n - d_t\xi^{n-1})/\Delta t$, $\partial_t \xi^n = (d_t\xi^n + d_t\xi^{n-1})/2$, we get the equality

$$2\Delta t \sum_{n=1}^{N-1} L_3^n = 2\lambda^2 (\Delta t)^3 (\|Dd_t\xi^{N-1}\|^2 - \|Dd_t\xi^0\|^2). \tag{3.7}$$

Now, we turn our attention to each R_i^n on the right side and obtain

$$2\Delta t \sum_{n=1}^{N-1} R_1^n \le K[(\Delta t)^4 + \|\eta\|_{L^2(L^2)}^2 + \|\eta_t\|_{L^2(L^2)}^2]$$

$$+ K\Delta t \sum_{n=1}^{N-1} \|\xi^n\|_1^2 + \epsilon\Delta t \sum_{n=1}^{N-1} \|\partial_t \xi^n\|^2. \tag{3.8}$$

$$|2\Delta t \sum_{n=1}^{N-1} R_2^n| \le K[\|\xi^1\|_1^2 + \|\xi^0\|_1^2 + \|\eta^1\|^2 + \|\eta^0\|^2 + \|\eta\|_{L^2(L^2)}^2 + \|\eta_t\|_{L^2(L^2)}^2]$$

$$+ K\Delta t \sum_{n=1}^{N-1} \|\xi^n\|_1^2 + \epsilon[\|\nabla \xi^N\|^2 + \Delta t \sum_{n=1}^{N-1} \|\partial_t \xi^n\|^2], \tag{3.9}$$

$$|2\Delta t \sum_{n=1}^{N-1} R_3^n| \le K[(\Delta t)^4 + \|\nabla \xi^0\|^2 + \|\nabla \xi^1\|^2]$$

$$+ K\Delta t \sum_{n=1}^{N-1} \|\nabla \xi^n\|^2 + \epsilon[\|\nabla \xi^N\|^2 + \|\nabla \xi^{N-1}\|^2]. \tag{3.10}$$

Combining relations (3.4)-(3.10) and noting $\lambda > \frac{1}{4}p^*$, then using Grownwall's lemma, we deduce

$$\Delta t \sum_{n=1}^{N-1} \|\partial_t \xi^n\|^2 + \|\xi^N\|_1^2 + \|\xi^{N-1}\|_1^2 + (\Delta t)^3 \|Dd_t\xi^{N-1}\|^2$$

$$\le K[(\Delta t)^4 + \|\eta\|_{L^\infty(L^2)}^2 + \|\eta_t\|_{L^2(L^2)}^2 + \|\xi^0\|_1^2 + \|\xi^1\|_1^2 + (\Delta t)^3 \|Dd_t\xi^0\|^2]. \tag{3.11}$$

Summarizing (3.11) and (3.2), we can obtain the following approximation result.

Theorem 1. *For $\lambda > \frac{1}{4}p*$, if*

$$\|\xi^0\|_1 + \|\xi^1\|_1 + (\Delta t)^{3/2} \|Dd_t\xi^0\| = O(h^{k+1} + (\Delta t)^2) \tag{3.12}$$

satisfies, then for the ADFE scheme (2.3), there is

$$\|(\hat{u} - \hat{U})_t\|_{L^2(L^2)} + \|\hat{u} - \hat{U}\|_{L^\infty(L^2)} + h\|\hat{u} - \hat{U}\|_{L^\infty(H^1)} = O(h^{k+1} + (\Delta t)^2),$$

e.g. discrete $L^\infty(L^2)$ and $L^\infty(H^1)$ approximation norm are all optimal on the computational domain.

Taking the test function $\hat{v} = \partial_t \hat{U}^n$ in (2.3), and using an analogous reasoning procedure, we get the stability result on the computational domain.

Theorem 2. *Under the same condition of theorem 1, we have*

$$\Delta t \sum_{n=1}^{N-1} \|\partial_t \hat{U}^n\|^2 + \|\nabla \hat{U}^N\|^2 + \|\nabla \hat{U}^{N-1}\|^2$$
$$\leq K[\|\hat{U}^0\|_1^2 + \|\hat{U}^1\|_1^2 + (\Delta t)^3 \|Dd_t \hat{U}^0\|^2 + \Delta t \sum_{n=1}^{N-1} \|f^n(\hat{U})\|^2].$$

From Theorem 1 and Theorem 2, we can conclude the ADFE scheme (2.3) is uniquely solvable and has optimal H^1 and L^2 norm convergence properties on the computational domain. Since the coordinate transformations are inverse between the computational domain and the practical domain, by using the equivalent norm property (Lemma 2.2 in [5]), we can obtain the same conclusion for the corresponding approximation solution U and the exact solution u of (2.1) on the practical domain.

4 Discussion

To start the procedure (2.3), we need to define perfect initial values to satisfy (3.12). In fact, this condition is easy to fulfill. For example, letting $\hat{U}^0 = \tilde{u}^0$, denoting $\hat{U}^{\frac{1}{2}} = \frac{\hat{U}^0 + \hat{U}^1}{2}$, and defining $\hat{U}^1$ as

$$\sum_{i=1}^{2}(p_i(\hat{U}^{\frac{1}{2}})\frac{\partial \hat{U}^{\frac{1}{2}}}{\partial y_i}, \frac{\partial \hat{v}}{\partial y_i}) + \sum_{i=1}^{2}(q_i(\hat{U}^{\frac{1}{2}})\frac{\partial \hat{U}^{\frac{1}{2}}}{\partial y_i}, \hat{v}) + \kappa(\hat{U}^{\frac{1}{2}}, \hat{v}) = (\hat{f}(\hat{U}^{\frac{1}{2}}), \hat{v}), \quad \forall \hat{v} \in \mu.$$

will be a feasible choice.

By transforming the domain with perfect coordinate transformations, the ADFE method can be applied to solve a class of moving boundary problems. The idea can be easily extended to 3D problems. For certain more complicated boundaries whose functions are dependent on both the temporal and the spatial variants with enough smoothness, this strategy can also be considered. Of course, after the coordinate transformations, any natural efficient numerical procedure, such as ADFD, etc., can also be discussed to solve moving boundary problems which will appear in the future paper. For more complex problems, more detailed considerations are necessary.

Acknowledgement

The project is supported by Hi-Tech Research and Development Program of China (863 Project)(2003AA4Z1010) and National Key Basic Research Project of China (973 Project)(2002CB312106).

References

1. Zhao, Y. and Forhad, A.: A general method for simulation of fluid flows with moving and compliant boundaries on unstructured grids. Computer Methods in Applied Mechanics and Engineering **192** (2003) 4439–4466

2. Cui, X.: The finite element methods for the parabolic integro-differential equation in 2-dimensional time-dependent domain. Numerical Mathematics, A Journal of Chinese Universities **21** (1999) 228–235
3. Douglas, J. and Dupont, T.: Alternating-direction galerkin methods on rectangles. In Hubbard, B., ed.: Proceedings of Symposium on Numerical Solution of Partial Differential Equation II. (1971) 133–214
4. Cui, X.: Adfe method with high accuracy for nonlinear parabolic integro-differential system with nonlinear boundary conditions. Acta Mathematica Scientia **22B** (2002) 473–483
5. Lesaint, P. and Touzani, R.: Approximation of the heat equation in a variable domain with application to the stefan problem. SIAM Journal on Numerical Analysis **26** (1989) 366–379
6. Ciarlet, P.: The Finite Element Method for Elliptic Problems. North-Holland Publishing Company, Amsterdam (1978)

Binomial-Tree Fault Tolerant Routing in Dual-Cubes with Large Number of Faulty Nodes

Yamin Li[1], Shietung Peng[1], and Wanming Chu[2]

[1] Department of Computer Science, Hosei University, Tokyo 184-8584 Japan
[2] Department of Computer Hardware, University of Aizu, Fukushima 965-8580 Japan

Abstract. A dual-cube $DC(m)$ has $m + 1$ links per node where m is the degree of a cluster (m-cube), and one extra link is used for connection between clusters. The dual-cube mitigates the problem of increasing number of links in the large-scale hypercube network while keeps most of the topological properties of the hypercube network. In this paper, we propose efficient algorithms for finding a nonfaulty routing path between any two nonfaulty nodes in the dual-cube with a large number of faulty nodes. A node $v \in DC(m)$ is called k-safe if v has at least k nonfaulty neighbors. The $DC(m)$ is called k-safe if every node in $DC(m)$ is k-safe. The first algorithm presented in this paper is an off-line algorithm that uses global information of faulty status. It finds a nonfaulty path of length at most $d(s,t) + O(k^2)$ in $O(|F| + m)$ time for any two nonfaulty nodes s and t in the k-safe $DC(m)$ with number of faulty nodes $|F| < 2^k(m + 1 - k)$, where $0 \le k \le m/2$. The second algorithm is an online algorithm that uses local information only. It can find a fault-free path with high probability in an arbitrarily faulty dual-cube with unbounded number of faulty nodes.

1 Introduction

As the size of computer networks increases continuously, the node failures are inevitable. Routing in computer networks with faults has been more important and has attracted considerable attention in the last decade. Hypercube is a popular network studied by researchers and adopted in many implementations of parallel computer systems, such as Intel iPSC, the nCUBE, the Connection Machine CM-2, and the SGI's Origin 2000 and Origin 3000. Previous research has shown that a hypercube can tolerate a constant fraction of faulty nodes. For example, Najjar et al[1] demonstrated that for the 10-cube, 33 percent of nodes can fail and the network can still remain connected with a probability of 99 percent. Gu and Peng[2] proposed off-line routing algorithms for a k-safe n-cube with up to $2^k(n - k) - 1$ faulty nodes, where a node u is k-safe if u has at least k nonfaulty neighbor nodes, and an n-cube is k-safe if all nodes in the cube are k-safe. Chen et al[3] also proposed a distributed routing algorithm in hypercube with large amount of faulty nodes based on local subcube-connectivity.

A dual-cube $DC(m)$[4,5] is an undirected graph on the node set $\{0,1\}^{2m+1}$ and there is a link between two nodes $u = (u_0u_1 \ldots u_m u_{m+1} \ldots u_{2m})$ and $v =$

$(v_0v_1 \ldots v_mv_{m+1} \ldots v_{2m})$ if and only if the following conditions are satisfied: 1) u and v differ exactly in one bit position i, 2) if $1 \leq i \leq m$ then $u_0 = v_0 = 1$ and 3) if $m+1 \leq i \leq 2m$ then $u_0 = v_0 = 0$. We use $(u:v)$ to denote a link connecting nodes u and v, and $(u \to v)$ or $(v_1 : v_2 : \ldots : v_r)$ to denote a path or a cycle. For a node $u = (u_1 \ldots u_n)$, $u^{(i)}$ denotes the node $(u_1 \ldots u_{i-1}\overline{u_i}u_{i+1} \ldots u_n)$, where $\overline{u_i}$ is the logical negation of u_i. The set of neighbors of a subgraph T in G is denoted as $N(T) = \{v | (w:v) \in E(G), w \in V(T), v \notin V(T)\}$.

2 Fault Tolerant Routing Algorithm for k-Safe Dual-Cube

We first briefly introduce the algorithm for the fault-tolerant routing in k-safe hypercubes[2]. Given a k-safe n-cube H_n, a set of faulty nodes F with $|F| < 2^k(n-k)$, and two nonfaulty nodes s and t, the idea of finding a fault-free path $s \to t$ is as follows. First, partition H_n along dimension i into two $(n-1)$-cubes, H_{n-1}^0 and H_{n-1}^1, such that s and t are separated, say $s \in H_{n-1}^0$ and $t \in H_{n-1}^1$. Assume that $|F \cap H_{n-1}^1| \leq |F|/2$. Then, we want to route s to H_{n-1}^1 by a fault-free path of length at most $k+2$. This can be done by first constructing a fault-free k-binomial tree with root s $T_k(s)$ in H_n. Since H_n is k-safe the $T_k(s)$ can be found. If $T_k(s) \cap H_{n-1}^1 \neq \emptyset$ or $u^{(i)}$ is nonfaulty, where $u \in T_k(s)$, then s is routed to H_{n-1}^1. Otherwise, since $|F| < 2^k(n-k)$ there exists a $u \in N(T_k(s))$ such that u and $u^{(i)}$ are nonfaulty. Therefore, we can route s to $s' \in H_{n-1}^1$. The fault-free path $s' \to t$ in H_{n-1}^1 can be found recursively since H_{n-1}^1 is $(k-1)$-safe and $|F \cap H_{n-1}^1| < 2^{k-1}((n-1)-(k-1))$. The recursion halts when $k = 0$. In this case, $|F| < n$ and a fault-free path $s \to t$ of length at most $d(s,t) + 2$ can be found in $O(n)$ time[6]. The fault-free path $s \to s'$ of length at most $k+2$ can be found in $O(|F|)$ time. Since at most half of the faulty nodes are involved in the next recursion. The time complexity of the algorithm $T(n) = \sum_{i=0}^{k-1} O(|F|/2^i) + O(n) = O(|F| + n)$. The length of the path, $L(k)$, satisfies the equation $L(k) \leq L(k-1) + (k+2)$ if $k > 0$, and $L(0) \leq d(s,t) + 2$. From this, $L(k) = d(s,t) + O(k^2)$. The algorithm described above is denoted as Hypercube_Routing(H_n, s, t, k, F, P), where P is the fault-free path $s \to t$ in H_n.

A dual-cube DC(m) is k-safe if every node in DC(m) is k-safe. We present an algorithm for finding a fault-free path $s \to t$ in a k-safe DC(m) with number of faulty nodes $|F| < 2^k(m-k+1)$. First, we describe two key techniques for the design of the algorithm. The first one is called *virtual cube*. Given two distinct clusters of the same class in DC(m), say C_s and C_t are of class 0, the virtual $(m+1)$-cube $VH_{m+1} = C_s \cup C_t \cup \{(u:v)|u \in C_s, v \in C_t, \text{ and } u_i = v_i \text{ for all } i, m+1 \leq i \leq 2m\}$.

We call the edge $(u:v)$ virtual edge. A virtual edge in VH_{m+1} corresponds to a path $(u \to v)$ in DC(m), and $(s \to t) = (s \to u : u' = u^{(0)} \to v' = v^{(0)} : v \to t)$, where $(u' \to v')$ is a path of length at most m in $C_{u'}$, a cluster of class 1. The 2^m paths in DC(m) corresponding to the 2^m virtual edges in VH_{m+1} are disjoint. The virtual edge $(u:v)$ is nonfaulty if nodes u, v, u', and v' are nonfaulty and

$|F \cap C_{u'}| < 2^{k-1}(m - k + 1)$. If the virtual edge is nonfaulty then since $C_{u'}$ is a $(k-1)$-safe m-cube, a fault-free path $u \to v$ in DC(m) can be found. Finding all faulty virtual edge takes at most $O(|F|)$ time.

The second one is a technique to find a fault-free path $s \to u : u' = u^{(0)}$ of length at most $k+2$, where path $s \to u$ is a path in C_s under the condition that DC(m) is k-safe and $|F| < 2^k(m-k+1)$. The path $s \to u$ can be found by constructing a fault-free k-binomial-tree $T_k(s)$ in C_s, and then considering the nodes in $N(T_k(s))$.

Algorithm 1 (Binomial_Tree_Routing(DC(m), s, F, P))
Input: DC(m), a nonfaulty node s, and a set of faulty nodes F
  with $|F| < 2^k(m - k + 1)$
Output: a fault-free path $P = (s \to u : u')$ of length at most $k+2$
begin
  $P = \emptyset$;
  find a fault-free $(k-1)$-binomial tree $T_{k-1}(s)$ in C_s;
  if there exists a nonfaulty u' for $u \in T_{k-1}(s)$
  then $P = P \cup (s \to u : u')$;
  else find a fault-free k-binomial tree
      $T_k(s) = T_{k-1}(s) \cup \{(u : u^{(i)}) | u \in T_{k-1}(s)\}$,
      where $u^{(i)}$ is nonfaulty, and $i \neq i_j, 1 \leq j \leq r$, the dimensions
      used for the path $s \to u$ in T_{k-1};
    find a node $u \in N(T_k(s)) \cap C_s$ such that u and u' are nonfaulty;
    $P = P \cup (s \to w : u : u')$, where $s \to w$ is a path in $T_k(s)$;
end

The details is depicted by Algorithm 1. The next lemma shows that Binomial_Tree_Routing algorithm is correct.

Lemma 1. *For $0 \leq k \leq m/2$, and a nonfaulty node s in a k-safe DC(m) with number of faulty nodes $|F| < 2^k(m - k + 1)$, the fault-free path $(s \to u : u')$ of length at most $k+2$ can be found in $O(|F| + m)$ time.*

Proof. From Binomial_Tree_Routing algorithm, since DC(m) is k-safe we know that either a fault-free $T_k(s)$ in C_s is found or there exists a nonfaulty node $w \in T_{k-1}(s)$ such that w' is nonfaulty. In the letter case, let $w = u$ and it is done. So, we assume that $T_k(s)$ is found and for every w in $T_k(s)$, w' is faulty. It was known that for $T_k(s)$ in an m-cube, we have $|N(T_k(s))| \geq 2^k(m-k)$ ([6]). Since C_s is an m-cube and there are at most $|F| - 2^k < 2^k(m^k + 1) - 2^k = 2^k m^k$ faulty nodes in C_s, there exists a node $u \in N(T_k(s))$ such that u and u' are nonfaulty.

The main idea of the proposed algorithm is to route s to C_t if s and t are in different clusters and $|F \cap C_s| \geq |F \cap C_t|$. This can be done using the similar idea of Algorithm 1 and is shown in Algorithm 2, Cluster_Routing. The next lemma shows that the algorithm is correct.

Algorithm 2 (Cluster_Routing(DC(m), s, t, F, P))
Input: a k-safe DC(m), nodes s and t, C_s and C_t are of the same class,
and a set of faulty nodes F with $|F| < 2^k(m - k + 1)$ and $|F \cap C_s| \geq |F \cap C_t|$
Output: a fault-free path $P = (s \to u \to v)$, where $v \in C_t$
and $u \to v$ is the path corresponding to virtual edge $(u : v)$
begin
    $P = \emptyset$;
    find a fault-free $(k - 1)$-binomial tree $T_{k-1}(s)$ in C_s
    if there is a $u \in T_{k-1}(s)$ such that u' is nonfaulty
    then if virtual edge $(u : v)$ is nonfaulty
        then $P = (s \to u \to v)$, $(u \to v)$ is a fault-free path in DC(m)
        else find a fault-free $(k - 1)$-binomial tree $T_{k-1}(u')$ in $C_{u'}$;
        find a nonfaulty $w \in N(T_{k-1}(u')) \cap C_{u'}$
        such that virtual edge $(u = w' : v)$ is nonfaulty;
        $P = (s \to w : u \to v)$, $u \to v$ is a fault-free path in DC(m)
    else find a fault-free k-binomial tree $T_k(s)$ by extending $T_{k-1}(s)$;
    find a nonfaulty $u \in N(T_k(s)) \cap C_s$
    such that virtual edge $(u : v)$ is nonfaulty;
    $P = (s \to u \to v)$, $u \to v$ is a fault-free path in DC(m);
end

Lemma 2. *For $0 \leq k \leq m/2$, and nonfaulty nodes s and t in a k-safe DC(m) with number of faulty nodes $|F| < 2^k(m - k + 1)$, the fault-free path $(s \to u \to v)$ can be found in $O(|F| + m)$ time, where $(s \to u)$ is a path in C_s and $(u \to v)$ is the path corresponding to virtual edge $(u : v)$.*

Proof. We divide the proof into two cases. Case 1: $T_k(s)$ is in C_s. Since $|N(T_K(s))| \geq 2^k(m - k + 1)$ we can find a nonfaulty virtual edge $(u : v)$, $u \in N(T_k(s))$. Then we are done. Case 2: there exists $u \in T_{k-1}(s)$ such that u' is nonfaulty and virtual edge is faulty. In this case, we should try to route u' to C_t using a fault-free $(k-1)$-binomial tree in $C_{u'}$. From Cluster_Routing algorithm, since the paths that route $u \in N(T_{k-1}(s))$ to C_t and the path that route $w \in N(T_{k-1}(u'))$ to C_t are disjoint and there are totally $2^k(m - k + 1)$ disjoint paths, we claim that a fault-free path that route s to C_t does exist.

The algorithm that constructs $s \to t$ is shown in Algorithm 3.

Theorem 1. *For $0 \leq k \leq m/2$, and two nonfaulty nodes s and t in in a k-safe DC(m) with number of faulty nodes $|F| < 2^k(m-k+1)$, the fault-free path $s \to t$ of length at most $d(s,t) + O(k^2)$ can be found in $O(|F| + m)$ time.*

Proof. The correctness of the DualCube_Routing algorithm follows easily from Binomial_Tree_Routing, Cluster_Routing, and Hypercube_Routing algorithms. The Length of the path $L(s,t)$ and time complexity of the algorithm $T(m)$ are shown below for Case 1. The other cases follows easily from the algorithm. $L(s,t) \leq (k+2) + (d(u',v') + 1) + (d(v,t) + O(k^2))$. Since $d(u',v') + d(v,t) \leq d(s,t) + d(s,u) = d(s,t) + k + 1$, we have $L(s,t) = d(s,t) + O(k^2)$. The time is $O(|F|)$ for Binomial_Tree_Routing; $O(m)$ for finding path $u \to v$; and $O(|F|+m)$ for Hypercube_Routing. Therefore, the running time $T(m) = O(|F| + m)$.

Algorithm 3 (DualCube_Routing(DC(m), s, t, k, F, P))
Input: DC(m), nonfaulty nodes s and t, and
 a set of faulty nodes F with $|F| < 2^k(m - k + 1)$
Output: a fault-free path $P = (s \to t)$
begin
 $P = \emptyset$;
 Case 1: $C_s \neq C_t$ and class_id(s) = class_id(t)
  Cluster_Routing(DC(m), s, F, P);
  Hypercube_Routing($C_t, v, t, k - 1, F \cap C_t, P'$);
  $P = P \cup P'$;
 Case 2: class_id(s) $\neq$ class_id(t)
  Binomial_Tree_Routing(DC(m), s, F, P);
  if $C_{u'} = C_t$
  then Hypercube_Routing($C_t, u', t, k - 1, P$);
  else find a fault-free path $P' = (u' \to t)$ as in Case 1;
   $P = P \cup P'$;
 Case 3: $C_s = C_t$
  if $|F \cap C_s| < 2^{k-1}(m - (k - 1))$
  then Hypercube_Routing($C_s, s, t, k - 1, P$);
  else Binomial_Tree_Routing(DC(m), s, F, P);
   Binomial_Tree_Routing(DC(m), t, F, P');
   find a fault-free path $P' = (u'_s \to u'_t)$ as in Case 1;
   $P = P \cup P' \cup (u'_s \to u'_t)$;
end

3 A Practical Fault-Tolerant Routing Algorithm

The proposed algorithm in the previous section requires that the dual-cube is k-safe. In reality, the chance that the dual-cube will not be k-safe increases when the number of faulty nodes grows. We propose an efficient routing algorithm for the fault-tolerant routing in dual-cube containing a large number of faulty nodes and dual-cube may not be k-safe. In general, without k-safe property, the fault-free k-binomial tree $T_k(s)$ might not exist and the routing algorithm might fail to find the fault-free path $s \to t$ although the fault-free path does exist. However, for practical reasons, it is interesting to design an efficient algorithm that find a fault-free path $s \to t$ using only local information of fault status in the dual-cube with a large number of faulty nodes.

The proposed algorithm for fault tolerant routing in an arbitrary faulty dual-cube is distributed and local-information-based. The algorithm is similar to algorithm 3. However, the new algorithm doesn't calculate F while route s to cluster C_t, and it uses binomial trees of increasing size starting with 0-binomial tree, a tree of single node. If the algorithm fails to find a fault-free path $s \to u \to v$ then it tries 1-binomial tree and so on until either a fault-free path is found or the k-binomial tree cannot be constructed. Due to the page limitation, we will not present the details of our algorithm in this draft. The simulations for this

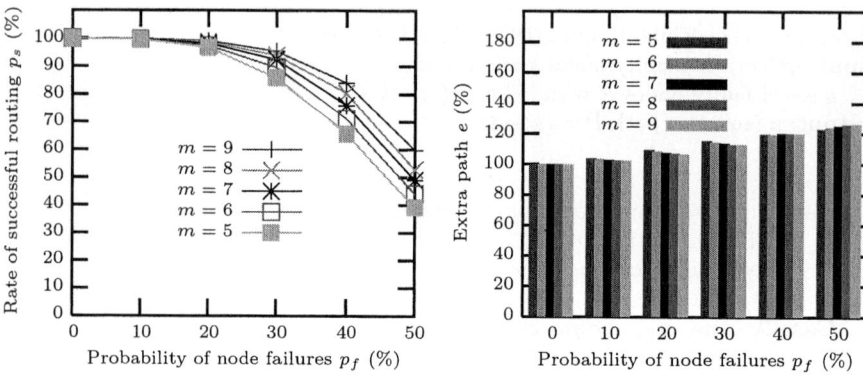

Fig. 1. Successful routing rate

Fig. 2. Path length (%)

algorithm have been conducted with uniformly distribution of faulty nodes in $DC(m)$ for $m = 5, 6, 7, 8$ and 9.

The results for successful routing v.s. the node failure rate are shown in Figure 1. It can be seen from the figure that the successful routing rate is very high ($> 90\%$) if the node failure rate is less than 30%. The successful routing rate drop more deeply when the node failure rate is beyond 30%. However, we can say that in most cases, the successful routing rates are still larger than 50% with the node failure rates up to 50%. As for the length of the routing path, we show the results in Figure 2. From the figure, we can say that the fault-free paths found by our algorithm are very close to the minimum paths in most of the cases. The experimental data show that the proposed algorithm performs well in an arbitrarily faulty dual-cubes with possible very large set of faulty nodes.

References

1. Najjar, W., Gaudiot, J.L.: Network resilience: A measure of network fault tolerance. IEEE Transactions on Computers **39** (1990) 174–181
2. Gu, Q.P., Peng, S.: Unicast in hypercubes with large number of faulty nodes. IEEE Transactions on Parallel and Distributed Systems **10** (1999) 964–975
3. Chen, J., Wang, G., Chen, S.: Locally subcube-connected hypercube networks: Theoretical analysis and experimental results. IEEE Transactions on Computers **51** (2002) 530–540
4. Li, Y., Peng, S.: Dual-cubes: a new interconnection network for high-performance computer clusters. In: Proceedings of the 2000 International Computer Symposium, Workshop on Computer Architecture, ChiaYi, Taiwan (2000) 51–57
5. Li, Y., Peng, S., Chu, W.: Efficient collective communications in dual-cube. The Journal of Supercomputing **28** (2004) 71–90
6. Gu, Q.P., Peng, S.: Optimal algorithms for node-to-node fault tolerant routing in hypercubes. The Computer Journal **39** (1996) 626–629

The Half-Sweep Iterative Alternating Decomposition Explicit (HSIADE) Method for Diffusion Equation

J. Sulaiman[1], M.K. Hasan[2], and M. Othman[3]

[1] School of Science and Technology, Universiti Malaysia Sabah, Locked Bag 2073, 88999 Kota Kinabalu, Sabah
[2] Faculty of Information Science and Technology, Universiti Kebangsaan Malaysia, 43600 Bangi, Selangor D.E.
[3] Department of Communication Tech. and Network, Faculty of Computer Science and Info. Tech., Universiti Putra Malaysia, 43400 Serdang, Selangor D.E., Malaysia
mothman@fsktm.upm.edu.my

Abstract. The primary goal of this paper is to apply the Half-Sweep Iterative Alternating Decomposition Explicit (HSIADE) method for solving one-dimensional diffusion problems. The formulation of the HSIADE method is also derived. Some numerical experiments are conducted that to verify the HSIADE method is more efficient than the Full-Sweep method.

1 Introduction

The half-sweep iterative method is introduced by Abdullah [1] via the Explicit Decoupled Group (EDG) iterative method to solve two-dimensional Poisson equations. The further application of this method to solve partial differential equations can be found via Ibrahim & Abdullah [4], Yousif & Evans [6], and Abdullah & Ali [2].

Based on this method, the concept of the half-sweep method is applied onto the Iterative Alternating Decomposition Explicit (IADE) method. This is an excellent technique for solving system of linear equations, and it is one of the two-step iterative methods. In this paper, combination of both methods will result in the Half-Sweep Iterative Alternating Decomposition Explicit (HSIADE) method. Other IADE methods include the Reduced Iterative Alternating Decomposition Explicit (RIADE) method (Sahimi & Khatim [5]).

To show the effectiveness of the HSIADE method, let us consider the one-dimensional diffusion equation as given by

$$\frac{\partial U}{\partial t} = \alpha \frac{\partial^2 U}{\partial^2 x}, \quad a \leq x \leq b, \text{ and } 0 \leq t \leq T, \tag{1}$$

subject to the initial condition and $U(x,t) = g_1(x)$, $a \leq x \leq b$ the boundary conditions

$$\left.\begin{array}{l}U(a,t) = g_2(t),\\ U(b,t) = g_3(t),\end{array}\right\} \quad 0 \le t \le T,$$

which α is a diffusion parameter.

Before describing formulation of the finite difference approximation equation in case of the full- and half-sweep iterative over the problem (1), we assume the solution domain (1) can be uniformly divided into $(n+1)$ and M subintervals in the x and t directions. The subintervals in the x and t directions are denoted by Δx and Δt respectively which are uniform and defined as

$$\left.\begin{array}{l}\Delta x = h = \frac{(b-a)}{m} \quad m = n+1,\\ \Delta t = \frac{(T-0)}{M}.\end{array}\right\} \tag{2}$$

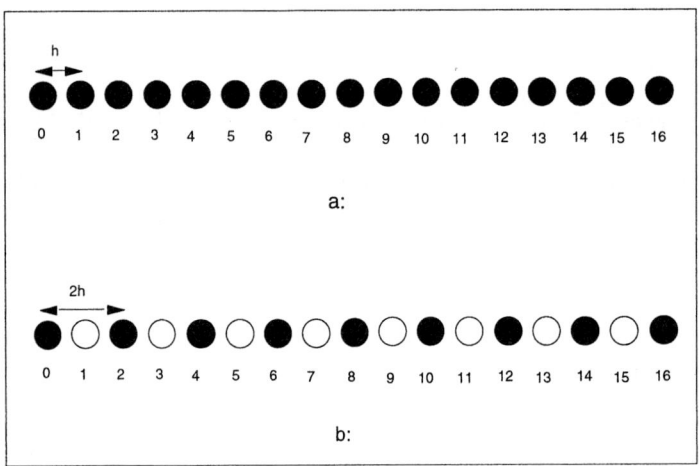

Fig. 1. a and b shows the distribution of uniformly points at any t level for the full- and half- sweep cases, respectively

2 The Half-Sweep Finite Difference Approximation

According to Figure 1 in the previous section, the finite grid networks provide a guide for formulation the full- and half-sweep finite difference approximation equations for the problem (1).

Here, implementations of the full- or half-sweep iterative involve the ● type of node points only. It is obvious that the implementation of the half-sweep iterative method just involves half of whole inner points as shown in Figure 1b compared with the full-sweep iterative method. Then the approximated solution at remaining points can be computed directly, see Abdullah [1], Ibrahim & Abdullah [4], Yousif & Evans [6].

By using the central difference and Crank-Nicolson approaches, the full- and half-sweep finite difference approximation equations can generally be stated as

$$-\beta_1 U_{i-p,j+1} + \beta_2 U_{i,j+1} - \beta_1 U_{i+p,j+1} = f_{i,j}^\beta \qquad (3)$$

where

$$\beta_1 = \frac{\alpha \Delta t}{2(ph)^2}, \quad \beta_2 = \left\{1 + \frac{\alpha \Delta t}{(ph)^2}\right\}, \quad \beta_3 = \left\{1 - \frac{\alpha \Delta t}{(ph)^2}\right\}$$
$$f_{i,j}^\beta = \beta_1 U_{i-p,j} + \beta_3 U_{i,j} + \beta_1 U_{i+p,j}$$

The values of p which correspond to 1 and 2 represent the full- and half-sweep cases respectively. The equation (3), considered at the $j+1$ time level, generates a system of linear equation as follows

$$A_\beta \underline{U}_{j+1}^\beta = f_j^\beta \qquad (4)$$

where coefficient matrix A_β is given by

$$\begin{bmatrix} \beta_2 & -\beta_1 & & & & \\ -\beta_1 & \beta_2 & -\beta_1 & & & \\ & -\beta_1 & \beta_2 & -\beta_1 & & \\ & & \ddots & \ddots & \ddots & \\ & & & -\beta_1 & \beta_2 & -\beta_1 \\ & & & & -\beta_1 & \beta_2 \end{bmatrix}_{(\frac{m}{p}-p)\times(\frac{m}{p}-p)}$$

Moreover, the full- and half sweeps Gauss-Seidel (FGS and HGS) methods acts as the control methods of numerical results.

3 The HSIADE Method

In developing of formulations of the full- and half-sweep IADE methods namely FSIADE and HSIADE respectively, let us consider the system of equation (4) in case of $p = 1, 2$. Hence, the scheme of the IADE method using the Mitchell-Fairweather variant (Sahimi & Khatim [5]) can be expressed as

$$\begin{aligned}(rI + G_1)U^{(k+\frac{1}{2})} &= (rI - gG_2)U^{(k)} + f \\ (rI + G_2)U^{(k+1)} &= (rI - gG_1)U^{(k+\frac{1}{2})} + gf \end{aligned} \qquad (5)$$

where r, k and I represent as an acceleration parameter, the k iteration and an identity matrix respectively and the relation of g and r is given by $g = (6+r)/6$. In fact formulations of the FSIADE and HSIADE methods are to assume that the coefficient matrix, A of the system (6) can be written generally as

$$A = G_1 + G_2 - \frac{1}{6}G_1 G_2 \qquad (6)$$

where

$$G_1 = \begin{bmatrix} 1 & 0 & & & & \\ a_{1p} & 1 & 0 & & & \\ & a_{2p} & 1 & 0 & & \\ & & \ddots & \ddots & \ddots & \\ & & & a_{\frac{m}{p}-3p} & 1 & 0 \\ & & & & a_{\frac{m}{p}-2p} & 1 \end{bmatrix}_{(\frac{m}{p}-1)\times(\frac{m}{p}-1)}$$ and

$$G_2 = \begin{bmatrix} e_{1p} & h_{1p} & & & & \\ 0 & e_{2p} & h_{2p} & & & \\ & 0 & e_{3p} & h_{3p} & & \\ & & \ddots & \ddots & \ddots & \\ & & & 0 & e_{\frac{m}{p}-2p} & h_{\frac{m}{p}-2p} \\ & & & & 0 & e_{\frac{m}{p}-1p} \end{bmatrix}_{(\frac{m}{p}-1)\times(\frac{m}{p}-1)}.$$

The scheme of the FSIADE and HSIADE onto the system (6) methods can be shown by determining values of G_1 and G_2 matrixes as follow

$$\begin{aligned} e_{1p} &= \frac{6}{5}(\beta_2 - 1) \\ h_i &= -\frac{6}{5}\beta_1, \quad a_i = \frac{-6\beta_1}{6 - e_i}, \quad e_i \neq 6 \\ e_{i+p} &= \frac{6}{5}(\beta_2 + \frac{1}{6}a_i h_p - 1) \end{aligned} \quad (7)$$

for $i = 1p, 2p, \ldots, \frac{m}{p} - 2p$. The FSIADE and HSIADE algorithms at level $k+1/2$ and level $k+1$ are explicitly implemented by using the equations (8) and (9) along points in interval [a, b] until the specified convergence criterion is satisfied.

1. At level $(k + 1/2)$

$$U_i^{(k+1/2)} = \frac{1}{d}\left(-a_{i-p}U_{i-p}^{(k+1/2)} + s_i U_i^{(k)} + w_i U_{i+p}^{(k)} + f_i\right) \quad (8)$$

for $i = 1p, 2p, \ldots, \frac{m}{p} - 1p$ where $d = 1 + r$, $a_0 = w_m = 0$, $s_j = r - ge_j$, $w_j = -gu_j$, $\forall j = 1p, 2p, \ldots, \frac{m}{p} - p$.

2. At level $(k + 1)$

$$U_i^{(k+1)} = \frac{1}{d_i}\left(v_{i-p}U_{i-p}^{(k+1/2)} + sU_i^{(k+1/2)} + gf_i - h_i U_{i+p}^{(k+1)}\right) \quad (9)$$

for $i = \frac{m}{p} - 1p, \frac{m}{p} - 2p, \ldots, 2p, p$ where $d_i = r + e_i$, $v_0 = u_m = 0$, $s = r - g$, $v_j = -ga_j$, $\forall j = \frac{m}{p} - 1p, \frac{m}{p} - 2p, \ldots, p$.

Table 1. Comparisons of no. of iteration, exe. time and max. absolute errors of the methods

	No. of iterations					
Methods	Grid size					
	64	128	256	512	1024	2048
FGS	204	691	2315	7491	22653	55466
FSIADE ($r = 0.6$)	40	152	537	1823	5965	18372
FSIADE (r=opt)	39	80	384	618	800	1528
HGS	62	204	691	2315	7491	22653
HSIADE ($r = 0.6$)	29	40	152	537	1823	5965
HSIADE (r=opt)	25	39	80	384	618	800
	Execution time (seconds)					
Methods	Grid size					
	64	128	256	512	1024	2048
FGS	0.22	1.32	9.23	63.88	456.76	3106.91
FSIADE ($r = 0.6$)	0.11	0.49	5.11	26.42	189.83	1455.54
FSIADE (r=opt)	0.06	0.39	3.73	12.96	34.99	136.71
HGS	0.06	0.22	1.48	9.40	63.83	459.67
HSIADE ($r = 0.6$)	0.05	0.11	0.50	6.54	24.88	336.20
HSIADE (r=opt)	0.00	0.11	0.44	5.27	14.50	38.56
	Maximum Absolute Errors					
Methods	Grid size					
	64	128	256	512	1024	2048
FGS	3.3e-7	4.7e-7	7.4e-7	1.7e-6	5.7e-6	2.2e-5
FSIADE ($r = 0.6$)	3.1e-7	4.0e-7	4.7e-7	6.6e-7	1.4e-6	4.4e-6
FSIADE (r=opt)	4.1e-7	4.1e-7	4.8e-7	7.9e-7	3.3e-7	4.7e-7
HGS	7.9e-9	3.3e-7	4.7e-7	7.4e-7	1.7e-6	5.7e-6
HSIADE ($r = 0.6$)	4.1e-9	3.1e-7	4.0e-7	4.6e-7	6.6e-7	1.4e-6
HSIADE (r=opt)	4.0e-9	3.1e-7	3.9e-7	4.1e-7	4.1e-7	4.1e-7

4 Numerical Experiments

To verify the efficiency of the implementation of the HSIADE scheme as derived in equation (5), which is based on the approximation equation (3), some numerical experiments are conducted to solve the one-dimensional diffusion equation as follows

$$\frac{\partial U}{\partial t} = \frac{\partial^2 U}{\partial^2 x}, \quad 0 \le x \le 1, 0 \le t \le 1. \tag{10}$$

The initial and boundary conditions and exact solution of the problem (10) is given by

$$U(x,t) = e^{-\pi^2 t}\sin(\pi x), \quad 0 \le x \le 1, 0 \le t \le 1.$$

All results of numerical experiments, which were gained from implementations of the FGS, HGS, FSIADE and HSIADE methods have been recorded in

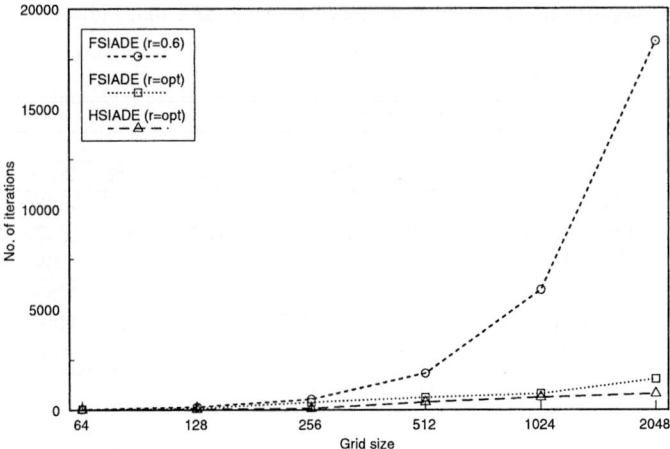

Fig. 2. Number of iterations versus grid size of the FSIADE and HSIADE methods

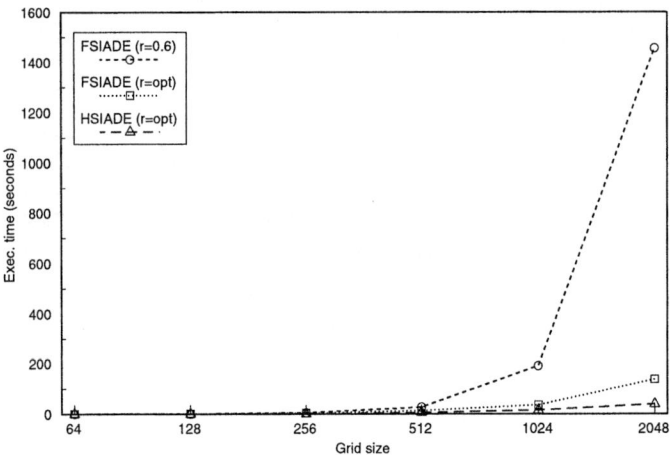

Fig. 3. Execution time (seconds) versus grid size of the FSIADE and HSIADE methods

Table 1. In implementations mentioned above, the convergence test considered the tolerance error $\epsilon = 10^{-10}$. Figures 2 and 3 show number of iterations and execution time against grid size respectively.

5 Summary

By referring onto Table 1 and it has shown in Figure 2 that number of iterations decreased by 37.50-95.64% and 2.50-91.68% respectively correspond to the HSIADE (r optimum) and FSIADE (r optimum) methods compared with the FSIADE ($r = 0.6$) method. In addition, the execution time against the grid size of both the HSIADE (r optimum) and FSIADE (r optimum) methods are much faster about 77.55-100% and 20.40-90.61% respectively than the FSIADE method, see Figure 3.

The whole results show that the HSIADE method is much better in terms of a number of iterations and the execution time than the FSIADE method. This is because the computational complexity of the HSIADE method is nearly 50% of the IADE method.

References

1. Abdullah, A.R.: The Four Explicit Decoupled Group ($\mathcal{EDG}$) Method: A Fast Poisson Solver. Intern. Journal of Computers Mathematics, **38** (1991) 61–70.
2. Abdullah, A.R., Ali, N.H.M.: A Comparative Study of Parallel Strategies for the Solution of Elliptic PDE's. Parallel Algorithms and Aplications, **10** (1996) 93–103.
3. Evans, D.J., Sahimi, M.S., : The Alternating Group Explicit iterative method (AGE) to solve parabolic and hyperbolic partial differential equations. Ann. Rev. Num. Fluid Mechanic and Heat Trans., **2** (1988) 283-389.
4. Ibrahim, A., Abdullah, A.R., : Solving the two-dimensional diffusion equation by the four point explicit decoupled group (EDG) iterative method. International Journal Computer Mathematics, **58** (1995) 253-256.
5. Sahimi, M.S., Khatim, M. : The Reduced Iterative Alternating Decomposition Explicit (RIADE) Method for the Diffusion Equation. Pertanika J. Sci. & Technol., **9(1)** (2001) 13-20.
6. Yousif, W. S., Evans, D. J. : Explicit De-coupled Group iterative methods and their implementations. Parallel Algorithms and Applications, **7** (2001) 53-71.

An Effective Compressed Sparse Preconditioner for Large Scale Biomolecular Simulations*

Dexuan Xie

Department of Mathematical Sciences, University of Wisconsin,
Milwaukee, WI 53201-0413, USA
dxie@uwm.edu

Abstract. The natural preconditioner defined by local potentials is effective in the truncated Newton method for minimizing large scale biomolecular potential energy functions. This paper extends its definition, and proposes an algorithm for generating the sparse pattern of the preconditioner from the primary structure of a molecular system with N atoms. It shows that the implementation of the new compressed sparse preconditioner requires only a linear order of N memory locations.

1 Introduction

The natural preconditioner M defined by local potentials [7] can significantly improve the convergence and performance of the CHARMM version of TNPACK (the truncated Newton program package [6]) for minimizing the potential energy function of a biomolecular system with N atoms [3, 8–10]. Here CHARMM is one widely-used molecular mechanics and dynamics program [1]. While the truncated Newton (TN) method is a second derivative minimization method [2], it can be simply modified into a first derivative method by using a finite difference approximation to the product of the Hessian matrix (i.e., the second derivative of the potential function) with a vector. Thus, the total memory requirement of TNPACK can be reduced to a linear order of N if the preconditioner M can be formulated in a compress format by a linear order of N memory locations. If so, the size of a biomolecular system that TNPACK can be applied to will be sharply increased. However, since the sparse pattern of M is extremely irregular, which varies with molecular systems, it is difficult to compress M without employing a full $3N \times 3N$ matrix. How to formulate a compressed M by a linear order of N memory locations has been an open problem for a long time.

This paper solves this problem and extends the definition of the preconditioner using the spherical cutoff approach to improve the efficiency of the preconditioner for a biomolecular system containing a large number of water molecules. Note that a sparse Hessian matrix has been programmed in CHARMM in a

---

* This work was supported by the National Science Foundation through grant DMS-0241236, and, in part, by the Graduate School Research Committee Award (343267-101-4) of the University of Wisconsin-Milwaukee.

special compress format (which will be referred to as the CHARMM compress format for clarity) [4]. Hence, a new scheme is first developed to generate the sparse pattern of the preconditioner from the primary structure of a biomolecular system, along with a given spherical cutoff condition, and express it in the CHARMM compress format. With such a sparse pattern, the preconditioner is then defined as a compressed sparse Hessian matrix so that the program routines for evaluating the preconditioner can be easily created by adapting the current CHARMM program routines. Finally, the compressed preconditioner is converted from the CHARMM compress format to the YSMP (Yale Sparse Matrix Package [5]) compress format, resulting in the compressed sparse preconditioner suitable for TNPACK. It is shown that the new compressed preconditioner can be formulated and evaluated in a linear order of N memory locations.

The remainder of this paper is organized as follows. Section 2 defines the new natural preconditioner. Section 3 expresses the compressed preconditioner in both CHARMM and YSMP formats. Section 4 describes a scheme for formulating the sparse pattern of the preconditioner directly from the primary structure of a molecular system. Section 5 presents a scheme for converting the CHARMM compress format to the YSMP compress format. Numerical results will be reported in the subsequent paper.

2 The New Definition of the Natural Preconditioner

Let $E(X)$ be a biomolecular potential energy function with $X = (X_1, X_2, \ldots, X_N)$, where $X_i = (x_i, y_i, z_i)$ denotes the position of atom i for $i = 1, 2, \ldots, N$. An important task in biomolecular simulations is to find a minimum point of $E(X)$ over the $3N$ dimensional space, which often leads to a feasible conformation of the molecular system. In general, E includes the local potential E_{local}, the van der Waals potential energy, and the Coulomb potential energy. E_{local} is defined as the sum of the bond length, bond angle, and torsional potential terms. See [1] for a detail description of the biomolecular potential energy function.

With a given E_{local} and a given nonnegative value of η, the sparse pattern set P is introduced as below:

$$P = \{(i,j) | \frac{\partial^2 E_{local}(X)}{\partial X_i \partial X_j} \neq 0 \text{ or } \|X_i - X_j\|_2 \leq \eta \text{ for } i < j, i, j = 1, 2, \ldots, N\}, \tag{1}$$

where $\|X_i - X_j\|_2 = \sqrt{(x_i - x_j)^2 + (y_i - y_j)^2 + (z_i - z_j)^2}$.

The natural preconditioner is a sparse $N \times N$ symmetric block matrix M defined by $M = (M_{ij})_{i,j=1}^{N}$ with block entry M_{ij} as given below:

$$M_{ij} = \begin{cases} \frac{\partial^2 E(X)}{\partial X_i \partial X_j} & \text{if } (i,j) \in P \\ \mathbf{0} & \text{otherwise} \end{cases} \text{ for } i \leq j, \text{ and } M_{ij} = M_{ji} \text{ for } i > j, \tag{2}$$

where $i, j = 1, 2, \ldots, N$, $\mathbf{0}$ denotes a 3×3 matrix of zero, and

$$\frac{\partial^2 E(X)}{\partial X_i \partial X_j} = \begin{bmatrix} \frac{\partial^2 E}{\partial x_i \partial x_j} & \frac{\partial^2 E}{\partial x_i \partial y_j} & \frac{\partial^2 E}{\partial x_i \partial z_j} \\ \frac{\partial^2 E}{\partial y_i \partial x_j} & \frac{\partial^2 E}{\partial y_i \partial y_j} & \frac{\partial^2 E}{\partial y_i \partial z_j} \\ \frac{\partial^2 E}{\partial z_i \partial x_j} & \frac{\partial^2 E}{\partial z_i \partial y_j} & \frac{\partial^2 E}{\partial z_i \partial z_j} \end{bmatrix}. \quad (3)$$

In [7], M was defined as the Hessian matrix of E_{local}. Clearly, replacing E_{local} by E and adding the cutoff condition $\|X_i - X_j\|_2 \leq \eta$ in the definition of M in (1) and (2) can significantly improve the approximation of the preconditioner to the Hessian matrix of E. Hence, the new preconditioner is expected to be more effective than the traditional one in [7] for TNPACK.

The preconditioner M is very sparse. It has only a linear order of N nonzero entries. In fact, each local potential term of E_{local} involves at most three bond connections, and the value of η is small (a default value of η in TNPACK is 4 Å). Hence, the total number of nonzero block entries on each row of M is a small fixed number. Thus, M has only a linear order of N nonzero entries.

3 The Preconditioner in Two Compress Formats

In this section, the preconditioner is compressed in the CHARMM and YSMP formats, respectively. For clarity, the ith row of the upper triangle part of M is assumed to have $n_i + 1$ nonzero block entries M_{ij} with column indices $j = i, \nu_{i1}, \nu_{i2}, \ldots, \nu_{i,n_i}$ satisfying

$$i < \nu_{i1} < \nu_{i2} < \ldots < \nu_{in_i} \text{ and } n_i \geq 0 \text{ for } i = 1, 2, \ldots, N. \quad (4)$$

Thus, the sparse pattern set P can be rewritten as

$$P = \{(i,j) | j = i, \nu_{i1}, \nu_{i2}, \ldots, \nu_{i,n_i} \text{ satisfying (4) for } i = 1, 2, \ldots, N\}. \quad (5)$$

The block entry M_{ij} is also denoted by

$$M_{ij} = \begin{bmatrix} m_{\mu,\nu} & m_{\mu,\nu+1} & m_{\mu,\nu+2} \\ m_{\mu+1,\nu} & m_{\mu+1,\nu+1} & m_{\mu+1,\nu+2} \\ m_{\mu+2,\nu} & m_{\mu+2,\nu+1} & m_{\mu+2,\nu+2} \end{bmatrix},$$

where $\mu = 3i - 2$ and $\nu = 3j - 2$ for $i, j = 1, 2, \ldots, N$. It is only needed to store the upper triangle part of M since M is symmetric.

In the CHARMM compress format, a sparse matrix is stored into two integer arrays $\mathcal{C}$ and $\mathcal{P}$ and one real array $\mathcal{M}$, where $\mathcal{M}$ holds the nonzero entries of M, $\mathcal{C}$ the column indices of the nonzero block entries in $\mathcal{M}$, and $\mathcal{P}$ the position of the column index ν_{i1} in $\mathcal{C}$ for $i = 1, 2, \ldots, N$. For the preconditioner M with the sparse pattern set P given in (5), the three arrays $\mathcal{M}$, $\mathcal{C}$ and $\mathcal{P}$ can be found as below:

$$\mathcal{C} = (\nu_{11}, \nu_{12}, \ldots, \nu_{1,n_1}, \nu_{21}, \nu_{22}, \ldots, \nu_{2,n_2}, \ldots, \nu_{N1}, \nu_{N2}, \ldots, \nu_{N,n_N}), \quad (6)$$

$\mathcal{P} = (\mathcal{P}(1), \mathcal{P}(2), \ldots, \mathcal{P}(N))$ with the ith component $\mathcal{P}(i)$ being defined by

$$\mathcal{P}(1) = 1 \quad \text{and} \quad \mathcal{P}(i) = 1 + \sum_{k=1}^{i-1} n_k \quad \text{for } i = 2, 3, \ldots, N, \tag{7}$$

and $\mathcal{M} = (\mathcal{R}(1), \mathcal{R}(2), \ldots, \mathcal{R}(N))$ with $\mathcal{R}(i)$ holding the $n_i + 1$ nonzero block entries of row i in the form

$$\mathcal{R}(i) = (\mathcal{D}(i), \mathcal{U}(i, \nu_{i1}), \mathcal{U}(i, \nu_{i2}), \ldots, \mathcal{U}(i, \nu_{in_i})).$$

Here $\mathcal{D}(i)$ holds the upper triangle part of M_{ii}, and $\mathcal{U}(i,j)$ the 9 entries of M_{ij} with $i < j$ in a row by row pattern.

In terms of $\mathcal{C}$ and $\mathcal{P}$, the column index ν_{ij} and number n_i are expressed as

$$\nu_{i,j} = \mathcal{C}(\mathcal{P}(i) + j - 1) \quad \text{and} \quad n_i = \mathcal{P}(i+1) - \mathcal{P}(i), \tag{8}$$

where $j = 1, 2, \ldots, n_i$, and $i = 1, 2, \ldots, N$.

In the YSMP compress format, a sparse matrix is stored into two integer arrays IA and JA and one real array A, where A holds the nonzero entries of the matrix in a row-wise fashion, JA the column indices of the entries of A, and IA is a pointer vector in which the ith entry $IA(i)$ points to the location of the first nonzero entry of the i-th row of the matrix in vector A. If $A(k) = m_{ij}$, where m_{ij} is the first nonzero entry of the i-th row, then $IA(i)$ and $JA(k)$ are defined by $IA(i) = k$ and $JA(k) = j$. Clearly, the difference $IA(i+1) - IA(i)$ indicates the total number of nonzero entries in the ith row of M, and the nonzero entries of the ith row of M can be expressed by

$$A(IA(i)), A(IA(i)+1), \ldots, A(IA(i+1)-1),$$

while their column indices by

$$JA(IA(i)), JA(IA(i)+1), \ldots, JA(IA(i+1)-1),$$

where $i = 1, 2, \ldots, n$, and $IA(n+1) = IA(n) + 1$.

For the natural preconditioner M, there are $n_i + 1$ nonzero block entries in the ith row of the upper triangle part of M, and each block entry is a 3×3 matrix. Hence, for each value of i for $i = 1, 2, \ldots, N$, the corresponding three rows of M have labelling indices $3i-2, 3i-1$ and $3i$, and contain $3n_i + 3, 3n_i + 2$ and $3n_i + 1$ nonzero entries, respectively. Thus, the compressed M by the YSMP compress format can be easily obtained.

4 Formulation of the Sparse Pattern Arrays $\mathcal{C}$ and $\mathcal{P}$

The primary structure of a molecular system is described as a list of pairs (ζ_k, η_k) for $k = 1, 2, \ldots, N_b$ in CHARMM, where ζ_k and η_k denote the labelling indices of the two end atoms of bond k, and N_b the total number of bonds. For a given bond list, the sparse pattern arrays $\mathcal{C}$ and $\mathcal{P}$ can be generated by using the following scheme:

It is clear that comparing the two atomic indices of each bond with the index of a given atom can identify all the atoms that share one bond with the given atom. Suppose that for $i = 1, 2, \ldots, N$, there exist l_i such atoms with labelling indices $\mu_{i,1}, \mu_{i,2}, \ldots, \mu_{i,l_i}$. Then, these indices can be held by vector $\mathcal{J}$ as below:

$$\mathcal{J}(\mathcal{I}(i) + k - 1) = \mu_{i,k} \quad \text{for} \quad k = 1, 2, \ldots, l_i, \text{ and } i = 1, 2, \ldots, N, \quad (9)$$

where $\mathcal{I}(1) = 1$, and $\mathcal{I}(i) = \sum_{k=1}^{i-1} l_k$ for $i = 2, 3, \ldots, N$. Since $\mathcal{I}(i)$ points to the position of $\mu_{i,1}$ in vector $\mathcal{J}$, l_i can be expressed by

$$l_i = \mathcal{I}(i) - \mathcal{I}(i-1) \quad \text{for } i = 1, 2, \ldots, N.$$

By using $\mathcal{J}$ and $\mathcal{I}$, the sparse pattern arrays $\mathcal{C}$ and $\mathcal{P}$ are formulated in four steps for each given atom, say atom i, where $1 \leq i \leq N$. In *Step 1*, find l_i atoms with indices $\mu_{i,1}, \mu_{i,2}, \ldots, \mu_{i,l_i}$ that share one bond with atom i from $\mathcal{J}$ and $\mathcal{I}$. If there exists $\mu_{i,j} > i$ with $1 \leq j \leq l_i$, then $\mu_{i,j}$ is selected as a new entry of $\mathcal{C}$. In *Step 2*, for each selected atom from Step 1, say atom k, find l_k atoms with indices $\mu_{k,1}, \mu_{k,2}, \ldots, \mu_{k,l_k}$ that share one bond with atom k from $\mathcal{J}$ and $\mathcal{I}$. If there exists $\mu_{k,j} > i$ with $1 \leq j \leq l_k$, then $\mu_{k,j}$ is selected as a new entry of $\mathcal{C}$. In *Step 3*, for each selected atom in Step 2, say atom t, similarly find the atoms that share one bond with atom t and have indices more than i from $\mathcal{J}$ and $\mathcal{I}$. Clearly, such selected atoms connect to atom i through three bonds. Hence, their indices become new entries of $\mathcal{C}$. Finally, in *Step 4*, the indices of atoms satisfying the cutoff condition are selected as new entries of $\mathcal{C}$ if $\eta \neq 0$. To ensure that the indices $\{\nu_{ij}\}$ satisfy (4), in each step of selecting the ith entry of $\mathcal{C}$, the repeated indices between $\mathcal{P}(i)$ and $\mathcal{P}(i+1) - 1$ are excluded.

The above scheme can be implemented in a linear order of N memory locations. In fact, both $\mathcal{I}$ and $\mathcal{P}$ are vectors of size $N + 1$, and the sizes of $\mathcal{J}$ and $\mathcal{C}$ can be found to be $\sum_{i=1}^{N} l_i$ and $\sum_{i=1}^{N} n_i$, respectively. Set $C_{max} = \max\{\max_{1 \leq i \leq N} n_i, \max_{1 \leq i \leq N} l_i\}$. Clearly, C_{max} is a constant much less than N. A default value of C_{max} is set as 180 in TNPACK. Hence, the sizes of $\mathcal{J}$ and $\mathcal{C}$ are estimated as $C_{max} N$. Therefore, the total number of integers required to formulate the sparse pattern arrays $\mathcal{C}$ and $\mathcal{P}$ are estimated as

$$2(N + 1 + C_{max} N) \approx 2(C_{max} + 1)N = O(N).$$

5 Switch the CHARMM Format to the YSMP Format

Since TNPACK uses the YSMP compress format to store the preconditioner M, a scheme is needed to switch the CHARMM format $(\mathcal{M}, \mathcal{C}, \mathcal{P})$ to the YSMP format (A, IA, JA). It is easy to obtain IA and JA from $\mathcal{C}$ and $\mathcal{P}$ since the involved numbers $\{\mu_{i,j}\}$ and $\{n_i\}$ can be produced directly from formula (8) when the sparse pattern arrays $\mathcal{C}$ and $\mathcal{P}$ are available.

Clearly, the sparse pattern of M remains the same in the whole minimization process because it is defined according to the bond list. Hence, the formulation of the sparse pattern (IA, JA) or $(\mathcal{C}, \mathcal{P})$ is carried out only at the initial step of TN iterations. After the pattern (IA, JA) is produced, the memory locations of the four integer arrays $\mathcal{I}, \mathcal{J}, \mathcal{C}$, and $\mathcal{P}$ can be released immediately.

To switch $\mathcal{M}$ to A, two additional pointer arrays, ξ and ζ, are introduced as below. Set their ith components $\xi(i)$ and $\zeta(i)$ to store the positions of $m_{\mu\mu}$ and $m_{\mu\nu}$ in the vector $\mathcal{M}$, respectively, where $\mu = 3i - 2$ and $\nu = 3\nu_{i,\tau} - 2$ for $\tau = 1, 2, \ldots, n_i$, and $i = 1, 2, \ldots, N$. $\xi(i)$ and $\zeta(i)$ can be found as below:

$$\xi(1) = 1, \quad \text{and} \quad \xi(i) = 9n_{i-1} + \xi(i-1) + 6 \quad \text{for} \quad i = 2, 3, \ldots, N, \qquad (10)$$

$$\zeta(i) = \xi(i) + 9(\tau - 1) + 6 \quad \text{for} \quad \tau = 1, 2, \ldots, n_i, i = 1, 2, \ldots, N. \qquad (11)$$

In terms of ξ and ζ, the nonzero entry array A can be produced directly from $\mathcal{M}$ by the following expressions:

$$A(IA(\mu) + k) = \mathcal{M}(\xi(i) + k), \quad A(IA(\mu + 1)) = \mathcal{M}(\xi(i) + 3),$$

$$A(IA(\mu + 1) + 1) = \mathcal{M}(\xi(i) + 4), \quad A(IA(\mu + 2)) = \mathcal{M}(\xi(i) + 5),$$

$$A(IA(\mu) + 3j + k + 1) = \mathcal{M}(\zeta(i) + k),$$

$$A(IA(\mu + 1) + 3j + k) = \mathcal{M}(\zeta(i) + k + 3),$$

$$A(IA(\mu + 2) + 3j + k - 1) = \mathcal{M}(\zeta(i) + k + 6),$$

where $k = 0, 1, 2$, $\mu = 3i - 2$, $j = 1, 2, \ldots, n_i$, and $i = 1, 2, \ldots, N$.

Both $\mathcal{M}$ and A have the same size, which can be found by $6N + 9\sum_{i=1}^{N} n_i$, and estimated by

$$(9 \max_{1 \leq i \leq N} n_i + 6)N \leq 9C_{max}N = O(N).$$

Hence, the formulation of A from $\mathcal{M}$ requires only a linear order of N memory locations.

References

1. Brooks, B. R., Bruccoleri, R. E., Olafson, B. D., States, D. J., Swaminathan, S. Karplus, M.: CHARMM: A program for macromolecular energy, minimization, and dynamics calculations. J. Comp. Chem., **4** (1983) 187-217.
2. Dembo, R. S., Steihaug, T.: Truncated-Newton algorithms for large-scale unconstrained optimization. Math. Prog., **26** (1983) 190-212.
3. Derreumaux, P., Zhang, G., Brooks, B., Schlick, T.: A truncated-Newton method adapted for CHARMM and biomolecular applications. J. Comp. Chem., **15** (1994) 532-552.
4. Perahia, D., Mouawad, L.: Computation of low-frequency normal modes in macromolecules: improvements to the method of diagonalization in a mixed basis and application to hemoglobin. Computers & Chemistry, **19** (1995) 241-245.

5. Schultz, M. H., Eisenstat, S. C., Sherman, A. H.: Algorithms and data structures for sparse symmetric Gaussian elimination. SIAM J. Sci. Statist. Comput., **2** (1981) 225-237.
6. Schlick, T., Fogelson, A.: TNPACK — A truncated Newton minimization package for large-scale problems: I. Algorithm and usage. ACM Trans. Math. Softw., **14** (1992) 46-70.
7. Schlick, T., Overton, M. L.: A powerful truncated Newton method for potential energy functions. J. Comp. Chem., **8** (1987) 1025-1039.
8. Xie, D., Schlick, T.: Efficient implementation of the truncated-Newton algorithm for large-scale chemistry applications. SIAM J. OPT., **10** (1999) 132-154.
9. Xie, D., Schlick, T.: Remark on Algorithm 702—The updated truncated Newton minimization package. ACM Trans. on Math. Software, **25** (1999) 108-122.
10. Xie, D., Schlick, T.: A more lenient stopping rule for line search algorithms. Optimization Methods and Software, **17** (2002) 683-700.

A Study on Lower Bound of Direct Proportional Length-Based DNA Computing for Shortest Path Problem

Zuwairie Ibrahim[1], Yusei Tsuboi[1], Osamu Ono[1], and Marzuki Khalid[2]

[1] Institute of Applied DNA Computing, Meiji University, 1-1-1 Higashi-mita, Tama-ku, Kawasaki-shi, Kanagawa-ken, Japan 214-8571
{zuwairie, ono, tsuboi}@isc.meiji.ac.jp
[2] Center for Artificial Intelligence and Robotics (CAIRO), Universiti Teknologi Malaysia, Jalan Semarak, 54100, Kuala Lumpur, Malaysia
marzuki@utmkl.utm.my

Abstract. Previously, we proposed a direct proportional length-based DNA computing approach for weighted graph problem. The approach has been proposed essentially to overcome the shortcoming of constant proportional length-based DNA computing approach. However, by using this approach, the minimum weight of edges that can be encoded is limited. Hence, in this paper, the lower bound, in term of minimum weight that can be encoded by direct proportional length-based DNA computing is analyzed. Also, the parameters contribute to the lower bound are investigated in order to identify the relation between those parameters and the lower bound of the direct proportional length-based DNA computing approach.

1 Introduction

A constant proportional length-based DNA computing specifically proposed for solving Traveling Salesman Problem (TSP) has been proposed by Narayanan and Zorbalas [4]. A constant increase of DNA strands is encoded according to the actual length of the distance. A drawback of this method is that, there is a possibility of an occurrence of concatenated DNA strands of two distances which could be longer than the DNA strand of the longest distance that has been encoded. This may lead to errors in computing the shortest path [3]. This scheme, however, has not been realized by any laboratory experiment.

Due to drawbacks in implementation, the constant proportional length-based DNA computing has not yet been implemented in any laboratory experiment. Thus, with the aim to solve the limitation of the constant-proportional length-based approach, by improving the previously proposed encoding style in [3], a direct proportional length-based DNA computing approach has been proposed [2]. The computation can be carried out *in vitro* by using the same computing architecture as Adleman [1]. In this approach, the cost of an edge is encoded as a direct-proportional length oligonucleotides, or oligos for short. As a result, during the computation, the important information is the length of the DNA duplex. Since this will result in numerous numbers of combinations, by using the standard bio-molecular laboratory operations, it is possible to extract the optimal combination which represents a solution to the problem.

However, we realized that there arises a drawback of direct proportional length-based DNA computing. The minimum weight of edges can be handled is limited. In other words, there is a lower bound in term of the weight of edges can be encoded by oligos. At first, the direct proportional length-based approach will be presented. At the end of this paper, the relation between those parameters and the lower bound, is studied. It is expected that this study will give some impacts and considerations when the laboratory experiment will be carried out soon, in future.

2 Direct Proportional Length-Based DNA Computing

As the shortest path problem is chosen as a benchmark, the input to this problem is a weighted, directed graph $G = (V, E, \omega)$, a start node u and an end node v. The output of the shortest path problem is a (u, v) path with the smallest cost. In the case given in Figure 1, if u is V_1 and v is V_5, the cost for the shortest path will be given as 27 and the optimal path is clearly shown as $V_1 - V_2 - V_5$.

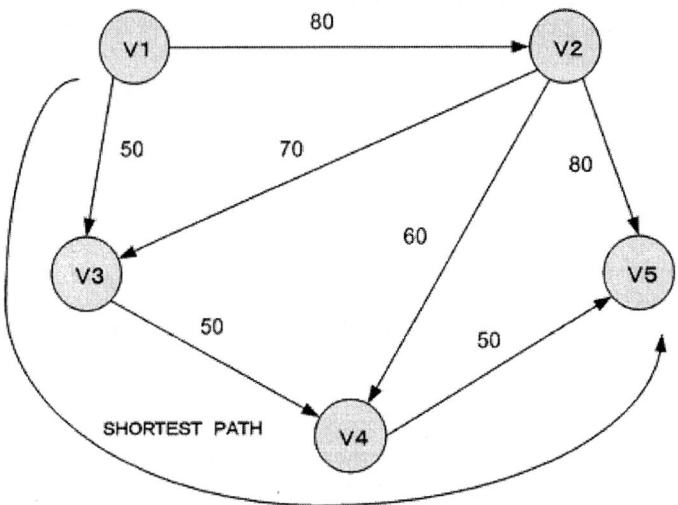

Fig. 1. Example showing a weighted directed graph $G = (V, E)$ with the shortest path shown as $V_1 - V_3 - V_4 - V_5$

If the directed graph shown in Figure 3 is considered. Let n be the total number of nodes in the graph. The DNA sequences correspond to all nodes and its complements are designed. Let V_i (i= 1, 2, ... , n) and $\overline{V_i}$ (i= 1, 2, ... , n) be the β-mer (β is an even number) DNA sequences correspond to the ith node in the graph and its complement respectively. If $\beta = 20$, by using the available software for DNA sequence design, DNASequenceGenerator [5], the DNA sequences V_i is designed and listed in Table 1.

The complement, the GC contents (GC%), and melting temperature (Tm) of each sequence are shown as well. In Table 1, V_i can be separated into V_{ia} and V_{ib} where V_{ia} is defined as half-5-end and V_{ib} is defined as half-3-end of V_i.

Table 1. DNA sequences for nodes. The melting temperature, Tm are calculated based on Sugimoto nearest neighbor thermodynamic parameter (Sugimoto et al., 1996)

Node (Vi)	20-mer sequences (5'-3')		20-mer complement sequences (5'-3')	GC%	Tm (°C)
	V_{ia}	V_{ib}			
V_1	TATGCTCATT	TGCATTTTGA	TCAAAATGCA AATGAGCATA	30	46.13
V_2	CAGGAGCGTC	TGAGAGCGAG	CTCGCTCTCA GACGCTCCTG	65	58.58
V_3	TACCGGATCG	ACCGGCTAAG	CTTAGCCGGT CGATCCGGTA	60	55.12
V_4	TCGTCCAAGG	GAGGCTTCTC	GAGAAGCCTC CCTTGGACGA	60	54.46
V_5	GCTATATTGC	GCTGGATGTG	CACATCCAGC GCAATATAGC	50	52.16

We introduce three rules to synthesize oligos for each edge in the graph as follows:

(i) If there is a connection between V_1 to V_j, synthesize the oligo for edge as

$$V_1(\beta) + W_{1j}(\omega\ \alpha - 3\beta/2) + V_{ja}(\beta/2) \quad (1)$$

(ii) If there is a connection between V_i to V_j, where $i \neq 1, j \neq n$, synthesize the oligo for edge as

$$V_{ib}(\beta/2) + W_{ij}(\omega\ \alpha - \beta) + V_{ja}(\beta/2) \quad (2)$$

(iii) If there is a connection between V_i to V_n, synthesize the oligo for edge as

$$V_{ib}(\beta/2) + W_{in}(\omega\ \alpha - 3\beta/2) + V_n(\beta) \quad (3)$$

The synthesized oligos consist of three segments. The number of DNA bases for each segment is shown in parenthesis, α is direct proportional factor, β is the length of the node sequences, and '+' represents the join. W_{ij} denotes the DNA sequences representing a cost between node V_i and V_j. As such, if $\alpha = 1$, for an edge from V_i to V_j, due to the synthesized rules, the DNA sequences for distance or costs, W_{ij} are designed using the DNASequenceGenerator [5] and the results are listed as shown in Table 2. Table 3 on the other hand lists all the synthesized oligos based on the proposed synthesis rules. The complement oligos of each node and cost are synthesized as well.

Table 2. DNA sequences for costs. Melting temperature, Tm are calculated based on Sugimoto nearest neighbor thermodynamic parameter (Sugimoto et al., 1996)

Cost, W_{ij}	DNA sequences (5'-3')	Length	GC%	Tm (°C)
W_{13}	GAAGTGTACGTTAGGCTGCT	20	50	51.59
W_{45}	AAAGGTCGTCTTTGAACGAG	20	45	50.34
W_{34}	AAAGGCCCTCTTTTAACGAA GTCCTGTACT	30	43	60.45
W_{24}	AAAGCCCGTCGGTTAAGCAA GTAGTTTACGCTGCGTCATT	40	48	69.47
W_{25}	GCGTTGTTGCGAGGCATGTGGAGAA TTGATCGCTTTCGTGCATAACTGGG	50	52	74.08
W_{12}	CAGCATCGTAGTAGAGCTAGTATCG AACTG ATAAGTAACG GAGGGGGCTC	50	50	72.19
W_{23}	AAAGCTCGTC GTTTAAGGAAGTAC GGTACTATGCGTGATTTGGAGGTGGA	50	46	70.91

Table 3. DNA sequences for edges

Edge	DNA sequences (5'-3')
$V_1 - W_{13} - V_{3a}$	TATGCTCATTTGCATTTTGAGAAGTGTAC GTTAGGCTGCTTACCGGATCG
$V_{4b} - W_{45} - V_5$	GAGGCTTCTCAAAGGTCGTCTTTGAACGA GGCTATATTGC GCTGGATGTG
$V_{3b} - W_{34} - V_{4a}$	ACCGGCTAAGAAAGGCCCTCTTTTAACGA AGTCCTGTACT TCGTCCAAGG
$V_{2b} - W_{24} - V_{4a}$	TGAGAGCGAGAAAGCCCGTCGGTTAAGCAA GTAGTTTACG CTGCGTCATT TCGTCCAAGG
$V_{2b} - W_{23} - V_{3a}$	TGAGAGCGAGAAAGCTCGTCGTTTAAGGAA GTACGGTACTATGCGTGATTTGGAGGTGGA TACCGGATCG
$V_{2b} - W_{25} - V_5$	TGAGAGCGAGGCGTTGTTGCGAGGCATGT GGAGAATTGATCGCTTTCGTGCATAACTGG GGCTATATTGCGCTGGATGTG
$V_1 - W_{12} - V_{2a}$	TATGCTCATTTGCATTTTGACAGCATCGTA GTAGAGCTAGTATCGAACTGATAAGTAACG GAGGGGGCTCCAGGAGCGTC

Then, all the synthesized oligos are poured into a test tube for initial pool generation. The generation of an initial pool solution is based on the hybridization/ligation method. In fact, the hybridization/ligation method for initial pool generation has been introduced firstly by Adleman (Adleman, 1994) in order to solve HPP. If the shortest path V_1 - V_3 - V_4 - V_5 is emphasized, Figure 2 clearly shows which kinds of oligos are important for the generation of this path. However, the unwanted combinations are also generated in the same manner.

At this stage, an initial pool of solution has been produced and it is time to filter out the optimal combinations among the vast alternative combinations of the problem. Unlike conventional filtering, this process is not merely throwing away the unwanted DNA duplex but rather copying the target DNA duplex exponentially by using the incredibly sensitive PCR process. This can be done by amplifying the DNA molecules that contain the start node V_1 and end node V_5 using primers. After the PCR operation is accomplished, there should be numerous number of DNA strands representing the start node V_1 and end node V_5 traveling through a possible number of nodes.

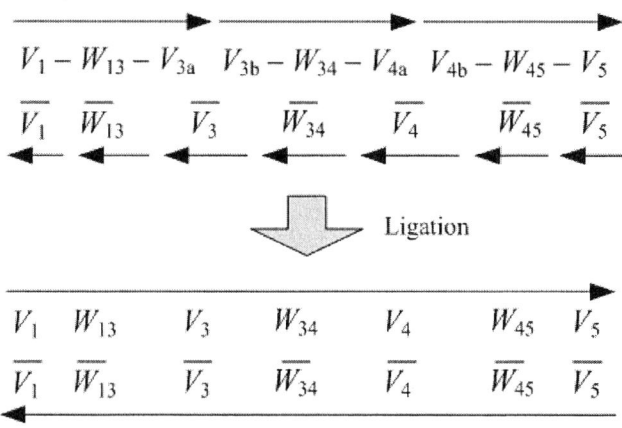

Fig. 2. DNA duplex for path V_1 - V_3 - V_4 - V_5. The arrowhead indicates the 3' end

The output solution of PCR operation is brought for gel electrophoresis operation. During this operation, the DNA molecules will be separated in term of its length and hence, the DNA molecules $V_1 - V_2 - V_5$ representing the shortest path starting from V_1 and end at V_5 would be extracted for sequencing.

3 Lower Bound Analysis

Basically, according to the DNA synthesis rules, the lower bound is achieved when:

$$\omega\alpha - \frac{3}{2}\beta = 0 \qquad (4)$$

At this time, the minimum weight, which can be encoded by oligos, ω_{min} is attained as:

$$\omega_{min} = \frac{3}{2}\frac{\beta}{\alpha} \qquad (5)$$

4 Conclusions

This paper is written to analyze a lower bound of direct proportional length-based DNA computing for graph problems. Importantly, the definition of lower bound is given. From the analysis, the lower bound is basically depends on the length of node sequences, α and the direct proportional factor, β. The main conclusion of this study is that lower bound can be kept as low as possible by controlling the length of the node sequences, β and the direct proportional factor, α. It is now clear that in order to keep the lower bound effect as low as possible, one should decrease the parameter β, whereas the parameter α, should be increased at the same time. However, from the experimental view, one can not simply decrease the parameter β and increase the parameter α because such parameters will greatly affect the performance of the computation. Therefore, the best value for the length of the node sequences and the direct proportional factor will be further investigated during experimental setup for an accurate direct proportional length-based DNA computing for weighted graph problems.

References

1. Adleman, L.: Molecular Computation of Solutions to Combinatorial Problems. Science, Vol. 266 (1994) 1021-1024
2. Ibrahim, Z., Ono, O., Tsuboi, Y., Khalid, M.: Length-Based DNA Computing for 1-Pair Shortest Path Problem. Proc. of the Ninth International Symposium on Artificial Life and Robotics (2004) 299-302
3. Lee, J.Y., Shin, S.Y., Augh, S.J., Park, T.H., Zhang, B.T.: Temperature Gradient-Based DNA computing for Graph Problems with Weighted Edges. Lecture Notes in Computer Science, Vol. 2568 (2003) 73-84
4. Narayanan, A., Zorbalas, S: DNA Algorithms for Computing Shortest Paths. Proceedings of Genetic Programming (1998) 718-723
5. Udo, F., Sam, S., Wolfgang, B., Hilmar, R.: DNA Sequence Generator: A Program for the Construction of DNA Sequences. In N. Jonoska and N. C. Seeman (editors), Proceedings of the Seventh International Workshop on DNA Based Computers (2001) 23-32

Key Management for Secure Multicast Using the RingNet Hierarchy

Guojun Wang[1,2], Lin Liao[2], Jiannong Cao[1], and Keith C. C. Chan[1]

[1] Department of Computing, Hong Kong Polytechnic University,
Hung Hom, Kowloon, Hong Kong
[2] School of Information Science and Engineering, Central South University,
Changsha, Hunan Province, P. R. China 410083

Abstract. We propose a novel multicast communications model using a RingNet hierarchy, called the RingNet model, which is a combination of logical trees and logical rings for multicast communications. The RingNet hierarchy consists of four tiers: Border Router Tier (BRT), Access Gateway Tier (AGT), Access Proxy Tier (APT), and Mobile Host Tier (MHT). Within the hierarchy, the upper two tiers are dynamically organized into logical rings with network entities. In this paper, based on the RingNet model, local group concept is proposed. For simple illustration, we choose each AG in AGT as the controller of each local group. Each local group has its own independent local group key. The member's join or leave in a local group only affects the local group, which makes multicast communications potentially scalable to very large groups. In this paper, we propose a novel key management scheme for secure multicast with the RingNet model.

1 Introduction

Multicast is an efficient service that provides delivery of data from a source to a group of receivers. It reduces transmission overhead and network bandwidth. Recently with the convergence of Internet computing and wireless communications, research on multicast communications in mobile Internet becomes more and more active and challenging. In mobile Internet, more concerns should be considered. Firstly mobile hosts such as laptop computers, PDAs, and mobile phones have severe resource constraints in terms of energy and processing capabilities. Secondly, wireless communications involves high error rate and limited bandwidth problems. Thirdly, the number of mobile hosts may be very huge and they may move very frequently. Thus in mobile Internet the security issues of multicast are more complicated than in traditional Internet.

The remainder of this paper is organized as follows. In section 2, we introduce some related works about key management for secure multicast. In section 3, we introduce our RingNet communications model, which is an efficient multicast communications model with the notion of a combination of tree and logical ring structures. In section 4, we describe the key management for secure communications using the RingNet model, and show how to implement secure multicast in RingNet and how it works when compared with others. The final section concludes this paper.

2 Related Works

In a group communications system, the basic requirement is that newly-joined users are not allowed to access the former group communications and thus the group key should be rekeyed when new members join; and departed users are not allowed to access the current and future group communications after the group key is rekeyed.

Conceptually, since every point-to-multipoint communications can be represented as a set of point-to-point communications, the current unicast technology can be extended in a straightforward way to do secure group communications. However, such an extension is not scalable to very large groups. As we previously mentioned, when a user joins a group, a new group session key (K_{GSK}) would be generated to replace the former one, which will be distributed to all current and joining users. A simple way is to multicast new K_{GSK} to current users encrypted by the old K_{GSK}. Accordingly the joining user will be informed of the new K_{GSK} by means of unicast. In case of user departure, since we should transmit the new K_{GSK} to all remaining users, we can use respective key materials to encrypt it individually. As a result, we have to take separate users into account for each leave and join. If extended to large dynamic group, the scalability problem is a potential bottleneck.

In order to resolve the scalability issue and improve efficiency in secure group communications, many solutions are proposed. The mainstream ideas are the *Iolus* system [1] and the *Key Graph* system [2]. Based on these schemes, different rekeying policies in terms of user departure, user join and traffic source rate are proposed, such as periodic rekeying [3], batch rekeying [4], and exposure-oriented rekeying [5].

Iolus is a novel framework for scalable secure multicast. A group is divided into many subgroups. Iolus uses independent keys for each subgroup and the absence of a general group key means membership change in a subgroup is treated locally. Thus scalability is achieved by having each subgroup to be relatively independent. Iolus uses entity-GSC (Group Security Controller) to manage the top-level subgroup, and entity-GSI (Group Security Intermediary) to act as proxy of GSC. However, it requires decryption and encryption of data packets while transmitted into different subgroups.

Except for Iolus, the other main idea is to organize users and server into a tree structure. In such a centralized system, there is only one central controller that doesn't rely on any auxiliary entity to perform secure access control. The typical scheme is the Key Graph system. Due to its verticality, the tree structure is easy to be destroyed in a single node especially for the group key. Furthermore, users' join and leave will cause rekeying and affect other users. By means of the tree structure, cost will be exaggerated when the number of users becomes huge.

3 The RingNet Communications Model

Researchers have proposed many mobile Internet architectures, such as Unified Wireless Networks Architecture [6], System Architecture for Mobile Communication System [7], All-IP Wireless/Mobile Network Architecture [8], and FIT-MIP Global System Architecture [9]. Based on them, we propose a multicast communications model called a RingNet hierarchy shown in Figure 1.

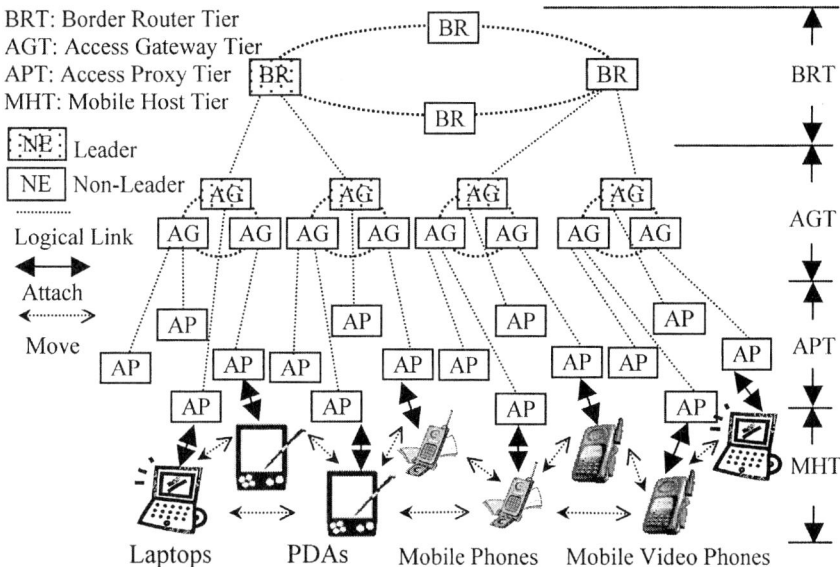

Fig. 1. The RingNet Hierarchy

The four tiers of the RingNet hierarchy are Border Router Tier (BRT), Access Gateway Tier (AGT), Access Proxy Tier (APT), and Mobile Host Tier (MHT). The higher two tiers are dynamically organized into logical rings. Each logical ring has a leader node, which is also responsible for interacting with upper tiers. Access Proxies (APs) are the Network Entities (NEs) that communicate directly with the Mobile Hosts (MHs). Access Gateways (AGs) are the NEs that communicate either between different wireless networks or between one wireless network and one wired network. Border Routers (BRs) are the NEs that communicate among administrative domains. Notice that only those NEs that are configured to run the proposed protocol will be involved in the hierarchy.

In order to form such a hierarchy, we require each AP, AG, and BR having some knowledge of its candidate contactors, either some candidate neighbor nodes through which it can join a logical ring, or some candidate parent nodes through which it can attach to an existing hierarchy. Multicast communications using the RingNet hierarchy is simple: Multicast Senders (MSs) send multicast messages to any of the BRs at the top logical ring. Then the multicast messages are transmitted along each logical ring, and downward to all the children nodes. Finally the MHs will receive multicast messages from their attached APs. Thus the multicast data are delivered to all users efficiently.

Ideally, there is only one RingNet hierarchy containing all the operational group members. The RingNet membership management works as follows. Each MH can join or leave a group at will, or fail at any time. The membership change message is firstly captured by the MH's attached AP node. Then it is propagated to the AP's parent AG node. If the AG happens to be a leader node, then it propagates such message to its parent node; if not, then it propagates it to the leader node in the logical ring where it resides. This process continues until the leader node in the top logical ring is reached.

Loosely speaking, the proposed RingNet hierarchy is a novel distribution vehicle that combines advantages of both logical tree and logical ring. Each logical ring is dynamically organized according to some criteria such as locality/proximity or Quality of Service (QoS). Due to the combination of ring and tree, fault tolerance is stronger than the tree structure that is fragile because of single node being attacked easily. What's more, due to the setting of logical rings, it's easy to ensure key packet transmitted along one vertical path even though some of the rings are destroyed. However, the latency incurred by logical ring is a drawback of RingNet. Therefore, in our model, we consider small logical rings. Based on these assumptions, we discuss some security issues.

4 Key Management Using the RingNet Model

In [10], we introduce the RingNet hierarchy for totally ordered multicast, which are proved to be scalable, reliable, and self-organizable in large dynamic group communications. Based on these works, we will discuss key management using the RingNet hierarchy.

In RingNet, the rekeying process is different from both the Iolus system and the Key Graph system. Due to the setting of logical rings, it ensures that non-leader AG-node receives key messages safely without being derived from leader AG-node through which the messages are forwarded. In order to solve the scalability problem, we borrow some idea from Iolus. We consider every AG node and its attaching AP/MH nodes as a local group. Then the multicast communications group is divided into many local groups and they are linked together by some rings. The local groups have their own local group keys (K_{LG}) and each AG serves as a trusted server proxy. Here we use "local group" other than "subgroup" to emphasize that, the local groups in our RingNet hierarchy are *dynamically* formed, while the subgroups in Iolus are *statically* configured.

In general, a global server in the top ring is used to control all the local groups, which are controlled by the AG nodes. Local groups obtain the rekeyed group session key K_{GSK} created by the global server. In each local group, if the members change within a local group, the local group changes its own K_{LG} and its members' individual keys. On the other hand, local group feedbacks its change to the global server. Each AP member only need to know its local group key and the server node has a record for each leader local group' K_{LG}.

In order to simplify the problem, we assume that initially the number of APs in every AG local group is L; the number of AGs under each BR is K; the number of nodes of each AG ring is R; the number of nodes in each BR ring is B. Each local group has a unique number to be distinguished from each other, with the label like LG_{ijk}, where $0 \le i \le B - 1, 0 \le j \le K - 1$, and $0 \le k \le R - 1$.

Multicast data sender firstly reaches the BRT and one of the BR nodes can be the server as shown in Figure 2. BR nodes will transmit the data packet to its next BR node and its children nodes almost at the same time, which will work in the same way as the server BR. Therefore, we put the other BR nodes aside because they are the same and act like duplicators of the BR server. In the following, we will describe the protocols in four situations.

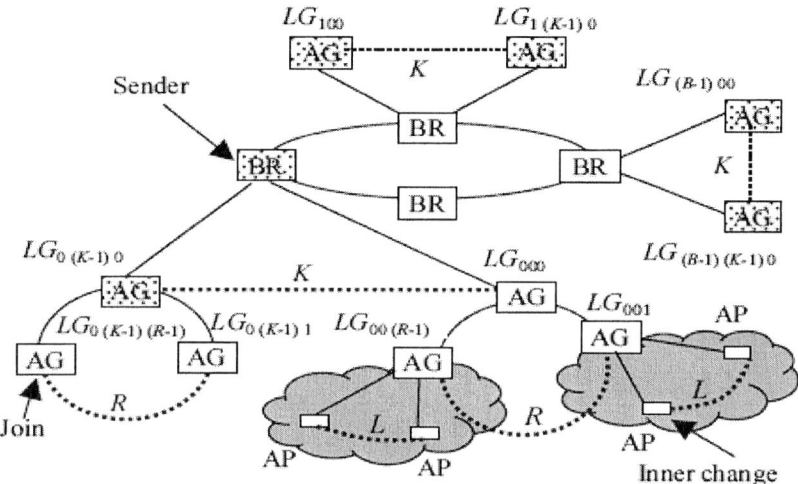

Fig. 2. Local Group Core Architecture in RingNet

4.1 An AG-Local Group Joins

When an AG-local group joins a multicast group, the APs and MHs involved in the AG-local group are not concerned in the rekeying directly. After the AG node and the server node have authenticated for each other, the server node creates a new K_{LG} of the local group. The server node renews the K_{GSK}, and multicasts it to all former AG nodes encrypted by old K_{GSK}. During this process, the multicast rekeying messages are firstly forwarded to the leader AG-nodes and then they forward the messages to the ring members as second level servers. The process is specifically expressed as follows assuming $LG_{0\,(k-1)(r-1)}$ wants to join.

1. $LG_{0\,(k-1)(r-1)}$ requests for joining a new group associated with a server node;
2. $LG_{0\,(k-1)(r-1)}$ and the server node will do authentication for each other;
3. The server node generates K_{LG} of $LG_{0\,(k-1)(r-1)}$ and renews K_{GSK} to K'_{GSK};
4. The server node sends key packet with $(K'_{GSK})_{K_{GSK}}$ to leader nodes of local groups from $LG_{0\,(k-1)\,0}$ to LG_{000};
5. Key packet with K'_{GSK} encrypted by K_{LG} of newly joining $LG_{0\,(k-1)(r-1)}$ is sent to $LG_{0\,(k-1)(r-1)}$ by the server node;
6. Key packet with $(K'_{GSK})_{K_{GSK}}$ is transmitted to the non-leader AG-nodes along the logical rings by each leader AG-node except for the newly joining node.

Notice that using K to encrypt K' is expressed as $(K')_K$. The key packet with $(K'_{GSK})_{K_{GSK}}$ firstly needs to be informed to the leader AG-nodes in step 4, and then the leader AG-nodes will transmit it in their AG-rings in step 6.

4.2 A Non-leader Local Group Leaves

A non-leader local group leaving means that all the attached MHs leave. After MHs leave, the link between the AG node and the adjacent nodes will detect that the AG is not active. Then the server sets the AG node as half-active. After a certain period of time, the server confirms that the AG is not connected. Then two neighbors of the unconnected AG establish a new link and the AG node disappears. After confirming the departure, the server node will re-create a new group session key K'_{GSK}. The following procedure will be a little different with a local group joining, for the departed AG node has known the old K_{GSK}. The new group session key K'_{GSK} will be multicast to all the remaining AGs encrypted by their respective K_{LG} grouped by the logical rings.

1. $LG_{0\,(k-1)(r-1)}$ requests for leaving a group associated with a server node;
2. The server node confirms the departure and makes the new ring work as well;
3. The server node renews K_{GSK} to K'_{GSK};
4. For each logical AG ring, the server node sends an individual key packet with K'_{GSK} encrypted by K_{LG}; then AG nodes in this ring can recover the K'_{GSK} using the key packet using its own K_{LG}.

4.3 A Local Group with Leader Node Leaves

Different from ordinary hierarchy, a local group leaving with leader node is quite different in RingNet, because leader node is the intermediary between two tiers. Once the leader node is requesting for leaving, the server node will establish a connection with an adjacent node, give it a leader certificate and update the membership information. Except for the above process, the remaining procedure is the same as a non-leader local group leaves.

4.4 Membership Changes Within a Local Group

If the membership changes within a local group, to some degree the local group hierarchy will hide such kind of changes, for these changes only affect within its own local group. As to the inner local group change, it firstly should be rekeyed like an independent group. We can use any existing rekeying approach such as tree key graph to rekey the local group and a new K_{LG} will be generated. Then the server node will receive the change message and update the membership information. Scalability is sufficed by means of local group hierarchy. The whole process can be represented as follows:

1. An MH of LG_{001} requests for leaving or another MH request for joining LG_{001};
2. LG_{001} is rekeyed and the leader AG-node of the group generates a new K_{LG}, i.e., K'_{LG};
3. Key packet with $(K'_{LG})_{K_{LG}}$ is propagated upward to the server node through leader node LG_{000}; but the leader AG–node doesn't obtain the packet because it doesn't know K_{LG} of LG_{001};
4. The server node gets the K'_{LG}, and updates the key information.

It's concluded that the server node can't overwrite the K_{LG} autonomously except the server node initiates the local group and the local group itself generates a new K_{LG}. The members of AG local group only need to know its K_{LG}, which becomes the bridge of members and the server node.

5 Conclusion

In this paper, we introduced the RingNet model and proposed a key management scheme based on the model. Complementary to the scalability, reliability and self-organizability of the RingNet model, we tackle the security issue in RingNet. Since RingNet has three kinds of network entities, theoretically we can generate at least three local group schemes. Furthermore, some combinations of these local group schemes can also be possible. However in this paper, we just consider every AG as a local group controller and the upper distribution structures rely on the RingNet inner structure. In order to take advantages of the local groups, the AG-tier local group scheme is probably the best choice for medium to large sized groups.

Acknowledgments

This work was supported by the Hong Kong Polytechnic University Central Research Grant *G-YY41*, the University Grant Council of Hong Kong under the CERG Grant PolyU *5170/03E*, and the China Postdoctoral Science Foundation (*No. 2003033472*).

References

1. Mittra, S., Iolus: A Framework for Scalable Secure Multicasting, *Proceedings of the ACM SIGCOMM '97 Conference on Applications, Technologies, Architectures, and Protocols for Computer Communication*, (1997) 227-288.
2. Wong, C.K., Gouda, M., Lam, S.S.: Secure Group Communications Using Key Graphs, *Proceedings of the ACM SIGCOMM '98 Conference on Applications, Technologies, and Protocols for Computer Communication*, (1998) 68-79.
3. Setia, S.J.S., Koussih, S., Jajodia, S., Harder, E.: Kronos: A Scalable Group Re-keying Approach for Secure Multicast, *Proceedings of the 2000 IEEE Symposium on Security and Privacy*, Berkeley, CA, (2000) 215-228.
4. Zhang, X.B., Lam, S., Lee, D.Y., Yang, Y.R.: Protocol Design for Scalable and Reliable Group Rekeying, *IEEE/ACM Transactions on Networking*, Vol.11, No.6, (2003) 908-922.
5. Zhang, Q., Calvert, K.L.: On Rekey Policies for Secure Group Applications, *Proceedings of the 12th International Conference on Computer Communications and Networks (ICCCN 2003)*, (2003) 559-564.
6. Lu, W.W.: Compact Multidimensional Broadband Wireless: The Convergence of Wireless Mobile and Access, *IEEE Communications Magazine*, Vol. 38, Issue 11, (2000) 119-123.
7. Otsu, T., Umeda, N., Yamao, Y.: System Architecture for Mobile Communication Systems beyond IMT-2000, *Proceedings of the IEEE Global Telecommunications Conference (GLOBECOM 2001)*, Vol. 1, (2001) 538-542.

8. Zahariadis, T.B., Vaxevanakis, K.G., Tsantilas, C.P., Zervos, N.A., Nikolaou, N.A.: Global Roaming in Next-Generation Networks, *IEEE Communications Magazine*, Vol. 40, Issue 2, (2002) 145-151.
9. Morand L., Tessier, S.: Global Mobility Approach with Mobile IP in all IP Networks, *Proceedings of the 2002 IEEE International Conference on Communication (ICC 2002)*, Vol. 4, (2002) 2075-2079.
10. Wang, G.J., Cao, J.N., Chan, K.C.C.: A Reliable Totally-Ordered Group Multicast Protocol for Mobile Internet, *Proceedings of the 2004 International Workshop on Mobile and Wireless Networking (MWN 2004), held in conjunction with the 33rd International Conference on Parallel Processing (ICPP 2004)*, Montreal, Quebec, Canada, August 15-18, 2004.

Open Middleware-Based Infrastructure for Context-Aware in Pervasive Computing

Xianggang Zhang, Jun Liao, and Jinde Liu

Micro Computer & Network Technology Research Institute,
College of Computer Science and Engineering,
University of Electronic Science and Technology of China,
Chengdu, Sichuan, 610054, P. R. China
csxgzhang@uestc.edu.cn

Abstract. Due to the lack of open infrastructure support, the development of context aware systems in pervasive computing environment is difficult and costly. To solve this problem, we have implemented an open middleware-based context-aware infrastructure for context aware in pervasive computing, named oca-Infrastructure. In oca-Infrastructure, (1) a generic layered model is proposed to specify the functional elements of context-aware computing and provides context-aware computing systems with robust separation of concerns. (2) Wireless CORBA is used to enable distributed components to communicate each other and support for wireless access and terminal mobility. (3) Context-aware supporting platform is created to mask the dynamic environment. (4) Context query is issued based on type-subject model. The context message is delivered in XML. (5) Knowledge-based access control is used to improve security. Finally, we have presented detailed examples to evaluate the feasibility of oca-Infrastructure.

1 Introduction

How to provide best-suited services to users in pervasive computing environment [1][2]? The context plays an important role. In the paper, we designed and implemented an open middleware-based context-aware infrastructure, called oca-Infrastructure, which uses CORBA to enable distributed components to find and communicate with one another and provides the functions to meet the essential requirements for context aware computing in pervasive computing, which are: (1) Supports for context-aware. This includes context acquisitions, context model and representation, context aggregation, context query, context interpretation and so on; (2) Supports for pervasive computing environment; (3) Supports for open systems, which includes portability, interoperability, scalability, security and privacy.

In the rest of the paper, section 2 is related to the related works. In section 3, we propose a layer model for context aware in pervasive computing. In section 4, the architecture of oca-Infrastructure is showed. In section 5, we explain the details of oca-Infrastructure. Then we build a location-aware system with oca-Infrastructure to evaluate it in section 6. Finally, section 7 is related to the further work and we summarize the paper.

2 Related Work

Dey and etc developed a set of abstractions, Context Toolkit [3], for sensors data processing in order to facilitate reuse and make context aware applications easier to build. The Stick-e notes system is a general framework for supporting a certain class of context-aware applications [6]. The Solar [4] system architecture proposed a graph-based abstraction for context aggregation and dissemination. The Gaia project [5] developed at the University of Illinois is a distributed middleware infrastructure that provides support for context aware agents in smart spaces.

To make a summary, the above-mentioned context aware architectures cannot provide a complete solution for all the essential requirements for context aware computing in pervasive computing.

3 A Layered Model for Context Aware in Pervasive Computing

In this section, we present a generic layered model as a design approach for context-aware computing in pervasive computing. Fig.1 shows the layer model.

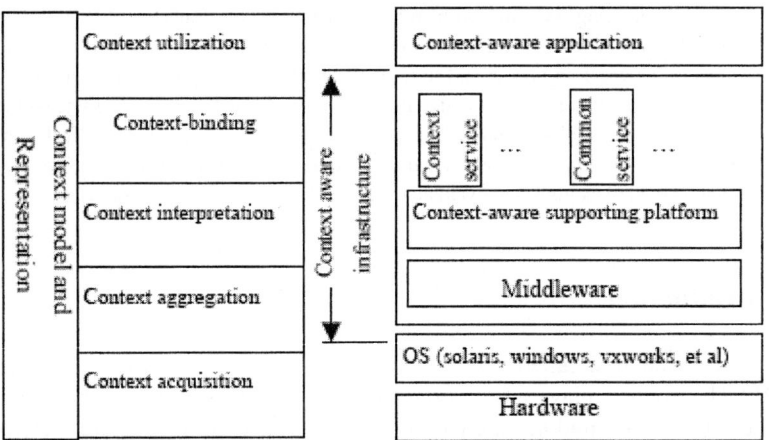

Fig. 1. A layer model for context-aware

Fig. 2. The architecture of oca-Infrastructure

Context model forms the foundation for expressive context representation and high-level context interpretation [8]. Context representation is the foundation for communication and interoperation. The lowest level is context acquisition layer, in which context in raw format may be acquired from a wide variety of ubiquitous context sensors. The context aggregation layer aggregates and relates contexts from distributed context sources to form a centralized context repository. The context interpretation layer leverages reasoning/learning techniques to deduce high-level, implicit context needed by intelligent services from related low-level, explicit context. The context-binding layer maps the context queries of the applications to the context services in order to adapt to the dynamic environment. Finally, on the uppermost level, context aware applications utilize context to adjust their behaviors.

4 Overview of oca-Infrastructure

Fig. 2 shows the architecture of oca-Infrastructure. From bottom up, the lowest layer is the hardware layer. The second layer is OS and driver layer. The third layer is Context-aware infrastructure. It is composed of, from bottom up, underlying middleware, context aware supporting platform and distributed services. The middleware layer provides the communication capability for distributed components. The middle layer is context aware supporting platform, which provides interface for context query and dynamically binds. In the upper layer, there is the service layer. The top layer is Context aware applications, which utilize both raw and abstract context to adjust their behaviors.

5 Implementation

5.1 Context Service and Taxonomy

A *Context Service* is a software (or software and hardware) component. It encapsulates the details of process and provides context information. Through the context service, context consumer is provided separation of concerns. All context services are implemented on top of CORBA. All of context services have a similar structure showed in fig.3. They can be encapsulated into higher Context services.

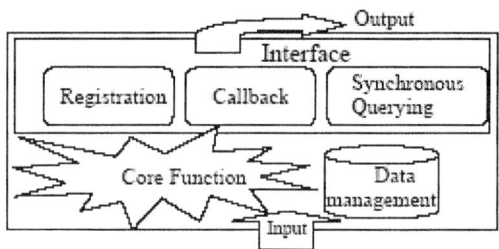

Fig. 3. General architecture of context service

The *Core Functional of a Context Service* implements context acquisition, aggregation, or the other functions. For example, GPS context service reads signals from the serial ports and analyzes signal stream, then derives the location information in form of (x, y, z). The *Uniform Interface* is composed of three components: Registration and de-registration, Callback, Synchronous querying. The *Registration and Callback Components* are adopted for the asynchronous delivery of events. The registration and de-registration interfaces allows an application to register and de-register some context event (represent some situation of circumstance), and context service uses callback interface to deliver context events to applications when the context events occur. The synchronous interface is provided to acquire context information synchronously. Using this, context information are returned immediately. The part of **Data Management** is provided to store context events, templates and history context information and so on.

Furthermore, we propose taxonomy for context services based on the functional requirements of context aware systems. The different taxonomy of the context service has a similar structure and the different core function.

- Elementary Context Service: An elementary context service offers a uniform way to encapsulate a physical sensor such as a thermometer, or any computational component that can be used as a source of physical data. For example, suppose we encapsulate a GPS device within an elementary context service. Then, core function acquires the signals of GPS, process the special-device details, output uniform context information through interface.
- Aggregation Service: The aggregation service aggregates and relates contexts from distributed context services to form a centralized context repository. Aggregation service helps to merges the required information which is related to a particular entity (e.g., 'user'), or which is of relevance of a particular context aware system (e.g., the entire context needed by the smart phone service), and then provides a basis for further interpretation over the related contextual knowledge. It also simplifies the query of distributed services through a centralized manner instead of distributed approach. In some degree, context aggregation by its nature is to provide the functionalities of a context Knowledge Base.
- Translation Service: A translation service performs type recasting but does not change the meaning, or the level of abstraction of the values received. For example, a translation service is used to transform input temperatures from one representation system to another (for example Celsius, Fahrenheit, Kelvin).
- Interpretation Service: The interpretation service receives multiple context Data, then derives higher level of abstraction than that of the input data types. Context can be interpreted with various kinds of reasoning and/or learning mechanisms to derive additional context.
- Abstract Context Services: So far, we have introduced the basic classes of context services. The Context services may be composed into abstract context services, whose internal composition is hidden to other system components.

5.2 Context Model and Representation

Context model forms the foundation for expressive context representation and high-level context interpretation [8]. A uniformed context models can facilitate context interpretation, context sharing and semantic interoperability.

Authors' assumptions are: (1) each context is anchored in an entity. (2) Context-awareness is always related to an entity. (3) Specifying context around the notion of context types provides a general mechanism for context sharing and semantic interoperability.

In oca-Infrastructure, we choose to use a general entity-oriented context model called type-subject model. In such context model, context information is structured around a set of entities, each describing a physical or conceptual object such as a person or an activity. Furthermore, the context set about an entity is classed by context type. For example, "room" has a location context, a temperature context and a light context.

In oca-Infrastructure, standard common representations for context has a form of 3-tuple as follows:

$$\text{(Subject, type, attribute\_set)} \quad (1)$$

Subject represents the entity name with which the context is anchored, for example: person1; Type describes the type of context, for example: location context, and so on; Attribute_set is a set of attributes about context, which has several elements or none. Each element of attribute_set has a form of 2-tuple as follows:

$$\text{(Attribute type, attribute value)} \quad (2)$$

For example: (userId-1, Location, (format, GPS; precision, 50m)) represent userId-1's location information, which has a data format of GPS and the precision of it is 50m.

Context event represent a situation of environment. Application can register oca-Infrastructure with some context events. When context event occur, oca-Infrastructure notifies application that the registered situation occur. Element context event has a form of 3-tuple:

$$\text{(Context, relater, referenced\_value)} \quad (3)$$

Context represents some context information, whose form is same as above (1). Referenced_value is a value associated with the context, Relater is something that relates the context and referenced_value. Relater can be a comparison operator (such as =, >, or <), verb, or preposition. For example, (userId-1, location, in, room321) refer to userId-1 is in room321. We can construct more complex context expressions by performing Boolean operations such as AND, OR, NOT over element context event.

In oca-Infrastructure, context Messages are delivered in XML. If the original data is in binary format, such as a frame of video, Base64 encoding is used to convert it to ASCII format.

5.3 Discovery Service and Context Aware Supporting Platform

Through the Context discovery Service, the context aware applications can find the context services needed in pervasive computing environment. There must be one such service in a single ubiquitous computing environment. Our discovery service is based on CORBA trader service, added the lease mechanism.

Context-aware supporting platform is used to mask the dynamic environment, and provide constant available context information through uniform interface. Its architecture is as fig. 4.

Context-aware supporting platform is composed of, from bottom up, discovery and rebinding function, core function, and context-aware interface. The context-aware

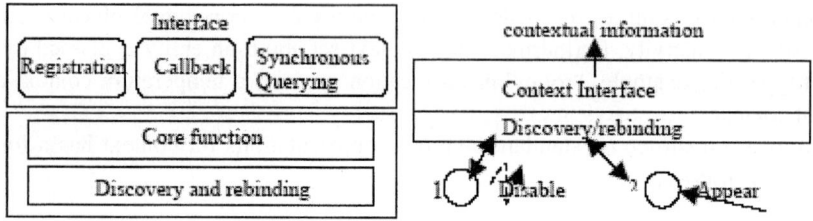

Fig. 4. Architecture of supporting platform **Fig. 5.** Illustration of rebinding

interface provides uniform interface for applications, which includes three kinds of interfaces (i.e. Registration and de-registration, Call-back, Synchronous querying). The middle layer is the core function layer, which includes data management, event filtering, query parsing and so on. The function of the bottom layer is discovery and rebinding.

Through discovery and rebinding layer allowed with the discovery service, context-aware supporting platform enables client applications to utilize the context services dynamically and remains available context services for client applications despite frequent changes of the environment. By the following scenario (fig. 5), we can understand how to use supporting platform to rebind services to client applications. For example, a client application uses a GPS context service to obtain location information about itself. When it enters the smart room, GPS context service is disabling. The supporting platform can rebind the active badges context service to the client application when the client application issues the same queries. In fig.5, "1" represents the GPS context service and "2" represents the active badges context service.

5.4 Middleware and Security

Our middleware for context-awareness does use wireless CORBA [9] to enable distributed components to communicate each other and support for Wireless Access and Terminal Mobility.

In pervasive computing environment, the security of information has become more important than ever. In order to form a perfected access control system, we put forward knowledge-based access control, which can use security-relevant contexts to discovery some rules and knowledge and combines these rules with traditional access controls to perfect the security system [7].

6 Evaluation

To evaluate oca-Infrastructure, we have implemented a location-aware system, called MOTS (mobile object trace system), as test bed. The mobile object taking a PDA (for example: a person taking a PDA) can monitor its location through the PDA. The PDA is equipped with GPS, and the smart building is equipped with an active badges location system. In the beginning, the mobile object is out of building. When the locations monitor application on PDA issues a question for the location of the mobile object, the oca-Infrastructure will bind the GPS context service to the application through Context-aware supporting platform and the discovery service (explain in section 5.4), because the GPS context service is usable and in local. So the returning result to issues from the application is GPS location information. The display on the PDA is as the left of fig.6. Then the mobile object enters the smart building. When the application on PDA issues the location of the mobile object, the oca-Infrastructure will bind the active badges context service to it, because the GPS device is disable and the active badges can be used. So the result is the location information in the form of the active badges system. The result on the PDA is as the right of fig.6. In spite of the changes of the location systems available, the MOTS remains available by binding the different context service to the application.

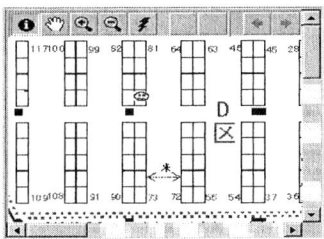

Fig. 6. Display of MOTS

This demonstration highlights the ability of oca-Infrastructure as follows: (1) makes it easier to acquire and deliver context to applications, and make it easier to building complex distributed context aware system; (2) Supporting Applications in Changing Environments; (3) Supporting shared access to context; (4) Supporting Application Extensibility, Portability and interoperability;

7 Further Work and Conclusion

Now we have started a study to make the oca-Infrastructure with the ability of syntactic and semantic interoperability. In addition, we extend the logical operators in order to create complex context event and to better simulate the complex interactions, which occur within the real world, such as, near, far, inside, opposite.

In the paper, we have implemented an open middleware-based context-aware infrastructure for context aware services, namely oca-Infrastructure, to support essential context aware functions. Furthermore, oca-Infrastructure needs be adapted to the pervasive computing environment, and hold the characteristics of open system. Through evaluation, we can prove that oca-Infrastructure will greatly simplify the implementation of context aware systems and meet the requirements for context aware computing in pervasive computing.

References

1. Garlan, D., Siewiorek, D., Smailagic, A., Steenkiste, P.: Project Aura: Towards Distraction-Free PervasiveComputing. IEEE Pervasive Computing, 1 (2):22-31,2002.
2. Weiser, M.: The computer for the twenty-first century, scientific american, pp. 94-10, September 1991.
3. Dey, A. K.: A Conceptual Framework and a Toolkit for Supporting the Rapid Prototyping of Context -Aware Applications. Anchor article of a special issue on Context -Aware Computing, Human-Computer Interaction (HCI) Journal, Vol. 16, 2001.
4. Chen, G., Kotz, D.: Solar: An Open Platform for Context –Aware Mobile Applications. In Proceedings of the First International Conference on Pervasive Computing (Pervasive 2002), Switzerland, June, 2002.
5. Ranganathan, A, Campbell, R. H.: A Middleware for Context -Aware Agents in Ubiquitous Computing Environments. In ACM/IFIP/USENIX International Middleware Conference, Brazil, June, 2003.
6. Brown, P. J.: The Stick-e Document: A framework for creating context-aware applications. In the *Proceedings of the Electronic Publishing '96*, pp. 259-272, Laxenburg, Austria, IFIP. September 1996.
7. Zhang, X., Liu, J.: EH_GRBAC: a knowledge-based access control prototype for pervasive computing. Computer sciences, vol 30 (9), 2003.9
8. Jiang, X., Landay, J. A.: Modeling Privacy Control in Context –Aware Systems. IEEE Pervasive Computing, Vol. 1, No. 3 July-September 2002.
9. Object Management Group, Needham, Massachusetts. Wireless Access and Terminal Mobility in CORBA Specification, June 2001. OMG document dtc/01-06-0

Boundary Integral Simulation of the Motion of Highly Deformable Drops in a Viscous Flow with Spontaneous Marangoni Effect

Wei Gu, Olga Lavrenteva, and Avinoam Nir

Department of Chemical Engineering, Technion, Haifa 32000, Israel

Abstract. We report on three dimensional boundary-integral code and on simulations of the motion of highly deformable drops in the presence of Marangoni effect. Our focus is on the case when the concentration gradients that cause the tangential stresses on the interfaces are not externally imposed but are induced spontaneously induced by the interfacial transfer of surfactants (spontaneous Marangoni effect). When drops move in viscous fluid under intensive external forcing, large deformations, cusps formation and breakup of the drops are typical. The results of the simulation of an initially deformed single drop in gravity field and in linear flow are presented. Our simulations show that even weak Marangoni effect may drastically change the deformation pattern in critical situations.

1 Introduction

The main advantage of the boundary integral equations (BIE) methods is the reduction of the dimension of linear problems as their implementation involves values of the variables only on the interfaces. These methods are extensively used for simulations of complex multiphase flows. In particular, numerous studies of the interaction of deformable drops in buoyancy-driven motion in creeping flow were performed making use of BIE methods and revealed a rich variety of deformation patterns depending on the Bond number and the initial configuration of the system (see e.g. Manga and Stone [1], Zinchenko *et al* [2] and the literature cited).

Most of the BIE simulation available in literature are devoted to multiphase problems with tangential stresses continuous across the interfaces that is typical for pure interfaces in isothermal fluids. In contrast to this, in processes accompanied with heat or mass transfer, surface tension that depends on temperature and concentration of surface-active substance is not constant. Surface tension gradients result in tangential stresses that, in turn, induce the so-called Marangoni flow in the vicinity of the interfaces. This flow may substantially alter the flow pattern, e.g. cause migration of drops in the absence of body forces. From the mathematical point of view, the boundary integral equation modeling of such flows contains an additional term with a tangential stress jump, which provides additional difficulties in the course of numerical solutions.

The goal of the present work was to develop a 3D boundary-integral code for the accurate calculations of the evolution of highly deformable drops in the presence of tangential stress jump and to employ it for simulation of various critical and near critical regimes such as breakup, coalescence and cusp formation in order to study the influence of the Marangoni effect in these processes. Our focus is on the case when the concentration gradients that cause tangential stress jump on the interfaces are not externally imposed but are spontaneously induced by the interfacial transfer of surfactants (spontaneous Marangoni effect).

2   Formulation of the Problem

Consider a number of drops of radii a_i, $i = 1,...,n$ moving in an immiscible viscous fluid submerged in an unbounded Newtonian fluid which has a uniform constant concentration of a weak surfactant C_0 at infinity. The initial concentration inside the drops, C_1, is constant and different from C_0. A Robin type boundary condition is assumed on the interfaces for the concentration in the outer fluid (see [3] for the detailed discussion of the applicability of this model of mass transfer). It is supposed further that the concentration does not affect any physical properties of the liquids in the bulk and at the interfaces except for the interfacial tension which depends linearly on concentration, $\sigma = \sigma_0 + \sigma_c(C - C_0)$, where $\partial\sigma/\partial C$ is constant and typically negative. We shall be interested in the case of small Reynolds and Peclet numbers, when the inertial and convective transport effects are negligible.

The following scaling is chosen: a_1 for length, $v_0 = \eta_0^{-1}|\sigma_c||C_1/k_e - C_0|$ for velocity, a_1/v_0 for time and $v_0\eta_0/a_1$ for the pressure. Here k_e is the phase distribution coefficient (the ratio of equilibrium concentration inside and outside the drops), η_0 is the viscosity of the continuous fluid and ρ_0 is its density. The governing dimensionless parameters of the problem are the Marangoni number, $Ma = |\sigma_c|/(a_1 g \Delta\rho)$, the capillary number, $Ca = \eta_0 v_0/\sigma_0$, the ratios of the material properties of the fluids, $\lambda_i = \eta_2/\eta_1$ and ρ_i/ρ_1, and the initial geometry of the system. The motion inside and outside the droplets is governed by the steady Stokes equations and satisfy the conventional conditions on the interfaces, while the dimensionless concentration $c(t, \mathbf{x}) = (C - C_0)/|C_1/k_e - C_0|$, $\mathbf{x} \in \Omega$ is harmonic outside of droplets, vanishes at infinity and satisfies

$$\partial c/\partial n = Sh(c - 1), \quad \mathbf{x} \in \partial\Omega, \tag{1}$$

where Ω denotes the domain occupied by the continuous phase and $\partial\Omega$ is the boundary separating it from the drops. $\mathbf{n}$ is an outer unit normal to a corresponding droplet interface and Sh is the Sherwood number. Following the classical potential theory the concentration problem is reduced to a boundary-integral equation

$$2\pi c(\mathbf{x}) + \oint_{\partial\Omega} \left[\frac{Sh}{|\mathbf{r}|} + \frac{\mathbf{r}\cdot\mathbf{n}(\mathbf{y})}{|\mathbf{r}|^3}\right] c(\mathbf{y}) dS_y = Sh \oint_{\partial\Omega} \frac{1}{|\mathbf{r}|} dS_y, \tag{2}$$

where $\mathbf{r} = \mathbf{x} - \mathbf{y}$ and $\mathbf{x} \in \partial\Omega$.

After the equation of concentration is solved, the values of concentration at each node are available to compute the jump of normal and tangential stress across the interface. The velocity on the interfaces, thus, can be determined solving another system of boundary integral equations

$$\mathbf{v}(\mathbf{x}) = \frac{2}{1+\lambda_i} \left\{ \mathbf{v}^\infty(\mathbf{x}) + \sum_{k=1}^{n}(1-\lambda_k) \oint_{\partial\Omega_k}^{PV} \mathbf{K}(\mathbf{r}) : \mathbf{v}(\mathbf{y})\,\mathbf{n}(\mathbf{y})\,dS_y \right.$$
$$\left. - \sum_{k=1}^{n} \oint_{\partial\Omega_k} \mathbf{J}(\mathbf{r}) \left[H\left(\frac{1}{Ca} - c(\mathbf{y})\right) \mathbf{n}(\mathbf{y}) - \phi(\mathbf{y})\mathbf{n}(\mathbf{y}) - Ma\nabla_s c(\mathbf{y}) \right] dS_y \right\}, \quad (3)$$
$$\mathbf{x} \in \partial\Omega_i \quad i = 1, ..., n,$$

where PV denotes the principal value, H is a mean curvature, $\nabla_s = \nabla - (\nabla\cdot\mathbf{n})\mathbf{n}$ is a surface gradient and $\phi(\mathbf{y})$ is a dimensionless body force potential. The kernels for the single and double layer potentials here are, respectively

$$\mathbf{J}(\mathbf{r}) = \frac{1}{8\pi}\left(\frac{\mathbf{I}}{|\mathbf{r}|} + \frac{\mathbf{r}\,\mathbf{r}}{|\mathbf{r}|^3}\right), \quad \text{and} \quad \mathbf{K}(\mathbf{r}) = -\frac{3}{4\pi}\frac{\mathbf{r}\,\mathbf{r}\,\mathbf{r}}{|\mathbf{r}|^5}. \quad (4)$$

3 Numerical Scheme and Result Discussion

3.1 Surface Discretization and Numerical Integration

The discretization of the surface is performed by means of uniform triangulation. An icosahedron, whose 20 triangular faces are inscribed into a sphere, is used as a basis for further triangulation. Each face of icosahedron is further divided into four triangles by connecting the edge midpoints. These vertices created by dividing triangles are projected radially on the sphere. Therefore, an umbrella neighborhood of 5 or 6 surrounding nodes and triangular elements is formed around each node. This process is repeated as many times as necessary. The total number N of triangular elements is determined by the mesh order k of the simulation, $N = 20 \cdot 4^k$.

A least square method is employed to calculate the normal vector, mean curvature and concentration gradient on the curved surfaces. A local coordinate system is built on each node to make local parametrization of this and the nodes in its neighborhood that is used to fit a paraboloid with least square method. However, the normal vector is not known a priori, and it is solved by iterative process. The works of Zinchenko et al [2, 4] provide the details of this procedure. The values of concentration and velocity on each node stand for the averaged values within the 1/3 of its umbrella neighborhood that is used in the course of the numerical integration.

Singular integration is the most difficult issue of boundary element method. In our code, the so-called singularity subtraction is required to relax the singularities of the kernels (see e.g. [2, 5] and [6] for details). The remaining singularities following this procedure are of the form $1/|\mathbf{r}|$ or $\mathbf{J}\cdot\mathbf{s}$ with $\mathbf{s}$ being a constant vector tangential to the interface. The singular integral of the first type can be analytically approximated as

$$\int_\Delta \frac{1}{|\mathbf{r}|} dS = \frac{2\Delta S}{b} Log \frac{a+b+c}{a-b+c}, \tag{5}$$

where ΔS is the area of the triangle for integration, while a, b and c denote the length of each side of the triangle. For the singular integral of the second type, an analytical approximation is derived taking advantage of coordinate transformation, which converts the 3D integration into 2D integration and reduces the integrand to the form $|\mathbf{r}|^{-3}(\mathbf{r}\cdot\mathbf{a})\mathbf{R}^{-1}\cdot\mathbf{r}$, where $\mathbf{R}^{-1}$ is the inverse transformation matrix. $\mathbf{a}$ in this integration denotes the tangential component of concentration gradient.

3.2 Mesh Propagation and Redistribution

Because of the strong tangential flow on the interface, the propagation of the nodes with calculated velocities would result in an irregular mesh. Thus, mesh redistribution should be carried out for the purpose of mesh stabilization. The effect of local curvature should be taken into account when the surface tension matters in the thermal capillary interfacial flows. A curvature spring model is developed for this purpose, in which the local curvature and the size of mesh are significant to the mesh distribution. The mechanism of our curvature spring model is described as $f_{ij} = k_{ij}^m l_{ij}^n$ (no summation), with f_{ij} being a spring force between nodes i and j; $k_{ij} = (H_i + H_j)/2$ is the elastic coefficient; $l_{ij} = |\mathbf{r}_{ij} + \Delta t \mathbf{v}_{ij}|$ is the length between each pair of nodes, in which $\mathbf{r}_{ij} = \mathbf{r}_i - \mathbf{r}_j$ and $\mathbf{v}_{ij} = \mathbf{v}_i - \mathbf{v}_j$. m is any positive number while n must be an odd integer. The density of the mesh is dynamically adjusted according to the distribution of curvature and the uniformity of the mesh. On one hand, increasing m will pull more grid points to the regions of high curvature; on the other hand, increasing n will make the mesh as uniform as possible.

The criterion of a "good" distribution of mesh nodes is that the spring force imposed on each node satisfies the force balance on the tangential plane, which could be expressed mathematically as $T\left[\sum_{j=1}^{N} f_{ij}\mathbf{e}_{ij}\right] = 0$, where T is the tangential projection operator, $\mathbf{e}_{ij}$ is the unit vector of the spring force, and N is the number of springs sharing the marked node i. For practical computations, an iterative algorithm is used to calculate the corrected tangential velocity. Our simulations revealed that higher values of m and n require more iterations.

3.3 Results of Simulations

Simulations were performed for the initially deformed single drop in a gravity field and in a linear flow as well as the pairwise interaction of the drops. It was revealed that even weak Marangoni effect may drastically change the deformation pattern in critical situations, e.g. it may cause the breakup or prevent the drop from breakup. In the two drop case, the Marangoni effect may promote or inhibit the coalescence.

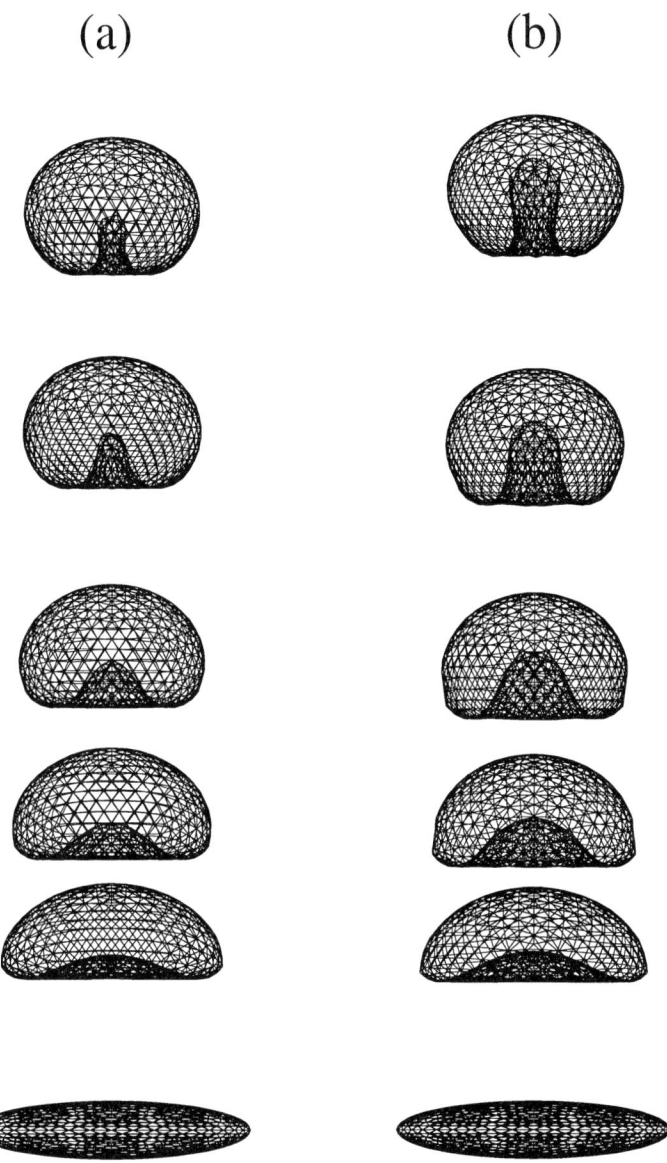

Fig. 1. Evolution of an initially oblate drop ($\lambda = 5$, $Bo = 2$, $Sh = 1$). (a) Buoyancy-driven motion. The drop will attain spherical shape. (b) Combined action of buoyancy and Marangoni effect. The drop will break up

Numerical simulation of the motion of an initially ellipsoidal drop with various aspect ratios and various inclinations with respect to the direction of gravity was carried out with and without the Marangoni effect. The equations of concentration and velocity are coupled when Marangoni effect is taken into account, otherwise, only the equation for the velocity is solved. Most calculations are performed for a mesh order $k = 3$, where the drop has 1280 triangular elements on the interface. Test computations with a mesh order 4 have demonstrated that $k = 3$ is sufficient to acquire a reasonable numerical accuracy except for the cases with extremely high deformations. Runge-Kutta method is employed to make mesh advancement with a time step determined by CFL algorithm. For mesh redistribution, the combination of $m = 0.6$ and $n = 5$ is found to be effective to stabilize the mesh and keep the mesh relative uniformity.

A sample result presented in figure 1 demonstrates the influence of the Marangoni effect on the motion and deformation of an initially oblate drop in a gravity field. Numerical simulation of the motion of an initially oblate drop with 1:4 aspect ratio with and without Marangoni effect is shown. The properties of the fluids are set to $\lambda = 5$, $\phi(\mathbf{y}) = Bo(\mathbf{g} \cdot \mathbf{y})$, where $Bo = 2$ is the Bond number, $\mathbf{g}$ is a unit vector in the direction of gravity and $Sh = 1$. $m = 1$ and $n = 5$ are chosen for the mesh refinement. The shape of the drop could be compared to each other at the same time step 2.575, 4.155, 6.06, 8.72 and 11.75. The left and right sequences are calculated in the absence and in the presence of Marangoni effect, respectively. With the passage of time the drop driven solely by gravity attains a spherical shape, while under the combined action of gravity and Marangoni effect, it breaks up. Thus, it is found that the tangential flow induced by concentration gradients influences the stability of the drop significantly.

Acknowledgement

The research was supported by ISF grant 74/01.

References

1. Manga, M., and Stone H. A.: Buoyancy-driven interaction between two deformable viscous drops. J. Fluid Mech. **256** (1993) 647-683.
2. Zinchenko, A., Rother M., and Davis, R.: Cusping, capture and breakup of interacting drops by a curvatureless boundary integral algorithm. J. Fluid Mech.**391** (1999) 249-293.
3. Golovin, A.A., Nir, A. and Pismen, L.P.: Spontaneous motion of two droplets caused by mass transfer. Ind. Eng. Chem. Res. **34** (1995) 3278-3288.
4. Zinchenko, A., Rother M., and Davis, R.: Phys. Fluids **9**, (1997) 1493-1511.
5. Berejnov, V., Lavrenteva O.M., and Nir, A.: Interaction of two deformable viscous drops under external temperature gradient. J. Colloid Interface Sci. **242** (2001) 202-213.
6. Lavrenteva O.M., Berejnov, V., and Nir, A.: Axisymmetric motion of a pair of deformable heavy drops in an upward temperature gradient. J. Colloid Interface Sci. **255** (2002) 214-217.

Solving Separable Nonlinear Equations with Jacobians of Rank Deficiency One

Yun-Qiu Shen and Tjalling J. Ypma

Department of Mathematics, Western Washington University,
Bellingham, WA 98225-9063, USA

Abstract. Nonlinear systems of equations of the separable form $A(y)z + b(y) = 0$, with only one nonlinear variable $y \in \mathbb{R}$, can be reduced to a single nonlinear equation in y. We develop a technique for the case in which $A(y)$ has rank deficiency one. The method requires only one LU factorization per iteration and is quadratically convergent. Numerical examples and applications are provided.

1 Introduction

Separable nonlinear equations arise in many applications, especially data fitting [5]. We focus here on equations of the form

$$A(y)z + b(y) = 0, \qquad (1)$$

where the $(n+1) \times n$ matrix $A(y)$ and the vector $b(y)$ are both continuous functions of the scalar variable y alone, and z is an n-dimensional vector. We refer to y as the nonlinear variable, and note immediately that once y is known then z can be computed from (1) by using a linear least squares technique.

The separation of linear and nonlinear variables in nonlinear systems was first investigated in [1]. The system (1) is a special case of the more general class of separable nonlinear systems that can be solved by first determining the value of the nonlinear variable(s), or equivalently by eliminating the linear variables [3]. The recent survey by Golub and Pereyra [5] outlines the historical development of various algorithms for separable systems and the myriad of applications in which such systems arise. A recent thesis [6] details the correspondence between our previous work concerning equations of the form (1) [7] and other work on the solution of separable systems.

Solving (1) may be reformulated as solving

$$min_{y,z}||b(y) + A(y)z||_2, \qquad (2)$$

which can be shown [3] to be equivalent to solving the optimization problem

$$min_y||[I - A(y)A^+(y)]b(y)||_2, \qquad (3)$$

where $A^+(y)$ is the Moore-Penrose pseudo-inverse of $A(y)$. Thus the linear variables have been eliminated. Standard variable projection methods such as

VARPRO [4], [5] are based on the use of a nonlinear least squares method to solve (3). Such iterative methods for solving (3) numerically involve either a QR or SVD decomposition in each step [2], [4], [5].

In [7] we gave an alternative algorithm for solving (1) under the condition that $A(y)$ has full rank n. That algorithm reduces (1) to a single equation in the nonlinear variable y, and requires only one LU factorization in each iterative step. The algorithm is based on appending a vector to the $(n+1) \times n$ full rank matrix $A(y)$ to produce a bordered, square, well-conditioned matrix, so that QR factorizations can be avoided. In this paper we extend that algorithm to the case that $A(y)$ has maximum rank deficiency one at y^*, i.e., $A(y)$ has minimum rank $n-1$ at the solution.

We present the method and its analysis in Section 2, and develop an efficient algorithm in Section 3. In Section 4 we give numerical examples of the method and its applications. An extension of the method to the case of higher rank deficiencies of $A(y)$ at y^* is developed in [8].

2 Reduction to One Nonlinear Equation in One Variable

Let (y^*, z^*) denote a solution of (1). When $A(y^*)$ is rank deficient there is no unique solution z^* of (1). Assuming the rank deficiency at y^* is one, which implies the rank deficiency is at most one in a neighborhood of y^*, we reformulate (1) as

$$\begin{bmatrix} A(y) \\ L^T \end{bmatrix} z + \begin{bmatrix} b(y) \\ 1 \end{bmatrix} = 0, \tag{4}$$

where L is any n-dimensional vector such that $L \notin Range(A^T(y^*))$.

Theorem 1. *Let $L \in \mathbb{R}^n$ and $R, S \in \mathbb{R}^{n+1}$ be vectors with $L \notin Range[(A^T(y^*)]$, $R \notin Range[A(y^*)]$ and $S \notin Range[\,A(y^*) \; R\,]$. Let $c(y) \in \mathbb{R}^{n+2}$ satisfy*

$$\text{(a)} \quad c^T(y) \begin{bmatrix} A(y) & R \\ L^T & 0 \end{bmatrix} = 0, \quad \text{(b)} \quad c^T(y) \begin{bmatrix} S \\ 0 \end{bmatrix} = 1. \tag{5}$$

Then (y^*, z^*) is a solution of (4) if and only if it is a solution of

$$\text{(a)} \quad c^T(y) \begin{bmatrix} b(y) \\ 1 \end{bmatrix} = 0, \quad \text{(b)} \quad \begin{bmatrix} z \\ 0 \end{bmatrix} = -\begin{bmatrix} A(y) & R \\ L^T & 0 \end{bmatrix}^+ \begin{bmatrix} b(y) \\ 1 \end{bmatrix}. \tag{6}$$

Proof. Note that the conditions on L and R imply that the matrix $\begin{bmatrix} A(y) & R \\ L^T & 0 \end{bmatrix}$ has full rank in a neighborhood of y^*.
($\Longrightarrow$) (4) implies

$$\begin{bmatrix} A(y) & R \\ L^T & 0 \end{bmatrix} \begin{bmatrix} z \\ 0 \end{bmatrix} + \begin{bmatrix} b(y) \\ 1 \end{bmatrix} = 0. \tag{7}$$

Multiplying (7) by $c^T(y)$ and $\begin{bmatrix} A(y) & R \\ L^T & 0 \end{bmatrix}^+$ we obtain (6a) and (6b) respectively, using (5a) and (5b) and noting that the full rank of the $(n+2) \times (n+1)$ matrix $\begin{bmatrix} A(y) & R \\ L^T & 0 \end{bmatrix}$ implies that

$$\begin{bmatrix} A(y) & R \\ L^T & 0 \end{bmatrix}^+ \begin{bmatrix} A(y) & R \\ L^T & 0 \end{bmatrix} = I_{n+1}. \tag{8}$$

($\Longleftarrow$) Multiply (6b) by $\begin{bmatrix} A(y) & R \\ L^T & 0 \end{bmatrix}$ and writing $P = \begin{bmatrix} A(y) & R \\ L^T & 0 \end{bmatrix} \begin{bmatrix} A(y) & R \\ L^T & 0 \end{bmatrix}^+$ gives

$$\begin{bmatrix} A(y) & R \\ L^T & 0 \end{bmatrix} \begin{bmatrix} z \\ 0 \end{bmatrix} + P \begin{bmatrix} b(y) \\ 1 \end{bmatrix} = 0. \tag{9}$$

Writing $\begin{bmatrix} b(y) \\ 1 \end{bmatrix}$ as a linear combination of the columns of $\begin{bmatrix} A(y) & R \\ L^T & 0 \end{bmatrix}$ and $\begin{bmatrix} S \\ 0 \end{bmatrix}$ with a coefficient vector α and scalar coefficient β respectively, (5–6a) imply that

$$0 = c^T(y) \begin{bmatrix} b(y) \\ 1 \end{bmatrix} = c^T(y) \left(\begin{bmatrix} A(y) & R \\ L^T & 0 \end{bmatrix} \alpha + \begin{bmatrix} S \\ 0 \end{bmatrix} \beta \right) = c^T(y) \begin{bmatrix} S \\ 0 \end{bmatrix} \beta = \beta.$$

Combined with (8) this implies that

$$(I_{n+2} - P) \begin{bmatrix} b(y) \\ 1 \end{bmatrix} = (I_{n+2} - P) \begin{bmatrix} A(y) & R \\ L^T & 0 \end{bmatrix} \alpha$$

$$= \left(\begin{bmatrix} A(y) & R \\ L^T & 0 \end{bmatrix} - \begin{bmatrix} A(y) & R \\ L^T & 0 \end{bmatrix} \begin{bmatrix} A(y) & R \\ L^T & 0 \end{bmatrix}^+ \begin{bmatrix} A(y) & R \\ L^T & 0 \end{bmatrix} \right) \alpha = 0. \tag{10}$$

Equations (9) and (10) give (7) which is equivalent to (4). □

Theorem 1 justifies our method: solve (6a) for y^*, then use (6b) to find z^*.

3 An Algorithm for the Nonlinear Equation

We solve (6a) for y using Newton's method. The following technical lemma is aimed at facilitating an efficient implementation of Newton's method in this particular context.

Lemma 1. *Assume that $A'(y)$ and $b'(y)$ are continuous in a neighborhood of y^*. Using the notation of Theorem 1, let*

$$M(y) = \begin{bmatrix} A(y) & R & S \\ L^T & 0 & 0 \end{bmatrix}, \tag{11}$$

and suppose $\mu(y) \in \mathbb{R}^{n+1}$ and $h(y) \in \mathbb{R}$ satisfy

$$M(y) \begin{bmatrix} \mu(y) \\ h(y) \end{bmatrix} = - \begin{bmatrix} b(y) \\ 1 \end{bmatrix}. \tag{12}$$

Then

(a) $h(y) = c^T(y) \begin{bmatrix} b(y) \\ 1 \end{bmatrix}$, (b) $h'(y) = c^T(y) \begin{bmatrix} [A'(y) \ 0]\mu(y) + b'(y) \\ 0 \end{bmatrix}$. (13)

Furthermore, if $\begin{bmatrix} A'(y^*)z^* + b'(y^*) \\ 0 \end{bmatrix} \notin Range\left(\begin{bmatrix} A(y^*) & R \\ L^T & 0 \end{bmatrix}\right)$, then $h'(y^*) \neq 0$.

Proof. The conditions on L, R, S imply that the $(n+2) \times (n+2)$ bordered matrix $M(y)$ defined in (11) is nonsingular in a neighborhood of y^*. Thus (12) has a unique solution in this neighborhood. Conditions (5a) and (5b) imply that

$$c^T(y)M(y) = -\begin{bmatrix} O_{1 \times (n+1)} & 1 \end{bmatrix}, \quad (14)$$

hence by (12)

$$h(y) = \begin{bmatrix} O_{1 \times (n+1)} & 1 \end{bmatrix} \begin{bmatrix} \mu(y) \\ h(y) \end{bmatrix} = -c^T(y)M(y) \begin{bmatrix} \mu(y) \\ h(y) \end{bmatrix} = c^T(y) \begin{bmatrix} b(y) \\ 1 \end{bmatrix} \quad (15)$$

which proves (13a). Differentiating both sides of (12) with respect to y, we obtain

$$M(y) \begin{bmatrix} \mu'(y) \\ h'(y) \end{bmatrix} + \begin{bmatrix} A'(y) & 0 & 0 \\ 0 & 0 & 0 \end{bmatrix} \begin{bmatrix} \mu(y) \\ h(y) \end{bmatrix} = -\begin{bmatrix} b'(y) \\ 0 \end{bmatrix},$$

which with (14) implies

$$h'(y) = -\begin{bmatrix} O_{1 \times (n+1)} & 1 \end{bmatrix} [M(y)]^{-1} \begin{bmatrix} [A'(y) \ 0]\mu(y) + b'(y) \\ 0 \end{bmatrix}$$

$$= c^T(y) \begin{bmatrix} [A'(y) \ 0]\mu(y) + b'(y) \\ 0 \end{bmatrix}$$

thus proving (13b). Since $h(y^*) = 0$, equations (6b), (12) and (8) imply that $\begin{bmatrix} z^* \\ 0 \end{bmatrix} = \mu(y^*)$, therefore $h'(y^*) \neq 0$ if the additional condition is satisfied. □

The idea is to solve (6a) by applying Newton's method to solve $h(y) = 0$ where $h(y)$ is defined through equation (12). By (13a) we know that this $h(y)$ is the appropriate function, while (13b) gives a convenient expression for its derivative.

The vectors L, R, S must be such that the matrix $M(y)$ of (11) is nonsingular. Any vectors produced by using a random number generator will almost certainly meet this requirement. However, it is desirable that $M(y)$ be well-conditioned, which demands more careful selection of L, R, S. For example, we can ensure that $L \notin Range[A^T(y)]$ by picking L with a large component in $Null[A(y)]$ so that L is largely orthogonal to $Range[A^T(y)]$. In practice we have found that computing a single SVD of the matrix $A(y^{(0)})$ at the starting point $y^{(0)}$ of the iteration is very effective, i.e., we select $L = v_n$, $R = u_n$ and $S = u_{n+1}$ from

$$A(y^{(0)}) = U\Sigma V^T = \begin{bmatrix} u_1 & \cdots & u_{n+1} \end{bmatrix} \begin{bmatrix} \sigma_1 & \cdots & 0 \\ 0 & \ddots & 0 \\ 0 & \cdots & \sigma_n \\ 0 & \cdots & 0 \end{bmatrix} \begin{bmatrix} v_1^T \\ \cdots \\ v_n^T \end{bmatrix}.$$

The above considerations result in the following algorithm for solving (1).

Algorithm 1.

1. Let $y^{(0)}$ be given. Choose appropriate vectors L, R, S. For $j = 0, 1, \ldots$ until convergence repeat Step 2 through Step 6:
2. Compute $A(y^{(j)}), b(y^{(j)}), A'(y^{(j)}), b'(y^{(j)})$.
3. Form $M(y^{(j)})$ as in (11).
4. Compute the LU factors of $M(y^{(j)})$; solve (14) for $c^T(y^{(j)})$ and (12) for $h(y^{(j)}), \mu(y^{(j)})$.
5. Compute $h'(y^{(j)})$ using (13b).
6. Compute $y^{(j+1)} = y^{(j)} - h(y^{(j)})/h'(y^{(j)})$.
7. If convergence has occurred, output $y = y^{(j+1)}, z = [I_n \; 0_{n\times 1}]\mu(y^{(j+1)})$.

4 Numerical Examples

The following illustrative example is adapted from [7]:

Example 1.
$$\begin{bmatrix} y & 0 \\ 1 & 1 \\ 1+y & 1-y^2 \end{bmatrix} z + \begin{bmatrix} y \\ 1+y \\ 1+y \end{bmatrix} = 0. \tag{16}$$

This equation has two solutions (y^*, z^*) at which the matrix $A(y^*)$ has full rank, namely $(1, (-1, -1))$ and $(-1, (-1, 1))$. Those solutions may be found by the method of [7]. At $y^* = 0$ the matrix $A(y^*)$ has rank 1 and any $z^* \in \mathbb{R}^2$ with the property that $z_1^* + z_2^* + 1 = 0$ satisfies the equation. The method of [7] fails to find this solution. We list in Table 1(a) the first five successive estimates of $y^* = 0$ generated by our new method starting from $y^{(0)} = 0.25$, generating L, R, S by performing one SVD of $A(y^{(0)})$. Quadratic convergence is evident.

The following example is a generalization of Example 5.6.2 of [9]:

Example 2. Consider a generalized eigenvalue problem for the matrices $\{M, N\}$:
$$(yN - M)z = 0, \tag{17}$$
$$(\alpha e_1^T + \beta e_2^T)z = 1, \tag{18}$$

where $M = \text{diag}\{\lambda I_2, -I_{n-2}\}$, $N = \text{diag}\{I_2, 0_{(n-2)\times(n-2)}\}$, e_1, e_2 are the first two unit vectors in the standard basis for $\mathbb{R}^n$, and α, β are real scalars. Clearly

Table 1.

	(a)		(b)
k	$y^{(k)}$	k	$e^{(k)}$
1	-1.8959×10^{-01}	1	1.4884×10^{-01}
2	-2.2961×10^{-02}	2	3.8405×10^{-02}
3	-4.9753×10^{-04}	3	2.6831×10^{-03}
4	-2.4866×10^{-07}	4	1.3416×10^{-05}
5	-6.2203×10^{-14}	5	3.3609×10^{-10}

$y^* = \lambda$ is the only generalized eigenvalue of this problem, and the matrix $A(y) = \begin{bmatrix} yN - M \\ \alpha e_1^T + \beta e_2^T \end{bmatrix}$ has rank deficiency one at $y = y^* = \lambda$ so long as $\alpha^2 + \beta^2 \neq 0$. The error $e^{(i)} = y^{(i)} - 1$ in the first five successive estimates of $y^* = 1$ generated by our method, for the case $\lambda = \alpha = \beta = 1, n = 5$ and $y^{(0)} = 0.75$ are given in Table 1(b). Quadratic convergence is again evident.

References

1. Drane, J.W., Schucany, W.R.: Another approach to nonlinear least squares. The 7th Annual Symposium on Biomathematics and Computer Science in the Life Sciences. Houston, Texas (1969)
2. Golub, G.H., LeVeque, R.J.: Extensions and uses of the variable projection algorithm for solving nonlinear least squares problems. Proceedings of the Army Numerical Analysis and Computing Conference. Washington DC (1979) 1–12
3. Golub, G.H., Pereyra, V.: The differentiation of pseudo-inverses and nonlinear least squares problems whose variables separate. SIAM J. Numer. Anal. **10**(2) (1973) 413–432
4. Golub, G.H., Pereyra, V.: The differentiation of pseudo-inverses and nonlinear least squares problems whose variables separate. Stanford Univ. CS Dept. Report. STAN-CS-72-261 (1972)
5. Golub, G.H., Pereyra, V.: Separable nonlinear least squares: the variable projection method and its applications. Inverse Problems **19** (2003) Topic Review, R1–R26
6. Lukeman, G.G.: Applications of the Shen-Ypma algorithm for separable overdetermined nonlinear systems. MSc Thesis, Dalhousie Univ., Halifax, NS (1999)
7. Shen, Y.-Q., Ypma, T.J.: Solving nonlinear systems of equations with only one nonlinear variable. J. Comput. Appl. Math. **30** (1990) 235–246
8. Shen, Y.-Q., Ypma, T.J.: Solving separable nonlinear equations with higher rank deficient Jacobians (in preparation)
9. Watkins, D.S.: Fundamentals of Matrix Computations. Wiley, New York (1991)

Optimal Capacity Expansion Arc Algorithm on Networks

Yuhua Liu[1,2], Shengsheng Yu[1], Jingzhong Mao[3], and Peng Yang[2]

[1] College of Computer Science and Technology, Huazhong
University of Science and Technology,
Wuhan 430074, China
ssyu@mail.hust.edu.cn
[2] Department of Computer Science, Central China Normal University,
Wuhan 430079, China
yhliu@mail.ccnu.edu.cn
[3] Department of Mathematics, Central China Normal University,
Wuhan 430079, China
jzmao@nail.ccnu.edu.cn

Abstract. The problem of the capacity expansion of a network by means of expanding the capacity of its arcs is elaborated in this paper. By using the theory and algorithm of the maximum flow, the minimum cut-set and the bottleneck of the network can be found, then the expansible arcs are sought according to whether there exists any augmenting path in the minimum cut-set. If the expansible arcs exist, the capacity of the network can be expanded by means of expanding the capacity of those arcs to increase the maximum flow. In order to fulfill the goal, the algorithms of the expansible arc and the optimal expansion arc are presented and the complexity analyses of those algorithms are given as well.

1 Introduction

When we plan a large size network topology structure or administrate a running network, the problem of congestion and bottleneck should be considered as the flow of the network increases. The problem of the traffic jam in a network can often be resolved by expanding the capacity of the network. Namely, new arcs or the capacity of the arcs in the bottleneck are added. All those methods belong to the category of capacity expansion of the network. In this paper the Ford-Fulkerson algorithm, which can figure out the maximum flow, is employed to obtain the minimum cut-set and the bottleneck part of the network, then the expansible arcs are sought according to whether there exists any augmenting path in the minimum cut-set. If the expansible arcs are found, the capacity of them can be expanded to destroy the saturation and increase the maximum flow so as to reach the goal of capacity expansion of the network. In order to fulfill this goal, the algorithms of the expansible arc and the optimal expansion arc are presented and the complexity analyses of the algorithms are given as well[1,2].

2 Basic Concept

In this section the mathematical concepts and symbols used throughout the paper is explained. The Ford-Fulkerson algorithm[3], which is employed to obtain the maximum flow-minimum cut-set, is described in detail.

2.1 Definitions and Symbols

Definition 1. Given a directed graph $D=(V,A,C)$ with vertex set V, arc set A and a capacity function $C: A \to R^+$, a vertex V_s and a vertex V_t in V are designated as the source and the sink respectively. All the other vertices are called intermediate vertices. For every arc $(v_i, v_j) \in A$, there exists a corresponding nonnegative value $c(v_i, v_j) \geq 0$ (labeled c_{ij}) called the capacity of the arc. The network $D=(V,A,V_s,V_t,C)$ is called a capacity-designated network or capacity network for short.

Definition 2. The maximum flow problem: In a capacity network $D=(V,A,V_s,V_t,C)$ with the source v_s and the sink v_t, search for a flow $f: A \to R = \{f_{ij}\}$, where $f_{ij}=f(v_i,v_j)$ denotes the flow that pass through the arc (v_i,v_j), which can make the flow value $V(f) = \sum_{(v_s,v_j) \in A} f(v_s,v_j) - \sum_{(v_i,v_s) \in A} f(v_i,v_s)$ reach its maximum and satisfies:

$$0 \leq f_{ij} \leq c_{ij} \quad (v_i,v_j) \in A \tag{1}$$

$$\sum_f f_{ij} - \sum_j f_{ji} = \begin{cases} V(f) & (i=s) \\ 0 & (i \neq s,t) \\ -V(f) & (i=t) \end{cases} \tag{2}$$

Such flow f is called the maximum flow with the flow value $V(f)$ and the flow $f = \{f_{ij}\}$, which meets the formula (1) and (2), is called a feasible flow.

Definition 3. Assume that f is a feasible flow of a capacity network D and (v_i,v_j) is an arc of it. If the flow value f_{ij} on (v_i,v_j) is zero, then (v_i,v_j) is called a zero arc of the flow f. If $f_{ij}>0$, then (v_i,v_j) is called a positive arc. If $f_{ij}<0$, then it is called a negative arc. Moreover if the flow value f_{ij} on the arc (v_i,v_j) equals c_{ij}, then the arc(v_i,v_j) is a saturated arc, otherwise it is a non-saturated arc.

2.2 The Maximum Flow and Minimum Cut-Set of a Network

The Ford-Fulkerson algorithm (1957) applies a labeling method to search for the augmenting path of an arbitrary feasible flow f_1 in a capacity network D with the source v_s and the sink v_t, then it adjusts f_1 to get a new flow f_2 whose capacity is bigger than f_1's. Again it searches for the augmenting path of f_2 in D and adjusts f_2. The process is repeated until there is no augmenting path with respect to the f_i in D. Finally the maximum flow capacity $V(f)$ and its corresponding flow f of the network is obtained. After the process above, there exist a labeled vertex set S and a non-labeled vertex set

$\overline{S}$ in D, such that $v_s \in S$, $v_t \in \overline{S}$, $S \cup \overline{S} = V$, $S \cap \overline{S} = \Phi$, then $R=(S, \overline{S})$ is a minimum cut-set of the network D. The sum of the capacity of the arcs in the minimum cut-set is called the capacity of this minimum cut-set, labeled $c(S, \overline{S}) = \sum f(v_i, v_j)$, where $\forall v_i \in S, v_j \in \overline{S}$. Then the maximum flow can be determined from the capacity of the minimum cut-set. This theory is called the maximum flow-minimum cut-set theorem.

Theorem 1. Given an arbitrary capacity network $D=(V,A,C)$, the flow value of the maximum flow from the source v_s to the sink v_t equals the capacity of the minimum cut-set $(S, \overline{S})$.

In sum, when the Ford-Fulkerson algorithm is applied to get the maximum flow of a network, a minimum cut-set of it is also found, and the arcs in the minimum cut-set are all saturated arcs and form the bottleneck part, which inhibits further increase of the flow value. Only when the saturation of those bottleneck arcs is destroyed, namely, the capacity of them is increased, can we control the congestion of the network. Based on the Ford-Fulkerson algorithm, the maximum flow and the minimum cut-set of a network can be obtained in finite steps. Thus the capacity of the saturated arcs in the minimum cut-set can be expanded to ameliorate the traffic ability of the whole network.

3 Algorithm of the Expansible Arc

From the discussion above we know that the key factor that affects the maximum flow of a network is the arcs in the minimum cut-set. If the capacity of one kind of those arcs is expanded, the feasible flow of the network can be augmented. Thus we have the following definition of an expansible arc:

Definition 4. If expanding the capacity of an arc in the minimum cut-set can lead to an increase of the capacity of the feasible flow, such kind of arc is called the expansible arc of the network.

How the expansible arc can be found? According to the concept of the augmenting path of the Ford-Fulkerson algorithm, the feasible flow is the maximum flow if and only if there is no augmenting path of f in the network. Thus the key of improving a flow into a bigger flow is to search whether there exists an augmenting path of f. In terms of this theory, an algorithm of finding the expansible arc of the network is presented.

Given a capacity network $D=(V,A,V_s,V_t,C)$, the algorithm of the expansible arc can be described as follows:

Step 1. Input the parameters of the network D;
Step 2. Use the Ford-Fulkerson algorithm to get a maximum flow f_0 of the network D and generate a minimum cut-set $(s, \overline{s})$;
Step 3. Initialize the set B, $B = \Phi$, reset the array δ, f, set counter k to 1;

3.1. If $(S,\overline{S})$ is not empty, then choose an arc (v_a, v_b) $(S,\overline{S})$, otherwise go to step 4;

3.2. Invoke the Ford-Fulkerson algorithm to search for the augmenting path starting at v_b. If there exists an augmenting path then $k=k+1$ and (v_a, v_b) is added to $B[k]$; otherwise delete (v_a, v_b) from $(S,\overline{S})$ and go to step 3.1;

3.3. Check all the arcs that are incident to v_b and $\notin (S,\overline{S})$;

Let $P = \{v_j \mid v_j \in V, (v_b, v_j) \in A\}$
$Q = \{v_i \mid v_i \in V, (v_i, v_b) \in A\}$

Let $g = \sum_{v_j \in p} f_{bj} - \sum_{v_i \in q} f_{ib}$, $\delta_b = g - c_{ab}$

3.4. If $v_a \neq v_s$ then let $\delta = \delta_b$, otherwise let $\delta = +\infty$;

3.5. Let $\delta = \min(\delta, \delta_b)$, $c_{ab} = c_{ab} + \delta$, $f = c_{ab}$,
$\delta[k] = \delta$, $f[k] = f$

Step 4. End.

When the maximum flow of the network is obtained at the step 1, a minimum cut-set is also obtained. The step 2 searches for the expansible arcs in the minimum cut-set one by one and deletes those arcs with no augmenting path, because the capacity expansion of those arcs do not lead to an increase of the feasible flow. The algorithm starts at v_b and searches for the augmenting path. If there exists a augmenting path, the capacity of the arc $(v_a v_b)$ is expansible. Step 3.3 calculates the increment of capacity, where P is the vertex set corresponding to the outgoing arcs of v_b and Q is the vertex set corresponding to the incoming arcs of v_b, g is the difference between the outgoing-flow sum and incoming-flow sum of v_b, δ is the capacity increment of the c_{ab} on the arc $(v_a v_b)$, thus the maximum flow of the new capacity network is $f' = f + \delta$. The step 3 repeatedly chooses one arc from the cut-set $(S,\overline{S})$ at a time, searches for its augmenting path and expands its capacity. Finally the set of the expansible arcs can be found in the cut-set.

Because the computational complexity of the Ford-Fulkerson algorithm depends on the capacity of the arcs in the network, this is a disadvantage from the perspective of theory and practical computation. After the advent of the Ford-Fulkerson algorithm, people used the labeling method in the Ford-Fulkerson algorithm for reference and proposed several improved methods, thus make the computational complexity of the maximum flow algorithm depend only on the number of vertices and arcs of the network, namely, the computational complexity depends only on the scale of the network, such as the Edmords-Kasp algorithm and so on. For the discussion convenience, the Ford-Fulkerson algorithm is employed to obtain the maximum flow-minimum cut-set, but this algorithm can be also taken place by other improved methods [4].

4 Algorithm of the Optimal Expansion Arc

In the practical application of a network, people always wish that the bandwidth and traffic ability of a network are sufficiently high, the same is true for the flow value. So we do not only want to seek the expansible arc in a network but also attend to the way of finding a specific kind of the arc so that the capacity expansion of it can lead to a maximum increase of the feasible flow[5].

Definition 5. In the set of the expansible arcs, there exists one arc whose capacity expansion can lead to a maximum increase of the maximum flow of a network. This arc is called the optimal expansion arc.

Based on the expansible arcs of a network worked out by the algorithm 1, it is easy to contrive an algorithm to seek the optimal expansion arc of the network, that is using some high efficient sorting algorithms to sort and search the array of the maximum flow obtained by the algorithm 1. This algorithm is described as follows:

Step 1. Input the parameters of the network D ;
Step 2. Invoke the algorithm 1 and work out the set of the expansible arcs B as well as the corresponding capacity expansion array δ and maximum flow array f;
Step 3. Let $max=f_0$, $i=1, j=0$;
Step 4. Choose the element $f[i]$ from the array f sequentially and carry out the maximum finding. If $f[i] > max$ then $max = f[i]$, $j=i$;
Step 5. Output $max = f[j]$, $B[j]$, $\delta[j]$;
Step 6. End.

$B[j]$ is the optimal expansion arc obtained by the algorithm. $\delta[j]$ is the capacity increment of the arc $B[j]$ and $f[j]$ is the maximum flow of the new network after expansion. Obviously the complexity of the algorithm depends on the scale of the network and the number of the expansible arcs, thus it can be finished in finite steps.

5 An Instance

Use the algorithm stated above to find an optimal expansion arc of a capacity network $D=(V, A, V_s, V_t, C)$, which is illustrated in figure 1 and calculate the maximum flow of the new network after the optimal capacity expansion.

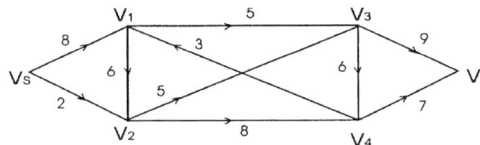

Fig. 1. Capacity network D

By using the algorithm of the optimal expansion arc, a conclusion can be drawn that the arc (v_s, v_2) is the optimal expansion arc, 6 is the capacity increment of it and the maximum flow of the new network D' is 16 after the capacity expansion, while the maximum flow of the network D before capacity expansion is only 10. The new network D' and its distribution of the maximum flow is showed in figure 2.

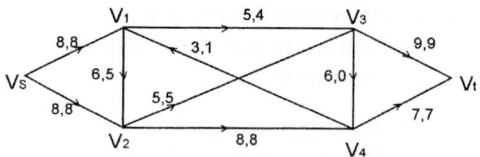

Fig. 2. The maximum flow of capacity network D

6 Conclusion

Through the capacity expansion of an arc, the algorithm of the optimal expansion arc presented in this paper can reach the goal of maximum increment of the maximum flow. If there is no expansible arc in the network, some general methods such as destroying the saturation of the arcs in the cut-set can be employed to generate the new maximum flow and the new minimum cut-set. Then the capacity expansion can be applied to the arcs in the new minimum cut-set and this process is repeated until there is no cut-set or the sink v_t has been already searched. This algorithm has been elaborated in another paper[6]. In the discussion of this paper, the capacity of the network is an integer. When the capacity of the network is a rational, it can be handled by converting the rational capacity into the integral capacity.

References

1. Balakrishnam, V. K.: Network Optimization. London:Chapman & Hall, (1995)
2. Ahujia,R.K., Magnanti,T.L., Orlin,J.B.: Network Flows:Theory, Algorithms and Applications. Englewood Cliffs, N J:Prentice-Hall, (1993)
3. Ford, L. R., Fulkerson, D. R.: Maximum flow through a network. Canadian Journal of Math, 1956, 8(5): 399-404
4. Andrew, V., Goldberg, Satish, R.: Flows in Undirected Unit Capacity Networks. SIAM Journal on Discrete Mathematics,1999,12: 1-5
5. Zhang, J., Yang C., Lin,Y.: A class of bottleneck expansion problems[J]. Computer Operate Research, 2001:286:505-519
6. Yu, S., Liu, Y., Mao, J., Xu, K.: Algorithm of capacity expansion networks on optimization. Chinese Science Bulletin, 2003, 48(10): 1047-1050

Solving Non-linear Finite Difference Systems by Normalized Approximate Inverses

George A. Gravvanis[1] and Konstantinos M. Giannoutakis[2]

[1] Department of Computer Science, Hellenic Open University, Patras, Greece
gravvanis@eap.gr
[2] Department of Informatics and Telecommunications, University of Athens,
Panepistimioupolis, GR 157 84 Athens, Greece

Abstract. A class of inner-outer isomorphic iterative procedures in conjunction with Picard/Newton methods based on normalized explicit approximate inverse matrix techniques for solving efficiently sparse non-linear finite difference systems is presented. Applications on characteristic non-linear boundary value problems in three dimensions are discussed and numerical results are given.

1 Introduction

Let us consider a class of non-linear boundary value problems defined by the non-linear elliptic partial differential equations in three space-variables, i.e.:

$$Lu = f(u), \quad (x, y, z) \in R, \tag{1}$$

$$\alpha u + \beta \frac{\partial u}{\partial \zeta} = \gamma, \quad (x, y, z) \in \partial R, \tag{2}$$

where L is a linear partial differential operator.

One may linearize the above problem, by the non-linear iterative method, namely the Picard method, viz.,

$$L\left[u^{(k+1)}\right] = f\left(u^{(k)}\right), \tag{3}$$

or the Newton method, viz.,

$$L\left[u^{(k+1)}\right] - f'\left(u^{(k)}\right)u^{(k+1)} = f\left(u^{(k)}\right) - u^{(k)}f'\left(u^{(k)}\right), \tag{4}$$

where f' is the Jacobian of f with respect to $u^{(k)}$, cf. [10]. The above iterative schemes, using the seven-point molecule, lead to the non-linear finite difference system, cf. [10], which can be compactly written as:

$$A_k u^{(k+1)} = s\left(u^{(k)}\right), \quad k > 0, \tag{5}$$

where A_k is a sparse, diagonally dominant, positive definite, symmetric $(n \times n)$ matrix of the following form:

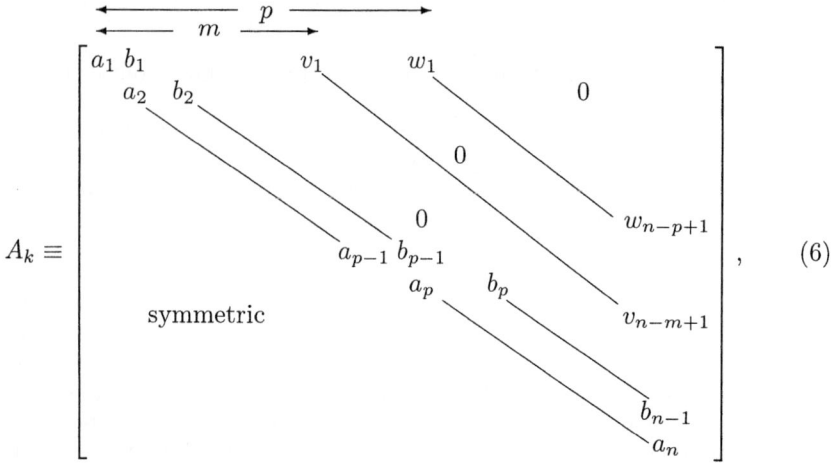

with $A_k = A$ for the Picard iterative method. The solution of 2D non-linear elliptic P.D.E's, using finite difference or finite element method, has been studied in [1, 6].

The performance and applicability of the new composite algorithmic schemes on 3D non-linear elliptic P.D.E. is discussed and numerical results are given.

2 Normalized Explicit Preconditioned Conjugate Gradient Methods

In this section we present a new class of normalized explicit preconditioned conjugate gradient-type schemes, based on the normalized approximate inverses. Let us now assume the normalized approximate factorization such that:

$$A_k \approx D_{r_1,r_2} T^t_{r_1,r_2} T_{r_1,r_2} D_{r_1,r_2}, \quad r_1 \in [1,\ldots,m-1), \quad r_2 \in [1,\ldots,p-1), \quad (7)$$

where r_1, r_2 are the "fill-in" parameters, D_{r_1,r_2} is a diagonal matrix and T_{r_1,r_2} is a sparse upper triangular matrix of the same profile as the coefficient matrix A_k, cf. [5]. The elements of the decomposition factors can be computed by the **M-NOBAR-3D** algorithm, cf. [5]. When the normalized approximate factorization or the incomplete Choleski factorization is used in conjunction with iterative schemes, a forward-backward substitution is involved, which does not parallelize easily, cf. [3, 11]. It has been shown that if we consider a restricted shared memory architecture concerning the number of processors available, the values of the relative speedup and efficiency are the best possible obtained, cf. [3].

Let $M^{\delta l}_{r_1,r_2} = (\mu_{i,j})$, $i \in [1,n]$, $j \in [\max(1, i - \delta l + 1), \min(n, i + \delta l - 1)]$, be the normalized approximate inverse of the coefficient matrix A_k, where δl is the "retention" parameter of elements in the approximate inverse, i.e.:

$$M_{r_1,r_2}^{\delta l} = \left(D_{r_1,r_2} T_{r_1,r_2}^t T_{r_1,r_2} D_{r_1,r_2}\right)^{-1} = D_{r_1,r_2}^{-1} \widehat{M}_{r_1,r_2}^{\delta l} D_{r_1,r_2}^{-1}. \tag{8}$$

Then, the elements of the normalized approximate inverse are computed by the **NORBAIM-3D** algorithm, cf. [4], solving recursively the following systems, cf. [4, 7, 8]:

$$\widehat{M}_{r_1,r_2} T_{r_1,r_2}^t = (T_{r_1,r_2})^{-1} \quad \text{and} \quad T_{r_1,r_2} \widehat{M}_{r_1,r_2} = \left(T_{r_1,r_2}^t\right)^{-1}. \tag{9}$$

Various families of normalized approximate inverses can be derived according to the requirements of accuracy, storage and computational work, cf. [4]. The parallel construction of similar approximate inverses has been studied and implemented in [2], and will be further investigated.

The **N**ormalized **E**xplicit **P**reconditioned **C**onjugate **G**radient (**NEPCG**) method, can be stated as follows:

Let u_0 be an arbitrary initial approximation to the solution vector u. Then,

form $r_0 = s - A_k u_0$, compute $r_0^* = D_{r_1,r_2}^{-1} \widehat{M}_{r_1,r_2}^{\delta l} D_{r_1,r_2}^{-1} r_0,$ (10)

set $\sigma_0 = r_0^*.$ (11)

Then, for $i = 0, 1, \ldots,$ (until convergence) compute the vectors $u_{i+1}, r_{i+1}, \sigma_{i+1}$ and the scalar quantities α_i, β_{i+1} as follows:

form $q_i = A_k \sigma_i,$ and $p_i = (r_i, r_i^*)$ when $i = 0$ only, (12)

evaluate $\alpha_i = p_i/(\sigma_i, q_i),\ u_{i+1} = u_i + \alpha_i \sigma_i,$ and $r_{i+1} = r_i - \alpha_i q_i.$ (13)

Then, form $r_{i+1}^* = D_{r_1,r_2}^{-1} \widehat{M}_{r_1,r_2}^{\delta l} D_{r_1,r_2}^{-1} r_{i+1},$ and $p_{i+1} = (r_{i+1}, r_{i+1}^*),$ (14)

evaluate $\beta_{i+1} = p_{i+1}/p_i,$ and $\sigma_{i+1} = r_{i+1}^* + \beta_{i+1} \sigma_i.$ (15)

The **N**ormalized **E**xplicit **P**reconditioned **C**onjugate **G**radient **S**quare (**NEPCGS**) method has been presented in [4] and can be similarly derived. The computational complexity of the **NEPCG** method is $\approx O[(2\delta l + 14)n$ mults $+3n$ adds$]\nu$ operations, while of the **NEPCGS** is $\approx O[(4\delta l + 27)n$ mults $+8n$ adds$]\nu$ operations, where ν is the number of iterations required for the convergence to a certain level of accuracy.

The parallel implementation and the performance of the Normalized Explicit Preconditioned Conjugate Gradient - type methods on distributed memory and symmetric multiprocessor systems has been studied in [4, 7, 8].

Let us consider the explicit iterative scheme for solving a boundary value problem:

$$u_{i+1} - u_i = \alpha_i \left(\Omega^*\right)^{-1}(s_i - \Omega u_i), \quad i \geq 0, \tag{16}$$

where the operator Ω^* is chosen to be "isomorphous", i.e. to agree closely in a quantitative sense, to the operator Ω, cf. [9]. The isomorphic relationship in the discretized operators Ω_h^*, and Ω_h, has certain similarities to the L_2-comparability between the same operators. Let us consider the following definition, cf. [9]:

Definition 1. The operator $\Omega_h^* \equiv \{w_{i,j}^*\}$, $i,j \in [1,n]$, is said to be δ_i - isomorphous to an operator $\Omega_h \equiv \{w_{i,j}\}$, $i,j \in [1,n]$ if the following relations hold:

$$w_{i,i}^* \equiv w_{i,i} + \sum_{k=i+1}^{n} w_{i,k} + \sum_{k=i+1}^{n} w_{k,i}, \quad i \in [1,n], \tag{17}$$

$$w_{i,j}^* \equiv \sum_{k=1}^{n-j+1} w_{i,j+k-1}, \quad i \in [1, n-1], j \in [2, \mu], \tag{18}$$

$$w_{i,j}^* \equiv \sum_{k=1}^{n-i+1} w_{i+k-1,j}, \quad i \in [2, \mu], j \in [1, n-1], \tag{19}$$

and $\Delta_k^*(\Omega_h^*) = \Delta_k(\Omega_h)$, $k \in [\mu+1, n]$, where Δ^* and Δ are the corresponding diagonal of operators Ω_h^* and Ω_h respectively.

An efficient solution of the linear system $Au = s$ can be obtained by considering a δ_i-isomorphic iterative scheme. Let us consider $\Omega_h^* u^* = s$, where Ω_h^* is a δ_i-isomorphous operator to A. Then assuming that the construction of Ω_h^* can be easily obtained the approximate solution u^* has to be proved that it is an acceptable approximate solution to the original system $Au = s$, cf. [9]. Further details on isomorphic iterative schemes can be found in [9].

3 Numerical Results

In order to demonstrate the applicability and efficiency of the proposed composite iterative schemata, we consider the following 3D non-linear elliptic partial differential equation subject to Dirichlet boundary conditions:

$$\partial^2 u/\partial x^2 + \partial^2 u/\partial y^2 + \partial^2 u/\partial z^2 = -1/2\,(1+e^u), \quad (x,y,z) \in R. \tag{20}$$

The linearized Picard and quasi-linearized Newton iterations are outer iterative schemes of the form:

$$L_h u^{(k+1)} = -\frac{1}{2}\left(1 + e^{u^{(k)}}\right), \tag{21}$$

and

$$L_h u^{(k+1)} + \frac{1}{2}e^{u^{(k)}} u^{(k+1)} = -\frac{1}{2}\left[\left(1 - u^{(k)}\right)e^{u^{(k)}} + 1\right], \tag{22}$$

respectively, with L_h denoting the finite difference operator.

The "fill-in" parameters were set to $r_1 = r_2 = 2$ while the initial guess was $u(0) = 0.1$. The termination criterion for the inner iteration was $\|r_i\|_\infty < 10^{-5}$,

Table 1. The convergence behavior of the composite "inner-outer" iterative scheme for the **NEPCG** method, using δ_i isomorphic schemes

n	m	p	δl	δ_0 isomorphic scheme				δ_3 isomorphic scheme			
				Picard method		Newton method		Picard method		Newton method	
				outer iter.	inner iter.	outer iter.	inner iter.	outer iter.	inner iter.	outer iter.	inner iter.
343	8	50	1	4	36	3	27	3	24	2	18
			2	4	33	3	25	3	18	2	14
			m	4	25	3	19	3	14	2	11
			$2m$	4	23	3	18	3	13	2	9
			p	4	17	3	13	3	12	2	8
729	10	82	1	4	44	3	34	3	26	2	19
			2	4	38	3	30	3	18	2	14
			m	4	29	3	22	3	15	2	11
			$2m$	4	27	3	21	3	14	2	10
			p	4	20	3	16	3	12	2	9
2744	15	197	1	4	59	3	46	3	28	2	21
			2	4	52	3	41	3	19	2	14
			m	4	39	3	29	3	15	2	11
			$2m$	4	36	3	28	3	13	2	10
			p	4	28	3	22	3	13	2	10
6859	20	362	1	4	72	3	59	3	27	2	21
			2	4	63	3	42	3	18	2	14
			m	4	47	3	34	3	15	2	12
			$2m$	3	39	3	31	3	13	2	11
			p	4	33	3	24	3	12	2	10
13824	25	577	1	3	79	3	65	3	28	2	22
			2	3	66	3	50	3	18	2	14
			m	3	51	3	37	3	14	2	12
			$2m$	3	48	3	35	3	13	2	11
			p	3	34	3	26	3	11	2	10

where r_i is the recursive residual and the criterion for the termination of the outer iteration was $\max_{i \in [1,n]} \left| \left(u_j^{(k+1)} - u_j^{(k)} \right) / \left(1 + u_j^{(k)} \right) \right| < 10^{-5}$, $j \in [1, n]$.

Numerical results are presented in Tables 1 and 2 for the composite "inner-outer" iterative scheme in conjunction with the **NEPCG** and **NEPCGS** method respectively, for several values of the order n, semi-bandwidths m and p, and the "retention" parameter δl, using a δ_0 and δ_3 isomorphic schemes.

Finally, it should be stated that the normalized explicit preconditioned conjugate gradient methods can be used efficiently for solving highly non-linear initial value problems.

Table 2. The convergence behavior of the composite "inner-outer" iterative scheme for the **NEPCGS** method, using δ_i isomorphic schemes

n	m	p	δl	δ_0 isomorphic scheme				δ_3 isomorphic scheme			
				Picard method		Newton method		Picard method		Newton method	
				outer iter.	inner iter.	outer iter.	inner iter.	outer iter.	inner iter.	outer iter.	inner iter.
343	8	50	1	3	23	4	18	3	10	3	10
			2	3	19	3	14	3	9	2	7
			m	4	17	3	12	3	6	2	5
			2m	4	16	3	13	3	6	2	5
			p	4	12	3	10	3	5	2	4
729	10	82	1	3	30	3	27	4	14	2	10
			2	3	23	3	17	3	10	2	8
			m	4	21	3	13	4	10	2	7
			2m	4	19	3	12	3	7	2	6
			p	4	15	3	11	3	7	2	6
2744	15	197	1	3	48	2	30	3	12	3	12
			2	3	33	2	20	3	9	2	8
			m	3	28	3	20	3	8	2	7
			2m	3	25	3	19	3	7	3	7
			p	3	16	4	13	3	7	2	6
6859	20	362	1	3	59	2	37	3	13	2	12
			2	3	40	2	24	3	9	2	8
			m	3	34	2	22	3	8	2	7
			2m	3	28	2	18	3	7	3	7
			p	3	21	4	16	3	7	2	6
13824	25	577	1	3	79	2	49	3	13	2	13
			2	3	53	2	32	3	9	2	8
			m	3	48	2	31	3	8	2	7
			2m	3	38	2	25	3	7	3	7
			p	3	24	2	15	3	7	2	6

References

[1] Gravvanis, G.A.: Generalized approximate inverse preconditioning for solving nonlinear elliptic boundary-value problems. I. J. Applied Math. **2(11)** (2000) 1363-1378.
[2] Gravvanis, G.A.: Parallel matrix techniques. In: K. Papailiou D. Tsahalis J. Periaux C. Hirsch M. Pandolfi eds. Computational Fluid Dynamics **I** (1998) 472-477 Wiley.
[3] Gravvanis, G.A., Bekakos, M.P., Efremides, O.B.: Parallel implicit preconditioned conjugate gradient methods for solving biharmonic equations. Sixth Hellenic-European Conf. on Computer Math. and its Applications (2004) (to appear).
[4] Gravvanis, G.A., Giannoutakis, K.M.: Normalized explicit approximate inverse preconditioning for solving 3D boundary value problems on uniprocessor and distributed systems. submitted for publication.

[5] Gravvanis, G.A., Giannoutakis, K.M.: On the rate of convergence and complexity of normalized implicit preconditioning for solving finite difference equations in three space variables. I. J. of Comput. Meth. World Scientific (to appear).
[6] Gravvanis, G.A., Giannoutakis, K.M.: Normalized finite element approximate inverse preconditioning for solving non-linear boundary value problems. In: Bathe KJ editor Computational Fluid and Solid Mechanics 2003 Elsevier Vol. **2** (2003) 1958-1962.
[7] Gravvanis, G.A., Giannoutakis, K.M., Bekakos, M.P.: Parallel finite element approximate inverse preconditioning on symmetric multiprocessor systems. Inter. Conf. on Parallel and Distributed Processing Techn. and Appl. 2004 (to appear).
[8] Gravvanis, G.A., Giannoutakis, K.M., Missirlis, N.M.: A Distributed Normalized Explicit Preconditioned Conjugate Gradient Method. Parallel Algorithms and Applications (to appear).
[9] Lipitakis, E.A., Evans, D.J.: Numerical solution of non-linear elliptic b.v.p. by isomorphic iterative methods. I. J. Comp. Math. **20** (1986) 261-282.
[10] Ortega, J.M., Rheinboldt, W.C.: Iterative solution of non-linear equations in several variables. Academic Press (1970).
[11] Saad, Y., van der Vorst, H.A.: Iterative solution of linear systems in the 20th century. J. Comp. Applied Math. **123** (2000) 1-33.

An Adaptive Two-Dimensional Mesh Refinement Method for the Problems in Fluid Engineering

Zhenquan Li

Department of Mathematics and Computing Science,
The University of the South Pacific, Suva, FIJI
li_z@usp.ac.fj

Abstract. Mesh generation is one of the key issues in Computational Fluid Dynamics. This paper presents an adaptive two-dimensional mesh refinement method based on the law of mass conservation. The method can be used to a governing system that includes the law of mass conservation (continuity equation) for incompressible or compressible steady flows. We show one example that demonstrates the streamlines constructed using the refined mesh is accurate.

1 Introduction

There are a large number of publications on mesh adaptive refinements and their applications. The Berger-Oliger method is one of the well-known adaptive mesh refinements [5]. The refinement criterion for this method is local truncation errors. As the solution progresses mesh points with high local truncation errors are flagged. Fine meshes are created such that all the flagged points are interior to some fine mesh. The method suits for solving hyperbolic partial differential equations on structured computational domains and the refinement factor is the same in both space and time. The method has been extended to other applications [e.g. 4, 3, 1]. The other common methods include h-refinement (e.g. [10]), p-refinement (e.g. [2]) or r-refinement (e.g. [12]), with various combinations of these also possible (e.g. [6, 7]). The overall aim of any adaptive algorithm is to allow a balance to be obtained between accuracy and computational efficiency.

Mass conservation is a key issue for accurate streamline construction of flow fields. In particular, failure to conserve mass can produce errors that cannot be eliminated by reducing the integration step and which can generate artificial effects, such as false spiraling [11]. Mass conservative streamline tracking methods for two-dimensional and three-dimensional CFD velocity fields have produced much more accurate streamlines than non-mass conservative methods [8, 9].

We describe an adaptive mesh refinement method based on the law of mass conservation for two-dimensional incompressible flows in this paper. We assume that f is a scalar function depending only on spatial variables such that its product with the linear interpolation of velocity fields at the nodes of a triangle satisfies the law of mass conservation on the triangle. The criteria for mesh refinement are the scalar functions f not equalling zero or infinity.

The method introduced in this paper can be used for compressible steady flows by replacing the vector fields by the momentum fields.

2 The Mass Conservative Conditions for Linear Interpolations of Vector Fields Over Triangle Domains

Assume that $V_l = AX + B$ is the linear interpolation of a vector field at the three vertexes in a triangle, where A is a 2-by-2 constant matrices, B and X are two-dimensional vertical vector. V_l is unique if the area of the triangle is not zero [8]. Mass conservation for an incompressible fluid means that

$$\nabla \cdot V_l = trace(A) = 0.$$

Let f be a scalar function depending only on spatial variables. We assume that $f V_l$ satisfies the law of mass conservation first and then calculate the expressions of f.

For incompressible flows, the mass conservation means

$$\nabla \cdot (f V_l) = 0,$$

or in Cartesian coordinates,

$$\frac{\partial (f V_x)}{\partial x} + \frac{\partial (f V_y)}{\partial y} = 0, \quad (1)$$

where $V_l = (V_x, V_y)^T$. If $V_l = AX + B$, then (1) can be written as

$$\frac{df}{dt} = -(a_{11} + a_{22}) f \quad (2)$$

where $\frac{df}{dt} = \frac{\partial f}{\partial t} + V_x \frac{\partial f}{\partial x} + V_y \frac{\partial f}{\partial y}$ is the material derivative. We can calculate the expressions of f from (2).

Table 1 shows the Jacobian of matrix A and corresponding expressions of f for all possible cases in which the linear interpolations of the vector fields over triangle domains do not hold the law of mass conservation.

The conditions (MC) that $f V_l$ satisfies the law of mass conservation in its triangle domains are the functions f in Table 1 not equalling zero or infinity in these triangles.

3 The Adaptive Mesh Refinement Method

This section presents how to refine a given mesh. The unstructured mesh is usually used in practice and the most of the elements are quadrilateral. The adaptive refinement method is for each element in a mesh. Fig.1 is a quadrilateral element of a mesh The conditions (MC) are for triangles only. The following process describes how to use the conditions (MC) to refine a quadrilateral element in a given mesh.

Table 1. Jacobian and expressions of f for all possible cases of a non mass conservative linear field

Case	Jacobian	f
1	$\begin{pmatrix} r_1 & 0 \\ 0 & r_2 \end{pmatrix} (0 \neq r_1 \neq r_2 \neq 0)$	$\dfrac{C}{\left(y_1 + \dfrac{b_1}{r_1}\right)\left(y_2 + \dfrac{b_2}{r_2}\right)}$
2	$\begin{pmatrix} r_1 & 0 \\ 0 & 0 \end{pmatrix} (r_1 \neq 0)$	$\dfrac{C}{y_1 + \dfrac{b_1}{r_1}}$
3	$\begin{pmatrix} r & 1 \\ 0 & r \end{pmatrix} (r \neq 0)$	$\dfrac{C}{\left(y_2 + \dfrac{b_2}{r}\right)^2}$
4	$\begin{pmatrix} \mu & \lambda \\ -\lambda & \mu \end{pmatrix} (\mu \neq 0, \lambda \neq 0)$	$\dfrac{C}{\left(y_1 + \dfrac{\mu b_1 - \lambda b_2}{\mu^2 + \lambda^2}\right)^2 + \left(y_2 + \dfrac{\lambda b_1 + \mu b_2}{\mu^2 + \lambda^2}\right)^2}$

where c is a constant.

The refinement process is as follows.

1) Subdivide the quadrilateral into two triangles and check if V_l satisfies the law of mass conservation on both triangles. If yes, no refinement for the quadrilateral is required. If no, go to Step 2.

Fig. 1. An element in a mesh

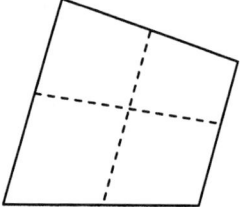

Fig. 2. Subdividing the element into four elements

2) Apply the conditions (MC) to both of the triangles. If the conditions (MC) are satisfied on both triangles, there is no need to subdivide the element. Otherwise, we need to subdivide the element into a number of small elements such that the lengths of all sides of the small elements are truly reduced (e.g. half). Fig. 2 is an example that subdivides an element into four small quadrilaterals by connecting the midpoints of its non-neighboring sides. Fig. 3 is an example that subdivides the element into nine small quadrilaterals.
3) Take the elements in the subdivided quadrilateral in Fig. 2 or Fig. 3 as new elements of the mesh by replacing the initial element in Fig. 1 and repeat these three steps until a pre-specified threshold number T is reached.

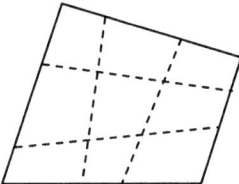 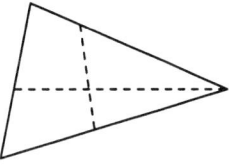

Fig. 3. Subdividing the element into nine elements **Fig. 4.** Subdividing the triangle into small elements

The threshold number T equals one in Fig. 2. A triangle is a special quadrilateral and can also be subdivided as the same as the element in Fig. 1. Fig. 4 shows a triangle can be subdivide into four elements.

It is the fact that every simple closed polygon with more than three vertices can be split into only triangles [13]. Thus the conditions (MC) can be used to any shape of elements in a mesh. Because this adaptive mesh refinement method is element based, it can be used to both structured and unstructured meshes.

The connectivity of the initially generated mesh is given while generating the mesh. The connectivity is easy to generate in the refined elements. Since the adaptive refinement method is element based, the connectivity is easy to work out for the refined mesh by making some appropriate changes to the initial connectivity and then inserting the connectivity of refined elements.

4 Example

One example is shown here to demonstrate the effectiveness of the adaptive mesh refinement method. In the example, exact velocity field is used to show that the values at the nodes of the refined mesh can present the field very well by comparing the exact streamlines with the constructed streamlines using the values. We use the exact streamline expressions for linear vector fields in streamline drawing processes [14]. The threshold number T can be infinity in the example but we can only choose T as an integer. The bigger the threshold number T (or the more number of refinements), the more accurate the streamlines can be constructed based on the values at the nodes of the refined mesh. We refined the mesh in the example for T=4. The red lines in the following figures are the exact streamlines, the green line is the streamline constructed using the refined mesh, and the blue line is the streamline constructed using the unrefined mesh. The example here is using a structured mesh and the subdivision schemes subdivide a square into four equal squares.

Example: Velocity Field

$$V = \left(26y + x^2 y, 16x - xy^2\right)^T.$$

The closed streamline can be used to check if a numerical method is accurate. We here use a closed streamline to check how well the refined mesh can be used to present the velocity field.

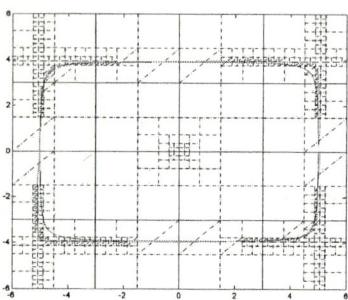
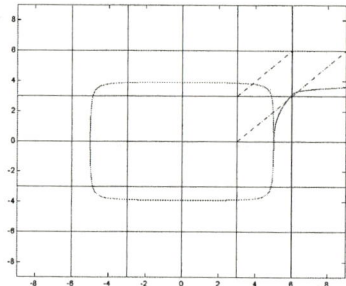

Fig. 5. Constructed streamline (green) using refined mesh with seed at (5, 0)

Fig. 6. Constructed streamline (blue) using initial mesh with seed at (5, 0)

The streamline with the same seed using the initial mesh is shown in Fig. 6. It is clear that the values of the velocity field at the nodes of the refined mesh present the exact velocity field very well by comparing the streamlines in the above latter two figures. The error is about 0.05 using the refined mesh, i.e., the x-coordinate of the intersection of the green streamline with the positive x-axis is about 5.05 with starting from seed point $(5, 0)^T$.

5 Discussion

The adaptive mesh refinement method given in this paper can find the accurate positions of some phenomena such as the core of vortex by refining the regions in which the structure of flows may be complicated. The applications of the adaptive mesh refinement method in practice will be further research topics.

References

1. Almgren, A., Bell, J., Colella, P., Howell, L., Welcome, M.: A conservative adaptive projection method for the variable density incompressible Navier-Stokes equations. J. Comp. Phys. 142 (1998) 1–46

2. Babuska, I., Szabo, B.A., Katz, I.N.: The p-Version of the Finite Element Method. SIAM. J. on Numer. Anal. 18 (1981) 515–545
3. Bell, J., Berger, M., Saltzman, J., Welcome, M.: Three dimensional adaptive mesh refinement for hyperbolic conservation laws. J. Sci. Comp. 15(1)(1994) 127–138
4. Berger, M.J., Colella, P.: Local adaptive mesh refinement for shock hydrodyamics. J. Comp. Phys. 82(1)(1989) 64–84
5. Berger, M.J., Oliger, J.: Adaptive Mesh Refinement for Hyperbolic Partial Differential Equations. J. Comp. Phys. 53(1)(1984) 484–512
6. Capon, P.J., Jimack P.K.: An Adaptive Finite Element Method for the Compressible Navier-Stokes Equations. In: Baines, M.J., Morton, K.W. (eds.): Numerical Methods for Fluid Dynamics 5. OUP (1995) 327–334
7. Demkowicz, L., Oden, J.T., Rachwicz, W., Hardy, O.: An h-p Taylor-Galerkin Finite Element Method for the Compressible Euler Equations. Comp. Meth. In Appl. Mech and Eng. 88(1991) 363–396
8. Li, Z.: A Mass Conservative Streamline Tracking Method for Two Dimensional CFD Velocity Fields. J. Flow Visual. And Image Proc. 9(3)(2002) 75–87
9. Li, Z.: A Mass Conservative Streamline Tracking Method for Three Dimensional CFD Velocity Fields. Proc. FEDSM'03, FEDSM2003-45526 (2003) 1–6
10. Lohner, R.: An Adaptive Finite Element Scheme for Transient Problems in CFD. Comp. Meth. in Appl. Mech. and Eng., 61(1987) 323-338
11. Mallinson, G: The Calculation of the Lines of a Three Dimensional Vector Fields. In: Davies, G. de Vahl, Fletcher, C. (eds.): Computational Fluid Mechanics, North Holland (1988) 525-534
12. Miller, K., Miller, R.: Moving Finite Elements. Part I. SIAM J. on Numer. Anal., 18(1981) 1019-1032
13. Nielson, G. M.: Tools for Triangulations and Tetrahedrizations and Constructing Functions Defined over Them. In: Nielson, G.M., Hagen, H., Mueller, H., (eds.): Scientific Visualization: Overviews, Methodologies, and Techniques. IEEE CS Press (1997) 429-526
14. 14 Nielson, G. M., Jung, I.-H., Srinivasan, N., Sung, J., Yoon, J.-B.: Tools for Computing Tangent Curves and Topological Graphs for Visualizing Piecewise Linearly Varying Vector Fields over Triangulated domains. In: Nielson, G.M., Hagen, H., Mueller, H., (eds.): Scientific Visualization: Overviews, Methodologies, and Techniques. IEEE CS Press (1997) 527-562

High Order Locally One-Dimensional Method for Parabolic Problems

Samir Karaa

Department of Mathematics and Statistics, Sultan Qaboos University,
P. O. Box 36, Al-khod 123, Muscat, Sultanate of Oman
skaraa@squ.edu.om

Abstract. We propose a high order locally one-dimensional scheme for solving parabolic problems. The method is fourth-order in space and second-order in time. It is unconditionally stable and provides a computationally efficient implicit scheme. Numerical experiments are conducted to test its high accuracy and to compare it with other schemes.

1 Introduction

Let Ω be a rectangular domain in $\mathbb{R}^2$, and let $J = (0, T]$ be a time interval, $T > 0$. Consider the parabolic problem

$$\frac{\partial u}{\partial t} - \nu \nabla^2 u + \boldsymbol{c} \cdot \nabla u + \eta u = 0, \quad \text{in } \Omega \times (0, T], \tag{1a}$$

$$u(x, y, t) = g(x, y, t), \quad (x, y) \in \partial\Omega, \ t \in J, \tag{1b}$$

$$u(x, y, 0) = u_0(x, y), \quad (x, y) \in \Omega, \tag{1c}$$

where the coefficients $\nu > 0$, η and the velocity vector $\boldsymbol{c} = (c_x, c_y)^T$ are constants, and g and u_0 are given functions of sufficient smoothness. Parabolic equations arise in various applications, including the study of flows in porous media and modeling of economic processes.

Split schemes such as alternating direction implicit (ADI) [2, 10, 3] and locally one-dimensional (LOD)[4, 13] methods have proved valuable in the approximation of the solutions of parabolic and elliptic problems in two and three variables. In this paper we are concerned with highly accurate numerical solution of (1) with split implicit schemes. A high order split formula was first obtained by Michell and Faireweather [8] for diffusion problems in two space dimensions. High order compact split schemes have been constructed for the wave equation in [7] and [1]. In [6], we derived a compact ADI method for solving unsteady convection-diffusion problems. In this paper, we propose a high order locally one-dimensional solution method for solving parabolic problems. The scheme is fourth-order accurate in space and second-order accurate in time and does not need much more work than the classical second-order LOD scheme.

2 High Order LOD Scheme

Let's first examine the one-dimensional parabolic equation

$$-\nu \frac{d^2 u}{dx^2} + c_x \frac{du}{dx} + \bar{\eta} u = f(x), \tag{2}$$

where $\nu > 0$, c_x and $\bar{\eta}$ are constants, and f is a function of x. The standard second-order central difference operators for the first and second derivatives of u are denoted by δ_x and δ_x^2, respectively. Using techniques from [12], it is easy to derive a three-point fourth-order scheme for (2) as

$$B_x u_i = L_x f_i + O(\Delta x^4),$$

where the finite difference operators B_x and L_x are respectively defined by

$$L_x = 1 + \frac{\Delta x^2}{12}\left(\delta_x^2 - \frac{c_x}{\nu}\delta_x\right),$$

$$B_x = \bar{\eta} - \left(\nu + \frac{c_x^2 \Delta x^2}{12\nu} - \frac{\bar{\eta}\Delta x^2}{12}\right)\delta_x^2 + \left(c_x - \bar{\eta}\frac{c_x \Delta x^2}{12\nu}\right)\delta_x.$$

A fourth-order semi-discrete approximation to the 2-D parabolic equation in (1) can now be easily obtained. If we let $\bar{\eta} = \eta/2$ and assume that (1a) is approximated on a uniform grid with mesh sizes Δx and Δy in the x- and y-direction, respectively, then we have the semi-discrete approximation

$$\frac{\partial u^n}{\partial t} = -(L_x^{-1} B_x + L_y^{-1} B_y) u^n + O(\Delta^4), \tag{3}$$

where u^n is the approximate solution at time $t^n = n\Delta t$, $n \geq 0$, and $O(\Delta^4)$ denotes the $O(\Delta x^4) + O(\Delta y^4)$ term. Here the meaning of the notations L_y and B_y is obvious. The Crank-Nicolson implicit discretization of (3) is

$$\frac{u^{n+1} - u^n}{\Delta t} = -\frac{1}{2}(L_x^{-1} B_x + L_y^{-1} B_y)(u^{n+1} + u^n) + O(\Delta^4) + O(\Delta t^2), \tag{4}$$

and after applying to both sides with $L_x L_y$, we have the (9,9) approximation

$$L_x L_y \frac{u^{n+1} - u^n}{\Delta t} = -\frac{1}{2}(L_y B_x + L_x B_y)(u^{n+1} + u^n) + O(\Delta^4) + O(\Delta t^2). \tag{5}$$

This approximation is obviously of order two in time and four in space. Linear systems arising from (5) are expensive to solve by direct methods based on Gaussian elimination. Krylov subspace methods, are generally efficient, however, they may be expensive to use at each time step, and especially for higher dimensional problems. A way around developing an efficient solution method to solve (3), is to use a locally one-dimensional scheme which is written as the pair of equations

$$\frac{1}{2}\frac{\partial u}{\partial t} = -L_x^{-1} B_x u^n, \tag{6a}$$

$$\frac{1}{2}\frac{\partial u}{\partial t} = -L_y^{-1} B_y u^n. \tag{6b}$$

In advancing a calculation from t^n to t^{n+1}, it is assumed that Equation (6a) holds from t^n to $t^{n+1/2}$, and Equation (6b) holds from $t^{n+1/2}$ to t^{n+1}. The Crank-Nicolson implicit discretization of (6a) is

$$\frac{1}{2}\frac{u^{n+1/2} - u^n}{\Delta t/2} = -L_x^{-1} B_x \left(\frac{u^{n+1/2} + u^n}{2}\right) + O(\Delta^4) + O(\Delta t^2). \tag{7}$$

Applying the operator $\Delta t L_x$ to both sides of (7), we obtain

$$L_x(u^{n+1/2} - u^n) = -\Delta t B_x \left(\frac{u^{n+1/2} + u^n}{2}\right) + O(\Delta t \Delta^4) + O(\Delta t^3). \tag{8}$$

After rearrangements and dropping the error terms, we have

$$\left(L_x + \frac{\Delta t}{2} B_x\right) u^{n+1/2} = \left(L_x - \frac{\Delta t}{2} B_x\right) u^n, \tag{9a}$$

$$\left(L_y + \frac{\Delta t}{2} B_y\right) u^{n+1} = \left(L_y - \frac{\Delta t}{2} B_y\right) u^{n+1/2}, \tag{9b}$$

where the second relation is derived from (6b) using the same manipulations. Eliminate the intermediate solution $u^{n+1/2}$ and compare the resulting scheme to (5), we verify that the difference is the term

$$\frac{\Delta t^2}{4} B_y B_x (u^{n+1} - u^n) = \frac{\Delta t^3}{4} B_y B_x \frac{\partial u^n}{\partial t} + O(\Delta t^4)$$
$$= O(\Delta t^3) + O(\Delta t^3 \Delta^2) + O(\Delta t^4).$$

Thus, we added to (5) an extra term which is of similar order to its truncation error. It then follows that the split scheme (9) is second-order in time and fourth-order in space.

To determine the stability of (9), we notice that, because of consistency, the lower order term ηu in (1a) contributes to the expression of the amplification factor of (9) only terms that are $O(\Delta t)$. It follows that the scheme is stable if and only if it is stable without this undifferentiated term [11]. The unconditional stability of (9) follows then from the results in [6].

3 Numerical Experiments

We conduct numerical experiments to test the proposed LOD scheme. We first consider the parabolic problem

$$\frac{\partial u}{\partial t} - \nabla^2 u - \pi^2 u = 0, \quad \text{in } \Omega \times (0, T], \tag{10}$$

where $\Omega = (0,1) \times (0,1)$. The exact solution for this problem is given by

$$u(x,y,t) = e^{-\pi^2 t} \sin(\pi x) \sin(\pi y).$$

The initial and Dirichlet boundary conditions are set to satisfy the exact solution. We solve the problem on uniform grids with different mesh sizes and compare the accuracy of the computed solutions from the present high order LOD scheme and the classical second-order LOD scheme. The quantity that we compare is the L^2-norm error of the computed solution with respect to the exact solution. We choose a time step $\Delta t = 0.001$ and $T = 1$.

Fig. 1 displays the L^2-norm errors at each time step in each case. The results show the superiority of the present LOD scheme over the classical scheme. The error obtained on a 10×10 grid is much smaller than the one obtained using the second-order LOD scheme on a 40×40 grid.

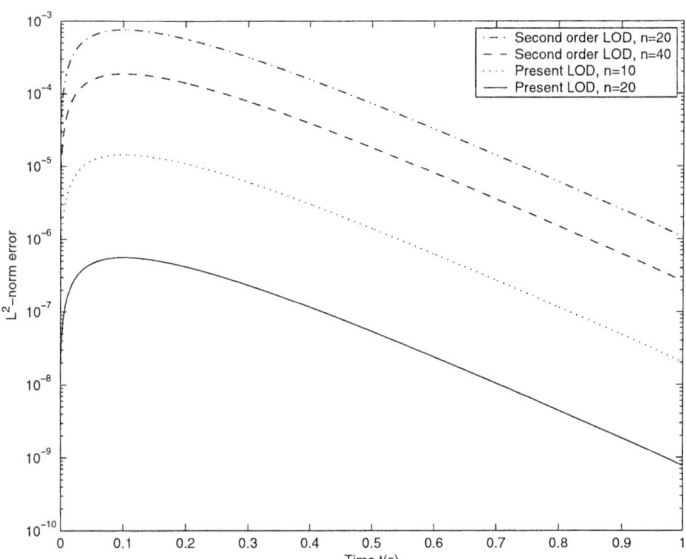

Fig. 1. Comparison of the L^2-norm errors produced by the pres-ent LOD scheme and the second-order LOD scheme at each time level

In the second test problem, we consider a convection-diffusion problem in the square region $(0,2) \times (0,2)$. We set $\eta = 0$, $\nu = 0.01$ and $c_x = c_y = 0.8$ as in [9]. An analytical solution for this problem is given by

$$u(x,y,t) = \frac{1.2}{4t+1} \exp\left[-\frac{(x-c_x t - 0.5)^2}{\nu(4t+1)} - \frac{(y-c_y t - 0.5)^2}{\nu(4t+1)}\right].$$

The initial condition is a Gaussian pulse centered at $(x, y) = (0.5, 0.5)$ with a pulse height 1.2. At $t = 1.25$, the pulse moves to a position centered at $(x, y) = (1.5, 1.5)$ with a pulse height $1/5$.

Table 1 lists the L^2-norm errors and the total elapsed time (CPU) in seconds delivered by the two LOD schemes, the spatial third-order nine-point compact scheme of Noye and Tan [9], and the fourth-order $(9,9)$ compact scheme of Kalita et al. [5]. The data show that the present LOD scheme provides the most accurate solution. From the table, we can also see that the two LOD schemes deliver very small CPU times. Almost seven times smaller than the one delivered by Noye and Tan scheme. Hence, the new LOD scheme is the most effective in terms of accuracy and time consumption.

Contour plots of the exact and numerically approximated pulses in the subregion $1 \leq x, y \leq 2$ are drawn in Fig. 2 for the two LOD schemes. The figure shows that, similar to what have been observed in [9] and [5], the present LOD scheme captures very well the moving pulse, yielding a pulse centered at $(1.5, 1.5)$ and indistinguishable from the exact one. The second-order LOD scheme produces however a pulse distorted in the x- and y-directions, owing to the fact the

Table 1. L^2-norm errors at $T = 1.25$ s and CPU times delivered by four different schemes, with $\Delta t = 0.00625$ and $\Delta x = \Delta y = 0.025$

Method	L^2-norm error	CPU time (s)
Noye and Tan	1.29×10^{-4}	29.5
Kalita et al.	1.23×10^{-4}	21.2
second-order LOD	2.43×10^{-3}	4.9
Present LOD	6.62×10^{-5}	3.9

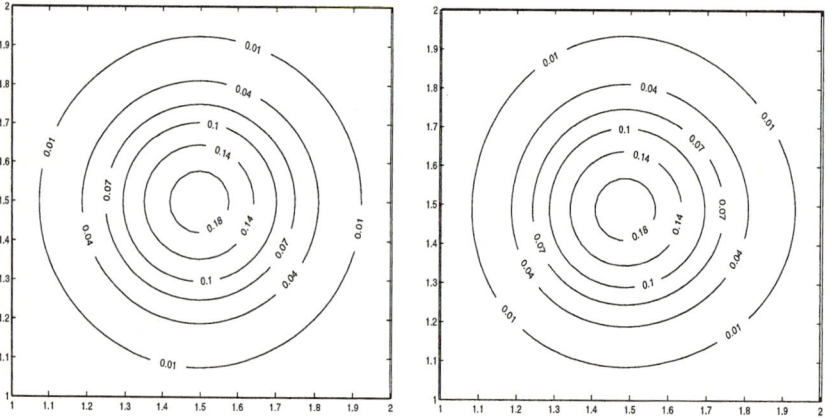

Fig. 2. Contour plots of the pulse in the subregion $1 \leq x, y \leq 2$ at $T = 1.25$ s, (a) Present LOD scheme, and (b) second-order LOD scheme, with $\Delta t = 0.00625$

second-order error terms of the method is related to the wave numbers in both directions, as is explained in [9]. We finally notice that, GMRES/ILU(0) was used to solve the sparse linear systems arising from Noye and Tan discretization (and the discretization of Kalita et al.). All programs are written in Fortran 77 and were run on a SunBlade 100 machine.

4 Concluding Remarks

We proposed a high order LOD scheme for solving parabolic problems. The scheme is of order 2 in time and order 4 in space. It is unconditionally stable and allows a considerable saving in computing time. Numerical tests are given to validate its high accuracy and to show its superiority over other schemes, in terms of accuracy and computational cost.

References

1. Ciment, M., Leventhal, S. H.: Higher order compact implicit schemes for the wave equation. Math. Comp. **29** (1975) 985–994
2. Douglas, J., Peaceman, D. W.: Numerical solution of two-dimensional heat flow problems. American Institute of Chemical Engineering Journal. **1** (1959 505–512
3. Douglas, J.: On the numerical integration of $\frac{\partial^2 u}{\partial x^2} + \frac{\partial^2 u}{\partial y^2} = \frac{\partial u}{\partial t}$ by implicit methods. J. Soc. Ind. Appl. Math. **3** (1959) 42–65
4. D'yakonov, E.: Difference schemes with splitting operators for multidimensional unsteady problems (English translation). URSS Comp. Math. **3** (1963) 581–607
5. Kalita, J. C., Dalal, D. C., Dass, A. K.: A class of higher order compact schemes for the unsteady two-dimensional convection-diffusion equation with variable convection coefficients. Int. J. Numer. Methods Fluids **38** (2002) 1111–1131
6. Karaa, S., Zhang, J.: High order ADI method for solving unsteady convection-diffusion problems. J. Comput. Phys. **198** (2004) 1–9
7. McKee, S.: High accuracy ADI methods for hyperbolic equations with variable coefficients. J. Inst. Maths. Applics. **11** (1973) 105-109
8. Mitchell, A. R., Fairweather, G.: Improved forms of the alternating direction methods of Douglas, Peaceman, and Rachford for solving parabolic elliptic equations. Numer. Math. **6** (1964) 285–292
9. Noye, B. J., Tan, H. H.: Finite difference methods for solving the two-dimensional advection-diffusion equation. Int. J. Numer. Methods Fluids **26** (1989) 1615–1629
10. Peaceman, D. W., Rachford Jr., H. H.: The numerical solution of parabolic and elliptic differential equations. J. Soc. Ind. Appl. Math. **3** (1959) 28–41
11. Strikwerda, J. C.: Difference Schemes and Partial Differential Equations, Wadsworth & Brooks/Cole, Pacific Grove, CA , 1989.
12. Spotz, W. F., High-Order Compact Finite Difference Schemes for Computational Mechanics. Ph.D. Thesis, University of Texas at Austin, Austin, TX, 1995.
13. Yanenko, N.: Convergence of the method of splitting for the heat conduction equations with variable coefficients (English tradition). USSR Comp. Math. **3** (1963) 1094–1100

Networked Control System Design Accounting for Delay Information*

Byung-In Park[1] and Oh-Kyu Kwon[2]

[1] FA Engineering Group 1, Samsung Electronics Co, Ltd.,
Hwasung, Gyeonggi-Do, 445-701, Korea
bi95.park@samsung.com
[2] School of Electrical Engineering, INHA University,
Incheon, 402-751, Korea
okkwon@inha.ac.kr

Abstract. In this paper a networked control system (NCS) design problem is discussed where a single digital controller and multiple plants interconnected by a common computer network with transfer time delays. The network-induced delays are assumed to be unknown constant in this paper, and a measuring method for delays in TCP/IP network is proposed to compensate the performance degradation due to the delay effect. An NCS is designed with LQR algorithm and applied to an inverted pendulum accounting for the multiple time delays as a benchmark plant to check its performance under time delays induced by the network.

1 Introduction

A control system communicating with sensors and actuators over a communication network is called the networked control systems (NCS). In these days, most of the industrial control systems are requiring decentralization of control, modularity and low costs, and so analysis, modeling and control of NCS have been investigated in the literature. There are many different network types in control systems: Ethernet bus, ControlNet, and DeviceNet, *etc*. These common-bus network architectures for distributed control systems offer several advantages in industrial area, and they have been a major trend in industrial fields [1-5].

By the way, the communication architecture change from point-to-point type to common-bus type introduces different forms of time delays between sensors, actuators, and controllers. Such time delays could be constant, bounded, or random depending on the network protocols and status. The network-induced delay can be larger than the sampling time, which makes the system performance degraded and even be a cause of instability. Therefore, the problem related to the time delays on NCS has been studied for several years. Analyses of time-delay systems have been investigated mainly in the frequency domain for single-delay cases. For multiple delay cases, LMI (Linear Matrix Inequality) or functional approach is used. In the functional approach, the controller

---

* This work is supported by Inha University under grant INHA30293.

design adopts robust or stochastic control approach. But unfortunately, these methods are too complicated to be implemented in real NCS [6-10].

This article proposes an NCS implementation method to solve the network-induced delay problem. Network-induced delays are composed of two parts. One is the delay from controller to plant, and another is from plant to controller. They are assumed to be unknown constant in this paper, and the plant and controller are discretized for the computer implementation. In this paper, networked control system with constant delays is designed for an inverted pendulum, which is an unstable systems frequently used as a benchmark plant. Therefore, time-delays are very important factors in the NCS to be designed. To implement NCS for inverted pendulum, a measuring method for the time delays is proposed and tested in the local area network environment. The control algorithm is based on the standard LQ regulator [5]. To apply the LQ regulator the NCS model is transformed to a standard model with delayed states as state variable.

2 NCS Configuration and System Description

Suppose that networked control systems are composed of two computers and two hubs shown in Fig. 1, where one computer is used as the controller and another as sensor and actuator. An inverted pendulum is to be taken as the controlled plant, which is known to be nonlinear and unstable, and so frequently used as a benchmark plant. So the cart position and the rod angle must be controlled at the same time. LQR algorithm has been used to control the inverted pendulum. The information of the cart position and the rod angle is converted by hardware such as AD/DA device. The linearized model of the inverted pendulum at an equilibrium point is described by the continuous state-space equation as follows:

$$\dot{x}(t) = A_p x(t) + B_p u(t) \qquad (1)$$
$$y(t) = C_p x(t),$$

where A_p, B_p, C_p are constant matrices.

The NCS configuration can be described by the block diagram of Fig. 2 [1]. This is a MIMO model with multi sensor input and multi actuator output. Each input/output node is assumed to have different delay scale. In order to construct a digital closed-loop control system, the system should be discretized considering the network delay. Taking the sampling time and actuator delay induced by the network into consideration, Eq. (1) can be discretized as follows:

$$x_{k+1} = A x_k + \sum_{j=0}^{1} B_k^j v_{k-j} \qquad (2)$$

$$A \equiv \exp(A_p T), \quad \Gamma(T,q) \equiv \exp(A_p(T-q))B_p,$$
$$B_m^1(k) \equiv \int_0^{a_m} \Gamma_m(T,q)dq, \quad B_m^0(k) \equiv \int_{a_m}^{T} \Gamma_m(T,q)dq,$$

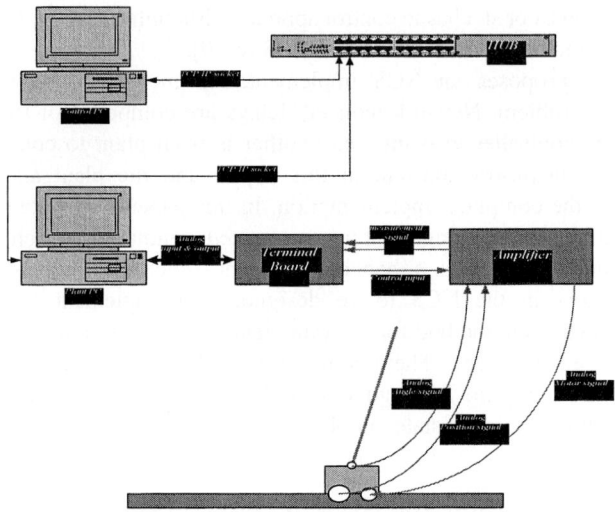

Fig. 1. NCS configuration for an inverted pendulum

where Γ_m and a_m denotes the m-th element of Γ and the m-th actuator delay, respectively. Considering sensor delay δ, Eq. (2) is converted to the equation as follows:

$$\mathbf{x}((k+1-\delta)T) = \mathbf{A}_\delta \mathbf{x}_k + \sum_{j=0}^{1} \mathbf{B}_{\delta k}^j \mathbf{v}_{k-j}, \tag{3}$$

where $\mathbf{A}_\delta = \exp(\mathbf{A}_p(T-\delta))$. Since the NCS problem requires a multivariable control law, the LQR algorithm [5,6] is adopted in this paper. In order to apply LQR algorithm to the above system, a new state variable is taken as follows:

$$\mathbf{z}(k) = \left[\mathbf{x}_s(k)^T, \mathbf{v}(k-1)^T, \mathbf{v}(k-2)^T\right]^T \tag{4}$$

$$\mathbf{z}(k+1) = \begin{bmatrix} \mathbf{A}_{x_s} & \mathbf{B}_{x_s}^1 & \mathbf{B}_{x_s}^2 \\ 0 & 0 & 0 \\ 0 & \mathbf{I} & 0 \end{bmatrix} \mathbf{z}(k) + \begin{bmatrix} \mathbf{B}_{x_s}^0 \\ \mathbf{I} \\ 0 \end{bmatrix} \mathbf{v}(k). \tag{5}$$

And the cost function for NCS and the optimal controller are as follows:

$$V = \sum_{k=0}^{N} \left\{ \mathbf{x}_s(k)^T \mathbf{Q}_s \mathbf{x}_s(k) + \sum_{t=0}^{2} \mathbf{v}(k)^T \mathbf{R}_t \mathbf{v}(k) \right\} \tag{6}$$

$$\mathbf{v}^*(k) = -\mathbf{K}_s(k)\mathbf{x}_s(k) - \sum_{i=1}^{2} \mathbf{K}_v^i(k)\mathbf{v}^*(k). \tag{7}$$

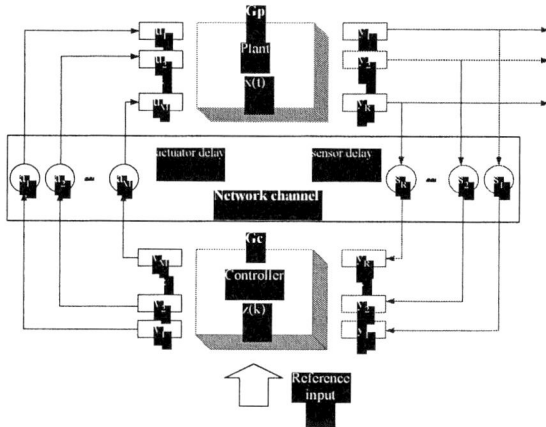

Fig. 2. Block diagram of NCS with time delays

This LQ optimal control algorithm utilizes the measured state and accounts for the estimated delay information.

3 Delay Measurement and NCS Simulations

An NCS design method is proposed to account for the network-induced delays. However, the size of delays is unknown, and so it is estimated by a test method in this paper. The test program is a modified ping program based on TCP/IP. The NCS under consideration is composed of two computers and two hubs in a building with local area network. Tested network route is from the sensor computer to the control computer through two hubs, where the distance between two hubs of 10 Mbps is about 30m. The test is performed by transmitting 40 bytes test data, which is similar size to that of real control signal in the NCS, and measuring the elapse time to receive the acknowledge signal. The measured time delays are almost about 16~20 ms, and they are seldom over 40ms. So delays larger than 40ms are not considered in this paper. Each measured value of network-induced delays is used to compute the control signal in Eq. (2)-(7).

In order to exemplify the performance of the proposed method, some network-induced delays are generated and tested in the NCS with inverted pendulum. The sampling time is taken as 40ms, and measured actuator and sensor delays are 10 and 16 ms, respectively. Then, the LQR control gain is obtained accounting for the delay. By Eq. (5), the state equation accounting for the time delays is given as follows:

$$z(k+1) = \begin{bmatrix} 1 & -0.0017 & 0.0236 & 0 & 0.0013 & 0 \\ 0 & 1.0202 & 0.0213 & 0.0302 & -0.0043 & 0 \\ 0 & -0.1069 & 0.6032 & -0.0017 & 0.0580 & 0 \\ 0 & 1.3209 & 1.3131 & 1.0202 & -0.1902 & 0 \\ 0 & 0 & 0 & 0 & 0 & 0 \\ 0 & 0 & 0 & 0 & 1 & 0 \end{bmatrix} z(k) + \begin{bmatrix} 0.0002 \\ -0.0005 \\ 0.0285 \\ -0.0819 \\ 1 \\ 0 \end{bmatrix} v(k),$$

where $\mathbf{R} = 0.0003$, $\mathbf{R}_1 = 0.0002$, $\mathbf{R}_2 = 0.00001$ and

$$\mathbf{Q} = \begin{bmatrix} 0.025 & 0 & 0 & 0 & 0 & 0 \\ 0 & 2 & 0 & 0 & 0 & 0 \\ 0 & 0 & 0 & 0 & 0 & 0 \\ 0 & 0 & 0 & 0 & 0 & 0 \\ 0 & 0 & 0 & 0 & 0.0002 & 0 \\ 0 & 0 & 0 & 0 & 0 & 0.00001 \end{bmatrix}.$$

Then, the optimal LQR gain of the NCS is computed as follows:

$$\mathbf{K} = \begin{bmatrix} -143213 & -737166 & -251758 & -104192 & 0.6429 & 0 \end{bmatrix}.$$

Fig. 3 shows the effect of the network-induced delay. It is shown that the non-NCS LQR works well with short time delays less than 5 ms, but that the tested delays of 10~16 ms in the actuator and the sensor make the controller useless as shown in Fig. 3-(a). On the other hand, the NCS LQR accounting for the delays works well even in the case of delays, which is shown in Fig. 3-(b).

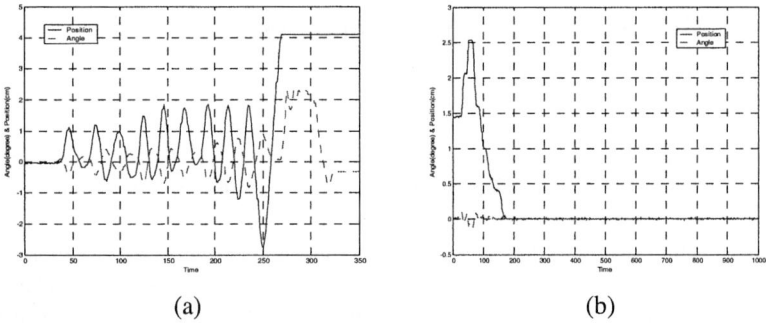

(a)   (b)

Fig. 3. Simulation result : (a) without compensation of delay, (b) accounting for delay (solid line : position, dashed line : angle)

4 Conclusions

In this paper, an NCS is designed and implemented using two computers and two hubs, and applied to an inverted pendulum as the benchmark plant. Since the network based control problems are to generate network-induced delays, NCS LQR technique is modified to account for the delays, and a simple method is proposed to measure the network-induced delays. Via some simulations, various time delays are generated to check the performance of the NCS proposed. Simulation results show that the NCS works well even in the case of various time delays induced by the network. It is noted that the permissible range of the network-induced delays depends on the sampling time and the plant dynamics in NCS, which requires further research.

References

1. Lian, F., Moyne, J., Tilbury, D.: Modeling and Optimal Controller Design of Networked Control Systems with Multiple Delays, Int. J. of Control, **76** (2003) 591-606
2. Kirk, D.E.: Optimal Control Theory, Prentice Hall (1989)
3. Hart, S., Vozdolsky, N., Djaferis, T.E.: A Class of Networked Control Systems, Proc. IEEE Conf. on Decision and Control, **41** (2002) 1643-1648
4. Shin, K.G., Chou, C.: Design and Evaluation of Real-time Communication for Fieldbus-based Manufacturing Systems, IEEE Trans. on Robotics and Automation, **12** (1996) 357-367
5. Burl, J.B.: Linear optimal Control, Addison Wesley (1999)
6. Nilson, J., Bernhardsson, B.: LQG Controller over a Markov Commu-nication Network, Proc. IEEE Conf. on Decision and Control, **36** (1997) 4586-4591
7. Zhen, W., Hong, L.C., Ying, X.J.: A Class of Networked Control Systems, Proc. IEEE Conf. on Decision and Control, **41** (2002) 1643-1648
8. Yoo, H.J.: Design of Network Control Systems with Time Delay, PhD. Dissertation, Inha University (2003)
9. Astrom, K.J.: Computer Controlled systems, 3rd edn. Prentice Hall (1997)
10. Nilson, J.: Real-Time Control Systems with Delays, Lund Institute (1992)

Eidolon: Real-Time Performance Evaluation Approach for Distributed Programs Based on Capacity of Communication Links*

Yunfa Li, Hai Jin, Zongfen Han, Chao Xie, and Minna Wu

School of Computer Science and Technology,
Huazhong University of Science and Technology, Wuhan, 430074, China
hjin@hust.edu.cn

Abstract. In this paper, we propose an approach for the real-time performance analysis of distributed programs, called Eidolon. The approach is based on the capacity of communication links. In Eidolon, some algorithms are designed to find all the transmission paths of each data file needed while the program executes, count the transmission time for each the transmission path, then get the minimum and the maximum transmission time by analyzing all the transmission paths of each data file, and at last, calculate the fastest and the slowest response time of distributed programs. A case study is given at the end of this paper to justify the validity of this approach.

1 Introduction

The performance of distributed systems includes the reliability, real-time characteristic, scalability and etc. Many methods are developed for performance evaluation purpose. For reliability, a useful notion called *Minimal File Spanning Tree* (MFST) is proposed and an algorithm to find MFST is developed [9]. To improve the MFST algorithms, there are some further improved algorithms [4][8].

There is little effort in studying effective algorithms to evaluate real-time performance of *distributed programs* (DPs) at present. In this paper, we propose a novel approach, called Eidolon, to evaluate the real-time performance of DPs.

The rest of this paper is organized as follows. We discuss the related work in section 2. In section 3, we present theoretic foundation about real-time performance analysis. We propose, in section 4, some algorithms about the real-time performance of DPs. In section 5, an evaluation case of the real-time performance of DPs is presented. Finally, the conclusions are drawn in section 6.

2 Related Works

Many approaches are given to analysis reliability of DPs. General factoring and reduction methods are developed for the DP reliability [1][3][5][10]. The algorithms SM and FM for computing the reliability of DP have been discussed in [7].

---

* This paper is supported by National Science Foundation of China under grant 60273076.

There are also many approaches to test real-time distributed systems. The approach for generalization of real-time distributed systems benchmarking is described in [2][6]. In [11][12], the algorithms EOG and GEX are proposed to test multitasking real-time systems and distributed real-time systems.

There are two factors that are similar between DPs and distributed systems. First, the distributed characteristic of DP is similar to that of distributed systems. Secondly, the scheduling strategy of the data required during the execution of DPs is similar to the strategy of the job scheduling in distributed systems. Considering these two factors, we use the basic principle of distributed systems real-time performance analysis for reference to analyze the real-time performance of DPs.

3 Real-Time Performance Analysis

Since the execution time of a DP usually comprises two parts: the data transmission time and the actual running time of the program. The DP usually runs on some *processing elements* (PEs) and each PE has powerful computing power. The actual running time of the program is so short compared with the data transmission time in the distributed system, therefore it can be even omitted.

3.1 Notations

A distributed system is composed of more than one node, in which many components are included, such as a series of PEs, available data files, and programs. Figure 1 is an example of a simple distributed system. There are six circles. Each circle represents a node and every node has a PE. Every node communicates with others through communication links. Here we use PA_i to denote the set of programs available on node x_i, FA_i the set of data files available for node x_i, FN_i the set of dada files needed for program P_i, F_j the data file j, and $x_{i,j}$ a link between PEs i and j.

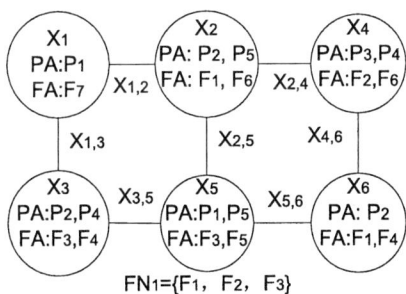

Fig. 1. A simple distributed system

In order to describe the analyzing process of the real-time performance evaluation of DPs, we use x_i to denote the node i, $S(F_j)$ the size of data file F_j, $B(P_i)$ the set of nodes in which P_i can be executed, R_i the transmission path i, $C_{i,j}$ the communication capacity of $x_{i,j}$, $T(R_i)$ the transmission time of data file through R_i, $T_{min}(F_j)$ the minimum transmission time of F_j, $T_{max}(F_j)$ the maximum transmission time of F_j, $T_{min}(P_i)$ the fastest response time of P_i, and $T_{max}(P_i)$ the slowest response time of P_i.

3.2 Theoretic Model

Suppose that program P_i is on node x_k, data file F_j ($F_j \in FN_i$) is on other nodes and these nodes can reach x_k through n(j) paths. These n(j) paths are of $R_1, R_2, ..., R_{n(j)}$ (n(j) represents a positive integer here). The transmission path R_i is composed of link x_{i_1,i_2}, x_{i_2,i_3}, ..., $x_{i_{(g-1)},i_g}$ (i_k represents the suffix of the node). The start node is x_{i_1} and the end node is x_{i_g} (namely node x_k). The capacity of communication link x_{i_1,i_2}, x_{i_2,i_3}, ..., $x_{i_{(g-1)},i_g}$ is C_{i_1,i_2}, C_{i_2,i_3}, ..., $C_{i_{(g-1)},i_g}$, respectively. The transmission time $T(R_i)$ for data file F_j through transmission path R_i is:

$$T(R_i) = \frac{S(F_j)}{C_{i_1,i_2}} + \frac{S(F_j)}{C_{i_2,i_3}} + ... + \frac{S(F_j)}{C_{i_{(g-1)},i_g}} \quad (1)$$

If data file F_j is contained in node x_k, we regard that C_{i_k,i_k} is ∞ and the transmission time of data file F_j is zero. We use π to represent a sorting operation. By sorting the transmission time of data file F_j from R_1 to $R_{n(j)}$, we get a transmission time list $(T_1(F_j), T_2(F_j), ..., T_{n(j)}(F_j))$.

$$(T_1(F_j), T_2(F_j), ..., T_{n(j)}(F_j)) = \pi(T(R_1), T(R_2), ..., T(R_{n(j)})) \quad (2)$$

From list $(T_1(F_j), T_2(F_j), ..., T_{n(j)}(F_j))$, $T_{min}(F_j)$ and $T_{max}(F_j)$ of data file F_j can be calculated as below:

$$T_{min}(F_j) = \min\{T_1(F_j), T_2(F_j), ..., T_{n(j)}(F_j)\} = T_1(F_j) \quad (3)$$

$$T_{max}(F_j) = \max\{T_1(F_j), T_2(F_j), ..., T_{n(j)}(F_j)\} = T_{n(j)}(F_j) \quad (4)$$

As all data files needed for the execution of program P_i can be transmitted concurrently, if the concurrence degree of data file transmission is the highest, and the transmission time of each data file transmitted to node x_k is the shortest, then the time that program P_i needs is the shortest. This is the optimal case that distributed programs can successfully execute. Thus, $T_{min}(P_i)$ can be calculated:

$$T_{min}(P_i) = \max(T_1(F_1), T_1(F_2), ..., T_1(F_j), ..., T_1(F_m)) \quad (5)$$

Similarly, if the concurrence degree of data file transmission is very low, even zero, and the transmission time of each data file transmitted to node x_k is the longest, then the time that program P_i needs is the longest. This is the worst case that DPs can successfully execute. Thus, $T_{max}(P_i)$ can be calculated:

$$T_{max}(P_i) = T_{n(1)}(F_1)+T_{n(2)}(F_2)+\ldots+T_{n(j)}(F_j)+\ldots+ T_{n(m)}(F_m) \tag{6}$$

4 Algorithms Description

Since distributed programs often use data files located on other nodes during their execution. The precondition of successful execution is that all the data files are schedulable and transmittable. The transmission path and the transmission time are two important factors for the performance evaluation.

4.1 Data File Transmission Path Generation Algorithm

At the beginning of the transmission, each data file is sent by broadcast. It is stochastic which transmission path a data file chooses. Furthermore, the probability of choosing each of the transmission paths should not be equal to zero at any time. In this section, we give a data file transmission path generation algorithm as follow:

Step 1: Initialize G the distributed system graph, Find_path = ϕ, B(P_i) = ϕ;

Step 2: Check all nodes in G to find out all the nodes where the distributed program P_i can be executed, and put these nodes in the set B(P_i);

Step 3: Randomly take a node x_i in the set B(P_i) as the root, search for the location of all data files needed for the execution of the program P_i in G;

Step 4: Find out all the possible transmission paths for each data file to the root by the depth-first search, and put these possible transmission paths in the set Find_path;

Step 5: Map the capacity of communication links to the transmission paths, and store the result in file File_Path.

4.2 Transmission Time Evaluation Algorithm of Data File

In this section, we give the transmission time evaluation algorithm of data file. In the algorithm, we first compute the transmission time by using equation (1) above. Then store the result into file Time_Path. The algorithm is described below:

Step 1: Open File_Path, get all the transmission paths $R_1, R_2, \ldots, R_{n(j)}$ of data file F_j;

Step 2: Initialize the transmission time of data file F_j through path R_i, namely $T(R_i)$ = 0 ($1 \leq i \leq n(j)$);

Step 3: Get all the links $x_{i_1,i_2}, x_{i_2,i_3}, \ldots, x_{i_{(g-1)},i_g}$ of path R_i;

Step 4: Calculate the transmission time $T(R_j)$ of data file F_j through path R_i;

Step 5: Store the transmission time $T(R_1), T(R_2), \ldots, T(R_{n(j)})$ of F_j into Time_Path.

4.3 Evaluation of Minimum and Maximum Transmission Time

In this section, we give the minimum and the maximum transmission time evaluation algorithm. In the algorithm, we first compute $T_{min}(F_j)$ and $T_{max}(F_j)$ by using equation (3) and equation (4) above. Then store the result into file Min_Max_Time by a triple (P_i, $T_{min}(F_j)$, $T_{max}(F_j)$). The algorithm is described below:

Step 1: Get data file F_j used by distributed program P_i during its execution;
Step 2: Get the transmission time $T(R_1)$, $T(R_2)$, ..., $T(R_{n(j)})$ of F_j from Time_Path;
Step 3: Sort the transmission time $T(R_1)$, $T(R_2)$, ..., $T(R_{n(j)})$ in increasing order;
Step 4: $T_{min}(F_j) = T(R_1)$, $T_{max}(F_j) = T(R_{n(j)})$;
Step 5: Store $T_{min}(F_j)$ and $T_{max}(F_j)$ of data file F_j in file Min_Max_Time by a triple $(P_i, T_{min}(F_j), T_{max}(F_j))$ (j = 1, 2, ..., m).

4.4 Real-Time Performance Evaluation Algorithm of DPs

In this section, we give the real-time performance evaluation algorithm of DPs. In the algorithm, we calculate the fastest and the slowest response time of distributed program P_i. The algorithm is shown below:

Step 1: Get the triple $(P_i, T_{min}(F_j), T_{max}(F_j))$ (j=1, 2, ..., m) from file Min_Max_Time and initialize $T_{min}(P_i) = 0$, $T_{max}(P_i) = 0$;
Step 2: Calculate $T_{min}(P_i)$ and $T_{max}(P_i)$ by using equation (5) and equation (6);
Step 3: Output $T_{min}(P_i)$ and $T_{max}(P_i)$.

5 Case Study

Suppose in Fig.1, the size of data file F_1, F_2, F_3, F_4, F_5, F_6, and F_7 is 70KB, 60KB, 58KB, 42KB, 54KB, 64KB, and 55KB, respectively, and the corresponding capacity of communication link $x_{1,2}$, $x_{1,3}$, $x_{2,4}$, $x_{2,5}$, $x_{3,5}$, $x_{4,6}$, and $x_{5,6}$ is 1.6Mb/s, 0.8Mb/s, 2.5Mb/s, 1.5Mb/s, 2Mb/s, 1.2Mb/s, and 0.5Mb/s, respectively. Distributed program P_1 needs to access data file F_1, F_2 and F_3 when it is executed. We use Eidolon to evaluate the real-time performance of distributed program P_1. The results are shown as below:

1. When distributed program P_1 is executed on x_1, the fastest response time is $T_{min}(P_1)$ =72.5 (ms) and the slowest response time is $T_{max}(P_1) = 876.616$ (ms).
2. When distributed program P_1 is executed on x_5, the fastest response time is $T_{min}(P_1)$ = 64(ms) and the slowest response time is $T_{max}(P_1) = 718.866$(ms).

In fact, $T_{min}(P_1)$ is the time for program P_1 successfully executed in the optimal case. $T_{max}(P_1)$ is the time for program P_1 successfully executed in the worst case.

6 Conclusions and Future Work

In this paper, based on the capacity of communication links, we investigate the transmission time of data files and the real-time characteristics of DPs. Because the reliability of communication links is very high and has little influence on the real-time performance of DPs, the little influence is always ignored in Evaluation. But, under some special circumstance, such as military control, even this kind of minute influence should not be ignored. So far in Eidolon, we don't take into consideration, the influence of the reliability of communication links. In particular, we plan to analyze and deduce a more detailed approach taking the influence of the reliability of communication links into consideration.

References

1. Chin, C. C., Yeh, Y. S., and Chou, J. S., "A Fast Algorithm for Reliability-oriented Task Assignment in a Distributed System", *Computer Communication*, Vol.25. 17 (2002), pp.1622-1630
2. Drummond, J., "Establishing a Real-Time Distributed Benchmark", *Proceeding of the 4th International Workshop on Parallel and Distributed Real-Time Systems*, 1996, pp.198-201
3. Lin, M. S., "The Reliability Analysis on Distributed Computing Systems", Ph D dissertation, National Chiao Tung University, Hsinchu, Taiwan, 1994
4. Lin, M. S., Chang, M. S., and Chen, D. J., "Efficient Algorithms for Reliability Analysis of Distributed Computing Systems", *Information Sciences*, 117 (1999), pp.89-106
5. Lin, M. S., and Chen, D. J., "General Reduction Methods for the Reliability Analysis of Distributed Computing Systems", *Computer Journal*, 36 (1993), pp.631-644
6. Kamenoff, N. I., "One Approach for Generalization of Real-Time Distributed Systems Benchmarking", *Proceeding of the 4th International Workshop on Parallel and Distributed Real-Time Systems*, 1996, pp.204-207
7. Ke, W. J., and Wang, S. D., "Reliability Evaluation for Distributed Computing Networks with Imperfect Nodes", *IEEE Transaction on Reliability*, 46 (1997), pp.342-349
8. Kumar, A., and Agrawal, D. P., "A Generalized Algorithm for Evaluating Distributed Program Reliability", *IEEE Transactions on Reliability*, Vol. 42, 3 (1993), pp.416-426
9. Kumar, V. P., Hariri, S., and Raghavendra, C. S., "Distributed program reliability analysis", *IEEE Trans Software Engineering*, 12 (1986), pp.42-50
10. Tanenbaum, A. S., *Distributed Operating Systems*, Vol.1, Prentice Hall PTR, 1995
11. Thane, H., "Monitoring, Testing and Debugging of Distributed Real-time Systems", Ph D dissertation, Royal Institute of Technology, KTH, Sweden, 2000
12. Thane, H., and Hansson, H., "Testing Distributed Real-time Systems", *Microprocessors and Microsystems*, 24 (2001), pp.463-483

Approximate Waiting Time Analysis of Burst Queue at an Edge in Optical Burst-Switched Networks*

SuKyoung Lee

Graduate School of Information & Communications
Sejong University, Seoul, Korea
sklee@sejong.ac.kr

Abstract. In this paper, we analyze the burst queue with burst assembler at an edge in optical burst-switched networks. It is significant from the fact that most of processing and buffering are concentrated at the edge in OBS networks. Modelling the burst queue in terms of IP packets, we aim to obtain the expected number of packets in burst queue the size of which influences the cost of constructing OBS networks. The analytical model is verified through simulation.

1 Introduction

In an Optical Burst Switching (OBS) network, multiple IP packets are aggregated into a burst that is then transmitted without the need for any type of buffering at the intermediate nodes. This novel approach is expected to provide an infrastructure of the next-generation Internet against the exponential growth of the demands. One of the key problems in the application of burst switching in an optical domain is the aggregation of several IP packets in a single optical burst and, hence, it is necessary to implement the assembly and disassembly functions from IP packets to the burst format and vice versa [1]. In particular, once IP packets are aggregated into much larger bursts, they should wait in the burst queue[2] till the enough wavelength is reserved, before transmission through the network. Even though there have been several works regarding burst queue [3]-[6], most of them have focused on how to assemble packets into a burst. Thus, in this paper, we focus on developing an analytical model of the burst queue with burst assembler in the light of IP packets. We investigate the mean waiting time in burst queue following the burst assembler at an edge because most of processing and buffering are concentrated at the edge in an OBS network. This investigation is meaningful in the aspect that the burst queue size could be decided on the basis of this analytical model and the tolerable delay limit of each traffic class.

---

* This work was supported in part by Korea Science and Engineering Foundation (KOSEF) through OIRC project.

2 Burst Queueing Architecture at an OBS Edge Node

An edge OBS node is connected to a number of users or Label Switched Paths (LSPs) [2]. A burst queue can be shared among several users or LSPs where the traffic arrives according to a burst arrival process at an edge node in OBS networks. If the traffic is composed of multiple classes, it is desirable to implement multiple burst queues with the same number of multiple burst assemblers, where incoming packets destined for the same (destination node, CoS: Class of Service) are sent to the same burst queue. That is, as soon as bursty IP traffic streams are assembled into bursts at burst assembler, these bursts are classified according to the priority of the traffic [3]. Each incoming packet for a given burst needs to be stored in the burst assembler until the last packet of the burst arrives. As soon as the last packet arrives, all packets of the burst are transferred to the burst queue.

Considering the burst arrival process to a given burst queue, the arrival process to the burst queue can be modelled as a superposition of N independent effective On/Off processes, each coming from N users or LSPs. If we assume that both On and Off periods are geometrically distributed with the same parameters α and β that denote the transition rates from Off to On and On to Off states, respectively, an Interrupted Bernoulli Process (IBP) can be used to model each burst assembler. Within the On state, at each time instant, there is a packet arrival with probability γ from each active user/LSP and generated packets during an On period form a single burst. If we denote π_i as the steady-state probability that the burst assembler contains i packets under a single On/Off process, it is possible to model the Markov chain for each burst assembler and to show that the steady-state probability, π_i is

$$\pi_i = \begin{cases} \frac{\beta}{\alpha+\beta}, & \text{for } i = 0 \\ \frac{\alpha}{\beta+\gamma(1-\beta)}\pi_0, & \text{for } i = 1 \\ \frac{\gamma(1-\beta)}{\beta+\gamma(1-\beta)}\pi_{i-1}, & \text{for } i \geq 2 \end{cases} \quad (1)$$

Assuming the event that the burst assembler is not empty, the conditional probability, Π_i that the burst assembler contains i packets becomes

$$\Pi_i = \frac{\pi_i}{1-\pi_0}, \quad \text{for } i \geq 1. \quad (2)$$

Let N_{ba} denote the number of packets in burst assembler under the assumption that the burst assembler is not empty. From the Eq. 2 and the fact that memoryless property of π_i leads Π_i to having the geometric distribution, we have

$$E[N_{ba}] = \frac{\beta+(1-\beta)r}{\beta} \quad (3)$$

3 Burst Queue Modelling

The data bursts in burst queue are assumed to be drained virtually in terms of packet. Burst queue is investigated using a discrete-time system where one time slot is equivalent to the duration a packet is transmitted on a wavelength. In OBS networks, it is known that the a burst in burst queue is transmitted to an output port after offset time when offset-based wavelength reservation scheme is being run on the network.

Thus, if t_c denotes an offset time for the bursts belonging to a service class c and L is the mean burst length, one packet service time becomes t_c/L. On the assumption that packets are not bound to arrive back-to-back, inter-arrival times of successive packets are i.i.d. (independent identically distributed)[4]. Often we think in terms of activity cycle, β^{-1} is the active period, which has units of the average number of seconds per an On duration and α^{-1} means the idle period. Then, overall one cycle T becomes $\alpha^{-1} + \beta^{-1}$. The probability of such a traffic source being in the On state, p_{on} is given by $\frac{\beta^{-1}}{T}$. Then, we get the probability that a single source transits from On state to Off state in a time slot p'_{on} as $p_{on}\beta$.

Let Γ_i denote the probability that i out of N active sources transit from On state to Off state in a time slot. Then, Γ_i can be expressed as

$$\Gamma_i = b(i; N, p'_{on}) \tag{4}$$

where $b(i; n, p)$ represents the binomial distribution with all n sources.

Let x be the burst queue probability distribution. We define x_i as the steady-state probability that the burst queue contains i packets under a total of N On/Off sources from users/LSPs. Let $\mathbf{x}_i$ denote the probability vector with N components, $x_{i,n}$ where $x_{i,n}$ is the probability that burst queue contains i packets and n sources are at On state. We can describe the aggregate arrival process by a discrete-time batch Markovian arrival process (D-BMAP). Subsequently, the algorithm for computing the vector $\mathbf{x}_i$ follows D-BMAP/D/1 model. However, it brings about the computational complexity in getting $\mathbf{x}_i$ to compute the transition probability that the number of On sources changes from i to j generating a total of k packets in D-BMAP/D/1 model.

In advance of computing the approximation of $\mathbf{x}_i$, the link density can be defined as

$$\rho = \frac{E[N_{ba}]N}{T} \tag{5}$$

Substituting Eq. 5 and $E[N_{ba}] = 1/\lambda$ into Eq. 4 and taking $N \to \infty$ and $\lambda\rho \ll 1$, we obtain the approximation of Eq. 4

$$\Gamma_i \approx \begin{cases} e^{-\lambda\rho} & \approx 1 - \lambda\rho, \; i = 0 \\ \lambda\rho e^{-\lambda\rho} & \approx \lambda\rho, \quad i = 1 \\ \frac{(\lambda\rho)^i}{i!} e^{-\lambda\rho} \approx 0, & \text{otherwise} \end{cases} \tag{6}$$

where the first approximation can be applied. From Eq. 6, we get the probability, β_i that i packets arrive to form a burst transiting from an On state to an Off state. This is expressed in the form of

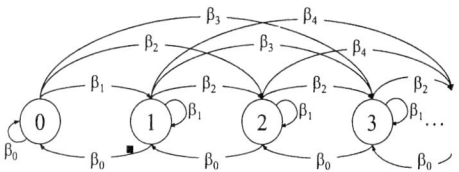

Fig. 1. State transition diagram of burst queue

$$\beta_i = \begin{cases} \Gamma_0 = 1 - \lambda\rho, & i = 0 \\ \Gamma_1 \Pi_i = \lambda^2 \rho (1-\lambda)^{i-1}, & i \geq 1 \end{cases} \quad (7)$$

where β_i also follows a geometric distribution. Fig. 1 depicts the state transition diagram of the burst queue which works on the basis of one time slot. According to the transition rules defined in this Figure, a system of difference equations may be derived for the approximation of x_i, $\tilde{x}_i$ as follows:

$$\tilde{x}_0(1-\beta_0) = \tilde{x}_1\beta_0 \quad (8)$$
$$\tilde{x}_1(1-\beta_1) = \tilde{x}_2\beta_0 + \tilde{x}_0\beta_1$$
$$\tilde{x}_i(1-\beta_1) = \tilde{x}_{i+1}\beta_0 + \tilde{x}_0\beta_i + \sum_{n=1}^{i-1}\tilde{x}_i\beta_{i-n+1}, \text{ for } i > 1$$

As usual, we solve the above equilibrium Eqs. 8 using the method of z-transform, thus we have

$$(1-\beta_1)\sum_{n=2}^{\infty}\tilde{x}_n z^n = \frac{\beta_0}{z}\sum_{n=2}^{\infty}\tilde{x}_{n+1}z^{n+1} + \tilde{x}_0\sum_{i=2}^{\infty}\beta_n z^n + \sum_{n=2}^{\infty}\sum_{i=1}^{n-1}\tilde{x}_i\beta_{n-i+1}z^n \quad (9)$$

From the Eq. 9, z-transform of $\tilde{x}_i$, $X(z)$ is derived as follows:

$$X(z) = \tilde{x}_0 \frac{zG(z) - G(z)}{z - G(z)}, \quad (10)$$

where $G(z)$, the z-transform of β_i is

$$1 - \lambda\rho + \lambda^2\rho \frac{z}{1-(1-\lambda)z}. \quad (11)$$

To eliminate $\tilde{x}_0$, $X(1) = 1$ is applied, that yields $\tilde{x}_0 = 1 - \rho$. We now make use of Eq. 11, $\tilde{x}_0 = 1 - \rho$ and $\tilde{x}_n = \frac{1}{n!}\frac{d^n X(z)}{dz^n}|_{z=0}$ to obtain $\tilde{x}_n$. Thus, we have

$$\tilde{x}_n = \frac{\lambda\rho(1-\rho)(1-\lambda)^{n-1}}{(1-\lambda\rho)^n}. \quad (12)$$

The expected number of packets in burst queue under N users/LSPs is derived as

$$E[n] = \sum_{n=0}^{\infty}n\tilde{x}_n = \sum_{n=1}^{\infty}n\tilde{x}_n = \frac{\rho(1-\lambda\rho)}{\lambda(1-\rho)} \quad (13)$$

Then, letting T_{bq} denote the waiting time in burst queue, the mean waiting time, $\bar{T}_{bq}$ becomes

$$\bar{T}_{bq} = \frac{E[n]}{\lambda} = \frac{\rho(1-\lambda\rho)}{\lambda^2(1-\rho)}. \tag{14}$$

This analytic model of burst queue, in combination with carriers' QoS mechanisms, could be used to allow the carriers to offer customized levels of burst assembly latency which is most part of each burst's end-to-end delay.

4 Performance Evaluation

To assess the accuracy of our model, we compare our analytical results with those obtained by means of simulations. In our simulation, burst sources are individually simulated with the on-off model. In order to capture the burstiness of data at the edge nodes, the traffic from a user to the edge node is generated by Poisson packet arrivals from 10 and 50 independent traffic sources that are either transmitting packets at a mean rate during the On period or is idle during the Off period. The average packet size is fixed to 100 in the Poisson case.

Fig. 2 plots the average waiting time in burst queue with burst assembly versus the traffic density for the simulation with different number of sources such as users or LSPs: $N = 10$, $N = 50$ and the analytical model.

As can be easily expected, the burst buffering in any OBS network strongly depends on the traffic statistics of arriving packets [7]. In other words, as the mean arrival rate increases, so does the average waiting time. From the graphs in Fig. 2, we observe that the error increases as the ρ increases, that is, the link density increases. There may be some concern that the analytical model may not work very well if there are just a few sources on the assumption that $N \to \infty$.

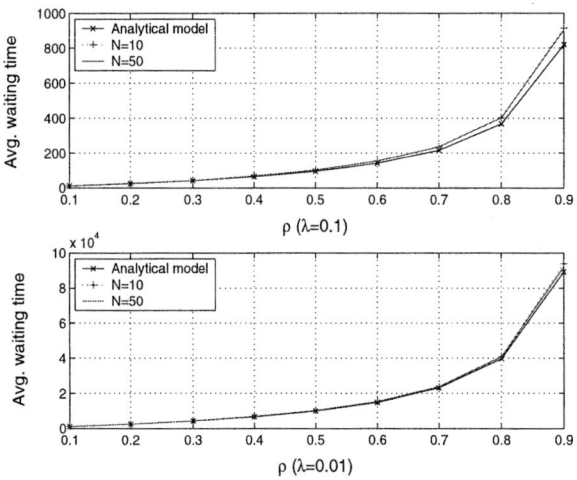

Fig. 2. Average waiting time in burst queue (λ=0.1 and 0.01)

But this concern falls off in the curves for both experiments with $N = 10$ and $N = 50$. The maximum error in difference between the analytical and simulation results is about 10% when there are 10 sources. As the link density increases, the error gets much smaller for $N = 50$ than for $N = 10$. Especially, at very high loads the analytical model performs better than that at low loads. Therefore, we know that the difference is further reduced as N increases.

As expected, the simulation results show that the analytical results behave to track the simulation results more closely as the traffic arrivals increase, because $\lambda \rho \ll 1$ is assumed in our analytical model. When the burst arrival is varied, we also see good agreement between the analytical and the simulation results. At last, from this analytical model, Internet Service Providers (ISPs) expect how tolerable delay limit could be provided for each service class.

5 Conclusion

OBS is especially economic and efficient for providing high-end applications with sessions requesting a high bit-rate and low latency. However, it is possible that the processing at an edge would be a bottleneck to those applications. Therefore, it is necessary to investigate burst queue at an edge in the aspect that most of processing and buffering are concentrated at the edge in an OBS network. In this paper, we investigated the burst queue with burst assembler at an edge in OBS networks. We observed that the analytical model performs as well as the simulations when there are large number of sources. It could be also known that as the burst arrivals increase, the analytical model approximates more to the simulation results, while the analytical model lightens the computation overhead.

References

1. Battestilli, T., Perros, H.: An Introduction to Optical Burst Switching. IEEE Commun. Mag., Vol. 41, No.8 (Aug. 2003) s10-s15.
2. Chaskar, H.M., Verma, S., Ravikanth, R.: Robust Transport of IP Traffic over WDM Using Optical Burst Switching. Optical Networks Mag. Vol.3, No.4 (Jul./Aug. 2001) 47-60.
3. Oh, S.Y., Kang, M.: A Burst Assembly Algorithm in Optical Burst Switching Networks. OFC, Anaheim CA USA (2002) 771-773.
4. XU, L., Perros, H.G., Rouskas, G.N.: A Queueing Network Model of an Edge Optical Burst Switching Node, IEEE NFOCOM, Vol.3, San Francisco CA USA, (2003) 2019-2029.
5. Vokkarane, V., Haridoss, K., Jue, J.: Threshold-Based Burst Assembly Policies for QoS Support in Optical Burst-Switched Networks. SPIE Optical Networking and Communication Conference, Vol. 4874, Boston MA USA (2002) 125-136.
6. Long, K., Tucker, R.S., Wang, C.: A New Framework and Burst Assembly for IP DiffServ over Optical Burst Switching Networks. IEEE Globecom, Vol.6, San Francisco CA USA (2003) 3159-3164.
7. Dueser, M., Bayvel, P.: Performance of a Dynamically Wavelength-Routed Optical Burst Switched Network. IEEE Photonics Technology Letters, Vol. 14, No. 2 (Feb. 2002) 239-241.

A Balanced Model Reduction for T-S Fuzzy Systems with Uncertain Time Varying Parameters[1]

Seog-Hwan Yoo and Byung-Jae Choi

School of Electronic Engineering, Daegu University,
15 Naeri Jinryang Gyeongsan Gyeongbuk, 712-714 Korea
{shryu, bjchoi}@daegu.ac.kr

Abstract. This paper deals with a balanced model reduction for a class of Takagi-Sugeno(T-S) fuzzy systems with uncertain time varying parameters. We define a generalized controllability Gramian and a generalized observability Gramian for quadratically stable T-S fuzzy systems with uncertainties. We introduce a balanced state space realization using the generalized controllability and observability Gramians and obtain a reduced model by truncating not only states but also time varying uncertain parameters from the balanced state space realization. We also present an upper bound of the approximation error. The generalized controllability and observability Gramians can be computed from solutions of linear matrix inequalities(LMI's).

1 Introduction

For linear finite dimensional systems with high orders, optimal control techniques such as linear quadratic gaussian and H_∞ control theory, usually produce controllers with the same state dimension as the model. Lower dimensional linear controllers are normally preferred over higher dimensional controllers in control system designs for some obvious reasons: they are easier to understand, implement and have higher reliability. Accordingly the problem of model reduction is of significant practical importance in control system design. Existing methods such as the balanced model reduction or the optimal Hankel norm model reduction are widely used to obtain a reduced model of linear time invariant systems [1]-[4].

Model reduction techniques have been developed for linear uncertain systems as well. Using LMI machinery, Beck et al. suggested a balanced truncation method for linear uncertain discrete systems with related error bounds [5]. Model reduction techniques for linear parameter varying systems were also reported by several researchers [6]-[8]. However, comparatively little work has been reported for the model reduction for nonlinear systems.

In recent years, a controller design method for nonlinear dynamic systems modeled as a T-S fuzzy model has been intensively addressed [9]-[11]. Unlike a single conventional model, this T-S fuzzy model usually consists of several linear models to describe the global behavior of the nonlinear system. Typically the T-S fuzzy model

---

[1] This work was supported in part by the 2004 Daegu university research fund.

is described by fuzzy IF-THEN rules. Based on this fuzzy model, many researchers use one of control design methods developed for linear parameter varying systems. In order to alleviate computational burden in design phase and simplify the designed fuzzy controller, the state dimension of the T-S fuzzy model should be low.

In this paper, using LMI machineries we develop a balanced model reduction scheme for T-S fuzzy systems with norm bounded time varying uncertain parameters. In section 2, we define the T-S fuzzy system with time varying uncertain parameters. A generalized controllability Gramian and a generalized observability Gramian are defined and a balanced model reduction of T-S fuzzy system using the generalized controllability and observability Gramians is also presented in section 3. Finally some concluding remarks are given in section 4.

The notation in this paper is fairly standard. R^n denotes n dimensional real vector space and $R^{n \times m}$ is the set of real $n \times m$ matrices. A^T denotes the transpose of a real matrix A. 0 and I denote zero matrix and identity matrix respectively. $M > 0$ means that M is a positive definite matrix. In a block symmetric matrix, $*$ in (i, j) block means the transpose of the matrix in (j,i) block. Finally $\|\cdot\|_\infty$ denotes the H_∞ norm of the system.

2 T-S Fuzzy System

We consider the following fuzzy dynamic system with uncertain time varying parameters.

Plant Rule i ($i = 1, \cdots, r$):
IF $\rho_1(t)$ is M_{i1} and $\cdots$ and $\rho_g(t)$ is M_{ig},
THEN

$$\begin{aligned} \dot{x}(t) &= A_i x(t) + F_i w(t) + B_i u(t) \\ z(t) &= H_i x(t) + J_i u(t) \\ y(t) &= C_i x(t) + G_i w(t) + D_i u(t) \\ w(t) &= \Theta(t) z(t), \end{aligned} \qquad (1)$$

where r is the number of fuzzy rules. $\rho_j(t)$ and M_{ij} ($j = 1, \cdots, g$) are the premise variables and the fuzzy set respectively. $x(t) \in R^n$ is the state vector, $u(t) \in R^m$ is the input, $y(t) \in R^p$ is the output variable and $w(t) \in R^q$, $z(t) \in R^q$ are variables related to uncertain parameters. $A_i, F_i, \cdots, D_i$ are real matrices with compatible dimensions. $\Theta(t)$ is an uncertain parameter matrix defined as follows:

$$\begin{aligned} \Theta(t) &= diag(\theta_1(t)I_{q1}, \theta_2(t)I_{q2}, \cdots, \theta_s(t)I_{qs}), \\ \Theta(t)^T \Theta(t) &\leq I, \quad q = q_1 + q_2 + \cdots + q_s. \end{aligned} \qquad (2)$$

Let $\mu_i(\rho(t))$, $i = 1, \cdots, r$, be the normalized membership function of the inferred fuzzy set $h_i(\rho(t))$,

$$\mu_i(\rho(t)) = \frac{h_i(\rho(t))}{\sum_{i=1}^{r} h_i(\rho(t))}, \tag{3}$$

where

$$h_i(\rho(t)) = \prod_{j=1}^{g} M_{ij}(\rho_j(t)), \quad \rho(t) = \begin{bmatrix} \rho_1(t) & \rho_2(t) & \cdots & \rho_g(t) \end{bmatrix}^T. \tag{4}$$

In this paper, assuming, for all i, $h_i(\rho(t)) \geq 0$ and $\sum_{i=1}^{r} h_i(\rho(t)) > 0$, we obtain

$$\mu_i(\rho(t)) \geq 0, \quad \sum_{i=1}^{r} \mu_i(\rho(t)) = 1. \tag{5}$$

For simplicity, by defining $\mu_i = \mu_i(\rho(t))$ and $\mu^T = [\mu_1 \cdots \mu_r]$, the uncertain fuzzy system (1) can be written as follows :

$$\begin{aligned} \dot{x}(t) &= \sum_{i=1}^{r} \mu_i (A_i x(t) + F_i w(t) + B_i u(t)) = A(\mu)x(t) + F(\mu)w(t) + B(\mu)u(t) \\ z(t) &= \sum_{i=1}^{r} \mu_i (H_i x(t) + J_i u(t)) = H(\mu)x(t) + J(\mu)u(t) \\ y(t) &= \sum_{i=1}^{r} \mu_i (C_i x(t) + G_i w(t) + D_i u(t)) = C(\mu)x(t) + G(\mu)w(t) + D(\mu)u(t) \\ w(t) &= \Theta(t)z(t). \end{aligned} \tag{6}$$

In a packed matrix notation, we express the fuzzy system (6) as follows:

$$G_\Theta = \left[\begin{array}{c|c|c} A(\mu) & F(\mu) & B(\mu) \\ \hline H(\mu) & 0 & J(\mu) \\ \hline C(\mu) & G(\mu) & D(\mu) \end{array} \right]. \tag{7}$$

3 A Balanced Model Reduction

In this section we present a balanced model reduction for the uncertain fuzzy system (7) using generalized controllability and observability gramians. First we define generalized controllability and observability gramians.

Lemma 1. (Generalized controllability and observability gramians)
1) Suppose that there exist $Q = Q^T > 0$ and $R = R^T > 0$, $R = diag(R_1, \cdots, R_s)$, $R_i \in R^{q_i \times q_i}$ ($i = 1, \cdots, s$) satisfying LMI (9), then the output energy is bounded above as follows:

$$\int_0^\infty y(t)^T y(t)dt < x(0)^T Q x(0) \text{ for } u(t) \equiv 0. \tag{8}$$

$$L_{oi} = \begin{bmatrix} A_i^T Q + Q A_i & * & * & * \\ R H_i & -R & * & * \\ F_i^T Q & 0 & -R & * \\ C_i & 0 & G_i & -I \end{bmatrix} < 0, \quad i = 1, \cdots, r. \tag{9}$$

2) Suppose that there exist $P = P^T > 0$ and $S = S^T > 0$, $S = diag(S_1, \cdots, S_s)$, $S_i \in R^{q_i \times q_i}$ $(i = 1, \cdots, s)$ satisfying LMI (11), the input energy transferring from $x(-\infty) = 0$ to $x(0) = x_0$ is bounded below as follows:

$$\int_{-\infty}^0 u(t)^T u(t)dt > x_0^T P^{-1} x_0. \tag{10}$$

$$L_{ci} = \begin{bmatrix} P A_i^T + A_i P & * & * & * \\ H_i P & -S & * & * \\ S F_i^T & 0 & -S & * \\ B_i^T & J_i^T & 0 & -I \end{bmatrix} < 0, \quad i = 1, \cdots, r. \tag{11}$$

Proof. The proof is omitted due to space limitation.

As in [6], we say Q and P, solutions of LMI's (9) and (11), are generalized observability Gramian and controllability Gramian respectively. While the observability and controllability Gramians in linear time invariant systems are unique, the generalized Gramians of the fuzzy system (7) are not unique. But the generalized Gramians are related to the input and output energy as can be seen in Lemma 1.

Using the generalized Gramians, we suggest a balanced realization of the uncertain fuzzy system (7). We obtain a transformation matrix T and W satisfying

$$\Sigma = diag(\sigma_1, \sigma_2, \cdots, \sigma_n) = T^T Q T = T^{-1} P T^{-T}, \quad \sigma_1 \geq \sigma_2 \geq \cdots \geq \sigma_n, \\ \Pi = diag(\Phi_1, \Phi_2, \cdots, \Phi_s) = W^T R W = W^{-1} S W^{-T}, \quad tr(\Phi_1) \geq tr(\Phi_2) \geq \cdots \geq tr(\Phi_s). \tag{12}$$

With T and W defined in (12), the change of coordinates in the fuzzy system (7) gives

$$G_{\Theta_b} = \begin{bmatrix} A_b(\mu) & F_b(\mu) & B_b(\mu) \\ H_b(\mu) & 0 & J_b(\mu) \\ C_b(\mu) & G_b(\mu) & D_b(\mu) \end{bmatrix} = \begin{bmatrix} T^{-1} A(\mu) T & T^{-1} F(\mu) W & T^{-1} B(\mu) \\ W^{-1} H(\mu) T & 0 & W^{-1} J(\mu) \\ C(\mu) T & G(\mu) W & D(\mu) \end{bmatrix}, \tag{13}$$

where $\Theta_b(t) = W^{-1} \Theta(t) W$.

We say that the realization (13) is the balanced realization of the fuzzy system (7) and Σ is the balanced Gramian. We assume that the fuzzy system (7) is already balanced and partitioned as follows:

$$G_\Theta = \sum_{i=1}^{r} \mu_i \begin{bmatrix} A_{i,11} & A_{i,12} & F_{i,11} & F_{i,12} & B_{i,1} \\ A_{i,21} & A_{i,22} & F_{i,21} & F_{i,22} & B_{i,2} \\ H_{i,11} & H_{i,12} & 0 & 0 & J_{i,1} \\ H_{i,21} & H_{i,22} & 0 & 0 & J_{i,2} \\ \hline C_{i,1} & C_{i,2} & G_{i,1} & G_{i,2} & D_i \end{bmatrix}, \quad (14)$$

$$w(t) = \begin{bmatrix} w_1(t) \\ w_2(t) \end{bmatrix} = \begin{bmatrix} \Theta_1(t) & 0 \\ 0 & \Theta_2(t) \end{bmatrix} \begin{bmatrix} z_1(t) \\ z_2(t) \end{bmatrix},$$

where $A_{i,11} \in R^{k \times k}$, $F_{i,11} \in R^{k \times (q_1 + \cdots q_v)}$ and the other matrices are compatibly partitioned. From (14) we obtain a reduced order model by truncating $n-k$ states and $s-v$ uncertain parameters as follows:

$$\bar{G}_{\Theta_1} = \sum_{i=1}^{r} \mu_i \begin{bmatrix} A_{i,11} & F_{i,11} & B_{i,1} \\ H_{i,11} & 0 & J_{i,1} \\ \hline C_{i,1} & G_{i,1} & D_i \end{bmatrix}, \quad w_1(t) = \Theta_1(t) z_1(t). \quad (15)$$

Theorem 2. The reduced order system (15) is quadratically stable and balanced. Moreover the model approximation error is given by

$$\|G_\Theta - \bar{G}_{\Theta_1}\|_\infty \leq 2 \left(\sum_{j=k+1}^{n} \sigma_j + \sum_{j=v+1}^{s} tr(\Phi_j) \right). \quad (16)$$

Proof. The proof is omitted due to space limitation.

In Theorem 2, we have derived an upper bound of the model reduction error. In order to get a less conservative model reduction error bound, it is necessary for $n-k$ smallest eigenvalues of Σ and $\sum_{j=v+1}^{s} tr(\Phi_j)$ of Π to be small. As an alternative, one can use a cost function defined as $J = tr(PQ) + \alpha tr(RS)$ for a positive constant α. Thus, one can minimize the non-convex cost function subject to the convex constraints (9) and (11). Since this optimization problem is non-convex, the optimization problem is very difficult to solve it so that a suboptimal procedure using an iterative method may be used to obtain a local minimum. So we suggest an alternative suboptimal procedure using an iterative method. We summarize the iterative method to solve a suboptimal problem.

Step 1. Set $i = 0$. Initialize P_i, Q_i, R_i and S_i such that $tr(P_i + Q_i) + \alpha tr(R_i + S_i)$ is minimized subject to LMI's (9) and (11).

Step 2. Set $i = i+1$.
  1) Minimize $J_i = tr(P_{i-1}Q_i) + \alpha tr(R_i S_{i-1})$ subject to LMI (9).
  2) Minimize $J_i = tr(P_i Q_i) + \alpha tr(R_i S_i)$ subject to LMI (11).

Step 3. If $|J_i - J_{i-1}|$ is less than a small tolerance level, stop iteration. Otherwise, go to Step 2.

4 Concluding Remark

In this paper, we have studied a balanced model reduction problem for T-S fuzzy systems with time varying uncertain parameters. For this purpose, we have defined generalized controllability and observability Gramians for the fuzzy system. This generalized Gramians can be obtained from solutions of the LMI problem. Using the generalized Gramians, we have derived the balanced state space realization. We have obtained the reduced model of the fuzzy system by truncating not only some state variables but also some uncertain parameters. We have also derived an upper bound of the model reduction error.

References

1. Moore B.C.: Principal component analysis in linear systems: Controllability, observability and model reduction. IEEE Trans. Automatic Contr., vol.26 (1981) 17-32
2. Pernebo L., and Silverman L.M.: Model reduction via balanced state space representations. IEEE Trans. Automatic Contr., vol.27 (1982) 382-387
3. Glover K.: All optimal Hankel-norm approximations of linear multivariable systems and their error bounds. Int. J. Control, vol.39 (1984) 1115-1193
4. Liu Y., and Anderson B.D.O.: Singular perturbation approximation of balanced systems. Int. J. Control, vol.50 (1989) 1379-1405
5. Beck C.L., Doyle J., and Glover K.: Model reduction of multidimensional and uncertain systems. IEEE Trans. Automatic Contr., vol.41 (1996) 1466-1477
6. Wood G.D., Goddard P.J., and Glover K.: Approximation of linear parameter varying systems. Proceedings of the 35th CDC, Kobe, Japan, Dec. (1996) 406-411
7. Wu F.: Induced L_2 norm model reduction of polytopic uncertain linear systems. Automatica, vol.32, No.10, (1996) 1417-1426
8. Haddad W.M., and Kapila V.: Robust, reduced order modeling for state space systems via parameter dependent bounding functions. Proceedings of American control conference, Seattle, Washington, June (1996) 4010-4014
9. Tanaka K., Ikeda T., and Wang H.O.: Robust stabilization of a class of uncertain nonlinear systems via fuzzy control: Quadratic stabilizability, control theory, and linear matrix inequalities. IEEE Trans. Fuzzy Systems, vol.4, no.1, Feb., (1996) 1-13
10. Nguang S.K., and Shi P.: Fuzzy output feedback control design for nonlinear systems: an LMI approach. IEEE Trans. Fuzzy Systems, vol.11, no.3, June (2003) 331-340
11. Tuan H.D., Apkarian P., Narikiyo T., and Yamamoto Y.: Parameterized linear matrix inequality techniques in fuzzy control system design. IEEE Trans. Fuzzy Systems, vol.9, no.2, April (2001) 324-332

Genetic Algorithms with Stochastic Ranking for Optimal Channel Assignment in Mobile Communications

Lipo Wang[1,2] and Wen Gu[2]

[1] College of Information Engineering, Xiangtan University Xiangtan, Hunan, China
[2] School of Electrical and Electronic Engineering Nanyang Technological University
Block S1, Nanyang Avenue, Singapore 639798
Elpwang@ntu.edu.sg

Abstract. Optimal channel assignment can enhance traffic capacity of a cellular mobile network and decrease interference between calls, thereby improving service quality and customer satisfaction. We combine genetic algorithms with stochastic ranking, to solve the problem of assigning calls in a cellular mobile network to frequency channels in such a way that interference between calls is minimized, while demands for channels are satisfied. Simulation results showed that this approach is able to further improve on the results obtained by some other techniques.

1 Introduction

There is a continuously growing number of mobile users. However, the number of usable channels (frequencies), which are necessary for the communication between mobile users and the base stations of cellular radio networks, is very limited.

The purpose of channel assignment problems is to assign channels efficiently in order that interference is minimized while the demand of each cell is satisfied. Our research is focused on static channel assignment problems since static assignment is the basis of dynamic assignment to a large extent.

One kind of SCA is to minimize the number of channels used while interference is precluded and the demand of each cell is satisfied. This kind of problem is defined by Gamst and Rave [16] and denoted as CAP1 in [12]. In most practical situations, the number of available channels may be not enough for an interference-free assignment. In such cases, we have to try to minimize the interference for a given set of channels while the demand is met, which can be called CAP2 [12]. Over recent years, many heuristic techniques have been applied to solve channel assignment problem [1-2]. In [12], demand satisfaction is treated as hard constraints and non-interference is treated as soft constraints. However, how to balance the constraints becomes a difficult problem.

In this paper, we apply a new constraint balance technique proposed by Runarsson and Yao [4], i.e., stochastic ranking, together with genetic algorithms, to balance the constraints in CAP2.

2 Channel Assignment Problem

Let us consider a network of N cells and M channels. The channel required (expected traffic) for cell i is given by D_i. The electromagnetic compatibility (EMC) constraints specify the minimum distance in the frequency domain by which two channels must be separated so that no interference exists. These minimum distances are stored in a symmetric compatibility matrix C which has a dimension of $N \times N$.

The mathematical formulation of CAP2 given by Smith and Palaniswami [12] is reviewed below.

$$X_{j,k} = \begin{cases} 1, & \text{if cell } j \text{ is assigned to channel } k, \\ 0, & \text{otherwise}, \end{cases} \quad (1)$$

for $j = 1, \cdots, N$ and $k = 1, \cdots, M$. One way to measure the degree of interference caused by such assignments is to weight each assignment by an element in a cost tensor $P_{j,i,m+1}$ where $m = |k - l|$ is the distance (in the channel domain) between channels k and l [12]. Now, the problem becomes how to minimize the total cost of all the assignments in the network.

Minimize

$$F(X) = \sum_{j=1}^{N} \sum_{k=1}^{M} X_{j,k} \sum_{i=1}^{N} \sum_{l=1}^{M} X_{i,l} P_{j,i,(|k-l|+1)}, \quad (2)$$

Subject to

$$\sum_{k=1}^{M} X_{j,k} = D_j, \quad \forall j = 1, \cdots, N, \quad (3)$$

The proximity factor tensor P described above can be generated according to the recursive relation

$$P_{j,i,m+1} = \max\{0, P_{j,i,m} - 1\}, \text{ for } m = 1, \cdots, M-1, \quad (4)$$

$$P_{j,i,1} = C_{ji}, \quad \text{for all } j, i \neq j, \quad (5)$$

$$P_{j,j,1} = 0, \quad \text{for all } j. \quad (6)$$

The cost function given by eqn. (2) is the objective function to be minimized and the demand vector D is the constraint that can be transformed to penalty function as shown below

$$p(x) = \sum_{j=1}^{N} \left[(\sum_{k=1}^{M} X_{j,k}) - D_j \right]^2. \quad (7)$$

3 Brief Review of Stochastic Ranking

3.1 Constrained Optimization

The general nonlinear programming problem can be formulated as: Find x to

$$\text{minimize } y(x), \quad x = (x_1, \cdots, x_n) \in R^n . \tag{8}$$

In eqn. (8), x is an n-dimensional vector and $x \in S \cap F$, where S defines the *search space* and F defines the feasible region. S is an n-dimensional space whose boundary is:

$$x_{\min} \leq x_i \leq x_{\max}, \quad i \in (1, \cdots, n), \tag{9}$$

where x_{min} is the lower bound of x_i and x_{max} is the upper bound of x_i. The *feasible region* F is given by

$$F = \{x \in R^n \mid c_j(x) \leq 0 \ \forall j \in \{1, \cdots, m\}\}, \tag{10}$$

where $c_j(x), j \in \{1, \cdots, m\}$ are constraints.

One way often used to deal with constrained optimization problems is to introduce a penalty term into the objective function to penalize constraint violations [4].

$$f(x) = y(x) + w_g p(c_j(x); j = 1, \cdots, m) . \tag{11}$$

In this equation, $p \geq 0$ is a real-valued function and it represents a "penalty". The "strength" of the penalty is controlled by a sequence of penalty coefficients. There are some forms of penalty functions. One often used is the following quadratic loss function [3]:

$$p(c_j(x); j = 1, \cdots, m) = \sum_{j=1}^{m} \max\{0, c_j(x)\}^2 . \tag{12}$$

Although for many problems, the penalty function approach may work quite well, another difficult optimization problem, i.e., how to determine the penalty coefficients arises.

3.2 Stochastic Ranking

In order to avoid setting penalty coefficients, Runarsson and Yao [4] proposed a novel constraint-handling technique for evolutionary algorithms. Using this new technique, the right balance between objective and penalty functions can be achieved stochastically.

In stochastic ranking [4], firstly we initialize n individuals randomly, $X_1, X_2, \cdots, X_n$. Each individual represents a potential solution and has an objective value $y(X_i)$ and a penalty value $p(X_i)$, $i = 1, 2, \cdots n$. Then, we rank the individuals according to either their objective function values eqn. (2) or penalty function values eqn. (7) in order to select m individuals to be parents. $m<n$. m/n is called truncation level. In order to balance the dominace of objective function and penalty function without setting w_g, a probability P_f is used. P_f decides whether we use objective function or penalty function to compare adjacent individuals. That is, given any pair of two

adjacent individuals (in order to determine which one is fitter) if both of them are feasible, the comparison is determined by the objective function; otherwise, the probability of comparing them according to the objective function is P_f. The one with lower values (in the case of minimizing fitness function) has higher position than those with larger values in the ranking list. Ranking can be achieved by a bubble-sort-like procedure [4]. After ranking, m individuals are chosen as parents to generate next generation. Then the new generation is ranked and parents are chosen. This procedure will go on until stop criteria are satisfied.

4 Experimental Studies

The evolutionary optimization algorithm described in this section is based on genetic algorithms (GA) and stochastic ranking. There have been several attempts in using GA to solve CAPs. For example, [5-9] used GA for CAP1. [10-11] and [13] used GA for CAP2.

The data set can be divided into three classes. The first class consists of the problems EX1 and EX2 [15]. The second class of test problems (Kunz1-Kunz4) was used by Kunz [14]. The third class of test problems (HEX1-HEX4) is based on the 21-cell regular hexagonal network used by Sivarajan et al. [15].

During simulations, several parameters, such as P_f, the crossover probability, mutation probability, and population size need to be set. These values were set by trial and error. Table 1 shows the parameters for various CAP2s. Table 2 compares the interference values obtained by different methods used in [12-13] and [18] with those obtained here using stochastic ranking. The minimum values and average values obtained by other algorithms listed in that table are calculated from 10 runs' results. In order to compare with them, we also run the program 10 times and calculate the average values. We note that standard deviations are not available in most methods shown in Table 2; however, we list the standard deviation for our method for comparisons with further work. "Min" represents the minimum objective function. "Ave" is the average objective value and "Sd" is the standard deviation.

Table 1. Parameters used in the simulations. (Ps is the population size; Pc is the crossover probability; Pm is the mutation probability)

Problem	P_s	P_c	P_m	P_f
EX1	40	0.75	0.3	0.45
EX2	60	0.85	0.2	0.46
HEX1	100	0.7	0.4	0.46
HEX2	120	0.65	0.35	0.465
HEX3	140	0.8	0.4	0.455
HEX4	140	0.85	0.35	0.46
KUNZ1	80	0.75	0.25	0.45
KUNZ2	80	0.7	0.2	0.445
KUNZ3	120	0.8	0.3	0.455
KUNZ4	140	0.75	0.4	0.46

Table 2. Comparison of the total interference obtained by different methods: Genetic algorithm with stochastic ranking (GASR), genetic algorithm without stochastic ranking (GA) [13], the self-organizing neural network (SONN) [12] and stochastic chaotic simulated annealing [18]. The values in bold indicate improved solutions over all other algorithms

Problem	GA-SR		GA		SONN		SCSA	
	Min	Ave±Sd.	Min	Ave	Min	Ave	Min	Ave±Sd.
EX1	0	0±0.0	0	0	0	0.4	0	0.0±0.0
EX2	0	0±0.0	0	0	0	2.4	0	0.0±0.0
HEX1	**46**	**47.7±0.4**	48	48.1	52	53.0	47	47.7±0.3
HEX2	**17**	**18.4±0.2**	19	19.3	24	28.5	18	18.5±0.2
HEX3	76	76.5±0.5	76	76.4	84	87.2	76	77.3±0.3
HEX4	16	17.5±0.6	17	17.2	22	29.1	16	17.2±0.4
KUNZ1	19	19.8±1.0	20	20.1	21	22.0	19	20.0±2.1
KUNZ2	29	29.4±0.3	29	29.4	33	33.4	30	30.3±0.2
KUNZ3	13	13.00	13	13.0	14	14.4	13	13.0±0.0
KUNZ4	0	0±0.0	0	0	1	2.2	0	0.0±0.0

We can see that for HEX1 and HEX2 we obtained lower interference values than other methods including genetic algorithm [13] without stochastic ranking. In [14], the search space is limited to feasible regions.

5 Conclusions and Discussions

In this paper, we have considered the problem of assigning channels to calls in a cellular mobile communication network. We have combined stochastic ranking proposed by Runarsson and Yao with GA and applied to CAP2. The simulations done on the benchmark problems showed that genetic algorithm using stochastic ranking can achieve desirable results compared to other heuristic approaches. For several problems, GA with stochastic ranking can obtain better results than using GA only.

References

1. Kunz, D.: Suboptimal Solutions Obtained by The Hopfield-Tank Neural Network Algorithm. Biological Cybernetics, Vol.65 (1991) 129-133
2. Duque-Anton, M., Kunz, D., Ruber, B.: Static and Dynamic Channel Assignment Using Simulated Annealing, Neural Networks in Telecommunications. B. Yuhas and N. Ansari, Eds. Boston, MA: Kluwer (1994)
3. Fiacco, A., McCormick, G.: Nonlinear Programming: Sequential Unconstrained minimization Techniques. New York: Wiley (1968)
4. Runarsson, T., Yao, X.: Stochastic Ranking for Constrained Evolutionary Optimization. IEEE Transactions on Evolutionary Computation, Vol.4 (2000) 284-294
5. Beckmann, D., Killat, U.: A New Strategy for The Application of Genetic Algorithms to The Channel Assignment Problem. IEEE Trans. Veh. Technol., Vol. 48 (1999) 1261-1269
6. Thavarajah, A., Lam, W.: Heuristic Approach for Optimal Channel Assignment in Cellular Mobile Systems. IEEE Proceedings Communications, Vol. 1463 (1999) 196-200

7. Chakraborty, G., Chakraborty, B.: A Genetic Algorithm Approach to Solve Channel Assignment Problem in Cellular Radio Networks. Proc. 1999 IEEE Midnight-Sun Workshop on Soft Computing Methods in Industrial Applications (1999) 34-39
8. Williams, M.: Making The Best Use of The Airways: An Important Requirement for Military Communications. Electronics & Communication Engineering Journal, Vol. 12 (2000) 75-83
9. Jaimes-Romero, F., Munoz-Rodriguez, D., Tekinay, S.: Channel Assignment in Cellular Systems Using Genetic Algorithms. IEEE 46th Vehicular Technology Conference, Vol. 2 (1996) 741-745
10. Lai, W., Coghill, G.: Channel Assignment Through Evolutionary Optimization. IEEE Transactions on Vehicular Technology, Vol. 45 (1996) 91-96
11. Ngo, C., Li, V.: Fixed Channel Assignment in Cellular Radio Networks Using A Modified Genetic Algorithm. IEEE Trans. Vehicular Technology, Vol. 47 (1998) 163-172
12. Smith, K., Palaniswami, M.: Static and Dynamic Channel Assignment using Neural Networks, IEEE Journal on Selected Areas in Communications, Vol. 15(1997) 238-249
13. Wang, L., Arunkumaar, S., Gu, W.: Genetic Algorithms for Optimal Channel Assignment in Mobile Communication. Proceeding of The 9th International Conference on Neural Information Processing (ICONIP'02), Vol.3 (2002) 1221-1225
14. Kohonen, T.: Self- Organized Formation of Topologically Correct Feature Maps. Biol. Cybern., Vol. 43 (1982) 59-69
15. Sivarajan, K., McEliece, R.: Channel Assignment in Cellular Radio. Proc. IEEE Veh. Technol. Conf (1989) 846-850
16. Gamst, A., Rave, W.: On Frequency Assignment in Mobile Automatic Tlephone Systems. Proc. GLOBECOM'82 (1982) 309-315
17. Smith, K.: A Genetic Algorithm for The Channel Assignment Problem. IEEE Global Technology Conference, Vol. 4 (1998) 2013-2018
18. Li, S., Wang, L.: Channel Assignment for Mobile Communications Using Stochastic Chaotic Simulated Annealing. Proceedings of the 2001 International Work-Conference on Artificial Neural Networks (IWANN2001, Granada, Spain, June 13-15,2001, Mira, Jose; Prieto, Alberto (Eds)), Part I, Lecture Notes in Computer Science, Vol. 2084 (2001) 757-764

A MPLS-Based Micro-mobility Supporting Scheme in Wireless Internet

SuKyoung Lee

Graduate School of Information & Communications
Sejong University, Seoul, Korea
sklee@sejong.ac.kr

Abstract. In this paper, we propose to exploit pre-established label switched paths with the aim to reduce handover processing latency which is one of the most critical performance metrics in micro-mobility environment. Thus, our proposed scheme has an advantage of offering micro-mobility to Mobile IP (MIP) with lower-latency handover as compared with the several existing MPLS-based MIP approaches. An analytic model of setting up LSPs for fast handover processing is presented. Further, we investigate the performance of our model.

1 Introduction

The IETF working group on Mobile IP (MIP) is proposing MIPv4 and MIPv6 as the main protocols for supporting IP mobility across different access networks. Additionally, some extensions have been made to MIP in order to support micro-mobility environments where the mobile nodes change their point of attachment so frequently that the MIP protocol would result in high handover latency as well as a high signaling load [1][2]. To support the micro-mobility, there have been several schemes to integrate MultiProtocol Label Switching (MPLS) into MIP. The development of these schemes was inspired by one of the main latest networking trends that a significant number of carriers are migrating towards a common background based on MPLS as the transport options for IP services. In addition, it is beneficial in terms of the Quality of Service (QoS) requested by the wireless Internet users to develop MIP based on MPLS as in [3]-[5] because the integration of MPLS and IP inherits the noted merits of MPLS in terms of QoS support, traffic engineering, and advanced IP services [4].

Therefore, in this paper, as proposed in [3]-[5], we also utilize MPLS to provide fast handover for the wireless Internet users with micro-mobility. In wireless Internet, it is important to maintain the QoS demands for real-time data traffic during handover. That means the handover processing delay should be as small as possible to ensure a deterministic QoS for real-time applications. Our model proposed in this paper makes very fast handovers workable by establishing Label Switched Paths (LSPs) between neighboring Base Stations (BSs) without the need of modifying the IP address. An analytic model is used to capture the behavior of a system in a concise mathematical formulation. Furthermore, we do not only investigate the performance of our model but also propose signaling so that our proposed scheme could be run on MPLS-based MIP system.

2 Reservation of LSPs

In our proposed system, the association change occurs between current BS and a new BS in order to complete mobile tunnel toward the new BS, where the BSs also function as edge Label Switching Router (LSR) similarly with [3]-[5], which is called BS-LSR from now on. To support fast handover in MPLS-based wireless Internet, our proposed scheme adopts a format of LSP extension between BS-LSRs. Several labels are devoted to connect between current and new BS-LSRs as soon as a handover occurs. The labels are not reserved between all BS-LSR pairs, but only for neighboring BS-LSRs are done so.

To analyze performance implication of label reservation, let us assume that the arrival rate of calls to a Mobile Host (MH) in a subnet or cell, covered by a single BS-LSR, be modelled as a Poisson process. In our traffic model, it is assumed that the call holding time, t_c be exponentially distributed with the density function, $f_c(t) = \mu e^{-\mu t}$ and a MH resides in a cell controlled by j^{th} BS-LSR for a time period, $t_j (j > 0)$. Let t_j be independent each other and identically distributed random variable with the distribution, $F_r(t_j)$ and the density function, $f_r(t_j)$ (mean is $1/\gamma$).

Let the label occupation time be denoted by t_l which implies the duration from the time when a label is assigned for a handover to the time when it is released. Then, as defined in [6], we have

$$t_l = \min(t_h, t_c), \qquad (1)$$

where $t_h = t_1 + t_2 + \cdots + t_K$ given that K is the number of cells in which the MH resides before the handover is blocked. From Eq. 1, it is then easy to see that t_h denotes the total LSP occupancy time before handover request is blocked. Given that K is the random variable with the geometric probability distribution, $Pr[K = k] = (1-p_f)^{k-1} p_f (k = 1, 2, \cdots)$ where p_f is the probability that a handover call is likely to be force-terminated in each cell.

Let $f_l(t)$ and $F_l(t)$ be the density function and the distribution of t_l, respectively. It now follows from the definition of t_l given in Eq. 1 that

$$F_l(t) = F_h(t) + (1 - F_h(t))F_c(t). \qquad (2)$$

Then, the density function $f_l(t)$ is

$$f_l(t) = \frac{d}{dt}[F_h(t) + (1 - F_h(t))F_c(t)] \qquad (3)$$

$$= \mu e^{-\mu t} + e^{-\mu t} f_h(t) - \mu e^{-\mu t} \int_0^t f_h(x)dx$$

Let $T_l^*(s)$, $T_h^*(s)$, and $T_r^*(s)$ denote the LST (Laplace-Stieltjes Transform) of the distributions for t_l, t_h and t_j, respectively. Under the assumption of $t_h = \sum_{j=1}^K t_j$, we get

$$T_h^*(s) = G_K[T_r^*(s)] \tag{4}$$

where $G_K[z] = \frac{p_f z}{1-(1-p_f)z}$ as the probability generating function of K. Using Eq. 3 and Eq. 4, we have

$$\begin{aligned}T_l^*(s) &= \int_0^\infty f_l(t)e^{-st}dt \\ &= \int_0^\infty \mu e^{-\mu t}dt + \int_0^\infty e^{-\mu t}f_h(t)e^{-st}dt - \int_0^\infty \mu e^{-\mu t}\int_0^t f_h(x)dx e^{-st}dt \\ &= \frac{\mu}{s+\mu} + G_K[T_r^*(s+\mu)] - \frac{\mu}{s+\mu}G_K[T_r^*(s+\mu)].\end{aligned} \tag{5}$$

From Eq. 5, the mean of t_l is given by

$$\begin{aligned}E[t_l] &= \frac{d}{ds}T_l^*(s)\mid_{s=0} \\ &= \frac{1 - T_r^*(\mu)}{\mu\{1 - (1-p_f)T_r^*(\mu)\}}.\end{aligned} \tag{6}$$

Let λ_o and p_o be the originating call arrival rate to a cell and the new call blocking probability, respectively. For the moment, we assume p_o (and also p_f) is known. Under the assumption that t_j is i.i.d, we get $E[t_l] = \frac{1}{\gamma}E[K]$. Thus, the handover arrival rate, λ_h can be expressed as

$$\begin{aligned}\lambda_h &= \lambda_o(1-p_o)E[K] \\ &= \frac{\lambda_o(1-p_o)(1-T_r^*(\mu))\gamma}{\mu\{1-(1-p_f)T_r^*(\mu)\}}.\end{aligned} \tag{7}$$

If $F_r(t)$ is an exponential distribution, $f_r^*(\mu) = \frac{\gamma}{\mu+\gamma}$. Then, Eq. 6 and 7 are rearranged as

$$E[t_l] = \frac{1}{\mu + \gamma p_f} \quad \text{and} \quad \lambda_h = \frac{\lambda_o(1-p_o)\gamma}{\mu + \gamma p_f}. \tag{8}$$

Now, we describe briefly the mobility model for MHs following the assumption in [7]. It is assumed that wireless Internet users are moving at an average velocity of v, and their direction of movement is uniformly distributed over $[0, 2\pi]$. Supposing that wireless Internet users are uniformly populated with a density of ρ, and the perimeter of a cell is length of L, the cell-crossing rate becomes $\frac{\rho v L}{\pi}$. Then, the active handover rate across a cell boundary is

$$\lambda_h' = \frac{\rho v L \lambda_h}{\pi} \tag{9}$$

given that each handover arrival rate to a cell is the same as the handover departure rate.

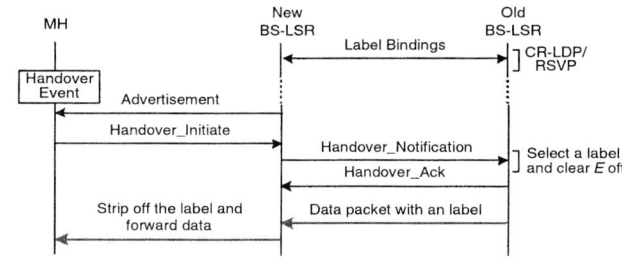

Fig. 1. Signaling

The handover blocking probability from lack of reserved labels can be analyzed numerically.

$$p_b^{(l)} = \frac{\lambda_h' E[t_l]^{N_l}/N_l!}{\sum_{n=0}^{N_l} \lambda_h' E[t_l]^n/n!}. \tag{10}$$

From the above Eq.10, we can obtain the optimum number of allocated LSPs, N_l to satisfy $p_b^{(l)} \leq \epsilon$, where ϵ is the threshold of the handover blocking probability requested by the wireless Internet users, especially who are using real-time applications.

The pre-allocation of multiple labels for handover is signaled as a part of the *Label Request* message in MPLS signaling such as CR-LDP or RSVP [8][9]. A list of groups of nodes to bind labels for handover is inserted into Explicit Route (ER) TLV/Object in *Label Request* Message. Thus, in the set up message, included is the information indicating if the LSP is primary or secondary for handover. A new flag E is introduced, which stands for "E"gress of the LSP. The flag is added to each entry of Label Forwarding Information Base (LFIB) at BS-LSR while all initialized with one during the time of allocating labels for handover. Fig. 1 shows the signaling procedure of pre-allocation of labels. Note that in the Fig. 1, when handover occurs, there is no need to exchange some MPLS signaling message related to label allocation, resulting in lowering the handover processing time.

3 Performance Evaluation

We now investigate the performance of our proposed scheme in comparison with that of LSP rerouting scheme [3]-[5]. The values used in numerical analysis for all the relevant parameters are shown in the Table 1. First, system performance is evaluated by considering processing overhead.

Whenever a MH moves into a new cell, the associated label mappings should be changed at every LSR related to rerouting (i.e. the LSRs over both new subpath and the subpath which should be released). Let N_m be the total number of label mapping changes in call duration. Let N_n and N_o denote the number of new LSRs over the new rerouted LSP and the number of LSRs over the old end-

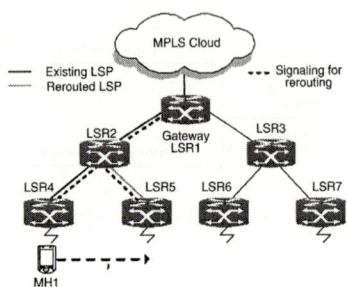

Fig. 2. A Sample Network Topology

Table 1. Parameter values

Parameter Name	Parameter Symbol	Value
Rate of new calls	λ_o	30/min
Average call duration	$1/\mu$	3, 5, and 10 min
Average cell sojourn time	$1/\gamma$	2 min
Blocking probability of a new call	p_o	0.01
Forced termination probability	p_f	0.001

to-end LSP but would not be necessary anymore over the new LSP, respectively. The average label mapping changes are derived as Fig. 2.

$$E[N_m] = \lambda'_h E[K] N_r \qquad (11)$$

where $N_r = N_n + N_o$. For example, a sample network is taken in Similar to LSP rerouting methods described in [3] and [5], the existing LSP (LSR1-LSR2-LSR4) for MH1 is rerouted to LSP (LSR1-LSR2-LSR5) as shown in Fig. 2. Thus, N_r becomes 3. On the other hand, our proposed scheme does not require any rearrangements of an existing LSP. The old BS-LSR and the handovered one alone change their own LFIB. Therefore, N_r becomes always 2. Table 2 shows the average number of label mapping changes. Table 2 shows that the proposed LSP extension method performs better than rerouting method in terms of label mapping change overhead.

Fig. 3 shows the probability of handover blocking due to the lack of LSPs for different mobilities and the average call duration in a system. From this figure, it is easy to see that the more LSPs are allocated, the smaller $p_b^{(l)}$ becomes. From the results shown in Fig. 3, we know that the optimal number of LSPs to satisfy $\epsilon = 10^{-3}$ and 10^{-4} are 10 and 12, respectively for 90 erlangs, while 14 and 16 for 150 erlangs.

Table 2. Average number of label mapping changes

Average call duration (min)	LSP Rerouting	LSP extension
3	100	67
5	278	110
10	1103	221

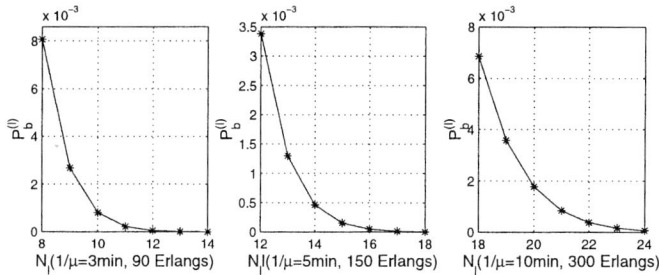

Fig. 3. Handover blocking probability versus number of LSPs under varying call duration time

4 Conclusion

In this paper, we proposed an efficient handover scheme to provide micro-mobility for MIP. We considered a MPLS architecture which was also taken into account in some existing schemes for wireless Internet since the MPLS-based wireless Internet inherits the native traffic engineering and QoS benefits provided by MPLS. The proposed scheme provides fast handover processing by taking advantage of LSPs for handovers. The reduction of handover processing latency makes our proposed scheme be well suited for micro-mobility. The numerical results indicated the handover blocking probability requested by users could be met by reserving some LSPs for handover.

References

1. Campbell, A.T., et al.: Comparison of IP Micro-mobility Protocols. IEEE Wireless Commun. Mag., Vol. 9, No.1 (Feb. 2002) 72-82.
2. Soliman, H., et al.: Hierarchical Mobile IPv6 Mobility Management. Internet Draft, draft-ietf-mipshop-hmipv6-00.txt, Work in progress (Jun. 2003).
3. Yang, T., Dong, Y., Zhang, Y., Makrakis, D.: Practical Approaches for Supporting Micro Mobility with MPLS, International Conference on Telecommunications, Beijing China (2002).
4. Chiussi, F.A., Khotimsky, D.A., Krishnan, S.: Mobility Management in Third-Generation All-IP Networks. IEEE Commun. Mag., Vol. 20, No. 9 (Sep. 2002) 124-135.
5. Vassiliou, V., Owen, H., Barlow, D., Soko, J., Huth, H., Grimminger, J.: M-MPLS: Micromobility-enabled Multiprotocol Label Switching. IEEE ICC (2003) 250-255.
6. Lin, Y.B.: Performance Modeling for Mobile Phone Networks. IEEE Network Mag., Vol. 11, No. 6 (Nov./Dec. 1997) 63-68.
7. Wu, W., Misra, A., Das, S.K., Das, S.: Scalable QoS Provisioning for Intra-Domain Mobility. IEEE Globecom (2003).
8. Awduche, D.O., et al.: RSVP-TE: Extensions to RSVP for LSP Tunnels. RFC3209 (Dec 2001).
9. Ash, J., et al.: LSP Modification Using CR-LDP. RFC3214 (Jan. 2002).

A Novel RBF Neural Network with Fast Training and Accurate Generalization

Lipo Wang[1,2], Bing Liu[2], and Chunru Wan[2]

[1] College of Information Engineering, Xiangtan University,
Xiangtan, Hunan, China
[2] School of Electrical and Electronic Engineering, Nanyang Technology University,
Block S1, Nanyang Avenue, Singapore 639798
{elpwang, liub0002, ecrwan}@ntu.edu.sg

Abstract. For the reason of all the centers and radii needed to be adjusted iteratively, the learning speed of radial basis function (RBF) neural networks is always far slower than required, which obviously forms a bottleneck in many applications. To overcome such problem, we propose a fast and accurate RBF neural network in this paper. First we prove the universal approximation theorem for RBF neural networks with arbitrary centers and radii. Based on this theory, we propose a new learning algorithm called fast and accurate RBF neural network with random kernels (RBF-RK). With the arbitrary centers and radii, our RBF-RK algorithm only needs to adjust the output weights. The experimental results, on function approximation and classification problems, show that the new algorithm not only runs much faster than traditional learning algorithms, but also produces better or at least comparable generalization performance.

Keywords: RBF, random kernel, fast, accurate

1 Introduction

Due to its structural simplicity, the radial basis function (RBF) has attracted an extensive research interest in the area of function approximation and pattern classification [1]. The standard RBF neural network is composed of three layers: an input layer, a hidden layer and an output layer. Very often, the RBF neural network is trained using two-step procedures. In the first step, the hidden units is fixed a priori based on some properties of the input data[2]. Then the weights are estimated by linear least squares method[2]. Several learning algorithms have been proposed in the literature for training RBF network. The classical training method is to apply cluster technique, such as vector quantization[3] and input-output clustering[4]. Besides clustering methods, the orthogonal forward selection algorithm[5], and support vector machine method [6] are the other two frequently used methods for RBF network training.

However, these traditional learning methods are usually far slower than required. In this paper, we proposed a new learning method called a RBF Neural Network with Random Kernels (RBF-RK) which arbitrarily chooses the centers and standard deviations and analytically determines the output weights of

RBF neural networks. Compared with traditional learning methods, our method is very fast. Usually, it is at least 20-30 times faster than traditional learning methods. Our experiments show that the proposed RBF-RK algorithm produces better or at least comparable generalization performance compared to other popular classification algorithms.

It was stated in [7] that a RBF neural network can be trained by fixing centers prior. However, in contrast to the present paper, there is no proof of universal approximation for such a RBF neural network with fixed kernels or no extensive simulations in [7]. Our present algorithm is also inspired by the recent work by Huang, Zhu, and Siew [8], who proposed an extreme learning machine (ELM) algorithm for multi-layer perceptron feedforward neural networks with arbitrary input weights, which shows fast training and excellent generalization.

This paper is organized as follows. To provide a theoretical base for our RBF-RK algorithm, we firstly propose the proof of universal approximation using RBF neural networks with arbitrary centers and standard deviations in Section 2. In Section 3 we propose the RBF-RK algorithm. Performance evaluation of the proposed RBF neural network is presented in Section 4. Finally, discussions and conclusions are given in Section 5.

2 Universal Approximation

We consider a neural network with n hidden neurons and one output defined as follows: $f_n(\boldsymbol{x}) = \sum_{i=1}^{n} \beta_i g_i(\frac{|\boldsymbol{x}-\boldsymbol{a}_i|^2}{2\sigma_i^2})$. Here β_i are adjustable weights connecting the hidden neurons with the network output. $g_i(y)$ is a continue function, where $y = \frac{|\boldsymbol{x}-\boldsymbol{a}_i|^2}{2\sigma_i^2}$, with centers $\boldsymbol{a}_i$ ($\boldsymbol{a}_i \in R^d$) and standard deviations σ_i ($\sigma_i \in R, \sigma_i \neq 0$). $\boldsymbol{x} \in R^d$ is the input vector. $|\cdot|$ denotes the Euclidean norm.

Theorem 1. *Given activation function $g_n(\boldsymbol{x}) \in L^2(R^d)$, $\int_{R^d} g_n^2(\boldsymbol{x})d\boldsymbol{x} = M_0$ where M_0 is constant values, $M_0 > 0$, for any $\boldsymbol{a}_i$ and σ_i chosen from any intervals of R^d and R, respectively, according to any continuous probability, then with probability 1, $\lim_{n \to \infty} \|f - f_n\| = 0$.*

The proof for the above theorem is given in [9]. When RBF networks choose other nonlinear continuous functions, such as Multiquadrics function[10], Inverse multiquadrics[10], sigmoid function, sine and cosine, we can similarly prove the incremental network f_n with different parameters, i.e. arbitrary parameters, converge to any continuous target function f.

3 RBF-RK Algorithm

Algorithm 1. *Given a training set $\Phi = \{(\boldsymbol{x}_i, \boldsymbol{t}_i) | \boldsymbol{x}_i \in R^d, \boldsymbol{t}_i \in R^m, i = 1, 2, ..., N\}$, a validation set $Z = \{(\tilde{\boldsymbol{x}}_i, \tilde{\boldsymbol{t}}_i) | \tilde{\boldsymbol{x}}_i \in R^d, \tilde{\boldsymbol{t}}_i \in R^m, i = 1, 2, ..., \tilde{N}\}$ and activation function $g(\boldsymbol{x})$. Let n denote the number of the radial basis kernels.*

- *Step 1:* Let $n = n_0$, where n_0 is a small value set a priori.
- *Step 2:* Assign arbitrary center $\boldsymbol{a}_i$ and standard deviation σ_i, $i = 1, 2, ..., n$. Calculate the output weight $\beta = H^+T = (H^TH)^{-1}H^T$, where $H = [\boldsymbol{p}_1, \boldsymbol{p}_2, ..., \boldsymbol{p}_n]$, $\boldsymbol{p}_i = [g_i(\boldsymbol{x}_1), g_i(\boldsymbol{x}_2), ..., g_i(\boldsymbol{x}_N)]^T$ and $T = [\boldsymbol{t}_1, \boldsymbol{t}_2, ..., \boldsymbol{t}_N]^T$.
- *Step 3:* With the output weights obtained by Step 2, calculate the validation error.
- *Step 4:* Repeat Steps 2 – 3 for a number of times, then compute the average validation error $E(n)$ for such n.
- *Step 5:* Let $n = n + \Delta n$, where $\Delta n > 0$, and repeat Steps 2–4 until $n > n_{max}$, where n_{max} is a large value set a priori.
- *Step 6:* Plot the average validation error $E(n)$ versus n. Find the optimal number of kernels n_{op}, where $E(n_{op})$ reaches a minimum.
- *Step 7:* Do Step 2 with the optimal number of kernels n_{op} obtained in Step 6 and compute test error using test data set.

According to [11], when $\beta = H^+T$, we have $\|H\beta - T\|_F = \|HH^+T - T\|_F = \min_\beta \|H\beta - T\|_F$, which means our RBF-RK algorithm reaches the minimum training error and avoids being trapped into a local minimum for any given set of kernel parameters. Here $\|\cdot\|_F$ means matrix $F-norms$ [11].

4 Experimental Results

In this section, the performance of our RBF-RK algorithm is compared with conventional learning algorithms for RBF neural networks. All the simulations are carried out in MATLAB 6.5 environment running in a Pentium 4, 2.4 GHZ CPU. To compare with RBF-RK, we use MATLAB functions NewGrnn and NewPnn to design generalized regression neural networks and probabilistic neural networks, respectively. The GRNN algorithm is used for function approximation problems, while the PNN algorithm is used for classification problems.

In our RBF-RK algorithm, the activation function was chosen as a standard Gaussian function, and all the centers and radii are randomly distributed in (0, 1) according to a uniform probability distribution function. All the input values are normalized to the range [0, 1]. The training and testing errors in function approximation problems are calculated by root mean squares (RMS). "Training Dev." and "Testing Dev." are the standard deviations in the training error (error rate) and test error (error rate), respectively. "Num. of Neurons" stands for the number of radial basis kernels. It should be noted that the number of radial basis kernels in the GRNN algorithm or the PNN algorithm is always equal to the number of training/target pairs and there is no validating process during the learning of the GRNN algorithm or the PNN algorithm.

We repeat Steps 2 – 3 of the proposed RBF-RK algorithm 20 times and then compute the average validation error (or error rate) $E(n)$ and stand deviations $STD(n)$ in the validation error. Using the polynomial method provided by MATLAB package, we plot $E(n)$ versus n and the error bar $STD(n)$ in one figure. When the optimal number of radial basis kernels n_{op} is selected out, we perform Step 7 for 50 times in all experiments.

California Housing Data Set. The California Housing data set is obtained from StatLib repository (www.niaad.liacc.up.pt/~ltorgo/Regression/cal_housing.html). There are 20640 cases, 8 continue inputs and one continue output in this data set. 5000 training data, 3000 validating data and 12640 testing data are chosen for RBF-RK algorithm, while 8000 training data and 12640 testing data are selected for GRNN algorithm. We plot $E(n)$ versus n and the error bar $STD(n)$ in Fig. 1. (a). Seen from Fig. 1. (a), when the number of neurons n is small, the average validation error $E(n)$ keeps decreasing as n increases. When $n > 15$, $E(n)$ is flat with n increasing. Therefore, in this experiment we choose the optimal $n_{op} = 20$.

10 trials were run for the GRNN algorithm, since it takes much longer to train the GRNN compared to the RBF-RK. The average results obtained by both algorithms are shown in Table 1, which shows that the learning speed of our RBF-RK algorithm is 617.9 times faster than GRNN algorithm and obtains a smaller testing error than that obtained by the GRNN.

Diabetes Data Set. The Diabetes data set [12] contains 768 cases, 8 attributes and 2 classes. 384 training data, 192 validating data and 192 testing data are

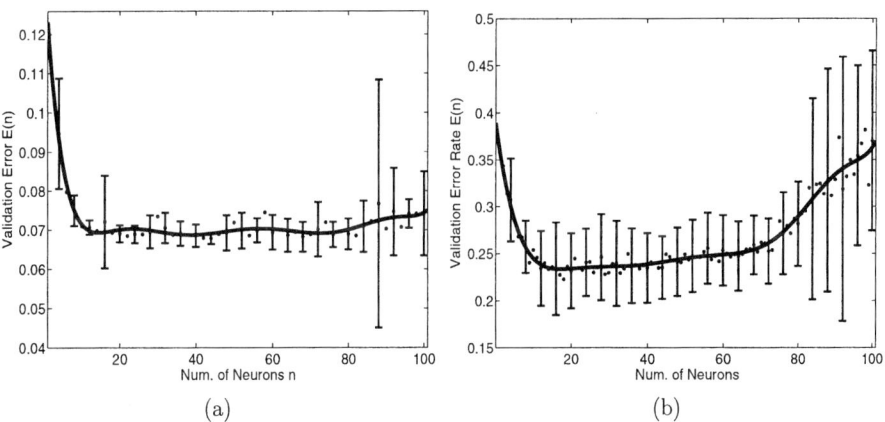

Fig. 1. Average validation error rate of California Housing and Diabetes data sets

Table 1. Comparison for California Housing and Diabetes data sets

Problems	Algorithms	Time (Sec.)	Training Error /Error Rate	Training Dev.	Testing Error /Error Rate	Testing Dev.	Num. of Neurons
California Housing	RBF-RK	0.5881	0.0614	0.0010	0.0802	0.0179	20
	GRNN	363.4	0.1122	0.0023	0.1159	0.0021	8000
Diabetes	RBF-RK	0.0203	21.49%	1.68%	22.73%	3.83%	20
	PNN	0.628	3.06%	0.22%	25.64%	4.01%	576

chosen for RBF-RK algorithm, while 576 training data and 192 testing data are selected for PNN algorithm. We plot $E(n)$ versus n and the error rate bar $STD(n)$ in Fig. 1. (b). Seen from Fig. 1. (b), when the number of neurons n is small, the average validation error rate $E(n)$ keeps decreasing as n increases. $E(n)$ reaches a minimum when n is between 15 and 25. When $n > 25$, $E(n)$ increases as n increases. Therefore, in this experiment we choose the optimal $n_{op} = 20$.

50 trials have been conducted for the PNN algorithms and the average results obtained by both algorithms are shown in Table 1. Seen from Table 1, our RBF-RK algorithm obtains a smaller testing error rate and runs 30.9 times faster than the PNN does. Compared with other methods' results in Table 2, the testing accuracy is comparable to that in [14][15] and higher than others.

Table 2. Comparison in accuracy for Diabetes data set

Algorithms	Testing Accuracy
RBF-RK	77.27%
SVM [13]	76.50%
SAOCIF [14] [15]	77.32%
Cascade-Correlation [14] [15]	76.58%
AdaBoost [16]	75.60%
C4.5 [16]	71.60%
RBF [17]	76.30%
Heterogeneous RBF [17]	76.30%

5 Discussion and Conclusion

In this paper, we firstly prove the universal approximation capability for a RBF neural network with arbitrary centers and standard deviations. Based on this theory, we proposed a new learning algorithm called RBF-RK. Compared with the conventional learning methods, the proposed RBF-RK algorithm showed some important advantages as follows:

1. The learning speed of RBF-RK is very fast. According to our results, the learning speed of RBF-RK is several to hundreds times faster than other popular learning methods for RBFNs.
2. At each experiment, the testing error in our RBF-RK algorithm is always smaller than or at least comparable to that in the traditional learning methods.

References

1. Poggio, T., Girosi, F.: Network for Approximation and Learning. Proc. IEEE, Vol. 78 (1992) 1481-1497

2. Wei, L.Y., Sundararajan, N., Saratchandran, P.: Performance Evaluation of a Sequential Minimal Radial Basis Function (RBF) Neural Network Learning Algorithm. IEEE Trans. Neural Networks, Vol. 9 (1998) 308-318
3. Kohonen, T.: Self-Organizing Maps. Springer-Verlag (1995)
4. Uykan, Z., Guzelis, C., Celebi, M.E., Koivo, H.N.: Analysis of Input-output Clustering for Determining Centers of RBFN. IEEE Trans. Neural Networks, Vol. 11 (2000) 851-858
5. Gomm, J.B., Yu, D.L.: Selecting Radial Basis Function Network Centers with Recursive Orthogonal Least Squares Training. IEEE Trans. Neural Networks, Vol. 11 (2000) 306-314
6. Schokopf, R., Sung, K.K., Burges, C.J.C., Girosi, F., Niyogi, P., Poggio, T., Vapnik, V.N.: Comparing Support Vector Machines with Gaussian Kernels to Radial Basis Function Classifiers. IEEE Trans. Signal Processing, Vol. 45 (1997) 2758-2765
7. Lowe, D.: Adaptive Radial Basis Function Nonlinearities, and the Problem of Generalisation. First IEEE International Conference on Artificial Neural Networks (1989) 29-33
8. Huang, G.B., Zhu, Q.Y., Siew, C.K.: Extreme Learning Machine. 2004 International Joint Conference on Neural Networks (IJCNN'2004) (2004)
9. Liu, B., Wan, C.R., Wang, L.: A Fast and Accurate RBF Neural Network with Random Kernels. Submitted for publication
10. Haykin, S.: Neural Networks-a Comprehensive Foundation. 2nd edn. Prentice Hall International,Inc. (1999)
11. Golub, G.H., Loan, C.F.V.: Matrix Computation. 3rd edn. Johns Hopkins University Press (1996)
12. Blake, C., Merz, C.: UCI Repository of Machine Learning Databases. http://www.ics.uci.edu/~mlearn/MLRepository.html, Department of Information and Computer Science, University of California, Irvine, USA (1998)
13. Ratsch, G., Onoda, T., Muller, K.R.: An Improvement of AdaBoost to Avoid Overfitting. Proceedings of the 5th International Conference on Neural Information Processing (ICONIP'1998) (1998)
14. Romero, E.: Function Approximation with Saocif: a General Sequential Method and a Particular Algorithm with Feed-forward Neural Networks. http://www.lsi.upc.es/dept/techreps/html/R01-41.html, Department de Llenguatgesi Sistemes Informatics, Universitat Politecnica de catalunya (2001)
15. Romero, E.: A New Incremental Method for Function Approximation Using Feedforward Neural Networks. Proc. INNS-IEEE International Joint Conference on Neural Networks (IJCNN'2002) (2002) 1968-1973
16. Freund, Y., Schapire, R.E.: Experiments with a New Boosting Algorithm. International Conference on Machine Learning (1996) 148-156
17. Wilson, D.R., Martinez, T.R.: Heterogeneous Radial Basis Function Networks. Proceedings of the International Conference on Neural networks (ICNN 96) (1996) 1263-1267

Basic Mathematical Properties of Multiparty Joint Authentication in Grids*

Hui Liu and Minglu Li

Department of Computer Science and Engineering,
Shanghai Jiaotong University, 200030 Shanghai, China
{liuhui, li-ml}@cs.sjtu.edu.cn

Abstract. This paper introduces the semantic of Multiparty Joint Authentication (MJA) into the authentication service, which is to find simplified or optimal authentication solutions for all involved principals as a whole instead of to authenticate each pair of principals in turn. MJA is designed to support multiparty security contexts in grids with a specified, understood level of confidence and reduce the time cost of mutual authentications. Graph theory model is employed to define MJA, and analyze its mathematical properties. Two algorithms to find an n-principal, n-order MJA solution are also presented and proved.

1 Introduction

The security solution for grids is of great importance. Such issues may exist through the whole lifecycle of designing, implementing, deploying and managing the grid systems. Security components are fundamental building blocks to bring grids into reality and are undergoing renovations along with the evolution of Web Services (WS) [1]. According to the GGF specification roadmap, future OGSA security architecture will leverage the existing and emerging WS security specifications as much as possible, fitting them into a layered grid security specification stack [2].

Grid security services comprise a series of web services [3]. In any case, the authentication service acts as a security entry for a grid application and different Single Sign-On (SSO) strategies built on top of it have different influences on the downstream activities, namely, authorizing, accounting, auditing, etc.

This paper introduces the semantic of Multiparty Joint Authentication (MJA) into the authentication service, which is to find simplified or optimal authentication solutions for all involved principals as a whole instead of to authenticate each pair of principals in turn. MJA supports multiparty security contexts in

---

* This research is supported by the National Grand Fundamental Research 973 Program of China (No.2002CB312002), China Grid Project, China Postdoctoral Science Foundation, and Grand Project of the Science and Technology Commission of Shanghai Municipality (No.03dz15027).

grids with a specified, understood level of confidence and reduces the time cost of mutual authentications.

The rest of this paper is organized as follows. Section 2 presents related work about GSI. New semantics for SSO in grids are extended in section 3, at the same time, the graph theory model of multiparty authentication and the definition of MJA are also proposed. In section 4, basic mathematical properties of MJA are analyzed in the form of theorems and proofs. Section 5 gives us two algorithms to find a possible MJA solution. Section 6 illustrates the potential efficiency of MJA and section 7 concludes this paper.

2 Related Work

GSI of GLOBUS is the de facto standard and infrastructure for current mainstream grid applications. GSI focuses on providing authentication and access control mechanism for the grid environment It proposes and implements a security architecture based on four interoperability protocols that are used to handle U-UP, UP-RP, RP-P and P-P interactions cooperatively [4]. The separation and linkage between the protocols and the underlying programming libraries is implemented with GSS-API that is oriented toward two-party security contexts.

Because GSS-API does not offer a clear solution to delegation, nor does it provide any support for group contexts, GSI employs proxy certificates and online credential repository to facilitate SSO. These techniques regard SSO as a series of authentications and authorizations, which can be expressed as: $Single\ Sign\text{-}On = \Sigma(authentication + authorization)$. According to this semantic, authentication and authorization are coupled too tightly to be separated from each other and a multiparty security context must be decomposed into a series of two-party security contexts. The intention of MJA is to break these two kinds of constraints.

3 Semantic of Multiparty Joint Authentication

3.1 Extended Semantics of Single Sign-On

According to the OASIS SAML architecture, authentication and authorization are interpreted as creating, exchanging and processing of security information expressed in the form of assertions about subjects within an XML based framework [5]. Therefore, authentication and authorization are separable, and the comprehension of SSO can be expressed as: $Single\ Sign\text{-}On = \Sigma authentication + \Sigma authorization$.

On the other hand, because a grid system intends to provide coordinated resources sharing, problem solving or services outsourcing in dynamic, multi-institutional virtual organizations [6][7], a typical grid application often spreads over multiple resource hosting sites and needs multiparty authentication. For example, the computational power providers of a computation job may need multi-party authentication, the idle resource providers of a cryptographic problem may need multiparty authentication, the search service providers of a parallel searching engine may need multiparty authentication, thousands of players participated in an online game may need multiparty authentication, a set of web services constituting an outsourcing workflow may need multiparty authentication, etc.

In these scenarios, it seems awkward for developers to regard SSO as a series of two-party authentications. Therefore, the semantic of SSO can be expressed as: *Single Sign-On = Σ two-party authentication + Σ multiparty authentication + Σ authorization*.

There are two main approaches to establish multiparty relationships: in the static manner, all the parties involved must be known and presented in advance of authentications; in the dynamic manner, old involved parties can quit the multiparty relationship while new parties involved can join the multiparty relationship. Obviously, the latter can be achieved by using the static multiparty approach together with the two-party approach repeatedly. This paper focuses on multiparty relationships formed in the static manner unless explicitly stated.

3.2 Graph Theory Model of Multiparty Authentication

Denote each principal to be authenticated by a vertex, and let the edge connecting a pair of vertices represent that two principals have confirmed the counterparty's identity mutually. Multiparty authentication that involves n principals can be modeled as a graph of order n. We denote such a graph by MAG_n. After authenticating each pair of distinct principals, MAG_n become a complete graph K_n with $n(n-1)/2$ edges.

A straightforward simplification is to choose one principal as a trusted third party and let it to mutually authenticate with the other principals in turn. This simplification changes MAG_n into a complete bipartite graph $K_{1,n-1}$ called a star, needing only $(n-1)$ mutual authentications. A further simplification is to distribute the responsibility of the trusted third party and to establish the MJA supposition as follows:

MJA Supposition. One principal can regard another principal as a trusted third party if either of the following conditions is satisfied:

1. Two principals have authenticated each other mutually.
2. Two principals have authenticated with one common trusted third party in advance.

Based on the MJA supposition, another simplification changes MAG_n into a Hamilton chain of the complete graph K_n, which also needs $(n-1)$ mutual authentications.

3.3 Defining Multiparty Joint Authentication

Definition 1. *Multiparty Joint Authentication (MJA) is to find a simplified or optimal authentication solution in a multiparty security context that involves n principals, which is based on three conditions below:*

1. If principal P_i, P_j have authenticated with one common trusted third party, then both P_i and P_j can confirm the counterparty's identity with a specified, understood level of confidence even without a real mutual authentication.
2. The relationship between a principal and its trust third party must satisfy the MJA supposition (or other feasible substitute).
3. There are $m(1 \leq m \leq n)$ principals to act as the trusted third parties to serve certain subsets comprising different principals. For short, let m be the order of the n-principal MJA, denote them by $n{:}m$.

The Hamilton chain of the complete graph K_n is one possible MJA solution, however, it is not the optimal answer to MAG_n if we take some practical constraints into account:

1. Different mutual authentications have different QoS and cost.
2. Users insist on performing mutual authentication for certain pairs of principals.
3. The MJA service provider can cache some mutual authentications performed by several principals and trusted third parties for a period of time.
4. The policies for a principal to become a trusted third party are of great varieties.
5. A principal may trust a trusted third party with different policies and security level.

These constraints indicate how to find an optimal MJA solution face lots of challenges. In this paper, we focus on n-principal, n-order MJA, i.e., $n{:}n$ MJA.

4 Basic Mathematical Properties of $n{:}n$ MJA

Theorem 1. *The spanning tree of a complete graph K_n is an $n{:}n$ MJA solution.*

Proof. According to the definition of MAG_n, the edge joining two vertices indicates that the corresponding two principals have confirmed the counterpart's identity with a specified, understood level of confidence.

According to the definition of MJA, if there has an edge joining P_i and P_j, then either P_i or P_j can regard its counterpart as a trusted third party. Without the loss of generality, let P_i regard P_j as a trusted third party. Thus, another edge joining P_j and P_k is equivalent to P_i and P_k has confirmed the counterpart's identity.

Therefore, if there has a walk that joins vertices P_i and P_j, then the corresponding two principals can confirm the counterpart's identity.

For every pair of distinct vertices in the spanning tree of K_n, there must be a walk joining them, therefore, every pair of distinct principals can confirm the counterpart's identity with a specified, understood level of confidence. That is, such a spanning tree stands for an $n{:}n$ MJA solution. □

Theorem 2. *The spanning tree of a complete graph K_n is one of the simplest $n{:}n$ MJA solution.*

Proof. Because each edge of the spanning tree is a bridge, removing any edge from the spanning tree leaves a disconnected graph of order n. Such a graph indicates that some principals have not confirmed their identities with a specified, understood level of confidence, and hence does not stand for any $n{:}n$ MJA solution.

On the other hand, because every principal can become a trusted third party, how to choose a trusted third party comes under no constraints, and hence such a spanning tree is one of the simplest $n{:}n$ MJA solution. □

Theorem 3. *The number of $n{:}n$ MJA solutions is n^{n-2}.*

Proof. Cayley asserts that the number of spanning trees of a completed graph K_n is n^{n-2} [8]. Therefore, the number of $n{:}n$ MJA solutions is n^{n-2}. □

This theorem indicates that searching the optimal solution of $n{:}n$ MJA by enumeration is an unsolvable problem. If we can design a nondeterministic Turing machine algorithm to find the optimal $n{:}n$ MJA solution with polynomial-time computation, this unsolvable problem become an NP problem.

Theorem 4. *Suppose it spends the time of T for each pair of principals to authenticate mutually, and each principal can carry out only one two-party authentication once a time. Then, the theoretical minimum time cost of an $n{:}n$ MJA solution is $2T$.*

Proof. According to theorem 2, a Hamilton chain of a complete graph K_n is an $n{:}n$ MJA solution.

Color all edges of the Hamilton chain as follows: choose either one of the two end vertices as a starting point; color each edge in the chain with the red and blue color in turn. After coloring, the Hamilton chain turns into a chain in the form of 'red-blue-red-blue-...'. Obviously, each pair of the edges that have the same color has no vertex in common, and all vertices except for the two end vertices are incident with one red edge and one blue edge simultaneously.

When carrying out MJA, let each pair of principals joined by red edge authenticate each other in parallel, it spends the time of T; then, let each pair of principals joined by blue edge authenticate each other in parallel, it also spends the time of T. Therefore, such a MJA solution spends the time of $2T$ in all. Of course, the order of the color has no influence on the final result. □

This theorem determines the infimum of the theoretical minimum time cost of an $n{:}n$ MJA solution. However, we can further reduce the time cost of MJA in a real program by means of multiple threads techniques.

Theorem 5. *Denote a spanning tree of a completed graph K_n by T_n. Suppose it spends the time of T for each pair of principals to authenticate mutually, and each principal can carry out only one two-party authentication once a time. Then, the theoretical minimum time cost of an $n{:}n$ MJA solution is mT if the maximum value of the degree sequence of T_n is $\deg(P_k) = m$.*

Proof. Because $deg(P_k) = m$, the corresponding principal P_k must carry out m mutual authentications with other principals. Color m edges which are incident with P_k with m different colors, construct a sub-tree.

Because each edge in the spanning tree T_n is a bridge, when inserting a new edge that is incident with any vertex P in the sub-tree, the other vertex joined by the new edge can not be any vertex in the sub-tree. This means the color of the new edge is determined by vertex P solely. Since the degree of a vertex in T_n is not more than m, it can be colored with m different colors so that all edges being incident with one common vertex have different colors. Order m different colors into a permutation, let each pair of principals joined by the same color edge authenticate each other in parallel, it spends the time of T. Therefore, such a MJA solution spends the time of mT in all. □

Theorem 6. *Suppose it spends the time of T for each pair of principals to authenticate mutually, and each principal can carry out only one two-party authentication once a time. Then, the maximum value of the theoretical minimum time cost of an n:n MJA solution is $(n-1)T$.*

Proof. Denote a spanning tree of a completed graph K_n by T_n. If the maximum value of the degree sequence of T_n is $(n-1)$, then the spanning tree is a star. According to theorem 5, the theoretical minimum time cost of such a solution is $(n-1)T$.

On the other hand, the degree of any vertex in T_n is not more than $(n-1)$, therefore, the maximum value of the theoretical minimum time cost of the n:n MJA solution is $(n-1)T$. □

Theorem 4, 5 and 6 indicate that the bigger the number of the trusted third party is, the smaller the theoretical minimum time cost is. If all principals can act as a trusted third party, the theoretical minimum time cost takes its minimum value; If only one principal act as a trusted third party, the theoretical minimum time cost takes its maximum value.

Theorem 7. *Let T be a spanning tree of a connected graph G. Let $\alpha = a, b$ be an edge of G which is not an edge of T. Then there is an edge β of T such that the graph T' obtained from T by inserting α and deleting β is also a spanning tree of G* [8].

5 Algorithms to Find *n:n* MJA Solution

If each pair of principals spends the same time to authenticate mutually (other factors can be converted into the equivalent time cost), a Hamilton chain of a completed graph K_n stands for one optimal n:n MJA solution. Algorithm 1 is to grow this kind of spanning tree.

Algorithm 1.

1. Number n principals with one of the numbers $1, 2, \ldots, n$.
2. Generate a permutation of natural number n, denoted by $p_1, p_2, \ldots, p_i, \ldots, p_n$, where $p_i \in N \wedge 1 \le p_i \le n$.
3. Join $p_1, p_2, \ldots, p_i, \ldots, p_n$ in turn, the chain in the form of $p_1 - p_2 - \ldots - p_i \ldots - p_n$ is an n-principal, n-order MJA solution.

Theorem 8. *If all mutual authentication spends the same time, algorithm 1 can find one optimal n:n MJA solution.*

Proof. The connected graph of order n, in the form of $p_1 - p_2 - \ldots - p_i - \ldots - p_n$, is a spanning tree of a complete graph K_n because it has exactly $(n-1)$ edges.

The walk from p_1 to p_n has distinct edges and distinct vertices, thus, it is a Hamilton chain of a complete graph K_n. According to theorem 2, this Hamilton chain stands for one optimal n:n MJA solution. □

Generally, different pair of principals spends different time to authenticate mutually (including converted factors). Therefore, we must design some algorithms to grow a weighted spanning tree from a weighted complete graph K_n. The most straightforward algorithm is the greedy algorithm, which can be used to grow a minimum-weight spanning tree. Let K_n be a weighted complete graph with weight function c, the greedy algorithm is shown as follows.

Algorithm 2.

1. Put $F = \phi, V = \phi$.
2. Put an edge α that has minimum weight in F, at the same time, put two vertices which are incident with α in V.
3. While there exists an edge α not in F such that $F \cup \{\alpha\}$ put only one new vertex in V, determine such an edge α of minimum weight and put α in F, at the same time, put the new vertex that is incident with α in V.
4. Put $T = \{V, F\}$, which is a minimum-weight spanning tree of K_n.

Theorem 9. *If different mutual authentication spends different time, algorithm 2 can find an n:n MJA solution that has the minimum-time-cost.*

Proof. Because step 2 put two vertices in V while step 3 put only one new vertex in V repeatedly, the algorithm terminates after $n-2$ iterations. On termination, T have $n-1$ edges in F and n vertices in V. While repeating step 3, there must exist a vertex in V that is incident with the new edge α, and hence T is connected and is a spanning tree.

Let the $n-1$ edges of F be $\alpha_1, \alpha_2, \ldots, \alpha_{n-1}$ in the order that they put in F. Let $T^* = (V, F^*)$ be a minimum-weight spanning tree which has the largest number of edges in common with T. Thus no minimum-weight spanning tree has more edges in common with F than F^* does. If $F^* = F$, then it follows that T is a minimum-weight spanning tree. Suppose to the contrary that $F^* \ne F$. Let

α_k be the first edge of F which is not in F^*. Thus the edges $\alpha_1, \alpha_2, \ldots, \alpha_{k-1}$ all belong to F^*. By theorem 7 there is an edge β of T^* such that the graph T^{**}, obtained from T^* by inserting α_k and deleting β, is a spanning tree of G. The weight of T^{**} is $c(T^{**}) = c(T^*) - c(\beta) + c(\alpha_k)$. Since T^* is a minimum-weight spanning tree we conclude that $c(\alpha_k) \geq c(\beta)$.

If $c(\alpha_k) > c(\beta)$, then in determining the kth edge to be put in F in carrying out the greedy algorithm, β is a possible choice. That is, choosing α_k is wrong.

If $c(\alpha_k) = c(\beta)$, then T^{**} is also a minimum-weight spanning tree. Since T^{**} has one more edge in common with T than T^* has, we contradict our choice of T^*. Suppose the weight stands for the time cost of each mutual authentication, then, T stands for an n:n MJA solution that has the minimum-time-cost. □

However, greedy algorithm may not find the optimal n:n MJA solution because the real time cost of an n:n MJA solution depends not only on the time cost for each pair of principals to authenticate mutually, but also on how many mutual authentications are to be performed by each principal (the degree sequence of the spanning tree).

This conclusion can be illustrated by one concrete example. Suppose it takes the time of T for principal P_i to authenticate mutually with all other $(n-1)$ principals, and it takes the time of $2T$ for each pair of principals, not including P_i, to authenticate mutually. Then, the minimum-weight spanning tree grown with greedy algorithm is a star whose center is P_i. The theoretical minimum time cost of such a MJA solution is $(n-1)T$. However, we can construct a Hamilton chain as follows: firstly, find a Hamilton chain for all $(n-1)$ principals, excluding P_i, with algorithm 1; secondly, join P_i with either end vertex of the previous Hamilton chain to form a new Hamilton chain. Obviously, the theoretical minimum time cost of this MJA solution is $4T$. Of course, the latter is better than the former.

6 Efficiency Analysis of MJA

The essential of MJA is to increase the efficiency of multiparty authentication by distributing the responsibility of the trusted third party. Because modeling and simulating the real performance of MJA is of great complexities, this paper only analyzes the theoretical cases to reveal some efficiency features of MJA.

At current stage, we suppose all mutual authentications spend the same time and denote this time cost by one unit. As shown in Fig.1, a traditional multiparty authentication needs $n(n-1)/2$ mutual authentications and spends $n(n-1)/2$ units of time. If only one principal acts as a trusted third party, the n:n MJA solution is indeed a star. If each principal can act as a trusted third party, the n:n MJA solution is a Hamilton chain found with algorithm 1. Because an n:n MJA needs only $(n-1)$ mutual authentications, the time cost reduces to $(n-1)$ units. Obviously, the more the principals involved in a multiparty authentication are, the greater the efficiency of MJA is. If there are $1,000$ principals, a traditional solution spends $499,500$ units of time and an n:n MJA solution spends only 999 units of time; if there are $1,000,000$ principals (e.g., online gaming), the traditional solution spends 499,999,500,000 units of time and an $n{:}n$ MJA solution spends only 999,999 units of time. On the other hand, parallel techniques can be used to speed up the processing and to make responses more efficiently. For example, if all principals can act as a trusted party and different principals perform mutual authentications in parallel, an $n{:}n$ MJA solution for 1,000,000 principals can be finished just in 2 units of time.

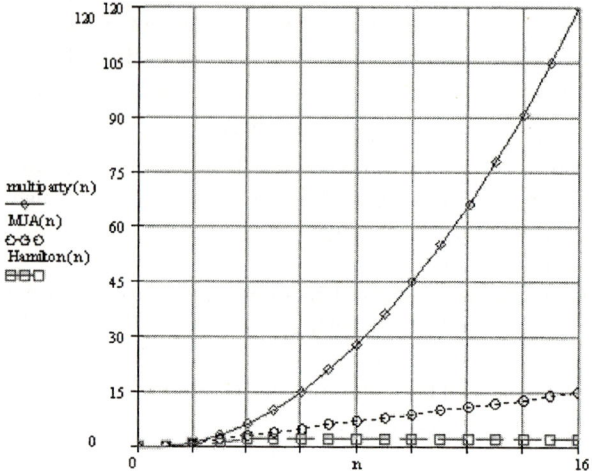

Fig. 1. This graph shows the time cost of the traditional multiparty authentication and two kinds of theoretical $n{:}n$ MJA solutions. The red solid line represents the time cost of traditional multiparty authentication, the blue dot line represents a general $n{:}n$ MJA solution, and the magenta dash line represents a Hamilton chain that is performed in a parallel way. These cases suppose all mutual authentications spend the same time

Of course, in a real grid system, the efficiency of MJA depends not only on the time used for mutual authentications, but also on the time used to find the MJA solution and to communicate. Therefore, a complete MJA infrastructure must take all these factors into account.

7 Conclusion

Along with the evolution of WS, security solution for grids is undergoing renovations. In any case, the authentication service is the security entry of a grid application. Traditional authentication service in GSI is two-party oriented. However, multiparty contexts can enrich semantics of grid security dramatically.

Multiparty Joint Authentication (MJA) is to find some simplified or optimal authentication solutions in multiparty contexts instead of to authenticate each pair of principals in turn. It is designed to support multiparty security scenarios in grids with a specified level of confidence and reduce the time cost of authentications.

By means of graph theory model, this paper reveals some mathematical properties of MJA and puts forward two algorithms to find an $n{:}n$ MJA solution. Future research topics on MJA includes: 1) MJA performance modeling and simulating; 2) MJA protocols; 3) MJA services; 4) MJA based authorization; 5) MJA based Leave/Join services; 6) Policy based MJA; 7) new algorithms to find MJA solutions with certain constraints; 8) MJA based applications, etc.

References

1. Ferreira, L., Berstis, V., Armstrong, J., Kendzierski, M., Neukoetter, A., Takagi, M., Bing-Wo, R., Amir, A., Murakawa, R., Hernandez, O., Magowan, J., Bieberstein, N.: Introduction to Grid Computing with Globus. IBM Corp. (2002)
2. Siebenlist, F., Welch, V., Tuecke, S., Foster, I., Nagaratnam, N., Janson, P., Dayka, J., Nadalin, A.: OGSA Security Roadmap. GGF OGSA Sec. Workgroup Doc. (2002)
3. Welch, V., Siebenlist, F., Foster, I., Bresnahan, J., Czajkowski, K., Gawor, J., Kesselman, C., Meder, S., Pearlman, L., Tuecke, S.: Security for Grid Services. In: Azada, D. (ed.): Proc. of 12th Intl. Symposium on H. Performance Distributed Computing. IEEE Press, Washington (2003) 48-57
4. Foster, I., Kesselman, C., Tsudik, G., Tuecke, S.: A Security Architecture for Computational Grids. In: G, Li, Reiter, M. (eds.): Proc. of the 5th ACM Conf. on Computer and Comm. Sec. ACM Press, New York (1998) 83-92
5. Website: http://www.oasis-open.org/
6. Foster, I., Kesselman, C.: The Grid: Blueprint for a New Computing Infrastructure. Morgan Kaufmann Publisher (1998) 259-278
7. Foster, I., Kesselman, C., Tuecke, S.: The Anatomy of the Grid, Enabling Scalable Virtual Organizations. Intl. J. of H. Performance Computing Applications 3 (2001) 200-222
8. Brualdi, R.: Introductory Combinatorics, Third Edition. Pearson Ed. Inc. (1999)

GA Based Adaptive Load Balancing Approach for a Distributed System

SeongHoon Lee[1] and DongWoo Lee[2]

[1] Division of Information and Communication Engineering,
Cheonan University, 115, Anseo-dong, Cheonan,
Choongnam, Republic of Korea
shlee@cheonan.ac.kr
[2] Department of Computer Science, WooSong University,
17-2, Jayang-dong, Dong-ku, Daejeon, Republic of Korea
dwlee@woosong.ac.kr

Abstract. In this paper, we propose a genetic algorithm(GA) based load balancing approach for a distributed system. Fisrt we develop two GA based algorithms to effectively reduce unnecessary request messages in sender-initiated approach and receiver-initiated approach respectively and then construct an adaptive algorithm for load balancing in a distributed system by combining the two GA based algorithms. The experiments show the proposed GA based adaptive algorithm performs efficiently in the variance of load of the distributed system.

1 Introduction

A distributed system consists of a collection of autonomous computers connected in networks. The primary advantages of the system are high performance, availability, and extensibility at low cost. To achieve the high performance of the distributed system, it is essential to keep the load of each processor evenly.

Dynamic load balancing algorithms are classified into three methods: sender-initiated, receiver-initiated, symmetrically-initiated. Basically our approach is sender-initiated and receiver-initiated. Under sender-initiated algorithms, when the distributed system becomes heavily loaded, it is difficult to find a suitable receiver because most processors have additional tasks to send. Similarly, Under receiver-initiated algorithms, when the distributed system becomes lightly loaded, it is difficult to find a suitable sender because most processors have a few tasks. Therefore, lots of unnecessary request and reject messages are repeatedly sent back and forth consuming a lot of time before real tasks can be executed. It results in consuming of the task processing time, low system throughput, and low CPU utilization.

To solve these problems in sender-initiated and receiver-initiated approaches, we propose a new genetic algorithm for both, and then combine both of them together to expand to a new adaptive load balancing algorithm.

2 GA-Based Method

2.1 Load Measure

We employ the CPU queue length as a suitable load index because it is known as most suitable one [1]. This measure means a number of tasks in CPU queue residing in a processor.

We use a 3-level scheme to represent a load state on its own CPU queue length of a processor. Table 1 shows the 3-level load measurement scheme.

Table 1. 3-level load measurement scheme

Load state	Meaning	Criteria
L-load	light-load	$CQL \leq T_{low}$
N-load	normal-load	$T_{low} < CQL \leq T_{up}$
H-load	heavy-load	$CQL > T_{up}$

(CQL : CPU Queue Length)

The transfer policy is triggered when a task arrives. A node is identified as a sender if a new task originating at the node makes the CPU queue length exceed T_{up}. A node is identified as a suitable receiver for a task acquisition if the node's CPU queue length will not cause to exceed T_{low}.

Each processor in a distributed system has its own population which genetic operators are applied to. There are many encoding methods; Binary encoding, Character and real-valued encoding, Tree encoding [2]. We use binary encoding method in this paper. So, a string in population can be defined as a binary-coded vector $< v_o, v_1, ..., v_{n-1} >$ which indicates a set of processors to which the request messages are sent off. If the request message is transferred to the processor $P_i(0 \leq i \leq$ n-1, where n is the total number of processors), then $v_i=1$, otherwise $v_i=0$. Each string has its own fitness value.

2.2 GA-Based Sender-Initiated Load Balancing Approach

In the sender-based load balancing approach using genetic algorithm, processors, which received a request message from a sender, send accept messages or reject messages depending on their own CPU queue length. In the case of more than two accept messages are returned, one is selected at random.

Each string included in a population is evaluated by the fitness function using following formula:

$$F_i = 1/((\alpha \times TMP) + (\beta \times TMT) + (\gamma \times TTP)) \qquad (1)$$

Here, α, β, and γ denote the weights for parameters, TMP, TMT, and TTP, respectively. The purpose of the weights is to be operated equally for each parameter to fitness function F_i.

TMP (Total Message Processing time) is the summation of the processing times for request messages to be transferred. TMT(Total Message Transfer time)

```
procedure Genetic_algorithm Approach
{
    Initialization()
    while (Check_load())
      if ( Load_i > T_up ) {
          Individual_evaluation();
          Genetic_operation();
          Message_evaluation();
      }
      Process a task in local processor;
}

Procedure Genetic_operation()
{
    Local_improvement_operation();
    Reproduction();
    Crossover();
}
```

Fig. 1. GA-based sender-initiated load balancing algorithm

means the summation of each message transfer times from the sender to processors corresponding to bits set '1' in selected string. The objective of this parameter is to select a string with the shortest distance eventually. TTP(Total Task Processing time) is the summation of the times needed to perform a task at each processor corresponding to bits set '1' in selected string.

The algorithm of the proposed sender-based load balancing is as shown in Fig. 1.

An Initialization module is executed in each processor. A population of strings is randomly generated without duplication. A Check_load module is used to observe its own processor's load by checking the CPU queue length, whenever a task is arrived at a processor. If the observed load is heavy, the load balancing algorithm performs the above modules. A String_evaluation module calculates the fitness value of strings in the population. A Genetic_operation module such as Local improvement, Reproduction, Crossover is executed on the population in such a way as above.

Local Improvement Operation. String 1 is chosen. A copy version of the string 1 is generated and part 1 of the newly generated string is mutated. This new string is evaluated by proposed fitness function.

If the evaluated value of the new string is higher than that of the original string, replace the original string with the new string. After this, the local improvement of part 2 of string 1 is done repeatedly. This local improvement is applied to each part one by one. When the local improvement of all the parts is finished, new string 1 is generated. String 2 is then chosen, and the above-mentioned local improvement is done. This local improvement operation is applied to all the strings in population.

```
procedure Genetic_algorithm Approach
{
   Initialization()
   while (Check_load())
      if ( Load_i ≤ T_low) {
         Individual_evaluation();
         Genetic_operation();
         Message_evaluation();
      }
      Process a task in local processor;
}
```

Fig. 2. GA-based receiver-initiated load balancing algorithm

Reproduction. The reproduction operation is applied to the newly generated strings. We use the "wheel of fortune" technique [3].

Crossover. The crossover operation is applied to the newly generated strings. These newly generated strings are evaluated.

Suppose that there are 5 parts in distributed systems. A boundary among the many boundaries(B_1, B_2, B_3, B_4) is determined at random as a crossover point. If a boundary B_3 is selected as a crossover point, crossover activity is generated based on the B_3. So, the effect of the local improvement operation in the previous phase is preserved through crossover activity.

The Genetic_operation selects a string from the population at the probability proportional to its fitness, and then sends off the request messages according to the contents of the selected string.

A Message_evaluation module is used whenever a processor receives a message from other processors. When a processor P_i receives a request message, it sends back an accept or reject message depending on its CPU queue length.

2.3 GA-Based Receiver-Initiated Load Balancing Approach

A GA-based receiver-initiated load balancing is shown by Fig 2.

Each string included in a population is evaluated by the fitness function using the formula 2.

$$F_i = 1/((\alpha \times TMP) + (\beta \times TMT)) + (\gamma \times TTP) \tag{2}$$

3 GA-Based Adaptive Load Balancing Approach

Combining the two proposed GA-based approaches, an algorithm of adaptive load balancing can be constructed, which is shown in Fig. 3.

---
if (load of distributed system < T_{low})
    if (a specific processor = sender)
        Use a conventional sender-initiated load balancing algorithm
    else
        Use a proposed a GA-based receiver-initiated load balancing algorithm
    if (a specific processor = receiver)
        Use a conventional receiver-initiated load balancing algorithm
    else
        Use a proposed a GA-based sender-initiated load balancing algorithm
---

Fig. 3. A GA-based Adaptive Load Balancing Algorithm

4 Experiments

Our experiments have the following assumptions. First, each task size and task type are the same. Second, the number of parts(p) in a string is five. In genetic algorithm, crossover probability(P_c) is 0.7, mutation probability(P_m) is 0.05. The values of these parameters P_c, P_m were known as the most suitable values in various applications [2, 3, 4, 5]. The table 2 shows the detailed contents of parameters used in our experiments.

Our experiments for sender-initiated load balancing approach have the following assumptions. The load rating over the systems is about 60 percent. The number of tasks to be performed is 3000. The weight for TMP is 0.025. The weight for TMT is 0.01. The weight for TTP is 0.02.

Experiment 1. We compared the performance of proposed method with a conventional method in this experiment by using the parameters on the table 2. The experiment is to observe the change of response time when the number of tasks to be performed is 3000(see Fig. 4).

Our experiments for receiver-initiated load balancing approach have the following assumptions. The load rating over the systems is about 20 percent. The number of tasks to be performed is 2000. The weight for TMP is 0.025. The weight for TMT is 0.01. The weight for TTP is 0.02.

Experiment 2. We compared the performance of proposed method with a conventional method in this experiment by using the parameters on the table 2. The experiment is to observe the change of response time when the number of tasks to be performed is 2000(see Fig. 5).

Table 2. experimental parameters

number of processor	30
P_c	0.7
P_m	0.05
number of strings	50

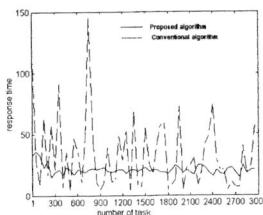 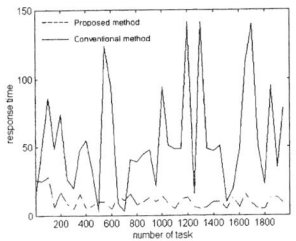

Fig. 4. Result of Experiment 1   **Fig. 5.** Result of Experiment 2

5   Conclusions

We propose a new adaptive load balancing scheme for a distributed system which is based on a genetic algorithm. The genetic algorithm is used to decide to suitable candidate senders or receivers which task transfer request messages should be sent off. Several experiments have been done to compare the proposed scheme with a conventional algorithm. Through the various experiments, performances of proposed scheme are better than those of conventional scheme on the response time and mean response time. The experimental results show the effectiveness of the proposed algorithm.

References

1. Kunz, T.: The Influence of Different Workload Descriptions on a Heuristic Load Balancing Scheme. *IEEE Trans on Software Engineering*, 17(7), July (1991)
2. Mitchell, M.: *An Introduction to Genetic Algorithms.* MIT Press (1996)
3. Grefenstette, J.: Optimization of Control Parameters for Genetic Algorithms. *IEEE Trans on SMC*, SMC-16(1), January (1986)
4. Miller, J.A., Potter, W.D., Gondham, R.V., and Lapena, C.N.: An Evaluation of Local Improvement Operators for Genetic Algorithms. *IEEE Trans on SMC*, 23(5), Sept (1993)
5. Srinivas, M., and Patnait, L. M.: Adaptive Probabilities of Crossover and Mutation in Genetic Algorithms. *IEEE Trans on SMC*, 24(4), April (1994)

A Novel Approach to Load Balancing Problem

Chuleui Hong[1], Wonil Kim[2,*], and Yeongjoon Kim[1]

[1] Software School, Sangmyung University, Seoul, Korea
{hongch, yjkim}@smu.ac.kr
[2] Dept. of Digital Contents, College of Electronics and Information Engineering,
Sejong University, Seoul, Korea
wikim@sejong.ac.kr

Abstract. In this paper, we propose a novel approach to the load balancing problem, which is an important issue in parallel processing. The proposed load balancing algorithm called Mean Field Genetic Algorithm (MGA) is a hybrid algorithm of Mean Field Annealing (MFA) and Simulated annealing-like Genetic Algorithm (SGA). The proposed MGA combines the benefit of rapid convergence property of MFA and the effective genetic operations of SGA. Our experimental results indicate that the composition of heuristic mapping methods improves the performance over the conventional ones in terms of communication cost, load imbalance and maximum execution time.

1 Introduction

The load balance mapping problem is assigning tasks to the processors in distributed memory multiprocessors [1–5]. Multiple tasks are allocated to the given processors in order to minimize the expected execution time of the parallel program. Thus, the mapping problem can be modeled as an optimization problem in which the interprocessor communication overhead should be minimized and computational load should be uniformly distributed among processors in order to minimize processor idle time. The load balancing is an importance issue in parallel processing.

The proposed Mean Field Genetic Algorithm (MGA) is a hybrid algorithm based on mean field annealing (MFA) [1, 4] and genetic algorithm (GA) [2, 5, 6]. MFA has the characteristics of rapid convergence to the equilibrium state while the simulated annealing (SA) [6, 7] takes long time to reach the equilibrium state. In the proposed method, the typical genetic algorithm is modified where the evolved new states are accepted by the Metropolis criteria as in simulated annealing. The modified Simulate annealing-like Genetic Algorithm is called SGA. The simulation results show that the new MGA is better than MFA and GA, as it reduces inter-processor communication time, load imbalance among processors and expected maximum execution time of the program.

---

* Author for correspondence : +82-2-3408-3795.

2 The Mapping Problem in Multiprocessors

The multiprocessor mapping problem is a typical load balancing optimization problem. A mapping problem can be represented with two undirected graphs, called the Task Interaction Graph (TIG) and the Processor Communication Graph (PCG). TIG is denoted as $G_T(V, E)$. $|V| = N$ vertices are labeled as (1, 2, ..., i, j, ..., N). Vertices of G_T represent the atomic tasks of the parallel program and its weight, w_i, denotes the computational cost of task i for $1 \leq i \leq N$. Edge E represents interaction between two tasks. Edge weight, e_{ij}, denotes the communication cost between tasks i and j that are connected by edge $(i, j) \in E$. The PCG is denoted as $G_P(P, D)$. G_P is a complete graph with $|P| = K$ nodes and $|D| = {}_KC_2$ edges. Vertices of the G_P are labeled as (1, 2, ..., p, q, ..., K), representing the processors of the target multicomputers. Edge weight, d_{pq}, for $1 \leq p,q \leq K$ and $p \neq q$, denotes the unit communication cost between processor p and q.

The problem of allocating tasks to a proper processor is to find a many-to-one mapping function $M: V \rightarrow P$. That is, each vertex of G_T is assigned to a unique node of G_P. Each processor is balanced in computational load (*Load*) while minimizing the total communication cost (*Comm*) between processors.

$$Comm = \sum_{(i,j) \in E, M(i) \neq M(j)} e_{ij} d_{M(i)M(j)} \cdot \quad (1)$$

$$Load_p = \sum_{i \in V, M(i)=p} w_i, \quad 1 \leq p \leq K \cdot \quad (2)$$

$M(i)$ denotes the processor to which task i is mapped, i.e. $M(i) = p$ represents that task i is mapped to the processor p. In Equation (1), if tasks i and j in G_T are allocated to the different processors, i.e. $M(i) \neq M(j)$ in G_P, the communication cost occurs. The contribution of this to *Comm* is the multiplication of the interaction amount of task i and j, e_{ij}, and the unit communication cost of different processors p and q, d_{pq}, where $M(i) = p$ and $M(j) = q$. $Load_p$ in Equation (2) denotes the summation of computational cost of tasks i, w_i, which are allocated processor p, $M(i) = p$.

A spin matrix consists of N task rows and K processor columns representing the allocation state. The value of spin element (i, p), s_{ip}, is the probability of mapping task i to processor p. Therefore, the range of s_{ip} is $0 \leq s_{ip} \leq 1$ and the sum of each row is 1. The initial value of s_{ip} is $1/K$ and s_{ip} converges 0 or 1 as solution state is reached eventually. $s_{ip} = 1$ means that task i is mapped to processor p.

The cost function, $C(s)$, is set to minimize the total communication cost of Equation (1) and to equally balance the computational load among processors of Equation (2).

$$C(s) = \sum_{i=1}^{N} \sum_{j \neq i} \sum_{p=1}^{K} \sum_{q \neq p} e_{ij} s_{ip} s_{jq} d_{pq} + r \sum_{i=1}^{N} \sum_{j \neq i} \sum_{p=1}^{K} s_{ip} s_{jp} w_i w_j \cdot \quad (3)$$

e_{ij} : The interaction amount of task i and j in TIG
w_i : The computational cost of task i in TIG

d_{pq} : The unit communication cost of processor p and q in PCG
s_{ip} : The probability of task i mapping to processor p
r : The ratio of communication to computation cost

The first term of cost function, Equation (3), represents interprocessor communication cost (IPC) between two tasks i and j when task i and j are mapped to different processor p and q respectively. Therefore the first IPC term is minimized as two tasks with large interaction amount are mapped to the same processors. The second term of Equation (3) means that the multiplication of computational cost of two tasks i and j mapped to the same processor p. The second computation term is also minimized when the computational costs of both processors are almost the same. It is the sum of squares of the amount of tasks in the same processor. The ratio r changes adaptively in the optimization process in order to balance the communication and computation cost. Changing the ratio r adaptively results in better optimal solution than fixing the ratio r. The optimal solution is to find the minimum of the cost function.

3 Load Balancing in Parallel Processing

The mean field annealing (MFA) is derived from simulated annealing (SA) based on mean field approximation method in physics [1]. Though SA provides the best solutions for various kinds of general optimization problems, it takes long time to achieve those. While SA changes the states randomly, MFA makes the system reach the equilibrium state very fast using the mean value estimated by mean field approximation.

MFA can be applied to the combinatorial optimization problem. It combines the collective computations of Hopfield neural network (HNN) and annealing properties of SA. HNN can efficiently solve the small size of problems, while it is hard to be applied to the large size of problems. MFA represents the states as the same way in HNN and repeats the annealing process to reach the thermal equilibrium.
In implementing MFA, the cooling schedule has a great effect on the solution quality. Therefore the cooling schedule must be chosen carefully according to the characteristics of problem and cost function.

Genetic algorithm (GA) is a powerful heuristic search method based on an analogy to the biological evolution model. GA provides an effective means for global optimization in a complex search space such as NP-complete optimization including mapping problem, function optimization and etc.

We modified GA such that the new evolved state is accepted with a Metropolis criterion like simulated annealing in order to keep the convergence property of MFA. The modified GA is called SGA. In order to keep the thermal equilibrium of MFA, the new configurations generated by genetic operations are accepted or rejected by the Metropolis Criteria which is used in SA. In the Equation (4), ΔC is the cost change of new state from old state which is made by subtracting the cost of new state from that of old one. T is the current temperature.

$$\Pr[\Delta C \text{ is accepted}] = \min\left(1, \exp\left(\frac{\Delta C}{T}\right)\right) \qquad (4)$$

A new hybrid algorithm called MGA combines the merits of mean field annealing (MFA) and simulated annealing-like genetic algorithm (SGA). MFA can reach the thermal equilibrium faster than simulated annealing and GA has powerful and various genetic operations such as selection, crossover and mutation.

First, MFA is applied on a spin matrix to reach the thermal equilibrium fast. After the thermal equilibrium is reached, the population for GA is made according to the distribution of task allocation in the spin matrix. Next, GA operations are applied on the population while keeping the thermal equilibrium by transiting the new state with Metropolis criteria. MFA and GA are applied by turns until the system freeze.

4 Simulation Results

The proposed MGA hybrid algorithm is compared with MFA and GA that use the same cost function as that of MGA. In this simulation, the size of tasks is 200 and 400 (only the results of task size of 400 are shown). The multiprocessors are connected with wrap-around mesh topology. The computational costs of each task are distributed uniformly ranging [1..10]. The communication costs between any two tasks ranges [1..5] with uniform distribution. The number of communications is set to 1, 2, or 3 times of the number of tasks. Each experiment is implemented 20 times varying the seed of random number generator and TIG representing the computational and communication cost.

The coefficient r in a linear cost function is for balancing the computation and communication cost between processors. The initially task allocations to processors are uniformly distributed and accordingly a spin matrix is generated. The initial ratio r is computed as in Equation (4). As the temperature decreases, the coefficient r varies adaptively according to Equation (5) in order to reflect the changed interprocessor communication cost. r_{old} is the ratio used at the previous temperature and r_{new} is the newly calculated ratio at the current temperature.

$$r_{init} = \frac{\text{Comm. cost}}{\text{\# of processors} \times \text{Comp. cost}} = \frac{\sum_{i=1}^{N}\sum_{j\neq i}\sum_{p=1}^{K}\sum_{q\neq p} e_{ij} S_{ip} S_{jq} d_{pq}}{K \times \sum_{i=1}^{N}\sum_{j\neq i}\sum_{p=1}^{K} S_{ip} S_{jp} w_i w_j} \quad (4)$$

$$r_{new} = \begin{cases} 0.9 \times r_{old} & \text{if } r_{new} < r_{old} \\ r_{old} & \text{Otherwise} \end{cases} \quad (5)$$

Table 1. The maximum completion times when the initial value of r is fixed or r varies adaptively according to Equation (5)

Problem Size (N = 400)		MFA		MGA	
|E|	K	fixed r	variable r	fixed r	variable r
400	16	627.1	279.05	256.4	222.75
800	16	1189.95	888.1	617.15	587
1200	16	1862.8	1557.4	971.9	987.5
400	36	410.5	152.85	143.85	128.65
800	36	834.45	617	410.9	385.15
1200	36	1376.8	1065.5	714.65	692.95

Table 1 compares the maximum completion times when the initial value of r is fixed or r varies adaptively according to Equation (5). The completion time of an arbitrary processor is the sum of computation costs of its processor and the communication cost with other processors. The maximum completion time is defined as the maximum value among the completion times of all processors. In Table 1, N is the number of tasks, $|E|$ is the total number of interprocessor communications, and K represents the number of processors. The averaged performance improvement, $Perf_r$ in Equation (6), is 33% in MFA, while it is 6% in MGA. The reason that the performance measure of MGA is less than that of MFA is that genetic operations of MGA have a great contribution to find the better solutions. Therefore, changing r has a less effect on the solution quality of MGA than that of MFA. $TMAX_{kr}$ and $TMAX_{cr}$ are the maximum completion times when the ratio r is fixed and varied respectively.

$$Perf_r = \frac{TMAX_{kr} - TMAX_{cr}}{TMAX_{kr}} * 100 \tag{6}$$

Table 2. Total interprocessor communication cost, Percent computational cost imbalance, and Execution Time for Problem size 400

		Total Comm. Time			Comp. Cost. Imbalance			Exec. Time (secs.)		
$|E|$	K	MFA	GA	MGA	MFA	GA	MGA	MFA	GA	MGA
400	16	994.3	2370.5	629.7	47%	37%	32%	26.2	61.7	95.6
800	16	8089.1	6714.6	4004.2	61%	37%	57%	40.7	70.3	150.9
1200	16	14677	11743	8348.4	49%	35%	54%	51.7	63.7	190.9
400	36	1062.7	3539.5	852.4	60%	59%	50%	94.9	77.6	212.4
800	36	10021	10360	5603	79%	62%	72%	122.9	76.4	260.6
1200	36	19937	17780	11868	75%	60%	70%	148.1	70.6	317.5

Table 2 displays average total interprocessor communication cost of each algorithm. The average performance improvement from MFA to MGA, which is the percent reduction of communication cost normalized by that of MFA, is 33%. It is 45% from GA to MGA. It also displays computational cost imbalance which is defined as the difference between maximum and minimum computational cost of processors normalized by the maximum cost. The computational cost imbalance of each algorithm displays a little difference, while total communication costs in Table 2 are much more different. This implies that the interprocessor communication cost has a greater effect on the solution quality than the computational cost. Finally the last column of Table 2 displays the execution time of each algorithm. The averaged execution time of MGA is 1.5 and 1.7 times longer than that of MFA and GA respectively. This is a trade-off between the solution quality and execution time. Figure 1 shows that MGA is more useful as the problem size increases.

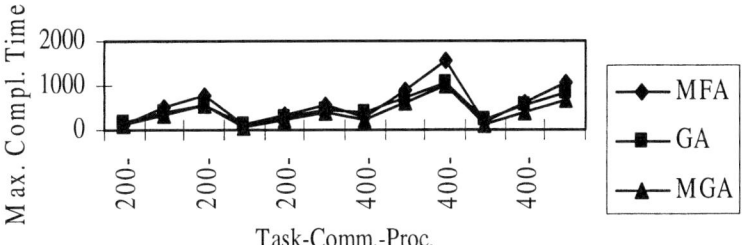

Fig. 1. Maximum completion time for various problem sizes

5 Conclusions

In this paper, we proposed a new hybrid algorithm called MGA. The proposed approach combines the merits of MFA and GA on a load balancing problem in parallel distributed memory multiprocessor systems. The solution quality of MGA is superior to that of MFA and GA while execution time of MGA takes longer than the compared methods. There can be the trade off between the solution quality and execution time by modifying the cooling schedule and genetic operations. MGA was also verified by producing more promising and useful results as the problem size and complexity increases. The proposed algorithm can be easily developed as a parallel algorithm since MFA and GA can be parallelized easily. This algorithm also can be applied efficiently to broad ranges of NP-Complete problems.

References

1. Bultan, T., Aykanat, C. : A New Mapping Heuristic Based on Mean Field Annealing. Journal of Parallel & Distributed Computing,16 (1992) 292-305
2. Heiss, H.-U., Dormanns, M. : Mapping Tasks to Processors with the Aid of Kohonen Network. Proc. High Performance Computing Conference, Singapore (1994) 133-143
3. Park, K., Hong, C.E. : Performance of Heuristic Task Allocation Algorithms. Journal of Natural Science, CUK, Vol. 18 (1998) 145-155
4. Salleh, S., Zomaya, A. Y.: Multiprocessor Scheduling Using Mean-Field Annealing. Proc. of the First Workshop on Biologically Inspired Solutions to Parallel Processing Problems (BioSP3) (1998) 288-296
5. Zomaya, A.Y., Teh, Y.W.: Observations on Using Genetic Algorithms for Dynamic Load-Balancing. IEEE Transactions on Parallel and Distributed Systems, Vol. 12, No. 9 (2001) 899–911
6. Hong, C.E.: Channel Routing using Asynchronous Distributed Genetic Algorithm. Journal of Computer Software & Media Tech., SMU, Vol. 2. (2003)
7. Hong, C., McMillin, B.: Relaxing synchronization in distributed simulated annealing. IEEE Trans. on Parallel and Distributed Systems, Vol. 16, No. 2. (1995) 189-195

Asynchronous Distributed Genetic Algorithm for Optimal Channel Routing

Wonil Kim[1], Chuleui Hong[2], and Yeongjoon Kim[2]

[1] Dept. of Digital Contents, College of Electronics and Information Engineering,
Sejong University, Seoul, Korea
wikim@sejong.ac.kr
[2] Software School, Sangmyung University,
Seoul, Korea
{hongch, yjkim}@smu.ac.kr

Abstract. This paper presents a distributed genetic algorithm for the channel routing problem in MPI environments. This system is implemented on a network of personal computers running Linux operating system connected via 10Mbps Ethernet. Each slave processor generates its own sub-population using genetic operations and communicates with the master processor in an asynchronous manner to form the global population. The experimental results show that the proposed algorithm maintains the convergence properties of sequential genetic algorithm while it achieves linear speedup as the nets of the channel routing and the number of computing processors increase.

1 Introduction

Known as an NP-Complete problem, the channel routing problem is very important in the automatic layout design of VLSI circuits and printed circuit boards. In the channel routing problem, a channel consists of two parallel horizontal rows of points which are called terminals. These terminals are placed at a regular interval and identified by the columns of the channel. A net consists of the same terminals that must be interconnected through certain routing paths without being overlapped with others. A layer is an interconnected routing area of the nets to connect the terminals. The channel routing problem is routing the given nets between terminals on the minimum multilayer channel areas.

Although many different algorithms have been proposed, the problem of finding globally optimal solution for the channel routing is still open [1, 2]. Parallel routing algorithms have been also proposed using knock-knee model, simulated annealing and neural networks [3, 4].

In this paper, we propose an asynchronous distributed genetic algorithm (ADGA) for the channel routing problems. This approach not only accommodates future advancement of VLSI technology flexibly but also can be applied to many NP-Complete problems efficiently. The distributed and asynchronous nature of the proposed algorithm effectively speeds up the processing time. This ADGA is verified practical in the solution quality and computation time.

2 GA Operations for Channel Routing

In the neural network optimization approach [3, 6] for a channel routing problem of n nets with m tracks and l layers, Funanbiki et. al. used $n \times m \times l$ processing elements each of which represents the output of the solution. However, we use a scalar value representation, a_i ranging 0 to $m \times l - 1$. When the ith net is assigned to the jth track of layer k, $a_i = j+(k-1) \times m$. Funabiki's algorithm needs more processing elements as m and l increases, while the proposed algorithm does not depend on the size of a track and layer. For example as in Figure 1, which is a 10-Net, 3-Track and 2-Layer problem, Funabiki's algorithm needs $10 \times 3 \times 2 = 60$ processing elements, while 400 processing elements are needed for a 10-Net, 10-Track and 4-Layer problem. However, our algorithm needs 10 integers for both cases.

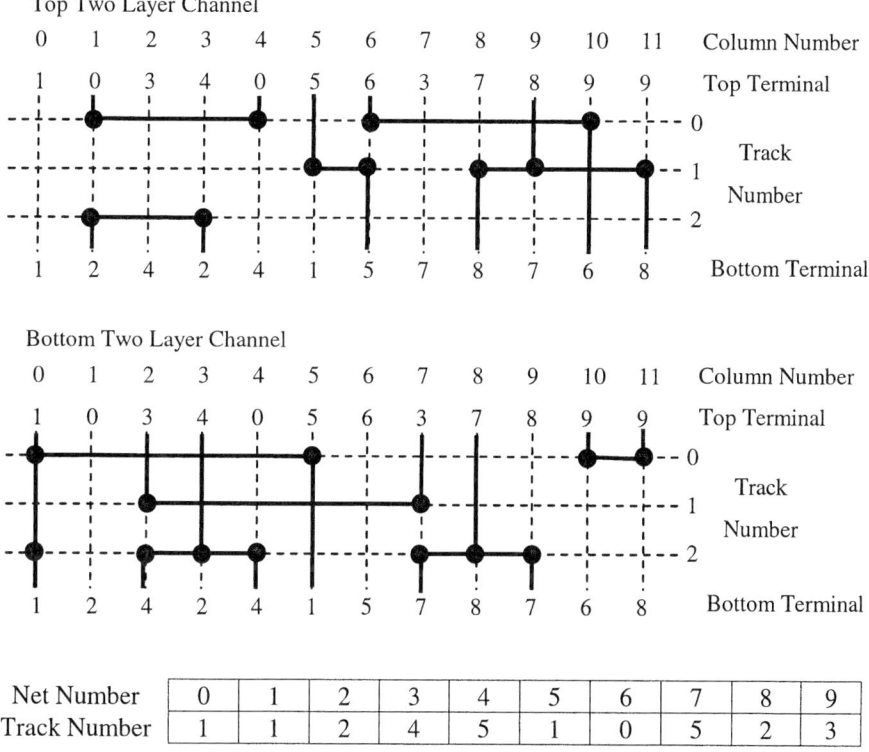

Fig. 1. Representation of Chromosome for 10-Net, 3-Track, 2-Layer Problem

2.1 Selection and Fitness Function

When a net is assigned to the track, it may overlap to other nets in the vertical and horizontal direction. Since the objective is to find the placement of nets free of overlaps, the fitness is the reciprocal of the total overlaps of a chromosome. The fitness function, f, is $a / (\beta + TOTAL\_C)$ where $TOTAL\_C$ is the total number of overlaps, a is the square of the total number of nets and β is the small positive real number forbidding division by zero.

For the selection of new population from the old population, a fitness proportionate section method called roulette scheme is used. An individual with higher fitness has more chances to be selected as the member of new population.

2.2 Crossover and Mutation

We adopted a novel crossover operator to our algorithm as shown in Figure 2. In order to generate new children, C1 and C2, from the parents, P1 and P2, a crossover position is selected randomly. The proposed crossover operator follows the two steps:

1. Simple Crossover: First step is copying the configuration of the parents, P1 and P2, to the children, C1 and C2, respectively. But unlike the standard simply crossover, only alleles of the larger part over the crossover position are copied.
2. Overlap-Free Crossover: Second step is filling the remaining alleles. For the undetermined alleles of C1, the candidate configuration is the remaining, or smaller, part of P2. In this step we check whether the candidate allele assignment generates vertical or horizontal overlaps with previous assignments of C1. If overlaps are found, we search for a new allele assignment which is not overlapped with the previous assigned alleles. We repeat this step until we fill in all the alleles of C1. For a child configuration, C2, we follow the same steps.

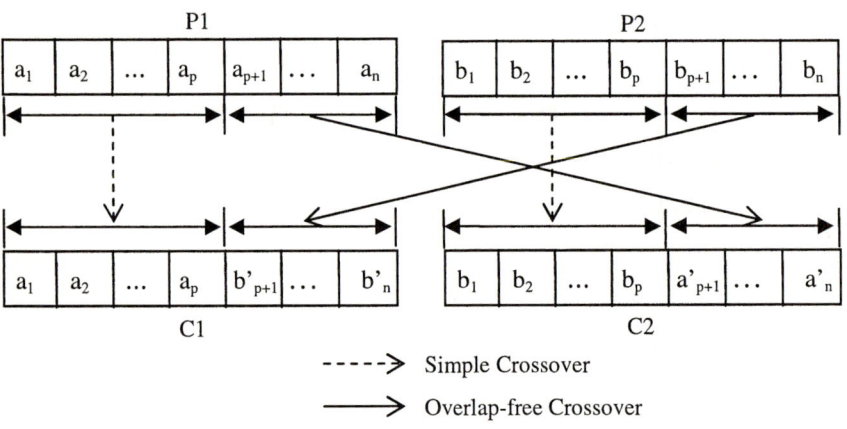

Fig. 2. Crossover Operation

Every two neighboring chromosomes involve in crossover operation with crossover probability, $P_{crossover}$. After the crossover operation, each of the alleles is changed randomly with mutation probability, $P_{mutation}$.

3 The Proposed Asynchronous Distributed Genetic Algorithm

There are many different distributed and parallel genetic algorithms. Previous researches have shown that the parallel genetic algorithm with punctuated equilibrium yields the better problem solutions and speed-up compared to sequential genetic approaches [7]. This scheme evolves independent subpopulation in an isolated manner and exchanges individuals when a state of equilibrium throughout all the subpopulation has been reached.

In our asynchronous distributed genetic algorithm, the master processor forms initial global population whose size is the multiplication of subpopulation size and number of slaves. And then the master randomly selects each subpopulation from the global population and distributes the selected subpopulation to each slave processor. Therefore, each slave processor has the same size of subpopulation which may or may not consist of the same individuals comparing with other subpopulation. Each slave executes the sequential genetic algorithm in parallel. Independent genetic operation such as crossover and mutation are implemented and evaluated independently to its subpopulation.

The duration of isolated evolution is called one epoch and the epoch length is the number of predefined generations for a slave processor before communication with the master. The epoch length is randomly selected between 1 and 2 times the population size. This random number reduces the possibility that more than two slaves communicate with the master simultaneously. After each epoch, the newly evolved subpopulation and their fitness values of each slave processor are sent to the master processor asynchronously. The master replaces the relevant subpopulation in a global population and selects subpopulation randomly. And then the master sends back it to the relevant slave. The processes of distributing and collecting the sub-populations to and from the slave processors iterates as many as predefined epoch or until solution is found.

4 Simulation and Evaluation

The simulation is implemented in MPI environments which are made up of 600Mhz personal computers running Linux operating system connected via 10Mbps Ethernet. The experiment shows that the crossover probability is optimal at 0.6 and the mutation probability has a great effect on the solution quality. When the mutation probability is large, it becomes quite a random search. While the mutation probability is small, it is difficult to escape from the local optimum configurations. We simulated the proposed algorithm using various crossover and mutation probabilities.

The inherent nature of a genetic algorithm makes it possible to be paralleled in multiprocessing environment. The population is almost equally divided to processors and genetic operations are implemented independently to the assigned subpopulation. After each generation, the newly evolved population is sent to the master and updated. The sequential genetic operation properties are maintained in the distributed version because selected genes of ADGA are as same as those of the sequential GA. Thus this distributed version keeps the solution quality as much as the sequential one with less computation time.

The parallel speedup generally increases proportional to the number of nets and the population size (Table 1, Figure 3). However, for the net-45 problem, it is much complicated and the computing time is extremely long, so the speedup is lager than other problems. This shows that proposed distributed algorithm is useful as the problem size increases.

Table 1. The Average Execution Time per one generation in msec

# of nets \ # of proc.	1	2	4	8	16	32
10	7	8	16	25	30	35
21	46	29	30	33	38	40
45	624	325	177	100	65	59
47	165	102	71	60	59	56
54	195	110	80	68	61	54
55	206	116	85	68	61	60
61	503	338	218	152	128	114
72	2211	1146	623	337	208	122

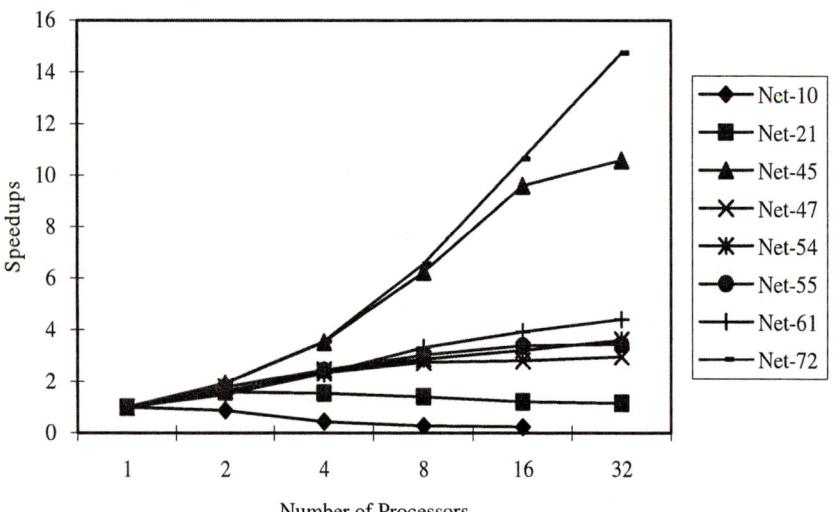

Fig. 3. Speedups of Different Nets

This distributed genetic algorithm was successful to find the solutions on several benchmark problems [8] as shown in Figure 4.

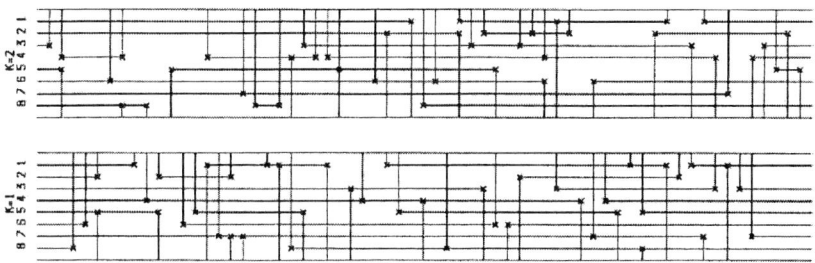

Fig. 4. The solutions of the benchmark problems.

5 Conclusions

In this paper, we proposed an asynchronous distributed genetic algorithm (ADGA) for the channel routing problems. This new asynchronous distributed algorithm was verified successful to the several benchmark routing problems. In addition, this algorithm may be easily modified to solve the partitioning and placement problems for the design automation. Considering the network latency in the distributed system, the results are promising. The proposed algorithm can be easily modified and applied to the broad range of NP-Complete problems efficiently such as chip placement, image processing, and traveling salesman problem.

References

1. Susmita, S., Bhargab, S., Bhattacharya, B.: Manhattan-diagonal routing in channels and switchboxes. ACM Trans on DAES, Vol. 09, No. 01. (2004)
2. Yang, C., Chen, S., Ho, J., Tsai, C.: Efficient routability check algorithms for segmented channel routing. ACM Trans on DAES, Vol. 05, No. 03. (2000)
3. Ning, X.: A Neural Network Optimization Algorithm For Channel Routing. Journal Of Computer Aided Design And Computer Graphics. Vol. 14, No. 1. (2002)
4. Chen, S.: Bubble-Sort Approach to Channel Routing. IEE Proceedings Computer and Digital Techniques. Vol. 147, No. 6. (2000)
5. Holland, J.: Adaptation in Natural and Artificial Systems. Univ. of Michigan Press. (1988)
6. Funabiki, N., Takefuji, Y: A Parallel Algorithm for Channel Routing Problem. IEEE Trans. of CAD, Vol. 11, No. 4. (1992) 464-474
7. Lienig, J., Thulasiraman, K.: A New Genetic Algorithm for the Channel Routing Problem. ACM Proc. of the 7th International Conference on VLSI Design. (1994)
8. Yoshimura, T., Kuh, E.S.: Efficient Algorithms for Channel Routing. IEEE Trans. of CAD, Vol. CAD-1, No. 1. (1982) 25-35

High-Level Language and Compiler for Reconfigurable Computing

Fu San Hiew and Kah Hoe Koay

Faculty of Engineering and Technology, Multimedia University,
Jalan Ayer Keroh Lama, 75450 Melaka, Malaysia
{fshiew, khkoay}@mmu.edu.my

Abstract. This paper presents a high-level, algorithmic, single-assignment programming language and its tailor-made optimizing compiler. The compiler is able to generate a synthesizable hardware description language for reconfigurable systems based on input instruction set. Simulated annealing and force-directed scheduling approaches were employed in this compiler for design speed and resource optimizations. The tasks of the designed compiler include control flow graph transformation, component selection, component scheduling, and VHDL transformation. Language features are introduced and the structure of the compiler is discussed.

1 Introduction

Designing digital system on Field Programmable Gate Arrays (FPGAs) by using hardware description languages is a time consuming task and heavily relied on designers' expert knowledge because algorithms are expressed in bits, registers and clock signals [1]. In order to shorten design time, an effective system design methodology is needed to reduce the designer's involvement in device-level considerations so that the designer can focus more on the architectural designs.

This paper presents a programming language and its compiler to change how reconfigurable systems are programmed from a circuit design paradigm to an algorithmic one. In our proposed method, system-level entry is a clear, simple, readable and C-like programming source code, which is a high-level, algorithmic and single-assignment language. This high-level source code is then transformed into a dataflow graph. Speed and area of the circuit are optimized in dataflow graph by using simulated annealing (SA) [2] and force-directed scheduling (FDS) [3] algorithms. Lastly, the optimized dataflow graph is transformed into synthesizable VHDL codes for implementation in FPGA. Fig. 1 illustrates the overview of the proposed system. The output of the compiler, VHDL, can be used as input to the FPGA synthesis tool for synthesizing reconfigurable computing modules and easily extended to ASICs, or any other devices that accept hardware description language as the design input.

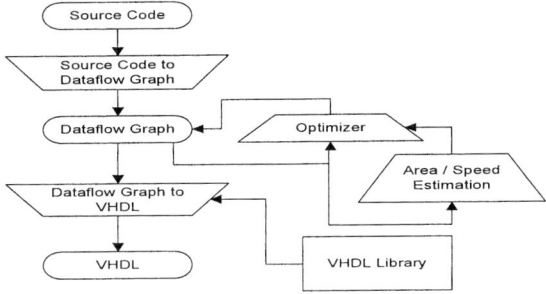

Fig. 1. System overview

2 The Proposed High-Level Language

The proposed language is high-level and algorithmic which can shorten the design time. All instructions are single-assignments and no pointer is involved for better compiler analysis and dataflow graph transformation. Besides, data types are same as used in VHDL. The proposed language's variable name, type, and bit-width are user-specified. Operation assignment is similar to C programming as shown in Fig. 2(a). The timing and parallelism are excluded during system level design for hiding the details and intricacies of low-level hardware design.

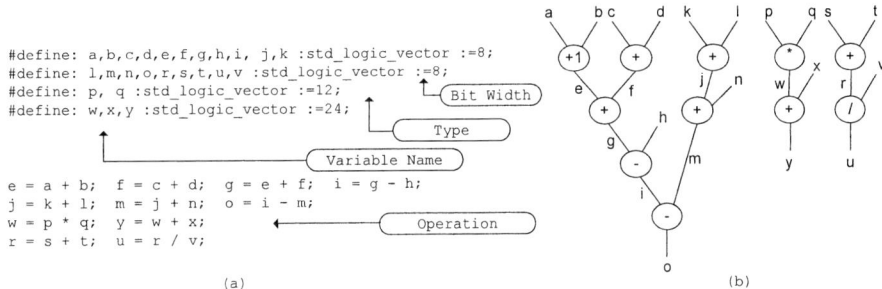

Fig. 2. (a) High-level source code (b) Control-data flow graph (CDFG)

3 Control Flow Transformations

The intermediate representation (IR) of the algorithm is based on a control-data flow graph (CDFG). CDFG is a directed graph and its nodes represent operations and arcs display dependencies between nodes. For the example shown in Fig. 2(b), operation of $e = a + b$, the add operator will become the node "$+_1$", a and b will become the input arcs and e will become the output arc. For those nodes with same arc's name will be connected and form a linked CDFG. CDFG generated during control flow transformations just reveals its nodes operation and dependencies between nodes. The timing and other information will be added on the CDFG in the subsequent processes.

4 Component Selection

In this section, each node is assigned a VHDL component, which is chosen from Zimmermann's arithmetic library [4], as shown in Table 1. The VHDL components are synthesized using Xilinx Webpack 5.2i to get its area (slice) and delay (ns). The VHDL component selection is based on user's optimization mode selection:

- *Speed (SD)*. The speed of the circuit is set to maximum regardless how large the area will be.
- Area (AR). The area of the circuit is set to minimum regardless how slow the speed will be.
- Speed with acceptable area (SAA). The speed of the circuit is set to as-high-as-possible with acceptable increase in area.
- *Area with acceptable speed (AAS)*. The area of the circuit is set to as-small-as-possible with acceptable decrease in speed.

If optimization mode is set for *SD*, VHDL component with smallest delay will be chosen for all nodes in a particular circuit; smallest area assignment for *AR*. SA performs speed (*SAA*) and area (*AAS*) optimizations in CDFG by randomly assigning various architecture of defined VHDL component to CDFG's nodes. The number of nodes (VHDL components) to be changed in each iteration is based on the concept of temperature cooling function in SA. This concept is called as *component reduction function* in our work. For *SAA* mode, the delay of the critical path is evaluated every iteration. SA will just accept new component assignment if the iteration causes decrease in critical path delay. An increase in delay is accepted if the Metropolis criteria are fulfilled. For *AAS* mode, new assignment will be accepted if there is a decrease in total circuit area. On the other hand, an increase in area will be accepted if the Metropolis criteria are fulfilled.

Table 1. Components Library

Component	Preferences	Properties	4 bit	8 bit	......	64 bit
AddC	Slow	Area	6.00	12.00		
AddC	Slow	Delay	19.66	23.49		
AddCFast	Slow	Area	8.00	17.00		
AddCFast	Slow	Delay	18.99	23.50		

5 Component Scheduling

Once timing constraint has been set, area is another factor that needs serious consideration. Force-directed scheduling (FDS) is chosen to minimize the area (number of slices) required [3]. This algorithm will uniformly distribute the operations of each type across a time-constrained schedule, thus resulting in higher functional unit usage. The result is shown in Fig. 3(b). From the figure, we can observe that the width of c-step is equal to the maximum delay (41.93ns) is caused by "*" node. All nodes are scheduled to their corresponding step with the least force. From this optimized CDFG, designer can easily partition the designed circuit according to clock cycle, for run-time reconfiguration into FPGA.

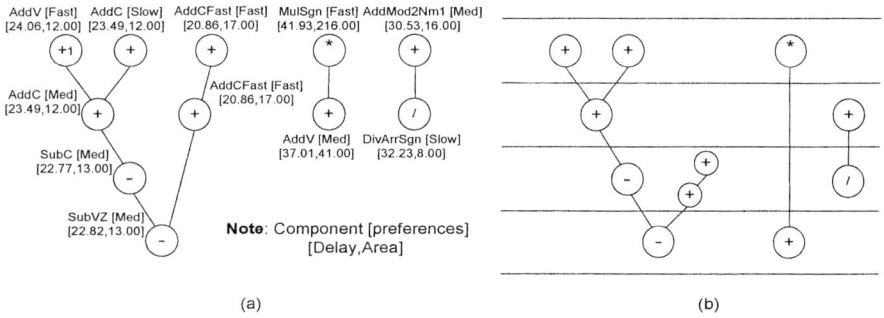

Fig. 3. (a) Nodes with VHDL component assignment (b) Optimized CDFG

6 VHDL Transformations

All nodes in CDFG are transformed into VHDL modules with input ports, output ports, clock, chip able, and reset signal, as shown Fig. 4(a). Note that, input ports in module are the nodes' input arcs and output port is the node's output arc in CDFG. In order for this module to perform the operation, the corresponding VHDL component from Zimmermann's arithmetic library will be instantiated into these modules. This process is repeated until each node is assigned with a module. Fig. 4(b) depicts the abstract view of the top module formed.

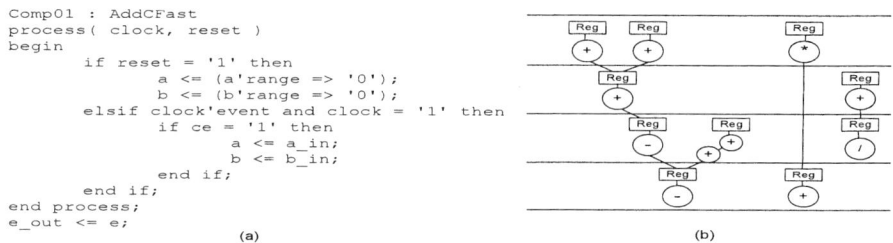

Fig. 4. (a) VHDL code for instruction $f = c + d$ (b) Top level module

7 Experimental Results

Experiments performed are based on the design illustrated in Fig. 5. This design was also used by Paulin and Knight [3].

```
#define: a, b, e, f, j, k, q, r :integer := 4;
#define: c, d, l, m, p, s, v, w, u, t, x, y :integer := 8;
#define: g, h, i, n, o :integer := 16;

c = a * b; d = e * f; g = c * d; i = g - h;
l = j * k; n = l * m; o = i - n;

p = q * r; s = p + t; u = v + w; x = u + y;
```

(a)   (b)

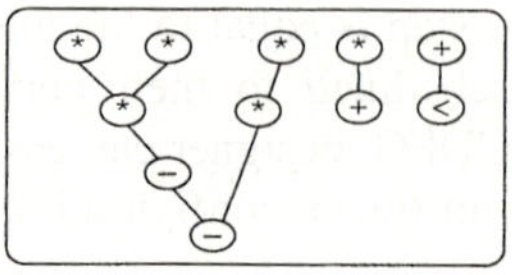

Fig. 5. Example (a) instructions set and (b) CDFG

A suitable *component reduction function* is adopted here so that SA can perform efficiently in our compiler. We have compared the results from various *component reduction functions* in our compiler, see formulas below.

$$T_1 = T_o - i\left((T_o - T_N)/N\right) \tag{1}$$

$$T_2 = T_o \left(T_N/T_o\right)^{i/N} \tag{2}$$

$$T_3 = T_o - i^A, A = \ln(T_o - T_N)/\ln(N) \tag{3}$$

$$T_4 = (1/2)(T_o - T_N)\left(1 - \tanh\left((10i)/N - 5\right)\right) + T_N \tag{4}$$

$$T_5 = (T_o - T_N)/\cosh\left((10i)/N\right) + T_N \tag{5}$$

Table 2 shows the delay time results of SAA for 10 trials with various *component reduction functions*. From the table, we can see that T_5 is superior as it obtains the highest probability of achieving minimum critical path delay. Besides, the delay time obtained in various trials is always less than 106.0 ns. Fig. 6 shows the delays obtained from our SAA experiments. It is shown that the critical path delay is decreasing along the temperature steps. Some increases in delay time are accepted to avoid being stuck at local minimum.

Table 2. Results (delay) of *SAA* for 15 trials with various component reduction functions

Trial	Eqn T1	Eqn T2	Eqn T3	Eqn T4	Eqn T5
1	106.3	103.7	**102.3**	104.3	103.9
2	112.0	106.6	103.9	103.7	106.0
3	106.9	107.1	112.8	108.0	**102.3**
4	113.5	104.3	105.8	104.5	104.5
5	108.9	103.7	106.3	104.9	**102.3**
6	112.7	**102.3**	106.0	108.5	104.3
7	106.5	106.6	103.9	103.9	**102.3**
8	114.4	106.9	109.1	**102.3**	**102.3**
9	108.5	107.0	**102.3**	106.0	103.7
10	104.5	106.9	107.8	106.5	106.0

In our subsequent experiment, the design is simulated based on T_5 as *component reduction function*. The simulated delay and area of the abovementioned optimization modes are compared. From Table 3, we can see that optimization with *SD* mode out-

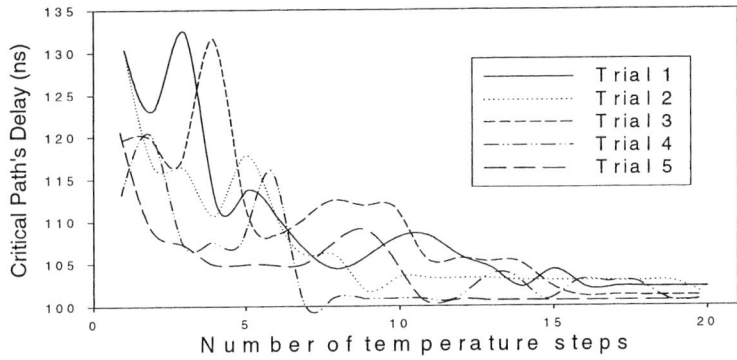

Fig. 6. Critical path's delay obtained in 5 *SAA* trials

performs the others in terms of speed with minimum delay time (102.3ns). Whilst, optimization with *AR* mode is superior in terms of area because area size (296 slices) obtained is the smallest among others. The simulation results for *SAA* and *AAS* modes lie in between *SD* and *AR* optimization modes' results.

Table 3. Simulation Results

Modes	*Speed (SD)*				*Area (AR)*			
Speed (ns)	102.3				118.9			
Area (slices)	329				**296**			
Modes	*Speed with acceptable area(SAA)*				*Area with acceptable speed(AAS)*			
Trials	1	2	3	4	1	2	3	4
Speed (ns)	104	106.0	104.5	104.3	114.0	110.4	111.4	114.4
Area (slices)	312	322	320	313	309	320	307	312

8 Concluding Remarks

In this paper, we have shown how a self-defined high-level language is transformed into a synthesizable VHDL. Simulated annealing and force-directed scheduling approaches are used for speed and resource optimization. A digital system design shown in Fig. 2 just requires less than 10 minutes design time. It is evident that we are able to move to a higher level of abstraction in reconfigurable systems design. Future effort will be concentrated on refinement of the presented techniques. Image processing component is our interest for future research and it will be added in our library.

References

1. Najjar, W. A., Bohm, W., Draper, B. A., Hammes, J., Rinker, R., Chawathe, M. and Ross, C.: High-level language abstraction for reconfigurable computing. Computer, vol. 36. issue. 8, (2003) 63-69

2. Kirkpatrick, S., Gelatt, C. D., Jr. and Vecchi, M. P.: Optimization by simulated annealing. Science, vol. 220, (1983) 671-680
3. Paulin, P.G and Knight, J.P.: Force-directed scheduling for the behavioral synthesis of ASICs. IEEE Transactions on Computer-Aided Design of Integrated Circuits and Systems, vol. 8, issue. 6, (1989) 661-679
4. Zimmermann, R.: VHDL Library of Arithmetic Units. in Proc. First Int. Forum on Design Languages (FDL'98), Lausanne, Switzerland, (1998)

A Parallel Algorithm for the Biorthogonal Wavelet Transform Without Multiplication

HyungJun Kim

Graduate School of Information Security, Korea University
1 5-ga Anam-dong Sungbuk-ku, Seoul 136-701, Korea
hyungjun@korea.ac.kr

Abstract. We present parallel algorithms based on biorthogonal wavelet transforms(BWTs). We have constructed processing elements(PEs) for the one-dimensional(1-D) and two-dimensional(2-D) reconstruction filter masks to minimize computational operations. The proposed architectures are efficient, scalable for dyadic filters of BWT. They may be applied to the implementation of codec in various image and video processing standards such as JPEG 2000 and MPEG-4.

1 Introduction

General wavelet transforms and subband filters have many taps and require floating point multiplication operations; thus, the derived compression systems often require longer processing time in comparison to JPEG. For image compression and real-time motion video display, it is critical that all phases of computation be performed as efficiently as possible to maximize image quality and frame rate. A technique commonly used to achieve this efficiency avoids multiplication and division operations wherever possible or implements them as simple shift operations. The wavelet coefficients chosen for decomposition and reconstruction processes of the BWT are intentionally defined to be as short as possible with good compression performance. Like the discrete Fourier transform, BWT is computation intensive and has many real-time applications. Therefore, a special parallel computation is desirable. By exploiting the characteristics of wavelet filter coefficients, we have designed parallel architectures which can compute the BWT efficiently. They can produce very high throughput because they possess high degrees of pipelining and parallel processing.

2 The Choice of Wavelet Filters

Perfect reconstruction conditions for two-channel filter banks with $H_0(z)$ and $G_0(z)$ as lowpass filters, and $H_1(z)$ and $G_1(z)$ as highpass filters are given as:

$$G_0(z)H_0(z) + G_1(z)H_1(z) = 2z^{-l}, \quad G_0(z)H_0(-z) + G_1(z)H_1(-z) = 0. \quad (1)$$

The aliasing is removed when the filters are properly chosen. Wavelet-based image compression requires wavelets with some degree of regularity for quality

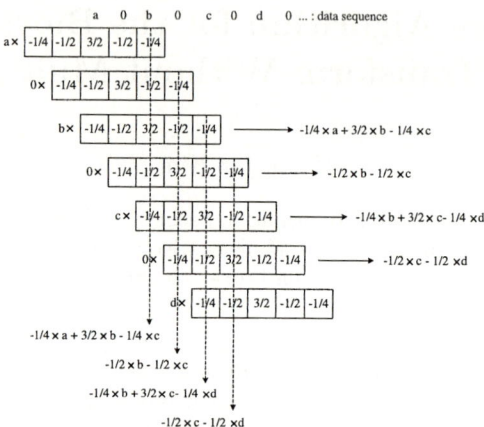

Fig. 1. An example of convolution and shift/superposition operations using a 5-tap highpass filter for the 1-D reconstruction process

reconstruction and filters with relatively short support for efficient processing. Usually, a more regular wavelet is preferred for use as the synthesis filter, since reconstruction is directly affected by its regularity[1]. One option is

$$H_0(z) = \frac{1}{4\sqrt{2}}(-z^{-2} + 2z^{-1} + 6 + 2z - z^2), \quad G_0(z) = \frac{1}{2\sqrt{2}}(z^{-1} + 2 + z). \quad (2)$$

The chosen filter set is the same as one of LeGall's factorized product filters[2], also known as Daubechies' generalized biorthogonal filters for $N=2$[3]. The default reversible transform of the JPEG 2000 still-image compression standard is also implemented by means of the Le Gall 5-tap/3-tap filter[4]. The highpass filters are then given as

$$H_1(z) = \frac{1}{2\sqrt{2}}(-z^{-1} + 2 - z), \quad G_1(z) = \frac{1}{4\sqrt{2}}(-z^{-2} - 2z^{-1} + 6 - 2z - z^2). \quad (3)$$

We avoid the factor $1/\sqrt{2}$ in the expressions by multiplying $1/\sqrt{2}$ in the analysis filters and $\sqrt{2}$ in the synthesis filters in order to create filter banks with dyadic rational coefficients for integer arithmetic operations.

3 Algorithm for the 1-D Reconstruction Process

In order to explain the proposed algorithm, we use an example of a regular convolution procedure using a 5-tap highpass filter for the reconstruction process. First, consider a data sequence given by $\{a, b, c, d, ..\}$ and a 5-tap highpass filter for reconstruction as $\{-\frac{1}{4}, -\frac{1}{2}, \frac{3}{2}, -\frac{1}{2}, -\frac{1}{4}\}$. Because of the interpolation characteristic of a wavelet transform, this results in a new data sequence $\{a, 0, b, 0, c, 0, d, 0, ..\}$ for convolution with the 5-tap filter. Excluding the incomplete boundary outputs, the first output is $-\frac{1}{4}a + \frac{3}{2}b - \frac{1}{4}c$ which is calculated as

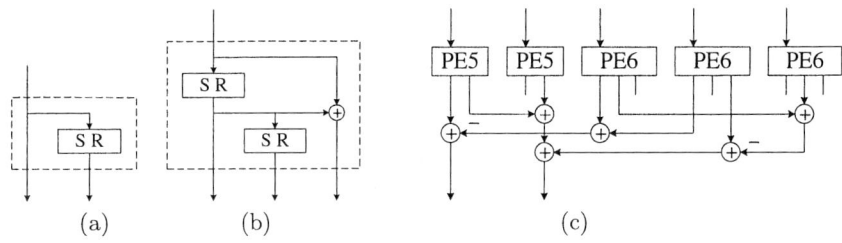

Fig. 2. (a) Lowpass filter(PE5), (b) Highpass filter(PE6), and (c) Systolic architecture for the 1-D multiplier-free reconstruction process of the BWT

the convolution of data sequence $\{a, 0, b, 0, c\}$ with filter coefficients as illustrated by the first horizontal arrow in Fig. 1. The same procedure can be repeated until the final data point is reached.

Alternatively, the computation may be performed simply by shift and superposition operations as follows. One can fetch one data point(a for example) which is located at the center of the filter, and then scale it. It looks like the center value spreads out toward both ends of the filter and is weighted according to the filter coefficients. At the next clock cycle, we do not need to calculate the intermediate data set since the input is zero, as shown in Fig. 1. At the third clock cycle, we perform similar shift operations. After five clock cycles, one superimposes all three intermediate data sets, except the two zero sets, to obtain a value of $-\frac{1}{4}a + \frac{3}{2}b - \frac{1}{4}c$, shown by the first vertical arrow at the position of the input data point b, which is the same result one would obtain using the regular convolution computation as explained previously. The same process may be repeated until one covers all the input data points in the sequence.

The benefits of shift and superposition operations include increases in processing speed and architectural efficiency. Whereas in the usual convolution computation, even if many data points are zeroes, such as in the highpass outputs of the wavelet decomposition, one must perform the convolution operation which is obviously a waste of time. Alternatively, in addition to its architectural simplicity, the shift and superposition operation case confers speed by allowing one to skip the whole shift and superposition operation if the center data value is zero. This increase becomes more pronounced in the 2-D case and will be very useful in accelerating the reconstruction process.

We constructed two PEs, each with lowpass and highpass filters for the 1-D multiplier-free reconstruction of the BWT as shown in Fig. 2. Parallel architecture for the 1-D reconstruction process of the BWT is very simple, as shown in Fig. 2, where *PE5* and *PE6* represent PEs for a lowpass and a highpass filter, respectively. The proposed architecture can compute the odd indexed and the even indexed outputs for both lowpass and highpass filterings simultaneously. If one set of lowpass data sequences for PE5s and another set of highpass data sequences for PE6s at a coarser level are fed into the PEs from right to left at every clock cycle T, the two PE5s of the systolic architecture generate the odd and the even indexed outputs which are reconstructed lowpass wavelet coefficients

Fig. 3. 2-D filter masks for the multiplier-free reconstruction of the BWT: (a) LL-filter, (b) LH-filter, and (c) HH-filter

on a finer scale. The three PE6s also generate the odd and the even indexed highpass wavelet coefficients simultaneously on a finer scale. Therefore, reconstructed data points can be obtained at a rate of two per clock period T. Note that the input sequence does not need to be interpolated, and the shift and superposition operation for zero values are automatically not processed. The same PE arrays can be used for lowpass and highpass data processings of additional levels. Note that reconstructed data points can be obtained simultaneously at the rate of two per clock period T.

4 Algorithm for the 2-D Reconstruction Process

We have developed a fast reconstruction algorithm, using 2-D filter masks, that takes advantage of the characteristics of the wavelet transformed data; i.e., there are many zero-valued pixels which need not be processed in highpass subbands at each level. Although processing by 2-D filters generally requires more time than processing by two 1-D directional filters, our special algorithm based on 2-D masks and shift-superposition operations provides a fast reconstruction process. The 2-D filters are constructed first by considering tensor products of two 1-D synthesis filters: $\{\frac{1}{2}, 1, \frac{1}{2}\}$ for lowpass and $\{-\frac{1}{4}, -\frac{1}{2}, \frac{3}{2}, -\frac{1}{2}, -\frac{1}{4}\}$ for highpass. Note that HL-filter is the transpose of LH-filter. All of the filter coefficients can be implemented using a combination of basic shift operations as shown in Fig. 3. The proposed algorithm uses 2-D filter masks, LL(3×3), LH(3×5), and HH(5×5), for carrying out shift and superposition operations to implement the convolution process and to minimize data access operations by processing only the nonzero pixels in the decomposed subbands.

Analogous to the 1-D case, the reference pixel value at the center of the mask spreads out toward neighboring pixels via shift and superposition operations using the 2D-filter masks of the BWT. For illustration, let us consider an image of size 3×3 at a coarse level having pixel values $\{a, b, c; d, e, f; g, h, i\}$ and the LL-filter mask for reconstruction as shown in Fig. 4. The pixel value at the center of the mask spreads to the eight neighboring pixels and is weighted by the filter coefficient values at the respective positions in the mask as illustrated in Fig. 4. The dark squares represent pixels at a coarser level, and the white squares show the interpolated pixels at the next finest level. Analogous to the

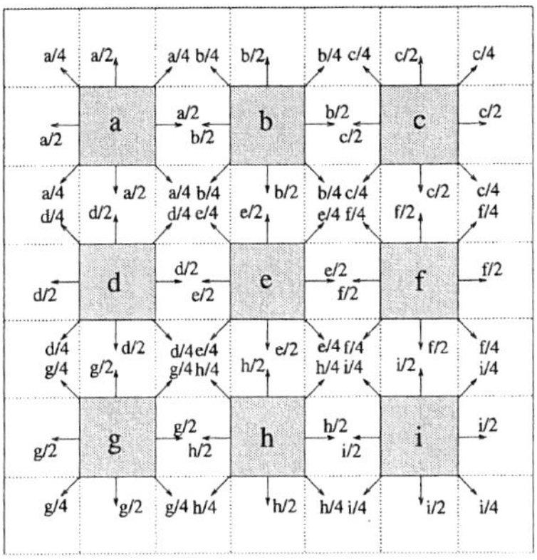

Fig. 4. An example of shift and superposition operations using the LL-filter mask in the 2-D reconstruction process for a 3×3 image data having pixel values $\{a, b, c; d, e, f; g, h, i\}$

reconstruction process with LL-filter mask, similar processes are performed using LH-, HL-, and HH-filter masks for the corresponding detail subbands at the coarser level. The reconstruction at one level is completed when all four subbands are superimposed. Reconstruction processing then continues at the next finest level.

For the architecture of the 2-D wavelet transform, we have developed a fast reconstruction architecture, using 2-D filter masks, that utilizes many intermediate states of shift operations. We constructed three PEs, each with LL-, LH-, and HH-filters for the 2-D multiplier-free reconstruction of the BWT as shown in Fig. 5. Note that PE for HL-filter mask is the same as PE for LH-filter mask since HL-filter mask is the transpose of LH-filter mask. Compared to two 1-D directional architectures which must process horizontally and then vertically(or vertically and then horizontally), mesh arrays of four PEs in the parallel architecture generate the reconstructed coefficients using LL-, LH-, HL-, and HH-filters simultaneously, and therefore, are applicable to high speed architectures. The reconstruction at one level is completed when all four reconstructed data sequences are superimposed. The reconstruction process then continues at the next finer level.

5 Conclusions

In this paper we present a parallel algorithm without multiplication for the efficient computation of 1-D and 2-D reconstruction processes. The proposed algorithm for the one-level 1-D wavelet reconstruction process utilizes only eight shift registers and nine 2-to-1 adders, i.e., about four shift registers and four 2-to-1 adders for each output. Note that reconstructed data points can be obtained simultaneously at the rate of two per clock period T. In the 2-D reconstruction

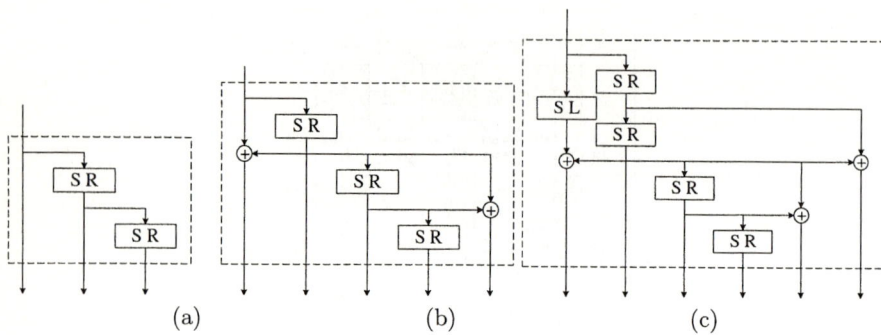

Fig. 5. PEs for the 2-D multiplier-free reconstruction of the BWT: (a) LL-filter(PE7), (b) LH-filter(PE8), and (c) HH-filter(PE9).

from a coarse level to the next finest level, the proposed architecture requires thirteen shift registers and twenty-three 2-to-1 adders for each pixel of four subbands. The proposed 2-D reconstruction process can be significantly accelerated in comparison to the usual two directional 1-D processes. It can produce very high speed throughput and process signals in real-time since it possesses high degrees of pipelining and parallel processing, and therefore, can be applied to the implementation of codec in various image and video processing standards, such as JPEG 2000 and MPEG-4.

The benefit of shift and superposition operations is shortening the processing time in convolution. Because there are many zero-valued pixels in the three highpass subbands at each decomposition level, if the pixel value at the center of the mask is zeroes, one may skip the whole shift and addition operation. In addition, the reconstruction process can be significantly accelerated in comparison to the usual tensor product of 1-D processing since all four reconstructed data sequences are processed simultaneously. Compared to the conventional wavelet transform method, the proposed parallel algorithm reduces the number of multiplications through bit-shift and addition operations by almost half.

References

1. Strang, G., Nguyen, T.: Wavelets and filter banks. Wellesley-Cambridge Press (1996)
2. LeGall, D., Tabatabai, A.: Sub-band coding of digital images using symmetric short kernel filters and arithmetic coding techniques. in IEEE Proc. Int. Conf. ASSP (1988) 761–764
3. Daubechies, I.: Ten lectures on wavelets. **CBMS-61** Philadelphia, PA: SIAM (1992)
4. Skodras, A., Christopoulos, C., Ebrahimi, T.: The JPEG 2000 still image compression standard. IEEE Signal Processing Magazine **Sep.** (2001) 36–58

Algorithms for Loosely Constrained Multiple Sequence Alignment[*]

Bin Song, Feng-feng Zhou, and Guo-liang Chen

Department of Computer Science and Technology,
University of Science and Technology of China, Hefei, Anhui, 230027, P. R. China
heather@mail.ustc.edu.cn, steam@ustc.edu, glchen@ustc.edu.cn

Abstract. For finding accurate and biologically meaningful multiple sequence alignment, it is required to consider a loosely constrained version of multiple sequence alignment. This paper is the first to define a loosely constrained multiple sequence alignment (LCMSA). And it also designs a complete algorithm for the loosely constrained pair-wise sequence alignment problem with the time complexity of $O((\alpha+1)n^{2(\alpha+1)})$ and an approximation algorithm for LCMSA running in time $O(k^2(\alpha+1)n^{2(\alpha+1)})$, k and n is the number and the maximum length of input sequences, and α is the length of constrained sequence. In fact, α is small compared to n and it is reasonable to regard it as a constant. So, the former two algorithms run in polynomial time. Finally, a LCMSA software system is built up.

1 Introduction

The multiple sequence alignment problem (MSA) is a fundamental and challenging problem in computational molecular biology [1]. Many algorithms for MSA have come out. Algorithm of Dynamic Programming (DP) is suitable for the problem only when the length of the sequences is short and the number of the sequences is small. Gusfield achieved a useful approximation algorithm with performance ratio of 2-2/k, k is the number of input sequences [2]. With the knowledge of the functionalities or structures of the input sequences, which is important for accurate and biologically meaningful alignment, Tang et al firstly gave the definition of the constrained multiple sequence alignment problem (CMSA) that required each character of the constrained sequence to appear in an entire column of the result alignment and they also presented some algorithms for CMSA [3]. Later, Chin et al improved the time complexity of the CMSA algorithms [4].

But, Tang and Chin did not consider the variations of the constrained sequence. In nature, it is frequent that there are some mutations and each character of the constrained sequence might not exactly appear in an entire column. For example, it is known that the three active-site residues His(H), Lyn(K), His(H) are significant for

---

[*] This work was supported by the National '863' High-Tech Program of China under the grant No.2001AA111041, 2002AA104560, and High-standard Universities Construction Project in CAS.

RNA degrading. And in the practical alignment, the residues might not appear in the entire columns for variations happen, though they are expected so. Then, it is required to define the loosely constrained multiple sequence alignment problem (LCMSA) that tries to find a loose version of CMSA, in which the dissimilarity between any two aligned sequences and the constrained sequence no more than the variation distance d and the score is optimal. Specially, when the number of the sequences is two, it is a loosely constrained pair-wise sequence alignment problem (LCPSA). More, this paper designs a complete LCPSA algorithm with the complexity of time $O((\alpha+1)n^{2(\alpha+1)})$, α is the length of the constrained sequence and n is the maximum length of the input sequences. And an approximation algorithm for LCMSA that runs in time $O(k^2(\alpha+1)n^{2(\alpha+1)})$ is presented. Since α is small compared to n and it is reasonable to regard it as a constant. Then, the former two algorithms run in polynomial time. Finally, a LCMSA software system is built up for biologists.

2 Preliminaries

In this section, some definitions are defined formally. The notations are similar to those in [1,3,4]. Let $S_i[x]$ denote the x-th character of sequence S_i.

Definition 1[1,4]: Given a set $S = \{S_1, S_2, ..., S_k\}$ of k sequences where $k \geq 2$, insert spaces into each sequence so that the resulting sequences S_i' ($i = 1, 2, ..., k$) have the same length n', and $\{S_1', S_2', ..., S_k'\}$ is called a multiple sequence alignment. The **Multiple Sequence Alignment Problem** (MSA) is to construct a multiple sequence alignment minimizing its score. In particular, when $k = 2$, MSA is the pair-wise sequence alignment problem.

The score of the multiple sequence alignment is defined as

$$\sum_i u(S_1'[i], S_2'[i], ..., S_k'[i])$$

and it reflects the degree of dissimilarity among characters in the columns of the multiple sequence alignment. With the different scoring definition, there are different scoring schemes. The following defines the most popular scoring scheme, SP-score. This paper only considers SP-score and the results in this paper can be easily extended to the other scoring schemes.

SP (Sum-of-Pairs)-Score:

$$u(S_1'[i], S_2'[i], ..., S_k'[i]) = \sum_{1 \leq p < q \leq k} u(S_p'[i], S_q'[i])$$

where $u(S_p'[i], S_q'[i])$ is the score of the opposing letters $S_p'[i]$ and $S_q'[i]$ in column i.

Definition 2[9,10]: Given a set $S = \{S_1, S_2, ..., S_k\}$ of k sequences and a constrained sequence $P = P[1]P[2]...P[\alpha]$ which is a common subsequence of $S_i \in \{S_1, S_2, ..., S_k\}$. The **Constrained Multiple Sequence Alignment Problem** (CMSA) is to find a multiple sequence alignment M with minimum score, such that each character in P appears in an entire column of M and also in the same order, i.e. there is a list of integers $\{c_1, c_2, ..., c_\alpha\}$, $1 \leq c_1 < c_2 < ... < c_l < ... < c_\alpha \leq n'$, and for all $1 \leq i \leq k$, $1 \leq l \leq \alpha$, let $S_i'[c_l] = P[l]$. Specially, when $k = 2$, CMSA is the **Constrained Pair-Wise Sequence Alignment Problem (CPSA)**.

Definition 3: In addition to the inputs of CMSA, a variation distance d is given. The **Loosely Constrained Multiple Sequence Alignment Problem (LCMSA)** is to find a multiple sequence alignment M minimizing its score, such that there is a list of integers $\{c_1, c_2, ..., c_\alpha\}$, $1 \leq c_1 < c_2 < ... < c_l < ... < c_\alpha \leq n'$, where $\sum_{1 \leq l \leq \alpha}\{u(S_i'[c_l], P[l]) + u(S_j'[c_l], P[l])\} \leq d$ and $S_i'[c_l]$ and $S_j'[c_l]$ are not spaces, for all $1 \leq i \leq k$, $1 \leq j \leq k$, $1 \leq l \leq \alpha$. Similarly, when $k = 2$, LCMSA is the **Loosely Constrained Pair-Wise Sequence Alignment Problem (LCPSA)**.

In Definition 3, d is introduced to evaluate the variation of the constrained sequence. And, the dissimilarity between any two aligned sequences, but not k sequences, and the constrained sequence is to avoid that the variations concentrate into one or two sequences. Specially, when the scoring scheme is metric and $d = 0$, LCMSA is CMSA and LCPSA is CPSA. If $d = \infty$, LCMSA is a general MSA.

3 Loosely Constrained Pair-Wise Sequence Alignment Algorithm

In this section, a complete algorithm for LCPSA is presented.

Algorithm 3.1 An Complete Algorithm for LCPSA

Input: Sequences S_1 and S_2, a constrained sequence $P = P[1]P[2]...P[\alpha]$ and a variation distance d.

Output: The optimal loosely constrained pair-wise sequence alignment of S_1 and S_2 with respect to P and d.

Step 1: For each list of positions $(c_1^1, c_2^1, ..., c_\alpha^1)$ of S_1 and $(c_1^2, c_2^2, ..., c_\alpha^2)$ of S_2, find all the positions that satisfy P and d, that is $\sum_{1 \leq l \leq \alpha}\{u(S_1[c_l^1], P[l]) + u(S_2[c_l^2], P[l])\} \leq d$.

Step 2: Consider all the positions gotten from Step 1. The position $(c_1^1, c_2^1, ..., c_\alpha^1)$ separates S_1 to $(\alpha+1)$ parts, such as $S^1{}_1, S^2{}_1, ..., S^{\alpha+1}{}_1$. And $(c_1^2, c_2^2, ..., c_\alpha^2)$ separates S_2 to $(\alpha+1)$ parts, such as $S^1{}_2, S^2{}_2, ..., S^{\alpha+1}{}_2$. Align $S^i{}_1$ and $S^i{}_2$ for all $1 \leq i \leq \alpha+1$ with the pair-wise sequence alignment algorithm of Dynamic Programming. Merge the $(\alpha+1)$ alignments, and a loosely constrained pair-wise sequence alignment comes.

Step 3: From all the loosely constrained pair-wise sequence alignments of Step 2, find the optimal alignment that has the minimum score.

Algorithm Analysis: The correctness of Algorithm 3.1 is easy to prove. The following analyses the time complexity. Step 1 need $C_n^\alpha C_n^\alpha = O(n^{2\alpha})$ time. Step 2 runs in time $O(n^{2\alpha})O((\alpha+1)n^2) = O((\alpha+1) n^{2(\alpha+1)})$ at most. Step 3 runs no more than $O(n^{2\alpha})$ time. Therefore, the total time complexity of Algorithm 3.1 is $O(n^{2\alpha}) + O((\alpha+1) n^{2(\alpha+1)}) + O(n^{2\alpha}) = O((\alpha+1) n^{2(\alpha+1)})$ and Theorem 3.1 follows. In fact, α is relatively small compared to n, and it is reasonable to regard it as a constant. So Algorithm 3.1 runs in polynomial time.

Theorem 3.1 The loosely constrained pair-wise sequence alignment problem can be solved in time $O((\alpha+1) n^{2(\alpha+1)})$.

4 Loosely Constrained Multiple Sequence Alignment Algorithm

Based on Gusfield's center-star approach [2], the following approximation algorithm for LCMSA is obtained. In order to describe it easily, the *star-sum score* of a loosely constrained multiple sequence alignment with a center sequence S_c is defined as the sum of SP-score of S_c with all other sequence $S_j \in S - \{S_c\}$.

Algorithm 4.1 An Approximation Algorithm for LCMSA

Input: Sequences $S=\{S_1, S_2, ..., S_k\}$, a constrained sequence $P = P[1]P[2]...P[\alpha]$ and a variation distance d.

Output: A loosely constrained multiple sequence alignment M with respect to P and d.

Step 1: For each S_i in S, treat S_i as the center sequence S_c. At first, find all the positions $(c_1, c_2, ..., c_\alpha)$ of sequence S_c that satisfy the constrained sequence P and the variation distance $d/4$, that is $\sum_{1 \leq l \leq \alpha} \{u(S_c[c_l], P[l])\} \leq d/4$. Then, perform the LCPSA algorithm between S_c and other S_j under $(c_1, c_2, ..., c_\alpha)$, using a slightly modified Algorithm 3.1. In details, treat S_c as S_1 and other S_j as S_2, step 1 of Algorithm 3.1 is changed to "for each list of positions $(c_1^2, c_2^2, ..., c_\alpha^2)$ of sequence S_2, find all the positions which is satisfied the constrained sequence P and the variation distance $d/4$, that is $\sum_{1 \leq l \leq \alpha} \{u(S_1[c_l], P[l]) + u(S_2[c_l^2], P[l])\} \leq d/4$" and the position $(c_1^1, c_2^1, ..., c_\alpha^1)$ in other steps of Algorithm 3.1 is changed to the position $(c_1, c_2, ..., c_\alpha)$. Therefore, the alignments between all $S_j \in S - \{S_c\}$ and S_c, under the position $(c_1, c_2, ..., c_\alpha)$ can be obtained.

Step 2: Consider each sequence S_i and a list of position $(c_1, c_2, ..., c_\alpha)$ in S_i where $1 \leq i \leq k$, gotten from Step1. Find the sequence S_c and its position $(c_1, c_2, ..., c_\alpha)$ that makes the star-sum score with respect to S_c minimum.

Step 3: Without loss of generality, suppose the center sequence of S is S_1 under a list of positions $(c_1, c_2, ..., c_\alpha)$, gotten from Step2. From Step1, there are $(k-1)$ loosely constrained pair-wise alignments, each of which is the alignment between S_1 with other S_j, for $2 \leq j \leq k$, noted $M_{1,j}$. In all $(k-1)$ $M_{1,j}$'s, let p_0 and p_n be the longest sequences of spaces before $S_1[1]$ and after $S_1[n_1]$ (n_1 is the length of S_1). Equally, for $1 \leq i \leq n_1-1$, p_i is the longest spaces' sequence between $S_1[i]$ and $S_1[i+1]$ in all $(k-1)$ $M_{1,j}$'s. Then $S_1' = p_0 \oplus S_1[1] \oplus p_1 \oplus S_1[2] \oplus ... \oplus p_{n-1} \oplus S_1[n] \oplus p_n$, where $\oplus$ denotes the string concatenation operation. Then, add S_j, $2 \leq j \leq k$, one at a time to M that initially contains only S_1', that is, in $M_{1,j}$, let S_1 and S_j be aligned to $(S_1^s, S_j^s)^{-1}$, insert spaces' columns to $(S_1^s, S_j^s)^{-1}$ until S_j^s is equal to S_1'.

So, M is the loosely constrained multiple sequence alignment for $\{S_1, S_2,...,S_k\}$ with respect to P and d since the insertion of spaces of Step 3 does not change the pair-wise alignment score between S_1 and S_j, $2 \leq j \leq k$, according to triangle inequality.

Algorithm Analysis: In Step 1, the loosely constrained pair-wise sequence alignment is obtained with P and $d/4$, not d. It is in order to get a correct loosely constrained multiple sequence alignment with d finally and quickly. The correctness and time complexity of Algorithm 4.1 is proved in Theorem 4.1 and Theorem 4.2.

In order to reduce the time complexity, parallelizing Algorithm 4.1, and using mainframe computer to solve LCMSA can cut down the time complexity greatly. For saving the length of paper, the details of parallel algorithm is not described and parallel algorithm is easy to draw from Algorithm 4.1.

Theorem 4.1 The loosely constrained multiple sequence alignment can be computed in Algorithm 4.1.

Proof. By Algorithm 4.1, suppose $S = \{S_1, S_2,...,S_k\}$ is aligned to $\{S_1', S_2',..., S_k'\}$ with the center sequence S_1 under the optimal position $(c_1, c_2, ..., c_\alpha)$, and each character of the constrained sequence P is aligned at the columns of $(c_1', c_2', ..., c_\alpha')$ in the result multiple sequence alignment.

For all $1 \leq i, j \leq n$

$$\sum_{1 \leq l \leq \alpha} \{u(S_i'[c_l'], P[l]) + u(S_j'[c_l'], P[l])\}$$
$$= \sum_{1 \leq l \leq \alpha} \{u(S_i[c_l^i], P[l]) + u(S_j[c_l^j], P[l])\} \quad \text{(Algorithm 4.1)}$$
$$\leq \sum_{1 \leq l \leq \alpha} \{u(S_i[c_l^i], S_1[c_l]) + u(S_1[c_l], P[l])$$
$$\quad + u(S_j[c_l^j], S_1[c_l]) + u(S_1[c_l], P[l])\} \quad \text{(triangle inequality)}$$
$$\leq \sum_{1 \leq l \leq \alpha} \{u(S_i[c_l^i], P[l]) + u(S_1[c_l], P[l]) + u(S_j[c_l^j], P[l])$$
$$\quad + u(S_1[c_l], P[l]) + u(S_1[c_l], P[l]) + u(S_1[c_l], P[l])\} \quad \text{(triangle inequality)}$$
$$\leq d/4 + d/4 + 2d/4$$
$$\quad (for\ all\ 1 \leq i \leq n, \sum_{1 \leq l \leq \alpha}\{u(S_1[c_l], P[l]) + u(S_i[c_l^i], P[l])\} \leq d/4)$$
$$= d$$

Theorem 4.2 Algorithm 4.1 runs in time $O(k^2(\alpha+1)n^{2(\alpha+1)})$.

Proof. The time for locating the center sequence S_c and its position $(c_1, c_2, ..., c_\alpha)$ is $O(k^2(\alpha+1)n^{2(\alpha+1)})$ for k sequences, with respect to P and $d/4$ according to Step 1 and 2. Then merging the $(k-1)$ loosely constrained pair-wise alignments need $O(kn)$ time. Therefore, Algorithm 4.1 runs in time $O(k^2(\alpha+1)n^{2(\alpha+1)})$ totally.

In fact, α is small compared to n and it is reasonable to regard it as a constant. So Algorithm 4.1 runs in polynomial time. Since the variation distance d is introduced, it becomes difficult to compute the performance ratio of Algorithm 4.1. However, from some experiments in Section 5, it is easy to see that the performance of Algorithm 4.1 is practical and useful.

5 Experimental Results

Based on Algorithm 4.1, a LCMSA tool is built up for some practical problems. All the experiments are conducted on a PC with Intel Celeron CPU 1.1 GHz and 256MB of memory. To guarantee the correctness of LCMSA, a metric scoring scheme is required. Now, a simple metric scoring scheme is assumed,

$$u(a,b) = \begin{cases} 0 & if\ a=b \\ 1 & otherwise \end{cases}$$

For saving the length of this paper, the detailed experiment results are not shown. However from experiments, it is easy to see that the LCMSA tool is practical and useful and biologists can adjust the variation distance d to get a biological meaningful multiple sequence alignment according to their factual requirements. Wish the LCMSA tool might direct them to find some achievements.

6 Conclusion

In this paper, the first definition of LCMSA is presented. And a complete algorithm for LCPSA is designed. Based on the LCPSA algorithm, an approximation algorithm for LCMSA is given. More, a software system is developed for LCMSA and biologists can get different results with different variation distance d. This helps them to find some useful and biologically meaningful results more easily. In addition, there are some open problems concerning this paper for the future work, such as, whether there is an approximation algorithm with guaranteed error bounds, and whether the time complexity of the LCMSA algorithm can be further improved.

References

[1] Jiang T., Xu Y., and Zhang M. Q.: Current Topics in Computational Molecular Biology. The MIT Press (2002)
[2] Gusfield D.: Efficient Methods for Multiple Sequence Alignment with Guaranteed Error Bounds. Bulletin of Mathematical Biology. 55 (1993) 141-154
[3] Tang C. Y., Lu C. L., Chang M. D-T., Tai Y-T., Sun Y-J., Chao K-M., Chang J-M., Chiou Y-H., Wu C-M., Chang H-T., and Chou W-I.: Constrained Multiple Sequence Alignment Tool Development and Its Application to RNase Family Alignment. In Proceeding of the first IEEE Computer Society Bioinformatics Conference. (2002) 127-137
[4] Francis Y.L. Chin, Ho N.L., Lam T.W., Prudence W.H. Wong, and Chan M.Y.: Efficient Constraint Multiple Sequence Alignment with Performance Guarantee. In Proceeding of the second IEEE Computer Society Bioinformatics Conference. (2003) 337-346

Application of the Hamiltonian Circuit Latin Square to the Parallel Routing Algorithm on 2-Circulant Networks

Yongeun Bae[1], Chunkyun Youn[2], and Ilyong Chung[1]

[1] Department of Computer Science, Chosun University, Kwangju, Korea
iyc@chosun.ac.kr
[2] Information Technology Division, Honam University, Kwangju, Korea
chqyoun@itc.honam.ac.kr

Abstract. Double-loop and 2-circulant networks are widely used in the design and implementation of local area networks and parallel processing architectures. In this paper, we investigate the routing of a message on circulant networks, that is a key to the performance of this network. We would like to transmit 2k packets from a source node to a destination node simultaneously along paths on G(n; $\pm s_1, \pm s_2, ..., \pm s_k$), where the ith packet will traverse along the ith path ($1 \leq i \leq 2k$). In oder for all packets to arrive at the destination node quickly and securely, the ith path must be node-disjoint from all other paths. For construction of these paths, employing the Hamiltonian Circuit Latin Square(HCLS), a special class of $(n \times n)$ matrices, we present $O(n^2)$ parallel routing algorithm on circulant networks.

1 Introduction

The intense interest in interconnection network used graph-theoretic properties for its investigations and produced various interconnection schemes. Many of these schemes have been derived to optimize important parameters such as degree, diameter, fault-tolerance, hardware cost, and the needs of particular applications. Double-loop[1] and 2-circulant networks(2-CN)[2] are widely used in the design and implementation of local area networks and parallel processing architectures. These networks are defined as follows. Let n, s_1, s_2 be positive integers such that $0 < s_1 < s_2 < n/2$. A double-loop network is a directed graph G(n; s_1, s_2), where n nodes labeled with integers modulo n, and 2 links per vertex such that each node i is adjacent to the 2 other nodes i+s_1, i+s_2. In the undirected case, which is known as a 2-circulant network and is denoted by G(n; $\pm s_1, \pm s_2$). It is well known that G(n; s_1, s_2) and G(n; $\pm s_1, \pm s_2$) are connected iff gcd(n, s_1, s_2) =1.

The routing of message is thus a key to the performance of such networks. There are routing algorithms using well-known methods, such as the Short-

---

* Corresponding Author : Ilyong Chung(iyc@chosun.ac.kr).

est Path Algorithm(the Forward Algorithm)[3], the Backward Algorithm[4], the Spanning Tree Algorithm[8]. These algorithms provide for only sequential transmission, from the source node to the desired node in a short time. We now look for algorithms that are capable of handling, multiple data items simultaneously transmitted from the staring(source) node to the destination node. There are a few algorithms on the n-dimensional hypercube network[5]-[7] that allow us to locate n disjoint paths such as the Hamiltonian path Algorithm [10], the Rotation Algorithm using Tree Structure[8], the Disjoint Path Algorithm[8], and the Routing Algorithms[9].

In this paper, we propose the algebraic approach to the routing of message on the $G(n; \pm s_1, \pm s_2)$. As described above, four packets are simultaneously transmitted from the starting(source) node to the destination node. In this case, the ith packet is sent along the ith path from the starting node to the destination node. In order for all packets to arrive at the destination node quickly and securely, the ith path must be node-disjoint from all other paths. To accomplish this, we employ the operations of nodes presented in Cayley Graph[13] and the special matrix called as Hamiltonian Circuit latin Square(HCLS)[10], which is used to find a set of node-disjoint paths on hypercube network.

2 Design of the Shortest Path

Let A and B be any two nodes on $G(n; \pm s_1, \pm s_2)$. The paper's objective is to find algorithms that will facilitate the transmission of data from node A to B in that network. In order for the data to traverse from node A to node B, it must cross, successively, intermediate nodes along a path.

Definition 1: The $G(n; \pm s_1, \pm s_2)$ is defined as follows: Let V = {0,1,2,...,n-1} as a set of nodes and E ={(v,w) | v$\pm s_i$ = w (mod n)} as a set of edges.

Researches on 2-CN are actively performed in graphic-theoretical area such as embedding and fault-tolerance. In this paper, we focus on parallel routing algorithm for $G(n; \pm s_1, \pm s_2)$. As mentioned earlier, a node on Cayley Graph can traverse to another node by performing a certain operation. We now employ these operations to CN.

Definition 2: The routing function R for $\pm s_i$ is as follows:

$$R(A) = A \pm s_i \pmod{n}, \text{ where A is node address}$$

Node A is physically connected to 2k neighboring nodes and these paths are node-disjoint. Data is transmitted from source node along the ith path, which is physically connected to. The path above is selected by the routing function described in Definition 2. To do this, the relative address of starting node and destination node can be obtained below.

Definition 3: The relative address r of nodes A and B on $G(n; \pm s_1, \pm s_2)$ is computed as the value of difference between A and B.

$$r = B-A$$

Let two addresses of node A and node B be 1 and 3. What is the relative address of two nodes? The value of the relative address is 2.

Definition 4: Let T(A,S) be the logical transmission path of data starting from node A to the destination node B, where S is a multiset and a sequence of operations, via which data can reach at the destination node. T(A,S) is determined by the order of the elements in the set S. between A and B.

Given the starting node A and a multiset S on G(16; ±2,±3), we would like to transmit to the destination node via intermediate nodes. Suppose that node A and a set S be 1 and <3,2,2,-3>, respectively. The traversal of the data along the path outlined by the sequence of nodes is $1 \rightarrow 4 \rightarrow 6 \rightarrow 8 \rightarrow 5$. The path from node A to a destination node is obtained from T(A,S) specified in Definition 4, that is (1,4,6,8,5). The path from node A to a destination node is obtained from T(A,S) specified in Definition 4, that is (1,4,6,8,5).

By applying operations obtained from the set, data can arrive at the destination node. However, since the operations are various, a number of paths are also made. Since the paper's objective is to find algorithms that will facilitate the fast transmission of data from a starting node to a destination node, those operations should be appropriate for this objective. Given S, the sequence should be reconfigured to a smallest size of sequence since routing distance is the same as the size of sequence. For example, supposed that S = <2,2,2,2,-3,-3,-3,2> on G(n; ±2,±3), S should be minimized to <2,-3,2> since <-3,-3> = <-2,-2,-2>.

3 Application of the Hamiltonian Circuit Latin Square to the Parallel Routing Algorithm on 2-Circulant Networks

The 2k packets are transmitted from a source node to a destination node on G(n; ±s_1,±s_2, ...,±s_k). In this section, we focus on G(n; ±2,±3) and construct a set of four node-disjoint and shortest paths in order to transmit these packets safely and quickly. First, four packets residing at a node on the network are sent to its four neighboring nodes along a set of disjoint paths. These paths are generated by employing four different operations at the beginning step and by performing four different operations at the last step in order to arrive at a destination node. The figure below illustrates the operations applied to generate 2k paths from a source node to a destination node.

The ith packet is transmitted along the ith path, the first intermediate node of which is obtained from applying the ith operation at a starting node and the last intermediate node transmits the packet to a destination node by applying the ith operation. In some cases, the two operations can be the same.

Definition 5: Let O^s be a set of operations occurring at a starting node when four packets are transmitted simultaneously and Let O^d be a set of operations

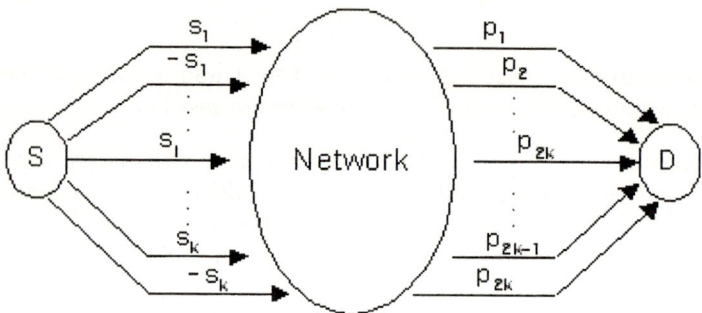

Fig. 1. The operations applied at the first step and the last step

occurring at a destination node when four packets arrive. These sets are defined as follows:

$O^s = \{s_1, -s_1, s_2, -s_2\}$
$O^d = \{p_1, p_2, p_3, p_4\}$
$O^s = O^d$

We now apply the HCLS(Hamiltonian Circuit Latin Square) to find a set of m shortest and node-disjoint paths. A latin square is a square matrix with m^2 entries of m different elements, none of the elements occurring twice within any row or column of the matrix. The integer m is called the order of latin square. The next definition describes the HCLS.

Definition 6: The HCLS M^1 is constructed as follows: Given distinct m points $a_0, a_2, \ldots, a_{m-2}, a_{m-1}$, a Hamiltonian circuit $a_i \to a_j \to \ldots \to a_k \to a_i$ is randomly selected. On the circuit each row of M can be obtained from the Hamiltonian path, starting at any position a_k ($0 \leq k \leq$ m-1), under the condition that no two rows begin at the same position. If a Hamiltonian path is $a_i \to a_j \to \ldots \to a_k$, then the row obtained from it is $[a_i, a_j, \ldots, a_k]$

From the definition of the HCLS given in Definition 6, the MHCM(Modified Hamiltonian Circuit Matrix) is constructed below.

Definition 7: Given the HCLS $M^1 = [a_{i,j}]$, the MHCM M^2 is constructed as follows: $M^2 = [A_{i,j}]$, $A_{i,j} = \{a_{i,0}, a_{i,1}, \ldots, a_{i,j-1}, a_{i,j}\}$, $0 \leq$ i,j $\leq$ m-1.

Referring to [13], the MHCM satisfies the conditions of the MGNDP(Matrix for Generating Node-Disjoint Paths), which is applied to the parallel routing on the hypercube network and a number of node-disjoint paths can be created. Since the HCLS belongs to a latin square, a set of elements in the first column is the same as that of the last column. On the G(n; $\pm s_1, \pm s_2$), an element in the HCLS is represented as an operation. Also, O^s and O^d in Definition 6 is described as a set of elements in the first column and a set of elements in the last column, respectively. We, intuitively, realize that a set of n shortest and node-disjoint paths is generated if the number of distinct sequences of operations for arriving at an arbitrary node in a short time is n(n≤m). The remaining operations excluding these distinct operations from O^s and O^d, should be performed. The process to find a set of a shortest and node-disjoint paths is described above. We now propose a parallel routing algorithm that generates a set of m minimum-distance and node-disjoint paths for the network. In this paper, we will use the term "distance" between two nodes in a interconnection network to refer to the number of routing steps(also called hopcounts) needed to send a message from one node to another.

CN_Routing_Algorithm

A ← an address of a starting node A
B ← an address of a destination node B
O^s ← a set of operations occurring at a starting node A
O^d ← a set of operations requisite for reaching to a destination node B

begin

(1) Compute the relative address R of nodes A and B; R = B-A
(2) Using the relative address R, a sequence S of operations to arrive at node B in a short time are produced
(3) In order to design a set of shortest and node-disjoint paths, find a set S_1 of distinct elements in S. A set of $|S_1|$ shortest and node-disjoint paths are generated. Each path of length is $|S|$,

    (3-1) Using the set S_1, (n×n) HCLS is constructed, where n = $|S_1|$.
    (3-2) Operations in the ith row of the HCLS are performed for traversal of the ith packet and the remaining operations in S should be executed at the point except the first and the last points.
    (3-3) $O^s \leftarrow O^s - S_1$ and $O^d \leftarrow O^d - S_1$.

(4) Construct two node-disjoint paths, each path has length $|S|+2$.

    (4-1) If $O^s = \phi$, the process is finished.
    (4-2) If a set of $\{s_i, -s_i\}$ is found in O^s , then these operations are performed at the first and the last steps of two paths newly designed, and operations in S at the middle steps of them, otherwise go to (5).
    (4-3) $O^s \leftarrow O^s - \{s_i, -s_i\}$, $O^d \leftarrow O^d - \{s_i, -s_i\}$ and go to (4-1).

(5) Generate the remaining paths.

    (5-1) If $O^s = \phi$, the process is finished.
    (5-2) Produce a sequence S_2 of minimum number of operations by reducing the size of $S_U \{-s_i, -s_i\}$, $s_i \in O^s$.
    (5-3) Operation g_i is performed at the first and the last steps at traversal and operations of S_2 are executed at the middle steps.
    (5-4) $O^s \leftarrow O^s - \{s_i\}$, $O^d \leftarrow O^d - \{s_i\}$ and go to (5-1).

end.

$CN\_Routing\_$ Algorithm is thus fairly straightforward. The time involved in performing Steps (1), (2) and (4) is small compared to the remaining steps. The first, second and fourth steps of this algorithm does not, therefore, contribute to an objectionable overhead.

4 Conclusion

In this paper, we present the algorithm that generates a set of 2k shortest and node-disjoint paths on G(n; $\pm s_1, \pm s_2, ..., \pm s_k$), employing the Hamiltonian Circuit Latin Square(HCLS). Even n and k are fixed values, the algorithm can be easily extended on arbitrary circulant networks. Important steps for determining time complexity requisite for the algorithm are two things. One is to design the HCLS, which needs O(n). The other is to execute Step (5) of $CN\_Routing\_$ Algorithm, which requires $O(n^2)$. Therefore, we can create $O(n^2)$ parallel routing algorithm for constructing 2k shortest and node-disjoint paths.

References

1. Bermond, J., Comellas, F., Hsu, D., "Distributed Loop Computer Networks: A Survey," J. Parallel and Distributed Computing, Academic Press, no. 24, pp.2-10, 1995.
2. Park, J., "Cycle Embedding of Faulty Recursive Circulants," J. of Korea Info. Sci. Soc., vol.31, no. 2, pp. 86-94, 2004.
3. Basse, S., Computer Algorithms : Introduction to Design and Analysis, Addition-Wesley, Reading, MA, 1978.
4. Stallings, W., Data and Computer Communications. Macmillan Publishing Company, New York, 1985.
5. Bae, M. and Bose, B., "Edge Disjoint Hamiltonian Cycles in k-ary n-cubes and Hypercubes," IEEE Trans. Comput., vol. 52, no. 10, pp. 1259-1270, 2003.
6. Thottethodi, M., Lebeck, A., and Mukherjee, S., "Exploiting Global Knowledge to Achieve Self-Tuned Congetion Control for k-ary n-cube Networks," IEEE Trans. Parallel and Distributed Systems, vol 15, no. 3, pp. 257-272, 2004.
7. Wu, J. and Huang, K., "The Balanced Hypercube:A Cube-Based System for Fault-Tolerant Application," IEEE Trans. Comput., vol. 46, no. 4, pp. 484-490, Apr. 1997.
8. Johnson, S.L. and Ho, C-T., "Optimum Broadcasting and Personalized Communication in Hypercube," IEEE Trans. Comput., vol. 38, no. 9, pp. 1249-1268, Seep. 1989.
9. Rabin, M.O., "Efficient Dispersal of Information for Security, Load Balancing, and Fault Tolerance," J. ACM, vol. 36, no. 2, pp. 335-348, Apr. 1989.
10. _____, "Application of the Special Latin Squares to the Parallel Routing Algorithm on Hypercube," J. of Korea Info. Sci. Soc., vol. 19, no. 5. Sep. 1992.
11. Gibbons, A., Algorithmic Graph Theory. Cambridge University Press, New York, 1985.
12. Denes, J. and Keedwell, A.D., Latin Square and Their Applications. Academic Press, New York, 1974.
13. Stone, H.S., Discrete Mathematical Structures and Their Applications. SRA, Chicago, IL., 1973.

A Distributed Locking Protocol

Jaechun No[1] and Sung Soon Park[2]

[1] Dept. of Computer Software,
College of Electronics and Information Engineering,
Sejong University, Seoul, Korea
jano@sejong.ac.kr

[2] Dept. of Computer Science & Engineering,
College of Science and Engineering,
Anyang University, Anyang, Korea
sspark@aycc.anyang.ac.kr

Abstract. Concurrent accesses to a file frequently occur in a distributed computing environment where a few number of network-attached servers are designated as a data storage pool and the clients are physically connected to the servers via network, like GigaEthernet or Fibre Channel. In such a distributed computing environment, one of the major issues affecting in achieving substantial I/O performance and scalability is to build an efficient locking protocol. We present a distributed locking protocol that enables multiple client nodes to simultaneously write their data to distinct data portions of a file, while providing the consistent view of client cached data, and conclude with an evaluation of the performance of our locking protocol on a Linux cluster.

1 Introduction

Concurrent accesses to a file frequently occur in a distributed computing environment where a few number of network-attached servers are designated as a data storage pool and the clients are physically connected to the servers via network, like GigaEthernet or Fibre Channel [1–3]. In such a computing environment, a critical issue affecting in achieving high I/O bandwidth and scalability is to build an efficient locking protocol. A locking protocol to support data consistency and cache coherency has a significant effect on generating high performance I/O. For example, many large-scale scientific applications use parallel I/O methods where multiple client nodes simultaneously perform their I/O operations. MPI-IO is among those parallel I/O methods. MPI-IO [4] is specifically designed to enable the optimizations that are critical for generating high-performance I/O. These optimizations include collective I/O and the ability to access noncontiguous data sets. However, in order to achieve high I/O performance using MPI-IO on a network-oriented distributed storage, the distributed file system which is running on top of the storage must provide the ability to lock a file per data section to have multiple concurrent writers. However, many of the locking protocols integrated with distributed file systems are based on a coarse-grained method

[1, 5, 6] where only a single client at any given time is allowed to write its data to a file. In this paper, we present a distributed locking protocol based on multiple reader/single writer semantics for a data portion to be accessed. In this scheme, a single lock is used to synchronize concurrent accesses to a data portion of a file. But, several nodes can simultaneously run on the district data sections in order to support data concurrency. We conclude our paper by discussing performance evaluation of our locking protocol on a Linux cluster.

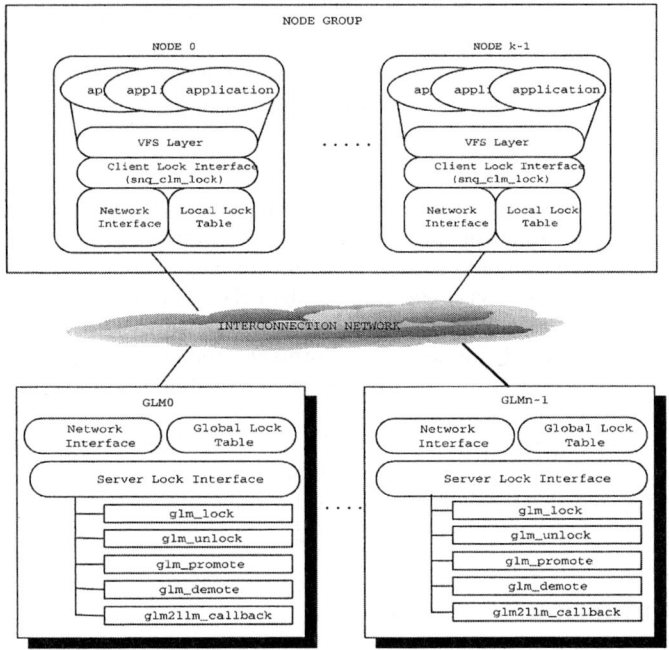

Fig. 1. A distributed lock interface

2 Implementation Details

2.1 Overview

Figure 1 illustrates the distributed lock interface that is integrated with distributed file systems. Applications issue I/O requests using local file system interface, on top of VFS layer. Before performing an I/O request, each client should acquire an appropriate distributed lock from GLM (Global Lock Manager) in order to maintain data consistency between the cached data on clients and the remote, shared data on servers. The lock request is initiated by calling the lock interface, *snq_clm_lock*.

Figure 2 represents a hierarchical overview of the locking construct with two client nodes and one GLM. The lock modes that we provide for are SHARED for multiple read processes and EXCLUSIVE for a single write process. The lock

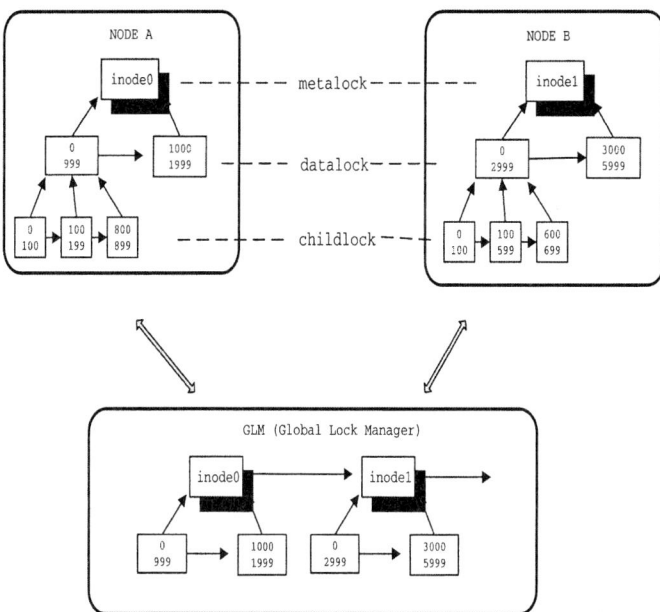

Fig. 2. A hierarchical overview of distributed locking protocol

structure consists of three levels: metalock, datalock, and childlock. The metalocks, inode0 on node A and inode1 on node B in Figure 2, synchronize accesses to files and the value of a metalock is an inode number of the corresponding file. Below the metalock is a datalock responsible for coordinating accesses to a data portion. For example, on node A, metalock inode0 is split into two datalocks associated with the data sections 0-999 and 1000-1999 in bytes and, on node B, two datalocks below inode1 are associated with the data sections 0-2999 and 3000-5999 in bytes. In order to grant a datalock, the lock mode of the higher lock (metalock) must be SHARED, meaning that a file is shared between multiple clients.

The lowest level is a childlock that is of a split datalock. Given that a datalock is granted, the datalock can be split further to maximize local lock services as long as the data section to be accessed by a requesting process does not exceed the data section of the datalock already held. In other words, in Figure 2, the datalock for the data portion 0-999 is split into three childlocks that control accesses to the data portions 0-100, 100-199, and 800-899, respectively. The childlock is locally granted and therefore the requesting process needs not communicate with GLM to obtain the childlock. However, the childlock is granted only when the lock mode of a childlock is compatible with that of the higher datalock. The datalock and childlock are found by comparing the starting file offset and data length being passed from the local file interface.

3 Performance Evaluation

We measured the performance of the distributed locking protocol on the machines that have Pentium3 866MHz CPU, 256 MB of RAM, and 100Mbps of Fast Ethernet. The operating system installed on those machines was RedHat 9.0 with Linux kernel 2.4.20-8 and four machines were configured as GLMs. On top of Linux kernel, we installed SANique cluster file system which is a Linux-based software solution in a SAN environment. However, at the time we measured the performance we didn't integrate our distributed locking protocol with SANique file system. Therefore, the performance results focused on the time to obtain locks by performing lock revoke, downgrade, and upgrade operations. The time to invalidate client cached data and to write dirty data to disk was not included in the evaluation.

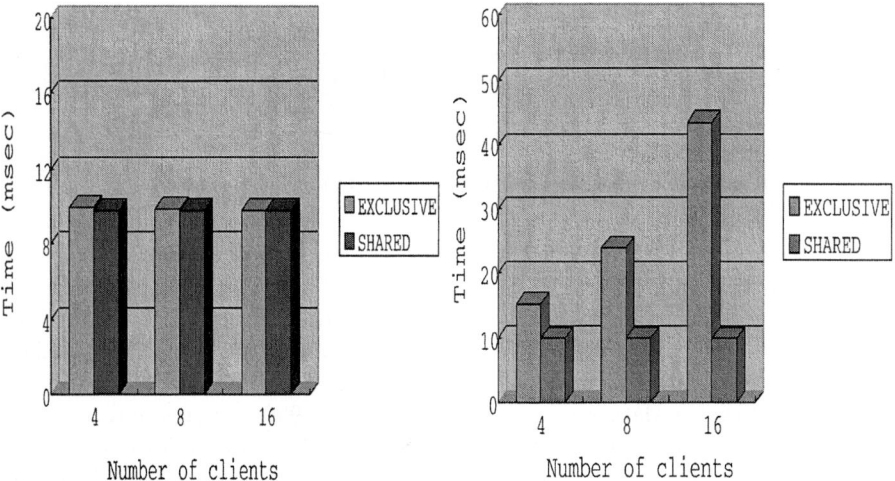

Fig. 3. Time overhead to acquire a distributed lock. Each client read or wrote 1Mbytes of data to the distinct section of the same file

Fig. 4. Time to acquire a distributed lock. A client's data section is shifted to the one given to the neighbor at the previous step

Figure 3 represents the time to obtain the locks with the exclusive mode in write operations and with the shared mode in read operations, as the number of clients increases from 4 to 16. All clients read or wrote 1Mbytes of data to the distinct portions of the same file. In this case, the lock requested by each client is newly created on GLM and returned to the requesting client, causing no callback message to be sent to the remote lock holder.

Figure 4 shows the time to obtain the locks with the exclusive mode and with the shared mode, while moving each client's data section to access to the one given to the neighbor at the previous step. For example, at the first step, the

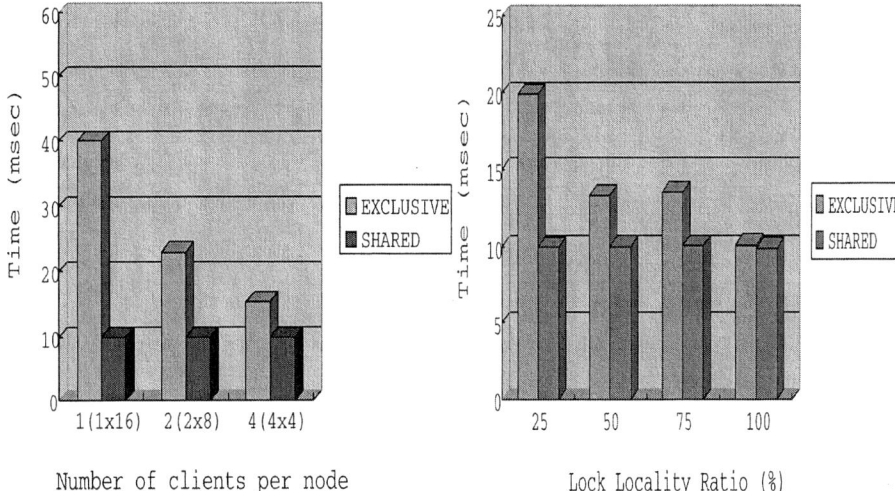

Fig. 5. Time overhead to acquire a distributed lock as a function of number of clients running on each node. A client's data access range is shifted right at each step

Fig. 6. Time to obtain a distributed lock as a function of lock locality ratio using four clients

first client accesses to the first 1Mbytes of data section of a file and the second client accesses to the second 1Mbytes of data section of the same file. At the second step, the first client's data section changes to the second 1Mbytes of data section for which the second client has already acquired the lock at the first step. Therefore, at the second step, the second client should yield the lock held to the first client, while taking a new lock from the third client.

Figure 4 illustrates that the overhead of the lock revocation is significant with the exclusive mode because only a single client is allowed to write to a data section at any given time. With the shared mode, there is no need to contact the remote lock holder since a single lock can be shared between multiple nodes. With the shared lock mode, GLM just increases a counter denoting the number of shared lock holders before granting the lock.

In order to figure out how much the network latency occurred at the lock negotiation step dominates the performance, in Figure 5, we changed the number of clients running on each node, while keeping the total number of clients as 16. Also, as did in Figure 4, we changed the data access range of each client to the one given to the neighbor at the previous step.

In Figure 5, with two clients running on the same node, the callback message is sent to the remote lock holder every two I/O operations to revoke a lock. With four clients, the callback message is sent to the remote lock holder every four I/O operations, resulting in the lock negotiation overhead decrement, compared to two clients on each node. According to this experiment, we could see that the dominating performance factor with 16 clients is the network overhead to contact the remote lock holder.

Figure 6 shows the effect of childlocks exploiting locality in the lock requests. The lock locality ratio means how often childlocks are taken. Figure 6 shows that, with the exclusive lock mode, the more childlocks are generated, the smaller time is taken to serve a lock request due to the drop in time to negotiate with GLM and remote lock holders. With the shared lock mode, however, the time to take a lock flattens out at about 9 msec because the remote shared lock holders needs not give up the lock requested, allowing to have multiple lock holders with the shared mode.

4 Conclusion

Concurrent accesses to the same file frequently occur in many scientific applications where allowing parallel write operations significantly improves I/O bandwidth. In this paper, we presented a distributed locking protocol with which several nodes can simultaneously write to the distinct data portions of a file, while guaranteeing a consistent view of client cached data. The distributed locking protocol has also been designed to exploit locality of lock requests to minimize communication overhead with GLM and remote lock holders. As a future work, we plan to integrate the locking scheme with a SAN-based cluster file system, called SANique, developed by MacroImpact company.

References

1. Preslan, K.W., Barry, A.P., Brassow, J.E., Erickson, G.M., Nygaard, E., Sabol, C.J., Soltis, S.R., Teigland, D.C., and O'Keefe, M.T.: A 64-bit Shared Disk File System for Linux. In Proceedings of Sixteenth IEEE Mass Storage Systems Symposium Seventh NASA Goddard Conference on Mass Storage Systems & Technologies, March 15-18, 1999
2. Schmuck, F., Haskin, R.: GPFS: A Shared-Disk File System for Large Computing Clusters. In Proceedings of the First Conference on File and Storage Technologies(FAST), pages 231–244, Jan. 2002
3. Braam, P.J.: The Lustre stroage architecture. Technical Report available at - http://www.lustre.org, Lustre, 2002
4. Gropp, W., Lusk, E., and Thakur, R.: Using MPI-2: A dvanced Features of the Message-Passing Interface, MIT Press, 1999, Cambridge, MA
5. Devarakonda, M., Kish, B., and Mohindra, A.: Recovery in the Calypso file system. ACM Transactions on Computer Systems, 14(3):287–310, August 1996
6. Thekkath, C.A., Mann, T., and Lee, E.K.: Frangipani: A Scalable Distributed File System. In Proceedings of the Symposium on Operating Systems Principles, 1997, pages 224–237

A Study on the Efficient Parallel Block Lanczos Method

Sun Kyung Kim and Tae Hee Kim

School of Computer and Information Technology,
Daegu University, Gyeongsan 712-714, Korea
skkim@daegu.ac.kr

Abstract. In order to use parallel computers in specific applications, algorithms need to be developed and mapped onto parallel computer architectures. Main memory access for shared memory system or global communication in message passing system deteriorate the computation speed. In this paper, it is found that the m-step generalization of the block Lanczos method enhances parallel properties by forming m simultaneous search direction vector blocks. QR factorization, which lowers the speed on parallel computers, is not necessary in the m-step block Lanczos method. The m-step method has the minimized synchronization points, which resulted in the minimized global communications and main memory accesses compared to the standard method.

1 Introduction

Memory contention on shared memory machines constitutes a severe bottleneck for achieving their maximum performances.[11] The same is true for communication costs on a message passing system.[10] It would be desirable to have methods for specific problems, which have low communication costs compared to the computation costs. This is interpreted as a small number of main memory accesses for the shared memory systems and a small number of global communications for the message passing systems. It also reduces the need for frequent synchronizations of the processors. Linear algebra algorithms, which are implemented efficiently on parallel computers, have also been studied.[3, 6, 8] Many important scientific and engineering problems require the computation of a small number of eigenvalues of symmetric large sparse matrices. One of the commonly used algorithm for solving a multiple eigenvalue problem is the block Lanczos algorithm. QR factorization process in the block Lanczos method is bottleneck when this method is implemented on parallel computers because QR factorization process requires many synchronization points.[1, 9] In this paper we introduce the m-step block Lanczos method that needs no QR factorization process. In the m-step method, m consecutive steps of the standard method are performed simultaneously. This means, for example, that the inner products needed during m steps of the standard method can be performed simultaneously.

2 The Parallel Block Lanczos Method

2.1 The Block Lanczos Method

The block Lanczos algorithm for computing extreme multiple eigenvalues of symmetric matrices is based on the block Lanczos recursion for the block tridiagonalization of a real symmetric matrix.[2, 4, 5] The block Lanczos is as follows:

Algorithm 2.1 The block Lanczos algorithm

$X_1 \in \mathbb{R}^{n \times p}$ given, with $X_1^T X_1 = I_p$.
$M_1 = X_1^T A X_1$
For $j=1$ **until** Convergence **Do**
$\quad C_j = AX_j - X_j M_j - X_{j-1} B_{j-1}^T (X_0 B_0^T = 0)$
$\quad X_{j+1} B_j = C_j$ (QR factorization process)
$\quad M_{j+1} = X_{j+1}^T A X_{j+1}$
EndFor

At the beginning of the j-th pass through the loop we have

$$A[X_1, ..., X_j] = [X_1, ..., X_j] T_j + C_j[0, ..., 0, I_j],$$

where $T_j = \begin{bmatrix} M_1 & B_1^T & & & \\ B_1 & M_2 & & & \\ & & \ddots & & \\ & & & \ddots & \\ & & & B_{j-1} & M_j \end{bmatrix}$

The eigenvalues of the block tridiagonal matrix T_j are called Rits values of large sparse matrix A. For many matrices and for relatively small j, several of the extreme multiple eigenvalues of A, that is several of the algebraically-largest or algebraically-smallest eigenvalues of A, are well approximated by eigenvalues of the corresponding matrices T_j. The steps to implement the algorithm 2.1 on parallel computer is as follows:

Select Q_1 of size $n \times p$ where p is at least as large as the number of eigenvalues desired and the columns of Q_1 are orthonormal.

For $j = 1$ **until** Convergence **Do**
  1. compute AQ_j
  2. compute (AQ_j, Q_j)
  3. $M_j = (AQ_j, Q_j)$: $p \times p$ matrix
  4. $C_j = AQ_j - Q_j M_j - Q_{j-1} R_{j-1}^T (Q_0 R_0^T = 0)$
  5. Apply modified Gram-Schmidt orthogonalization to the columns of C_j
     (that is, $Q_{j+1} R_j = C_j$:QR factorization)
EndFor

Next, compute the p algebraically-largest eigenvalues of T_j. The Ritz vector $Q_j y(= Z)$ obtained from an eigenvector y of a given T_j is an approximation to a corresponding eigenvector of A.

The above block Lanczos procedure with no reorthogonalization has minimal storage requirements and can therefore be used on very large matrix A, if only approximations to eigenvalues are required. However, the inner products cannot be performed simultaneously because of steps 2, 5 and inner products require global communications among all processors on message passing systems. These force several accesses of vectors Q_j, C_j, AQ_j per one iteration of above procedure. For shared memory systems, main memory accessing may be slow. So in the section 2.2, the m-step block Lanczos method is proposed to perform simultaneously all the inner products needed for m iterations of above procedure. For QR factorization, $p(p+1)/2$ inner products should be separately repeated $2p-1$ times and so communication time rate is high for this QR factorization process. In the next section we introduce m-step block Lanczos method which has much less synchronization points and needs no QR factorization process.

2.2 m-Step Block Lanczos Method

One way that can lead to build a new parallel block Lanczos algorithm is to perform m steps of the standard algorithm simultaneously in parallel using a set of $m \times p$ linearly independent vectors. In this case, the set of $m \times p$ linearly independent vectors $\overline{V}_k$ is spanned by $[V_k^1, AV_k^1, ..., A^{m-1}V_k^1, V_k^1 \in \mathbb{R}^{n \times p}]$.

Remark 1. If $\overline{V}_i$ and $\overline{V}_j$ are orthogonal for $i \neq j$, $V_k = [\overline{V}_i, \overline{V}_2, ..., \overline{V}_k]$ can be decomposed into $Q_k \times R_k$, where $Q_k = [\overline{Q}_1, \overline{Q}_2, ..., \overline{Q}_k]$ and $R_k = \text{diag}[\overline{R}_1, \overline{R}_2, ..., \overline{R}_k]$.

Remark 2. If T_j is a symmetric tridiagonal matrix generated by the standard block Lanczos algorithm and $\overline{T}_k = R_k^{-1} T_j R_k$ where $j = m \times k$, $\overline{T}_k$ becomes a non-symmetric matrix similar to T_j as follows:

$$\overline{T}_k = \begin{bmatrix} G_1 & E_1 & & & \\ F_1 & G_2 & E_2 & & \\ & & \ddots & & \\ & & & \ddots & E_{k-1} \\ & & & F_{k-1} & G_k \end{bmatrix}$$

where $G_i = [G_i^1, ..., G_i^m], G_i^j \in \mathbb{R}^{mp \times p}$ and $E_i = [E_i^1, ..., E_i^m], E_i^j \in \mathbb{R}^{mp \times p}$ for $i = 1, ..., k$, and $j = 1, ..., m$. Here F_i is a specially shaped matrix of which the block in the upper right corner is a triangular matrix with zero elements otherwise. This upper triangular matrix arises from the QR factorization of V_i^m, together with orthonormal vectors.

The following is an example of F_i for $p = 3$ and $m = 2$:

$$F_i = \begin{bmatrix} 0 & 0 & 0 & * & * & * \\ 0 & 0 & 0 & 0 & * & * \\ 0 & 0 & 0 & 0 & 0 & * \\ 0 & 0 & 0 & 0 & 0 & 0 \\ 0 & 0 & 0 & 0 & 0 & 0 \\ 0 & 0 & 0 & 0 & 0 & 0 \end{bmatrix}$$

Since $\overline{T}_k$ is similar to T_j with $j = k \times m$, they have the same eigenvalues. Note that V_i^j is not orthogonal, but it does not induce any problem in constructing the algorithm. The reason is that, as given in Remark 1, $\overline{V}_i$ and $\overline{V}_j$ are orthogonal for $i \neq j$ and $\overline{V}_i = [V_i^1, V_i^2, ..., V_i^m]$ is a set of linearly independent vectors that span $[V_i^1, AV_i^1, ..., A^{m-1}V_i^1, V_i^1 \in \mathbb{R}^{n \times p}]$. Based on the above analysis, a new parallel block Lanczos algorithm can be constructed as follows:

Algorithm 2.2 The m-step block Lanczos algorithm

$\overline{V}_0 = 0, \overline{V}_1 = [V_1^1, AV_1^1, A^2V_1^1..., A^{m-1}V_1^1]$

For $k = 1$ **until** Convergence **Do**
  Select G_k, E_{k-1} so that $\overline{V}_k$ is orthogonal to $\overline{V}_{k-1}$.
  This gives $V_{k+1}^1 = AV_k^m - \overline{V}_{k-1}E_{k-1}^m - \overline{V}_k G_k^m$
  Select C_k^j so that $\overline{V}_{k-1}$ is orthogonal to $[V_k^1, AV_k^1, ..., A^{m-1}V_k^1]$
  which gives $V_{k+1}^j = A^{j-1}V_{k+1}^1 - \overline{V}_k C_k^j$ for $j = 2, ..., m$.
EndFor

We will present the reduction matrix generated by m-step block Lanczos method for the special case of $m = 2$ and $p = 2$:

$$\overline{T}_k = \begin{bmatrix} * & * & * & * & | & * & * & * & * & | & & \\ * & * & * & * & | & * & * & * & * & | & & \\ * & * & * & * & | & * & * & * & * & | & & \\ * & * & * & * & | & * & * & * & * & | & & \\ - & - & - & - & | & - & - & - & - & | & - & - \\ & 1 & 0 & & | & * & * & * & * & | & \cdot & \\ & & 1 & & | & * & * & * & * & | & \cdot & \\ & & & & | & & * & * & * & | & \cdot & \\ & & & & | & & * & * & * & | & \cdot & \\ & & & & & \cdot & \cdot & \cdot & \cdot & \cdot & \cdot & \end{bmatrix}$$

Next, we demonstrate how to determine the parameters G_k, E_{k-1}, C_k in the above process:

Scalar 1: $\overline{V}_k^T A \overline{V}_k = \overline{V}_k^T \overline{V}_k G_k,$
$\overline{V}_{k-1}^T A \overline{V}_k = \overline{V}_{k-1}^T \overline{V}_{k-1} E_{k-1}$
Scalar 2: $0 = \overline{V}_k^T A^{j-1} V_k^1 - \overline{V}_k^T \overline{V}_k C_k^j$

The inner products $(\overline{V}_k, \overline{V}_k), (\overline{V}_k, A\overline{V}_k), (\overline{V}_{k-1}, A\overline{V}_k), (\overline{V}_k, A^{j-1}V_k^1)$ can be reduced to $(V_k^1, V_k^1), (V_k^1, AV_k^1), ..., (V_k^1, A^{2m-1}V_k^1)$ in a similar way to the s-step Lanczos method.[7] We now formulate the m-step block Lanczos algorithm.

Select V_1^1
For $k = 1$ until Convergence **Do**
   1. Compute $AV_k^1, A^2V_k^1, ..., A^mV_k^1$
   2. Compute $2p^2m$ inner products
   3. Compute Scalars
        C_{k-1}^j for $j = 2, ..., m$
        E_{k-1}^j, G_k^i for $i = 1, ..., m$
   4. Compute
        $V_k^j = A^{j-1}V_k^1 - \overline{V}_{k-1}C_{k-1}^j$
        (with $\overline{V}_0 = 0, C_0^j = 0$)
   5. Compute AV_k^m
   6. Compute
        $V_{k+1}^1 = AV_k^m - \overline{V}_{k-1}E_{k-1}^m - \overline{V}_k G_k^m$
        (with $E_0^m = 0$)

In the above algorithm all the inner products are performed at once in the step 2. Matrix-vector operations are performed in the step 1 and 5, and the other steps are for vector update operation.

3 Analysis of m-Step Iterative Method

During the procedure of the standard block Lanczos algorithm in the Section 2.1, $p^2 + p(p+1)/2$ inner products should be separately repeated at $2p$ times. Thus on message passing system, lots of times are needed for global communications, which in turn reduces the efficiency of the parallel system. However, the new parallel algorithm introduced in Section 2.2 can perform all the needed inner products at once during m iterations of the corresponding standard method, the

Table 1. Comparison of the numbers of the vector operations and the data communications for block Lanczos method

	standard algorithm	m-step algorithm
inner product	$[p^2 + p(p+1)/2]/m$	$2p^2m$
vector update	$(2p+1)pm$	$pm(m+1)$
matrix-vector multiplication	pm	$p(m+1)$
global communication	$2pm$	1
local communication	m	2

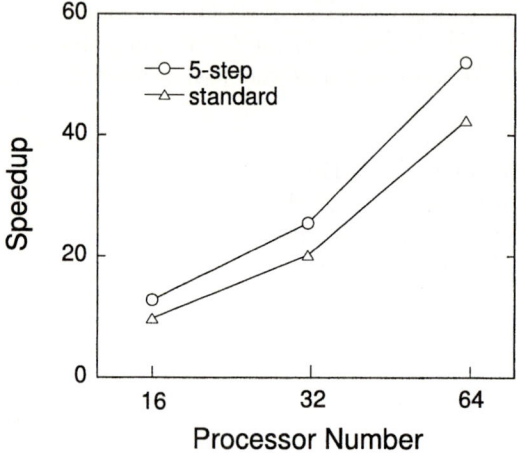

Fig. 1. Performance of the standard and the m-step block Lanczos algorithm in Cray T3E

time necessary for inner products is reduced by a factor of $1/(2pm)$ compared to the standard method. The main cause for the increased memory latency time for the shared memory system is too many synchronizing point. In a new parallel algorithm 2.2 such a bottleneck phenomena can be reduced. Table 1 shows the number of vector operations and the data communications during a single iteration of the m-step block Lanczos algorithm and m iterations of the corresponding standard method. As shown in Table 1, the communication cost can be greatly decreased by introducing more effective algorithm in parallel process.

Figure 1 shows the efficiency of the 5-step block Lanczos algorithm implemented on MP parallel computer Cray T3E. All the inner products needed for m iterations of the standard method are performed at the same time in the m-step method, so that only one global communication is required and the communication cost is decreased on a message passing system like Cray T3E. But the m-step block Lanczos method needs $p(m+1)$ times of matrix-vector operations compared to pm times in the standard methods. But in the case of vector updates, the number is slightly smaller in the m-step method. The m-step algorithm is also effective in the matrix-vector operations since the local communication time between the adjacent processors of Cray T3E can be reduced. While the efficiency of parallelism increases with increasing m and p, but a loss of accuracy for eigenvalues has been observed for large $m > 5$.

4 Conclusions

Parallel processing systems equipped with from a few to many thousand processors are now being used in many areas. As the number of the processors involved in the parallel system is increased, the relative importance of the communication cost grows. In this paper, we proposed a new m-step iterative method suitable to reduce the communication cost. The m-step block Lanczos algorithm utilizes a reduced matrix similar to that in the standard block Lanczos method with the same eigenvalues, but m-step method is more effective in the parallel system because a large amount of the inner products can be done at once. This process can reduce the data communication time in a message passing system. The m-step methods also help to reduce the memory latency time in shared memory systems by reducing the synchronization point showing a better data locality. The new m-step method has the better performance compared to the standard method in MP parallel computer Cray T3E.

Acknowledgment

This work is supported by the research grant of Daegu University, 2004.

References

1. Bendtsen, C., Hansen, P., Madsen, K., Nielsen, H., Pinar, M.: Implementation of QR up- and downdating on a massively parallel computer. Parallel Computing **21** (1995) 49–61
2. Cullum, J., and Willoughby, R.: Lanczos Algorithms for Large Symmetric Eigenvalues Computation, Birkhauser Boston, Inc. (1985)
3. Dave, A. and Duff, I.: Sparse Matrix Calculations on the CRAY-2. Parallel Computing **5** (1987) 55–64
4. Demmel, J.: Applied Numerical Linear Algebra. the Society for Industrial and Applied Mathematics press. (1997)
5. Golub, G. and Van Loan C.: MATRIX Computations. Johns Hopkins University Press. (1996)
6. Gutheil, I. and Krotz-Vogel, W.: Performance of a Parallel Matrix Multiplication Routine on Intel iPSC/860. Parallel Computing **20** (1994) 953–974
7. Kim, S. and Chronopoulos, A.: A Class of Lanczos Algorithms Implemented on Parallel Computers. Parallel Computing **17** (1991) 763–778
8. Mathur, K. and Johnsson, S.: Multiplication of Matrices of Arbitrary Shape on a Data Parallel Computers. Parallel Computing **20** (1994) 919–951
9. Matstoms, P.: Parallel sparse QR factorization on shared memory architectures. Parallel Computing **21** (1995) 473–486
10. Ranka, S., Won, Y., and Sahni, S.: Programming the NCUBE Hypercube. Tech. Rep. Csci No bf 88-13 Univ. of Minnesota, (1988)
11. Saylor, P.: Leapfrog Variants of Iterative Methods for Linear Algebraic Equations. Journal of Computational and Applied Mathematics bf 24 (1988) 169–193

Performance Evaluation of Numerical Integration Methods in the Physics Engine

Jong-Hwa Choi[1], Dongkyoo Shin[1], Won Heo[2], and Dongil Shin[1]*

[1] Department of Computer Science and Engineering, Sejong University,
98 Kunja-Dong Kwangjin-Gu, Seoul, Korea
com97@gce.sejong.ac.kr, {dshin, shindk}@sejong.ac.kr
[2] eSum Technologies Inc., Sungbo Yeoksam Bldg, 833-2, Yeoksam-Dong,
Kangnam-Gu, Seoul, Korea
heowon@esumtech.com

Abstract. A physics engine in computer games takes charge of the calculations simulating the physical world. In this paper, we evaluate the performance of three numerical integral methods: Euler method, Improved Euler method, and Runge-Kutta method. We utilized a car moving game for the simulation experiments logging fps (frame per second). Each numerical integral was evaluated under two different settings, one with collision detection and the other without it. The simulation environment without collision detection was divided into two sections, a uniform velocity section and a variable velocity section. The Euler method was shown to have the best fps in the simulation environment with collision detection. Simulation with collision detection shows similar fps for all three methods and the Runge-Kutta method showed the greatest accuracy. Since we tested with rigid bodies only, we are currently studying efficient numerical integral methods for soft body objects.

1 Introduction

Nowadays texture-mapped 3D games are the norm and abundant CPU powers can be utilized for realistic and smart games. Physics simulation is a key technology that makes a game world solid and real. A physics engine is a library of codes that takes charge of the calculations for physical phenomenon occurring between game objects [1]. The physics engine needs very many mathematical calculations and numerical integration is a major part of it. There has been much research on the implementation of physics engines, but not on the performance evaluations of numerical integral modules for physics engines.

In this paper, we compare the performance of three numerical integral methods and suggest which one of them is most suitable for computer games. Section 2 presents previous studies related to implementing physics engines. Section 3 explains the structure of the physics engine and the numerical integral method used for the experiment. Section 4 presents the experimental data with comments. Section 5 presents our conclusions.

---

* Correspondence author.

2 Related Works

The study of the performance level of the physics engine has contributed much to our understanding of the efficient structure of the engine itself. The study that led to a physical model for the solution of physical problems in the real world led also to an algorithm of engine structure [1]. The interaction of multiple parts in the working of the physiotherapy engine developed from the study of collision detection; one such study led to *voxel's* efficient structure to improve the speed of collision detection [2]. Research that compares the performance of collision detection produced an algorithm described as *boxtree* that is defined by the structure of the object [3].

ODE (Open Dynamics Engine) is a .well-known open source based physics engine [4]. Also Math Engine [5], Havok [6] and Meqon [7] are widely used commercial physics engines.

3 The Physics Engine and Numerical Integration Methods

3.1 The Structure of the Physics Engine

Fig. 1 shows the structure of an ODE physics engine that is used in this paper [4]. The left side of Fig. 1 shows the structure of simulation contents that are created by the ODE physics engine.

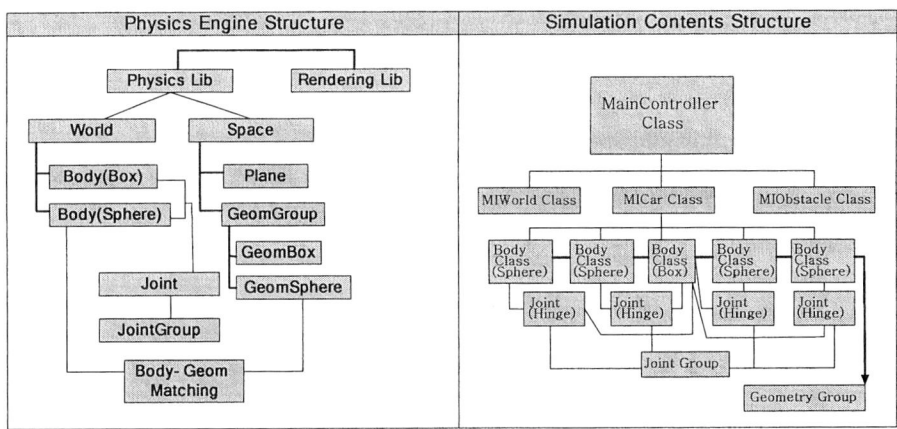

Fig. 1. Structures of the ODE engine and the simulator

The main functions of the ODE physics engine shown are a collision detection module and an object joint module. The physics engine calculates the joint value of the inside object and the collision detection, and the changed position of the object and rotation value. Newton's equation of motion is applied in this case.

$$d^2 X(t)/dt^2 = dV(t)/dt = A(t) = F(t)/m \qquad (1)$$

Fig. 2. Screen shot of the simulator

Numerical integral method is used to determine the object's position by calculating X(t) which express the position as a function of time in equation (1). Fig. 2 is the car's simulation content that is used in an experiment. Simulation content consists of car and obstacle that only are represented by rigid bodies.

3.2 Numerical Integration Methods

In this paper, we used a car simulation as simulation components for an experiment, and tested the Euler method [8], Improved Euler method, and the Runge-Kutta method [9] using numerical integration for a comparison experiment. The ODE physics engine used in one experiment employed the Euler method as numerical integral method. We used an Improved Euler and Runge-Kutta methods in ODE's inside module for another experiment.

3.2.1 Euler Method and Improved Euler Method

The Euler method approximates y(x) ignoring more than the first differential coefficient in the process by Taylor series [10] when differential equation and initial condition were given in the Eq. (2).

$$\frac{dy}{dx} = f(x, y), y(x_0) = y_0$$

$$\frac{dy}{dx} \cong \frac{y(x_0 + \Delta x) - y(x_0)}{\Delta x} \cong f(x_0, y_0)$$

$$f(x_i, y_i) \cong \frac{y(x_{i+1}) - y(x_i)}{\Delta x} \tag{2}$$

$$y(x_{i+1}) \cong y(x_i) + hf(x_i, y_i)$$

$$h = \Delta x, x_i = x_0 + ih, i = 1,2,3 \cdots, n$$

The improved Euler method approximates to the second differential coefficient. The following way shows the process of calculating to the second differential coefficient in Eq. (3).

$$y_{i+1} = y_i + (1-b)k_1 + bk_2, (b = \frac{1}{2} \text{ or } 1)$$
$$k_1 = hf(x_i, y_i) \quad (3)$$
$$k_2 = hf(x_i + \frac{h}{2b}, y_i + \frac{k_1}{2b})$$

3.2.2 Runge-Kutta Method

The Runge-Kutta method approximates to the fourth differential coefficient in the Taylor series. The following method shows the process that calculates to the fourth differential coefficient in Eq. (4).

$$y_{i+1} = y_i + \frac{1}{6}(k_1 + 2k_2 + 2k_3 + k_4)$$
$$k_1 = hf(x_i, y_i)$$
$$k_2 = hf(x_i + \frac{h}{2}, y_i + \frac{k_1}{2}) \quad (4)$$
$$k_3 = hf(x_i + \frac{h}{2}, y_i + \frac{k_2}{2})$$
$$k_4 = hf(x_i + h, y_i + k_3)$$

Runge-Kutta methods achieve the accuracy of a Taylor series approach without requiring the calculation of higher derivatives.

4 Experiments and Evaluations

An experimental environment is constructed using a Pentium-4 1.7G CPU, 512 MB RAM and GeForce4 Ti 4200 VGA card. Each numerical integral was evaluated under two different settings, one with collision detection and the other without it. Simulation environment without collision detection was divided into a uniform velocity section and a variable velocity section.

4.1 Simulation with Collision Detection

Simulation with collision detection shows irregular results according to collision frequency with obstacles. Euler method shows the best fps as shown in Table 1 and Fig. 3.

Table 1. Measurement value of Simulation Environment with Collision Detection

Sec • Method •	1	2	3	4	5	6	7	8	9	10	11	12	13	14	15
Euler	30.2	30.8	30.9	31.2	31.2	31	31	30.9	30.8	30.6	30.5	30.9	31.2	31.3	31.3
Improved-Euler	30.3	30.5	30.8	30.8	30.9	3.6	31	30.8	30.8	30.9	31.1	30.9	31	30.2	29.9
Runge-kutta	30.2	30.5	30.7	30.8	30.9	30.6	30.9	30.7	30.7	31	31	30.4	30.7	30.5	29.8

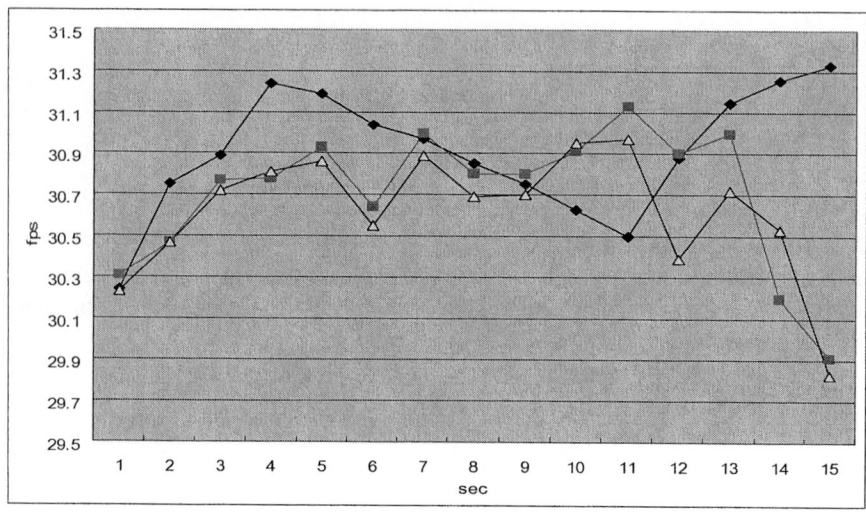

Fig. 3. Graph of Simulation Environment with Collision Detection

4.2 Simulation Without Collision Detection

Simulation without collision detection was divided into uniform velocity section and variable velocity section. Because Objects in simulation without collision detection do not collide with other objects, as computational complexity is less, fps (frame per second) shows high results. Table 2 and Fig. 4 show the sequence.

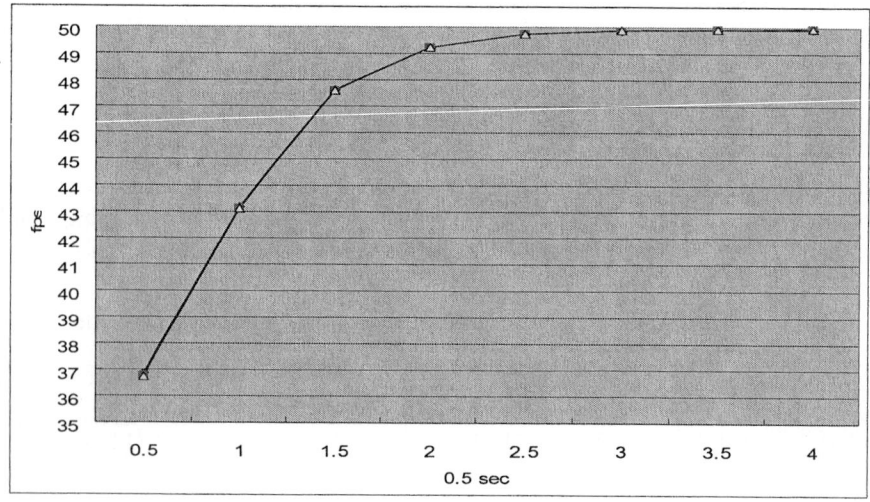

Fig. 4. FPS Graph in Variable Velocity Section

Table 2. FPS Measurement Value in Variable Velocity Section

method \ sec	0.5	1	1.5	2	2.5	3	3.5	4
Euler	36.85	43.16	47.62	49.24	49.74	49.87	49.91	49.92
Improved-Euler	36.79	43.13	47.6	49.24	49.73	49.87	49.91	49.93
Runge-Kutta	36.73	43.11	47.59	49.24	49.73	49.88	49.91	49.93

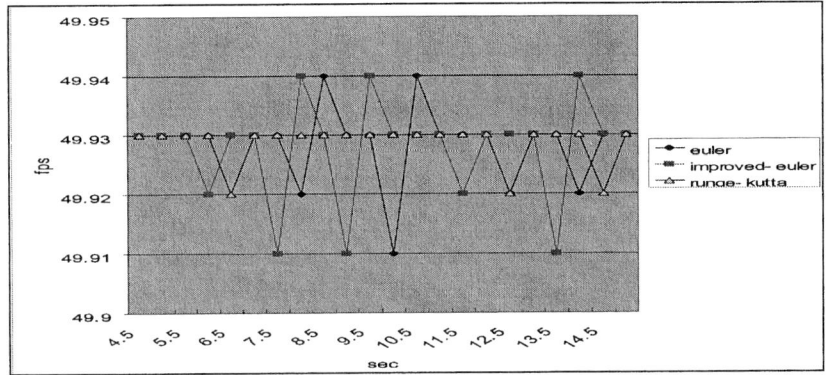

Fig. 5. FPS Graph in Uniform Velocity Section

Fig. 5 shows the result of fps value in uniform velocity section. Interval per calibration of graph is 0.01.

The results of the experiment show that where collision detection was not present, all methods produced similar fps, but the Runge-Kutta method showed the best performance in terms of accuracy.

5 Conclusions

In this paper, we compared the performance of three numerical integral methods that are used in the physics engine, and demonstrated an effective numerical integral method for computer games. We used car simulation for the experiment. Euler method, Improved Euler method, and Runge-Kutta method were evaluated. Each numerical integral was evaluated under two different settings, one with collision detection and the other without it. The simulation environment without collision detection was divided into a uniform velocity section and a variable velocity section.

The Euler method was shown to produce the best result in terms of fps in a simulation environment with collision detection. Simulation with collision detection shows similar fps for all three methods and the Runge-Kutta method shows the greatest accuracy. Since we tested with rigid bodies only, we are currently studying efficient numerical integral methods for soft body objects.

References

[1] Kook, H.J, Novak, G. S., Jr.: Representation of models for solving real world physics problems. Proceedings of the Sixth Conference on Artificial Intelligence for Applications, (1990) 274-280
[2] Lawlor, O.S., Kalee, L.V.: A Voxel-based Parallel Collision Detection Algorithm. Proceedings of the 6th international conference on Supercomputing, June (2002) 285-293
[3] Zachmann, G.: Minimal Hierarchical Collision Detection. Proceedings of the ACM symposium on Virtual reality software and technology, November (2002)
[4] Open Dynamics Engine, http://ode.org
[5] Math Engine, http://www.mathengine.com
[6] Havok, http://havok.com
[7] [Meqon, http://www.meqon.com
[8] Jameson, A, Baker, T.J: Improvements to the Aircraft Euler Method. AIAA, January (1987)
[9] Munthe-Kaas, H.: High order Runge-Kutta methods on manifolds. Journal of Applied Number Math, (1999) 115-127
[10] Moller, T., Machiraju, R., Mueller, K., Yagel, R.: Evaluation and design of filters using a Taylor series expansion: IEEE transactions on Visualization and Computer Graphics, Vol.3, No.2, (1997) 184-199

A Design and Analysis of Circulant Preconditioners

Ran Baik[1] and Sung Wook Baik[2]

[1] Dept. of Computer Engineering, Honam University, Gwangju 506-090, Korea
[2] College of Electronics and Information Engineering, Sejong University, Seoul 143-747, Korea
baik@honam.ac.kr, sbaik@sejong.ac.kr

Abstract. We propose a new type of preconditioners for symmetric Toeplitz system $Tx = b$. When applying iterative methods to solve linear system with matrix T, we often use some preconditioner C by the preconditioned conjugate gradient (PCG) method[3]. If T is a symmetric positive definite Toeplitz matrix, two kinds of preconditioners are investigated: the "optimal" one , which minimizes $\|C - T\|_F$, and the "superoptimal" one , which minimize $\|I - C^{-1}T\|_F$[8]. In this paper, we present a general approach to the design of Toeplitz preconditioners based on the optimal investigating and also preconditioners C with preserving the characteristic of the given matrix T. Fast all resulting preconditioners can be inverted via fast transform algorithms with $O(NlogN)$ operations. For a wide class of problems, PCG method converges in a finite number of iterations independent of N so that the computational complexity for solving these Toeplitz systems is $O(NlogN)$[2].

1 Introduction

Toeplitz systems arise in a variety of practical applications in mathematics and engineering. For instance, in signal processing , solutions of Toeplitz systems are required in order to obtain the filter coefficient in the design of recursive digital filters, time-series analysis also involves solutions of Toeplitz systems for the unknown parameters of stationary auto gressive models. There are a number of specialized fast direct methods for solving Toeplitz systems [2]. For an n by n Toeplitz system $Tx = b$, the algorithm requires $O(n^2)$ operations to solve it. The matrix T has the Toeplitz property: Down each diagonal its entries are constant. The i,j entry is t_{i-j}, and we assume symmetry and positive definite. Toeplitz matrices also arise directly from constant coefficient partial differential equations, and from integral equations with a convolution kernel, when those equations are made discrete. With periodicity, these problems can be solved quickly by a Fourier transform. The convolution becomes a multiplication and deconvolution is straight-forward. In the nonperiodic case, which is analogous to a problem on a finite interval (or on a bounded region in the multidimensional case), this direct solution is lost. The inverse of a Toeplitz matrix is not Toeplitz, because of the presence of a boundary and the absence of periodicity. Algorithms that exploit the Toeplitz property are much faster than the operations of symmetric elimination , and direct methods based on the Levinson recursion formula are in constant use. A number of superfast methods have been created in the last ten years, and an implementation by Ammar and Gragg is given excellent results. The iterative method uses a preconditioner. The Toeplitz matrix is replaced by a circulant matrix. It retains the Toeplitz property and adds periodicity. Thus, we denote M_n, T_n by the set of n by n matrices and Toeplitz matrices, respectively:

$$T_n \equiv \left\{ \begin{bmatrix} t_0 & t_1 & \cdots & t_{n-1} \\ t_{-1} & \ddots & \ddots & \vdots \\ \vdots & \ddots & \ddots & \vdots \\ t_{-(n-1)} & \cdots & t_{-1} & t_0 \end{bmatrix} \in M_n \right\} \subseteq M_n.$$

We denote by T_n^R the set of all real symmetric Toeplitz matrices:

$$T_n^R \equiv \left\{ \begin{bmatrix} t_0 & t_1 & \cdots & t_{n-1} \\ t_1 & \ddots & \ddots & \vdots \\ \vdots & \ddots & \ddots & \vdots \\ t_{n-1} & \cdots & t_1 & t_0 \end{bmatrix} \right\} \subseteq T_n(\mathbf{R}) \tag{1}$$

and by T_n^H the set of all hermitian Toeplitz matrices :

$$T_n^H \equiv \left\{ \begin{bmatrix} t_0 & t_1 & \cdots & t_{n-1} \\ \bar{t}_1 & \ddots & \ddots & \vdots \\ \vdots & \ddots & \ddots & \vdots \\ \bar{t}_{n-1} & \cdots & \bar{t}_1 & t_0 \end{bmatrix} \right\} \subseteq T_n(\mathbf{C}). \tag{2}$$

2 Analysis of Circulant Preconditioners

A Toeplitz matrix $C = [c_{ij}] \in T_n$ is called a circulant if $c_{ij} = c_{(n+j-i) \bmod n}$, $0 \leq i, j \leq n-1$, i.e.,
$$C = \begin{bmatrix} c_0 & c_1 & \cdots & c_{n-1} \\ c_{n-1} & c_0 & \ddots & \vdots \\ \vdots & \ddots & \ddots & c_1 \\ c_1 & \cdots & c_{n-1} & c_0 \end{bmatrix} \in T_n.$$

From the structure of the circulant matrix, we see that $C = c_0 I + c_1 J + \cdots + c_{n-1} J^{n-1}$ where $J = \begin{bmatrix} 0 & 1 & 0 & \cdots & 0 \\ \vdots & 0 & 1 & 0 & \vdots \\ 0 & \ddots & \ddots & \ddots & 1 \\ 1 & 0 & \cdots & \cdots & 0 \end{bmatrix} \in M_n$. Therefore, every circulant matrix is generated by a simple permutation matrix J.

Note that $J = FWF^* = Fdiag(1, \omega, \omega^2, \cdots, \omega^{n-1})F^*$ where $\omega = e^{2\pi i/n} = \cos\frac{2\pi}{n} + i\sin\frac{2\pi}{n}$ and F is a Fourier transformation. Thus, $C = F(c_0 I + c_1 W + \cdots + c_{n-1} W^{n-1})F^* = Fdiag(\sum_{j=0}^{n-1} c_j, \sum_{j=0}^{n-1} c_j w^j, \cdots, \sum_{j=0}^{n-1} c_j w^{j(n-1)})F^*$ where $\sum_{j=0}^{n-1} c_j w^{j(k-1)}$ is the kth eigenvalue and $\frac{1}{\sqrt{n}}(1, w^{(k-1)}, \cdots, w^{(k-1)(n-1)})^T \in \mathbf{C}^n$ is the corresponding kth eigenvector of C for $k = 1, \cdots, n$. We denote by C_n the set of all circulant matrices. Because all circulant matrices are generated by J, it is easy to verify that C_n forms a commutative ring over real or complex field. Also it should be noted that C_n is completely characterized by the diagonalizability under F−unitary similarity : $C \in C_n$ if and only if $C = Fdiag(\alpha_1, \cdots, \alpha_n)F^*$, $\alpha_i \in \mathbf{C}$. Thus, C_n is a very special commutative subclass of the normal Toeplitz matrices.

We denote by C_n^R the set of real symmetric circulant matrices:

$$C_n^R \equiv \begin{cases} T_n^R(c_0, c_1, \cdots, c_{k-1}, c_k, c_{k-1}, \cdots, c_1)\} & \text{for an even } n \\ T_n^R(c_0, c_1, \cdots, c_{k-1}, c_k, c_k, \cdots, c_1)\} & \text{for an odd } n, \end{cases} \quad (3)$$

where $k = [\frac{n}{2}]$ and T_n^R in (1).

This choice relies on the fact that $||A||_F^2 \geq \sum_{j=0}^{n} \lambda_j^2(A)$, $A \in M_n$ is hermitian. Let $T \in T_n^R$. Choose $C_0 \in C_n^R$ such that $||T - C_0||_F = \min_{C \in C_n^R} ||T - C||_F$. It is known that that if $T = T_n^R(t_0, t_1, \cdots, t_{n-1}) \in T_n^R$ then

$$C_0 = T_n^R(c_0, c_1, \cdots, c_k, \cdots, c_1) \in C_n^R \text{ such that}$$
$$c_j = \frac{jt_{(n-j)} + (n-j)t_j}{n}, \quad j = 0, \cdots, k. \quad (4)$$

Thus, (4) gives the formula for the initial symmetric circulant matrix choice for this case. Note that $|\lambda_j - \alpha_j|^2 \leq \sum_{j=0}^{n}(\lambda_j - \alpha_j)^2 \leq ||T - C_0||_F^2$ where λ_j and α_j are the eigenvalues of the matrices T and C_0, respectively. Thus, if $||T - C_0||_F$ is small then the eigenvalues of C_0 are close to the corresponding eigenvalues of T, and we have a good choice of the preconditioner.

Let C_n^R be the set of real symmetric circulant matrices in (3). We denote by C_n^H the set of hermitian circulant matrices:

$$C_n^H \equiv \begin{cases} T_n^H(c_0, c_1, \cdots, c_{k-1}, c_k, \bar{c}_{k-1}, \cdots, \bar{c}_1) & \text{for n even} \\ T_n^H(c_0, c_1, \cdots, c_{k-1}, c_k, \bar{c}_k, \cdots, \bar{c}_1)\} & \text{for n odd}, \end{cases} \quad (5)$$

where $k = [\frac{n}{2}]$ and T_n^H in (2).

Note that C_n^H is the $F-$ real diagonalizable class of matrices: $C \in C_n^H$ if and only if $C = Fdiag(\alpha_1, \cdots, \alpha_n)F^*$ where $\alpha_i \in \mathbf{R}$ for $i = 1, \cdots, n$. It can be verified easily that $C_n^R \subseteq C_n^H$ has the following finer spectral characteristic:

$C \in C_n^R$ if and only if $C = F diag(\alpha_1, \cdots, \alpha_n) F^*$ where $\alpha_i \in \mathbf{R}$ for $i = 1, \cdots, n$ such that the algebraic multiplicity of each α_i must be greater than or equal to 2, that is, where is no simple eigenvalue for real symmetric circulant matrices.

A Toeplitz matrix $K = [k_{ij}] \in T_n$ is called a skew circulant matrix if

$$k_{ij} = \begin{cases} c_{j-i} & \text{for } 0 \leq i \leq j \leq n-1 \\ -c_{-(i-j)} & \text{for } 0 \leq j \leq i \leq n-1. \end{cases}$$

Thus, $K = k_0 I + k_1 L + k_2 L^2 + \cdots + k_{n-1} L^{n-1}$ where $L = \begin{bmatrix} 0 & 1 & 0 & \cdots & 0 \\ \vdots & \ddots & 1 & \ddots & \vdots \\ & & \ddots & \ddots & \\ 0 & \ddots & & \ddots & 1 \\ -1 & 0 & \cdots & \cdots & 0 \end{bmatrix} \in$

M_n. Note that $P^* L P = \theta J$ where $P = diag(1, \theta, \theta^2, \cdots, \theta^{n-1})$, $\theta = e^{(\pi i)/n}$, $\theta^n = -1$. Therefore, $P^* K P = k_0 I + \theta k_1 J + \theta^2 k_2 J^2 + \cdots + \theta^{n-1} k_{n-1} J^{n-1} \in C_n$, and hence $F^* P^* K P F$ is a diagonal matrix. It is easy to identify that $P^{s^*} L P = \theta^s J$ for any odd number $s = 1, 3, \cdots$ and K is diagonalizable under $P^s F$-unitary similarity for $s = 1, 3, \cdots$. Note that $J = FWF^*$ and $J^T = F^* W F$ where $P^2 = W = diag(1, \omega, \omega^2, \cdots, \omega^{n-1})$, $\omega = e^{(2\pi i)/n}$, hence $F^* P F$ must be a square root of J^T. Note that we need only to consider for $s = 1, \cdots, 2n - 1$, because of $P^{(2n+1)} = W^n P = P$. We denote by $K_n \subseteq T_n$ the set of all skew circulant matrices, and by $K_n^R \subseteq K_n$ the set of all real symmetric skew circulant matrices.

Lemma 1. *The following are equivalent. (i) $K \in K_n^R$. (ii) K is $P^s F$-real diagonalizable for $s = 1, 3, 5, \cdots, 2n - 1$ i.e., $K = F^* P^{s^*} diag(\alpha_1, \cdots, \alpha_n) P^s F$, $\alpha_i \in \mathbf{R}$. (iii) $P^{s^*} K P^s \in C_n^H$ for $s = 1, 3, \cdots, 2n - 1$.*

Thus suppose $K \in K_n^R$ is a given real symmetric skew circulant matrix.

Consider the set $G = \{P^{s^*} K P^s\}_{s=1,3,\cdots,2n-1} = \{W^{i^*} P^* K P W^i\}_{i=0,\cdots,n-1} = \{W^{i^*} C^{(0)} W^i\}_{i=0,\cdots,n-1}$ where $C^{(0)}$ is a hermitian circulant matrix, $P = diag(1, \theta, \cdots, \theta^{n-1})$ and $W = diag(1, \omega, \cdots, \omega^{n-1})$, $\theta = e^{i\pi/n}$.

Since $FWF^* = J$, $G = \{F^*(FW^{i^*} F^* F C^{(0)} F^* F W^i F^*) F\}_{i=0,\cdots,n-1} = \{F^*(J^{i^*} diag(\alpha_1, \cdots, \alpha_n) J^i) F\}_{i=0,\cdots,n-1} = \{C^{(i)}\}_{i=0,\cdots,n-1} \subseteq C_n^H$, where $\{\alpha_1, \cdots, \alpha_n\}$ are the eigenvalues of K.

We present the following general result about hermitian matrices. We denote by H_n the set of all n by n hermitian matrices. Suppose $A \in H_n$, $A = V diag(\lambda_1, \lambda_2, \cdots, \lambda_n) V^*$, $\lambda_i \in \mathbf{R}$ and $V \in M_n$ is unitary. By $\mathcal{D}$, we denote the set of all real diagonal matrices, $\mathcal{D} = \{diag(\alpha_1, \cdots, \alpha_n) / \alpha_i \in R\}$.

For $a, b \in \mathbf{R}$, $a \geq b$, we let $\mathcal{D}(a,b) = \{diag(\alpha_1, \cdots, \alpha_n) / b \leq \alpha_i \leq a$ for all $i = 1, \cdots, n\} \subseteq \mathcal{D}$. A subset $\mathcal{D}^s(a,b) = \{diag(\alpha_1, \cdots, \alpha_n) / \sum_{i=1}^n \alpha_i = s$ and $b \leq \alpha_i \leq a\}$.

Theorem 1. *Suppose* $A = V \begin{bmatrix} \lambda_1 & & 0 \\ & \ddots & \\ 0 & & \lambda_n \end{bmatrix} V^* \in H_n$. *Let* $s = \sum_{i=1}^{n} \lambda_i$. *Then* $\min_{\Sigma \in \mathcal{D}} ||A - \Sigma||_2 = \min_{\Sigma \in \mathcal{D}(\lambda_{max}, \lambda_{min})} ||A - \Sigma||_2 = \min_{\Sigma \in \mathcal{D}^s(\lambda_{max}, \lambda_{min})} ||A - \Sigma||_2$.

As a simple consequence of Theorem 1, we have the following.

Corollary 1. *Let* $K_0 \in K_n^R$ *be given. Then* $\min_{C \in C_n^H} ||K_0 - C||_2 = \min_{C \in \mathbf{C}_0(G)}$ $||K_0 - C||_2$ *where* $\mathbf{C}_0(G)$ *is the convex hull of* $G = \{W^{i^*} P^* K_0 P W^i P\}_{i=0,\cdots,n-1}$ $= F^*(J^{i^*} diag(\alpha_1, \cdots, \alpha_n) J^i) F\}_{i=0,\cdots,n-1} = \{C^{(i)} \in C_n^H\}_{i=1,\cdots,n}$.

3 A Design of Circulant Preconditioners

Now suppose $T = T_n^R$. We define T^C, the complement of T by $T^C = T_n^R(t_0, t_{n-1}, \cdots, t_1) \in T_n^R$. Then notice that $T = \frac{T+T^C}{2} + \frac{T-T^C}{2}$ where $\frac{T+T^C}{2} \in C_n^R$ and $\frac{T-T^C}{2} \in K_n^R$. Thus we have $T = C_T + K_T$ where $C_T \in C_n^R$ and $K_T \in K_n^R$, and it is easy to verify that the decomposition is unique. Note that $\min_{C_0 \in C_n^H} ||T - C_0||_2 = \min_{C_0 \in C_n^H} ||C_T + K_T - C_0||_2 = \min_{C \in C_n^H}$
$||C_T + K_T - (C_T + C)||_2 \leq \min_{C \in \mathbf{C}_0(\mathbf{G})} ||K_T - C||_2$ where the last equality is from the Corollary 3. Since the object of this section is to obtain a preconditioner hermitian circulant matrix, it is clear from above equality that we need to only consider a hermitian circulant matrix from the convex hull of $\{P^* K_T P, \cdots, P^*(W^{n-1^*} K_T W^{n-1}) P\}$. We use the following steps to choose the preconditioner hermitian circulant matrix for $T \in T_n^R$. Set $S(\alpha, \beta) = \alpha P^* K_T P + \beta P^* W^T K_T W P$, $\alpha, \beta > 0$, $\alpha + \beta = 1$. Compute $\gamma(\alpha, \beta) = \frac{||K_T - S(\alpha,\beta)||_2}{||C_T + S(\alpha,\beta)||_2}$ for $\alpha = 0, 0.1, \cdots, 0.9, 1$. Then the preconditioner S of given T is the hermitian circulant matrix $C_0 = C_T + S(\alpha, \beta)$ with the minimum values of $\gamma(\alpha, \beta)$ where $T = C_T + K_T$.

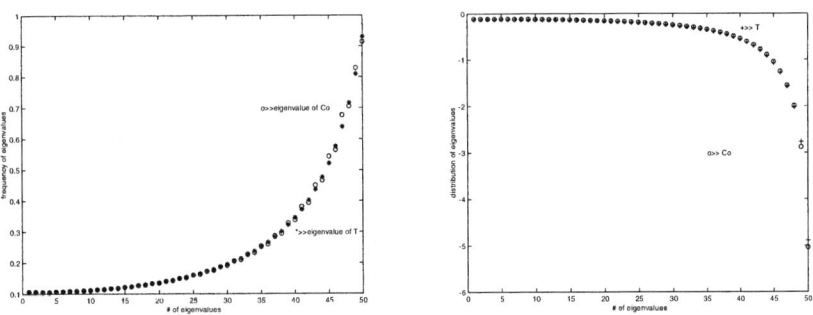

Fig. 1. Eigenvalue Distribution of C_0 and T of Example 4(case(a),(b))

Example 1. We choose α and β: $\alpha + \beta = 1$ such that $\gamma(\alpha, \beta) = \frac{\|K_T - S(\alpha,\beta)\|_2}{\|C_T + S(\alpha,\beta)\|_2}$ is the minimum. Suppose $\gamma(\alpha_0, \beta_0) = \min_{0 \leq \alpha,\beta \leq 1} \gamma(\alpha, \beta)$, where $\alpha + \beta = 1$. Then $C_0 = C_T + S(\alpha_0, \beta_0)$. We provide the following two examples.

(a) $T \in T_n^R(t_1, \cdots, t_n)$ such that $t_i = \frac{1}{(i+1)^2}$ for $i = 0, \cdots, 49$. $\alpha_0 = 0.7$, $\beta_0 = 0.3$.

(b) $T \in T_n^R(t_1, \cdots, t_n)$ such that $t_i = \frac{(-1)^i}{i+1}$ for $i = 0, \cdots, 49$. $\alpha_0 = 1.0$, $\beta_0 = 0.0$.

4 Conclusion

From those experiments, we see a new preconditioner C_0 instead of the given matrix T by the iterative method for linear systems. We expect that all eigenvalues of the given Toeplitz matrices are close to all eigenvalues of new circulant preconditioners[Fig 1]. Also the distributions of eigenvalues of $C_0^{-1}T$ are between 0.6 and 1.2 [Fig2] and are between 0.5 and 1.3 [Fig2] as shown in section 3(Example 4). i.e, we have the property that all eigenvalues of $C_0^{-1}T$ are very close to 1 excluding the extreme eigenvalues. It supports to reduce the iterations for the iterative method on the linear system. The objective matrix for case (a) is a symmetric positive matrix and $\|I - C_0^{-1}T\|_F^2 \approx 1.1591$. The objective matrix for case (b) is a symmetric matrix and $\|I - C_0^{-1}T\|_F^2 \approx 8.3689$. Our circulant preconditioners are much stronger on a symmetric positive Toeplitz systems.

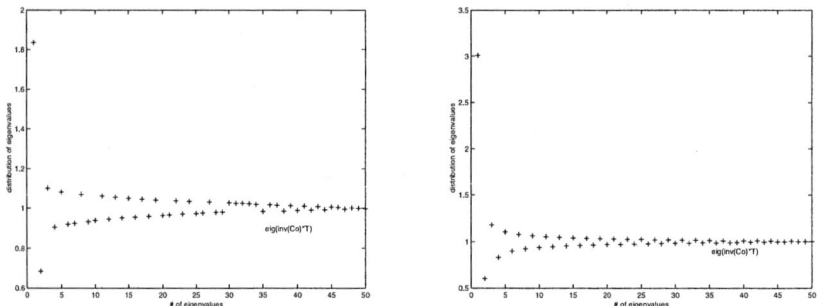

Fig. 2. Eigenvalue Distribution of $C_0^{-1}T$ of given Example 4(case(a),(b))

References

1. Baik, R., Datta, K., Hong, Y. : An application of homotopy method for eiegnproblem of a Symmetric Matrix. Iterative Methods in Linear Algebra, (1995) 367–376
2. Brent, R., Gustavson, F., Yun, D. : Fast solution of Toeplitz systems of equations and computations of Pade approximations, J. Algorithms **1**(1980) 259–295
3. Chan, R. : Circulant preconditioner for Hermition Toeplitz systems, SIAM. J. Matrix Anal. Appl. **10** (1989) 542–550

4. Chan, T. : an Optimal circulant preconditioners for Toeplitz systems, SIAM J. Sci. Numer. Anal. **29**(1992) 1093-1103
5. Davis, P. : Circulant Matrices, Wiley. New York.(1979)
6. Golub, G. H., F. Van Loan, C. : Matrix Computations. JOHN HOPKINS.**3** (1995)
7. Ortega, J. : Numerical Analysis, A Second Course. SIAM Series in Classical in Applied Mathematics. Philadelphia, SIAM Publications. (1990)
8. Tyrtshnikov, E. : Optimal and super-optimal circulant preconditioners. SIAM J. Matrix Anal. Appl.**13** (1992) 459–473

An Approximation Algorithm for a Queuing Model with Bursty Heterogeneous Input Processes

Sugwon Hong, Tae-Sun Chung, Yeonseung Ryu, Hyuk Soo Jang, and Chung Ki Lee

Department of Computer Software, Myongji University,
San 38-2 Namdong, Yongin, Gyeonggi-do, 449-728 Korea

Abstract. We show a discrete-time queuing model to represent a single server queue with constant service rate and N bursty different input processes. The arrival process is modeled by an Interrupted Bernoulli Process. We propose an approximation algorithm to solve this queuing system. The algorithm is based on aggregation technique of input processes, reducing the state space of the Markov Chain. We validate the algorithm by comparing the results with simulation results. In addition, we explain briefly a continuous-time queuing model of the same case.

1 Introduction

Many systems in a real world are modeled by a deterministic single server queue with N arrival streams. The queue has a finite capacity and each arrival stream is assumed to be different each other and a bursty process, which is realistic assumption in a real application.

The analysis of such a queuing system is quite complex due to the large number of arrival processes. One way of analyzing approximately this queue is to first obtain the superposition of all the arrival processes, and then analyze the queue with a single arrival process. This approach reduces the dimensionality of the problem. However, it requires the construction of the superposition process which is fairly complex problem in itself. One approach for obtaining the superposition process focuses on an approximating it by a renewal process[1][2][3]. A Markov Modulated Poisson Process(MMPP) is used to superpose approximately voice sources[4]. An alternative way to analyze this queuing system involves the so-called fluid-flow approximation[5][6].

Most of the methodologies reported above have assumed identical arrival processes. However, this assumption is unrealistic in the real environment such as packet arrivals in communication networks. In this paper, we analyze a discrete-time queuing system assuming different bursty arrival sources and propose the approximation algorithm to obtain the performance measures such as queue length distribution and delay distribution. In addition, we show the continuous-time version of the same queuing model.

In the following section, we describe the discrete-time queuing model and in section 3, we give the approximation algorithm, and derive the queue length distribution probability and the delay distribution probability. In section 4, we validate the approximation algorithm by comparing it with simulation results. In section 5, we show a continuous-time queuing model for the same system briefly. Finally, the conclusion is given in section 5.

2 Discrete-Time Queuing Model

We analyze a discrete-time queuing system consisting of a FIFO single server queue with N different input processes which is modeled by a n-IBP/D/1 queuing system. The server is assumed to be slotted, and the service time is constant equal to 1 slot. Each arrival process is also slotted with a slot equal to a service slot. If a customer arrives when the system is empty, it waits until the beginning of the next service slot. The customer departs one service slot later.

The capacity of the queue including the customer in service is finite and it is equal to M. Each arrival process is modeled by a Interrupted Bernoulli Process(IBP). An IBP process may find itself either in the busy state or in the idle state. Arrivals occur in a Bernoulli fashion only when the process is in the busy state. No arrivals occur if the process is in the idle state. Let us assume that at the end of slot i the process is in the idle (or busy) state. Then in the next slot i+1 it will remain in the idle (or busy) state with probability q (or p), or it will change to busy (or idle) with probability 1-q (or 1-p). If during a slot the process is in the busy state, then with probability α a customer will arrive during the slot. In IBP, $\alpha=1$. That is, at every busy slot there is an arrival.

Let t be the inter-arrival time of a customer. Then, it can be shown that the mean arrival time E[t] and the squared coefficient of variation of the inter-arrival time C^2 are as follows:

$$E[t] = \frac{2-p-q}{\alpha(1-q)}$$

and

$$C^2 = \frac{Var(t)}{E[t]^2} = 1 + \alpha(\frac{(1-p)(p+q)}{(2-p-q)^2} - 1) \tag{1}$$

The link utilization, ρ, i.e., the probability that any slot contains a customer is as follows:

$$\rho = \frac{\alpha(1-q)}{2-p-q} \tag{2}$$

3 Approximation Algorithm

The state of the queuing model is completely described immediately after the beginning of each service slot by the vector $(\underline{w};n) = (w_1,...,w_N;n)$, where w_i is the state of ith IBP arrival process. $w_i=0$ or $w_i=1$, and n is the number of customers in the queue, n=0,1,2,...,M where M is the queue capacity. If $w_i=0$, the IBP arrival process is in the idle state, and if $w_i=1$, the process is in the busy state. Based on this state description, we can compute the transition probability P_{ij} from state i to state j, and subsequently the matrix of transition probabilities $P=[P_{ij}]$. Finally, solving the steady state equation $\pi P=\pi$ where $\underline{\pi}$ is the steady state vector, we can obtain the exact probabilities of all states numerically. The total number of states of this Markov chain is $2^N(M+1)$.

In this paper, we analyze the Markov chain associated with the random variable $(w_1,...,w_N; n)$ using aggregation. The state of arrival processes $(w_1,..., w_n)$ is described by random variable k which represents the number of the heterogeneous arrival processes in the busy state, $k=0,1,...,N$. The state of the queuing system under study is represented approximately by the state (k,n). If it is found in state (k,n) at a certain slot, then during the subsequent slot k arrivals will occur and at the end of the slot one customer will depart. Finally, the queue-length distribution of the number of customers in the queue is obtained by solving numerically the reduced Markov Chain (k,n). In this way, we significantly reduce the state space of the original Markov Chain to $(N+1)(M+1)$, while at the same time through the aggregate random variable k we can maintain enough information regarding the state of the N arrival processes.

Let $r(\underline{w}\rightarrow\underline{w}')$ be the transition probability from state $\underline{w}$ to state $\underline{w}'$. Define S_i to be the set of all states $\underline{w}$ in which there are exactly i arrival processes in the busy state. Using this definition we can lump the state space into $N+1$ sets of states S_i, $i=0,1,...,N$.

Let R_{ij} be the transition probability from lump i to lump j. Then we have

$$R_{ij} = \sum_{\underline{w}\in S_i}\Pr[\underline{w}|S_i][\sum_{\underline{w}'\in S_j}r(\underline{w}\rightarrow\underline{w}')]$$

where

$$\Pr[\underline{w}|S_i]=\frac{\Pr[\underline{w}]}{\Pr[S_i]}, \quad \Pr[S_i]=\sum_{\underline{w}\in S_i}\Pr[\underline{w}]$$

Seeing that the N arrival processes are independent from each other, we have $\Pr[\underline{w}]=\Pr[w_1]\cdot\Pr[w_2]\cdots\Pr[w_N]$, where

$$\Pr[w_i=0]=\frac{1-p_i}{2-(p_i+q_i)}., \quad \Pr[w_i=1]=\frac{1-q_i}{2-(p_i+q_i)}$$

Thus we have

$$\Pr[\underline{w}]=[\prod_{i\in W_0}\frac{1-p_i}{2-(p_i+q_i)}][\prod_{i\in W_1}\frac{1-q_i}{2-(p_i+q_i)}]$$

where for a given state $\underline{w}$, W_0 and W_1 are the state of the arrival processes in the idle state and the busy state respectively.

The generation of the aggregate transition matrix is the most time consuming operation. The reason is that we first have to generate the transition matrix associated with the N arrival processes and subsequently calculate the transition probabilities R_{ij}, $i,j=0,1,2,...,N$.

The computational effort for the generation of the aggregate matrix A can be significantly reduced by observing that the parameters p and q of a bursty IBP process are very likely to be close to 1. It is also highly unlikely that a bursty source will change its current state at the next slot, seeing that $1-p\approx 0$ and $1-q\approx 0$. Therefore, we can neglect all the transitions where more than two arrival processes change their states in the next slot.

The transition probabilities $p_{(k,n)(k',n')}$ between two states are obtained using the transition probabilities R_{ij} of the aggregate matrix A. We have

$$P_{(k,n)(k',n')} = \Pr[(k,n) \rightarrow (k', \max(0, n-1)+)]$$
$$= R_{kk'}$$

We can now solve numerically the system of linear equations $\underline{\pi}P=\underline{\pi}$ where $P=[P_{(k,n)(k',n')}]$ in order to obtain the steady state probabilities $\pi(k,n)$.

Once we have obtained the stationary vector, $\pi(k,n)$, we can easily compute the loss probability by dividing the expected number of lost customers by the expected number of arriving customers. We have

$$P_{loss} = \frac{\sum (k+n-1-M)\pi(k,n)}{\sum_{k=1}^{N} k \sum_{n=0}^{M} \pi(k,n)}$$

where the summation in the numerator runs for all (k,n) such that k+n-1>M.

The probability distribution of the number of the time slots that a customer waits in the queue before the transmission can be also computed. Let C_r denote the expected number of customers which wait exactly r slots and C_E denotes the expected number of customers that entered the queue. Then C_E is expressed as

$$C_E = \sum_{n=0}^{M} \sum_{k=1}^{N} \tilde{k}\,\pi(k,n)$$

where $\tilde{k} = \min(k, M-(n-1))$. And C_r can be obtained as follows:

$$C_r = \sum_{n=0}^{r+1} \sum_{k=r+1-\tilde{n}}^{N} \pi(k,n)$$

where $\tilde{n} = \max(0, n-1)$. The probability that a customer waits exactly r slots before transmission can now be obtained by dividing C_r by C_E.

4 Validation

The approximation algorithm is validated by comparing it with simulation results for eight sources and sixteen sources. Figures 1, 2 and 3 show the results for the queue length distributions and delay distributions. The results are obtained assuming a buffer capacity M equals to 24 for 8 inputs and 32 for 16 inputs. The 8 arrival processes are set as follows: $\rho_i=0.12$, i=1,2,...,8, and $C_1^2 = C_2^2 = 50$, $C_3^2 = C_4^2 = 100$, $C_5^2 = C_6^2 = 200$, $C_7^2 = C_8^2 = 500$ where C^2 is the squared coefficient of variation of the inter-arrival time. The parameters p and q of each arrival process are then obtained from expressions (1) and (2). The 16 arrival processes are set as follows: $\rho_i = 0.03125$, i = 1,2,...,16, and

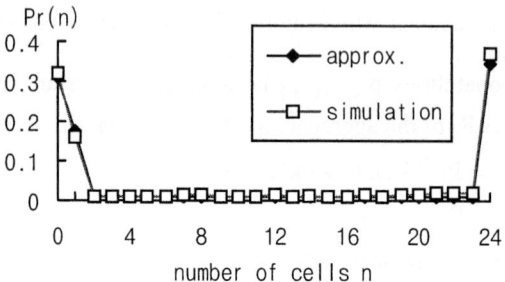

Fig. 1. Queue-length distribution (N=8)

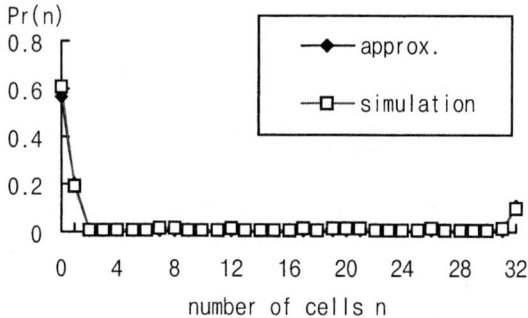

Fig. 2. Queue-length distribution (N=16)

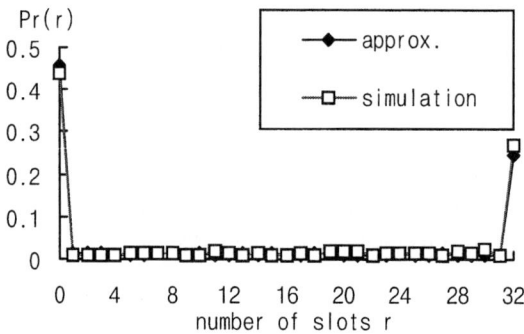

Fig. 3. Delay distribution (N=16)

$C_1^2=C_2^2=C_3^2=C_4^2=50$, $C_5^2=C_6^2=C_7^2=C_8^2=100$, $C_9^2=C_{10}^2=C_{11}^2=C_{12}^2=200$, $C_{13}^2=C_{14}^2=C_{15}^2=C_{16}^2=500$.

We observe that there is a good agreement between the approximation and simulation results. We note that the queue-length distribution in the figures is bi-modal. The two peaks correspond to the case where the queue is empty and the case where the queue is full. This pattern is due to the fact that the overall arrival process to the queue is highly bursty. The accuracy of the algorithm seems to be affected by the variability in the C^2 values of arrival processes. C^2 is used in this paper as a measure of burstiness of an arrival process.

5 Continuous Queuing Model

In a continuous model, a deterministic single server is modeled by an Erlang distribution with r stages. The arrival process to each input queue is represented by Interrupted Poisson Process(IPP). That is, two separate exponentially distributed periods occur alternatively. During the active period, arrivals occur in a Poisson fashion. It can be shown that the time between two successively arrivals has a hyperexponential distribution with two stages, which in turn is equivalent to a Coxian distribution with two stages(C_2). In view of this, the arrival process to the ith input queue is described by a C_2 distribution with parameters $\lambda_{i,1}$, $\lambda_{i,2}$, and a_i for i=1,2,...N, where a_i is the probability of going from stage 1 to stage 2. We can apply the similar approximation algorithm to this continuous model by aggregation as we do to the discrete case[7].

6 Conclusion

In this paper we present an approximation algorithm to analyze the n-IBP/D/1 queuing model. The Markov chain associated with these N arrival processes is first aggregated to a small matrix A, which characterizes approximately the input processes. Subsequently, the queue is analyzed numerically assuming a single arrival stream described by matrix A. Comparisons with simulation data show that the approximation algorithm has a good accuracy. The accuracy of the algorithm is affected by the range of C^2 values of the arrival processes.

References

1. Albin S.L.: Approximating a point process by a renewal process, 2: Superposition arrival processes to queues. Oper. Res. **32**(1984) 1133-1162
2. Sriram K., White W.: Characterizing superposition arrival processes n packet multiplexers for voice and data. IEEE J. SAC **4**(1986) 833-846
3. Perros H.G., Onvural R.: On the superposition of arrival processes for voice and data. Fourth Int. Conf. On Data Communication Systems and Their Performance (1990) 341-357
4. Heffs H., Lucantoni D.M.: A Markov modulated characterization of packetized voice and data traffic and related statistical multiplexer performance. IEEE J. SAC **4**(1986) 856-868
5. Anick D., Mitra D., Sondhi M.M.: Stochastic theory of a data-handling system with multiple sources. Bell Sys. Tech. J. **61**(1982) 1871-1894
6. Tucker R.: Accurate method for analysis of a packet-speech multiplexer with limited delay. IEEE Tran. Comm. **36**(1988) 479-483
7. Hong S.: Continuous-Time Queuing Model and Approximate Algorithm of a Packet Switch under Heterogeneous Bursty Traffic. J. of KISS **30**(2003) 416-423

Improved Adaptive Modulation and Coding of MIMO with Selection Transmit Diversity Systems

Young-hwan You[1], Min-goo Kang[2], Ou-seb Lee[2], Seung-il Sonh[2],
Tae-won Jang[2], Hyoung-kyu Song[1], Dong-oh Kim[3], and Hwa-seop Lim[3]

[1] uT Communication Research Institute, Sejong University, 143-747 Seoul, Korea
{songhk, yhyou}@sejong.ac.kr
[2] Dept. of Inform. & Telecomm., Hanshin University, 447-791 Kyongggi-do, Korea
{kangmg, ouseblee, saisonh}@hs.ac.kr, tecdong@hitel.net
[3] Kaon Media Co. Ltd., Office No.901, Ssangyoung IT Twin Tower A-Dong, 442-17 Sangdaewon-Dong, Jungwon-Gu, Sungnam City, Kyunggi-Do, 462-120, Korea
{renno.kim, hslim}@kaonmedia.com

Abstract. In this paper, Adaptive Modulation and Coding (AMC) is combined with Multiple Input Multiple Output (MIMO) multiplexing to improve the throughput performance of AMC. In addition, a system that adopts Selection Transmit Diversity (STD) in the AMC-MIMO multiplexing system is proposed. The received SNR is improved by adopting STD techniques. And it increases probability of selecting MCS (Modulation and Coding Scheme) level that supports higher data rate. This leads to an increased throughput of the AMC-MIMO multiplexing system. STD in our simulation selects 2 transmission antennas from 4 antennas and AMC-MIMO multiplexing process operates with the selected antennas. The computer simulation is performed in flat Rayleigh fading channel. The results show that the proposed system achieves a gain of 1Mbps over the AMC-MIMO multiplexing system with the same number of antennas at 15dB SNR.

1 Introduction

In order to fulfill the need for ultra-high speed service, researches about Multiple Input Multiple Output (MIMO) system have been in progress. The Bell-lab LAyered Space-Time (BLAST) is representative of the MIMO multiplexing scheme [1]. Also, in order to improve throughput performance, together with MIMO multiplexing system, Adaptive Modulation and Coding (AMC) has drawn much attention [2][3]. The AMC scheme adapts coding rate and modulation scheme to channel condition, resulting in improved throughput and guarantees transmission quality. For this advantage, the High Speed Downlink Packet Access (HSDPA) and 1x-Evolved high-speed Data Only/Data and Voice (1xEV-DO/DV) for high-speed data packet transferring adopts the AMC in the current 3rd wireless communication systems. The reason of the selecting STD from the other TD techniques is as follows. First, closed-loop TD is considered since the AMC system basically contains feedback information. Second, instead of open-loop TD like Space-Time Block Coding (STBC), higher diversity gain is achievable by using closed-loop TD. Finally, the feedback information of STD is simpler than that of the other closed-loop TD like Transmit Adaptive Array (TxAA). In the combination of MIMO multiplexing system and AMC, the Vertical BLAST (V-BLAST) is considered for lower complexity [6].

2 Proposed AMC-MIMO with Selection Transmit Diversity

The system configuration of the AMC-MIMO multiplexing system and the proposed system is shown in this section.

2.1 Transceiver of the AMC-MIMO Multiplexing System

AMC with a single antenna makes use of a single received SNR to select MCS level. But, in the AMC-MIMO multiplexing system, the received SNR can not be determined at each receive antenna because these signals contain signals from the other transmit antennas. Therefore, the SNR in receiver is achieved after nulling. The calculation of SNR from each transmit antennas can be summarized as following. If the V-BLAST system has M transmit antennas and N receive antennas, MIMO channel will have a $M \times N$ channel response matrix as $\mathbf{H}^{M \times N} = \{h_{ij}\}$, where h_{ij} means the channel response from the i-th transmit antenna to the j-th receive antenna. Then the received signal, $\mathbf{X}$ can be written as

$$\mathbf{X} = \sqrt{\frac{E_s}{M}} \mathbf{H} \mathbf{s}_n + \mathbf{V}_n \qquad (1)$$

With AMC-MIMO, the received SNR of signal from each transmission layer needs to be calculated for the MCS level selection. But the received SNR cannot be directly obtained from Eq. (1). After the nulling process, the SNR becomes visible. The SNR of each transmitted layer is determined as the following equations [7].

$$\tilde{\mathbf{s}}_n = \mathbf{G}\mathbf{X}_n = \sqrt{\frac{E_s}{M}} \mathbf{G}\mathbf{H}\mathbf{s}_n + \mathbf{G}\mathbf{V}_n \qquad (2)$$

with

$$\mathbf{G} = \begin{cases} \mathbf{H}^+, & \text{for ZF} \\ (\mathbf{H}^*\mathbf{H} + N_0/E_s \mathbf{I}_M)^{-1} \mathbf{H}^*, & \text{for MMSE} \end{cases} \qquad (3)$$

and

$$\mathrm{SNR}_k = \begin{cases} \dfrac{E_s/MN_0}{[\mathbf{H}^*\mathbf{H}]^{-1}_{kk}}, & \text{for ZF} \\ \dfrac{E_s/MN_0}{\left[\mathbf{H}^*\mathbf{H} + \frac{MN_0}{E_s \mathbf{I}_M}\right]^{-1}_{kk}} - 1, & \text{for MMSE} \end{cases} \qquad (4)$$

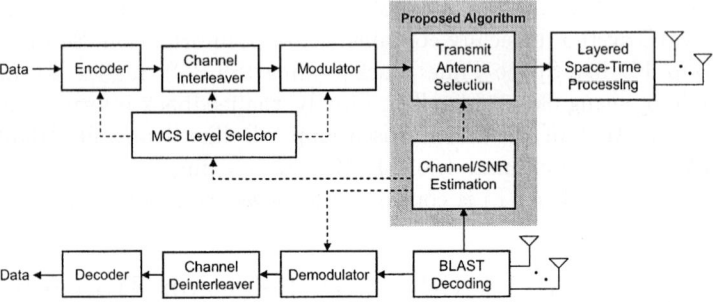

Fig. 1. Transmitter and receiver structure of the proposed AMC-MIMO multiplexing system

where N_0 is a noise variance, $\mathbf{G}$ is a nulling matrix, $\mathbf{V}_n$ is a noise vector at each receive antennas, $\mathbf{s}_n$ is a transmitted symbol vector, and $\tilde{\mathbf{s}}_n$ means an estimated symbol vector. Notation $(\cdot)^*$ means a hermitian matrix, $(\cdot)^+$ denotes the pseudo inverse, and $[\cdot]_{kk}$ represents the element at the kth-row and the kth-column ($k = 1, 2, \cdots, M$).

The results from Eq. (4) are used for the selection of MCS level. If channel encoding/decoding, interleaving/de-interleaving and modulation/de-modulation scheme is separately applied to each layer, it will contribute to the throughput improvement. In this paper, the simulation is performed using the same MCS level for all antennas in consideration of the simple feedback information.

2.2 Combining AMC-MIMO Multiplexing with Selection Transmit Diversity

The AMC-MIMO multiplexing system attains an improved throughput compared to a conventional AMC system. The Space-Time Block Coding (STBC) or the STD can be one of the many choices. In the simulation, STD is applied because it achieves higher diversity gain than that of STBC.

Figure 1 shows the transmitter and receiver structure of the proposed system. Figure 2 shows the transmit antenna selection algorithm when STD is applied to AMC-MIMO multiplexing system where M_t is the number of transmit antenna which is possible to select and M is the number of selected transmit antenna for MIMO multiplexing.

The SNR of each layer is obtained as a result of nulling. For each possible combination, SNR for each of the M selected transmit antennas can be obtained. The layer that has the minimum SNR in each set of antennas limits the throughput of the system. Consequently, the selection of the set of antennas aims at the maximization of the minimum SNR. First, SNR values that are the minimum in each combination are compared. The selected set of antennas is one that has the maximum value among the compared SNRs. Figure 2 illustrates the transmit antenna selection algorithm in AMC-MIMO multiplexing systems.

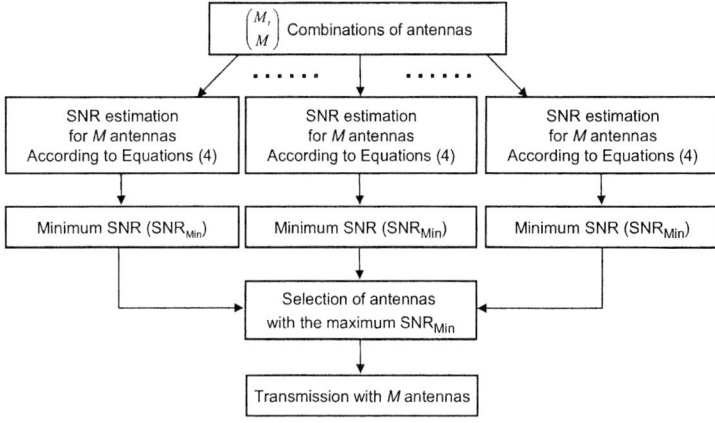

Fig. 2. The transmit antenna selection algorithm in AMC-MIMO multiplexing systems

3 Adaptive Modulation and Coding with Transmit Diversity

Figure 3 shows the AMC-STTD structure in CDMA systems. The information bits are encoded by turbo coding and interleaved. After modulation, the complex valued symbols are STTD encoded, then the Walsh modulation and scrambling are to be done. The structure of AMC-STD is described in Figure 4. The structure of AMC-STD is similar to AMC-STTD case. The difference is that transmit antenna selection part exists and selection information is included in the feedback. The weight vector **W** for transmit antenna selection is given by

$$\mathbf{W} = \{W_0, W_1, \cdots, W_{N-1}\} \quad (5)$$

with

$$W_i = \begin{cases} 1, & \text{if } i = n \\ 0, & \text{otherwise} \end{cases} \quad (6)$$

where W_i is weight of i-th transmit antenna, n is an index of the selected antenna, and N is the number of transmit antennas.

The concept of the estimated SNR for proposed STTD and STD with two transmit antennas is illustrated in Figs. 3 and 4. In STTD, the transmission power is normalized by the number of transmit antennas [4]. Therefore, the estimated SNR from STTD decoding can be seen as an averaged SNR. On the other hand, STD with two transmit antennas does not suffer 3dB transmit power penalty that appears in STTD case because it selects only one transmit antenna.

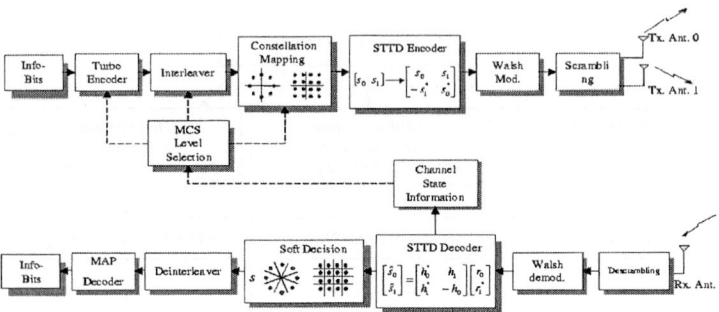

Fig. 3. AMC-STTD structure

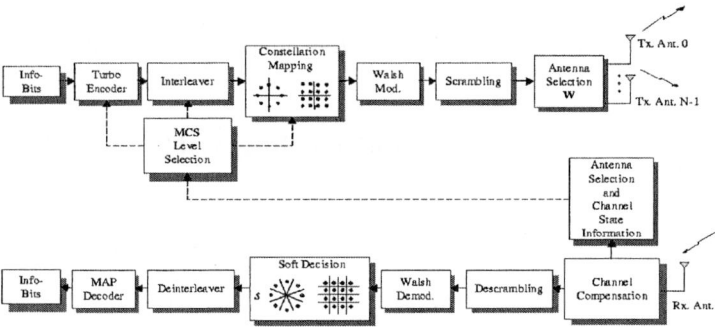

Fig. 4. AMC-STD structure

4 Simulation Results

4.1 MCS Levels and Simulation Parameters

Table 1 and Table 2 show simulation parameters and the MCS level selection used in the letter, respectively. As shown in Eq. (7), the throughput has much relation to frame error ratio E_{FER}.

$$P_{Throughput} = (1 - E_{FER}) \times R \tag{7}$$

Data rate of MCS levels is shown in Table 2. In AMC-MIMO multiplexing system, data transmission rate of each MCS level increases proportional to the number of the transmit antennas. There are many references in the selection of the MCS level selection threshold. In this paper, since the main emphasis is on data transmission rate, the thresholds that maximize throughput are selected. Accordingly, each MCS level selection threshold is defined as the throughput performance cross point in the considered Rayleigh fading environment. MCS level selection thresholds decided by the cross points are 3.25dB, 8.70dB, and 9.55dB, respectively.

Table 1. Simulation parameters

Parameter	Value
Bandwidth	1.2288MHz
Slot length	1.67msec
Channel model	Flat Rayleigh fading channel
	Two-path Rayleigh fading channel
Modulation	QPSK, 8PSK, 16QAM
Channel coding	Turbo coding (1/2 or 2/3 rate)
# of Tx. antennas	1, 2, 4
Spreading factor	16
Doppler frequency	50Hz

Table 2. MCS levels

MCS level	Data rate (Kbps)	Bits/frame	Code rate	Modulation
1	614.4	1024	1/3	QPSK
2	1228.8	2048	2/3	QPSK
3	1843.2	3072	2/3	8PSK
4	2457.6	4096	2/3	16QAM

4.2 Performance of the Proposed AMC Systems with Multiple Antennas

Figure 5 shows throughput performance of the AMC-MIMO multiplexing system when MMSE nulling is adopted. The AMC-STD systems that select one antenna from 4 candidate antennas, AMC-STD (4Tx), and AMC+STBC (2Tx) are also compared. The result verifies that maximum throughput is increased by applying MIMO multiplexing to AMC. In the low SNR, approximately under 1dB, the 2×2 AMC-MIMO (2 transmit and 2 receive antennas) multiplexing system shows a poor performance. AMC-STBC scheme hardly gets a considerable diversity gain in the low SNR since STBC just averages the channel gains of two branches.

Figure 6 depicts the throughput improvement for the proposed AMC-STD-MIMO multiplexing system. The same experiment environment of Table 1 and 2 is applied. 4-2×2 means the proposed AMC-MIMO multiplexing system that selects 2 transmit antennas from 4 candidate antennas and the number of receive antennas is two. 3-2×2 indicates the proposed system which selects 2 antennas from 3 candidate antennas. The result shows the proposed system achieves better throughput performance than the AMC-MIMO multiplexing system in the total SNR range. In case of applying the MMSE nulling method, the proposed achieves the gain of 4.5dB SNR compared to the AMC-MIMO multiplexing system at 3Mbps throughput. 2 × 4 MMSE AMC-MIMO multiplexing system obtains the receiver diversity gain. Accordingly, 2 × 4 MMSE AMC-MIMO multiplexing system can be seen as an upper bound of the other schemes in the simulation. The result demonstrates that the performance of the proposed scheme approaches that of 2 × 4 MMSE AMC-MIMO multiplexing system up to 2dB for the same throughput.

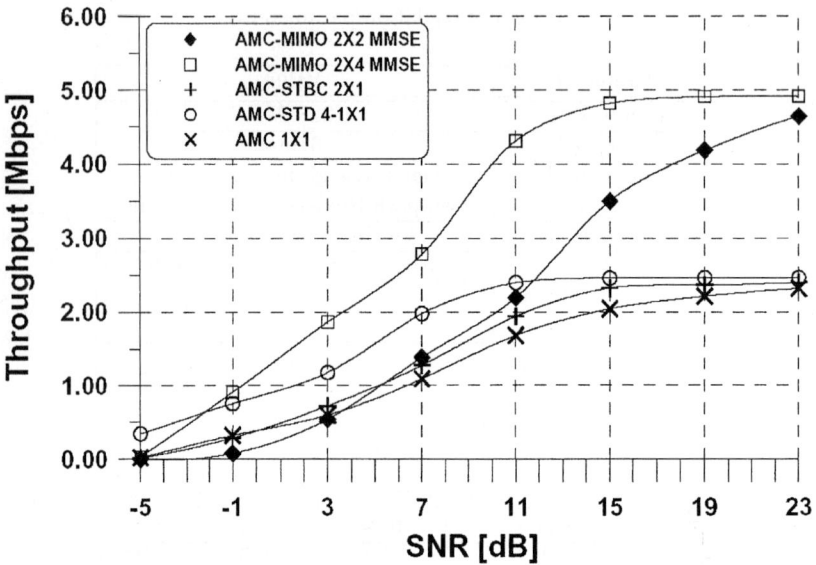

Fig. 5. The throughputs (Mbps) performance of AMC-MIMO multiplexing systems and AMC-TD systems

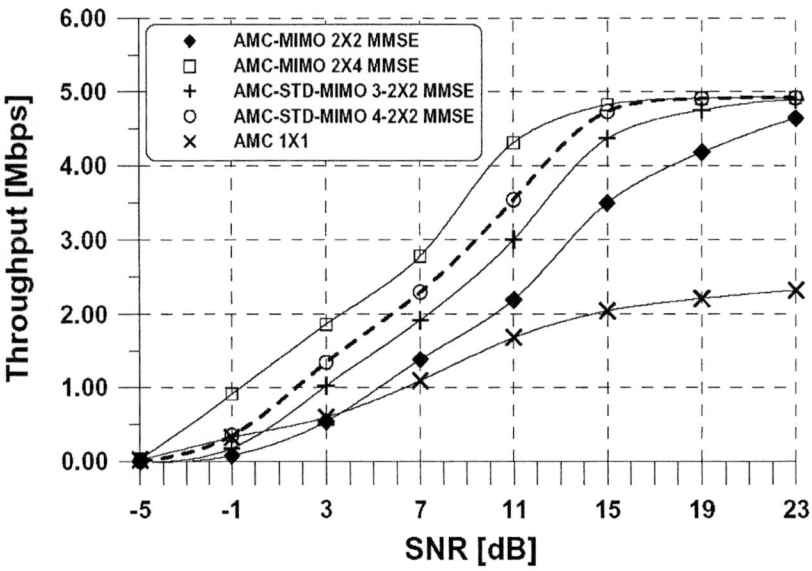

Fig. 6. The throughputs (Mbps) performance of the proposed AMC-STD-MIMO multiplexing systems

5 Conclusion

In this paper, we have investigated the performance of AMC, AMC-STTD, and AMC-STD schemes with 2 or 4 transmit antennas. Using the diversity gain, the effectiveness of the received SNR is improved so that the probability of selecting MCS level with higher data transfer rate is increased.

And the AMC-MIMO multiplexing system is implemented and its performance is shown from the computer simulations. The results prove that the AMC-MIMO multiplexing system contributes to the throughput enhancement. On the other hand, it hardly improves the performance in a low SNR range. To mitigate this problem, the AMC-STD-MIMO multiplexing system is proposed.

The improvement of the received SNR by STD allows the AMC-MIMO multiplexing system to select MCS level that supports higher data rate. Moreover, STD makes the error probability lower in the relatively low SNR and ultimately the throughput performance of system is improved. From simulation results, it is shown that the proposed system provided the maximum throughput improvement and stable throughput increase in the whole SNR range

References

1. Bender, P., Black, P., Grob, M., Padovani, R., Sindhushayana, N., Viterbi, S.: CDMA/HDR : A bandwidth-efficient high-speed wireless data service for nomadic users. IEEE Comm. Magazine. **7** (2000) 70-77
2. Goldsmith, A., Chua, S.: Variable-rate variable-power MQAM for fading channels. IEEE Trans. on Comm. **10** (1997) 1218-1230
3. Goldsmith, A., Chua, S.: Adaptive coded modulation for fading channels. IEEE Trans. on Comm. **5** (1998) 595-602
4. Alamouti, S.: A simple transmit diversity technique for wireless communications. IEEE J. Select. Areas Comm. **8** (1998) 1451-1458
5. Foschini, G., Golden, G., Valenzuela, R., Wolniansky, P.: Simplified processing for high spectral efficiency wireless communication employing multi-element arrays. IEEE J. Select. Areas Comm. **17** (1999) 1841-1852
6. Naguib, A., Seshadri, N., Calderbank, A.: Increasing data rate over wireless channels. IEEE Sig. Proc. Magazine. **17** (2000) 76-92
7. Sandell, M.: Analytical analysis of transmit diversity in WCDMA on fading multipath channels. PIMRC'99. (1999) 946-950
8. 3GPP2 C.P9010: Draft baseline text for the physical layer portion of the 1X EV specification. (2000) 9-78
9. Hwang, I, Jang, T., Kang, M., Hong, S., Hong, D., Kang, C.: Adaptive modulation and coding combined with transmit diversity in MIMO systems. International Conference on Multimedia Technology and Its Applications. (2003) 331-334

Design of a Cycle-Accurate User-Retargetable Instruction-Set Simulator Using Process-Based Scheduling Scheme

Hoonmo Yang and Moonkey Lee

Dept. EE., Yonsei University, 134, Shincho-Dong Sudaemoon-ku, Seoul, Korea

Abstract. We designed a cycle-accurate user-retargetable instruction-set simulator (UR-ISS) based on architecture description language (ADL) which is suitable for system-on-chip (SoC) design. It uses a new scheduling method based on process control. It is effective for scheduling multi-cycle instructions and asynchronous events to a pipeline such as interrupts and exceptions frequently found in SoCs. The proposed UR-ISS consists of a byte-code compiler (BCC) and a virtual machine (VM); The BCC translates ADL semantics into byte-codes and the VM executes them. We have investigated that the UR-ISS is 5.5 times faster than HDL models and 2.5 times faster than System-C models on average. We also applied the UR-ISS for CALMRISC32^{TM} during its development and obtained good results for functional validation.

1 Introduction

Instruction-set simulators (ISS) are indispensable tools for architecture exploration, early system verification and pre-silicon software development. In the past years, ISSs were just used for analyzing dynamic characteristics of software so that they had no consideration of timing accuracy and retargetability. However, circumstances are quite different in SoC design. SoCs are very large-scale systems consisting of standardized components such as processors and peripheral devices and the processors are in the form of pipeline structure for high performance. Accordingly, ISSs should take consideration into retargetability in order to fit in with their modular property and also become cycle-accurate in order to obtain accurate measurements for complex timed operations due to pipeline structure.

A number of publication on machine models and simulation methods have been issued on the topic of how to meet trade-offs between retargetability and performance. The trends can be divided into 2 categories. The first uses general programming languages such as C/C++ but they makes up for the lack of semantics for hardware simulation with standardized software libraries. System-C[1] is the typical example. The second attempts to supply a special language named architecture description language (ADL) for describing machine models. The noteworthy ADLs are ISDL, Rapide, LISA, etc. [2-4]. They are able to obtain good results in generating cycle-accurate UR-ISS to some degree but they mostly adopt restricted timing models based on a reservation table[4]. Embedded processors found in SoCs usually have multi-cycle instructions containing internal iteration such as multiple load/store or Booth multiplication. There also exist asynchronous events to pipeline flow such as hardware interrupts and exceptions. Such timed operations are very complex and hard to schedule accurately by the existing methods. In this paper, we thus present a cycle-accurate user-retargetable ISS (UR-ISS) which is based on ADL and suitable for SoC design. It uses a new scheduling method based on process control which provides generic enough timing model to cover the cases the previous ones cannot handle. A tool set of the proposed UR-ISS consists of a byte-code compiler (BCC) and a virtual machine (VM); The BCC translates ADL semantics into byte-codes and the VM executes them.

This paper is organized as follows. Section 2 outlines the proposed ADL. Section 3 explains the timing model and simulative mechanism of pipeline flow. Section 4 outlines the proposed VM. Section 5 investigates the proposed UR-ISS by presenting experimental performance evaluation and practical application case. Section 6 draws conclusions.

2 Architecture Description Language

The proposed ADL consists of components as shown in Fig. 1. Prototypes are the highest hierarchical components. They can be instantiated and their instances communicate among themselves through ports. Each prototype represents entire structure of an independent pipeline. They possess components named stage, which represent pipeline stages.

The ADL has syntax of high-level language like C which is useful for modeling processors in a top-down way. It supplies flow control statements for loop and selection, and composite data types such as array, reference and record. It also supplies semantics for scheduling operations.

3 Timing Model

The proposed UR-ISS is based on a process, which is the smallest unit of concurrent behavior and represents a thread in pipeline flow as shown in Fig. 2-A. Processes are controlled by 5 basic operations as summarized in Table 1.

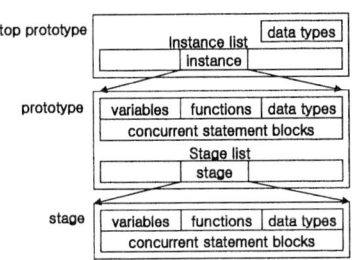

Fig. 1. Components of the proposed ADL

Table 1. Process control operations

Name	Behavior
SPLIT	creates a child process from a current process
STEP	delays a current process
ENTER	makes a current process enter the designated pipeline stage
WAIT	makes a current process wait until correspondent events have occurred
EXIT	destroys a current process

Processes are created by split operation, run through pipeline stages and are destroyed by exit operation. The proposed UR-ISS does not initiate iteration of pipeline flow by itself. Instead, it obligates users to explicate iteration in the ADL by using split operation as shown in example 1. for the sake of retargetability.

Example 1 (ADL specification of pipeline flow).

```
stage Ife : 1 {
  act {
    read_code_from_memory( );
    split {
      enter Ife; // create a child process
    }
    enter Dec; // a parent process advances toward Dec stage
  }
};
```

Such policy is especially effective on multi-cycle instructions containing internal iteration of pipeline flow such as multiple load/store and Booth multiplication. In the cases of multi-cycle instructions, split operation is executed at other pipeline stages besides the first one in order to initiate internal iteration as shown in Fig. 2-B.

The scheduler reserves processes at a time wheel as shown in Fig. 3-A. The time wheel is a circular queue of time slots. A time slot represents a single clock cycle and the simultaneous processes occupy the same time slot. The scheduler selects and executes processes from a current time slot one by one. If there remain no more processes, it moves to the next time slot, progresses simulative time and

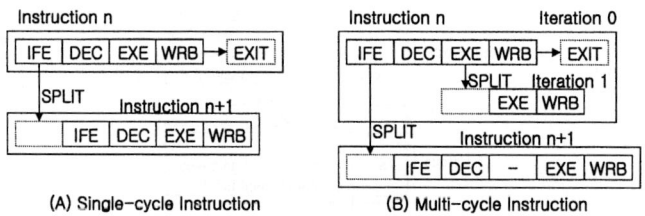

Fig. 2. Initiation of pipeline flow

repeat the same procedure. A time slot has a busy variable indicating whether there still remain any reserved processes. The busy time slots are formed into a linked list and the scheduler slides faster along time slots by visiting only busy slots directed by the linked list.

A process circulates through states as shown in Fig. 3-B during its life time. In active state, a process is currently running or reserved in the time wheel. A process becomes idle and moves to a wait queue when its pipeline stage stalls or wait operation is executed. If the causes are resolved, the idle process will be turned active back and moved to the time wheel.

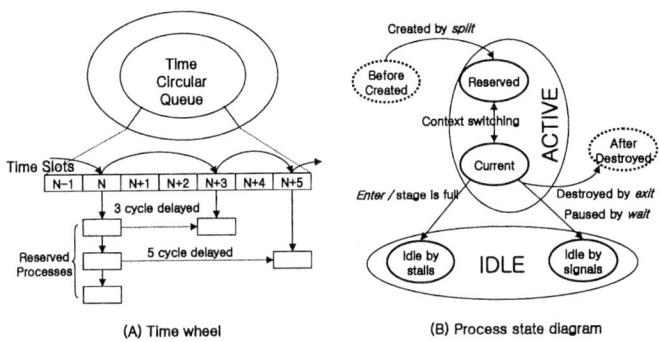

Fig. 3. Structure of a time wheel and process state diagram

When step operation is executed, a current process stops and moves to the $latency^{th}$ time slot from there. If designated latency is bigger than the total number of time slots, a current process will be pushed into the priority queue along with its simulative time. Otherwise, it will be moved to a destination time slot. An index of the destination time slot is calculated as below.

$$\text{index} = (\text{current time slot index} + \text{latency}) \% \text{total number of time slots} . \quad (1)$$

After moving to the destination time slot, the scheduler examines its busy variable. If the time slot is not busy, it will join into the linked list as shown in Fig. 4. Processes in the priority queue are sorted in increasing order by simulative time. Whenever the scheduler advances to a new time slot, it moves every process with simulative time of which difference from the current one is less than total number of time slots from priority queue to the new time slot.

3.1 Pipeline Flow Model

Every pipeline stage has its own capacity for accepting processes all at once. Capacities are specified along with their own stage declarations. These values are 1 in the models of scalar processors and they may be over 1 in the models of superscalar or VLIW processors. If enter operation is executed, the scheduler tries releasing one stalled process from wait queue of the current pipeline stage;

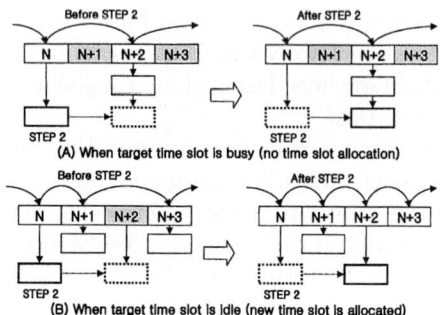

Fig. 4. Procedure of inserting new time slot into the busy time slot list

It decrements current pipeline stage count by 1 unless processes exist. It then changes the current pipeline stage into the destination one and checks the destination pipeline stage count; If the count does not exceed its capacity, it will move the current process into the time slot after 1 clock cycle or otherwise, it reserves the current process at the wait queue of the destination pipeline stage. Split and exit operations also affect pipeline stage count; Newly created processes enter the pipeline stages of their parent and destroyed processes leave their current pipeline stage. Procedure of pipeline flow in the time wheel is exemplified in Fig. 5.

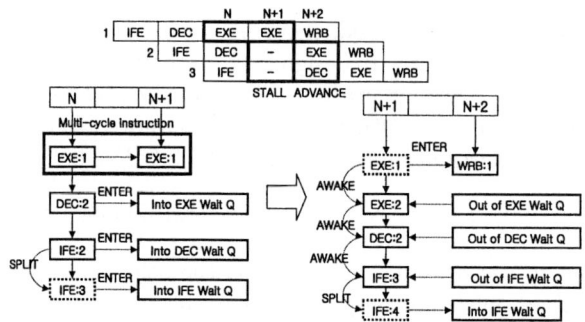

Fig. 5. Procedure of pipeline flow with pipeline stall

3.2 Event-Driven Scheme

Event-driven scheme based on signal variables is mainly used for specifying events asynchronous to pipeline flow such as interrupts/exceptions and data transactions through ports. If wait operation is executed, a current process is stopped and moved to one of signal wait queues. A signal wait queue can be discerned by a pair of P:a code-address where a byte-code for wait operation is located and Q:a prototype instance to which signal variables of the sensitivity list belong. Signal variables contain their own pointers indicating elements in the signal table as shown in Fig. 6. The element contains a list of signal events which specify types of signal sensitivity and value of P. Whenever a signal variable is written, the scheduler searches the signal table by the pointer the signal variable contains and checks whether any signal events have occurred. If so, the scheduler retrieves its P and Q and select the correspondent waiting processes and moved them into the time wheel.

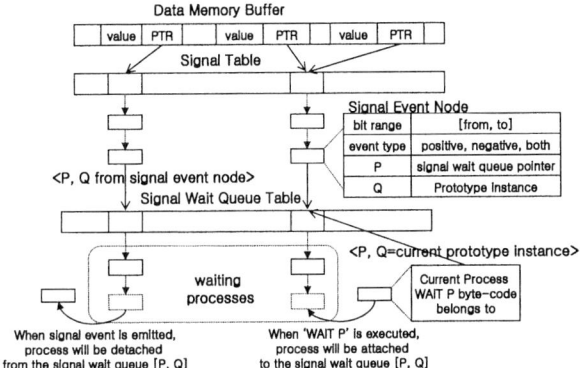

Fig. 6. Structure of the signal table

4 Virtual Machine

We have chosen virtual machine scheme for the proposed UR-ISS because it is more independent of platform and easier to modify and expand than direct syntactic translation. The VM comprises a byte-code processing routine, a process scheduler and a memory manager as shown in Fig. 7. The byte-code processing routine fetches byte-codes, decodes them and calls their correspondent call-back routines. The call-back routines use an evaluation stack for saving operands and a context buffer for looking up the information about a current process. The byte-codes for process control cause context switching by which the context buffer is substituted. The memory manager allocates dynamic memory space for a process during process creation.

5 Simulation Results

The design flow of the proposed UR-ISS is as follows; we first defined syntax of the ADL, checked ambiguity of its grammar and generated its parser by LEX & YACC. We implemented the BCC and the VM for the ADL in C++. In order to evaluate the proposed UR-ISS, we selected samples among existing embedded processors and modeled them in the ADL. We accomplished simulation of the models by the proposed UR-ISS. We used biased random vectors as

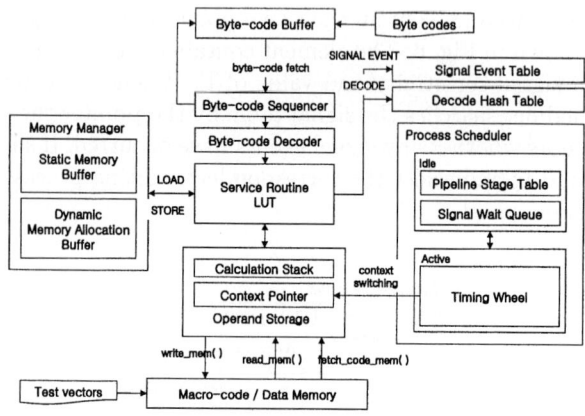

Fig. 7. Block diagram of the virtual machine

Table 2. Comparison of the proposed UR-ISS with other models in performance

			(# of codes / sec)
	ARM7TDMI	SPARC	CALMRISC32
Cycle-accurate model	19,000	20,000	22,000
Verilog HDL model	90	120	150
System-C model	260	330	370
The proposed UR-ISS	520	630	719

workload[6]. We also ran other models of the same processors to compare the proposed UR-ISS. The other models were classified into 3 types; Cycle-accurate C models, Verilog^{TM} models and System-C models. Simulation was carried out in SUN SPARC60^{TM} workstation. It has been verified that the proposed UR-ISS is 5.5 times faster than Verilog^{TM} HDL models and 2.5 times faster than System-C models on average as shown in Table 2. Although the UR-ISS is slower than cycle-accurate C models; it is still favorable in that architectural changes are frequent in the stage of architecture decision and high retargetability overwhelms time loss caused by slow simulation speed.

We also applied the UR-ISS for the validation of CALMRISC32^{TM}[5]. Its validation environment compares simulative results between a target machine and a reference one step by step on the basis of clock cycle[7]. We substituted the proposed UR-ISS for the reference machine and attained statistics of detected errors as shown in Table 3. We found that errors mostly crowded around the case of hardware interrupts and exceptions. CALMRISC32^{TM} also contains multi-cycle instructions containing internal iteration of pipeline flow such as multiple load/store and Booth multiplication. Such case is hard to schedule precisely by other than the proposed scheduling scheme.

Table 3. Distribution of errors detected in validation of CALMRISC32

Instruction	Number	Hardware	Number
Exception/Interrupt	13	Status register	12
Multi-cycle instruction	2	Code bus	6
Single-cycle instruction	5	Data dependency	1
Compound cases	1	Misc.	2
Total	21	Total	21

6 Conclusion

We have designed a cycle-accurate UR-ISS based on ADL which is suitable for SoC design. It uses a new scheduling method based on process control. It is very effective for scheduling multi-cycle instructions and asynchronous events to a pipeline such as interrupts and exceptions frequently found in SoCs. We have investigated that the UR-ISS is 5.5 times faster than HDL models and 2.5 times faster than System-C models on average. We also applied the UR-ISS for CALMRISC32^{TM} in the development and obtained good results for functional validation. Future work will concentrate on software/hardware co-design methodology by adding 2 new features; The one is assistance of software development tools including a compiler, a linker and a debugger and the other is assistance of hybrid integration with other simulative models including HDL models.

References

1. Liao, S., Tjiang, S., Gupta, R.: An Efficient Implementation of Reactivity for Modeling Hardware. DAC'97 (1997)
2. Hadjiyiannis, G., Hanono, S., Devadas, S.: ISDL : an Instruction Set Description Language for Retargetability. Proc of the 34th Design Automation Conference (1997), 299-302
3. Luckham, D.: Rapide: a Language and Toolset for Simulation of Distributed Systems. DIMACS Partial Order Methods Workshop IV, Princeton University (1996)
4. Z.,V., et al.: LISA - Machine Description Language and Generic Machine Model for HW/SW Co-design. IEEE Workshop on VLSI Signal Processing (1996)
5. Cho, S., et al.: CALMRISC32 : A 32bit Low-power MCU Core. IEEE Proceeding of AP-ASIC2000, Cheju, Korea (2000)
6. Chang, T.: A Biased Random Instruction Generation Environment for Architectural Verification of Pipelined Processor. Journal of Electronic Testing (2000)
7. Lee, C., Yang, H., et al.: Efficient Random Vector Verification Method for an Embedded 32bit RISC Core. Proc. of AP-ASIC2000, Cheju, Korea (2000), 291-294

An Authentication Scheme Based Upon Face Recognition for the Mobile Environment

Yong-Guk Kim and Taekyoung Kwon

School of Computer Engineering, Sejong University, Seoul 143-747, Korea
{ykim, tkwon}@sejong.ac.kr

Abstract. It is well known that humans are far better in recalling person's face than his name. One of the promising approaches for devising a secure password scheme would be using a set of images as passwords rather than conventional PIN numbers. We have investigated such potential using an experimental paradigm by which security and usability for three different categories of images (i.e. landscapes scenes, random faces, familiar faces) can be compared. The results suggest that performance of the subjects was reliably higher for the "familiar faces" case than for other cases. Issues such as "Known-face attack" and "Camera Attack" were discussed. We propose a more secure, and yet usable, visual password system by exploiting human's innate capability of fast face identification against serial images of faces and implement it on a PDA for using within the pervasive and mobile environment.

1 Introduction

In modern days, user authentications are so ubiquitous that we almost could not carry out our business and private life properly without passing them. However, it could be true that you sometime found yourself stumbling in front of an ATM machine, because the PIN number was recalled incorrectly or not recalled. Often it is getting worse when you have to manage several numbers for different bank accounts. You may think about choosing his PIN number that is easy to remember, based upon a phone number or birthday. But such choice is vulnerable and can be very risky [2,5,6,7]. The rationale of introducing the visual password system is to utilize the best of two worlds [9]: (1) human information processing power for discriminating visual images, which is considered to be a hard task for the computer; (2) fast serial processing power of the digital computer to minimize human errors in interacting with the computer. There are several attempts in this line of thinking. For example, in the early graphical system designed by Blonder [3], a bunch of objects are distributed in the main image. To login successfully, a user needs to touch those objects in a designated sequence. Recently such a system has been further improved by Passlogix^{TM} [9]. In this system, for instance, the time has to be put in when the selected object is a clock, and a telephone number has to be dialed for the telephone object. Secondly, Perrig and Song [11] proposed the hash visualization technique. Here, a user chose the images, as his keys, among other images that were generated by random-art algorithm, and memorized them

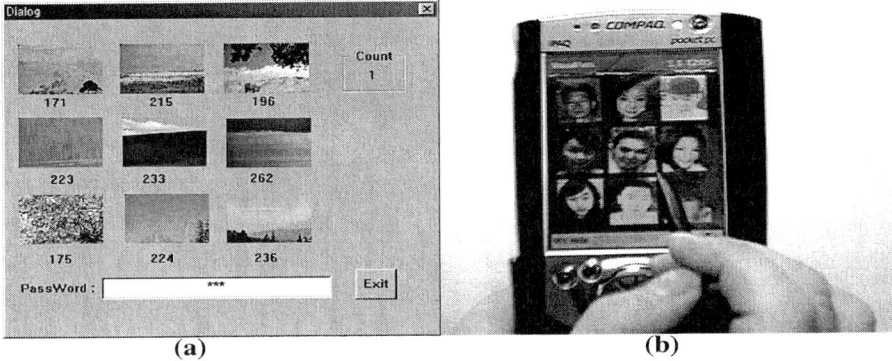

Fig. 1. The visual stimuli consisting of 3x3 landscapes (a) and faces (b)

through a session of training. In a login session, he has to pick up the known image among others in a trial and goes through a few trials to the end. Previous works have shown that humans have remarkable capacity in memorizing visual scenes for a long time span. This scheme intends to exploit such human capability for the authentication purpose. Thirdly, RealUser has designed passfaces^{TM} system [12,13], in which initially a user needs to choose five face images within a large face database and then consolidates those five images through the training session. In a testing trial, he is asked to select one face that he already knows among 9 face images embedded in a 3x3 grid, and carries on 4 more trials to successfully pass the login session. However, like the hash visualization technique, this also requires a training session to memorize the key faces. The aim of this study is to compare human visual performances for three different images groups: landscape scenes, randomly chosen faces and known faces. We reason that this could be used as a guideline for designing the better visual password system [1].

2 Experiments

2.1 Methods

Three groups of images are prepared: (1) landscape scenes, (2) random faces and (3) known faces. The "landscape scenes" are typically photographs taken in the outdoor, such as mountains, trees, loads and seaside views. Face images are all color photographs that were gathered from the private collections and the diverse web sites. For the "known faces", each subject was asked either to bring his (or her) images that could be his family members, friends, or favorite entertainers, or to choose from our face database. Some of them chose the key images from both sources. Each subject was instructed to choose the most memorable or favorite face images. There were two formal sessions in order: key viewing session and testing session. During either (1) or (2) viewing session, he was asked to view the key images (n=5, 10 or 15 in the different session, respectively) for three

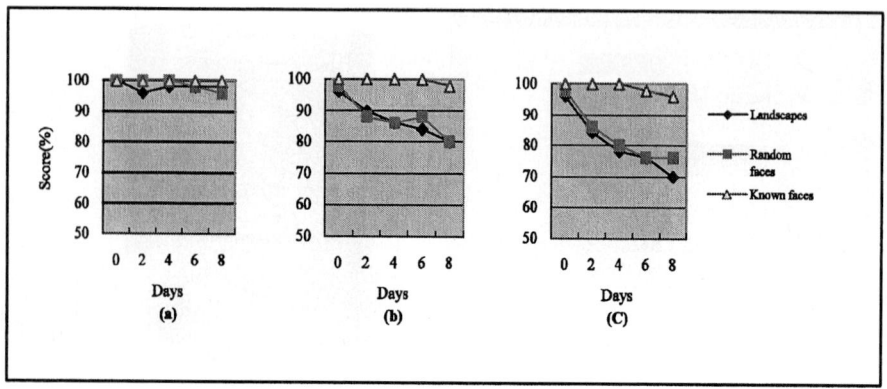

Fig. 2. Scores for three different key cases (5, 10, 15 keys)

seconds for each images. However, there was not any viewing session for the (3) case. During the testing session, the images were presented on a grid that contains embedded images: 2x2 (4 images), 3x3 (9 images) and 4x4 (16 images), respectively, in the different testing sessions. The subject had to choose a key image among decoy images. So that, depending upon the numbers of the key images, the subject would view either 5, 10 or 15 visual stimuli in each testing session. The viewing period for each stimulus was three seconds, but when the subject clicked a face (or landscape scene), the next stimulus came out. The subjects could choose either a mouse or a keyboard during their testing. For the cases of choosing the keyboard for entering the number of a chosen image, a number was designated just beside each image as shown in Fig. 1. Twenty subjects were recruited for these experiments.

2.2 The Size of the Key Images

In the first experiment, we tested how the size of the key images affects the performances of the subjects. The three cases were: 5, 10 and 15 for three different images groups (i.e. landscape scenes, random faces and known faces). During this experiment, the number of the embedded images within each visual stimulus was fixed as 9 (3x3). The task of the subjects was to choose a key image among 8 decoy images in each stimulus. Data were collected at every two days for the 8 days period. Results are shown in Figure 2. Although subjects' performance for the known faces was the best as shown in Figure 2(a), they also showed equally good performances for the other two cases, suggesting that as long as the size of the key images was 5, most subjects could successfully pass this authentication system. However, when the size of the key images was increased to 10 as shown in Fig. 2(b), the performances for the "landscape scenes" and "random faces" were deteriorated severely as passed especially. Notice, however, that the subjects maintained high performance for the "known faces" case. Moreover, as the size of the key was further increased as 15, the contrast between two cases became bigger as shown Fig. 2(c).

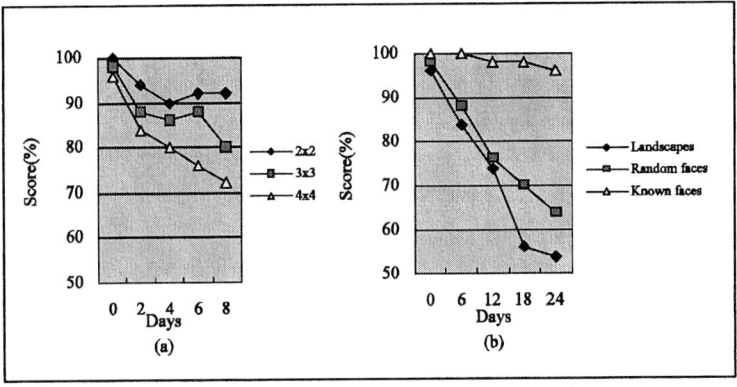

Fig. 3. The effect of size (a) and time (b)

2.3 The Size of the Embedded Images

In the second experiment, we tested whether the size of the embedded images within each stimulus could affect the performance of the subjects. Since the subject has to choose one among many detractors (or decoy images), this is closely related to the well-known "visual search" problem. Previous study demonstrated that there are two kinds of searches: serial and parallel searches. In a parallel search, search time does not increase by the size of the distracters, whereas it does when it is a serial search. Our result as shown in Fig. 3 suggests that the subjects did a serial search in detecting a key face among other distracting faces. This is an interesting finding, although we could reason that the subject may have to watch each face carefully to detect a known face. So that, his performance became worse as the size of the embedded images was increased shown in the same figure.

2.4 The Time Effect

In the third experiment, we tested how performance of the subjects was changed as the testing time was extended as 24 days or more. Here, we fixed the size of the embedded images at 9(3x3) and tested again for three image groups. Figure 4 shows that basically the performance for the "known faces" group maintained at the high rate of success, whereas those for two other groups were poor as days went by.

3 Potential Attacks

As for the security of the proposed scheme, we should examine two possible attacks that we call a known face attack and a camera attack, respectively, in this paper. The known face attack means that an adversary attempts to steal the authenticity of a known party of which the key face images could be guessed easily. This attack seems feasible by gaining a benefit from known face images, for example, from the related persons or one's favorite movie stars.

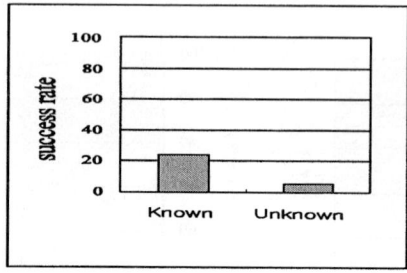

Fig. 4. Known-face attacks for the known and unknown groups

3.1 "Known-Face Attack"

To test such a case in the laboratory, a group of subjects (n=5), who are friends each other and so they know others' relatives and favorites, was recruited. The second group (n=4) was recruited as the control subjects, and so the first group's members did not know well the members of the second group. Each subject (n=9) prepared his own 10 key images and participated at the attacks. There were two sessions: one for 5 keys and 10 keys tests. The experiment was focused on how each member of the first group attacked his group members as well as other group members. Unlike the experiment 1 where the subjects needed to make a decision within 5 second for each visual stimulus, here the stimulus stayed until the subject made a choice to make him guess other's key as possible as he could. For the 5 keys sessions, 5 stimuli were given and 10 stimuli were given for the 10 key sessions. No feedbacks (i.e. answers) were given to the subjects. Each subject attacked each member of both group 3 times.

First of all, no subjects succeeded to pass any single session completely throughout the attack tests, even during the 5 keys sessions. However, in each session the subject hit zero or more key images. Fig. 5 shows the percentage graph of hit ratio averaged over the sessions. We could find from the experimental results that the knowledgeable case does not seem to have *notable* advantages for succeeding the referred attacks, although the hit ratios for the known cases were relatively higher than the unknown cases.

3.2 "Camera Attack"

The camera attack means that an adversary takes a picture of the given authentication images and then attempts to make a pair of the key images among the pictures taken. The adversary should impersonate the victim party, more than two times, in order to have the pictures taken. This attack seems possible because the redundant images should be chosen at random from a large set of images, i.e., the key image is only repeatedly emerging. However, we designed our system very carefully in a way to resist the camera attack. For example, the key images do not appear to the authenticating parties in order. Rather it appears at random from the one's key image set. Even with this disordered appearance of the key images, the authenticating user is able to provide correct answers easily. This is the given strength of the visual passwords. Also the continuous failure is restricted by five times as it is in a conventional password system. This was originally engaged to resist on-line guessing attacks [7] in our system but happened to be useful for preventing the referred specific attacks. The adversary who failed to provide correct answers in end, could not gather all key image pairs in five times. So, it is not easy for the adversary to make a complete set of key pairs on the given images.

4 Conclusions and Discussion

We have proposed the new visual password scheme for using within the pervasive and mobile environment as an authentication purpose. Using the experimental paradigm, we have tested the security and usability of three different image groups (i.e. landscape scenes, random faces, familiar faces), and found that the last one is the best for our purpose. Two potential attacks such as "Known-face attack" and "Camera attack" are tested and discussed. The present method is human-friendly and yet usable compared to the conventional password systems since it aims to utilize human's innate capability in checking security.

References

1. Adams, A., Sasse, MA., and Lunt, P.: Users are not enemy: Why users compro-mise computer security mechanisms and how to take remedial measures. Communication of the ACM, (1999) Vol. 42, 40-46
2. Bellovin, SM., and Merritt, M.: Encrypted Key Exchange: Password- Based Protocols Secure Against Dictionary Attacks. In Proceedings of the I.E.E.E. Symposium on Research in Security and Privacy, (1992) Oakland, May.
3. Blonder, G.: United States Patents, United States Patents 55599961 (1996)
4. Dhamija, R. and Perrig, A.: Deja Vu: A User Study. Using Images for Authentication, USENIX Security Symposium (2000)
5. Gong. L., Lomas, MR., Needham, R., and Saltzer, J.: Protecting Poorly Chosen Secrets from Guessing Attacks. I.E.E.E. Journal on Selected Areas in Communications, Vol. 11, No. 5 (1993) 648-656
6. Jablon, D.: Strong Password-Only Authenticated Key Exchange. Computer Communication Review, ACM SIGCOMM, vol. 26, no. 5, (1996) 5-26
7. Kwon, T.: Authentication and Key Agreement via Memorable Passwords. ISOC NDSS 2001
8. Nielson, J,: Usability Engineering. Academic Press (1993)
9. Passlogix. http://www.passlogix.com
10. Paulson, LD.: Taking a graphical approach to the password. IEEE Computer Magazine, vol. 19 (2002)
11. Perrig, A., and Song, D.: Hash visualization: A new technique to improve real-world security. In Proceeding of the 1999 International Workshop on Cryptographic Techniques and E-Commerce (1999)
12. Real User, The science behind passfaces. September (2001) Security Symposium (1999)

A Survey of Load Balancing in Grid Computing

Yawei Li and Zhiling Lan

Department of Computer Science,
Illinois Institute of Technology, Chicago, IL 60616
{liyawei, lan}@iit.edu

Abstract. Although intensive work has been done in the area of load balancing, the grid computing environment is different from the traditional parallel systems, which prevents existing load balancing schemes from benefiting large-scale parallel applications. This paper provides a survey of the existing solutions and new efforts in load balancing to address the new challenges in grid computing. We classify the surveyed approaches into three categories: resource-aware repartition, divisible load theory and prediction based schemes.

1 Introduction

Due to its critical role in high-performance computing, the load balancing issue has been studied extensively in recent years and a number of load balancing toolkits are available for parallel computing [4, 9, 11]. Most of these works focus on traditional distributed systems and are no longer suitable for the emerging Grid [6, 8]. The Computation Grid evolves high performance computing platforms to heterogeneous, dynamic and shared environments, which present obstacles for applications to harness conventional load balancing techniques directly. Thus how to migrate or adapt load balancing schemes to Grid becomes the focus of this research area.

The rest of this paper is organized as follows. Section 2 briefly describes the load balancing problem and gives taxonomy of traditional load balancing work. Section 3 elaborates and compares the current load balancing approaches in Grid environment. Section4 draws a concise conclusion.

2 Background of Load Balancing

For parallel applications, load balancing attempts to distribute the computation load across multiple processors or machines as evenly as possible with objective to improve performance. Generally, a load balancing scheme consists of three phases: information collection, decision making and data migration. During the information collection phase, load balancer gathers the information of workload distribution and the state of computing environment and detects whether there is a load imbalance. The decision making phase focuses on calculating an optimal data distribution, while the data migration phase transfers the excess amount of workload from overloaded processors to underloaded ones.

Based on different policies used in the three phases, load balancing schemes can be classified in three dimensions as follows:

1. *Static Schemes Versus Dynamic Schemes*: Static schemes [7] assume *priori* knowledge of both application and system state. Static load balancing is advantageous in terms of low overhead as the decision is only made once before computation; however, it cannot adapt to fluctuation of application computation requirement and system state. Dynamic load balancing (DLB) is proposed to make decision based on the changes of application and system environment.
2. *Centralized Schemes Versus Distributed Schemes:* In centralized schemes, a central manager takes charge of collecting load information and making decision based on global knowledge. Centralized schemes usually entail global synchronization to obtain global information at the cost of high synchronization cost. Distributed load balancing allows every processor maintains its local view of workload distribution and makes decision based on the partial knowledge. Typical distributed load balancing schemes are neighbor-based, such as diffusion method [2]. The lack of global knowledge slows down the convergence rate of global balancing, e.g., diffusion algorithm requires quadratic time complexity.
3. *Application-Level Schemes Versus System-Level Schemes:* Application-level load balancing focuses on minimizing the makespan of a parallel application. Here, makespan is defined as the completion time of the entire job. System-level load balancing, also known as distributed scheduling, is to maximize process throughput or the overall utilization rate of the machines.

3 Load Balancing Schemes for Grid Computing

Recently, there are some efforts in load balancing attempting to cope with one or more of the challenges brought by Grid: heterogeneity, resource sharing, high latency and dynamic system state. We classify these schemes into three categories as listed below.

3.1 Resource-Aware Repartition Based Schemes

Graph repartition is the most dominant dynamic load balancing approach in scientific computing area [11]. The repartition schemes model the problem domain as a graph consisting of weighted vertices and communication edges. Consequently, the load balancing problem is transformed into one with goal to reduce the imbalance among subdomains as well as minimize the communication edge-cut. Figure 1 illustrates the typical interactions between partition-bases algorithms and parallel applications.

Unfortunately, traditional partitioners share the drawback that stems from the *priori* knowledge of the workload and system state, and cannot adapt to system changes. Meanwhile, they mainly focus on tightly-coupled systems consisting of homogenous processors thus cannot handle the heterogeneity in Grid. Furthermore, these algorithms only consider the imbalance of subdomains and the edge-cut communication, while neglect the data movement cost that can be a crucial performance bottleneck due to the high communication latency in Grid.

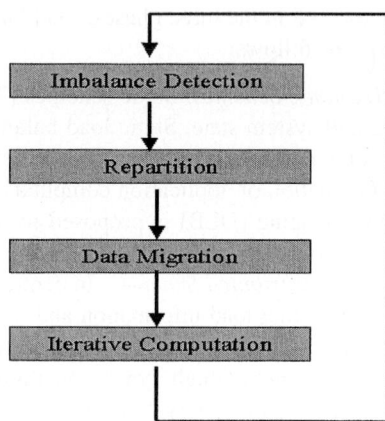

Fig. 1. Interactions between repartition algorithms and applications

To address the resource heterogeneity, the DRUM [5] project proposes a resource-aware partition model cooperating with the Zoltan toolkit [4] to perform dynamic load balancing operation. As indicated in formula (1), DRUM uses a linear model to calculate a scalar sum of node power, which uses both static benchmark data and dynamic performance data collected by monitor agents.

$$power_n = w_n^{comm} c_n + w_n^{cpu} p_n, w_n^{comm} + w_n^{cpu} = 1 \tag{1}$$

Here, p_n is the processing power on node n determined by CPU utilization rate, idle time and benchmark rating; c_n is the communication power on node n that reflects the dynamic information on each communication channel; w_n^{cpu} and w_n^{comm} are the weight factors for processing and communication power respectively; $power_n$ denotes the synthetic node power. Based on the node power above, the partition algorithm calculates a non-uniform workload distribution in heterogeneous machines. However, in this scheme, only CPU contention is considered, the impacts of bandwidth and memory contention are not addressed yet.

Similar to DRUM, GRACE [12] utilizes the NWS [13] package, a Grid resource performance monitoring toolkit, to obtain runtime state, e.g. the CPU available time, end-to-end TCP network bandwidth and free memory. Then the proposed system-sensitive partitioner calculates a new workload partition in proportional to the combined capacity metric synthesizing all the above runtime information. Although this approach can harnesses heterogeneous resources, currently it simply assumes that each resource has equal weight, which cannot be adjusted according to the application characteristics. Thus this partitioner cannot reflect the adaptive requirements of applications.

Different from the above two partitioners, the MinEX [3] partitioner proposes a latency-tolerant algorithm that emphasizes minimizing the data movement cost incurred by load balancing operation. The rationale behind MinEX is to initiate the communication as early as possible. Therefore it can take advantage of overlapping the computation of internal data with the communication of incoming data as much as possible. The inter-cluster experiments show that this approach can efficiently hide the communication latency in Grid.

3.2 Divisible Load Theory (DLT) [10] Based Schemes

In DLT, both the computation and communication load of applications can be arbitrarily divided. Furthermore, there is no special relation among data, which allows workload to be assigned to multiple processors or machines in a linear way.

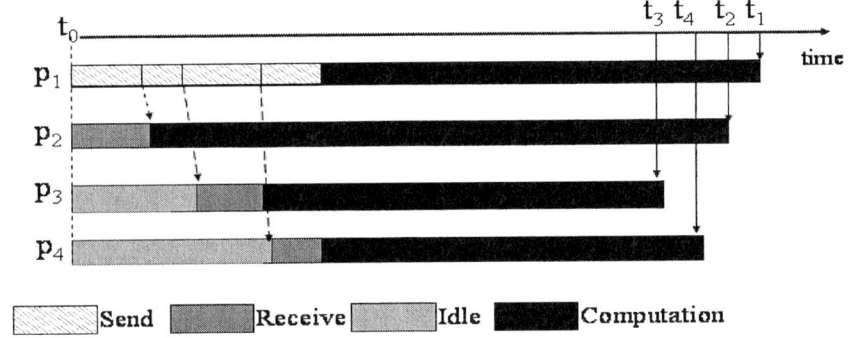

Fig. 2. A load balancing scattering operation

In [7], the MPI_Scatterv primitive in MPI is enhanced to support static load balancing data scattering in Grid environment. The authors attempt to find an optimal data distribution based on the DLT theory. For example, as indicated in figure 2, a Gantt-char-like timing diagram illustrates the communication and computation phases of divisible load. The application is running on four processors denoted by Pi. Root processor P1 scatters data to other processors at the initial time T0 then begins to process its own workload. Other processors start computation after receiving workload from root processor. The computation on each processor stops at the time Ti respectively. According to the DLT, the objective is to find out an optimized data distribution that minimizes the makespan of the application so that all processors can stop at the same time. The problem is modeled as a linear optimization problem under the subject to the communication latency and computation time, and a guaranteed heuristic solution is proposed to ensure all the sub-tasks end simultaneously. This method takes full consideration of the communication latency in Grid and makes an optimized load distribution decision. However, it suffers from the inability to dynamic changes of the processor utilization rate and fluctuations of network channel, which limits this approach to static load balancing.

Cyril Banino and Olivier Beaumont [1] recently propose a master-slave scheme for divisible load in heterogeneous platform. This scheme assumes parallel applications will always enter a steady state in which they spend a fix time fraction on computation and communication respectively. Therefore, instead of minimizing the makespan of the application, this scheme targets on maximizing the throughput of steady-state applications in unit time so as to avoid NP-complete complexity of makespan optimization. The objective of steady state optimization is to find out an optimal solution of time fractions application spent in computing, sending and receiving tasks. Another advantage of their work is that this model is very flexible and can extend to arbitrary platform topologies or application domains with different computation and communication requirement. But the validity of the steady state assumption still needs further verification, especially for adaptive applications.

3.3 Prediction Based Schemes

Load balancing schemes require accurate estimation of the future computation time and communication cost to establish the performance evaluation model. Simply using linear expression to calculate those values may be insufficient in Grid as unexpected system state changes and resource contention make the future system state quite different from the current.

For the Cactus application [14], Yang et al. propose a conservative time balancing scheme to calculate the load distribution of the Cactus application by utilizing the prediction value of host load in Grid. They use an aggregated time series prediction technique and give a simple but effective way to estimate the processor computation power during the period when the application will run. The work is allocated in a way so that the execute time of application is minimized. The experiments show that this scheme works well for short-time task. However, the performance model only focuses on the Cactus application, thus cannot directly apply to other applications, especially long-term running jobs.

Charm++ [9] also integrates prediction strategies into the load balancing framework. Charm++ uses multiple build-in prediction strategies to forecast the future system load so that the load balancer can make partition decision more accurately. It also exposes a callback interface that allows user implements prediction functions and gives load prediction hint to load balancing framework, which makes the predicted value more adaptive to the application at the cost of transparence.

4 Conclusions

These works make progress in addressing one or some challenging issues in load balancing for grid computing, but no solution is fully adaptive to the characteristics of Grid. Repartition methods focus on calculating data distribution in a heterogeneous way, but do not address the issue related to the cost incurred by data movement in Grid. DLT-based schemes can effectively model both computation and communication; however, the validation of these schemes for adaptive applications is needed. Prediction based schemes need further investigation in case of long-term applications.

References

1. Banino, C., Beaumont, O., Carter, L.: Scheduling Strategies for Master-Slave Tasking on Heterogeneous Processor Platforms, IEEE Transactions on Parallel and Distributed Systems, Vol. 15, No. 4, (2004) 319-330
2. Cybenko, G.: Dynamic load balancing for distributed memory multiprocessors, Journal of Parallel and Distributed Computing, Vol. 7, Issue 2. (1989) 279 - 301
3. Dasa, S.K., Harvey D.J., Biswas, R.: MinEX: A latency-tolerant dynamic partitioner for grid computing applications, Future Generation Computer Systems, Vol. 18, No. 4,(2002) 477--489
4. Devine, K., Hendrickson, B., Boman, E., John, M.S., Vaughan, C.: Design of Dynamic Load-Balancing Tools for Parallel Applications, Proceedings of the International Conference on Supercomputing, Santa Fe, (2000) 110 - 118
5. Faik, J., Gervasio, L. G., Flaherty, J. E., Chang, J. , Teresco,: A model for resource-aware load balancing on heterogeneous clusters, Tech. Rep. CS-03-03, Williams College Department of Computer Science, http://www.cs.williams.edu/drum/
6. Foster, I., Kesselman, C.: The Grid: Blueprint for a New Computing Infrastructure, Morgan Kaufmann Publishers, San Francisco, California, (1999).
7. Genaud, S., Giersch, A., Vivien, F.: Load-Balancing Scatter Operations for Grid Computing, International Parallel and Distributed Processing Symposium, Nice, France, (2003)
8. Johnston, W., Gannon, D., Nitzberg, B.: Grids as Production Computing Environments: The Engineering Aspects of NASA's Information Power Grid. IEEE Computer Society Press, (1999).
9. Laxmikant V. K., Krishnan, S.: CHARM++: a portable concurrent object oriented system based on C++, Proceedings of the eighth annual conference on Object-oriented programming systems, languages, and applications, Washington, D.C., (1993) 91-108
10. Thomas G. R.: Ten Reasons to Use Divisible Load Theory. IEEE Computer Vol. 36, No. 5, (2003) 63-68
11. Schloegel, K., Karypis, G..: Multilevel Diffusion Schemes for Repartitioning of Adaptive Meshes. Journal of Parallel and Distributed Computing, Vol.47, Issue 2, (1997) 109-124
12. Sinha, S., Parashar, M.: Adaptive System-Sensitive Partitioning of AMR Applications on Heterogeneous Clusters, Cluster Computing: The Journal of Networks, Software Tools, and Applications, Kluwer Academic Publishers, Vol. 5, Issue 4, (2002) 343-352
13. Wolskiy, R.: Dynamically Forecasting Network Performance Using the Network Weather Service, UCSD Technical Report TR-CS96-494, (1998)
14. Yang, L., Schopf, J. M., Foster, I.: Conservative Scheduling: Using Predicted Variance to Improve Scheduling Decisions in Dynamic Environments, Supercomputing, Phoenix, Arizona, (2003)

Fractal Tiling with the Extended Modular Group*

Rui-song Ye, Yu-ru Zou, and Jian Lu

Department of Mathematics, Shantou University,
Shantou, Guangdong, 515063, P. R. China
rsye@stu.edu.cn

Abstract. Automatic generation of fractal tiling patterns with the symmetry of the extended modular group is considered. Thanks to the unique one to one relationship between points in the upper half complex plane and the corresponding points in the fundamental region of the extended modular group, we can generate fractal tiling with the extended modular group by repeating the fractal pattern created in the fundamental region to the other tiling regions. We also produce such a kind of tiling in the unit disk by conformal mapping fractal tiling in the upper half complex plane. The method provides a novel approach for devising exotic fractal tiling patterns with symmetries.

1 Introduction

The perception of symmetry in art and nature has long captured man's attention. Symmetrical patterns arising naturally, whether they are stripes and spot formations in a biological context or X-ray diffraction patterns, are intriguing because they suggest both randomness and hidden patterns [1]. In some cases, tiling of abstract groups is rather unusual. The simultaneous sense of symmetry and the bizarre was used by the Dutch artist M. C. Escher, who often took advantage of planar crystallographic groups and hyperbolic groups [2], [3], [4]; in deed, he had significant interaction with mathematicians who studied group theory, for example, he kept contacting with Coxeter, a mathematician in the same age.

Symmetry can be understood more precisely by means of group theory [4], [5]. Actually a symmetrical tiling is related to a particular group. For instance, a repeated pattern in the Euclidean plane is related to one of the seventeen wallpaper or crystallographic groups [6], whereas a repeated pattern in the hyperbolic plane is related to one particular hyperbolic group [7]. Modern considerations with the aid of computers have become prolific, see for example [8], [9]. In recent years, automatic generation of symmetrical tiling patterns by means of computers attracts many mathematicians' interest. Wallpaper repeating patterns [10]

---

* Supported by Tianyuan Foundation, National Nature Science Foundation of China (No. A0324649).

and hyperbolic repeating patterns [11] are generated by visualizing dynamical systems' chaotic attractors.

The modular group Γ is the most widely studied of all discrete groups, the importance of which lies in its many connections with other branches of mathematics, especially with number theory. The symmetry of Γ can be visualized in the unit disk or the upper half complex plane. Although the symmetry has long been thoroughly studied, the creation of artistic patterns with such symmetry is rare. The reason may be due to the complicated geometric construction and numerical calculation. Chung et al. [12] proposed an algorithm to generate tiling patterns with the modular group from a dynamical system's point of view. By constructing some particular dynamical systems which satisfy the symmetry of the modular group elaborately, the authors created many exotic patterns. But the construction of dynamical systems seems rather complicated. As a result, the proposed method is not easy to manipulate and generalize.

In this paper, we present an alternative method to produce fractal tiling patterns. We propose a fast algorithm for the generation of fractal tiling patterns with the symmetry of the extended modular group Γ^*, which contains Γ as a subgroup of index 2. Such patterns are of infinitely fine structure and similarity. It is well known that fractal patterns are easy to generate and they are also very intriguing even by iteration of some simple complex dynamical systems [13], [14]. Though we needn't spend much time to construct some special dynamical systems associated with the extended modular group, we could generate large numbers of fractal tiling patterns with intoxicated artistic appearance. The method presented in this paper provides a novel approach for devising tessellation patterns with symmetries.

2 The Extended Modular Group

Let Z and C be the set of integers and complex numbers respectively. The group Γ of fractional linear substitution transformations, for which $\gamma \in \Gamma$ if

$$z \longmapsto z' = \gamma z = \frac{az+b}{cz+d} \quad (a,b,c,d \in Z, z, z' \in C)$$

with $ad - bc = 1$, is called the modular group. The modular group acts on the upper half complex plane $H = \{z \in C | Im(z) > 0\}$ onto itself. As is well-known,

$$D = \left\{ z \in H | -\frac{1}{2} \leq Re(z) \leq \frac{1}{2}, |z| \geq 1 \right\}$$

is a fundamental region for this action in a very precise sense that its images by Γ tessellate H and two distinct tiles have disjoint interiors. D is a hyperbolic triangle with vertices at $\rho = \exp(\pi i/3), \rho^2$ and ∞. This tessellation is called the modular tessellation.

Let D^* defined as follows:

$$D^* = \{z \in H | 0 \leq Re(z) \leq \frac{1}{2}, |z| \geq 1\}$$

which is the hyperbolic triangle with vertices at $i = \sqrt{-1}, \rho$ and ∞. Then D^* is a fundamental region for the extended modular group Γ^* which is the group generated by reflection along the edges of D^*. A tessellation of the upper half complex plane induced by Γ^* is shown in figure 1(a).

The elements of Γ^* map H onto itself. Γ^* has three generators κ, α and β such that

$$\kappa z = -\bar{z}, \alpha z = z + 1, \beta z = -\frac{1}{z}, \forall z \in H.$$

It is well know that α and β are also the generators for the modular group Γ. We note that every element γ of Γ can be expressed as a finite product of the generators' power including negative power:

$$\gamma = \alpha^{n_m} \beta \alpha^{n_{m-1}} \beta \cdots \beta \alpha^{n_1},$$

and every element γ of Γ^* can be expressed as

$$\gamma' = \kappa^s \alpha^{n_m} \beta \alpha^{n_{m-1}} \beta \cdots \beta \alpha^{n_1},$$

where $s \in \{0, 1\}$ and $n_i (i = 1, \cdots, m) \in Z$.

For a given point $z = x + yi$, there exist an element $\gamma \in \Gamma$ and a point $w \in D$ such that $w = \gamma z$. Similar conclusion is also hold for the extended modular group Γ^*, i. e., there exist $\gamma' \in \Gamma^*$ and $w' \in D^*$ such that $w' = \gamma' z$.

γ' and w' can be computed numerically as follows: if $x \notin I = (-\frac{1}{2}, \frac{1}{2}]$, the point z will be translated horizontally to z_1 by

$$z_1 = \alpha_{n_1} z = x - \lfloor x + \frac{1}{2} \rfloor + iy,$$

where $-n_1 = \lfloor x + \frac{1}{2} \rfloor$ is the largest integer smaller than $x + \frac{1}{2}$. As $Re(z_1) \in I$, it remains to see whether $|z_1| \geq 1$. If this is not the case, then z_1 will be transformed by β. Note that the imaginary part of βz_1 is strictly greater than y. The process is repeated by checking if $Re(\beta z_1) \in I$. Each time when β is applied, the imaginary part of the image is getting larger and eventually the final image will fall within D. As the image falls into D, we then decide whether we should apply κ to map the image into the fundamental region D^* of the extended modular group Γ^*, and this is easy to do by checking the real part of the image, i. e., if the real part of the image is less than zero, then we transform it by κ; if it is not the case, then we don't need to do anything.

Tiling with the extended modular group can also be visualized elegantly in the Poincaré model which contains all the points of the unit disk $D_0 = \{z \in C | |z| < 1\}$. The upper half complex plane is related to the Poincaré model by conformal mappings. We could choose suitable conformal mappings such as

$$\Pi_1(z) = \frac{zi + 1}{z + i}, \quad \Pi_2(z) = e^{zi}.$$

Tessellations of the Poincaré model with the above two conformal mappings are shown in figures 1(b-c).

3 Fractal Tiling with the Extended Modular Group

We have shown in Section 2 that every point in the upper half complex plane will be mapped to one corresponding point in the fundamental region of the extended modular group. Therefore, if we want to create tiling with the symmetry of the extended modular group, we just only need to generate fundamental patterns in the fundamental region. We here produce fractal patterns as fundamental patterns, then we map them by the extended modular group. As a result, we will obtain fractal tiling with the extended modular group.

It is well known that escape-time algorithm can be used to generate Mandelbrot sets and Julia sets of complex dynamical systems. The escape-time algorithm is one of the earliest coloring algorithms, and in many programs it is still the only option available. Its simplicity makes it a favorite with those learning to create fractal software.

In order to produce the basic fractal image in the fundamental region of the extended modular group, we choose the following complex dynamical systems:

$$\text{(I)} \quad z = \exp(z^\alpha + c),$$
$$\text{(II)} \quad z = \cos(\frac{2\pi}{z^2 - 0.33\pi}),$$
$$\text{(III)} \quad z = \cos(z^2 + c).$$

The corresponding Mandelbrot and Julia fractal tiling patterns with the extended modular group are shown in figures 2-4.

4 Conclusion

Thanks to the advancement of computer graphic techniques, one can easily create fascinating and beautiful fractal images by iterating some simple complex dynamical systems and even create fractal like tiling with symmetries. A fast and new algorithm has been proposed to automatically generate colored fractal tiling with the symmetry of the extended modular group. Compared with the algorithm proposed in [12] for producing tessellation associated with the modular group, the algorithm presented here is more easy and convenient to manipulate without requiring to construct complicated dynamical systems which satisfy the symmetry of the extended modular group. We could produce a variety of fractal tiling patterns by just changing the complex systems easily and could also produce different kinds of fractal tiling both in the upper half complex plane and the unit disk by choosing different kinds of conformal mappings. The fractal tiling patterns reveal more artistic appearance and are of similarity and infinitely fine structure. The method provides a novel approach for devising fractal tiling patterns with the symmetry of the extended modular group.

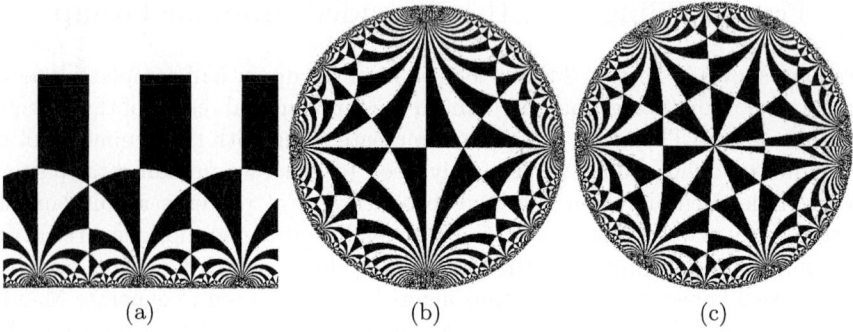

Fig. 1. (a) Tiling pattern with the extended modular group in the upper half complex plane. (b-c) Two kinds of tiling patterns with the extended modular group in the unit disk

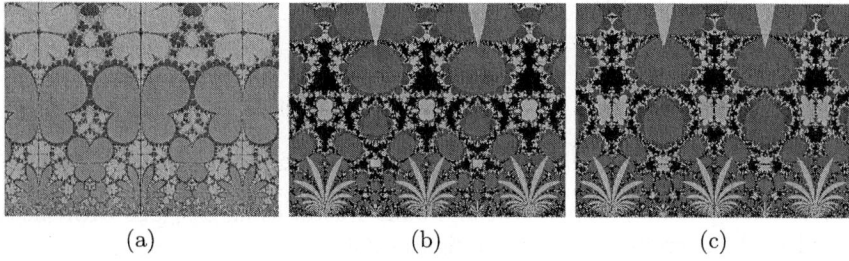

Fig. 2. (a) Julia fractal tiling in the upper half complex plane by complex system (II). (b) Julia fractal tiling in the upper half complex plane by complex system (III) with $c = (-0.64, 0.59)$. (c) Julia fractal tiling in the upper half complex plane by complex system (III) with $c = (-0.44, 0.59)$

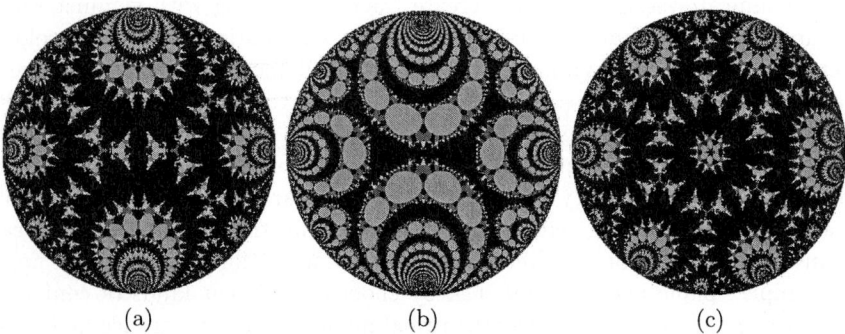

Fig. 3. (a) Mandelbrot fractal tiling in the unit disk by complex system (I) with $\alpha = 2$ and conformal mapping Π_1. (b) Mandelbrot fractal tiling in the unit disk by complex system (I) with $\alpha = 4$ and conformal mapping Π_1. (c) Mandelbrot fractal tiling in the unit disk by complex system (I) with $\alpha = 2$ and conformal mapping Π_2

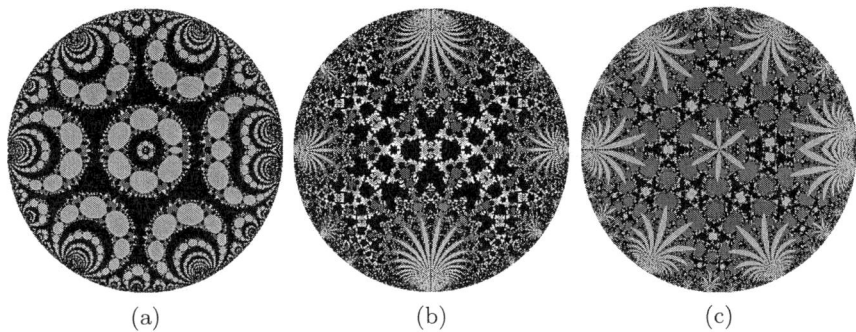

Fig. 4. (a) Mandelbrot fractal tiling in the unit disk by complex system (I) with $\alpha = 4$ and conformal mapping Π_2. (b) Julia fractal tiling in the unit disk by complex system (III) with $c = (-0.64, 0.59)$ and conformal mapping Π_1. (c) Julia fractal tiling in the unit disk by complex system (III) with $c = (-0.64, 0.59)$ and conformal mapping Π_2

References

1. Zhu, M., Murray, J.D.: Parameter domains for spots and stripes in mechanical models for biological pattern formation. Journal of Nonlinear Science, **5** (1995) 317–336
2. Escher, M.C.: Escher on Escher–Exploring the Infinity. Harry N. Abrams, New York (1989)
3. Schattschneider, D.: Visions of Symmetry: Notebooks, Periodic Drawings and Related Works of M.C. Escher. Freeman, New York (1990)
4. Coxeter, H.S.M., Moser, W.O.J.: Generators and Relations for Discrete Groups. 4th edition, Springer-Verlag, New York (1980)
5. Weyl, H.: Symmetry. Princeton University Press, Princeton (1952)
6. Armstrong, M.A.: Groups and Symmetry. Springer-Verlag, New York (1988)
7. Dunham, D.: Hyperbolic symmetry. Computers & Mathematics with Applications, **12B** (1986) 139–153
8. Pickover, C.A.: Computers, Pattern, Chaos and Beauty. Alan Sutton Publishing, Stroud Glouscestershire (1990)
9. Field, M., Golubistky, M.: Symmetry in Chaos. Oxford University Press, New York (1992)
10. Carter, N., Eagles, R., Grimes, S., Hahn, A., Reiter, C.: Chaotic attractors with discrete planar symmetries. Chaos, Solitons and Fractals, **9** (1998) 2031–2054
11. Chung, K.W., Chan, H.S.Y., Wang, B.N.: Hyperbolic symmetries from dynamics, Computers & Mathematics with Applications, **31** (1996) 33–47
12. Chung, K.W., Chan, H.S.Y., Wang, B.N.: Tessellations with the modular group from dynamics. Computers & Graphics, **21** (1997) 523–534
13. Barnsley, M.F.: Fractals Everywhere. Academic Press, San Diego (1988)
14. Ye, R.: Another choice for orbit traps to generate artistic fractal images. Computers & Graphics, **26** (2002) 629–633

Shelling Algorithm in Solid Modeling

Dong-Ming Yan, Hui Zhang, Jun-Hai Yong,
Yu Peng, and Jia-Guang Sun

Tsinghua University, Beijing 100084, China
yandm@cg.cs.tsinghua.edu.cn

Abstract. Shelling is a powerful modeling operation in current CAD/CAM systems. In this paper, we present a new shelling algorithm for regular B-Rep solids. We first generate an initial offset solid, then correct the self-intersecting offset solid if necessary. We obtain the resultant shelling solid through a regularized Boolean operation finally. The algorithm has been implemented in a solid modeling system, and some examples are given to illustrate it.

1 Introduction

Shelling and offsetting are two related modeling operations widely used in CAD/CAM systems. Offsetting is a procedure for adding or removing a constant thickness from a solid model. Shelling is a procedure for turning a solid model into a thin-walled shell of constant thickness or variable thickness. This "shelled" solid is a hollowed version of the original model.

The research on offsetting has been carried out for a long time. Lee et al. [8] classified it into two main categories: offset geometry and offset topology. In the last decade or so, numerous researchers focused on offsetting curves or surfaces [5]. Generating offset solids belongs to the area of offset topology [1, 2, 3]. Satoh et al. [4] developed a Boolean operation algorithm on open sets. They also presented an algorithm for offset solid generation using their Boolean operation method. Lee et al. [8] proposed an algorithm for non-manifold offsetting operation. Recently, Ravi [10] developed an algorithm for the automatic offset of a NURBS B-Rep solid.

But very few studies focus on shelling a solid. Forsyth [6] presented several algorithms for offsetting and shelling operations on B-Rep solids, but it must be ensured that the offset solid does not self-intersect. Zuo [9] also developed a shelling algorithm in a solid modeling system, but his method has the same drawback as Forsyth's.

We present a solid shelling algorithm in this paper, which is more robust than Zuo's and Forsyth's algorithm. Firstly, we compute an initial offset solid of the original solid, then turn the self-intersecting offset solid into 2-manifold, if necessary. Finally we compute the shelling body through a Boolean operation.

2 Definition

In this paper, we assume that the solid is 2-manifold and in B-Rep form. Given a 2-manifold B-Rep solid $O \subseteq \mathbf{R}^3$ (three-dimensional Euclidean space). The boundary ∂O of the body O is the union of a set of faces, denoted by $F(O)$. Each face in $F(O)$ is a trimmed surface, which includes an underling surface to represent its geometric shape and a set of loops to define its boundary. A face must have a counterclockwise outer loop and some clockwise inner loops. All the loops of a face don't intersect with each other.

Generally, shelling operation needs to specify one or more pierce faces and the shelling direction. The offset distances of pierce faces are zero. Each of the non-pierce faces of the body offsets a user-defined nonzero distance, and the distances can be varied. If the shelling direction is inward, the offset direction is opposite to the normal of the face, otherwise the offset direction is along the normal of the face. The definition of shelling is illustrated in Fig. 1.

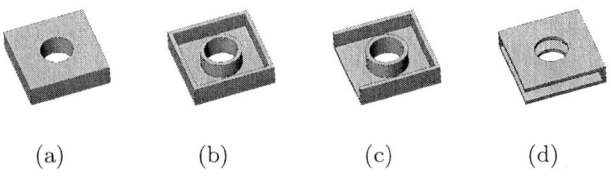

(a)   (b)   (c)   (d)

Fig. 1. Illustration of inward shelling operation. (a) The original solid. (b) Only the top face is a pierce face. (c) Both the top face and the left side face are pierce faces. (d) The left side, the front side and the cylindrical face are pierce faces

3 Methodology

At the beginning of the shelling operation, user should select the pierce face(s), set the offset distances of the non-pierce faces(the distances can be varied), and set the direction of shelling, i.e. inward or outward.

3.1 Initial Offset Solid Generation

In this step, we compute an initial offset solid, which may be topologically irregular in geometric substitution. We will solve the issue in the next step.

Firstly, we use Forsyth's method [6] to create an offset face F' of each face F. An offset surface is defined as the locus of points, which are at a constant distance d along the normal from the base surface [10]. Secondly, offset curve for each edge E is obtained by intersecting two offset surfaces of E's two adjacent faces. And then we use offset surfaces of E's two section faces to trim the intersecting curve. Thus the trimmed curve is between the two section faces. We denote edge E's offset edge as E', and attach the trimmed curve to the related offset edge E'. Then the offset edges of each offset face are connected to form offset loops L', which have the same direction as their original corresponding loops L, respectively, and the offset loops are attached to their corresponding offset faces. Finally we sew all the offset faces together to obtain an initial offset solid.

3.2 Facial Error Management

In the following step, we should judge whether the faces of the offset solid is valid. Several cases may occur on a single face. First of all, we judge whether a loop of face is self-intersecting, then we detect whether some loops of a face intersect with each other, and finally we judge whether a face has a clockwise outer loop. The offset solid becomes illegal if one or more cases occur.

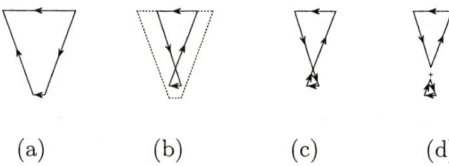

    (a)   (b)   (c)   (d)

Fig. 2. Illustration of loop self-intersection

Fig. 2 illustrates the loop self-intersection problem in 2D. Fig. 2(a) is the original loop. The offset loop in Fig. 2(b) is self-intersecting(The dashed denotes the original loop). It is detected by intersecting every two edges of the loop. If the intersection point is not the endpoint of an edge, the loop is self-intersecting. We split the intersected edges at the intersection points, as shown in Fig. 2(c), and partition the loop into several sub-loops, as shown in Fig. 2(d). If the original loop is counterclockwise, we create a new face whose direction is reversed from the original face for each clockwise sub-loop. Otherwise, the original loop is clockwise, and we create a new reversed face for each counterclockwise sub-loop.

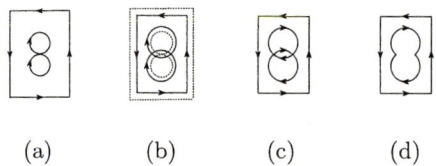

    (a)   (b)   (c)   (d)

Fig. 3. Illustration of inner-inner loop intersection

Two or more loops of an offset face may intersect with each other if the face has any inner loops. Fig. 3 and 4 illustrate two cases of loop intersection. One is that several inner loops intersect with each other, and the other is that one or more inter loops intersect with the outer loop.

Firstly, we compute the intersection points among all inner loops, and then spilt the intersected edges with those points, as shown in Fig. 3 and 4. At last, we merge the intersected inner loops into one inner loop(for example, the inner loops in Fig. 3 and a 3D example in Fig. 6).

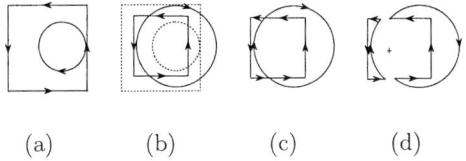

(a) (b) (c) (d)

Fig. 4. Illustration of inner-outer loop intersection

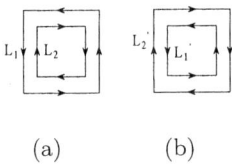

(a) (b)

Fig. 5. Illustration of loop direction error

After merging all the intersected inner loops, we compute the intersection between the outer loop and inner loops, and then split edges similar to the above case. We partition the intersected loops into several sub-loops, which don't intersect with each other. Then create a new face for the clockwise loop with the reversed direction of the original face. Fig. 7 gives a 3D example.

Each face must contain a counterclockwise outer loop and some clockwise inner loops. The loops don't intersect with each other. Fig. 5(a) shows an original face with an outer loop L_1 and an inner loop L_2. Fig. 5(b) shows the offset face of Fig. 5(a), where L'_1 and L'_2 are the offset loops of L_1 and L_2, respectively. The offset face is topologically incorrect. We deal with this problem by simply reversing the direction of the face. Fig. 8 gives an example for this case in 3D.

3.3 Solid Correction and Shelling

The cases discussed above only deal with the single face of the offset solid. Moreover we should judge whether the faces of the offset solid intersect with each other. Therefore, we apply a face-face intersection procedure in this step which partitions the intersection faces by intersection curves.

The offset solid may be non-manifold now. We should eliminate the illegal parts to obtain a 2-manifold solid. For each face of offset solid, we get an arbitrary reference point **p** on the face, which does not lie on the boundary of the face. The normal of the face at the point **p** is denoted by **N**. We create a ray line by using **p** as start point and **N** as its direction, and then calculate the intersection points between the ray line and all other faces. We mark the current face as undesired if the number of intersection points is odd, otherwise mark the face as reserved. We delete the undesired faces and the open faces (one or more edges of this face have only one adjacent face). The remaining faces will be sewed together to form the final offset solid. The offset solid will be $NULL$ if no face is left.

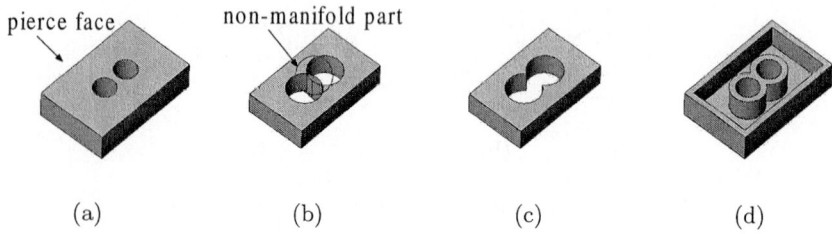

Fig. 6. Example of Inner-Inner loop intersection

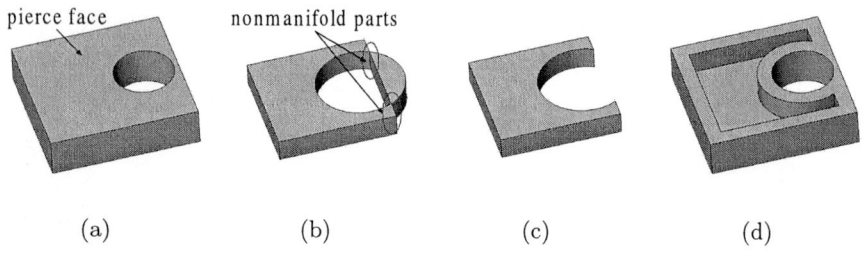

Fig. 7. Example of Outer-Inner loop intersection

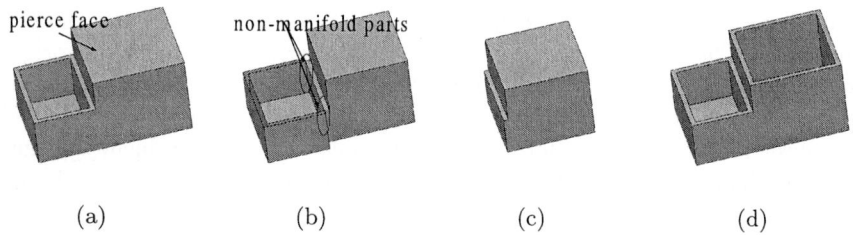

Fig. 8. Example of facial direction error

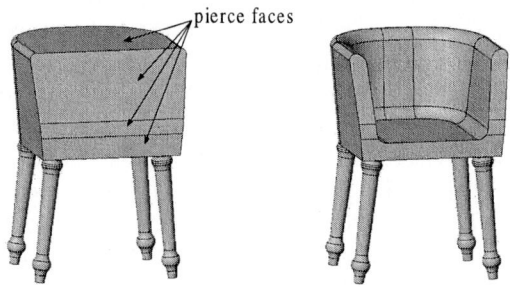

Fig. 9. Chair example

We denote the offset body after correction by O'. The shelling result of solid O is formed by the following regularized Boolean operation rules:

$$Shell(O) = \begin{cases} O, & \text{if } O' = NULL, \\ O - O', & \text{if } inward\ shelling, \\ O' - O, & \text{if } outward\ shelling. \end{cases}$$

4 Results and Conclusion

Fig. 6 - 8 give some examples of our algorithm. In each example, (a) shows the original solid, (b) shows the initial offset solid after loop splitting operations, (c) shows the offset solid after correction, and (d) shows the final shelling result. Fig. 9 gives a chair example using our shelling algorithm.

We present a new shelling algorithm for B-Rep solids. The loop operation used in our algorithm is similar to Gardan's algorithm [7], which reduces the complexity of the algorithm from 3D to 2D. Our algorithm has been implemented in a solid modeling system TiGems.

Acknowledgements

The research was supported by Chinese 973 Program (2002CB312106) and Chinese 863 Program (2003AA4Z3110).

References

1. Farouki, R.T.: Exact Offset Procedures for Simple Solids, Computer Aided Geometric Design. **2** (1985) 257-279
2. Rossignac, J. R., Requicha, A. A. E.: Offsetting Operations in Solid Modelling, Computer Aided Geometric Design. **3** (1986) 129-148
3. Saeed, S. E. O., de Pennington, A. and Dodsworth, J. R.: Offsetting in geometric modeling, Computer-Aided Design, **20** (1988) 50-62
4. Satoh, T., Chiyokura, H.: Boolean Operations on Sets Using Surface Data, ACM SIGGRAPH: Symposium on Solid Modeling Foundations and CAD/CAM Applications. (1991) 119-127
5. Pham, B.: Offset Curves and Surfaces: a Brief Survey, Computer-Aided Design. **24** (1992) 223-229
6. Forsyth, M.: Shelling and offsetting bodies, Proceedings of the third ACM symposium on Solid modeling and applications. (1995) 373-381
7. Gardan, Y., Perrin, E.: An algorithm reducing 3D Boolean operations to a 2D problem: concepts and results. Computer Aided Design, **28** (1996) 277-287
8. Lee, S. H.: Offsetting operations on non-manifold boundary representation models with simple geometry. Proceedings of the fifth ACM symposium on Solid modeling and applications, (1999) 42-53
9. Zuo, Z., Hu, S. M. and Sun, J. G.: A shell algorithm for solid on boundary representation. Journal of Software. **10** (1999) 761-765 (In Chinese)
10. Ravi Kumar, G. V. V., Shastry, K. G. and Prakash, B. G.: Computing constant offsets of a NURBS B-Rep. Computer-Aided Design. **35** (2003) 935-944

Load and Performance Balancing Scheme for Heterogeneous Parallel Processing

Tae-Hyung Kim

Hanyang University, Dept. of Computer Sci. & Eng., Kyunggi-Do 426-791, Korea
tkim@cse.hanyang.ac.kr

Abstract. Dynamic loop scheduling methods are suitable for balancing parallel loops on shared-memory multiprocessor machines. However, their centralized nature causes a bottleneck for relatively small number of processors in internet-wise parallel processing using computational grids because of the order-of-magnitude differences in communications overheads. Moreover, improvements of basic loop scheduling methods have not dealt effectively with irregularly distributed workloads in parallel loops, which commonly occur in large applications. In this paper, we present a decentralized balancing method, which tries to balance load and performance simultaneously, for parallel loops on computational grids.

1 Introduction

Recently, the computational *grids* [1] have emerged as a universal source of computing power, where many heterogeneous computing nodes can be utilized for a large application. In grid computation, a simple policy like equally distributing workloads to multiple processors may lead to a *parallelization anomaly*. That is, the execution time of the given workload may take longer even if the number of computing nodes is increased. Suppose there are n processors $\{P_1, \ldots, P_n\}$, and T identical tasks. Let τ_i be the number of tasks per unit time that P_i can process. In equal distribution, each processor has T/n numbers of tasks. The execution time of the program is determined by the critical processor that has the smallest τ_i value; let's say it is τ_{min}. Then the execution time is $\frac{T/n}{\tau_{min}} = \frac{T}{n\tau_{min}}$. Now, let's add a new processor of τ_{new} to the cluster for the application. Each processor will have $T/(n+1)$ tasks. Therefore, if $\tau_{new} < \frac{n}{n+1}\tau_{min}$, the execution time of $(n+1)$-processors cluster is $\frac{T}{(n+1)\tau_{new}}$, which is longer than that of n processors!

One may want to get around this problem by allocating tasks according to the known computing power of each processor [2]. However, their methods were static, thus of limited usefulness. Dynamic loop scheduling methods can deal with more general cases, but the centralized nature of the methods — the central processor that generates sub-tasks has to manage all other processors — may cause a bottleneck in a network of many workstations. Since there are many "embarrassingly parallel" applications, a decentralized form of load balancing scheme is called for.

Moreover, load balancing gets more complicated in grid computation environments, because of their nature of heterogeneity. *Self-scheduling* (SS) [3] is the simplest dynamic solution in heterogeneous parallel computing. It assigns a new iteration to a processor only when the processor becomes available. Thus, it conveys a perfect balance in whatever the circumstances. However, this method requires tremendous synchronization overhead. While improvements have been made on SS to deal with the overhead as reported in [4, 5, 6], they cannot deal with heterogeneity. Some researchers tried to develop a parallel loop scheduling method in a network of heterogeneous workstations [7]. They considered three aspects of heterogeneity — loop, processor, and network — and developed algorithms for generating optimal and sub-optimal schedules of loops. But their algorithm is basically static and the loop heterogeneity model is linear. As a result, it requires a predictable form of parallel loops to be effective.

In this paper, we present a performance-aware load balancing method for parallel loops of more general patterns including irregular ones, since many large applications often do not carry conventional regular loop patterns. Such a new scheme may be quite suitable for in grid computation, since the unpredictable loop patterns are natural in their applications. Note that the previous improvements [4, 5, 6] on SS cannot perform well but sometimes even worse for those parallel loops of irregular form.

To come up with such a scheme, we may remember that parallel tasks and their working platforms are two ingredients in parallel processing. Nonetheless, only the "tasks" part has been the focus of load balancing. There has been nothing wrong in this because the platforms part has been a fixed architecture. However, in heterogeneous grid computation, the conventional global load balancing and dynamic loop scheduling methods can become problematic as well because they do not consider the performance distribution among participating nodes. One of the key issues in dynamic load balancing is how to reduce the accompanying overheads. The main idea of our approach is balancing the performance as well to facilitate balancing the parallel tasks; i.e. we construct a special migration topology based on relative processor speeds in order to reduce migration overheads. Basically our method adopts demand-driven migrations the same way that dynamic loop scheduling methods do while well-constructed topology reduces unnecessary migrations.

2 Execution Model: Loop and Processor

In this section, we classify four typical parallel loop patterns that affect performance of load balancing schemes based on workload distribution in an iteration space. Next, we discuss our processor model to deal with those diverse patterns, especially if the processors involved are heterogeneous.

Fig. 1 shows four typical parallel loops where $L(i)$ represents the execution time of the i-th iteration. The workload may be uniformly distributed over an iteration space as shown in Fig. 1 (a). It may also be non-uniform but *linearly* distributed as in Figs. 1 (b) and (c); this kind of distribution is often contained in

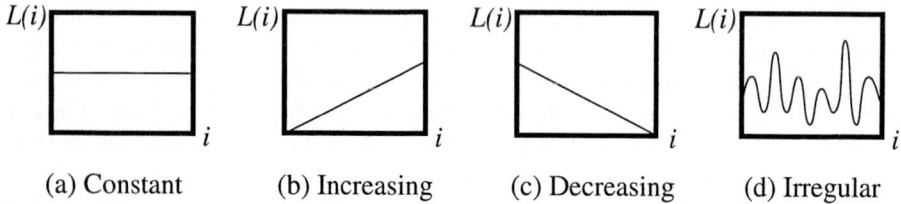

Fig. 1. Four typical parallel loops

scientific programs. Finally, as in Fig. 1 (d), the workload may be quite irregular. Many non-scientific applications carry parallel loops of this type. The first three cases have been specially considered by conventional loop scheduling methods [5, 6, 7] in order to improve the basic self-scheduling method.

Traditional loop scheduling methods [3, 4, 5, 6] are based on centralized processing. The main processor prepares a set of tasks and allocates them to each computing nodes whenever they demand. Since the scheduling process is dedicated to the main processor, its chance of creating a bottleneck rises as the number of processors present on the network increases. In our model, the entire workloads are statically distributed to all computing nodes according to the relative performance data in the computation nodes, and then task migration is performed in a decentralized fashion. In this way, the dynamic balancing overheads can be minimized.

3 Performance and Load Balancing Method

Two important components of dynamic load balancing schemes are transfer policy and location policy [8]. The transfer policy determines whether a task should be processed locally or remotely by transferring it at a particular load state. The location policy determines which process initiates the migration and its source or destination. These are for global load balancing from the OS's viewpoints. In our scheme, we aim to balance parallel loops in an application. A simple 'demand' message is enough to initiate load migration rather than load state exchange or random polling of candidate processors because the only load vector is the number of sub-tasks in a processor. The transfer policy then becomes simple: if a processor receives a request message for transfer from a processor that is running out of sub-tasks to work on, it migrates some of its sub-tasks to that processor.

Likewise, the location policy is now modified by the problem of establishing proper task migration paths. It may be assumed that any two point-to-point communication overheads are equal, but identifying the optimal sender and receiver pair is essential. Considering all possible candidates for sender (or receiver) to migrate the excess load causes high overhead, but it is avoidable. The key is how to identify the busy and the idle processors in the middle of computations.

Since the relative processing speeds of workstations in a cluster are known in advance, the possible senders and receivers of migrations are known — momentary overload by other activities is the reason for uncertainty.

In this section, we present how to construct such a task migration network. Once the network is constructed, load balancing is pursued through task migration on it. The task migration is performed between the paired processors; whenever the faster processor (P_j) depletes its workload, it demands that its pre-determined partner P_i, and P_i migrates γ_{ij} of its current workload to P_j.

A cluster is a bipartite form of (C_s, C_f), in which C_s is slower than C_f (i.e. $\tau_s < \tau_f$). An entire computation cluster is defined as follows:

Definition 1. The *cluster tree* (CT) of N processors $\{P_1, \ldots, P_N\}$ is a binary tree $CT = (V, E_{left} \cup E_{right})$, where

- The vertices V represent *clusters*. A distinguished vertex 'root' represents an entire cluster, and the right sub-cluster is faster than (or equal to) the left sub-cluster.
- E_{left} is a set of edges to the left sub-trees. E_{right} is a set of edges to the right sub-trees.
- If $(c, v) \in E_{left}$ and $(c, w) \in E_{right}$, a load migration path exists from v to w. When v and w are not terminal nodes, the path is established from the fastest node in cluster v, which is the rightmost terminal in the subtree of v, to the slowest node in cluster w, which is the leftmost terminal in the subtree of w.

In this definition, terminal vertices are individual processors. Each terminal v is associated with its throughput τ_v. Throughput of non-terminal node $C = (v, w)$ is defined by $(\tau_v + \tau_w)$. In a cluster, note that the migration overhead is proportional to the number of migrations rather than the size of workload itself due to higher communication latency in internet-wise communication. Thus, the balance ratio can be defined by the ratio between combined throughput and its throughput difference. In the extreme case that τ_1 is equal to τ_2, the *balance ratio* is zero; thus load is perfectly balanced. Likewise, in the other extreme in which τ_2 is much greater than τ_1, the ratio is asymptotically 1. The *balance ratio* in a cluster can be related to the amount of load migration. When the components in a cluster are equally loaded initially, if the cluster is perfectly balanced, then no intra-cluster migration is necessary. In other words, the more balanced a cluster is, the less migration is needed.

The process of constructing a cluster tree from a set of processors is done in recursive bitonic fashion. First, processors in the set $\{P_1, \ldots, P_n\}$ become terminal nodes in the tree. They are sorted in ascending order by their throughputs. Let the sorted set be $\{P'_1, \ldots, P'_n\}$. The fastest one (P'_n) is coupled with the slowest one (P'_1), the second fastest one (P'_{n-1}) is coupled with the second slowest one (P'_2), and so forth. The couples come to have parents in the tree, i.e. $\{c_1 = (P'_1, P'_n), \ldots, c_{n/2} = (P'_{n/2}, P'_{n/2+1})\}$, which are likewise sorted by their throughputs. Again, they are coupled in bitonic fashion, i.e. $\{cc_1 = (c'_1, c'_n), \ldots, cc_{n/4} = (c'_{n/4}, c'_{n/4+1})\}$. This process continues until it reaches a

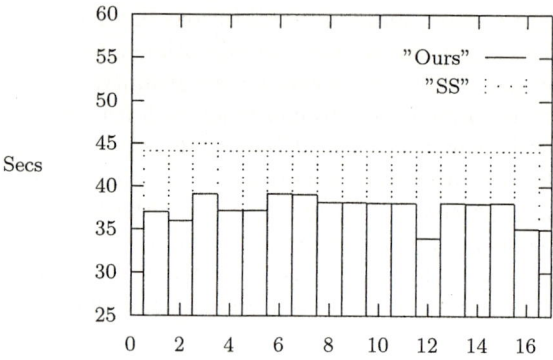

Fig. 2. Execution time: Mandelbrot set computation on [0.5,-1.8] to [1.2,-1.2]

single cluster. Notice that the cluster of the two identical components still needs an intra-cluster migration because an equal distribution is not always possible. Once such a tree is constructed, the task migration topology is determined as follows:

For all clusters (non-terminal nodes) c in CT
    For two children v and w such that $(c,v) \in E_{left}$ and $(c,w) \in E_{right}$
        if $(v, w$ are terminals) then $CONNECT\ v\ TO\ w$
        else $CONNECT$ RightmostTerminal(v) TO LeftmostTerminal(w)

4 Experimental Results

To demonstrate the performance of our method, we conducted our experiment on a cluster of 16 heterogeneous PCs. Specifically, the fastest PC is about 10 times more powerful than the slowest one. The example program was Mandelbrot set computation on $[0.5, -1.8]$ to $[1.2, -1.2]$ using a 800 × 800 pixel window. Mandelbrot program contains unpredictably irregular loops, thus it is almost impossible to determine an ideal load distribution statically among multiple processors.

We have initially distributed those tasks in a round-robin style. The results by the 16-node cluster are given by Fig. 2. The dotted boxes represent the finish times of each participating PC under the pure self-scheduling method, which substantiate the expected good load balance. The result by our method is seemingly imbalanced but note that the actual finish time is much improved. Perfect balance may be good but the evaluation should be finally made based on how much its overheads negate its resulting benefits as well.

The following table summarizes the size of each migration and its frequency that are counted in our experimentation. For example, the single-task migration occurred 13 times, and the 30-tasks migration occurred once, etc, during the entire task migration attempts. In the table, we can compute the total occurrences of migrations by summing all frequencies up. Thus, the latency time (α) can be drastically reduced to 40α, rather than 147α in self-scheduling case, which can illustrate where our improvement comes from; load balance is not for free, and if the price is too high, then we should refrain from buying it.

taskcount	1	2	3	4	5	6	7	10	11	12	17	19	20	30
frequency	13	6	3	3	3	1	4	1	1	1	1	1	1	1

5 Conclusion

We have presented a performance-aware load balancing method for parallel tasks that is suitable for computational grid environments. We discussed why the conventional global dynamic load balancing methods are not adequate to this new computing paradigm. Loop scheduling schemes that have been useful under shared-memory multiprocessor machines cause a bottleneck because the communication overheads are higher in order of magnitude scale. The predefined path for load migration based on performance of each node has not been considered important heretofore because normally the parallel processing architecture is given in a hard-wired form or it is meaningless where distributed load patterns cannot be assumed to be known in advance [8]. We have shown experimentally that our scheme performs better eventually, although it does not reach a perfect balance like SS, since it reduces the unnecessary migrations. Since the grid environment provides us very flexible form of parallel computation, we may need to investigate more interesting topologies in the future.

References

1. Foster, I., Kesselman, C., Nick, J., Tuecke, S.: The Physiology of the Grid: An Open Grid Services Architecture for Distributed Systems Integration, Open Grid Service Infrastructure WG, Global Grid Forum. (2002)
2. Grimshaw, A., Wulf, A.: The Legion Vision of a Worldwide Virtual Computer. Communications of the ACM, Vol. 40, Issue. 1 (1997)
3. Tang, P., Yew, P.-C.: Processor Self-Scheduling for Multiple Nested Parallel Loops. Proceedings of International Conference on Parallel Processing(1986)
4. Kruskal, C. P., Weiss, A.: Allocating Independent Subtasks on Parallel Processors. IEEE Transactions on Software Engineering, Vol. 11. 1001–1016 (1985)
5. Polychronopoulos, C. D., Kuck, D. J.: Guided Self-scheduling: A Practical Scheduling Scheme for Parallel Supercomputers. IEEE Transactions on Computer, Vol. C-36, No. 12, 1425–1496 (1987)
6. Tzen, T. H., Ni, L. M.: Dynamic Loop Scheduling for Shared-Memory Multiprocessors, Proceedings of International Conference on Parallel Processing (1991)
7. Cierniak, M., Li, W., Zaki, M. J.: Loop Scheduling for Heterogeneity, Proceedings of International Symposium on High Performance Distributed Computing. (1995)
8. Krueger, P., Shivaratri, N.: Adaptive Location Policies for Global Scheduling. IEEE Transactions on Software Engineering, Vol. 20, No. 6 (1994)

A Nonlinear Finite Difference Scheme for Solving the Nonlinear Parabolic Two-Step Model

Weizhong Dai[1] and Teng Zhu[2]

[1] Mathematics & Statistics, College of Engineering & Science,
Louisiana Tech University, Ruston, LA 71272, USA
dai@coes.latech.edu
[2] Department of Physics, Grambling State University,
Grambling, LA 71270, USA
zhut@gram.edu

Abstract. A nonlinear finite difference scheme is developed by obtaining an energy estimate for solving the parabolic two-step model with temperature-dependent thermal properties in a double-layered micro thin film irradiated by ultrashort-pulsed lasers. The method is applied to investigating the heat transfer in a gold layer on a chromium layer.

1 Introduction

Ultrashort-pulsed lasers with pulse durations of the order of sub-picosecond to femtosecond domain possess exclusive capabilities in limiting the undesirable spread of the thermal process zone in the heated sample. They have been widely applied in structural monitoring of thin metal films, laser micromachining and patterning, structural tailoring of microfilms, and laser synthesis and processing in thin-film deposition as well as in physics, chemistry, biology, medicine, and optical technology [1].

For a ultrashort-pulsed laser, the heating involves high-rate heat flow from electrons to lattices in the picosecond domains. This stage is termed non-equilibrium heating due to the large difference of temperatures in electrons and lattices [2]. The energy equations describing the continuous energy flow from hot electrons to lattices during non-equilibrium heating can be written as:

$$C_e(T_e)\frac{\partial T_e}{\partial t} = \nabla[k_e(T_e, T_l)\nabla T_e] - G(T_e - T_l) + S, \quad (1)$$

$$C_l\frac{\partial T_l}{\partial t} = G(T_e - T_l), \quad (2)$$

where $C_e(T_e) = A_e T_e$, $k_e(T_e, T_l) = k_0(\frac{T_e}{T_l})$, T_e is electron temperature, T_l lattice temperature, k_0 thermal conductivity in thermal equilibrium, C_e and C_l volumetric heat capacity, G electron-lattice coupling factor, S laser heating source, and ∇ the gradient operator. Here, A_e, C_l, k_0 and G are all positive. In the classical theory of diffusion, $T_e = T_l$ because thermal equilibrium between the electrons and lattices is reached. Thus, the above two equations can be reduced to the classical heat conduction equation. However, for sub-picosecond pulses and sub-microscale conditions, $T_e > T_l$ during non-equilibrium heating. The significance of the heat transport equations (1) and (2) as opposed to the classical heat conduction equations has been discussed in [4].

The above coupled Eqs. (1) and (2), often referred to as parabolic two-step micro heat transport equations have been widely applied for thermal analysis of thin metal films exposed to picosecond thermal pulses. Most of researches are limited to the constant thermal property case. For the temperature-dependent thermal property case, a recent report shows that using temperature-dependent conductivity instead of constant conductivity leads to better agreement between temperature predictions and corresponding measurements for short-pulse laser heating [5]. Since Eq. (1) is highly nonlinear, it must be solved numerically. In this study, we consider the parabolic two-step model with temperature-dependent thermal properties in a double-layered micro thin film irradiated by ultrashort-pulsed lasers. By obtaining an energy estimate for the system, we develop a nonlinear finite difference scheme which has a discrete analogue of the energy estimate. It should be pointed out that double(or multi)-layered metal thin-films are widely used in engineering applications since a single metal layer often cannot satisfy all mechanical, thermal and electronic requirements. For example, chromium can be used to act as a heat sink which significantly reduces the temperatures rise in the top gold layer.

2 Governing Equations and an Energy Estimate

Consider a double-layered thin film with thickness L of the order 0.1 μm, which is subjected to a subpicosecond-pulse irradiation. Based on Eqs. (1) and (2), the governing equations for the double-layered thin film can be written as follows:

$$C_e^{(m)} \frac{\partial T_e^{(m)}}{\partial t} = \frac{\partial}{\partial x}(k_e^{(m)} \frac{\partial T_e^{(m)}}{\partial x}) - G^{(m)}(T_e^{(m)} - T_l^{(m)}) + S^{(m)}, \qquad (3)$$

$$C_l^{(m)} \frac{\partial T_l^{(m)}}{\partial t} = G^{(m)}(T_e^{(m)} - T_l^{(m)}), \qquad (4)$$

where $C_e^{(m)} = A_e T_e^{(m)}$, $k_e^{(m)} = k_0 \frac{T_e^{(m)}}{T_l^{(m)}}$, and $0 \leq x \leq \frac{L}{2}$ when $m = 1$, and $\frac{L}{2} \leq x \leq L$ when $m = 2$. The heat source for both layers is chosen to be [4]

$$S^{(m)}(x,t) = 0.94 J \left[\frac{1-R}{t_p \delta}\right] e^{-\frac{x}{\delta}} I(t), \qquad (5)$$

where $I(t)$ is light intensity of the laser beam, δ the penetration depth of laser radiation, J laser fluence, R the radiative reflectivity of the sample to the laser beam, t_p the full-width-at-half-maximum pulse duration. The interfacial equations are assumed to be

$$k_e^{(1)} \frac{\partial T_e^{(1)}}{\partial x} = k_e^{(2)} \frac{\partial T_e^{(2)}}{\partial x}, \quad T_e^{(1)} = T_e^{(2)}, \quad x = \frac{L}{2}. \tag{6}$$

The initial and boundary conditions are assumed to be

$$T_e^{(m)}(x,0) = T_l^{(m)}(x,0) = T_0 (= 300K), \tag{7}$$

and

$$\frac{\partial T_e^{(1)}(0,t)}{\partial x} = \frac{\partial T_l^{(1)}(0,t)}{\partial x} = 0, \quad \frac{\partial T_e^{(2)}(L,t)}{\partial x} = \frac{\partial T_l^{(2)}(L,t)}{\partial x} = 0. \tag{8}$$

Such boundary conditions arise from the fact that there are no heat losses from the film surfaces in the short time response [4].

Assume that $T_e^{(m)} \geq T_0$ and $T_l^{(m)} \geq T_0$. We seek an energy estimate for the above problem. To this end, we first introduce the L^P-norm:

$$\left\| u^{(m)} \right\|_{L^P} = \left(\int_{I^{(m)}} \left| u^{(m)} \right|^p dx \right)^{\frac{1}{p}}, \quad 1 < p < +\infty, \tag{9}$$

where $I^{(m)}$ represents the intervals $[0, \frac{L}{2}]$ when $m = 1$ and $[\frac{L}{2}, L]$ when $m = 2$, respectively. The following theorem has been obtained.

THEOREM 1. Assume that the solutions of Eqs. (3)-(8) are smooth, and $T_e^{(m)} \geq T_0$ and $T_l^{(m)} \geq T_0$, where $T_0 = 300K$. Then the solutions satisfy an energy estimate as follows:

$$\sum_{m=1}^{2} \left[\frac{1}{3} A_e^{(m)} \left\| T_e^{(m)}(t) \right\|_{L^3}^3 + \frac{1}{2} C_l^{(m)} \left\| T_l^{(m)}(t) \right\|_{L^2}^2 \right]$$

$$\leq e^t \{ \sum_{m=1}^{2} [\frac{1}{3} A_e^{(m)} \left\| T_e^{(m)}(0) \right\|_{L^3}^3 + \frac{1}{2} C_l^{(m)} \left\| T_l^{(m)}(0) \right\|_{L^2}^2]$$

$$+ \int_0^t \sum_{m=1}^{2} \frac{2}{3\sqrt{A_e^{(m)}}} \left\| S^{(m)}(s) \right\|_{L^{\frac{3}{2}}}^{\frac{3}{2}} ds \}. \tag{10}$$

3 Finite Difference Scheme and Its Energy Estimate

We denote $(T_e^{(m)})_j^n$ and $(T_l^{(m)})_j^n$ as the numerical approximations of $(T_e^{(m)})(j\Delta x, n\Delta t)$ and $(T_l^{(m)})(j\Delta x, n\Delta t)$, respectively, where Δx and Δt are the x directional spatial and temporal mesh sizes, respectively, $0 \leq j \leq N+1$ so that $(N+1)\Delta x = \frac{L}{2}$. A nonlinear finite difference scheme for solving Eqs. (3)-(8) can be written as follows:

$$A_e^{(m)} \frac{\left|(T_e^{(m)})_j^{n+1}\right|^3 - \left|(T_e^{(m)})_j^n\right|^3}{3\Delta t \frac{(T_e^{(m)})_j^{n+1}+(T_e^{(m)})_j^n}{2}} = \nabla_{\bar{x}}[(k_e^{(m)})_{j-\frac{1}{2}}^{n+\frac{1}{2}} \nabla_{\bar{x}} \frac{(T_e^{(m)})_j^{n+1}+(T_e^{(m)})_j^n}{2}]$$

$$-G^{(m)}[\frac{(T_e^{(m)})_j^{n+1}+(T_e^{(m)})_j^n}{2} - \frac{(T_l^{(m)})_j^{n+1}+(T_l^{(m)})_j^n}{2}]$$

$$+(S^{(m)})_j^{n+\frac{1}{2}}, \qquad (11)$$

$$C_l^{(m)} \frac{(T_l^{(m)})_j^{n+1} - (T_l^{(m)})_j^n}{\Delta t} = G^{(m)}[\frac{(T_e^{(m)})_j^{n+1}+(T_e^{(m)})_j^n}{2} - \frac{(T_l^{(m)})_j^{n+1}+(T_l^{(m)})_j^n}{2}], \qquad (12)$$

where $m=1,2$, and $(k_e^{(m)})_{j-\frac{1}{2}}^{n+\frac{1}{2}} = \frac{1}{2}k_0 \left|\frac{(T_e^{(m)})_j^{n+1}+(T_e^{(m)})_j^n}{(T_l^{(m)})_j^{n+1}+(T_l^{(m)})_j^n}\right| + \frac{1}{2}k_0 \left|\frac{(T_e^{(m)})_{j-1}^{n+1}+(T_e^{(m)})_{j-1}^n}{(T_l^{(m)})_{j-1}^{n+1}+(T_l^{(m)})_{j-1}^n}\right|.$

The discrete interfacial equations are chosen to be

$$(k_e^{(1)})_{N+\frac{1}{2}}^{n+\frac{1}{2}} \nabla_{\bar{x}} \frac{(T_e^{(1)})_{N+1}^{n+1}+(T_e^{(1)})_{N+1}^n}{2} = (k_e^{(2)})_{\frac{1}{2}}^{n+\frac{1}{2}} \nabla_{\bar{x}} \frac{(T_e^{(2)})_1^{n+1}+(T_e^{(2)})_1^n}{2}, \qquad (13a)$$

$$(T_e^{(1)})_{N+1}^n = (T_e^{(2)})_0^n, \qquad (13b)$$

for any time level n. The initial and boundary conditions are

$$(T_e^{(m)})_j^0 = (T_l^{(m)})_j^0 = T_0, \qquad (14)$$

and

$$\nabla_{\bar{x}}(T_e^{(1)})_1^n = \nabla_{\bar{x}}(T_e^{(2)})_{N+1}^n = 0, \quad \nabla_{\bar{x}}(T_l^{(1)})_1^n = \nabla_{\bar{x}}(T_l^{(2)})_{N+1}^n = 0, \qquad (15)$$

for any time level n. It can be seen that the order of the truncation error of Eq. (11) is $O(\Delta x^2 + \Delta t^2)$.

THEOREM 2. Suppose that $(T_e^{(m)})_j^n$ and $(T_l^{(m)})_k^n$ are the solutions of the scheme, Eqs. (11)-(25). Then the solutions satisfy

$$\sum_{m=1}^{2}[\frac{1}{3}A_e^{(m)}\left\|(T_e^{(m)})^n\right\|_{l^3}^3 + \frac{1}{2}C_l^{(m)}\left\|(T_l^{(m)})^n\right\|_{l^2}^2]$$

$$\leq e^{\frac{3}{2}n\Delta t}\{\sum_{m=1}^{2}[\frac{1}{3}A_e^{(m)}\left\|(T_e^{(m)})^0\right\|_{l^3}^3 + \frac{1}{2}C_l^{(m)}\left\|(T_l^{(m)})^0\right\|_{l^2}^2]$$

$$+n\Delta t \max_{0\leq k\leq n}\sum_{m=1}^{2}\frac{2}{3\sqrt{A_e^{(m)}}}\left\|(S^{(m)})^{k-\frac{1}{2}}\right\|_{l^{\frac{3}{2}}}^{\frac{3}{2}}\}. \qquad (16)$$

It can be seen that Eq. (16) is the discrete analogue of Eq. (10).

Here, the l^p-norm and inner product for the mesh functions u_j and v_j are

$$(u,v) = \sum_{j=1}^{N} u_j v_j, \qquad \|u\|_{l^p} = (\sum_{j=1}^{N} |u_j|^p \, dx)^{\frac{1}{p}}, \qquad 1 < p < +\infty.$$

The first-order forward and backward finite difference operators are defined as follows:

$$\nabla_x u_j = \frac{u_{j+1} - u_j}{\Delta x}, \qquad \nabla_{\bar{x}} u_j = \frac{u_j - u_{j-1}}{\Delta x}.$$

4 Numerical Example

To test the applicability of the above numerical scheme, we investigated the temperature rise in a double-layered thin film, namely a gold layer on a chromium padding layer, where the gold layer and the chromium padding layer are 0.05 μm in thickness. The parameters for the heat source were chosen to be $t_p = 0.1$ps, $\delta = 15.3$ nm, $R = 0.93$, and J is the laser fluence.

Figures 1 and 2 plot the electron temperature profiles and lattice temperature profiles with $J = 500.0 \frac{J}{m^2}$, respectively. Results are no significant differences from those obtained by the parabolic two-step model (see Figure 6 in [3]).

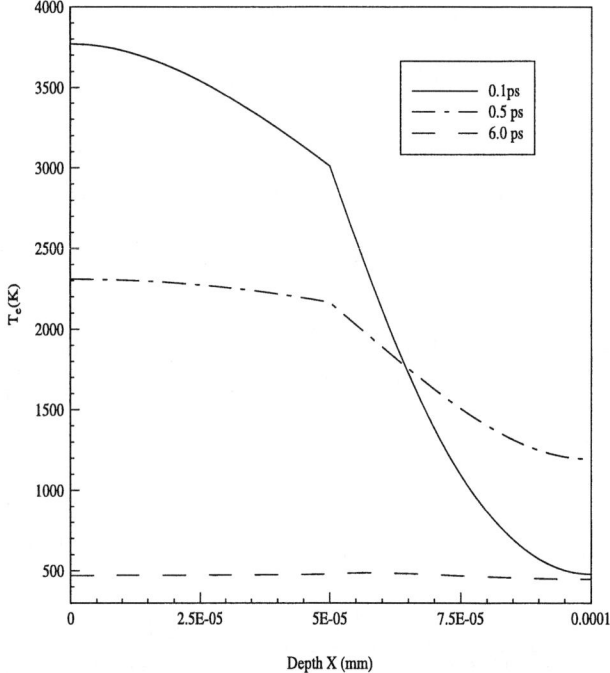

Fig. 1. Calculated electron temperature profiles for a 100 nm gold and chromium thin film irradiated with a 0.1 ps laser pulse at a fluence of 500.0 J/m²

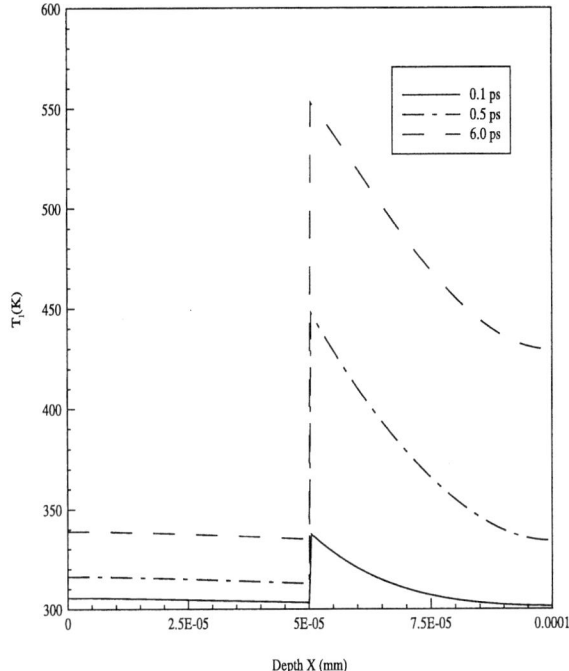

Fig. 2. Calculated latice temperature profiles for a 100 nm gold and chromium thin film irradiated with a 0.1 ps laser pulse at a fluence of 500.0 J/m^2

Acknowledgments

This research is supported by a Louisiana Educational Quality Support Fund (LEQSF) grant. Contract No: LEQSF (2002-05)-RD-A-01.

References

1. Tzou, D.Y., Chen, J.K., Beraun, J.E.: Hot-electron Blast Induced by Ultrashort-pulsed Lasers in Layered Media. Int. J. Heat Mass Transfer 45 (2002) 3369-3382
2. Chen, J.K., Beraun, J.E.: Numerical Study of Ultrashort Laser Pulse Interactions with Metal Films. Numerical Heat Transfer A 40 (2001) 1-20
3. Qiu, T.Q., Tien, C.L.: Femtosecond Laser Heating of Multi-layer Metals-I. Analysis. Int. J. Heat Mass Transfer 37 (1994) 2789-2797
4. Tzou, D.Y.: Macro To Micro Heat Transfer, Taylor & Francis, Washington DC (1996)
5. Antaki, P.J.: Importance of Nonequilibrium Thermal Conductivity During Short-pulse Laser-induced Desorption from Metals. Int. J. Heat Mass Transfer 45 (2002) 4063-4067

Analysis on Network-Induced Delays in Networked Learning Based Control Systems

Li Lixiong, Fei Minrui, and Zhou Xiaobing

School of Mechatronical Engineering & Automation, Shanghai University,
200072 Shanghai, P.R. China
lilix@sh163.net

Abstract. Local controller and remote learning device are connected through communication network in Networked Learning based Control Systems (NLCSs). Network-induced delays are inevitable during data transmission, and will deteriorate the real-time transmission of learning result from learning device to controller and even destabilize the entire system. This paper deals with the delays in NLCSs. The sources of delays are discussed and the possibility of reducing all delays into an equivalent delay is proposed. Finally, experimental measurements for that equivalent delay on Ethernet are conducted.

1 Introduction

Over past two decades, with the fast development of information technologies, it is very convenient and cost-efficient to connect different computers or devices with communication network. Consequently, communication network has been introduced into the control systems, and Networked Control Systems (NCSs) whose control loops are closed over communication network is becoming more and more common. Based on NCSs, we proposed a new control architecture: Networked Learning based Control Systems (NLCSs) in literature [3]. It consists of a local controller and a remote learning device, and they are connected through communication network. Obviously, the NLCS setup provides many advantages like NCS. For instance, it can reduce cost of installation and cabling and offer modularity and flexibility in system design, and the entire system is easy to maintain. And more importantly, it can implement remote monitoring and tuning of control systems.

It is well-known that network-induced delay caused by communication network is a crucial problem in NCSs, which will affect the performance even destabilize the control systems. As a result, there is a growing interest in the field of network-induced delays' effect on control systems [1, 2]. In this paper, the main intention is to study the network-induce delays in the learning loop of NLCSs. The delays occur in the learning loop of NLCSs are different from those in the control loop of NCSs and they will make different influence to the control system. It is clear that the research on network-induced delays is a solid foundation for the following research on performance analysis, stability analysis and compensation strategies for delays in NLCSs.

2 Networked Learning Based Control Systems

In NCSs, sensors, controllers and actuators are connected through communication networks (see Fig. 1). And network-induced delays occur during the data transmission between above control devices. In this case, the most popular and efficient networks in practical application proved to be control networks with different incarnations: DeviceNet, ControlNet, Profibus, Modbus, etc. Therefore, the network-induced delays in control networks are approximately fixed, i.e. the delays are deterministic.

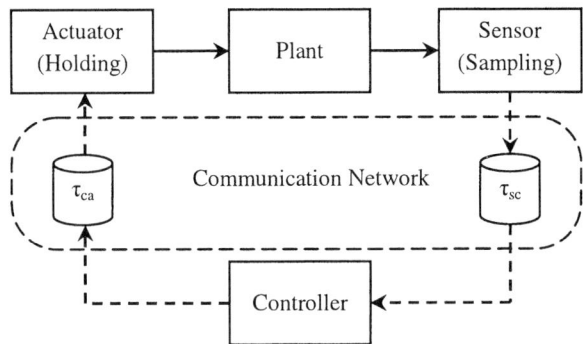

Fig. 1. Setup of NCSs. A typical NCS is with two delays: one is between sensor and controller; another is between controller and actuator

With the help of above-mentioned control networks, the efficiency and capability of the entire system can be highly improved in NCSs. In the meantime, many controllers with learning algorithm emerge to improve control performance. At the very start, the learning algorithm was implemented by the controller, i.e. the learning algorithm and control algorithm were carried out by same computer. However, considering the increasing complexity of learning algorithm, the controller with limited computation resources is no longer suitable for implementing the learning algorithm. Therefore, the possibility of implementing the learning algorithm by another computer or device is considered.

Therefore, the concept of NLCSs was proposed by our research group in literature [3], where local controller is simple and easy to implement, and complex learning algorithm is implemented by another remote computer via communication network (shown in Fig. 2). As a result, a low cost, high performance control system can be established with communication network.

In the learning loop of NLCSs, resources of other computers can be shared to operate complicated learning algorithm with communication link. Local Area Networks (LANs) even Wide Area Networks (WANs) can be used as a communication link and network-induced delays are inevitable just like NCSs. They include delay from controller to learning device τ_{cl} and delay from learning device to controller τ_{lc}. Longer and time-variant delays are expected in NLCSs because of LANs or WANs, and it will pose more challenges.

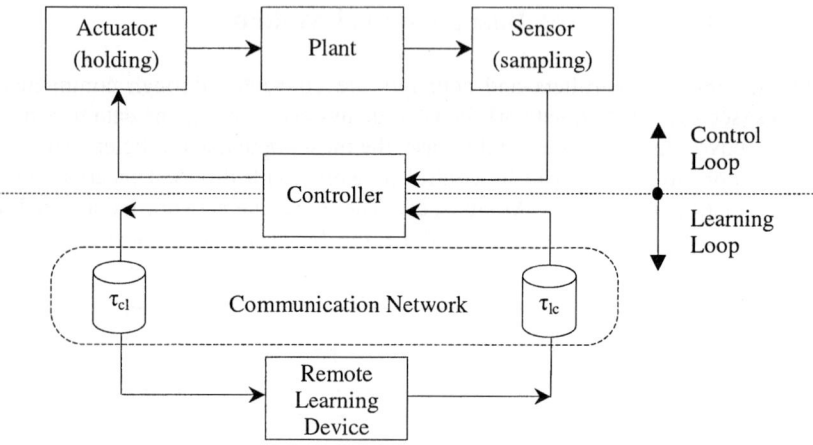

Fig. 2. Setup of NLCSs. There are two loops: a control loop without network connection, and a learning loop with network connection

3 Delays in the Learning Loop of NLCSs

3.1 Sources of Network-Induced Delays

The network-induced delays occur in the learning loop of NLCSs, and the communication network and the devices on the network have great influence on the characteristic of the delays (see literature [4]). The network-induced delay caused by network and devices can be expressed by

$$T_{delay} = T_{pre} + T_{wait} + T_{tx} + T_{post} \tag{1}$$

where T_{pre} and T_{post} are device delays, and T_{wait} and T_{tx} are link delays. Here, T_{pre}, T_{post} and T_{tx} can be considered as fixed when devices and link are fixed. And T_{wait} is due to network congestion, it is nondeterministic in general.

It is obvious that network-induced delay is mainly influenced by two aspects. 1) The waiting delay T_{wait} is decided by communication protocol of network; 2) The device delays are decided by the devices on the network.

3.2 Reducing All Delays into an Equivalent Delay

Considering the learning loop in Fig. 2, local controller and remote learning device are connected via communication network. The delay from controller to learning device is τ_{cl}, the delay from learning device to controller is τ_{lc}, while the computational delay of learning device is τ_c. A delay matrix can be define as

$$\tau = [\tau_{1,1}, \tau_{1,2}, \tau_{2,1}, \tau_{2,2}] \tag{2}$$

where $\tau_{i,i}$ (i=1,2) is computational delay of *ith* device, $\tau_{i,j}$ (i, j=1,2; i≠j) is network-induced delay between *ith* and *jth* device. Here, controller is Device 1 and learning device is Device 2. We assume for simplicity that control algorithm is so simple that controller's computational delay can be ignored, i.e. $\tau_{1,1}=0$; the leaning device has a fixed computational delay, i.e. $\tau_{2,2}=\tau_c$; and the time-variant network-induced delays $\tau_{1,2}=\tau_{cl}$, $\tau_{2,1}=\tau_{lc}$.

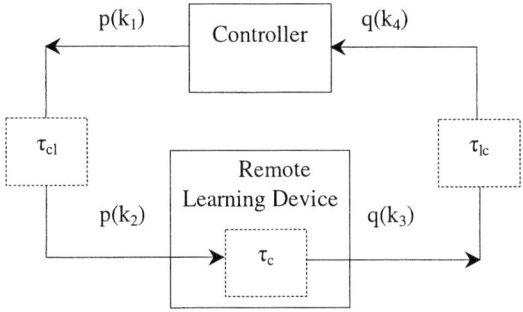

Fig. 3. Delays in the leaning loop. Assuming the controller is clock-driven and the remote learning device is event-driven. The controller sends plant output $p(k_1)$ to learning device at time k_1. After network-induced delay $\tau_{cl}(k_1)$, the learning device receives the data at time k_2. Processed by learning algorithm, the learning result can be downloaded back to controller at time k_3 while the computational delay of learning device is τ_c. After network-induced delay $\tau_{lc}(k_3)$, the controller receives the learning result $q(k_4)$ at time k_4

Next, the possibility of reducing three delays into one equivalent delay is considered. From literature [5], the end-to-end delay can be described as

$$\tau(k_1) = k_4 - k_1 = \tau_{cl}(k_1) + \tau_c + \tau_{lc}(k_3) \tag{3}$$

where $k_2=k_1+\tau_{cl}(k_1)$, $k_3=k_2+\tau_c$, and $k_4=k_3+\tau_{lc}(k_3)$, then three delays can be reduced into an equivalent delay, this will make the analysis on delay and entire system much simpler and easier.

4 Experimental Measurements for the Equivalent Delay

An experiment platform shown in Fig. 4 is established to conduct experimental measurements for the equivalent delay in previous section. The experiments through Ethernet are under different network loads since the computer 3 also sends data to the computer 1 and computer 2 periodically as network loads. With this platform, equivalent delay can be measured at different time, and the distribution of delay can be illustrated like Fig. 5. The experiment results show that Ethernet is suitable for closing the learning loop for it has almost no delay at low network loads. But Ethernet is also a nondeterministic protocol and message collision is a major problem at high network loads.

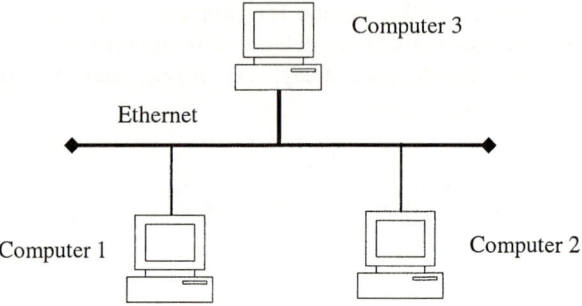

Fig. 4. Experimental setup through Ethernet. In order to realize the network-induced delays shown in Fig. 2, the clock-driven Computer1 sends 8 bytes data to Computer 2 at fixed interval (100ms); and Computer 2 is event-driven, that means it starts its activity when an event occurs

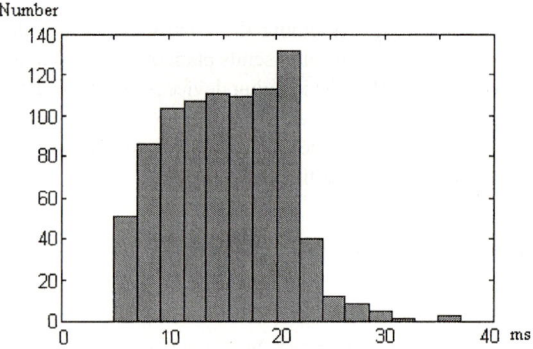

Fig. 5. Experimental results. The time-variant equivalent delay between computer 1 and computer 2 is measured for 1000 times. Its distribution looks like a Gamma distribution

From the perspective of devices, the equivalent delay is discrete when the controller is clock-driven, and digital control theory can be used to analyze and design the NLCSs. In addition, the equivalent delay can be shorter when learning device is event-driven. Therefore, in NLCSs clock-driven controller and event-driven learning device is a popular choice.

5 Conclusions and Future Work

In this paper, the network-induced delays in NLCSs are studied in detail. As the major sources of delays, communication link and control devices are discussed. The possibility of reducing all delays into one equivalent delay is proposed. And some experimental measurements for the equivalent delay are conducted. Experimental results show that clock-driven local controller plus event-driven remote learning device with Ethernet connection is suitable for NLCSs in general.

The above work on network-induced delays is the very beginning of the delay-related performance and stability analysis for NLCSs. With the knowledge of delays, our future efforts will focus on compensation approaches to guarantee system stability and promote control performance.

Acknowledgements

This work was partially supported by National Natural Science Foundation of China under grant 60274031; and Key Project of Science & Technology Commission of Shanghai Municipality under grant 025111052.

References

1. Nilsson, J.: Real-Time Control Systems with Delays, PhD thesis. ISRN LUTFD2/TFRT -1049-SE. (1998)
2. Zhang, W., Branicky, M.S., Phillips, S.M.: Stability of Networked Control Systems. IEEE Control Systems Magazine. Vol. 21, No. 1 (2001) 84-99
3. Fei, M.R., Li, L.X.: Networked Learning based Control Systems (in Chinese). in Proc. CIAC 2003, Hong Kong (2003)
4. Lian, Feng-Li, Moyne, J.R., Tilbury, D.M.: Performance Evaluation of Control Networks: Ethernet, ControlNet, and DeviceNet. IEEE Control Systems Magazine. Vol. 21, No. 1 (2001) 66-83
5. Bauer, P., Sichitiu, M., Premaratne, K.: On the Nature of the Time-Variant Communication Delays. In Proc. IASTED Conference Modeling, Identification and Control (MIC 2001). Innsbruck, Austria. (2001) 792-797

A New Boundary Preserval and Noise Removal Method Combining Gibbs Random Field with Anisotropic-Diffusion*

Guang Tian and Fei-hu Qi

Department of Computer Science and Engineering, Shanghai Jiao Tong University,
Shanghai, China
{Tianguang, Qifeihu}@sjtu.edu.cn

Abstract. In this paper, we present a new filter model which combines Gibbs random field with anisotropic-diffusion. The Gibbs random field is used to determine the boundaries of the objects in image according to the spatial information of image. Then the anisotropic-diffusion propagates different energy at different orientation with respect to conduction coefficient, and stops diffusing at the boundaries of the objects in image. We also provide the numerical implementation of the proposed method. The numerical experimental results show that our method has a high performance.

1 Introduction

During imaging, the formed images usually involve a lot of noise because of the effects of the environment and the limitations of imaging equipments. The general noise is white Gaussian noise. Only if the noise of image is filtered from the original image, the consequent image processing can be efficiently carried out. Currently, the 2-D Gaussian filter is widely introduced to remove white Gaussian noise in image processing, which convolves the original image $I(x, y)$ with Gaussian kernel $G(x, y, t)$. However, the symmetric Gaussian kernel smoothes image in isotropic way, which results in losing the details of the image. For example, the boundaries of the objects in image are blurred, and the angles of the objects in image are lost. Therefore the result of image processing does more harm to the consequent image processing than the original noisy image. Several advanced Gaussian filter models have been introduced to overcome these problems, which use anisotropic diffusion to remove the noise and preserve the details of the image ([1], [2] and [4]). However, their performances of removing noise are not satisfactory, so we present our method through modifying the partial differential equation (PDE) that describes the filtering process to reach a high performance.

The rest of this paper is organized as follows: in Section 2 we simply introduce several current main models based on anisotropic diffusion and point out their

---

* Our work is supported by the National Natural Science Founds of China (N.o. 60072029).

limitations. In Section 3 we present a new method combining Gibbs random field with anisotropic-diffusion. In Section 4 we propose the numeric implementation of our method. Meanwhile, we give our experimental results and compare our method with the previous methods. In Section 5, we draw a conclusion about our work.

2 The Previous Methods

Let u be the representation of the reconstructed image. u can be defined as the function of $\Omega \subset \mathbb{R}^2 \to \mathbb{R}$, which associates the pixel $(x,y) \in \mathbb{R}^2$ to its grey-levels $u(x,y)$. Ω is the image support. $u(x,y,t)$ is the reconstructed image at time t, which is equal to $I(x,y) * G(x,y,t)$. It is also the solution to the isotropic diffusion equations,

$$u_t = c \triangle u, \quad (x,y) \in \Omega, \ t \in \mathbb{R}^+. \qquad (1)$$
$$u(x,y,0) = I(x,y), \quad (x,y) \in \Omega. \qquad (2)$$

where c is the constant conduction coefficient. Malik and Perona put forth the following anisotropic diffusion equations in [1].

$$u_t = div(g(||\nabla u||) \nabla u), \quad (x,y) \in \Omega, \ t \in \mathbb{R}^+. \qquad (3)$$
$$u(x,y,0) = I(x,y), \quad (x,y) \in \Omega. \qquad (4)$$

Where the conduction coefficient, $g(||\nabla u||)$, is the function of $||\nabla u||$ at a pixel. ∇ is gradient operator, and $g(||\nabla u||)$ is a non-increasing function whose value range is (0,1]. $g(0) = 1$ when $||\nabla u|| = 0$, i.e., smoothing in a homogeneous grey-level region. The diffusion is very large in these regions. $g(||\nabla u||) \to 0$ when $||\nabla u||$ is very large, i.e., the pixel at the boundary of the region. Therefore their method can detect the boundary of region, and will almost not diffuse the energy at the boundary. So the exact location of object boundary will be preserved when an image is smoothed.

However, the $g(||\nabla u||)$ can't do well if the image is too noisy. Since the quantity of gradient will be very large and the value of the $g(||\nabla u||)$ is close to zero almost everywhere. As a result, almost all noise in image will remain when we use this model to smooth the corrupted image.

Alvarez, Lions, and Morel proposed that the diffusion process becomes more intense along the edges and weaker along the perpendicular orientation of the edges [2]. They proposed the following model for image recovery:

$$u_t = g(||\nabla u * G||) ||\nabla u|| div(\frac{\nabla u}{||\nabla u||}), \quad (x,y) \in \Omega, t \in \mathbb{R}^+. \qquad (5)$$
$$u(x,y,0) = I(x,y), \quad (x,y) \in \Omega. \qquad (6)$$
$$\left. \frac{\partial u}{\partial n} \right|_{\partial \Omega \times \mathbb{R}^+} = 0, \quad (x,y) \in \partial \Omega, \ t \in \mathbb{R}^+, \quad \frac{\partial u}{\partial n} = n \cdot \nabla u. \qquad (7)$$

The numerical experimental results show that this method obtain a superior performance on smoothing noise and enhance the edges of the region in image, but it still makes sharp corners round. In this paper we will refer to this model as ALM model.

Nordström [3] modified the diffusion (3) through adding a forcing term, and obtained the following model:

$$u_t = div(g(||\nabla u||)\nabla u) + I - u, \quad (x,y) \in \Omega, \; t \in \mathbb{R}^+. \quad (8)$$

$$u(x,y,0) = I(x,y), \quad (x,y) \in \Omega. \quad (9)$$

Because of adopting the new forcing term, the (8) has the feature of making $u(x,y,t)$ close to the original image. The quantity of $g(||\nabla u||)$ is very low at the boundaries of region, and the value of $div(g(||\nabla u||)\nabla u)$ is very low as well. Then u_t is mainly determined by the forcing term, $I - u$, and this term can reduce the degenerative effects of the diffusion to an acceptable levels at the boundary. However, this model is similar to (3) whose performance of eliminating noise is not satisfying. In this paper we will refer to this model as Nordström model.

3 A New Method

In this section, we provide a new method combining Gibbs random field with anisotropic diffusion. In 2-D image, the grey-levels of the pixels can be considered as random variables. The correlation between the grey-levels of the neighboring points can be described by the Markov Random Field (MRF), i.e., $P\{X(i,j)|X(k,l),(k,l) \neq (i,j)\} = P\{X(i,j)|X(k,l),(k,l) \in N(i,j)\}$. Every MRF can be described by Gibbs distribution. The Gibbs distribution effectively expresses the local interaction between pixels. If the differences of $||\nabla u||$ between a pixel and its neighboring pixels are small, the $P\{\cdot\}$ will be large. This shows that those pixels belong to a same region and locate interior region. Otherwise, the pixel must be at the boundary of a certain region.

3.1 The Modified Diffusion Equation

In order to preserve image's details, such as boundaries and sharp corners, we should stop diffusing along perpendicular direction of edges and use the forcing term to make new image close to the original image. In the region of an image, we should enforce the intensity of diffusion to reduce the noise of its interior region. Our idea is to add a boundary detector on both terms of the right side of (8). Our diffusion equations is proposed as follows:

$$u_t = P\{x,y,t\}div(C(x,y,t)\nabla u) + (1 - P\{x,y,t\})(I - u),$$
$$(x,y) \in \Omega, \; t \in \mathbb{R}^+. \quad (10)$$

$$u(x,y,0) = I(x,y), \quad (x,y) \in \Omega. \quad (11)$$

$$\left.\frac{\partial u}{\partial n}\right|_{\partial\Omega \times \mathbb{R}^+} = 0, \quad (x,y) \in \partial\Omega, \; t \in \mathbb{R}^+, \quad \frac{\partial u}{\partial n} = n \cdot \nabla u. \quad (12)$$

The right side of (10) consists of two terms: the first is the diffusion term and the second is the forcing term. $P\{\cdot\}$ is a local conditional probability with respect to the neighboring $||\nabla u||$ of a pixel. $C(x, y, t)$ is the conduction coefficient at pixel (x, y) at time t. As described above, $P\{\cdot\}$ is used as the boundary detector, which controls the diffusing speed. The conduction coefficient, $C(x, y, t)$, is the function of $||\nabla u||$, $C(x, y, t) = g(||\nabla u^t||)$, which controls diffusion intensity along propagating orientation. In addition, a good-designed $C(x, y, t)$ can not only preserve the boundaries of image, but also enhance the boundaries of image, i.e., making boundaries steeper. We use the following exponential function for $C(x, y, t)$,

$$C(x, y, t) = g(||\nabla u^t||) = exp\{-\frac{||\nabla u||^2}{K^2}\} . \tag{13}$$

Where K is a parameter. $C(x, y, t)$ is a non-increasing negative exponential function, which can privilege high-contrast edges over low-contrast ones. The forcing term can prevent the diffusion function from resulting in the degeneration of image, which only acts at the boundaries of regions. Equations (10) and (12) ensure the diffusion zero along the normal direction of edges. Equation (12) is a boundary value condition, and (11) is a initial value condition. $(1 - P\{\cdot\})$ is a moderation coefficient, which balances the forcing term and the diffusion term with respect to the neighboring gradient of the smoothed pixel.

3.2 Computational Expression of Boundary Detector

Gibbs distribution is an exponent function of Gibbs potential, which is described in [5]. The derivation of the local conditional probability density function (pdf), $P\{\cdot\}$, is shown in the following. Starting with the Bayes rule,

$$P\{(||\nabla u^t_{(i,j)}||)|(||\nabla u^t_{(k,l)}||, (k,l) \in N_{(i,j)}\}$$
$$= P\{(||\nabla u^t_{(i,j)}||)|(||\nabla u^t_{(k,l)}||, (i,j) \neq (k,l)\}$$
$$= \frac{1}{Q_{(i,j)}} exp(-\frac{1}{T} \sum_{(i,j) \in C} V_c(||\nabla u^t_{(i,j)}||, ||\nabla u^t_{(k,l)}||, (k,l) \in C)) . \tag{14}$$

where

$$Q_{(i,j)} = \sum_{||\nabla u^t_{(i,j)}|| \in \Gamma} exp(-\frac{1}{T} \sum_{(i,j) \in C} V_c(||\nabla u^t_{(i,j)}||, ||\nabla u^t_{(k,l)}||, (k,l) \in C)) . \tag{15}$$

In (14) and (15), C is clique, and $V_c(\cdot)$, called the clique potential, depends on the differences of $||\nabla u||$ between a pixel and its neighboring pixels. The parameter T, known as the temperature, is used to control the peaking of the distribution.

In this paper, we study only first-order clique and 4-Neighborhood, so

$$V_c(||\nabla u^t_{(i,j)}||, ||\nabla u^t_{(k,l)}||, (k,l) \notin C)) = 0 . \tag{16}$$

We use a contaminated Gaussian model [6], which is a robust penalty method.

$$V_c(||x||) = -log((1-\epsilon_p)exp(-\frac{||x||^2}{2\delta_p^2}) + \epsilon_p) . \tag{17}$$

3.3 Mathematical Validity

The determining solution problem of PDE consists of (10), (11) and (12). Equation (10) is a nonlinear parabolic equation with degeneration possibilities, so we discuss the mathematical validity of our model based on the theory of diffusion equation's solution. Only if the solution to our equations is existing, unique, and stable, the producing image is significant.

Theorem 1 (Maximum Principle). *The maxima and minima of the diffusion equation only belong to either the initial value condition or the boundary of the region.*

Uniqueness of Solution. Given u^I, u^{II} are two solutions to our equations, and we denote $u = u^I - u^{II}$, then u should be the solution to the following equations in terms of the theory of the PDE.

$$u_t = P\{x,y,t\}div(C(x,y,t)\nabla u) - P\{x,y,t\}u,$$
$$(x,y) \in \Omega, \ t \in \mathbb{R}^+. \tag{18}$$

$$u(x,y,0) = 0, \quad (x,y) \in \Omega. \tag{19}$$

$$\left.\frac{\partial u}{\partial n}\right|_{\partial\Omega \times \mathbb{R}^+} = 0, \quad (x,y) \in \partial\Omega, \ t \in \mathbb{R}^+, \ \frac{\partial u}{\partial n} = n \cdot \nabla u. \tag{20}$$

Equation (18) is a homogeneous diffusion equation whose initial- boundary value conditions are equal to zero, so we can deduce $u \leq 0$ in the interior region. In another case, i.e., $u = u^{II} - u^I$, we also can deduce $u \leq 0$ in the interior region. Therefore $u \equiv 0$, and $u^I = u^{II}$. This shows that the solution to our equations is unique.

Stability of Solution. Suppose u^I is the solution to our equations in one condition, and u^{II} is the another solution to our equations in another condition. We can also conclude that the $u = u^I - u^{II}$ is the solution to (18), (19) and (20). Set $\forall \varepsilon > 0, \exists \eta > 0$

$$|u^I - u^{II}|_\Gamma < \eta, |u^I - u^{II}|_\mathbb{R} < \varepsilon. \tag{21}$$

$\mathbb{R}$ denotes the interior region and Γ is the initial-boundary value conditions. According to the Maximum Principle, we can deduce

$$|u^I - u^{II}|_\mathbb{R} \leq |u^I - u^{II}|_\Gamma < \eta. \tag{22}$$

Let $\eta = \varepsilon$, $\forall \varepsilon > 0$, $\exists \eta = \varepsilon$, $|u^I - u^{II}|_\mathbb{R} < \varepsilon$ when $|u^I - u^{II}|_\Gamma < \eta$. This shows that the solution to our equations is stable for the fluctuation of the initial-boundary value conditions.

4 Numerical Implementation and Experimental Result

The images used in our experiments are represented by 256×256 matrices. We denote $u(x_i, y_j, t_n)$ by $u_{i,j}^n$, where $t_n = n \triangle t$.

Firstly, a numerical scheme is presented for our equations and Nordströn equations. A 4-neighborhood discretization of their diffusion terms can be used:

$$div(C(x,y,t)\nabla u) \sim (C_{-N} \cdot d_N u + C_{-S} \cdot d_S u + C_{-W} \cdot d_W u + C_{-E} \cdot d_E u)_{i,j}^t. \quad (23)$$

The notations in this numerical formula of the diffusion equation are as following:

$$d_W u_{i,j} = u_{i-1,j} - u_{i,j}, \quad d_E u_{i,j} = u_{i+1,j} - u_{i,j},$$
$$d_N u_{i,j} = u_{i,j-1} - u_{i,j}, \quad d_N u_{i,j} = u_{i,j+1} - u_{i,j}. \quad (24)$$

We denote $C(x,y,t)$ by $g(||\nabla u(x,y,t)||)$. Then the conduction coefficients are modified at every iteration as a function of the grey-level gradient.

$$C_{N_{i,j}}^t \sim g(||\nabla d_N I_{i,j}^t||), \quad C_{S_{i,j}}^t \sim g(||\nabla d_S I_{i,j}^t||),$$
$$C_{W_{i,j}}^t \sim g(||\nabla d_W I_{i,j}^t||), \quad C_{E_{i,j}}^t \sim g(||\nabla d_E I_{i,j}^t||). \quad (25)$$

Secondly, we also provide a numerical recipe for ALM method. The derivative of u with respect to the time t, i.e., u_t calculated in (x_i, y_j, t_n) is approximated by Euler's method, that is, $u_t \sim (u_{i,j}^{n+1} - u_{i,j}^n)/\triangle t$.

$$||\nabla u||div(\frac{\nabla u}{||\nabla u||}) = ||\nabla u||(\frac{\partial}{\partial x}(\frac{u_x}{||\nabla u||}) + \frac{\partial}{\partial y}(\frac{u_y}{||\nabla u||}))$$
$$= \frac{1}{||\nabla u||}(||\nabla u||u_{xx} - \frac{\partial}{\partial x}(||\nabla u||u_x) - \frac{\partial}{\partial y}(||\nabla u||u_y))$$
$$= \frac{1}{u_x^2 + u_y^2}(u_y^2 u_{xx} - 2u_x u_y u_{xy} + u_x^2 u_{yy}). \quad (26)$$

The computational complexities of the models presented in this paper are determined by the diffusion term which is the term of greater order in those PDEs that define them. All of the models are equal on computational complexity because the orders of those diffusion terms are equal. The iterative formula of our equations is given by:

$$u_{i,j}^{n+1} = u_{i,j}^n - \triangle t \cdot [P(x,y,t)(C_{-N} \cdot d_N u$$
$$+ C_{-S} \cdot d_S u + C_{-W} \cdot d_W u + C_{-E} \cdot d_E u)_{i,j}^t$$
$$- (1 - P(x,y,t))(u(x,y,t) - I(x,y))]. \quad (27)$$

Thirdly, we use two images in our experiments. One is very complex that contains two houses. The other is relatively simple that contains a rectangle and a triangle. Each image has several sharp corners and at least one square object. We add 12db noise to the former and 20db noise to the later. In the experiments of ALM model, we set a initial value of K in $g(x)$ in terms of the scale-space theory. Then we adjust the parameter K based on initial K until getting the best experimental results. We get $K=0.03$ at last. In the experiments of Nordströn method, we get a best value of K that is 0.02. In the experiments of our method, we set $K = 0.025$, $T = 3$, $\delta_p = 4$, and $\varepsilon_p = 0.01$ [6].

Our experimental results are showed in Fig. 1 and Fig. 2. As introduced in the previous sections, the ALM method has good performance in removing noise, but destroys the corners of the objects of the new image. The Nordströn method can preserve the boundary of the region, but is not good at smoothing noise. Our method has better performance. It not only does well in removing noise, but also has a high performance in preserving boundary. We also discover that the value of K in $g(x)$ affects the processing results of the previous methods. The previous methods will approximate to the traditional Gaussian filter when K changes from 0 to 0.001. It will result in best effects when K changes among 0.001 and 0.9. It will produce a lot of noise when K change from 1 to $+\infty$ as well.

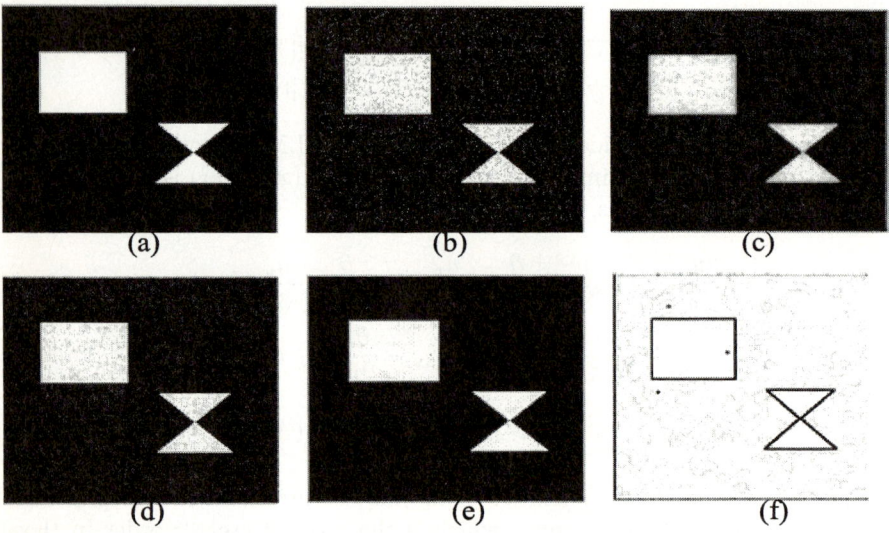

Fig. 1. (a) Original and (b) noisy image – SNR = -20 db. (c) Image reconstructed by ALM model. (d) Image reconstructed by Nordströn model. (e) Image reconstructed image by our model and (f) its segmentation

5 Conclusion

In this paper, we present a new method to remove noise and preserve boundary in 2-D image plain. We prove its validity in mathematical way. We also give the numerical scheme for our equations and use this scheme to do a lot of numerical experiments. The experimental results show that our method has a better performance in all aspects than the other presented methods. Compared with the other presented methods, our method obtains the producing images that have a high quality of eliminating noise, preserving boundary and segmentation. We can conclude that our method is much useful for those cases where the boundary of region must be involved.

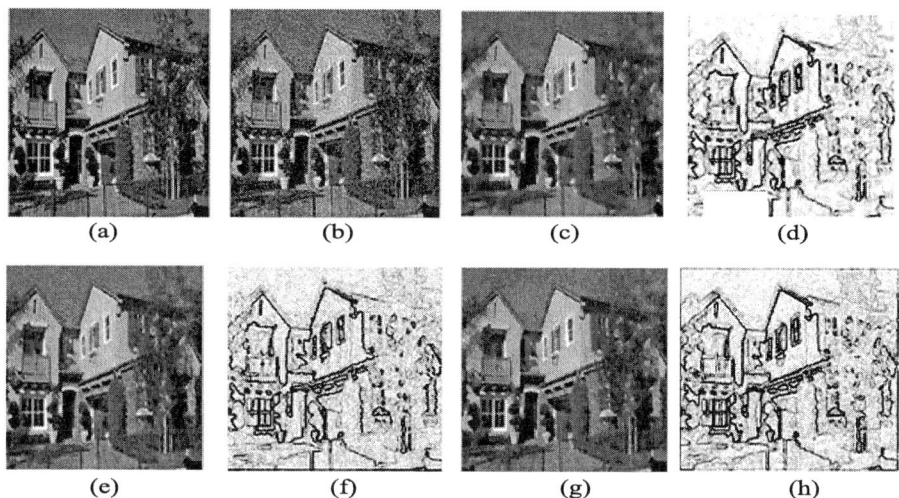

Fig. 2. (a) Original image and (b) noisy image – SNR=-12.5db. (c) Image reconstructed by ALM model and (d) its segmentation. (e) Image reconstructed by Nordströn model and (f) its segmentation. (g) Image reconstructed by our model and (h) its segmentation

References

1. Malik, J. and Perona, P.: Scale-space and Edge Detection Using Anisotropic Diffusion. IEEE Trans. Pattern Anal. Machine Intell. **12(7)** (1990) 629–639
2. Alvarez, L. Lions, P.L. and Morel, J.M.: Image Selective Smoothing and Edge Detection by Nonlinear Diffusion. SIAM J. Numer. Anal. **29(3)** (1992) 845-866
3. Nordströn, K.N.: Biased Anisotropic Diffusion: a Unified Regularization and Diffusion Approach to Edge Detection. Image Vis. Comput. **8** (1990) 318-327
4. Barcelos, C.A.Z. and Chen, Y.: Heat Flows and Related Minimization Problem in Image Restoration. Comput. Math. Applicat. **39** (2000) 81-97
5. Kuang, J. and Zhu, J.: On Markov Random Field Models for Segmentation of Noisy Images. Journal of Electronics. **13(1)** (1996) 31-39
6. Scharstein, D. and Szeliski, R.: Stereo Matching with Nonlinear Diffusion. Int. J. Comput. Vision. **28(2)** (1998) 155-174

The Geometric Constraint Solving Based on Mutative Scale Chaos Genetic Algorithm

Cao Chunhong and Li Wenhui

College of Computer Science and Technology, Jilin University,
Changchun 130012, P.R. China
chunhongcao_li@163.com, liwh@public.cc.jl.cn

Abstract. The constraint problem can be transformed to an optimization problem. In this paper a new hybrid algorithm—MSCGA was introduced which mixes genetic algorithm with chaos optimization method. The character of this new method is that the mechanism of the GA was not changed but the search space and the coefficient of adjustment was reduced continually and this can lead generation to evolve to the next generation in order to produce better optimization individuals. It can improve the performance of the GA and get over the disadvantage of the GA. The examination indicates that this algorithm can show a better performance than the normal GA and other hybrid methods in solving a geometric constraint and acquires a satisfying result.

1 Introduction

Machine design as a science technology impulses the social and economical development in the human production activities in a long time. The shortage of this process is that the cycle is very long, the accuracy and reliability is limited by this method, and the quality and optimization design can not be reached. The quality of the design can not be evaluated until the sample machine was made. In order to get over those disadvantages, the geometric constraint solving is a popular problem of the study of constraint design. Many scholars worked over the constraint solving by numeric computing approach, artificial intelligence approach, degree of freedom approach and graph-based approach. To conclude there are whole solving, sparse matrix solving, joint analysis solving, constraint diffuse solving, symbolic algebra solving and guides solving. [1]

Geometric constraint problem is equivalent to the problem of solving a set of nonlinear equations substantially. In constraint set every constraint corresponds to one or more nonlinear equations, all the unattached geometric parameters in the geometric element set constitute the variable set of nonlinear equations. Nonlinear equations can be solved by classical Newton-Raphson algorithm. When the size of geometric constraint problem is large, the scale of equation set is very large. It will be difficult for the efficiency and stability to meet the requirement of interactive design. Furthermore, the geometric information in the geometric system will not be treated properly and we can't deal with under- and over-constrained design systems well when all the equations must be solved.

In recent years the Genetic Algorithm simulating the evolution process attracts many researchers' attention in different areas. Because of its many advantages such as simplicity, less strict need on objective function, this algorithm is applied into many areas, such as image processing, artificial intelligence and so on. The main trait of GA is that it can ensure that the colony evolves continually, and the solutions change continually. Its disadvantage is that it may probably lack strong capacity of producing best individuals and cause the speed slow when it is near to the global best solution and sometimes it may run into local best solution. The other disadvantages such as slow convergence, prematurity convergence are also correlative with the former disadvantages. In order to get over these shortages of this algorithm, the hybrid method was introduced. Usually there are three kinds of hybrid algorithms[2].

(1) The two algorithms solve the problem respectively. One of them uses the solution of the other, but one doesn't take part in the other's search process. The ordinary method is that once the GA has found a good solution, we can change to the other algorithm quickly.
(2) This idea of this kind of method is that one algorithm is added into another algorithm as an accessory element.
(3) The algorithm associates the two algorithms to solve the problem. One algorithm acts as a necessary element, and participates in another's search process directly. When the taboo searching method is mixed with GA[3], we can consider the taboo search method as a variation operator of the GA, and the algorithm mixing the Simulated Annealing Algorithm with the GA is the same as that.

The chaos is a general phenomenon in non-line system, which has some special characters, such as acquiring all kinds of states in a self-rule in a certain range. And it has such sensitivity that a tiny change of initial condition can lead to a big change of the system. Based on the two advantages of the chaos, some chaos optimization algorithms were proposed that can prevent the algorithm from getting into the local optimization solution and have a high efficiency of calculation. In this paper a new hybrid algorithm—MSCGA[4] was introduced which mixes genetic algorithm with chaos optimization method. The character of this new method is that the mechanism of the GA was not changed but the search space and the coefficient of adjustment was reduced continually, and this can lead generation to evolve to the next generation in order to produce better optimization individuals, which can improve the performance of the GA and get over the disadvantage of the GA. The examination indicates that this algorithm shows a better performance than the normal GA and other hybrid methods in solving a geometric constraint and acquires a satisfying result.

2 The Geometric Constraint Solving

The machine design is a constraint-satisfied problem from the point of artificial intelligence. That's to say that we should find the design details of the object to satisfy these constraints given such as function, structure, material, and manufacture. The traditional CAD system without management and technical support of the geometric constraint only memorizes the coordinate of location-shape, and doesn't memorize information of the constraint. That is to say this system only maintains the final solution of the design and abandons the constraint information in all the process. This kind of system cannot satisfy the needs of modern manufacturing to the design. The research and exploiture of new CAD is based on geometric and model. The design model associates the geometric model with design information. It can communicate the intention of design better and perfectly. In all the design constraints, geometric constraint is the base of all, which is the base of other constraints, and also is the preferential problem in constraint management and solving technology.

The constraint problem can be formalized as (E, C)[5], here E=(e_1, e_2, ..., e_n), it expresses geometric elements, such as point, line and circle etc; C=(c_1, c_2, ... ,c_m), c_i is the constraint set in these geometric elements. Usually a constraint corresponds to an algebraic equation, so the constraint can be expressed as the following:

$$\begin{cases} f_1(x_0, x_1, x_2, ..., x_n) = 0 \\ \quad ... \\ f_m(x_0, x_1, x_2, ..., x_n) = 0 \end{cases} \quad (1)$$

X=(x_0, x_1, ..., x_n),
X_i are some parameters, for example, planar point can be expressed as (x_1, x_2). Constraint solving is to get a solution x to satisfy formula (1).

$$F(X_j) = \sum_1^m |f_i| \quad (2)$$

Apparently if X_j can satisfy F (X_j) =0, then X_j can satisfy formula (1). Then the constraint problem can be transformed to an optimized problem and we only need to solve min (F (X_j)) < ε. ε is a threshold. In order to improve the speed of ant algorithm, we adopt the absolute value of f_i not the square sum to express constraint equation set.

3 Mutative Scale Chaos Optimization Algorithm

Considering Logistic Mapping
$$x_{n-1} = \mu x_n(1-x_n)$$

Here, μ is the controlling parameter, n=0, 1, 2, ... When μ =4, we can get n chaos variables whose tracks are different for discretionary n initialization values which have a bit difference (they can not be the immobility points of x), $x_n(0) \in [0, 1]$.

The basic steps of Mutative Scale Chaos Optimization Algorithm[6] are as follows:

1) The initialization of the algorithm: suppose k=1, k'=1, $x_i(k)=x_i(0)$, $x_i^* = x_i(0)$, $f^* =f(0)$, $a_i(k') = a_i$, $b_i(k') = b_i$.

2) Map the chaos variable $x_i(k)$ to the choosing value area, search in an ordinary chaos optimization method; Let k=k+1, $x_i(k)=4\ x_i(k)(1- x_i(k))$, until f^* maintains invariable in a certain steps.
3) Change the search scale of the chaos variable (here adjusting coefficient $\gamma \in$ (0, 0.5), mx_i^* is the current best solution.

$$a_i(k'+1) = mx_i^* - \gamma (b_i(k') - a_i(k')) \qquad (3)$$
$$b_i(k'+1) = mx_i^* + \gamma (b_i(k') - a_i(k'))$$

4) Revert optimization variable

$$x_i^* = \frac{mx_i^* - a_i(k'+1)}{b_i(k+1) - a_i(k'+1)} \qquad (4)$$

5) Do the chaos search by the iteration of step 2) ~step 4) using the new chaos variable $y_i(k) = (1-\alpha)\ x_i^* + \alpha\ x_i(k)$ (α is a small number); Let $k' = k'$ +1, until maintains invariable in a certain steps.
6) Stop after repeat 3) and 4) some times, now mx_i^* is the best variable, f^* is the best solution acquired by the algorithm.

4 Mutative Scale Chaos Genetic Algorithm (MSCGA)

Because the structure of GA is open and not related with the problem, it is easily mixed with other kinds of algorithms to produce a new algorithm in order to improve the capability of local search and increase the optimization quality and the searching efficiency, such as GA mixed with SA, GA mixed with climbing method, GA mixed with the gradient method, and so on. Based on the foreign and domestic research, we can find that these hybrid methods can overcome some of the shortages of the GA, but there still are some disadvantages. For example the algorithm of GA mixed with SA shows a good performance in solving the geometric constraint problem than other methods and reduces the calculation time. But the parameter decision of the SA is very difficult and needs many times tries. So it is very inconvenient to use. The chaos is a general phenomenon in non-line system, which has some special characters, such as getting all kinds of states, randomcity and disciplinarian. It can acquire all kinds of states in its rule in a certain scale. So it has a high feasibility of getting the optimization solution.

To the problem described in formula 1, the best solution is corresponding to the lowest point of the system's energy, so the points near the best solution also have lower energy. If we use the chaos optimization method to search the individuals that have a higher adaptation value after they have completed an operation of generation process, then the speed of calculation and the performance of the GA will be improved rapidly. In this paper, we construct a new algorithm MSCGA that mixes the GA with the Mutative Scale Chaos Optimization Algorithm. The steps are as follows:

1) Initialize the colony.
2) Finish a genetic operation by the normal GA to get a better colony.
3) Search the individuals that have a higher adaptation value by Mutative Scale Chaos Optimization Algorithm so that it can lead the colony to evolve rapidly.

Repeat 2) and 3) until it can satisfy the stop condition, then the algorithm ends.

5 Application Instance and Result Analysis

We can realize from the above figures that once a user defines a series of relations, the system will satisfy the constraints by selecting proper state after the parameters are modified.

From the table 1, we can see that MSCGA proposed in this paper can have a convergence to the global best solution for figure 1 and figure 2 by the percent of 100%, SA+GA can finish this goal in most conditions, while SGA cannot finish this goal. Seen from the feasible solution number that it has been searched when it has found the global best solution, the searching times of MSCGA is fewer than SGA and SA+GA. This reduces the running time of the program and improves the search speed.

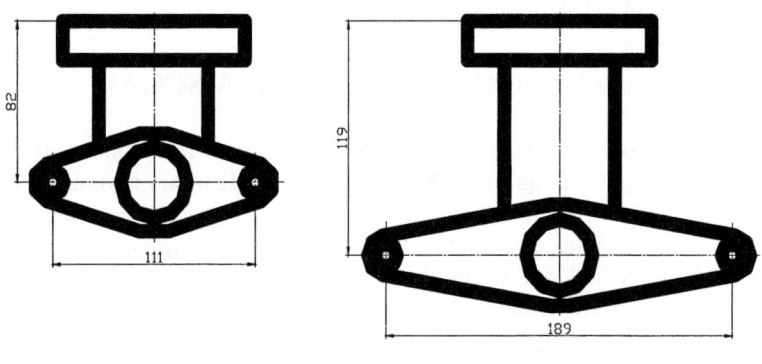

Fig. 1. (a) A design instance   (b) solving result

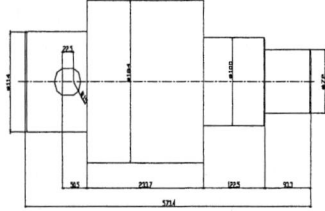
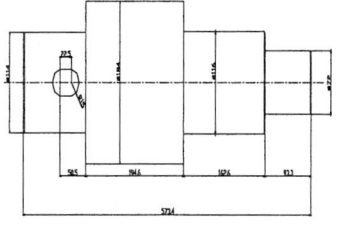

Fig. 2. (a) A design instance   (b) solving result

Table 1. The capability comparison between MSCGA and SGA, SA+GA

Testing figure	optimization method	the ratio of convergence to the global best solution	the number of feasible solutions when convergence to the global best solution
Figure 1	MSGA	100%	2074
	SGA	71%	4003
	SA+GA	100%	4351
Figure 2	MSGA	100%	986
	SGA	34%	1737
	SA+GA	76%	2896

6 The Conclusion

MSCGA operates searching by the rule of chaos variable itself and reduces the space continually in the process of optimization. It can improve the search precision and is easy to avoid the local best solution, so it has high search efficiency. In this paper we introduce Mutative Scale Chaos Optimization Algorithm into GA when not changing the search mechanism of Genetic Algorithm. It can quicken the evolvement speed of the generation and can improve the capability of GA. The algorithm has a fast search speed and high calculation precision. Its structure is simple and convenient. It can be indicated that its integrated performance is superior to SGA and other hybrid GA.

References

1. Yuan, B.: Research and Implementation of Geometric Constraint Solving Technology, doctor dissertation of Tsinghua University (1999): 1-8
2. Deming, L.: A hybrid genetic algorithm using chaos for globally optimal solution [J]. System Engineering and Electronics, 1999, 21(12): 81-84
3. Li, D.: An improved hybrid genetic algorithm. Information and Control, 1997, 26 (6): 449-454
4. Wang, M., Liu, J., and Sun, Y.: The hybrid genetic algorithm based on the mutative scale chaos optimization. Control and Decision, 2002, 17 (6): 958-960
5. Liu, S., Tang, M. and Dong J.: Two Spatial Constraint Solving Algorithms. Journal of Computer-Aided Design & Computer Graphics, 2003, 15(8): 1011-1029
6. Tong, Z., Wang, H., and Wang, Z.: Mutative scale chaos optimization algorithm and its application [J]. Control and Decision, 1999, 14 (3): 285-288

Genetic Algorithm Based Neuro-fuzzy Network Adaptive PID Control and Its Applications

Dongqing Feng[1,2], Lingjiao Dong[2], Minrui Fei[1], and Tiejun Chen[2]

[1] School of Mechatronical Engineering and Automation, Shanghai University,
200072 Shanghai, China,
dqfeng@zzu.edu.cn, mrfei888@x263.net
[2] Institute of Information and Control, Zhengzhou University,
450002 Zhengzhou, China
{dqfeng, dlj7905, tchen}@zzu.edu.cn

Abstract. It is difficult to satisfy most of the performance targets by using the PID control law only, if the plants are the processes with uncertain time-delay, varying parameters and non-linearity. For this reason a genetic algorithm based neuro-fuzzy network adaptive PID controller is proposed in this paper. The neuro-fuzzy network is used to amend the parameters of the PID controller online, the global optimal parameters of the network are found with a high speed, and the improved genetic algorithm is introduced to overcome the local optimum defect of the BP algorithm. Finally, the simulation experiment of the control method on the tobacco-drying control process is performed. The simulation results demonstrate that this kind of control method is effective.

1 Introduction

Tobacco drying is a process with uncertain time-delay, varying parameters and non-linearity. It is difficult to control the whole process effectively only by applying a group of fixed PID parameters. To meet the control performance targets this paper has presented a neuro-fuzzy network adaptive PID controller based on genetic algorithm. It integrates the merits of neural network, fuzzy control and PID control and is easy to be realized. It works according to the factual feedback information and uses the neuro-fuzzy network to amend the parameters of the PID controller online. The global optimal parameters of the network can be found at a high speed by using an improved genetic algorithm [1][2].

2 Problem Description

Tobacco-drying process is to refabricate the pre-drying tobacco. It's a key process to ensure the quality of the tobacco. During the whole drying process tobaccos exchange heat with the heat transfer medium having a certain temperature, stress and flow rate. It includes the processes of drying, cooling, dampening and so on. The variation of the temperature and stress in each zone will affect the parameters of the following zones. The critical task of tobacco drying is to control the temperature and humidity, which are coupled mutually. To decouple them, the compensation method can be used [3]. After that both the temperature and the humidity can be controlled separately and the controlled process can be treated as an object with concentrated parameters. So a first-order inertial object with pure delay is adopted to describe the whole process, given by

$$G(s) = \frac{Ke^{-\tau s}}{TS+1}$$

Where the static gain K, the pure delay and the constant time T of the object vary during different stages.

3 The Neuro-fuzzy Adaptive PID Controller

3.1 The Whole Control System Design

This paper presents an adaptive PID control system based on neuro-fuzzy network shown in Fig.1. It consists of the traditional PID controller, neuro-fuzzy network (FNN), and the self-learning device. The neuro-fuzzy network has two inputs, which we call respectively error e and error variance ration ec. It has three outputs K_p, K_i and K_d, which are corresponded to the variance of the proportion coefficient, integral coefficient and differential coefficient separately. In the control process they vary according to different inputs e and ec to meet the different needs. With the neuro-fuzzy network the parameters of the PID controller can be self-adjusted online. The learning device is to find the optimum parameters of the neuro-fuzzy network.

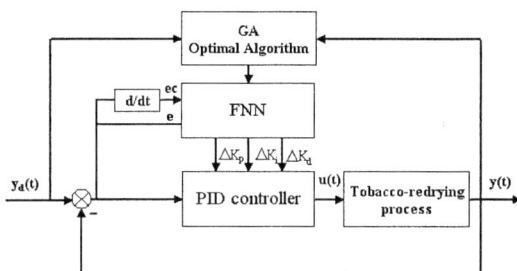

Fig. 1. Neuro-fuzzy Network Adaptive PID Controller based on Genetic Algorithm

3.2 The Common Used PID Controller

The routine PID control algorithm is given by

$$u(t) = K_p [e(t) + \frac{1}{T_i} \int_0^t e(t)dt + T_d \frac{de(t)}{dt} dt] \tag{1}$$

Where e (t) is the error of the system; K_p, K_i, K_d are the proportion coefficient, integral coefficient and differential coefficient separately; $K_i = K_p/T_i$, $K_d = K_p T_d$.

The corresponding incremental digital PID control algorithm is the following:

$$u(t) = u(t-1) + k_p[e(t) - e(t-1)] + k_i e(t) + k_d[e(t) - 2e(t-1) + e(t-2)] \quad (2)$$

In this paper the latter algorithm is adopted and the optimal PID parameters of given control law can be found by FNN self-learning.

3.3 The Neuro-fuzzy Network

If we describe the two-dimension fuzzy logic controller by using a kind of linking mechanism then the structure of the FNN can be shown as Fig. 2.

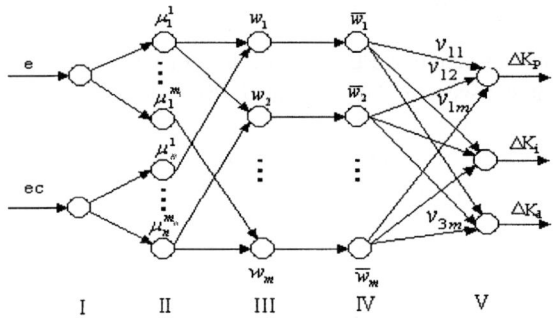

Fig. 2. The neuro-fuzzy network (FNN)

It consists of five layers and each of them has an explicit meaning. The first is the input layer. The second is the fuzzification layer. Each node of this layer represents a language variable, for example NB, PS and so on. It computes the membership function of each component μ_i^j. If we select gauss function as the membership function, then

$$\mu_i^j = e^{-\frac{(x_i - c_{ij})^2}{\sigma_{ij}}} \quad i=1,2, \ j=1,2,\cdots,m_i \quad (3)$$

Where m_i is the total number of partition of the x_i, c_{ij} and σ_{ij} representing center and width of the membership function separately. The third layer is to match the first component of the fuzzy control rule. Each node of this layer represents a fuzzy control rule. The fitness degree of every rule can be calculated. The forth layer executes a normalization process. The fifth is a defuzzification layer, with which we can achieve an explicitly result, given by

$$y_i = \sum_{l=1}^{m} v_{il}\overline{w}_l \quad i=1,2,3 \quad l=1,2,\cdots,m \quad (4)$$

Considered the network in Fig.2, we know that the parameters needing optimization are σ_{ij}, m_{ij} and v_{il}. This paper has improved the genetic algorithm and uses it to find the global optimum values of the parameters. The detailed introduction is as follows.

4 Optimization Algorithms

To optimize the FNN is to find the optimal parameters under a given performance index function. This paper adopts ITAE as the performance index, defined by

$$J(ITAE) = \int_0^\infty t|e(t)|dt \quad (5)$$

The optimal problem in this paper can be described by

$$J^* = \min J(ITAE) \quad (6)$$

If the parameters of the FNN are selected properly then the performance index will reach an optimal value J^*. It can be optimized without knowing the accurate mathematical model of the controlled process and the input and output datum of the process in advance.

To improve the optimization efficiency the genetic algorithm should be improved. This paper has presents some new strategies based on the research of the original genetic algorithm [4].

(1) It presents a new scaling method to solve the minimum optimum problem. The scaling processes mainly involves two steps:

a. Arrange the individuals by a descending order. In another word, the least fit individual, having the highest fitness value, put to the first place while the fittest individual, having the lowest fitness value, put to the end.

b. Assign fitness value to each individual based on its position. The scaling formula is written as

$$f(i) = 2\times(i-1)/N, \quad 1 \le i \le N \quad (7)$$

Where i means the position of the individual and N is the scale of the population.

(2) This paper presents an improved optimal selection method. In the first step, it arranges the individuals of the population in terms of fitness value. Then it keeps the individuals with higher level and let them retain in the next generation directly. In the following step it selects the other individuals based on the arranging-selection mechanism [5] The main reason for this is to extend the searching space without spoiling the best individual, so that the evolution process is in the optimal direction.

(3) This paper introduces a self-adaptive mutation strategy. With it the mutation behavior of the individual is related to the solution quality and the searching space is adjusted adaptively. The individuals with higher fitness value are searched in a small scale while the less fit individuals are hunted in a larger range. With the assumption that $\{x_i\}$ is the parent individual that is to be mutated, if some variables of it are mutated then the offspring are produced according to the self-adaptive mutation rule,

$$y_k = \begin{cases} x_k + \Delta(T, b_k - x_k) & \mu = 0 \\ x_k - \Delta(T, x_k - a_k) & \mu = 1 \end{cases} \quad (8)$$

$$\Delta(T, z) = z(1 - r^{T^\lambda}) \quad (9)$$

Where μ is created at random which is 0 or 1. r is chosen uniformly at random in $[0,1]$; λ typically is in $[2,5]$; mutation temperature $T = 1 - f/f_{max}$. f is the fitness value and f_{max} the highest fitness value.

5 Simulation Example

The mathematical model of the whole tobacco-drying process is given by

$$G(s) = \frac{k(t)}{T(t)s + 1} e^{-\tau(t)s}$$

Where $k(t) = a/(b+t)$, $T(t) = c + dt$, $\tau(t) = e + ft$.

From above we can know that the parameters of the process are nonlinear and time varying. In this simulation experiment a, b, c, d, e and f are selected as:

a=b=-30, c=0.7, d=0.03, e=0.35, f=0.1.

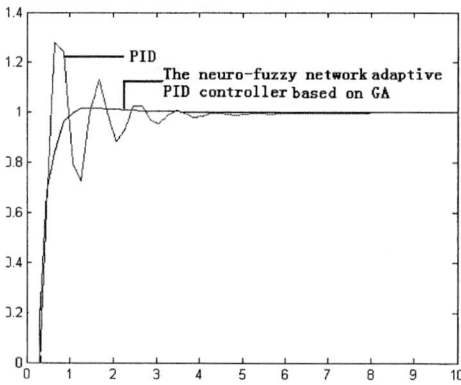

Fig. 3. System response to the unit step input with different types of controller

Using the common PID controller and the neuro-fuzzy network adaptive PID controller based on genetic algorithm separately, the responses of the whole system to the unit step input are shown in Fig.3. Viewing Fig.3, we can know the control effects under the two controllers clearly. With the common used PID controller, the system oscillated. It is for the parameters of the process vary while the parameters of the common used PID controller are fixed during the whole control process. With the neuro-fuzzy network adaptive PID controller based on genetic algorithm, the steady-state error and the overshoot are shown to be zero and the rising time is very short.

6 Conclusion

This paper presents an improved GA to optimize the FNN, without knowing the accurate mathematical model of the controlled process and gaining the input and output datum of the process in advance. The simulation results have demonstrated that the optimized FNN can improve the control performance of the common used PID controller greatly, by amending its parameters online.

Acknowledgment

This project is supported in part by the National Natural Science Foundation of China under grant no. 60274031 and Henan Province Natural Science Foundation of China under grant no. 0311011300.

References

1. Feng, D., Xie, S.: Fuzzy and Intelligence Control. The Publishing of Chemical Industry (1998)
2. Li, S.: Intelligence Control Theory and Application. The Publishing of Ha Er Bi Bin University (1996)
3. Chen, J.: Tobacco-drying Intelligence Control System. Kun Ming University of Science and Technology, Vol.6 (2001)
4. Chen, G., Wang, X.: Genetic Algorithm and Application. The Publishing of Ren Min Post Office (1996)
5. Baker, Z.: Adaptive Selection Methods for Genetic Algorithms Proc. 1st Int'1. Conf On Genetic Algorithms. Lawrence Earlbaum Associates, Hilladale, NJ, (1985) 110–111

Formalizing the Environment View of Process Equivalence*

Yuxi Fu and Xiaoju Dong

BASICS**, Department of Computer Science,
Shanghai Jiaotong University, Shanghai 200030, China

Abstract. The notion of program equivalence plays a fundamental role in the understanding of programming related issues in the framework of concurrent/distributive/mobile/global/grid computing. Many observational equivalences have been proposed in the literature. These equivalences are based on the intuition that different classes of environments have different observing powers. The paper provides a formalization of the observations of environments. This formalization leads to an equivalence relation called global bisimulation. We examine this relation in some well known computing models.

1 Observation by Environment

The notion of program equivalence plays a fundamental role in the understanding of programming related issues in concurrent/distributive/mobile/global/grid computing. Many observational equivalences have been proposed in the literature. These equivalences are based on the intuition that different classes of environments have different observing powers. From the observational point of view, two processes are equivalent if no environments can detect any observable operational difference. There are two important issues: One is what it takes to observe. The other is what can be observed. The general view is that an observation is made by an interaction between the environment and the process being observed. From a game theoretical point of view, an equality tells what environments cannot do whereas an inequality indicates what environments can do. In other words, an equality is the failure of the environments while an inequality is the success of the environments. Here the basic assumption is that the environments are always malicious. According to this view it is clear that the environments play a crucial role in the theory of concurrent/distributive/mobile/global/grid computing. Let's look at the issue in some detail. Suppose P and Q are processes and $C[\_]$ is a context. We would like to test the behavior of P and Q using the context $C[\_]$. Now suppose that

$$C[P] \stackrel{\tau}{\longrightarrow} C'[P'] \tag{1}$$

---

* The work is supported by The National Distinguished Young Scientist Fund of NNSFC (60225012) and The National 973 Project (2003CB316905).
** Laboratory for Basic Studies in Computing Science.

is caused by an interaction between P and $C[\_]$. Here P' and $C'[\_]$ are the residuals, or descendants, of P and $C[\_]$ respectively. Note that we are using the word 'descendant' in an informal way. So the environment $C[\_]$ has made an observation on P by communicating with it. The environment has been changed by the observation. So has the process P. Now what does it take to say that $C[\_]$ can make the same observation on Q? If Q and $C[\_]$ interact then one would have the following

$$C[Q] \stackrel{\tau}{\Longrightarrow} C''[Q'] \qquad (2)$$

where Q' and $C''[\_]$ are the residuals. Now (1) admits the situation

$$C[P] \stackrel{\tau}{\longrightarrow} C[P']$$

where the context has made a vacuous observation and the tau action is caused by an internal communication of P. This internal action of P could be simulated by *zero* or a finite number of internal actions of Q. So (2) should be improved to the following

$$C[Q] \Longrightarrow C''[Q'] \qquad (3)$$

An immediate question is whether we should make it a condition that $C''[\_]$ and $C'[\_]$ are syntactically the same. Here are some comments:

- Assume that $C[\_]$ is $(\bar{a}a.\mathbf{0}+\bar{a}a.(b)\bar{b}b) \mid \_$. It is perfectly possible that $C'[\_]$ is $\mathbf{0} \mid \_$ and $C''[\_]$ is $(b)\bar{b}b \mid \_$. So in general $C'[\_]$ and $C''[\_]$ are different in syntax. However one might as well take $C'''[\_]$ to be $\mathbf{0} \mid \_$. It seems to us that for the first order calculi one does not lose any generality by assuming that $C'[\_]$ and $C''[\_]$ are syntactically the same.
- Suppose that we are working in the higher order π-calculus. Let us assume that $C[\_]$ is $a(X).C_1[\_]$ and that P and Q are $\bar{a}\langle \mathbf{0}\rangle.A$ and $\bar{a}\langle (b)\bar{b}b\rangle.A$. In this case, $C'[\_]$ and $C''[\_]$ must be syntactically distinct if $C_1[\_]$ contains X.
- It is possible that in (3) the context has made more than one observation. If we assume that $C'''[\_]$ is the same as $C'[\_]$ then at most one observation can be made.
- If one takes the liberal view that $C'[\_]$ and $C''[\_]$ are distinct then one calls for some notion of equivalence of contexts! But it is conceivable that the equivalence of contexts and the equivalence of processes are inter-dependent. The general assumption does not always pay off.

In summary assuming $C'[\_] \equiv C''[\_]$ has many advantages, although it is not a general assumption. Under this assumption (3) becomes

$$C[Q] \Longrightarrow C'[Q'] \qquad (4)$$

The paper provides a formalization of the observational equivalence based on the idea that (4) simulates (1). This formalization leads to an equivalence relation called global bisimilarity. We take a look at this relation in some well known models of computing and prove some coincidence results.

2 Formalizing Observations

It is necessary to have a process calculus in order to formalize the idea of observation. The first language we use to test the idea is the χ-calculus. This language is a variant of the π-calculus ([7]) that has a uniform treatment of names. See [2–6] for more on the language. The syntax of the calculus is define as follows:

$$P := 0 \mid \alpha x.P \mid P|P' \mid (x)P \mid [x{=}y]P \mid !P$$

Due to space limitation we will not explain the standard programming constructs, nor do we define or state any well-known terminologies and properties. The notion of context is important since it formalizes the notion of environment.

Definition 1. *Contexts are defined as follows: (i) $[\,]$ is a context; (ii) If $C[\,]$ is a context then $\alpha x.C[\,]$, $C[\,]\mid P$, $P\mid C[\,]$, $(x)C[\,]$ and $[x{=}y]C[\,]$ are contexts.*

Throughout this paper, whenever we write $C[P] \xrightarrow{\tau} C'[P']$, we do mean that $C[\_]$ has made an observation on P and that $C'[\_]$ and P' are the descendants of $C[\_]$ and P' respectively. Unless this is formally defined, confusion would arise. The approach we adopt is to define the observation like $C[P] \xrightarrow{\tau} C'[P']$ in the same labelled transition system that defines the operational semantics of the χ-calculus. In the following semantic rules, one thinks of P as a process and of $C[P]$ as an observation pair. We use V to range over the set of all processes union the set of all observation pairs.

Sequentialization

$$\frac{}{\alpha x.P \xrightarrow{\alpha x} P}\ Sqn \qquad \frac{P \xrightarrow{\lambda} P'}{[P] \xrightarrow{\lambda} [P']}\ Cnxt$$

Composition

$$\frac{V \xrightarrow{\gamma} V'}{V|Q \xrightarrow{\gamma} V'|Q}\ Cmp_0 \qquad \frac{V \xrightarrow{y/x} V'}{V|Q \xrightarrow{y/x} V'|Q\{y/x\}}\ Cmp_1$$

$$\frac{V \xrightarrow{(y)/x} V'}{V|Q \xrightarrow{(y)/x} V'|Q\{y/x\}}\ Cmp_2$$

Communication

$$\frac{V \xrightarrow{a(x)} V' \quad Q \xrightarrow{\bar{a}y} Q'}{V|Q \xrightarrow{\tau} V'\{y/x\}|Q'}\ Cmm_0 \qquad \frac{V \xrightarrow{a(x)} V' \quad Q \xrightarrow{\bar{a}(x)} Q'}{V|Q \xrightarrow{\tau} (x)(V'|Q')}\ Cmm_1$$

$$\frac{V \xrightarrow{ax} V' \quad Q \xrightarrow{\bar{a}y} Q'}{V|Q \xrightarrow{y/x} V'\{y/x\}|Q'\{y/x\}}\ Cmm_2 \qquad \frac{V \xrightarrow{ax} V' \quad Q \xrightarrow{\bar{a}(y)} Q'}{V|Q \xrightarrow{(y)/x} V'\{y/x\}|Q'\{y/x\}}\ Cmm_3$$

Localization

$$\frac{V \xrightarrow{\lambda} V' \quad x \notin n(\lambda)}{(x)V \xrightarrow{\lambda} (x)V'} Loc_0 \qquad \frac{V \xrightarrow{\alpha x} V' \quad x \notin \{\alpha, \bar{\alpha}\}}{(x)V \xrightarrow{\alpha(x)} V'} Loc_1$$

$$\frac{V \xrightarrow{y/x} V'}{(x)V \xrightarrow{\tau} V'} Loc_2 \qquad \frac{V \xrightarrow{y/x} V'}{(y)V \xrightarrow{(y)/x} V'} Loc_3$$

$$\frac{V \xrightarrow{(y)/x} V'}{(x)V \xrightarrow{\tau} (y)V'} Loc_4 \qquad \frac{V \xrightarrow{x/x} V'}{(x)V \xrightarrow{\tau} (x)V'} Loc_5$$

Condition

$$\frac{V \xrightarrow{\lambda} V'}{[x=x]V \xrightarrow{\lambda} V'} Cnd$$

Replication

$$\frac{V \mid !V \xrightarrow{\lambda} V'}{!V \xrightarrow{\lambda} V'} Rpl$$

Using the above semantic rules, it is easy to see that

$$(x)(ax.R \mid Q) \xrightarrow{\tau} R\{y/x\} \mid Q\{y/x\} \qquad (5)$$

Now suppose $C[\_]$ is $(x)(ax.R \mid \_)$ and $C'[\_]$ is $R\{y/x\} \mid \_$. Then we have the following observation

$$C[\bar{a}y.Q] \xrightarrow{\tau} C'[Q\{y/x\}] \qquad (6)$$

It is important to distinguish between (5) and (6).

A lot of process equivalences have been studied. Some of them are specific to a particular calculus; others are applicable to a wide range of languages. The barbed bisimilarity of Milner and Sangiorgi is of the second nature ([8]). This equivalence is based on the notion of barb.

Definition 2. *A process P is strongly barbed at a, notation $P\downarrow a$, if $P \xrightarrow{\alpha(x)} P'$ or $P \xrightarrow{\alpha x} P'$ for some P' such that $a \in \{\alpha, \bar{\alpha}\}$. P is barbed at a, written $P\Downarrow a$, if some P' exists such that $P \Longrightarrow P'\downarrow a$. A binary relation $\mathcal{R}$ is barbed if $\forall a \in \mathcal{N}.P\Downarrow a \Leftrightarrow Q\Downarrow a$ whenever $P\mathcal{R}Q$.*

Two processes are barbed bisimilar if they can simulate each other while maintaining the same barbedness.

Definition 3. *Let $\mathcal{R}$ be a barbed symmetric relation on $\mathcal{C}$. The relation $\mathcal{R}$ is a barbed bisimulation if whenever $Q\mathcal{R}P \xrightarrow{\tau} P'$ then $Q \Longrightarrow Q'\mathcal{R}P'$ for some Q'. The barbed bisimilarity $\approx_b$ is the largest barbed bisimulation closed under context.*

The next lemma describes an important properties of the barbed bisimilarity. Its proof has to be omitted in this short paper.

Lemma 1. *Suppose $P \approx_b Q$ and $C[P] \xrightarrow{\tau} C'[P']$ for a context $C[\_]$. Then Q' exists such that $C[Q] \Longrightarrow C'[Q']$ and $P' \approx_b Q'$.*

3 Bisimulation Via Environment

We are now in a position to formally define what we mean by observational equivalence. Intuitively, two processes P and Q are observationally equivalent if for every context $C[\_]$ whenever $C[\_]$ has made an observation on P in the following manner

$$C[P] \xrightarrow{\tau} C'[P']$$

then $C[\_]$ can make the *same* observation on Q in the following manner

$$C[Q] \Longrightarrow C'[Q']$$

such that P' and Q' are also observationally equivalent. This leads to the following definition.

Definition 4. *A relation $\mathcal{R}$ satisfies the* Global Bisimulation Property *if whenever $\langle P, Q \rangle \in \mathcal{R}$ then for each context $C[\_]$ the following properties hold:*

(i) If $C[P] \xrightarrow{\tau} C'[P']$ then $C[Q] \Longrightarrow C'[Q']$ for some Q' such that $\langle P', Q' \rangle \in \mathcal{R}$.
(ii) If $C[Q] \xrightarrow{\tau} C'[Q']$ then $C[P] \Longrightarrow C'[P']$ for some P' such that $\langle P', Q' \rangle \in \mathcal{R}$.

A global bisimulation is a relation that satisfies the Global Bisimulation Property. The global bisimilarity, written $\approx_g$, is the largest global bisimulation.

Lemma 2. *The global bisimilarity is closed under all the combinators.*

Proof. It is easy to show that if $\mathcal{R}$ is a global bisimulation then the relation

$$\{(C[P], C[Q]) \mid \langle P, Q \rangle \in \mathcal{R} \text{ and } C[] \text{ is a context}\}$$

is also a global bisimulation. Closure under replication is also easy to prove. □

4 Global Bisimulation in χ-Calculus

This section serves to characterize the global bisimilarity of the χ-calculus. The main result here is that the global bisimilarity coincides with the barbed bisimilarity.

Theorem 1. *In the χ-calculus, the global bisimilarity and the barbed bisimilarity are the same.*

Proof. The inclusion $\approx_b \subseteq \approx_g$ is supported by Lemma 1.

Now suppose $\mathcal{S} = \{(P,Q) \mid P \approx_b Q\}$. We prove that $\mathcal{S}$ is a global bisimulation. The following is the case analysis:

1. Suppose $C[P] \xrightarrow{\tau} C'[P']$. There are eight subcases. We examine one of them.

   (a) $C[P] \xrightarrow{\tau} C'[P']$ is induced by $C[] \xrightarrow{a(x)} C''[]$ and $P \xrightarrow{\bar{a}(x)} P'$, where $C'[] \equiv (x)C''[]$. It follows from $P \approx_b Q$ and

   $$P \mid a(x).bx \mid \bar{b}x \xrightarrow{\tau} (x)(P' \mid bx) \mid \bar{b}x \xrightarrow{\tau} P' \mid \mathbf{0} \mid \mathbf{0}$$

   where b is fresh, that $Q \mid a(x).bx \mid \bar{b}x \Longrightarrow Q' \mid \mathbf{0} \mid \mathbf{0}$ for some Q' and $Q' \approx_b P'$. Then $Q \xRightarrow{\bar{a}(x)} \xRightarrow{(y_1)/x\ (y_2)/y_1} \cdots \xRightarrow{(y_n)/y_{n-1}} Q''$ for some $Q'', y_1, \ldots, y_n$ and $Q' \equiv Q''\{x/y_n\}$, where $n \geq 0$. Therefore $C[Q] \xrightarrow{\tau} C'[Q']$ and $Q' \mathcal{S} P'$.

2. Suppose $(x)C[P] \xrightarrow{\tau} C'[P']$ is induced by $C[P] \xrightarrow{y/x} C'[P']$. Then there are four subcases. We look at one of them.

   (a) $C[P] \xrightarrow{y/x} C'[P']$ is induced by $C[] \xrightarrow{\bar{a}y} C''[]$ and $P \xrightarrow{ax} P''$, where $C'[] \equiv C''[]\{y/x\}$ and $P' \equiv P''\{y/x\}$. By $P \approx_b Q$ and

   $$(x)(P \mid \bar{a}z.bx) \mid \bar{b}z \xrightarrow{\tau} P''\{z/x\} \mid bz \mid \bar{b}z \xrightarrow{\tau} P''\{z/x\} \mid \mathbf{0} \mid \mathbf{0}$$

   for some fresh b and z, $(x)(Q \mid \bar{a}z.bx) \mid \bar{b}z \Longrightarrow Q'' \mid \mathbf{0} \mid \mathbf{0}$ for some Q'' and $Q'' \approx_b P''\{z/x\}$. Then either

   $$Q \xRightarrow{(y_1)/x\ (y_2)/y_1} \cdots \xRightarrow{(y_n)/y_{n-1}} \xRightarrow{ay_n} Q'$$

   and $Q'' \equiv Q'\{z/y_n\}$ for some $Q', y_1, \ldots, y_n$, where $n \geq 0$, or

   $$Q \xRightarrow{(y_1)/x\ (y_2)/y_1} \cdots \xRightarrow{(y_n)/y_{n-1}} \xRightarrow{a(z)} \xRightarrow{(z_1)/y_n\ (z_2)/z_1} \cdots \xRightarrow{(z_m)/z_{m-1}} \xRightarrow{z/z_m} Q''$$

   for some $y_1, \ldots, y_n, z_1, \ldots, z_m$, where $n \geq 0, m \geq 0$. In the former case, $(x)C[Q] \xRightarrow{\tau} C'[Q'\{y/y_n\}]$ and $P'\mathcal{S}Q'\{y/y_n\}$. In the latter case, $(x)C[Q] \xRightarrow{\tau} C'[Q'\{y/z\}]$ and $P'\mathcal{S}Q'\{y/z\}$.

3. Suppose $(x)C[P] \xrightarrow{\tau} C'[P']$ is induced by $C[P] \xrightarrow{(y)/x} C''[P']$, where $C'[] \equiv (y)C''[]$. Then there are four subcases. We look at one of them.

   (a) $C[P] \xrightarrow{y/x} C'[P']$ is induced by $C[] \xrightarrow{\bar{a}(y)} C''[]$ and $P \xrightarrow{ax} P''$, where $C'[] \equiv C''[]\{y/x\}$ and $P' \equiv P''\{y/x\}$. By $P \approx_b Q$, either

   $$Q \xRightarrow{(y_1)/x\ (y_2)/y_1} \cdots \xRightarrow{(y_n)/y_{n-1}} \xRightarrow{ay_n} Q''$$

   and $Q''\{x/y_n\} \approx_b P''$ for some $Q'', y_1, \ldots, y_n$, where $n \geq 0$, or

$$Q \xRightarrow{(y_1)/x} \xRightarrow{(y_2)/y_1} \cdots \xRightarrow{(y_n)/y_{n-1}} \xRightarrow{a(z)} \xRightarrow{(z_1)/y_n} \xRightarrow{(z_2)/z_1} \cdots \xRightarrow{(z_m)/z_{m-1}} \xRightarrow{z/z_m} Q''$$

and $Q''\{x/z\} \approx_b P''$ for some $Q'', y_1, \ldots, y_n, z_1, \ldots, z_m$, where $n \geq 0, m \geq 0$. So either $(x)C[Q] \xRightarrow{\tau} C'[Q''\{y/y_n\}]$ and $P'SQ''\{y/y_n\}$, or $(x)C[Q] \xRightarrow{\tau} C'[Q''\{y/z\}]$ and $P'SQ''\{y/z\}$.

This completes the proof. □

5 Global Bisimulation in π-Calculus

Having seen Theorem 1, one is tempted to think that in the π-calculus, the global bisimilarity and the barbed bisimilarity are the same. This is however not the case. It is surprising to know that the global bisimilarity in the π-calculus relates to the open bisimilarity of Sangiorgi. The open bisimulations ([9]) are not as general as the barbed bisimulations. But they have established their authentic role in the calculi of mobile processes. We assume that the reader is familiar with the π-calculus. So without further ado, we state and prove the interesting result.

Definition 5. *Suppose S is a symmetric relation on the π-processes. It is called a rigid open bisimulation if PSQ and for every substitution σ whenever $P\sigma \xrightarrow{\alpha} P'$, then Q' exists such that $Q\sigma \xRightarrow{\widehat{\alpha}} Q'SP'$. The rigid open bisimilarity $\approx_o^r$ is the largest rigid open bisimulation.*

The rigid open bisimilarity is of the open style but is different from the open bisimilarity defined in [9].

The definition of the global bisimilarity for the π-calculus is the same as that for the χ-calculus.

Theorem 2. *In the π-calculus, the global bisimilarity and the rigid open bisimilarity are the same relation.*

Proof. Let S be $\{(P,Q) \mid P \approx_o^r Q\}$. Then S is a global bisimulation. Suppose PSQ and, for a context $C[\,]$, $C[P] \xrightarrow{\tau} C'[P']$. Consider two cases:

1. $C[P] \xrightarrow{\tau} C'[P']$ is induced by $C[\,] \xrightarrow{x(w)} C''[\,]$ and $P \xrightarrow{\overline{x}(w)} P'$, where $C'[\,] \equiv (w)C''[\,]$. It follows form $P \approx_o^r Q$ that $Q \xRightarrow{\overline{x}(w)} Q' \approx_o^r P'$ for some Q'. Therefore $C[Q] \xRightarrow{\tau} C'[Q']$ and $P'SQ'$.

2. $C[P] \xrightarrow{\tau} C'[P']$ is induced by $C[\,] \xrightarrow{\overline{x}(w)} C''[\,]$ and $P \xrightarrow{x(w)} P'$, where $C'[\,] \equiv (w)C''[\,]$. It follows from $P \approx_o^r Q$ that $Q \xRightarrow{x(w)} Q' \approx_o^r P'$. Therefore $C[Q] \xRightarrow{\tau} C'[Q']$ and $Q'SP'$.

For the reverse inclusion, let S be $\{(P,Q) \mid P \approx_g Q\}$. Then S is a rigid open bisimulation. Suppose $P\sigma \xrightarrow{\alpha} P'$ for some substitution σ. Consider one case:

– α is $\bar{x}(w)$. Using the fact that $P \approx_g Q$ and $P\sigma \mid x(w).\bar{b}w \xrightarrow{\tau} (w)(P' \mid \bar{b}w)$ for some fresh b, it is immediate to see that Q' exists such that $Q\sigma \mid x(w).\bar{b}w \Longrightarrow (w)(Q' \mid \bar{b}w)$ and $Q' \approx_g P'$. It then follows that $Q\sigma \xrightarrow{\bar{x}(w)} Q'$ and $P'\mathcal{S}Q'$.

This completes the proof. □

6 Global Bisimulation for Higher Order Calculi

As we have discussed in the introduction, for higher order calculi where the content of a communication can be a process, it is too strong to assume that equivalent processes exert the *same* effect on the environments. What one can safely assume is that equivalent processes exert *similar*, or *equivalent*, effect on the environments. This means that we need to have a notion of equivalence on contexts. But usually an equivalence on the contexts is based on an equivalence on the processes. We seem to be in a circle. In order to break the circle, we introduce the notion of context algebra.

Definition 6. *A context algebra is a tuple $(\mathcal{C}, \asymp)$ such that $\mathcal{C}$ is a set of contexts closed under composition and $\asymp$ is an equivalence relation on $\mathcal{C}$. We often write $\mathcal{C}$ for $(\mathcal{C}, \asymp)$.*

Using this notion, we can define the observational equivalence with respect to a context algebra as follows.

Definition 7. *Suppose $\mathcal{R}$ is a symmetric binary relation on the set of processes and $(\mathcal{C}, \asymp)$ is a context algebra. It is a global bisimulation with respect to $\mathcal{C}$ if whenever $P\mathcal{R}Q$ then the following property holds for each pair of contexts $C_1[], C_2[] \in \mathcal{C}$ satisfying $C_1[] \asymp C_2[]$:*

If $C_1[P] \longrightarrow C_1'[P']$ then $C_2[Q] \Longrightarrow C_2'[Q']$ for some process Q' and context $C_2'[] \in \mathcal{C}$ such that $P'\mathcal{R}Q'$ and $C_1'[] \asymp C_2'[]$.

The global bisimilarity with respect to $\mathcal{C}$, noted $\approx_{\mathcal{C}}$, is the largest global bisimulation with respect to $\mathcal{C}$.

Lemma 3. *For each context algebra $\mathcal{C}$ the bisimilarity $\approx_{\mathcal{C}}$ with respect to $\mathcal{C}$ is an equivalence relation.*

The problem with $\approx_{\mathcal{C}}$ is that it refers to a context algebra. What we seek is a definition that starts with a trivial context algebra. Now given a context algebra $(\mathcal{C}_0, \asymp_0)$, we can define a new context algebra $(\mathcal{C}_{i+1}, \asymp_{i+1})$ from $\approx_{\mathcal{C}_i}$ as follows:

– For every i, $\mathcal{C}_{i+1}$ is the same as $\mathcal{C}_i$;
– For every i, $C[] \asymp_{i+1} C'[]$ if and only if $\forall P \in \mathcal{P}.C[P] \approx_{\mathcal{C}_i} C'[P]$.

In this way we can define a sequence of bisimilarities:

$\mathcal{C}_0, \mathcal{C}_1, \mathcal{C}_2, \ldots, \mathcal{C}_i, \ldots$

Let $\mathcal{C}_0$ be $(\mathcal{C}, \equiv)$, where $\mathcal{C}$ is the set of all contexts and $\equiv$ is the syntactical equality. We then have a sequence of concrete observational equivalences.

There are alternative definitions of the global style bisimilarities that define the equivalence of processes and the equivalence of contexts at one go. As a future work we will investigate these ideas by applying them to the Ambient Calculus ([1]), a higher order calculus for global computing.

7 Remark

We have given one formalization of observation and proposed a general definition of bisimulation equivalence on top of the formalization. We have proved that in the χ-calculus this is the same as the barbed bisimilarity and in the π-calculus it gives rise to a variant of the open bisimilarity. This is a little surprising because in most languages the barbed bisimilarity is the weakest bisimulation equivalence while the open bisimilarity is the strongest. The implication of these facts need be further investigated before making any useful comments. But we believe that the approach used in this paper is both interesting and useful. It is interesting because it formalizes our original intuition about observational equivalence. It is useful because it applies to a wide range of models and suggests a natural bisimulation equivalence for every one of them. The importance of the global bisimilarity is that it is conceptually clear and simple, even though it is less tractable.

The global bisimilarity assumes that the environments are global. A different view is that they are local. The latter view suggests the following definition.

Definition 8. *A relation $\mathcal{R}$ satisfies the* Local Bisimulation Property *if whenever $\langle P, Q \rangle \in \mathcal{R}$ then for each context $C[\_]$ the following properties hold:*

(i) If $C[P] \xrightarrow{\tau} C'[P']$ then $C[Q] \Longrightarrow C'[Q'] \mathcal{R} C'[P']$ for some Q'.
(ii) If $C[Q] \xrightarrow{\tau} C'[Q']$ then $C[P] \Longrightarrow C'[P'] \mathcal{R} C'[Q']$ for some P'.

A local bisimulation *is a relation that satisfies the Local Bisimulation Property. The* local bisimilarity, *written $\approx_l$, is the largest local bisimulation.*

The investigation of this equivalence has to be carried out in another occasion.

References

1. Cardelli, L. and Gordon, A.: Mobile Ambient, *Theoretical Computer Science*, **240** (2000) 177-213.
2. Fu, Y.: A Proof Theoretical Approach to Communications, *ICALP '97*, Lecture Notes in Computer Science 1256 (July 7th-11th, Bologna, Italy, Springer-Verlag, 1997) 325–335.

3. Fu, Y.: Variations on Mobile Processes, *Theoretical Computer Science*, **221** (1999) 327–368.
4. Fu, Y.: Bisimulation Congruence of Chi calculus, *Information and Computation*, **184** (2003) 201–226.
5. Fu, Y. and Yang, Z.: Chi Calculus with mismatch, *CONCUR 2000*, Lecture Notes in Computer Science 1877 (Springer, 2000) 596–610.
6. Fu, Y. and Yang, Z.: Understanding the Mismatch Combinator in Chi Calculus, *Theoretical Computer Science*, **290** (2002) 779-830.
7. Milner, R., Parrow, J. and Walker, D.: A Calculus of Mobile Processes, *Information and Computation*, **100** (1992) 1–40 (Part I), 41–77 (Part II).
8. Milner, R. and Sangiorgi, D.: Barbed Bisimulation, *ICALP '92*, Lecture Notes in Computer Science 623 (Springer-Verlag, 1992) 685–695.
9. Sangiorgi, D.: A Theory of Bisimulation for π-Calculus, *Acta Informatica*, **3** (1996) 69–97.

A Scalable and Reliable Mobile Agent Computation Model

Yong Liu, Congfu Xu, Zhaohui Wu, and Yunhe Pan

College of Computer Science, Zhejiang University
Hangzhou 310027, China
cckaffe@yahoo.com.cn

Abstract. This paper presents a high performance service based mobile agent computation model. The scalability and reliability of this model is secured through the service clone policy and access privilege policy. With the introduction of service density of group, we can further decrease resource waiting and balance the service occupancy for the whole network.

1 Introduction

Mobile Agents are programs that can be migrated and executed between different network hosts. There are tow types of mobile agents classified by the migration ability of the agents. They are strong migration mobile agents and weak migration mobile agents. When using the strong migration policy, the mobile agent system needs to record all the states and related data in each position of the agent, which will spend a huge amount of time and space for the transport, and will lead to low efficiency. When using weak migration policy, the transportation of data will decrease greatly, however, the abilities of adapting the complicated network topology will decrease too. Therefore, how to design a reliable and high-efficiency work pattern for the mobile agent is one of the most important problems.

In fact, the Grid[3] technology has provided a powerful platform for the mobile agent. The WSDL[4] provides a web service description language for service disposing. How dose the service be composed together can be described by the WSFL[6]. And the UDDI (Universal Description, Discovery and Integration)[5] is used to enable online registry and the publishing and dynamic discovery of Web services offered by businesses. The RDF[7], which is recommended by W3C and can support a domain-independent metadata description, is used to describe the resources.

The mobile agent computation model that is built upon these above techniques is convergence and hard to scalability. In order to solve this problem, a scalable and reliable service based mobile agent computation model is proposed based on decentralized virtual organization architecture.

J. Zhang, J.-H. He, and Y. Fu (Eds.): CIS 2004, LNCS 3314, pp. 346–352, 2004.
© Springer-Verlag Berlin Heidelberg 2004

2 Overview of MACM and Fabric Architecture

Our computation model is established upon the virtual organization architecture, namely virtual organization (VO or group), which can greatly reduce the mobile agent size during migration, and this architecture can provide more intelligent support for mobile agent. The basic elements of virtual organization are nodes, which can be PC, PDA, laptop, and other devices connected by network. The nodes are grouped in virtual, and they can join and leave the group dynamically. The groups are virtually hierarchical, which means that the groups can be classified into root-layer, middle-layers and leaf virtual layers. The virtual group based fabric architecture is the platform of the mobile agent migration. By this way, the mobile agent can discovery and move more effective and it can also greatly decrease the mobile agent size during migration.

2.1 Formal Definition of Infrastructure

The VO based architecture mentioned in this paper is a structure similar to the fabric layer in [1]. Some definitions are given as the following:

Definition 1. Node, the minimized devices that can load and execute the mobile agents in network are denoted as R_i. Each node in VO can be a service provider or a service consumer.

Definition 2. Key Node is a kind of nodes that deals with the remote communications, denoted as R_i^0. Each group must contain a key node. And the key nodes are always positioned in two or more than tow groups. Normally the contiguity key nodes form a kernel group.

Definition 3. Group, the set includes one node or several nodes, denoted as $G_a^i = \{R_a^0, R_a^1, R_a^2, ..., R_a^n\}$, where a is the name of the group. Each group has a layer identifier i, which means that the group is the ith layer in VO. The node can join in more than one group in VO, which means that each node including key node can belong to two or more than two groups. R_j^i means that node belongs to group G_j. Group is a comparatively stable organization; the nodes belonging to certain group can leave this group and join in another group dynamically. The login and logout of nodes adopt the GGMP (Grid Group Management Protocol)[1], which is similar to the IGMP.

Definition 4. Service, in VO architecture, service is a kind of dealing process that provided by a certain nodes in VO. It can be formally defined as four-tuple: $S(GUIDS, Privilege, Content, Operation)$, where $GUIDS$ is the global unique ID of the service. And $privilege$ is the privilege defined by the service provider, which include public and private privileges. $Content$ is the process of service. $Operation$ is a kind of special operations associated with service, which includes clone, serve, migration etc.

Definition 5. Service Management Node, is a kind of nodes that contains the services information of current group, denoted as R_s. Normally, all these contiguity service management nodes form a service management group automatically.

Definition 6. Virtual Organization, VO is a fabric structure that composed of nodes and is established by a serial of protocols. It is a hierarchical tree structure constituted by virtual groups. There exist a root group (denoted as RG, and it is the first layer) and other middle layer groups, denoted as G_j^i, i means that the group is the ith layer in the virtual organization. L is the leaf group, which is positioned at the bottom of the virtual tree, shown as figure 1. Normally, the group contains resembling and adjacent nodes. There is a key node R_i^0 in each group G_i. The functions of the key node in a group are similar to the gateway in a LAN, which communicates with other nodes outside the group. A protocol called GGMP (Grid Group Management Protocol)[2] has been used to determine the key node. Among all the nodes and groups, the key nodes constitute a virtual group called *kernel group*, G_k. It is the most important portion that serves for other nodes. It deals with communication and seeking etc. Among all of the nodes in virtual organization topology, the root (first layer) group in the tree structure of virtual organization is always chosen from one of the kernel groups.

2.2 Safe Service Management Protocol

Service is disposed upon the virtual organization. Compared with the UDDI[5], the virtual organization based service is more dispersed. It is not necessary to

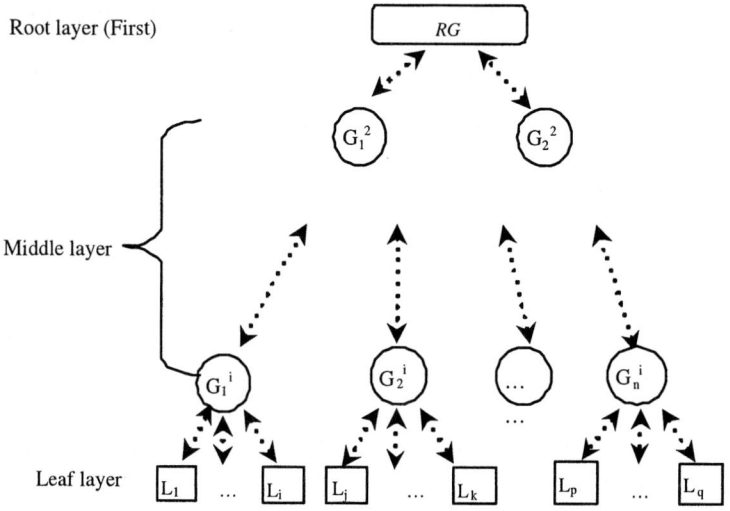

Fig. 1. Tree-Structure of the Virtual Organization

establish a universal service information server in virtual organize, on the contrary, the service information is distributed in normal group, there is a service management node in each group. The available service information is stored in these service management nodes. For safe service management protocol, once the service has been released by the original node, it can be replicated to the same group nodes and others, and this operation is called *clone*. However, there may exist some services that would not share them with all the other nodes in VO. So the service disposed in the virtual organization has a privilege attribute. For the virtual tree is organized in hierarchy, the privilege of each layer is different, and in general the higher privilege is assigned to the higher layer group in VO.

Table 1. Privileges defined by the SSMP

Privilege Types	Privilege Sub-types	Notes
Private privileges	Self private	The node, which releases the service, engrosses the service.
	Group private	The service is privately belonged to the released node and its group.
Public privileges	Below access	The service is public to the groups, which are the children of the original service released group.
	Same level access	The service is public to the same layer group nodes in VO, which means the service released by the ith layer group node, and it can be cloned to the same ith layer group nodes.
	Full tree access	The service is public to all the tree nodes.
	Special level access	The service can specify the public layer from ith layer to jth layer, $i < j$.

Table 1 lists the privileges that are defined by the SSMP. The privilege can be classified into two categories: private service and public service. This category division is based on the concept of the group, that is to say, whether the service is private or public to other groups. A farther category in private privilege can be classified into self-private and group-private, the difference between these two categories is whether the service can be cloned among the original group. There are four kinds of farther public privilege categories, they are below access, same level access, full tree access, and special level access respectively. This kind of public privilege category means that the services can be cloned out of their released groups, and the access degree out of the original groups is controlled by this privilege category.

2.3 Service Density

Before the service dispose protocol is presented, a tolerance named *Service Density* should be introduced firstly. It is one of the most important factors in our mobile agent computation model.

Definition 7. Service Density of Group, is a function, $f_s(A_{G_a^i}, N_{G_a^i})$, which symbolizes a certain available service in groups. Where s is the service name. $A_{G_a^i}$ is the number of nodes in group G_a^i, which contains available service of s. $N_{G_a^i}$ is the node number in group G_a^i.

In our mobile agent computation model, the service density of group (SDG) is defined as formulation (1). In SSMP, the service density of current group is stored in this service management node. Other groups can request for service density by sending message to this management node.

$$\eta_{G_a^i}^S = \frac{A_{G_a^i}}{N_{G_a^i}} \times 100\% \tag{1}$$

3 Service Based Mobile Agent Computation Model

Definition 8. Service based mobile agent computation model, is a six-tuple $MACM = (R, S, M, \phi, v, E)$,

Where, R is the node set, S is the finite state set of the mobile agent. S does not include the state of the agent migration Λ and the null state ε. Here, migration state Λ means that the mobile agent starts to move to another node to execute new state; null state means that the mobile agent does not perform any action (execution and migration).

$M \subset S$ is the set of all the message operation states for mobile agent. $M = \{M_s, M_a\}$, M_s is the state of sending message, and M_a is the state of receiving message.

$v \in R$ is the initial node that the mobile agent has been produced, a mobile agent's service firstly comes from the node v, and then cycles are driven by the finite states.

$E \subset R$ is the set of final node for the mobile agent, only in the final node the mobile agent can be destroyed and the service ends.

ϕ, the transition relation, is a finite subset of $(R \times (S \cup \{\Lambda, \varepsilon\})) \longrightarrow R$: $\phi : (R \times (S \cup \{\Lambda, \varepsilon\})) \longrightarrow R$, Where

(1) To all the $R_i, R_j \in R$, if $\phi(R_i, \varepsilon) = R_j$, then $R_i = R_j$,
(2) To all the $R_i, R_j \in R$, if $\phi(R_i, \Lambda) = R_j$, then $R_i \neq R_j$,
(3) To all the $R_i, R_j \in R$, $S_k \in S$, if $\phi(R_i, S_k) = R_j$, then $R_i = R_j$,
(4) To all the $R_i \in R$, if $\phi(R_i, M_a) = R_i$, then the next transition state relation is $\phi(R_i, M_s) = R_i$,
(5) To $VO = \{G_{a_m}^n, n = 1, 2, 3, ..., m = 1, 2, 3, ...\}$, $\eta_{G_{a_m}^n}^{S_r}$ is the service S_r's density of group $G_{a_m}^n$. When mobile agent begins to switch to S_r service, there will be:

$$\phi_{S_r}(R_i, \Lambda) = R_j$$

Here R_i is the current position node of the mobile agent. To the target node R_j, we have:

$$(R_j \in SN^{S_r}_{G^q_{a_k}}) \cap (SN^{S_r}_{G^q_{a_k}} \subseteq G^q_{a_k}) \cap (G^q_{a_k} \in G^n_{a_m}) \cap (\eta^{S_r}_{G^q_{a_k}} = max(\eta^{S_r}_{G^n_{a_m}}, n, m = 1, 2, 3, ...))$$

where $SN^{S_r}_{G^q_{a_k}}$ is the available service node in $G^q_{a_k}$ of service S_r.

(6) After the migration in (5) has finished, the service S_r's density of $G^q_{a_k}$ will change to:

$$\eta^{S_r}_{G^q_{a_k}} = \frac{A^{S_r}_{G^q_{a_k}} - 1}{N^{S_r}_{G^q_{a_k}}} \times 100\% \qquad (2)$$

In this computation model, the migration state Λ is established by the communication of the nodes in VO. By adopting the service fabric communication protocol, mobile agent can move from the original node to the destination node efficiently. The state transition and message communication are all implemented by this protocol. The service density has been treated as an important tolerance to dispatch the mobile agent's migration.

4 Conclusions

In this paper, a safe service based mobile agent computation model is introduced. With the policy of service clone in a tree structure hierarchy group and the policy of flexible service access, the model is more effective and reliable compared with traditional methods. Furthermore, agent can use group service density to monitor the whole network service occupancy in groups. It can balance the service's serving time and decrease the waiting list of service.

Acknowledgements. This paper was sponsored by China Defense Ministry, and was partially supported by the Aerospace Research Foundation (No. 2003-HT-ZJDX-13). Thanks for Dr. Zhang Qiong's advice to this paper.

References

1. Huang, L.C., Wu, Z.H., and Pan, Y.H.: Virtual and Dynamic Hierarchical Architecture for E-Science Grid. International Journal of High Performance Computing Applications, 17 (2003)329-350
2. Huang, L.C., Wu, Z.H., and Pan, Y.H.: A Scalable and Effective Architecture for Grid services' Discovery. First Workshop on Semantics in Peer-to-Peer and Grid Computing. Budapest, Hungary, 20 May 2003: 103-115
3. Foster, I., Kesselman, C., and Tuecke, S.: The Anatomy of the Grid: Enabling Scalable Virtual Organizations. International J. Supercomputer Applications, 15(3), 2001

4. Christensen, E., Curbera, F., Meredith, G. and Weerawarana., S.: Web Services Description Language (WSDL) 1.1. W3C, Note 15, 2001, http://www. w3. org /TR/wsdl
5. UDDI http://www.uddi.org/
6. Web Services Flow Language (WSFL) Version 1.0, http://www4.ibm.com/ software/ solutions/ Webservices/pdf/WSFL.pdf
7. Brickley, D., and Guha,R. V.: Resource Description Framework (RDF) Schema Specification 1.0, W3C Candidate Recommendation 27 March 2000.

Building Grid Monitoring System Based on Globus Toolkit: Architecture and Implementation

Kejing He, Shoubin Dong, Ling Zhang, and Binglin Song

Guangdong Key Laboratory of Computer Network, South China University of Technology,
Guangzhou, 510641, P.R. China
{kjhe, sbdong, ling, blsong}@scut.edu.cn

Abstract. Computational Grids, coupling geographically distributed resources, have emerged as a next generation computing platform for solving large-scale problems in scientific arena. Monitoring these Grid computing resources, which not only helps end-users monitor the status of Grid resources, but also provides quick and comprehensive information for resource scheduling and management, becomes extremely important for efficiently using Grid Computing resource. This paper presents architecture and implementation of a Grid Monitoring System based on Globus Toolkit. The Grid Monitoring System presented in this paper is flexible, extensible and, most important, secure. The architecture and implementation presented in this paper has important practical value for the monitoring of China Education and Research Grid.

1 Introduction

Computational Grid has emerged as a next generation computing platform for solving large-scale problems. Computational Grid often consists of large and diverse sets of distributed resources, which are geographically distributed [8]. Grid Monitoring, which is aimed at making information about the status of a Grid available to users and to other components of the Grid, is necessary for building Computational Grid [3]. First, administrators and users want to see the dynamic information of the computational grid. Second, other parts of Grid components need know the Grid status for scheduling or allocating. How to monitor those large amounts of resources becomes a very key question in the computational grid. The monitor should be scalable to a large number of resources and events. The start-up of the monitor and collection of data from remote sites are more complex and difficult than in a single cluster [10].

Some systems are currently available and others are being developed for collecting and forwarding this data. For example, Ganglia [9] is a distributed monitoring system that may span multiple clusters, but if those clusters belong to different organization, exporting all information to collectors is not a secure behavior. Current version of ganglia don't have any delegation and certification mechanism for safety management,

---

* This research was supported by China Education and Research Grid (China Grid) Project under grant CG2003-CG005 and CG2003-GA002.

so it is not very suitable for Grid-level monitoring respect to security, which is the most important factor in Grid monitoring [4].

The goal of this study is to examine the representative work related to Grid monitoring and presents a flexible, extensible and, most important, secure Grid monitoring architecture based on Globus Toolkit Version 3. Section 2 describes the challenges and architecture of a Grid Monitoring System. Section 3 shows the implementation mechanisms of Grid Monitoring Architecture.

2 Design of Grid Monitoring System

The ability to monitor and manage distributed computing resources is critical for enabling high-performance distributed Grid computing. Monitoring data is needed to determine the source of performance problems and to tune the system and application for better performance.

Large distributed systems such as Computational and Data Grid require a substantial amount of monitoring data be collected for a variety of tasks such as fault detection, performance analysis, performance tuning, performance prediction, and scheduling.

2.1 Design Principles

Grid monitoring system is differentiated from cluster monitoring system in that it must be scalable across wide-area networks, include a large number of heterogeneous resources, and be integrated with other Grid middleware in terms of scheduling and security issues.

The following is a list of features required in the Grid Monitoring System:

1. Functionality: Gathering the static and dynamic information about the distributed resources. Static information covers parameters like number of computing elements, total storage capacities, total memory etc. whereas dynamic information covers parameters like free memory, free storage space, load average etc.

2. Security: When the case comes from Cluster to Grid, security will be a significant factor for building a safe monitoring system. Clusters be monitored often belong to different organizations; thus, delegation and certification mechanism will be needed respect to security.

3. Applicability: Support of multiple, existing monitoring systems.

4. Extensibility: The framework should be extensible to incorporate other monitoring components and the format of resource information data should be extensible.

2.2 Architecture

Based on the design challenges and requirements analyzed above, the Grid monitoring Architecture is presented as below.

As figure shown above, each small VO logically represents a cluster in computational concept and each leaf node represents a computational node. Each VO is an Autonomous System that can run independently. This is similar to the real situation of China Education and Research Grid. Every computational node runs a Resource Information Provider, which responsible for provider host information to Monitoring Service. Monitoring Service, responsible for aggregate computational nodes' Resource Information and push that Resource Information to Index Service, contained in the cluster's management node.

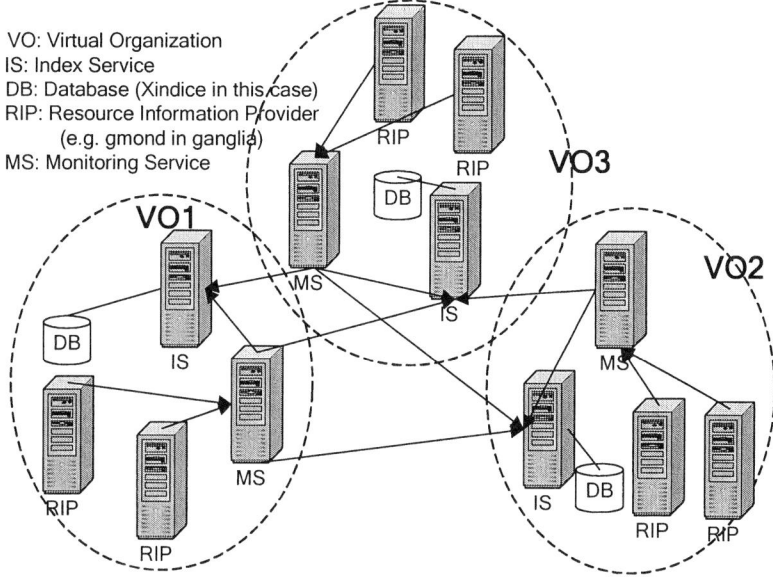

Fig. 1. Flexible Grid Monitoring Architecture

Public key based X.509 identity certificates are a recognized solution for cross-VOs identification of users. When the certificate is presented through a secure protocol such as SSL (Secure Socket Layer), the server side can be assured that the connection is indeed to the legitimate user named in the certificate. That solution overcomes the weakness of Ganglia and makes the cross-VOs monitoring is security and safety. For example, as shown in Fig.1, VO1's MS only push its Resource Information to VO2's IS, VO2's MS push its Resource Information to VO2's IS and VO3's IS, VO3's MS push its Resource Information to VO1's IS, VO2's IS and VO3's IS.

In the architecture presented above, each VO can run its Monitoring System autonomously. Among VOs, there are only Resource Information Flows between Monitoring Services and Index Services. Interactions between Monitoring Services and Index Services are under the protection of X.509 identity certificates, which provide Grid Monitoring Architecture with security.

3 Implementation Mechanisms

The key middleware supporting computational Grid is the Globus Toolkit. The Globus Toolkit version 3.0 (GT3) based on an Open Grid Service Architecture

(OGSA) [2] mechanism includes a set of core services such as security, communication, managing distributed applications, and information. In the Open Grid Service Architecture (OGSA), everything is represented by a Grid service and expresses its state in a standardized way as Service Data. The Grid service is a web service that provides a set of well-defined interfaces and that follows specific conventions. The interfaces address discovery, dynamic service creator, lifetime management, notification, and manageability; the conventions address naming and upgradeability. Ganglia is a scalable distributed monitoring system used in cluster monitoring, but doesn't suitable for Grid monitoring respect to security.

The architecture presented above provides clear delineation between cluster monitoring and Grid monitoring. Cluster administrator can use Ganglia or other monitoring system for cluster monitoring and does not need care of security problem and the communication among clusters. Globus Toolkit version 3.0 (GT3), which includes a set of core services such as security, communication, managing distributed applications, and information, is responsible for the security problem brought by spanning clusters, aggregating host information from clusters and indexing host information for quick querying. The Grid Monitoring architecture is flexible, because it is a modularization approach. Administrator can choose suitable cluster monitoring system, resource information provider and the Grid Services will be aggregated by index service.

3.1 Resource Information Provider Mechanism

Resource Information Provider mechanism consist of a Monitoring Service and one or more Plug-in Resource Information Providers, which are regularly executed by the Monitoring Service. The Resource Information Provider mechanism provides a standard mechanism for dynamics generation of service data via external programs. External provider programs can be the core providers that are part of GT3 or can be

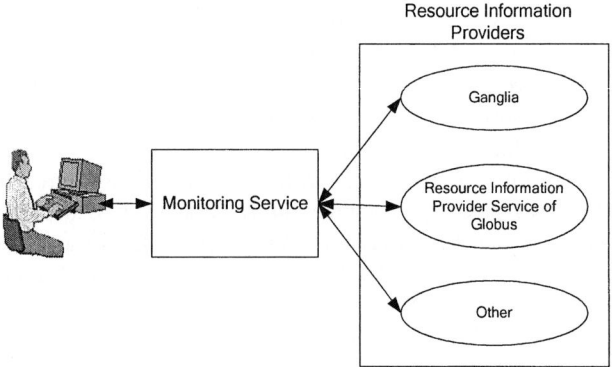

Fig. 2. Resource Information Provider mechanism

user-created providers, which means, user can use his own Java provider to generate dynamic information. But, of course, the user-defined provider must conform to some regulation. First, the provider must be composed of a Java class which implements at least one of predefined Java interfaces. Second, the provider must generate a compatible form of XML output as the result of its execution.

3.2 Information Aggregation Mechanism

The data aggregation mechanism provides a reusable mechanism for handling subscription, notification, and updating of locally stored copies of service data that is generated by Resource Information Providers. The diagram above show the aggregator mechanism used in other system.

While resource information received by the Index Service is represented externally as Service Data in typical OGSI fashion, the components at the Collective Layer represent Index Service enhancements to the standard OGSI behavior. There is typically one Monitoring Service per Virtual Organization or resource site. When there is a large VO that consists of multiple large sites, then often each site will run its own Index Service that will index the various resources available at that site. Then each of those Monitoring Services would be included in the VO's Index Service.

So, query the Virtual Organization's index service can get all the organization's host information.

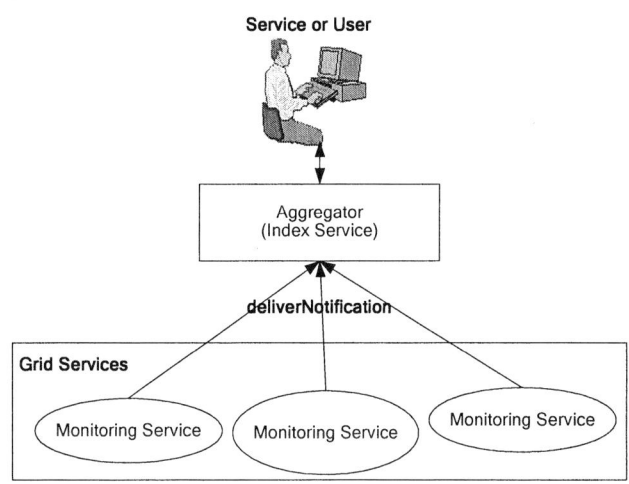

Fig. 3. Service data aggregation mechanism

4 Conclusions and Future Work

In this paper we present an Extensible Grid Monitoring System based on Globus and Ganglia. The Extensible Grid Monitoring System presented in this paper take the advantages of Globus Toolkit and ganglia into account. Ganglia is a scalable distributed monitoring system and often used in cluster monitoring, but it is not very suitable for Grid-level monitoring, because ganglia don't have any delegation and certification mechanism for safety management. Globus Toolkit version 3.0 (GT3) based on an

Open Grid Service Architecture (OGSA) mechanism includes a set of core services such as security, communication, managing distributed applications, and information. Integrating Globus with ganglia brings many advantages include stability, security, flexibility and so on. Further enhancements and optimization of this architecture are currently under investigation.

References

1. Gerndt, M., Wismueller, R., Balaton, Z., Gombas, G., Kacsuk, P., Nemeth, Z., Podhorszki, N., Truong, H., Fahringer, T., Bubak, M., Laure, E., Margalef, T.: Performance Tools for the Grid: State of the Art and Future, Technical report, Research Report Series, Lehrstuhl fuer Rechnertechnik und Rechnerorganisation (LRR-TUM) Technische Universitaet Muenchen, Vol. 30, Shaker Verlag (2004)
2. Foster, I., Kesselman, C., Nick, J., Tuecke, S.: The Physiology of the Grid: An Open Grid Services Architecture for Distributed Systems Integration, Open Grid Service Infrastructure WG, Global Grid Forum (2002)
3. Czajkowski, K., Fitzgerald, S., Foster, I., Kesselman, C.: Grid Information Services for Distributed Resource Sharing, Proceedings of the Tenth IEEE International Symposium on High-Performance Distributed Computing (HPDC-10), IEEE Press (2001)
4. Welch, V., Siebenlist, F., Foster, I., Bresnahan, J., Czajkowski, K., Gawor, J., Kesselman, C., Meder, S., Pearlman, L., Tuecke, S.: Security for Grid Services, Twelfth International Symposium on High Performance Distributed Computing (HPDC-12), IEEE Press (2003)
5. Andreozzi, S., De Bortoli, N., Fantinel, S., Ghiselli, A., Tortone, G., Vistoli, C.: GridICE: a monitoring service for the Grid, in Proceedings of the Third Cracow Grid Workshop (2003) 220-226
6. Andreozzi, S., Sgaravatto, M., Vistoli, C.: Sharing a conceptual model of Grid resources and services, Conference for Computing in High Energy and Nuclear Physics (CHEP 2003) (2003)
7. Sacerdoti, F.D., Katz, M.J., Massie, M.L., Culler, D.E.: Wide Area Cluster Monitoring with Ganglia, In Proceedings of the IEEE Cluster 2003 Conference (2003)
8. Plale, B., Jacobs, C., Jensen, S., Liu, Y.: Charlie Moad, Rupali Parab, and Prajakta Vaidya, Understanding Grid Resource Information Management through a Synthetic Database Benchmark/Workload to appear in 4th IEEE/ACM International Symposium on Cluster Computing and the Grid (CCGrid2004) (2004)
9. Massie, M.L., Chun, B.N., Culler, D.E.: The Ganglia Distributed Monitoring System:Design, Implementation, and Experience, submitted for publication (2003)
10. Balaton, Z., Kacsuk, P., Podhorszki, N.: From cluster monitoringto grid monitoring based on grm and prove, Technical Report LPDS-1/2000, Laboratory of Parallel and Distributed System, Hungary (2000)

History Information Based Optimization of Additively Decomposed Function with Constraints

Qingsheng Ren[1], Jin Zeng[2], and Feihu Qi[1]

[1] Department of Computer Science and Engineering,
Shanghai Jiaotong University, Shanghai 200030, P.R. China
{ren-qs, fhqi}@cs.sjtu.edu.cn
[2] Department of Mathematics,
Shanghai Jiaotong University, Shanghai 200030, P.R. China
zengjin@sjtu.edu.cn

Abstract. In this paper, we propose a modified estimation of distribution algorithm HCFA (History information based Constraint Factorization Algorithm) to solve the optimization problem of additively decomposed function with constraints. It is based on factorized distribution instead of penalty function and any transformation to a linear model or others. The history information is used and good results can be achieved with small population size. The feasibility of the new algorithm is also given.

1 Introduction

The constraint problem we discuss here is defined as the following:

$$max f(X) = \sum_i f_i(S_i) \qquad (1)$$
$$s.t. C_i(S_i)$$

where $X = \{x_1, \cdots, x_n\}$ and the value of the ith variable belongs to the set $\{x_{i,1}, \cdots, x_{i,n_i}\}$. $C_i(S_i)$ stands for the ith constraint function (it may be equality or inequality) and the variable set $S_i \subseteq X$ ($i = 1, \cdots, l$). The function $f(x)$ is called additively decomposed function (ADF). This class of functions is of great theoretical and practical importance. Optimization of an arbitrary function in this space is NP complete.

Evolutionary computation (EC) is widely used for constraint optimization [1]. This method is loosely based on the mechanics of artificial selection and genetic recombination. Two crucial factors of the EC success, a proper growth and mixing of good building blocks, are often not achieved [2]. The problem of building block disruption is often referred to as the linkage problem [3]. Various attempts to prevent the disruption of important partial solutions have been done. One way to generate new individuals is to use the information extracted from the entire set of promising solutions. A general scheme of the algorithms based on this principle is called the estimation of distribution algorithm (EDA)[4]. The main problems of EDA are how to estimate the distribution and how to get new individuals by this distribution. EDA is used successfully not only for unconstrained optimization but also for constrained optimization [5].

Although EDA is used successfully in real application, it also has the shortcoming of large population size. One of the reasons is that EDA only uses the information of the selected population and does not use any information of the population before selection. In this paper a history information based algorithm is proposed to solve additively decomposed function with constraints. We called this new algorithm as HCFA (History information based Constraint Factorization Algorithm). The main advantage of HCFA is that it can get good results with small population size. It can always produce feasible solutions if the individuals of the initial population are all feasible by using a factorization of the distribution of selected points. It can handle not only linear constraints but also nonlinear constraints. The paper is organized as follows. Section 2 gives the frame of the algorithm. In section 3, some important topics, including initialization and feasibility are discussed. Some numerical results are given in section 4 to show the efficiency of the algorithm.

2   The Frame of the Algorithm

2.1   Factorization

To solve the problem, we must get a factorization of the probability of distribution at first. For convenience, just suppose $X = \bigcup_{i=1}^{l} S_i$. Define

$$d_i = \bigcup_{j=1}^{i} S_j, \ b_i = S_i \setminus d_{i-1}, \ c_i = S_i \cap d_{i-1}$$

Set $d_0 = \phi$ and then we get the factorization as

$$P(X) = \prod_{i=1}^{l} P(x_{b_i}|x_{c_i})$$

Let's see an example. Suppose $X = \{x_1, x_2, x_3\}$, $x_i \in \{0,1\}$ and the constraint functions are:

$$x_1 + x_2 \leq 1$$
$$x_2 + x_3 \leq 1$$

Then we have

$$S_1 = \{x_1, x_2\}, \ S_2 = \{x_2, x_3\}$$

So the factorization is

$$P(X) = P(x_1, x_2) P(x_3|x_2)$$

If
$$b_i \neq \phi, \forall i = 1, \cdots l; d_l = X$$
$$\forall i \geq 2, \exists j < i \text{ such that } c_i \subseteq S_j$$

we say the factorization satisfy the *running intersection property*. At this condition, $P(x_1, \cdots, x_n) = \prod_{i=1}^{l} P(x_{b_i}|x_{c_i})$ really holds [6].

If the running intersection property is violated the factorization might not be exact. But we will show the running intersection property is not necessary by the numerical examples.

2.2 Algorithm

We assume the factorization of the probability distribution is given. The following is the frame of the algorithm HCFA to solve the constraint problem:

1. Get initial feasible population;
2. Selection;
3. Compute the probabilities $P^s(x_{b_i}|x_{c_i}, t-1)$ using the selected individuals;
4. Generate a new population according to

$$P(x,t) = \prod_{i=1}^{l} [\lambda P^s(x_{b_i}|x_{c_i}, t-1) + (1-\lambda) P^s(x_{b_i}|x_{c_i}, t-2)] \quad (2)$$

where $0 < \lambda \leq 1$
5. If the termination criterion is met, finish;
6. Add the best point of the previous generation to the generated points;
7. $t = t+1$, go to 2).

This algorithm can be run with any popular selection methods.

3 Further Discussion

3.1 Feasibility

Before the further discussion, we should know if the algorithm HCFA is feasible. This means the final solution should be feasible. The following theorem gives us the guarantee.

Theorem 1. *If the former generation is feasible, the new generation will be feasible too.*

Proof: If $\exists x \in Population(t)$ and x doesn't satisfy the kth constraint $C_k(S_k)$. Then

$$0 \neq P(x,t) = \prod_{i=1}^{l} [\lambda P^s(x_{b_i}|x_{c_i}, t-1) + (1-\lambda) P^s(x_{b_i}|x_{c_i}, t-2)]$$

$$\rightarrow P^s(x_{b_k}|x_{c_k}, t-1) \neq 0 \text{ or } P^s(x_{b_k}|x_{c_k}, t-2) \neq 0$$
$$\rightarrow P^s(x_{S_k}, t-1) \neq 0 \text{ or } P^s(x_{S_k}, t-2) \neq 0$$

Then $\exists \tilde{x} \in Population\,(t-1)$, $\tilde{x}_{S_k} = x_{S_k}$ or $\exists \hat{x} \in Population\,(t-2)$, $\hat{x}_{S_k} = x_{S_k}$. Therefore $\tilde{x}$ or $\hat{x}$ does not satisfy $C_k(S_k)$

And it is impossible because we suppose the former generation is feasible.

3.2 How to Get Feasible Initial Population

Although Theorem 1 guarantees that we can get a feasible population from a feasible former, we need a feasible initial population. The easiest way is to create a solution randomly and then check the feasibility. When $t = 1$, we will generate new population according to $P(x,1) = \prod_{i=1}^{l} P^s(x_{b_i}|x_{c_i}, 0)$.

4 Numerical Results

In the following examples, the experiment results are expressed by how many runs it gets the global optimum among 100 independent runs. The max generation is 100 and the population size is denoted as N.

Example 1.

$$max \sum_{i=1}^{n} x_i$$
$$s.t. \, 2 \leq x_{2j-1}^2 + x_{2j}^2 + x_{2j+1}^2 \leq 8$$

where $n = 2m+1$, $x_i \in \{0, \pm 1, \pm 2\}$, $i = 1, \cdots n$, $j = 1, \cdots, m$.

This example is a nonlinear programming problem with concave domain. The table 1 shows the results. When $\lambda = 1$, it is the algorithm used in [5]. From this table we can see that the new algorithm can get better results with smaller population size.

Example 2.

$$max \sum_{i=1}^{n} x_i$$
$$s.t. \, x_{2j-1} + x_{2j} + x_{2j+1} \leq 2$$
$$x_1 + x_{n-1} + x_n \leq 2$$

where $n = 2m$, $x_i \in \{0, 1\}$, $i = 1, \cdots n$, $j = 1, \cdots, m-1$

Table 1. Results of Example 1

N	1000		2000		3000		4000	
λ	[5]	HCFA	[5]	HCFA	[5]	HCFA	[5]	HCFA
	1	0.5	1	0.5	1	0.5	1	0.5
n=101	0	3	46	74	80	97	97	100
n=201	0	0	0	5	14	66	55	93
n=299	0	0	0	0	5	50	37	89

Table 2. Results of Example 2. Not satisfy the running intersection property

N	200		500		700		1000	
λ	[5]	HCFA	[5]	HCFA	[5]	HCFA	[5]	HCFA
	1	0.5	1	0.5	1	0.5	1	0.5
n=200	1	12	81	88	99	100	99	100
n=300	0	0	45	80	79	97	99	100
n=398	0	0	41	85	88	100	98	100
n=502	0	0	8	58	60	98	95	100

The table 2 shows the results. When $\lambda = 1$, it is the algorithm used in [5].

This example is a linear programming problem. In order to achieve the same convergence ratio, the population size needed by the new algorithm is much smaller than that of [5]. At the same time the factorization we used is

$$P(x_1, \cdots, x_n) = P(x_1 x_2 x_3) P(x_4 x_5 | x_3) \cdots P(x_{n-2} x_{n-1} | x_{n-3}) P(x_n | x_1 x_{n-1})$$

Obviously it does not satisfy the running intersection property. But we can still get the global optimum. We can also use the following factorization which satisfies the running intersection property.

$$P(x_1, \cdots, x_n) = P(x_1 x_2 x_3) P(x_4 x_5 | x_1 x_3) \cdots P(x_{n-2} x_{n-1} | x_1 x_{n-3}) P(x_n | x_1 x_{n-1})$$

Table 3 shows the result with the above factorization. We get much better results just because we change the factorization. This example shows the running

Table 3. Results of Example 2. Satisfy the running intersection property

N	200		500		700		1000	
λ	[5]	HCFA	[5]	HCFA	[5]	HCFA	[5]	HCFA
	1	0.5	1	0.5	1	0.5	1	0.5
n=200	27	78	95	100	99	100	100	100
n=300	5	34	80	98	95	100	100	100
n=398	6	51	97	99	100	100	100	100
n=502	0	27	86	99	98	100	100	100

intersection property is not necessary. In our research we do find that some examples can not converge to the global optimum just because the factorization does not satisfy the running intersection property. We still need further study.

5 Conclusion

In this paper we proposed a new algorithm based on factorized distribution to solve constraint optimization problems of additively decomposed function. Future research work will focus on the running intersection property, converge theory, population size, etc.

Acknowledgement

This project was supported by the National Natural Science Foundation of China.

References

1. Coello, C., Carlos, A.: A Survey of Constraint Handling Techniques Used with Evolutionary Algorithms. Technical Report Lania-RI-9904. Laboratorio Nacional de Informtica Avanzada, Xalapa, Veracruz, Mexico(1999)
2. Thierens, D.: Analysis and Design of Genetic Algorithms. PhD thesis. Katholieke Universiteit Leuven, Leuven, Belgium(1995)
3. Harik, G., Goldberg, D. E.: Learning Linkage. In: Richard, K., Michael, D.(eds.): Foundations of Genetic Algorithms 4. Morgan Kaufmann, San Francisco (1997) 247-262
4. Muehlenbein, H., Paaβ G.: From Recombination of Genes to the Estimation of Distributions I: Binary Parameters. Lecture Notes in Computer Science, Vol. 1411. Springer-Verlag, Berlin Heidelberg New York (1996) 178-187
5. Ren, Q., Zeng, J., Qi, F.: Optimization of Additively Decomposed Function with Constraints. Journal of Engineering and Electronics. 4(2002)85-90
6. Muhlenbein, H., Mahnig, T. and Rodriguez, A.O.: Schemata, Distribution and Graphical Models in Evolutionary Optimization. Journal of Heuristics. 5(1999) 215-247

An Efficient Multiple-Constraints QoS Routing Algorithm Based on Nonlinear Path Distance

Xiaolong Yang[1,2], Min Zhang[1], and Keping Long[1,2]

[1] Chongqing Univ. of Posts and Telecommunications, Chongqing 400065, China
[2] Univ. of Electronic Science and Technology of China, Chengdu 610054, China
{yangxl, zhangmin}@cqupt.edu.cn

Abstract. This paper proposes a QoS routing algorithm (called MIS-LB), which extends the definition of nonlinear path distances into the link-sharing and load balancing. Based on the extension, it can find feasible paths by the shortest-first criterion, but also adjust the link-sharing and the loads among multiple feasible paths. Finally, the simulation results show that it outperforms other multiple-constraints routing, such as TAMCRA and H_MCOP.

1 Introduction

In IP networks, the QoS routing under multiple constraints is a NP-complete problem [1]. Currently, many algorithms proposed by the literatures (e.g., [3]-[6]) are based on the linear path distance, and it is inefficient and difficult for them to search feasible solutions. However, [3] presents another concept, i.e., the nonlinear path distance. Because it exactly contains the information of the individual and collective QoS requirements, the multiple constraints can be converted into a uniform definition.

However, [3] does not consider the link-sharing interference and load balancing. In order to further improve the performance of QoS routing, this paper extends the concept of nonlinear distance, and brings the link sharing and load balancing into the nonlinear path distance. Then we propose a novel QoS routing, called MIS-LB (Min-Sharing Interference and Load Balancing QoS Routing), which can not only find feasible paths, but also adjust the link-sharing, and balance the network loads among multiple feasible paths. Hence, it outperforms TAMCRA and H_MCOP.

2 The Extension of Nonlinear Path Distance

2.1 The Basic Processing of QoS Constraints

Usually, the QoS requirements of an application can be described by some constraints, which would be additive, multiplicative, or concave [1]. Firstly, we discuss the additive and multiplicative constraints, e.g. delay and reliability. For a QoS-aware application, the delay of its feasible paths must be less than an upper bound T. Because of its additivity, the relationship between the individual link delay and the path delay can be represented as follows

$$\sum_i delay_i \leq T \quad (1)$$

where $delay_i$ denotes the individual delay of a link i within the feasible path.

Similarly, the reliability of its feasible paths must be no less than an upper bound P. Because of its multiplicativity, the relationship between the individual link reliability and the path reliability can be represented as follows

$$\prod_i prob_i \geq p \quad (2)$$

where $prob_i$ denotes the individual reliability of a link i belonging to the feasible path. (2) can convert to another form by the logarithm, as shown the following

$$\sum_i \log prob_i \geq \log p \quad (3)$$

In this way, any multiplicative constraint can convert into the additive one.

For the concave constraint, such as the bandwidth, we can express it as follows

$$Min(BW_i) \geq BW \quad (4)$$

where BW_i denotes the individual available bandwidth of a link i within the feasible path. If an application has some concave QoS constraints, we firstly prune away the links that dissatisfy the concave constraints, and then process the remaining links [5].

2.2 The Representation of Nonlinear Path Distance

Assumed that an application has m types QoS constraints; and one of path between source S and destination D contains h intermediate nodes, e.g., a, b...etc. lc_i, pc_i and L_i denote i-th type metric of link, i-th type metric of path, and i-th type QoS constraint. Hence, the QoS metrics of the link a-b can be denoted as follows

$$Link(a\text{-}b) = [lc_1(a\text{-}b), lc_2(a\text{-}b), \ldots, lc_m(a\text{-}b)] \quad (5)$$

Similarly, the QoS metrics of the path S-D can be denoted as follows

$$Path(S\text{-}D) = [pc_1(S\text{-}D), pc_2(S\text{-}D), \ldots, pc_m(S\text{-}D)] \quad (6)$$

Expression (5) and (6) hold the following relationship

$$pc_i(S-D) = lc_i(S-a) + lc_i(a-b) + \ldots + lc_i(r-D)lc_i(a-b) \quad i=1,2,\cdots,m \quad (7)$$

Under the linear distance, the QoS routing can be formulated as the following:

$$\begin{cases} Min(Length(S-D)) = Min(\sum_i w_i \cdot pc_i(S-D)) \\ Subject\ to: pc_i(S-D) \leq L_i \end{cases} \quad (8)$$

where w_i denotes the weight, which is the contribution proportion of each type QoS metric to the path distance. As known, the problem (8) is a linear programming [2], which can be simply solved by illustrated method. From the operation of illustrated method [2] and the results in [3], the feasible solution is better than any other ones outside the QoS constraint space (i.e., the distance of path satisfying constraints is less than that of path not satisfying constraints in the maximum likelihood) if the weight and the QoS constraint of each type satisfy the following relationship:

$$w_1 : w_2 : \ldots : w_m = \frac{1}{L_1} : \frac{1}{L_2} : \ldots : \frac{1}{L_m} \tag{9}$$

Therefore, the problem (8) can approximately convert to another form with implicit constraint condition, which is rewritten as follows

$$Min(Length(S-D)) = Min \sum_i \frac{pc_i(S-D)}{L_i} \tag{10}$$

In fact, the expression (10) must subject to the following connotative condition

$$\frac{pc_i(S-D)}{L_i} \leq 1 \tag{11}$$

However, if the equilength line of object function is similar to the boundary line of the QoS constraint space, the search space of feasible solution can be reduced, and so the nonlinear path distance can be rewritten as follows [3]

$$Length(S-D) = \left[\sum_i \left(\frac{pc_i(S-D)}{L_i} \right)^q \right]^{1/q} \tag{12}$$

where q denotes the nonlinear degree of path distance, and $q \geq 1$. When q is infinity, the nonlinear path distance can be ideally expressed as follows

$$Length(S-D)\big|_{p \to \infty} \approx Max_i \left(\frac{pc_i(S-D)}{L_i} \right) \tag{13}$$

Hence, the problem (8) translates the following simple expression:

$$Min(Length(S-D)) = Min \left[Max_i \left(\frac{pc_i(S-D)}{L_i} \right) \right] \tag{14}$$

2.3 The Distance Definition of Link-Sharing Interference

Because the link-sharing interference may influence the QoS guarantee of each path, Ref. [5]-[6] proposed new QoS routing algorithms based on it. However, its definition is more complex. Here, it is simply defined as the increment of nonlinear path distance. Assumed that n paths, which have m types QoS constraints, share the link a-b. For k-th path, $h^{(k)}$, $lc^{(k)}_i$, and $pc^{(k)}_i$, $L^{(k)}_i$, $Link^{(k)}(a-b)$, and $Path^{(k)}(S-D)$ ($i=1,2,\ldots,m$; $k=1,2\ldots,n$) denote the hop count, i-th type link QoS metric, i-th type path QoS constraint, i-th type QoS requirement of application, the link metrics vector, and the path metrics vector. According to (5)-(7), the link-sharing vector is expressed as follows

$$Share\_link(a-b) = [Share\_lc_1(a-b), Share\_lc_2(a-b)], \ldots, Share\_lc_m(a-b) \tag{15}$$

where $Share\_lc_i(a-b)] = \sum_k Share\_lc_i^{(k)}(a-b)]$. Analogously, the sharing-path vector is expressed as follows

$$Path^{(k)}(S-D) = [pc_1^{(k)}(S-D), pc_2^{(k)}(S-D), \ldots, pc_m^{(k)}(S-D)] \tag{16}$$

where
$$pc_i^{(k)}(S-D) = lc_i^{(k)}(S-a) + Share\_lc_i^{(k)}(a-b) + ... + lc_i^{(k)}(r-D)$$
$$= lc_i^{(k)}(S-a) + \sum_k lc_i^{(k)}(a-b) + ... + lc_i^{(k)}(r-D)$$

Note that (16) is meaningless if $pc_i^{(k)}(S-D) \leq L_i^{(k)}$ does not hold, which means that the sharing link is bottleneck, and obviously the path distance should be defined infinite.

2.4 The Distance Definition of Load Balancing

It is important to distribute some traffic from the congesting link to others, i.e., the load balancing. This paper implements load balancing based on nonlinear path distance. Here, we simply define a non-descending function with respect to path distance and traffic load, which maybe have the following form:

$$\Delta l = \begin{cases} f(v) & 0 \leq v < C \\ \infty & v \geq C \end{cases} \quad (17)$$

where Δl, v and C denote the increment of path distance, the input traffic load, and the maximal link capacity, respectively.

If so, traffic load can be adaptively distributed among several feasible paths. In order to consider simultaneously the sharing link and load balancing during the selection of feasible routing path, the definition of the nonlinear path length can be rewritten as follows according to the representations (13), (16) and (17)

$$Length(S-D) = Max_i\left(\frac{pc_i(S-D)}{L_i}\right) + \Delta l \quad (18)$$

3 MIS-LB: The Multiple-Constraints QoS Routing Algorithm

According to the above extension of the nonlinear path length with sharing link and load balancing, this section presents the implementation of our QoS routing algorithm. The algorithm works as follows. At first, it translates the QoS constraints into the nonlinear path distance, which contains the information of the link sharing

```
Input :
        G= (V , E )  //A graph with vertex set V and edge set E //
        (S-D)        //A source-destination pair: s: source node, d: destination node //
        m            //The type number of QoS constraints //,

Implementation:
1.      for each LinkE
2.      if Link does not satisfy the concave QoS constraint L_i" then G'(V,E')=Prune(Link) //Prune these links not satisfying the concave QoS constraint //
3.      if K feasible paths have been searched, or there exists none of feasible path in G'(V,E'), then stop
4.      else Path[k]← Tamcra(G'(V,E'),m types QoS constraints, S-D) // To invoke Tamcra function in the literature [3] to record the links of a feasible routing path //
5.      if the intersection between Path[k] and Path[i] (i<k) is empty, or there is no traffic load in the path then k← k+1
6.      else to compute the distance with sharing link according to representation (15) and (18),
7.           then to retune the row 4
8.      endif
9.      endif
Output: K feasible paths
```

Fig. 1. The pseudo code implementation of MIS-LB

interference and the traffic load distribution. Then under the shortest-path first routing criterion, while searching feasible routing paths from solution space, it adaptively regulates the sharing of link and the distribution of traffic load. Its implementation pseudo code is shown as Fig. 1.

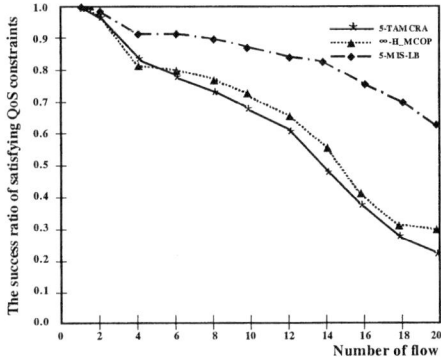

Fig. 2. The performance comparison of QoS support of three algorithms (the bandwidth requirement of each flow is 20Mbps)

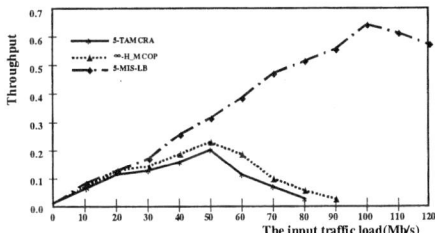

Fig. 3. The throughput comparison of three algorithms (3 flows)

4 Numerical Simulation Results

In the section, we evaluate the performance of our proposed QoS routing algorithm, i.e. MIS-LB. The simulations are on the basis of Waxman-based random graph with 50 nodes, where there exists edge between any two nodes according to a probability expressed by [7]. Assumed that the link bandwidth uniformly distributes in [0, 155] Mb/s, and the flows have three types QoS constraints: delay, packet loss probability (PLP), and bandwidth. Delay and PLP of link, both of which are independent to each other, also uniformly distribute in [1, 50] ms and [10^{-6}, 0.3], respectively. The delay and PLP constraint of application are assumed to respectively distribute in [10, 80] ms and [10^{-4}, 0.4] while bandwidth requirement is variable. For simplicity, the function with respect to path distance and traffic load is defined as follows

$$\Delta l = f(v) = \frac{v}{C-v} \tag{19}$$

Here, we compare *MIS-LB* with TAMCRA [3] and H_MCOP [4] from two sides. One is the ratio that can find feasible paths to satisfy the QoS requirements; the other is the throughput. Simply, *MIS-LB* is assumed to only find 5 shortest feasible paths at the ideal situation, and short for 5-MIS-LB in Fig.2 and Fig.3. Similarly, the tunable parameter of TAMCRA can also take 5 according to [3] (short for 5-TAMCRA) while the path index of H_MCOP is just infinite according to [4] (short for ∞-H_MCOP).

Usually, MIS-LB can support multi-path. However, the multi-path support capacities of TAMCRA and H_MCOP are indeed weaker, and the traffic is carried though the unique best routing path. Hence, some parts of network is possibly overloading in TAMCRA and H_MCOP. As illustrated in Fig.3, MIS-LB obviously outperforms the other two algorithms in throughput. This phenomenon can be explained by (17)-(19).

5 Conclusion

This paper extends the definition of nonlinear path distance represents, and integrates the additive and multiplicative QoS metrics, the link sharing and load balancing into the definition. Under the shortest-path first routing criterion, the path distance acts as the unique factor of QoS routing decision-making. So, it can effectively simplify the process of QoS routing, but its performance does not degrade. Contrarily, compared with 5-TAMCRA and ∞-H_MCOP, MIS-LB can adaptively regulate the sharing of link and the distribution of traffic load.

References

1. Wang, Z., and Crowcroft, J.: Quality-of-service routing for supporting multimedia applications, IEEE JSAC, 14(1996) 1288–1234
2. Fu, Y., Cheng,X., and Tang,Y.: Optimization theory and method, Press of UESTC (1996)
3. Neve, H., and Mieghem, P.: TAMCRA a tunable accuracy multiple constraints routing algorithm, Computer Communications, 23(2000) 667-679
4. Korkmaz, T., and Krunz, M.: Multi-Constrained Optimal Path Selection, IEEE Infocom'01, vol.2, Alaska, USA (2001) 834-843
5. Kodialam,M., and Lakshman, T.: Minimum Interference Routing with Applications to MPLS Traffic Engineering, IEEE Infocom'00, vol.2, Tel-Aviv, Israel (2000) 884–893
6. Nelakuditi, S., Zhang, Z., and Tsang, R.: Adaptive Proportional Routing: A Localized QoS Routing Approach, IEEE/ACM ToN, 10(2002) 790-804
7. Calvert, K., Doar, M., and Zegura, E.: Modeling Internet Topology, IEEE Communications Magazine, 35(1997) 160-163

The Early and Late Congruences for Asymmetric $\chi^{\neq}$-Calculus*

Farong Zhong

Department of Computer Science,
Shanghai Jiaotong University, Shanghai 200030, China

Abstract. The paper investigates into the early and late congruences on the asymmetric $\chi^{\neq}$-processes. The two bisimilarities are defined and their closure properties are given. Sound and complete equational systems are constructed for both congruences.

1 Introduction

The π-calculus [1] is a successful paradigm of the mobile process calculi. The χ-calculus [3–5] is an addition to the family of the mobile process calculi. Many aspects of the χ-calculus have been studied [6–9]. These investigations improve our understanding of the π-like mobile process calculi. From the operational viewpoint, the χ-calculus is obtained from the π-calculus by removing the asymmetry of the latter. This modification results in significant change in observational properties. Assume $x \neq y$, one has the following reductions in the χ-calculus:

$$(x)(ax.P \mid \overline{a}y.Q \mid R) \xrightarrow{\tau} (P \mid Q \mid R)\{y/x\} \qquad (1)$$

$$(x)(\overline{a}x.P \mid ay.Q \mid R) \xrightarrow{\tau} (P \mid Q \mid R)\{y/x\} \qquad (2)$$

$$(x)ax.P \mid \overline{a}y.Q \xrightarrow{\tau} P\{y/x\} \mid Q \qquad (3)$$

$$(x)\overline{a}x.P \mid ay.Q \xrightarrow{\tau} P\{y/x\} \mid Q \qquad (4)$$

In $(x)(ax.P \mid \overline{a}y.Q \mid R)$ the name x is bound under a localization operator (x). Processes $ax.P$ and $\overline{a}y.Q$ are in prefix form, in which x and y are free. The interactions between $ax.P$ and $\overline{a}y.Q$ cause the bound name x to be replaced by y throughout the term over which the localization operator (x) applies. In (3) and (4) the interactions do not affect Q as it is not bound by (x).

The asymmetric χ-calculus, which disallows (2) and (4), is a variant of the χ-calculus. It can be regarded as an extension of the π-calculus. The asymmetric χ-calculus with mismatch operator is written as the asymmetric $\chi^{\neq}$-calculus. In this paper, the early and late open bisimilarities are defined and their closure

---

* The work is supported by The National Distinguished Young Scientist Fund of NNSFC (60225012) and The National 973 Project (2003CB316905).

properties under composition and restriction are given. The sound and complete equational systems are constructed for both the early and late open congruences. The completeness theorem is presented for each of the open congruences.

2 Asymmetric $\chi^{\ne}$-Calculus

The asymmetric $\chi^{\ne}$-calculus can be seen as obtained from π-calculus by unifying the two classes of bound names. The approach is to unify the input prefix and the output prefix. In the asymmetric $\chi^{\ne}$-calculus, a prefix takes the form of $\alpha x.P$, where α indicates either a name a or co-name $\bar{a}$.

Let $\mathcal{N}$ be the set of names ranged over by the lower case letters and $\overline{\mathcal{N}}$ the set $\{\bar{x} \mid x \in \mathcal{N}\}$ of co-names. The Greek letter α ranges over $\mathcal{N} \cup \overline{\mathcal{N}}$. For $\alpha \in \mathcal{N} \cup \overline{\mathcal{N}}$, $\bar{\alpha}$ is defined as a if $\alpha = \bar{a}$ and as $\bar{a}$ if $\alpha = a$. The asymmetric $\chi^{\ne}$-processes are defined by the following abstract grammar:

$$P := \mathbf{0} \mid \alpha x.P \mid P|P \mid (x)P \mid [x{=}y]P \mid [x{\ne}y]P \mid P{+}P$$

The set of the asymmetric $\chi^{\ne}$-processes will be denoted by $\mathcal{C}$. Let λ range over the set of the transition labels $\{\tau\} \cup \{ax, \bar{a}x, a(x), \bar{a}(x), y/x, (y)/x \mid a, x, y \in \mathcal{N}\}$ and γ over $\{\tau\} \cup \{ax, \bar{a}x, a(x), \bar{a}(x) \mid a, x \in \mathcal{N}\}$. In $(y)/x$, x and y must be different. A name in λ is bound if it appears as x in $\alpha(x)$ or $(x)/y$; it is free otherwise. Let $bn(\lambda)$, respectively $fn(\lambda)$, denote the set of bound, respectively free, names occurred in λ; and let $n(\lambda)$ denote the set of names in λ. The sets $bn(\gamma)$, $fn(\gamma)$ and $n(\gamma)$ are defined accordingly. The followings are the operational rules:

$$\frac{}{\alpha x.P \xrightarrow{\alpha x} P} Sqn \qquad \frac{P \xrightarrow{\gamma} P' \quad bn(\gamma) \cap fn(P) = \emptyset}{P|Q \xrightarrow{\gamma} P'|Q} Cmp_0$$

$$\frac{P \xrightarrow{y/x} P'}{P|Q \xrightarrow{y/x} P'|Q\{y/x\}} Cmp_1 \qquad \frac{P \xrightarrow{(y)/x} P' \quad y \notin fn(Q)}{P|Q \xrightarrow{(y)/x} P'|Q\{y/x\}} Cmp_2$$

$$\frac{P \xrightarrow{a(x)} P' \quad Q \xrightarrow{\bar{a}y} Q'}{P|Q \xrightarrow{\tau} P'\{y/x\}|Q'} Cmm_0 \qquad \frac{P \xrightarrow{ax} P' \quad Q \xrightarrow{\bar{a}(x)} Q'}{P|Q \xrightarrow{\tau} (x)(P'|Q')} Cmm_1$$

$$\frac{P \xrightarrow{ax} P' \quad Q \xrightarrow{\bar{a}y} Q' \quad x \ne y}{P|Q \xrightarrow{y/x} P'\{y/x\}|Q'\{y/x\}} Cmm_2 \qquad \frac{P \xrightarrow{ax} P' \quad Q \xrightarrow{\bar{a}x} Q'}{P|Q \xrightarrow{\tau} P'|Q'} Cmm_3$$

$$\frac{P \xrightarrow{ax} P' \quad Q \xrightarrow{\bar{a}(y)} Q' \quad y \notin fn(P)}{P|Q \xrightarrow{(y)/x} P'\{y/x\}|Q'\{y/x\}} Cmm_4 \qquad \frac{P|!P \xrightarrow{\lambda} P'}{!P \xrightarrow{\lambda} P'} Rpl$$

$$\frac{P \xrightarrow{\lambda} P' \quad x \notin n(\lambda)}{(x)P \xrightarrow{\lambda} (x)P'} Loc_0 \qquad \frac{P \xrightarrow{\alpha x} P' \quad x \notin \{\alpha, \bar{\alpha}\}}{(x)P \xrightarrow{\alpha(x)} P'} Loc_1$$

$$\frac{P \xrightarrow{y/x} P'}{(x)P \xrightarrow{\tau} P'}Loc_2 \qquad \frac{P \xrightarrow{y/x} P'}{(y)P \xrightarrow{(y)/x} P'}Loc_3 \qquad \frac{P \xrightarrow{(y)/x} P'}{(x)P \xrightarrow{\tau} (y)P'}Loc_4$$

$$\frac{P \xrightarrow{\lambda} P'}{[x=x]P \xrightarrow{\lambda} P'}Mtch \qquad \frac{P \xrightarrow{\lambda} P' \quad x \neq y}{[x \neq y]P \xrightarrow{\lambda} P'}Mismtch \qquad \frac{P \xrightarrow{\lambda} P'}{P+Q \xrightarrow{\lambda} P'}Sum$$

Now some notations need be fixed. Let $\Longrightarrow$ be the reflexive and transitive closure of $\xrightarrow{\tau}$. We write $\xLongrightarrow{\lambda}$ for $\Longrightarrow \xrightarrow{\lambda} \Longrightarrow$. The notation $\xLongrightarrow{\widehat{\lambda}}$ stands for $\xLongrightarrow{\lambda}$ if $\lambda \neq \tau$ and for $\Longrightarrow$ otherwise.

Suppose Y is a finite set $\{y_1, \ldots, y_n\}$ of names. The notation $[y \notin Y]P$ signifies $[y \neq y_1] \ldots [y \neq y_n]P$. We write ϕ and ψ, called conditions, to stand for sequences of match and mismatch combinators concatenated one after another, and δ for a sequence of mismatch operators. Consequently we write ψP, μP and δP. When the length of ψ (μ, δ) is zero, ψP (μP, δP) is just P. The notation $\phi \Rightarrow \psi$ says that ϕ logically implies ψ and $\phi \Leftrightarrow \psi$ that ϕ and ψ are logically equivalent.

In what follows we will often use a substitution that draws a particular relationship with a condition. Some of these relationships are made precise in the following definition.

Definition 1. *A substitution σ respects ψ if $\psi \Longrightarrow x=y$ implies $x\sigma=y\sigma$ and $\psi \Longrightarrow x \neq y$ implies $x\sigma \neq y\sigma$. Dually ψ respects σ if $x\sigma=y\sigma$ implies $\psi \Longrightarrow x=y$ and $x\sigma \neq y\sigma$ implies $\psi \Longrightarrow x \neq y$. The substitution σ agrees with ψ, and ψ agrees with σ, if they respect each other. The substitution σ is induced by ψ if it agrees with ψ and $n(\sigma) \subseteq n(\psi)$.*

A sequence of names $x_1, \ldots, x_n$ will be abbreviated as $\widetilde{x}$; and consequently $(x_1)\ldots(x_n)P$ will be abbreviated to $(\widetilde{x})P$. When the length of $\widetilde{x}$ is zero, $(\widetilde{x})P$ is just P.

Definition 2. *Let V be a finite set of names. ψ is complete on V if $n(\psi)=V$ and for each pair x,y of names in V it holds that either $\psi \Longrightarrow x=y$ or $\psi \Longrightarrow x \neq y$.*

For convenience of characterizing the equivalence relations of this paper, we need to internalize the labels of the transition system. For that purpose we introduce some auxiliary prefix combinators. In the following definition a is fresh:

$$\langle y/x \rangle.P \stackrel{\text{def}}{=} (a)(\overline{a}y \mid ax.P)$$
$$(y/x).P \stackrel{\text{def}}{=} (y)\langle y/x \rangle.P, \text{ where } x \neq y$$
$$\tau.P \stackrel{\text{def}}{=} \langle a/a \rangle.P$$
$$\alpha(x) \stackrel{\text{def}}{=} (x)\alpha x$$

The prefix $\langle y/x \rangle$, first introduced in [3, 11], is called an update. Correspondingly the prefix (y/x) is called a bound update.

Contexts are defined inductively as follows:

- $[\,]$ is a context;
- If $C[\,]$ is a context then $\alpha x.C[\,]$, $C[\,]\,|\,P$, $P\,|\,C[\,]$, $(x)C[\,]$ and $[x{=}y]C[\,]$ are contexts.

Full contexts are those contexts that satisfy additionally:

- If $C[\,]$ is a context then $C[\,]{+}P, P{+}C[\,]$ and $[x{\neq}y]C[\,]$ are contexts.

3 Early and Late Open Bisimilarities

In this section we define the early and late open bisimilarities and establish the closure properties of them.

Definition 3. *Let $\mathcal{R}$ be a binary symmetric relation on $\mathcal{C}$. It is called an early open bisimulation if it is closed under substitution and whenever $P\mathcal{R}Q$ then the following properties hold:*

1. *If $P\xrightarrow{\tau}P'$ then some Q' exists such that $Q\Longrightarrow Q'\mathcal{R}P'$.*
2. *If $P\xrightarrow{(y)/x}P'$ then some Q' exists such that $Q\overset{(y)/x}{\Longrightarrow}Q'\mathcal{R}P'$.*
3. *If $P\xrightarrow{y/x}P'$ then some Q' exists such that $Q\overset{y/x}{\Longrightarrow}Q'\mathcal{R}P'$.*
4. *If $P\xrightarrow{\alpha(x)}P'$ then for every y, some Q',Q'' exist such that $Q\Longrightarrow\xrightarrow{\alpha(x)}Q''$ and $Q''\{y/x\}\Longrightarrow Q'\mathcal{R}P'\{y/x\}$.*
5. *If $P\xrightarrow{\alpha x}P'$ then for every y, some Q',Q'' exist such that $Q\Longrightarrow\xrightarrow{\alpha x}Q''$ and $Q''\{y/x\}\Longrightarrow Q'\mathcal{R}P'\{y/x\}$.*

The early open bisimilarity $\approx^o_e$ is the largest early open bisimulation.

The clauses (1), (4) and (5) are similar to the counterparts of the weak early bisimilarity for the π-calculus. In the asymmetric $\chi^{\neq}$-calculus free input action can also incur name updates in suitable contexts. Suppose $P\xrightarrow{ax}P''$. Then $(x)(P\,|\,\bar{a}y.Q)\xrightarrow{\tau}P''\{y/x\}\,|\,Q\{y/x\}$.

Definition 4. *Let $\mathcal{R}$ be a binary symmetric relation on $\mathcal{C}$. It is called a late open bisimulation if it is closed under substitution and whenever $P\mathcal{R}Q$ then the following properties hold:*

1. *If $P\xrightarrow{\tau}P'$ then some Q' exists such that $Q\Longrightarrow Q'\mathcal{R}P'$.*
2. *If $P\xrightarrow{(y)/x}P'$ then some Q' exists such that $Q\overset{(y)/x}{\Longrightarrow}Q'\mathcal{R}P'$.*
3. *If $P\xrightarrow{y/x}P'$ then some Q' exists such that $Q\overset{y/x}{\Longrightarrow}Q'\mathcal{R}P'$.*
4. *If $P\xrightarrow{\alpha(x)}P'$ then there exists some Q'' such that $Q\Longrightarrow\xrightarrow{\alpha(x)}Q''$ and for every y some Q' exists such that $Q''\{y/x\}\Longrightarrow Q'\mathcal{R}P'\{y/x\}$.*
5. *If $P\xrightarrow{\alpha x}P'$ then there exists some Q'' such that $Q\Longrightarrow\xrightarrow{\alpha x}Q''$ and for every y some Q' exists such that $Q''\{y/x\}\Longrightarrow Q'\mathcal{R}P'\{y/x\}$.*

The late open bisimilarity $\approx^o_l$ is the largest late open bisimulation.

Obviously $\approx_l^o \subseteq \approx_e^o$. The following example shows that inclusion is strict: $ax.[x{=}y]+ax.[x{\neq}y]\tau.P \approx_e^o ax.[x{=}y]\tau.P+ax.[x{\neq}y]\tau.P+ax.P$. But it doesn't hold that $ax.[x{=}y]+ax.[x{\neq}y]\tau.P \approx_l^o ax.[x{=}y]\tau.P+ax.[x{\neq}y]\tau.P+ax.P$, since the transition $ax.[x{=}y]\tau.P+ax.[x{\neq}y]\tau.P+ax.P \xrightarrow{ax} P$ cannot be matched up by any transition $ax.[x{=}y]\tau.P+ax.[x{\neq}y]\tau.P \xrightarrow{ax} P$ in the late approach.

Lemma 1. $\approx_e^o$ and $\approx_l^o$ are closed under restriction and composition operations.

4 Axiomatization

The relations $\approx_e^o$ and $\approx_l^o$ are closed neither under the mismatch operation nor under the summation operation. To obtain the largest congruence contained in the corresponding bisimilarity, we now use the standard approach as follows:

Definition 5. Two processes P and Q are early open congruent, written $\simeq_e^o$, if $P \approx_e^o Q$, and for every substitution σ, the following condition must be satisfied:

- If $P\sigma \xrightarrow{\tau} P'$ then Q' exists such that $Q\sigma \Longrightarrow Q'$ and $P' \approx_e^o Q'$.
- If $Q\sigma \xrightarrow{\tau} Q'$ then P' exists such that $P\sigma \Longrightarrow P'$ and $P' \approx_e^o Q'$.

The late open congruence $\simeq_l^o$ is defined in the same way.

An important problem about a congruence relation is whether there is a finite set of equation schemes and inference rules such that all congruent pairs can be derived from these equation schemes and rules. In [2] Sangiori presents a complete system for strong congruence on π-processes. Parrow and Victor consider in [11] two axiomatization systems for strong bisimilarity on Update processes, one with mismatch operator and one without mismatch operator. The latter is of similar style to Sangiorgi's system. The strong bisimilarity of Fusion processes has also been axiomatized [12]. In [8] Fu and Yang present a complete system for the weak barbed congruence on the symmetric χ-processes. In Figure 1 a conditional equational system is defined. Let AS denote the system consisting of the rules and laws in Figure 1 plus the following expansion law:

$$P \mid Q = \sum_{i \in I} \psi_i(\widetilde{x})\pi_i.(P_i \mid Q) + \sum_{\substack{\pi_i = a_i x_i \\ \pi_j = \overline{b_j} y_j}} \psi_i \varphi_j(\widetilde{x})(\widetilde{y})[a_i{=}b_j]\langle y_j/x_i \rangle.(P_i \mid Q_j)$$

$$+ \sum_{j \in J} \varphi_j(\widetilde{y})\pi_j.(P \mid Q_j) + \sum_{\substack{\pi_i = \overline{a_i} x_i \\ \pi_j = b_j y_j}} \psi_i \varphi_j(\widetilde{x})(\widetilde{y})[a_i{=}b_j]\langle x_i/y_j \rangle.(P_i \mid Q_j)$$

where P is $\sum_{i \in I} \psi_i(\widetilde{x})\pi_i.P_i$ and Q is $\sum_{j \in J} \varphi_j(\widetilde{y})\pi_j.Q_j$ and $\{\widetilde{x}\} \cap \{\widetilde{y}\} = \emptyset$. The second component in the right hand of the above equality captures the idea that whenever π_i is of the form $a_i x_i$ for some $i \in I$ and π_j is of the form $\overline{b_j} y_j$ for some $j \in J$ then there is a summand $\psi_i \varphi_j(\widetilde{x})(\widetilde{y})[a_i{=}b_j]\langle y_j/x_i \rangle.(P_i \mid Q_j)$.

We will write $AS \vdash P{=}Q$ to mean that the equality $P = Q$ can be inferred from AS. When $R_1 \cdots R_n$ are the major axioms used to derive $P = Q$, we write $P \stackrel{R_1 \cdots R_n}{=} Q$.

C1	$\alpha x.P = \alpha x.Q$	$P = Q$
C2	$(x)P = (x)Q$	$P = Q$
C3a	$[x{=}y]P = [x{=}y]Q$	$P = Q$
C3b	$[x{\neq}y]P = [x{\neq}y]Q$	$P = Q$
C4	$P+R = Q+R$	$P = Q$
C5	$P_0 \mid P_1 = Q_0 \mid Q_1$	$P_i = Q_i,\ i \in \{1,2\}$
L1	$(x)\mathbf{0} = \mathbf{0}$	
L2	$(x)\alpha y.P = \mathbf{0}$	$x \in \{\alpha, \overline{\alpha}\}$
L3	$(x)\alpha y.P = \alpha y.(x)P$	$x \notin \{y, \alpha, \overline{\alpha}\}$
L4	$(x)(y)P = (y)(x)P$	
L5	$(x)[y{=}z]P = [y{=}z](x)P$	$x \notin \{y, z\}$
L6	$(x)(P+Q) = (x)P+(x)Q$	
L7	$(x)[x{=}y]P = \mathbf{0}$	
L8	$(x)\langle y/x \rangle.P = \tau.P\{y/x\}$	
L9	$(x)\langle y/z \rangle.P = \langle y/z \rangle.(x)P$	$x \notin \{y, z\}$
L10	$(x)\tau.P = \tau.(x)P$	
M1	$\psi P = \varphi P$	if $\psi \Leftrightarrow \varphi$
M2	$[x{=}y]P = [x{=}y]P\{y/x\}$	
M3a	$[x{=}y](P+Q) = [x{=}y]P+[x{=}y]Q$	
M3b	$[x{\neq}y](P+Q) = [x{\neq}y]P+[x{\neq}y]Q$	
M4	$[x{=}y]P+[x{\neq}y]P = P$	
M5	$[x{\neq}x]P = \mathbf{0}$	
S1	$P+\mathbf{0} = P$	
S2	$P+Q = Q+P$	
S3	$P+(Q+R) = (P+Q)+R$	
S4	$P+P = P$	
U1	$\langle y/x \rangle.P = \langle y/x \rangle.[x{=}y]P$	

Fig. 1. Axioms for Strong Bisimilarity on Chi Processes

Definition 6. *A process P is in normal form if it is of the form*

$$\sum_{i \in I_1} \psi_i \tau.P_i + \sum_{i \in I_2} \psi_i \alpha_i x_i.P_i + \sum_{i \in I_3} \psi_i \alpha_i(x).P_i + \sum_{i \in I_4} \psi_i \langle z_i/y_i \rangle.P_i + \sum_{i \in I_5} \psi_i (y/x_i).P_i$$

such that neither x nor y appears free in P, and P_i is in normal form for each $i \in I_1 \cup I_2 \cup I_3 \cup I_4 \cup I_5$. Here I_1, I_2, I_3, I_4 and I_5 are pairwise disjoint finite indexing sets.

The depth of a process measures the maximal length of nested prefixes in the process. The structural definition goes as follows:

- $d(\mathbf{0}) = 0$;
- $d(\alpha x.P) = 1 + d(P)$;
- $d(P \mid Q) = d(P) + d(Q)$;
- $d((x)P) = d(P)$;
- $d([x{=}y]P) = d(P)$;
- $d([x{\neq}y]P) = d(P)$;

- $d(P+Q) = max\{d(P), d(Q)\}$;
- $d(\langle y/x\rangle.P) = 1 + d(P)$.

In this definition we take the update operation as primitive.

Lemma 2. *For each process P there is some P' in normal form such that $AS \vdash P = P'$ and $d(P') \leq d(P)$.*

The complete system for the weak relation is obtained from AS by adding some tau laws. The tau laws used in the asymmetric $\chi^{\neq}$-calculus are given in Figure 2.

T1	$\alpha x.\tau.P = \alpha x.P$
T2	$\alpha x.(P+\delta\tau.Q) = \alpha x.(P+\delta\tau.Q) + [x \notin n(\delta)]\delta\alpha x.Q$
T3	$P+\tau.P = \tau.(P+\psi\tau.P)$
T4	$TT4 = TT4+\tau.(P+\delta[x\notin n(\delta)]ax.P)$

Fig. 2. The Tau Laws

Some explanations of these tau laws are as follows:

- T1 is Milner's first tau law.
- T2 is a nontrivial extension of Milner's third tau law.
- T3 is purely for tau prefix. If let ψ be false, we get the Milner's second tau law $P+\tau.P = \tau.P$, from which $TD3b$ can be inferred immediately.
- In T4 the shorthand notation $TT4$ stands for

$$\sum_{y \in Y} ax.(P_y+\delta[x=y]\tau.Q)+ax.(P+\delta[x\notin Y]\tau.Q)$$

- In TD4 the abbreviation D4 is for

$$\sum_{y \in Y} a(x).(P_y+\delta[x=y]\tau.Q)+a(x).(P+\delta[x\notin Y]\tau.Q)$$

To state the completeness results we define the equational systems AS^o_e and AS^o_l as in Figure 3.

system	rules and laws
AS^o_e	$AS \cup \{T1, T2, T3, T4\}$
AS^o_l	$AS \cup \{T1, T2, T3\}$

Fig. 3. Two systems

The proof of the completeness results is divided into two steps. First a promotion is established:

Theorem 1. *The following properties hold:*

- *If $P \simeq_l^o Q$ then $AS_l^o \vdash \tau.P = \tau.Q$.*
- *If $P \simeq_e^o Q$ then $AS_l^o \vdash \tau.P = \tau.Q$.*

Then in the second step the full completeness is proved using the weak completeness.

Theorem 2 (Completeness). *The following properties hold:*

- *AS_l^o is sound and complete for $\simeq_l^o$.*
- *AS_e^o is sound and complete for $\simeq_e^o$.*

References

1. Milner, R., Parrow, J., and Walker, D.: A Calculus of Mobile Processes, *Information and Computation*, **100**(1992) 1-77
2. Sangiorgi, D.: A Theory of Bisimulation for π-calculus, *Acta Informatica* **3** (1992)69–97
3. Fu, Y.: A Proof Theoretical Approach to Communications, *ICALP '97*, Lecture Notes in Computer Science 1256 (July 7th-11th, Bologna, Italy, Springer-Verlag, 1997) 325–335
4. Fu, Y.: Bisimulation Lattice of Chi Processes, *ASIAN '98*, Lecture Notes in Computer Science 1538 (December 8th-10th, Manila, The Philippines, Springer-Verlag, 1998) 245–262
5. Fu, Y.: Variations on Mobile Processes, *Theoretical Computer Science*, **221** (1999) 327–368
6. Fu, Y.: Open Bisimulations of Chi Processes, *CONCUR'99*, Lecture Notes in Computer Science 1664 (Springer, 1999) 304–319
7. Fu, Y.: Bisimulation Congruence of Chi calculus, *Information and Computation*, **184** (2003) 201–226
8. Fu, Y. and Yang, Z.: Chi Calculus with mismatch, *CONCUR 2000*, Lecture Notes in Computer Science 1877 (Springer, 2000) 596–610
9. Fu,Y. and Yang, Z.: The Ground Congruence for Chi Calculus, *FST&TCS 2000*, Lecture Notes in Computer Science 1974 (Springer, 2000) 385-396
10. Fu, Y. and Yang, Z.: Understanding the Mismatch Combinator in Chi Calculus, *Theoretical Computer Science*, **290** (2003) 779-830
11. Parrow, J. and Victor, B.: The Update Calculus, *AMAST '97*, Lecture Notes in Computer Science 1119 (Springer-Verlag,1997) 389–405
12. Parrow, J. and Victor, B.: The Tau-Laws of Fusion, *CONCUR '98*, Lecture Notes in Computer Science 1466(Springer,1998) 99–114

Improvement of the Resolution Ratio of the Seismic Record by Balanced Biorthogonal Multi-wavelet Transform*

Wenzhang He[1,2], Aidi Wu[2], and Guoxiang Song[1]

[1] Department of Applied Mathematics, Xi'dian University, Xian,
710071, P. R. China
Hewenzhang@tute.edu.cn
[2] Department of Mathematics and Information Science,
Tianjin Universityof Technology and Education, Tianjin, 300222, P. R. China

Abstract. This paper discusses an attempt to improve the resolution of a seismic record by using balanced biorthogonal multi-wavelet transform. Wavelet transform offers a lot of local information of seismic signals relevant to time, space, frequency and wave number. Compensation for the loss of high frequency information allows an improvement of the resolution ratio of the seismic record. Numerical results show that the suggested method has several advantages over single wavelet in feasibility and effectiveness and the practice shows some satisfactory results.

1 Introduction

At present, wavelet transform has found a wider application in the processing of seismic data and the application has yielded some desired results. It is well-known that orthogonal property preserves energy in the practical application of signal processing, and symmetric property may preserve linear phase, and make signals easy to process at the edge, so analyzing tools with the two properties simultaneously are important. But in real domain, no extraordinary wavelet can be found, which has compactly supported set, orthogonal and symmetric property at the same time. To make up for the drawbacks mentioned above, Goodman et al.(1994) proposed the concept of multi-wavelet [1],the basic idea of which is that the single scaling function is expanded to multiscaling function to achieve the greater degree of freedom. And Geronimo et al.(1994)adopted fractal interpolation method and worked out the celebrated CHM wavelets, which not only keeps the super local character in time domain and frequency domain, but also overcome the drawback of single wavelet and combine perfectly the important smooth property, compactly supported property, symmetric property and orthogonal property in practical application.

---

* This work is supported by Science and Technology Development Foundation of Tianjin Higher Education (20040708), and partly by both the Key Laboratory of Geophysical Exploration, China National Petroleum Corporation (GPKL0204).

In processing the high resolution of the seismic data, the first step is to eliminate or reduce the noise as much as possible and to increase the signal-noise rate of the seismic data for the reason that the lack of higher signal noise rate means the failure to realize a higher resolution. The second step is to compensate the decreased high frequency elements of seismic waves flowing along the ground. As the result of wavelet transform, a binary seismic signal can be changed into a 4-ary function relevant to time, frequency, space and wave-number, which provides the local information involving time, frequency, space and wave-number of the seismic signal. The lost frequency elements and decreased energy are absorbed by the layers of the earth, and to some extent, the seismic signals can be compensated by compensating the different frequency elements in different frequency ranges or in the different periods of time, thus increasing the resolution ratio of the seismic data. This paper deals with the method directed at improving the resolution ratio of the seismic records by making use of the plentiful information provided by the balanced biorthogonal multi-wavelet transform.

2 Balanced Biorthogonal Multi-wavelet

Single wavelet is connected to one scaling function while multi-wavelet is connected to multiple corresponding scaling functions. These scaling functions can be denoted by one vector $\phi(x) \equiv [\phi_1(x), \phi_2(x), \cdots, \phi_r(x)]^T$, where $\phi(x)$ is called multiscaling vector function. Similar case can be defined as multi-wavelet vector function $\psi(x) \equiv [\psi_1(x), \psi_2(x), \cdots, \psi_r(x)]^T$. They can be produced by multi-resolution analysis of multiplicity r. In case of r = 1, ordinary wavelet, namely single wavelets occur. The study of multi-wavelet mainly focuses on the case of r = 2. Biorthogonal wavelets have two groups of scaling functions, which can satisfy the following 2-scale function equations:

$$\phi(x) = \sum_{k=0}^{N} h_k \phi(2x - k),$$

$$\tilde{\phi}(x) = \sum_{k=0}^{N} \tilde{h}_k \tilde{\phi}(2x - k),$$

where $\{h_k\}$ and $\{\tilde{h}_k\}$ are 2×2 matrices.

Two groups of multi-wavelet can be denoted by multi-scaling functions

$$\psi = \sum_k g_k \phi(2 \cdot -k),$$

$$\tilde{\psi} = \sum_k \tilde{g}_k \tilde{\phi}(2 \cdot -k),$$

where $\{g_k\}$ and $\{\tilde{g}_k\}$ are 2×2 matrices.

These filter matrices satisfy double orthogonal conditions:

$$\sum_k h_k \tilde{h}_{k+2m}^T = 2\delta_{m,0} I, \quad m \in Z,$$

$$\sum_k h_k \tilde{g}_{k+2m}^T = 0, \quad m \in Z,$$

$$\sum_k g_k \tilde{g}_{k+2m}^T = 2\delta_{m,0} I_r, \quad m \in Z.$$

Let $f(t)$ be given signal, $f(t) \in V_0$, then $f(t)$ can be decomposed into the following form:

$$f(t) = \sum_k v_{0,k}^T \phi(t-k).$$

Here exists a discrete multi-wavelet transform:

$$v_{j,k} = \sum_m h_{m-2k} v_{j-1,m},$$

$$w_{j,k} = \sum_m g_{m-2k} v_{j-1,m}, \qquad j = 1, 2, 3 \cdots N.$$

And the inverse discrete multi-wavelet transform:

$$v_{j-1,k} = \sum_m \tilde{h}_{k-2m}^T v_{j,k} + \sum_m \tilde{g}_{k-2m}^T w_{j,m}, \qquad j = N, N-1, \cdots 1.$$

Vector filters $\{h_k\}, \{\tilde{h}_k\}$ and $\{g_k\}, \{\tilde{g}_k\}$ determine the performance of transform, where $\{h_k\}, \{g_k\}$ are named as the analysis filters, and $\{\tilde{h}_k\}, \{\tilde{g}_k\}$ are named as the reconstruction filters. Compared with scalar filters, vector filters offer a greater degree of freedom in structure. A full use of these degrees of freedom can increase the performance of multi-wavelets, satisfy orthogonality and symmetry, and increase approximation order or optimum areas of time frequency.

If the polynomial, which is no more than order p treated by vector filters in multi-wavelet transform, remains unchanged, this kind of multi-wavelet can be defined as balanced multi-wavelet of order p. Thus constant vector signal passing low pass filters remains constant signal and vector signal which is lower than order p polynomial passing low pass filters remains polynomial signal. The application of the property plays a considerably important part in the processing seismic data. Lebrun et al.(2001) created a method of working on high-order balanced orthogonal multi-wavelet and offered a coefficient matrix value of balanced orthogonal multi-wavelet filter with length 1-4[2]. The use of balanced orthogonal multi-wavelet illustrates that multi-wavlet transform can be made possible simply by resolving original signal into odd and even parts by subscribing and combining them into elementary vectors without any need of performing complex pre-processing.

Hwee et al.(1999) worked out two ways to structure symmetric and antisymmetric biorthogonal multiwavelets, optimize the parameters and obtain filter system with high properties and the frequently used filter coefficient matrix of the biorthogonal multi-wavelet: BSA6/16,BSA7/19,BSA8/16[3].

3 Improvement of Resolution Ratio

Two-dimension wavelet transform offers the local information of signals relevant to time, space, frequency, and wave number. And various noises can be suppressed efficiently by making use of the information. Obviously, the narrower the frequency band and the wave number band are, the bigger the calculating amount is, thus lowering the calculating speed. The improvement of the resolution ratio can be accomplished by two-dimensional f – k analysis of the signals to make clear the distribution range of the frequency and wave number of the signals as well as by two-dimension resolution of the signals and treatment of the decreased noise in the two-dimension wavelet domain, according to positional difference in time and space of the signals and noise and the difference in frequency and wave number of the signals and noises.

It is known that the earth layers absorb the seismic waves so that the resolution ratio of the seismic record reduces. There are two results of the earth layer absorbing. One is that the energy decreases as the spread time increases and the other is that the high-frequency elements will be lost to a larger extent than the low-frequency elements with the spread time increasing, which makes the wave form longer, contributing to the reduction of the resolution ratio recorded. The change in the rock of the earth results in the different resolution ratio recorded in every time layer. This means compensation for the loss of information on different time layers and different frequency, especially on the same axis. By contrast, wavelets transform gives the information of seismic section on different time layers and different frequency bands and compensates the lost energy and frequency. The resolution ratio of the seismic record is improved by processing the seismic records with multi-wavelet transform.

LL	LH
HL	HH

L_1L_1	L_1L_2	L_1H_1	L_1H_2
L_2L_1	L_2L_2	L_2H_1	L_2H_2
H_1L_1	H_1L_2	H_1H_1	H_1H_2
H_2L_1	H_2L_2	H_2H_1	H_2H_2

(a) Single wavelet resolution   (b) Multi-wavelet resolution

Fig. 1. Two-dimensional wavelet resolution

The use of single wavelet in the case of single resolution of wavelet transform entails decomposing seismic data into 4 blocks as shown in Fig.1(a). For example, the data in LH is obtained by performing high pass filtering of the rows and then performing low-pass filtering of the columns. This paper introduces the use of multi-wavelet of multiplicity r = 2 because the occurrence of two low pass filters and two high pass filters results in decomposing seismic data into 16 blocks, as shown in Fig.1 (b). The structure resembles the resolution of single wavelet packets. For example, the data in $L_1 H_2$ is obtained by acting the second high pass filter on the rows and acting the first low pass filter on the columns. Multi-wavelet resolution of low frequency data leads to a pyramid data structure.

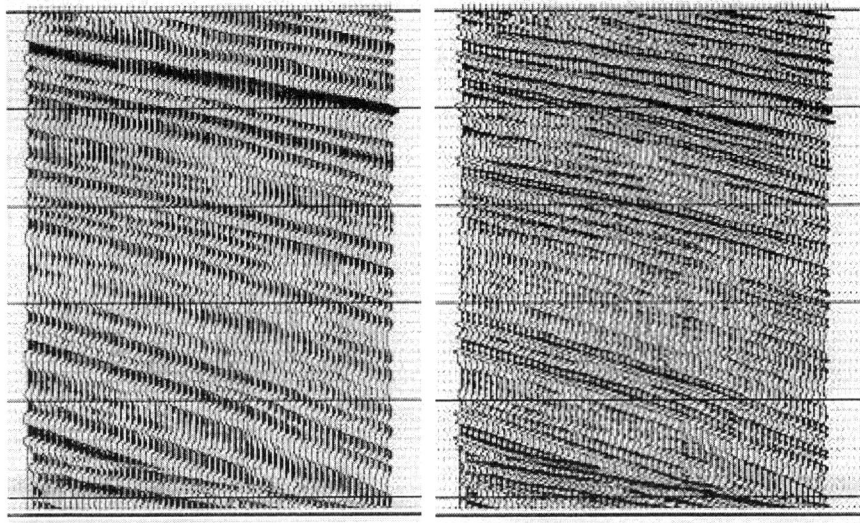

Fig. 2. Seismic section of some oil field   Fig. 3. Result of improved resolution ratio

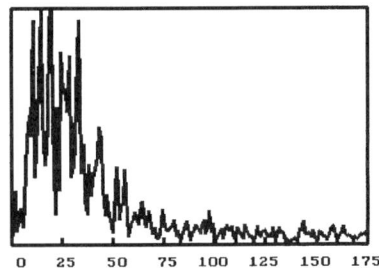
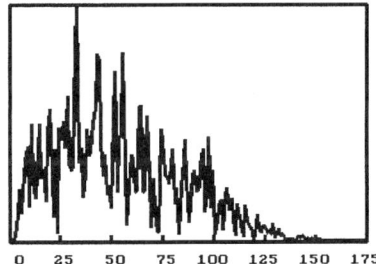

Fig. 4. The frequency-spectrum graph of the first seismic record of Fig. 2

Fig. 5. The frequency-spectrum graph of the first seismic record of Fig. 3

Multi-wavelet coefficients can be decomposed respectively into vertical, horizontal, and diagonal tree, in each of which, every parental node has four corresponding sub-nodes except the lowest frequency. If the value of the parental node is very small, the values of the formal sub-nodes are generally very small. Conversely, the value of the parental node is very big and so are the values of the sub-nodes. The big wavelet coefficients demonstrate the edge of image as well as the reflected intensity of the seismic records. The energy of big frequency is increased by compensating the wavelet coefficients of all the corresponding sub-nodes of the parental node with big value to improve the resolution ratio.

To show the feasibility and the practical effect, the practical seismic section of an oil field is chosen and processed. And the data of seismic section was decomposed into 4 layers with balanced biorthogonal multi-wavelet BSA6/6. Fig. 2. is the seismic section of an oil field and Fig. 3. is the improved result of resolution ratio by multi-wavelet transform, which shows great improvement compared with Fig. 2. Fig. 4. is the frequency-spectrum graph of the first seismic record of Fig. 2., and Fig. 5. is the frequency-spectrum graph of the first seismic record of Fig. 3, which clearly shows that the dominant frequency and frequency width of the processed section has been increased greatly, which is 1.5 times as many as the original seismic section and consequently the resolution ratio has been greatly improved.

4 Conclusion

Multi-wavelet transform overcomes the drawbacks of scalar wavelet and enables to satisfy orthogonal and symmetric property simultaneously. Two-dimensional multi-wavelet transform offers the local information of seismic signals relevant to time, space, frequency and wave number. The dominant frequency and the frequency width of the seismic sections have increased roughly 2 times as many as the original seismic section by compensating the high frequency of the reflected signals. It is revealed clearly that this method proves both feasible and effective in practical application.

References

1. Goodman, T.N.T., Lee, S.L.: Wavelets of Multiplicity r. Trans. Amer. Math.Soc. ,342 (1994) 307–324.
2. Lebrun, J., Vetterli, M.: High-order Balanced Multiwavelets: Theory, Factorization and Design. IEEE Trans SP, 9 (2001) 1918–1929.
3. Hwee,H.T.:New Biorthogonal Multiwavelets for Image Compression. Signal Processing, 79 (1999) 45–65.
4. Shaprio, J.M.: Emedded Image Coding Using Zerotrees of Wavelet Coefficients. IEEE Trans SP, 12 (1993) 3445–3462.
5. Lixin,S., Hwee,H.T.: On a Family of Orthonormal Scalar Wavelets and Related Balanced Multiwavelets.IEEE Trans SP,7(2001)1447–1453.

Formally Specifying T Cell Cytokine Networks with B Method

Shengrong Zou

Institute of Information Technology, Yangzhou University,
225000, Yangzhou, China
Zoushengrong@vip.sina.com

Abstract. In our work, We have specified aspects of T-cell cytokine networks using B method. With this model, we are able to run verification with B-toolkit and allow us to compare the dynamic behavior of the model to actual experimental data from College of Animal Science and Veterinary Medicine. Our results show that the use of B method can help confront open questions in immunology and probably in other fields of biology, which, because of their complexity, cannot be addressed by standard laboratory techniques alone.

1 Introduction

The immune system protects our organisms against pathogens and aberrant cells. It achieves this capability by processing and evaluating a wide variety of signals many of which are generated by its own agents. An illustrative example for the current situation in immunological research is the case of cytokine networks. Cytokines are small protein or glycoprotein messenger molecules that convey information from one cell to another. Various aspects of the immune response are regulated by cytokine networks[1]. More than 200 cytokines have now been identified. Yet, although many details of particular cytokine interactions have been elucidated, " practically nothing is known about the behavior of the network as a whole" [2].

The transition from identifying building blocks to integrating the parts into a whole will have to use the language of mathematics and algorithmics. We need a language that is legible both to biologists and computers, and that is faithful to the logic of the biological system of interest.

2 B Method

The B abstract Machine Notation specification language[3], based on mathematical notation: predicate logic, symbols denoting sets, sequences, functions and other abstract data types, originally developed in the early and mid 1980s by J.R.Abrial and by research groups at BP Research, Matra and GEC alsthom, is currently attracting increasing interest in both industry and academia. It is one of the few "formal methods" which has robust, commercially available tool support for the entire development lifecycle from specification through to code generation[4].

B, Z and VDM are formal methods[5] based on the construction of models (as opposed to those based on an algebraic approach like LARCH, OBJ).

3 Specifying T Cell Cytokine Networks

In this paper, we specified aspects of T-cell cytokine networks using the B method[6][7][8]. With this specification, we are able to run verification with B-toolkit and allow us to compare the dynamic behavior of the model with actual experimental data from College of Animal Science and Veterinary Medicine . Our results show that the use of B method can help confront open questions in immunology and probably in other fields of biology, which, because of their complexity, cannot be addressed by standard laboratory techniques alone[9].

3.1 T Cell Cytokine Networks Specification

Due to space limitations, we omitted in this paper many details of cytokine network that we used for constructing the specification. We present here the B machine that describes the dynamic behavior of the T helper cell[10][11] .

The Machine of the T helper cell describes three components of the behavior of a T cell.

---Its immunological state- whether the T cell is active or not.
---Its phase in the cell cycle.
---Its anatomical location.

Immunological State. The immunological state component contains two super-states: active and nonactive.

The default transition is conditioned: if the number of cell cycle through which the T cell has gone is zero(i.e., the cell is naive),then the cell will enter the nonactive super-state. If, however, the number of cell cycles is greater than zero, implying that this newly born cell is the progeny of an active T cell ,it enters the active state.

Control of proliferation is carried out within the cell cycle component of the machine. In the current model, we do not get too deeply into the differentiation aspect of T cell activation. However, a basic sketch appears in the Active sub-machine.

As helper T cells differentiate, they are thought to go through an intermediate stage known as Th0.Therefore,the default way in the Active sub-machine leads to the Th0 state. Further differentiation into Th1 or Th2 armed effector T cells is dependent on two basic factors:

 – The cytokine pattern sensed by the T cells, Two key cytokines that control this process are IL-12 and IL-4,which drive the cells towards the Th1 and Th2 pathways, respectively.

 – The number of cell cycles these cells have gone through.Th0 cells fail to respond to IL-4 unless they have traversed at least 3 cell cycles, while the responsiveness of Th0 cells to Il-I2 was found to increase as cell proliferation proceeds: "A naive cell can become a Th1 cell with increasing likelihood after each successive cell cycle but can only become a Th2 cell when eight siblings have arisen". This dependency on cell cycle progression was found to be related to cell cycle dependent changes in chromatin structure, which gradually 'expose' these cytokine genes to transcription.

Cell Cycle Control. As mentioned above ,naive T cells can live for many years without replicating, implying that during this time period they stay in the G0 state.The initial encounter with a specific antigen in the presence of the required co-stimulatory signal triggers the transition from G0 to the G1 phase of the cell cycle. In our model, the transition from the G0 state to G1 state is driven by the 'EnterCellcycle' triggered operation, which is initiated during the transition from the standby state to the IL2production state. Further progression through the cell cycle is dependent on IL-2 binding by the IL-2 receptor, an event that triggers both the transition from the G1 state to the S state in the cell-cycle-control component and the transition from the nonactive state to the active state in the immunological state component. If no death signal is induced while being in the S phase of the cell cycle, the cell will go through the rest of the cell cycle, depending only on timeout events. Once entering the M state, a series of actions is carried out, resulting in the generation of two new instances of the Thelpercell class, representing the two newly born daughter cells.

3.2 B Machine of T Cell Cytokine Network

For example, we show the framework of the T helper cell machine.
```
Machine ThelperReceptor_class
Extends T_class
Sets
  Receptor;
  Thehelper-
cell_state={cellcycle,g0,inblood,inecm,inlymph,nonactive,active}
Variables
  Thelpercell_state,
  Place,
  Initialize,
  Ts
Invariant
  Thelpercell_stage:receptor→thelpercell_state&
  Place:receptor→omstring&
  Initialize:receptor→omboole&
  Ts:receptor→pow(t)
Operations
      Entercellcycle(thisreceptor)=
        Pre
          Thisreceptor:receptor
        Then
          Select thelpercell_state(thisreceptor)=g0
          Then                                    thelper-
cell_state(thisreceptor)  :=cellcycle
          end
        end;
```

```
      Bindligand(thisreceptor) =
        Pre
          Thisreceptor:receptor
        Then
          skip
        end;
      Tm(24000)(thisreceptor) =
        Pre
          Thisreceptor:receptor
        Then
          Select                                              thelper-
cell_state(thisreceptor)=inlymph
          Then                                                thelper-
cell_state(thisreceptor):=inblood
          end
        end;
      Freeligand(thisreceptor) =
        Pre
          Thisreceptor:receptor
        Then
          skip
            end;
      Evil2rbound(thisreceptor) =
        Pre
          Thisreceptor:receptor
        Then
          Select                                              thelper-
cell_state(thisreceptor)=nonactive
          Then thelpercell_state(thisreceptor):=active
          end
        end;
      Exitcellcycle(thisreceptor) =
        Pre
          Thisreceptor:receptor
        Then
          Select                                              thelper-
cell_state(thisreceptor)=cellcycle
          Then thelpercell_state(thisreceptor):=g0
          end
        end;......
```

3.3 Refine and Implement

The formal modeling technique that we use for our application does not consist of directly constructing the software algorithm[12]. Instead, we shall obtain it in a rigorous way, from the main requirements using the refinement mechanism. We consider that the system dynamics are expressed through asynchronous events.

The development is made in several refinements: each step must refine its preceding level. The transition to the next level is characterized by a transformation of the state abstract data to more concrete data and by the refinement of the events.

Usually, the model development begins with an abstract model featuring very few events. At the beginning of the model development, we attempt to add events. These events provide an implementation of the algorithm.

The refinement principle presented above will be followed in the following steps, when we present each refinement step of the actual B model for the transaction mechanism (particularly at the first refinement step). The mathematical assumptions which must be true for one B model to be a correct refinement of another are completely defined in the B theory [13](as stated in the B-Book).

3.4 Verification

As the refinement consists of a simplification of the previous model, the resulting proof obligations are either obvious or automatically demonstrated by the prover (in the tool we use, the "obvious" proof obligations are the ones eliminated by a basic prover included in the proof obligation generator, for instance when the goal in literally found in the hypotheses).

We have given a specification of cytokine network, validated it against the informal requirement, refined this machine. Now if we can prove the refinement relation from the specification toward the implementation, then our verification is done. We have done this with B-toolkit in our project and have described our proof in detail in technical report. Because of space limit, we do not present the proof here.

4 Conclusion

In this work, we examined the feasibility of B method in bridging the gap between analysis and synthesis by modeling a well-studied immunological research termed T cell cytokine network using the language of B/AMN, as implemented in the software engineering.

The use of the B method for the model of the biological system has been greatly beneficial for the following reasons:

- formal description of the functional requirements, together with the environment assumptions,
- formal validation of the detailed design and the pseudo-code algorithm,
- reduction of the validation task load,
- 100% of the proof obligations of the project (over 2000 in total) have been demonstrated in interactive mode.
- It shall be able to generate source code,

We believe that these goals may be reached. The compactness of the code may be achieved by including in the last stages of the B development all the size optimization tricks. The flexibility and the accordance to programming style guides may be achieved in the same way. The automatic translator would then be a very simple and literal one, potentially defined and proved using B. Then we can test whether the formal representation of the model fulfills existing biological data by model simulation and formal verification methods.

References

1. Jordan, J.D., Landac, E.M., Iyengar, R.: Signaling Networks: The Origins of Cellular Multasking. Cell. 103(2000)193-200
2. Callard, R., George, A.J., Stark, J.: Cytokines, Chaos, and Complexity. Immunity. 11(1999)507-513
3. Kevin Lano: The B Language and Method. Springer-Verlag, Berlin Heidelberg New York (1996)
4. Walden, M., Sere, K.: Refining Action Systems within B-tool. In: Gaudel, M.C., Woodcock, J.(eds.): FME'96:Industrial Benefit and Advances in Formal Methods, LNCS , Vol. 1051(1996)85-104
5. Satpathy, M., Snook, C., Harrison, R., Butler, M., Krause, P.A.: Comparative Study of Formal and Informal Specification through an Industrial Case Study. Proc IEEE/IFIP Workshop on Formal Specification of Computer Based Systems (2001)
6. Morpurgo, D.: Modelling Thymus Functions in a Cellular Automation. Int'l Immunology. 7(1995)505-516
7. Steven, H., Kleinstein, Philip, E., Seiden: Simulating The Immune System. Computing in Science & Engineering. (2000)69-77
8. Zou Shengrong: Modeling T cell Cytokine Network with B method. Proc The Second International Forum on Post-Genome Technologies(2'IFPT). (2004)136-139
9. Peter, D., Karp: Pathway Databases: A Case Study in Computational Symbolic Theories. Science. 293(2001)2040-2044
10. Endy, D., Brent, R.: Modeling Cellular Behavior. Nature. 409(2001)391-395
11. Naaman kam, David Harel, Irun, R., Cohen: Model Biological Reactivity: Statecharts vs. Boolean logic. Proc. Working Conf. on Advanced Visual Interfaces (AVI'02), Trento, Italy(2002)345-353
12. Zou Shengrong: Modeling Distributed Algorithm using B. Proceeding of the International Grid and Cooperative Computing Conference. Lecture Notes in Computer Science, Vol 3033. Springer (2004)683-689
13. Abrial, J.R., Mussat, L.: Introducing Dynamic Constraints. in Bert, B.D. (Eds.): Proceedings of the Second International B Conference B'98: Recent Advances in the Development and Use of the B Method. Springer. 4(1998)

Three-Dimensional Motion Analysis of the Right Ventricle Using an Electromechanical Biventricular Model

Ling Xia[1] and Meimei Huo[2]

[1] Department of Biomedical Engineering,
Zhejiang University, Hangzhou, 310027, China
[2] Department of Computer Science, Zhejiang University City College,
Hangzhou, 310015, China

Abstract. Previous mechanical heart models were mainly constructed for analysis of the left ventricle behavior, and the right ventricle was almost not included because of its complex geometry structure. The motion of the healthy right ventricle and its alteration due to disease are not currently well understood. In this paper, a 3-D finite element biventricular model with real geometric shape and fiber structure has been constructed and the right ventricular wall motion and deformation during systole phase have been simulated. The results show that: 1) The right ventricular free wall moves towards the septum, and at the same time, the base and middle of the free wall move towards the apex, which reduce the volume of right ventricle; 2) The minimum principal strain is largest at the apex, then at the middle of free wall, and its direction is in the approximate direction of the epicardial muscle fibers. The results are in good accordance with solutions obtained from MR tagging images.

1 Introduction

Heart diseases remain a major killer in the world. Any asynchrony of the electrical activation and the disorder of the mechanical properties can lead to abnormalities in heart function. Accurate estimates of heart motion and deformation are of importance for evaluating normal and abnormal cardiac physiology and mechanics. The measurement of regional myocardial injury due to ischemic heart diseases, for example, is an important clinical problem.

At present, many forms of cardiac imaging and image analysis methods, including MRI, echocardiography and so on, aim to measure the regional function of the left ventricle (LV) [1]–[3]. However, the complexity of LV motion, the absence of internal landmarks in the myocardium and the influence of the right ventricle (RV) on the LV imply that the true motion trajectories of tissue elements are, at best, difficult to obtain from image analysis.

Compared with above measurement methods, computer modeling of the heart is a useful theoretical approach for the investigation of cardiac motion. Many scientific and medical researchers have developed cardiovascular models, hoping to better understand physiological and pathological properties of the heart. Among the previous heart models, there were many mechanical heart models that mainly focussed on LV, the right ventricle, however, was almost not included because of its complex geometry structure. The motion of the healthy RV and its alteration due to disease are not currently well understood. Consider that the RV also plays an important role in normal and abnormal hemodynamics, a biventricular model should be constructed and used to study the mechanical property of the heart. In this report, a 3-D finite element biventricular model with real geometric shape and fiber structure has been set up. It contracts with modulation from electrical excitation, and the resultant wall motion and deformation of the right ventricle during systole phase have been simulated.

2 Methods

2.1 Electromechanical Biventricular Model Construction

As Fig. 1 showed, the biventricular model was constructed on the basis of our previous electrical heart model [4]. Based on CT image data of the human body and heart sections, the electrical heart model was reconstructed as a 3-D array of approximately 65000 cell units spaced 1.5mm apart. It has been proved the computational method in this model is efficient with reasonable accuracy [4]. It is well known that the myocardial fiber structure of the heart plays a critical role in electrical propagation and force production. Myocardial electrical propagation is anisotropic, with the spread of current greatest in the direction of the long axis of the fiber. Fiber orientation is also

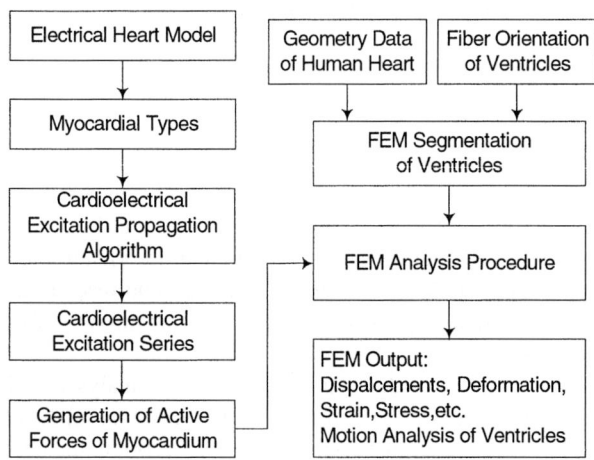

Fig. 1. The construction flow chart of biventricular model

an important determinant of myocardial stress and strain. According to the literature reports [5], we considered the fiber orientation in the biventricular model as follows: in the left and right ventricular free wall, fiber orientation typically varies from -60^0 at the epicardium to $+90^0$ at the endocardium, whereas in the septal wall the fiber angle ranges from approximately -90^0 at the right ventricular endocardium to around $+80^0$ at the left ventricular endocardium. At the junction of ventricular free wall and the septum, certain interpolation function was considered to make the junction smoothly.

It is known that the stiffness in the fiber direction is much greater than that of in the transverse direction, and the along-fiber speed of electric stimulus transmission is also greater than that of other directions. As far as mechanical properties are concerned, at any point in the myocardium the material may be considered as transversely isotropic.

Because ventricles are composite structures with very thick cross sections, 2-D plate analyses are inadequate. 3-D finite element (FE) analysis is necessary in order to calculate the stresses and strains accurately. Since material properties vary from layer to layer due to the change of fiber orientation, FE modeling for ventricles becomes extremely difficult and expensive. The normal 3-D FE methods (based on one layer per element) are not efficient enough for the analysis [10]. Hence, a simple but sufficiently accurate FE analysis is needed for the analysis of ventricles. Based on paper [10], we proposed an isoparametric method to discrete the ventricles. The major feature of the proposed analysis is that each element in the analysis can accommodate several layers having different ply orientations. When discretizing the ventricles into finite elements in radial direction, we group 10 layers per element, and it may be legitimate to assume that all the fibers within a layer are in the same orientation. We adopted 3-D 8- nodes isoparametric element as the basic element, and there were total 13-layer nodes, 1489 brick elements, and 5937 degrees of freedom in this model as shown in Figure 2.

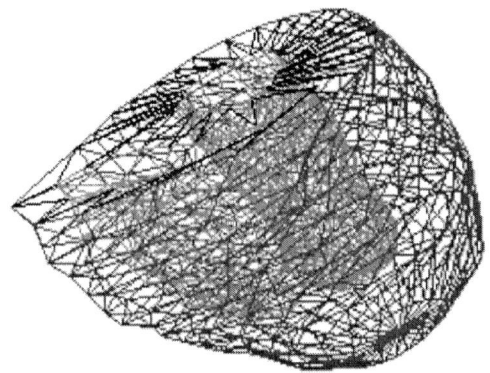

Fig. 2. The mesh graph of ventricles

In order to analysis the motion of ventricles, we needed to calculate the following equation (1) at time t.

$$[K]\{\delta\} = \{F_f\} \tag{1}$$

Here, $[K]$ is the total stiffness matrix, $\{\delta\}$ is the volume vector of nodal displacement, and $\{F_f\}$ is the total vector of the active forces.

$$K = \sum_e [K]^e; \quad F_f = \sum_e [F]^e \tag{2}$$

As equation (2) expressed, $[K]^e$ is the stiffness matrix of one element, $[F]^e$ is nodal force vector of one element.

$$[k]^e = \sum_{l=1}^{L_e} \int_{\xi_{l-1}}^{\xi_l} \left(\sum_{m=1}^{M_e} \int_{\eta_{m-1}}^{\eta_m} \int_{-1}^{1} [B]^T [C]_{lm} [B] |J| d\xi d\eta \right) d\zeta \tag{3}$$

As equation (3) expressed, $[B]$ is the matrix associated with the derivatives of the shape functions, $[C]_{lm}$ is elasticity matrix of layer l and segment m, m/M_e stand for the number /total numbers of segments in layer l in circumferential direction respectively. After calculation of equation (3) and (5), through equation (1) we can get $\{\delta\}$. A 3-D FE program was developed during the investigation for calculating the strains and deformations of ventricles.

2.2 Myocardium Active Forces Computation

After the cardiac electrical excitation series are determined based on the electrical simulation, we could compute the resultant active forces of myocardium [6].

In the fiber-coordinate system (one axis is chosen to coincide with the local muscle fiber direction, another one is determined by the epicardium surface normal vector), the active force along the fiber can be expressed as:

$$\sigma_e' = \begin{cases} \sigma_o' \sin(\dfrac{t-\tau}{T_e}\pi), & 0 \le t \le T_e \\ 0, & t > T_e \end{cases} \tag{4}$$

where t is the transient time; τ is the time lag of active stress, T_e is the activation period. σ_o' is the active force of a myocardial fiber and which is a function of fiber length, time t, fiber directions, etc.

In the calculation, the stress and strain vector is expressed with respect to two coordinate frames of reference: global coordinate system X-Y-Z and local coordinate system ξ-η-ζ. The nodal force equivalent to the active contractile force is calculated based on equation (4) and the isoparametric transformation:

$$\{F_f\}^e = -\sum_{l=1}^{L_e} \int_{\xi_{l-1}}^{\xi_l} \int_{-1}^{1} \int_{-1}^{1} [B]^T T\{0,0,\sigma_e',0,0,0\}_l^T |J| d\xi d\eta d\zeta \tag{5}$$

where $[B]$ is geometric matrix of element, l, L_e stand for the number and total number of layers in an element respectively, T is the transformation matrix between the fiber coordinate and global coordinate. $\{0,0,\sigma_e',0,0,0\}_l^T$ is the active force row vector in the fiber-coordinate system, |J| is the determinant of the Jacobian matrix. ξ, η, ζ are the local coordinate system with the magnitudes ranging from −1 to 1; ξ^{l} and ξ^{l-1} represents individually the coordinate values along ξ direction on the front-side and backside of layer l in an element. $\xi^{l-1}=-1$ when $l=1$ and $\xi^{l}=1$ when $l=L_e$.

2.3 Deformation Calculation

Strain analysis is a method to describe the internal deformation of a continuum body. It is an appealing tool to study and quantify myocardial deformation. The Lagrangian strain E is used to describe systolic deformation in a region surrounding a point in the heart wall relative to its initial position at end-diastole. E was calculated from the relationship:

$$E = 1/2(F^T F - I) \tag{6}$$

where the superscript T represents the matrix transpose, F represents deformation gradient tensor, and I, the identity matrix. The deformation gradient tensor F can be calculated for the relationship:

$$F_{pq} = \partial x_p / \partial X_q \tag{7}$$

where X and x represent the initial and final position of a point in the material respectively, the subscripts p and q range from 1 to 3 and denote one of the 3D Cartesian coordinates. The tensor, F, includes both the rotation and deformation around a point in the material.

3 Results

Based on the model we constructed above, the motion of ventricles was analyzed. At first, on the basis of myocardial physiological property as well as some references [11, 12], the material constants are employed as follows: Young's modulus $E_L: 50\,Kpa, E_T: 11\,Kpa$; shear modulus $G_{LT}: 10\,Kpa$; Poisson's ratio: $\mu_{LT} = \mu_{TT} = 0.49$; The initial value of $\sigma_0' = 53\,Kpa$. In the simulation, the time incremental step of simulation is 7 ms. Activation period $T_e = 353\,ms$. The active force and material constants are adjusted slightly in order to get proper results. We did not consider the blood pressure in the simulation.

Fig. 3 shows the displacements of endocardium of RV at four different times during the period of systole, the color bar denotes the length of displacements (mm). The white dots mean the position of nodes at end-diastole, the red dots mean the position of nodes at different times through the systole, and the red line between the white dots and red dots mean the direction of nodes' displacements. We can see that there is a considerable displacement of the free wall towards the septum and a large displacement of the base towards the apex in the free wall, which contributed to a smaller RV cavity. The results are similar to those obtained from the Tagged MRI [7].

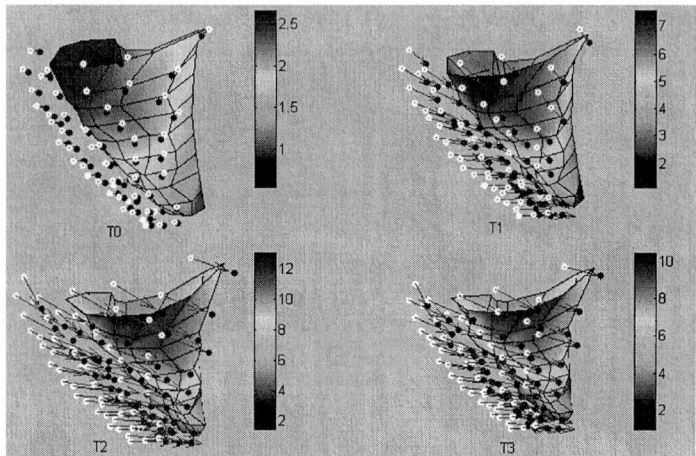

Fig. 3. The displacements of endocardium of RV at four different instants (T0: end-isovolumetric contraction T1: middle-maximum ejection T2: end-maximum ejection T3: end-ejection)

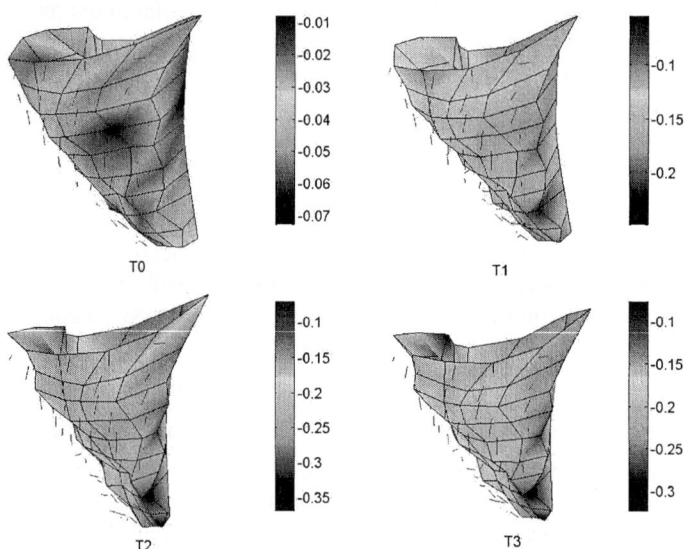

Fig. 4. The negative minimum principal strains and their directions of endocardium of RV at four different instants (see Fig. 2)

As the RV has complicated shape/geometry, it is difficult to find a single axis about which to calculate twisting and radial contraction like those of LV. Here we use coordinate-system-independent principal strains and their directions to describe the deformation of RV. These quantities are derived by finding the eigenvalues and eigenvectors of the Lagrangian strain tensor, respectively. The eigenvalues of E are the principal strains, where a positive value indicates extension while a negative value indicates compression. The minimum principal strain, E3, is the smallest eigenvalue of E, and along with the minimum principal strain directions, supplies potentially useful information about the magnitude and direction of contraction in the heart [8].

Fig. 4 shows the distribution of E3 at the endocardium of RV. The short red lines denote the directions of E3 of the central point of finite elements at the endocardium of RV. It can be seen that the minimum principal strain is not uniform at the endocardium. A maximum value of –0.35 is at the apex and the directions of E3 mostly towards the apex, which means that the maximum contraction occurs towards the apex and in the approximate direction of the muscle fibers. These results are also correlated well with those from the Tagged MRI [7, 9].

4 Conclusion

The right ventricular wall motion and deformation caused by the cardiac electrical activation have been analyzed based on a biventricular model and some general parameters, such as displacements, minimum principal strain, which are accepted in clinic, were used to describe the motion and deformation of the heart. The simulation results are in good accordance with solutions obtained from MR tagging images that reported in the literature. The future work will investigate the inter-ventricular dependence and ventricular wall relative motion within this framework.

Acknowledgments

This work is supported by the National Natural Science Foundation of China (30170243, 30370400), a Special Grant to Authors of Excellent Doctoral Dissertations by the Ministry of Education (1999941), and 973 National Key Basic Research & Development Program (2003CB716106).

References

1. Papademetris, X., Sinusas, A.J., Dione, D.P., Constable, R.T. and Duncan, J.S.: Estimation of 3-D Left Ventricular Deformation From Medical Images Using Biomedical Models. IEEE Trans. Med. Imag. 21 (2002) 786-800
2. Azhari, H., Oliker, S., Rogers, W.J., Weiss, J.L., and Shapiro, E.P.: Three-Dimensional Mapping of Acute Ischemic Regions Using Artificial Neural Networks and Tagged MRI. IEEE Trans. Biomed. Eng. 43 (1996) 619-626
3. Linares, P., Torrealba, V., Montilla, G., Bosnjak, A., Jimenez, C., and Barrios, V.: Deformable Model Application on Segmentation in 3-D Echocardiography. Computers in Cardiology 23 (1996) 413-416
4. Lu, W. and Xia, L.: Computer Simulation of Epicardial Potentials Using a Heart-Torso Model with Realistic Geometry. IEEE Trans. Biomed. Eng. 43 (1996) 211-217

5. Scollan, D.F., Holmes, A., Winslow, R.L.: Histological Validation of Myocardial Micro-Structure Obtained from Diffusion Tensor Magnetic Resonance Imaging. Am. J. Physiol. 275 (1998) H2308-H2318
6. Liu, F., Lu, W., Xia, L. and Wu, G.: The Construction of Three-Dimensional Composite Finite Element Mechanical Model of Human Left Ventricle. JSME Int. J. C44 (2001) 125-133
7. Haber, I., Metaxas, D.N., and Axel, L.: Using Tagged MRI to Reconstruct A 3D Heartbeat. IEEE Comput. Sci. Eng. 2 (2000) 22-34
8. Haber, I., Metaxas, D.N., and Axel, L.: Three-Dimensional Motion Reconstruction and Analysis of the Right Ventricle Using Tagged MRI. Med. Imag. Ana. 4 (2000) 335-255
9. Young, A.A., Fayad, Z.A., and Axel, L.: Right Ventricular Midwall Surface Motion and Deformation Using Magnetic Resonance Tagging. Am. J. PhysioL. 271 (1996) H2677-H2688
10. Chang, F.K., Perez, J.L., and Chang, K.Y.: Analysis of Thick Laminated Composites. J. Compos. Mater. 24 (1990) 801-822
11. Tokuda, M., Sekioka, K., Ueno, T., Hayashi, T., Havlicek, F. and Sawaki, Y.: Numerical Simulator for Estimation of Mechanical Function of Human Left Ventricle (Study of Basic System). JSME Int. J. A37 (1994) 64-70
12. Beyar, R. and Sideman, S.: A Computer Study of the Left Ventricular Performance Based on Fiber Structure, Sarcomere Dynamics, and Transmural Electrical Propagation Velocity. Circ. Res. 55 (1984) 358-375

Growing RBF Networks for Function Approximation by a DE-Based Method

Junhong Liu, Saku Kukkonen, and Jouni Lampinen

Lappeenranta University of Technology, Department of Information Technology,
PO Box 20, 53851 Lappeenranta, Finland
{junhong.liu, Saku.Kukkonen, jlampine}@lut.fi

Abstract. The Differential Evolution (DE) algorithm is a floating-point encoded Evolutionary Algorithm for global optimization. It has been demonstrated to be an efficient, effective, and robust optimization method especially for problems containing continuous variables. The paper concerns applying a DE-based method to perform function approximation using Gaussian Radial Basis Function (RBF) networks with variable widths. This method selects centres and decides weights of the networks heuristically, then uses the Differential Evolution algorithm for local and global tuning iteratively to find the widths of RBFs. The method is demonstrated by training networks that approximate a set of functions. The Mean Square Error from the desired outputs to the actual network outputs is applied as the objective function to be minimized. A comparison of the net performances with other approaches reported in the literature has been performed. The proposed approach effectively overcomes the problem of how many radial basis functions to use. The obtained initial results suggest that the Differential Evolution based method is an efficient approach in approximating functions with growing radial basis function networks and the resulting network generally improves the approximation results reported for continuous mappings.

Keywords: Radial Basis Function, Neural Network, Differential Evolution, Evolutionary Algorithm.

1 Introduction

Radial Basis Functions (RBFs) emerged as a variant of artificial neural networks in the late 80's. RBFs are embedded in a three-layer network, i.e., the input layer, the hidden layer, and the output layer. Each unit in the hidden layer implements a radial activation function. The output units implement a weighted sum of hidden unit outputs. Approximation capabilities of RBFs have been studied in [10, 8]. Due to their nonlinear approximation properties, RBF networks are able to model complex mappings.

The performance of a trained RBF network depends on variable factors, like the number of the RBFs as well as their locations, shapes, weights, and the method used for learning the input-output mapping. Different approaches

[1, 2, 4, 6, 9, 11, 15, 16] for training RBF networks have been developed and can be divided into three categories: (*i*) learning the centres and widths in the hidden layer; (*ii*) learning the connection weights from the hidden layer to the output layer; (*iii*) learning the network structure; and (*iv*) hybrid learning: learning the centres, widths, weights, or the network structure together.

Among these approaches, evolutionary techniques become outstanding, because they do not require high order derivatives when applied to nonlinear optimization problems, especially those with stochastic, temporal, or chaotic components. One of such technique is the Differential Evolution (DE) algorithm introduced by Price and Storn [12], a floating-point encoded Evolutionary Algorithm (EA) for global optimization. This paper introduces a DE-based method for finding the set of Gaussian RBFs with variable widths to provide the best function approximation.

The DE-based method, characterized as a growing method, can be classified in Category (*iv*), starting with a single hidden node network and iteratively growing the network. A centre plus its weight is decided heuristically based on the difference between the desired outputs and the actual outputs of the existing RBF network. The adaptation of widths is conducted using DE locally and globally.

The paper is structured as follows: the formulation of an optimization problem is explained in Sect. 2; Sect. 3 describes the optimization system of RBF networks using the DE-based method—how to combine iteratively the selection and training procedures to learn the network structure and the widths in order to achieve an overall efficient method; experimental results are shown in Sect. 4; and the conclusion is given in Sect. 5.

2   Optimization Problem Formulation

2.1   Gaussian Radial Basis Function Networks

RBFs have their origin in the solution of the multivariate interpolation problem [3]. An arbitrary function $g(\mathbf{v}) : \Re^d \to \Re$ can be approximate by mapping, using a RBF network with a single hidden layer of p units:

$$\widehat{g}(\mathbf{v}, \mathbf{x}) = \sum_{j=1}^{p} w_j r_j(\mathbf{v}, \boldsymbol{\sigma}_j, \mathbf{c}_j) = \sum_{j=1}^{p} w_j \phi_j(\boldsymbol{\sigma}_j, \|\mathbf{v} - \mathbf{c}_j\|), \qquad (1)$$

where $\mathbf{v} \in \Re^d$; $\mathbf{x}$ is the vector of all variables including w_0, w_j, $\boldsymbol{\sigma}_j$, and $\mathbf{c}_j$; p denotes the number of basis functions; $\mathbf{w} = (w_1, w_2, ..., w_p)^T$ are the weight coefficients; $r_j(\cdot)$ represents the d-dimensional activation function (also known as the radial basis function) from $\Re^d$ to $\Re$; $\mathbf{c}_j$, $j = 1, 2, ..., p$, are the centres of the basis functions; $\boldsymbol{\sigma}_j$, $j = 1, 2, ..., p$, are the widths, which are called scaling factors for the radii $\|\mathbf{v} - \mathbf{c}_j\|$, $j = 1, 2, ..., p$, of the basis functions, respectively; and $\phi(\cdot)$ is a non-linear function that monotonically decreases (or increases) as $\mathbf{v}$ moves away from $\mathbf{c}_j$.

In order to simplify the notation, coordinate axes-aligned Gaussian RBF functions are used. When a 1-dimensional Gaussian radial basis function is centred at the centroids c_j, it follows from (1) that

$$\widehat{g}(\mathbf{v},\mathbf{x}) = \sum_{j=1}^{p} w_j e^{-\frac{\|\mathbf{v}-\mathbf{c}_j\|^2}{2\sigma_j^2}} = \sum_{j=1}^{p} w_j e^{-\frac{(v-c_j)^2}{2\sigma_j^2}}, \qquad (2)$$

where $\mathbf{x}$ is the vector of all variable factors and can be written as

$$\mathbf{x}^T = (w_1, ..., w_j, ..., w_p, \sigma_1, c_1, ..., \sigma_j, c_j, ..., \sigma_p, c_p)^T. \qquad (3)$$

The network can be trained to approximate an unknown function $g(\mathbf{v})$ by finding the optimal vector $\mathbf{x}$ given (possibly noisy) training set $V = \{(\mathbf{v}_n, y_n) \mid n = 1, 2, ..., N, \mathbf{v}_n \in \Re^d, y_n \in \Re\}$.

2.2 Objective Function

The objective function is defined as to minimize the Mean Square Error (MSE):

$$f(\mathbf{x}) = \frac{1}{N} \sum_{n=1}^{N} \left(y_n - \widehat{g}(\mathbf{v}_n, \mathbf{x}) \right). \qquad (4)$$

2.3 Brief Description of the Differential Evolution Algorithm

The optimization target function is of the form

$$f(\mathbf{x}) : \Re^D \to \Re. \qquad (5)$$

The optimization objective is to minimize the value of the target function by finding the optimal values of vector $\mathbf{x}$, which is composed of D objective function parameters. The parameters of the target function are also subject to lower and upper boundary constraints $\mathbf{x}^{(L)}$ and $\mathbf{x}^{(U)}$: $x_k^{(L)} \leq x_k \leq x_k^{(U)}$, $k = 1, 2, ..., D$.

DE is a parallel direct search method with three control parameters, which remain constant during the optimization process (mutation amplification, crossover operator, and the number of population members) that utilizes D-dimensional parameter vectors, $\mathbf{x}_{i,G}, i = 1, 2, ..., N_{pop}$, as a population for generation G.

3 Optimization of Gaussian Radial Basis Function Networks by a Differential Evolution Based Method

3.1 Control Parameters' Setting of DE

The control parameters' setting affects the performance of DE, its values were chosen based on discussions in [13, 5] and are given in Table 1.

Table 1. Control parameters' setting of DE

Variable	Value
Strategy	DE/$rand$/1/bin
Crossover operator: C_r	0.9
Mutation amplification: F	0.9
Number of population members: N_{pop}	100

3.2 Differential Evolution Based Method

Training. The entire sample points in training set V are used for training. $\mathbf{v}_n \in \Re^d$, $n = 1, 2, ..., N$, possible centre candidates are in set V_e and each can only be chosen once. Initially the number of hidden nodes, p, is zero.

1. (i) p is increased by 1; and (ii) the sample pair $(\mathbf{v}_i, y_i)$ from set V_e, for which the function approximation is the worst—having the highest difference between the desired output and the actual output of the existing RBF network (i.e., before adding the new node), is chosen to be taken away from V_e as a new node centre, i.e., $\|y_i - \hat{g}_{p-1}(\mathbf{v}_i)\| = \max\{\|y_n - \hat{g}_{p-1}(\mathbf{v}_n)\|, (\mathbf{v}_n, y_n) \in V_e, n = 1, 2, ..., N - p + 1\}$, where $\hat{g}_{p-1}$ is the existing Gaussian RBF network.
2. Growing: (i) the chosen node $\mathbf{v}_i$ is added to centres' set $C = \{c_{ji} \mid j = 1, 2, ..., p, i = 1, 2, ..., d\}$, where p is the number of hidden nodes present in the RBF network; (ii) the weight of the node is set as $w_p = y_i - \hat{g}_{p-1}(\mathbf{v}_i)$, i.e., the error between the desired output and the existing RBF network output.
3. Local tuning: find the width of this newly added node after G_{local} generations, using DE. Each newly generated parameter vector mentioned in Sect. 2.3 of a population is based on the boundaries of the discussed function, $g(v)$, i.e., $x_{ji} \in (0, \frac{v_i^{(U)} - v_i^{(L)}}{2}]$, where $v_i^{(U)}$ and $v_i^{(L)}$ are the upper and lower limits of the defined region of the computed function.
4. Global tuning: the widths of the entire RBF network are learned after G_{global} generations, using DE, setting dimensionality $D = p$.
5. Criterion: if $f(\mathbf{x}) \geq \epsilon_f$ and $V_e \neq \emptyset$, where ϵ_f is a threshold, go to Step 1. Otherwise, stop the process.

Testing. For each function, testing set V_t is different from V and contains the same number of testing samples as in the training set, but are uniformly distributed random values in the function domain.

4 Experimental Results and Analysis

The proposed approach is applied to single-input and single-output test functions in Table 2 without any disturbance.

Table 2. Test functions

Index	Test function
1	A Hermite polynomial: $g_1(v) = 1.1 \cdot (1 - v - 2v^2) \cdot e^{-v^2/2}, v \in [-4, 4]$
2	$g_2(v) = \sin(12v), v \in [0, 1]$
3	$g_3(v) = \sin(20v^2), v \in [0, 1]$
4	$g_4(v) = 1 + (v + 2v^2)\sin(-v^2), v \in [-4, 4]$
5	$g_5(v) = \sin(v) + \sin(3v) + \sin(5v), v \in [-0.35, 3.5]$
6	$g_6(v) = 0.5 \cdot e^{-v} \cdot \sin(6v), v \in [-1, 1]$
7	$g_7(v) = \sin(v) + \sin(3v) + \sin(6v), v \in [-0.35, 3.5]$
8	$g_8(v) = \sum_{n=1}^{4} \left(n \cdot e^{-n^2 \cdot (v-4+2n)^2} \right), v \in [-4, 4]$
9	$g_9(v) = v, v \in [0, 1]$

For a continuous function $g(v)$, the input is considered as data set V, containing N data points in the function's defined region: $V = \{(\mathbf{v}_n, y_n) \in \Re^d \times \Re, 1 \leq n \leq N, y_n = g(\mathbf{v}_n)\}$, where d denotes the dimension of the discussed function $g(v)$, i.e., $d = 1$ for a single input function, and $\mathbf{v}_n$, $n = 1, 2, ..., N$, are N uniformly spaced noiseless points in the defined region of the computed function.

The functions in Table 2 were tested. The approach performed with the control parameters set in Table 1 and $N = 101$. Key experimental results are:

1. Fig. 1 shows the examples of the net functions in Table 2, the approximated functions as well as their individual composing RBFs, and the convergence histories: (i) g_8 can be represented by a network of four hidden nodes with the objective function value $1.8868 \cdot 10^{-5}$; and (ii) g_9 can be represented by a network of eight hidden nodes with the objective function value $7.6906 \cdot 10^{-6}$.
2. Table 3 lists the numerical results of the experiments.
3. As can be seen from Fig. 1 and Table 3, the DE-based approach is effective with respect to the objective function in approximating the original functions with growing RBF networks and the proposed method is efficient since most of the functions in Table 2 can be represented by a certain number of RBFs with very small error, for example, for g_1, the maximum approximation error is $2.24 \cdot 10^{-2}$ and the standard deviation is $5.4 \cdot 10^{-3}$ and the maximum testing error is $1.64 \cdot 10^{-2}$ with the standard deviation $5.3 \cdot 10^{-3}$, for g_2, the maximum approximation error is $8.24 \cdot 10^{-2}$ and the standard deviation is $9.4 \cdot 10^{-3}$ and the maximum testing error is $3.01 \cdot 10^{-2}$ with the standard deviation $5.5 \cdot 10^{-3}$ and so on.
4. Table 4 shows the comparisons between the DE-based method (short as DE) and the method (short as EA) proposed by Esposito et al. [4] for functions g_5, g_6, g_8, and g_9. Esposito et al. proposed an incremental algorithm for growing RBF networks by means of an evolutionary strategy and showed their strategy can achieve better performance than a greedy algorithm [14]

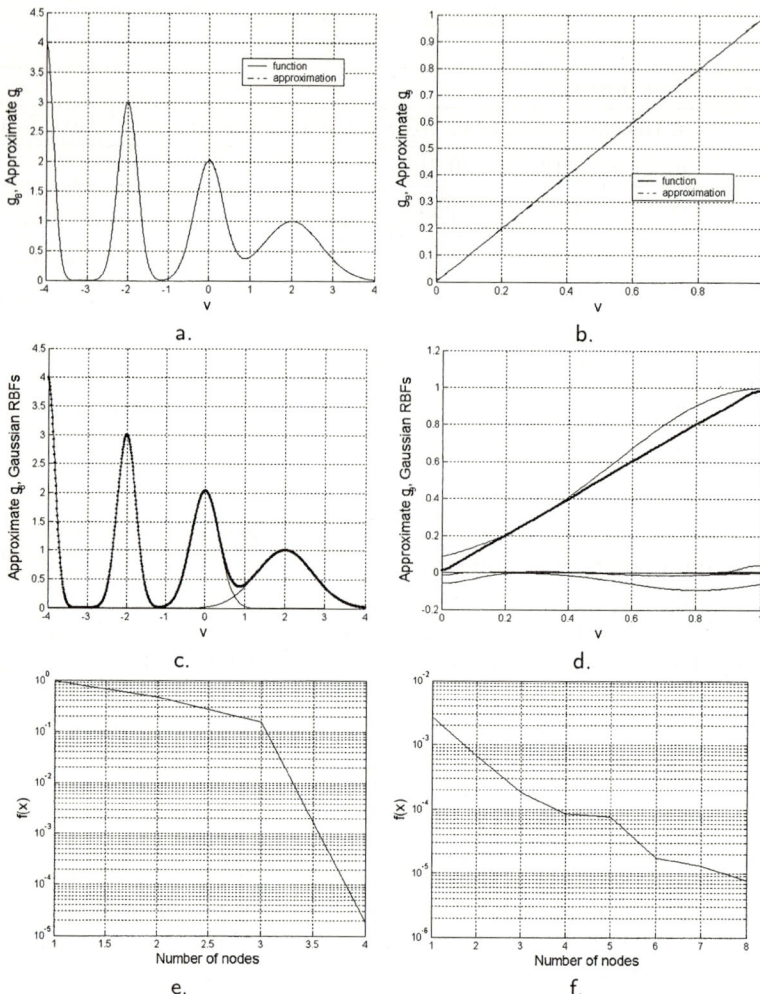

Fig. 1. Examples of original functions, approximations as well as the decomposition of individual RBF networks, and convergence histories. (i) a and b: net functions g_8 and g_9; (ii) c and d: the decomposition of respective RBF networks approximating functions g_8 and g_9; (iii) e and f: respective convergence histories (Figure legends in c and d: "·" network output; "–" individual RBF)

and Wavelet Neural Network [17] both in terms of net size and in terms of computation time. Setting the sizes of training set and testing set the same as reported by the authors for each function, the DE-based method got worse results than their method for g_5 but better solutions for the rest three.

Table 3. Experimental results

Test Function	p	Training			Testing		
		Training error (10^{-5})	Maximum error (10^{-2})	std (10^{-3})	Testing error (10^{-5})	Maximum error (10^{-2})	std (10^{-3})
$g_1(v)$	16	9.4648	2.24	5.4	13.142	1.64	5.3
$g_2(v)$	10	18.412	8.24	9.4	11.533	3.01	5.5
$g_3(v)$	14	86.318	7.07	17.9	77.32	6.97	16.8
$g_4(v)$	22	190260	813.61	105.79	181710	708.31	990.7
$g_5(v)$	10	49.735	5.13	13.2	52.135	5.16	1.31
$g_6(v)$	7	11.842	4.42	6.7	7.1339	2.08	5
$g_7(v)$	10	190	11.74	27.5	160	11.74	25.5
$g_8(v)$	4	1.8868	1.88	3.7	1.4703	1.88	3.2
$g_9(v)$	8	0.76906	1.51	2	0.71346	1.02	1.8

Note: *std* stands for standard deviation.

Table 4. Comparisons

Function	Algorithm	N	p	Training		Testing	
				Training error(10^{-5})	Maximum error (10^{-2})	Testing error (10^{-5})	Maximum error (10^{-2})
$g_5(v)$	DE	40	9	420	11.89	1010	36.25
$g_5(v)$	EA	40	6	500		400	18
$g_6(v)$	DE	100	9	2.1354	2.16	1.9292	1.53
$g_6(v)$	EA	100	25	0.11		0.42	
$g_8(v)$	DE	400	5	15.668	9.190	12.326	3.85
$g_8(v)$	EA	400	6	22		24	
$g_9(v)$	DE	500	6	0.87505	1.55	0.94226	1.32
$g_9(v)$	EA	500	6			1.4	

5 Conclusions

The DE-based method was applied to growing Gaussian RBF networks. The method starts from a single-hidden-node network and finally achieves a minimum network through an iterative procedure of local tuning and global tuning. The choice of the optimal network parameters corresponds to the minimum Mean Square Error between the desired outputs and the actual network outputs. A set of continuous functions were used to demonstrate the approach. The proposed approach effectively overcomes the problems of how many RBFs to use. The results obtained from this initial investigation suggest that the DE-based method is efficient and effective in approximating functions with growing RBF networks, and further studies on complicated and high dimensional data are therefore motivated and needed.

References

1. Alexandridis A., Sarimveis H., and Bafas G.: A new algorithm for online structure and parameter adaptation of RBF networks. *Neural Networks* **16** (2003) 1003-1017
2. Billings S. A. and Zheng G. L.: Radial basis function networks configuration using genetic algorithms. *Neural Networks* **8**(6) (1995) 877-890
3. Broomhead D. S. and Lowe D.: Multivariable functional interpolation and adaptive networks. *Complex Systems* **2** (1988) 321-355
4. Esposito A., Marinaro M., Oricchio D., and Scarpetta S.: Approximation of continuous and discontinuous mappings by a growing neural RBF-based algorithm. *Neural Networks* **13** (2000) 651-665
5. Lampinen J. and Zelinka I., On stagnation of the differential evolution algorithm. *Proc. 6th Int'l Conf. on Soft Computing (MENDEL 2000), Brno, Czech Republic, June 7-9 2000* (2000) 76-83
6. Leonardis A. and Bischof H.: An efficient MDL-based construction of RBF networks. *Neural Networks* **11** (1998) 963-973
7. Musavi M. T., Ahmed W., Chan K. H., Faris K. B., and Hummels D. M.: On the training of radial basis function classifiers. *Neural Networks* **5** (1992) 595-603
8. Park J. and Sandberg J. W.: Universal approximation using radial-basis-function networks. *Neural Computation* **3** (1991) 246-257
9. Plagianakos V. P. and Vrahatis M. N.: Neural network training with constrained integer weights. *Proc. 1999 Congress on Evolutionary Computation, Washington, DC USA, 6-9 July 1999* **3** (1999) 2007-2013
10. Poggio T. and Girosi F.: Networks for approximation and learning. *Proc. IEEE, MIT, Cambridge, MA, USA September 1990* **78**(9) (1990) 1481-1497
11. Schmitz G. P. J. and Aldrich C.: Combinatorial evolution of regression nodes in feedforward neural networks. *Neural Networks* **12** (1999) 175-189
12. Storn R. and Price K.: Differential Evolution a simple and efficient heuristic for global optimization over continuous spaces. *Journal of Global Optimization* **11**(4) (December 1997): 341-359
13. Storn R. and Price K.: On the usage of Differential Evolution for function optimization. *1996 Biennial Conference of the North American Fuzzy Information Processing Society (NAFIPS), Berkeley, CA USA, 19-22 June* (1996) 519-523
14. Vinod V. and Ghose S.: Growing nonuniform feedforward networks for continuous mapping. *Neural computing* **10** (1996): 55-69
15. Wang Z. and Zhu T.: An efficient learning algorithm for improving generalization performance of radial basis function neural networks. *Neural Networks* **13** (2000) 545-553
16. Zhu Q., Cai Y., and Liu L.: A global learning algorithm for a RBF network. *Neural Networks* **12** (1999) 527–540
17. Yong F. and Chow T.: Neural network adaptive wavelets for function approximation. *Intern report.* Department of Electrical Engineering. City University of Hong Kong (1996)

Dual-Source Backoff for Enhancing Language Models

Sehyeong Cho

MyongJi University, Department of Computer Science,
San 38-2 Yong In, KyungGi, Korea
shcho@mju.ac.kr

Abstract. This paper proposes a method of combining two n-gram language models to construct a single language model. One of the corpora is constructed from a very small corpus of the right domain of interest, and the other is constructed from a large but less adequate corpus. This method is based on the observation that a small corpus from the right domain has high quality n-grams but suffers from sparseness problem, while a large corpus from another domain is inadequately biased, but easy to obtain bigger size. The basic idea behind dual-source backoff is basically the same with Katz's *backoff*. We ran experiments with 3-gram language models constructed from newspaper corpora of several millions to tens of millions words together with models from smaller broadcast news corpora. The target domain was broadcast news. We obtained significant improvement by incorporating a small corpus around one thirtieth size of the newspaper corpus.

1 Introduction

Language modeling is an attempt to capture the regularities and make predictions, and is an essential component of automatic speech recognition. Statistical language models are constructed from large text, or corpus. However, it is not easy to collect enough spoken language text. Other corpora are readily available, but the difference in spoken and written language leads to a poor quality.

This granted, what we need is a way of making use of existing information to help lower the perplexity of the language model. However, simply merging two corpora will not help much, as we shall see later in the next section.

2 Related Work

Linear combination is probably the simplest way of combining two language models:

$$P_{combined}(w \mid h) = \sum_{k=1..n} \lambda_k P_k(w \mid h)$$

. Linear interpolation has the advantage of extreme simplicity. It is easy to implement, easy to compute. Linear combination is consistent as far as n-gram models are concerned.

Maximum entropy method[1] can be used to incorporate two information sources. For instance, suppose we had an *n*-gram model probability and a trigger pair model probability: $P(bank \mid in, the)$ and $P(bank \mid loan \in history)$. When both conditions are satisfied, then maximum entropy method can find a solution without sacrificing the consistency by imposing that the constraints are satisfied *on the average*.

However, if we had the same event space, then Maximum entropy method will result in trouble, because $E_{h \text{ ends in 'in the'}} [P_{combined}(bank \mid h)] = P_1(bank \mid in, the)$ and $E_{h \text{ ends in 'in the'}}[P_{combined}(bank \mid h)] = P_2(bank \mid in, the)$ should both be true. However they contradict because $P_1(bank \mid in, the) \neq P_2(bank \mid in, the)$.

Akiba [2] proposed using selective backoff. Their approach is similar to ours in that they use backoff with two different models. One of the models is probabilistic model and the other is a grammar network. The aim of their combination is to delete probabilities of all unnecessary *n*-grams, that is, those that are not possible word sequences according to the simpler grammar-based transition network.

Adaptation([3], for example) is a dynamic switching of language models based on the present situation. While adaptation focuses on dynamically detecting the shift among domains or topics, our problems deals with constructing a language model *per se* by using information from two models.

3 Proposed Approach

Define a *primary corpus* as a corpus from a domain of interest. A *secondary corpus* is a (relatively larger) corpus, from another domain. A *primary language model*, then, is a language model constructed from a primary corpus. A *secondary language model* is a language model constructed from a secondary corpus. C_1 is the primary corpus, and C_2 is the secondary corpus. P_1 denotes the probability obtained by maximum likelihood estimation from the primary corpus. $\overline{P_1}$ denotes a discounted primary probability. P_2 and $\overline{P_2}$ are likewise defined.

We measured perplexity[4] of spoken language models and written language models against broadcast script. With the same 3-gram hit ratio, the perplexity of the latter is almost twice higher (e.g., 17.5% / 560 / 970). Conversely, with similar perplexity, hit ratio of the primary model is lower. In other words, a small primary model has quality 30gram statistics, while a large, secondary model has simply more 3-grams. Then it would be nice if we had a way of combining the two.

Therefore given appropriate sizes, we may be able to take advantage of *n*-gram probabilities in both models. We assumed that the secondary corpus is at least one order of magnitude larger than the primary corpus. We also assumed primary 3-gram statistics best reflect the actual probability. Next are 3-gram(secondary), 2-gram(primary), 2-gram(secondary), 1-gram(primary), and 1-gram (secondary), in order.

The basic idea is this: the probability of a trigram x,y,z is calculated from primary 3-gram, if any. If it doesn't exist, use probability of a trigram x,y, z from secondary 3-grams, then primary 2-gram, and so on. In order to be consistent (i.e., sum of probabilities equal to 1), we do as follows, based on Katz's idea[5].

We first discount[6] the MLE probabilities of the non-zerotons. Let $\beta = 1 - \sum_{xyz \in C_1} \overline{P_1}(z \mid x, y)$. Then we redistribute the mass to zeroton 3-grams (i.e., the 3-gram xyz's, such that $xyz \notin C_1$), in proportion to either secondary 3-gram probability or primary 2-gram, as in equation 1.

$$\overline{P}(z \mid xy) = \begin{cases} \overline{P_1}(z \mid xy) & \text{if } xyz \in C_1 \\ \alpha_{xy} \overline{P_2}(z \mid xy) & \text{if } xyz \notin C_1, xyz \in C_2 \\ \alpha_{xy} \overline{P}(z \mid y) & \text{otherwise} \end{cases} \quad (1)$$

In the above formula, α_{xy} is a normalizing constant, defined by equation 2.

$$\alpha_{xy} = \frac{\beta}{\sum_{\substack{xyz \notin C_1 \\ xyz \in C_2}} \overline{P_2}(z \mid xy) + \sum_{\substack{xyz \notin C_1 \\ xyz \notin C_2}} \overline{P}(z \mid y)} \quad (2)$$

Unlike Katz's coefficients, there is no simple computation procedure for α_{xy}, and thus repeated summation is required, which took hours in a machine with two Xeon 2GHz processors. Fortunately, the calculation needs to be done only once and it need not be calculated in real-time.

The 2-gram probability $\overline{P}(z \mid y)$ and 1-gram probability are defined in a similar manner, as in equation 3 and equation 4, respectively.

$$\overline{P}(z \mid y) = \begin{cases} \overline{P_1}(z \mid y) & \text{if } yz \in C_1 \\ \alpha_y \overline{P_2}(z \mid y) & \text{if } yz \notin C_1, yz \in C_2 \\ \alpha_y \overline{P}(z) & \text{otherwise} \end{cases} \quad (3)$$

$$\overline{P}(z) = \begin{cases} \overline{P_1}(z) & \text{if } z \in C_1 \\ \alpha_0 \overline{P_2}(z) & \text{if } z \notin C_1, z \in C_2 \\ \alpha'_0 & \text{otherwise} \end{cases} \quad (4)$$

4 Results

We used CMU-Cambridge toolkit to construct secondary models in ARPA-format from newspaper corpora (Dong-A Ilbo news) from 4 million to 8 million words. We also constructed 4 primary models from SBS broadcast news corpora (100K to 400K words). Test corpus was a separate SBS broadcast news text of 10K size.

By simply mixing up primary and secondary models, we obtained 10 to 17 percent decrease in perplexity. With optimal mixing ratio by linear interpolation, additional 5 to 6% decrease was seen [7]. The result of the dual-source experiment showed around 30% decrease in perplexity. Considering that 20% decrease in perplexity shows notable increase in the accuracy of the speech recognizer, this can be regarded a meaningful result.

Table 1. Resulting Perplexity of interpolated model and dual-source backoff model

Size(1-ary/2ndary)	Linear Interpolation (1:1)	dual-source backoff
100K/4M	377	242
200K/5M	359	244
300K/6M	333	230
400K/8M	300	206

5 Discussion

The experiment clearly showed some improvement. However, it is not certain if this is indeed the optimal. As we discussed earlier the relative quality of the primary and the secondary n-grams depend on the corpora sizes. For instance, if the size of the primary corpus is very small compared to the secondary model, the secondary 2-gram probability may prove to be more reliable than the primary 3-gram. Lastly, the algorithm needs to be generalized to n-gram models of arbitrary n values. Theoretically, it is a straightforward generalization of equation 4 into equation 5:

$$\overline{P}(w_m | w_{1..n-1}) = \begin{cases} \overline{P_1}(w_m | w_{1..n-1}) & \text{if } w_{1..n} \in C_1 \\ \alpha_{w_{1..n-1}} \overline{P_2}(w_m | w_{1..n-1}) & \text{if } w_{1..n} \notin C_1, w_{1..n} \in C_2 \\ \alpha_{w_{1..n-1}} \overline{P}(w_m | w_{2..n-1}) & \text{otherwise} \end{cases} \quad (5)$$

However, the real problem is in determining the order of applications. This is not merely a theoretical problem, but a practical one, since it may well depend on the sizes of the corpora – relative or absolute – and also on the similarity among primary, secondary, and the test corpora.

Unigram case is a little more complicated. Actually in the experiment, dual-source backoff is not applied to unigrams (i.e., equation 4), because we found that it has negative impact. The reason is this. Usually, when unknown words appear in a text, they are mapped to a single symbol, sometimes referred to as UNK.(stands for unknown). For practical reasons in speech recognition, this is acceptable. If the speech recognizer hears a word which it doesn't know, it should say that it doesn't know it. However, for the purpose of comparing the perplexity – the measure of the quality of a language model – it is not fair. Replacing all "infrequent" words by a single symbol makes the probability of the word higher, thus making the perplexity lower that it actually is.. If one model recognizes the word and the other model doesn't, then the former should be deemed better.

In order to do justice with the complexity, we compare the perplexity of a language model using "complete" vocabulary, or a sufficiently large vocabulary. Given a language model, we assume the probability of an unknown word to be the total probability of unknown word divided by the total number of unknown word types. In effect, we are assuming the language model is using a backoff, by discounting the unigram model.

Now, the problem is in determining the discount ratio. The "exact" discount ratio is practically impossible to obtain, but we can obtain an approximate value by using a relatively large corpus. Since we are dealing with very small spoken language corpora, say, some tens or hundreds of thousands of words, a few million words will be enough for obtaining the probability of unknown words.

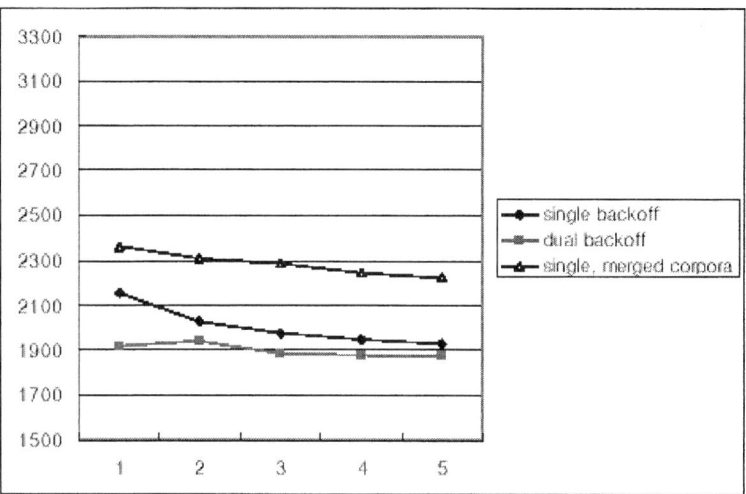

Fig. 1. Perplexity measures in unigram language models. X-axis indicates the size of the corpora in 100,000 words. The secondary is a 10M newspaper corpus

Given that, α' is computed as the ratio of appearance of words in the "purportedly infinite" corpus which were not in C_1, nor in C_2, divided by such word types. The result of applying this to a unigram-only dual-source backoff is shown in Figure 1 below[8]. The result in Table 1 does not count for the enhancement that can be obtained through applying 1-gram dual-source backoff. It is clear from Table 1 and Figure 1 that the combination of 1,2,3-gram dual-source backoff will bring further improvement, but the extent of the improvement needs to be seen.

We are still working on continued enhancement. Among the plans are: 1) combining 1,2, and 3-gram dual-source models, and 2) dual-source backoff with 4 and higher n-gram models.

Acknowledgement. This work was supported by grant R05-2003-000-11830-0 from the Basic Research Program of the Korea Science and Engineering Foundation.

References

1. Rosenfeld, R.: Adaptive Statistical Language Modeling: A Maximum Entropy Approach, Ph.D. dissertation, Carnegie-Mellon University (1994)
2. Akiba, T., Itou, K., Fujii, A., and Ishikawa, T.: Selective Backoff Smoothing for Incorporating Grammatical Constraints into the n-gram Language Model. In: Proc. International Conference on Spoken Language Processing (2002) 881-884
3. Chen, S. F., Seymore, K., and Rosenfeld, R.: Topic Adaptation for Language Modeling using Un-normalized Exponential Models. In: Proc. ICASSP'98, Vol. 2. (1998) 681-684
4. Jelinek, F., Mercer, R. L., Bahl, L. R., and J, K. B.: Perplexity – A Measure of the Difficulty of Speech Recognition Tasks. Journal of the Acoustics Society of America, 62, S63. Supplement 1, (1977)
5. Katz, S.M.: Estimation of Probabilities from Sparse Data for the Language Model Component of a Speech Recognizer. IEEE Transactions on Acoustics, Speech and signal Processing, vol. ASSP-35 (1987) 400-401
6. Good, I.J.: The Population Frequencies of Species and the Estimation of Population Parameters. Biometrica, vol.40, parts3,4 (1952) 237-264
7. Cho, S.: Experimental Results of Applying Dual-source Backoff. Technical Memo TM-CS-2004-9, Department of Computer Software, MyongJi University (2004)
8. Cho, S.: A Dual-course Back-off for Exploiting Large Text of Non-spoken Language Corpora in Constructing a Language Model for Speech Recognition: the Case of 1-gram. Tech memo TM-CS-2004-3, Department of Computer Software, MyongJi University (2004)

Use of Simulation Technology for Prediction of Radiation Dose in Nuclear Power Plant

Yoon Hyuk Kim and Won Man Park

School of Advanced Technology, Kyung Hee University,
Yongin-si, Gyeonggi-do, 449-701, Korea
yoonhkim@khu.ac.kr, muhaguy@naver.com
http://web.khu.ac.kr/~vbeslab

Abstract. In this study, a simulation program based on virtual environment concerning radiation work at nuclear power plants was developed. VRML and Java Applet were used to develop this simulation program. Radiation exposure levels were predicted through radiation work in a virtual reality environment, and high dose danger zones were visually represented by graphically visualizing dose rates so that it is possible to predict exposure rates of virtual workers. As a result of the simulation it turned out that the work traffic line that moved to the area where the dose rate concentric circle was light-shaded receives the lowest exposure dose radiation exposure level. According to the conventional method, the shortest work time or the shortest work path has the lowest radiation, but the result of the simulation showed that the exposure dose not depended on only work time, but also the distance between the worker and the radiation source.

1 Introduction

In this study the authors developed a simulation program based on virtual environment concerning radiation work at nuclear power plants. VRML and Java Applet were used to develop this simulation program. Radiation exposure levels were predicted through radiation work in a virtual reality environment, and high dose danger zones were visually represented by graphically visualizing dose rates so that it was possible to predict exposure rates of virtual workers.

2 Materials and Methods

For the 3D graphic work in this study a commercial graphic program and VRML 97 were used, which is the latest version of the VRML protocol. For modeling control of the VRML language Java Applet was used. VRML enables control through Java EAI. Both VRML and Java Applet are Web languages, and they are executed on the Inter-

---

This work was financially supported by MOCIE through IERC program.
Corresponding author : Yoon Hyuk Kim, School of Advanced Technology, Kyung Hee University, Korea, (Tel) +82-31-201-2028, (e-mail) yoonhkim@khu.ac.kr

net browser. The environment of the latest Java Applet program development is JDK(Java™ Development Kit) 1.4.2, but for interoperability with VRML through the Applet, JDK 1.1.8 was used in this study.

In the simulation program developed in this study, graphic simulation was performed through VRML, and the data resulting from simulations was displayed through the Java Applet instead of the VRML panel (Fig. 1).

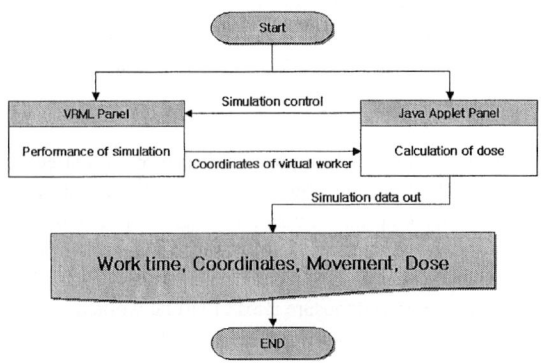

Fig. 1. Flow chart of virtual work simulation program

For first-person work VRML's 'NavigationInfo' was used, and for computation of the distance between the worker and the radiation source VRML's 'ProximitySensor' was inserted in the virtual work space, thus enabling the computation of the coordinates of the virtual worker. Java receives the coordinates obtained in VRML and calculates the exposure level, and then displays it in the Applet window. After the work is finished, it is outputted to a data file (Fig. 1). In the simulation program the 'Choice' command was used to apply 8 radioactive materials such as ^{192}Iridium, ^{51}Chromium, ^{60}Cobalt and ^{226}Radium, and 8 values such as 5, 10, or 40 to the radiation source and radioactivity concentration, and the program was developed in such a way that these values are applied simultaneously to two point isotropic radiation sources.

In case Java applet accesses a file in the computer, e.g. executing interconnected programs inside the simulation program and saving the simulation results, 'Security exception' occurs. This is a mechanism for maintaining the stability of the computer system. In this study the authors made arbitrary certificates to take care of the 'Security exception'. If a certificate is made and inserted in the HTML file, a window for accepting the certificate is displayed when the simulation program is executed through the Web browser. Acceptance of the certificate means that Java Applet is allowed to access files in the computer, thereby making it possible to save simulation results in a file and execute interconnected programs.

The work space for virtual work simulation was set up as illustrated in Fig. 2. Two identical point isotropic radiation sources (Source Point 1, Source Point 2 in Fig. 2) were assumed in a space 31m wide and 15m high. In a movement starting from the start point (Point 1 in Fig. 2), passing through the two points (Point 2 and Point 3 in

Fig. 2) and returning to the start point, the traffic line was varied to compare radiation exposure levels. Here, the assumption is that the worker is a point with his/her volume ignored, and the worker's movement velocity is constant.

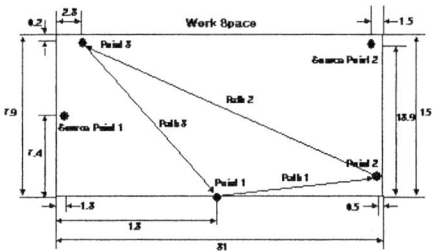

Unit [m]

Fig. 2. Virtual work space with 2 sources on 3 predefined paths

In computing the radiation exposure level in motion by given path is calculated by integrating the dose rate (E) dependent on the distance between the radiation source and the worker with respect to time.

3 Results and Discussion

The simulation program consists of two VRML panels and two Java Applet panels (Fig. 3). For the two VRML panels, the 'ViewPoint' of VRML is set up differently in each panel by means of one VRML file so that it is displayed in a different screen. The start and end of simulation is controlled by the 'Button Event' of the Java Applet, and results such as the work time for the performance of work simulation, the distance of movement, the current location, and the dose rate of the current location, and the exposure level are displayed in the Java Applet panel on a real-time basis, and they are saved in a data file after the completion of the simulation. To solve the 'Security Exception' that occurs at this time, the Java Singed Applet was used.

Movements in the virtual work space were executed by the arrow keys on the keyboard and the mouse. The 'NavigationInfo' of the VRML used to develop the simulation program enables forward movement and backward movement by means of the '↑ and ↓' buttons on the keyboard, and left and right rotation by means of the '← and →' buttons, thereby making it possible for the simulation performer to move about in the virtual work space. 'NavigationInfo' also supports movement by the drag of the mouse. In addition, 'TouchSensor' was inserted in the door to the virtual work space so that the simulation performer can click on the door to open it and enter the work space.

When a radiation source is selected, the radiation exposure level per unit time due to a certain gamma ray can be calculated, and accordingly the dose rate is visualized as a concentric circle shown in the Minimap (panel C) in Fig. 4 (b), (c), (d) so that the simulation performer can visually identify the location where the dose rate is low. Then, it was possible to find the path with a relatively low radiation exposure level, since the radiation exposure level was expressed by the integral of the dose rate with respect to time. After simulation is completed, the resulting data file is entered in such programs as Microsoft® Excel(Microsoft Corp. USA) so that the radiation exposure level per hour according to the worker's location and work traffic line can be reexamined. In calculating the radiation exposure level during simulation, the assumption was that the walls surrounding the work space and the door to the work space are completely shut off, with only the radiation exposure in the work space taken into consideration.

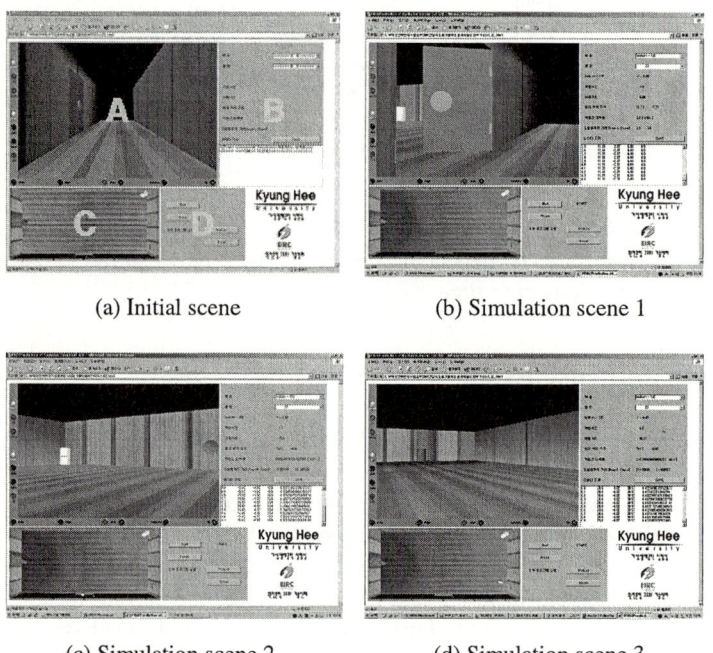

(a) Initial scene (b) Simulation scene 1

(c) Simulation scene 2 (d) Simulation scene 3

Fig. 3. Developed simulation program for calculation & visualization of dose. Panel A : simulation panel. Panel B : data input/output panel. Panel C : Minimap and dose rate distribution display panel. Panel D : simulation control panel

In this simulation 20 Ci ^{192}Ir was selected as the radiation source, and the worker entered the virtual work space through the door. Along the 4 paths such as the work path connecting the areas in low concentration on the Minimap, and the shortest work path, the worker passed the two points (Point 2 & Point 3) and exited the work space through the door.

The graphs in Fig. 4 represent the worker's movement path(left) during simulations with different traffic lines and the radiation exposure level and dose rate(right) between jobs.

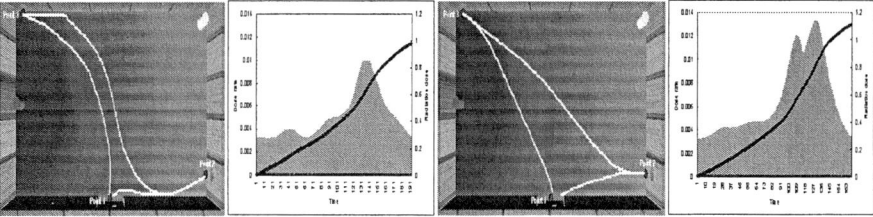

(a) Path 1 (The path through the areas with dose rates in low concentration on the Minimap)

(b) Path 2 (The shortest path)

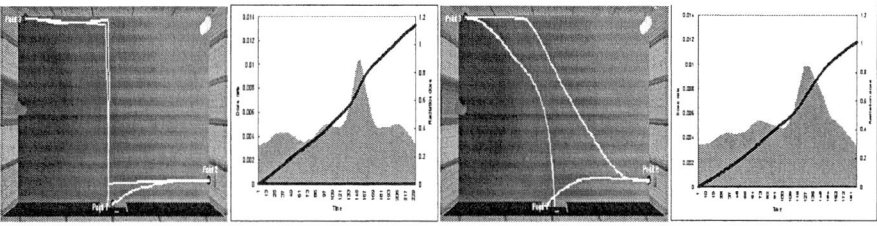

(c) Path 3 (Rectangular path)

(d) Path 4 (Hybrid path)

Fig. 4. Simulation result about dose variation by paths

Table 1. Radiation dose during movement along four different paths in simulation

Path No.	Time [sec]	Movement [m]	Velocity [ms^{-1}]	Dose(E) [mSv]	E/E$_{path1}$
1	19.1	69.9	3.66	$9.81*10^{-1}$	1.00
2	16.7	65.7	3.93	$1.11*10^{0}$	1.13
3	23.0	80.3	3.49	$1.14*10^{0}$	1.16
4	18.6	70.2	3.77	$1.01*10^{0}$	1.03

According to the previous method assuming that the dose rate of the work space is constant, the shortest work path is 13% (2.4 seconds) shorter than the work traffic line that passes the area where the concentric circle is light-shaded, so the exposure level must be lower, but the result of the simulation showed that the exposure level was 13% higher on the contrary. In the shortest movement (path 2), it was possible to reduce the distance of movement and work time, but as it approached the radiation source during movement, it recorded a relatively higher radiation exposure level than Path 1. In addition, path 3 with focus on the distance from the radiation source showed a high exposure dose due to increased work time.

The exposure dose received during work is a value determined by a combination of factors such as work time, and the distance between the worker and the radiation source according to the movement path, so reduction of work time alone cannot decrease the exposure dose. As the result of the simulation showed, the work path is an important factor that affects radiation exposure. It is an essential consideration when a work plan is formulated for optimal radiation protection. Therefore, the authors expect that it will be possible to secure a work traffic line capable of ensuring the best radiation protection through selection of an appropriate work path that takes the distance from the radiation source into consideration.

4 Conclusions

In this study the virtual reality technology was used to develop a radiation exposure level prediction and dose rate visualization simulation program. The simulation program based on VRML and Java Applets was developed to calculate the radiation exposure level of the worker during radiation work, and the dose rate was visualized by means of a concentric circle around the radiation source. To this end, the radiation exposure level computation expression was derived through numerical analysis, and Java EAI and Java Signed Applet were used.

By means of the dose rate distribution visualized by the concentric circle around the radiation source through the virtual work simulation program, high-dose danger zones were visually represented so that it was possible to predict work paths more favorable in terms of radiation protection. In addition, work simulation was performed to compute radiation exposure levels, and through graphic representation utilizing the file in which the results of simulation of work time, movement distance and radiation exposure levels are saved, work paths were visually verified, and problems of each path were identified. The results of simulation performance showed that exposure levels varied depending on work paths.

References

1. VRML 97, International Specification ISO/IEC IS 147772-1, http://www.vrml.org (1997)
2. Java Development Kit 1.4.2 , http://java.sun.com/j2se/1.4.2
3. Java Development Kit(JDKTM) 1.1.x-Signed Applet, http://java.sun.com/security/
4. ICRP, Recommendations of the ICRP, ICRP pub. 60, Pergamon Press (1991)
5. Fumizawa, M., Kameda, A., Nakagawa, T., Wu, W., Yoshikawa, H.: Development of Simulation-based Evaluation System for Iterative Design of Human-machine Interface in Nuclear Power Plant-Application for Reducing Workload. Nuclear Technology **141** (2003) 78-87
6. Hornaes, A., Hulsund, J. E., Végh, J., Major, C., Horváth, C., Lipcsei, S., Kapocs, G.: The EOP Visualization Module Integrated into the Plasma On-Line Nuclear Power Plant Safety Monitoring and Assessment System. Nuclear Technology **135** (2001) 123-130

A Numerical Model for Estimating Pedestrian Delays at Signalized Intersections in Developing Cities

Qingfeng Li, Zhaoan Wang, and Jianguo Yang

School of Electrical Engineering, Xi'an Jiaotong University, Xi'an 710049, China
qfli78@yahoo.com.cn
{zawang, yjg}@xjtu.edu.cn

Abstract. Considering the traffic situations in developing cities like Xi'an, China, a numerical model is developed to estimate pedestrian delays at signalized intersections. In the model, signal cycle is divided into a series of subphases, and each subphase lasts 1 second. The model estimates the average conflicting vehicle flow rate for each subphase firstly, and next estimates the average delay of pedestrians arriving during each subphase using probability theory and gap acceptance theory, and then estimates the overall average delay by aggregating the average delay of pedestrians arriving during each subphase. Finally, field data collected from a crosswalk at a signalized intersection in Xi'an are used to validate the numerical model, and the validation results indicate that the model is able to estimate pedestrian delays accurately.

1 Introduction

Pedestrian delay, which is defined as the additional travel time experienced by pedestrians, is the key performance index to evaluate intersections' level-of-service (LOS) for pedestrians [1]. Therefore, basing on the studies conducted in developed cities, several models for estimating pedestrian delays at signalized intersections have been developed [1-6]. However, at signalized intersections in developing cities like Xi'an, China, traffic situations are significantly different. Though vehicles usually comply with traffic signals, the signal noncompliance rates of pedestrians arriving during nongreen phases are usually quite high. And since vehicles start and run slower, and drivers get used to pedestrians' signal noncompliance, pedestrians usually can cross successfully during non-green phases.

Therefore, in this paper a numerical model is proposed to estimate pedestrian delays at signalized intersections in developing cities like Xi'an. The model divides signal cycle into a series of subphases, and each subphase lasts 1 second. It estimates the average conflicting vehicle flow rate for each subphase firstly, and then estimates the average delay for pedestrians arriving during each subphase using probability theory and gap acceptance theory, and finally obtains the overall average pedestrian delay by aggregating the average delay of pedestrians arriving during each subphase. And the model is validated using the field data collected from a crosswalk at a signalized intersection in Xi'an.

2 A Numerical Model

At signalized crosswalks, usually there are 3 kinds of phases for pedestrians: green phases, clearance phases and red phases. In China, a clearance phase usually includes two parts: the signal is flashing green during the first part, and the signal is all-red during the second part, i.e., both pedestrians and pedestrians' conflicting vehicles receive red signals.

At the beginning, pedestrians are divided into two groups: complying pedestrians and opportunistic pedestrians. Complying pedestrians are not willing to enter crosswalks during non-green phases even if there are acceptable gaps. Opportunistic pedestrians are willing to enter crosswalks during non-green phases if there are acceptable gaps. The average delay of complying pedestrians can be estimated accurately using the model proposed by Braun and Roddin [2], and then it can be integrated into the overall average delay if the percentage of complying pedestrians is known. Therefore, the key problem the numerical model needs to solve is the estimation of the average delay of opportunistic pedestrians.

And considering that walking directions might influence pedestrian delays at signalized intersections, the numerical model estimates respectively the delays of downward-to-upward (D2U) pedestrians, who encounter the downward vehicle flow firstly, and the delays of upward-to-downward (U2D) pedestrians, who encounter the upward vehicle flow firstly. Here the downward vehicle flow is the vehicle flow leaving an intersection, and the upward vehicle flow is the vehicle flow entering an intersection.

For the purpose of simplification, it is assumed that pedestrians do not influence each other; the walking speeds and the critical acceptable gaps of pedestrians arriving during the same subphase are the same; pedestrians decide wait or walk according to the vehicle flow faced, but the other vehicle flow does not influence pedestrians' decisions; pedestrians do not influence vehicles; the control scheme adopted by intersections is pre-timed two-phase control, which is the most popular control scheme in Xi'an; and there is no vehicle queue overflow.

2.1 Estimation of Conflicting Vehicle Flow Rates

Let q_U be the average upward vehicle flow rate and q_D be the average downward vehicle flow rate. And the algorithm for q_U and q_D estimation is as follows.

Firstly, the signal cycle is divided into two parts. For the upward vehicle flow, the first part includes the green phase, the flashing green phase, the all-red subphases, and the subphases when vehicles recognize signal, start and arrive at the crosswalk. And all the other subphases are included in the second part. For the downward vehicle flow, the first part includes the medium and latter subphases of the green phase, the flashing green phase, the all-red subphases, and the subphases when the vehicles from the opposite approach recognize signal, start and arrive at the crosswalk. And the second part includes the other subphases of the red phase and the former subphases of the green phase, when the remaining vehicles from the opposite approach cross the crosswalk.

Secondly, q_U and q_D for each subphase during the first part of the cycle are estimated. For the upward vehicle flow, if right-turn-on-red (RTOR) is permitted, and there is an exclusive lane for right-turn, right-turn vehicles cross the crosswalk uniformly during the first part of the cycle; or else, q_U is zero. For the downward vehicle flow, right-turn and left-turn vehicles from the other two approaches cross the crosswalk uniformly during the first part of the cycle.

Thirdly, q_U and q_D for each subphase during the second part of the cycle is estimated. For the upward vehicle flow, the second part of the cycle begins with saturation flow, and then q_U decreases gradually. The basic idea of the algorithm is to estimate the distribution of queue length, and then determine the probability of no queue during each subphase, and finally determine q_U for each subphase.

Assume vehicles arrive randomly with uniform distribution, let n be the number of vehicles arriving within a cycle, then n satisfies Poisson distribution:

$$P\{n = k\} = \frac{\lambda^k \exp(-\lambda)}{k!}, \quad k = 0, 1, 2, \ldots, \tag{1}$$

where λ is the mean of n.

Assume the newly arriving vehicle is also queued if there is a queue, then,

$$l = R_E \frac{n}{C} + \frac{nR_E}{Cs}\frac{n}{C} + \frac{n^2 R_E}{C^2 s^2}\frac{n}{C} + \cdots = \frac{snR_E}{Cs - n}, \tag{2}$$

where l is the queue length, s is the saturation flow rate, R_E is the effective red time for conflicting vehicles, and C is the signal cycle length.

And then, according to the correspondence between l and n, the probability distribution of l can be determined. Assume the queue dissipates with saturation flow rate, the probability of no queue during each subphase can also be determined. Next, assume the flow rate equals saturation flow rate when there is queue, and the flow rate equals arrival flow rate when there is no queue (if arrival flow rate is not greater than saturation flow rate), q_U for each subphase can be determined.

The algorithm to estimate q_D for each subphase during the second part of the cycle is a little bit different. At the beginning of the second part of the cycle, the through vehicles from the opposite approach cross the crosswalk, and the flow rate is the saturation flow rate of the opposite approach multiplied by the percentage of though vehicles at the opposite approach. And if RTOR is permitted, and there is an exclusive right-turn lane, the right-turn vehicles from another approach cross the crosswalk uniformly.

2.2 Estimation of Pedestrian Delays

After the determination of q_U and q_D for each subphase, the probability of no vehicle during each subphase can be estimated. Since the average headway is about 2 seconds for saturation vehicle flow at a lane, and a subphase lasts only 1 second, it is reasonable to assume that at most one vehicle crosses the crosswalk during a subphase at a lane. And then, according to the classical probability model,

$$p = (1 - q/N)^N, \quad q/N < 1, \tag{3}$$

where p is the probability of no vehicle within the subphase, N is the number of lanes for the vehicle flow, q is the vehicle flow rate.

Next, the average delay for pedestrians arriving during each subphase can be estimated. Pedestrian critical acceptable gap and walking speed are two key parameters here. During the latter part of green phases, flashing green phases, and the former part of red phases, to avoid big delays, pedestrian critical acceptable gaps are smaller, and walking speeds are higher. These subphases are named special subphases.

For pedestrians arriving during the m-th subphase, according to the classical probability model, the probability distribution of delays received during the crossing of one vehicle flow is,

$$P\{d=0\} = \frac{p(m)p(m+1)...p(m+[t_c]-1)}{(1-(t_c-[t_c])+(t_c-[t_c])p(m+[t_c]))},$$
$$P\{d=1\} = \frac{(1-P\{d=0\})p(m+1)p(m+2)...p(m+[t_c])}{(1-(t_c-[t_c])+(t_c-[t_c])p(m+[t_c]+1))},$$
$$P\{d=2\} = \frac{(1-P\{d=0\}-P\{d=1\})p(m+2)p(m+3)...p(m+[t_c]+1)}{(1-(t_c-[t_c])+(t_c-[t_c])p(m+[t_c]+2))},$$
$$......$$
(4)

where t_c is the critical acceptable gap, and $[t_c]$ is the integral part of t_c. And similar method can also be used to determine the probability distribution of the delays received during the crossing of the other vehicle flow. And finally the average delay for the pedestrians arriving during the subphase can be determined.

Finally, the overall average delay of D2U and U2D pedestrians can be estimated respectively, by averaging the average delay of pedestrians arriving during each subphase. Non-random arrivals can be considered if pedestrian arrival patterns are known. And if the percentage of complying pedestrians is known, their delays can also be considered in the overall average delay estimation.

3 Model Validations

The data used in model validations were collected from the crosswalk on the eastern leg of the intersection of West Xianning Road and South Xingqing Road in Xi'an, China. The phase plan adopted by the intersection is pre-timed two-phase control, and RTOR is permitted at the intersection. The signal cycle length is 123 seconds. For the crosswalk, there are 45 seconds for the green phase, 2 seconds for the flashing green phase, and 76 seconds for the red phase. And the red phase for conflicting vehicles begins 2 seconds ahead of the green phase for pedestrians, but ends 2 seconds later.

To decrease data amount requirements, the subphase in the field study is longer than the subphase in the numerical model. The pedestrian green phase is divided into 5 subphases (PG1-PG5), 10 seconds each, except that 7 seconds for PG5 (the 2-second flashing green phase is included in PG5). The pedestrian red phase is divided into 8 subphases (PR1-PR8), 10 seconds each, except that 6 seconds for PR1.

In field data collection, firstly 5 hours were chosen to take videos in the tall buildings near the intersection. The hours were selected under the condition that the weather is sunny or cloudy, during 3:00 p.m. to 5:00 p.m., from July 7 to July 10, 2003 (all the days are ordinary workdays). And then, playing back the videos, the delay and the arriving subphase of each pedestrian were measured. Totally 1050 D2U pedestrians and 634 U2D pedestrians were recorded.

The major parameters of the model are estimated as follows. For the upward vehicle flow, the first part of the cycle includes subphase 1-55, the second part includes subphase 56-123; for the downward vehicle flow, the first part includes subphase 11-65, and the second part includes subphase 66-123 and subphase 1-10. The critical acceptable gap is 4.2 seconds for D2U pedestrians, 4.5 seconds for U2D pedestrians during usual subphases, and 3 seconds for both during special subphases. And the time needed to cross one vehicle flow is 6 seconds during usual subphases, and 3 seconds during special subphases. For U2D pedestrians, the special subphases are subphase 43-52; for D2U pedestrians, the special subphases are subphase 43-57. Besides, all pedestrians are assumed to be opportunistic.

Model inputs are mainly average vehicle flow rates and saturation flow rates, which are set using the data collected within about 30 minutes. As the upward vehicle flow is concerned, the average vehicle flow rate is 0.32 veh/sec, and no vehicle crosses the crosswalk during green phases. And the saturation flow rate for the upward vehicle flow is 0.64 veh/sec. As the downward vehicle flow is concerned, the average through vehicle flow rate from the opposite approach is 0.19 veh/sec, the average left-turn vehicle flow rate is 0.03 veh/sec, and the average right-turn vehicle flow rate is 0.12 veh/sec. And the highest vehicle flow rate from the opposite approach is 0.44 veh/sec.

And then, after implementing the numerical model using MATLAB, estimations of pedestrian delays are obtained by executing the model, and then compared with field measurements to perform validations, as show in Fig. 1 and Fig. 2. Since each subphase lasts only 1 second in the numerical model, but 6-10 seconds in the field study, the estimated average delay used in comparison is actually the mean of estimated average delays of pedestrians arriving during several subphases.

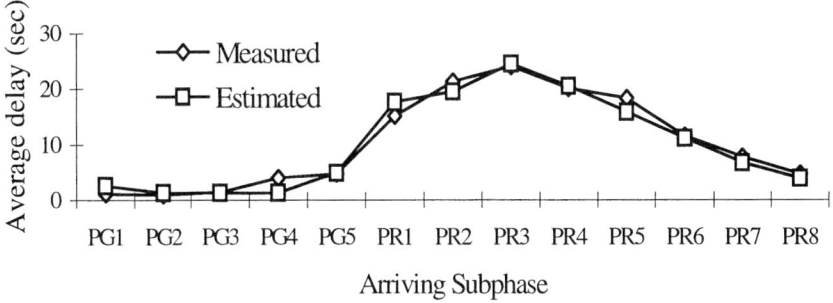

Fig. 1. Comparison of measured and estimated average delay of D2U pedestrians

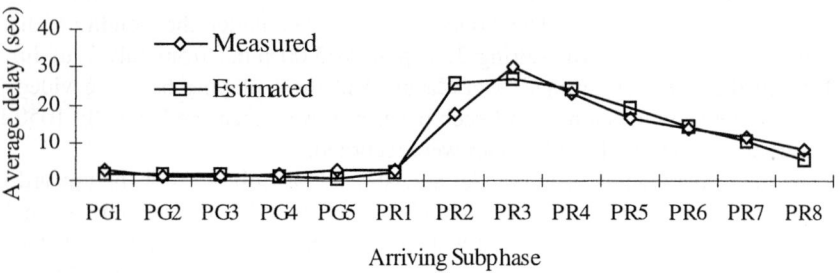

Fig. 2. Comparison of measured and estimated average delay of U2D pedestrians

The comparison results indicate that the numerical model is able to estimate the average delay of pedestrians arriving during each subphase with satisfactory accuracy. Therefore, if pedestrian arrival patterns are known, the numerical model is also able to estimate overall average delay accurately.

4 Conclusions

Considering the traffic situations in developing cities like Xi'an, a numerical model is proposed to estimate pedestrian delays at signalized intersections. And the model is validated using the field data collected from a crosswalk at a signalized intersection in Xi'an, and the validation results indicate that the model is able to estimate pedestrian delays accurately.

References

1. Transportation Research Board: Highway Capacity Manual. National Research Council, Washington, D.C. (2000)
2. Braun, R., Roddin, M.: Quantifying the Benefits of Separating Pedestrians and Vehicles. National Research Council, Washington, D.C. (1978)
3. Rouphail, N., Hummer, J., Milazzo II, J., Allen, P.: Capacity Analysis of Pedestrian and Bicycle Facilities: Recommended Procedures for the "Pedestrians" Chapter of the Highway Capacity Manual. U.S. Department of Transportation, Washington, D.C. (1998)
4. Virkler, M.: Pedestrian Compliance Effects on Signal Delay. Transport. Res. Rec. 1636 (1994) 88-91
5. Pretty, R.: The Delay to Pedestrians and Vehicles at Signalized intersections. ITE J. 49 (1979) 20-23
6. Hunt, J.G., Al-Neami, A.H.K.: Effectiveness of Pedestrian Facilities at Signal Controlled Junctions. P. I. Civil Eng.-Transp. 111 (1995) 268-277

Feature Selection with Particle Swarms

Yu Liu[1], Zheng Qin[1,2], Zenglin Xu[1], and Xingshi He[3]

[1] Department of Computer Science, Xian JiaoTong University,
Xian 710049, P.R. China
liuyu@mailst.xjtu.edu.cn
http://www.psodream.net
[2] School of Software, Tsinghua University, Beijing 100084, P.R. China
[3] Department of Mathematics, Xian University of Engineering Science and
Technology, Xian 710048, P.R. China

Abstract. Feature selection is widely used to reduce dimension and remove irrelevant features. In this paper, particle swarm optimization is employed to select feature subset for classification task and train RBF neural network simultaneously. One advantage is that both the number of features and neural network configuration are encoded into particles, and in each iteration of PSO there is no iterative neural network training sub-algorithm. Another is that the fitness function considers three factors: mean squared error between neural network outputs and desired outputs, the complexity of network and the number of features, which guarantees strong generalization ability of RBF network. Furthermore, our approach could select as small-sized feature subset as possible to satisfy high accuracy requirement with rational training time. Experimental results on four datasets show that this method is attractive.

1 Introduction

Feature selection is widely used as the first stage of classification task to reduce the dimension of problem, decrease noise, improve speed and relieve memory constraints by the elimination of irrelevant or redundant features. Basically, there are two types of approaches for feature selection [1,2,3] : filter approach where feature selection process occurs before and is independent to learning algorithms, and wrapper approach where feature selection is seen as a subroutine and feature subsets are evaluated by the learning algorithms. The wrapper approach is generally better because the search of optimum is guided by the learning algorithm in order to get good measures of accuracy and efficiency [4]. Genetic algorithms are often used to select feature subsets. Using genetic algorithm to perform feature selection requires many iterations. If the learning algorithm is an iterative one, the computational cost of the whole process would be very expensive, since in each iteration learning algorithm is used to evaluate the fitness for every individual. Most learning algorithms for neural networks are iterative methods. Therefore, it is infeasible to directly adopt these learning algorithms in feature selection for neural networks because of too expensive computational cost. The combination of feature selection and neural network classifier as wrapper approach is a challenge task. Yang et. al [5] used a fast constructive neural network learning algorithm called DistAL, which is not iterative, to build the neural network classifier for the evaluation of feature selection. But the no-iterative learning method [6] can just train multiple layer networks of threshold logic units. Brill et. al [7] employed nearest-neighbor error estimation for feature selection due to spending less time compared with the corresponding neural network estimation. Subsequently, a counterpropagation network (CPN) was constructed with only the selected features as inputs. The success of this method strongly depends on the close relationship of nearest-neighbor classifier and counterprogagation neural network. The above two methods can just work on special neural networks. This paper proposes a novel method to resolve the problem of feature selection for neural network classifier as wrapper approach. The new method is that feature selection and neural networks training are processed at the same time by Particle Swarm Optimization (PSO), which do not depends on special neural networks. This method is simple and practical because there is no iterative sub-algorithm in each iteration of PSO algorithm.

The rest of this paper is organized as follows: Section 2 gives the description of PSO algorithm.Section 3 presents our PSO solution to the combination of feature selection and neural network training. Section 4 presents and analyzes the experimental results on 4 UCI data sets. Finally, Section 5 gives conclusions.

2 Particle Swarm Optimization

Particle Swarm Optimization (PSO), a new population-based evolutionary computation technique inspired by social behavior simulation, was first introduced in 1995 by Eberhart and Kennedy [8]. PSO is an efficient and effective global optimization algorithm, which has been widely applied to nonlinear function optimization, neural network training, and pattern recognition. In PSO, a swarm consists of N particles moving around in a D-dimensional search space.The position of the ith particle at t iteration is represented by $X_i^{(t)} = (x_{i1}, x_{i2}, \ldots, x_{iD})$ that are used to evaluate the quality of the particle. During the search process the particle successively adjusts its position toward the global optimum according to the two factors: the best position encountered by itself (pbest) denoted as $P_i = (p_{i1}, p_{i2}, \ldots, p_{iD})$ and the best position encountered by the whole swarm (gbest) denoted as $P_g = (p_{g1}, p_{g2}, \ldots, p_{gD})$. Its velocity at t iteration is represented by $V_i^{(t)} = (v_{i1}, v_{i2}, \ldots, v_{iD})$. The position at next iteration is calculated according to the following equations:

$$V_i^{(t)} = \lambda(\omega * V_i^{(t-1)} + c_1 * rand() * (P_i - X_i^{(t-1)}) + c_2 * rand() * (P_g - X_i^{(t-1)})) \quad (1)$$

$$X_i^{(t)} = X_i^{(t-1)} + V_i^{(t)} \quad (2)$$

where c_1 and c_2 are two positive constants, called cognitive learning rate and social learning rate respectively; $rand()$ is a random function in the range $[0, 1]$;

w is $inertia factor$; and λ is $constriction factor$. In addition, the velocities of the particles are confined within $[Vmin, Vmax]^D$. If an element of velocities exceeds the threshold $Vmin$ or $Vmax$, it is set equal to the corresponding threshold.

3 Feature Selection with PSO

3.1 Feature Selection for Neural Network

Neural network offers an attractive framework for the design of trainable pattern classifiers for real-world pattern classification tasks. In order to remove the redundant or irrelevant features, feature selection is often used as the first stage of classification task for neural network classifiers. In a neural network classifier, the number of features determines the number of neurons in input layer. Therefore, if different features are selected, subsequent neural network must be trained according with the corresponding training set determined by the selected features.

In this paper, we propose that feature selection and neural network training are done simultaneously in each iteration. We formulated the combination of feature selection and neural network as an optimization problem and employed PSO algorithm to resolve it. RBF neural network was employed as classifier in this paper. PSO algorithm is a population based iterative global optimization algorithm. In each iteration, each particle consists of two components: feature subset and neural network configuration. The former encodes the feature subset and determines the training set and the number of neurons in input layer of RBF network, while the latter determines the number of neurons in hidden layer and corresponding parameters of RBF neural network. The resulting RBF neural network is measured on the training set formed by feature selection to evaluate the fitness of a particle. Fitness function of PSO takes into account three measures: mean squared error (MSE) between network outputs and desired outputs, the complexity of networks, and the number of features. Therefore, the swarm evolves toward the following configuration: minimal number of features and the corresponding RBF that has not only minimal MSE but also appropriate network architecture. Thus the resulting RBF network has strong generalization ability. Moreover, in each iteration of PSO algorithm, there is no iterative process to train the RBF neural network, so it saves a great deal of time. Depending on the powerful evolutionary mechanism of PSO, feature subset and RBF neural network evolve to better configuration in next iteration simultaneously.

3.2 Encoding Feature Selection and Neural Network

Traditional methods just encode features, while we encode features and subsequent neural network at the same time. Therefore, the position of a particle is represented by concatenation of feature subset and neural network as shown in Figure 1. Here $SFlag_i$ ($1 \leq i \leq fmax$) indicates whether or not the i-th feature is selected. If $SFlag_i > 0$, then the i-th feature is selected, else not selected. The encoding neural network is described in detail. RBF network is three-layer

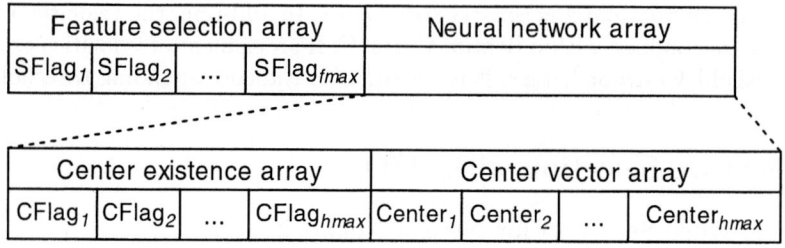

Fig. 1. Structure of a particle

feedback network. The number of neurons in hidden layer determines the complexity of networks. Too few hidden neurons lead to low classification accuracy due to limited flexibility, while too many result in poor generalization ability since the noise in the training data is fitted. The number of input neurons is variable, which is determined by selected features, so it is impossible to specify the number of hidden neurons in advance. The hidden layer is evolved during the search. $CFlag_i$ ($1 \leq i \leq hmax$) indicates whether or not the i-th hidden neuron is involved into the network. If $CFlag_i > 0$, the i-th hidden neuron is included in the network. Otherwise the i-th hidden neuron is removed from the network. Since the performance of RBF networks mainly depends on the centers of hidden neurons, we just encode the centers into the particle for stochastic search. Then the widths of hidden neurons and weights between hidden layer and output layer are determined by heuristic methods and analytic methods, respectively [9].

Designing Fitness Function. In the way mentioned above, in every iteration particles can be interpreted into networks with variable number of hidden neurons. According to the following formula, the resulting network is measured on the training set that is determined by feature selection.

$$Fitness = k_1 \cdot \frac{NSelectedFeature}{NmaxFeature} + k_2 \cdot \frac{1}{PN} \sum_{p=1}^{PN} \|t_p - o_p\|^2 + k_3 \cdot \frac{Nhidden}{Nmaxhidden} \quad (3)$$

where k_i ($i = 1, 2, 3$) are constant and indicate the weight of each component respectively; $NSelectedFeature$ indicates the number of features selected and $NmaxFeature$ indicates the total number of features; PN input-target patterns are given; t_p and o_p are the desired output and network output for pattern p respectively; $Nhidden$ is the number of hidden units involved in networks; $Nmaxhidden$ is predefined as the max number of hidden units.

4 Experiments

Four data sets from UCI Repository of machine learning databases [10] were used in our experiments. These data sets are image segmentation, page-blocks, ionosphere and wine. The properties of these four data sets were described in Table 1. Two training schemes were compared. One was PSO algorithm without feature selection [9], and the other was PSO algorithm with feature selection, which was proposed in this paper. The parameters of PSO algorithm were set as follows: weight w decreasing linearly between 0.9 and 0.4, learning rate $c_1 = c_2 = 2$, and $Vmax = 1$. Max iterations were set to 2000 iterations. All the experiments were conducted 10 runs. In each experiment, each data set was randomly divided into two parts: 2/3 as training set and 1/3 as test set. As each feature was normalized as the following formula, normalized values were in range [-1,1].

$$Normalized\ value = \frac{Original\ value}{|\ max(orinial\ values\ of\ all\ instances)\ |} \qquad (4)$$

Table 1. The information of data sets and results of two training schemes for different data sets

	Wine	Ionosphere	Image segmentation	Page-blocks
Data sets information				
Instances	178	351	300	5473
Features	13	34	30	10
Feature type	Numeric	Numeric	Numeric	Numeric
Classes	3	2	7	5
RBF Without feature selection				
Hidden neurons	15	42	40	15
TrainCorrect	0.9995	0.9605	0.9661	0.9474
TestCorrect	0.9688	0.9012	0.9086	0.9121
RBF With feature selection				
Selected features	4	8	5	2
Hidden neurons	12	20	32	5
TrainCorrect	0.9815	0.9707	0.9736	0.9255
TestCorrect	0.9488	0.9209	0.9116	0.9073

The results were listed in Table 1. TrainCorrect and TestCorrect refer to mean correct classification rate averaged over 10 runs for the training and test set, respectively. The number of selected features and hidden neurons achieved for each data set were averaged over 10 runs, and then were round off to integer. As can be seen from Table 1, the percentage of the number of selected features to the total number of features in each dataset is relatively small. Although, many features are removed from the original feature set, the mean classification rate still remains high. Moreover, feature selection improves the correct classification rate on date sets, Ionosphere and Image segmentation.

5 Conclusions

In this paper, we have proposed a new feature selection method, in which PSO is employed. RBF neural network is used as the classifier to evaluate the accuracy of selected features. The PSO algorithm simultaneously considers three measures as the fitness function: MSE, the complexity of networks and the number of features, guaranteeing strong generalization ability of RBF network. Experimental results on the datasets from UCI repository show that this method effectively selected features with high accuracy as well as small feature subset size. Furthermore, this method is efficient because it encodes not only feature subset but also neural network configuration in each iteration of PSO algorithm, and hence avoiding time-consuming process to train neural network in each iteration.

References

1. Blum, A., Langley, P.: Selection of relevant features and examples in machine learning. Artificial Intelligence **97** (1997) 245–271
2. John, G.H., Kohavi, R., Pfleger, K.: Irrelevant features and the subset selection problem. In: International Conference on Machine Learning. (1994) 121–129
3. Kohavi, R., John, G.H.: Wrappers for feature subset selection. Artificial Intelligence **97** (1997) 273–324
4. Martin-Bautista, M.J., Villa, M.: A survey of genetic feature selection in mining issues. In Angeline, P.J., Michalewicz, Z., Schoenauer, M., Yao, X., Zalzala, A., eds.: Proceedings of the Congress on Evolutionary Computation. Volume 2., IEEE Press (1999) 1314–1321
5. Yang, J., Honavar, V.: Feature subset selection using a genetic algorithm. In Koza, J.R., Deb, K., Dorigo, M., Fogel, D.B., Garzon, M., Iba, H., Riolo, R.L., eds.: Genetic Programming 1997: Proceedings of the Second Annual Conference, Stanford University, CA, USA, Morgan Kaufmann (1997) 380
6. Yang, J., Parekh, R., Honavar, V.: Distal: An inter-pattern distance-based constructive learning algorithm. Intelligent Data Analysis **3** (1999) 55–73
7. Brill, F.Z., Brown, D.E., Martin, W.N.: Fast genetic selection of features for neural network classifiers. IEEE Transactions on Neural Networks **3** (1992) 324–328
8. Kennedy, J., Eberhart, R.: Particle swarm optimization. In: Proceeding of IEEE International Conference on Neural Networks (ICNN'95). Volume 4., Perth, Western Australia, IEEE (1995) 1942–1947
9. Liu, Y., Qin, Z., Shi, Z., Chen, J.: Training radial basis function networks with particle swarms. In: Proceeding of the IEEE International Symposium Neural Networks. Lecture Notes in Computer Science, Springer (2004)
10. Blake, C., Merz, C.J.: UCI repository of machine learning databases, http://www.ics.uci.edu/~mlearn/MLRepository.html (1998)

Influence of Moment Arms on Lumbar Spine Subjected to Follower Loads

Kyungsoo Kim[1] and Yoon Hyuk Kim[2]

[1] Impedance Imaging Research Center, Kyung Hee University,
Yongin-si, Gyeonggi-do, 449-701, Korea
azureton@kaist.ac.kr
[2] School of Advanced Technology, Kyung Hee University,
Yongin-si, Gyeonggi-do, 449-701, Korea
yoonhkim@khu.ac.kr

Abstract. In this paper, an influence of moment arms on the lumbar spine subjected to the follower load was investigated. A two-dimensional finite element model of the lumbar spine including idealized trunk muscles was developed to evaluate the muscle force activation generating the follower load when the lengths of moment arms varied. The follower forces, the shear forces and the resultant joint moments were also calculated to confirm the generation of the follower load. Finally, the dependence of the lengths of moment arms on the deformed shape of the lumbar spine model concerning the spinal stability was examined.

1 Introduction

It has been shown by several past literatures e.g. [1-3] that the human lumbar spine can maintain its stability when external loads are transmitted along the path that approximates the tangent to the curve of the spinal column, called the follower load path. The resultant compressive loads described above are called the follower loads. The trunk muscles were assumed to have an essential role in generating the follower loads. However, it is quite difficult to measure the magnitudes of muscle forces experimentally, so that the computational investigations are attempted [4-6]. It was verified with simplified anatomical data that there were activations of the human trunk muscles generating the follower loads and maintaining the spinal stability when physiologically meaningful external forces and moments were applied to the spine. Thus, it is necessary to study the dependence of the simplified anatomical data on the response of the lumbar spine, especially the stability.

In this paper, the influence of the distance as a moment arm between the body center and the attachment point of a muscle in each vertebra on the lumbar spine subjected to the follower load was investigated.

This study was partly supported by RRC program of MOST and KOSEF.
Corresponding author : Yoon Hyuk Kim, School of Advanced Technology, Kyung Hee University, Korea, (Tel) +82-31-201-2028, (e-mail) yoonhkim@khu.ac.kr

2 Materials and Methods

A two-dimensional finite element model of a human lumbar spine in the frontal plane with a laterally flexed standing posture was developed. The response of the whole lumbar spine (L1-S1) subjected to the follower load and the activations of muscles on the lumbar spine were investigated as in [5, 6]. A global coordinate system in the frontal plane was defined, with the origin at the center of the first sacral vertebra, the x-axis pointing horizontally to the right, and the y-axis pointing upwards. Five node points of finite elements identified with the vertebral body centers of L1-L5 using geometric data in [4-6] (Fig. 1). Each finite element was assumed to have elastic stiffness properties given in [7, 8].

As shown in Fig. 1, a vertical load (P) was applied at the top body center (L1) to simulate an external load including an external force and an external moment. (F1-F10) denoted by unknown muscle forces applied at five nodes along the idealized lines of action of muscles whose fascicles had a distal attachment in the sacrum/pelvis region and a proximal attachment near the vertebral body centers. Muscles were assumed to be activated statically. The distances between the vertebral body centers and the proximal attachment points of ten muscles were regarded as perpendicular moment arms. In this model, it was supposed that all the length of the moment arms were identical while L denoted the length. The length L of moment arms were assumed to vary from 0mm to 30mm in order to examine the influence of the length of moment arms on the lumbar spine model subjected to the follower load.

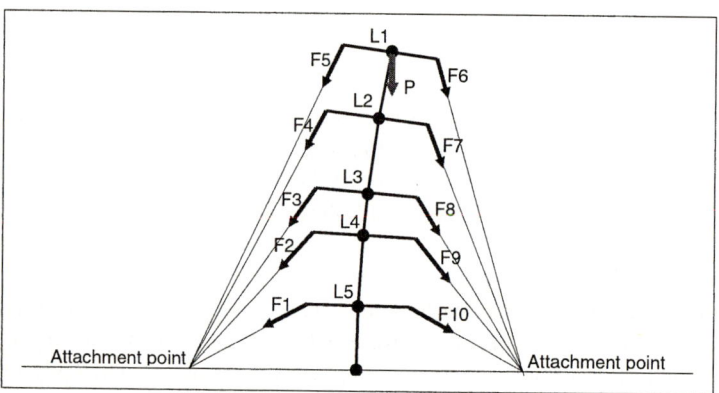

Fig. 1. Two-dimensional finite element model of the lumbar spine

A follower load path was defined as having a direction parallel to the bisection of two adjacent segments at each node. A local coordinate system at each node in the frontal plane could then be defined with the origin at the node that is the vertebral body center. The x-axis is in the opposite direction to the follower load path, i.e., the follower direction, and the y-axis is in the perpendicular direction to the x-axis, or the shear direction. At each node, the resultant joint force could be decomposed into two force components, one in the follower direction and one in the shear direction. The former is called as the follower force and the latter the shear force.

The lumbar spine model is in static equilibrium when all the forces and moments at each node equal zero. The forces and moments used in the analysis consist of muscle forces and moments derived by muscle forces, motion segment forces and moments by the stiffness property, as well as applied external forces and moments as in [5, 6, 8]. The total static equilibrium equations at each vertebral segment are as follows:

$$\mathbf{F}_m - \mathbf{K} \cdot \mathbf{d} + \mathbf{F}_{ext} = 0 \quad (1)$$

where $\mathbf{F}_m$ represents the muscle forces and the moments activated by muscles, $\mathbf{F}_{ext}$ the external forces and moments acting on the five lumbar vertebrae, $\mathbf{K}$ the global stiffness matrix of motion segments in the lumbar spine, and $\mathbf{d}$ the translations and the rotations at each node, respectively. In this analysis two translations and a rotation at each vertebra and ten muscle forces were considered as state variables. The number of unknowns that consisted of magnitudes of muscle forces and displacements exceeded that of equations describing the equilibrium state of the lumbar spine model so that the optimization technique is necessary to solve this redundant problem. In this study, the quadratic sequential programming method was utilized to solve this optimization problem using MATLAB®(MathWorks Inc., USA).

The optimization problem minimizing the sum of follower forces was formulated allowing no shear forces as follows:

$$\text{Minimize } f = \sum_{i=1}^{n_e} \left| F_f^i \right| \quad (2)$$

$$\text{subject to } \left| F_s^i \right| = 0 \text{ , } i = 1, \ldots, n_e$$

$$\text{and } \mathbf{F}_m - \mathbf{K} \cdot \mathbf{d} + \mathbf{F}_{ext} = 0$$

where F_f^i denoted the follower force at the i-th node, $i = 1, \ldots, n_e$, F_s^i the shear force at the i-th node, $i = 1, \ldots, n_e$, n_e the number of elements. Here $n_e = 5$. In the optimization problem, the sum of follower forces was minimized in order to reduce the intervertebral disk stresses. Two loading cases were tested for various values of L, L=0, 10, 20, 30 mm: (a) Case 1: P=350N and 3.5Nm (the clockwise direction is positive), (b) Case 2: P=350N and 0.0Nm.

In this paper, the muscle force activations generating the follower load were calculated for the variation of the length L of moment arms under both loading cases. The corresponding resultant joint forces (the follower forces and the shear forces) and the resultant joint moments, and the deformed shapes of the model were also derived. Finally, tilts of elements, which were angles between the vertical axis and the elements, were investigated for each L and their changes with the variation of L were examined in order to verify the influence of the length of moment arms on the deformed shapes.

3 Results and Discussion

Muscle force activations satisfying constraints of the formulated optimization problem could be obtained for various L under both loading cases. Five muscle forces (F1, F2, F3, F4, F6) were zero for every L in each loading case and the others were presented in Fig. 2. It was verified that all the shear forces were eliminated at every node regardless of the length L and the load case. This indicated that the calculated muscle force activations generated the follower load. While the length L of moment arms changed, the magnitudes of activated muscle forces did not vary so much at each vertebra. The follower forces corresponding to those muscle force combinations also had little variations (Fig. 3). These observations implied that muscle force activations and the follower forces at all vertebrae were less dependent on the length of the moment arms.

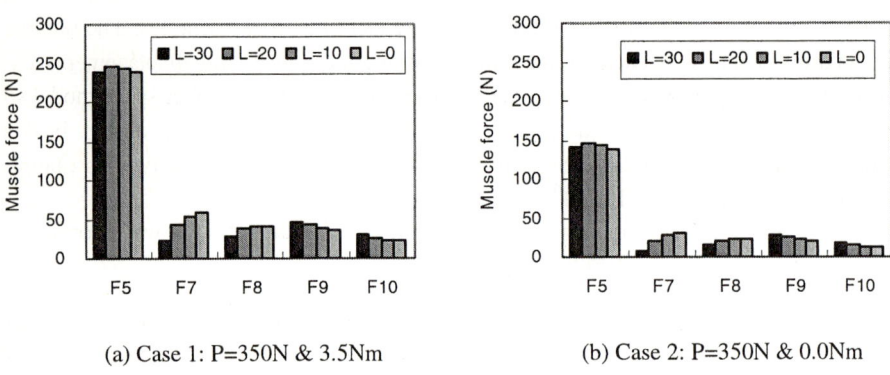

(a) Case 1: P=350N & 3.5Nm   (b) Case 2: P=350N & 0.0Nm

Fig. 2. Nonzero muscle force activations for various L

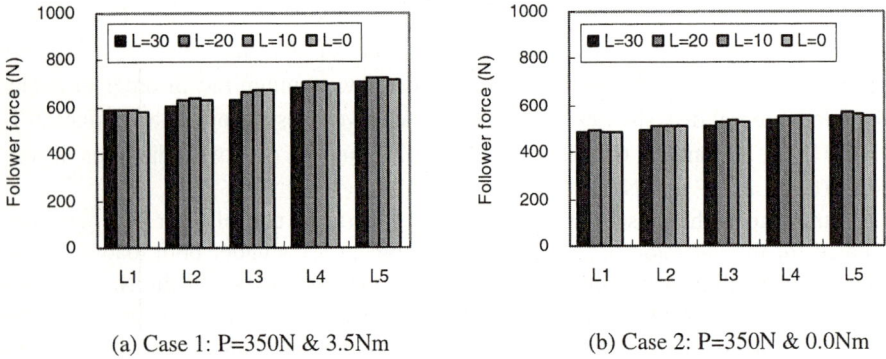

(a) Case 1: P=350N & 3.5Nm   (b) Case 2: P=350N & 0.0Nm

Fig. 3. Follower forces subjected to the follower load for various L

However, the resultant joint moments at L1 were affected by the length of moment arms in both loading cases though joint moments at the other vertebrae changed a little (Fig. 4). There were tendencies that the amounts of joint moments decreased as the length of the moment arms reduced. These phenomena would be observed since the moment arm is directly used to calculate moments in opposite to forces. Thus, attachment points of muscles need to be investigated carefully in order to estimate the resultant joint moment precisely.

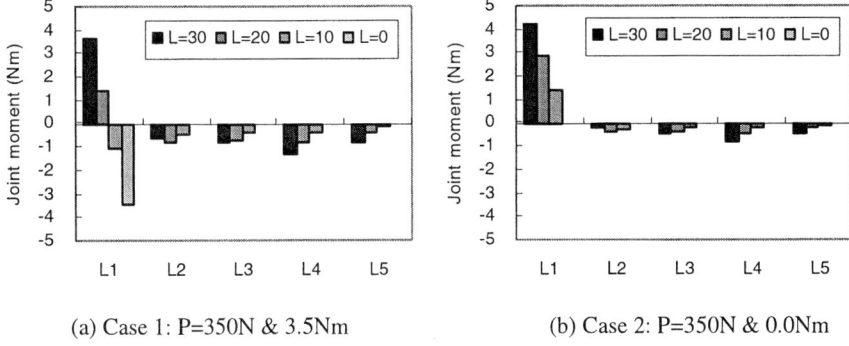

Fig. 4. Resultant joint moments subjected to the follower load for various L

The deformed shapes of the finite element model were close to the initial shape regardless of the length L of moment arms and the loading case. This meant that the stability of the model could be preserved under the follower load as shown in the previous researches. However, the tilt of each element decreased as the lengths of moment arms increased for both loading cases. Figure 5 represented tilts of L1-L2 for various lengths L. From these results, it was verified that the length of moment arms was an influencing factor on the deformed shape while the follower load maintained the spinal stability.

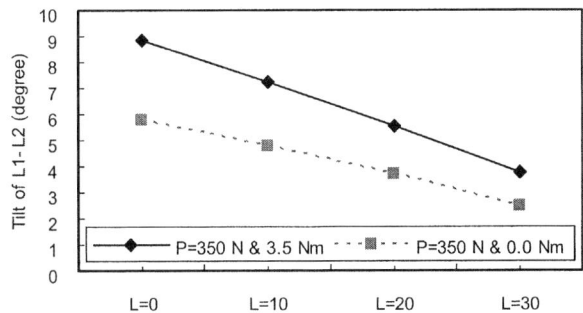

Fig. 5. Tilts of L1-L2 subjected to the follower load for various L

4 Conclusion

The influence of the moment arm on the lumbar spine subjected to the follower load was investigated by using a two-dimensional finite element model of the lumbar spine including the idealized trunk muscles in the frontal plane. The distance between the body center and an attachment point of a muscle in each vertebra was regarded as moment arm in this study. The deformed shape of the finite element model and the tilt of each element as well as the resultant joint moment at each vertebral body center were dependent on the length of the moment arm while the muscle force activations to generate the follower load, and the corresponding follower force and shear force at each vertebra were influenced little by the moment arm. In this manner, the length of the moment arm obtained by the anatomical data, such as the positions of muscle attachment points and vertebral body centers, could be an important factor in studying computational works in biomechanics fields. Therefore, proper and precise knowledge of the human anatomy can contribute to the development and improvement of computational researches on biomechanics, biomedical engineering and so on.

References

1. Patwardhan, A. G., Havey, R. M., Meade, K. P., Lee, B., Dunlap, B.: A follower load increases the load-carrying capacity of the lumbar spine in compression, Spine 24 (1999) 1003-1009
2. Patwardhan, A. G., Havey, R. M., Ghanayem, A. J., Diener, H., Meade, K. P., Dunlap, B., Hodges, S. D.: Load-carrying capacity of the human cervical spine in compression is increased under a follower load: Spine 25 (2000) 1548-1554
3. Rohlmann, A., Neller, S., Claes, L, Bergman, G., Wilke, H.-J.: Influence of a follower load on intradiscal pressure and intersegmental rotation of the lumbar spine. Spine 26 (2001) E557-E561
4. Patwardhan, A. G., Meade, K. P., Lee, B.: A frontal plane model of the lumbar spine subjected to a follower load: Implications for the role of muscles. J. Biomech. Eng. 123 (2001) 212-217
5. Kim, Y. H., Kim, K.: Musculoskeletal modeling of lumbar spine under follower loads, Lecture Notes in Computer Science Vol. 3044. Springer-Verlag, Berlin Heidelberg New York (2004) 467-475
6. Kim, Y. H., Kim, K.: Numerical analysis on quantitative role of trunk muscles in spinal stabilization, JSME Int. (submitted)
7. Gardner-Morse, M. G., Laible, J. P., Stokes, I. A. F.: Incorporation of spinal flexibility measurements into finite element analysis. J. Biomech. Eng. 112 (1990) 481-483
8. Stokes, I. A. F., Gardner-Morse, M. G.: Lumbar spine maximum efforts and muscle recruitment patterns predicted by a model with multijoint muscles and joints with stiffness. J. Biomech. 28 (1995) 173-186

Monte Carlo Simulation of the Effects of Large Blood Vessels During Hyperthermia

Zhong-Shan Deng and Jing Liu

Cryogenics Laboratory, P.O. Box 2711, Technical Institute of Physics and Chemistry,
Chinese Academy of Sciences, Beijing 100080, P.R. China
{zsdeng, jliu}@cl.cryo.ac.cn

Abstract. Large blood vessels can produce localized cooling in heated tissues during tumor hyperthermia. This study has developed a Monte Carlo algorithm to simulate the cooling effect of large blood vessels during hyperthermia. Similar to previous formulations, heat transfer coefficients are used to calculate heat transfer of large blood vessels, and several typical vascular geometries are applied. The corresponding thermal model combines the Pennes bioheat transfer equation for perfused tissues and the energy equation for blood vessels with a constant Nusselt number. Using the Monte Carlo algorithm, numerical analyses are performed to test the influences of large blood vessels to the temperature distributions of tissues. The results indicate that during hyperthermia the presence of large vessels can be an important source of temperature non-uniformity and possible under-dosage. Reduction of blood flow through the large vessels is found to be an effective approach of reducing the localized cooling.

1 Introduction

Hyperthermia treatment is the induction of temperatures above 42 °C in order to treat diseases. The clinical application of hyperthermia is mainly as a combined therapy to radiotherapy and/or chemotherapy against cancer [1]. In hyperthermia, heating is usually limited to the tumor and a small margin of the surrounding healthy tissue. Because the temperatures in the rest of the body remain normal, the blood that supplies the tumor will be relatively cold. As a result, tissue in the treatment volume that is close to a large blood vessel may remain below the desired treatment temperature. In fact, both theoretical and clinical studies have demonstrated that during hyperthermia large blood vessels can produce steep temperature gradients around the vessels with associated thermal under-dosage near the vessels [2,3]. The problems in reaching therapeutic temperatures in all of the treatment volume are hindering the use of hyperthermia in the clinics.

One effective strategy for preventing regions of thermal under-dosage is by compensation of power deposition. More heat must be deposited where there is localized cooling caused by the larger vessels. For this purpose, the power deposition patterns must be carefully devised to effectively heat the tumor without overheating normal tissues. Consequently, there has been an increasing amount of work dedicated to the effects of large blood vessels on temperature distributions during hyperthermia [4,5].

In these works, many vascular models were developed, which account for the convective effects of large blood vessels. The complexity makes the analysis of heat transfer of large vessels rather hard, and only numerical approaches work well.

Considering that the conventional numerical methods such as finite difference method, finite element method, and boundary element method must solve the temperatures at all mesh points simultaneously, and that in hyperthermia only the temperatures at specific area are useful for determination of power deposition pattern, the newly developed Monte Carlo algorithm [6] was extended in this study to investigate the effects of large blood vessels on the temperature distributions during hyperthermia. The particularly attractive feature of Monte Carlo method lies in that the temperature at any desired point can be obtained independently from solutions at the other points, which is an asset when temperature are needed only at some specific area [6]. Aiming at developing a new highly flexible method for the thermal analysis of large vessels, only several typical vascular models were studied in this article.

2 Models and Algorithm Development

In this study, several vascular models with typical geometrical configurations are applied to simulate the cooling effect of large blood vessels during hyperthermia. The vascular models are shown in Fig. 1 (the bioheat transfer equation model, BHTE, is not presented here for brevity). The boundary conditions are a constant temperature of 37 °C on all surfaces of the parallelepiped, and the initial condition is defined as uniform temperature of 37 °C over the whole area. Similar vascular models were firstly developed by Chen and Roemer [2]. In these models, the whole tissue domain consists of perfused tissue and large blood vessel domains. For the perfused tissue, thermal equation is described by the Pennes bioheat transfer equation

$$\frac{\partial T(\mathbf{X},t)}{\partial t} = \alpha \nabla^2 T(\mathbf{X},t) - \frac{\omega_b \rho_b c_b}{\rho c} T(\mathbf{X},t) + \frac{Q(\mathbf{X},t)}{\rho c}, \quad \mathbf{X} \in \Omega_1 . \tag{1}$$

where $Q(\mathbf{X},t) = Q_m(\mathbf{X},t) + Q_r(\mathbf{X},t) + \rho_b c_b \omega_b T_a$; $\alpha = k/\rho c$ is the thermal diffusivity of tissue; ρ, c, k are the density, the specific heat, and the thermal conductivity of the tissue respectively; ρ_b, c_b denote density and specific heat of blood; ω_b the blood perfusion; T_a the arterial temperature; Q_m is the metabolic heat generation, and Q_r the distributed volumetric heat source due to spatial heating; $\mathbf{X}$ contains the Cartesian coordinates x, y and z, and Ω_1 denotes the perfused tissue domain.

For the large blood vessel, the thermal balance is governed by the convective heat transfer equation [7]

$$\rho_b c_b \frac{\partial T}{\partial t} = \frac{hP}{S}(T_w - T) - \rho_b c_b v \frac{\partial T}{\partial z} + Q_r(\mathbf{X},t), \quad \mathbf{X} \in \Omega_2 . \tag{2}$$

where $h = Nu \cdot k_b /D$ is the convective heat transfer coefficient between the blood and tissue, D is the diameter of the vessel, P is the perimeter of the vessel, S is the cross-sectional area of the vessel, v the mean blood velocity along the vessel, Nu the Nusselt number, T_w the wall temperature of the vessel, and Ω_2 denotes the large blood vessel domain. Like the study of Chato [8], conduction inside the vessel in the z direction (flow direction) is neglected for large flow rate, and a constant Nusselt number is assumed. The sign of blood velocity v is assigned as positive, i.e., the velocities for bloods in artery and vein are positive and minus respectively.

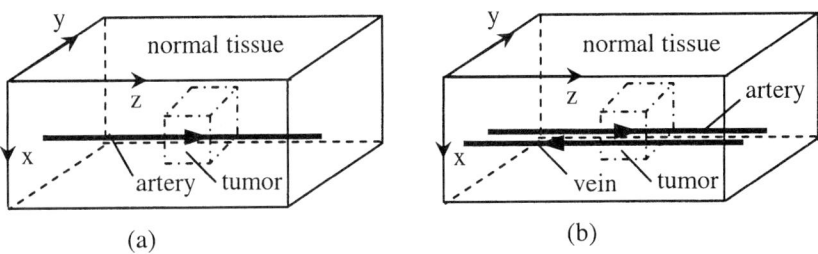

Fig. 1. Illustration of the typical vascular models. The central region is tumor and the outer is normal tissue. (a) a model with a single artery transiting the tumor, SATT; (b) a model with counter-current flow vessels transiting the tumor, CVTT

The probability model for Eq. (1) had been presented in our previous work [6]. The detailed derivation is not repeated here for brevity. If the large blood vessel is present, the probability model in blood vessel domain needs special treatment. In order to avoid the convective numerical instabilities, the up-wind scheme is used to discretize the convective term in equation (2). Then the probability models for the vessels can be written as

$$T_{i,j,k}^{n+1} = p_1 T_{i,j,k}^n + p_2 T_{i+1,j,k}^n + p_3 T_{i-1,j,k}^n + p_4 T_{i,j+1,k}^n + p_5 T_{i,j-1,k}^n + p_6 T_{i,j,k-1}^n + Q_r/F_1 . \tag{3}$$

where $p_1 = a_1/F_1$, $p_2 = a_2/F_1$, $p_3 = a_3/F_1$, $p_4 = a_4/F_1$, $p_5 = a_5/F_1$, $p_6 = a_6/F_1$, $a_1 = \rho_b c_b/\Delta t$, $a_2 = 1/(S \cdot R_{i+1/2})$, $a_3 = 1/(S \cdot R_{i-1/2})$, $a_4 = 1/(S \cdot R_{j+1/2})$, $a_5 = 1/(S \cdot R_{j-1/2})$, $a_6 = \rho_b c_b |v|/\Delta z$, $F_1 = a_1 + a_2 + a_3 + a_4 + a_5 + a_6$. The subscripted R's are the thermal resistance between the center vessel node (i,j,k) and the four neighboring tissue nodes outside the vessel. These thermal resistances are composed of two parts, the convective and conductive resistances. The total thermal resistances between the two nodes, (i,j,k) and $(i-1,j,k)$, (i,j,k) and $(i+1,j,k)$, (i,j,k) and $(i,j-1,k)$, (i,j,k) and $(i,j+1,k)$, are respectively [2, 7]

$$R_{i-1/2} = R_{i+1/2} = \frac{4}{h\pi D} + \frac{2}{\pi k}\ln\frac{\Delta x}{D/2} , \quad R_{j-1/2} = R_{j+1/2} = \frac{4}{h\pi D} + \frac{2}{\pi k}\ln\frac{\Delta y}{D/2} . \tag{4}$$

The thermal resistance between the artery and the vein is

$$R_{a-v} = \frac{8}{h\pi D} + \frac{4}{\pi k}\ln\frac{\Delta y}{D} . \tag{5}$$

For the four neighboring tissue nodes outside the vessel, the probability models are different from those for the other tissue nodes, which are not listed for brevity (the finite difference formulations can be found in [7]).

The probabilistic interpretations of the above equations are similar to that given in [7], and detailed description is omitted here for brevity.

The computer code compiled in this article is revised from the code developed in our previous study [6], which had been validated through comparing the numerical results with the one-dimensional exact solution.

3 Results and Discussion

The tissue domain is prescribed in a rectangular geometry with 10×10×20 cm in the x, y and z directions respectively, and the dimensions of the tumor (which is located at the center of the tissue region) are 3.2×3.2×3.2 cm for all configurations. The blood flow velocity is set as 10 cm/s, and the vessel diameter is set as 1 mm, based on the data compiled by Chato [8]. The constant Nusselt number is taken as $Nu=4$ [2, 8]. The typical tissue properties for normal and tumor tissues are applied in this study, which can be found in [9]. The power deposition pattern used in this article is: $Q_r = 12000 W/m^3$, $X \notin \Omega_t$, $Q_r = 120000 W/m^3$, $X \in \Omega_t$.

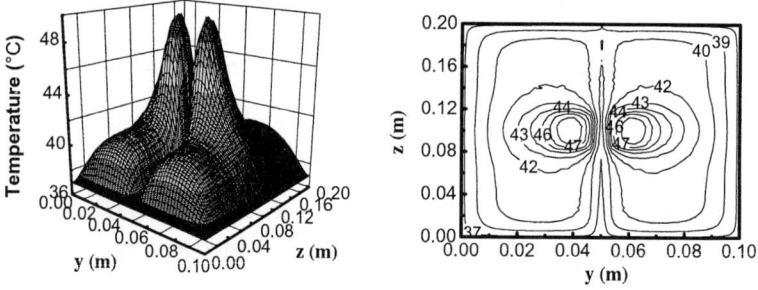

Fig. 2. Temperature distribution at section $x=0.05m$ when t=2000s for SATT model

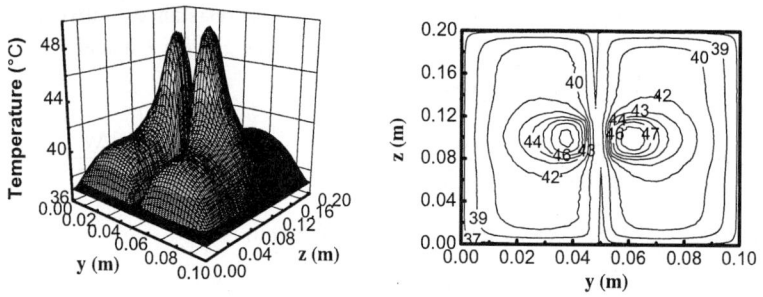

Fig. 3. Temperature distribution at section $x=0.05m$ when t=2000s for CVTT model

Figure 2 depicts the result for SATT model, in which the central line of the single artery is at (x=0.05m, y=0.05m). Figure 3 depicts the result for CVTT model, in which the central line of the counter-current flow vessels are at (x=0.05m, y=0.048m) and (x=0.05m, y=0.05m), respectively. In order to better illustrate the effects of blood vessel on the temperature distributions of tissue during hyperthermia, similar calculation for BHTE model is also performed, and the corresponding result is depicted in Fig. 4. It can be seen from these figures that blood vessel has produced significant temperature non-uniformity around the vessel. Due to the cooling behavior of the flowing blood in large blood vessel, the temperatures of tissues near the vessel for the case of SATT and CVTT models are much lower than that for the case of BHTE model. This may result in thermal under-dosage in the tumor tissue near the large vessels during hyperthermia. For effective killing of tumor, the power deposition around large vessels must be carefully devised to compensate the cooling of vessels.

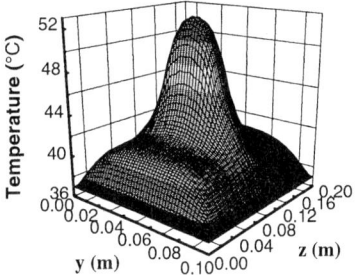 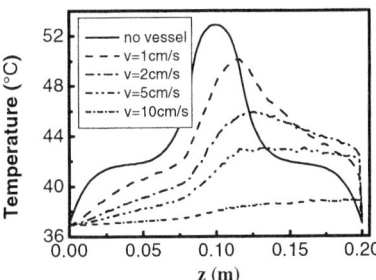

Fig. 4. Temperature distribution at section x=0.05m when t=2000s for BHTE model

Fig. 5. Comparison of the cooling effects of different blood flows

In all the above calculations, the blood flow velocities were taken as 10 cm/s. To see the cooling effects of large vessels under different velocities, Fig. 5 depicts the blood temperatures along the single artery under different blood flow velocities for SATT model. As comparison, the tissue temperatures at the same area for BHTE model are also shown in Fig. 5 (depicted by solid curve, i.e., the case of no vessel). It is clearly seen that the larger the flow velocity, the stronger the cooling effect. Therefore, reduction of blood flow through the large vessel may possibly be served as an effective approach of reducing the localized cooling. Besides, it is also seen that the temperature at some area for SATT model is much higher than that at the same area for BHTE model. This indicates that the blood in large vessel may carry away deposition energy from the tumor to the backward normal tissue along the vessel, possibly causing undesired damage to healthy tissues. Compared with the cooling behavior of large vessel, this feature has received relative few attentions up to now, and must be carefully considered during devising power deposition pattern for hyperthermia.

To prevent cold regions in tumors around large vessel and to obtain improved hyperthermia temperature fields, optimal heating protocols must be found. Since this research involves three dimensional inhomogeneous models with different types and configurations of blood vessels, a complete optimization study would involve many parameters and be very complicated. Therefore, an exhaustive optimization study is beyond the scope of the present work and should be done in the near future.

4 Conclusions

This study has developed a Monte Carlo algorithm for solving three-dimensional heat transfer problems of biological tissues embedded with large blood vessels. In the algorithm, both the convective mechanism for the blood flow in the vessels and the thermal effects for the perfused blood are considered. Using this algorithm, the effects of large blood vessels to the tissue temperature distributions during hyperthermia are investigated. The computation results indicate that the presence of large vessels can produce large temperature non-uniformity around the vessel and thus possibly result in thermal under-dosage in the tumor during hyperthermia, and that reduction of blood flow can reduce the cooling effect of the large vessel. It concludes that to improve the tumor killing effect, the power depositions must be carefully devised to compensate the cooling of large vessels.

Acknowledgment

This work was partially supported by the National Natural Science Foundation of China under Grant 50325622.

References

1. Roemer, R.B.: Engineering aspects of hyperthermia therapy. Annu. Rev. Biomed. Eng. 1 (1999) 347-376
2. Chen, Z.P., Roemer, R.B.: The effects of large blood vessels on temperature distributions during simulated hyperthermia. ASME J. Biomech. Eng. 114 (1992) 473-481
3. Levin, W., Sherar, M.D., Cooper, B., Hill, R.P., Hunt, J.W., Liu, F.F.: The effect of vascular occlusion on tumor temperatures during superficial hyperthermia, Int. J. Hyperther. 10 (1994) 495-505
4. Crezee, J., Lagendijk, J.J.W.: Temperature uniformity during hyperthermia: the impact of large vessels. Phys. Med. Biol. 37 (1992) 1321-1337
5. Koliost, M.C., Sherar, M.D., Hunt, J.W.: Large blood vessel cooling in heated tissues: a numerical study. Phys. Med. Biol. 40 (1995) 477-494
6. Deng, Z.S., Liu, J.: Monte Carlo method to solve multidimensional bioheat transfer problem. Numer. Heat Tran. B. 42 (2002) 543-567
7. Deng, Z.S., Liu, J.: Effects of large blood vessels on 3-D temperature distributions during simulated cryosurgery. submitted to ASME J. Biomech. Eng. (2004)
8. Chato, J.C.: Heat transfer to blood vessels. ASME J. Biomech. Eng. 102 (1980) 110-118
9. Chato, J.C.: Selected thermophysical properties of biological materials. In: Shitzer, A., Eberhart, E.C. (eds.): Heat Transfer in Medicine and Biology, Vol.1. Plenum Press, New York, (1985) 413-418

A Delimitative and Combinatorial Algorithm for Discrete Optimum Design with Different Discrete Sets

Lianshuan Shi[*] and Heng Fu

Computer Department, Tianjin University of Technology and Education,
Tianjin 300222, P.R. China
shilianshuan@263.net

Abstract. The sequential delimitative and combinatorial algorithm for optimum design of structure with discrete variables is popularized to the optimum design, in which, the number of elements of the allowable discrete set of design variables is the same, but the allowable discrete sets of design variables are different. Firstly, the optimal problem is converted into several sequential sub-problems with lower dimensions and similar structures by recurrent method. Secondly, the algorithm generates all combinations according to a certain order of the magnitude of objective function by using a multi-level generating method of proceeding for higher place. In the procedure of delimiting, the objective function and constraint functions are converted into new functions by the linear transforms. And, these new functions are used to delimit.

1 Introduction

Consider a linear programming problem with discrete variables,

$$P \min f(X) = \sum_{i=1}^{n} c_i x_i$$

$$s.t. \; g_i(X) = \sum_{j=1}^{n} a_{ij} x_j \geq b_i$$

$$i = 1, 2, ..., m$$

$$x_j \in S_j(p) = \{s_{j1}, s_{j2}, ..., s_{jp}\}$$

$$s_{j1} < s_{j2} <, ..., < s_{jp}$$

$$j = 1, 2, ..., n$$

$$c_1 \leq c_2 \leq, ..., \leq c_n, \; a_{ij} \geq 0.$$

Where, $f(X)$ is the linear combinations of designing variables $x_i (i=1,2,...,n)$, $g_i(X)$ are the linear combinations of designing variables $x_i (i=1,2,...,n)$, $S_j(p)$

---

[*] The author is a visiting professor at Dept of Computer Science, University of Otago, Dunedin, New Zealand.

is the allowable discrete set of x_j, which is composed of m different discrete values.

Chai S., et al [1,2] given the delimitative and combinatorial algorithms for (0,1) programming problem with the same discrete sets. Shi L.,et al [5] popularized this algorithm to discrete optimum design problem, in which, the number of the design variables is arbitrary, and a sequential delimitative and combinatorial algorithm is given. At the sequential delimitative and combinatorial algorithm, the primal problem is converted into several sequential sub-problems with lower dimensions and similar structures by recurrent method. In the process of optimizing, the multi-level generating method is used to generate the new combinations, and the multilevel delimitative and combinatorial algorithm is used to search optimum solution. In the procedure of delimitation at every level, for the first step, the objective function and constraint functions are converted into some new functions by the linear transforms, then, these new functions are used to delimit by the delimitative algorithm [1]. Both the constraint functions and the unified constraint function are used to delimit so that the computational efficiency is very higher than only using the unified constraint function. A few examples show that the algorithm has higher accuracy and the algorithm is used to optimum design of structure with discrete variables. However, in real engineering design, the discrete sets of design variables of a lot of mathematical model are different. It is necessary to solve the problem with different discrete set for different design variables. In this paper, the sequential delimitative and combinatorial algorithm is popularized, in which, the numbers of elements of the allowable discrete sets of design variables are the same, but the allowable discrete sets of design variables are different. The objective function and constraint functions are converted into new functions by the linear transforms, and these new functions are used to delimit.

2   The Multi-level Generating Algorithm of Combinations

In order to generate all combinations of the model P conveniently, a multi-level generating algorithm proceeding for higher place is given. The total combinations of the design variables of the mathematical model P are p^n. According to the equality $p^n = \sum_{r=0}^{n} C_n^r (p-1)^{n-r}$, these p^n combinations can be divided into $n+1$ groups. $C_n^r(p-1)^{n-r}$ denotes the total number of combinations which r variables taken $s_{jp-(r-1)}$ and other $n-r$ design variables taken values from set $S_j(p-(r-1)) = \{s_{j1}, s_{j2}, ..., s_{jp-(r-1)}\}$. The combination that is taken $s_{jp-(r-1)}$ is generated at the rth level corresponding to the rth group. Other $n-r$ design variables taken values from set $S_j(p-(r-1)) = \{s_{j1}, s_{j2}, ..., s_{jp-(r-1)}\}$ will be generated at the rth level to the $(n+1)th$ level. In order to generate the supperscript combination of the design variables that are taken $s_{jp-(r-1)}$ at the rth level, the algorithm proceeding for higher place is used here.

3 Determining the Lower Bound According to the Constraint Functions

The constraint conditions of the mathematical model P are

$$\sum_{j=1}^{n} a_{ij} x_j \geq b_i.$$

Let

$$x_j = (1 - y_j) s_{j(p-1)} + y_j s_{jp},$$

then

$$x_j = s_{j(p-1)} + y_j (s_{jp} + s_{j(p-1)});$$

If $y_j = 0$, then $x_j = s_{j(p-1)}$; If $y_j = 1$, then $x_j = s_{jp}$. Then constraint condition

$$\sum_{j=1}^{n} a_{ij} x_j \geq b_i$$

is converted into

$$\sum_{j=1}^{n} a_{ij} ((1 - y_j) s_{j(p-1)} + y_j s_{jp}) \geq b_i,$$

or

$$\sum_{j=1}^{n} a_{ij} s_{jp} y_j \geq b_i - \sum_{j=1}^{n} a_{ij} s_{j(p-1)}.$$

Let

$$a_{ij} s_{jp} = k_{ij}, \quad b_i - \sum_{j=1}^{n} a_{ij} s_{j(p-1)} = b'_i,$$

then the constraint condition of the mathematical model P is converted into

$$\sum_{j=1}^{n} k_{ij} y_j \geq b'_i, \quad S'_j = \{s'_{j1}, s'_{j2}, ..., s'_{j(p-2)}, 0, 1\}, \quad k_{ij} \geq 0.$$

Because the elements in the set S_j are ordered according to the increasing order, and the function

$$x_j = s_{j(p-1)} + y_j (s_{jp} - s_{j(p-1)})$$

is a linear and monotonous increasing function of y_j, the elements in the set S'_j are arranged according to the increasing order, namely there are the relation

$$s'_{j1} < s'_{j2} <, ..., < s'_{j(p-2)} < 0 < 1$$

between the element of the S'_j. Therefore, the minimum r that meets constraint conditions can be obtained by using the delimitative and combinatorial algorithm [1]. Then, the mathematical model P can be changed into

$$P_1 \; \min f(X) = \sum_{i=1}^{n-r} c_i^{(1)} x_i$$

$$\text{s.t.} \; g_i^{(1)}(X) = \sum_{j=1}^{n^{(1)}} a_{ij}^{(1)} x_j \geq b_i^{(1)}, i = 1, 2, ..., m$$

$$x_j \in S_j(p-1) = \{s_{j1}, s_{j2}, ..., s_{j(p-1)}\}$$
$$s_{j1} < s_{j2} <, ..., < s_{j(p-1)}$$
$$c_1 \leq c_2 \leq, ..., \leq c_n, a_{ij}^{(1)} \geq 0$$
$$j = 1, 2, ..., n^{(1)}.$$

Where, $b_i^{(1)} = b_i - \sum_{x_j = s_p} a_{ij} x_j$, $c_i^{(1)}$ and $a_{ij}^{(1)}$ are the coefficients of the objective and constraint functions and are reordered according to the increasing order of the coefficients. The model P_1 is the same as P, so it can be solved according to the above method and the recursive algorithm can be used to search optimum solution.

4 Determining the Upper Bound According to the Objective Functions

The objective function is

$$f(X) = \sum_{i=1}^{n} c_i x_i,$$

Where

$$x_j \in S_j(p) = \{s_{j1}, s_{j2}, ..., s_{jp}\},$$

and

$$c_1 \leq c_2 \leq, ..., \leq c_n.$$

Let $x_j = \alpha_j s_{jp}$, then the objective function is changed into

$$f(\alpha) = \sum_{i=1}^{n} \alpha_i c_i x_i,$$

Where

$$\alpha_j \in S_j(p) = \{\frac{s_{j1}}{s_{jp}}, \frac{s_{j2}}{s_{jp}}, ..., \frac{s_{j(p-1)}}{s_{jp}}, 1\}.$$

Let $k_j = c_j s_{jp}$, then $f(\alpha) = \sum_{j=1}^{n} k_j \alpha_j$. After the coefficients of $f(\alpha)$ is reordered according to the increasing order and it is denoted as

$$f'(\alpha) = \sum_{j=1}^{n} k_j \alpha'_j,$$

then the delimitative method as follow can be used.

Definition 1. *Let $u_1 u_2 ... u_r$ is the combination of the magnitude of design variables $\alpha_{u_1} \alpha_{u_2} ... \alpha_{u_r}$, $F = (k'_{u_1} \alpha'_{u_1} + k'_{u_2} \alpha'_{u_2} + ... + k'_{u_r} \alpha'_{u_r})_{u_i = 1, i = 1, 2, ..., r}$. F^* is the current optimum solution and $F > F^*$, then there are four delimitative cases:*

Lemma 1. *(1) If $u_r = u_{r-1} + 1$, then all combinations, which have not been searched in this one-order peak cannot have their objective function values smaller than F^*, and it is not necessary to search for them, Turn to the search for the combinations of the two-order peak.*

(2) If $u_{r-k+1} = u_{r-k} + 1, u_{r-k+2} = u_{r-k+1} + 1, ..., u_r = u_{r-1} + 1$, all combinations which have not been searched in $k+1$ peak must have their objective function values not less than F^*. Turn to the search for the combinations of the $k+2$-order peak.

(3) If $u_2 = u_1 + 1, ..., u_k = u_{k-1} + 1, ..., u_r = u_{r-1} + 1$, the objective function values of all the remaining combinations in (n, r) combinations must not be less than F^*. Turn to $(n, r+1)$ combinations until (n, n) combinations.

(4) If $u_2 = u_1 + 1, ..., u_k = u_{k-1} + 1, ..., u_r = u_{r-1} + 1$, and $u_1 = 1$, the values of objective function of all the remaining combinations must not be less than F^*. The optimum design ends and the current optimum solution is the optimum solution of the problem.

By delimiting of the objective function, a series of non-optimum combinations are eliminated.

5 Examples

Example 1.

$$\min \quad F = x_1 + 2x_2 + 2x_3 + 3x_4 + 4x_5 + 5x_6$$
$$\text{s.t.} \quad 4x_1 + 3x_2 + 2x_3 + 3x_4 + 4x_5 + 6x_6 \geq 510$$
$$3x_1 + 2x_2 + 3x_3 + 4x_4 + 2x_5 + 3x_6 \geq 400$$
$$x_i \in \{1, 2, 3, 4, 8, 9, 11, 14, 18, 20, 21, 27\}, i = 1, ..., 5$$
$$x_6 \in \{3, 4, 8, 9, 11, 14, 18, 20, 21, 27, 28, 29\}.$$

The optimum solution is $(27, 27, 27, 27, 3, 29)$, the searching times are 708, the optimum value is $V = 373$. The total combinations are $12^6 = 2985984$, the searching times is 0.01249 % of the total combinations.

Example 2.

$$\min \quad F = 2x_1 + 3x_2 + 5x_3 + 6x_4 + 8x_5 + 10x_6$$
$$\text{s.t.} \quad 8x_1 + 2x_2 + 4x_3 + x_4 + 5x_5 + 3x_6 \geq 550$$
$$2x_1 + 10x_2 + 5x_3 + 4x_4 + 2x_5 + 7x_6 \geq 650$$
$$4x_1 + 3x_2 + 10x_3 + 5x_4 + 5x_5 + 2x_6 \geq 500$$
$$3x_1 + 3x_2 + 4x_3 + 7x_4 + x_5 + 3x_6 \geq 400$$
$$x_1 \in \{1, 3, 4, 6, 7, 8, 9, 10, 14, 26\}$$
$$x_2 \in \{2, 3, 4, 5, 7, 8, 9, 12, 14, 22\}$$
$$x_3 \in \{1, 3, 5, 6, 7, 8, 9, 12, 15, 26\}$$

$$x_4 \in \{3, 4, 5, 7, 8, 9, 10, 14, 26, 27\}$$
$$x_5 \in \{2, 3, 6, 7, 8, 10, 11, 14, 26, 28\}$$
$$x_6 \in \{6, 7, 8, 9, 10, 11, 13, 14, 27, 29\}.$$

The searching times are 43, the optimum solution is $X^* = \{26, 22, 26, 27, 26, 13\}$, and the objective value is $F^*(X) = 748$. The total combination is $10^6 = 1000000$. The searching time is 0.00043% of the total combinations.

6 Conclusions

The method is an accurate algorithm. The information of the objective function and constraint functions is used fully. The efficiency of this algorithm depends on the given model. By testing a lot of examples generated randomly, the multilevel delimitative and combinatorial algorithm has higher calculating efficiency for a class of linear programming problems with discrete variables. The idea of the delimitative and combinatorial algorithm for discrete optimum design with different discrete sets can also be popularized to the discrete optimum design with reciprocal design variables and structural optimum design with discrete variables. But it is also an exponential algorithm, the bigger dimension of problem is, the more the computational times is.

Acknowledgments

This work was supported by the TianJin Natural Science Foundation of China under the grant No.02360081 and the Education Committee Foundation of Tianjin under the grant No.20022104.

References

1. Chai, S., Sun H.: An application of delimitative and combinatorial algorithm to optimum design of structures with discrete variables. Struc. Optim. **11** (1996) 151-158
2. Chai, S., Sun H.: A two-level delimitative and combinatorial algorithm for the discrete optimization of structures. Strut. Optim. **13** (1997) 250-257
3. Lu, K.: Combinatorial mathematics.(second edition). The Press of Qinghua University, Beijing(In China), (1991)
4. Sun, H., et al: Discrete Optimum design of structure. The Press of Dalian University of Technology, DaLian(in China) (2002)
5. Shi, L. et al.: An application of a sequential delimitative and combinatorial algorithm to the discrete optimum design of structures. Journal of Dalian University of Technology (in China), **39** (1999) 592-596
6. Shi, L. et al: A sequential delimitative and combinatorial algorithm for a class of integer programming. Proceedings of The Fourth International Conference on Parallel and Distributed Computing, Applications and Technologies. (2003) 801-804

A New Algebraic-Based Geometric Constraint Solving Approach: Path Tracking Homotopy Iteration Method

Wenhui Li[1], Chunhong Cao[1], and Wan Yi[2]

[1] College of Computer Science and Technology, Jilin University,
Changchun 130012, P.R. China
liwh@public.cc.jl.cn, chunhongcao_li@163.com
[2] College of Mechanical Science and Technology, Jilin University,
Changchun, 130025, P.R. China
pcnite@sohu.com

Abstract. Geometric constraint problem is equivalent to the problem of solving a set of nonlinear equations substantially. Nonlinear equations can be solved by classical Newton-Raphson algorithm. Path tracking is the iterative application of Newton-Raphson algorithm. The Homotopy iteration method based on the path tracking is appropriate for solving all polynomial equations. At every step of path tracking we get rid of the estimating tache, and the number of divisor part is less, so the calculation efficiency is higher than the common continuum method and the calculation complexity is also less than the common continuum method.

1 Introduction

Geometric constraint solving approaches are made up of three approaches: algebraic-based solving approach, rule-based solving approach and graph-based solving approach. Once a user defines a series of relations, the system will satisfy the constraints by selecting proper states after the parameters are modified. The idea is named as model-based constraints[1]. In the practical project application, there are three basic needs for the geometric constraint problems: real-time quality (the speed must be fast), integrality (all the solutions must be gained) and stability (a small change from the solutions to the problems can't result in a large change).

Geometric constraint problem is equivalent to the problem of solving a set of nonlinear equations substantially. In constraint set every constraint corresponds to one or more nonlinear equations, all the unattached geometric parameters in the geometric element set constitute the variable set of nonlinear equations. Nonlinear equations can be solved by Newton-Raphson algorithm. When the size of geometric constraint problem is large, the scale of the equation set is very large. Furthermore, the geometric information in the geometric system will not be treated properly and we can't deal with under- and over-constrained design system well when all the equations must be solved. Accordingly, we can differentiate the geometry constraint method based on algebra method, the method based on numerical value and the method based on symbol. Symbol method can solve closed formal analytical solution of the problem. It is certainly the best method, but sometimes, its difficulty is also very high. Along with the accretion of the problem's mathematical model, the process of eliminating variable is more and more complicate. [2]

The most widely used method is Newton Iteration method. For example, the method is proposed by Gossard[3]. Generally, the method can get the solution quickly, but it needs a nice initialization value and it is quite sensitive to the initialized value. Using the method, when the initialization value changes a little, it may lead to emanative or converge to the solution that the user doesn't want. Homotopy can overcome these disadvantages[4-6] to a certain extent. Homotopy is an effective numerical iterative method; it has a strong whole convergence and it can solve all the solutions or a set of the isolated solutions of the equations reliably. For the big scale and under-constraint problem, the calculating efficiency of solving also restricts the practical application of the method. [7-12]

Based on the spirit of the Homotopy continuum, we put forward a new numerical iterative method that integrates the Homotopy function with the traditional iteration method. We define it as Homotopy iteration method. Homotopy iteration method can solve all numerical solution of the non-linear equations without choosing appropriate initialization value effectively. For the under-degree equation, it not only needn't to predispose such as homogeneous Homotopy or coefficient Homotopy, but also has a higher calculation efficiency and reliability of getting all the solutions.

2 Homotopy Iteration Method

2.1 The Summarization of the Homotopy Method

Homotopy itself is the conception in the Analisissitus. In 1976, Kellogg and Yorke. Li, solved the global astringency problem of the Homotopy algorithm by differential coefficient topology tools, and proved the theorem's constructive character of Brouwer fixed point. From then on scientists began to restudy the Homotopy method. In several decades, this algorithm became a successful algorithm and is applied to the Economics, the design of the electronic circuitry, auto control, computer aided design, and computer aided manufacture and many other areas and so on.

The main spirit of the Homotopy is that: firstly, choose a simple equation f(x) =0, whose solution is known, then construct the Homotopy mapping H(x, t) which contains parameters, so that H (x, 0) = f (x), H (x, 1) = g (x). In a given condition, the solution of the H (x, t) = 0 can define the curve x (t) which starts from the solution of f (x) = 0, x (0) = x^0, when t approximates 1, the curve achieves the solution x*=x (1) of the g(x) = 0, this is just Homotopy method.

Homotopy method has two steps: Step 1. Insert the equations P (x) into a cluster of the equations H (x, t). This is just the reason why we call it as Homotopy. Step 2. Carry numerical track to the solution curve by the Homotopy-Continuation Method.

Definition1. Homotopy H (x, t) = 0 is defined as a set of equations as follows:

$$H(x, t) = c(1-t)^k f(x) + t^k g(x) = 0, k \in N, c \in C\setminus\{0\}, t \in [0, 1],$$ here t is a homotopy parameter.

When K=1, we usually call it as Protruding Linear Homotopy; When K>1 the every solution path just starts or almost ends and then the lengths of step are both quite short. But the Homotopy mapping H (x, t): $R^n \times [0, 1] \to R^n$ has original equation f (x) = 0, its solution is called as original solution, then g(x) = 0 is called as target equation. For any t R, every H (x, t) is a polynomial equation set about x.

Homotopy method may also constringe to an unexpected solution. But because of its property of explaining itself the terminate state can be found in the process of the solving and the user can easily find out what is wrong at his initial hypothesize. In this way they then can backtrack to former state and correct the mistake.

2.2 Homotopy Iteration Method

Common continuum method's calculation efficiency is quite low when it is applied in solving big under-constraint equations. Although efficient method can preclude many emanative paths, it leaves large numbers of emanative paths at most conditions; furthermore, its preprocessing course needs specific proposal for specific problem and sometimes some skills are needed. Coefficient Homotopy method is a quite highly effective method when solving the polynomial equations which has the same structure and different coefficients, but when it is applied to solving common initial equations, it still needs to adopt the common continuum method or efficient method, thus it come back to the problem of getting solutions using this two methods.

Based on calculation practice, [13],[14] brought forward a new method solving under-constraint polynomial equations and the main spirit is: (1) To construct Homotopy function $H(t, x)=(1-t)F(x)+t \gamma G(x)$ for the equation F(x)=0 which is to be solved, here G (x)is an initial function whose solutions are known. It isn't required that G (x) and F (x) are of the same efficient characters, commonly $G_i(x)=x_i^{di}-1$, in which di is the degree of $F_i(x)$, t is a Homotopy parameter, γ is a random constant plural number whose imaginary part is not zero. (2) For every solution to the G (x) =0, by the Newton iteration (or any traditional iteration), calculate directly the solution of the H (t, x) =0 when t=0.5; (3) Adopt the solution as an initial value which is solved at the former step, using Newton iteration, calculate directly the solution to the H(t, x)=0 when t=0 . If the iteration emanative, then preprocess the above course from another solution to the G (x) =0 again; if the iteration is convergent, then we can get the solution to the F (x) =0. Repeat the process described above until getting all the solutions to the G (x) =0. The common Homotopy method is 10-20 times slower than the Newton-Raphson method when it comes to non-linear equations, but in practice, the essence of following tracking precept is the application of Newton method (or other traditional iterations) many times. Most traditional iteration requires choosing an appropriate initial value to ensure its local convergence; however, choosing an initial value is usually of blindfold character which blocks the efficiency of the method. Now by the Homotopy function, we can resolve the problem well. The Homotopy iteration method based on the following tracking precept is appropriate for solving all polynomial equations. At every step of path tracking we get rid of the estimating tache and the number of divisor part is less, therefore the calculation efficiency is higher than the common continuum method and the calculation complexity is also less than the common continuum method.

2.3 The Solving of Geometrical Constraint Based on Path Tracking Homotopy Iteration Method

The steps of the algorithm based on the Path Tracking Homotopy Iteration method are as the following:

Step 1. Suppose the constraint equations are $F(x)=0$, for polynomial equations $F(x)=0$, we can construct aiding equations $G(x)=0$ whose full solutions are known or can be got easily, generally adopting $G_i(x)=x_i^{d_i}-1$, in which d_i is the degree of $F_i(x)$;

Step 2. Construct Homotopy function $H(t, x) = (1-t)\bar{t} + t\gamma G(x)$, in which t [0, 1], $\gamma=e^{ii}$ is a random constant plural number whose imaginary part is not zero;

Step 3. Choose and input the iteration controlling parameter, which includes the iteration degree k, for a given precision d and emanative condition number, select the increment Δt of Homotopy parameter t or the interval number n (=1/Δt), and $\bar{t} = 1 - \Delta t$;

Step 4. Select a solution $X^{(0)}$ as an initial value to the $G(x)=0$;

Step 5. Adopt Newton method (or another traditional numerical iteration method), to get a solution $X^{(1)}$ of the $H(\bar{t}, x) = 0$;

Step 6. If $\sum \|H(\bar{t}, X^{(1)})\| < d$, then turn to the step 9;

Step 7. If $\sum \|H(\bar{t}, X^{(1)})\| > f$, then turn to the step 10;

Step 8. If the time of the iteration is less than k, then let $X^{(0)} = X^{(1)}$, turn to the step 5;

Step 9. Set $\bar{t} = \bar{t} - \Delta t$, if 0, then adopt $X^{(0)} = X^{(1)}$, turn to the step 5; if $\bar{t} < 0$, we can get the solution $X^* = X(1)$ of $F(x)=0$, turn to the step 11;

Step 10. The process of the iteration is emanative;

Step 11. If there is any solution to the $G(x)=0$, then turn to the step 4; Else the calculation is end.

To make the above algorithm be processed quickly and efficiently, we should select appropriate Homotopy parameter increment Δt (or the interval division number n), the control parameter k, and the parameters d and f. Especially, the size of the interval division number n can affect directly the speed and precision of the calculation, so the choosing principle is given as follows: If the DOD of the polynomial equations is larger, then n becomes larger too, in which DOD is under-degree of equations; DOD= (TD-the number of the equations) × 100%/TD.

3 Result

For the two examples in the Figure 1, the results by different methods are indicated in Table 1. The comparison of the result indicates that the solving efficiency can be improved greatly by our method.

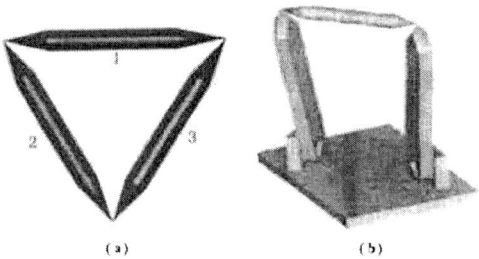

Fig. 1. The example of Tracking Homotopy Iteration method

Table 1. The comparison of result by different methods

	Newton Method	ordinary homotopy method	Path Tracking Homotopy Iteration method
Fig.1 (a) convergence	not convergent	convergent	convergent
time of solving(s)	-----	<100	<5
Fig.1(b) convergence	not convergent	convergent	convergent
time of solving(s)	<1 (if convergent)	<60	<3

We can see that when DOD=0 (all the solutions to the equation to be solved are 0), if n is larger, the area is set off more detailed, the time spent on tracking every route is more, the precision of tracking is higher, so the probability of getting all of the solutions is higher. For the under-system of DOD>0 if we set a smaller n value and the division is larger, then we can judge many emanative routes and decrease the calculating time. For the hyper-under system (DOD$\geq$90%), n can be the minimum $n_{min}=2$, corresponding $t_{max}=1/ n_{min}=0.5$. This is the homotopy iteration method of two-step tracking mentioned in [13, 14]. The number of emanative condition f will have an effect on the calculating efficiency and the probability of getting all the solutions will increase. The experiment indicates in most conditions the homotopy iteration method will get the same solution many times, though it can improve the reliability of getting all the solutions. In order to decrease unnecessary iteration computation the appropriate scope of value of f is 10^3-10^{10}, which should make about half of the routes converge. When there is shortage of experiment, we can select $f=10^7$, 10^8 firstly, then adjust by computing some routes. The selection of d and k will affect computation efficiency, precision and the probability of getting all the solutions, but they are less important than the parameters n and f. Usually the value of d is about 10^{-6}-10^{-4}, and the value of k is small about 10-20 in the beginning of tracking (for example two-step tracking t=0.5), and the value of k is large about 200-500 before the tracking ceases (for example two-step tracking t=0).

We can see the advantages of Homotopy Iteration method are: (1) it has a good universality. We can adopt a uniform solving method to the nonlinear equations; (2) it can deal with the complex constraint solving; (3) when the figure has a great change, it can also achieve convergence. (4) It can have a not high demand for the iteration initialized value. The shortcomings are: (1) Because of iteration solving the speed is still slow. If the number of equations is large, it is not convenient to realize. (2) It can't choose a solution using geometric information when there are many solutions.

References

1. Yuan, B.: Research and Implementation of Geometric Constraint Solving Technology, Doctor Dissertation, Tsinghua University (1999)
2. Chen, Y., Li, Y., Zhang, J., and Yan, J.: An iteration method based on the homotopy function—homotopy iteration method, the numerical calculation and computer appliance. 2: 6 (2001), 149-155
3. Light, R.: Modification of geometric models through variational geometry[J], Computer Aided Design, vol.14. (1982): 209-214
4. He, J.: Homotopy perturbation technique, Computer Methods in Applied Mechanics and Engineering 178 (1999): 257-262
5. He, J.: A coupling method of homotopy technique and perturbation technique for nonlinear problems, International J. of Nonlinear Mechanics, vol.35 (2000): 37-43
6. He, J.: Homotopy perturbation method: a new nonlinear analytical technique, Applied Mathematics and Computation, 135 (2003): 73-79
7. Wang, K., Gao, T.: The introduction of homotopy method, Chongqing Press (1990)
8. Tsai, L., Morgan, A.: Solving the Kinematics of the Most General Six-and-Five-Degree of Freedom Manipulators by Continuation Methods, ASME Journal of Mechanisms, Transmissions and Automation in Desig, 107 (1985), 189-200
9. Morgan, A.: Solving Polynomial Systems Using Continuation for Engineering and Scientific Problems, Prentice-Hall (1987)
10. Li, T., Yorke, J.: The Random Product Homotopy and deficient Polynomial Systems, Numerische Mathemetik, 51(1987), 481-500
11. Li, T., Yorke, J.: The Cheater's Homotopy: An efficient Procedure for Solving System of Polynomial Equations, SIAM, Journal of Numerical Analysis, 18: 2 (1988), 173-177
12. Morgan, A.: Coefficient Parameters Polynomial Continuation, Appl. Math. Comput.29 (1989), 123-160
13. Chen, Y., Yan, J.: An iteration Method Based on Homotopy Function For Solving Polynmial Systems and Application to Mechanisms Problems, Journal of Southwest Jiaotong University, 5: 1(1997), 36-41
14. Chen, Y., Yan, J.: The Homotopy Iteration and its application in the a general 6-SPS Parallel Robert Position Problem, Mechanism Science and Technology, 16: 2(1997), 189-194

A BioAmbients Based Framework for Chain-Structured Biomolecules Modelling*

Cheng Fu, Zhengwei Qi, and Jinyuan You

Dept. of Computer Science and Engineering, Shanghai Jiao Tong Univ., China
{fucheng, qi-zw, you-jy}@cs.sjtu.edu.cn

Abstract. This paper presents a formal extension to describe simple chain-like biomolecular structures and related operations based on the calculus of BioAmbients which serves as a basic framework to model compartments. In our extension, we represent these biomolecules by means of an abstract chain structure possibly with paired molecules at each node, including the effect between molecules, the intra interaction and movement inside linked molecules, and the inter interaction and movement between linked molecules and general compartments. Moreover, An example is given to model the one of the major phases during the process of protein synthesis.

1 Introduction

The calculus of BioAmbients [1] was proposed to serve as a basic framework to abstract compartments as hierarchical organizations in biomolecular systems. As an extension to *Mobile Ambients* [2], it varies that all ambients are anonymous (or with comments) in BioAmbients where handshaking takes place between the action and corresponding co-action through the same channel name, and that communication takes place not only between local parallel systems, but also between sibling ambients and between parent and child ambients. As a kind of bioware biolanguage [3], this abstraction successfully handled dynamic rearrangements of the hierarchical organization in biomolecular systems, including movement of molecules between compartments. As for related works, BioSpi simulation system [4, 5, 1] has been developed to execute bioambient programs and pi-calculus [6] has been used to model chemical reactions [5, 7].

In this paper, we present an extension of BioAmbient, called *BioAmbients with Chains* (BAC), which is suitable to express the static chain structures of biomolecules, the dynamic change of these structures, including binding and pairing operations, the dynamic rearrangements between molecules and compartments, and the intra and inter communications inside and among those chains.

The paper is organized as follows: Section 2 presents a review of BioAmbients. Section 3 presents our extension of BioAmbients. Section 4 shows how the

---

* Supported by the National Natural Science Foundation of China (No. 60173033).

resulting *BioAmbients with Chains* (BAC) models transcription phase during the process of protein synthesis inside the eukaryotic cell. In section 5 goes the final conclusion.

2 BioAmbient Review

The calculus of BioAmbients [1] is proposed to properly model compartments of biological systems. With some improvements, communication channel names take the place of the functioning of original ambient names, and handshaking always occurs through the same channel name. The BNF-like syntax and reduction rules of BioAmbients is summarized below:

Systems $P ::= C \parallel P|P \parallel \lambda.P$ chain/par/pre
$\parallel P+P \parallel (n)P \parallel !P$ choice/res/rpl
Actions $\lambda ::=$ **enter** $n \parallel$ **enter**$^\circ n \parallel$ **exit** $n \parallel$ **exit**$^\circ n$ (co-)enter/(co-)exit
$\parallel$ **merge** $n \parallel$ **merge**$^\circ n \parallel$ **p2c** $n\$$ (co-)merge/parent-child comm
$\parallel$ **c2p** $n\$ \parallel$ **s2s** $n\$ \parallel$ **local** $n\$$ sibling/local comm
Comm $\$ ::= ?n \parallel !n$ input/output Chains $C ::= \beta$ single ambient
Ambient $\beta ::= \mathbf{0} \parallel [P]$ empty/ambient(molecule)
par $P \longrightarrow P' \Longrightarrow P|Q \longrightarrow P'|Q$ str $P \equiv P', P' \longrightarrow Q', Q' \equiv Q \Longrightarrow P \longrightarrow Q$
res $P \longrightarrow Q \Longrightarrow (n)P \longrightarrow (n)Q$ amb $P \longrightarrow Q \Longrightarrow [P] \longrightarrow [Q]$
enter $[(R + \mathbf{enter}\ ch.P)\,|\,S]\,|\,[(R' + \mathbf{enter}^\circ ch.Q)\,|\,S'] \longrightarrow [[P\,|\,S]\,|\,Q\,|\,S']$
exit $[[(R + \mathbf{exit}\ ch.P)\,|\,S]\,|\,(R' + \mathbf{exit}^\circ ch.Q)\,|\,S']] \longrightarrow [P\,|\,S]\,|\,[Q\,|\,S']$
merge $[(R + \mathbf{merge}\ ch.P)\,|\,S]\,|\,[(R' + \mathbf{merge}^\circ ch.Q)\,|\,S'] \longrightarrow [P\,|\,S\,|\,Q\,|\,S']$
sibling $[(R + \mathbf{s2s}\ ch!n.P)\,|\,S]\,|\,[(R' + \mathbf{s2s}\ ch?m.Q)\,|\,S'] \longrightarrow [P\,|\,S]\,|\,[Q\{m \leftarrow n\}\,|\,S']$
local $[(R + \mathbf{local}\ ch!n.P)\,|\,(R' + \mathbf{local}\ ch?m.Q)\,|\,S] \longrightarrow [P\,|\,Q\{m \leftarrow n\}\,|\,S]$
p2c $[(R + \mathbf{p2c}\ ch!n.P)\,|\,S\,|\,[(R' + \mathbf{c2p}\ ch?m.Q)\,|\,S']] \longrightarrow [P\,|\,S\,|\,[Q\{m \leftarrow n\}\,|\,S']]$
c2p $[(R + \mathbf{p2c}\ ch?n.P)\,|\,S\,|\,[(R' + \mathbf{c2p}\ ch!m.Q)\,|\,S']] \longrightarrow [P\{n \leftarrow m\}\,|\,S\,|\,[Q\,|\,S']]$

All ambients are anonymous. $n, m, k, \ldots$ are names or channels. Linked names and chains will be gradually extended in the next section. Reduction semantics remains identical to the original BioAmbients. Structural congruence rules for BAC are defined as follows:

S-Par	$P \equiv Q \Longrightarrow P\,\|\,R \equiv Q\,\|\,R$	S-RplNil	$!\mathbf{0} \equiv \mathbf{0}$
S-Sym	$P \equiv Q \Longrightarrow Q \equiv P$	S-ResNil	$(n)\mathbf{0} \equiv \mathbf{0}$
S-Trans	$P \equiv Q, Q \equiv R \Longrightarrow P \equiv R$	S-ParNil	$P\,\|\,\mathbf{0} \equiv P$
S-ParAssoc	$(P\,\|\,Q)\,\|\,R \equiv P\,\|\,(Q\,\|\,R)$	S-Rpl	$!P\,\|\,P \equiv P$
S-ResPar	$(n)(P\,\|\,Q) \equiv (n)P\,\|\,Q\quad n \notin fn(Q)$	S-ResRes	$(n)(m)P \equiv (m)(n)P$
S-PairNil	$C \frown \beta \frown D \equiv C \frown \beta \times \mathbf{0} \frown D$	S-ResAmb	$(n)[P] \equiv [(n)P]$
S-Prefix	$P \equiv Q \Longrightarrow \lambda.P \equiv \lambda.Q$	S-BindNil1	$C \frown \mathbf{0} \equiv C$
S-BindNil3	$C \frown \mathbf{0} \frown D \equiv C \frown D$	S-BindNil2	$\mathbf{0} \frown C \equiv C$
S-BindAssoc	$(C \frown D) \frown E \equiv C \frown (D \frown E)$		
S-ChainSym	$[P_1] \asymp [Q_1] \frown \ldots \frown [P_k] \asymp [Q_k] \equiv [Q_1] \asymp [P_1] \frown \ldots \frown [Q_k] \asymp [P_k]\quad k \geqslant 1$		

For precedence rules in BioAmbients: $\lambda.P\,|\,Q$ stands for $(\lambda.P)\,|\,Q$, $\lambda.P + Q$ stands for $(\lambda.P) + Q$, $!\lambda.P$ stands for $!(\lambda.P)$, $!P\,|\,Q$ stands for $(!P)\,|\,Q$, and $!P + Q$ stands for $(!P) + Q$. For other operators in the extension, the related precedence rules will be introduced section by section.

3 Extensions to the BioAmbients

Chains with Bind/Unbind Actions. Chain structure is one of the abstract models for those large bio-molecules inside the cell. For example, a strand of DNA is composed of four kinds of nucleotides by linking one by one. Moreover, during the gene translation, single amino acid is bound to the sequence of amino acids by means of transfer RNA.

Actions $\lambda, \mu ::= ...\ \|\ \textbf{bind}\ ch\ \|\ \textbf{bind}^\circ ch$ extended with synch (co-)bind
     $\|\ \textbf{unbind}\ ch\ \|\ \textbf{unbind}^\circ ch$ synch (co-)unbind
Chain $C, D ::= ...\ \|\ C \frown D$ extended with ambient sequence
bind $C \frown [(R + \textbf{bind}^\circ ch.P)\,|\,P']\,|\,[(R' + \textbf{bind}\ ch.Q)\,|\,Q'] \frown D$
     $\longrightarrow C \frown [P\,|\,P'] \frown [Q\,|\,Q'] \frown D$
unbind $C \frown [(R + \textbf{unbind}\ ch.P)\,|\,P'] \frown [(R' + \textbf{unbind}^\circ ch.Q)\,|\,Q'] \frown D$
     $\longrightarrow C \frown [P\,|\,P']\,|\,[Q\,|\,Q'] \frown D$

Pairs and Paring Actions. Pair structure forms essentially by means of the same effect between biomolecules as chain structure. But to simply express the special behavior in DNA pairing operation, pair structure and *pair/unpair* actions are abstracted separately from chain structure and *bind/unbind* actions.

Actions $\lambda, \mu ::= ...\ \|\ \textbf{pair}\ u\ \|\ \textbf{pair}^\circ u\ \|\ \textbf{unpair}\ \|\ \textbf{unpair}^\circ$ (co-)pair/(co-)unpair
Pair $\rho ::= \beta \asymp \beta$ ambient pair   Chain $C ::= ...\ \|\ \rho$ single/multi pairs
pair $C_1 \frown [(R + \textbf{pair}^\circ u.P)\,|\,P'] \frown C_2\,|\,[(R' + \textbf{pair}\ u.Q)\,|\,Q']$
     $\longrightarrow C_1 \frown [P\,|\,P'] \asymp [Q\,|\,Q'] \frown C_2$
unpair $C_1 \frown [(R + \textbf{unpair}^\circ.P)\,|\,P'] \asymp [(R' + \textbf{unpair}.Q)\,|\,Q'] \frown C_2$
     $\longrightarrow C_1 \frown [P\,|\,P'] \frown C_2\,|\,[Q\,|\,Q']$

Chain Splitting Action. In the end of DNA transcription, mRNA is separated from the DNA strand which leads to the *split* action. *Split* action allows a chain with each node having an ambient pair to split into two chains.

Actions $\lambda, \mu ::= ...\ \|\ \textbf{split}$ extended with synch split
split $[(R_1 + \textbf{split}.P_1)\,|\,P_1'] \asymp [(R_1' + \textbf{split}.Q_1)\,|\,Q_1'] \frown ... \frown [(R_k + \textbf{split}.P_k)\,|\,P_k']$
     $\asymp [(R_k' + \textbf{split}.Q_k)\,|\,Q_k'] \longrightarrow [P_1\,|\,P_1'] \frown ... \frown [P_k\,|\,P_k']\,|\,[Q_1\,|\,Q_1'] \frown ... \frown [Q_k\,|\,Q_k']$

Mobility Inside Chain. A set of move capabilities are added to the Action. *moven/movep* allows a ambient inside current node to move into the next or previous node. Both of them must synchronize on the same channel specified by action arguments.

Actions $\lambda, \mu ::= ...\ \|\ \textbf{moven}\ ch\ \|\ \textbf{moven}^\circ ch$ extended with synch right (co-)move
     $\|\ \textbf{movep}\ ch\ \|\ \textbf{movep}^\circ ch$ synch left (co-)move
rmove $C_1 \frown [[(R + \textbf{moven}\ ch.P')\,|\,P'']\,|\,P_1] \asymp \beta_1 \frown [(R' + \textbf{moven}^\circ ch.P_2)\,|\,P_2'] \asymp \beta_2$
     $\frown C_2 \longrightarrow C_1 \frown [P_1] \asymp \beta_1 \frown [P_2\,|\,P_2'\,|\,[P'\,|\,P'']] \asymp \beta_2 \frown C_2$
lmove $C_1 \frown [(R + \textbf{movep}^\circ ch.P_1)\,|\,P_1'] \asymp \beta_1 \frown [[(R' + \textbf{movep}\ ch.P')\,|\,P'']\,|\,P_2] \asymp \beta_2$
     $\frown C_2 \longrightarrow C_1 \frown [P_1\,|\,P_1'\,|\,[P'\,|\,P'']] \asymp \beta_1 \frown [P_2] \asymp \beta_2 \frown C_2$

Intra Communications of Chain. Now we add some synchronized communication capabilities to the actions where p2n/n2p takes the responsibility of interaction between the two neighboring ambients on the same side, while b2b is used to communicate with the paired ambient on the other side.

Actions	$\lambda, \mu ::= ... \parallel \mathbf{p2n}\ ch\$ \parallel \mathbf{n2p}\ ch\$ \parallel \mathbf{b2b}\ ch\$$ n-synch/p-synch
pton	$C_1 \frown [(R + \mathbf{p2n}\ ch!n.P_1) \mid P_1'] \asymp \beta_1 \frown [(R' + \mathbf{n2p}\ ch?m.P_2) \mid P_2'] \asymp \beta_2 \frown C_2$
	$\longrightarrow C_1 \frown [P_1 \mid P_1'] \asymp \beta_1 \frown [P_2\{m \leftarrow n\} \mid P_2'] \asymp \beta_2 \frown C_2$
ntop	$C_1 \frown [(R + \mathbf{p2n}\ ch?m.P_1) \mid P_1'] \asymp \beta_1 \frown [(R' + \mathbf{n2p}\ ch!n.P_2) \mid P_2'] \asymp \beta_2 \frown C_2$
	$\longrightarrow C_1 \frown [P_1\{m \leftarrow n\} \mid P_1'] \asymp \beta_1 \frown [P_2 \mid P_2'] \asymp \beta_2 \frown C_2$
btob	$C_1 \frown [(R + \mathbf{b2b}\ ch!n.P) \mid P'] \asymp [(R' + \mathbf{b2b}\ ch?m.Q) \mid Q'] \frown C_2$
	$\longrightarrow C_1 \frown [P \mid P'] \asymp [Q\{m \leftarrow n\} \mid Q'] \frown C_2$

Chain Transforming. So far, there is no global behavior on chains, and it is difficult to express a chain to pair with another chain or even a segment of it. Therefore we devised a virtual structure, called *Linked Name*, for a chain or part of it, which can exercise pairing operations like an ambient.

Action $\lambda, \mu ::= \mathbf{trfm}\ u \parallel \mathbf{trfm}°n$ (co-)transform   Link $u, v ::= n \parallel u \cdot v$ l-names

int/sep   if $u = n_1 \cdot ... \cdot n_k$ and for each $n_i \notin fn(P_1) \cup ... \cup fn(P_k)$ where $k \geqslant 2$ then
$$C \frown [P_1] \frown ... \frown [P_k] \frown D$$
$$\longleftrightarrow C \frown (n_1)...(n_k)[[P_1 \mid \mathbf{trfm}°n_1] \mid ... \mid [P_k \mid \mathbf{trfm}°n_k] \mid \mathbf{trfm}\ u] \frown D$$

tf-pair   if $u = n_1 \cdot n_2 \cdot ... \cdot n_k$, $u' = n_1' \cdot n_2' \cdot ... \cdot n_k'$, $k \geqslant 2$ and $\psi \in \{\mathbf{pair}, \mathbf{pair}°\}$ then
$C \frown (n_1)...(n_k)[[(T_1 + \psi n_1'.P_1) \mid Q_1 \mid \mathbf{trfm}°n_1] \mid ... \mid [(T_2 + \psi n_k'.P_k) \mid Q_k \mid \mathbf{trfm}°n_k]$
$\mid \mathbf{trfm}\ u] \frown D \longrightarrow C \frown (n_1)...(n_k)[[P_1 \mid Q_1 \mid \mathbf{trfm}°n_1] \mid ...$
$\mid [P_k \mid Q_k \mid \mathbf{trfm}°n_k] \mid \mathbf{pair}\ u' \mid \mathbf{trfm}\ u] \frown D$

tf-λ if $u = n_1 \cdot n_2 \cdot ... \cdot n_k$, $u' = n_1' \cdot n_2' \cdot ... \cdot n_k'$, $k \geqslant 2$ and $\lambda \notin \{\mathbf{pair}\ n, \mathbf{pair}°n\}$ then
$C \frown (n_1)...(n_k)[[(T_1 + \lambda.P_1) \mid Q_1 \mid \mathbf{trfm}°n_1] \mid ... \mid [(T_k + \lambda.P_k) \mid Q_k \mid \mathbf{trfm}°n_k]$
$\mid \mathbf{trfm}\ u] \frown D \longrightarrow C \frown (n_1)...(n_k)[[P_1 \mid Q_1 \mid \mathbf{trfm}°n_1] \mid ...$
$\mid [P_k \mid Q_k \mid \mathbf{trfm}°n_k] \mid \lambda \mid \mathbf{trfm}\ u] \frown D$

4   DNA Transcription Example

During the process of transcription, one DNA strand stands for a template, or a pattern, for the construction of mRNA, then free floating RNA nucleotides travelling into the nucleus, pair with the complementary bases on the DNA template strand. After that, adjacent RNA nucleotides join to form a precursor mRNA. In this model, a sample DNA pattern T-A-C-C-G-T-T-A-A-A-T-T is given where T, A, C, G represent thymine, adenosine, cytosine and guanine respectively. As RNA nucleotides pair with DNA bases, U from RNA pairs with A on the DNA strand, A from RNA pairs with T on DNA strand, and C pairs with G. Different free floating RNA nucleotides are identified through their communication channel names. The ambient name just appears as a comment to the sideward ambient.

$RP_x \triangleq \mathbf{pair}\ x.RS_x$ $Nucleotides \triangleq !U[RP_{ua}] \mid !A[RP_{at}] \mid !C[RP_{cg}] \mid !G[RP_{gc}]$
$DP_x \triangleq \mathbf{pair}\ x.DS_x$ $NucleusEnv \triangleq Nucleotides \mid Others$
$Pattern \triangleq T[DP_{at}] \frown A[DP_{ua}] \frown C[DP_{gc}] \frown C[DP_{gc}] \frown G[DP_{cg}] \frown T[DP_{at}]$
$\quad \frown T[DP_{at}] \frown A[DP_{ua}] \frown A[DP_{ua}] \frown A[DP_{ua}] \frown T[DP_{at}] \frown T[DP_{at}]$
$Sys \triangleq Nc[Pattern \mid NucleusEnv] \longrightarrow^* Sys_1 \triangleq Nc[\ T[DSp_{at}] \asymp A[RS_{at}] \frown$
$\quad A[DSp_{ua}] \asymp U[RS_{ua}] \frown \cdots \frown T[DS_{at}] \asymp A[RS_{at}] \mid NucleusEnv]$

For the above DNA pattern, process $Pattern \mid NucleusEnv$ exercises 12 steps to form a precursor mRNA. After that, the precursor mRNA strand will detach from the DNA template strand:

$RS_x \triangleq \mathbf{split}.RE_x \quad RS_x \triangleq \mathbf{split}.RE_x \quad RS_x \triangleq \mathbf{split}.RE_x \quad DS_x \triangleq \mathbf{split}.DP_x$
$mRNA \triangleq A[RE_{at}] \frown U[RE_{ua}] \frown G[RE_{gc}] \frown G[RE_{gc}] \frown C[RE_{cg}] \frown A[RE_{at}]$
$\quad \frown A[RE_{at}] \frown U[RE_{ua}] \frown U[RE_{ua}] \frown U[RE_{ua}] \frown A[RE_{at}] \frown A[RE_{at}]$
$Sys_1 \longrightarrow Sys_2 \triangleq Nc[Pattern \,|\, mRNA \,|\, NucleusEnv]$

5 Conclusion

In this paper, we briefly introduced an extension of the calculus of BioAmbients, called *BioAmbients with Chains* (BAC)[1], with the additional constructs of chain and pair structures. By comparing with the original BioAmbients, BAC delves more attention into the structure of the biomolecules and the effects among them in a more realistic way. By comparing with *Brane Calculi* (BC)[8], BAC emphasizes that capabilities are carried inside the ambient, but in BC, actions are on-brane; moreover, multisets of molecules in brane calculi are non-structured where chemical reaction are defined as transcendental rules, while, in BAC, within the limit of chain structured molecules, those chemical reactions can be coded by related capabilities.

References

[1] Regev, A., Panina, E.M., Silverman, W., Cardelli, L., Shapiro, E.: Bioambients: An abstraction for biological compartments. Theoretical Computer Science, to appear (2003)
[2] Cardelli, L., Gordon, A.D.: Mobile ambients. In Nivat, M., ed.: Proc. FoSSaCS'98. Volume 1378 of Lecture Notes in Computer Science., Springer Verlag (1998) 140–155
[3] Cardelli, L.: Bioware languages. Computer Systems: Theory, Technology, and Applications - A Tribute to Roger Needham (Springer, 2003)
[4] Priami, C., Regev, A., Shapiro, E.Y., Silverman, W.: Application of a stochastic name-passing calculus to representation and simulation of molecular processes. **80(1)** (2001) 25–31
[5] Regev, A., Silverman, W., Shapiro, E.Y.: Representation and simulation of biochemical processes using the pi-calculus process algebra. In Altman, R.B., amd L. Hunter, A.K.D., Klein, T.E., eds.: Pacific Symposium on Biocomputing, World Scientific Press (2001) 459–470
[6] Milner, R.: Communicating and Mobile Systems: The π-Calculus. Cambridge University Press (1999)
[7] Curti, M., Degano, P., Baldari, C.T.: Causal pi-calculus for biochemical modelling. In: CMSB 2003. Volume 2602 of Lecture Notes in Computer Science., Springer (2003) 21–33
[8] Cardelli, L.: Brane calculi. (Draft paper, available at www.luca.demon.co.uk, 2003)

Stability of Non-autonomous Delayed Cellular Neural Networks*

Qiang Zhang[1,2], Dongsheng Zhou[2], and Xiaopeng Wei[1]

[1] Center for Advanced Design Technology,
University Key Lab of Information Science & Engineering,
Dalian University, Dalian, 116622, China
[2] School of Mechanical Engineering, Dalian University of Technology,
Dalian, 116024, China
zhangq30@yahoo.com

Abstract. The stability of non-autonomous delayed cellular neural networks is studied in this paper. By applying a delay differential inequality, a new sufficient condition which guarantees the global asymptotic stability is established. Since the condition does not impose differentiability on delay, it is less conservative than some established in the earlier references.

1 Introduction

Recently, the dynamics of autonomous delayed cellular neural networks have been extensively investigated and many important results on the global asymptotic stability and global exponential stability of equilibrium point have been given, see, for example,[1]-[11] and references cited therein. However, to the best of our knowledge, few studies have considered stability for non-autonomous delayed cellular neural networks [12]. In this paper, by using differential inequalities, we analyze the global asymptotic stability of non-autonomous delayed cellular neural networks and obtain a new sufficient condition. We do not require the delay to be differentiable. For this reason, the result obtained here improves and generalizes the corresponding results given in the earlier literature.

2 Preliminaries

The dynamic behavior of a continuous time non-autonomous delayed cellular neural networks can be described by the following state equations:

$$x'_i(t) = -c_i(t)x_i(t) + \sum_{j=1}^{n} a_{ij}(t)f_j(x_j(t)) + \sum_{j=1}^{n} b_{ij}(t)f_j(x_j(t - \tau_j(t))) + I_i(t). \quad (1)$$

---

* The project supported by the National Natural Science Foundation of China and China Postdoctoral Science Foundation.

where n corresponds to the number of units in a neural network; $x_i(t)$ corresponds to the state vector at time t; $f(x(t)) = [f_1(x_1(t)), \cdots, f_n(x_n(t))]^T \in R^n$ denotes the activation function of the neurons; $A(t) = [a_{ij}(t)]_{n \times n}$ is referred to as the feedback matrix, $B(t) = [b_{ij}(t)]_{n \times n}$ represents the delayed feedback matrix, while $I_i(t)$ is an external bias vector at time t, $\tau_j(t)$ is the transmission delay along the axon of the jth unit and satisfies $0 \leq \tau_i(t) \leq \tau$.

Throughout this paper, we will assume that the real valued functions $c_i(t) > 0, a_{ij}(t), b_{ij}(t), I_i(t)$ are continuous functions. The activation functions $f_i, i = 1, 2, \cdots, n$ are assumed to satisfy the following hypothesis

$$|f_i(\xi_1) - f_i(\xi_2)| \leq L_i |\xi_1 - \xi_2|, \forall \xi_1, \xi_2. \tag{2}$$

This type of activation functions is clearly more general than both the usual sigmoid activation functions and the piecewise linear function (PWL): $f_i(x) = \frac{1}{2}(|x+1| - |x-1|)$ which is used in [4].

The initial conditions associated with system (1) are of the form

$$x_i(s) = \phi_i(s), \ s \in [-\tau, 0], \ \tau = \max_{1 \leq i \leq n} \{\tau_i^+\} \tag{3}$$

in which $\phi_i(s)$ are continuous for $s \in [-\tau, 0]$.

Lemma 1. *Assume $k_1(t)$ and $k_2(t)$ are nonnegative continuous functions. Let $x(t)$ be a continuous nonnegative function on $t \geq t_0 - \tau$ satisfying inequality (4) for $t \geq t_0$.*

$$x'(t) \leq -k_1(t) x(t) + k_2(t) \bar{x}(t) \tag{4}$$

where $\bar{x}(t) = \sup_{t-\tau \leq s \leq t} \{x(s)\}$. If the following conditions hold

$$\begin{aligned} &1) \ \int_0^\infty k_1(s) ds = +\infty \\ &2) \ \int_{t_0}^t k_2(s) e^{-\int_s^t k_1(u) du} ds \leq \delta < 1. \end{aligned} \tag{5}$$

then, we have $\lim_{t \to \infty} x(t) = 0$.

Proof. It follows from (4) that

$$x(t) \leq x(t_0) e^{-\int_{t_0}^t k_1(s) ds} + \int_{t_0}^t k_2(s) e^{-\int_s^t k_1(u) du} \bar{x}(s) ds, \ t \geq t_0 \tag{6}$$

For $t \geq t_0$, let $y(t) = x(t)$, and for $t_0 - \tau \leq t \leq t_0$, $y(t) = \sup_{t_0 - \tau \leq \theta \leq t_0} [x(\theta)]$. From (6), we obtain

$$x(t) \leq x(t_0) + \delta \sup_{t_0 - \tau \leq \theta \leq t} [x(\theta)], \ t \geq t_0 \tag{7}$$

then, we can get

$$y(t) \leq x(t_0) + \delta \sup_{t_0 - \tau \leq \theta \leq t} [y(\theta)], \ t \geq t_0 - \tau \tag{8}$$

Since the right hand of (8) is nondecreasing, we have

$$\sup_{t_0-\tau \leq \theta \leq t} [y(\theta)] \leq x(t_0) + \delta \sup_{t_0-\tau \leq \theta \leq t} [y(\theta)], \; t \geq t_0 - \tau \tag{9}$$

and

$$x(t) = y(t) \leq \frac{x(t_0)}{1-\delta}, \; t \geq t_0 \tag{10}$$

By condition 1), we know that $\lim_{t \to \infty} \sup x(t) = x^*$ exists. Hence, for each $\varepsilon > 0$, there exists a constant $T > t_0$ such that

$$x(t) < x^* + \varepsilon, \; t \geq T \tag{11}$$

From (6) combining with (11), we have

$$\begin{aligned} x(t) &\leq x(T)e^{-\int_T^t k_1(s)ds} + \int_T^t k_2(s)e^{-\int_s^t k_1(u)du} \bar{x}(s)ds \\ &\leq x(T)e^{-\int_T^t k_1(s)ds} + \delta(x^* + \varepsilon), \; t \geq T \end{aligned} \tag{12}$$

On the other hand, there exists another constant $T_1 > T$ such that

$$\begin{aligned} x^* - \varepsilon &< x(T_1) \\ e^{-\int_T^{T_1} k_1(u)du} &\leq \varepsilon \end{aligned} \tag{13}$$

therefore,

$$x^* - \varepsilon < x(T_1) \leq x(T)\varepsilon + \delta(x^* + \varepsilon) \tag{14}$$

Let $\varepsilon \to 0^+$, we obtain

$$0 \leq x^* \leq \delta x^* \tag{15}$$

this implies $x^* = 0$. This completes the proof.

3 Global Asymptotic Stability Analysis

In this section, we will use the above Lemma to establish the asymptotic stability of system (1). Consider two solutions $x(t)$ and $z(t)$ of system (1) for $t > 0$ corresponding to arbitrary initial values $x(s) = \phi(s)$ and $z(s) = \varphi(s)$ for $s \in [-\tau, 0]$. Let $y_i(t) = x_i(t) - z_i(t)$, then we have

$$\begin{aligned} y_i'(t) = &-c_i(t)y_i(t) + \sum_{j=1}^n a_{ij}(t)\left(f_j(x_j(t)) - f_j(z_j(t))\right) \\ &+ \sum_{j=1}^n b_{ij}(t)\left(f_j(x_j(t-\tau_j(t))) - f_j(z_j(t-\tau_j(t)))\right) \end{aligned} \tag{16}$$

Set $g_j(y_j(t)) = f_j(y_j(t) + z_j(t)) - f_j(z_j(t))$, one can rewrite Eq.(16) as

$$y'_i(t) = -c_i(t)y_i(t) + \sum_{j=1}^{n} a_{ij}(t)g_j(y_j(t)) + \sum_{j=1}^{n} b_{ij}(t)g_j(y_j(t - \tau_j(t))) \quad (17)$$

Note that the functions f_j satisfy the hypothesis (2), that is,

$$|g_i(\xi_1) - g_i(\xi_2)| \leq L_i |\xi_1 - \xi_2|, \forall \xi_1, \xi_2. \quad (18)$$
$$g_i(0) = 0$$

By Eq.(17), we have

$$D^+|y_i(t)| \leq -c_i(t)|y_i(t)| + \sum_{j=1}^{n} |a_{ij}(t)|L_j|y_j(t)| + \sum_{j=1}^{n} |b_{ij}(t)|L_j|y_j(t-\tau_j(t))| \quad (19)$$

Theorem 1. Let $k_1(t) = \min_i \left\{ c_i(t) - \frac{\alpha_j}{\alpha_i} \sum_{j=1}^{n} |a_{ji}(t)|L_i \right\} > 0$,

$k_2(t) = \max_i \left\{ \frac{\alpha_j}{\alpha_i} \sum_{j=1}^{n} |b_{ji}(t)|L_i \right\}$, then Eq.(1) is globally asymptotically stable if

$$\begin{aligned} &1)\ \int_0^\infty k_1(s)ds = +\infty \\ &2)\ \int_{t_0}^{t} k_2(s) e^{-\int_s^t k_1(u)du} ds \leq \delta < 1. \end{aligned} \quad (20)$$

Proof. Let $z(t) = \sum_{i=1}^{n} \alpha_i |y_i(t)|$. Calculating the upper right derivative of $z(t)$ along the solutions of Eq.(17), we have

$$D^+ z(t) \leq \sum_{i=1}^{n} \alpha_i D^+ |y_i(t)|$$

$$\leq \sum_{i=1}^{n} \alpha_i \left\{ -c_i(t)|y_i(t)| + \sum_{j=1}^{n} |a_{ij}(t)|L_j|y_j(t)| + \sum_{j=1}^{n} |b_{ij}(t)|L_j|y_j(t-\tau_j(t))| \right\}$$

$$\leq \sum_{i=1}^{n} \alpha_i \left\{ -c_i(t)|y_i(t)| + \sum_{j=1}^{n} |a_{ij}(t)|L_j|y_j(t)| + \sum_{j=1}^{n} |b_{ij}(t)|L_j|\bar{y}_j(t)| \right\}$$

$$= -\sum_{i=1}^{n} \alpha_i \left[c_i(t) - \frac{\alpha_j}{\alpha_i} \sum_{j=1}^{n} |a_{ji}(t)|L_i \right] |y_i(t)|$$

$$+ \sum_{i=1}^{n} \alpha_i \left[\frac{\alpha_j}{\alpha_i} \sum_{j=1}^{n} |b_{ji}(t)|L_i \right] |\bar{y}_i(t)|$$

$$\leq -k_1(t)z(t) + k_2(t)\bar{z}(t) \quad (21)$$

Using the above Lemma, if the conditions 1) and 2) are satisfied, we have $\lim_{t\to\infty} z(t) = \lim_{t\to\infty} \sum_{i=1}^{n} \alpha_i |y_i(t)| = 0$, this implies that $\lim_{t\to\infty} y_i(t) = 0$. The proof is completed.

Remark 1. Note that the condition obtained here is independent of delay and the coefficients $c_i(t), a_{ij}(t)$ and $b_{ij}(t)$ may be unbounded.

4 Conclusion

We analyze global asymptotic stability of non-autonomous delayed cellular neural networks by establishing a delay differential inequality. A new stability condition is presented. Compared with the Lyapunov functional method which is widely used in the stability analysis, our method is simpler and more straightforward. Since the condition does not impose the differentiability on delay, it is less restrictive than some previously given in the earlier publications.

References

1. Arik, S.: An Improved Global Stability Result for Delayed Cellular Neural Networks. IEEE Trans. Circuits Syst. I. **49** (2002) 1211–1214
2. Arik, S.: An Analysis of Global Asymptotic Stability of Delayed Cellular Neural Networks. IEEE Trans. Neural Networks. **13** (2002) 1239–1242
3. Cao, J., Wang, J.: Global Asymptotic Stability of a General Class of Recurrent Neural Networks with Time-Varying Delays. IEEE Trans. Circuits Syst. I. **50** (2003) 34–44
4. Chua, L.O., Yang, L.: Cellular Neural Networks: Theory and Applications. IEEE Trans. Circuits Syst. I. **35** (1988) 1257–1290
5. Huang, H., Cao, J.: On Global Asymptotic Stability of Recurrent Neural Networks with Time-Varying Delays. Appl. Math. Comput. **142** (2003) 143–154
6. Roska, T., Wu, C.W., Chua, L.O.: Stability of Cellular Neural Network with Dominant Nonlinear and Delay-Type Templates. IEEE Trans. Circuits Syst. **40** (1993) 270–272
7. Zeng, Z., Wang, J., Liao, X.: Global Exponential Stability of a General Class of Recurrent Neural Networks with Time-Varying Delays. IEEE Trans. Circuits Syst. I. **50** (2003) 1353–1358
8. Zhang, J.: Globally Exponential Stability of Neural Networks with Variable Delays. IEEE Trans. Circuits Syst. I. **50** (2003) 288–290
9. Zhang, Q., Ma, R., Xu, J.: Stability of Cellular Neural Networks with Delay. Electron. Lett. **37** (2001) 575–576
10. Zhang, Q., Ma, R., Wang, C., Xu, J.: On the Global Stability of Delayed Neural Networks. IEEE Trans. Automatic Control **48** (2003) 794–797
11. Zhang, Q., Wei, X.P. Xu, J.: Global Exponential Convergence Analysis of Delayed Neural Networks with Time-Varying Delays. Phys. Lett. A **318** (2003) 537–544
12. Jiang, H., Li, Z., Teng, Z.: Boundedness and Stability for Nonautonomous Cellular Neural Networks with Delay. Phys. Lett. A **306** (2003) 313–325

Allometric Scaling Law for Static Friction of Fibrous Materials

Yue Wu[1], Yu-Mei Zhao[1], Jian-Yong Yu[1], and Ji-Huan He[2,3,*]

[1] College of Textiles, Donghua University, Shanghai, China
[2] College of Science, Donghua University, 1882 Yan'an Xilu Road,
Shanghai 200051, People's Republic of China
jhhe@ dhu.edu.cn
[3] Key Lab of Textile Technology, Ministry of Education, Shanghai, China

Abstract. Scaling relationship between the friction and its normal force for fibrous materials is discussed, the scaling exponent is about 2/3 for fibre-fibre contact.

1 Introduction

Friction is a vital important factor in the application of fibrous materials to engineering. Fibers are assembled into various textile products and structures largely to say the least through friction mechanisms. However, due to the complex nature of the inter-fiber friction, our understanding of this fundamental phenomenon is scarce and primitive.

1.1 Historical Development

As it is well-known that the static friction can be written in the form [1]

$$F = \mu N \quad (1)$$

where F is the frictional force, μ friction coefficient, N normal force.

The frictional behavior probably was first understood by Leonardo da Vinci, who found that the frictional force is independent of the area of contact between the two faces and is proportional to the normal force between them. The phenomena were re-discovered by Amontons in 1699, and were verified by Coulomb in 1781. He also pointed out the distinction between static friction and kinetic friction.

The Amontons' law for a yarn passing round a guide can be expressed in the form[1]

$$T_2 / T_1 = e^{\mu \theta} \quad (2)$$

where T_1 and T_2 denote incoming tension and leaving tension respectively, μ the coefficient of friction, and θ angle of contact.

---

* Corresponding author. College of Science, Donghua University, Shanghai 200051, China.
  jhhe@dhu.edu.cn

Many experimental data show that the ratio of frictional force F to normal load N for fibres is found to decrease as the load is increased. In other words, Coulomb's law or Amontons' law is not obeyed. The study of fibrous friction between soft materials or between soft material and metal has largely been the experimental observation of departures from these laws, the reasons for such departures, and their consequences. Among the various mathematical relations that have been used to fit the experimental data are the following[1]:

$$F = \mu N + \alpha S \qquad (3)$$

$$F / N = A - B \log N \qquad (4)$$

$$F = aN + bN^c \qquad (5)$$

$$F = aN^c \qquad (6)$$

where S is the area of contact, $\mu, \alpha, A, B, a, b,$ and c are constants determined from experiment data.

In this paper, we will apply allometric approach to the discussed problem, and the technology has emerged as an interesting and fascinating mathematical tool to the search for the scaling relations of various real-life problems[2~8].

2 Allometric Scaling in Nature

Scaling and dimensional analysis actually started with Newton[5], and allometry exists everywhere in our daily life and scientific activity. Generally, the form of an allometric relation can be expressed as

$$Y \sim X^b \qquad (7)$$

where Y and X are two measures describing an organism or an event, such as the weight of the antlers X and total weight of a deer Y.

When $b=1$ the relationship is isometric and when $b \neq 1$ the relationship is allometric. The exponent is critically important, Ref.[3] points out that and the exponent is relevant to space dimension for allometry, so (7) can be re-written in the form

$$Y \sim X^{D/(D+1)} \qquad (8)$$

or

$$Y \sim X^{k/(D+1)} \qquad (9)$$

where D is dimension of the discussed problem, k is an integral.

Allometry is widely applied in biology, the most fruitful achievement is the allometric scaling relationship relating metabolic rate (B) to organ mass (M)[6,7,8]

$$B \sim M^{D/(D+1)} \tag{10}$$

where D is the dimension of the organism construction, for example $D=2$ for a leaf, and $D=3$ for whole body of an animal.

Similar phenomenon arises from our daily life observation. The best known example is the simple pendulum, with the period:

$$T \sim R^{D/(D+1)}, D=1. \tag{11}$$

Allometry is very effective to quantitative analysis. Conductive textile is widely used as a new kind of intelligent material[3], the classic Ohm law is not valid for calculation of the resistance for intelligent textile.

The Ohm's resistance formulation for metals reads

$$R \sim \frac{L}{A} \tag{12}$$

where R is the resistance, L length of the metal wire, A its section area.

This scaling (12) is valid only for the case of plentiful free electrons in the conductor. For non-metal material, (12) should be modified as

$$R \sim \frac{L}{A^{D/(D+1)}} \tag{13}$$

Here $D=2$ for case when moving charges are distributed on a section, and $D=1$ if the moving charges (for example, in electrospinning[3]) are distributed only on its surface.

For conductive textile we have ($D=2$)

$$R \sim \frac{L}{A^{2/3}} \tag{14}$$

If we want to design an electrochemical cell constructed of two electrodes, which are made of knitted, woven or non-woven conductive textile material, the allometry leads to the following formulation

$$R = k \frac{L}{cA^{2/3}} \tag{15}$$

where k is a constant, L the distance between the electrodes, A the surface area of the electrodes, and c the concentration of the electrolyte solution.

3 Allometric Scaling Law for Fibrous Static Friction

3.1 Solid-Solid Friction

Coulomb's law is valid for metal-metal friction, the scaling relation between frictional force F and the normal force N can be written in the form

$$F \sim N^1 \tag{16}$$

The frictional metal-metal force is independent of the area (A) of the contact between the two surfaces:

$$F \sim A^0 \tag{17}$$

These two scaling relations (16) and (17) are invalid for non-metals.

3.2 Viscous Friction for Newtonian Flow

We consider the viscous friction of Newtonian flow. The viscous friction F is proportional to its area:

$$F \sim A^1 \tag{18}$$

The relations (17) and (18) are valid for two extreme conditions respectively, the former for stability contact, and the later for instability contact.

3.3 Friction for Soft Materials

The frictional force between soft materials, e.g. cotton flow, scales as

$$F \sim A^\alpha, \ 0 < \alpha < 1 \tag{19}$$

where the scaling exponent α represents the mechanical character for the frictional partners. When $\alpha = 0$, the friction is of solid-solid type, and when $\alpha = 1$, it turns out to be viscous friction. Soft material behaves in neither the manner of Newtonian flow, nor solid contact, so the value of α lies between 0 and 1. According to He Chengtian's interpolation(see Appendix), the value of α can be approximated as

$$\alpha(m,n) = \frac{n}{m+n} \tag{20}$$

where m and n are integers. When n is fixed, $m \to \infty$ corresponds to solid-solid contact; when m is fixed, $n \to \infty$ corresponds to viscous flow. The most successful relation is $\alpha(1,1) = 1/2$, which can be used for cotton flow.

3.4 Fibre-Fibre Friction

Now we consider fibre-fibre (e.g. nylon-nylon, nylon-acetate)friction, we have

$$F \sim A^\beta, \ 0 < \beta < \alpha \tag{21}$$

In case $\beta = 0$, the frictional partners can be considered as rigid bodies, and $\beta = \alpha$ for flexible bodies. Most textile productions has a fixed surface similar to a solid surface, but the fibrous flexibility leads to a similar phenomenon as flow, especially as cotton flow, so the value of β generally lies between 0 and 1/2, i.e.,

$$0 < \beta < \frac{1}{2} \tag{22}$$

By He Chengtian's interpolation[9], we have

$$\beta(m,n) = \frac{m+n}{2m+n} \tag{23}$$

If we choose $\beta(1,1) = 2/3$, then we have the following scaling relation $F \sim A^{2/3}$. Tab.1 shows the experimental value δ and our prediction. In practice, we fix $m=1$ for fibre-fibre contact, the more soft of the martial, the larger value of n.

For fibre-fibre friction, the relation (16) is modified as

$$F \sim N^{\delta} \tag{24}$$

or in general form

$$F \sim N^{D/(D+1)} \tag{25}$$

Here D is the dimension of the contact area, $D=2$, so for fibre-fibre friction we have $F \sim N^{2/3}$.

Table 1. Experimental value of δ and prediction

	acetate	nylon	Viscose rayon	Terylene polyester fibre	wool
acetate	0.94 $\delta(1,11) = 0.93$	0.89 $\delta(1,6) = 0.88$	0.90 $\delta(1,7) = 0.90$	0.86 $\delta(1,4) = 0.85$	0.92 $\delta(1,10) = 0.91$
nylon	0.86 $\delta(1,4) = 0.85$	0.81 $\delta(1,2) = 0.80$			
Viscose rayon	0.89 $\delta(1,6) = 0.88$	0.88 $\delta(1,6) = 0.88$	0.91 $\delta(1,10) = 0.92$	0.88 $\delta(1,6) = 0.88$	0.87 $\delta(1,6) = 0.88$
Terylene polyester fibre	0.88 $\delta(1,6) = 0.88$				
wool	0.88 $\delta(1,6) = 0.88$	0.86 $\delta(1,4) = 0.85$	0.92 $\delta(1,10) = 0.92$	0.86 $\delta(1,4) = 0.85$	0.90 $\delta(1,7) = 0.90$

4 Conclusions

We have proposed a theoretical model dealing with for the time fibre-fibre friction, which is critical importance for textile industry. The model is able to describe an allometric form requiring less empirical or semi-empirical input. Of course the authors understand that no matter how rigorous, some experimentally verification is needed to validate the model. A thorough such experimental work is under way and the results will be reported in a separate paper.

Acknowledgement

The work is supported by grant 10372021 from National Natural Science Foundation of China.

References

1. Morton,W.E., Hearle,J.W.S.: Physical properties of textile fibres. The Textile Institute and Heinemann, London, 1975
2. Wan,Y.Q., Guo,Q., Pan,N.,: Thermo-electro-hydrodynamic model for electrospinning process. *International Journal of Nonlinear Sciences and Numerical Simulation,* **5** (2004),5-8
3. He, J.H., Wan,Y.Q., Yu,J.Y.: Allometric Scaling and Instability in Electrospinning. *International Journal of Nonlinear Sciences and Numerical Simulation*, **5**(3), 2004, 243-252
4. He,J.H.: Mysterious Pi and a Possible Link to DNA Sequencing, *International Journal of Nonlinear Sciences and Numerical Simulation* 5(3), 2004, 263-274
5. West, B.J.: Comments on the renormalization group, scaling and measures of complexity. Chaos, Solitons and Fractals, **20**(2004): 33-44
6. He JH, Chen, H.: Effects of Size and pH on Metabolic Rate. *International Journal of Nonlinear Sciences and Numerical Simulation*, **4** (2003): 429-432
7. Kuikka JT.: Scaling Laws in Physiology: Relationships between Size, Function, Metabolism and Life Expectancy. *International Journal of Nonlinear Sciences and Numerical Simulation,* **4** (2003): 317-328
8. Kuikka JT.: Fractal analysis of medical imaging. *International Journal of Nonlinear Sciences and Numerical Simulation,* **3** (2002): 81-88
9. He,J.H.: He Chengtian's inequality and its applications. *Applied Math. Computation,* **151**(2004), 887-891

Flexible Web Service Composition Based on Interface Matching*

Shoujian Yu[1], Ruiqiang Guo[1,2], and Jiajin Le[1]

[1] College of Information Science and Technology, Donghua University,
1882 West Yan'an Road, Shanghai, China 200051
{Jackyysj, grq}@mail.dhu.edu.cn, lejiajin@dhu.edu.cn
[2] College of Mathematics and Information Science, HeBei Normal University,
Shijiazhuang, China 050016

Abstract. Composition of Web services has received much interest in the field of business-to-business or enterprise application integration. In our work, we propose a composition method based on service interface matching. Compared to previous work, our approach only uses the information that is already available in service interface definitions. It does not require that service providers describe their interfaces with semantic markup. We propose data type matching and service composition algorithm. The similarity score calculated by our algorithm allows for flexible Web service composition. Users can define different score threshold for either relaxed or accurate service composition. The experimental results show that the matching algorithm has good precision and recall.

1 Introduction

The emergence of Web service has led to more research into Web services composition, which enables the ability to combine existing services form new services [1]. The business world has developed a number of XML-based composition languages, such as BPEL (Business Process Execution Language) [2], WSCL (Web Services Conversation Language) [3]. These languages only support static composition. On the other side, the Semantic Web community uses semantic annotations to define preconditions and effects of services using terms from pre-agreed ontologies, e.g. DAML-S [4]. But commonly, there aren't universal ontologies to which all partners will prefer to refer. Our approach does not require that service providers describe their interfaces using semantic markup. Instead we attempt to perform composition based solely on the information that is already available in the service descriptions. The rest of this paper is structured as follows. Section 2 gives a detailed discussion of Web service interface matching. In section 3, we discuss the Web service composition algorithm based on interface matching. In section 4, we conduct experiments on interface matching algorithm evaluation. Section 5 concludes this paper and gives some future work.

---

* This work has been partially supported by The National High Technology Research and Development Program of China (863 Program) under contract 2002AA4Z3430.

2 Web Service Interface Matching

WSDL (Web Services Description Language) is the current standard for Web service description [5]. The syntax of WSDL is defined in XML Schema. WSDL document describes its available operations, their associated messages and data types as well as the format of their result values. Automatic services composition relies on the automatic matching of inputs/outputs of operations in Web services, i.e. interface matching. It is in fact the comparison of the messages contained in Web services. Two messages similarity degree is based on how similar their parameter lists are, in terms of the data types they contain and their organization.

2.1 Data Type Matching

The data types of Web service specified in WSDL can be primitive, simple or complex. A simple data type is associated with element name and value type. The element name describes the actual semantic of the element. The value type should be one of the primitive data types. Thus, the simple data type matching can be done from element name and its value type aspects. Various name and string matching algorithms, like *NGram*, semantic matching and abbreviation expansion, are used for element name matching. The *NGram* algorithm calculates the similarity by considering the number of sub string that the names have in common [6]. Equation 1 explains all these cases.

$$ElementNameMatch = \begin{cases} 1 & \text{if } (ms_1 \cup ms_2 \cup ms_3 = 1) \\ ms_2 & \text{if } ((0 < ms_2 < 1) \cap (ms_1 = ms_3 = 0)) \\ 0 & \text{if } (ms_1 = ms_2 = ms_3 = 0) \\ (ms_1 + ms_2 + ms_3)/3 & \text{if } (ms_1 \neq 1, ms_2 \neq 1, ms_3 \neq 1) \end{cases} \quad (1)$$

Where ms_1=MatchScore (*NGram*), ms_2=MatchScore (*SemanticMatching*), ms_3=MatchScore (*Abbreviation*).

The *SemanticMatching* algorithm uses WordNet [7] to find semantic matching between two element names based on the semantic similarity. If two words are identical or synonymous, they are assigned a maximum score of 1 and 0.8 respectively. Otherwise, if two words are in a hierarchical semantic relation, we count the number of semantic links between these words along their shortest path in WordNet hierarchy. The similarity score between two such terms is calculated by dividing 0.6 by the number of links found between them.

$$ValueTypeMatch = \begin{cases} 0 & \text{if ElementNameMatch} = 0 \\ 1 & \text{if two datatypes have compatible value type} \\ 0.8 & \text{if two datatypes have semi-compatible value type} \\ 0 & \text{if two types are incompatible} \end{cases} \quad (2)$$

For each simple data type, there is a primitive type for its values. Equation 2 illustrates the use of value type matching. If two element names return a zero match, their value type matching score should also be zero. The matching score of two simple data types is calculated by combination of *ElementNameMatch* and *ValueTypeMatch*: $S_{simpleDatatype} = w \times S_{ElementNameMatch} + (1-w) \times S_{valueTypeMatch}$, where the constant w is the weight of element name matching. It is in the range of 0 to 1.

$$ComplexDatatypeMatch = \frac{\sum_{i=1}^{n} S(subSimpleDatatype)}{n} \qquad (3)$$

Where n is the number of sub simple data types in the source complex data type.

Complex data types are user-defined data types that contain child elements or attributes. We treat attribute as sub element for simplicity. We match all sub elements in it to find corresponding data types pairs with the maximal sum matching score. If a complex data type also includes another complex data type, the matching process should be proceeded recursively. In order to be with the same measure of matching score as simple data types, the sub elements matching score of complex data types is computed as the average score of its all sub simple data types, as shown in Equation 3.

2.2 Web Service Interface Matching

After evaluating the data type matching scores, Web service interface matching can be proceeded by comparing source service messages against the target service message. The objective of interface matching is to identify the parameter correspondence that maximizes the sum of their individual data type matching score.

```
1    int matchInterface (sourceList(m), targetList(n)){
2    matrix = construct a m⊗n matrix;
3    //exhaustive matching
4    for (int i=0; i<m; i++) {
5      for (int j=0; j<n; j++) {
6        sourceType = sourceList(i); targetType = targetList(j);
7        if (both sourceType and targetType are primitive)
8          matrix[i, j] = matchSimpleTypes(sourceType, targetType);
9        else if (either sourceType or targetType is complex) {
10         newSourceList = getCompositeDataElements(sourceType);
11         newTargetList = getCompositeDataElements(targetType);
12         matrix[i, j] = matchInterface (newBaseList, newTargetList);}}}
13   return the matches with the maximum score;}
```

Fig. 1. Web service interface matching algorithm

Several data types can be enclosed in a message. As can be seen in Fig. 1, the algorithm takes as input two lists of data types: *sourceList*, which contains *m* data types, and *targetList*, which contains *n* data types. Using these two lists, it constructs a m⊗n matrix, whose rows correspond to the source data types and columns correspond to the target data types. For each two data type from the source list and the target list, if they are both simple data types, the procedure *matchSimpleTypes* uses the algorithm in

Sect. 2.1 to compute their matching scores and the score is stored in the corresponding cell of the matrix. If one (or both) of the data types being compared is (are) complex, the procedure *getCompositeDataElements* collects all sub elements of the complex data type(s) to form new data type list(s) to be further matched recursively. After the matrix is filled, the algorithm forms all possible matches between the two lists represented by the matrix and returns the highest matching score between the two lists of data types.

3 Web Service Composition

A service contains several operations. The input/output of each operation tells us what type of parameter needs to be provided, as well as the types that will be returned. One of the possible composition mechanisms is to combine multiple services, such that the execution result of the former service is passed to the next service and accepted as its input, or combination of results from several services is passed to the next service and accepted as its input. This process can be implemented automatically and recursively. In order to express service requestor' need, the service requestor may provide some local information. The *inputList* represents the local information that the requestor provides. *Goal* represents the desirous output that the requestor wants. For simplicity, the operations candidate (*canOperList*) of services is represented by two sets of parameters: *canInList* represents parameter lists of input messages in operations, and *canOutList* represents those of output messages.

Definition. Let DT_1 $(O_1, O_1, ..., O_n)$ and DT_2 $(S_1, S_2, ..., S_m)$ represent two sets of data types. If $\forall O_i \in DT_1$, $\exists S_j \in DT_2$, that match score of $(O_i, S_j) \geq$ threshold, we call DT_1 **satisfies** DT_2.

```
1  comSerList serCompose (inputList, goal, canInList (k), canOutList (k)){
2  for (int i= 0; i <k; i++) {
3      outputMatchScore(i) = matchInterface (goal, canOutList (i));}
       //reorders the canOutList with descending matching score and returns DescOutList
4  DescOutList = descendSort (canOutList);
5  for (i= 0; i <l; i++){ //l≤k, some messages are omitted with a matching score threshold
6      {if (DescOutList (i) satisfies goal)
7         if (inputList satisfies canInList (i))
8             comSerList.add (DescOutList (i));
9         else//get the unmatched parts
10            unMatched = getLeftParts(canInList (i), inputList);
11        comSerList.add serCompose (inputList, unMatched, canInList (k), canOutList (k));
12     else
13        unMatched = getLeftParts(goal, DescOutList (i));
14     comSerList.add serCompose (inputList, unMatched, canInList (k), canOutList (k));}}}
```

Fig. 2. Web service composition algorithm

Now we discuss the algorithm *serCompose* in Fig. 2. First, the algorithm takes *goal* as input to be achieved and calculates the interface matching score between the *goal* and each candidate output message (*canOutList*) (line 3) with the interface matching algorithm described in Sect. 2. The output message lists are reordered with descending matching score, which is *DescOutList* (line 4) (step 1). Second, the algorithm checks each output message if it *satisfies* the *goal* (line 6). If not, this means that individual operation is not enough for the *goal* and composition is necessary. Then the algorithm searches the outputs of other operations for the left unmatched part of the goal (line 13, 14). This is implemented by the procedure calling itself recursively (step 2). Third, if the output message satisfies the goal, whether the input data types list *satisfies* that output message' input part should also be checked (line 7). If not, the procedure *getLeftParts* gets the unmatched data types in that input message to form new data types to be further composed recursively (line 10, 11). This procedure proceeds recursively the same as in step 2 (step 3).

After the matching between the goal and each candidate output message in step 1, some output message with small matching score can be omitted (line 5). User can use a matching score threshold to filter some irrelevant messages, and the system burden is alleviated. Also user can use different threshold for flexible service composition. If one of the output message *satisfies* the *goal* and the input data types list *satisfies* that message' input part, that operation for this message is added to the composite services list (*comSerList*) (line 8). In the end, the algorithm returns the composite services list *comSerList*.

4 Experiment and Evaluation

We evaluate our composition method by evaluating the interface matching algorithm. We find a collection of Web services from SALCentral.org and XMethods.com. We extract 394 pairs of input messages and output messages from their WSDL descriptions. To evaluate the quality of the matching algorithm, we compare the manually determined real matches for those messages (R) with the matches returned by our matching algorithm (P). We determine the correctly identified matches as (I). The following quality measures are computed.

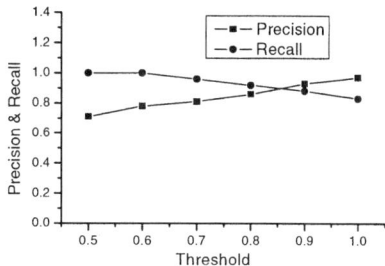

Fig. 3. Web service interface matching algorithm evaluation

- $\Pr ecision = \frac{|I|}{|P|}$ estimates the reliability of the match predictions;
- $\operatorname{Re} call = \frac{|I|}{|R|}$ specifies the share of real matches that is found.

We vary the threshold form 0.5 to 1; the experimental result is shown in Fig. 3. We can see that the matching algorithm has good precision and recall. The threshold of 0.86, where the precision and recall interconnect, should be used for service composition.

5 Conclusions and Future Work

In this paper, we have proposed a novel Web service composition method based on interface matching. The intuition behind our work is that the semantic information is already implicitly captured in the current WSDL interface specification. We have proposed data type matching algorithm for both simple types and complex types. The interface matching score calculated by our algorithm allows for flexible Web service composition. Users can use different score threshold for both relaxed and accurate service composition. Experiments conducted to evaluate the effectiveness of our data type matching algorithm shows that it has good precision and recall.

While good results are obtained using our method, there is room for improvement. First, the weight value used in data type matching can be determined experimentally by constantly changing its value to get the best precision and recall. Second, only sequential composite services can be constructed by interface matching. We will develop workflow techniques facilitated complex Web services composition in the future work.

References

1. Benatallah, B., Dumas, M., Fauvet, M. C., Rabhi, F.: Towards Patterns of Web Services Composition. Patterns and Skeletons for Parallel and Distributed Computing. Springer-Verlag, UK (2003) 265-296
2. Curbera, F., Goland, Y., Klein, J., Leymann, F., Roller, D., Thatte, S., Weerawarana, S.: Business Process Execution Language for Web Services, Version 1.0. http://www-106.ibm.com/developerworks/webservices/library/ws-bpel/ (2002)
3. Banerji, A., Bartolini, C., Beringer, D.: Web Services Conversation Language (WSCL) 1.0. http://www.w3.org/TR/wscl10/ (2002)
4. Ankolekar, A., Burstein, M., Hobbs, J. R., et al: DAML-S: Web Service Description for the Semantic Web. In: Horrocks, I., Hendler, J. (eds.): The Semantic Web - ISWC 2002. Lecture Notes in Computer Science, Vol. 2342. Springer-Verlag, Berlin Heidelberg New York (2002) 348–363
5. Chinnici, R., Gudgin, M., Moreau, J. J., Schlimmer, J., Weerawarana, S.: Web Services Description Language (WSDL) Version 2.0 Part 1: Core Language. http://www.w3.org/TR/2004/WD-wsdl20-20040326/ (2004)
6. Angell, R. C., Freund, G. E., Willett, P.: Automatic spelling correction using a trigram similarity measure. Information Processing and Management. 19 (1983) 255–261
7. WordNet 2.0. http://www.cogsci.princeton.edu/~wn/

Representation of the Signal Transduction with Aberrance Using Ipi Calculus*

Min Zhang[1,**], Guoqiang Li[1], Yuxi Fu[1], Zhizhou Zhang[2], and Lin He[2]

[1] BASICS, Department of Computer Science and Engineering,
Shanghai Jiao Tong University, Shanghai 200030, China
[2] BDCC, College of Life Science and Biotechnology,
Shanghai Jiao Tong University, Shanghai 200030, China
{zhangmin, liguoqiang, yxfu, zhangzz, helin}@sjtu.edu.cn

Abstract. The pi calculus has been applied to modelling biochemical networks. In these applications, the modelling is done without considerations to exceptions. The *Ipi calculus*, the *Interference pi calculus*, is introduced to describe the signal transduction with aberrance. The calculus is obtained by adding two aberrant actions into the pi calculus and a tag system to check existing aberrance. A model of the signal transduction, RTK-MAPK, with the aberrant Ras is highlighted to illustrate the expressive power of the Ipi calculus.

1 Introduction

In recent years, various approaches from computer science have been adapted for the research of biochemical processes. These include boolean networks, petri nets, and object-oriented databases, to name a few. The pi calculus [1, 6] is an alternative way to model biochemical processes. In the pi calculus approach, molecules and their individual domains are treated as computational processes, where their complementary structural and chemical determinants correspond to the communication channels. Chemical interaction and subsequent modification coincide with communication and channel transmission. There are some related research about modelling various biochemical systems based on the pi calculus. Such systems cover for instance the signal transduction (ST for short) [4, 5, 2, 3]. However the biochemical systems considered so far are restricted in the sense that one assumes that there are no exceptions when they evolve.

ST, a process linking the detection of certain kinds of external events to biochemical responses on the part of the cell is a very important biochemical process in biology. An aberrant ST is the cause of many diseases challenged by modern medicine, including cancer, inflammatory diseases, cardiovascular disease and neuropsychiatric disorders. In the search for treatments, cures and preventions

---

* The work is supported by the Young Scientist Research Fund (60225012) and BDCC (03DZ14025).
** The author is also supported by PPS, University Paris 7, France.

of these diseases, in-depth understanding of the biochemical processes of ST is crucial.

In order to describe more complex biochemical systems like the aberrant ST, we develop a calculus called the *Interference pi calculus* by extending the pi calculus. The calculus is obtained by adding two aberrant actions into the pi calculus and a tag system to check existing aberrance. We also illustrate our system using a model of the ST, RTK-MAPK, with the aberrant Ras (a signal that can lead to cancer) which will result to pathological changes.

In this short paper, we have to assume that the reader is familiar with the pi calculus and has some knowledge of biochemical systems.

2 The Interference Pi Calculus

In this section we define the syntax and the semantics of the *Interference pi calculus*.

2.1 Syntax

Processes evolve by performing actions. In process algebra action capabilities are introduced by prefix capabilities. In our calculus we define prefix as a pair, where each capability has a label we call *interference coefficient*, regulating whether it can act in a normal way, or in an aberrant manner. So we define our *interference coefficient* primitive before we introduce the prefix.

We assume an infinite countable set $\mathcal{N}$ of *names* and an infinite countable set $\mathcal{V}$ of *values*. We let $x, y,\ldots$ range over the names. Let σ, ρ be functions from $\mathcal{N}$ to $\mathcal{V}$. One can think of σ as an interference function and that $\sigma(x)$ as the interference degree of x. The function ρ is a critical function and that $\rho(x)$ is the critical value of the interference degree of x. The interference coefficient can be defined below:

Definition 1 (Interference Coefficient). *For $x \in \mathcal{N}$, let i_x be $|\rho(x) - \sigma(x)|$. We say that i_x is the interference coefficient of x.*

Intuitively, when i_x is equal to zero, we take that x is in an aberrant state; when i_x is not zero, we think that x is still in a normal state. For convenience of representation, when i_x is equal to zero, we write 0 as the interference coefficient of x. Otherwise we write i_x as the interference coefficient of x.

We also define two symbols, § and ♯, to represent the aberrance capability. Here § represents the killer capability and ♯ the propagation capability. When a process has the killer capability, it will terminate immediately. And when a process has the propagation capability, it will duplicate himself infinitely.

We define the prefix of Ipi calculus as follows:

Definition 2 (Prefix). *The prefixes of the Ipi calculus are given below:*

$$\langle i_\pi, \pi \rangle := \langle i_x, \overline{x}(y) \rangle \mid \langle i_x, x(y) \rangle \mid \langle i_x, \overline{x} \rangle \mid \langle i_x, x \rangle$$
$$\pi_i := \langle i_\pi, \pi \rangle \mid \langle 0, \S \rangle \pi_i \mid \langle 0, \sharp \rangle \pi_i$$

The capabilities of $\langle i_\pi, \pi \rangle$ is the same as in the π-calculus. Here are some explanations of the π_i capabilities: $\langle 0, \S \rangle$ has the capability that kills all the processes; $\langle 0, \sharp \rangle$ has the capability that duplicates processes infinitely; $\langle 0, \S \rangle \pi_i$ and $\langle 0, \sharp \rangle \pi_i$ are the substitution capabilities: they are respectively the capabilities $\langle 0, \S \rangle$ and $\langle 0, \sharp \rangle$ if the interference coefficient i_π of the π is zero.

We now define the processes of the Ipi calculus. The expression of a process is also a pair $\langle I_P, P \rangle$ where I_P is the tag of the process P. Using the tag in processes we can know the existence of aberrance. If $0 \in I_P$ we say that P has aberrance. If for any $i \in I_P$ one has that $i \neq 0$ then we say that P is normal. The value of tag is produced recursively.

Definition 3 (Process). *The Ipi processes are defined as follows:*

$$\langle I_P, P \rangle := \langle I_0, 0 \rangle \mid \pi_i.\langle I_P, P \rangle \mid \pi_i.\langle I_P, P \rangle + \pi'_i.\langle I_{P'}, P' \rangle \mid \langle I_P, P \rangle | \langle I_{P'}, P' \rangle \mid$$
$$(\nu x)\langle I_P, P \rangle \mid \langle I_P, P \rangle; \langle I_{P'}, P' \rangle$$

The syntax of the tags are defined inductively by the following rules, where the symbol $\uplus$ means disjoint union: $\biguplus_{n=1}^{\infty} I_P \triangleq I_P \uplus I_P \uplus ...$:

$$\frac{}{I_0 = \emptyset} \text{ 0-t} \qquad \frac{\langle I_P, P \rangle = \langle i_\pi, \pi \rangle.\langle I_Q, Q \rangle}{I_P = \{i_\pi\} \uplus I_Q} \text{ t-N}$$

$$\frac{\langle I_P, P \rangle = \langle 0, \S \rangle \langle i_\pi, \pi \rangle.\langle I_Q, Q \rangle}{I_P = \{0\}} \text{ §-t} \qquad \frac{\langle I_P, P \rangle = \langle 0, \sharp \rangle \langle i_\pi, \pi \rangle.\langle I_Q, Q \rangle}{I_P = \biguplus_{n=1}^{\infty} (\{0\} \uplus \{i_\pi\} \uplus I_Q)} \text{ ♯-t}$$

$$\frac{\langle I_P, P \rangle = \langle i_{\pi_i}, \pi_i \rangle.\langle I_Q, Q \rangle + \langle i_{\pi'_i}, \pi'_i \rangle.\langle I_R, R \rangle}{I_P = f\langle \{i_{\pi_i}\} \uplus I_Q, \{i_{\pi'_i}\} \uplus I_R \rangle} \text{ sum-t}$$

$$\frac{\langle I_P, P \rangle = \langle I_Q, Q \rangle | \langle I_R, R \rangle}{I_P = I_Q \cup I_R} \text{ com-t} \qquad \frac{\langle I_P, P \rangle = (\nu x)\langle I_Q, Q \rangle}{I_P = I_Q} \text{ res-t}$$

$$\frac{\langle I_P, P \rangle = \langle I_Q, Q \rangle; \langle I_R, R \rangle}{I_P = I_Q \uplus I_R} \text{ seq-t}$$

In the above definition, $\langle I_P, I_Q \rangle$ is a pair, f is the projection, and $f_{P,Q}\langle I_P, I_Q \rangle$ represents the tag of the process which has the operator "sum". I_P and I_Q are nondeterministically chosen as the process P or Q is chosen to act.

Intuitively the constructs of the Ipi processes have the following meaning: $\langle I_0, 0 \rangle$ is the inert process. The prefix process $\pi_i.\langle I_P, P \rangle$ has a single capability imposed by π_i, that is, the process $\langle I_P, P \rangle$ cannot proceed until that capability has been exercised. The capabilities of the sum $\pi_i.\langle I_P, P \rangle + \pi'_i.\langle I_{P'}, P' \rangle$ are those of $\pi_i.\langle I_P, P \rangle$ plus those of $\pi'_i.\langle I_{P'}, P' \rangle$. When a sum exercises one of its capabilities, the other is rendered void. In the composition process $\langle I_P, P \rangle | \langle I_Q, Q \rangle$, the components $\langle I_P, P \rangle$ and $\langle I_Q, Q \rangle$ can proceed independently and can interact via shared channels. In the restriction process $(\nu x)\langle I_P, P \rangle$, the scope of the name x is restricted to $\langle I_P, P \rangle$. The sequential process $\langle I_P, P \rangle; \langle I_{P'}, P' \rangle$ can run the process $\langle I_{P'}, P' \rangle$ after the process $\langle I_P, P \rangle$.

2.2 Semantics

The structural congruence $\equiv$ is the least equivalence relation on processes that satisfies the following equalities:

$$\langle I_P, P\rangle \mid \langle I_Q, Q\rangle \equiv \langle I_Q, Q\rangle \mid \langle I_P, P\rangle$$
$$(\langle I_P, P\rangle \mid \langle I_Q, Q\rangle) \mid \langle I_R, R\rangle \equiv \langle I_P, P\rangle \mid (\langle I_Q, Q\rangle \mid \langle I_R, R\rangle)$$
$$\langle I_P, P\rangle + \langle I_Q, Q\rangle \equiv \langle I_Q, Q\rangle + \langle I_P, P\rangle$$
$$(\langle I_P, P\rangle + \langle I_Q, Q\rangle) + \langle I_R, R\rangle \equiv \langle I_P, P\rangle + (\langle I_Q, Q\rangle + \langle I_R, R\rangle)$$
$$(\nu x)\langle I_0, 0\rangle \equiv \langle I_0, 0\rangle$$
$$(\nu x)(\nu y)\langle I_P, P\rangle \equiv (\nu y)(\nu x)\langle I_P, P\rangle$$
$$((\nu x)\langle I_P, P\rangle) \mid \langle I_Q, Q\rangle \equiv (\nu x)(\langle I_P, P\rangle \mid \langle I_Q, Q\rangle) \text{ if } x \notin FN(Q)$$

Let I_P, I_Q be the tags of the processes P and Q. We define

$$I_P = I_Q \Leftrightarrow \langle I_P, P\rangle \equiv \langle I_Q, Q\rangle$$

So we have defined an equivalence on the tags in terms of the structural equivalence.

The reaction relation, introduced initially by Milner [1], is a concise account of computation in the pi calculus. A reaction step arises from the interaction of the adjacent process with $\overline{m}\langle M\rangle.P$ and $m(x).Q$, which is also included in the Ipi calculus. Besides this rule, our reaction relation also include two new rules representing reactions with aberrance. All the rules react with their tags reacting simultaneously. We define them and their tag reaction rules blow:

$$\frac{}{\langle 0, \S\rangle\langle i_\pi, \pi\rangle.\langle I_P, P\rangle \longrightarrow \langle \emptyset, 0\rangle}; \quad \frac{}{\{0\} \setminus \{0\} = \emptyset}\text{pre-}\S;$$

$$\frac{}{\langle 0, \sharp\rangle\langle i_\pi, \pi\rangle.\langle I_P, P\rangle \longrightarrow \langle i_\pi, \pi\rangle.\langle I_P, P\rangle; \langle 0, \sharp\rangle\langle i_\pi, \pi\rangle.\langle I_P, P\rangle};$$

$$\frac{}{\biguplus_{n=1}^{\infty}(\{0\} \uplus \{i_\pi\} \uplus I_P) \setminus \{0\} = \{i_\pi\} \uplus I_P \uplus \biguplus_{n=1}^{\infty}(\{0\} \uplus \{i_\pi\} \uplus I_P)}\text{pre-}\sharp;$$

$$\frac{}{\langle i_x, \overline{x}(z)\rangle.\langle I_Q, Q\rangle \mid \langle i_x, x(y)\rangle.\langle I_P, P\rangle \longrightarrow \langle I_Q, Q\rangle \mid \langle I_P, P\rangle\{z/y\}};$$

$$\frac{}{(\{i_x\} \uplus I_Q) \cup (\{i_x\} \uplus I_P) \setminus \{i_x\} = I_Q \cup I_P}\text{com-N};$$

$$\frac{\langle I_P, P\rangle \longrightarrow \langle I_{P'}, P'\rangle}{\langle I_P, P\rangle + \langle I_Q, Q\rangle \longrightarrow \langle I_{P'}, P'\rangle}; \quad \frac{I_P \setminus \{i_y\} = I_{P'}}{f_P\langle I_P, I_Q\rangle \setminus \{i_y\} = I_{P'}};$$

$$\frac{\langle I_P, P\rangle \longrightarrow \langle I_{P'}, P'\rangle}{\langle I_P, P\rangle \mid \langle I_Q, Q\rangle \longrightarrow \langle I_{P'}, P'\rangle \mid \langle I_Q, Q\rangle}; \quad \frac{I_P \setminus \{i_y\} = I_{P'}}{I_P \cup I_Q \setminus \{i_y\} = I_{P'} \cup I_Q};$$

$$\frac{\langle I_P, P\rangle \longrightarrow \langle I_{P'}, P'\rangle \quad x \neq y}{(\nu x)\langle I_P, P\rangle \longrightarrow (\nu x)\langle I_{P'}, P'\rangle}; \quad \frac{I_P \setminus \{i_y\} = I_{P'}}{I_P \setminus \{i_y\} = I_{P'}};$$

$$\frac{\langle I_Q, Q\rangle \equiv \langle I_P, P\rangle \quad \langle I_P, P\rangle \longrightarrow \langle I_{P'}, P'\rangle \quad \langle I_{P'}, P'\rangle \equiv \langle I_{Q'}, Q'\rangle}{\langle I_Q, Q\rangle \longrightarrow \langle I_{Q'}, Q'\rangle};$$

$$\frac{I_Q = I_P \quad I_P \setminus \{i_x\} = I_{P'} \quad I_{P'} = I_{Q'}}{I_Q \setminus \{i_x\} = I_{Q'}}.$$

The first and the second rules deal with reactions with aberrance: the former says that the resulting process is terminated and its tag also changes to empty set. The latter declares that the resulting process duplicates itself infinitely and its tag also duplicates itself. The third reaction rule deals with the interaction in which one sends a message with a channel while the other receives a message with the same channel so that they have an interactive action. This is quite common in the pi calculus. Each of the reduction rules also includes its tag reductions, which means that all the reaction rules are closed in the summation, composition, restriction and structural congruence.

3 An Example in ST Pathway with the Aberrance

In this section we take a look at an example of ST pathways using the new calculus.

3.1 The RTK-MAPK Pathway

In biology pathways of molecule interactions provide communication between the cell membrane and intracellular endpoints, leading to some change in the cell.

We focus our attention on the well-studied RTK-MAPK pathway. The RTK-MAPK pathway is composed of 14 kinds of proteins. A protein ligand molecule (GF), with two identical domains, binds two receptor tyrosine kinase (RTK) molecules on their extracellular part. The bound receptors form a dimeric complex, cross-phosphorylate and activate the protein tyrosine kinase in their intracellular part. The activated receptor can phosphorylate various targets, including its own tyrosines. The phosphorylated tyrosine is identified and bound by an adaptor molecule SHC. A series of protein-protein binding events follows, leading to formation of a protein complex (SHC, GRB2, SOS, and Ras) at the receptor intracellular side. Within this complex the SOS protein activates the Ras protein, which in turn recruits the serine/threonine protein kinase, Raf, to the membrane, where it is subsequently phosphorylated and activated. A cascade of phosphorylations/activations follows, from Raf to MEK1 to ERK1. This cascade culminates in the activation of the threonine and tyrosine protein kinase, ERK1. Activated ERK1 translocates to the nucleus, where it phosphorylates and activates transcription factors.

Within the framework of Ipi calculus, we set some principles for the correspondence. Firstly, we choose the functional signaling *domain* as our primitive *process*. This captures the functional and structural independence of domains in signaling molecules. Secondly, we model the component *residues* of domains as communication *channels* that construct a process. Finally, molecular interaction and modification is modelled as communication and the subsequent change of channel names.

3.2 Representation for the Ras and the Aberrant Ras

Fig.1 gives an example Ras Activation of the ST pathway, RTK-MAPK. In this ST, Ras Activation is the part of the pathway. At the normal state, the protein-to-protein interactions bring the SOS protein close to the membrane, where Ras can be activated. SOS activates Ras by exchanging Ras's GDP with GTP. Active Ras interacts with the first kinase in the MAPK cascade, Raf. GAP inactivates it by the reverse reaction. Aviv Regev and his colleagues have given the representation of normal RTK-MAPK using the pi calculus.

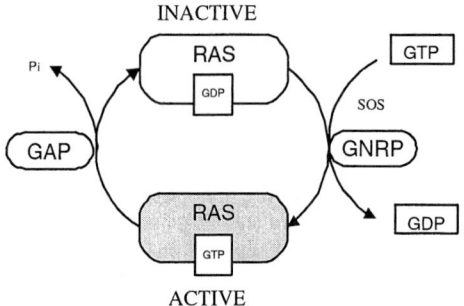

Fig. 1. Ras Activation

The interpretation of Ras in the Ipi calculus can be done in the following manner:

The system defined in (1) is a collection of concurrently operating molecules, seen as processes with potential behavior. Here the operator | is the concurrent combinator:

$$\langle I_S, SYSTEM \rangle ::= \langle I_{RS}, RAS \rangle \mid \langle I_S, SOS \rangle \mid \langle I_G, GAP \rangle \mid \\ \langle I_{RF}, RAF \rangle \mid \cdots \quad (1)$$

A protein molecule is composed of several domains, each of which is modelled as a process as well. In (2) through (5) the detailed Ipi calculus programs for the proteins Ras, SOS, Raf and GAP are given:

$$\langle I_{RS}, RAS \rangle ::= \langle I_{ISI}, INASWI\_I \rangle \mid \langle I_{ISII}, INASWI\_II \rangle \quad (2)$$
$$\langle I_S, SOS \rangle ::= \langle I_{SSB}, S\_SH3\_BS \rangle \mid \langle I_{SG}, S\_GNEF \rangle \quad (3)$$
$$\langle I_{RF}, RAF \rangle ::= \langle I_{RN}, R\_Nt \rangle \mid \langle I_{RACB}, R\_ACT\_BS \rangle \mid \langle I_{RMB}, R\_M\_BS \rangle \\ \mid \langle I_{IRC}, INA\_R\_Ct \rangle \mid \langle I_{RATB}, R\_ATP\_BS \rangle \quad (4)$$
$$\langle I_G, GAP \rangle ::= \langle i_s, sg \rangle (c\_ras).\langle i_c, \overline{c\_ras} \rangle (gdp).\langle I_G, GAP \rangle \quad (5)$$

The molecules (or domains) interact with each other based on their structural and chemical complementarity. Interaction is accomplished by the motifs and residues that constitute a domain. These are viewed as channels or communication ports of the molecule:

$$\langle I_{ISI}, INASWI\_I \rangle ::= \langle i_b, \overline{bbone} \rangle.\langle I_{ASI}, ACTSWI\_I \rangle \qquad (6)$$

$$\langle I_{ISII}, INASWI\_II \rangle ::= \langle i_{sg}, \overline{sg} \rangle (rs\_1).\langle i_{rs1}, rs\_1 \rangle (x).\langle i_b, bbone \rangle.$$
$$\langle I_{ASII}, ACTSWI\_II \rangle \qquad (7)$$

$$\langle I_{SG}, S\_GNEF \rangle ::= \langle i_b, bbone \rangle.\langle I_{SG}, S\_GNEF \rangle + \langle i_{sg}, sg \rangle (c\_ras).$$
$$\langle i_{cr}, \overline{c\_ras} \rangle (gtp).\langle I_{SG}, S\_GNEF \rangle \qquad (8)$$

The following interactions are possible:

$$\langle I_{ISI}, INASWI\_I \rangle \mid \langle I_{SG}, S\_GNEF \rangle \longrightarrow$$
$$\langle I_{ASI}, ACTSWI\_I \rangle \mid \langle I_{SG}, S\_GNEF \rangle \qquad (9)$$

$$\langle I_{ISII}, INASWI\_II \rangle \mid \langle I_{SG}, S\_GNEF \rangle \longrightarrow$$
$$\langle i_b, bbone \rangle \langle I_{ASII}, ACTSWI\_II \rangle \mid \langle I_{SG}, S\_GNEF \rangle \qquad (10)$$

The interaction (9) shows that the domain $INASWI\_I$ of Ras is activated by the domain of S_GNEF of SOS. The interaction (10) shows that the domain $INASWI\_II$ of Ras is activated by the domain S_GNEF of SOS.

The detailed Ipi programs for activated domains, $ACTSWI\_I$, $ACTSWI\_II$ of the protein Ras and the domain R_Nt of Raf are defined in (11) through (13):

$$\langle I_{ASI}, ACTSWI\_I \rangle ::= \langle i_s, \overline{s} \rangle (rs\_2).\langle i_{is2}, \overline{rs\_2} \rangle.\langle I_{ASI}, ACTSWI\_I \rangle$$
$$+ \langle i_b, \overline{bbone} \rangle.\langle I_{ISI}, INASWI\_I \rangle \qquad (11)$$

$$\langle I_{ASII}, ACTSWI\_II \rangle ::= \langle i_{sg}, \overline{sg} \rangle (r\_swi\_1).\langle i_{rs1}, r\_swi\_1 \rangle (x).$$
$$\langle i_b, \overline{bbone} \rangle \langle I_{ASII}, ACTSWI\_II \rangle \qquad (12)$$

$$\langle I_{RN}, R\_Nt \rangle ::= \langle i_s, \overline{s} \rangle (c\_ras).\langle i_{cr}, c\_ras \rangle.\langle I_{ARN}, ACTR\_Nt \rangle \qquad (13)$$

The processes so defined have the following interactions:

$$\langle I_{ISI}, ACTSWI\_I \rangle \mid \langle I_{RN}, R\_Nt \rangle \longrightarrow^*$$
$$\langle I_{ISI}, ACTSWI\_I \rangle \mid \langle I_{ARN}, ACTR\_Nt \rangle \qquad (14)$$

$$\langle I_{ISII}, ACTSWI\_II \rangle \mid \langle I_G, GAP \rangle \longrightarrow^*$$
$$\langle i_b, bbone \rangle \langle I_{ISII}, INASWI\_II \rangle \mid \langle I_G, GAP \rangle \qquad (15)$$

$$\langle i_b, \overline{bbone} \rangle.\langle I_{ASII}, ACTSWI\_II \rangle \mid \langle I_{ASI}, ACTSWI\_I \rangle \longrightarrow$$
$$\langle I_{ISII}, INASWI\_II \rangle \mid \langle I_{ISI}, INASWI\_I \rangle \qquad (16)$$

The interaction (14) shows that the active domain $ACTSWI\_I$ of Ras interacts with the domain R_Nt of Raf. (15) shows that GAP inactivates the domain $ACTSWI\_II$ of Ras. (16) says that the domains of Ras interact with each other and that Ras rollbacks to the initial inactivated state.

When Ras mutates aberrantly, it does not have any effect on the Ras's binding with GTP and GDP but will reduce the activity of the GTP hydrolase of Ras and lower its hydrolysis of GTP greatly; in the meantime Ras will be kept in an active state; they keep activating the molecule, induce the continual effect of signal transduction, and result in cell proliferation and tumor malignancy.

(17) defines the Ipi representation of GAP in the aberrant state. (18) shows that GAP loses its function and does nothing, meaning that it can not inactivate the domain $ACTSWI\_II$ of Ras.

$$\langle I_G, GAP \rangle ::= \langle 0, \S \rangle \langle i_s, sg \rangle (c\_ras).\langle i_c, \overline{c\_ras} \rangle (gdp).\langle I_G, GAP \rangle \qquad (17)$$
$$\langle I_G, GAP \rangle \longrightarrow \langle \emptyset, 0 \rangle \qquad (18)$$

But then the interaction (16) will not occur whereas the interaction (14) will occur infinitely. Now observe that

$$\langle 0, \sharp \rangle \langle I_{ASI}, ACTSWI\_I \rangle \longrightarrow \langle I_{ASI}, ACTSWI\_I \rangle; \langle 0, \sharp \rangle \langle I_{ASI}, ACTSWI\_I \rangle$$

It reaches in an abnormal state with exceptions. The pi calculus could not easily describe this aberrant case. Ipi calculus, on the contrary, can describe it quite precisely. In fact, when the aberrance occurs, it will be marked into the tag. So we can check the existing aberrance, using the tag system of Ipi calculus.

4 Future Prospects

The Ipi calculus is an extension of the pi calculus. The desirable outcomes and properties of a biochemical process can be formally proven in the framework of the Ipi caluclus. The theory of process calculus allows us to formally compare two programs using *bisimulation*. We can also define different levels of bisimilarity for verifying different properties of biochemical processes.

The research opens up new possibilities in the study of biochemical systems with exceptions. Designing and implementing an automatic tool will be our next work.

References

1. Milner, R., Parrow, J., Walker, D.: A Calculus of Mobile Processes, parts I and II. In: Information and Computation. (1992) 1-77
2. Priami, C., Regev, A., Silverman, W., and Shapiro,E.: Application of a stochastic name passing calculus to representation and simulation of molecular processes. In: Information Processing Letters. **80**(2001) 25-31
3. Regev, A.: Representation and simulation of molecular pathways in the stochastic pi calculus. In: Proceedings of the 2nd workshop on Computation of Biochemical Pathways and Genetic Networks. (2001)
4. Regev, A., Silverman, W., and Shapiro, E.: Representing biomolecular processes with computer process algebra: pi calculus programs of signal transduction pathways. In: http://www.wisdom.weizmann.ac.il/~aviv/papers.htm (2000)

5. Regev, A., Silverman, W., and Shapiro, E.: Representation and simulation of biochemical processes using the pi calculus process algebra. In :Proceedings of the Pacific Symposium of Biocomputing. **6**(2001) 459-470
6. Sangiorgi, D., and Walker, D.: The pi calculus: a Theory of Mobile Process. In: Cambridge University Press. (2001)

The Application of Nonaffine Network Structural Model in Sine Pulsating Flow Field

Juan Zhang

College of Science, Donghua University, 1882 Yan'an Road West,
Shanghai 200051, China
zhangjuan@dhu.edu.cn

Abstract. In order to describe the rheological behavior of polymer melts in a practical vibration force field, the nonaffine network structural model is generalized from a simple oscillatory shear force field to a sine pulsating flow field. Through experiments and theoretic analyses, we prove that this model can be described preferably the viscous and elastic behavior of polymer melts in sine pulsating flow field, and the change amplitude of the shear stress increases with the vibration frequency or amplitude increasing.

1 Generalized Nonaffine Transient Network Structural Model

Recently, from the author's research [1-3], it is obvious that the accurate precision of the nonaffine transient network structural model is higher than that of the affine network model, which is made by Giacomin and coworkers [4-7], in the large amplitude oscillatory shear flow field. But we know that the large amplitude oscillatory shear is only a simple oscillatory shear force field. In fact, the practical one is more complicated, which usually includes both oscillatory shear force field and steady shear flow field. Therefore in this paper, we study the application of the nonaffine transient network structural model in the parallel superposition of sine vibration upon steady shear flow field (also named sine pulsating flow field).

The nonaffine transient network structural model is:

$$\tau = \sum_i \tau_i . \tag{1}$$

$$\frac{\tau_i}{G_i} + \lambda_i \frac{\delta}{\delta t}[\frac{\tau_i}{G_i}] = 2\lambda_i D . \tag{2}$$

where τ is the extra stress tensor, τ_i is the i-th spectral component of the extra tensor, t is the time, λ_i is the relaxation time of the structure dependent relaxation spectra (G_i, λ_i), and G_i is the relaxation module for the relaxation time, $D = (\nabla v + \nabla v^T)/2$ denotes the rate of deformation tensor with velocity gradient tensor ∇v, and the upper convected derivative is:

$$\frac{\delta \tau_i}{\delta t} = \frac{d\tau_i}{dt} - \nabla v \tau_i - \tau_i \nabla v^T + \zeta[\tau_i D + D\tau_i]. \tag{3}$$

where ζ measures how nonaffine the network deformation is.

In former research, the kinetic rate equation in the large amplitude oscillatory shear is:

$$\frac{dx_i}{dt} = \frac{k_1(1-x_i)}{\lambda_i} - k_2 \gamma_0 \omega |\cos \omega t| x_i = \frac{k_1(1-x_i)}{\lambda_{0i}(\frac{x_i+\alpha}{1+\alpha})^{1.4}} - k_2 \gamma_0 \omega |\cos \omega t| x_i. \tag{4}$$

$$\lambda_i = \lambda_{0i}(\frac{x_i+\alpha}{1+\alpha})^{1.4}, G_i = G_{0i}(\frac{x_i+\alpha}{1+\alpha}). \tag{5}$$

where k_1 is the kinetic rate constant for thermal regeneration of entanglements, k_2 is the kinetic rate constant for the destruction of entanglements, x_i denotes entanglement density, which is a set of scalar structural variables, each ranging from 0 to 1, γ_0 is strain amplitude, ω is circular frequency, α is a constant parameter, and (G_{0i}, λ_{0i}) is a discretization of the linear relaxation spectrum.

Now in sine pulsating flow field, the imposed shear rate is $\dot{\gamma} = \dot{\gamma}_m + \lambda_0 \omega \cos \omega t$, where $\dot{\gamma}_m$ is the shear rate that is no longer a function of time in steady state. Similarly, using the kinetic theory, we can obtain the new kinetic rate equation is:

$$\frac{dx_i}{dt} = \frac{k_1(x_{mi}-x_i)}{\lambda_i} - k_2 \Delta \dot{\gamma} x_i. \tag{6}$$

where $\Delta \dot{\gamma} = \gamma_0 \omega \cos \omega t$ is the net shear rate after imposing parallel superposition vibration force field upon steady shear flow, x_{mi} is the entanglement density in the steady state.

Combining equations (1)~(3) and (5)~(6), the generalized nonaffine model is made. Using Runge-Kutta fifth-order ordinary differential equation solver can obtain the numerical solution of this model.

2 Experiment

In order to obtain the accurate data, we use a capillary dynamic rheometer [8,9], in which a sine vibration force with high frequency and small amplitude is introduced into all the process of polymer melt extrusion through capillary. Here the experimental material is LDPE, the temperature of the melt is 150 degrees centigrade, and $k_1 = 0.32$, $k_2 = 0.1$, $\zeta = 0.02$, $\alpha = 0.1$.

Fig.1 describes the velocity of polymer melts in capillary dynamic rheometer. Where L is the length of the capillary, R is the radius of the capillary, ΔP is the pressure difference between the two ports.

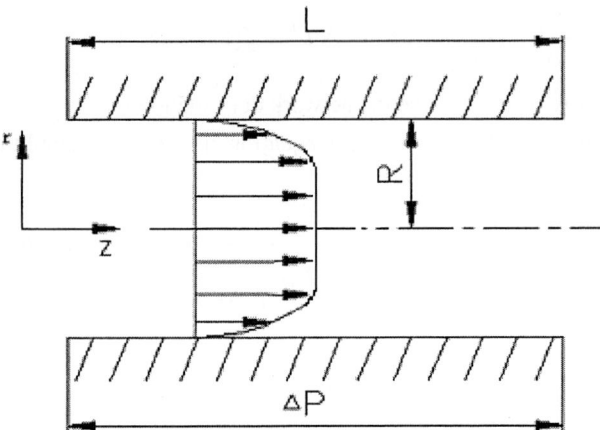

Fig. 1. Velocity of polymer melts in capillary dynamic rheometer (fully development section)

In capillary the velocity and the shear stress of the melt respectively are:

$$V = [V_z \quad 0 \quad 0]. \tag{7}$$

$$\tau = \begin{bmatrix} \tau_{zz} & \tau_{rz} & 0 \\ \tau_{rz} & \tau_{rr} & 0 \\ 0 & 0 & \tau_{\theta\theta} \end{bmatrix}. \tag{8}$$

where boundary conditions are:

$$V_z(r,t)\big|_{r=R} = 0. \tag{9}$$

$$\tau_{rz}\big|_{r=0} = const. \tag{10}$$

The displacement of the piston rod in capillary is:

$$S = \bar{v}_0 t + A \sin \omega t. \tag{11}$$

where $\bar{v}_0$ is the steady average velocity of the piston rod, A is the vibration amplitude, ω is the vibration circle frequency ($\omega = 2\pi f$, f is vibration frequency). Through analyzing the movement equation and using conservation of mass, the average velocity in the capillary section can be described as:

$$\bar{v}_z = (\frac{R_0}{R})^2 (\bar{v}_0 + A\omega \cos \omega t). \tag{12}$$

when $r = R$, there is $\bar{v}_z = 0$; and when $r = 0$, there is $\bar{v}_z = (\frac{R_0}{R})^2 (\bar{v}_0 + A\omega \cos \omega t)$ and

$$\frac{d\bar{v}_z}{dr} = \lim_{\Delta r \to 0} \frac{\Delta v_z}{\Delta r} \approx \frac{0 - \bar{v}_z}{R} = -\frac{R_0^2}{R^3}(\bar{v}_0 + A\omega\cos\omega t). \tag{13}$$

So the shear rate of the capillary wall can be described as:

$$\dot{\gamma} = \frac{R_0^2}{R^3}(\bar{v}_0 + A\omega\cos\omega t). \tag{14}$$

where R_0 is the radius of the cylinder, $\bar{v}_0$ is the steady average velocity of the piston rod.

Supposed:

$$\dot{\gamma}_m = -\frac{R_0^2}{R^3}\bar{v}_0. \tag{15}$$

$$\gamma_0 = -\frac{R_0^2}{R^3}A. \tag{16}$$

So the shear rate of the melt in the capillary wall can be described as:

$$\dot{\gamma} = \dot{\gamma}_m + \gamma_0 \omega \cos\omega t. \tag{17}$$

Obviously, there is $\Delta\dot{\gamma} = \gamma_0\omega\cos\omega t$.

From the direction-z of the movement equation, it can be gained such as:

$$0 = -\frac{\partial p}{\partial z} + \frac{1}{r}\frac{\partial}{\partial r}(r\tau_{rz}). \tag{18}$$

Integrating above equation and predigesting it, then the shear stress of the capillary wall in Z-direction can be approximately described as:

$$\tau_R(t) \approx \frac{R\Delta P(t)}{2L} - \frac{\rho R_0^2 A \omega^2}{6R} \cdot \sin\omega t = \frac{RP(t)}{2L} - \frac{2\rho\pi^2 R_0^2 A f^2}{3R} \cdot \sin\omega t. \tag{19}$$

With the on-line collection system, the capillary entry pressure can be measured instantaneously. So the shear stress of the capillary wall in Z-direction can be obtained.

3 Analysis

Figure 2 is a plot of comparison of the experimental values and the theoretical values of the shear stress with different vibration frequency in one period. It is clear that the change amplitude of the shear stress increases with the vibration frequency increasing. Because the frequency is larger, the vibration force is larger and the ability of overcoming viscoelastic resistance is more. So the change amplitude of the shear stress increases, too. At the same time, the effects of vibration frequency on the experimental shear stress and the theoretical shear stress are accordant.

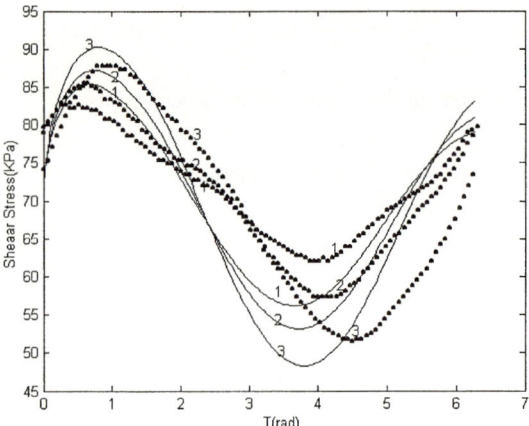

Fig. 2. Comparison of the experimental values and the theoretical values of shear stress with different vibration frequency in one period, $A = 0.1$ mm, $T = \omega t$, $\overline{v}_0 = 15$ (mm/s), 1: $f = 6$ Hz, 2: $f = 8$ Hz, 3: $f = 12$ Hz real line denotes theory values, broken line denotes experiment values

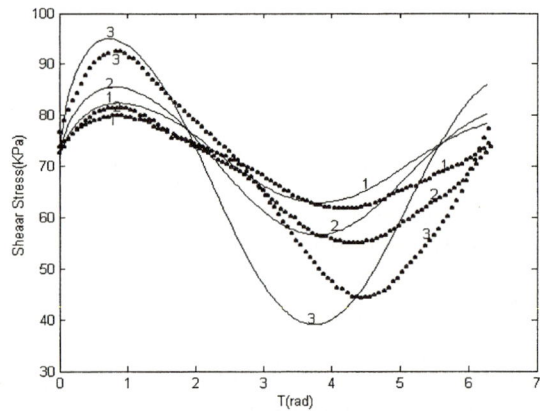

Fig. 3. Comparison of the experimental values and the theoretical values of shear stress with different amplitude in one period, $T = \omega t$, $f = 10$ Hz, $\overline{v}_0 = 15$ (mm/s), 1: $A = 0.05$ mm, 2: $A = 0.075$ mm, 3: $A = 0.15$ mm, real line denotes theory values, broken line denotes experiment values

Figure 3 is a plot of comparison of the experimental values and the theoretical values of the shear stress with different amplitude in one period. It is clear that the change amplitude of the shear stress increases with the amplitude increasing. Obviously, the shear stress change trend of the amplitude effect is as same as that of the vibration frequency effect. Similarly, the effects of amplitude on the experimental shear stress and the theoretical shear stress are also accordant.

4 Conclusion

Conclusionally, this generalized nonaffine transient network theory can be used for the interpretation of the rheological behavior of molten polymer in sine pulsating flow field. At one time, it is clear that the change amplitude of the shear stress increases with the vibration frequency or amplitude increasing, and it is the theoretical instruction for the engineering processing.

References

1. Zhang, J.: Constitutive Equations of Polymer Melts Under Vibration Force Fields: A Review, International Journal of Nonlinear Sciences and Numerical Simulation, 5(1)(2004) 37-44
2. Zhang, J.: Primary Research on Normal Stress Difference for Polymer Melts in Vibration Force Field, International Journal of Nonlinear Sciences and Numerical Simulation, 5(1)(2004) 97-98
3. Zhang, J., and Qu, J.P.: Nonaffine Network Structural Model for Molten Low-Density Polyethlene and High-Density Polyethylene in Oscillatory Shear, Journal of Shanghai University, 6(4)(2002) 292-296
4. Giacomin, A. J., and Jeyaseelan, R.S.: A Constitutive Theory for Polyolefins in Large Amplitude Oscillatory Shear, Polymer Engineering and Science, 35(1995) 768-777
5. Giacomin, A.J., and Oakley, J.G.: Structural Network Models for Molten Plastics Evaluated in Large Amplitude Oscillatory Shear, J.Rheol., 36(8)(1992)1529-1546
6. Giacomin, A.J., Samurkas, T., and Dealy, J.M.: A Novel Sliding Plate Rheometer for Molten Plastics, Polym Eng. Sci., 29(1989) 499-504
7. Giacomin, A. J., and Jeyaseelan, R. S.: How Affine is the Entanglement Network of Molten Low-Density Polyethylene in Large Amplitude Oscillatory Shear? Transactions of the ASME, 116(1994) 14-18
8. Liu, Y.J.: Dissertation of Doctoral Degree, South China University of Technology, (2002) 50-65
9. Zhang, J.: Dissertation of Doctoral Degree, South China University of Technology, (2003) 50-73

Microcalcifications Detection in Digital Mammogram Using Morphological Bandpass Filters

Ju Cheng Yang, Jin Wook Shin, Gab Seok Yang, and Dong Sun Park

Dept. of Infor. & Comm. Eng., Chonbuk National University,
Jeonju, Jeonbuk, 561-756, Korea
yangjucheng@hotmail.com

Abstract. Microcalcifications in digital mammogram images can be an early sign of breast cancer. It is very challenging, however, to detect all microcalcifications since they appear as slightly brighter spots than their backgrounds. In this paper, a new method is proposed to efficiently detect all microcalcifications using morphological bandpass filters. Morphological bandpass filters, each with two different structuring elements, are tuned to isolate frequency components of microcalcifications in this method. Experimental results show that the proposed method with bandpass filters can recognize microcalcifications with a higher visibility and more accurate positions and sizes comparing to the well-known wavelet transform method.

1 Introduction

The breast cancer is still one of main mortality causes in women, but the early detection can increase the chance of cure [1]. Microcalcifications are small-sized structures, which may indicate the presence of cancer since they are often associated to the most different types of breast tumors. In a mammogram, they are appeared as very small dots showing various shapes and gray levels with a limited resolution due to the X-ray systems limitations. This fact can be another severe constraint in case that a mammogram presents poor contrast between microcalcifications and the tissues around them.

Many approaches have been explored to detect microcalcifications. Some morphological methods for detecting microcalcifications are based on top-hat transform to elaborate all the bright blobs of small size which are the suspected microcalcifications [1, 2]. However, the top-hat transform enhances not only the microcalcifications but also other type of bright narrow elongated areas due to the connective tissue of the breast. There are other researches using multiresolution approaches based on wavelet transforms [3-5]. Since wavelet transform can decomposing images into subimages using a subband decomposing filterbank. In [4], low-high(LH), high-low(HL) and high-high(HH) subbands at full size are obtained, which correspond with horizontal, vertical and diagonal details. Reconstructing the sum of these bands (LH+HL+HH) generates an image containing small microcalcifications detected in each direction. But this method has a disadvantage of locating narrow, elongated objects in one direction. This would lead to false detection. In [5], this approach can have enhanced noises remaining in the subbands as well as microcalcifications.

To resolve the drawbacks of the multiresolution approaches, in this paper, we describe a method to detect the microcalcifications using four discrete steps including a morphological decomposing bandpass filtering process. The nonlinear morphological bandpass filter is implemented by applying the morphological multilevel opening operators twice to a mammogram image using two different structuring elements, and subtract one from another to obtain a bandpass filtered image [6]. Two parameters, sizes of the structure elements of the morphological bandpass filter and a threshold value, are determined by a tuning process. By this method, noises in the high frequency subbands and breast tissues with low frequency bands are removed by the morphological bandpass filter and microcalcifications, however, are remained in the frequency subbands for microcalcifications.

This paper is organized as follows: In Section 2, we describe the morphological bandpass filter and Section 3 explains the proposed detection method. Experimental results are explained in Section 4 and Section 5 shows the conclusion.

2 Morphological Bandpass Filter

Morphological filters perform nonlinear signal transformations that locally modify geometric features of signals. Basic morphological operations for binary images or multilevel images are dilation and erosion. By combining the basic operators, more complex operators, such as opening and closing, can be derived. Since a mammography contains gray level pixels, the basic multilevel morphological operations based on the umbra(U[]) and top(T[]) functions [8] are used in this paper.

From this definition, a multilevel dilation can be expressed in terms of a maximum operation and a set of addition operations as in Eq. 1.

$$(f \oplus s)(x) = \max[f(x-z) + s(z)] \quad \text{for all } z \in s \text{ and } x-z \in F \quad (1)$$

In this equation F and S are the domains of functions f and s, respectively. Similarly, the multilevel erosion is defined by Eq. 2.

$$(f \ominus s)(x) = \min[f(x+z) - s(z)] \quad \text{for all } z \in s \text{ and } x+z \in F \quad (2)$$

The opening of gray scale image can be expressed by a dilation of an erosion using the same structuring element. A morphological opening operator can eliminate protruding elements from images and the sizes of the protruding elements, which may be removed, depend solely on the size of the given structuring element. An opening operation with a smaller-size structuring element eliminates smaller protruding elements and we can consider the multilevel opening operation as a type of low-pass filter.

A morphological band-pass filter can be defined as the difference of two multilevel opening operations with two different structuring elements [6]. Outputs of a band-pass filter can be varied depending on the structuring elements of two opening.

Morphological bandpass filters can be used to find microcalcifications with certain sizes depending on structuring elements used and to eliminate scattered salt-and-pepper noises. Fig.1 shows an example. In Fig.1 (a), a structuring element with a smaller size is used for the opening operation. It removes small isolated noises. By applying an opening operator with a larger structuring element to the same original image, we can remove all protruding parts in the image including the microcalcifications as in Fig.1 (b). And Fig.1(c) shows the resulting, band-passed, difference image of previous two images filtered with opening operators. Fig.1 (d) shows the labeled binary image after a thresholding operation.

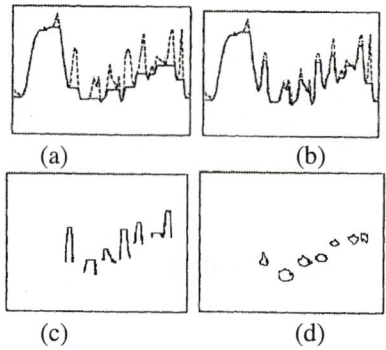

Fig. 1. (a) Opening operation with a smaller structuring element (b) with a larger element (c) Bandpass filtered image and (d) Labeled binary image after thresholding

3 Detection Method of Microcalcifications

The detection using morphological method consists of preprocessing process and ROI detection process.

3.1 Preprocessing

The preprocessing is performed using the five blocks. At first, the histogram equalization is used to enhance the contrast of the mammogram image and then, the next step is to determine whether the image contains a left or a right breast. A preliminary segmentation method is used to find the initial rough boundary in the third block. Since the boundary regions are rough and much distorted due to breast's own characteristics and noises, a breast curve tracking with a decision-making is used to find the precise boundaries of the breast.

3.2 ROI Detection

The proposed ROI detection algorithm, in this paper, consists of four steps as in Fig. 2: segmentation, bandpass filtering processing, labeling and post-processing. The segmentation step is to split the breast area into 256x256 segments. For each segmented subimage, a morphological bandpass filtering is applied to find the possible regions of microcalcifications. Two structuring elements for the bandpass filter are determined through a tuning process to generate the best result. We used square structuring elements filled with 1s.

After the filtering process, the third step is to find regions of interest (ROI) which possibly contain microcalcifications through a thresholding operation. This labeling operation with a threshold changes each subimage into a binary image. The threshold is also determined through a series of simulation study. The final step of the proposed detection algorithm is the post-processing to eliminate tiny isolated points using binary morphological closing and opening operators [9].

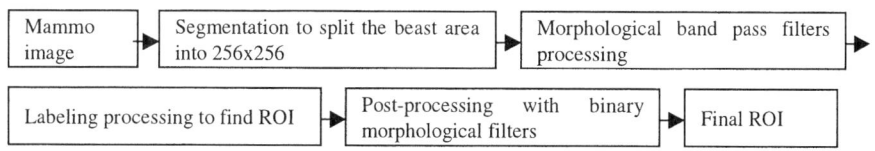

Fig. 2. Proposed four-step detection process

4 Experimental Results

The digital mammogram database used in this work is the MIAS (Mammographic Image Analysis Society [7]) database. Each pixel is 8-bits deep and at a resolution of 50μm x 50μm. And regions of microcalcifications have marked by the veteran diagnostician. Thirty 256x256 segments from eleven images (5 with benign and 6 with malignant microcalcifications) are selected for this experiment.

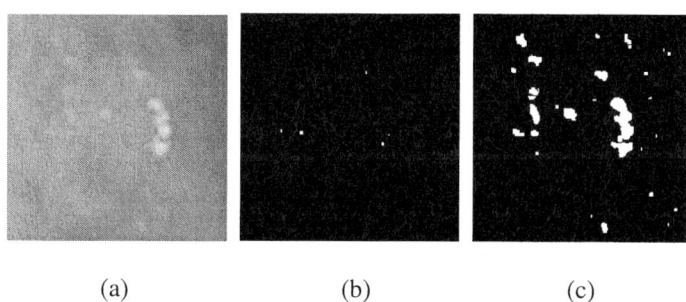

(a) (b) (c)

Fig. 3. (a)An original subimage. Simulation results using different sizes of structuring elements (b) 5X5 (f) 11X11 with a threshold T=10

In order to examine the effect of using different structuring elements and thresholds, a series of experiments are performed varying the size of structuring elements and threshold values. Fig.3 shows an example of simulation results using different sizes of structuring elements. From the experiments, we found that a 3x3 and 11x11 structuring elements are optimal for the bandpass filter to detect microcalcifications.

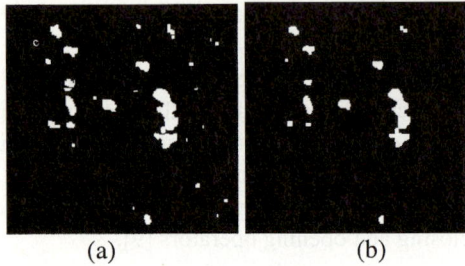

(a)   (b)

Fig. 4. (a) A bandpass filtered image (b) Post-processed image (3X3 and 11X11 structuring elements, threshold=10)

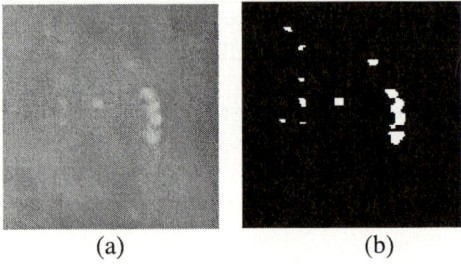

(a)   (b)

Fig. 5. (a)An original subimage (b)Simulation result using thewavelet transform

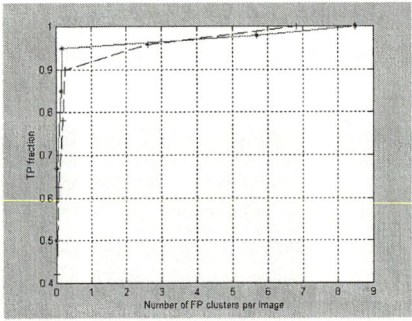

Fig. 6. FROC curve of MBF approach (solid curve) and DWT method (dashed curve)

The image processed by morphological bandpass filter 3X3 subtract 11X11, with a threshold 10 are showed as Fig. 4(a). And post-processing with binary morphological close-opening operation is used to acquire the final microcalcifications ROI as shown in Fig. 4(b).

To compare the performance of the proposed algorithm to an existing algorithm, we tested the well-known wavelet transform method introduced by Wang and Karayiannis in [5]. The final results were obtained using a nonlinear threshold method to enhance the histogram of the resulting mammograms. The results are shown in Fig.5. We can see that the detected microcalcifications are not accurate in terms of size, comparing to the proposed algorithm.

The performance of the proposed algorithm was evaluated by a free-response receiver operating characteristic (FROC) [10] in terms of TP (true-positive) fraction for a given number of FP (false-positive) clusters per image. Experimental results for 80 subimages were shown as in Fig. 6.

In the experiments, we used threshold values varying from 8 to 13 to draw the FROC curve. Seeing from Fig. 6, we can conclude that the MBF method is superior to DWT method. And a threshold of 10 can obtain a 95% TP rate with FP rate of 0.17 clusters per image when using MBF with 3X3 and 11X11 structuring elements.

5 Conclusion

In this paper, we describe a new four-step method for the detection of microcalcifications including morphological bandpass filters. The appropriate sizes of two structuring elements for the bandpass filters are determined through a number of experiments. To obtain regions of interest (ROI), a threshold value for the labeling process is also determined by a series of experiments. The proposed detection method generates accurate positions, with almost same sizes, of microcalcifications. A threshold of 10 can obtain a 95% TP rate with FP rate of 0.17 clusters per image when using MBF with 3X3 and 11X11 structuring elements. Further experiments have proceeded to obtain more objective results.

References

1. Dengler, J., *et al.* "Segmentation of microcalcifications in mammograms," IEEE, Trans. on Med. Imag., vol. 12, pp.634-64, 1993.
2. Astey, S., Hutt, I., Miller, P., Rose, P., Boggis, C. , Adamson, S., Valentine, T. , Davies, J. and Rmstrong, J., "Automation in mammography: computer vision and human perception, " Int. J. of pattern Recog. And Art. Intt., Vol 7, No 6,pp 1313-1338,m1993.
3. Yoshida, H., Doi, K. and Nishikawa, R.M., "Automated detection of clustered microcalcifications in digital mammograms using wavelet transform techniques," in proc. SPIE, 1994, vol.2167, pp.868-886.
4. Strickland, R.N. and Hahn, H.I., "Wavelet transform for detecting microcalcifications in mammograms," IEEE Trans. Med. Imag., vol.15, pp.218-229, Apr.1996.
5. Wang, T.C., Karayiannis, N.B., "Detection of microcalcifications in digital mammograms using wavelets," Medical Imaging, IEEE Transactions, Vo. 17 Issue: 4 , pp. 498 -509, Aug. 1998.
6. Saniie, J. and Mohamed, M.A.,"Ultrasonic Flaw Detection Based on Mathematical Morphology," IEEE Trans. on Ultrasonic, Ferroelectrics, and control, vol.41, No. 1, January 1994.
7. http://www.wiau.man.ac.uk/services/MIAS.
8. Haralick, R.M., Sternberg, S.R., and Zhuang, X., "Image analysis using mathematical morphology," IEEE Trans.Pattern Anal.Mach,Intell., vol.9,pp.532-550,july 1987.
9. Serra, J., Image analysis and Mathematical Morphology, Academic press,1988.
10. Swets, J. A. and pickette, R. M., Evaluation of diagnostic systems---Methods from signal detection theory, Acdemic,1982.

Peptidomic Pattern Analysis and Taxonomy of Amphibian Species

Huiru Zheng[1], Piyush C Ojha[1], Stephen McClean[2], Norman D Black[1], John G Hughes[1], and Chris Shaw[2]

[1] Faculty of Engineering, University of Ulster at Jordanstown,
Newtownabbey BT37 0QB, Northern Ireland
[2] School of Biomedical Sciences, University of Ulster at Coleraine,
Coleraine BT52 1SA, Northern Ireland
{h.zheng, pc.ojha, s.mcclean, nd.black, jg.hughes, c.shaw}@ulster.ac.uk

Abstract. In taxonomy, frog is a member of the Anura order under the class of Amphibia with many species. Some species of frog share a common biological function with other species, some have unique biological activities. Peptides contained in the frog species play an important role in these biological functions. In this paper, we investigate the degree of similarity of skin secretion peptide profiles between species. After the ESI-MS spectra from a sample of amphibian skin secretions are interpreted and deconvoluted by a heuristic deconvolution algorithm [1], a mass spectral profile is established. An overlap comparison algorithm is proposed and applied for the comparison of the peptide profiles between species. Results from the interspecies comparison and intraspecies comparison show that the peptide profiles of frog skin secretion can reflect their taxonomy classification.

1 Introduction

In taxonomy, frog is a member of the Anura order under the class of Amphibia. With so many different habitats ranging from arid desert regions to mountainous regions to swamps to tropical rainforests, it is not surprising that there are so many species of frogs. Some species of frogs share a common biological function with other species, some have unique biological activities, such as antibiotic, antimicrobial, or anticancer properties [2–4]. The comparison of peptides/proteins profiles of different species can help in searching for common peptides. In the view of taxonomy, specimens within the same species have higher similarity than different species from the same genus. It is a reasonable assumption that the degree of similarity of peptide profile between species is a reflection of their biological similarity.

Samples from skin or skin secretions (venom) of frogs are collected and analysed by HPLC/MS to generate mass spectral data by the Pharmaceutical Biotechnology Research Group (PBRG) in the University of Ulster at Coleraine (UUC). In order to study the taxonomy of species by using peptides mass spectral profiles, in this paper, the intraspecies and cross-species comparisons of peptides mass spectral profiles between different specimens or species have been made. In this study, two different comparisons, intraspecies and interspecies, were made to the skin secretion peptide profiles from three species *aurea, caerulea* and *infrafrenata* of the genus *Litoria*, and the species *capito* of the genus *Rana*.

The comparison of mass spectral peptide profiles of different samples of frog skin secretion indicates that the peptide profiles from same species but different specimens are significantly similar, peptide profiles from different species but same genus have some similarity while peptide profiles from different genera have very low similarity.

The remainder of this paper is organised as follows: In section 2 the establishment of peptides mass spectral profiles is described. Section 3 introduces the proposed overlap algorithm for the comparison of peptide profiles. The results of interspecies and intraspecies comparison of peptides mass spectral profiles are presented in Section 4 and Conclusions are presented in Section 5.

2 Establishment of Peptides Mass Spectral Profiles from Frog Skin Secretions

As an example of the process, analysis of *Litoria infrafrenata* 80 minute run mass spectra is presented in detail. The mass-charge ratio (m/z) in raw spectra is from 50 to 2000 m/z. There are many peaks in a raw spectrum and it is hard to determine manually the number of compounds and their molecular weights in the sample mixture.

After preprocessing of raw data, the total ion current (*TIC*) is calculated by the the algorithm described in [5]. The *TIC* gives the whole picture of retention time windows within which the compounds in the sample are separated. The appearance of peaks reflects the separation and detection of compounds at the relevant retention times. Higher peaks indicate the occurrence of more significant compounds, i.e. more abundant compound. Lower peaks, if above the noise threshold, show less significant compounds. These compounds are less significant only in their abundance but may have significant biological function. Overlapping peaks imply that more than one compound has been separated and detected at the same time or within a narrow time range.

In the second stage, time windows are automatically identified. Each time window starts from a local minimum, ends at another local minimum and centers at a local maximum.

At this point in the process, a heuristic deconvolution algorithm [1] is applied to identify the molecular weight of every compound in each time window determined above. Because of the complexity of the mass spectra of biological samples, the number of peptides is unknown in different mixtures, and differs across different time windows.

Finally, the peptides identified from each time window are put together to establish the peptide profile of the sample. For a sample S_i, the above analysis yields a peptide profile of the following form:

$$P(S_i) = \{(m_{i1}, t_{i1}, a_{i1}), (m_{i2}, t_{i2}, a_{i2}), ..., (m_{ij}, t_{ij}, a_{ij}), ..., (m_{in}, t_{in}, a_{in})\}$$

i.e. a set of n 3-tuples, each corresponding to one peptide; m_{ij}, t_{ij} and a_{ij} are the molecular weight, retention time (in the HPLC column) and relative abundance of the peptide respectively. The value of m (molecular weight) is the most important feature of the peptide. Relative abundance reflects the concentration of each peptide in the mixture. Retention time depends on the experimental conditions for liquid chromatography (flow rate, acetonitrile gradient etc) and to that extent it is less objective than the other two parameters. Its main use is to distinguish between two peptides which have identical or similar masses.

3 Overlap Algorithm for the Comparison of Peptide Profiles

We propose an overlap algorithm, inner product measurement, to measure of similarity between two peptide profiles.

Assuming two samples S_1 and S_2, their peptides mass spectral profiles are denoted as:

$$f(m,t,a) = \{(m_1,t_1,a_1),(m_2,t_2,a_2),...,(m_p,t_p,a_p),...,(m_m,t_m,a_m)\}$$

and

$$g(m',t',a') = \{(m'_1,t'_1,a'_1),(m'_2,t'_2,a'_2),...,(m'_q,t'_q,a'_q),...,(m'_n,t'_n,a'_n)\}$$

where m, t and a are the mass, retention time and relative abundance respectively, and the profiles have been normalised so that

$$\sum_{i=1}^{m} a_i^2 = 1,$$

$$\sum_{j=1}^{n} a'_j{}^2 = 1,$$

and m, n are the number of peptides respectively.

The overlap of two profiles $f(m,t,a)$ and $g(m',t',a')$ is then defined as the following:

$$overlap(f,g) = \sum_{i=1}^{m}\sum_{j=1}^{n} a_i a'_j \quad (1)$$

for all i and j that meet the following condition:

$$|m_i - m'_j| \le \Delta m, \ |t_i - t'_j| \le \Delta t$$

$\Delta m \approx 0.5$ Da and $\Delta t \approx 1$ minute represented the uncertainty in mass and retention time respectively to ensure that only similar/common peptides are taken into account.

Clearly, when $f(m,t,a)$ and $g(m',t',a')$ are the same, $overlap(f,g)$ equals 1, and when $f(m,t,a)$ and $g(m',t',a')$ are totally different, $overlap(f,g)$ equals 0, i.e., the higher similar of $f(m,t,a)$ and $g(m',t',a')$, the larger overlap $overlap(f,g)$ is, and the more abundant peptide makes greater contribution to $overlap(f,g)$ than the less abundant one.

4 Results

We have applied the above overlap algorithm to quantitatively measure the degree of similarity between two peptide profiles, and the results of the study of *Rana Capito*, *Litoria aurea*, *Litoria caerulea* and *Litoria infrafrenata* are presented.

4.1 Comparison of *Rana Capito*

Three venom samples of the species *Rana capito* are studied. The first sample is from a pool of female frogs, the second from male frogs, and the third one is from metamorphs[1].

Results show that the degree of similarity between female and male *R.captio* is 0.74, between metamorph and female is 0.67 and between metamorph and male is 0.40. These indicate the similarity between femal and male based on mass spectral peptide profiles is significant and metamorph is closer to female than to male. These agree with the known biological fact that samples from the same species have significant similarity and the metamorph is closer to the female frog while different sexual samples also show a difference. Peptide profiles are thus meaningful not only for the taxonomic classification, but also for the study of commonality and difference between female and male frogs and sexual possibility of metamorph.

4.2 Comparison of 9 Specimens of *L. infrafrenata*

Comparisons are carried out to measure the degree of similarity between nine specimens, denoted as *infr01*, *infr02*, ... , to *infr09*, of species *Litoria infrafrenata*. Table 1 shows the similarities between nine specimens vary from 0.19 to 0.72.

Table 1. The overlap measure of the degree of similarity for pairs of 9 specimens of *L.infrenata*

similarity (simple)	infr01	infr02	infor03	infr04	infr05	infr06	infr07	infr08	infr09
infr01	1.0	0.54	0.77	0.59	0.47	0.72	0.31	0.48	0.71
infr02		1.0	0.52	0.56	0.57	0.56	0.38	0.29	0.66
infr03			1.0	0.66	0.55	0.68	0.29	0.40	0.65
infr04				1.0	0.74	0.46	0.19	0.20	0.61
infr05					1.0	0.40	0.22	0.21	0.52
infr06						1.0	0.47	0.62	0.72
infr07							1.0	0.36	0.41
infr08								1.0	0.38
infr09									1.0

---

[1] Metamorph is a young amphibian that is newly transformed to the adult stage.

4.3 Comparison of Three Species of Genus *Litoria*

Samples from the pool of skin secretion of three species of genus *Litoria*: *L.aurea*, *L.caerulea* and *L.infrafrenata* are injected separately into HPLC/ESI-MS and the peptide profiles are established. The degree of the similarities of these three species are 0.021 (*aurea* vs *caerulea*), 0.094 (*aurea* vs *infrafrenata*) and 0.11 (*caerulea* vs *infrafrenata*) and are lower than the similarities between nine specimens of *L.infrafrenata*. The results also show that *L.aurea* is closer to *L.careulea* than to *L.infrafrenata*, and very few similarity between *L.caerulea* and *L.infrafrenata*.

Further comparisons were carried out to measure the degree of overlap similarity between the nine specimens (*infr01* to *infr09*) of *L.infrafrenata* and other pooled samples (see Table 2). Results indicate that the similarities between different species are less significant; the pooled sample of *L.caerulea* may have some similarity with some specimens of *L.infrafrenata* while very low similarity with some other specimens of *L.infrafrenata*; there is high similarity among the specimens from the same species and very low similarity between different genera.

Table 2. Overlap measure of degree of similarity between one individual specimen of *L.infrafrenata* and other pooled samples. "aur80", "cer80" and "infr80" are generated by 80 minute runs from the pooled sample of *L.aurea*, *L.cearulea* and *L.infrafrenata* respectively

similarity (simple)	infr01	infr02	infor03	infr04	infr05	infr06	infr07	infr08	infr09
aur80	0.043	0.13	0.080	0.071	0.017	0.056	0.012	0.006	0.015
cer80	0.15	0.20	0.15	0.053	0.16	0.052	0.015	0.033	0
infr80	0.48	0.37	0.45	0.31	0.35	0.54	0.46	0.38	0.52
CapitoFem	0	0.0053	0.0062	0	0.0020	0.012	0.0028	0.0082	0
CapitoMal	0.0045	0.0022	0.0023	0	0.0010	0.024	0.0053	0.011	0
CapitoMet	0.005	0.0048	0.0009	0	0.0028	0.022	0.0051	0.024	0

4.4 Similarities Between Different Species of *Litoria* and *Rana* Genus

Study has also been carried out to measure the similarity of three *Litoria* frogs with the frog *Rana capito* and the degree of similarities are 0.027(*aurea* vs *CapitoFem*), 0.013(*caerulea* vs *CapitoFem*), 0.012(*caerulea* vs *CapitoMal*), 0.0082(caerulea vs CapitoMet) and 0 (other pairs).

5 Conclusions

In conclusion, the comparisons of mass spectral peptide profiles of different samples of frog skin secretions show that:

1. the samples from different specimens of the same species are significantly similar;
2. the samples from different species but same genus also show some similarity; and
3. the samples from different genera have very low similarity.

These results reflect the taxonomic relationship. It suggests that it is potentially meaningful to apply the peptides mass spectral profiles in biological taxonomy. From the results obtained in this paper,

1. if two samples have degree of similarity greater than 0.30, then these two can be the same species;
2. if the degree of similarity is less than 0.01, then they are very likely from different genera; and,
3. otherwise, they may be from the same genus but different species.

However, the determination of a more accurate criterion would require analysis of a large number of species and genera.

References

1. Zheng, H., McClean, S., Ojha, P.C, Black,N.D., Hughes, J.G., Shaw, C.: Heuristic Charge Assignment For Deconvolution Of Electrospray Ionization Mass Spectra. Rapid Communications in Mass Spectrometry **17** (2003) 1 – 8
2. Pierre, T.N., Seon, A.A., Amiche, M., Nicolas, P.: Phylloxin, A Novel Peptide Antibiotic Of The Dermaseptin Family Of Antimicrobial/Opioid Peptide Precursors. European Journal of Biochemistry **267** (2000) 370–378
3. Rollins-Smith, L.A., Doersam, J.K., Longcore, J.E., Taylor, S.K., Shamblin, J.C., Carey, C., Zasloff, M.A.: Antimicrobial Peptide Defenses Against Pathogens Associated With Global Amphibian Declines. Developmental and Comparative Immunology **26** (2000) 63–72
4. Rozek, T., Wegener, K.L., Bowie, J.H., Olver, I.N., Carver, J.A., Wallace, J.C., Tyler, M.J.: The Antibiotic And Anticancer Active Aurein Peptides From The Australian Bell Frogs Litoria Aurea And Litoria Raniformis - The Solution Structure Of Aurein 1.2. European Journal of Biochemistry **267** (2000), no. 17, 5330–5341
5. Zheng, H., McClean, S., Ojha, P.C., Black, N.D., Hughes, J.G., Shaw, C.: Analysis And Interpretation Of Complex HPLC/MS Spectra For Drug Discovery. Technology and Health Care **9** (2001), no. 1,2, 95–97

Global and Local Shape Analysis of the Hippocampus Based on Level-of-Detail Representations

Jeong-Sik Kim[1], Soo-Mi Choi[1],
Yoo-Joo Choi[2], and Myoung-Hee Kim[2]

[1] School of Computer Engineering, Sejong University,
Seoul, Korea
jskim@sju.ac.kr, smchoi@sejong.ac.kr
[2] Dept. of Computer Science and Engineering, Ewha Womans University,
Seoul, Korea
{choirina, mhkim}@ewha.ac.kr

Abstract. Both volume and shape of the organs within the brain such as hippocampus indicate their abnormal neurological states such as epilepsy, schizophrenia, and Alzheimer's diseases. This paper proposes a new method for the analysis of hippocampal shape using an integrated Octree-based representation, consisting of meshes, voxels, and skeletons. Initially, we create multi-level meshes by applying the Marching Cube algorithm to the hippocampal region segmented from MR images. Then, we convert the polygonal model to intermediate binary voxel representation by a depth-buffer based voxelization, which makes it easier to extract a 3-D skeleton as well as relate to original MR images. As a similarity measure between the shapes, we compute L_2 norm and *Hausdorff* distance for each sampled mesh by shooting the rays fired from the extracted skeleton. It also allows an interactive analysis because of the octree-based data structure. Moreover, it increases the speed of analysis without degrading accuracy by using a hierarchical level-of-detail approach.

1 Introduction

The hippocampus in the brain plays an important role in memory encoding and retrieving. It is known that its morphological change is involved in neurological diseases such as epilepsy, schizophrenia, and Alzheimer's diseases [1]. Because of such tight relation between the shape and its function of the hippocampus, many researchers have been tried to investigate this issue.

In general, 3-D shape can be analyzed based on the skeleton-based description, e.g. spherical harmonic basis function (SPHARM) [3]. The M-rep is a local shape description that provides location, orientation and thickness of a shape, whereas the SPHARM is a global description using 3-D parametric surface. The former focuses on only large-scale shape differences since it is based on a coarse grid of medial atoms, and the latter captures both small- and large-scale shape differences. The combination of two methods can provide a coarse-to-fine shape analysis, but this cannot discriminate local shape differences interactively. Shenton *et al.* [4] proposed a spherical harmonic based method for the analysis of hippocampal left-right asymmetry. Styner *et al.* [5] developed a combined boundary and medial shape analysis method to estimate various hippocampal shape changes, such as bending, expansion and reduction.

In order to analyze local shape deformation efficiently, it is required to develop a hierarchical level-of-detail (LOD) representation consisting of hybrid information such as boundary surfaces, internal skeletons, and original medical images. We propose a new method for the analysis of hippocampal shape using an integrated Octree-based representation, consisting of meshes, voxels, and skeletons. As it allows us to analyze 3-D shape by changing the level-of-detail, it is possible to increases the speed of analysis without degrading accuracy.

The rest of this paper is organized as follows. Section 2 describes an integrated octree-based shape representation and Section 3 describes hippocampal shape analysis using the proposed representation. Experimental results and discussion are given in Section 4 and some conclusions and future works are given in Section 5.

2 Octree-Based Shape Representation

In this section, we describe how to represent the shape of hippocampal structure using a hierarchical LOD approach. Initially, multi-level surface meshes can be reconstructed by applying the Marching Cube algorithm [6] to the hippocampal region segmented from MR images. Then, we convert the polygon model to intermediate binary voxel representation by a depth-buffer based voxelization, which makes it easier to extract a skeleton as well as relate to original medical images. The extracted skeleton is used for sampling the meshes and computing a similarity measure between the shapes.

The octree is a data structure to represent objects in 3-D space, automatically grouping them hierarchically and avoiding the representation of empty portion of the space. This hierarchical structure is suitable for generating a local LOD representation. Fig. 1 shows an algorithm for octree-based shape representation. Three different types of shape information (i.e. meshes, voxels, and skeletons) are integrated into a single octree data structure. The function *Octree_Construction* stores the shape information into the octree data structure in a hierarchical way. The *root* node of the octree contains all of the surfaces meshes and intermediate level nodes represent subdivisions of the object space along the x, y, and z directions. The octree is successively refined by further subdivision of leaf nodes by inserting new vertices to them.

Binary voxel representation can be generated by the function *Depth_Buffer_Based_Voxelization*. The function computes the bounding box surrounding the surface mesh in the canonical coordinate frame and then specifies the buffer resolution n. In addition, it generates six buffer images by parallel-projecting the object onto the face of the box. For each pair of faces of the box along an axis, we obtain a minimum and maximum distance for an object. When a certain voxel is located inside all three distances, then the voxel belongs to the object. In this case, the computing time of voxelization is independent of the shape complexity and is proportional to the buffer resolution.

The function *Slice_Skeletonization* extracts 3-D skeleton based on the binary voxel representation. Here, the skeletal points are computed by the center of the object in each slice image or by 3-D distance transformation in more general case. The gaps between neighboring skeletal points are linearly interpolated. Fig. 2 shows how to integrate three types of shape information into a single octree data structure. Here, shape information in different nodes is labeled with different colors.

Depth_Buffer_Based_Voxelization( )
    Set up depth buffers for a 3-D model ($[x_1, x_2]$, $[y_1, y_2]$, $[z_1, z_2]$).
    Specify the buffer resolution n (n^3 voxels).
    For each voxel $v(i,j,k)$
        Compute corresponding buffer values. $x_1(j,k)$, $x_2(j,k)$, $y_1(k,i)$, $y_2(k,i)$, $z_1(j,i)$, $z_2(j,i)$
        Decide $v(i,j,k)$ belongs to the object or not.
Slice_Skeletonization( )
    Divide 3-D voxel space into n slices along y *axis*.
    For each slice S_i, Compute a center point C_i from the object boundary.
    For each pair of skeletal points: (C_i, C_j) // $j=i+1$ ($0 \leq i \leq n$)
        If (Euclidean_distance(C_i, C_j) > *threshold*) Interpolate (C_i, C_j) using *threshold*.
Octree_Construction(NODE *root*)
    If ((*meshes_count* > *max_triangles*) and (*curr_subdivison* < *max_subdivision*))
        Subdivide *root* node into eight child nodes.
        For each child node
            Set the child node to *current_node*.
            Find polygons that are included in *current_node*.
            Octree_Construction(*current_ node*)
    Else Store the polygons to *current_ node*.

Fig. 1. The algorithm for the octree-based shape representation

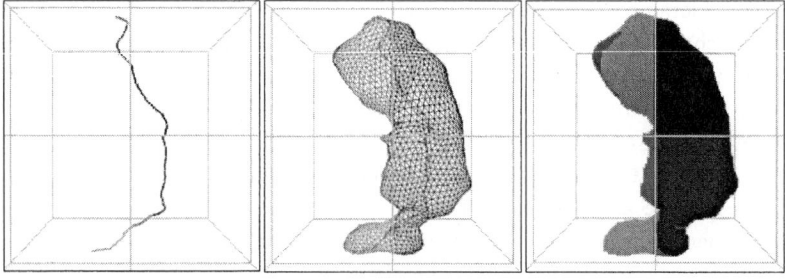

Fig. 2. An integrated octree-based representation of the right hippocampus: (left) Labeled skeleton; (middle) Labeled meshes; (right) Labeled voxels

3 Hippocampal Shape Analysis Using LOD Representations

In this section, we describe how to analyze the shape of the hippocampus based on the proposed octree-based shape representation. Given a reconstructed surface mesh, it has to be placed into a canonical coordinate system, where the position, orientation and scaling are normalized.

The position normalization is accomplished by placing the origin of the coordinate system at the center of the mesh model. For the rotation normalization, we apply the principal component analysis (PCA) to all vertices of the surface mesh. The PCA identifies the most informative directions called the principle components of the distribution, and are given by the eigenvectors of its covariance matrix. The scale of the surface mesh is normalized by three scaling factors that represent average distance if points form the origin along x, y, and z axes, respectively.

The feature vectors are computed for each sampled mesh by shooting the rays fired from the extracted skeleton. As a similarity measure between the shapes, we compute L_2 norm and *Hausdorff* distance for the sampled meshes. The L_2 norm is the metric to compute the distance between two 3-D points by Eq. (1), where x and y represents the centers of corresponding sample meshes. *Hausdorff* distance measures the extent to which $h(A, B)$ is the directed *Hausdorff* distance from shape A to shape B.

$$L_2(x, y) = \left(\sum_{i=0}^{k=3} |x_i - y_i|^2 \right)^{1/2} \quad (1)$$

$$H(A,B) = \max(h(A,B), h(B,A))$$
$$\text{where } A = \{a_1, \cdots, a_m\},\ B = \{b_1, \cdots, b_n\},\ h(A,B) = \max \min \|a - b\| \quad (2)$$

Mesh_Sampling( )
  Specify the number of skeletal points n and the number of rays m.
  For each skeletal point p_i
    Construct a circle map of p_i and shoot rays.
    For each ray r_j, Compute intersection between the ray and all polygon meshes.
Distance_Computation(*reference, target*)
  For each pair of sampled meshes   // $m_{reference}$, m_{target}
    Compute the centers of $m_{reference}$ and m_{target}.
    Compute the L_2 norm or *Hausdorff* distances between the two centers.

Fig. 3. The algorithm for measuring L_2 norm and *Hausdorff* distances

4 Results and Discussion

Our method was applied to analyze the 3-D hippocampal structure extracted from MR brain images. Fig. 4 (left) shows the results of comparison between the hippocampal structures of a normal subject (orange) and a patient with epilepsy (magenta). Fig. 4 (middle) and (right) show how to compare two hippocampal shapes based on the proposed octree scheme. It is possible to reduce the computation time in comparing two 3-D shapes by picking a certain skeletal point (Fig. 4(middle)) or by localizing an octree node (Fig. 4(right)) from the remaining parts. It is also possible to analyze the more detailed region by expanding the resolution of the octree, since it has a hierarchical structure. The result of shape comparison is displayed on the surface of the target object using color-coding.

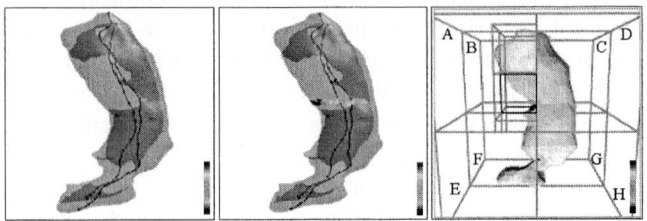

Fig. 4. Local and hierarchical shape analysis: (left) super-imposition view; (middle) skeletal point picking based local shape analysis; (right) octree-based hierarchical shape analysis

Table 1. The result of global shape analysis

	L_2 norm	*Hausdorff* distance	Volume difference	Rank
N_L:T1	1.220	1.441	94.3%	1
N_L:T2	1.554	1.664	109.3%	2
N_L:T3	2.420	2.573	88.8%	3

Table 2. The result of local shape analysis based on the octree structure

	A	B	C	D	E	F	G	H
P_L:T4	0.15	0.77	0.84	3.15	0.00	0.00	0.00	0.15
P_L:T5	1.20	0.00	0.00	0.00	3.12	2.00	1.00	1.44
N_R:T6	0.06	1.02	0.06	0.00	0.00	0.12	0.00	0.00
N_R:T7	0.00	0.00	0.00	0.00	1.54	1.31	1.313	1.54

Table 1 gives the results of global shape difference between the normal left hippocampus (N_L) and three deformed targets (T1, T2, and T3) in the upper area. From Table 1, it is founded that the volume of T1 and T3 are 5.7% and 11.2% smaller than N_L respectively, whereas T2 is 9.3% larger than N_L. Table 2 summarizes the result of local shape differences by comparing the 3-D shapes between the reference models (P_L and N-R) and deformed targets (T4~T7), respectively. P_L is an abnormal left hippocampus in epilepsy and N_R is a normal right hippocampus. T4~T7 are deformed targets at specific region (i.e. upper-front-right, bottom-front-left, upper-back-left, and the bottom region, respectively). In Table 2, we can easily observe that the similarity error at deformed region is higher than at other regions. As shown in Table 1 and 2, our method is able to discriminate the global shape difference and is also able to distinguish a certain shape difference at a specific local region in a hierarchical fashion.

5 Conclusions and Future Works

This paper presents a new method for the analysis of hippocampal shape using an integrated octree-based scheme where three different representations, i.e. meshes, voxels, and skeletons are combined in a hybrid fashion. In addition, the point-picking interface allows users to access a specific local area within the organ and then compare 3-D shapes interactively. The present system also increases the speed of analysis without degrading accuracy by using a hierarchical level-of-detail approach. Although the present study only addresses the issue of comparison between two 3-D anatomical objects, it can be extend, in the future, to a multiple comparison case, where it is possible to categorize abnormal and normal organs into several distinct groups using a multiple classifier such as Support Vector Machine.

Acknowledgements

We would like to thank Prof. Seung-Bong Hong and Woo-Suk Tae in Samsung Medical Center for giving useful comments. This work was supported in part by the Korean Ministry of Science and Technology under the National Research Laboratory program. This work was also supported in part by grant No. (R04-2003-000-10017-0) from the Basic Research Program of the Korea Science & Engineering Foundation.

References

1. Dean, D., Buckley, P., Bookstein, F., Kamath, J., Kwon, D., Friedman, L., Lys, C.: Three dimensional MR-based morphometric comparison of schizophrenic and normal cerebral ventricles. Vis. In Biom. Computing, Lecture Notes in Comp. Sc., (1996) 363-372
2. Pizer, S., Fritsch, D., Yushkevich, P., Johnson, V., Chaney, E.: Segmentation, registration, and measurement of shape variation via image object shape. IEEE Trans. Med. Imaging, Vol. 18. (1999) 851-865
3. Brechbühler, C., Gerig, G., Kübler, O.: Parameterization of closed surfaces for 3-D shape description. Computer Vision, Graphics, Image Processing, Vol. 61. (1995) 154-170
4. Shenton, ME., Gerig, G., McCarley, RW., Szekely, G., Kikinis, R.: Amygdala-hippocampal shape differences in schizophrenia: the application of 3D shape models to volumetric MR data. Psychiatry Research Neuroimaging, Vol. 115. (2002) 15-35
5. Styner, M., Lieberman, J.A., Gerig, G.: Boundary and Medical Shape Analysis of the Hippocampus in Schizophrenia. MICCAI, No. 2. (2003) 464-471
6. Lorensen W.E., Cline, H.E.: Marching cubes: A high resolution 3D surface construction algorithm. Computer Graphics, Vol. 21. No. 4. (1987) 163-169
7. Karabassi, E.A., Papaioannou, G., Theoharis, T.: A Fast Depth-Buffer-Based Voxelization Algorithm. Journal of Graphics Tools, ACM, Vol. 4. No.4. (1999) 5-10
8. Möller, T., Trumbore, B.: Fast, minimum storage ray-triangle intersection. Journal of Graphics Tools, Vol. 2. No. 1. (1997) 21-28

Vascular Segmentation Using Level Set Method

Yongqiang Zhao[1], Lei Zhang[2], and Minglu Li[1]

[1] Department of Computer Science and Engineering,
Shanghai Jiaotong University, Shanghai 200030, China
{zhao-yq, li-ml}@cs.sjtu.edu.cn
[2] Department of Radiology, Tongji Hospital of Tongji University,
Shanghai 200065, China
zhanglei4302@hotmail.com

Abstract. In this paper, we propose a two-stage level set segmentation framework to extract vascular tree from magnetic resonance angiography(MRA). First, we smooth the isosurface of MRA by anisotropic diffusion filter. Then this smoothed surface is treated as the initial localization of the desired contour, and used in the following geodesic active contours method, which provides accurate vascular structure. Results on cases demonstrate the effectiveness and accuracy of the approach.

1 Introduction

Accurate description of vasculature structure plays an important role in many clinical applications, e.g., for quantitative diagnosis, surgical planning, and monitoring disease progress, etc. However, due to the complex nature of vasculature structure, depicting it proves to be a difficult task. With the advancement of MRA acquisition technology, rapid and noninvasive 3D mapping of the vascular structures are available. A variety of methods have been developed for segmenting vessels within MRA [1, 3]. The most common method is maximum intensity projection(MIP); it was generated by selecting the maximum value along an optical ray that corresponds to each pixel of the 2-D MIP image. It is an easy and fast way for visualization of angiography; meanwhile, MIP can be obtained irrespectively from any direction of transverse. However, it loses the 3-D information, and it is not helpful for finding stenosis [1]. Another kind of popular methods is multiscale methods[4], which are based on the assumption that the centerlines of the vessels often appear brightest in the image. These methods firstly detect the intensity ridges of the image as the centerlines, then determine the width of vessels by multiscale response function.

The level set method for capturing moving fronts was introduced by Osher and Sethian [5]. It has proven to be a robust numerical device for this purpose in a diverse collection of problems. The advantages of the level set representation are that it is intrinsic (independent of parameterization) and that it is topologically flexible. The surface evolution in level set algorithm can be regarded as an initial value partial differential equation (PDE) in higher dimension. To get the good segmentation result, the proper initialization is necessary and important. Hossam used a level set based segmentation algorithm to extract the vascular tree from phase contrast MRA. The algorithm initializes level sets in each slice using automatic seed initialization and then each level set approaches the steady state and contains the vessel or non-vessel area iteratively [8]. The approach proved to be fast and accurate.

In this paper, we present a two-stage level set framework for vascular segmentation from time of flight (TOF)-MRA, including initialization and level set surface evolution. We observe that isosurfaces of MRA can closely approximate vessel shape although they remain subjective in nature. The isosurface can be refined to provide a smoother surface that is more closely aligned with image edges in the gradient magnitude of the image. In this framework, the smoothed isosurface is regarded as the initialization of the level set method. Then, we choose the speed function proposed by Caselles to evolve the initial contour.

2 Level Set Method

The basic idea of level set method is to start with a closed curve in two dimensions (or a surface in three dimensions) and allow the curve to move perpendicular to itself at a prescribed speed.

Consider a closed moving interface $\Gamma(t)$ in R^n with co-dimension 1, let $\Omega(t)$ be the region (possible multiply connected) that $\Gamma(t)$ encloses. We associate with $\Gamma(t)$ an auxiliary scalar function $\varphi(x,t)$, which is known as the level set function. Over the image area, it is a surface which is positive inside the region, negative outside and zero on the interfaces between regions. The evolution equation for the level set function $\varphi(x,t)$ takes the following formula:

$$\varphi_t + F|\nabla\varphi| = 0 \qquad (1)$$

The evolution of the level set function is determined by the speed function F. A number of numerical techniques make the initial value problem of Eq.(1) computationally feasible. The two of the most important techniques are "up-wind" scheme which addresses the problem of overshooting when trying to integrate Eq.(1) in time by finite differences and "narrow band" scheme that solves Eq.(1) in a narrow band of voxels near the surface.

Typically, in segmentation applications, the speed function may be made up of several terms. Eq.(1) is always modified as follows:

$$\varphi_t = F_{prop}|\nabla\varphi| + F_{curv}|\nabla\varphi| + F_{adv} \cdot \nabla\varphi \qquad (2)$$

where F_{prop}, F_{curv} and F_{adv} are speed terms those can be spatially varying. F_{prop} is an expansion or contraction speed. F_{curv} is a part of the speed that depends on the intrinsic geometry, especially the curvature κ of the contour and/or its derivatives. F_{adv} is an underlying velocity field that passively transports the contour [6].

Level set segmentation relies on a surface-fitting strategy, which is effective for dealing with both small-scale noise and smoother intensity fluctuations in volume data. The level set segmentation method creates a new volume from the input data by solving an initial value PDE with user-defined feature extracting terms. Given the local/global nature of these terms, level set deformations alone are not sufficient; they must be combined with powerful initialization techniques in order to produce successful segmentations [7]. Meanwhile, the speed term F also plays a key role in medical image segmentation. F depends on many factors including the local properties of the curve, such as the curvature, and the global properties, such as the shape and the position of the front.

3   Our Proposed Algorithm

Level set method contains many good mathematical properties which make it an accurate description for front propagation. For image segmentation, the level set method has the ability to handle objects with topology changes from the initial contour. This paper presents a framework which is composed of two major stages: initialization and level set surface evolution. Each stage is equally important for generating a correct segmentation.

3.1   Initialization

In MR blood imaging techniques, TOF MRA utilizes the in-flow effect. By using a very short repetition time during data acquisition, the static surrounding tissue become saturated, resulting in low signal intensity in the acquired images. In contrast, the replenished flowing spins are less saturated, providing a stronger signal, which allows the vessel to be differentiated from the surrounding tissues [2]. On the view of this principle, we can conclude that the most area of vasculature can be thought as on one isosurface.

Meanwhile, 3-D prefiltering is an important step before 3-D segmentation can proceed. In our level set framework, we propose an anisotropic diffusion process for filtering images. It overcomes the major drawbacks of conventional filtering methods and is able to remove noise while preserving edge information. It is also simple in implementation. The method has been used to enhance variety of medical images.

From above descriptions, in the first step of our level set framework, we firstly choose the value of isosurface in TOF MRA data sets according to MIP, then use an anisotropic diffusion filter to smooth this isosurface; the fourth order level set method is implemented to solve anisotropic diffusion on the normal map of the surface; finally, the surface deforms to fit the smoothed normal. After that, we get the initial contour of level set framework.

3.2   Level Set Surface Evolution

In second stage of framework, geodesic active contour is adopted. Caselles et al [10] used an energy minimization formulation to design the speed function. This leads to the following speed function formulation:

$$\varphi_t = c(\kappa + V_0)|\nabla \varphi| + \nabla c \cdot \nabla \varphi \qquad (3)$$

where κ denotes the front curvature; V_0 is constant, positive V_0 shrinks the curve, and negative V_0 expands the curve. In practice, as the initialization is already close to the boundaries of vasculature tree, we always set V_0 to be small value even zero. The curve evolution is coupled with the image data through a multiplicative stopping term $c = \frac{1}{1+|\nabla(G_\sigma*I)|}$. The expression $G_\sigma * I$ denotes the image convolved with a Gaussian smoothing filter whose characteristic width is σ. I is the intensity of image. The term c makes V_0 and κ near zero around the boundary. The term $\nabla c \cdot \nabla \varphi$ can pull back the contour if it passes the boundary. So the initial contour does not have to lie wholly within the shape to be segmented [9]. The initial contour is allowed to overlap the shape boundary, which makes our framework more robust.

4 Results

We have run the proposed framework on various TOF-MRA images. In Fig.1, the brain MRA data set consists of 60 slices. Each slice has a size of 256*256. Fig.1(a) and (b) are the MIP representations of this series from different direction; Fig.1(c) and (e) are the results of manual initialization level set method and our level set segmentation framework respectively, visualized by marching cube algorithm. We found that many small vessels could not be extracted in Fig.1(c). In this method, the initial contour is manually defined by placing some points at the positions of the larger vessels. As the human brain vascular system is a complex anatomical structure and MRA is of low contrast and resolution, it is hard to provide a proper initialization for level set segmentation. In contrast, our method treats smoothed isosurface as the initial contour so that it can extract mostly exact vascular tree (Fig.1(e)). Fig.1(d) and (f) are the first stage and final results respectively, also visualized by marching cube algorithm. Comparing these two results with MIP(Fig.1(b)), Fig.1(d) and Fig.1(f) extract almost the same vascular tree as Fig.1(b), at the same time, they provide the doctor with the 3D spatial information; moreover, the final result shows the better connectivity of the vessels.

However, isosurfaces have limited values for shape analysis since the image intensity of an object and the background are usually inhomogeneous. In MRA, other soft tissues may share the same range of grey levels as the vessels. In such condition, our proposed method cannot provide an accurate segmentation of an entire anatomical structure. The MRA series of Fig.2 is 256*256*50. Although the final result (Fig.2(b)) provides more 3D information and more details of vascular tree than in MIP(Fig.2(a)), other tissue or organs those are not of interest can obscure visualization of our interest. As a result, some vessels are hard to be distinguished in Fig.2(b).

5 Conclusion

In this paper, we present a two-stage level set framework to extract the vascular tree from TOF-MRA images. This framework provides a good initialization and

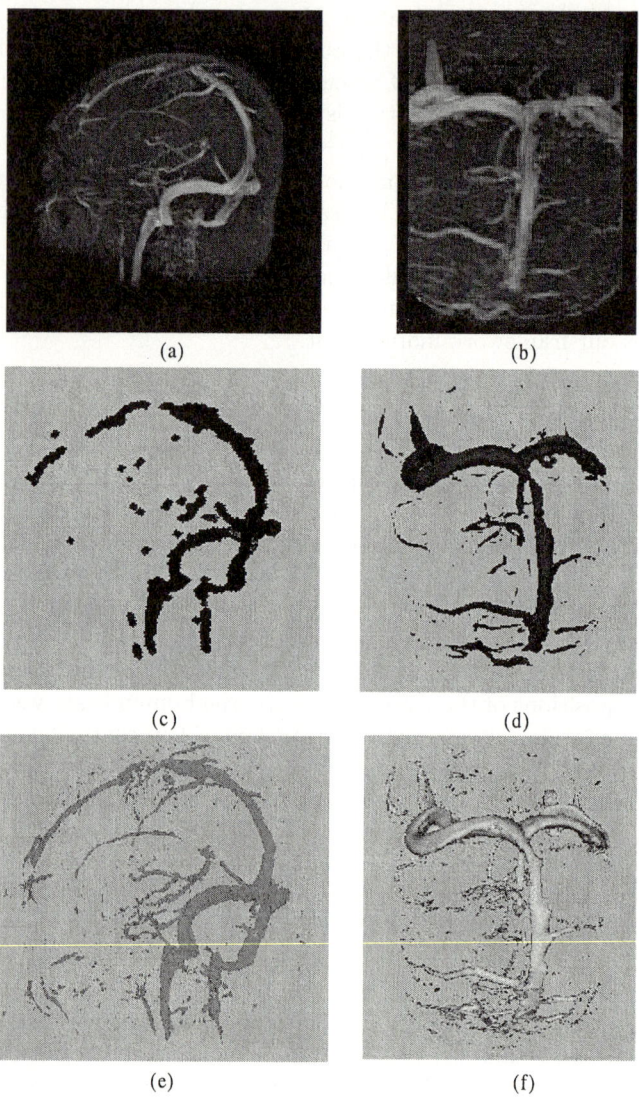

Fig. 1. (a) (b) MIP from different direction (c) manual initialization level set method result (e) our level set method result (d) first stage result of our method (f) final result of our method

makes the initial model reach the real boundary quickly. As we have said, blood vessels are especially difficult to segment. We are still far away from achieving the robust segmentation in real time. Fuzzy connectedness and differential geometry features of images can be added to initialization and the speed function to enhance the segmentation results and make the framework more robust.

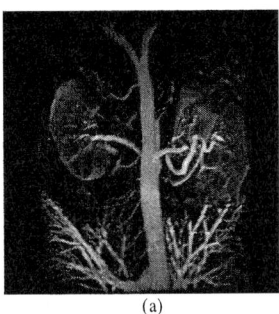
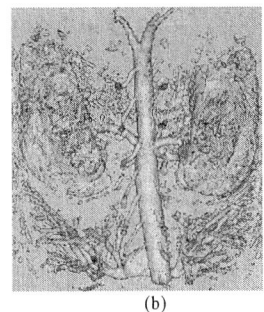

(a)   (b)

Fig. 2. (a) MIP (b) proposed method

References

1. Jasjit, S., Kecheng, L., Laura, R. and Swamy, L.: A Review on MR Vascular Image Processing: Skeleton Versus Nonskeleton Approaches:Part II. IEEE Trans.Information technology in Biomedicine, vol.6, no.4, pp. 338-350, Dec. 2002.
2. Jasjit, S., Kecheng, L., Laura, R. and Swamy, L.: A Review on MR Vascular Image Processing Algorithms: Acquisition and Prefiltering: Part I. IEEE Trans.Information technology in Biomedicine, vol.6, no.4, pp. 324-337, Dec. 2002.
3. Kirbas, C., Quek, F.: Vessel Extraction Techniques and Algorithms: A Survey. IEEE Conference Bio-Informatics and Bio-Engineering (BIBE), 2003, pp. 238-245.
4. Krissian, K., Malandain, G., Ayache, N., Vaillant, R. and Trousset, Y.: Model Based Detection of Tubular Structures in 3D Images Computer Vision and Image Understanding, vol. 80, no. 2, pp. 130-171, Nov. 2000.
5. Osher, S., Sethian, J. A.: Fronts propagating with curvature dependent speed: Algorithms based on Hamilton-Jacobi formulations, Journal of Computational Physics 79, pp. 12-49, 1988.
6. Han, X., Xu, C. and Prince, J. L.: A Topology Preserving Deformable Model Using Level Sets, Proc. IEEE Conf. CVPR 2001, vol. II, pages 765–770, Kauai, HI, Dec 2001.
7. Ross, W., David, B., Ken, M. and Neha, S.: A Framework for Level Set Segmentation of Volume Datasets, Proceedings of ACM International Workshop on Volume Graphics. pp. 159-168, June 2001.
8. Hossam, E. D., Abd, E. M.: Cerebrovascular segmentation for MRA data using level sets. International Congress Series Volume: 1256, pp. 246-252 June, 2003
9. Xu, C., Pham, D. L. and Prince, J. L.: Medical Image Segmentation Using Deformable Models, pp. 129-174, SPIE Press, May 2000.
10. Caselles, V., Kimmel, R. and Sapiro, G.: Geodesic active contours, Proc. 5th Int'l Conf. Computer Vision, pp. 694-699, 1995.

Brain Region Extraction and Direct Volume Rendering of MRI Head Data

Yong-Guk Kim[1], Ou-Bong Gwun[2], and Ju-Whan Song[3]

[1] School of Computer Engineering, Sejong University,
Seoul, Korea
ykim@sejong.ac.kr
[2] Division of Electronics and Information Engineering,
Chonbuk National University,
Jeonju, Jeonbuk, Korea
obgwun@chonbuk.ac.kr
[3] School of Liberal Art, Jeonju University,
Jeonju, Jeonbuk, Korea
jwsong@jj.ac.kr

Abstract. This paper proposes a new 3D visualization method for MRI head data based upon direct volume rendering. Surface rendering has difficulties in displaying speckles due to information loss during the surface construction procedure, whereas direct volume rendering does not have this problem, though managing MR head image data is not an easy task. In our method, brain structures are extracted from MR images, and then embedded back into the remaining regions. To extract the brain structure, we use a combination of thresholding, morphology and SNAKES operations. Experimental results show that our method makes it possible to simultaneously visualize all the anatomical organs of human brains in three dimensions.

1 Introduction

Direct volume rendering is usually used to explore the internal structure. But it is often difficult to visualize MRI head data directly, due to fact that the classification methods adopted in direct volume rendering cannot discriminate anatomical organs of MR head image from each other[3]. Although it is essential for radiologists to have a reliable MRI classification technique, there are too many different factors to consider, such as image contrast, image resolution, the ratio of signal to noise, and the capacity of data. The widely-used approaches in classifying anatomical objects from MR head images include: thresholding[1], statistical segmentation[2], and region growing[5]. The hierarchical approaches have achieved great success in recent years. In this paper we propose a hierarchical brain segmentation method that combines thresholding, morphology operation, SNAKES operations and a direct volume rendering method for MRI head data.

2 Overview of the Present Method

Fig. 1 shows the distinctive outcomes for the data acquired from different image modalities. The image from CT head data[7] can display skin and skull as shown in Fig. 1(a), whereas the image from MRI[8] cannot display skin and brain as shown in Fig. 1(b). Drevin et al. classify the anatomical parts of CT data assuming that air, fat, soft tissue, and bone all have different voxel values[3]. But the MRI head data has voxels with the same intensity that belong to different anatomical organ of the head. To resolve this problem, we have developed a three-stage pipeline for direct volume rendering MR head image data as illustrated in Fig. 2. In the first stage, the head region of MR head image slices is divided into the brain region and the remaining region. In the second stage, the values of voxels of the brain region are increased and the voxels of the brain region are recombined with the remaining region, i.e. skin, skull, and scarp, using the window-level transfer function. In the third stage, all slices of the recombined MR head image data is drawn using direct volume rendering.

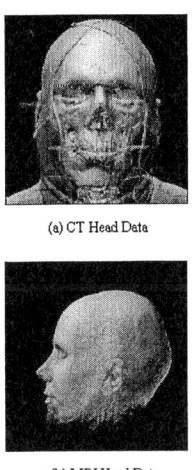

(a) CT Head Data

(b) MRI Head Data

Fig. 1. Direct volume rendering of 3D tomography

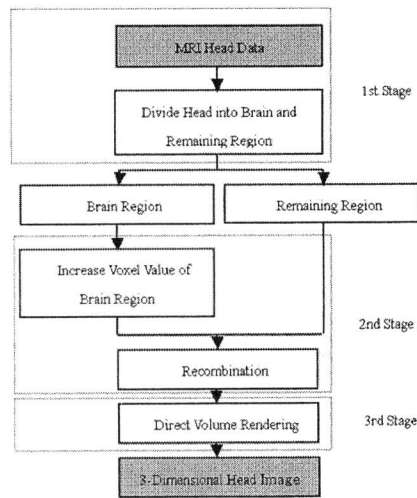

Fig. 2. Direct volume rendering pipeline for MRI head data

3 Brain Segmentation and Recombination

3.1 Head Segmentation

We segment the head region from an MR head image slice using Brummer's method[1] together with the morphology operations. Brummer assumed that the histogram of the noise voxels within such data confirms to a Rayleigh function. Rayleigh distribution function $r(x)$ is calculated by best curve fitting in the low grey-value range of the histogram. Equation (1) represents a grey level histogram of the rest of the data. The boundary threshold between background region and the head region with a minim error is found at the voxel and the value t minimizes equation (2).

$$g(x) = h(x) - \gamma(x) \tag{1}$$

$$\varepsilon_t = \sum_{x=0}^{t-1} \alpha(x) + \sum_{x=t}^{\infty} \gamma(x) \tag{2}$$

The thresholding procedure with the value of voxel t generates a set of binary slice masks. However, the salt and pepper noises may occur on the binary slice mask, because there are some voxels having higher values in the back ground noise (lower values in the head) than the value of voxel t. These noises can be removed from the binary slice mask using a median filter. In some cases, the noises of MRI slices generate lumps attached near the head region. The lumps can be removed by morphology opening operation. Edge detection procedure is operated on the slice mask for finding the head boundary. The largest edge among the edges on the slice becomes the head boundary. The head mask is acquired by filling the head boundary with the largest gray value.

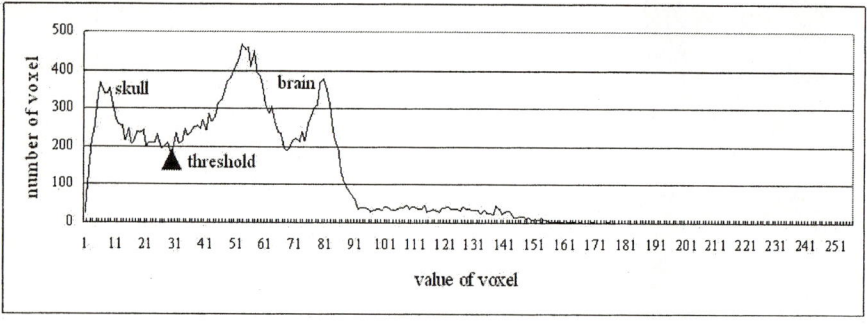

Fig. 3. Histogram of MR head image slice

3.2 Approximate Brain Segmentation

The approximate brain region is acquired by applying histogram thresholding and morphology operation. It is assumed that the histogram of the brain region has a normal distribution for the white and grey matter, since the brain consists of homogeneous material occupying the most part of the head. The threshold of the brain region and the remaining region (skull) becomes the first local minimum of the lower grey-value range of histogram as drawn in Fig. 3, where it shows the boundary threshold between the brain region and the remaining region for the histogram of NO 35 MRI Head slice acquired from University of North Carolina. The thresholding procedure with the first local minimum generates a binary slice mask. Since the remaining regions are usually connected to the brain part, those are separated from the brain region using morphology operation. And edge detection procedure is applied to the slice mask to find the approximate brain boundary.

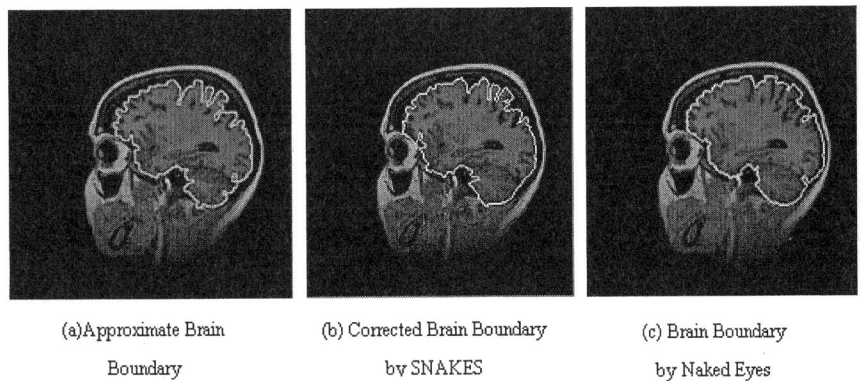

(a) Approximate Brain Boundary

(b) Corrected Brain Boundary by SNAKES

(c) Brain Boundary by Naked Eyes

Fig. 4. Brain boundaries detected by proposed method and by naked eyes

3.3 Exact Brain Region Detection

We correct the approximate brain boundary of section 3.2 into the exact brain boundary using the KWT SNAKES algorithm[4]. The brain boundary is represented by the collection of n points on the image plane. SNAKES operations finds the boundary by solving an energy minimum problem and moving the boundary into the area containing lower total energy. In SNAKES, the energy function is represented as equation (3).

$$E_i = \alpha_i E_{count,i} + \beta_i E_{curve,i} + \gamma_i E_{img,i} \qquad (3)$$

Continuous energy $E_{count,i}$ makes the boundary continuous, and curvature energy $E_{curve,i}$ makes the boundary smooth. Energy $E_{img,i}$ is called the image energy that can be defined for two types of features. The first type is I and the second type is $-|\nabla I|$. The first type makes the boundaries moving into the dark area, and the second does it moving into the large gradient area. α_i, β_i, γ_i are the relative weight parameters of energy function $E_{count,i}$, $E_{curve,i}$, $E_{img,i}$, respectively.

It is necessary to make use of the brain structure and SNAKES operations features in correcting the brain boundary. The starting brain boundary has to be inside the approximate brain boundary. SNAKES operations is apt to move the boundary to outside of it, since it tries to keep a boundary to be continuous and smooth. The brain boundary can be positioned a little more inside by making the brain threshold a little larger than the approximate brain threshold calculated from section 3.2. Fig. 4(a) shows the brain boundary detected by the approximate thresholding, and the brain boundary corrected by SNAKES operations shown in Fig. 4(b), and the brain boundary detected by naked eyes is shown in Fig. 4(c).

3.4 Recombining the Brain the Remaining Regions

Fig. 3 shows that the voxels of MR head image are concentrated in the lower grey-level area of the histogram. It is necessary for the voxels of brain region to be moved into the higher grey-level area of the histogram by adding some values to the original voxel values of the brain region. In the voxel moving process, if the brain region and the remaining region are overlapped, a window level transfer processing is necessary[6]. We recombine the brain and the remaining region as follows:

1. Draw the histograms of the voxels of the brain region and the remaining region respectively.
2. Find the minimum and maximum voxel values from the histogram of the brain region.
3. Calculate the control value α by subtracting the maximum voxel value of the brain region from the largest voxel value within the histogram. Add the control value α to all the values of the brain region.
4. Draw the new histogram of the brain region using the values acquired by procedure 3.
5. Investigate if the voxel range of the brain and the voxel range of the remains are overlapped or the maximum voxel value of the brain region exceeds the voxel limits.
6. If overlapping or exceeding happens, the voxel values of the brain region and the remaining region are adjusted with window level transfer function.

4 Implementation and Results

We have implemented the proposed method using Visual C++ on a personal computer equipped with Pentium III 550MHz. MRI Head data of University of North Carolina is used as the benchmark data. Fig. 4 shows the brain boundaries detected by the proposed method and by manually detected, respectively. The brain boundary by the approximate brain segmentation is positioned a little more inside than the brain boundary of the naked eyes. SNAKES operations corrects the problem.

Fig. 5 shows the typical 5 brain boundaries detected by the proposed method. We compare the brain boundary by the proposed method to the brain boundary

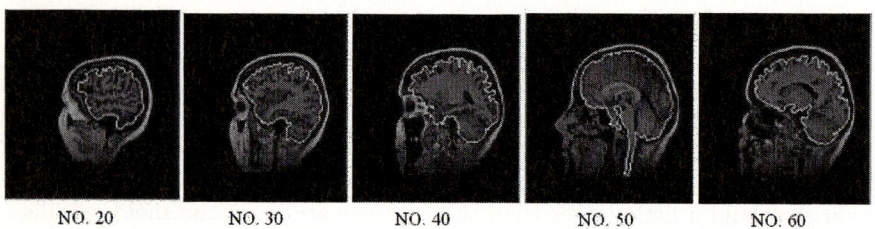

Fig. 5. Brain boundary detected by the proposed method

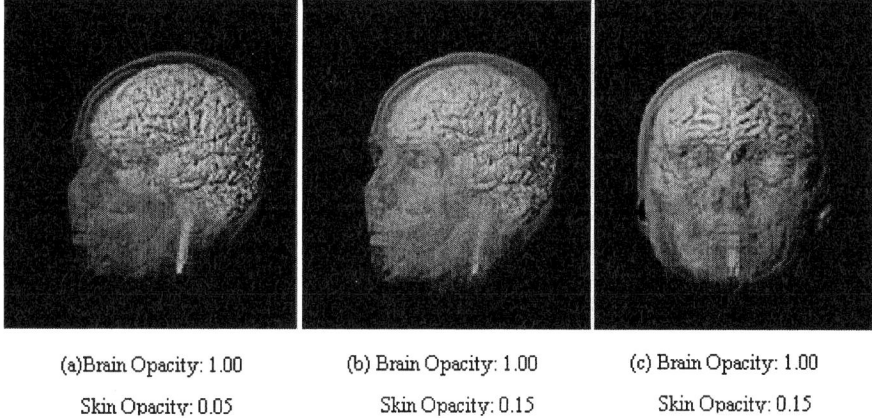

(a) Brain Opacity: 1.00　Skin Opacity: 0.05
(b) Brain Opacity: 1.00　Skin Opacity: 0.15
(c) Brain Opacity: 1.00　Skin Opacity: 0.15

Fig. 6. 3D visualization of MRI Head data by direct volume rendering

by the naked eyes using Zijdenbos similarity index[10]. The index is within 0.91-0.95 and the index goes near 0.95 as the slice becomes inside slice.

We direct volume rendered the recombined brain and rest organs using ray casting algorithm[9]. Fig. 6 shows the results that visualize all the organs of the head in 3 dimensions. The results show that the proposed method can visualize all the anatomical organs in the MRI Head Data simultaneously.

5 Discussion

In this paper, we proposed a new method for direct volume rendering of MRI head data without segmenting all the anatomical organs in a head. The proposed method directly volume renders MRI head data by segmenting head into the brain and the remains i.e. skin, skull, scarp, etc., and then recombining them with window transfer function. The segmentation is processed in two stages. The first stage extracts the approximated brain boundary by thresholding and morphology operation. The second stage corrects the approximated brain boundary by SNAKES operations. The proposed method is relatively simple and yet exact, since we are able to direct volume render the MRI head data by separating a head into only two types of material without complicating leveling of the material.

References

1. Brummer., M. E., Mersereau, R. M., Eisner, R. L., Lewine, R. R. J.: Automatic Detection of Brain Contours in MRI data sets. IEEE Transaction on Medical Imaging **12**(1993) 153–166
2. Cline, H. E., Lorensen, W. E., Kikinis, R., Jolesz, F.: Three-dimensional Segmentation of MR images of the Head using Probability and Connectivity. J. Comput. Assist. Tomogr. **14** (1990) 1037–1045

3. Drebin, R. A., Carpenter, L., Hanrahan, P.: Volume Rendering. Computer Graphics **22** (1988) 51–58
4. Kass, M., Witkin, A., Terzopoulos, D.: Snakes Active Countor Models. International Journal of Computer Vision **1** (1987) 321–331
5. Pannizzo, F., Stallmeyer, M. J. B., Whalen, J. P., Cahill, P. T: Quantitative MRI Studies for Assessment of Multiple Sclerosis. Magn. Reson. Med. **24** (1992) 90–99
6. Schroder, W., Martin, K., Lorenson, B.: The Visualization Toolkit 2nd Edition. Prentice Hall
7. http://www.nlm.nih.gov/research/visible/visible_human.html. NLM the Visible Human Project
8. http://www.siggraph.org/education/materials/volviz/volume_vis-ualization_data _sets.htm.SIGGRAPH
9. Song, J., Gwun, O., Jeong, H.: Operation Level Acceleration for Volume Rendering. Proceedings of SPIE:Visualization and Data Analysis 2002 **4665** (2002) 154–164
10. Zijdenbos, A. P., Dawant, B. M., Margolin, R. A., Palmer, A. C.: Morphometric Analysis of White Matter Lesions in MR images: Method and Validation. IEEE Trans. Med. Img. **13** (1994) 716–724

Text Retrieval Using Sparsified Concept Decomposition Matrix*

Jing Gao and Jun Zhang

Laboratory for High Performance Scientific Computing and Computer Simulation,
Department of Computer Science, University of Kentucky, 773 Anderson Hall,
Lexington, KY 40506-0046, USA
jzhang@cs.uky.edu
http://www.cs.uky.edu/~jzhang

Abstract. We examine text retrieval strategies using the sparsified concept decomposition matrix. The centroid vector of a tightly structured text collection provides a general description of text documents in that collection. The union of the centroid vectors forms a concept matrix. The original text data matrix can be projected into the concept space spanned by the concept vectors. We propose a procedure to conduct text retrieval based on the sparsified concept decomposition (SCD) matrix. Our experimental results show that text retrieval based on SCD may enhance the retrieval accuracy and reduce the storage cost, compared with the popular text retrieval technique based on latent semantic indexing with singular value decomposition.

1 Introduction

Many popular text retrieval techniques are based on the vector space model. Each document is represented as a vector of certain weighted word frequencies. A text dataset is modeled as a term-document matrix whose rows are terms (words) and columns are document vectors. A user's query of a database can be represented as a document vector. Relevant documents are found by querying the database. In other words, query matching is to find the documents that are most similar to the query in some sense. A measure of similarity, e.g., the cosine value of the angles between the vectors, is used to select the documents that are most relevant to the query vector. For real-life term-document matrices, both the number of terms and the number of documents are large, which result in high dimensionality of the database.

A standard strategy in dealing with high dimensional databases is dimensionality reduction. In text retrieval community, dimensionality reduction using latent semantic indexing (LSI) based on singular value decomposition (SVD)

---

* The research work of the authors was supported in part by the U.S. National Science Foundation under grants CCR-0092532, and ACR-0202934, and in part by the U.S. Department of Energy Office of Science under grant DE-FG02-02ER45961.

of the term-document matrix is popular [3]. The truncated SVD low-rank approximation to the vector space representation of the database can capture the semantics of the documents [2,3,7], and is used to estimate the structure in word usage across the documents [2]. Experimental results show that text retrieval performance is improved by SVD. However, the truncated SVD matrices are dense, and usually consume more storage space than the original sparse term-document matrix [1,6]. Several strategies have been proposed to reduce the memory cost of SVD, including the matrix sparsification strategies [6].

In this paper, we examine the advantages of text retrieval strategies using the sparsified concept decomposition (CD) matrix. We propose strategies to make query processing more efficient with CD. In particular, we propose to compute the inverse of the normal matrix of the concept matrix explicitly in order to facilitate real-life parallel query processing. We further propose to sparsify the dense inverse matrix to reduce storage cost. Our experimental results indicate that, compared with SVD, the new retrieval technique reduces storage cost and improves the performance of text retrieval.

Concept decomposition was proposed and analyzed in [4]. Text retrieval using concept projection from the PDDP clustering algorithm was experimented in [9]. Our contribution is mainly in advocating text retrieval using the CD matrix (not the concept projection as in [9]), the explicit computation procedure, and the sparsification of the inverse matrix.

This paper is organized as follows. In Section 2 we review the concept decomposition method and propose our sparsification strategy. Section 3 presents experimental results and comparisons. We summarize this paper in Section 4.

2 Concept Decomposition and Query Retrieval

A large data collection can be divided into a few smaller subcollections, each contains data that are close in some sense. This procedure is called clustering, which is a common operation in data mining. In information and text retrieval, clustering is useful for organization and search of large text collections, since it is helpful to discover obscured words in sets of unstructured text documents.

One of the best known clustering algorithms is the k-means clustering [8]. In our study, we use a standard k-means algorithm to cluster a document collection into a few tightly structured ones. These subclusters may have a certain mutual similarity behavior, i.e., documents of similar classes are grouped into the same cluster. The centroid of a tightly structured cluster can usually capture the general description of the documents in that cluster.

2.1 Document Clustering Based on k-Means Algorithm

Let the term-document matrix of dimension $m \times n$ be $A = [a_1, a_2, \ldots, a_i, \ldots, a_n]$, where a_i is the ith document in the collection. We would like to partition the documents into k subcollections $\{\pi_i, i = 1, \ldots, k\}$ such that

$$\bigcup_{j=1}^{k} \pi_j = \{a_1, a_2, \ldots, a_n\} \text{ and } \pi_j \cap \pi_i = \phi \text{ if } j \neq i.$$

For the jth cluster, the centroid vector of π_j is computed as

$$\tilde{c}_j = \frac{1}{n_j} \sum_{a_s \in \pi_j} a_s,$$

where $n_j = |\pi_j|$ is the number of documents in π_j. We normalize the centroid vectors such that

$$c_j = \frac{\tilde{c}_j}{\|\tilde{c}_j\|_2}, \quad j = 1, 2, \ldots, k.$$

If the clustering is good enough and each cluster is compact enough, the centroid vector may represent the abstract concept of the cluster very well.

2.2 Concept Decomposition

After the clustering, we have $A' = [\pi_1, \pi_2, \ldots, \pi_k]$, and a collection of the centroid vectors of the clusters $C = [c_1, c_2, \ldots, c_k]$. The matrix C is called the concept matrix [4]. Based on this concept matrix, we may use a straightforward query procedure as the following. For a given query vector q, we can find the closest matching clusters by computing and comparing the cosine similarity (inner product) values $q^T C = [q^T c_1, q^T c_2, \ldots, q^T c_k]$. We may obtain the closest matching documents by computing and comparing all or part of $q^T \pi_1, q^T \pi_2, \ldots, q^T \pi_k$.

To have better retrieval accuracy, the LSI technique with truncated SVD can be applied to each individual clusters [5]. Although retrieval performance is improved, the clustered SVD strategy still consumes much CPU time and storage space [5]. To further alleviate the problems associated with SVD, we examine text retrieval strategies using the concept matrix [9].

According to [4], the basic idea of concept decomposition (CD) is to project a high dimensional matrix into a lower-rank concept space. The concept vector c_i obtained by normalizing the centroid vectors provides the concept projection. Without loss of generality, we assume that the matrix C is of full rank k. We can project the document matrix onto the concept space spanned by the column vectors of the concept matrix $\tilde{A}_k = C M^\star$, such that $M^\star$ is a $k \times n$ matrix from the least squares problem $M^\star = \arg\min_M \|A - C M\|_F^2$, where $\|\cdot\|_F$ is the matrix Frobenius norm. The closed form solution of this problem is $M^\star = (C^T C)^{-1} C^T A$ [4]. The CD of the document matrix is $\tilde{A}_k = C (C^T C)^{-1} C^T A$, where C is of dimension $m \times k$, and $k \ll \{m, n\}$.

2.3 Retrieval Procedure

Applying a query vector q on the CD matrix, we obtain

$$q^T \tilde{A}_k = q^T C (C^T C)^{-1} C^T A.$$

This procedure can be carried out in a few steps. First, we can compute $q_1^T = q^T C$. Then, we obtain

$$q_2^T = q_1^T (C^T C)^{-1}, \tag{1}$$

and $q_3 = C q_2$. Finally, the query procedure is computed as $q^T \tilde{A}_k = q_3^T A$. The concept matrix is largely sparse, although denser than A. The entire procedure can be considered as expanding the query q to q_3 and then retrieval on the original matrix by $q_3^T A$.

All steps can be carried out straightforwardly, except Step (1), which requires the inverse of the normal matrix of the concept matrix. Although not directly applied to text retrieval, [4] suggests to compute the QR factorization of the concept matrix as $C = QR$, where Q is an orthogonal matrix, i.e., $Q^T Q = I$, and R is an upper triangular matrix of rank k. Because

$$C^T C = (Q^T R)^T Q^T R = R^T Q Q^T R = R^T R,$$

Step (1) becomes $q_2^T = q_1^T (R^T R)^{-1}$. Since $(R^T R)^{-1}$ is symmetric, we have $q_2 = (R^T R)^{-1} q_1$. This leads to $(R^T R) q_2 = q_1$, which can be solved to get q_2 by a forward elimination and back substitution as in standard Cholesky factorization of a symmetric positive definite matrix.

The QR factorization is fast, straightforward, and is used in the results reported in [9]. However, in realistic large scale text retrieval, such as those used in the online search engines, the query operations are usually computed by many processors with shared or distributed memory simultaneously. The forward elimination and back substitution procedure is inherently sequential and thus may present a bottleneck in the response time of a text retrieval system.

We prefer to invert the matrix $C^T C$ explicitly, so that Step (1) can be evaluated as a matrix vector operation, which can be carried out efficiently on parallel platforms. Since k is usually far smaller than both m and n, the computational and storage cost of computing $D = (C^T C)^{-1}$ explicitly is acceptable. The storage cost can be reduced by sparsification as we discuss below.

2.4 Sparsification Strategy

The matrix $(C^T C)^{-1}$ is of dimension $k \times k$ and dense. It may have many small size (or zero) entries. They play less important role during the query operations (inner product). Therefore, it is possible to replace these small size entries in the approximation matrix. If enough zeros are found in the matrix, some sparse matrix storage formats may be used to reduce the unnecessary storage space.

Following our previous work on sparsifying the SVD matrices, we propose a sparsification strategy to sparsify the matrix $(C^T C)^{-1}$. Given a threshold parameter ϵ, for any entry d_{ij} in $(C^T C)^{-1}$, if $|d_{ij}| < \epsilon$, we set $d_{ij} = 0$.

In our tests of sparsified SVD matrices, we found that the sparsification strategy may sometimes improve the performance of the SVD based text retrieval system, in addition to reduce storage cost [6]. The same effect will be demonstrated with the sparsified concept decomposition (SCD) technique.

3 Results and Discussion

To evaluate the performance of the text retrieval technique based on the concept decomposition (CD) matrix, we apply it to three popular text databases: CRAN, MED, and CISI [6]. The term-document matrices of these three databases are downloaded from http://www.cs.utk.edu/~lsi/.

A standard way to evaluate the performance of an information retrieval system is to compute the precision and recall. The precision is the proportion of the relevant documents in the set returned to the user; the recall is the proportion of all relevant documents in the collection that are retrieved by the system. We average the precision of all queries at fixed recall values such as $0, 10\%, 20\%, \ldots, 90\%$. The precision values that we report are the average precision over the number of queries at a given recall value. In precision-recall pair, the higher curves indicate better performance of an information retrieval system.

We first use the k-means algorithm to divide the original term-document matrix into 32, 64, 128, 256, and 500 small data clusters. For each data cluster, we compute the concept vector, and form a concept matrix collectively. To study the performance of the text retrieval technique based on concept decomposition with or without sparsification, we compare the query results with those of SVD. For SVD, we choose the reduced rank $k = 100$, as previous tests show good results with LSI by choosing the reduced rank at about 1/10 of the documents in the collection [2, 6].

We tested the sparsification of $D = (C^T C)^{-1}$ with the threshold values 0.01, 0.02, 0.03, and 0.04. For all three databases divided into 256 clusters, if $\epsilon = 0.04$, more than 60% of the nonzero entries would be set to zero in D.

Fig. 1 shows the precision-recall curves from CD and SCD with different number of clusters (64, 128, 256, and 500), compared with SVD. For the method of CD without sparsification, we just provide the best query results for every database.

When the number of cluster is 128, CD without sparsification reaches its best results for MED. For CISI and CRAN, the best results are obtained when the number of clusters is 500.

After sparsification with $\epsilon = 0.03$, the best results for MED are obtained when the number of clusters is 128 and 86% of the entries in D are dropped. For both CRAN and CISI, the best number of clusters is 256, but the best ϵ values are 0.02 and 0.03, respectively, under which 56% and 64% of the entries in D are dropped.

For MED and CISI, CD with and without sparsification have better performance than SVD, especially at the low recall range. For example, SCD has around 30% improvement over SVD at recall=10% for CISI. For CRAN, the precisions of SCD are significantly better than that of SVD and CD. Compared with SVD and CD, the sparsification procedure not only reduces the CPU time and storage space, but also improves the query precision.

We also compare the storage cost and the total CPU time for the query procedure between SVD and SCD. The results are shown in Table 1. The CPU times are counted for all queries for each database. Since SCD has a small rank and

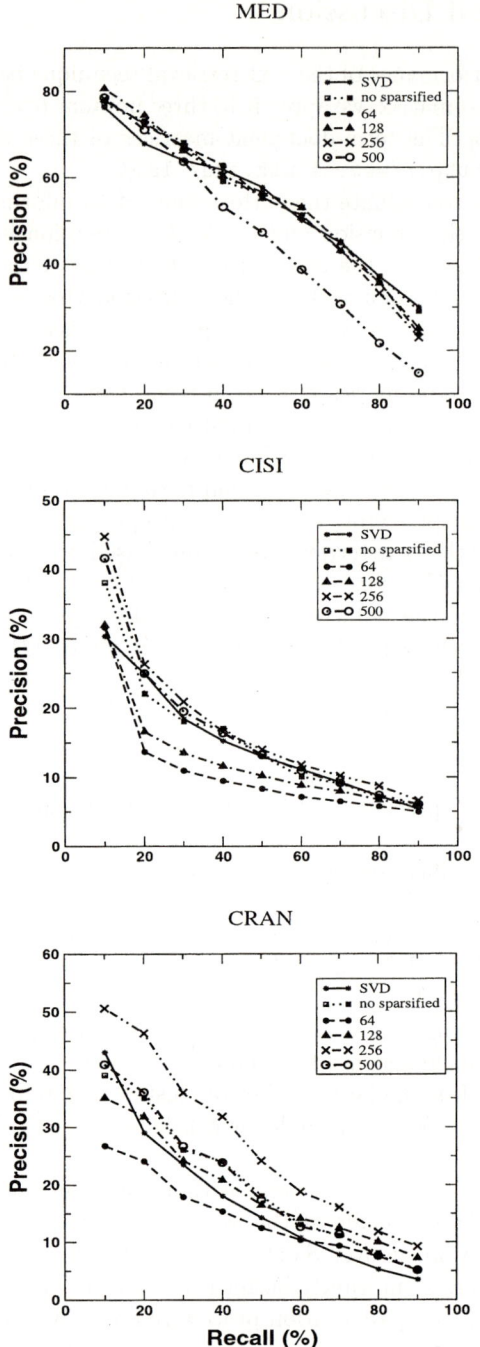

Fig. 1. Comparisons of precision-recall results using SVD, CD, and SCD

Table 1. Comparisons of CPU time and storage cost for SVD and SCD

Databases	MED		CISI		CRAN	
Techniques	SVD	SCD	SVD	SCD	SVD	SCD
CPU time (s)	4.5×10^2	0.3×10^2	6.2×10^2	2.1×10^2	1.4×10^3	5.5×10^2
Memory (MB)	1.3	0.26	1.6	0.31	1.4	0.29

the sparsification strategy further removes many small entries, the query time and storage cost of SCD are significantly smaller, compared to SVD. The experiments were carried out in MATLAB on a SUN Ultra 10 Workstation running at 500 MHz with 128 MB memory.

4 Summary

We proposed to use a sparsification technique to enhance the performance of text retrieval using the concept decomposition matrix, which was proposed for dimensionality reduction of vector space information retrieval model [4]. We give experimental results of query performance on three well-known databases using our sparsified concept decomposition technique and compare it with the SVD and the concept decomposition technique without sparsification. We found that the sparsified concept decomposition technique usually has better performance in most cases. The query time and the storage cost of the sparsified concept decomposition technique are substantially smaller than that of SVD.

References

1. Bassu, D., Behrens, C.: Distributed LSI: scalable concept-based information retrieval with high semantic resolution. In *Proceedings of the 2003 Text Mining Workshop*, San Francisco, CA (2003) 72–82
2. Berry, M.W., Drmac, Z., Jessup, E.R.: Matrix, vector space, and information retrieval. SIAM Rev. **41** (1999) 335–362
3. Deerwester, S., Dumais, S.T., Furnas, G., Landauer, T., Harshman, R.: Indexing by latent semantic analysis. J. Amer. Soc. Infor. Sci. **41** (1990) 391–407
4. Dhillon, I.S., Modha, D.S.: Concept decompositions for large sparse text data using clustering. Machine Learning **42** (2001) 143–175
5. Gao, J., Zhang, J.: Clustered SVD strategies in latent semantic indexing. Technical Report No. 382-03, Department of Computer Science, University of Kentucky, Lexington, KY (2003)
6. Gao, J., Zhang, J.: Sparsification strategies in latent semantic indexing. In *Proceedings of the 2003 Text Mining Workshop*, San Francisco, CA (2003) 93–103
7. Husbands, P., Simon, H., Ding, C.: On the use of singular value decomposition for text retrieval. In *Computational Information Retrieval*, SIAM, Philadelphia, PA (2001) 145–156
8. Jain, A., Dubes, R.C.: *Algorithms for Clustering Data*. Prentice Hall (1998)
9. Sasaki, M., Kita, K.: Information retrieval system using concept projection based on PDDP algorithm. In *Pacific Association for Computational Linguistics* (PACLING2001) (2001) 243–249

Knowledge-Based Search Engine for Specific 3D Models

Dezhi Liu[1] and Anshuman Razdan[2]

[1] College of Computer Science, Zhejiang University,
Hangzhou, Zhejiang Province 310027, China
liudezhi@zjuem.zju.edu.cn
http://nmlab.zju.edu.cn/en/teachers/ldz.htm
[2] Partnership for Research in Spatial Modeling, Arizona State University,
Tempe, AZ 85282, USA
razdan@asu.edu

Abstract. A search engine is implemented to support querying specific 3D models – digitized Indian pottery on Internet for scientists. The engine is knowledge-based. I.e. not only shape information but also shape feature information (knowledge) of 3D models can be retrieved via the engine. Shape information of 3D models is collected from Lasers Scanner and/or geometric modeling techniques. Feature information is generated from shape information via feature extracting techniques. All information is organized according to a prior defined XML schema. Matching algorithm, which is the key challenge of search engine design, is also presented. One of applications of this search engine for 3D objects is a retrieval system of Native American ceramic vessels for archaeologists.

1 Introduction

There is growing consensus among computational scientists that observational data, result of computation and other forms of information produced by an individual or a research group need to be shared and used by other authorized groups across the world through the entire life cycle of the information [1]. On the other side, the Web has revolutionized the electronic publication of data. It has relied primarily on HTML that emphasizes a hypertext document approach. More recently, *Extensible Markup Language (XML)*, although originally a document mark-up language, is promoting an approach more focused on data exchange. XML is a set of rules for defining semantic tags that break a document into parts and identify the different parts of the document. It is a meta-markup language that defines a syntax used to define other domain-specific, semantic, structured markup languages [2]. XML makes the sharing of scientific data on Internet more feasible.

As an important part of scientific data, shape information of 3D objects plays very key roles in scientific research. For example, Archaeologists study the 3D form of Native American pottery to characterize the development of cultures. Quantitative methods of reasoning about the shape of a vessel are becoming far more powerful than was possible when vessel shape was first given a mathematical treatment by G. Birkhoff [3]. Conventionally, mathematical treatment of vessels is done manually by an expert and is subjective and prone to inaccuracies. Recent theoretical and technological breakthroughs in mathematical modeling of 3D data and data-capturing techniques present the opportunity to overcome such inaccuracies in mathematical treatment of vessels. Our research involves obtaining shape information from the scanned three-dimensional data of archaeological vessels, using 2D and 3D geometric models to represent scanned vessels, extracting features from geometric models and storing original and modeled information in database for Web-based retrieval. In this paper we introduce the design of a Web based Visual Query Interface (VQI) for 3D archaeological vessels. This paper is structured as fellows. Part two describes archaeological vessel features from the point of view of archaeologists, and develops an XML-based information model for vessels. Part three introduces interface design on client site and server design on server site for the Web-based VQI. Conclusions and further research directions can be found in part four.

2 Information Model for 3D Vessels

The information model is used to organize and store shape and shape feature information. The useful information is gathered by several offline software tools. A powerful information model for archaeological vessels should include raw data, curve/surface information, and higher-level information – feature information. Raw data of archaeological vessels are 3D triangulated meshes composed of points, edges and triangles. They are collected from scanning vessels by 3D Laser Scanners. Curve and surface information is generated from raw data via geometric modeling techniques. Feature information is extracted from geometric models, and is organized according to formal description in 2.3. Figure 1 describes the hierarchical relationship among raw data, curve/surface models and feature models.

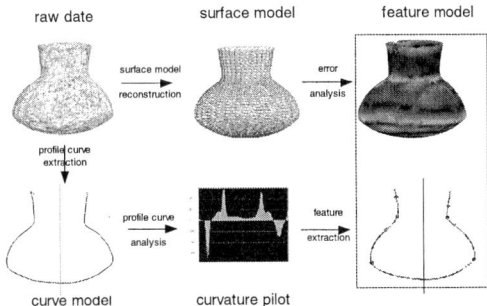

Fig. 1. Information models for vessels

2.1 Geometric Models

After scanning an archaeological vessel via a 3D laser scanner (Cyberware 3030), we can get a polygonal mesh that is constituted of faces, edges and vertices. The polygonal mesh is used as a raw data for further analysis. A Polygon mesh M is a 3-tuple, i.e. $M = (V, E, F)$ where V is vertex set, E is edge set, and F is face set.

Raw data run short of geometric information and cannot be used directly to get shape information of vessels. For example there is a lack of curvature information in raw data. Curve and surface models are generated from raw data via geometric modeling techniques and used to provide more geometric information. One of representing or modeling curves/surfaces is via parametric forms such as B-Spline or NURBS. Surface models are generated by fitting points of polygonal meshes with least squares approximation. We use such representation to enable us to rebuild models, analyze properties such as curvatures, make quantitative measurements as well "repair" incomplete models. A NURB surface can be represented as

$$\vec{P}(u,v) = \frac{\sum_{i=0}^{m}\sum_{j=0}^{n} w_{i,j} \vec{d}_{i,j} N_{i,k}(u) N_{j,l}(v)}{\sum_{i=0}^{m}\sum_{j=0}^{n} w_{i,j} N_{i,k}(u) N_{j,l}(v)} \quad (1)$$

where $\vec{d}_{i,j}$, $i = 0, 1, \ldots, m$; $j = 0, 1, \ldots n$ are control points, $w_{i,j}$ are weights, $N_{i,k}(u)$ and $N_{j,l}(v)$ are B-Spline basis functions. When weights equal 1.0, it reduces to a non-uniform B-Spline surface.

Because contour shape information plays an important role in analysis of archaeological vessels, we use 2D NURB curves to represent profile curves of archaeological vessels. Using 2D geometric models can make problem simple, and reduce 3D problems to 2D problems.

In order to get a 2D profile curve from a vessel, archaeologists use a cutting plane to intersect the vessel (polygonal mesh) and can get intersection points, then connect all the points according to some order, and get the chain code.

NURB curves are generated by fitting points of chain codes with least squares approximation. Since curvature has useful information such as convexity, smoothness, and inflection points of the curve needed by vessel analysis, we adopt cubic NURB curves to approximate profile curves of vessels

$$\vec{P}(u) = \frac{\sum_{i=0}^{n} w_i \vec{d}_i N_{i,k}(u)}{\sum_{i=0}^{n} w_i N_{i,k}(u)} \quad (2)$$

where $\vec{d}_i$, $i = 0, 1, \ldots, n$ are control points, w_i are weights, $N_{i,k}(u)$ are B-Spline basis functions. Details about generating 2D curve models and 3D surface models are in [4, 5].

2.2 Feature Models

Mostly archaeological vessels are (approximately) surfaces of revolution, and studying contour shape will suffice to gather shape information about the whole object. According to archaeological definition [3] there are four kinds of feature points on profile curves to calculate dimensions and proportions of vessels, see figure 2. They are *End Points (EPs), Points of Vertical Tangency (VTs), Inflection Points (IPs)* and *Corner Points (CPs)* found on the vertical profile curve of a vessel:

- *End Points* - points at the rim (lip) or at the base (i.e. top and bottom of vessels).
- *Points of Vertical Tangency* - points at the place that is the maximum diameter on spheroidal form or minimum diameter on hyperbolic form.
- *Inflection Points* - points of change from concave to convex, or vice versa.
- *Corner Points* - points of sharp change on a profile curve. See figure 2.

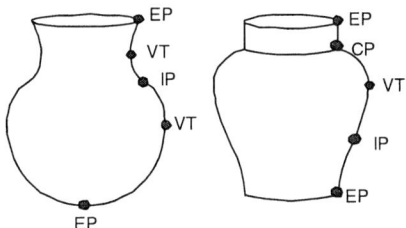

Fig. 2. Feature points of vessel profile curves

Next four features in figure 3 are common to all vessels:

- *Orifice* - the opening of the vessel, or the minimum diameter of the opening, may be the same as the rim, or below the rim.
- *Rim* - the finished edge of the top or opening of the vessel. It may or may not be the same as the orifice. It may have a larger diameter.
- *Body* - the form of the vessel below the orifice and above the base.
- *Base* - the bottom of the vessel, portion upon which it rests, or sits on a surface. The base may be convex, flat, or concave, or a combination of these.

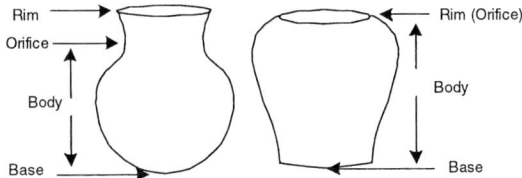

Fig. 3. Common features for vessels

2.3 XML-Based Information Models

From above definition for characteristic points and common features for all vessels, we can formalize feature representation of vessels as the following.

<Point Feature>:=<End Point Feature> |< Point of Vertical Tangency Feature> |< Inflection Point Feature> | <Corner Point Feature>;
<Curve Feature>:= <Rim Curve Feature> |< Orifice Curve Feature> |< Base Curve Feature>;
<Rim Curve Feature>:=<End Point Feature>< End Point Feature>;
<Orifice Curve Feature>:=<Corner Point Feature> <Corner Point Feature>;
<Base Curve Feature>:= <End Point Feature> |<End Point Feature>< End Point Feature>
<Region Feature>:=< Neck Region Feature> |< Body Region Feature> |< Base Region Feature>;
<Neck Region Feature> := <Rim Curve Feature><Orifice Curve Feature>;
<Body Region Feature>:=<Orifice Curve Feature>< Base Curve Feature>;
<Base Region Feature>:= <Base Curve Feature>;
<Volume Feature>:=< Unrestricted Volume Feature> |< Restricted Volume Feature>.

We use XML to represent information models of vessels. We design an XML schema to represent geometric information, feature information and measured value of archaeological vessels. Feature information is extracting from geometric information and is organized according to the feature formalism in the XML schema. Also feature information is used to index vessels stored in a database. The purpose of using XML to represent information is that we can develop a distributed and web based query interface for archiving and searching 3D archeological vessels. Embedding data in XML adds structure and web accessibility to the inherent information of archeological vessels.

3 Content-Based 3D Retrieval for Vessels

We offer users contend-based retrieval tools on the client site. Originally content-based retrieval system was mainly designed for still image libraries [6]. However there are lots of unsolved problems remained in the content-based retrieval of 3D models. Our research tries to implement the search engine for 3D models, especially 3D ceramics. We combine MS IIS, Tomcat and MS Access to design the web server for the VQI. Tomcat supports JSP and XML/XSL. MS Access serves as database to store information models of vessels. A Netscape/IE plug-in was developed using C++ and OpenGL, and allows users draw profile curves on screen. Drawn curves with other retrieval parameters, such as height, diameter, area, and volume are submitted to Web server, and process a content-based retrieval. Figure 4 demonstrates the retrieval procedure.

The query process in VQI combines a sketch-based interface and searches by traditional text and metric data. Representative vessel shapes can be selected from the supplied palette and modified, or a freeform profile sketch can be created in the interface window. Text and numeric fields support parallel query of descriptive and derived data within the databases. Query results from database are stored in XML format, and are visualized via a pre-designed Extensive Stylesheet Language (XSL) file.

A hierarchical indexing structure for the database design is used to speed up the query procedure. The structure includes compactness value of 3D solid objects, feature points and profile curves of 3D ceramics.

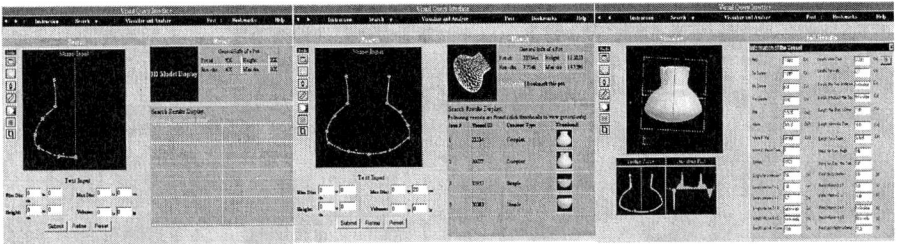

Fig. 4. (L to R) Query interface screen with sketch–based, numeric, and text-based input fields. Initial response screen with resulting thumbnail images and summary data, and wire frame of first matching vessel. Detail individual display screen with 2D, 3D, and descriptive vessel data

3.1 Compactness Computing

The initial database search field is the indexing of compactness. The basic descriptive properties of rigid solids are the enclosing surface area and volume. A measure of compactness for solids relates the enclosing surface area with the volume. Thus, a classical measure of compactness can be defined by the ratio $C = (area^3)/(volume^2)$, which is dimensionless and minimized by a sphere [7].

For a sphere, its area is equal to $4\pi r^2$ and volume $(4/3)\pi r^3$. Therefore, $C = 36\pi$ is the minimum compactness of a solid, since the sphere encloses maximum volume for a constant surface area. We can define regular compactness $C_{reg} = C_{min}/C$, where $C_{min} = 36\pi$, and C is classical compactness. Several compactness values of classic vessels can be found in figure 5.

The compactness value is a real number. Totally different vessels may have the same compactness values. But these values can narrow the search range, and speed up the search procedure. A solid 3D model of a submitted 2D sketch curve is generated by CGI via surface of rotation modeling techniques. Then the compactness value of this solid model is used to search the database. Tens of similar vessels are returned as the initial search result by comparing the compactness values.

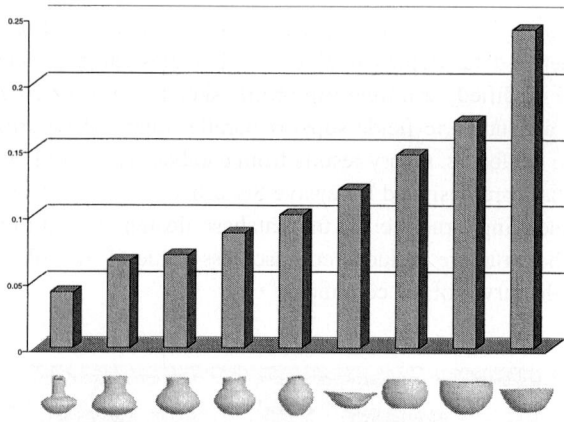

Fig. 5. Regular compactness values of some vessels

3.2 2D Sketch Matching

From the initial search result, a curve match process is called to perform the best matching of the submitted sketch curve and profile curves of 3D vessels. A curve-based Iterative Closest Point (ICP) algorithm [8] is implemented for this. The ICP algorithm can be stated as following. A "data" shape P is moved (registered, positioned) to be in best alignment with a "model" shape X. The data and model shape may be represented in any of the allowable forms (point sets, line segment set, implicit/parametric curves, triangle sets, implicit/parametric surfaces, in our project, we will use triangle sets). The number of points in the data shape will be denoted N_p. Let N_x be the number of points, line segments, or triangles involved in the model shape (line segments in our case after the parametric curves are discretized).

There are three basic computational components for ICP algorithm.

Step 1. Computing the closest points (costs: $o(N_p N_x)$ worst case, $o(N_p \log N_x)$ average;
Step 2. Computing the registration (costs: $o(N_p)$);
Step 3. Applying the registration (costs: $o(N_p)$)

The three basic computational components make up an iterative process of ICP algorithm, and the ICP algorithm always converges monotonically to the nearest local minimum of a mean square distance metric. Experience shows that the rate of convergence is rapid during the first few iterations. A trick to accelerate the ICP algorithm is that we use feature points on profile curves to calculate the initial position estimation.

4 Conclusions and Further Work

We present a method for archiving and searching 3D objects, Native American ceramic vessels on Web. We have (i) modeled raw data of 3D archaeological vessels with parametric curves and surfaces, (ii) extracted features to raise the level of abstraction of data, (iii) organized vessel data based on XML and (iv) developed a visual query interface on the web for sharing information.

During our research we have found some problems that need to be solved in the future. Firstly hundreds of vessels have been scanned and processed as our data set. The proposed VQI works well on this small data set. It is our further work to scan and process more vessels to test our system. Secondly we just implemented a 2.5 dimension search engine for 3D ceramics. If the shape of 3D objects is freeform, the 2D sketch-based query is not enough. To develop a general search engine for 3D objects is another future research direction.

References

1. Williams R.: Interfaces to Scientific Data Archives. Workshop Report, California Institute of Technology. Pasadena (1998)
2. Extensible Markup Language (XML) 1.0: http://www.w3.org/TR/REC-xml. (2004)
3. Birkhoff, G.: Aesthetic Measure. Harvard University Press (1933)
4. Farin, G.: Curve and Surface for Computer Aided Geometric Design. Academic Press, Boston, fourth edition (1996)
5. Bae, M.: Curvature and Analysis of Archaeological Shapes. MS Thesis, Arizona State University (1999)
6. University of California, Berkeley, Digital Library Project, Image Retrieval by Image Content: http://galaxy.cs.berkeley.edu/photos/blobworld/. (2004)
7. Bribiesca, E.: A Measure of Compactness for 3D Shape. Computers & Mathematics with Applications 40 (2000) 1275-1284
8. Besl, P.J., Mckay, N.D.: A Method for Registration of 3-D Shapes, IEEE Trans. Pattern Analysis and Machine Intelligence 2 (1992), 239-256

Robust TSK Fuzzy Modeling Approach Using Noise Clustering Concept for Function Approximation

Kyoungjung Kim[1], Kyu Min Kyung[1], Chang-Woo Park[2], Euntai Kim[1], and Mignon Park[1]

[1] Department of electrical and electronic engineering, Yonsei University,
134, Shinchon-dong, Sudaemoon-gu, Seoul, Korea
{kjkim01, mediart, etkim, mignpark}@yonsei.ac.kr
http://yeics.yonsei.ac.kr
[2] Korea electronics technology institute,
401-402 B/D 192, Yakdae-Dong, Wonmi-Gu, Buchon-Si, Kyunggi-Do, Korea
drcwpark@keti.re.kr

Abstract. This paper proposes the algorithm that additional term is added to an objective function of noise clustering algorithm to define fuzzy subspaces in a fuzzy regression manner to identify fuzzy subspaces and parameters of the consequent parts simultaneously and obtain robust performance against outliers.

1 Introduction

Many researches on the robust fuzzy modeling technique to apply when noise and outliers exist have been reported [2], [3], [6], [7], [9], [10]. The approach using an objective function with loss function with fuzzy regression [2], [7] can identify fuzzy subspaces and parameters of the consequent part, but it is complex and cannot use the prior knowledge of the given system. The approach to clustering noise [3], [6] is rather simple to implement, but it does not have a capability to identify fuzzy subspaces and parameters of the consequent parts. In this paper, we propose TSK fuzzy modeling algorithm that has a capability to identify fuzzy subspaces and parameters of the consequent parts simultaneously and is rather simple. The proposed algorithm adopted fuzzy regression and noise clustering concept [3], [6], [13], [14]. The proposed algorithm represents robust performance against outliers and can obtain fuzzy subspaces and parameters of the consequent parts simultaneously. The simulation results show that the performance of the proposed algorithm is superior to other approaches, especially in high order system.

2 TSK Fuzzy Modeling

In this paper, we deal with TSK fuzzy model. The TSK fuzzy model can present static or dynamic nonlinear systems. The TSK fuzzy model has consequent parts consisting of linear functions and can be viewed as expansion of piecewise linear partition. It has the form as

$$R^i : \text{If } x_1 \text{ is } A_1^i(\vec{\theta}_i^i) \text{ and } x_2 \text{ is } A_2^i(\vec{\theta}_2^i), ..., x_n \text{ is } A_n^i(\vec{\theta}_n^i) \quad (1)$$

$$\text{then } h^i = f_i(x_1, x_2, ..., x_n; \vec{a}^i) = a_0^i + a_1^i x_1 + ... + a_n^i x_n$$

For $i = 1, 2, ..., C$, where C is the number of rules, $A_j^i(\vec{\theta}_j^i)$ is the fuzzy set of the i th rule for x_j with the adjustable parameter set $\vec{\theta}_j^i$, and $\vec{a}^i = (a_0^i, ..., a_n^i)$ is the parameter set in the consequent part.

The predicted output of the fuzzy model is inferred as

$$\hat{y} = \frac{\sum_{i=1}^{C} h^i w^i}{\sum_{i=1}^{C} w^i} \quad (2)$$

where h^i is the output of the i th rule, $w^i = \min_{j=i,...,n} A_j^i(\vec{\theta}_j^i; x_j)$ is the i th rule's firing strength, which is obtained as the minimum of the fuzzy membership degrees of all fuzzy variables.

Let e_{ij} is the error between the j th desired output of the modeled system and the output of the i th rule with the j th input data.

$$e_{ij} = y_j - f_i(\vec{x}(j); \vec{a}^i), \; i = 1, 2, ..., C, \; j = 1, 2, ..., N \quad (3)$$

where y_j is j th desired output, C is the number of fuzzy rules and N is the number of the training data. The noise is considered to be a separate class and is represented by a prototype that has distance δ. The membership u_{*j} of a point x_j in the noise cluster is defined as in [2] to be

$$u_{*j} = 1 - \sum_{i=1}^{C} u_{ij} \quad (4)$$

We propose the objective function considering both error measure and distance measure. The objective function of the proposed algorithm is defined as

$$J = \alpha \sum_{i=1}^{C} \sum_{j=1}^{N} u_{ij}^2 e_{ij}^2 + \sum_{j=1}^{N} \delta^2 (1 - \sum_{i=1}^{C} u_{ij})^2 + \sum_{i=1}^{C} \sum_{j=1}^{N} u_{ij}^2 d^2(x_j, \beta_i) \quad (5)$$

where $d(x_j, \beta_i)$ is the distance from a feature point x_j to prototype β_i, u_{ij} is the grade of membership, δ is noise distance and α is a free parameter to be selected.

$$\frac{\partial J}{\partial u_{ij}} = 2\alpha u_{ij} e_{ij}^2 - 2\delta^2 (1 - \sum_{i=1}^{C} u_{ij}) + 2u_{ij} d^2(x_j, \beta_i) = 0 \quad (6)$$

Then the following membership update equation is derived

$$u_{ij} = \cfrac{1}{\displaystyle\sum_{k=1}^{C} \cfrac{\alpha e_{ij}^2 + d^2(x_j, \beta_i)}{\alpha e_{kj}^2 + d^2(x_j, \beta_k)} + \cfrac{\alpha e_{ij}^2 + d^2(x_j, \beta_i)}{\delta^2}} \quad (7)$$

To obtain the parameter vector $\vec{a}^i$ for the consequent part of the ith rule, differentiate the objective function with respect to $\vec{a}^i$, then is obtained as follows

$$\frac{\partial J}{\partial \vec{a}^i} = \sum (u_{ij})^2 \frac{\partial e_{ij}^2}{\partial \vec{a}^i} = 0 \quad (8)$$

$$\frac{\partial e_{ij}^2}{\partial \vec{a}^i} = 2e_{ij} \frac{\partial e_{ij}}{\partial \vec{a}^i} = 2[y_i - f_i(x_j; \vec{a}^i)] \frac{\partial e_{ij}}{\partial \vec{a}^i} \quad (9)$$

Substituting (9) into (8), we obtain

$$\sum_{j=1}^{N} (u_{ij})^2 y_i \frac{\partial e_{ij}}{\partial \vec{a}^i} - \sum_{j=1}^{N} (u_{ij})^2 f_i(x_j; \vec{a}^i) \frac{\partial e_{ij}}{\partial \vec{a}^i} = 0 \quad (10)$$

We can see the following from (3)

$$\frac{\partial e_{ij}}{\partial \vec{a}^i} = x_j \quad (11)$$

Now, we define $X \in R^{N \times (n+1)}$ is a matrix with x_k as its $(k+1)$th row (entries in the first row of X are all 1), $Y \in R^N$ is a vector with y_k as its kth element and $D_i \in R^{N \times N}$ is a diagonal matrix with u_{ik}^2 as its kth diagonal element. Then we can rewrite (11) as a matrix form as follows

$$X^T D_i Y - (X^T D_i X) \cdot \vec{a}^i = 0 \quad i = 1, 2, \ldots, C \quad (12)$$

The parameter vector $\vec{a}^i$ for the consequent part of the ith rule is obtained as

$$\vec{a}^i = [X^T D_i X]^{-1} X^T D_i Y \quad i = 1, 2, \ldots, C \quad (13)$$

The proposed algorithm is described in the following.

Step 1) Set the number of clusters C, and the noise distance δ.
Step 2) Initialize the prototypes and the grade of membership.
Step 3) Compute the consequent parameter sets $\vec{a}^i$ and the distance between prototypes and each data points.
Step 4) Compute the error e_{ij}, and update the grade of membership u_{ij}.
Step 5) Stop if iteration exceeds the maximum iteration number or the stop criterion is reached. Otherwise go to step 3.

We can obtain the membership functions of the fuzzy system by using the grade of the membership u_{ij} obtained in the above procedure. Assume that Gaussian membership functions are used in the premise parts, that is
$A_j^i(\theta_{j1}^i, \theta_{j2}^i) = \exp\{-(x_j - \theta_{j1}^i)^2 / (2(\theta_{j2}^i)^2)\}$, where θ_{j1}^i and θ_{j2}^i are two adjustable parameters of the jth membership function of the ith fuzzy rules. Then they can easily be obtained from u_{ij} as

$$\theta_{j1}^i = \frac{\sum_{k=1}^{N}(u_{ik})^2 x_j(k)}{\sum_{k=1}^{N}(u_{ik})^2} \tag{14}$$

$$\frac{\theta_{j2}^i}{\sqrt{2}} \theta_{j2}^i = \sqrt{\frac{\sum_{k=1}^{N}(u_{ik})^2 (x_j(k) - \theta_{j1}^i)^2}{\sum_{k=1}^{N}(u_{ik})^2}} \tag{15}$$

3 Simulation Results

To verify the validity of the proposed algorithm, two simple examples in [2] are considered. We use RMSE as a performance index.

First, the function to verify the algorithm is defined as

$$y = \begin{cases} x, & 0 \le x \le 3 \\ 7.5 - 1.5x, & 3 < x < 5 \end{cases} \tag{16}$$

201 training data are generated and the gross error model is used for modeling outliers as in [2]. The number of cluster in this example is selected as 2. RMSEs of this algorithm and other algorithms are given in Table 1.

Second, the function to test is defined as

$$y = x^{2/3}, \quad -2 \le x \le 2 \tag{17}$$

201 input-output data are used. The number of cluster in this example is selected as 2. RMSEs of this example are given in Table 2.

Table 1. Comparison of Performance using equation (16)

Example 1	
Algorithms	RMSE
RFRA with robust learning algorithm	0.0272
SONFIN with BP algorithm	0.0515
FCRM with BP learning algorithm	0.0607
The proposed algorithm	0.0029

Table 2. Comparison of Performance using equation (17)

Example 2	
Algorithms	RMSE
RFRA with robust learning algorithm	0.0662
SONFIN with BP algorithm	0.0814
FCRM with BP learning algorithm	0.0844
The proposed algorithm	0.0036

Next, we verify the performance when the given system is high order. The function to test is given as follows

$$y = 2x^3 + 12x^2 - 20x + 8.5, \qquad 0 \leq x \leq 4. \tag{18}$$

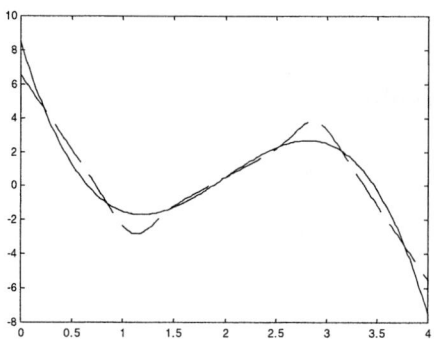

Fig. 1. The simulation result using the proposed algorithm with 3 clusters

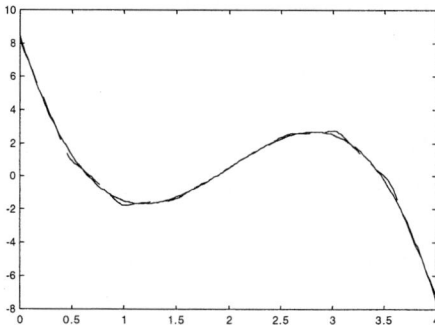

Fig. 2. The simulation result using the proposed algorithm with 7 clusters

Figure 1 presents the performance of the proposed algorithm when the number of cluster is 3. The performance of the proposed algorithm is better than other algorithm when the number of cluster is same. Figure 2 presents the performance of the proposed algorithm when the number of clusters is 7. Considering computation load and level of accuracy, we can select the number of clusters.

As seen in above three examples, the proposed algorithm shows superior performance to other algorithms and it has a capability to approximate the given high order system exactly with minimum number of fuzzy rules. If the proposed algorithm has the same rule as other algorithm, its performance is better than others.

4 Conclusion

We propose a new fuzzy modeling algorithm and describe its efficiency by comparison with various algorithms. Adopting noise clustering concept and fuzzy regression, the proposed algorithm has a robust capability against outliers and furthermore fuzzy subspaces and parameters of the consequent part simultaneously. The proposed algorithm is very simple and easy to implement and its capability to describe a given system is excellent. Especially, when the given system is high order the proposed algorithm shows better performance with minimum number of fuzzy rule compared with other algorithms. The proposed algorithm is verified by some examples and its performance is compared with other approaches.

References

1. Takagi T. and Sugeno, M.: Fuzzy Identification of Systems and Its applications to Modeling and Control. IEEE Trans. Systems, Man, Cybernetics, Vol. smc-15 (1985) 116-132
2. Chuang, C. -C., Su, S. –F. and Chen, S. –S.: Robust TSK Fuzzy Modeling for Function Approximation with Outliers. IEEE Tans. Fuzzy Systems, Vol. 9 (2001) 810-821
3. Dave, R. N. and Krishnapuram, R.: Robust clustering Methods: A Unified View. IEEE Trans. Fuzzy Systems, Vol. 5 (1997) 270-293
4. Kim, E., Park, M., Ji, S. and Park, M.: A New Approach to Fuzzy Modeling. IEEE Trans. Fuzzy Systems, Vol. 5 (1997) 328-337
5. Wang, L. X.: A Course in Fuzzy Systems and Control. Prentice Hall (1997)
6. Dave, R. N. and Sen, S.: Robust Fuzzy Clustering of Relational Data. IEEE Trans. Fuzzy Systems, Vol. 10 (2002) 713-727
7. Frigui, H. and Krishnapuram, R.: A robust competitive Clustering Algorithm with Applications in Computer Vision. IEEE Trans. Pattern Analysis and Machine Intelligence, Vol. 21 (1999) 450-465
8. Krishnapuram, R. and Keller, M.: A Possibilistic Approach to Clustering. IEEE Trans. Fuzzy systems, Vol. 1 (1993) 98-110
9. Chen, D. S. and Jain, R. C.: A robust back-propagation learning algorithm for function approximation. IEEE Trans. Neural Networks, Vol. 5 (1994) 467-479
10. Connor, J. T., Martin, R. D. and Atlas, L. E.: Recurrent neural networks and robust time series prediction, IEEE Trans. Neural Networks, Vol. 5 (1994) 240-254

Helical CT Angiography of Aortic Stent Grafting: Comparison of Three-Dimensional Rendering Techniques

Zhonghua Sun[1] and Huiru Zheng[2]

[1] Faculty of Life and Health Sciences, University of Ulster,
Newtownabbey BT 37 0QB, Northern Ireland
[2] Faculty of Engineering, University of Ulster at Jordanstown,
Newtownabbey BT37 0QB, Northern Ireland
{zh.sun, h.zheng}@ulster.ac.uk

Abstract. We aim to compare various 3D reconstruction techniques in the visualization of abdominal aortic aneurysms (AAA) treated with suprarenal stent grafts. 28 patients with AAA undergoing suprarenal fixation of aortic stent grafts were included in the study. Volumetric CT data were postprocessed with Analyze V 5.0 and four different 3D visualization methods were generated including shaded surface display (SSD), maximum intensity projection (MIP), volume rendering (VR) and virtual endoscopy (VE). Selection of CT threshold, time required for generation of 3D images as well as demonstration of 3D relationship between aortic stents and ostium were assessed in each method. Results showed that VR was found to be the most efficient 3D technique in the visualization of aortic stent grafts as well its relationship to the renal artery. VE was found to be particularly useful in demonstrating the encroachment of stent struts to the renal arteries.

1 Introduction

Abdominal aortic aneurysm (AAA) primarily affects elderly males (sex ratio 4:1) with prevalence up to 5% and accounts for over 11,000 admissions per year in England, requiring approximately 3000 elective operations and 1,500 emergency procedures [1]. The ultimate goal in the treatment of aortic aneurysm is to exclude the aneurysm from the aortic bloodstream without interfering with limb and organ perfusion. Conventional surgical repair of AAA is an invasive procedure and carries the overall operative mortality of 4% but can be as high as 8.4% [2]. Therefore, a less invasive alternative has been considered to treat AAA. Currently, endovascular aortic repair has been reported to be an effective alternative to open surgery and has been widely used in clinical practice for more than a decade [3]. With experience gathered, it has been found that nearly 30% to 40% of patients with AAA are unsuitable for the common treatment of infrarenal stent grafting due to suboptimal aneurysm necks [4]. Method of dealing with this has been investigated and placement of an uncovered suprarenal stents above the renal arteries has been reported to be a modified alternative to treat patients with suboptimal aneurysm necks [5]. Although short-medium term results of suprarenal stent grafting is satisfactory without significantly compromising renal function, long-term effect of suprarenal stents on the renal arteries is still unknown and not fully understood [6].

Unlike open repair, the success of endovascular stent grafting cannot be ascertained by means of direct examination and thus relies on imaging results. Helical CT angiography (CTA) is established as an important non-invasive imaging modality in the assessment of AAA both pre-and post-stent grafting [7, 8]. CT volumetric data acquisition has been complemented by the parallel development of image processing and visualisation methods to create high quality images including 3D representations of anatomical structures. These are shaded surface display (SSD), maximum intensity projection (MIP), volume rendering (VR) and virtual endoscopy (VE). However, it is important to know whether or not these postprocessing methods add additional information in comparison with axial CT images, as the rendering of 3D reconstructions can be very time consuming. Therefore, the purpose of this study was to generate and compare various 3D rendering techniques in patients with AAA following aortic stent grafting.

2 Materials and Methods

2.1 Patients Data and CT Scanning

28 patients with AAA (24 males and 4 females, age range: 66-87 years with a median of 76 years) who underwent endovascular repair were included in the study. CTA was performed on a Philips AV-E1 CT scanner in a single breathhold technique with slice thickness of 5 mm, pitch 1.0 (table speed 5 mm/second) and reconstruction interval of 2 mm. Uniphasic contrast enhancement was given to all patients to a total volume of 100 ml, flow rate of 2ml/s and scan delay of 30 seconds. All patients were treated with a Zenith AAA endovascular stent graft (Cook Europe, Denmark) with uncovered suprarenal struts placed above the renal arteries for obtaining proximal fixation.

All CTA datasets were burned into CD and transferred to a PC and processed using Analyze V 5.0 (www. Analyzedirect.com, Mayo Clinic, USA). The DICOM (Digital Imaging and Communications in Medicine) images were converted into volume data and resampled into 256 x 256 matrix with a pixel size of 0.82 mm to expedite the rendering process.

2.2 Generation of 3D Reconstructions

Shaded Surface Display. In SSD, user-selected upper and lower thresholds were used to define a specific range of Hounsfield unit (HU) to be displayed. As we aimed to demonstrate the stent-covered renal arteries, the range of threshold was determined by measuring the region of interest in the level of renal artery on sequential axial CT images. Therefore, voxels lower than the selected threshold were invisible in final images. SSD images were generated with edited images to remove osseous structures.

Maximum Intensity Projection. MIP does not require selection of any thresholds as it displays only the brightest voxel along every ray of sight. As a result, all darker voxels in front of or behind a brighter voxel were not displayed. Contrast-enhanced abdominal aorta and its branches together with high-density stent wires were clearly displayed in MIP images. Manual editing was applied to remove high-density structures such as bone, calcified plaque and veins.

Volume Rendering. In contrast to SSD and MIP, VR uses all of the information contained inside a volume dataset, thus allowing production of more meaningful images. By assigning a specific color and opacity value of every attenuation value of the CT data, groups of voxels were selected for display. In our study, we first measured the CT attenuation of aortic branches, aortic stents and bony structures. Then, an object map was created and seeded region grow technique was used to identify the object of interest. This allows providing unique visualization of segmented volume with connected components.

Virtual Endoscopy. The CT data was prepared for VE by removing the contrast-enhanced blood from the aorta using a CT number thresholding technique. A CT number threshold range of aortic lumen and stent was identified using region of interest measurements in the level of renal artery. The two individual virtual endoscopic images of lumen and stent wire were added numerically together producing a combination of aortic stent and aortic lumen and a direct encroachment of the renal ostium by a stent wire was clearly displayed on 3D VE images. The details of generation of VE images have been described elsewhere [9].

Processing time and threshold selection required for generation of 3D images were recorded and compared in each patient and a p value less than 0.05 was considered statistically significant difference.

2.3 Results

All of these 3D reconstructions were successfully generated in 28 patients. The median time required for generation of SSD, MIP, VR and VE was 2 min (range: 2-2.5 min), 27.5 min (range: 22-33 min), 4.5 min (range: 4-5.5 min) and 25 min (range: 20-30 min), respectively. The median threshold selected for generation of abdominal aorta and its branches on SSD, VR and VE was 105 HU (range: 80-130 HU), 105 HU (range: 80-130 HU), and 90 HU (range: 73-145 HU), respectively. The median threshold selected for generation of aortic stents on VR and VE was 300 HU (range: 300-400 HU) and 525 HU (range: 350-650 HU), respectively. There was significant difference between SSD and VR, MIP and VE, in terms of the time required for generation of 3D images ($p < 0.05$). No significant difference was found in the median threshold selection for generation of aortic branches among SSD, VR and VE ($p > 0.05$). However, significant difference was found in the median threshold selection for generation of aortic stents between VR and VE with a p value less than 0.001.

All of the stent-covered renal arteries remained patent on axial CT images and were visualized clearly on SSD, MIP, VR and VE in 27/28, 28/28, 27/28 and 28/28, respectively. Stent position relative to the renal artery/ostium was clearly visualized in nearly all of the MIP, VR and VE images and only displayed in 12 out of 28 patients on SSD. Image quality was more affected in SSD than other 3D methods due to the interference of artifacts caused by stent wires, which limited the observation of stent position relative to the renal artery in more than half of the patients (Fig1). The number of stent wires crossing the renal ostium was best demonstrated on VE images as intraluminal views were clearly demonstrated regarding the degree of encroachment to the ostium (Fig1D).

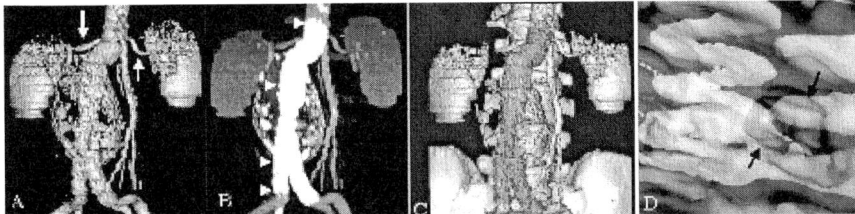

Fig. 1. Various 3D reconstruction techniques in a patient with AAA after aortic stent grafting. Renal arteries were shown to be patent on SSD (A), however, stent wires relative to the renal arteries was difficult to visualize. High-density stent wires and their relationship with renal arteries were easily identified on MIP (arrowheads in B) and VR (red color) images. Left renal ostium was found to be covered by two stent wires peripherally as shown in VE (D). White arrows indicate the renal arteries, while black arrows point to the left renal ostium

3 Discussion

Helical CTA has been recognized as the preferred imaging modality for evaluating the abdominal vasculature as well as stent grafts [7,8]. There is growing awareness of the necessity to interpret 3D reconstructions in conjunction with the axial source images [10]. CTA-generated 3D postprocessing methods have been reported to complement conventional axial images and enhance our understanding of the effect of aortic stent grafting [11]. Among these four 3D reconstructions presented in our study, SSD, MIP and VR are most commonly used in aortic stent grafting, while VE is a relatively new technique which allows intraluminal views of aortic lumen as well as stent wires [11]. Reliable recognition of the advantages and disadvantages of these 3D techniques will aid clinicians to make efficient use of 3D reconstructions and evaluate the treatment of aortic stent grafting, especially the effect of suprarenal stent on the renal arteries.

Although SSD makes images appear very three-dimensional and intuitive, it has serious limitations. As only gray-scale information is used for surface estimation, a large proportion of the available data is lost and not represented. It is very sensitive to changes in threshold and it is often difficult to determine which threshold range results in the most accurate depiction of the actual anatomic structures. This was observed in our study which showed that visualization of stent relative to the renal artery was significantly affected in more than half of the cases. MIP images are most commonly used in vascular imaging as they provide excellent differentiation between vascular and nonvascular structures. They clearly demonstrate contrast-enhanced arteries as well as aortic stents as shown in our study. However, MIP images lack depth information and MIP is a time-consuming technique due to manual editing of volume data. The median time of nearly 30 min required for generation of MIP images in our study prevents it from being accepted by clinicians as a routine imaging technique in clinical practice.

VR was found to be the most efficient reconstruction in aortic stent grafting, according to our study. VR is often combined with depth and surface shading and rotation to give the viewer a realistic sense of three-dimensionality. The use of varying degree of transparency combined with color coding of different structures allows simultaneous display of superficial and deep structures in an image, as shown in Fig1. 3D relationship between suprarenal stent wires and renal arteries was also clearly visualized by VR images regarding the encroachment or coverage of stent wires to the arterial branches. The final image quality depends on the segmentation of the volume data that are given to the rendering algorithm, which is the main limitation of VR.

VE was found to be particularly useful in the demonstration of stent wires relative to the renal ostium due to its capability of providing intraluminal views. We consider VE as a valuable technique in the follow-up of patients with AAA after aortic stent grafting as it provides clinicians with additional information when compared to other reconstruction methods regarding the effect of suprarenal stent wires on the renal ostium.

The uniphasic contrast enhancement and scanning parameters inherent in single slice CT used in our study resulted in inhomogeneous image quality with minor artifacts in 50% of cases when the contrast reached the iliac arteries. This limitation can be overcome using biphasic or multiphasic contrast enhancement, which has been reported to produce homogenous enhancement along the abdominal aorta [12, 13]. With the emergence of multislice CT in most clinical centres, we believe that image quality of CT volume data acquired on multislice CT will definitely improve with subsequent improvement of 3D reconstruction methods on account of its fast scanning, thinner collimation and high-spatial resolution [14].

4 Conclusion

We found in our study that VR is the most efficient technique in aortic stent grafting and VE is particularly useful in demonstration of the encroachment to the renal ostium by suprarenal stent wires. Further study of validation of diagnostic value of these 3D rendering techniques in a large cohort of patients deserves to be investigated.

References

1. Government Statistical Service: Hospital episode statistics. 1. Finished Consultant Episodes By Diagnosis, Operation and Speciality. England: Financial Year 1992-1993. London: HMSO (1995)
2. Lawrence, P.F., Gazak, C., Bhirangi, L, Jones, B, Bhirangi, K, Oderich, G, Treiman, G.: The Epidemiology Of Surgically Repaired Aneurysms In The United States. J. Vasc. Surg. **30** (1999) 632-640
3. Parodi, J.C., Palmaz, J.C., Barone, H.D.: Transfemoral Intraluminal Graft Implantation For Abdominal Aortic Aneurysms. Ann. Vasc. Surg. **5** (1991) 491-499
4. Parodi, J.C., Barone, A., Piraino, R., Schonholz, C: Endovascular Treatment Of Abdominal Aortic Aneurysms: Lessons learned. J. Endovasc. Surg. **4** (1997) 102-110
5. Malina, M., Lindh, M., Ivancev, K., Frennby, B., Lindblad, B., Brunkwall, J: The Effects Of Endovascular Aortic Stents Placed Across The Renal Arteries. Eur. J. Vasc. Endovasc Surg. **13** (1997) 207-213
6. Lau, L.L., Hakaim, A.G., Oldenburg, W.A.,Neuhauser, B, McKinney, J.M., Paz-Fumagalli, R, Stockland, A.: Effect Of Suprarenal Versus Infrarenal Aortic Endograft Fixation On Renal Function And Renal Artery Patency: A Comparative Study With Intermediate Follow-up. J. Vasc. Surg. **37** (2003) 1162-1168
7. Broeders, IAMJ., Blankensteijn, J.D., Olree, M., Mali, W., Eikelboom, B.C.: Preoperative Sizing Of Grafts For Transfemoral Endovascular Aneurysm Management: A Prospective Comparative Study Of Spiral CT Angiography, Arterial Angiography And Conventional CT Imaging. J. Endovasc. Surg. **4** (1997) 252-261
8. Armerding, M.D., Rubin, G.D., Beaulieu, C.F., Slonim, S.M., Olcott, E.W., Samuels, S.L., Jorgensen, M.J, Semba, C.P, Jeffrey, R.B., Jr, Dake, M.D.: Aortic Aneurysmal Disease: Assessment Of Stent-Grafted Treatment-CT Versus Conventional Angiography. Radiology **215** (2000) 138-146
9. Sun, Z., Winder, J., Kelly, B., Ellis, P., Hirst, D.: CT Virtual Intravascular Endoscopy Of Abdominal Aortic Aneurysms Treated With Suprarenal Endovascular Stent Grafting. Abdom. Imaging. **28** (2003) 580-587
10. Zeman,R.K., Silverman , P.M., Vieco, P.T., Costello, P.: CT Angiography. A. J. Roentgenol. **165** (1995) 1079-1088
11. Sun, Z, Winder, J, Kelly, B,Ellis, P, Kennedy, P, Hirst, D.: Diagnostic Value Of CT Virtual Intravascular Endoscopy In Aortic Stent Grafting. J. Endovasc. Ther. **11** (2004) 13-25
12. Choe, Y.H., Phyun, L.H., Han, B.K.: Biphasic And Discontinuous Injection Of Contrast Material For Thin-Section Helical CT Angiography Of The Whole Aorta And Iliac Arteries. Am. J. Roentgenol. **176** (2001) 454-456
13. Bae, K.T., Tran, H.Q., Heiken, J.P.: Multiphasic Injection Method For Uniform Prolonged Vascular Enhancement At CT Angiography: Pharmacokinetic Analysis And Experimental Porcine Model. Radiology. **216** (2000) 872-880
14. Hu, H., He, H.D., Foley, W.D., Fox SH.: Four Multidetector-Row Helical CT: Image Quality And Volume Coverage Speed. Radiology **215** (2000) 55-62

A New Fuzzy Penalized Likelihood Method for PET Image Reconstruction*

Jian Zhou, Huazhong Shu, Limin Luo, and Hongqing Zhu

Lab of Image Science and Technology,
Department of Biological Science and Medical Engineering,
Southeast University, Nanjing 210096, China
zjseu@hotmail.com, {shu.list, luo.list, hqzhu}@seu.edu.cn

Abstract. In positron emission tomography (PET) image reconstruction, classical regularization methods are usually used to overcome the slow convergence of the expectation maximization (EM) methods and to reduce the noise in reconstructed images. In this paper, the fuzzy set theory was employed into the reconstruction procedure. The observations of emission counts were viewed as Poisson random variables with fuzzy mean values. And the fuzziness of these mean values was modelled through choosing an appropriate fuzzy membership function with several adjustable parameters. Coupled with this fuzzy method, the new fuzzy penalized likelihood expectation maximization (FPL–EM) method was proposed for PET image reconstruction. Simulation results showed that the proposed method might perform better in both the image quality and the convergence rate compared with the classical maximum likelihood expectation-maximization (ML–EM).

1 Introduction

Statistical iterative methods for PET reconstruction have become popular since Shepp and Vardi introduced the ML–EM method [1]. However, the ML–EM method is of slow convergence and low resolution due to the ill-conditioned problem of PET. One way to overcome these problems is to use the regularization (or penalized) method.

Statistical regularization methods are thought as an appropriate way to solve the maximum of likelihood function for reconstruction with considering several additional penalty terms. In this paper, we proposed a new penalty term through investigating the fuzziness of the observations' mean values. According to fuzzy set theory (see [2] and [3]), in PET the observations can be modelled as random variables with noncognitive uncertainty known for the Poisson distribution and cognitive uncertainty embedded in their mean values. We addressed this cognitive uncertainty with an appropriate membership function that can be in-

---

* This work was supported by National Basic Research Program of China under grant No. 2003CB716102.

terpreted as a degree of belief for some parameters. Guided by this way, we presented our new fuzzy penalized method for PET reconstruction.

The rest of paper is organized as follows: The new reconstruction method is presented in Sec. 2; The experimental results are given in Sec. 3; Finally, Sec. 4 concludes the paper.

2 Method

2.1 Statistical Model and the ML Estimation

In the Poisson model for PET, the photon counts are represented as a spatial, inhomogeneous Poisson process with unknown intensity [1]. Suppose that the reconstruction regions are subdivided into N_p pixels, such that the emission counts in data collection time at different locations are independent and at location j constitutes a Poisson variable with mean x_j. Then the photon counts collected by N_d detector pairs, denoted y_i, $i = 1, ..., N_d$, are also independent Poisson random variables with mean value ξ_i,

$$\xi_i = (\boldsymbol{A}\boldsymbol{x})_i = \sum_{j=1}^{N_p} a_{ij} x_j \tag{1}$$

where the element a_{ij} in the system matrix $\boldsymbol{A}$ is the probability that a photon emitted at location j is detected by the detector pair i.

For the Poisson random variable Y_i, we have the following probability density function

$$P(Y_i = y_i) = \exp(-\xi_i) \cdot \frac{\xi_i^{y_i}}{(y_i)!}. \tag{2}$$

Then the likelihood function can be given by the multiplication of the probability density function of each Y_i, $i = 1, ..., N_d$. In PET reconstruction, the equivalent log-likelihood function [1]

$$\Phi(\boldsymbol{x}) = \sum_{i=1}^{N_d} \left\{ -(\boldsymbol{A}\boldsymbol{x})_i + y_i \cdot \log(\boldsymbol{A}\boldsymbol{x})_i \right\} \tag{3}$$

is often used (terms independent to $\boldsymbol{x}$ are ignored).

Therefore, the maximum likelihood (ML) estimation for image $\boldsymbol{x}$ can be computed by solving the following nonlinear optimization problem,

$$\hat{\boldsymbol{x}}_{ML} = \arg\max \Phi(\boldsymbol{x}) \quad \text{subject to: } \boldsymbol{x} \geq 0.$$

2.2 Fuzzy Reconstruction Method

We first make the following assumption: *For each $i = 1, ..., N_d$, the mean value ξ_i may take its value near around a ξ_i^*, where ξ_i^* is a properly selected value.* In most cases, this assumption holds because the system model usually has limited accuracy which leads to a bounded error for any ξ_i given the emission image intensities.

Under this assumption, for each $i = 1, ..., N_d$, we define a fuzzy subset Ω_i,

$$\Omega_i = \left\{ \omega | \xi_i^* - \delta \leq \omega \leq \omega_i^* + \delta, \omega \in \mathcal{Z}^+ \right\}$$

where $\mathcal{Z}^+$ is the set of positive integer and $\Omega_i \subset \mathcal{Z}^+$. Since the emission counts must be nonnegative, each element in Ω_i should be nonnegative integer. δ ($\delta \leq \xi_i^*, \delta \in \mathcal{Z}^+$) is an adjustable parameter that controls the cardinality of the fuzzy set.

For each $i = 1, ..., N_d$, we define a membership function $\alpha(\xi_i)$, $\forall i \in \mathcal{Z}^+$. For any given ξ_i, $\xi_i \in \Omega_i$, we may set a relatively greater degree of belief $\alpha(\xi_i)$ on Ω_i. If simply let $\alpha(\xi_i) = 1.0$ on Ω_i, then one can obtain a rectangle membership function shown in Fig. 1(a). Generally, for any fuzzy set Ω_i, we can define a Gaussian membership function (GMF) shown in Fig. 1(b) which is given by

$$\alpha(\xi_i) = \exp \left\{ -\frac{(\xi_i - \xi_i^*)^2}{2\sigma_i^2} \right\}. \tag{4}$$

In (4), the shape of function is adjusted through the parameter σ_i, The smaller the σ_i is, the smoother the curve will be. In this paper, our method is mainly developed by using this GMF.

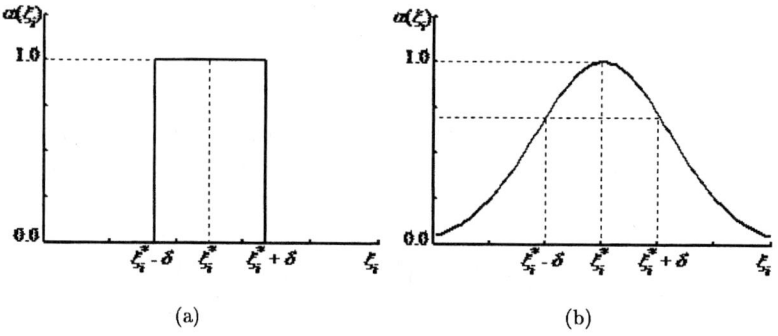

Fig. 1. Fuzzy membership function. (a) Rectangle membership function. (b) GMF

The function $\alpha(\xi_i)$ can also be viewed as the probability density function for the random variable Ξ_i if we normalize it by dividing its summation over the set $\mathcal{Z}^+$. The probability function $P(\Xi_i = \xi_i)$ for random variable Ξ_i is given by

$$P(\Xi_i = \xi_i) = \frac{\alpha(\xi_i)}{\sum_{\eta_i \in \mathcal{Z}^+} \alpha(\eta_i)} \rightarrow \frac{1}{C_i} \cdot \exp \left\{ -\frac{(\xi_i - \xi_i^*)^2}{2\sigma_i^2} \right\} \tag{5}$$

where C_i is the summation of $\alpha(\xi_i)$ over the set $\mathcal{Z}^+$.

The probability function $P(Y_i = y_i)$ in (2) can then be rewritten to be the conditional probability $P(Y_i = y_i|\Xi_i = \xi_i)$. Hence, the joint probability density function $P(Y_i = y_i, \Xi_i = \xi_i)$ is defined by the multiplication of $P(Y_i = y_i|\Xi_i = \xi_i)$ and $P(\Xi_i = \xi_i)$. Since the observations are assumed to be independent, the joint probability density can be represented as

$$\prod_{i=1}^{N_d} P(Y_i = y_i, \Xi_i = \xi_i) = \prod_{i=1}^{N_d} \exp(-\xi_i) \cdot \frac{\xi_i^{y_i}}{(y_i)!} \cdot \frac{1}{C_i} \cdot \exp\left\{-\frac{(\xi_i - \xi_i^*)^2}{2\sigma^2}\right\}. \quad (6)$$

Substituting (1) into (6) and taking logarithm, we obtain

$$\Psi(\boldsymbol{x}) = \sum_{i=1}^{N_d} \left\{-(\boldsymbol{Ax})_i + y_i \cdot \log(\boldsymbol{Ax})_i\right\} + \sum_{i=1}^{N_d} \left\{-\frac{[(\boldsymbol{Ax})_i - \xi_i^*]^2}{2\sigma_i^2}\right\}. \quad (7)$$

Notice that any observed photons y_i are realizations for Poisson random variables which, in a statistical sense, approximate the mean values. Hence, every observation yi should be fallen into our predefined fuzzy set Ω_i. Moreover, we can set the greatest degree of belief at the point of y_i, i.e., $\alpha(\xi_i^*) = \alpha(y_i) = \max \alpha(\xi_i)$, $\forall \xi_i \in \Omega_i$, which is equivalent to set ξ_i^* to y_i, $(i = 1, ..., N_d)$. The σ_i is a shape parameter controlling the degree of belief of each ξ_i. For the sake of simplicity, we apply the same shape parameter to each membership function, i.e., for each $i = 1, ..., N_d$, we let $\sigma_i = \sigma$, where σ is a constant. Thus we simplify (7) into the following function

$$\Psi(\boldsymbol{x}) = \sum_{i=1}^{N_d} \left\{-(\boldsymbol{Ax})_i + y_i \cdot \log(\boldsymbol{Ax})_i\right\} + \sum_{i=1}^{N_d} \left\{-\frac{[(\boldsymbol{Ax})_i - y_i]^2}{2\sigma^2}\right\}. \quad (8)$$

Since the observations are usually viewed as an incomplete data space, we can solve the maximum optimization problem of (8) by using the EM method [1], [4], and then obtain our FPL–EM method in which the update for each pixel is as follows

$$x_j^{k+1} = \frac{x_j^k \cdot \sum_{i=1}^{N_d} a_{ij} \frac{y_i}{(\boldsymbol{Ax}^k)_i}}{\sum_{i=1}^{N_d} a_{ij}\{1 + \sigma^{-2} \cdot [(\boldsymbol{Ax}^k)_i - y_i]\}} \quad (9)$$

where $k = 0, 1, 2, ...$, indicates the iteration number.

3 Experimental Evaluation

In our experiments, we used the Shepp-Logan phantom shown in Fig. 2. The projection space was assumed to be 160 radial bins and 192 angles evenly spaced over 180°. The final reconstructed images were set to a size of 64 × 64 pixel matrices.

Fig. 2. Phantom

The sinogram were globally scaled to a mean sum of $1,000,000$ true events. And the pseudo-random Poisson variants were drawn through the model in subsec. 2.1. Data were also studied with 5% uniform Poisson distributed background events, representing the range of random coincidences in PET scans. The initial estimator x_0 was set to a strictly positive vector.

Fig. 3 shows the reconstructed images by using ML–EM and FPL–EM. Different shape parameter σ's in FPL–EM were used to test its influence on the reconstructed image.

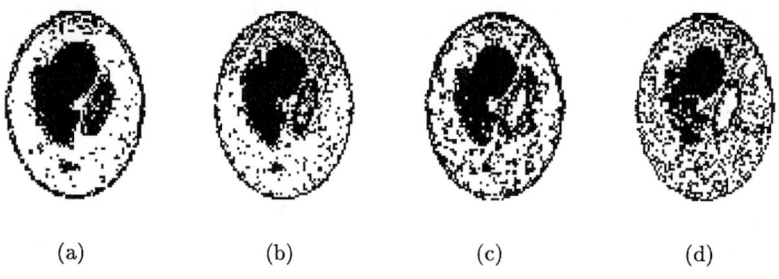

(a)   (b)   (c)   (d)

Fig. 3. Reconstructed images. (a) ML–EM reconstruction (20 iterations). (b), (c) and (d) FPL–EM reconstructions. (20 iterations for $\sigma^{-2} = 0.002, 0.004, 0.006$, respectively)

To evaluate the reconstructed images, the mean square error(MSE) between each iteration result x^k and the true phantom x^* was computed using the following definition

$$\text{MSE}(k) = \frac{\|x^k - x^*\|}{\|x^*\|}, \quad k = 1, 2, \ldots \tag{10}$$

where $\|x\|$ is the Euclidean length of a vector x

Fig. 4(a) shows the mean square error comparison between ML–EM and FPL–EM. Clearly, small mean square errors can be achieved by choosing an appropriate shape parameter σ in FPL–EM. In Fig. 4(a), we can observe that the FPL–EM method yields the best performance when $\sigma^{-2} = 0.004$.

Fig. 4(b) shows the log-likelihood values versus iteration number. The fastest convergence rate was also yielded with $\sigma^{-2} = 0.004$ in FPL–EM. Fig. 4(c) shows the line plot of row 32 of reconstructed images. It can be seen that the FPL–EM method has better performance than the ML–EM method.

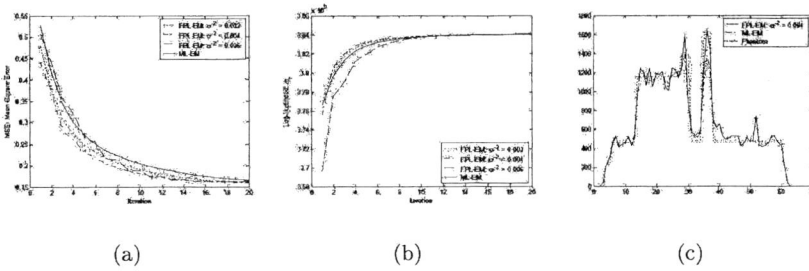

Fig. 4. Evaluation results. (a) Mean square errors versus iteration. (b) The log-likelihood values versus iteration. (c) Line plot of row 32 at the twentieth iteration

4 Conclusions

We have proposed a new fuzzy penalized likelihood method for PET image reconstruction. The simulations proved its better performance than the ML–EM method in both image resolution and convergence. The future work will concentrate on finding the optimal shape parameter for GMF. In addition, further studies on the image prior information, such as the Gibbs distribution, are currently under investigation.

References

1. Shepp, L. A., Vardi, Y.: Maximum likelihood reconstruction for emission tomography, IEEE Trans. Med. Imag. **MI-1(2)** (1982) 113–122.
2. Zadeh, L. A.: Fuzzy Sets. Information and Control, **8** (1965) 338–353.
3. Chao, R. J., Ayyub, B. M.: Distributions with Fuzziness and Randomness, Proceedings of ISUMA - NAFIPS '95 The Third International Symposium on Uncertainty Modeling and Analysis and Annual Conference of the North American Fuzzy Information Processing Society (1995) 668–673.
4. Lange, K., Carson, R.: EM reconstruction algorithm for emission and transmission tomography, J. Comput. Assist. Tomog. **8** (1984) 306–316.

Interactive GSOM-Based Approaches for Improving Biomedical Pattern Discovery and Visualization

Haiying Wang, Francisco Azuaje, and Norman Black

School of Computing and Mathematics, University of Ulster at Jordanstown,
Newtownabbey, Co. Antrim, BT37 0QB, N. Ireland, UK
{hy.wang, fj.azuaje, nd.black}@ulster.ac.uk

Abstract. Recent progress in biology and medical sciences has led to an explosive growth of biomedical data. Extracting relevant knowledge from such volumes of data represents an enormous challenge and opportunity. This paper assesses several approaches to improving neural network-based biomedical pattern discovery and visualization. It focuses on unsupervised classification problems, as well as on interactive and iterative methods to display, identify and validate potential relevant patterns. Clustering and pattern visualization models were based on the adaptation of a self-adaptive neural network known as *Growing Self Organizing Maps*. These models provided the basis for the implementation of hierarchical clustering, cluster validity assessment and a method for monitoring learning processes (cluster formation). This framework was tested on an *electrocardiogram beat data set* and *data consisting of DNA splice-junction sequences*. The results indicate that these techniques may facilitate knowledge discovery tasks by improving key factors such as predictive effectiveness, learning efficiency and understandability of outcomes.

1 Introduction

Advances in biological and medical sciences have been reflected in an explosive growth of diverse data sources. Due to its inherent complexity and the lack of comprehensive theories at the molecular and physiological levels, extracting relevant knowledge from such volumes of data represents an enormous challenge and opportunity.

It has been demonstrated that *artificial neural networks* (ANNs) are well suited to such domains. Ideker *et al.* [1], for example, used *Kohonen Feature Maps* (SOMs) [2] to support the study of a cellular pathway model, in which they identified 997 messenger RNAs (mRNAs) responding to 20 systematic perturbations of the yeast galactose utilization pathway. Azuaje [3] applied a *Simplified Fuzzy ARTMAP* (SFAM) model to analyse gene expression data. In this study, the SFAM-based system not only distinguished normal from diffuse large B-cell lymphoma (DLBCL) patients, but also it identified the differences between patients with molecularly distinct forms of DLBCL without previous knowledge of those subtypes.

While successfully addressing key biomedical pattern classification problems, most of the existing ANN-based solutions exhibit several limitations that limit its applicability. For example, a fundamental problem of ANNs is that the information that they encode cannot be easily understood by humans. The internal structure and the way in which an ANN derives an output value from a given feature vector have been traditionally concealed from the user. It has become apparent that, without some form of explanation and interpretation capability, the potential of ANNs-based pattern discovery may not be fully achieved [4, 5].

Based on *self-adaptive neural networks* (SANNs) together with graphical and statistical tools, this paper explores several approaches to support pattern discovery and visualization. The remainder of this paper is organized as follows. Section II describes the key component in this study: *Growing Self Organizing Maps* (GSOM), followed by a description of the data sets under study. The results are presented in Section IV. Discussion and Conclusions are included in the last section.

2 The GSOM

The GSOM [6] belongs to the family of SANNs, which follows the basic principle of the SOM but adapts its shape and size according to the structure of the input data. The GSOM learning process, which is started by generating an initial network composed of four neurons, includes three stages: *Initialization, growing* and *smoothing* phases. Like other SANNs, each input presentation involves two basic operations: (a) finding the best matching neurone for each input vector and (b) updating the winning neurone and its neighbourhood. The reader is referred to [6] and [7] for a detailed description of the learning dynamics of the GSOM.

In the context of pattern discovery and visualization, the GSOM exhibits two important features:

- The GSOM keeps a regular, 2D grid structure at all times. The resulting map reveals patterns hidden in the data by its shape and attracts attention to relevant areas by branching out. This provides the basis for a user-friendly and flexible pattern visualization and interpretation platform.
- The user can provide a spread factor, SF, $[0, 1]$ to specify the growth degree of the GSOM. This provides a straightforward way to measure and control the expansion of the networks. Based on the selection of different values of SF, hierarchical and multiresolution clustering may be implemented.

3 The Data Sets and Prediction Problems

Two data sets were used to test the approaches proposed. The first one is an ECG beat data set obtained from the MIT/BIH arrhythmia database [8]. Based on a set of descriptive measurements for each beat, the problem defined in this data set is to determine whether a beat is a ventricular ectopic beat (Class V) or a normal beat (Class N). Each beat is represented by 9 features. A detailed description of this data set can be found at http://www.cae.wisc.edu/~ece539/data/ecg/.

The second data set is a DNA sequence data set. The problem posed is to decide, given a DNA sequence, whether it is an 'intron ⟶exon' boundary (IE), 'exon ⟶ intron' boundary (EI), or neither (N). The data set consists of 2000 sequences taken from [9], in which 23.2% belong to Class EI, 24.25% are IE samples and 52.55% are N samples. Each sequence, which originally includes 60 nucleotides, is represented by 180 binary attributes. Each nucleotide is replaced by a three-digit binary variable as follows: A⟶100; C⟶010; G⟶001; T⟶000. Each class is represented by a number: EI⟶1; IE⟶2; N⟶3. The full data description is given in [9].

4 Results

4.1 Pattern Discovery and Visualization Using GSOM

One of the key advantages of the GSOM is that it can develop into different shapes to reveal key patterns hidden in the data. Such visualization capabilities may highlight the inherent data structure in a more meaningful way. Based on the relevant resulting maps, the cluster structure of the underlying data can be easily interpreted. The GSOM maps with $SF = 0.001$ for the ECG data set shown in Fig. 1(a), for example, has branched out in two directions, representing the two main classes of the data set. An analysis of the sample distribution over each branch indicates that there is a dominant class represented in each branch. Based on a majority voting strategy, the label map(1⟶N and 2⟶V) is shown in Fig. 1(b). Similar data are successfully clustered. Ambiguous areas, such as

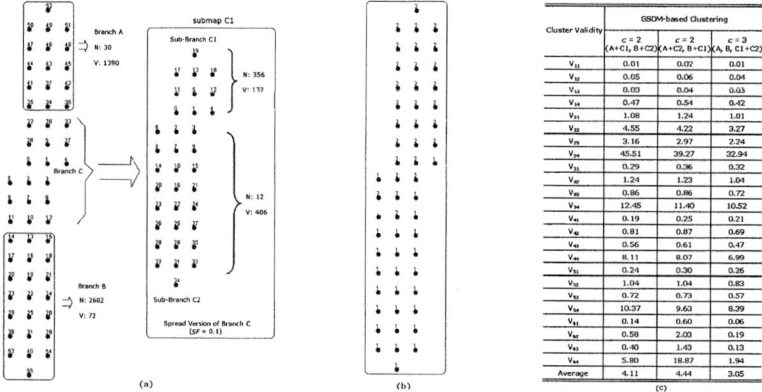

Fig. 1. GSOM-based data visualization for ECG data. (a) The resulting map. The numbers shown represent the order in which each neuron was created during the growth phase. (b) The label map. (c) Twenty-four Dunn's indices for the three partitions

the borders of the cluster regions, can be further analyzed with a higher *SF* value (see submap C1). This submap has clearly been dispersed in two directions (Sub-Branches C1 and C2). To identify the relationship between Branches A, B and Sub-Branches C1 and C2, a clustering identification and evaluation framework based on twenty-four Dunn's indices [7, 10] were implemented.

Fig. 1(c) shows twenty-four Dunn's validity indices for the three possible partitions suggested by visualizing the resulting map illustrated in Fig. 1(a) and (b). Good partitions are associated with larger value of the Dunn's indices. Fifteen validity indices indicate that the second 2-clusters partition is the optimum partition, which indicates that the Sub-Branch C1 is closer to Branch B and Sub-Branch C2 may be associated with Branch A. This coincides with the sample distribution in these areas.

4.2 Visualization of the Learning Process

The visualization of a learning process and its outcomes can help users to gain a better understanding of the predictions made by an ANN-based approach. Nevertheless, the study of advanced visualization techniques for ANNs has not received the attention it deserves [11]. The GSOM incrementally grows new neurons to achieve a better representation of the input data. Thus, graphical tools to visualize the phases of a GSOM learning process may provide a better understanding of the inter- and intra-cluster relationships of the data [12], as illustrated in Fig. 2 for the DNA data.

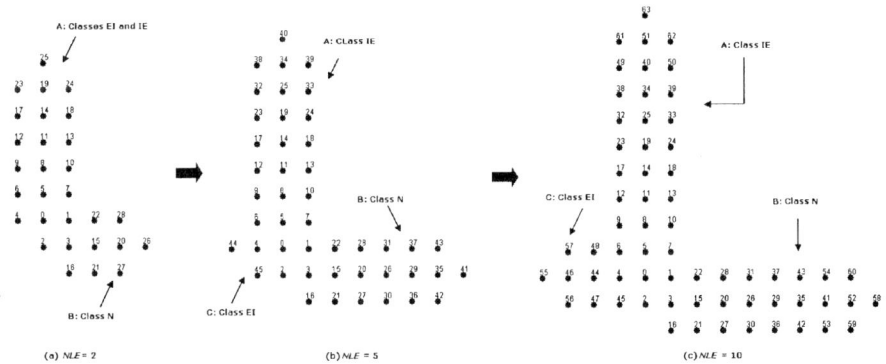

Fig. 2. The resulting GSOM map for DNA data at different learning stages: (a) *NLE* = 2, (b) *NLE* = 5; (c) *NLE* = 10. *NLE* stands for the number of learning epochs

At an early stage (*NLE* = 2, shown in Fig. 2(a)) the map branches out into two directions, indicating that the data set may be firstly clustered into two major categories: splice (Classes EI and IE) and non-splice (Class N) junctions. As the learning process progresses, the GSOM gradually produces a new branch (Branch C in Fig. 2(b)) to represent the underlying data structure, suggesting that the samples (splice junctions) assigned to Branch A (shown on Fig. 2(a)) could be further classified into two different classes (Classes EI and IE). The GSOM map shown on Fig. 2(c) ($NLE = 10$) support this argument.

4.3 Hierarchical Clustering with the GSOM

Based on the selection on different SF values, the GSOM can be used to perform hierarchical clustering. The results obtained show that the GSOM may highlight significant groups in the data through its shape even for lower SF values. A higher resolution view can be obtained with a larger SF value in the areas of interest. The map shown in Fig. 3(b) was produced with $SF = 0.01$, indicating the two significant groups in the DNA data set (splice and non-splice junctions). A higher SF value is then applied ($SF = 0.1$) to the Branch A (Fig. 3(a)). The new map has been developed in two directions, suggesting that there may be two subgroups (Classes EI and IE) in this branch. This is consistent with the sample distribution over these two sub-branches (Branches A1 and A2). Based on this observation together with the results obtained in the last section, we argued that the hierarchical structure shown in Fig. 3(c) perhaps reflects the inherent data structure in a more meaningful and precise way.

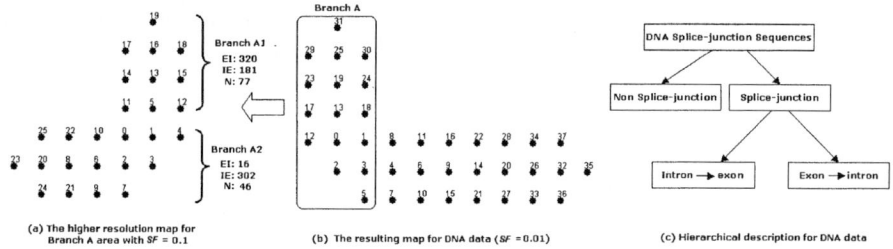

Fig. 3. The hierarchical clustering of DNA data based on different SF values. (a) The higher resolution map of Branch A with $SF = 0.1$. (b) The resulting map with $SF = 0.01$. (c) A hierarchical description for DNA data inferred by GSOM-based technique

5 Discussion

This study indicates that a GSOM-based system may facilitate two important biomedical pattern discovery and visualization problems. For instance, instead of using a static grid representation or long lists of numbers to describe patterns revealed, this research developed a graphical environment to dynamically depict a networks' evolution during learning. Such a facility may help scientists achieve a better understanding of the dynamics of the ANN-based model. The GSOM attempts to build a structure to represent the input data as accurately as possible. By monitoring its learning process, users can better understand how a model generates its decision. By introducing a SF, users have more control on the growth of the resulting map. Thus, hierarchical and multi-resolution clustering analysis may also be implemented. A robust framework for assessing the quality of clustering outcomes, which has been implemented based on the combination of cluster validity indices, provides statistical indicators that further enhance the understanding of the results generated by ANNs. The approaches discussed in this paper hold promise to deliver an iterative and interactive pattern visualization and interpretation platform.

Due to the generation of new neurons in all free neighbouring positions, the GSOM inevitably generates irrelevant neurons, which sometimes can severely degrade the visualization ability of GSOM-based models. Incorporation of pruning algorithms into GSOM-based learning techniques deserves further investigation.

Another problem that deserves further research is GSOM-based feature relevance assessment. In GSOM, a SF is independent of the dimensionality of the data. Thus, one can compare different maps by using the same SF and different combinations of features to identify their contributions to the cluster structure. Such a method can help the data miner to study the effect of each feature on the clustering and their relationships.

References

1. Ideker, T., Thorsson, V., Ranish, J. A., Christmas, R., Buhler, J., Eng, J. K., Bumgarner, R., Goodlett, D. R., Aebersol, R., Hood, L.: Integrated Genomic and Proteomic Analyses of a Systematically Perturbated Metabolic Network. Science, 292 (2001) 929-933
2. Kohonen,T.:Self-Organizing Maps. Heidelberg. Germany:Springer-Vin erlag (1995)
3. Azuaje, F.: A Computational Neural Approach to Support the Discovery of Gene Function and Classes of Cancer. IEEE Trans. on Biomedical Engineering, 48(3), (2001) 332-339
4. Baldi, P., Brunak, S.: Bioinformatics: The Machine Learning Approach. London: The MIT Press (2000)
5. Dybowski, R., Gant, V.: Clinical Applications of Artificial Neural Networks. London: Cambridge University Press (2001)
6. Alahakoon, D., Halgamuge, S. K., Srinivasan, B.: Dynamic Self-Oorganizing Maps with Controlled Growth for Knowledge Discovery. IEEE Transactions on Neural Networks 11(3), (2000) 601-614
7. Wang, H., Azuaje, F., Black, N.: Improving Biomolecular Pattern Discovery and Visualization with Hybrid Self-Adaptive Networks. IEEE Transactions on Nanobioscience 1(4), (2002) 146-166
8. Mark, R., Moody, G.: MIT-BIH Arrhythmia DataBase Directory. Cambridge: MIT (1988)
9. Brazdil, P., Gama, J.: DNA-Primate Splice-Junction Gene Sequences, with Associated Imperfect Domain Theory. Available at: http://porto.niaad.liacc.up.pt/niaad/statlog/datasets/dna/dna.doc.html (June 2002)
10. Bezdek, J. C., Pal, N. R.: Some New Indexes of Cluster Validity. IEEE Transactions on Systems, Man, and Cybernetics-Part B: Cybernetics, 28(3), (1998) 301-315
11. Haykin, S.: Neural Networks: A Comprehensive Foundation. New Jersey: Prentice Hall (1994)
12. Wang, H., Azuaje, F., Black, N.: An Integrative and Interactive Framework for Improving Biomedical Pattern Discovery and Visualization. IEEE Transactions on Information Technology in Biomedicine, 8(1), (2004) 16-27

Discontinuity-Preserving Moving Least Squares Method

Huafeng Liu[1,2] and Pengcheng Shi[2]

[1]State Key Laboratory of Modern Optical Instrumentation,
Zhejiang University, Hangzhou, China
[2]Department of Electrical and Electronic Engineering,
Hong Kong University of Science and Technology, Hong Kong
{eeliuhf, eeship}@ust.hk

Abstract. We present a discontinuity-preserving moving-least-squares (MLS) method with applications in curve and surface reconstruction, domain partitioning, and image restoration. The fundamental power of this strategy rests with the moving support domain selection for each data point from its neighbors and the associated notion of compactly supported weighting functions, and the inclusion of singular enhancement techniques aided by the data-dependent singularity detection. This general framework meshes well with the multi-scale concept, and can treat uniformly and non-uniformly distributed data in a consistent manner. In addition to the smooth approximation capability, which is essentially the basis of the emerging meshfree particle methods for numerical solutions of partial differential equations, MLS can also be used as a general numerical method for derivative evaluation on irregularly spaced points, which has a wide variety of important implications for computer vision problems.

1 Introduction

Data fitting and approximation is one of the central problems in computer vision, with wide range of applications in object reconstruction and recognition, image restoration, and domain partitioning. The fundamental needs come from the disparity between the continua of visual phenomena and the discreteness of measurement data that are usually corrupted by noises. A typical strategy to tackle this type of problems is to introduce a *smoothness* constraint:

$$\min \sum_{i=1}^{N} (S(t) - S_i)^2 + \int \alpha (S')^2 + \beta (S'')^2 \, dt \tag{1}$$

where $S(t)$ is the unknown smooth function and S_i is the measurement data. This actually become the regularization paradigm in computer vision, with a wide range of variations and applications developed in the past twenty years [1, 3, 4, 9].

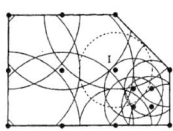 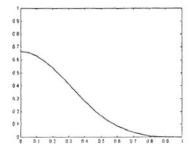 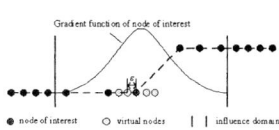

Fig. 1. Influence domains (left), cubic spline weighting function (middle) and The addition of gradient-dependent *virtual nodes* (right)

It has been suggested that the least squares approximation ideas can be applied to generate discontinuity-preserving interpolants by introducing the notion of *moving least squares* approximation together with appropriate singularities in the weights used in such approximations [6]. However, for computer vision problems, the question of where to enforce the interpolation and where to perform the smoothing approximation remains unanswered. Of course, thresholding of data gradient may do the trick of detecting the singularities under certain situations, which can then be followed by the interpolating MLS procedure. Nevertheless, the selection of proper thresholds introduce a whole new set of questions.

We present a MLS algorithm which performs smoothing approximation and discontinuity-preserving interpolation in a consistent fashion. Instead of direct manipulation of the MLS weight functions and/or the basis functions in order to create the singular effect at any particular data point [5, 6], we make changes to the data point distribution at the support domain around the point of interest, which in turn alters the weight distribution to achieve the desired interpolating or smoothing effect. We demonstrate the accuracy and robustness of the method with synthetic curve and surface data fitting, and present the experiment results with real images.

2 Moving Least Squares Approximation

It is assumed that a function $f : \bar{D} \to \mathbf{R}$ is to be approximated and that its values $f_i = f(\mathbf{x}_i)$ at sampling nodes $\mathbf{x}_i \in \bar{D}, i = 1, 2, ..., N$ are given, where domain $\bar{D}$ is assumed to be the closure of a simply connected subset $D \in \mathbf{R}^n$. We aim to develop an *smooth approximation* Gf to f on the basis of the available information about f at those scattered nodal locations, which are often noise corrupted and not necessary uniformly distributed.

2.1 MLS Formulation

At each point $\hat{\mathbf{x}} \in \bar{D}$, a local approximant $L_{\hat{\mathbf{x}}} f$ of the global approximating function Gf is defined in terms of some basis functions $\{p^{(i)}\}_{i=1}^n, n \leq N$ and a local L^2-norm:

$$L_{\hat{\mathbf{x}}} f := \sum_{i=1}^{n} a_i(\hat{\mathbf{x}}) p^{(i)}(\hat{\mathbf{x}}) \qquad (2)$$

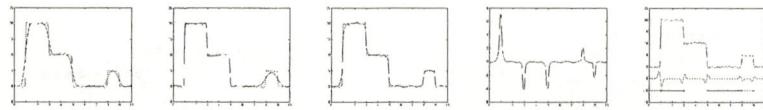

Fig. 2. Noisy curve fitting with MLS, GTMLS, GWMLS, 1st derivative (left) and 2nd derivative (green) with segment partitioning at zero-crossings (right) for the 20dB GWMLS curve

where the set of coefficients $a_i(\hat{\mathbf{x}}), i = 1, 2, ..., n$ are determined such that Equation (2) is, in the least squares sense, the best approximation of f. The global approximant Gf is then defined on the entire domain for any $\hat{\mathbf{x}} \in \bar{D}$

$$Gf(\hat{\mathbf{x}}) = L_{\hat{\mathbf{x}}} f(\hat{\mathbf{x}}) = \sum_{i=1}^{n} a_i(\hat{\mathbf{x}}) p^{(i)}(\hat{\mathbf{x}}) \qquad (3)$$

Commonly used basis functions in two-dimensional cases include the linear one $\mathbf{p}_{(n=3)}^T = \{1, x, y\}$ and the quadratic one $\mathbf{p}_{(n=6)}^T = \{1, x, y, x^2, xy, y^2\}$.

2.2 Determination of MLS Coefficients

Based on the least-squares criteria, we can calculate the coefficients $a_i(\hat{\mathbf{x}})$ using the equation, with $\mathbf{a}(\hat{\mathbf{x}}) = \{a_1(\hat{\mathbf{x}}), a_2(\hat{\mathbf{x}}), ..., a_n(\hat{\mathbf{x}})\}^T$:

$$\mathbf{A}(\hat{\mathbf{x}}) \mathbf{a}(\hat{\mathbf{x}}) = \mathbf{B}(\hat{\mathbf{x}}) \mathbf{f} \qquad (4)$$

where

$$\mathbf{A}(\hat{\mathbf{x}}) = \mathbf{P}(\hat{\mathbf{x}}) \mathbf{W}(\hat{\mathbf{x}}) \mathbf{P}(\hat{\mathbf{x}})^T \qquad (5)$$
$$\mathbf{B}(\hat{\mathbf{x}}) = \mathbf{P}(\hat{\mathbf{x}}) \mathbf{W}(\hat{\mathbf{x}}) \qquad (6)$$

where $\mathbf{f} = \{f_1, f_2, ..., f_n\}^T$ and $\mathbf{P}(\hat{\mathbf{x}})$ is an $n \times N$ matrix whose j^{th} row is $\{p^{(j)}(\mathbf{x}_1), p^{(j)}(\mathbf{x}_2), ..., p^{(j)}(\mathbf{x}_N)\}$. $\mathbf{W}(\hat{\mathbf{x}}) = \text{diag}(w_1(\hat{\mathbf{x}}) \ w_2(\hat{\mathbf{x}}) \ ... \ w_N(\hat{\mathbf{x}}))$ is an $N \times N$ diagonal matrix of weight functions. Each weight functions $w_i(\hat{\mathbf{x}})$, define for each $\hat{\mathbf{x}} \in \bar{D}$, is non-negative and has the form $w_i(\hat{\mathbf{x}}) = w(|\hat{\mathbf{x}} - \mathbf{x}_i|)$ where $|\hat{\mathbf{x}} - \mathbf{x}_i|$ is the distance between the point $\hat{\mathbf{x}}$ and $\mathbf{x}_i$.

In our implementation, we have chosen to use the following cubic spline function (see Fig. 1) which is dependent on the distance $d_i = |\mathbf{x} - \mathbf{x}_i|$:

$$w(r) = \begin{cases} \frac{2}{3} - 4r^2 + 4r^3 & \text{for } r \leq \frac{1}{2} \\ \frac{4}{3} - 4r + 4r^2 - \frac{4}{3}r^3 & \text{for } \frac{1}{2} < r \leq 1 \\ 0 & \text{for } r > 1 \end{cases} \qquad (7)$$

in which $r = d_i / d_{m_i}$ is a normalized radius. The support size d_{m_i} of the ith node is determined by

$$d_{mi} = d_{max} c_i \qquad (8)$$

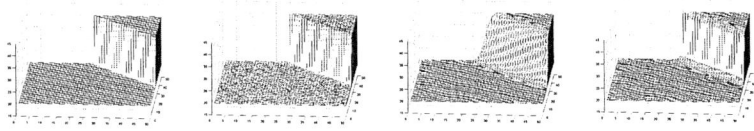

Fig. 3. Noisy surface (20dB) fitting with MLS (0.03 ± 1.60) and GWMLS (0.01 ± 0.37)

where d_{max} is a scaling parameter, and the distance c_i is determined by searching for enough neighbor nodes for the matrix $\mathbf{A}$ in Equation (5) to be regular, i.e. invertible at every point in the domain. Together, d_{max} and c_i ensure that the system has a unique solution, and they also serve similar role as the *scale* as in scale-space theory which allow multi-resolution MLS approximations of the function f.

Since the basis functions are designed to be linearly independent, together with the positive-definiteness of $\mathbf{W}(\hat{\mathbf{x}})$, it implies that $\mathbf{A}(\hat{\mathbf{x}})$ is positive-definite. Hence, the coefficients $a_i(\hat{\mathbf{x}})$ can be uniquely determined as

$$\mathbf{a}(\hat{\mathbf{x}}) = \mathbf{A}^{-1}(\hat{\mathbf{x}})\mathbf{B}(\hat{\mathbf{x}})\mathbf{f} \tag{9}$$

3 Interpolating MLS

It should be noted that MLS approximations do not satisfy the Kronecker delta criterion and result in $Gf(\mathbf{x}_i) \neq f_i$, i.e. the nodal data value f_i are not the nodal values of the MLS approximation $Gf(\mathbf{x}_i)$. Therefore, the approximation of the function at the ith node $Gf(\mathbf{x}_i)$ depends not only on the nodal data f_i but also on the nodal data of all the nodes within the influence domain of node i. However, for many situations in computer vision such as image denoising and partitioning, the edges or discontinuities need to be preserved while the noisy data being smoothed. In these cases, one would prefer that the approximation actually performs *exact interpolation* (EIMLS) or *near interpolation* (NIMLS) at selected nodal points, usually *the discontinuities*, and performs *approximation* for the rest.

With inspiration from Shepard [8], Lancaster and Salkauskas has proposed the basic principle of making weight function w_i infinity or a suitable singular at the nodal point $\mathbf{x}_i$ if Gf is to interpolate there [6]. This can be achieved by selecting proper weight function $w_i(\mathbf{x}) = w(d_i)$ such that $w_i \to \infty$ as $d_i \to 0$. It has been suggested to use a function in the form of

$$w(d_i) = e^{-d_i^2}/(d_i^2) \tag{10}$$

which clearly has the correct asymptotic behavior at $d_i = 0$ and $d_i = \infty$.

Exact interpolating MLS is, however, possible, and a scheme is developed originally intended for imposition of essential boundary conditions in the mesh-free particle methods [5, 7]. The singularities introduced by the singular weight functions in Equation (10) are removed through algebraic manipulations and all the resulting terms are well defined.

In order to achieve that goal, a new set of basis functions is generated. The first term of the new basis function $u^{(1)}(\mathbf{x})$ is obtained by normalizing the first term of the original basis functions $p^{(1)}(\mathbf{x})$:

$$u^{(1)}(\mathbf{x}) \equiv \frac{p^{(1)}}{|p^{(1)}|} \quad (11)$$

The remaining new basis functions are required to be orthogonal to the first basis function $u^{(1)}(\mathbf{x})$. Using the Gramm-Schmidt orthogonalization, the modified basis functions $u^{(i)}(\mathbf{x})$ are obtained through

$$u^{(i)}(\mathbf{x}) = p^{(i)}(\mathbf{x}) - Sp^{(i)}(\mathbf{x}) \quad (12)$$

where the operator S is defined as

$$Sp^{(i)}(\mathbf{x}) = \sum_{j=1}^{N} p^{(i)}(\mathbf{x}_j) v^{(j)}(\mathbf{x}) \quad (13)$$

$$v^{(j)}(\mathbf{x}) = \frac{w_j(\mathbf{x})}{\sum_{l=1}^{N} w_l(\mathbf{x})} \quad (14)$$

Now, the modified local and global approximant of Equation (2) and (3) becomes:

$$\hat{G}f(\mathbf{x}) = \hat{L}_\mathbf{x} f := Sf(\mathbf{x}) + \sum_{i=2}^{n} \alpha_i(\mathbf{x}) u^{(i)}(\mathbf{x}) \quad (15)$$

where $\alpha_i(\mathbf{x}), i = 2, 3, ..., n$ represent the modified unknown MLS coefficients to be determined from the nodal data.

Since the first basis function $u^{(1)}(\mathbf{x})$ is orthogonal to the rest of the basis functions by definition, the equation for the coefficient $\alpha_1(\mathbf{x})$ can be solved separately, and resulting in $Sf(\mathbf{x})$ in Equation (15). The other coefficients $\alpha^T(\mathbf{x}) = \{\alpha_2(\mathbf{x}), \alpha_3(\mathbf{x}), ..., \alpha_n(\mathbf{x})\}$ can be computed through

$$\hat{\mathbf{A}}(\mathbf{x})\alpha(\mathbf{x}) = \hat{\mathbf{B}}(\mathbf{x})\varphi(\mathbf{x}) \quad (16)$$
$$\alpha(\mathbf{x}) = \hat{\mathbf{A}}^{-1}(\mathbf{x})\hat{\mathbf{B}}(\mathbf{x})\varphi(\mathbf{x}) \quad (17)$$

where

$$\hat{\mathbf{A}}(\mathbf{x}) = \mathbf{U}(\mathbf{x})\mathbf{W}(\mathbf{x})\mathbf{U}(\mathbf{x})^T \quad (18)$$
$$\mathbf{U}(\mathbf{x}) = \hat{\mathbf{P}}(\mathbf{I} - \mathbf{V}(\mathbf{x}))^T \quad (19)$$
$$\hat{\mathbf{B}}(\mathbf{x}) = \mathbf{U}(\mathbf{x})\mathbf{W}(\mathbf{x}) \quad (20)$$

$\mathbf{U}(\mathbf{x})$ is a $(n-1) \times N$ matrix, $\hat{\mathbf{P}}$ is $(n-1) \times N$ matrix that contains row 2 through n of matrix $\mathbf{P}$, $\mathbf{V}(\mathbf{x})$ is a $n \times N$ matrix with each row being $\{v^{(1)(\mathbf{x})}, v^{(2)(\mathbf{x})}, ..., v^{(n)(\mathbf{x})}\}$, and $\varphi^T(\mathbf{x}) = \{\varphi_1(\mathbf{x}), \varphi_2(\mathbf{x}), ..., \varphi_N(\mathbf{x})\}$ with $\varphi_k(\mathbf{x}) = f_k - Sf(\mathbf{x}), k = 1, 2, ..., N$.

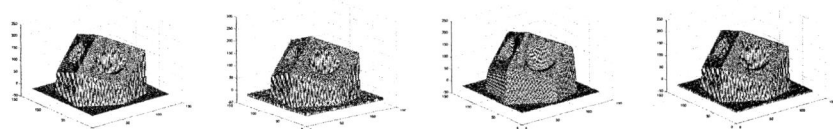

Fig. 4. Denoising of the range data: original, noisy, MLS, and GWMLS

Since the weight functions $w_i(\mathbf{x})$ are singular at the respective sampling nodes $\mathbf{x}_i$ by design, special cares must be taken in computing the terms in matrix $\hat{\mathbf{A}}$ and vector $\hat{\mathbf{Y}} = \hat{\mathbf{B}}\varphi$. After some algebraic manipulations [5], we have

$$\hat{\mathbf{A}}_{ij}(\mathbf{x}) = \sum_{k=1}^{N} \hat{a}_{ij}^{(k)}(\mathbf{x}) \quad i,j = 1, 2, ..., (n-1) \tag{21}$$

$$\hat{\mathbf{Y}}_i(\mathbf{x}) = \sum_{k=1}^{N} \hat{y}_i^{(k)}(\mathbf{x}) \quad i = 1, 2, ..., (n-1) \tag{22}$$

The values of $\hat{a}_{ij}^{(k)}(\mathbf{x})$ and $\hat{y}_i^{(k)}(\mathbf{x})$ depend on whether the point $\mathbf{x}$ coincides with a sampling node $\mathbf{x}_k, k = 1, 2, ..., N$ or not:

$$\hat{a}_{ij}^{(k)}(\mathbf{x}) = \begin{cases} s(\mathbf{x})v^{(k)}(\mathbf{x})c_k^{(i)}(\mathbf{x})c_k^{(j)}(\mathbf{x}), & \mathbf{x} = \mathbf{x}_k \\ u^{(i)}(\mathbf{x})w_k(\mathbf{x})u^{(j)}(\mathbf{x}), & \mathbf{x} \neq \mathbf{x}_k \end{cases} \tag{23}$$

$$\hat{y}_i^{(k)}(\mathbf{x}) = \begin{cases} v^{(k)}(\mathbf{x})c_k^{(i)}(\mathbf{x})\varphi_k(\mathbf{x}), & \mathbf{x} = \mathbf{x}_k \\ u^{(i)}(\mathbf{x})w_k(\mathbf{x})\varphi_k(\mathbf{x}), & \mathbf{x} \neq \mathbf{x}_k \end{cases} \tag{24}$$

where

$$s(\mathbf{x}) = \frac{1}{\sum_{j=1}^{N} w_j(\mathbf{x})} \tag{25}$$

$$c_k^{(i)}(\mathbf{x}) = \sum_{\substack{m=1 \\ m \neq k}}^{N} w_m(\mathbf{x})[p^{(i)}(\mathbf{x}) - p^{(i)}(\mathbf{x}_m)] \tag{26}$$

Note that there are no singularities in this set of equations if $\mathbf{x} \neq \mathbf{x}_k$. Since they contain the weight function $w_k(\mathbf{x})$, on surface the singularities may occur for functions $s(\mathbf{x})$ and $v^{(j)}(\mathbf{x})$. However, it can be shown that $\lim_{\mathbf{x} \to \mathbf{x}_k} s(\mathbf{x}) = 0$ and $\lim_{\mathbf{x} \to \mathbf{x}_k} v^{(j)}(\mathbf{x}) = 0$. The function $c_k^{(i)}(\mathbf{x})$ is not singular at $\mathbf{x} = \mathbf{x}_k$ because it does not include the weight function $w_k(\mathbf{x})$. Now, it clear that singularities for $\mathbf{x} = \mathbf{x}_k$ are removed and all terms needed for $\hat{\mathbf{A}}$ and $\hat{\mathbf{Y}}$ are well defined for all locations of the domain.

4 Data-Driven MLS for Interpolation and Approximation

NIMLS and EIMLS provide means to perform interpolations at selected nodal points and performs smoothing approximation for the rest. However, for practical problems, the questions of where to enforce the interpolation and where to

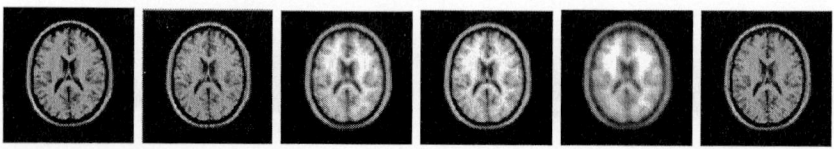

Fig. 5. Denoising of the noisy brain image: original, noisy, Gaussian filtered, anisotropic diffusion, MLS, and GWMLS

perform the smoothing approximation remain unanswered. Of course, for computer vision applications, we can always perform discrete derivative operations, such as gradient, on the image or measurement data first. Nevertheless, the selection of proper thresholds introduces a whole new set of questions which are difficult to handle.

We have developed a simple, yet powerful and flexible, alternative data-driven MLS process which treats the *near interpolating* at discontinuities and the smoothing approximation elsewhere in a coherent manner. Philosophically, this approach is very different from the NIMLS and EIMLS. Instead of direct manipulation of the weight functions $w_i(\mathbf{x})$ or the basis functions $p^i(\mathbf{x})$ in order to create the singular effect at any particular nodal point, we make changes to the nodal point distribution at the influence domain around the point of interest, which in turn alters the weight distribution to achieve the desired interpolating or smoothing effect.

Assuming that Gf is to interpolate at node $\mathbf{x}_i$, the influence domain of $\mathbf{x}_i$ contains K nodal data points $\mathbf{x}_{ik}, k = 1, 2, ..., K$, and we have the weight function $w_i(r)$ as defined in Equation (7). As discussed in the last paragraph of Section 2.2, in practice only the nodes $\mathbf{x}_{ik}$ within the influence domain of $\mathbf{x}_i$ are used for the MLS approximation $Gf(\mathbf{x})$, realized through the careful design of the weight function $w_i(\mathbf{x})$ and through Equation (3). In order to interpolate at node $\mathbf{x}_i$, we add a number of *virtual nodes* $\mathbf{x}_i \pm j\epsilon, j = 0, 1, 2, ..., M$ to the influence domain, where ϵ is a position vector with *very small* magnitude (see Fig. 1 for illustration). These virtual nodes distribute closely to node $\mathbf{x}_i$, and have identical value as $\mathbf{x}_i$:

$$f(\mathbf{x}_i \pm j\epsilon) = f(\mathbf{x}_i) \tag{27}$$

The number $2 \times M$ of virtual nodes determines if the process is smoothing approximation (no virtual node, $M = 0$) or near/exact interpolation (the more the virtual nodes, the closer the process converges to exact interpolating MLS). This way, without any attempts to modify the weight function itself, we can manipulate the *actual relative weight* for each of the node inside the influence domain relative to node $\mathbf{x}_i$, and perform approximation or interpolation in exactly the same manner. The main advantage of this strategy is that it affords a purely data-driven scheme to enforce interpolation or approximation at various nodal points. In our applications of image denoising and curve/surface fitting, we have used the gradient/derivative magnitude at the point of interest to determine the number of virtual points being added for the MLS procedure.

5   Applications

We denote the classic MLS approximation as MLS, the gradient thresholding detected singularities following by exact interpolating MLS as GTMLS, and the data-driven gradient-weighted MLS as GWMLS.

Curve fitting is performed on a set of noisy curve data generated from clean curve with $SNR = 20dB$. The reconstructed curves are shown in Fig. 2. In these tests, the GWMLS does substantially better reconstruction than the other two. Further, the derivatives of GWMLS are used to partition the curve into different segments, and the results are shown in Fig. 2. Similarly, we apply the MLS and GWMLS to synthetic surface data as shown in Fig. 3. And once again, the GWMLS does much better job in re-producing the original surface.

Using a well-known range data from the OSU/WSU/MSU Range Database, we test the ability of the GWMLS procedure on range data processing. Fig. 4 shows the wireframe plots of the range image. From these figures, it is obvious that the GWMLS does a wonderful job in reconstructing the original data. Finally, we add noise to the brain image ($SNR = 20dB$) and then perform Gaussian filtering (variance is 0.8), anisotropic diffusion [2], MLS, and GWMLS on the noisy brain image. The resulting images are shown in Fig. 5. Here, the GWMLS has slightly better performance than anisotropic diffusion, and much better than Gaussian filter and MLS.

Acknowledgments. This work is supported by HKRGC-CERG HKUST6151/03E and National Basic Research Program of China (No: 2003CB716104).

References

1. Bertero, M., Poggio, T.A., Torre, V. : Ill-posed problems in early vision, Proceedings of IEEE **76** (1988) 869–889
2. Black, M.J., Sapiro, G., Marimont, D., Heeger, D.J.: Robust anisotropic diffusion. IEEE TIP **7** (1998) 421-432
3. Chin, R.T., Yeh, C.L.: Quantitative Evaluation of Some Edge Preserving Noise Smoothing Techniques, CVGIP **23** (1983) 167–191
4. Geman, D., Reynolds, G.: Constrained Restoration and the Recovery of Discontinuities. IEEE PAMI **14** (1992) 367–383
5. Kaljevic, I. Saigal, S.: An improved Element Free Galerkin Formulation. Internaltion Journal for Numerical Methods in Engineering **40** (1997) 2953–2974
6. Lancaster, P. Salkauskas, K.: Surface Generated by Moving Least Squares Methods. Mathematics of Computation **37** (1981) 141–158
7. Liu, H., Shi, P.: Meshfree Particle Method. IEEE International Conference on Computer Vision. (2003) 289–296
8. Shepard, D.: A two-dimensional interpolation function for irregular spaced points. ACM National Conference (1968) 517–524
9. Terzopoulos, D.: The computation of visible-surface representation. IEEE PAMI **10** (1991) 417–438

Multiscale Centerline Extraction of Angiogram Vessels Using Gabor Filters

Nong Sang[1], Qiling Tang[1], Xiaoxiao Liu[2], and Wenjie Weng[2]

[1] Institute for Pattern Recognition and Artificial Intelligence,
Huazhong University of Science and Technology,
Wuhan 430074, PR China
nsang@hust.edu.cn, tqlinn@163.com

[2] Key Laboratory of Ministry of Education for Image Processing and Intelligent Control,
Huazhong University of Science and Technology,
Wuhan 430074, PR China
liu_xiaoxiao@263.net, hustwengwenjie@tom.com

Abstract. In this paper, we propose a new automated approach to extract the centerlines from 2-D angiography. The centerline extraction is the basis of 3-D reconstruction of the blood vessels, so the accurate localization of centerlines counts for much. The characteristics of multiscale Gabor even filter, flexible frequency bands and enhancement effects, are fully utilized to detect centerlines of the blood vessels in various size.

1 Introduction

Most angiogram analysis techniques, interactive or not, initially extract the 2-D centerlines of tubular object, especially in quantitative coronary angiography. Vessel centerlines have been used in three-dimensional vessel tree segmentation and reconstruction [1], computing edge gradients and searching for border positions [2], for calculation of lesion symmetry.

Medial representation of grayscale images has been well developed during the past several decades: multiscale approach has gained most attention [3], and others new methods include PDM technique by Jang [4], and snake-like algorithms [5].

Considering the curvilinear features and the complexity of the X-ray imaging artifacts, most of the reported image vessel features detectors are conventional ones: ridge detector based on level-set theory, morphological processing [6]. Many of them have high computational cost or poor robustness due to the complexity of the angiograph imaging.

To cope with close proximity or crossing of vessels and varying vessel widths including large stenoses and sever imaging artifacts, we develop a new automated method for accurate detection of centerlines in angiogram images. A characteristic of multiscale Gabor filters is its ability to tune to specific frequency, which allows conducting noise suppression and centerline enhancement. This is very beneficial to accurately extract centerlines of blood vessels in poor contrast and noisy background. The resulting centerline structure can be used for 3-D reconstruction.

2 Computational Models

2.1 Gabor Functions

Gabor functions were first defined by Gabor, and later extended to 2-D by Daugman [7]. A Gabor function is a Gaussian modulated by a complex sinusoid, as the following equation illustrates:

$$h(x, y) = g(x', y') \exp(j2\pi F x') \qquad (1)$$

where $(x', y') = (x\cos\theta + y\sin\theta, -x\sin\theta + y\cos\theta)$, denoting θ the orientation of the filter. Any desired orientation can be achieved via a rigid rotation in the x-y plane. F is the spatial center frequency, $g(x, y)$ is the 2-D Gaussian function:

$$g(x, y) = \frac{1}{2\pi\sigma_x\sigma_y} \exp\left\{-\frac{1}{2}\left[\left(\frac{x}{\sigma_x}\right)^2 + \left(\frac{y}{\sigma_y}\right)^2\right]\right\} \qquad (2)$$

where σ_x and σ_y correspond to the horizontal and vertical spatial extent of the filter. σ_y/σ_x, called the aspect ratio, gives a measure of the filter's asymmetry. σ_x and σ_y are related to the filter's frequency bandwidth and orientation bandwidth in the way illustrated in formula (3) and (4):

$$\sigma_x = \sqrt{\frac{\ln 2}{2}} \frac{1}{\pi F} \frac{2^{B_F} + 1}{2^{B_F} - 1} \qquad (3)$$

$$\sigma_y = \sqrt{\frac{\ln 2}{2}} \frac{1}{\pi F} \frac{1}{\tan(B_\theta/2)} \qquad (4)$$

B_F is the spatial frequency bandwidth and B_θ is the orientation bandwidth. In this paper, the parameters B_F and B_θ are set as 1.0 and 45°, respectively.

2.2 Centerline Response of Gabor Even Filters

The accurate localization of centerlines requires a high resolution in the spatial domain. An important property of multiscale Gabor filters is that they have optimal joint localization or resolution, in both the spatial and the spatial-frequency domains [8]. These results are extensively used for edge detection, texture classification; optical flow estimation and image compression. In this paper, the Gabor even filter,

$$h(x, y) = g(x', y') \cos(2\pi F x') \qquad (5)$$

the real component of Gabor functions, is proposed as the centerline extraction operator of blood vessels.

Here we demonstrate that a 1-D Gabor even filter with suitable center frequency can give maximal response at the center of tube-profile.

The tube-profile centered at $x = 0$ with height 1 and width τ can be expressed by the step-edge function $\varepsilon(x)$ as follows:

$$f_\tau(x) = \varepsilon\left(x + \frac{\tau}{2}\right) - \varepsilon\left(x - \frac{\tau}{2}\right) \qquad \varepsilon(x) = \begin{cases} 1 & x > 0 \\ 0 & else \end{cases} \qquad (6)$$

While the impulse response of a 1-D Gabor even filter is:

$$h(x) = \frac{1}{\sqrt{2\pi}\sigma} \exp\left(-\frac{x^2}{2\sigma^2}\right) \cos(2\pi F x) \qquad (7)$$

Here we define $w = 1/2F$ as the width of the central excitatory region of the Gabor even filters, as shown in Fig. 1.

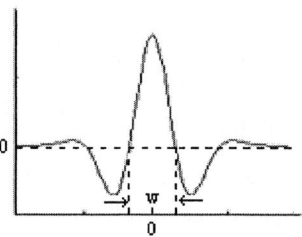

Fig. 1. 1-D Gabor even filter

The response of the filter to the tube-profile centered at $x = a$ is given by convolution

$$r(x) = f_\tau(x - a) * h(x) \qquad (8)$$

Then, the zero crossing point of the first derivative of response $r(x)$ corresponds to the maximum of the response map.

$$r'(x) = \frac{1}{\sqrt{2\pi}\sigma} \left\{ \exp\left[-\frac{1}{2}\frac{(x - a + \tau/2)^2}{\sigma^2}\right] \cos[2\pi F(x - a + \tau/2)] \right.$$

$$\left. - \exp\left[-\frac{1}{2}\frac{(x - a - \tau/2)^2}{\sigma^2}\right] \cos[2\pi F(x - a - \tau/2)] \right\} \qquad (9)$$

From formula (9), we get $r'(x) = 0$ at $x = a$, which is the center point we try to figure out.

Additionally, if $x = a$ is the maximum point, $r'(a_-) > 0$ and $r'(a_+) < 0$ are necessary. With $0 < \tau \le 2w$, that is $0 < \tau \le 1/F$, both the inequations can be met. Consequently, under $0 < F \le 1/\tau$, the maximum of response $r(x)$ correspond to the center of tube-profile. And we also prove that when τ is fixed, and $w \ge \tau/2$ is satisfied, the $r(a)$ get its maximum while $\tau = w$, that is to say, the center point got its maximum response when the width of the bar-profile signal is equal to the width of the central excitatory region of the Gabor even filter.

If we take a single scale to detect the vessel centerlines, the quality of centerline extraction largely related to the value of w, the width of the central excitatory region of the filter. If too wide, that is, the center frequency is too small, the spatial extent of the filter will become so extensive that the centerline responses are seriously blurred. On the contrary, the centerline may be wrongly localized at the edges. While applying the filter in 2D image, therefore, we take multiscale approach to detect vessels in various widths by choosing the maximum response among the different scales, also different orientations.

2.3 Centerline Segmentation

Local maxima extraction is then carried out on the maximal response map of Gabor even filters. After that, we apply a hysteresis thresholding, which retains connected components with most points that have values above a low threshold and with at least one point with a value above a high threshold. The two thresholds are computed as constant quantile values (at 90% and 98%) of the histogram of response map. The output of segmentation is a list of tagged chains, which are the collection of vessel segments.

3 Results

We use Gabor even filters of twelve different orientations, $\theta = (k-1)\pi/K$, for $k = 1, 2, \cdots, K$, $K = 12$. Furthermore, multiple scales are taken to adapt to the variety of the vessel size. In the original image of Fig.2, the width of the blood vessels varies from about 4 to 20 pixels. We choose five scales respectively corresponding to five center frequencies, $F \in \{0.125, 0.100, 0.075, 0.050, 0.025\}$. The maximum response from all 12 orientations and 5 scales for each position is taken as the output of the filter, as shown in Fig. 3.

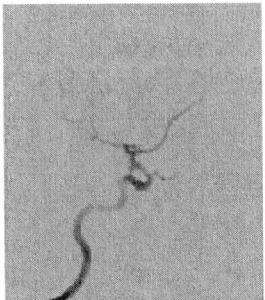
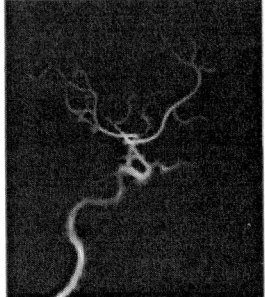

Fig. 2. 2-D angiography image of the brain   **Fig. 3.** the maxima from all orientations and scales

The maximum response map (Fig. 3) shows the prominent advantage of our method is Gabor even filters' ability to reduce noise and to enhance centerlines. Especially, some weak ends of the blood vessels are well intensified. After filtering, we adopt non-maximum suppression to detect local maxima and hysteresis thresholding to extract the final centerline. The result is shown in Fig. 4 and Fig. 5. In Fig. 6, the centerlines are overlaid on top of the original image. We can see that the centerlines are correctly localized.

The multiscale Gabor even filters manifest their good capability for accurate localization. To have a clearer view about the scales of the filters, we give the following two experiments to illustrate how scales influence the centerline localization. If a large scale is taken while detecting comparatively thin vessels, like in Fig. 7, it will cause serious faintness to the image, which therefore result in the deflection. However, if the scale is relatively small, the maxima will occur at the edge of the blood vessel, which can be utilized to detect edges, but not the centerlines we need (Fig. 8).

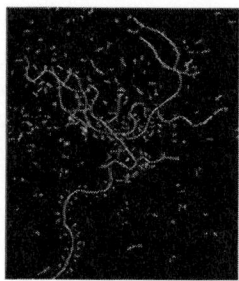
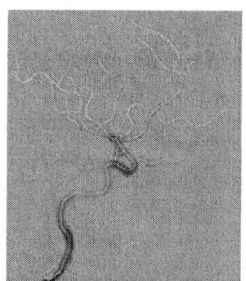

Fig. 4. The result of non-maximum suppression

Fig. 5. The centerlines after hysteresis thresholding

Fig. 6. The centerlines overlain on top of the Fig.2

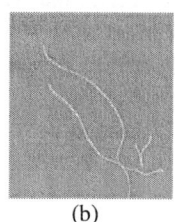
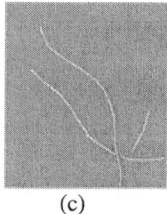

(a)   (b)   (c)

Fig. 7. (a) original image (b) proper scale corresponding to center frequency $F = 0.125$ (c) too big scale corresponding to center frequency $F = 0.025$

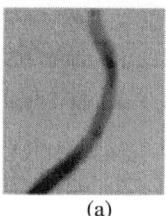
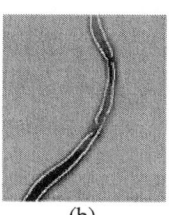
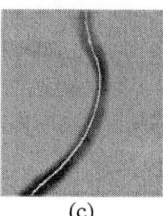

(a)   (b)   (c)

Fig. 8. (a) original image (b) too small scale corresponding to center frequency $F = 0.125$ (c) proper scale corresponding to center frequency $F = 0.025$

4 Conclusions

We introduce an automated new multiscale method based on Gabor even filters to extract vessel centerlines of angiogram images, whose flexible and effective enhancement on the centerline of the tubular objects made the results satisfying and encouraging.

Acknowledgements

This work was supported by the 973 Program of China (No. 2003CB716105).

References

1. Shechter, G., Devemay, F., Quyyumi, A.: Three-Dimensional Motion Tracking of Coronary Arteries in Biplane Cineangiograms. IEEE Transactions on Medical Imaging. Vol 22, No.4 (2000) 321-337.
2. Andress, K.M.: Evidential Reconstruction of Vessel Trees from Rotational Angiograms. Proc. IEEE Conf. Image Processing. Vol.3 (1998) 385-389.
3. Lindeberg, T.: Edge Detection and Ridge Detection with Automatic Scale Selection. International Journal of Computer Vision, Vol. 30, No. 2 (1998) 117-154.
4. Jang, J.H., Hong, K.S.: A Pseudo-Distance Map for the Segmentation-Free Skeletonization of Gray-Scale Images, Proc. IEEE Conf. Computer Vision. Vol. 2 (2001) 7-14.
5. Golland, P., Grimson, W.E.L.: Fixed Topology Skeletons. Proc. IEEE Conf. Computer Vision and Pattern Recognition. Vol. 1.(2000) 13-15.
6. Maglaveras, N., Haris, K.: Artery Skeleton Extraction Using Topographic and Connected Component Labeling. IEEE Computers in Cardiology. Vol.28 (2001) 17-20.
7. Dugman, J.G.: Uncertainty Relation for Resolution in Space, Spatial Frequency, and Orientation Optimized by Two-dimensional Visual Cortical Filters. J. Opt. Soc Am. A, Vol.2, No. 7 (1985) 1160~1169

Improved Adaptive Neighborhood Pre-processing for Medical Image Enhancement

Du-Yih Tsai and Yongbum Lee

Department of Radiological Technology, School of Health Sciences, Niigata University,
2-746, Asahimachi-dori, Niigata City, Niigata, 951-8518, Japan

Abstract. This paper presents an improved adaptive neighborhood contrast enhancement (ANCE) method for improvement of medical image quality. The ANCE method provides the advantages of enhancing or preserving image contrast while suppressing noise. However, it has a drawback. The performance of the ANCE method largely depends on how to determine the parameters used in the processing steps. The present study proposes a novel method for optimal and automatic determination of threshold-value and neighborhood-size parameters using entropy. To quantitatively compare the performance of the proposed method with that of the ANCE method, computer-simulated images are generated. The output-to-input SNR ratio and the mean squared error are used as comparison criteria. Results demonstrate the superiority of the proposed method. Moreover, we have applied our new algorithm to echocardiograms. Our results show that the proposed method has the potential to become useful for improvement of image quality of medical images.

1 Introduction

In recent decades a number of computer-aided diagnosis (CAD) schemes have been developed to aid in the interpretation of the increasing amounts of medical image data and clinical information [1]. In general the performance of CAD schemes largely depends on the image database employed. Therefore, in order to improve the performance of the schemes, enhancement of image quality of the image set is of importance.

Image quality is usually characterized by contrast, resolution, and signal to noise ratio. Adaptive neighborhood contrast enhancement (ANCE) is a recent approach to contrast enhancement [2, 3]. An adaptive neighborhood is constructed for each pixel, this pixel being called a seed pixel of the neighborhood [4]. In the ANCE, a variable shape and size neighborhood is defined using local characteristics of the image. Recently, Guis et al. [5] reported a novel ANCE technique. This ANCE method consists of computing a local contrast around each pixel using a variable neighborhood whose size depends on the statistical properties around the given pixel. The obtained contrast image is then transformed into a new contrast image using a contrast enhancement function. Finally, a contrast-enhanced image is obtained by applying inverse contrast transform to the previous step. This technique provides the advantages of enhancing or preserving image contrast while suppressing noise. However, it has a drawback. The performance of the ANCE method largely depends on how to determine the parameters used in the processing steps.

The present study proposes a method for optimal and automatic determination of threshold-value and neighborhood-size parameters using entropy. To quantitatively compare the performance of the proposed method with that of the ANCE method, computer-simulated images are generated. The output-to-input SNR ratio and the mean squared error are used as comparison criteria. Moreover, medical images obtained from various modalities are also used for performance comparison.

2 Methods

2.1 ANCE Method

Figure 1 shows the flowchart of the ANCE method proposed by Guis et al. [4]. Basic steps of the ANCE method are as follows:

1) Each pixel (i,j) is assigned an upper window W_{max} centered on it, whose size is $N \times N$ (N is an odd number). Let $I(i,j)$ be the gray level of pixel (i,j) in image I, and let T be a given threshold.

2) Pixel (k,l) within W_{max} is assigned a binary mask value 0 if $|I(k,l)-I(i,j)|>T$, else it is assigned a binary mask value 1. This results in constructing a binary image.

3) The percentage P_0 of zeros is computed over the region between the external $(c+2) \times (c+2)$ and the inner $(c \times c)$ areas (c is an odd number). The process stops if this percentage is greater than 60% or if the upper window W_{max} is reached. Let c_0 be the upper c value beyond which the percentage P_0 is greater than 60%. The pixel (i,j) is assigned the window $W=(c_0+2) \times (c_0+2)$. The set of pixels having the mask value 1 is defined as "center", and the set of pixels having both the mask value o and which are eight-neighborhood connected at least to a pixel 1 is defined as "background".

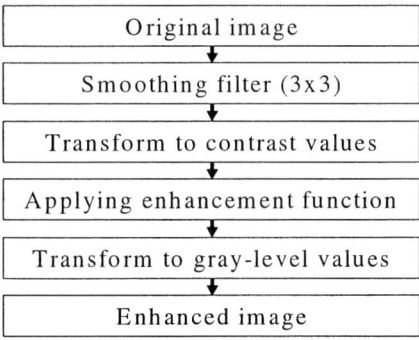

Fig. 1. Flowchart of the adaptive neighborhood contrast enhancement method proposed by Guis et al. [4]

4) A local contrast image is computed from

$$C(i,j) = \frac{|M_c(i,j) - M_b(i,j)|}{\max[M_c(i,j), M_b(i,j)]} \qquad (1)$$

where, $M_c(i,j)$ and $M_b(i,j)$ are the mean values in image I of pixels labeled as the center and as the background regions around pixel (i,j), respectively.

5) The local contrast image C is then transformed into a new image C' using

$$C'(i,j) = F[C(i,j)] \qquad (2)$$

where F is a contrast-enhancement function that depends on the features to be detected. For example, the sigmoidal function or the trigonometric function is used.

6) A new image E is obtained by the process of inverse contrast transform using

$$E(i,j) = M_b(i,j)[1 - C'(i,j)] \quad \text{if} \quad M_b(i,j) \geq M_c(i,j) \qquad (3)$$

$$E(i,j) = \frac{M_b(i,j)}{1 - c'(i,j)} \quad \text{if} \quad M_b(i,j) < M_c(i,j) \qquad (4)$$

7) Repeat step 1 to step 6 for each pixel in the image I.

2.2 Our Proposed Method for Parameter Determination

Two of the most important parameters used in the ANCE method are the threshold value T and the percentage P_0 of zeros computed over the region between the external and the inner areas. Guis et al. empirically used $T=5$ for thresholding and $P_0=60\%$ for determining neighborhood size in their study [4].

In this current study, we use a method for optimal and automatic determination of threshold value and neighborhood size from the viewpoint of information amount. Namely, the two parameters are determined when the entropy of the image I is at its maximum. The detail of determination process is described as follows.

1) Determination of the threshold value T: Let d be the difference between the maximum and minimum pixel values in the region of interest (ROI) whose size is $W_{max} \times W_{max}$. The value of T is then in the range of $0 \leq T \leq d$. When the maximum entropy in the ROI is obtained by varying threshold value, this threshold value is regarded as T. The entropy of the ROI is given by

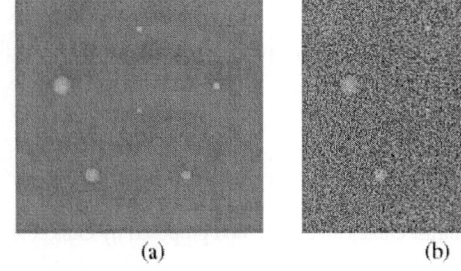
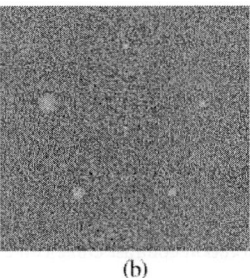
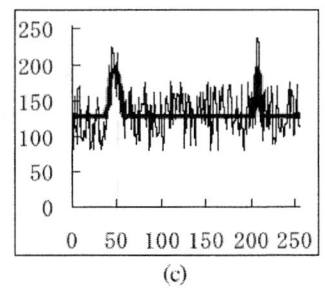

Fig. 2. Computer-simulated image of breast microcalcifications: (a) microcalcification noise-free image with a 30% contrast level, (b) noisy image with a SNR=18dB, (c) horizontal profile of both images (a) and (b) passing through the two different sizes of microcalcifications

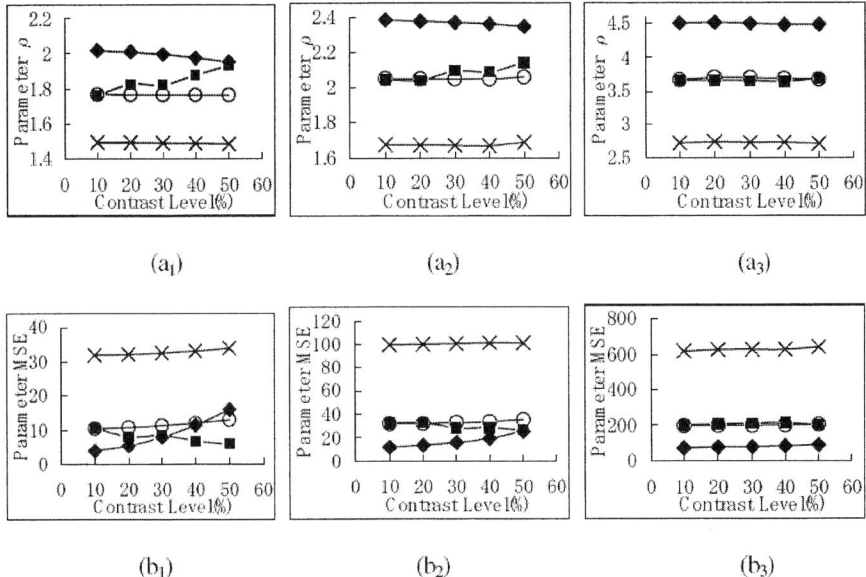

Fig. 3. Parameters ρ and MSE versus contrast level for computer simulated microcalcifications: (a_1), (a_2), and (a_3) results for parameter ρ at SNR=23dB, 18dB, and 10dB, respectively; (b_1), (b_2), and (b_3) results for parameter MSE at SNR=23dB, 18dB, and 10dB, respectively, where ■-conventional method , □-improved method , ○-5×5 smoothing filter, ×-5×5 median filter

$$ENT(t) = -p_{0t} \log_2 p_{0t} - p_{1t} \log_2 p_{1t} \quad (0 \leq t \leq d) \quad (5)$$

where $ENT(t)$ is the entropy of the binary image obtained using a threshold value t, p_{0t} and p_{1t} are the probability of pixel value=0 and that of pixel value=1 in corresponding binary image, respectively. All the values of $ENT(t)$ in the range of ($0 \leq t \leq d$) are computed. The value of t is considered as T when $ENT(t)$ is at its maximum.

2) Determination of the neighborhood size: The entropy in $c \times c$ area is calculated, where $0 \leq c \leq N$. The value of c is used as the neighborhood size when entropy is at its maximum.

3 Performance Assessment

In order to demonstrate the effectiveness of the improved method, computer-simulated images of breast calcifications were used for quantitative evaluation. Five different contrast levels (from 10% to 50%; with a step size of 10%) and three noise levels (signal to noise ratios =10dB, 18dB, and 22dB) for each contrast level were generated. Therefore a total of 15 compute-simulated images were employed. The images consist of 256×256 pixels. The images were coded on 256 gray levels and the background level was set at gray level of 128. Figure 2 shows an example of computer-simulated image related to breast calcifications.

Performance comparison was made among the proposed improved method, Guis's ANCE method, 5×5 smoothing filter, and 5×5 median filter. Two criteria, namely, output-to-input SNR and the mean-squared-error (MSE), were used to quantitatively evaluate the four algorithms on computer simulated images. The output-to-input *SNR* parameter (called ρ) is defined as the ratio

$$\rho = \frac{SNR_{out}}{SNR_{in}}, \qquad (6)$$

where SNR_{out} and SNR_{in} are the SNR after and before processing, respectively.

The MSE is calculated between the noise-free image f and the result $\hat{g}$ of the enhancement process on the input noisy image g:

$$MSE = \frac{\sum_{i=1}^{a}\sum_{j=1}^{b}[f(i,j) - \hat{g}(i,j)]^2}{a \times b}, \qquad (7)$$

where a and b are the numbers of pixels on the horizontal and vertical directions, respectively.

It is noted that ρ is higher when much more noise is removed, whereas the MSE value is smaller when the image is denoised and the structure is preserved. Figure 3(a_1), 3(a_2), and 3(a_3) shows the results of ρ versus contrast at SNR=23dB, 18dB, and 10dB, respectively. The improved method gives best results. Figure 3(b_1), 3(b_2), and 3(b_3) shows the results of MSE versus contrast at SNR=23dB, 18dB, and 10dB, respectively. Similarly, the improved method gives best results. Figure 4 shows the images and the corresponding profiles obtained after applying the proposed method, Guis's ANCE method, 5×5 smoothing filter, and 5×5 median filter. Visual observation demonstrates the superiority of the proposed method. Figure 5 shows the results obtained after applying four different methods to an ultrasonic image. It is noted from visual evaluation that the images processed using the improved method give the best results.

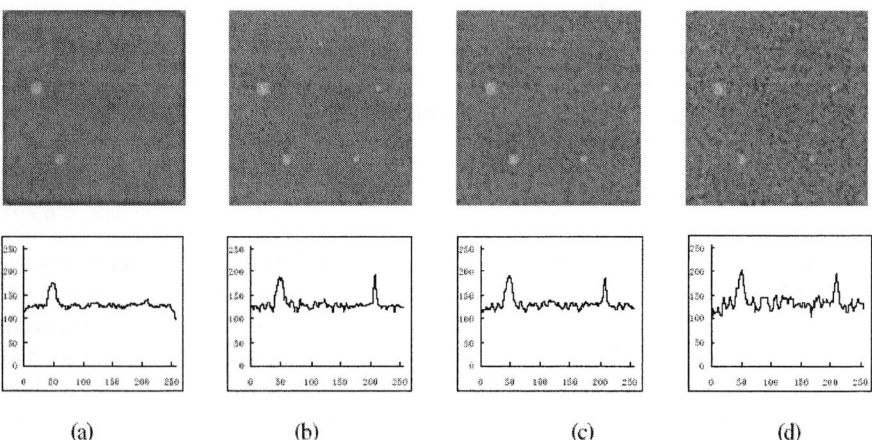

Fig. 4. Results obtained on the microcalcification image shown in Fig. 2 and corresponding horizontal profiles using (a) the proposed method, (b) Guis's ANCE method, (c) 5×5 smoothing filter, and (d) 5×5 median filter

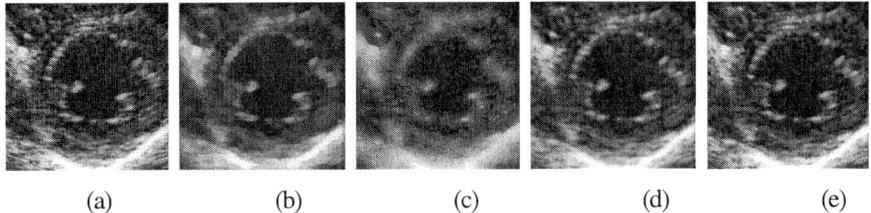

Fig. 5. Original and processed echocardiograms. (a) original end-diastole image, (b) image obtained using the proposed method, (c) image obtained using Guis's ANCE method, (d) image obtained using 5×5 smoothing filter, and (e) image obtained using 5×5 median filter

4 Conclusion

In this paper we have described an improved ANCE method for enhancement of medical image quality. The improved method was based on the algorithm proposed by Guis et al. The feature of the improved method is to automatically determine the optimal threshold-value and neighborhood-size parameters using entropy. Computer-simulated images were generated to quantitatively evaluate the effectiveness of the proposed method in terms of output-to-input SNR and the mean-squared error. The proposed method was also applied to clinical echocardiograms and CT phantom images. Results show that our proposed method performed well and clinically useful.

References

[1] Giger, M.L.: Computer-aided diagnosis in radiology. Acad. Radiol. 9 (2002) 1-3
[2] Dhawan, A.P., Buelloni, G., Gordon, R: Enhancement of mammographic features by optimal adaptive neighborhood image processing. IEEE Trans. Med. Imaging. 5 (1986) 8-15
[3] Paranjape, R.B., Rabie, T.F., Rangayyan, R.M.: Image restoration by adaptive neighborhood noise subtraction. Appl. Opt. 33 (1994) 2861-2869
[4] Jiang, M.: Digital Image Processing, lecture notes, Department of Information Science, School of Mathematics, Peking University (2002)
[5] Guis, V.H., Adel, M., Rasigni, M., Rasigni, G., Seradour, B., Heid, P.:Adaptive neighborhood contrast enhancement in mammographic phantom images. Opt. Eng. 42 (2003) 357-366

On the Implementation of a Biologizing Intelligent System

Byung-Jae Choi[1], Paul P. Wang[2], and Seog Hwan Yoo[1]

[1] School of Electronic Engineering, Daegu University,
Naeri Jillyang Gyungsan Gyungbuk, 712-714 Korea
{bjchoi, shryu}@daegu.ac.kr
[2] Department of Electrical & Computer Engineering, Duke University,
Durham NC, 27708 USA
ppw@ee.duke.edu

Abstract. According to the progress of an information-oriented society, more human friendly systems are required. Such systems can be implemented by a kind of much more intelligent algorithms. In this paper we propose the possibility of the implementation of an intelligent algorithm from gene behavior of human beings, which has some properties such as self organization and self regulation. The regulation of gene behavior was widely analyzed by Boolean network. Also the SORE (Self Organizable and Regulating Engine) is one of those algorithms. We here describe the concepts of the implementation of an intelligent algorithm through the analysis of both gene regulatory network.

1 Introduction

Many complex processes are difficult to control using existing techniques because they are highly interconnected nonlinear systems that operate over a wide range of conditions. Some novel control techniques are required in order to cope with increasing demands on convenience, comfort, and high performance. Some pioneering researchers such as James Albus have even undertaken the task of constructing a road map for the engineering of the mind [1-2]. A biocontrol system is recently considered as one of most important issues in control fields. This can also play an important role in the design of a kind of human friendly systems.

The development of control theory has a long history due to fairly intensive research efforts for at least half a century. Some issues were thoroughly investigated and have reached a very mature status, while others were left nearly untouched. In an article [3], John L. Casti coined the word "Biologizing" to reflect the recognition of so called "reliability and survivability" as topics of primary concern for engineers. A correct value judgment for a control system was thus finally determined. A blue-ribbon panel of 52 experts has nearly created a large set of research problems that lie ahead for control research engineers [4].

The concept of "feedback" in control theory is so far only known as a special trivial case of "homeostasis" - the function of an organism to regulate and to keep a

---

This research was partially supported by the Daegu University Research Grant, 2003.

constant "internal environment". These topics on biocontrol systems nevertheless bring much excitement and vision [5-6] to the research community.

Some models for a gene regulatory network were reported: The Boolean network (BN) model had a considerable amount of attention, which was originally introduced by Stuart A. Kauffman [7]. This is also called by NK-network. In a genetic network, the total number of genes is represented by N and K is the largest number of genes which regulates any one of the N genes in the genetic network.

SORE (Self Organizable and Regulating Engine) was first discovered as a classifier [8]. More interesting and robust properties subsequently emerged [9-12]. SORE is by far the most general mathematical structure of a family of automata theories listed as follows: SORE $\supset$ Boolean Network $\supset$ Cellular Automata $\supset$ Linear Automata. Based upon the theory of Stuart A. Kauffman, a biological genetic network usually has a much smaller K to more realistically model a biological setting. However, it is the condition of K=N in which the network assumes its full strength for the best possible performance in a "biologizing" intelligent system. This K=N condition is what distinguishes SORE from the standard Boolean network [9].

We discuss Boolean network and SORE in gene regulatory networks and a concept of colonization in Section 2 and 3, respectively. In Section 4 we describe some properties for a biologizing intelligent system. Discussions and conclusions are presented in Section 5.

2 Gene Regulatory Networks

The roots of the Boolean network lie in the automata theory which was a subject of Turing and von Neumann work. A Boolean network is a system of N interconnected binary elements (nodes). It is widely used to model gene regulatory networks and was originally introduced by Stuart A. Kauffman in 1969. It is frequently called by NK-network, where N and K are the number of genes and the maximum connectivity of a Boolean network, respectively. The number of inputs K may vary or be the same for all nodes in the network.

A Boolean network G(V, F) is defined by a set of nodes (genes) V=$\{x_1, x_2, \ldots, x_n\}$ and a set of Boolean functions F=$\{f_1, f_2, \ldots, f_n\}$. $x_i \in \{0, 1\}$ is a Boolean variable, where $i=1, 2, \ldots, n$. We write simply $x_i=1$ to denote that the i_th node is expressed ($x_i=0$ denotes that the i_th gene is not expressed). Each Boolean function $f_i(x_{i1}, \ldots, x_{ik})$ with K specific input nodes is assigned to node x_i and is used to update its value. The values of all the nodes in V are then updated synchronously. In general, there are $2^{\wedge}2^K$ possible Boolean activation functions for a node with K inputs.

Regulation of nodes is defined by the set F of Boolean functions. In detail, given the value of the nodes V at time t, the Boolean functions are used to update the value of the nodes at time $t+1$. In the NK network model for gene regulation, a symbolic logic function is used to describe the self-organization and self-regulation of genes' expressions. The Boolean functions are made up of the logical connectors AND, OR, and NOT, forming a complete logic set. The HIGH's and LOW's of gene expressions for the next state in the Boolean network are controlled by these Boolean functions. In order words, these logic functions become the rules that govern the HIGH's or LOW's of the genes' expressions in the next instant of time. The synchronous update process is then repeated making the network dynamic. In order to capture the dynamic nature of the network, it may be useful to consider the wiring diagram which gives an explicit way of implementing the updating procedure.

During the operation of the network, a sequence of states will converge to a limit cycle or an attractor (a limit cycle with a length of one). Each specific initial condition will converge to a specific attractor or limit cycle. For a genetic network of N genes, there are exactly 2^N possible initial states.

For an example, consider the lambda bacteriophage. It is a virus that invades E. coli bacteria and 2 distinct modes of operation: 1) it can become integrated into the host cell DNA, and be replicated automatically each time the bacterium divides. 2) it can multiply in the cytoplasm of bacterium, eventually killing its host. There are 2 proteins: lambda repressor and cro protein. The lambda repressor blocks the expression of the gene for the cro protein, and vice versa. So, it can be modelled by a one-input Boolean network with 2 nodes (lambda repressor and cro protein) as shown in Fig. 1. It shows that both of the nodes are INVERSE properties because cro protein blocks the expression of the lambda repressor and vice versa. Two of four states are 10 and 01, and they correspond to the exclusive expression of each protein. There are another cycle of period 2 (11 → 00 → 11) which is not a behavior observed in the lambda bacteriophage. This shows that the simple model is not a complete description of the lambda bacteriophage system.

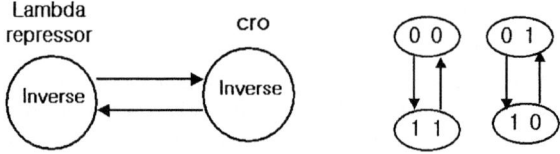

Fig. 1. The Boolean network model for the lambda bacteriophage

SORE is the most general Boolean network of which Stuart A. Kauffman's NK-network is a special case [9]. The NK-network imposes the severe restriction of K<N, which closely approximate the biological reality of K<N or K<<N. Since SORE uses the most general NN-network, it can be used in applications other than genetic networks alone.

Theoretically speaking, the whole family of the Boolean network and the cellular automata can best be explained by using the modern control theoretical language of state space. The N genes' expression levels naturally constitute $N\times1$ state vectors in a vector space. This discrete state space will consist of precisely 2^N state vectors. Unlike continuous differential dynamic systems, the discrete space can be completely displayed as long as N is not too large. For example, we can visualize a three-dimensional cube for $N=3$. An autonomous solution subject to an initial state vector is of primary concern because it gives the sign of life in a cell. In other words, a zero-state response does not even exist, and only a zero-input response has to be dealt with. For all NK-network and all SORE, the whole state vector space can always be partitioned into I independent subspaces $\{S_1, S_2, ..., S_I\}$, where I is the total number of attractors or limit cycles. If α_{Sj} and α_{Sk} represent the elements of subspace S_j and S_k

respectively, $\alpha_{Sj} \cap \alpha_{Sk} = \Phi$ and $S_j \cap S_k = \Phi$ for all j≠k. All isolated islands (subgraphs that describe the subspaces) are called "basins of attractors".

3 Concept of Colonization

Usually the complexity of a network is designated by the number of genes. Even with a few genes, the "colonizing action" of a network can create a much more complex colony in a short amount of time. As the "biologizing" intelligent system becomes more complex, so do the fundamental issues of reachability, controllability, and observability.

To illustrate only one example, let us assume that there are two known genetic networks and that both are made up of three genes (Fig. 2 (a) and (b)). There is only one attractor for each network where the existence of two three-gene networks would act autonomously. However, once the states of the two networks are connected (we call this a serial cascade), the behavior changes dramatically. Fig. 2 (c) shows that no less than five limit cycles now exist. The existence of these limit cycles carries special meaning when considered in reference to living cells. It is precisely this constant cyclic behavior that is prevalent in many forms of life. The number of limit cycles observed is also significant. A large number of them will exhibit the biological characteristics which Stuart A. Kauffman called "the edge of chaos." This chaotic behavior is triggered by the time-varying rules of Boolean logic.

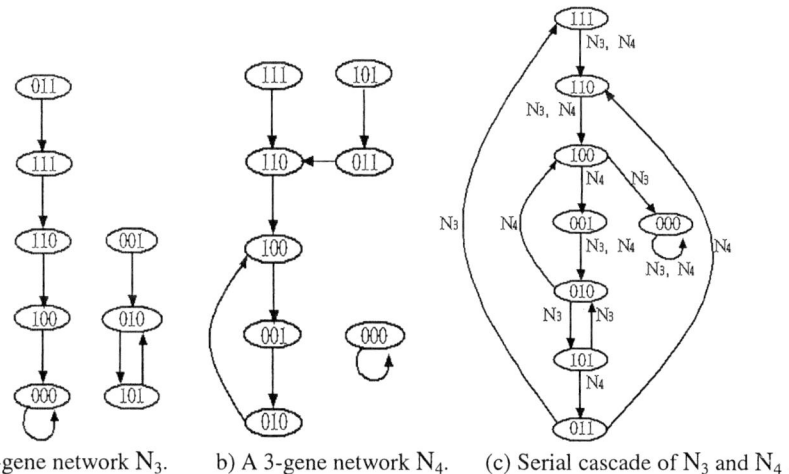

(a) A 3-gene network N_3.   b) A 3-gene network N_4.   (c) Serial cascade of N_3 and N_4.

Fig. 2. Two 3-gene networks and their serial cascade network

The complexity in chaotic orders is due to the time-varying nature of the rules of Boolean logic. However, all of the complexity is due to autonomous activities. So far, there exists no zero-state response. The equivalence of such a response does exist, but it takes the form of dynamic systems expansion and enlargement; one may call it "colonization". Each gene is made up of a combination of the same molecules used in molecular genomics G, A, T, and C. When a new gene is introduced into genetic networks, a larger network emerges. For example, if a new gene is added to a four-gene network N_4, and a five-gene network N_5, a single much larger ten-gene network N_{10} emerges. The new genetic network will have 2^{10} total state vectors, which is 2^6 or 64 times larger than the N_4 genetic network and 2^5 or 32 times larger than the N_5 genetic network. Why does one gene possess so much power? This single new gene may be viewed as a combination of big molecules, or it may represent a change in the chemical environment, such as temperature, pressure, or other changes.

The modeling of various biological phenomena within the general framework of automata has been documented for quite some time. However, the renewed interest in modeling gene regulations in molecular biology in recent years has caught the attention of a much larger set of researchers. Since a generalized example of ten genes demonstrating colonization requires much more space to document, its salient features will simply be mentioned without the presentation of its computer simulations.

4 Other Properties for an Intelligent System

The "biologizing" intelligent theory states that reliability and survivability are the top priorities for a biocontrol system. How does one achieve this goal? One example is a system component like the Boolean network or SORE which can deliver this requirement of reliability and survivability.

What method of control should be exercised in an autonomous manner? Redundancy with a simple switching action after a decision is not a good design with respect to control and economics. Self organization, regulation, and control are basic requirements for a "biologizing" intelligent system. Smooth control operation only be expected from a self organizable system. Self organization and regulation is a main capability of a gene behavior.

Classification lies in the heart of any intelligent and autonomous control system. For a "biologizing" intelligent system to be the most effective, some classification must be made before the best choice can be selected. In reference [10], the simplest case of SORE, the two gene network, was investigated and proven to be the best possible classifier suitable for any kind of data structure. Unfortunately, for more complicated networks, more time and research is needed because the problem becomes a synthesis problem for discovering Boolean logic functions.

Decision making capability under uncertainty, namely the approximation reasoning of fuzzy logic, is only a special case of SORE. The use of Boolean functions in Boolean networks allows each node or each gene to take any combination of logical connectors. Most of the methodology of fuzzy logic employed so far is a simple set of "IF ... THEN ..." rules, which is less general than SORE.

Brain wave bio potentials have been used in experiments for mobile robot control [13]. This is just one of the many possible brain-inspired applications. Intelligent computational analysis of the human genome will drive medicine for at least the next half century [14]. Will research into bioinformatics result in useful information well beyond what is currently known? Perhaps the manipulation of brain information may be governed by the device such as SORE.

5 Concluding Remarks

We discussed that a model for gene regulatory network can be a generic building block for a "biologizing" intelligent system. There has been a lot of research into control systems for many years. Everything from the relationships between intelligent systems and the fundamental rules of biology to system configurations and road maps for the future of control systems have been studied. On the other hand, we present a generic building block for control systems, because without effective system blocks, there can be no successful system.

The evolutionary processes of Boolean networks are also very important. SORE demonstrates evolutionary development through its adaptation and learning capabilities. In colonization, the input of one gene (logically connected) to two genetic networks of N_4 and N_5 will produce a colony of N_{10} which is many times larger. The biological phenomena of self-reproduction can also be explained.

As a model for a gene regulatory network can be considered as a generic building block for a biologizing intelligent system, a gene behavior can be studied as a new idea for the systemization of an intelligent algorithm.

References

1. Albus, J. S.: Outline for a Theory of Intelligence, IEEE Trans. SMC **21(3)** (1991) 473-509
2. Albus, J. S.: The engineering of mind", Information Sciences **117** (1999) 1-18
3. Casti, J. L.: "Biologizing" Control Theory: How to Make a Control System Come Alive, Complexity **7(4)** (2002) 10-12
4. Report of the workshop Held at the University of Santa Clara on September 18-19, 1986, Challenges to Control: A Collective View", IEEE Trans. Automatic Control **32(4)** (1987) 275-285
5. Lathrop, R.: Intelligent Systems in Biology: Why the Excitement?, IEEE Intelligent Systems, (2001) 8-13
6. Leith, J. *et al.*: Toward More Intelligent Annotation Tools: A Prototype, IEEE Intelligent Systems (2001) 42-51
7. Kauffman, S. A.: Metabolic Stability and Epigenesis in Randomly Constructed Genetic Nets, J. of Theoretical Biology **22** (1969) 437-467
8. Wang, P. P., and Cheng, H. D.: SORE: Self Organizable and Regulating Engine, A Research Proposal submitted to US NSF, File No.: 03248-27, Fastlane (2003)
9. Wang, P. P., and Robinson, J.: What is SORE?, The 7th Proceedings of the Joint Conferences on Information Sciences (2003)
10. Wang, P. P., *et al.*: A Study of the Two-Gene Network - The Simplest Special Case of SORE, The 7th Proceedings of the Joint Conferences on Information Sciences (2003)
11. Wang, P. P., and Tao, H.: A Novel Method of Error Correcting Code Generation based upon SORE, The 7th Proceedings of the Joint Conferences on Information Sciences (2003)
12. Wang, P. P., and Yu, J.: SORE - A Powerful Classifier, The 7th Proceedings of the Joint Conferences on Information Sciences (2003)
13. Choi, K. H., and Sasaki, M.: Brain-Wave Bio Potentials based Mobile Robot Control: Wavelet-Neural Network Pattern Recognition Approach, IEEE Int. Conference on SMC **1** (2001) 322-328

14. Altman, R. B.: Challenges for Intelligent Systems in Biology, IEEE Intelligent Systems, (2001) 14-18
15. Evans, J. M. *et al.*: Knowledge Engineering for Real Time Intelligent Control, Proceedings of the 2002 IEEE Int. Symposium on Intelligent Control (2002)

Computerized Detection of Liver Cirrhosis Using Wave Pattern of Spleen in Abdominal CT Images

Won Seong[1], June-Sik Cho[2], Seung-Moo Noh[3], and Jong-Won Park[1]

[1] Dept. of Information and Communications Eng.,
Chungnam National University, Korea
{wseong, jwpark}@crow.cnu.ac.kr
[2] Dept. of Diagnostic Radiology, Chungnam National University, Korea
[3] Dept. of General Surgery, Chungnam National University, Korea
{jscho, smnoh}@cnuh.co.kr

Abstract. We examined the wave pattern of the spleen by using abdominal CT images of a patient with liver cirrhosis, and found that they are different from those of a person with a normal liver. This paper suggests a method to diagnose liver cirrhosis by using the wave pattern of the spleen in abdominal CT images on the basis of the two principles. It tells us that we can judge if the liver has liver cirrhosis, without the test of the ratio of caudate lobe to right lobe, only with the spleen.

1 Introduction

The following two methods are typically used by radiologists to diagnose liver cirrhosis in abdominal CT images. One method is to diagnose by measuring the ratio of the caudate lobe to the right lobe of the liver[1][2][3]. The other method is to diagnose indirectly based on the fact that the spleen of patients with liver cirrhosis is hypertrophied[4-10]. However, the ratio of the caudate lobe to the right lobe of an abnormal liver with liver cirrhosis isn't that different from a normal liver, and, in some cases, patients with liver cirrhosis don't show hypertrophy of the spleen[11]. On the contrary, even the size of the patient's spleen with a normal liver may be large by nature and not abnormally hypertrophied. Therefore, calculating the ratio of caudate lobe to right lobe or measuring hypertrophy of the spleen is not a reliable method, and with only one method it is hard to diagnose correctly. For this reason, on the basis of their own standards, radiologists make a final diagnosis if the liver is normal or abnormal by examining the size and shape of the spleen and the ratio of the caudate lobe to the right lobe of the liver at the same time.

In this study, we examined the wave pattern of the spleen by using abdominal CT images of a patient with liver cirrhosis, and found that they are different from those of a person with a normal liver. In the abdominal CT image of the patient with liver cirrhosis, there is a deep wave part on the left side of the spleen. In the case of the normal liver, there are waves on the left side, but they aren't deep. Therefore, the total area of waving parts of the spleen with liver cirrhosis is found to be greater than that of the spleen with the normal liver. Moreover, when examining circularity by abstracting the waves of the spleen from the image with liver cirrhosis, we found they are more circular than those of the spleen accompanied by a normal liver.

The purpose of this study is to suggest a new automatic effective method to diagnose liver cirrhosis by using the wave pattern of the spleen in abdominal CT images on the basis of the above-mentioned principles. First of all, we segment and abstract the abnormal liver and the spleen by reinforcing the utilization of place information and using an angle line technique. It can be applied to the image with the normal liver and used for segmenting and abstracting the normal liver and the spleen. After segmenting the spleen, we perform two diagnosis tests for liver cirrhosis.

2 Material and Methods

This scheme was developed in the Intel P4 processor with a Linux operating system, GNU C, X window and XV for image viewing. The general computerized procedure of this study is shown in Figure 1. Here, Pre-processing steps consist of image equalization and removing background and removing muscle from abdominal CT images.

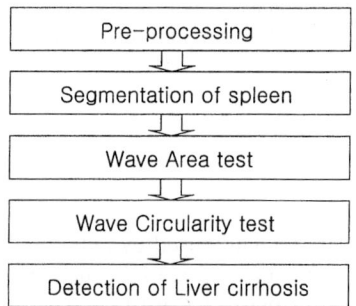

Fig. 1. Schematic Diagram of general process

2.1 Experiment Image

This study uses the abdominal CT images of 64 cases in total: 32 cases with normal livers and 32 cases judged to have liver cirrhosis. Each case consists of 17 slice images, and the resolution of each slice is 512 × 512 pixels. From the Chungnam National University Hospital, we obtained the CT image file in the DICOM (Digital Imaging and Communication in Medicine) format. In general, a normal person's liver with no lesion, such as liver cancer or liver cirrhosis (hereinafter referred to as a normal liver), shows even brightness values in an abdominal CT image.

2.2 Automatic Segmentation of the Spleen

Since a spleen which is with an abnormal liver with liver cirrhosis shows irregular brightness values, we can't segment properly only with the brightness value information. For this reason, we segment the spleen by using an angle line method. The angle line method applied to the spleen is the same as that applied to the liver, but only the place of its starting point of the search is different. The spleen is located opposite to the liver, on the right of the abdominal CT image, and accordingly the starting point can be set up at a point around that area. Figure 2 is an example of computerized segmentation of spleen.

(a)   (b)

Fig. 2. Example of computerized segmentation of the spleen and 3 distinctive wave in spleen. (a) is an example of a computerized segmentation using the angle line method in an abdominal CT image. (b) is shows 3 distinctive waves in abstracted spleen

2.3 Wave Area Test

In the test, we can calculate the sum of areas with the number of pixels by abstracting wave parts of the abstracted spleen and then judge if the liver is normal or abnormal by the standard value taken from the experiment in advance.

With the segmented spleen we will follow the procedure to abstract its wave parts as shown in figure 3(a). First, we draw a line which starts from the highest middle point of the spleen and links a vertex (K) of the first wave. It is to set a standard line for calculating the area of wave (A). The procedure to get wave vertex (K) is as follows. First, we get a tangential line tangent to the spleen from the highest middle point of the wave. Second, we find a tangential point (K) satisfying the conditions of tangential points out of the points that make the tangential line. In order to get the tangential line of the wave, we have only to take an angle line satisfying the conditions that the line should be tangent to the surface of the spleen, shooting lines of variety of angles from the starting point of the wave below the spleen.

After getting the vertex (K) of the first wave as figure 3(b), we calculate an area (A) of the first wave with the standard of a linking line, starting from the highest middle point (S) of the spleen and ending in the vertex (K). In order to get the area (A), we have only to add the number of points with the brightness values of the spleen, examining the composing points of all the lines with a smaller angle than angle line (a) which links the starting points (S) and (K).

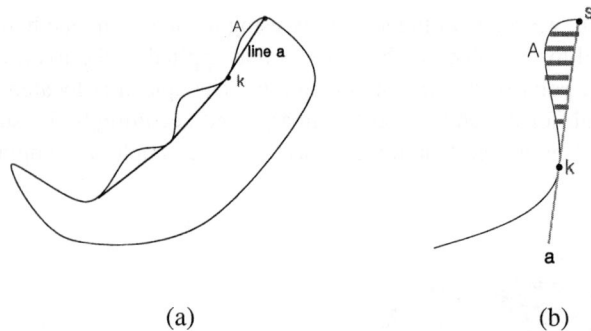

Fig. 3. Abstraction of wave parts and getting an area (A) of the first wave

After getting the area of the first wave, we calculate the area of the second wave. In order to get the area of the second wave, we have only to set the vertex (K) of the first wave as a starting point (S) of the second wave and repeat the procedure previously used for getting the area of the first wave. For the third wave, we repeat the procedure.

2.4 Circularity Test of Wave

From the example of the first wave we can learn the following mechanism for calculating circularity. After we get an area which takes an angle line linking (S) and (K) for the standard, we perform a circularity test before the next wave. The angle line linking (S) and (K) will be a standard line and form the ROI. The line becomes a basic diameter to calculate circularity. As shown in figure 4(a), we can draw a

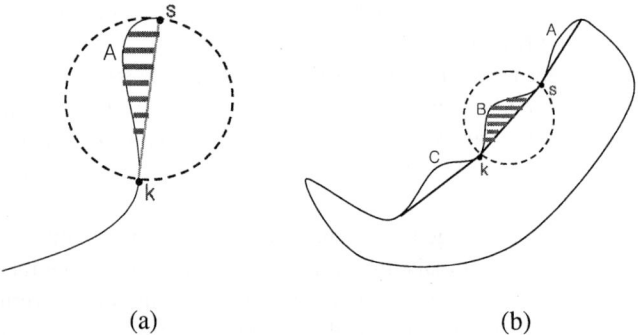

Fig. 4. Circle taking the line linking (S) and (K) as a diameter and calculating the degree of circularity of the second wave

hypothetical circle taking the line linking (S) and (K) as its diameter. We can get a radius and an area of the circle With the circle area gained above we can calculate the degree of circularity of the first wave. Since the wave area calculated in the first wave is included in a half circle, we should calculate the degree of an actual circularity after multiplying the double. Expression (1) shows how to calculate the degree of circularity of a wave.

$$\text{Degree of circularity} = (\text{area of wave A}) \times 2 \times 100 \ / \ \text{area of circle} \quad (1)$$

By calculating the degree of circularity we can reduce mistakes considerably (which result from individual difference in the size of the spleen) in judging if the liver is normal only with wave area. In other words, a person with a naturally big spleen can be regarded as a patient with an abnormal liver owing to a large wave area, although he/she has a normal liver. However, the waves that originate from abnormal livers regardless of individual size differences usually indicate a circular shape regardless of their size. Therefore, we can make an effective judgment by measuring the absolute area value of wave and calculating the degree of circularity of the waves at the same time.

Once we finish calculating the degree of circularity of the first wave, we should get the tangential point, the tangential line and the area of the second wave. Then we calculate the circularity of the second wave as figure 4(b). For the third wave, we repeat the procedure. The calculated values in order are used to judge if the liver is normal or abnormal according to criterion.

3 Experimental Results

This section describes experimental results with abdominal CT images of 32 cases of normal liver and 32 cases of liver cirrhosis. For the abdominal CT images of the 64 cases used in the experiment, radiologists have judged if they are liver cirrhosis in advance.

From the results of this experiment, we can find a difference between images with liver cirrhosis and those with normal livers. For the images with liver cirrhosis, the total sum of wave areas of the spleen is more than 1000 pixels, and the circularity average of waves is over 40% if the sum is 700 to 900 pixels. For the images of the normal liver, however, the total sum of valid wave areas of the spleen is not over 1000 pixels. If the area is over 700 pixels, the circularity average is less than 25%. Therefore, sorting out three with the large spleen among many slice images composing a case, we can judge liver cirrhosis if the sum total of valid wave areas of the spleen is more than 700 pixels and circularity average is over 35%.

Table 1 shows the results of comparing 32 liver cirrhosis cases under three categories like the ratio of caudate lobe to right lobe, size of the spleen, and wave pattern test. Through the experiment in this study, we can find that, all the cases of liver cirrhosis judged by the ratio of caudate lobe to right lobe can also be judged with the wave pattern of the spleen. This means we can judge liver cirrhosis using only the segmentation of the spleen in abdominal CT images.

Table 1. Comparison of three diagnoses in the image finally judged liver cirrhosis

Methods of diagnosis	Satisfied cases	Rates	Probability
Rate of caudate lobe	12	12 / 32	37.5%
Spleen size	18	18 / 32	56.2%
Wave pattern tests	30	30 / 32	93.7%

4 Conclusions

We have developed a computerized technique for the diagnosis and detection of liver cirrhosis in abdominal CT images. We examined the wave pattern of the spleen by using abdominal CT images of a patient with liver cirrhosis, and found they are different from those of a person with a normal liver. This paper suggests a new effective method to diagnose and detect liver cirrhosis by using the wave pattern of the spleen in abdominal CT images on the basis of the two principles. Therefore, this system effectively diagnoses and detects the liver cirrhosis by two tests when applied to the abdominal CT images which is difficult to detect the cirrhosis accurately.

References

1. Wegener, O.: The Liver. In: Whole body computed tomography. 2nd edn. Boston Blackwell Scientific Publications. (1993) 243-275
2. Hitomi, A., Donald, G., Tamotsu, K., George, H., Katsuyoshi, I., Tsuneo M.: Cirrhosis: Modified Caudate-Right Lobe Ratio, Radiology 224 (2002) 769-774
3. Torres, E., Whitmire, L., Gedgaudas, M., Bernardino, M.: Computed tomography of hepatic morphologic changes in cirrhosis of the liver. J. Comput. Assist. Tomogr. 10(1) (1986) 47-50
4. Lamb, P., Lund, A., Kanagasabay, R., Martin, A., Webb, J., Reznek, R.: Spleen size: how well do linear ultrasound measurements correlate with three-dimensional CT volume assessments?. Br. J. Radiol. 75(895) (2002) 573-577
5. Prassopoulos, P., Cavouras, D.: CT assessment of normal splenic size in children. Acta Radiol. 35(2) (1994)152-154
6. Schlesinger, A., Hildebolt, C., Siegel, M., Pilgrim, T.: Splenic volume in children: simplified estimation at CT. Radiology.193(2) (1994) 578-580
7. Rosenberg, H., Markowitz, R., Kolberg, H., Park, C., Hubbard, A., Bellah, R.: Normal splenic size in infants and children: sonographic measurements. Am. J. Roentgenol. 157(1) (1991) 119-121
8. Sheth, S., Mani, S., Tamhankar, H., Mehta, P.: Spleen size in health and disease: a sonographic assessment. J. Assoc. Physicians India. 43(3) (1995) 182-184
9. Frank, K., Linhart, P., Kortsik, C., Wohlenberg, H.: Sonographic determination of spleen size: normal dimensions in adults with a healthy spleen. Ultrachall Med. 7(3) (1986) 134-137
10. Loftus, W., Metreweli, C.: Normal splenic size in a Chinese population. J. Ultrasound Med. 16(5) (1997) 345-347
11. 11.Arkles, L., Gill, G., Molan, M.: A palpable spleen is not necessarily enlarged or pathological. Med. J. Aust. 145(1) (1986) 15-17

Automatic Segmentation Technique Without User Modification for 3D Visualization in Medical Images

Won Seong[1], Eui-Jeong Kim[2], and Jong-Won Park[1]

[1] Dept. of Information and Communications Eng.,
Chungnam National University, Korea
{wseong, jwpark}@crow.cnu.ac.kr
[2] Dept. of Computer Education, Kongju National University, Korea
ejkim@kongju.ac.kr

Abstract. It is necessary to analyze an image from CT or MR and then to segment an image of a certain organ from that of other tissues for 3D (Three-Dimensional) visualization. There are many ways for segmentation, but they have a somewhat ineffective problem because they are combined with manual treatment. In this study, we developed a new segmenting method using a region-growing technique and a deformable modeling technique with control points for more effective segmentation. As a result, we try to extract the image of liver and identify the improved performance by applying the algorithm suggested in this study to two-dimensional CT image of the stomach that has a wide gap between slices.

1 Introduction

The CT or MR image through advanced computer technology is usually processed two-dimensionally, since it is expressed as a series of two-dimensional slices. however, There must be a limit in processing the two-dimensional slices, because the organs in the body are a three-dimensional structure. If we reorganize the slices with three-dimensional image technology, Visualization, operation, and analysis of the organs can be done practically. For example, we can make an plan for effective operation completely in advance, by reorganizing and operating the real shape of organs in a surgical operation[1].

On the other hand, we need a segmentation process for separating the specific organ of interest from other tissues in order to analyze the image from CT or MR and visualize it three-dimensionally. It will be ineffective to take much time to process by manual treatment or user modification or semiautomatically. It is said that there is no perfectly automatic segmentation method[1]. Therefore, we need a method to segment a targeting organ from other organs or structures without person's intervention as much as possible. The existing general segmentation methods are thresholding, edge detection, morphological filtering, and deformable model techniques [2-8]. According to the existing techniques, they pursue too much detailed boundaries of the organ which lead to wrong segmentation results. This often requires a user modification or user intervention in the middle of processing segmentation in order to get a convincing results. After all, however, the segmentation with a user modification or intervention cannot be said to be a complete automatic segmentation.

Therefore, in this study we devised a new algorithm, on the basis of a region-growing technique and a deformable model technique using control points, to develop an automatic segmentation method applicable to the three-dimensional medical image.

2 Material and Methods

This scheme was developed in the Intel P4 processor with a Linux operating system, GNU C, X window and XV for image viewing. The general computerized procedure of this study is shown in Figure 1. Here, Pre-processing steps consist of image equalization and removing background and removing muscle from abdominal CT images.

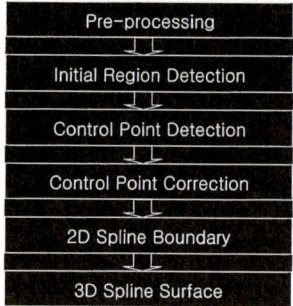

Fig. 1. Schematic diagram of general process

2.1 Experiment Image

This study uses the 850 abdominal CT images of 50 cases in total. Each case consists of 17 slice images, and the resolution of each slice is 512 × 512 pixels with 1 byte/pixel, and the total size was 256kb. From the Chungnam National University Hospital, we obtained the CT image file in the DICOM (Digital Imaging and Communication in Medicine) format.

2.2 Initial Region Detection

The region-growing method is performed from a random coordinate within the region according to the following process as shown in figure 2. Figure 2(b) is the result after applying the region-growing algorithm using gray-level thresholding.

2.3 Control Point Detection

As for a model detecting a control point within the boundary, it is good to choose a large model starting as from an inter center as possible. After inputting a possible coordinate of the central part, we make the greatest circle which is meeting at the boundary of segmentation centering round the coordinate. In Figure 3(a), there is possibility of a control point on false boundary around A to be detected, since initial coordinate is not placed on the center of segmentation in a model A. But in B we can detect control point evenly on each boundary.

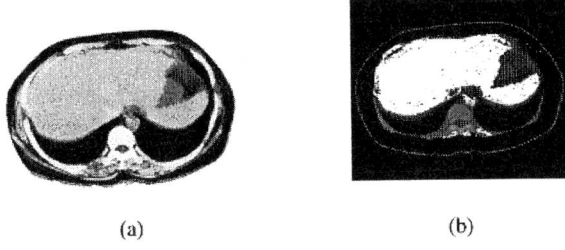

(a) (b)

Fig. 2. Example of initial region detection. (a) Equalized original image. (b) Initial region with region growing technique from figure 2(a)

As shown in Figure 3(b), the control points needed to create B-Spline Surface are selected from the center of model to the point which reaches the original boundary first. In this procedure, we can confirm the effect that a part with a great change in the angle is excluded in the control point, when sampling some boundary with a sharp inclination.

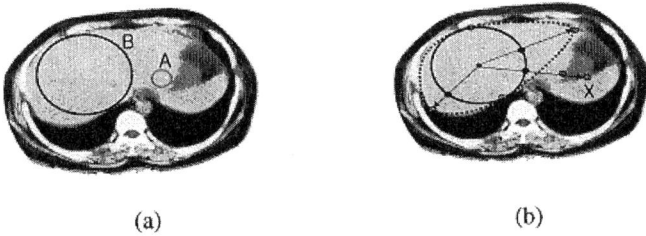

(a) (b)

Fig. 3. Control point detection. (a) Initial model setup. (b) Angle-line method for control point detection. In here, point X is excluded in the control point set

2.4 Control Point Correction

As Since the detected control points may be set up on some false boundary, if either sampling value of the points or that of neighbor control points doesn't satisfy continuity force related to the angle, we should move the unsatisfactory control point to a similar place to its neighbor. When finding control points due to irregular boundaries of the liver, the following cases of folded control points as shown in figure 4 can be often found.

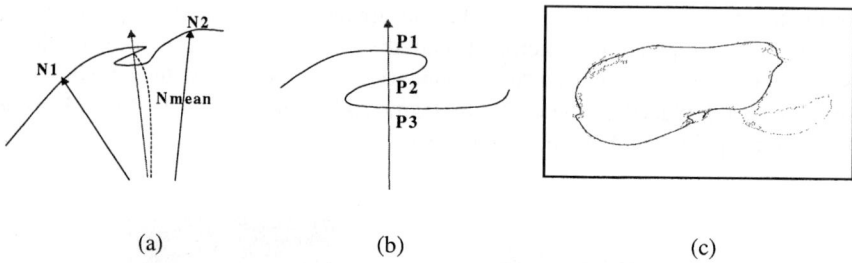

(a)　　　　　　　　　(b)　　　　　　　　　(c)

Fig. 4. Example of computerized segmentation of the spleen and 3 distinctive wave in spleen. (a) is an example of a computerized segmentation using the angle line method in an abdominal CT image. (b) is shows 3 distinctive waves in abstracted spleen

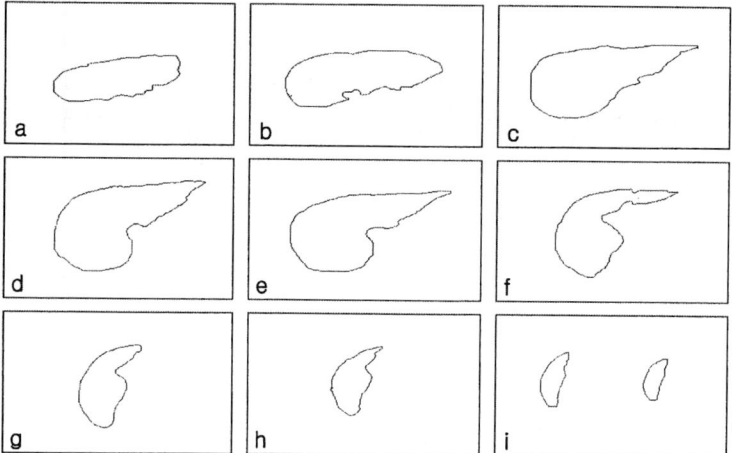

Fig. 5. Example of slices series(a-i) extracted from one case

In this case, only the adequate control point needs to be selected out of P1,P2, and P3. Since the boundaries as the first result have to be simple.

The steps for the selection are as follows ; Figure 5 shows the segmentations of a liver within CT through those process.

1. Nmean, the average of the distances N1 and N2 from the center, should be obtained
2. Among the values of $|\text{Nmean-P1}|$, $|\text{Nmean-P2}|$ and $|\text{Nmean-P3}|$, the smallest value is selected.
3. The Px which has been selected as the smallest value at the second step, is decided to be the control point for the folded section of the boundary. In other words, the nearest point from adjacent control point is selected as a control point

2.5 2D Spline Boundary and 3D Spline Surface Creation

After deciding the control points, They are applied to 2D spline function so that the boundaries can be resulted in a smooth line. The creation of 2D spline boundaries is applied to all consecutive slices S_i, S_{i+2},..., S_N. After the creation of the boundaries of all the slices, They are applied to 3D spline surface function[9] so that the entire shape of the organ, liver can be formed. At this point, The acquired 3D spline surface can be showed through 3D graphic viewer, GeomView. In addition, we performed a volume calculation of extracted organ. By using the values of pixel spacing, and slice thickness, the volume of extracted organ was calculated.

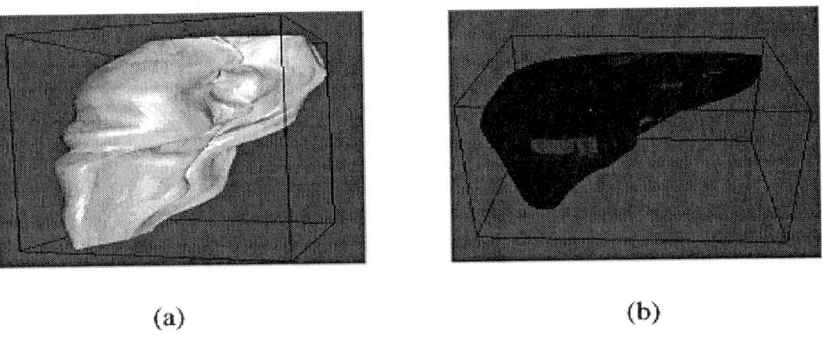

(a)   (b)

Fig. 6. Example of 3D surface visualization. Especially, figure 6(a) shows the result using figure 5

3 Experimental Results

This method has been applied to the abdominal CT images of 50 normal people. The following table shows the compared results between the volume calculation proposed by this new method and the volume calculation segmented by two radiologists' manual. In this case, the two radiologists extracted the proper organ directly from each slices by drawing the boundaries.

Table 1 represents that the difference between this new method and the manual method by radiologists is less than 4%. This percentage explains that this proposed method can be regarded as showing the entire shape of the organ smoothly.

Figure 6 shows the slices after segmentation processing as a three-dimensional image through a three-dimensional graphic data viewer.

Table 1. Comparison of volume calculation

Number of cases	Average: Manual Volume	Average: Automatic Volume	Difference (Automatic to manual)
50	1.135439	1.129095	3.410%

4 Conclusions

We have developed a automatic segmentation technique for 3D visualization in medical images. The existing segmentation techniques pursue too much detailed boundaries of the organ which lead to wrong segmentation results. This often requires a user modification or user intervention in the middle of processing segmentation in order to get a convincing results. After all, however, the segmentation with a user modification or intervention can not be said to be a complete automatic segmentation. Therefore, a complete automatic segmentation technique without a user modification or user intervention is required. For this reason, we focused on the literally automatic segmentation technique. The benefit of new technique is a complete automatic segmentation method without a user modification though it is not supposed to contain too much detailed boundaries of the organ.

References

1. Jayaram, K., Gabor, T. : 3D Imaging in Medicine. 2nd edn. CRC Press (2000)
2. Yoo, S. : A Study on the Segmentation of Liver and Spleen From the CT Image. Master thesis of Industrial Education, Chungnam National University in korea (1999)
3. Luomin, G., David, G., Brian, S., Elliot, K. : Automatic Liver Segmentation Technique for Three-dimensional Visualization of CT Data. Radiology. Vol. 201 (1996) 359-364
4. Marc, J., Bernard, M. : Three-Dimensional Segmentation and Interpolation of Magnetic Resonance Brain Images, IEEE Transactions on Medical Imaging. Vol. 12(2) (1993)(2) 573-577
5. Michael, B., Karl, H., Ulf, T., Martin, R. : 3-D Segmentation of MR images of the Head for 3-D Display. IEEE Transactions on Medical Imaging. Vol 9(2) (1990) 152-164
6. Karl, H., William, A. : Interactive 3D Segmentation of MRI and CT Volumes using Morphological Operations. J. Com. Assist. Tomogr. (1992) 578-580
7. Zahid, H. : Digital Image Processing. Ellis Horwood, (1991)
8. Parker, J. : Algorithms For Image Processing and Computer Vision. John Wiley & Sons (1997)
9. Bourke, F. : Spline curves in 3D. http://astronomy.swin.edu.au/pbourke/curves/spline

Adaptive Stereo Brain Images Segmentation Based on the Weak Membrane Model

Yonghong Shi and Feihu Qi

Department of Computer Science and Engineering,
Shanghai Jiao Tong University, Shanghai, 200030, China
{Shi-yh, fhqi}@cs.sjtu.edu.cn

Abstract. This paper presents a new method for automatically segmenting brain parenchyma and cerebrospinal fluid in routine single-echo magnetic resonance (MR) images. Our method is based on the weak membrane model. Weak membrane models can model intensity measurement at each voxel site to implement piecewise smoothness constraint, and at the same time model discontinuities to control the interaction between each pair of the neighboring pixel. Segmentation is obtained by seeking for the maximum a posteriori estimation of the regions and the boundaries by using Bayesian inference and neighborhood constraints based on Markov random fields (MRFs) or Gibbs random fields (GRFs) models. Our approach has the following desirable properties: (1) brain voxels can be accurately classified into white matter, grey matter and cerebrospinal fluid (CSF), and (2) relatively insensitive to noise and intensity inhomogeneity.

1 Introduction

Image segmentation is an important step for many image analysis systems. Its goal is to extract meaningful features from an input image and divide the image into separate regions based on the discontinuities existing in the feature space [1]. In medical imaging, segmentation often involves the delineation of anatomical structures or the detection of organs or tissue types from CT or MR images. But the issues such as biological variations of tissue and intensity inhomogeneities and poor contrast make the accurate segmentation a difficult task. In addition, many pixels in real images are factually ambiguous and cannot be classified reliably and consistently based on feature attributes alone. In medical images, voxels belonging to a same object usually have similar feature values, which reflect the inherent spatial consistence existing in medical images. Therefore, spatial information can be used, in principle, to lead to a more meaningful classification.

Based on the idea of integrating spatial consistence information into the classification procedure, Markov random fields (MRFs) have become a popular solution [2, 3]. The MRFs of a prior contextual dependent pattern is a powerful method for modeling spatial continuity or piecewise continuity, which provides useful information for the segmentation process. However, the prior smoothness assumption does not include, or actually excludes the abrupt changes between neighboring regions, which happens frequently in reality.

In this paper, we propose a modification method to the popular MRFs prior smoothness assumption by introducing an MRFs prior discontinuities assumption. These two prior assumptions make up a model, which we call 'weak membrane'. The method is to derive the maximum a posteriori estimate of the regions and the boundaries [2, 7] by using Bayesian inference and neighborhood constraints based on Markov random fields (MRFs) or Gibbs random fields (GRFs) models. In section 2, we give a more detailed description of the weak membrane models. And it is followed by a description of the adaptive segmentation algorithm, which includes a discussion of the parameter estimation and adaptation technique. Finally, we present the results of some experimental studies and conclusions.

2 Weak Membrane Model

When we establish a weak membrane model for our magnetic resonance images segmentation, two Markov random fields (MRFs) prior models need to be considered. Firstly, intensity measurement with respect to various tissues is modeled by a finite Gaussian mixture, and piecewise smoothness with respect to voxel values is modeled with MRF [2-5, 7]. Secondly, discontinuities or abrupt changes in certain image properties between neighboring areas, which are edges, are modeled by line process (LP) [4,9,10] in which each label takes a value on $\{0,1\}$ to signal the occurrence of edges.

The two-coupled MRFs are defined on two spatially interwoven sets. One set is the lattice for the intensity field and the other one is the dual lattice for the introduced edge field. A possible edge may exist between each pair of the neighboring pixel sites.

The objective is to find the global minimum of the energy function described as

$$E = \sum_{l=1}^{K}\sum_{i=1}^{N_l}\left[\frac{1}{2}\left(\frac{x_i - \mu_{l,i}}{\sigma_{l,i}}\right)^2 + \log(\sigma_{l,i})\right] + \sum_{\{i,i'\}\in C_2}[(y_i^P - y_{i'}^P)^2(1 - y_{i,i'}^E) + \gamma y_{i,i'}^E]. \quad (1)$$

Where we denote a 3D cubic lattice of the voxel sites by $S^P \subset N^3$ and the dual lattice by $S^E = \{(i,i') | i,i' \in S^P, i \leftrightarrow i'\}$ where $i \leftrightarrow i'$ means i and i' are neighbors. Let $x = (x_i, i \in S^P)$, be the observed intensities where x_i denotes the image intensity at the voxel site indexed by i. We assume that the total number of tissue classes in the image is K and the label set is $L^P = \{1,2,...,K\}$. The segmentation task is to determine the optimal classification denoted by $y = (y_i^P, i \in S^P)$, $y_i^P \in L^P$. We write $y_i^P = l$ to indicate that the tissue class l is assigned to the voxel site i. Let $y_{i,i'}^E, (i,i') \in S^E$, be an edge label taking a value in $L^E = \{0,1\}$, with 0 and 1 representing the absence and presence of an edge, respectively. Let $\mu_{l,i}$ and $\sigma_{l,i}$ represent the underlying image intensity of the class l at site i and the standard deviation of the noise signal for the tissue class l at site i, respectively. Let N_l denote the number of all voxel sites belonging to tissue class l and $\sum_{l=1}^{K} N_l = |S^P|$. Let $|S^P|$ denote the total number of voxels sites in the image. $C_2 = \{(i,i') | i' \in NB_i, i \in S^P\}$ is the clique representing a pair of neighboring sites in our prior model. NB_i is the neighbor set of i and defined as $NB_i = \{i' \in S^P \mid [dist(i,i')]^2 \leq r, i \neq i'\}$, Where $dist(i,i')$ denoting the Euclidean distance between i and i', r taking an integer value, and γ controlling the interaction between neighboring regions. To minimize the second term on the right of (1) alone, the intensity labels and edge labels are determined as following [4]:

If $(y_i^P - y_{i'}^P)^2 < \gamma$, then set $y_{i,i'}^E = 0$; otherwise, set $y_{i,i'}^E = 1$.

The first term on the right of (1) represents the binding between the image data and the segmentation, and the second term indicates the smoothness and discontinuity constraints in the weak membrane model.

3 Segmentation and Estimation

First, the parameters set $\Theta = \{(\mu_{l,i}, \sigma_{l,i}) \mid l \in L^P, i \in S^P\}$ related with the observed data x can be unknown, and a part of the parameters need to be estimated. The parameters set $\Phi = \{\gamma\}$ for MRFs also need to be estimated. These parameters should be estimated only by using realizations of MRFs. Second, to perform good segmentation, the parameter values of the MRFs have to be available. This is a dilemma between segmentation and estimation. In order to meet the optimal segmentation requirement, we should segment the image and estimate parameters simultaneously. We rewrite (1) as

$$E_i(y_i^P \mid x_i, y_{NB_i}^{P(n)}) = \frac{1}{2}\left(\frac{x_i - \mu_{l,i}}{\sigma_{l,i}}\right)^2 + \log(\sigma_{l,i}) + \sum_{i' \in NB_i}(y_i^P - y_{i'}^P)^2(1 - y_{i,i'}^E) + \gamma y_{i,i'}^E. \quad (2)$$

Therefore, the segmentation and estimation procedure can be implemented simultaneously in (2). We choose Iterative conditional model (ICM) [4] to find a local suboptimal solution y^{P^*} for (2).

The ICM algorithm uses the greedy strategy in the iterative local minimization. Given the data x_i and the other segmentation labels $y_{NB_i}^{P(n)}$, the algorithm sequentially updates each $y_i^{P(n)}$ into $y_i^{P(n+1)}$, where n is the time of cycle, by minimizing $E_i(y_i^P \mid x, y_{NB_i}^P)$. $E_i(y_i^P \mid x_i, y_{NB_i}^{P(n)})$ is evaluated with each $y_i^P \in L^P$ and the label causing the lowest value $E_i(y_i^P \mid x_i, y_{NB_i}^{P(n)})$ is chosen as the value for $y_i^{P(n+1)}$. When equation (2) is applied to each i in turn, the above defines an updating cycle of ICM. Then let $n \leftarrow n+1$ and continue next cycle, until converge. Convergence is decided when the number of changes of the tissue classes at the voxel sites drops below a certain threshold. We adaptively adjust parameters $\mu_{l,i}, \sigma_{l,i}$ in the ICM algorithm before we use equation (2) to determine the label at voxel i. We consider a neighborhood at each voxel i, and the ML estimates of the parameters $\mu_{l,i}, \sigma_{l,i}$ are determined by taking means and standard deviations of the intensities of the neighborhood voxels [2].

The result obtained by ICM depends very much on the initial estimator $y^{P(0)}$, as widely reported. Currently, a natural choice for $y^{P(0)}$ is to derive the maximum likelihood estimation of $y^{P(0)}$ [4]. Since the intensity measurement is assumed to be a finite Gaussian mixture, we can use EM algorithm to compute initial parameter Θ. To avoid intensive computation involved in the EM algorithm, K-mean algorithm is used to give a starting configuration for EM [2, 6, and 8]. K-mean's parameters are initialized as follows: First, compute the mean value μ_0 and standard deviation σ_0 of the image data; Second, set K-mean's initial parameters. The mean value for white matter, gray matter and CSF are $\mu_0 + \delta\sigma_0, \mu_0 - \delta\sigma_0, \delta\sigma_0$, and the standard deviations of all tissue chasses are $\delta\sigma_0$, where $0 < \delta < 1$. Therefore the initial parameter Θ used in ICM can be computed through K-mean and EM algorithm, and the initial estimator $y^{P(0)}$ is also derived. And the parameter γ controlling the interaction between neighboring voxels can be adaptively determined by setting it as the weighted mean value or median value within the neighborhood NB_i.

Whenever a cycle visiting all voxels in the image finished, the parameters Φ associated with MRFs prior models must be estimated using the newly segmentation result $y^{P(n)}$ again. Then the newly estimated prior parameters are used to continue the next cycle.

4 Experimental Result

The data used in this paper is an MRI brain scan. The brain consists of white matter (WM), grey matter (GM) and CSF. Illustrations of the segmentation are shown in Fig. 1. White matter, grey matter and CSF in an image are accurately segmented and also shown in the figure.

In order to compare our segmentation result with other popular methods, a brain image is segmented using two schemes: (M1) based on the assumption of parameters space invariant, the ML estimations of parameters and segmentation are derived by using K-mean and EM algorithm; (M2) first, use k-mean and EM algorithm to derive the initial configuration of the weak membrane model. Then, make use of our model to adaptively segment the input image and estimate the parameters. The two segmentation results are shown in Fig. 2. It's obvious that our approach gives a consistent segmentation result.

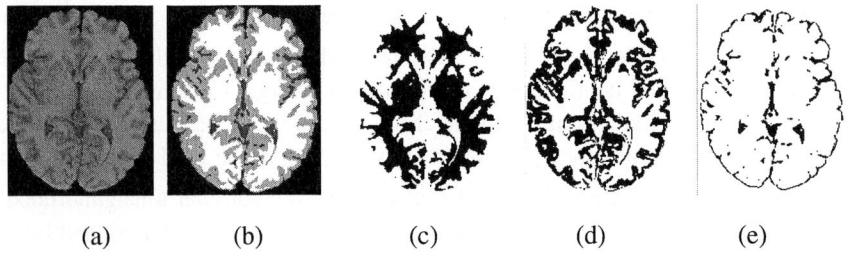

(a)　　　　　(b)　　　　　(c)　　　　　(d)　　　　　(e)

Fig. 1. Illustration axial segmentation (a) original image (b) segmentation result (c) WM (d) GM (e) CSF

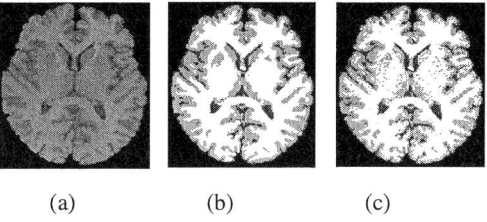

Fig. 2. Comparison of two methods (a) original image (b) result using M2 (c) result using M1

In order to demonstrate the effect of noise on the performance of our method, we added varying amounts of zero mean Gaussian noise to a 3D brain image to obtain noisy images. The pictures in Fig. 3 from second column to fifth column in the first row show cross sections of four noisy images that were corrupted by Gaussian noise with standard deviation of 5, 10, 15 and 20, respectively. The performance of our algorithm becomes worse with the increases of noise levels. And the method using K-mean and EM algorithm based on the spatial invariant parameters has much worse performance. Overall, our method is more robust in handling noise.

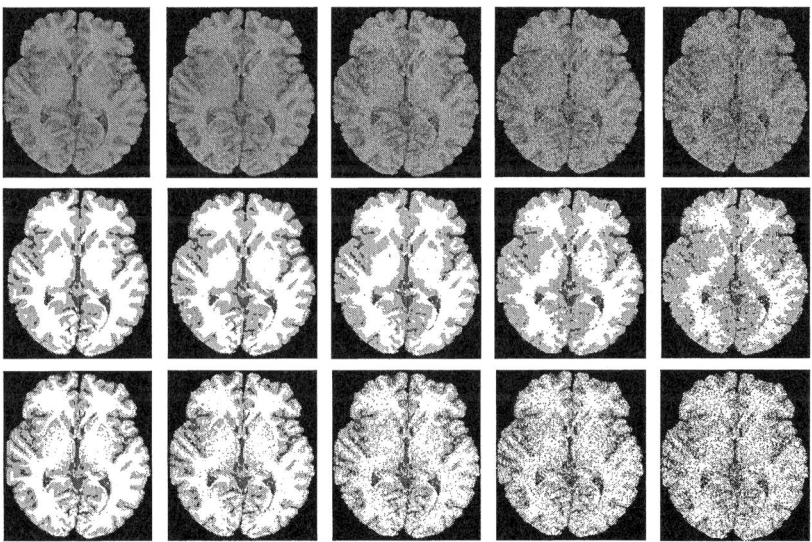

Fig. 3. Images used for demonstrating robustness of our methods. The images in the first row are the original image and noisy images. The ones in the second row are the segmented results using our method. The ones in the third row are the segmented results using K-mean and EM

5 Conclusion

Based on weak membrane model, this paper describes a stereo MRI brain image segmentation algorithm. Piecewise smoothness with respect to the intensity measurement and discontinuity with respect to the edges or boundaries are modeled respectively.

Due to the improvements brought by these two spaces interwoven models, brain images are accurately segmented into white matter, grey matter and CSF. Now we are trying to make a further improvement to our model in order to reduce the misclassification of voxels.

References

1. Pham D.L., Xu C.Y., Prince J.L.: A survey of current methods in medical image segmentation. Annual Review of Biomedical Engineering, 1 (1998) 19
2. Rajapakse J.C., Giedd J.N., Rapoport J.L.: Statistical approach to segmentation of single-channel cerebral MR Images. IEEE Tran Med. Imag., Vol.16, 2 (1997) 176–186
3. Wilson R., Li C.T.: A class of discrete multiresolution random fields and its application to image segmentation. IEEE Tran pattern analysis and machine intelligence, Vol. 25, 1 (2002) 42–55
4. Li S.Z., Markov Random Field Modeling in Computer Vision, Springer, 1995
5. Laidlaw D.H., Fleischer K.W., Barr A.H.: Partial-Volume Bayesian classification of material mixtures in MR volume data using voxel histograms. IEEE Tran Med. Imag., Vol.17, 1 (1998) 74–86
6. Li W.Q., Attikiouzel Y.: Initialization of Clustering Algorithms for Unsupervised Segmentation of Multi-echo MR Images. Third Australian and New Zealand Conference on Intelligent Information Systems, Perth, IEEE Australia and New Zealand Council. 1 (1995) 88–92
7. Geman S., Geman D.: Stochastic relaxation, Gibbs distribution, and the Bayesian restoration of images. IEEE Tran Pattern Anal. Machine Intel. Vol. PAMI-6, 1984, 721–741
8. Pappas T.N.: An adaptive clustering algorithm for image segmentation. IEEE Tran Signal Processing, Vol. 40, 4 (1992) 901–914
9. Blake A.: The least disturbance principal and weak constraints. Pattern Recog. Lett, 1 (1983) 393–399
10. Kubota T., Huntsberger T.L.: Adaptive anisotropic parameter estimation in the weak membrane model. Lecture Notes in Computer Science (Proc. Intern. Workshop on Energy Minimization Methods), Vol. 1223, Venice, Italy, 5 (1997) 179–194

PASL: Prediction of the Alpha-Helix Transmembrane by Pruning the Subcellular Location

Young Joo Seol[1], Hyun Suk Park[2], and Seong-Joon Yoo[1]

[1] School of Computer Engineering, Sejong University, 98 Gunja-Dong,
Gwangjin-gu, Seoul, 143-747, Korea
sjyoo@sejong.ac.kr
[2] Ewha Womans University, Seoul, Korea

Abstract. We have developed a software tool, called **PASL**, which predicts the transmembrane region and its topology by pruning the subcellular location. The main virtues of **PASL** are that it discriminates the integral proteins of the plasma membrane from the intracellular membranes, and it eliminates the possibility of misrecognition of the signal peptide as a transmembrane region. The transmembrane region prediction algorithm, which is based on the Hidden Markov Model, and the ER signal peptide detection architecture, which is based on neural networks, have been used for the actual implementation of a prototype. This paper mainly describes the prototype and how it works.

1 Introduction

Proteins embedded in the bilayer (a transverse hydrophobic region) are called transmembrane proteins. The transmembrane region of proteins has two secondary structures: alpha-helix and beta-barrel regions. This paper describes a software tool that improves the accuracy of predicting alpha-helical transmembrane regions. It is difficult for transmembrane-region-finding programs to distinguish between receptors and enzymes that are located in the plasma membrane, since intracellular membranes have similar structures.

A variety of tools for transmembrane prediction have been implemented based on approaches such as hydropathy values, biological rules and machine learning approaches. Among these, TMHMM[1] and HMMTOP[2], based on the Hidden Markov Model, have been reported to have the best performance[3]. These models predict the location of membrane-spanning regions and their topology. Currently available software, however, does not distinguish between plasma membrane spanning and intracellular membranes. Moreover, the existence of the N-terminal peptide sequence complicates the prediction of transmembrane regions.

This paper describes a prototype system, **PASL** (Prediction of Alpha-Helix Transmembrane by Pruning Subcellular Location), which implements a prediction model proposed in [4]. **PASL** predicts the plasma membrane-spanning region and its topology by separating subcellular locations through preprocessing. We summarize the model used in **PASL**, its performance, and the characteristics of the prototype.

2 Related Works

This section describes previous efforts in developing software tools for predicting the transmembrane spanning region and its topology. Although the researchers' primary interest lies in the plasma membrane proteins, currently available software does not discriminate plasma membrane spanning intracellular membranes. Moreover, the existence of the N-terminal peptide sequence complicates the prediction of the transmembrane regions. The N-terminal signal peptide can be regarded as a transmembrane region due to the hydrophobic nature of signal peptide [5]. Once assigned as a transmembrane region, it affects other topology predictions by frame shift.

Hydrophobicity and positively charged amino acid compositions were used in early models for predicting membrane-spanning regions. Neural network, matrix, multiple alignments, or the hidden markov model were also used in previous prediction tools.

Most subcellular location prediction methods are grouped into two approaches: one is based on the detection of the N-terminal signal peptide, and the other is based on full-length amino acid composition. SignalP [6] and TargetP [7] are prediction servers based on N-terminal sequence approaches. These programs also supply the information of mature protein by cleavage of signal peptides. The N-terminal signal peptides, however, are limited in some intracellular organelles, therefore preventing predictions from taking on a general approach in covering all kinds of parts in a cell.

Machine learning methods, such as neural network [8] and support vector machines [9], were also used to predict the subcelluar location of proteins. They are motivated by the variation of amino acid compositions of full-length protein according to the protein's location in the cell [10]. These machine learning methods show improved performance and more subdivided categorization compared to the N-terminal-sequences-based approach.

3 Prediction Model and Its Performance

A prediction system, called **PASL,** which integrated modules of pruning the subcellular location - the ER signal peptide and cleavage site detection - and of predicting the transmembrane region, was developed. The modules are based on the neural network model and on the Hidden Markov Model, respectively. For the recognition of the cleavage site, the approach based on N-terminal signal peptide detection was used. Preprocessing of the ER signal peptide can improve the prediction performance by blocking the frame shift caused by misinterpreting the signal peptide as a transmembrane region. The ER-targeted proteins will be transported to the plasma membrane through the vesicle transport by membrane fusion. For this reason, **PASL** makes it possible to discriminate plasma membrane proteins more accurately from other intracellular transmembrane-spanning proteins.

PASL is composed of two modules: the **Pruning Module (PM)** and the **Trans-Membrane region and topology prediction module (TM).** These modules are described above and shown in Figure 1. If the ER signal peptide and cleavage site exist in the **PASL** query sequences, the signal peptide is removed by the **PM** before entering the **TM** module. The output of **PASL** will be sequences labeled as i (inner), M (membrane-spanning region), and o (outer region of the membrane). Each module is described in the following sections.

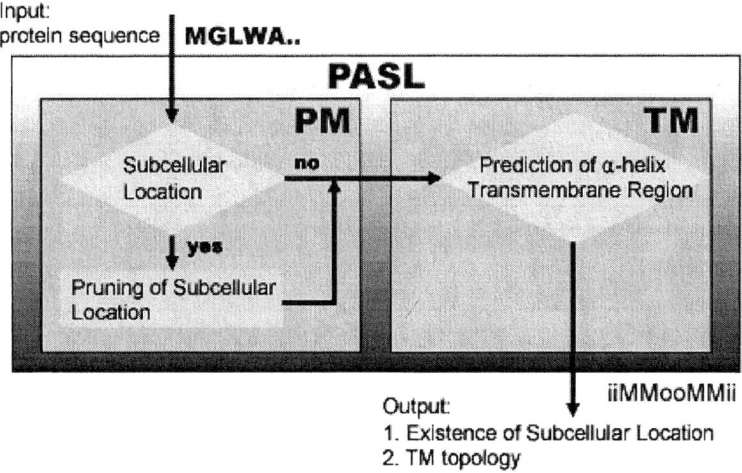

Fig. 1. Data Flow Model of PASL

To test the performance of **PASL**, 24 previously reported protein sequences with experimentally determined transmembrane topology were selected. Test data were selected from the Moller's TM data set [11] and SWISS-PROT [12]. The accuracy of each method is shown in Table 1. The accuracy of the TM topology was predicted as the same as or below the accuracy of the segments and the position. Therefore, the correct identification of the number of TM segments is a base of TM topology prediction. Confusing the signal peptide as a transmembrane region affects not only the number of TM segments, but also the TM topologies. **PASL** has an advantage in the detection of the transmembrane segment. For this reason, **PASL** performs better than any other prediction tool in the case of signal-peptide-contained queries.

Table 1. Performance test by ER-signal-peptide-sequence-contained set

Method	TM Segments and Position (%)	TM topology (%)
MEMSAT2	43.5	39.1
TMHMM 2.0	69.6	69.6
HMMTOP 2.0	47.8	47.8
PASL	82.6	82.6

4 Prediction with PASL System

The PASL prototype provides a web-based environment since it is implemented with JSP on a LINUX server. Fig. 2 shows the entering window of the PASL system. It is composed of five windows: the sequence input window, the prediction result window that shows the subcellular location and the transmembrane region, and three windows displaying the topology of the transmembrane region. Biologists or bioinformaticians can insert the protein sequence in FASTA or text format into the sequence input window. Once the "execute" button is pressed, the prediction algorithm runs and the result will be displayed in the result window, as shown in Fig. 3.

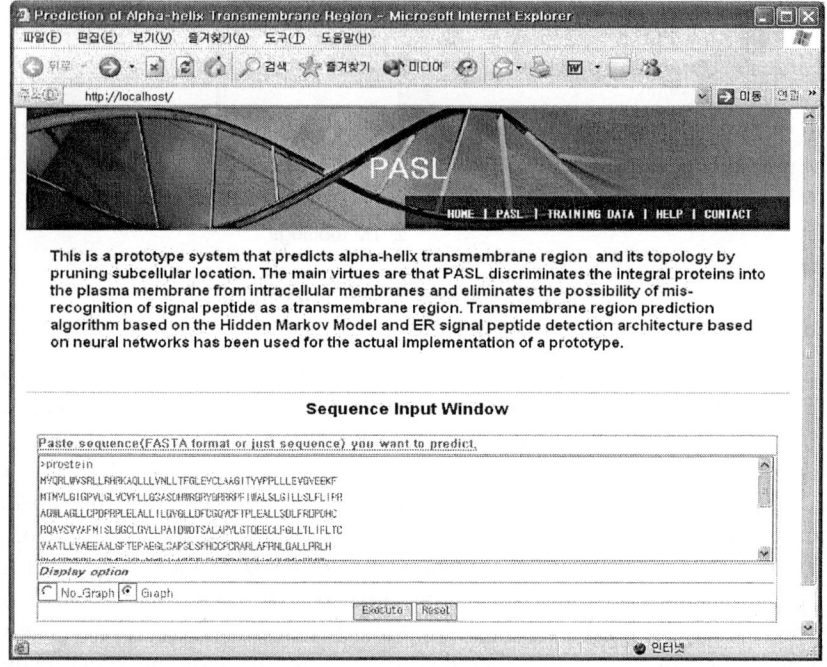

Fig. 2. Sequence Input Window of the PASL Prototype System

Fig. 3 shows that the input protein sequence "prostein" consists of 553 amino acids. The subcellular location module first found 33 amino acids composing a signal peptide. The original sequence with these 33 amino acids pruned is fed into the prediction of the alpha-helix region module in Fig. 1. The topology of the transmembrane region is another issue biologists are interested in. This can be displayed in a textual form as well as in a graphical form. An example of the graphical window is depicted in Fig. 4. This figure shows the posterior probability of the location of the example sequence. The 204th amino acid of the given sequence is highly probable to be in the transmembrane region. Also users can see the 250th amino acid is in inside of the membrane.

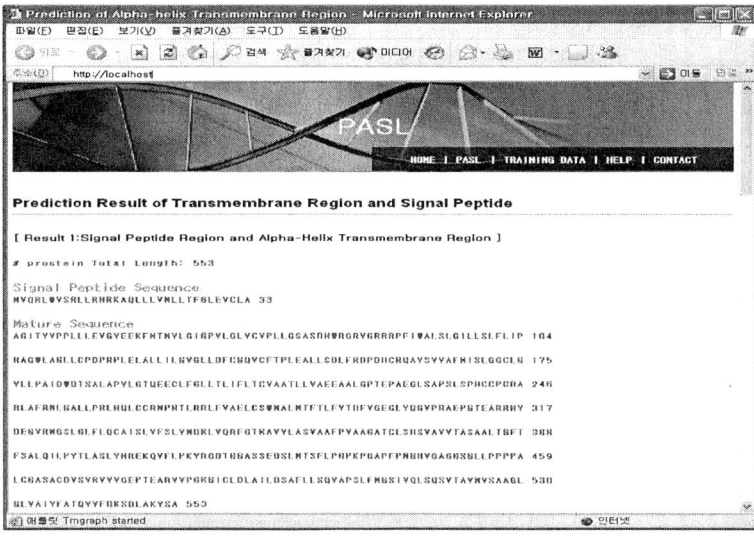

Fig. 3. Prediction Result of Signal Peptide and Transmembrane Protein Regions

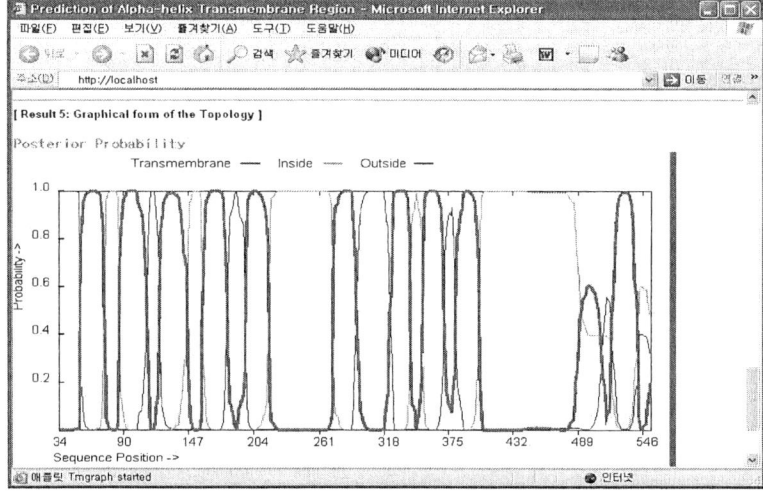

Fig. 4. A Window for the Graphical View of the Topology of the Predicted Protein

5 Conclusion and Future Work

We have presented the prediction model of PASL and how this system works. PASL integrated the location prediction module and the transmembrane region and topology prediction module (http://dblab.sejong.ac.kr:8080/pass/home.jsp). The main virtues of **PASL** are that it discriminates the integral proteins in the plasma membrane from intracellular membranes, and it eliminates the possibility of misrecognition of the signal peptide as a transmembrane region.

Many subcellular location prediction servers are now available. For fine resolution, however, such as mitochondrial inner space or mitochondrial membrane, some comprehensive remedial steps are required, e.g., the combined use of transmembrane prediction with location prediction.

References

1. Krogh, A., Larsson, B., von Heijne, G., Sonnhammer, E.L.: Predicting Transmembrane Protein Topology with a Hidden Markov Model: Application to Complete Genomes. J. Mol. Biol. Vol. 305 (2001) 567-580.
2. Tusnady, G.E., Simon, I.: The HMMTOP Transmembrane Topology Prediction Server. Bioinformatics. Vol. 17 (2001) 849-850.
3. Moller, S., Croning, M.D., Apweiler, R.: Evaluation of Methods for the Prediction of Membrane Spanning Regions. Bioinformatics. Vol. 17 (2001) 646-653
4. Kim, M.K., Park, H.S., Park, S.H.: Prediction of Plasma Membrane Spanning Region and Topology Using Hidden Markov Model and Neural Network, To appear in KES2004, 20-24, September (2004)
5. Lao, D.M., Arai, M., Ikeda, M., Shimizu, T.: The Presence of Signal Peptide Significantly Affects Transmembrane Topology Prediction. Bioinformatics. Vol. 18 (2002) 1562-1566.
6. Nielsen, H., Engelbrecht, J., Brunak, S., von Heijne, G.: Identification of Prokaryotic and Eukaryotic Signal Peptides and Prediction of their Cleavage Sites. Protein Eng. Vol. 10 (1997) 1-6.
7. Emanuelsson, O., Nielsen, H., Brunak, S., von Heijne, G.: Predicting Subcellular Localization of Proteins based on their N-terminal Amino Acid Sequence. J. Mol. Biol. Vol. 300 (2000) 1005-1016.
8. Reinhardt, A., Hubbard, T.: Using Neural Networks for Prediction of the Subcellular Location of Proteins. Nucleic Acids Res. Vol. 26 (1998) 2230-2236.
9. Park, K.J., Kanehisa, M.: Prediction of Protein Subcellular Locations by Support Vector Machines using Compositions of Amino Acids and Amino Acid Pairs. Bioinformatics. Vol. 19(2003) 1656-1663.
10. Nakashima, H., Nishikawa, K.: Discrimination of Intracellular and Extracellular Proteins using Amino Acid Composition and Residue-Pair Frequencies. J. Mol. Biol. Vol. 238 (1994) 54-61.
11. Moller, S., Kriventseva, E.V., Apweiler, R.: A Collection of Well-Characterised Integral Membrane Proteins. Bioinformatics. Vol. 16 (2000) 1159-1160.
12. Boeckmann, B., Bairoch, A., Apweiler, R., Blatter, M.C., Estreicher, A., Gasteiger, E., Martin, M.J., Michoud, K., O'Donovan, C., Phan, I., Pilbout, S., Schneider, M.: The SWISS-PROT Protein Knowledgebase and its Supplement TrEMBL in 2003. Nucleic Acids Res. Vol. 31 (2003) 365-370.

Information Processing in Cognitive Science

Sung-Kwan Je[1], Jae-Hyun Cho[2], and Kwang-Baek Kim[3]

[1] Dept. of Computer Science, Pusan National University,
[2] Dept. of Computer Information, Catholic University of Pusan,
[3] Dept. of Computer Engineering, Silla University, Korea
jimmy374@pusan.ac.kr

Abstract. At the retinal level, the strategies utilized by biological visual systems allow them to hypotheses machine vision systems that are difficult to apply to the real world, and they simply imitate a coarse form of the human visual system. Starting from research on the human visual system, we analyze a mechanism that processes input information when information is transferred from the retina to ganglion cells. In this study, a model for the characteristics of ganglion cells in the retina is proposed after considering the structure of the retina and the efficiency of storage space. The results of this study show that the proposed recognition model is the efficiency of storage space can be improved by constructing a mechanism that processes input information.

1 Introduction

Along commercially available machine vision sensors are beginning to approach the photoreceptor densities found in primate retinas, they are still outperformed by biological visual systems in terms of dynamic range, and strategies of information processing employed at the sensor level. Ongoing vision studies, however, imitate part of the function of human visual system and develop machine vision systems that can be supplied to machines. Recently, many developed countries are actively conducting research to maximize the performance of computer vision technology and to develop artificial vision through the modeling of human visual processing [3][4][13][14]. Artificial vision is to develop information processing procedures of the human visual system based on the biological characteristics. Compared with the machine vision technology, it can be effectively applied to industry. By storing images without processing, the efficiency of storage space is lowered. Inefficiency of storage space can influence the maintenance of the system. Therefore, it is necessary to have a mechanism that can process input information like the human visual system.

This study is to propose a model that can implement a mechanism of information processing that is actually occurring in the human visual system. We will review information processing procedures of the retina in Chapter 2, construct a model based on the mechanism in Chapter 3, and present the results of the experiment and conclusions in Chapter 4.

2 Retinal Cells

The retina is a highly structured network of neurons that extracts and pre-processes visual information from the image projected upon it by the optics of the eyes. A schematic diagram of the retinal cell structure is shown in Fig. 1. The biological retina is more than just a simple video camera. It not only converts optical information to electrical signals but performs considerable processing on the visual signal itself before transmitting it to higher levels. Various local adaptation mechanisms extend the retina's dynamic range by several orders of magnitude. In order to meet the transmission bottleneck at the optic nerve, the retina extracts only those features required at later stages of visual processing while employing many data reduction strategies such as information processing.

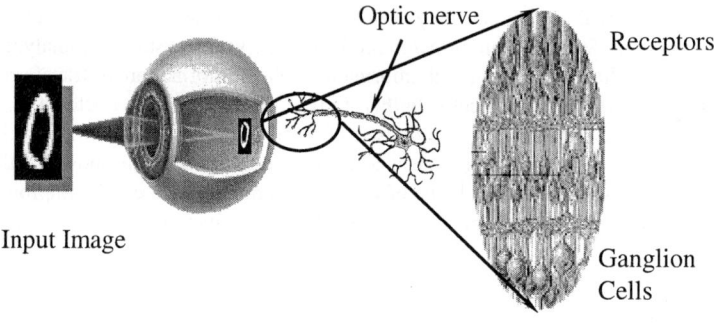

Fig. 1. Retinal Cells [13][14]

As for the human visual system, a light energy becomes a shadow with a reversed phase in the retina, and it transmits a displayed neural signal to ganglion cell and, as for this image, it is through a cognition process on delivery with primary visual cortex through optic nerve later by the retina. The retina located behind a pupil is complicatedly composed of ten vertical layers. However, since ganglion cell is composed with only about 1,000,000 in one eye, a lot of visual information is compressed for transferring from 125,000,000 receptors to ganglion cell. Actually it is turned into a neural signal by an operation of the retina, but the sampling image is transmitted to primary visual cortex of a brain through optic nerve of the retina which listened to this signal in 125,000,000 rod cells and 6,000,000 con cells.

Ganglion cells can be mapped into P-cells and M-cells. P-cells contain major information of images 'what', whereas M-cells contain edge information of images 'where'[4]. The information is decomposed into i) processes to recognize objects and ii) processes to recognize the location of objects. It depends on a mechanism that can minimize the loss of critical information when transmitting information and can maximally minimize the amount of total information. That is, information related to perceiving 'What' is transmitted to P-cells; and P-cells comprises 80% of total ganglion cells and minimize the loss during transmission. Whereas, information related to 'Where' is sent to M-cells; and M-cells comprises 20% of total ganglion cells [13][14]. Therefore, human vision is sensitive to 'What' information, but insensitive to 'Where' information. It values information about objects, but compresses information about background maximally. Even if a lot of information is compressed, human beings do not displaying a particular problem in recognizing images [5][6].

The inefficiency of storage space can influence greatly the maintenance of systems. Therefore, a mechanism like the human visual system is needed to process input information. Using the characteristics of the human visual system, this study implement a mechanism that processes input information when transmitting information from the retina to ganglion cells, and compares it to existing visual models and their performance.

3 Proposed Algorithm

Most of the current computer vision theories are based on hypotheses that are difficult to apply to the real world, and they simply imitate a coarse form of the human visual system. As a result, they have not been showing satisfying results. In the human visual system, there is a mechanism that processes information due to memory degradation with time and limited storage space. Starting from research on the human visual system, this study analyzes a mechanism that processes input information when information is transferred from the retina to ganglion cells. In this paper, the response of the computer retina model to various simple visual stimuli is presented and compared to responses in biological retinas in order to validate the model. In addition, experiments with more complex images are used to illustrate what the outputs of the primate retina might look like if such bulk recordings were possible. Currently study of these models is attempted actively each nation in the world [1]-[4]. Computer vision is emulating a rough form of human visual information processing, and the information processing is showing them the form that is different from a human information processing [13][14]. Due to this problem, machine visual systems cannot adapt to the changes of the environment, and have a lot of problems in real world application. In this paper, a model is proposed to implement a mechanism that processes visual information, occurring in the recognition process of the real human visual system. A flowchart of the general algorithm is shown in Fig. 2. In the retina, information is transmitted by minimizing the loss of important information in ganglion cells and by executing a mechanism that minimizes total amount of information maximally.

A mapping process of parvo-cellular of ganglion cell (P-cell) and the mechanism are very alike in receptor of the retina. Also, information of high-band of wavelet transform has edge information of original image compared to low-band which has a little information. These mechanisms are very similar to a mapping process of magno-cellular of ganglion cell (M-cell). Therefore, P-cell of ganglion cell deals with main information of an image like low-band, and M-cell is dealing with edge information of an image like the high-band. In this paper, we used wavelet transform in a compression process of this visual information and composed the model. We implemented the image recognition model are shown in Fig. 2. To process human visual information, letter recognition is a necessary problem to solve.

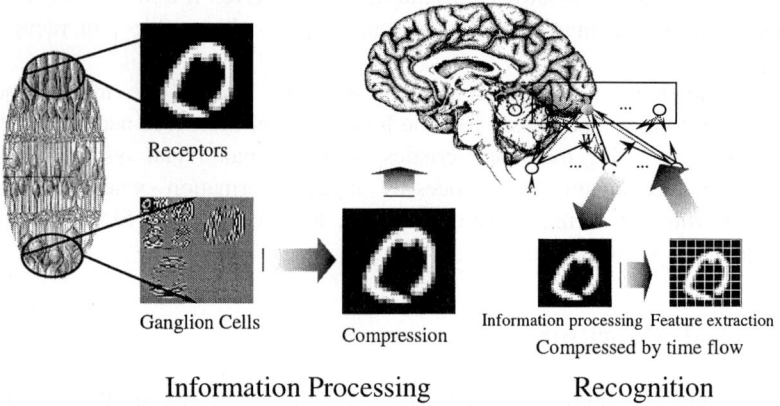

Fig. 2. Proposed Information Processing Model

Currently many algorithms have already been developed. Each algorithm has its own characteristic application areas. Generally, in case of classifying given data into clusters that have similar characteristics, the most commonly used methods are: Adaptive Resonance Theory (ART), Self-Organizing feature Map (SOM), and Fuzzy-ART. Among these methods, the algorithms of Grossberg's ART2 and Kohonen's SOM are chosen for this study. They are unsupervised learning algorithms that they learn on their own even if correct answers are not given to input patterns, which is similar to the biological recognition system. In this study, therefore, self-learning model is used as a recognizer and evaluates the performance of a model that implements a mechanism occurring in the retina.

4 Experimental Results

In this study, a proposed image model, which is based on the human visual information process, is implemented by using Visual C++ 6.0 in the environment of Pentium 1.7GHz, 256MB memory, and Window XP. We used MNIST database of the AT&T Corp. which it was used in a lot of paper on handwritten off-line number at this paper and tested. It is MNIST database newly compounding database of original NIST (National Institute of Standards and Technology). This is database using handwritten digit 0-9 in learning and recognition. NIST database is the binary image which was normalized with 20×20 sizes while keeping horizontal vertical ratio. The training data set selected 5,000 in 60,000. And, as for the test data set, a random sampling selected 5,000 in 10,000. And the person who wrote a number is not same. A 28×28 digit image was compressed with Daubechies (9,7) filter that is often used for the loss compression. For an experiment on the recognition process, the most commonly used neural network algorithms, ART2 and SOM, were used to suggest a recognition model.

The memory of human beings fades away with time. Human recognition ability also decreases over time, and the accuracy of recognition becomes low. In addition, due to the limited capacity of the brain, a lot of images cannot be remembered, so human beings compress and recollect past memory. In existing machine vision models, for mistaken recognition and data management, input data were stored in DB. If we store all the data to DB and backup them, it will cost a lot of money. Therefore the size of images should be considered for maintenance. The size of input data is 28×28, and a necessary storage space to recognize 5,000 input data is 3,920,000 bytes. If this data are processed over time, a necessary storage space shown in Fig. 3 will be increased exponentially.

In the experiment of this study, different compression rates were used over time by taking into consideration of human reflection functions. As recognition rates gradually decreases with compression rates, whereas the efficiency of storage space gets better, as shown in Fig. 3. Therefore, in a recognition system considering human vision, the use of storage space is efficient as input data are processed according to the time flow.

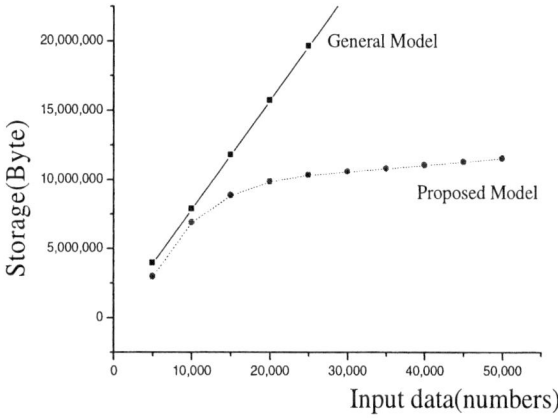

Fig. 3. Changes of storage space for proposed models

5 Conclusion

This study implemented mechanism information processing from the retina to ganglion cells, based on the human visual information processing. Many of the properties of the human visual system are ultimately limited by the fidelity of image sampling within the retina and the strategies used to process the sampled information before its transmission to the cortex. The same may be said of machine vision systems. They are ultimately limited by properties of the sensor. Even at the retinal level, the performance of the human visual system far exceeds the capabilities of commercially available visual sensors and machine vision systems, and thereby motivates the study of human or primate retinas. Decreased efficiency of storage space for the general model over time must be considered. This study proposed an image recognition model considering the characteristics of the human visual system, not a strategic model applying an existing recognition model. As human reflection capability decreases over time, human beings need mechanism information processing to overcome the limited use of storage space. Therefore, after implementing this mechanism and comparing it with a general model, the two were found to be not different in terms of recognition rate, but the efficiency of storage space for the mechanism was found to be excellent.

Starting from research on the human visual system, this study analyzes a mechanism that processes input information when information is transferred from the retina to ganglion cells. If the modeling from ganglion cells to the primary visual cortex as well as the modeling from primary visual cortex to cognition process is completed in the future, real recognition performance will be further improved. The proposed model can be applied to the tracking of image recognition.

References

[1] Firsxhler, M. F., Firschein, O.: Intelligence: The eye, the brain and the computer. Addison-Wesley. (1987)
[2] Arnheim, R.: Visual thinking. University of California Press. (1997)
[3] Jae-Hyun, C.: Research on data construction and recognition of artificial neural networks by fractal coefficients. Dissertation at the Pusan National University. (1998)
[4] Brain Science Research Center.: Research on Artificial Audiovisual System based on the Brain Information Processing. Research Paper by the Korea Advanced Institute of Science and Technology. Department of Science Technology. (2001)
[5] In-Sik, L.: Human Beings and Computer. Gachi Press. (1992)
[6] Dae-Soo, K.: Neural Networks: Theory and Application (1). Hi-Tech Infor-mation. (1992)
[7] Shapiro, J. M.: Embedded Image coding using zerotrees of wavelet coefficients. IEEE Trans. on Signal Processing. **41(12)** (Dec. 1993) 3445-3462
[8] Haykin, S.: Neural Networks: A Comprehensive Foundation. MacMillan. (1994)
[9] Gonzalez, R. C, Woods, R. E.: Digital image processing. Second edition. Prentice Hall. (2001)
[10] Dobelle, W. H.: Artificial Vision for the Blind by Connecting a Television Camera to the Visual Cortex. ASAIO journal. (2000) 3-9
[11] Mallat, S. G.: A Wavelet tour of Signal Processing. Academic Press. (1998)
[12] Kohonen, T.: Self-Organizing Maps. Berlin: Springer-Verlag. First edition was (1995). second edition (1997)
[13] Shah, S., Levine, M. D.: Information Processing in Primate Retinal Cone Pathways: A Model. TR-CIM-93-18. Centre for Intelligent Machines. McGill University. Montreal. (Dec. 1993)
[14] Shah, S., Levine, M. D.: Information Processing in Primate Retina: Experiments and results. TR-CIM-93-19. Centre for Intelligent Machines. McGill University. Montreal. (Dec. 1993)

Reconstruction of Human Anatomical Models from Segmented Contour Lines

Byeong-Seok Shin

Inha University, Department of Computer Science and Engineering,
253 Yonghyeon-Dong, Nam-Gu, Inchon, 402-751, Korea
bsshin@inha.ac.kr

Abstract. In order to make 3D anatomical models, a method that reconstructs geometric models from contour lines on 2D slices is commonly used. We can obtain a set of contours by acquiring CT or MR images and segmenting them. Previous methods divide entire contour line into simple matching regions and clefts. Since long processing time is required for clefts, performance might be degraded when manipulating complex structures. We propose a fast reconstruction method, which generates a triangle strip with single tiling operation for simple region that does not contain branches. If there are branches in contour lines, it partitions the contour lines into several sub-contours by considering the number of vertices and their spatial distribution. We implemented a human anatomical model reconstruction system by applying our method.

1 Introduction

Tomographic images produced by CT and MR are generally used for identifying pathological structures in human body. Although experienced doctors can infer the 3D shape of target structures using only the 2D images based on their prior knowledge, it is very hard for ordinary person and students to understand 3D anatomical structures intuitively. Therefore creating 3D anatomical models is useful for accurate diagnosis and education since we can recognize the size, position and orientation of pathologies.

Surface reconstruction is to restore original geometry by extracting contours of a specific organ from 2D slice images and connecting the geometric primitives. We can reconstruct detailed anatomical models for complicated and elastic structures. Reconstructed models can be used for animation, simulation and deformation as well as image display. Reconstruction process is composed of three components: correspondence determination, tiling, and branch processing [1]. Correspondence determination is to identify the contour lines that should be connected to a specific contour. Tiling is to produce triangle strips from a pair of contour lines. Branch processing is to determine the correspondence when a contour line corresponds to multiple contours.

When a contour is sufficiently close to its adjacent contour(s) in z-direction, it is easy to determine the exact correspondence using only the consecutive contours. Otherwise, we have to exploit prior knowledge or global information about overall shape of target objects [2]. Commonly used surface reconstruction method proposed by

Barequet et al. divides a pair of corresponding contour lines into simple matching region and clefts [3, 4]. It reconstructs triangle meshes by applying their tiling method to a simple region, and triangulates clefts with dynamic programming method [5]. Although it has an advantage of generating considerably accurate models, it is not efficient while manipulating contour lines containing a lot of clefts. Bajaj et al. defined three constrains for surface to be reconstructed and derived exact correspondence and tiling rules from those constraints [1]. Since it uses multi-pass algorithm, complicated regions are reconstructed after all the other regions are processed. Cline et al. proposed a method to generate accurate surface model using 2D distance transformation and medial axis [6]. Meyers et al. simplified multiple-branch problem into one-to-one correspondence by defining composition of individual contour lines [7]. Bajaj and Barequet presented methods for tiling branches that have complex joint topology and shape.

We propose fast reconstruction method of 3D anatomical models as an extension of Barequet's method. It generates a triangle strip with single tiling operation for simple matching region that does not contain branches. If there are branches, it partitions the contour line into several sub-contours by considering the number of vertices and their spatial distribution. Each sub-contour can be processed by using the same method applied to matching regions.

In section 2, we explain our method in detail. In section 3, we show some experimental results. Lastly, we summarize and conclude our work.

2 Anatomical Model Reconstruction Method

We propose a high-speed terrain model reconstruction algorithm that does not directly handle clefts. Each slice contains several contour lines for organs. Let a pair of consecutive slices be $<S_n, S_{n+1}>$ and sets of contours belong to the slices be C_i^n and C_i^{n+1}. A contour line is composed of a set of vertices $\{v(x,y,n)\}$.

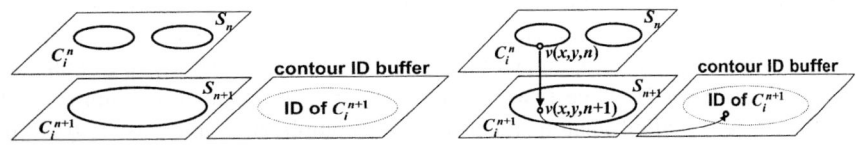

Fig. 1. Correspondence determination between two adjacent contours (left) fills up interior of C_i^{n+1} as its ID's (right) projects vertices of C_i^n onto S_{n+1} and checks the ID of those points

When a projected vertex of a contour belongs to another contour on its adjacent slice, two contour lines are regarded as corresponding to each other. It allocates a 2D-buffer that has the same size of a slice and fills up interior pixels of a C_i^{n+1} as its identifier with boundary-fill algorithm. Then it projects vertices $v(x,y,n)$ of C_i^n onto a slice S_{n+1}, and checks whether the buffer value is identical to the contour ID or not at the projected position $v(x,y,n+1)$ (see figure 1). If the number of corresponding contour is more than two, the contour lines may produce branches.

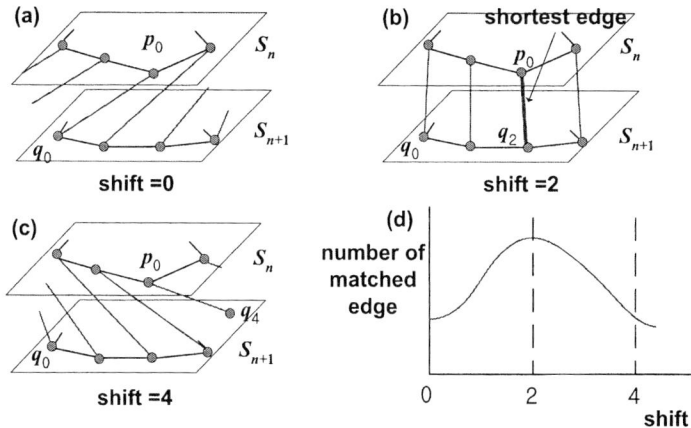

Fig. 2. An example of determining optimal shift value and the shortest edge (a) shift=0 (b) shift=2 (c) shift=4 (d) change the number of matched edge according to the shift value

After determining correspondence of adjacent contours C_i^n and C_i^{n+1}, it produces triangle strips connecting them. Since there is no topological information between two vertices on different contours, we have to determine the corresponding vertex on the adjacent contour for a specific vertex. Normally, topologically adjacent vertices have the smaller distance in comparison to the remaining vertices.

Assume that the vertex lists of $<C_i^n, C_i^{n+1}>$ are $(p_0,\ldots,p_{l-1})$ and $(q_0,\ldots,q_{l-1})$. When the length of a slice chord $e(p_i, q_i)$ is less than pre-defined threshold ε, the slice chord is called a *matched edge*. It changes the index of starting vertex j from 0 to l-1 for C_i^{n+1} while fixing the index of starting vertex as 0 for C_i^n. It determines the value of j that maximizes the number of matched edges among slice chords $e(p_k, q_{(j+k) \bmod l})$ defined by two vertex list $(p_0,\ldots,p_{l-1})$ and $(q_j,\ldots,q_{l-1},q_0,\ldots, q_{j-1})$. Here, j is the relative offset of two lists and called a *shift*. When applying the shift value that maximizes the number of matched edges, a slice chord that has the minimum length is regarded as the shortest edge. Figure 2 shows an example of determining the optimal shift value. While increasing the index j from 0 to l-1, in the case of figure 2(a) and (c), length of each edge $e(p_k, q_{(j+k) \bmod l})$ becomes longer and the number of matched edge is decreased. When the shift value is 2, the number of matched edges is maximized and $e(p_0, q_2)$ is the shortest edge. A triangle strip is generated from the shortest edge.

In order to produce visually pleased triangle strips, each triangle should be similar to equilateral triangle. When we assume that a contour is composed of equally distant vertices, we have only to minimize the length of slice chord while inserting a new vertex into a triangle strip. Let the most recently produced edge of the strip be $e(p_i, q_i)$, a vertex to be added is one of the candidate vertex p_{i+1} and q_{j+1}. When $|e(p_i, q_{i+1})| > |e(p_{i+1}, q_i)|$, p_{i+1} becomes the next vertex, otherwise q_{i+1} is selected.

A branch occurs when a contour C_i^{n+1} corresponds to N contours $\{C_i^n,\ldots,C_{i+N-1}^n\}$. Our method partitions C_i^{n+1} into N sub-contours $\{\overline{C}_i^{n+1},\ldots,\overline{C}_{i+N-1}^{n+1}\}$, determines which of $\{C_i^n,\ldots,C_{i+N-1}^n\}$ corresponds to each sub-contour, and tiles the contour pairs using the method described above. When two contours C_0^n and C_1^n on a slice S_n correspond to a contour C_0^{n+1} on adjacent slices S_{n+1}, surface reconstruction process is as follows:

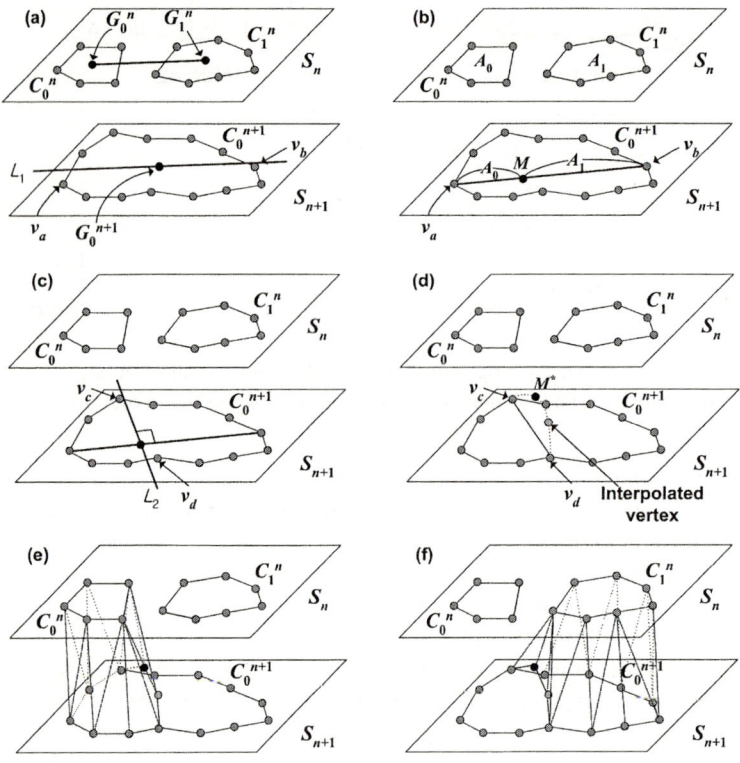

Fig. 3. An example of tiling a contour pair that contains branches

(1) Derives the center-of-gravity G_0^n, G_1^n, and G_0^{n+1} from C_0^n, C_1^n, and C_0^{n+1}.
(2) Computes a straight line L_1 parallel to the line from G_0^n to , G_1^n, and its projected line passes by the G_0^{n+1}, and determines two vertices v_a and v_b from the vertices of G_0^{n+1} that are mostly close to L_1 (figure 3 (a)).
(3) Partitions the line segment $\overline{v_a v_b}$ according to ratio of the area of C_0^n and C_1^n (A_0, A_1) and denotes the point as M. Area of contours can be computed by counting the number of pixels while calculating the center-of-gravity (figure 3 (b)).
(4) Computes a straight line L_2 perpendicular to the line segment $\overline{v_a v_b}$ and passes by the point M, and determines two vertices v_c and v_d from the vertices of G_0^{n+1} that are mostly close to L_2 (figure 3 (c)).
(5) Inserts shared vertices along with the line segment $\overline{v_c v_d}$. Interval of shared vertices is identical to that of ordinal vertices of original contour lines. Let the coordinates of points v_c, v_d, and M be (x_c, y_c, n), (x_d, y_d, n) and (x_M, y_M, n). Z-values of shared vertices should be interpolated along with a curve from v_c to v_d via new points M^* of which the coordinates is $(x_M, y_M, n+0.5)$ (figure 3 (d)).

(6) Partitions a contour line C_0^{n+1} into two sub-contours $\overline{C}_0^{n+1}$ and $\overline{C}_0^{n+1\,*}$ by duplicating the shared vertices, and reconstructs contour pairs $<C_0^n, \overline{C}_0^{n+1}>$ and $<C_1^n, \overline{C}_0^{n+1\,*}>$ using the tiling method (figure 3 (e) (f)).

In the case of $N \geq 3$, it partitions entire contour into several sub-contours and tiles the contour pairs by recursively applying the above method until all of the sub-contours have one-to-one correspondence with adjacent contour lines.

3 Experimental Results

In order to show how efficient our method, we implement 3D anatomical model reconstruction system which segments contour lines of organs from 2D slices, and produces

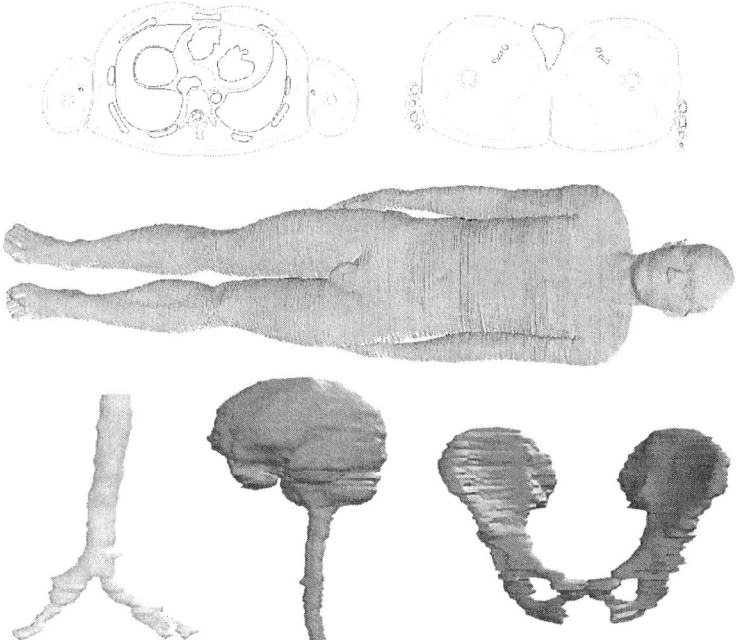

Fig. 4. Examples of reconstructed models and their shaded images with our method (top) segmented images for inner organs (middle) skin model (bottom) trachea, meninges, and pelvis models from left to right

Table 1. Time to be taken for reconstructing several anatomical structures using our method

	# of contours	# of polygons	Reconstruction time (sec)
Liver	50	11642	2.21
Heart	37	6770	1.82
Pelvis	161	21294	4.52

three dimensional models by tiling adjacent contours. It is implemented on a PC equipped with Pentium IV 2.2GHz CPU, 1GB main memory, and NVIDIA GeForce4 graphics accelerator. 613 horizontal MRIs of the entire body for young Korean male with standard body were scanned and 60 anatomical structures were segmented.

Figure 4 shows images produced by rendering 3D anatomical models with our method. It can reconstruct accurate models from contour lines even when we dealing with considerably complex region. It produces geometric model fairly well not only for simple matching area but also for branches.

Table 1 shows time to be taken for reconstruction of geometry according to the size of data and complexity. Experimental result shows that our method can reconstruct arbitrary shaped models using contour lines on slices within short time.

4 Conclusion

Human body modeling is an important technology for medical applications. We propose fast reconstruction method of 3D anatomical model from contour lines. It generates a triangle strip with single tiling operation for simple matching region that does not contain branch structures. If there are some branches, it partitions the contour line into several sub-contours by considering the number of vertices and their spatial distribution. Each sub-contour can be processed by using the same method applied to matching region. Experimental result shows that our method can reconstruct arbitrary shaped models using contour lines on 2D slices within short time.

Acknowledgement

This research was supported by University IT Research Center Project.

References

1. Bajaj, C., Coyle E. and Lin, K. : Arbitrary Topology Shape Reconstruction for Planar Cross Sections. Graphical Models and Image Processing, Vol. 58, No. 6 (1996) 524-543
2. Soroka, B. : Generalized Cones from Serial Sections. Computer Graphics and Image Processing, Vol. 15, No. 2 (1981) 154-166
3. Barequet, G. and Sharir, M. : Piecewise-Linear Interpolation between Polygonal Slices. Computer Vision and Image Understanding, Vol. 63, No. 2 (1996) 251-272
4. Barequet, G., Shapiro, D. and Tal, A. : Multilevel Sensitive Reconstruction of Polyhedral Surfaces from Parallel Slices. The Visual Computer, Vol. 16, No. 2 (2000) 116-133
5. Klincsek, G. : Minimal Triangulations of Polygonal Domains. Annals of Discrete Mathematics, Vol. 9 (1980) 121-123
6. Klein, R., Schilling, A. and Strasser, W. : Reconstruction and Simplification of Surface from Contours. IEEE Computer Graphics and Applications, (1999) 198-207
7. Meyers, D., Skinner, S. and Sloan, K. : Surfaces from Contours. ACM Transactions on Graphics, Vol. 11, No. 3 (1992) 228-258

Efficient Perspective Volume Visualization Method Using Progressive Depth Refinement

Byeong-Seok Shin

Inha University, Department of Computer Science and Engineering,
253 Yonghyeon-Dong, Nam-Gu, Inchon, 402-751, Korea
bsshin@inha.ac.kr

Abstract. In recent volume visualization, high-speed perspective rendering method is essential. Progressive refinement is well-known solution for real-time image generation under limited computing environment. We propose an accelerated rendering method that exploits *progressive depth refinement*. While performing progressive refinement of volume ray casting, it computes depth values for sub-sampled pixels and determines the minimum value of four neighboring pixels. Original ray casting starts ray traversal from its corresponding pixel. However, rays in our method jump as the amount of the minimum z-value calculated in the previous stage. Experimental results show that our method reduces rendering time in comparison to conventional volume ray casting.

1 Introduction

The most important issue in recent volume visualization is to produce perspective-projection images in real time. It is highly related to hardware-based approaches [1],[2]. Although they achieve 30 *fps* without preprocessing, it is too expensive and difficult to manipulate large volume data due to limitation of dedicated memory size.

Volume ray casting is the most famous software rendering algorithm [3]. Although it produces high-quality images, it takes long time due to unnecessary sampling of empty regions. Several optimized methods have been proposed for speed-up [4],[5],[6]. They have mainly concentrated on skipping over transparent regions using coherent data structures [7],[8],[9],[10]. However they require long preprocessing time and extra storage. There are some extended algorithms that require less computation for preprocessing such as min-max octree [11] and min-max block [12]. They are still insufficient for real-time classification or rendering of time-varying dataset.

Progressive refinement is regarded as simple but effective solution for real-time image generation under limited computing resources [13],[14]. Since image quality has trade-off for rendering time, it negotiates image quality against update speed. When an object or a camera rapidly changes its position or direction, it generates low-quality images by sub-sampling pixels with regular interval and casts rays only for those pixels. Colors for remaining pixels are computed by simple linear interpolation. On the contrary, the images are progressively refined for several stages when the object or camera stops moving for a moment.

In this paper, we present an acceleration method that produces images faster than conventional progressive refinement of volume ray casting. While casting rays in some pixels, it calculates depth values (z-values) for those pixels without extra cost and determines the minimum depth for rectangular regions (grids) defined by four neighboring samples. The minimum values computed in current stage are used to accelerate ray casting in the next stage of refinement. Our method moves forward the starting point of ray traversal as the amount of the minimum depth value. It doesn't need preprocessing or additional data structures. Experimental results show that our method is much faster than conventional progressive refinement of ray casting.

In Section 2, we present our progressive refinement algorithm in detail. Experimental results and remarks are shown in Section 3. Lastly, we conclude our work.

2 Volume Ray Casting Using Progressive Depth Refinement

Assume that a view plane P has the resolution of $N \times N$ for simplicity. While performing progressive refinement, we compute z-values d_{ij} for sub-sampled pixels q_{ij}, and estimate *potentially empty space* (*PES*) by comparing depth values of four neighboring samples. A PES is defined as the area between image plane ($z=0$) and a plane apart from view plane as the amount of minimum depth value ($z=d^{min}$).

Fig. 1 shows our refinement procedure not only for colors but also for depth values, where G_{uv} is a grid and d_{uv}^{min} means the minimum depth value of G_{uv}. Assume that initial sampling interval of progressive refinement is S ($S=2^i$ for simplicity). In the first stage (Fig. 1 (top)), it performs full range ray traversal to determine color values at four corner pixels of $2S \times 2S$ sized grid to derive initial value of d_{uv}^{min}. Then it casts rays at the corner pixels (depicted as gray squares) of $S \times S$ sized grid from the points apart from view plane as the amount of d_{uv}^{min}. Color values for remaining $N^2 - (N/S)^2$ pixels are interpolated from the pre-computed pixel values. At the same time, it stores their z-values into the depth buffer. A square stands for a pixel whose color and z-value are already calculated in the previous stage. A circle means an interpolated pixel. A small black circle is a starting point of ray traversal.

In the following stages, the minimum depth values calculated in the previous stages are used for speedup. While a ray fires from its corresponding pixel in conventional ray casting, all the rays in our method fired from refined pixels in a grid G_{uv} start to traverse at the points apart from their corresponding pixels as the value of d_{uv}^{min} stored in the depth buffer (Fig. 1 (bottom)). Consequently, we can reduce the rendering time as the amount of PES (gray regions). Additional samples for k-th stage ($k=0,\ldots,i$) is $3(kN/S)^2$. While computing color values for additional samples, depth values are also computed to refine the depth map more precisely.

Let the average time to compute color and z-value for each pixel be t_{rt}, bilinear interpolation time be t_{int}, and the cost for traversing a ray for our method be t_{minrt}. Rendering time of original ray casting t_{old} and our method t_{new} for each stage can be defined as Eq. (1). Since t_{minrt} is less than or equal to t_{rt} in any case, t_{new} is always smaller than t_{old} in all the stages of refinement.

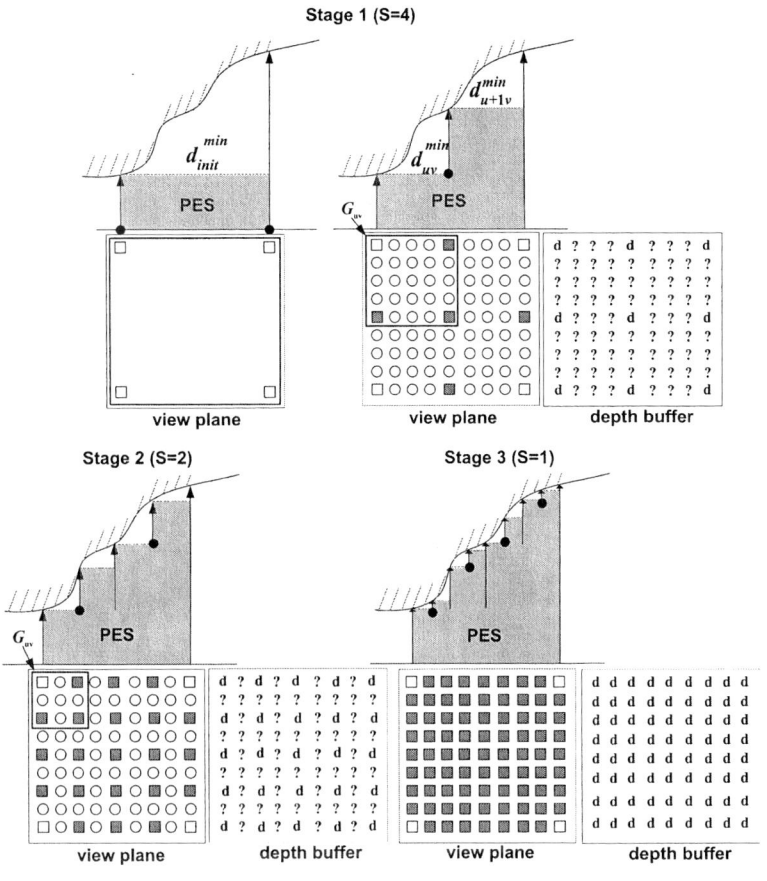

Fig. 1. A procedure of proposed refinement method

$$t_{old} = \begin{cases} \dfrac{1}{S^2} N^2 t_{rt} + \left(1 - \dfrac{1}{S^2}\right) N^2 t_{int} & (k=0) \\[2mm] \dfrac{(2^{2k} - 2^{2(k-1)})}{S^2} N^2 t_{rt} + \left(1 - \dfrac{2^{2k}}{S^2}\right) N^2 t_{int} & (1 \le k \le i-1) \\[2mm] \dfrac{(2^{2k} - 2^{2(k-1)})}{S^2} N^2 t_{rt} & (k=i) \end{cases}$$

$$t_{new} = \begin{cases} \dfrac{1}{(2S)^2} N^2 t_{rt} + \dfrac{3}{(2S)^2} N^2 t_{\min rt} + \left(1 - \dfrac{1}{S^2}\right) N^2 t_{int} & (k=0) \\[2mm] \dfrac{(2^{2k} - 2^{2(k-1)})}{S^2} N^2 t_{\min rt} + \left(1 - \dfrac{2^{2k}}{S^2}\right) N^2 t_{int} & (1 \le k \le i-1) \\[2mm] \dfrac{(2^{2k} - 2^{2(k-1)})}{S^2} N^2 t_{\min rt} & (k=i) \end{cases} \quad (1)$$

The most important factor is the sampling interval S. When the value of S is too large, image quality is degraded since spatial frequency of the volume data is much higher than the sampling rate. It might incorrectly estimate minimum depth values and a lot of visual artifacts may occur in final images (see Fig. 2). When S is too small, sufficient speed-up cannot be achieved. We have to choose optimal value of S to achieve maximum performance without deteriorating of image quality.

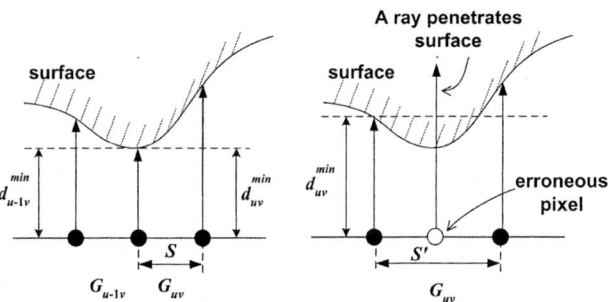

Fig. 2. An example of generating erroneous images when the sampling interval is too large in comparison to spatial frequency of volume data

3 Experimental Results

Virtual endoscopy is a non-invasive diagnosis method based on computer processing of 3D data sets in order to visualize inner structures of the organ cavity [15]. It is good example to verify the performance enhancement since it requires real-time generation of high-quality perspective images and interactive classification.

We compare the rendering time and image quality of original progressive refinement of volume ray casting and our method. In order to show that our method is still efficient even when it is combined with other optimization method, we take into account the min-max block method. Although it requires preprocessing to build up min-max block, it is inherently simple and fast since it does not require complicated operations. All of these methods are implemented on a PC equipped with Pentium IV 3.06GHz CPU, 2GB memory, and Radeon9800 GPU. Volume dataset is obtained by scanning a human abdomen with a MDCT of which resolution is 512×512×541.

We measure the rendering time in colon cavity under fixed viewing conditions. Fig. 3 shows comparison of rendering time of original ray casting (RC), our method (PDR), acceleration using min-max block (MINMAX), and combination of our method and min-max block method (PDR+MINMAX) in several refinement stages.

As the value of S decreases, rendering time gets longer since RC fires rays at more pixels. PDR produces the same quality images in much less time than RC. Rendering time of our method is only about 35% (S=1) ~ 55% (S=4) of that of RC. Rendering time of MINMAX is almost the same or longer than that of PDR. Rendering time of PDR+MINMAX is about 25% (S=1) ~ 36% (S=4) of that of RC. High-speed flythough can be possible since it provides 5.0 ~ 6.2 *fps* when the value of S is 4.

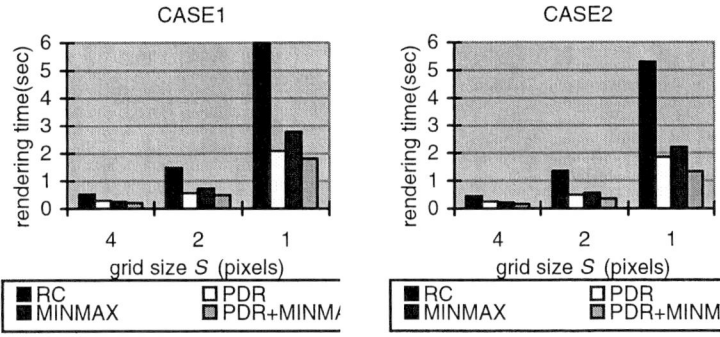

Fig. 3. Rendering time of RC, PDR, MINMAX, and PDR+MINMAX for CASE 1 (left) and CASE 2 (right) in the same condition. Image size is 512x512

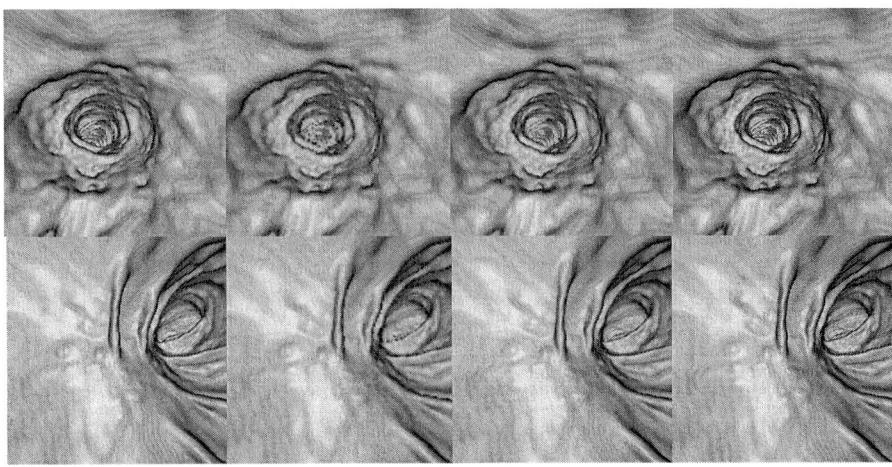

Fig. 4. A comparison of image quality of virtual colonoscopy in different regions: Leftmost images are produced by RC and remaining ones are obtained by our method in CASE1 (top row) and CASE2 (bottom row) region. Intervals between two consecutive sub-sampled pixels decrease as 4 (2nd column), 2 (3rd column) and 1 (rightmost column)

In order to examine the influence of local surface geometry on rendering speed, we measure the rendering time in a place where surface's tangent is almost parallel to viewing direction (CASE1) and another place where surface is perpendicular to viewing direction (CASE2). When we apply our method to those regions, rendering time in CASE2 is 15% ~ 20% shorter than that in CASE1. This implies that local surface geometry is related to rendering speed.

Fig. 4 shows the quality of images produced by RC and PDR as an image is progressively refined under fixed viewing condition. It is very hard to recognize the difference between images from the two methods.

4 Conclusion

The most important issue in volume visualization is to produce high quality images in real time. We propose an efficient progressive refinement of ray casting that reduces the rendering time in comparison to the conventional algorithms in any situation without loss of image quality. Using depth-refinement scheme, our method moves forward the starting point of ray traversal as the amount of the minimum depth value calculated in the previous refinement stage. It can be applied to generate endoscopic image for any kind of tubular-shaped organs and flight-through image sequence of large-scale terrain. Experimental result shows that it normally produces high-quality images as in ray casting and takes less time for rendering.

References

1. Meissner, M., Hoffmann, U., Strasser, W. : Enabling Classification and Shading for 3D Texture Mapping based Volume Rendering using OpenGL and Extensions. Proceedings of IEEE Visualization '99. San Francisco, CA (1999) 207–214
2. Pfister, H., Hardenbergh, J., Knittel, J., Lauer, H., Seiler, L. : The VolumePro Real-Time Ray-Casting System. SIGGRAPH 99 Proceedings. Los Angeles, CA (1999) 251–260
3. Levoy, M. : Display of Surfaces from Volume Data. IEEE Computer Graphics and Applications, Vol. 8, No. 3 (1988) 29-37
4. Lacroute, P., Levoy, M. : Fast Volume Rendering Using a Shear-Warp Factorization of the Viewing Transformation. SIGGRAPH 94 Proceedings. Orlando, Florida (1994) 451-458
5. Yagel, R., Shi, Z. : Accelerating Volume Animation by Space-Leaping. Proceedings of IEEE Visualization 93 (1993) 62-69
6. Udupa, J. K., Odhner, D. : Shell Rendering. IEEE Computer Graphics and Applications, Vol. 13, No. 6 (1993) 58–67
7. Levoy, M. : Efficient Ray Tracing of Volume Data. ACM Transactions on Graphics, Vol. 9 (1990) 245-261
8. Avila, R., Sobierajski, L., Kaufman, A. : Towards a Comprehensive Volume Visualization System. Proceedings of IEEE Visualization 92 (1992) 13-20
9. Cohen, D., Sheffer, Z. : Proximity Clouds. An Acceleration Technique for 3D Grid Traversal. The Visual Computer, Vol. 11, No. 1 (1994) 27-28
10. Devillers, O. : The Macro-Regions: An Efficient Space Subdivision Structure for Ray-Tracing, Proceedings of Eurographics 89 (1989) 27-38
11. Lacroute, P. : Fast Volume Rendering Using a Shear-Warp Factorization of the Viewing Transformation. Doctoral Dissertation, CSL-TR-95-678. Stanford University (1995)
12. Kim, T., Shin, Y. : Fast Volume Rendering with Interactive Classification. Computers & Graphics, Vol. 25 (2001) 819-831
13. Gotsman, C., Reisman, A., Schuster, A. : Parallel Progressive Rendering of Animation Sequences at Interactive Rates on Distributed-Memory Machines. Journal of Parallel and Distributed Computing, Vol. 60 (2000) 1074-1102
14. Roerdink, J. : Multiresolution Maximum Intensity Volume Rendering by Morphological Adjunction Pyramids. IEEE Trans. on Image Processing, Vol.12, No. 6 (2003) 653-660
15. Hong, L., Muraki, S., Kaufman, A., Bartz, D., He, T. : Virtual Voyage: Interactive Navigation in the Human Colon. SIGGRAPH 97 Proceedings (1997) 27-34

Proteomic Pattern Classification Using Bio-markers for Prostate Cancer Diagnosis

Jung-Ja Kim[1], Young-Ho Kim[2], and Yonggwan Won[2]

[1] Research Institute of Electronics and Telecommunications Technology, Chonnam National University, 300 Yongbong-Dong Buk-Gu Kwangju, Republic of Korea
j2kim@chonnam.ac.kr
[2] Department of Computer and Information Engineering, Chonnam National University, 300 Yongbong-Dong Buk-Gu Kwangju, Republic of Korea
{ykwon, melchi}@grace.chonnam.ac.kr

Abstract. Decision trees (DTs) and multi-layer perceptron (MLP) neural networks have long been successfully used to various pattern classification problems. Those two classification models have been applied to a number of diverse areas for the identification of 'biologically relevant' molecules. Surface enhanced laser desorption/ionization time-of-flight mass spectrometry (SELDI-TOF MS) is a novel approach to biomarker discovery and has been successfully used in projects ranging from rapid identification of potential maker proteins to segregation of abnormal cases from normal cases. SELDI-TOF MS can contain thousands of data points. This high dimensional data causes a more complex neural network architecture and slow training procedure. In the approach we proposed in this paper, a decision tree is first applied to select possible biomarker candidates from the SELDI-TOF MS data. At this stage, the decision tree selects a small number of discriminatory biomarker proteins. This candidate mass data defined by the peak amplitude values is then provided as input patterns to the MLP neural network which is trained to classify the mass spectrometry patterns. The key feature of this hybrid approach is to take advantage of both models: use the neural network for classification with significantly low-dimensional mass data obtained by the decision tree. We applied this bioinformatics tool to identify proteomic patterns in serum that distinguish prostate cancer samples from normal or benign ones. The results indicate that the proposed method for mass spectrometry analysis is a promising approach to classify the proteomic patterns and is applicable for the significant clinical diagnosis and prognosis in the fields of cancer biology.

1 Introduction

The study of the cell's proteome presents a new horizon for biomarker discovery. The discovery, identification, and validation of proteins associated with a particular disease are difficult and laborious task. Presently, disease biomarker discovery is generally carried out using two dimensional polyacrylamide gel electrophoresis (2D-PAGE) to separate and detect differences in protein expression [1][2]. Advances have also been made in mass spectrometry to achieve high-throughput separation and analysis of proteins [3]. One of the recent advances is surface enhanced laser desorption/ionization time-of-flight mass spectrometry (SELDI-TOF MS) which provides a rapid protein expression profile from a variety of biological and clinical samples [4].

SELDI-TOF MS can contain thousands of data points, which are mass/intensity spectra. Molecular weight (mass) and intensity are presented on the x-axis and y-axis, respectively. Due to the high dimension of data that is generated from a single analysis, it is essential to use algorithms that can detect expression patterns from such large volumes of data correlating to a given biological phenotype from multiple samples. The algorithm should serve to classify the samples into patient or normal case according to their molecular expression profile [5][6].

DT (decision tree) and MLP (multi-layer perceptron) have been widely used for many pattern classification problems successfully [7]. Decision trees generally run significantly faster in training and give better expressiveness, and the MLPs are often more accurate at classifying novel examples in the presence of noisy data. They were applied to classify human cancers and identify the potential biomarker proteins using SELDI-TOF mass spectrometry [5][6][8].

In this paper, we propose a DT and MLP hybrid system. In the proposed approach, a decision tree is first applied to select possible biomarker candidates from the SELDI-TOF MS data. At this stage, the decision tree selects a small number of discriminatory biomarker proteins. This candidate mass data defined by the peak amplitude values is then provided as input patterns to the MLP neural network which is trained to classify the mass spectrometry patterns. The key feature of this hybrid approach is to take advantage of both models: using the neural network for classification with significantly low-dimensional mass data obtained by the decision tree.

We demonstrate proof of principle that our hybrid approach can more accurately discriminate prostate cancer from patients with benign prostate hyperplasia and healthy men. Data set of SELDI-TOF MS used for our work is available from [6], and materials and methods to obtain the spectrometry data were described in detail. Our results show that the hybrid approach outperforms the decision tree approach used in [6]. The result suggests that this hybrid approach can identify molecular ion patterns that strongly associate with certain disease and potentially be applicable for significant clinical diagnosis or prognosis for the cancer biology.

2 SELDI-TOF MS Data

Mass spectrometry data we used for this study was obtained from the Virginia Prostate Cancer Center which is available at http://www.evms.edu/vpc/seldi [6]. There are two sets of mass spectrometry data: one is the training set for constructing a classification model and the other is the test set for validating the model. For our study, the training data set were only used, since the test set does not include the true class index. Details for the biological/chemical processes and the pre-processing procedure are described in [6].

SELDI-TOF MS is able to fast analyze on the basis of protein molecular weight in comparison with MALDI-TOF MS, finding biomarkers without pre-processing. SELDI-TOF MS experimentation must take protein chip array composition by biological characteristic and chemical characteristic. Protein samples become to bind the same characteristic spot in protein chip array. Here, we separated and quantified proteins using Surface-Enhanced Laser Desorption/Ionization Time-of-Flight mass spectrometry (SELDI-TOF-MS) (Ciphergen, Fremont, CA), in which proteins were bound to active surfaces, ionized by a laser, moved to protein chip reader through energy law $e = mv^2$ *, and recorded by MS(mass spectrum) [9].

Specimens from three groups of patients were used to obtain the SELDI mass spectrometry data: 82 healthy men, 77 patients with BPH(Benign Prostate Hyperplasia) and 167 patients with PCA(Prostate Cancer). SELDI protein profiling is the process that prepares the protein chip array with the patient's serum samples after appropriate biological or chemical treatment, reads the array by the SELDI mass spectrometry reader, and generates the time-of-flight spectra for the samples. Chip reading and TOF spectra generation are automated by the Protein Biological System II which is the SELDI mass spectrometry reader (Ciphergen Biosystems, Inc.).

3 A Hybrid System for Proteomic Pattern Classification

3.1 Potential Biomarkers: Feature Selection by Decision Tree

Decision tree classifies a pattern through a sequence of questions, in which the next question asked depends upon the answer to the current question [7]. Such a sequence of questions forms a node connected by successive links or branches downward to other nodes. The questions asked at each node concern a particular property of the spectrum patterns such as 3,017 <= 12.14, where the first number indicates the molecular weight and the second one does the peak intensity value. With the decision tree, classification of a particular pattern begins at the root node, following the appropriate link based on the answer to the question at the node.

At each node of the tree, the question concerns the property of the spectrum that makes the data reach the next descendent nodes as "pure" as possible. For this purpose, a "cost" function is computed that reflects the heterogeneity of each descendent node. The most popular measurement for the "cost" is the entropy impurity defined as:

$$i(N) = -\sum_j P(w_j) \log_2 P(w_j) \quad (1)$$

where $P(w_j)$ is the fraction of the spectrum data patterns at node N that is in class category w_j. Thus, peaks selected by this method to form the classification rules are the ones that achieve the maximum reduction of the "cost" in the next descendent nodes.

Each node selects the best feature (i.e., a mass value in the spectrum) with which heterogeneity of the data in the next descendent nodes is as pure as possible. However, with the data set in the presence of noise, this decision scheme is not always adequate. For instance, classification rules from a decision tree constructed by a set of pattern data are often different from those obtained by another set of pattern data collected

---

* e = energy, m = molecular weight, v = velocity.

from the same problem domain. This phenomenon mostly happens when the noise is involved. This is because the decision tree training algorithm considers only the *best* feature at each node. Unfortunately, the *best* feature could be the second or the third best one if the noise is not involved.

To overcome this drawback of the decision tree, we used the decision tree as a feature selection system, not the classification system. Suppose we have a possible test with n outcomes that partitions the training set T into subsets $T_1, T_2, ..., T_n$. With the decision tree model C4.5 [10], we computed *gain ratio* at each node which is defined by

$$gain\ ratio(X) = gain(X) / split\ info(X). \tag{2}$$

where X is the test feature. This expresses the proportion of information generated by the split that is useful for classification. In this equation, *split info* is defined by

$$split\ info(X) = -\sum_{i=1}^{n} \left(\frac{|T_i|}{|T|} \times \log_2 \left(\frac{|T_i|}{|T|} \right) \right), \tag{3}$$

and the *gain* measures the information that is gained by partitioning T in accordance with the test feature X, which is defined by

$$gain(X) = info(T) - info_X(T). \tag{4}$$

The first term in this equation *info* is also known as the *entropy* shown in the equation 1, and the second term is the expected information requirement that can be found as the weighted sum over the subsets, as

$$info_X(T) = \sum_{i=1}^{n} \frac{|T_i|}{T} \times info(T_i). \tag{5}$$

C4.5 decision tree algorithm selects a feature that maximizes the ration defined in the equation 2. Instead of a single best feature, we selected some of the top-ranked best multiple features as the inputs to the MLP classifier. In terms of biomarker discovery, those features are considered as the potential maker proteins with which the abnormal cases (patients) can be distinguished from the normal cases (healthy men).

From the protein profiling procedure described in the section 2, we obtained 326 pattern vectors from three classes (healthy, BPH, PCA) with 779 dimensions (molecular weight values). With the training patterns of 779 dimensions, we first constructed the decision tree using C4.5 [10]. During the training procedure, we selected the 5 best features at each branch node. The term 'feature' represents the molecular mass in this study. In one of our simulation studies, a decision tree generated 22 nodes that perfectly classified the training patterns. From 110 mass values (22 nodes * 5 features), the unique mass values were 54. In other words, the 779 feature dimensions were reduced down to 54 dimensions. In the figure 1, some of selected molecular weights are shown, and we can intuitively see that the features selected by the decision tree have better discriminatory power than those not selected.

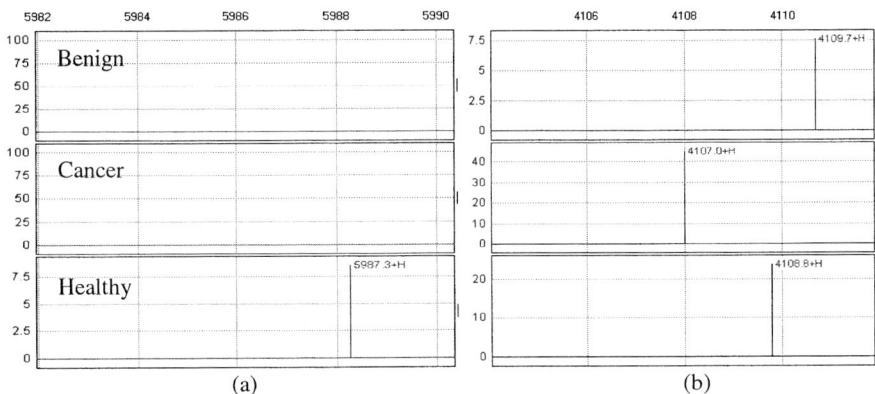

Fig. 1. Relative intensity values of molecular weights (a) not selected and (b) selected by the decision tree. Note that the peaks in (b) represent the same protein molecular weight after they are aligned

3.2 Diagnosis by MLP Neural Network

Multi-layer perceptron neural networks have one input layer, one output layer, and one or more hidden layers between the input and output layers. Each layer has several processing units, called nodes, neurons, or units which are generally non-linear units except the input units that are simple bypass buffers. The unit operation is characterized by the equation

$$O_j = f(net_j). \tag{6}$$

The input to the unit net_j and the activation function f are respectively given by

$$net_j = \sum_i w_{ij} O_i + bais_j \tag{7}$$

and

$$f(net_j) = 1/(1 + e^{-net_j}) \tag{8}$$

where i denotes the units in the next lower layer, w_{ij} is the connection strength, generally called the weight, and $bias_j$ is a weight whose input is always 1. The activation function f is a sigmoid or logistic function which is usually differentiable and non-decreasing.

Multi-layer perceptron neural networks have been widely employed to solve many classification problems with the development of the effective error back-propagation learning rule. This Learning rule is an iterative gradient descent algorithm that updates the weights to minimize the error defined by

$$E = \frac{1}{2} \sum_j (t_j - O_j)^2 \tag{9}$$

where t_j is the target output and O_j is the actual output. By applying the chain rule and the partial derivative on E with respect to each weight, one can obtain the amount of weight change for a single input, described as

$$\Delta w_{ji} = -\eta \frac{\partial E}{\partial w_{ji}} = \eta \delta_j O_i \quad (10)$$

where $\delta_j = f'(net_j) \sum_k \delta_k w_{kj}$ for the hidden units and $\delta_j = (t_j - O_j) f'(net_j)$ for the output units.

4 Experiments and Result Analysis

We first randomly collected 20% of patterns in each class for the test data set; the rest 80% were used to construct a decision tree and MLP classifier. Table 1 shows the number of patterns for each class.

Table 1. Number of training and testing patterns for each class. Test set was ramdom collected from the data for constructing a decision tree [6]

	BPH	PCA	Healthy	No. Samples
Training set	62	134	66	262
Test set	15	33	16	64

New 54-dimensional pattern vectors were then used as the inputs to the multi-layer perceptron neural network trainable by back-propagation learning algorithm. The multi-layer perceptron neural network had 54 input units, 20 hidden units and 3 output units. Each output unit corresponds to one of the tree classes: benign, prostate cancer, and healthy men.

Table 2 shows one of the classification results for the test set obtained from the MLP network trained with the features selected by the decision tree. For this table, we used the MAX rule "Pattern p is classified as the class j if O_j is the maximum value among the values of the output nodes." As seen in the table, our approach outperforms the decision tree classification that the most previous works used. We ran several simulations with the different training sets. For each simulation, 20% of the total samples were randomly collected as the test set and the rest 80% were used as the training set. Other results also showed the similar results.

Another attractive property of our approach is its rejection capability. Most of the misclassified patterns can be rejected as "unknown" or "ambiguous" as shown in the table 3, if we apply an appropriate decision making rule. An example will be one named as the DIFF rule "Pattern p is classified into the class j, if O_j is larger than 0.7 and $(O_j - O_k)^2$ is larger than 0.09 for all $k \neq j$." With this rule, the patterns misclassified by the MAX rule can be easily classified as "unknown", and some of correctly classified ones can also become "unknown". This property can not be expected by the rules generated from the decision tree algorithm.

Table 2. A classification result on the test set by the decision tree only and the MLP network with the features selected by the decision tree

	Decision Tree Only				MLP with features from DT			
Sample	BPH	PCA	Healthy	Error	BPH	PCA	Healthy	Error
BPH	11	4	0	4(27%)	12	2	1	3(20%)
PCA	1	27	5	6(18%)	0	32	1	1(3%)
Healthy	2	2	12	4(25%)	1	0	15	1(6%)

Table 3. Output vectors for the misclassified and the ambigious patterns. The misclassified patterns with the MAX rule can be easily classified as *unknown* or *anbigious* pattern, while some of correctly classified one can also become *unknown*

Output vector	True class	By MAX	By DIFF
[0.121342 0.117438 0.524633]	BPH	Healthy	Unknown
[0.624093 0.733536 0.010744]	BPH	PCA	Unknown
[0.732484 0.748367 0.034827]	BPH	PCA	Unknown
[0.442034 0.132481 0.228364]	BPH	BPH	Unknown
[0.027561 0.324827 0.736442]	PCA	Healthy	Healthy
[0.302657 0.035846 0.336854]	Healthy	Healthy	Unknown
[0.628723 0.323544 0.021657]	Healthy	BPH	Unknown

5 Conclusion

We proposed a hybrid approach to classify the serum samples using the SELDI-TOF mass spectrum. The hybrid system is composed of two classification systems: DT(decision tree) and MLP(multi-layer perceptron). DT was used as a feature selection system. It selected some features (i.e., mass values) at each branch node that have better discriminatory power. Discriminatory power for each feature was computed by the gain ratio [10]. Those selected features then composed the pattern vectors as the input to the MLP trained by the back-propagation algorithm. The key feature of our hybrid approach was to take advantages of both systems: use the MLP for better accurate classification with the significantly reduced number of features obtained by DT that significantly faster in training.

We applied the proposed bioinformatics tool to identify the proteomic patterns in serum samples to distinguish prostate cancer cases using SELDI-TOF mass spectrometry data. The serum samples were collected from three groups: healthy men, patients with benign prostate hyperplasia, and patients with prostate cancer. They were processed to obtain the SELDI-TOF mass spectrometry data [6]. The hybrid system outperformed the decision tree approach that has been widely used for biomarker discovery studies. It produced significantly better classification accuracy and demonstrated a rejection capability for the suspicious samples.

Finally, it is suggested that this bioinformatics approach could lead to the prediction for the patient's responsiveness to particular forms of therapy and diagnosis for a certain disease.

Acknowledgement

This study was supported by a grant of Gwangju Research Center for Biophotonics 2004, Republic of Korea (BPRC-2004-02-001).

References

1. Sirinivas, P.R., Srivastavas, S., Hanash, S., and Wright, G.: Proteomics in early detection of cancer, Clinical Chemistry, 47 (2001) 1901-1911
2. Adam, B., Vlahou, A., Semmes, O., and Wright, G.: Proteomic approaches to biomarker discovery in prostate and bladder cancers, Proteomics, 1 (2001) 1264-1270
3. Keough, T., Lacey, M.P., Fieno, A.M., et al: Tandem mass spectrometry methods for definitive protein identification in proteomics research. Electrophoresis, 21. (2000) 2252-2265
4. Fung, E.T., and Enderwick, C.: ProteinChip Clinical Proteomics: Computational Challenges and Solutions. Computational Proteomics Supplement, 32. (2002) 34-41
5. Petricoin III, E., Ardekani, A., Hitt, B, et al.: Use of proteomic patterns in serum to identify ovarian cancer. The Lancet, 359 (2002) 572-577
6. Adam, B., Qu, Y., Davis, J, et al: Serum protein fingerprinting coupled with a pattern-matching algorithm distinguishes prostate cancer from benign prostate hyperplasia and healthy men. Cancer Research, 62 (2002) 3609-3614
7. Duda, R., Hart, P, Stork, D.: Pattern classification. Wiley-interscience, New York (2001)
8. Ball, G., Mian, S., Holding, F., et al: An integrated approach utilizing artificial neural networks and SELDI mass spectrometry for the classification of human tumors and rapid identification of potential biomarkers. Bioinformatics, Vol. 18, No. 3 (2002) 395-404
9. Merchant, M., and Weinberger, S.R.: Recent advancements in surface-enhanced laser desorption/ionization time-of-flight mass spectrometry. Electrophoresis, 21 (2000) 1164-1177
10. Quinlan, J.R.,C4.5: Programs for machine learning, Morgan Kaufmann Publisher (1993)

Deterministic Annealing EM and Its Application in Natural Image Segmentation

Jonghyun Park,[1] Wanhyun Cho,[2] and Soonyoung Park[3]

[1] Department of Computer Science, Chonbuk National University, S. Korea
jhpark@dahong.chonbuk.ac.kr
[2] Department of Statistics, Chonnam National University, S. Korea
whcho@chonnam.ac.kr
[3] Department of Electronic Engineering, Mokpo National University, S. Korea
sypark@mokpo.ac.kr

Abstract. In this paper, we present a color image segmentation algorithm based on a finite mixture model and examine its application to natural scene segmentation. Gaussian mixture model (GMM) is first adopted to represent the statistical distribution of multi-colored objects. Then a deterministic annealing Expectation Maximization (DAEM) formula is used to estimate the parameters of the GMM. The experimental results show that the proposed DAEM can avoid the initialization problem unlike the standard EM algorithm during the maximum likelihood (ML) parameter estimation and natural scenes containing texts are segmented more efficiently than the existing EM technique.

1 Introduction

Color image segmentation is to partition a given image into homogeneous regions or pattern classes and label each region by a pattern type. Color image segmentation occurs frequently in many image processing applications such as intelligent robot vision systems, medical diagnosis and retrieval in large image databases.

In recent years, statistical model-based segmentation algorithms have been proposed and successfully applied to various areas [1-5]. McLachlan et al. [1] considered the GMM approach to the clustering of multispectral magnetic resonance images and X. Chen et al. [5] used GMM to characterize color distributions of the foreground and background of a sign from natural scenes. The EM algorithm is generally employed to fit the mixture model to the observed data and then ML parameter estimate of the mixture model is obtained. However, the estimates of parameters obtained by the EM algorithm are strongly dependent upon their initial conditions. To overcome this problem, the concepts of the DAEM have been proposed. Ueda and Nakano [6] proposed a new posterior parameter that was called as 'temperature' derived by using the principal of maximum entropy and used this parameter for controlling the annealing process. Hofmann et al. [7] formulated the texture segmentation as a data clustering problem based on spare proximity data. M.A.T. Figueiredo [3] proposed an unsupervised algorithm for learning a finite mixture model, which is capable of selecting the number of clusters and avoids the initialization problem.

In this paper, we are going to consider how to use the parametric probability models to segment the text from natural scenes. We employ the GMM to statistically represent the color distributions of multi-objects in natural scenes and use the DAEM to estimate the parameters in a mixture model.

2 Natural Image Segmentation Using DAEM Algorithm

When a natural image is given, assuming the image has N pixels, we use $\mathbf{y}_i$ to denote the observation at the i-th incomplete data. The whole samples in the image form data set $\mathbf{S} = \{\mathbf{y_i}\}_{i=1}^{N}$, assuming that $\mathbf{y}_i$ is a sample from a finite mixture distribution. In the following we briefly review the Gaussian finite mixture model with DAEM algorithm and its application in natural image segmentation.

2.1 Deterministic Annealing EM Algorithm

We let $\mathbf{y}$ denote the incomplete data consisting by the feature vector extracted from a given image, and we define a indicate vector $\mathbf{z}$ to denote all of hidden vector to represent the particular object or region that each feature vector belongs to. And also we let $p(\mathbf{y}|\mathbf{z};\theta)$ denote the conditional density function of the random vector Y given $\mathbf{z}$, $p(\mathbf{z};\pi)$ and denote the probability function of Z, where π is a parameter vector of prior probabilities. Then the complete data vector is defined by $\mathbf{x} = (\mathbf{y},\mathbf{z})$, and the log likelihood function that can be formed on the basis of the complete data $\mathbf{x}$ if we adopt the GMM for an observed feature vector, is given by

$$\log L_C(\Theta|\mathbf{x}) = \log p(\mathbf{y}|\mathbf{z};\pi) + \log p(\mathbf{z};\pi)$$
$$= \sum_{k=1}^{K}\sum_{i=1}^{N} Z_{ik}\log(N(\mu_k,\Sigma_k)) + \sum_{k=1}^{K}\sum_{i=1}^{N} Z_{ik}\log(\pi_k) \quad (1)$$

where Θ is the vector containing the elements of θ and π, $N(\cdot)$ is Gaussian probability distribution with mean vector and covariance matrix.

The problem of maximum likelihood estimation of Θ given the observed vector $\mathbf{y}$ can be solved by applying the EM algorithm for the incomplete data. However the EM algorithm has two kinds of disadvantages. The first is hard to avoid unfavorable local maximum of the log-likelihood and the second is overfitting problem. Thus we have to think about the new method that is able to improve the EM algorithm. It is known as DAEM algorithm. This is to use the principle of maximum entropy to estimate the parameter [6]. We consider the complete data log likelihood $\log L_C(\Theta|\mathbf{x})$ as a function of the hidden variable $\mathbf{z}$ for fixed parameter vector Θ, and define a cost function on the hidden variable space Ω_z as follows:

$$H(\mathbf{z};\mathbf{y},\Theta) = -\log L_C(\Theta|\mathbf{x}) \quad (2)$$

Then we need to minimize $E(H(\mathbf{z};\mathbf{y},\Theta))$ with respect to probability distribution $p(\mathbf{z};\pi)$ over the distribution space subject to a constraint on the entropy. It yields a quantity, which is known as the generalized free energy in statistical physics. Introducing a Lagrange parameter β, we arrive at the following objective function :

$$\vartheta(P_{\mathbf{z}}^{(t)},\Theta) = E_{P_{\mathbf{z}}^{(t)}}(H(\mathbf{z};\mathbf{y},\Theta)) + \beta \cdot E_{P_{\mathbf{z}}^{(t)}}(\log P_{\mathbf{z}}^{(t)}) \quad (3)$$

The solution of the minimization problem associated with the generalized free energy in $\vartheta(P_{\mathbf{z}}^{(t)},\Theta)$ with respect to probability distribution $p(\mathbf{z};\pi)$ with a fixed parameter is the following Gibbs distribution:

$$P_\beta(\mathbf{z}|\mathbf{y},\Theta) = \frac{1}{\sum_{\mathbf{z}\in\Omega_z} \exp(-\beta H(\mathbf{z}'))} \cdot \exp(-\beta H(\mathbf{z})) \quad (4)$$

Hence we can obtain a new posterior distribution, $P_\beta(\mathbf{z}|\mathbf{y},\Theta)$ parameterized by β. Next, we should find the minimum of $\vartheta(P_{\mathbf{z}}^{(t)},\Theta)$ with respect to Θ with fixed posterior $P_\beta(\mathbf{z}|\mathbf{y},\Theta)$. It means finding the $\Theta^{(t)}$ that minimizes $\vartheta(P_{\mathbf{z}}^{(t)},\Theta)$. The generalized free energy, $\vartheta(P_{\mathbf{z}}^{(t)},\Theta)$ can be written by the following form :

$$\vartheta(P_{\mathbf{z}}^{(t)},\Theta) = Q_\beta(\Theta) + \beta \cdot E_{P_{\mathbf{z}}^{(t)}}(\log P_{\mathbf{z}}^{(t)}) \quad (5)$$

Since the second term on the right hand side of the generalized free energy is independent of Θ, we should find the value of Θ minimizing the first term

$$Q_\beta(\Theta) = E_{P_{\mathbf{z}}^{(t)}}(H(\mathbf{z};\mathbf{y},\Theta)) \quad (6)$$

To achieve this purpose, we can add a new β-loop, which is called annealing loop, to the original EM-algorithm and replace the original posterior with the new posterior distribution, $P_\beta(\mathbf{z}|\mathbf{y},\Theta)$ parameterized by β. Finally, after finishing fully iteration, we can obtain the conditional expectation of the hidden variable, Z_{ik} given the observed feature data from E-step. This is given by

$$\tau_k(\mathbf{y_i}) = E(Z_{ik}) = \frac{\hat{\pi}_k N(\mathbf{y}_i; \hat{\mu}_k, \hat{\Sigma}_k)}{\sum_{j=1}^{K} \hat{\pi}_j N(\mathbf{y}_i; \hat{\mu}_j, \hat{\Sigma}_j)} \quad (7)$$

And we can obtain the estimators of mixing proportions, a component mean vector and covariance matrix from M-step. These are respectively given as

$$\hat{\pi}_k = \frac{1}{N}\sum_{i=1}^{N} \tau_k(\mathbf{y}_i), k=1,\cdots,K \quad (8)$$

$$\hat{\mu}_k = \frac{\sum_{i=1}^{N} \tau_k(\mathbf{y}_i)\mathbf{y}_i}{\sum_{i=1}^{N} \tau_k(\mathbf{y}_i)}, k=1,\cdots,K \quad (9)$$

$$\hat{\Sigma}_k = \frac{\sum_{i=1}^{N} \tau_k(\mathbf{y}_i)(\mathbf{y}_i-\hat{\mu}_k)(\mathbf{y}_i-\hat{\mu}_k)^t}{\sum_{i=1}^{N} \tau_k(\mathbf{y}_i)}, k=1,\cdots,K. \quad (10)$$

2.2 Application in Natural Image Segmentation

Suppose that a natural image consists of a set of the K distinct objects or regions. We usually segment a natural image to assign each pixel to some region or objects. To do this, we need a posterior probability of the i-th pixel belonging to k-th region. These probabilities have already estimated by DAEM algorithm in last section. That is, given the observed data $\mathbf{S} = \{\mathbf{y_i}\}_{i=1}^{N}$ and a knowledge of estimated parameter vector $\hat{\Theta}$, the probability of the i-th pixel belongs to k-th region is given by

$$\tau_k(\mathbf{y_i}) = E(Z_{ik}) = \frac{\hat{\pi}_k N(\mathbf{y}_i; \hat{\mu}_k, \hat{\Sigma}_k)}{\sum_{j=1}^{K} \hat{\pi}_j N(\mathbf{y}_i; \hat{\mu}_j, \hat{\Sigma}_j)} \quad (11)$$

Next, we try to find what the component or region has the maximum value of probabilities among the estimated posterior probabilities. This is define as

$$\hat{Z}_i = \arg\max_{1 \leq k \leq K} \tau_k(\mathbf{y}_i), i = 1, \cdots, N. \quad (12)$$

Then, we can segment a natural image by assigning each pixel to the region or object having the maximum a posterior probability.

3 Experimental Results

In order to assess the performance of the proposed algorithm, we have conducted the experiment using data obtained from natural images containing text regions. Figure 1 (a) is an original image that has 400×300 pixels with 3byte RGB color levels and consists of multi-colored objects and a background. The conversion of the RGB color to HSI model is carried out and the hue and saturation components are only used as values of the feature vectors. Figure 1 (b) shows color distributions in HS-space and Figure 1(c) shows the initial positions being selected as the centers of the circles. Here, the points on the circles have the unit Mahalanobis distance from each center.

We have applied the EM and the DAEM algorithms to the natural image in Figure 1(a) to examine the initialization effects on the parameter estimation.

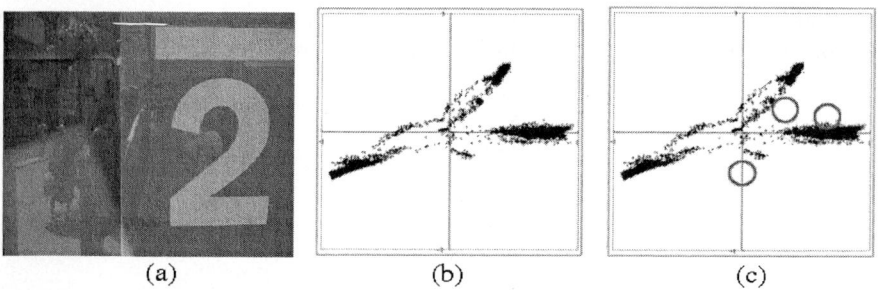

Fig. 1. Natural image used in the experiment: (a) natural image, (b) color distributions in HS-space, (c) initial position for ML parameter estimate

Figure 2 (a) and (b) show the segmentation and convergence results for the EM and DAEM, respectively. We can observe that the DAEM tends to accurately estimate the parameters of each cluster rather than the standard EM regardless of initial positions. The segmentation results for three other natural images using

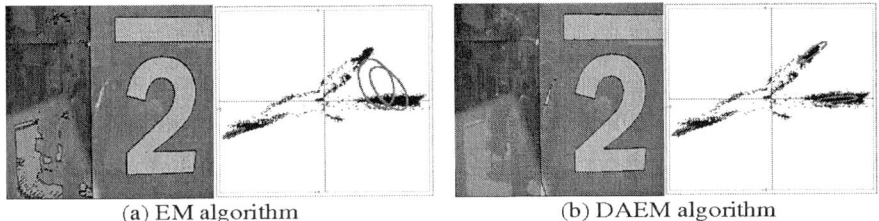

(a) EM algorithm (b) DAEM algorithm

Fig. 2. Segmentation and convergence results of applying EM and DAEM

Fig. 3. Segmentation result by EM and DAEM: (a) natural image, (b) EM algorithm result, (c) DAEM algorithm result

the EM and DAEM algorithms are shown in figure 3. We can see that the DAEM provides a subjectively superior segmentation in all cases. This is also given from the fact that the text and background are partitioned into the same region accurately and the fine structure is preserved.

4 Conclusion

In this paper, we have proposed a DAEM algorithm for the natural color image segmentation. The proposed algorithm was derived using the principle of maximum entropy to overcome the local maximal problem associated with the conventional EM algorithm.

We conclude from the experiments that the DAEM algorithm is robust to initial conditions and it provides a global optimal solution for the ML parameter estimator. It has also provided a superior segmentation in that each text is partitioned into the same homogeneous region accurately.

References

1. McLachlan, G.J., Nguyen, S.k., Galloway, G.J., and Wang, D.: Clustering of magnetic resonance images. Technical Report, Department of Mathematics, University of Queensland, (1998)
2. Deng, Y., and Manjunath, B.S.: Unsupervised segmentation of color-texture regions in images and video. IEEE Transactions on Pattern Analysis and Machine Intelligence, vol. 23, no. 8 (2001) 800-810
3. Figueiredo, M.A.T. and Jain, A.K.: Unsupervised learning of finite mixture models. IEEE Transactions on Pattern Analysis and Machine Intelligence, vol. 24, no. 3 (2002) 381-396
4. Permuter, H., Francos, J., and Jermyn, I.H.: Gaussian mixture models of texture and colour for image database retrieval. Proc. of the IEEE International Conference on Acoustics, Speech and Signal Processing, vol. III (2003) 569-572
5. Chen, X., Yang, J., Zhang, J., and Waibel, A.: Automatic detection and recognition of signs from natural scenes. IEEE Transactions on Image Processing, vol. 13, no. 1 (2004) 87-99
6. Ueda, N. and Nakano, R.: Deterministic annealing EM algorithm. Neural Networks, vol. 11 (1998) 271-282
7. Hofmann, T. and Buhman, J.M.: Pairwise data clustering by deterministic annealing. IEEE Transactions on PAMI, vol. 19, no. 1 (1998) 1-13

The Structural Classes of Proteins Predicted by Multi-resolution Analysis[1]

Jing Zhao[1], Peiming Song[1], Linsen Xie[3], and Jianhua Luo[1,2,3]

[1] School of Life Science & Technology, Shanghai Jiaotong University, Shanghai 200240
[2] Shanghai Research Center of Bioinformatics and Technology, Shanghai 200002
[3] Lishui College, Lishui, Zhejiang 32300
zjane_cn@yahoo.com.cn

Abstract. Prediction of protein structural class from primary structure is studied in this paper. Wavelet packet transform is used to decompose the corresponding numerical signal of protein into several sub-signals at different resolution scales. The auto-correlation functions based on the sub-signals are used as feature vectors of the protein. The Bayes decision rule is used as classification algorithm. Experiments show that for the same datasets, the prediction accuracy is improved compared with the existed methods.

1 Introduction

From the work of Levitt and Chothia, proteins can be generally classified into four different structural classes, all-α, all-β, α/β and α+β, according to the content and topology of α-helices and β-strands [1]. Since it is generally accepted that the protein structure is determined by its primary sequence, the prediction of protein structure from amino acid sequence is one of the major goals of bioinformatics.

Actually, many efforts have been made in this field. The existed prediction algorithms were mostly based on the amino acid composition of proteins. Although they are reasonable approximate approaches and did yield some encouraging results, the prediction quality will be most probably improved if the information from residue order along the primary structure of protein can also be incorporated into the prediction algorithm.

Wavelet packet transform is a signal processing method efficient for multi-resolution analysis and local feature extraction [2]. In this paper, by making use of wavelet packet transform, the information hidden in the primary sequences of proteins can be measured at different resolution scales based on a space-scale analysis. The feature vectors of proteins are extracted based on wavelet packet decomposition and auto-correlation analysis. Then we use Bayes decision rule as classifying algorithm. Our experiments demonstrate this is a successful method in protein structural class prediction.

---

[1] This work is supported by 863 Project of China (No.2002AA234021) and 973 Project of China (No. 2002CB512800).

2 Wavelet Transform and Wavelet Packet Transform

The wavelet transform (WT) splits a signal S at different frequency bands with different resolutions by decomposing the signal into a coarse approximation A_1 and detail information D_1. The approximation A_1 is then itself split into a second-level approximation A_2 and detail D_2. The decomposition process can be iterated, with successive approximations being decomposed in turn, so that one signal is broken down into many lower-resolution components [2].

The wavelet packet transform (WPT) is a generalization of wavelet decomposition that provides a complete level-by-level decomposition to a signal S. Instead of dividing only the approximation spaces to construct detail spaces and wavelet bases in wavelet decomposition, wavelet packet split the details as well as the approximations. As a result, the wavelet packet decomposition enables the extraction of features from signals that combine stationary and nonstationary characteristics with an arbitrary time-frequency resolution. The wavelet packet decomposition tree at level 3 is shown in Figure 1. Suppose the frequency range of the original signal S is from 0 to 1, then the frequency range of each reconstructed component sub-signal is shown in Fig.1. Reference [3] gives the detail of WPT algorithm.

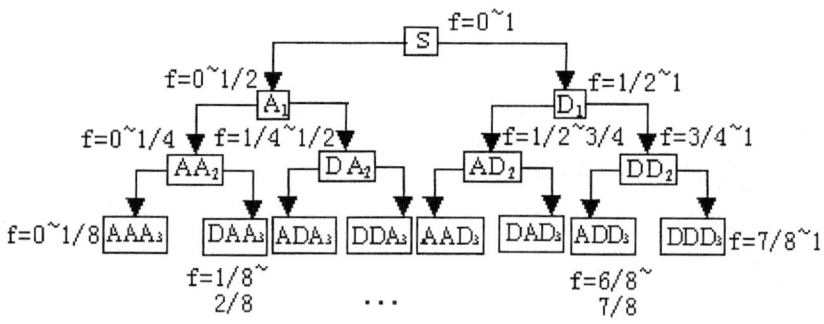

Fig. 1. Wavelet Packet Decomposition Tree at Level 3

3 Feature Vectors and Classifying Algorithm

3.1 Perform Wavelet Packet Decomposition

In order to study amino acid sequence with wavelet packet transform, we make every amino acid of one sequence correspond to its amino-acid index. An amino acid index is a set of 20 numerical values representing any of the different physicochemical properties of the 20 amino acids. Here the index of Oobatake and OOi, i.e. the average nonbonded energy per residue [4], is used, which is listed in Table 1. In this way, one amino acid sequence x is changed to a numerical sequence S which can be seen as a one-dimensional numerical signal. Then S is normalized to zero mean and unit standard deviation. For simplicity, we still denote it by S.

Table 1. Amino Acids Index of Oobatake and OOi

Residue	Value	Residue	Value	Residue	Value	Residue	Value
G	-7.592	C	-12.210	Y	-20.232	K	-12.366
P	-11.893	W	-26.166	H	-17.55	I	-15.608
F	-20.504	D	-12.144	L	-15.728	S	-10.518
E	-13.815	V	-13.867	T	-12.369	N	-12.480
A	-9.475	M	-15.704	Q	-13.689	R	-16.225

For every numerical signal S, we compute its wavelet packet decomposition at level 3 and get its 8 component sub-signals $S^{(j)}$ at different frequency bands, $j=1,2,3,4,5,6,7,8$, i.e.,

$$S = \sum_{j=1}^{8} S^{(j)} \qquad (1)$$

3.2 Compute the Auto-correlation Functions

For a numerical sequence:

$h_1, h_2, ..., h_N,$

its auto-correlation function r_n is defined by [5]:

$$r_n = \frac{1}{N-n} \sum_{i=1}^{N-n} h_i h_{i+n} \qquad n=1,2,...,N-1 \qquad (2)$$

Bu et al. [6] have studied the prediction of protein structural classes by auto-correlation functions based on the amino acid index of Oobatake and OOi. In their study, the feature vector of a protein is defined as:

$$X = (r_1, r_2, ... r_m)^T, \qquad (3)$$

where r_n is defined by equation (2), h_i is the amino acid index for the i-th residue and m is an integer to be determined by the optimum prediction.

From the auto-correlation function of one numerical signal, we can derive the average information about the correlation inside the data. Similarly, when we consider the auto-correlation functions of the wavelet packet decomposition sub-signals, we are able to distinguish the correlation properties of the original signal at different scales or different frequency bands. Because of this, we define 9 feature vectors for each protein as following,

$$X^{(0)} = (r_1, r_2, ... r_{m0})^T \qquad (4)$$

$$X^{(j)} = (r_1^{(j)}, r_2^{(j)}, ..., r_{mj}^{(j)})^T \qquad j=1, 2, ..., 8 \qquad (5)$$

where, feature vector $X^{(0)}$ is the same as that defined by Bu et al, $r_n^{(j)}$ ($n=1,2,...,mj$) is the auto-correlation function of sub-signal $S^{(j)}$ and mj is an integer to be determined by the optimum prediction. The feature vectors $X^{(j)}$ describe the correlation of the signal with a multi-resolution point of view.

Integer mj (j=0, 1, ..., 8) means the optimal number of the auto-correlation functions used in the feature vector and leads to optimal prediction accuracy. It is dependent on the training dataset and can be determined by respectively using the classifying algorithm based on different length of feature vectors.

3.3 Classifying Algorithm

We use Bayes decision rule as classifying algorithm [7].

Suppose the classification training set T is the union of 4 subsets, i.e.

$$T = T_1 \cup T_2 \cup T_3 \cup T_4$$

where the subset T_1, T_2, T_3 and T_4 respectively consist of all-α, all-β, α/β and α+β proteins.

One frequently used discriminant function is following,

$$g_k(X) = -(X - \mu_k)^T \Sigma_k^{-1} (X - \mu_k) - \ln|\Sigma_k| \qquad (6)$$

where X is the feature vector of the query protein, μ_k is the mean vector of all the feature vectors in training subset T_k, Σ_k is the covariance matrix of all the feature vectors in training subset T_k, $|\Sigma_k|$ is the determinant of the square matrix Σ_k, Σ_k^{-1} is the inverse of the square matrix Σ_k.

Then the Bayes Decision Rule could be expressed as:

If $g_p(X) = \max\{g_k(X) \mid k=1,2,3,4\}$, then X is decided as class T_p. (7)

Obviously, this classification algorithm is equivalent to the component-coupled algorithm of Chou and Maggiora [8].

According to the discuss above, in our study, any protein corresponds to 9 feature vectors $X^{(j)}$, j=0,1,2,...8. We advance the classifying algorithm as following,

(1) Respectively use Bayes Decision Rule based on the 9 feature vectors
(2) Apply majority voting on the 9 results and get the final prediction result

4 Experiments

4.1 Datasets

In this research, the datasets used are following:

(1) The 138 domains and 253 domains constructed by Chou et al [8]
(2) PDB40-B constructed by Brenner et al [9]

In the dataset of PDB40-B, the pairwise sequence identities are less than 40%. This means it is non-redundant dataset. It is available at http://scop.mrc-lmb.cam.ac.uk/scop/parse/index.html. In this dataset, there are 1323 domain sequences in seven structural classes, where 1112 belong to the first four major structural classes, i.e. 85% sequences in PDB40-B are in all-α, all-β, α/β and α+β. After excluding the sequences with unknown residues, a dataset including 1050 sequences is obtained, which is labeled as PDB40-b. In this dataset there are 219 domain sequences of all-α class, 308 of all-β class, 284 of α/β class, and 239 of α+β class.

4.2 Accuracy Measure and Results

We use jackknife test for cross-validation. Comparing with resubstitution test or independent data set test, the jackknife test is thought to be more rigorous and reliable [10]. In the jackknife test, each protein in the dataset is singled out in turn as a testing set and all the remaining $n-1$ domains make up a training set. So during the process of jackknife analysis, both the training and testing datasets are actually open, and a protein will in turn move from each to the other. The prediction quality was evaluated by the overall prediction accuracy and prediction accuracy for each class.

In this research, we perform the jackknife test for the datasets described above and compare the accuracy with those of Chou's component-coupled algorithm (AAC) [8] and Bu's auto-correlation function-based approach (ACF) [6]. In the work of Chou, amino acid composition is used as feature vector and the component-coupled algorithm which is equivalent to Bayes decision rule is used as classifying algorithm. Bu's work uses auto-correlation functions based on amino acid index sequence of protein as feature vector and component-coupled algorithm as classifying algorithm. The results of jackknife tests are summarized in Table 2.

Table 2. Predicted results for 138 domains, 253 domains and PDB40-b by jackknife tests

Dataset	Methods	all-α	all-β	α/β	α+β	Overall accuracy
138 domains	AAC	77.78%	55.17%	28.12%	85.37%	63.77%
	ACF	63.9%	41.4%	53.1%	65.9%	57.2%
	This Study	100%	65.52%	68.75%	100%	85.51%
253 domains	AAC	84.13%	79.31%	70.49%	81.69%	79.05%
	ACF	88.89%	74.14%	80.33%	88.73%	83.40%
	This Study	100%	65.52%	77.05%	100%	86.56%
PDB40-b	AAC	54.3%	58.4%	75.4%	30.1%	55.7%
	ACF	41.1%	32.14%	69.01%	26.78%	42.76%
	This Study	100%	28.57%	91.9%	26.78%	60.19%

5 Discussion

From the experiment, we can see that for the data sets we used, the total accuracy is obviously higher than the result got from ACF and AAC algorithm.

Instead of using the amino acid composition as feature vector, ACF algorithm and the method in this paper lay special emphasis on extracting feature vectors from the residue order along the primary structure of protein. But the performance of ACF doesn't markedly better than that of AAC. Comparing with ACF algorithm, our methods extracts more information about the correlation properties of the corresponding numerical signal of a protein on different scales, i.e., it allows to derive some interesting information about the correlation properties of different periodicities present in the numerical signal. So it has better performance than ACF algorithm. From Table 3, we can see that the feature vectors based on sub-signals get higher prediction accuracy than the feature vector based on original numerical signal. By majority voting, we get the final prediction result 60.19%, which is better than all of the results in Table 3.

Table 3. The total accuracy of jackknife test for PDB40-b by using different feature vectors

Feature vector	$X^{(0)}$	$X^{(1)}$	$X^{(2)}$	$X^{(3)}$	$X^{(4)}$	$X^{(5)}$	$X^{(6)}$	$X^{(7)}$	$X^{(8)}$
Total accuracy	42.76%	55.33%	56.86%	55.71%	56.76%	53.33%	56%	56.1%	56%

References

1. Levitt, M., Chothia, C.: Structural Patterns in Globular Proteins. Nature 261(1976)552-557
2. Mallat, S.G.: A Theory for Multirosolutin Signal Decomposition: A Wavelet Representation. IEEE Tnans Pattern Anal. Mach. Intell. 11(1989)674-693
3. Coifman, R.R., Wickerhauser, M.V.: Entropy-Based Algorithms for Best Basis Selection. IEEE Trans. Information Theory 38(1992)713-718
4. Oobatake, M.. Ooi, T.: An Analysis of Non-Bonded Energy of Proteins. J. Theor. Biol. 67(1977)567-584
5. Cornette, J.L., Cease, K.B., Margali, H., Spouge, J.L., Berzofsky, J.A., DeLisi, C.: Hydrophobicity Scales and Computational Techniques for Detecting Amphipathic Structures in Proteins. J. Mol. Biol. 195 (1987) 659-685
6. Bu, W.S., Feng, Z.P., Zhang, Z.D., Zhang, C.T.: Prediction of Protein (Domain) Structural Classes Based on Amino-Acid Index. Eur. J. Biochem. 266 (1999)1043-1049
7. Richard, O. D., Peter, E. H., David G. S.: Pattern Classification, 2nd edn. John Wiley & Son, Inc. (2001)
8. Chou, K.C., Maggiora, G.M.: Domain Structural Class Prediction. Protein Eng. 11 (1998) 523-538

9. Brenner, S.E., Chothia, C., Hubbard, T.: Assessing Sequence Comparison Methods with Reliable Structurally Identified Distant Evolutionary Relationships. Proc. Natl Acad. Sci. USA 95(1998)6073–6078
10. Mardia, K.V., Kent, J.T., Bibby, J.M.: Multivariate Analysis. Academic Press. London (1979) 322- 381

A Brief Review on Allometric Scaling in Biology

Ji-Huan He

College of Science, Donghua University, 1882 Yan'an Xilu Road, Shanghai 20051, China
jhhe@dhu.edu.cn

Abstract. A brief review on allometric scaling in biology is given. Dueling theories aim at explaining the mystery of how an animal's metabolic rate is related to its size, but no convincing result is so far obtained, so the allometric scaling in biology still keeps an intriguing and enduring problem. The contention throughout this body of literature focuses on the arguments for and against the Rubner law or Kleiber law.

1 Introduction

The biological processes underlying allometric scaling relations for metabolic rates in biology have been a topic of long-standing interest and speculation. The relation between metabolic rate of an animal (B) and its mass (M) has the following allometric form

$$B \sim M^a, \tag{1}$$

where a is the scaling exponent. The scaling exponent is a point of contention throughout the open literature, investigators have either contested 3/4-law or attempted it.

In 1883, Rubner[1] found that the metabolic rate depended on the body's surface-to-volume ratio, or $M^{2/3}$. The Rubner law can be obtained by a simple geometry analysis. This law was misunderstood as a "surface law", which correlates the metabolic rate with the body surface. Actually the metabolic cost does not depend upon only the heat exchange through the body surface. So the Rubner 2/3-law, resulting in a Euclidean surface-to-volume relationship, was rebutted by many researchers due to its theory pitfall, but a thorough re-examination of large data sets has concluded that the null hypothesis of a 2/3 power relation can not be rejected[2,3]. He and Chen[4] suggested that the Rubner's law is valid for 2 dimensional organisms.

In 1932, Max Kleiber [5] plotted the first accurate measurements of body size versus log metabolic rate for mammals and birds (Fig.1) and discovered that the exponent was approximately 3/4. The curve was extended by Brody [6] in 1945, and it is now called the mouse-to-elephant curve [7]. It has since been extended, controversially, to a wide range of organisms from mycoplasma ($\sim 10^{-13}$ g) to the blue whale ($\sim 10^8$ g), and it is considered as a ubiquitous law in biology [8,9]. The 3/4-law is relevant in medical imaging [10], life expectancy [11], and many other parameters. A convincing unitary explanation for the allometric scaling laws would have broad implications for our understanding of developmental ontogeny, regeneration, population growth and evolutionary processes, but there is no generally agreed mechanistic model.

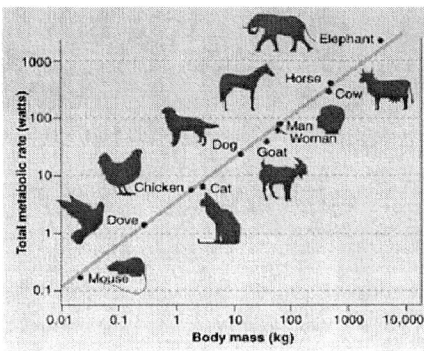

Fig. 1. Kleiber's Mouse-to-Elephant curve

There is no obvious reason for the exponent to have this value, and dueling theories[2,3, 8,9] aim at explaining the mystery of how an animal's metabolic rate is related to its size, but no convincing result is so far obtained, so the allometric scaling in biology still keeps an intriguing and enduring problem. The contention throughout this body of literature focuses on the arguments for and against the Rubner 2/3 law [1] or Kleiber 3/4 law [5].

Taylor et al. found that the maximal metabolic rate induced by exercise scales with $M^{0.86}$ rather than $M^{0.75}$ [12]. No theory can explain this phenomenon till now.

Wang et al. [13] reconstructed Kleiber's law at the organ-tissue level, and they found that 4 metabolically active organs, brain, liver, kidneys and heart, have high specific resting metabolic rates when compared with the remaining less-active tissues, such as skeletal muscle, adipose tissue, bone and skin. Brain, liver, kidneys and heart together account for ~60% of resting energy expenditure in humans, even though the 4 organs represent <6% of body mass. It is still a conundrum of why different organs in mammals have different allometric cascades, and how basal metabolic rate and maximal metabolic rate take different scaling forms continues to exercise biologists.

2 Fractal Networks

West et al.[14] proposed a general model(Fig.2) that describes how essential materials are transported through space-filling fractal networks of branching tubes. The theory, as pointed out by Weibel[15], is indeed an attractive model, a simple unified theory as the solution for a complex system—almost too good to be true. This theory has been questioned by many authors[16,17]. Dawson[16] points out that the assumption that the capillary dimensions do not vary with mammal size is contrary to his early work and to the basic cardiovascular design and physiological processes of mammals. The most pitfall of the theory is to apply the well-known Poiseuille formula to the calculation of the viscous resistance of a capillary, which is not valid for blood flow in capillary, the viscosity depends strongly upon the radius of the capillary. And the assumed steady laminar flow of a Newtonian fluid in capillary does not lead to a minimal viscous resistance, this is contrary to their second assumption that energy

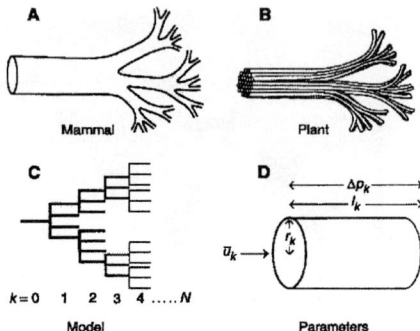

Fig. 2. West et al.'s fractal networks

dissipated is minimized. Dodds et al.[17] also pointed out the contradiction existed in West et al.'s theory, and the optimization procedure based on Poiseuille formula is seen to yield $a \approx 1$ rather than $a=3/4$.

3 Multiple-Cause Cascade Model

Recently Darveau et al. suggested a multiple-cause cascade model of metabolic allometry[18]. The basic equation of the cascade model is

$$MR = a\sum_i c_i M^{b_i}, \qquad (1)$$

where MR is the metabolic rate in any given state, M is body mass, b_i the scaling exponent of the process i, and c_i the control coefficient of the process i. The model was considered as a new perspective on comparative integrative physiology and scaling relationships, but the theory has been refuted elsewhere[19,20]. West et al.[19] pointed out that this scaling equation is based on technical, theoretical and conceptual errors, including misrepresentations of their publications. Furthermore, West et al.[19] concluded that the scaling exponent b depends upon the units of mass: for the basal rate $b \approx 0.76$ when M is in kilograms, and $b \approx 1.08$ when M is in picograms. Banavar et al.[20] showed that this cascade model is flawed and is therefore meaningless both for control of metabolic rate in an organism of a given size and for scaling of the metabolic rate.

4 Cytofractalogy

He and Chen[4] predicted 1/2-law, 2/3-law, and 3/4-law for, respectively, one-dimensional, two-dimensional, and three-dimensional organisms $B \sim M^{D/(D+1)}$, where D is the dimension of space, B is the basal metabolic rate of an organism, M is the mass of the organism. Based on cell fractal geometry(*cytofractalogy*)[21], which is quite different from the fractal-like distribution networks proposed by West et al.[14], we can obtain both Rubner's 2/3 law for 2-dimensional organisms and Kleiber's 3/4 law for 3-dimensional organisms rather than a ubiquitous law in biology. *Cytofractalogy* can successfully explain Taylor et al.'s phenomenon[12] and Wang's experimental observation[13].

The basic idea of *cytofractalogy* is based the fractal dimension of a cell. A cell's metabolic rate (*b*) scales as

$$b \sim r^\delta, \tag{2}$$

where r is the character radius of the cell, δ is the fractal dimension of the cell's surface. When an animal in motion, each cell has to enlarge its surface in order to obtain more energy from its environment, leading to increase the value of δ, and obtaining maximal metabolic rate when in motion $B_{max} \sim M^{0.9}$, which agrees reasonably with Taylor et al.'s data, $B_{motion} \sim M^{0.86}$ [12].

5 4-Dimentional Life

In 1977, Blum[22] suggested that the 3/4-law can be understood by a four dimensional approach. In $D+1$ dimensions, the "area" A of the hypersurface enclosing a D+1 dimensional hypervolume scales like $A \sim V^{D/(D+1)}$, where D is the spatial dimension of the organism. When $D=3$, we have $A \sim V^{3/4}$, a four dimensional construction. The fourth dimension of life have caught much attention since West et al.'s work[23]. But so far no physical explanation is given for the fourth dimension of life. According to *cytofractalogy*, We have proved that $n \sim r$ for three dimensional organisms, so the number of cells in an organ endows the fourth dimension of life, which exists only in three dimension organs, such as heart and kidneys. If a cell is isolated from a heart of a rat, no 4th dimension is endowed. However, if enough number of heart cells are cultivated and accumulated together in three dimensions, the isolated cardiac cell begins to beat[24], and the 4th dimension of life is endowed.

Diversity of motion patterns in nature suggests different scaling patters in biology by constraints. Constraining a particle motion will naturally change its motion patterns, similarly artificial or natural energetic constraint of cell growth may alter expected scaling patters. If heart cells are cultivated on a plane, the isolated cardiac cells can not beat[24].

We have also illustrated $n \sim r^0$ for two dimensional lives, so the 4th dimension of life does not exist in leaves or barks of plants. Some animal organs such as liver and cornea behave in 2-dimensional forms. This can explain a destroyed leaf can still make photosynthesis, though the output depend linearly on the total number of cells in the leaf.

6 5-Dimentional Human Brain

In Ref. [25], we will show that human brain has a further dimension. Brain cells are not approximately spherical, in contrast to other human cells. The brain is made up of many cells, including neurons and glial cells. The branching dendrites and the axon of a neuron make the brain cell more geometrically complex than spherical cells.

Nature selection leads most cells to spheres, which have maximal surface so that the energy exchange between cell and its environment reaches maximum. In the other hand, snowflake is a nature phenomenon, endowing an optimal fractal dimension in nature, human brain cell has a similar construction of snowflake, so that it can receive as much as possible information from its environment. Recalling that fractal dimensions of a snowflake is about 2.26. We, therefore, assume that $\delta \approx 2.26$ for human brain. Note that the human brain is best developed system among animals, the fractal dimension for vertebrate brain or spinal cord, or an insect brain, or the nerve net of a coelenterate can not arrive at this extremum.

Assuming that there are n basal cells with characteristic radius r in human brain, the overall metabolic rate, B_{brain}, of the brain scales as

$$B_{brain} \sim M_{brain}^{3/(6-\delta)}. \tag{36}$$

The fractal dimension of a cell in human brain is $\delta \approx 2.26 \approx 9/4$. Substituting this into Eq.(36) results in a 4/5-law for human brain:

$$B_{brain} \sim M_{brain}^{4/5}. \tag{37}$$

Nature endows humans uniquely with a 5th dimension!
Wang et al. [13] obtain the following relations for human brain:

$$B_{brain} = 20.5 m^{0.62} \text{ (kJ/d)}, \tag{38}$$

$$M_{brain} = 0.011 m^{0.76}, \tag{39}$$

where m is the body mass. From (38) and (39), we have

$$B_{brain} \sim M_{brain}^{0.62/0.76} = M_{brain}^{0.81}, \tag{40}$$

which is in good agreement with our prediction, Eq.(37).

The 0.81 power relationship implies an optimal fractal figure of brain cells, and the fractal shape rather than sphere leads to maximal heat exchange due to its extreme surface, this can explain why brain consumes large amounts of energy, and its metabolic rate can be arrived at near "maximal" even under general BMR conditions in the body, the maximal scaling exponent as illustrated in the above section reaches 0.9.

We might call the 4th dimension of life the life-dimension, and the 5th dimension of the brain as the thought-dimension; it originates from the fractal structures of brain cells.

7 Allometric Scaling Laws for Different Organs

In[21], we rebuild an allometric scaling relationship for metabolic rates for organs in the form $B_{organ} \sim T_{organ}^{(D+N/6)/(D+1)}$, where B_{organ} is the metabolic rate of an organ,

T_{organ} its mass, D is the dimension of the studied organ, N is cell's degree of freedom in the discussed organ, for example, $N=4$ for kidneys, and $N=3$ for heart. This prediction agrees quite well with the experiment data for brain, liver, heart, and kidneys, and can also explain very well the reason why the maximal metabolic rate induced by exercise scales with $M^{0.86}$ rather than $M^{0.75}$.

It is mysterious why kidney is of 3 dimensional construction with 4 freedoms of motion($D=3, N=4$), while liver acts like a rest leaf of plant ($D=2$, $N=0$), the microstructure evidences of histology in kidney and liver successfully support our points.

References

1. Rubner, M.: Zeitschrift fur Biologie 19 (1883) 536-562
2. Riisgard, H.U.: No foundation of a "3/4 power scaling law" for respiration in biology. Ecology Let. 1 (1998) 71-73.
3. White, C.R., Seymour, R.S.: Mammalian basal metabolic rate is proportional to body mass 2/3. Proc. Natl. Acad. Sci., USA 100(2003) 4046-4049
4. He, J.H., Chen, H.: Effects of Size and pH on Metabolic Rate. International Journal of Nonlinear Sciences and Numerical Simulation. 4 (2003) 429-432
5. Kleiber, M.: Body size and metabolism. Hilgardia 6(1932) 315-353
6. Brody, S.: Bioenergetics and Growth, with Special Reference to the Efficiency Complex in Domestic Animals, Reinhold, New York(1945)
7. Mackenzie, D.: New clues to why size equals destiny. Science. 284(1999)1607-1609
8. West, G., Woodruff, W.H., Brown, J.H.: Allometric scaling of metabolic rate from molecules and mitochondria to cells and mammals. Proc. Natl. Acad. Sci., USA. 99(2002) 2473-2478
9. Damuth, J.: Scaling of growth: plants and animals are not so different. Proc. Natl. Acad. Sci., USA. 98(2001) 2113-2114
10. Kuikka, J.T.: Scaling Laws in Physiology: Relationships between Size, Function, Metabolism and Life Expectancy. International Journal of Nonlinear Sciences and Numerical Simulation. 4(2003) 317-328
11. Kuikka, J.T.: Fractal analysis of medical imaging. International Journal of Nonlinear Sciences and Numerical Simulation 3(2002) 81-88.
12. Taylor, C. R. et al.: Respir. Physiol. 44(1981)25-37.
13. Wang, Z. M., O'Connort, T.P., Heshka, S., Heymsfield, S. B.: The reconstruction of Kleiber's law at organ-tissue level. Journal of Nutrition 131(2001)2967-2970
14. West, G.B., Brown, J.H., Enquist, B.J.: A general model for origin of allometric scaling laws in biology. Science. 276 (1997) 122-126
15. Weibel, E. R.: The pitfalls of power laws. Nature. 417(2002) 131-132
16. Dawson, T. H.: Allometric scaling in biology. Science. 281(1998) 751-751
17. Dodds, P. S., Rothman, D. H.,Weitz, J. S.: Re-examination of the 3/4-law of metabolism. Journal of Theoretical Biology. 209(2001) 9-27.
18. Darveau, C. A., Suarez, R. K. , Andrews, R. D., Hochachka, P. W.: Allometric cascade as a unifying principle of body mass effects on metabolism. Nature. 417 (2002) 166-170.
19. West, G. B., Savage, V. M., Gillooly, J., Enquist, B. J., Woodruff, W. H. & Brown, J. H. Why does metabolic rate scale with body size. Nature. 421(2003) 713.
20. Banavar, J. R., Damuth, J., Maritan, A., Rinaldo, A.: Allometric cascades. Nature. 421 (2003) 713-714.

21. Blum, J. J.: On the geometry of four-dimensions and the relationship between metabolism and body mass. J. Theor. Biol. 64(1997) 599-601
22. He,J.H., Huang,Z.: A novel model for allometric scaling laws for different organs . Submitted.
23. West, G. B., Brown, J. H., Enquist, B. J.: The Fourth Dimension of Life: Fractal Geometry and Allometric Scaling of Organisms. Science. 284 (1999) 1677-1679
24. NASA microgravity research highlights, Advancing heart research, reprinted from the Spring 2000 issue of Microgravity News. Detailed information is published in American Journal of Physiology 277(Heart Circ. Physiol.46), H433-H444 (1999) and Biotechnology and Bioengineering 64(1999) 580-589.
25. He,J.H., Zhang,J.: Fifth dimension of life and the 4/5 allometric scaling law for human brain. Submitted to Cell Biology International.

On *He Map* (*River Map*) and the Oldest Scientific Management Method

Ji-Huan He

College of Science, Donghua University, 1882 Yan'an Xilu Road, Shanghai 20051, China
jhhe@dhu.edu.cn

Abstract. In ancient China, preliterate societies whose past is hard to study showed excellent management through *Luo Writing* and *He Map*(*River Map*), which have millennia history. It is intriguing that the recovered optimal management is based on Pi, the value of Pi was approximated by 3 in *He Map*. If we choose Pi=22/7 or Pi=335/113, another reasonable management model appears.

1 What is He Map (River Map)

Before introduction of the *He Map(Ho Map, River Map)*, we first introduce the oldest ancient Chinese classics named *Yi Jing(* the *I Ching*, book of changes , in lit. theory of changes, or science for changes)[1], it is a book of divination practice expound with profound and peculiar philosophical theory, and it has a history of several millennia. The classics might appear before preliterate times, many commentaries were made by many famous ancient philosophers(e.g. *Confucius)* and mathematicians (Zu Chongzhi[2]).

The dates of ancient Chinese works[2,3,4,5] are very unclear. A given book may have been produced as a single work, or it may be a compilation of several earlier works. In either case, the material may be centuries older than the book and the preserved version of the book may be several centuries and editions later than its original form.

The 64 double *Hexagrams* can be obtained by duplicating the 8 *Hexagrams,* each of which is composed by 3 linear signs (▬▬▬ continued line called *Yang* or ▬ ▬ dashed line called *Yin),* see Fig.1. In Chinese, *Yang* denotes Sun, man, strong, positive, etc., and *Yin* Moon, woman, weak, negative, etc..

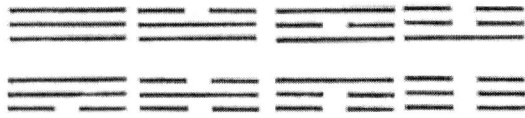

Fig. 1. Eight Hexagrams representing respectively Heaven, water, fire, Sun, wind, Moon, mountain, and Earth

Ancient Chinese used these simplest signs ('━━━━ and ━━ ━━) to record the mostmysterious philosophic ideas and evens, for example, ☰ records the activity of dragon, from its initial, to powerful stage, to utmost, then to its comedown, it can be considered as the best stenography in world. In 1701, Joachim Bouvet, a Jesuit missionary, described the binary ordering of the *I Ching* hexagrams to famous mathematician G.W. Leibniz, reviving Leibniz's interest in binary in *Yin-Yang* system[1]. Fig.2 is a *Taiji* image which is of *Yin-Yang* original.

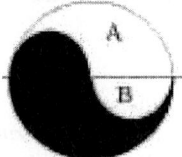

Fig. 2. *Taiji* image, the ratio of the area A to the area B is 3, which was used as Pi in China millennia years ago

The mystery in the *I Ching* has been searched for millennia, but we can not still explain it reasonably, either does the *He Map*(See Fig.3), which also has a history of several millennia, it is said that the *He Map* is obtained from a *River God*. The really meaning of the *He Map* is still an open problem.

2 Mystery in He Map

Why was the magical image called *He Map*(lit. a schematic for river)? Why were the white and black stones so arranged? What did the mysterious meaning for each stone? Among others, these are open questions till now.

It is astonishing that the *He Map* is the oldest blueprint for water engineering, and it is even more amazing that *He Map* is relative to pi, and the constructer of the *He Map* is similar to that of G-C and A-T pairs in DNA sequencing, it is really beyond belief [6].

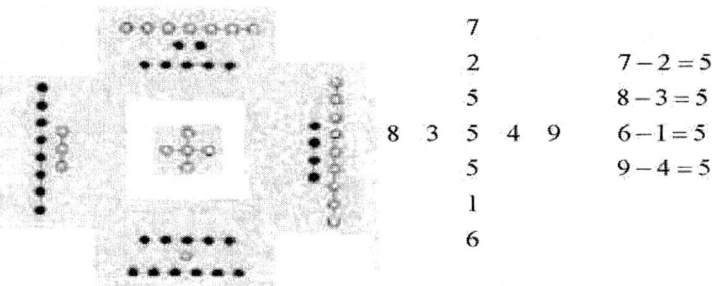

Fig. 3. The *He Map* (The *River Map*),which was preliterate, so it has a history of at least 5 millennia

2.1 The Number 5

Taoism places great importance on the number 5. There were 5 elements: earth, fire, water, wood, metal; five directions: north, east, south, west, center; 5 colours: yellow, blue, red, white, black; five planets: Venus, Jupiter, Mercury, Mars, and Saturn.

In the *He Map,* only the number 5 was repeated, and it was placed the center of the map, illustrating the importance of the number 5, which is also located at center of the *Luo Writing,* as illustrated in Fig. 4.

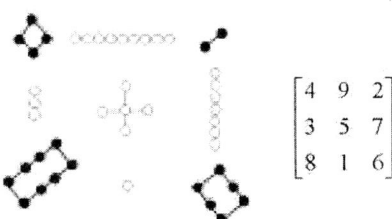

Fig. 4. The *Luo Writing* and numbers in the image *of Luo Writing,* which might be a defence pattern, the center white stone denotes the most important position

The mystery in *Luo Writing* is not solved yet. If it is a number magic as illustrated in Fig. 4b, then why does it need white and black stones? Why are the black stones arranged in such a strange form?

Understanding the regulation of number 5 in *He Map* would have broad implications on furthering our knowledge of this mysterious image. The white stone in *Yin*-form should be of woman, weakness, and the black stone in *Yang*-form should be of man, mightiness. The so-called *He Map* might be an oldest blueprint for water engineering used to establish a riverbank. The white stone stands for woman, while black stone man or young boy. So the present author explains the *He Map* as illustrated in Fig. 5.

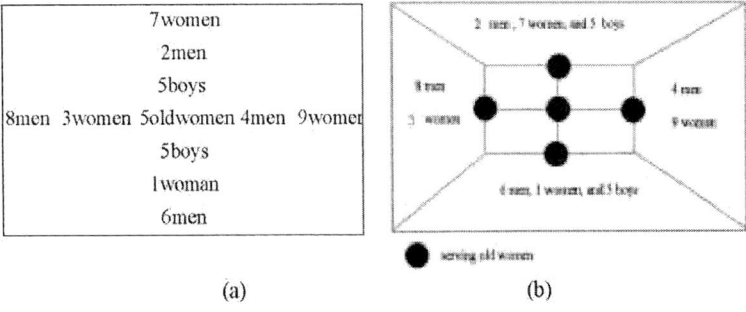

Fig. 5. Is the *He River* used for riverbank establishment ? (a) Stones in *He Map* denote the number of differential people working together in a water engineering. It is an oldest blueprint in world, and it is best managed even in modern view. (b) The labour force is equivalently distributed in the 4 parts of the riverbank, so that the progress is well matched. There 5 serving persons, so that each side can be best served

2.2 He Map and Pi

If the a man, a woman, and a young boy can dig, respectively, x, y, and z m^3 earth per hour, as illustrated in Fig. 5b, each side has same labor force, so we have the following identity.

$$8x + 3y = 4x + 9y = 2x + 7y + 5z = 6x + y + 5z, \quad (1)$$

leading to the following results

$$x=3n,\ y=2n,\ z=2n. \quad (2)$$

Eq. (2) replies that the ratio of labor ability of woman to man is 2: 3

$$\text{Woman : Man} = 2:3 \approx 0.66. \quad (3)$$

It is interesting that 2/3 is closed to the optimal number(pi/2)[6], so we write $\pi/2 \approx 3/2$. It is amazing that in preliterate times, ancient Chinese use pi=3, and it was considered as a ruler that the perimeter of circle is 3 when diameter equals to 1. And ratio 3:1 or 3:2 was widely used in arts and construction design. Standard cards are 2 by 3, creating a ratio of 0.666, slightly larger than a perfect golden section of 0.618. In *Taiji* image(Fig.2), the ratio of the area A to the area B is 3.

It should be also emphasized that $y:z=1:1$, that means a woman and a boy contribute same in the engineering, so in the *He Map*, so boys and men are assigned in different line same signs.

In order to validate our deduction, we enlarge the *He Map* as follows (Fig.6).

		$11p$		
		$7y$		
		$2x$		
		$5z$		
$12p$ $8x$ $3y$		5		$4x$ $9y$ $13p$
		$5z$		
		$1y$		
		$6y$		
		$10p$		

Fig. 6. A modification of He Map, leading to the interesting result: pi/2=22/14. Here x,y,z, and p represent, respectively the labor ability for man, woman, young man, old man

If a man, a woman, a young man, and an old man can dig, respectively, x, y, z, and p m^3 earth per hour, then we have the following identity.

$$8x + 3y + 12p = 4x + 9y + 13p = 2x + 7y + 5z + 11p = 6x + y + 5z + 10p \qquad (5)$$

Solving (5), we have

$$x: y: z : p = 22 : 14 : 16 : 4. \qquad (6)$$

It is really very astonishing to find that

$$\frac{\pi}{2} \approx \frac{x}{y} = \frac{22}{14}. \qquad (7)$$

The result (7) is really very excited, and $y: z = 14:16 \approx 1:1$ agrees well the previous assumption.

Now we further add more 62 old women to do water engineering, as a result, the *He Map* can be modified as follows (Fig. 7).

			15q			
			11p			
			7y			
			2x			
			5z			
16p	12q	8x 3y	5	4x 9y	13p	17q
			5z			
			1y			
			6x			
			10q			
			14p			

Fig. 7. A modification of He Map, leading to an astonishing result: pi/2=335/226. Here x,y,z, p, and q represent, respectively the labor ability for man, woman, young man, old man, and old woman.

If a man, a woman, and a young man, an old man and an old woman can dig, respectively, $x, y, z, p,$ and q m^3 earth per hour, then we have the following identity

$$8x + 3y + 16p + 12q = 4x + 9y + 13p + 17q = 2x + 7y + 5z + 11p + 15q$$
$$= 6x + y + 5z + 11p + 10q. \qquad (8)$$

Solving (8), we have

$$x : y : z : p : q = 8875 : 5650 : 7986 : 3200 : 2240. \qquad (9)$$

It is amazing to find the remarkable result for pi:

$$\frac{\pi}{2} \approx \frac{8875}{5650} = \frac{335}{226}. \qquad (10)$$

If we add more 79 young girls, the result is astounding

$$\frac{\pi}{2} \approx \frac{3927}{2500}.$$
(11)

We have now the very reason that x/y converges fast to pi/2 if more labors take in the establishment.

3 Conclusion

China is one of the four countries with an ancient civilization. The ancient Chinese civilization had made greatest contributions to the development of human culture. The *He Map* might be the oldest preserved blueprint in the world.

References

1. Book of Changes, Hunan PublishingHouse, Changsha (1993) (in Chinese with EnglishTranslation by Canon McClatchie).
2. He,J.H.: Zu-Geng's axiom vs Cavalieri's theory, Applied Math. Computation, 152(2004) 9–15
3. He,J.H.: Solution of nonlinear equations by an ancient Chinese algorithm, Applied Math.Computation. 151(2004)293–297
4. He,J.H.: Some interpolation formulas in Chinese ancient mathematics, Applied Mathematics and Computation. 152(2004)367–371
5. He,J.H. He Chengtian's inequality and its applications, Applied Math. Computation,151(2004): 887-891
6. He, J.H.: Mysterious Pi and a Possible Link to DNA Sequencing, International Journal ofNonlinear Sciences and Numerical Simulation 5(2004)263–274

A Novel Feature Selection Approach and Its Application

Gexiang Zhang[1,2], Weidong Jin[2], and Laizhao Hu[1]

[1] National EW Laboratory, Chengdu 610036 Sichuan, China
dylan7237@sina.com
[2] School of Electrical Engineering, Southwest Jiaotong University,
Chengdu 610031 Sichuan, China

Abstract. Feature selection is a satisfactory optimization problem. Most feature selection methods did not consider the cost of feature extraction and the automatic decision of feature subset dimension. So a novel approach called satisfactory feature selection method (SFSM) was proposed. SFSM integrated feature extraction with feature selection and considered classification performance of feature samples, the dimension of feature subset and the complexity of feature extraction simultaneously. Experimental results show that SFSM selects more satisfying feature subset than sequential forward selection using distance criterion (SFSDC) and the method presented by Tiejun Lü (GADC). Also, SFSM achieves higher accurate recognition rate than original feature set, SFSDC and GADC, which verifies the validity of the proposed method.

1 Introduction

Feature selection is a typical NP-hard constrained combination optimization problem. Besides exhaustive search method, no approach can obtain the optimal solution. [1-4] In fact, feature selection is a satisfactory optimization problem essentially and the solutions obtained using feature selection are satisfying results. Few feature selection methods in existing literatures [1-4] consider the cost of feature extraction and automatic decision of the dimension of feature subset. So it is necessary that the principles of multi-objective satisfactory optimization (SAOP) are introduced into feature selection to consider many factors simultaneously and to evaluate the feature subset selected. SAOP was presented to solve two difficult optimization problems: no optimal solution problem, and too much cost for obtaining the optimal solution or no method for achieving the best solution. [4-7] SAOP has good application potentiality. Jin [5] presented principles of satisfactory solution for fuzzy-neural computing. Xi [6] proposed satisfactory control of complex industrial process. Jin [7] studied the principles and method of SAOP systematically and gave two models of SAOP to optimize parameters of control systems and train operation. Zhao [8] and Zhang [9] did further research on SAOP. Some remarkable fruits have been achieved [4-9]. The core of SAOP is to emphasize *satisfaction* instead of *optimum*. In this paper, the principles of

---

[1] This work was supported by the National Defence Foundation (No.51435030101ZS0502).

SAOP are introduced into pattern recognition and satisfactory feature selection method (SFSM) is proposed to select the most satisfying feature subset with small dimension, low complexity and strong discrimination from a large number of features.

2 Satisfactory Feature Evaluation and Selection

Satisfactory rate (SR) is the most important and fundamental concept in SAOP. The definition of SR is different in different applications. [4-7,10] This paper gives a new definition of SR as follows.

Definition 1. Suppose that there are n classes, SR of a feature t is defined as

$$s_t = s(J_t), \ J_t = J(J_{ij}, 1 \leq i < j \leq n), \ J_{ij} = JT(X_i, X_j), \ s_t \in [0,1] \quad (1)$$

Where $s(J_t)$ is feature satisfactory rate function (FSRF) and is a monotone and nondecreasing function; J_t is class-separability criterion of feature t; $J(\cdot)$ is a synthesis function of class-separability. $JT(\cdot)$ is class-separability criterion function of the ith class and the jth class. J_{ij} is the value of function $JT(\cdot)$. X_i and X_j are feature sample vector of the ith class and the jth class, respectively. If the number of feature samples is M, $X_i = [x_{i1}, x_{i2}, \cdots, x_{iM}]$, $X_j = [x_{j1}, x_{j2}, \cdots, x_{jM}]$.

FSRF is used to evaluate one feature. The bigger SR of a feature is, the stronger discriminatory capability of the feature is. The following description gives the definition of feature set SR (FS²R).

Definition 2. Suppose that there is a feature set $T = [t_1, t_2, \cdots, t_Z], Z(Z \geq 2)$ is the dimension of T. FS²R is defined as

$$S_{ts} = \varphi(s_d, s_c, s_m), S_{ts} \in [0,1] \quad (2)$$

Where $\varphi(\cdot)$ is FS²R function; s_d is the SR of the dimension of feature set T; s_c is the SR of complexity of feature set T and s_m is the SR of class-separability of feature set T. s_d, s_c and s_m are respectively

Definition 3. If the dimension of a feature set is d, s_d is defined as

$$s_d = f(d) = (d_{max} - d)/(d_{max} - d_{min}), s_d \in [0,1] \quad (3)$$

Where d_{max} is the dimension of the feature set and d_{min} is the minimal dimension of expected feature subset. s_d reflects on SR of the dimension of selected feature subset for a designer. The purpose of considering s_d is to decide automatically the dimension of a satisfying feature subset.

Definition 4. Suppose that there is a feature set $T = [t_1, t_2, \cdots, t_Z]$, Z is the dimension of T. Complexity SR of T is defined as

$$s_c = g(C) = (C_{sum} - C)/(C_{sum} - C_{min}), s_c \in [0,1] \quad (4)$$

Where C_{min} is the minimal complexity in original feature set and C_{sum} is total complexity of all features in original feature set, i.e. $C_{sum} = \sum_{i=1}^{L} C_i$, Where L is the dimension of original feature set and C_i is complexity of the ith feature. In (4), C is complexity of feature set T, i.e. $C = \sum_{j=1}^{Z} C_j$. Introduction of complexity SR can consider the cost of feature extraction.

Definition 5. Class-separability SR of a feature set is used to evaluate satisfying degree of discriminatory capability of feature set $T = [t_1, t_2, \cdots, t_Z]$ and is defined as

$$s_m = \psi(J^m), J^m = JM(J_{ij}^m, 1 \leq i < j \leq n), J_{ij}^m = JTM(X_i^m, X_j^m), s_m \in [0,1] \quad (5)$$

Where $\psi(J^m)$ is feature-set-class-separability satisfactory rate function and is a monotone and nondecreasing function; J^m is class-separability criterion of feature set T; $JM(\cdot)$ is a synthesis function of feature-set class-separability. $JTM(\cdot)$ is class-separability criterion function of the ith class and the jth class in the feature set T. J_{ij}^m is the value of function $JTM(\cdot)$. X_i^m and X_j^m are feature sample matrix of the ith class and the jth class, respectively. If the number of feature samples is M, there are

$$X_i^m = \begin{bmatrix} x_{i1}^{t_1} & x_{i2}^{t_1} & \cdots & x_{iM}^{t_1} \\ x_{i1}^{t_2} & x_{i2}^{t_2} & \cdots & x_{iM}^{t_2} \\ \vdots & \vdots & \cdots & \vdots \\ x_{i1}^{t_Z} & x_{i2}^{t_Z} & \cdots & x_{iM}^{t_Z} \end{bmatrix} \quad X_j^m = \begin{bmatrix} x_{j1}^{t_1} & x_{j2}^{t_1} & \cdots & x_{jM}^{t_1} \\ x_{j1}^{t_2} & x_{j2}^{t_2} & \cdots & x_{jM}^{t_2} \\ \vdots & \vdots & \cdots & \vdots \\ x_{j1}^{t_Z} & x_{j2}^{t_Z} & \cdots & x_{jM}^{t_Z} \end{bmatrix} \quad (6)$$

According to the above evaluation criterion, the detailed algorithm of selecting a satisfying feature subset from original feature set is described as follows.

Step 1 Deciding the number n of classes and the dimension L of original feature set.

Step 2 Deciding the parameters d_{max} and d_{min}. Computing the complexity of each feature in original feature set and deciding the parameters C_{sum} and C_{min}.

Step 3 Quantum genetic algorithm (QGA) [11] is used for finding automatically the most satisfying feature subset.

Step 4 Computing SRs of the dimensions of all feature subsets in terms of the dimensions of all feature subsets using (3).

Step 5 Computing SRs of the complexities of all feature subsets in terms of the complexities of all feature subsets using (4).

Step 6 According to (5), SRs of discriminatory capabilities of all feature subsets are calculated. In calculating SRs, several functions are respectively

$$\psi(J^m) = (J^m)^2, JM(J_{ij}^m, 1 \le i < j \le n) = \sum_{i=1}^{n-1}\sum_{j=i+1}^{n} w_{ij} J_{ij} \tag{8}$$

$$JTM(X_i^m, X_j^m) = 1 - \frac{\int_z \cdots \int f(x_1, \cdots x_z) \cdot g(x_1, \cdots x_z) dx_1 \cdots dx_z}{\sqrt{\int_z \cdots \int f^2(x_1, \cdots x_z) dx_1 \cdots dx_z} \sqrt{\int_z \cdots \int g^2(x_1, \cdots x_z) dx_1 \cdots dx_z}} \tag{9}$$

In (8), all w_{ij} ($1 \le i < j \le n$) are chosen as the same values. Z is the dimension of feature subset. $f(\cdot)$ and $g(\cdot)$ are distribution functions of feature samples of the ith class and the jth class in the feature subset, respectively. In general, Gaussian function with the parameters of means μ_i, μ_j and variances σ_i, σ_j is chosen as the distribution functions. $\mu_i = E(X_i^m), \mu_j = E(X_j^m), \sigma_i = D(X_i^m), \sigma_j = D(X_j^m)$. From (9), if the distribution functions of feature samples of two classes are completely separable in multi-dimensional feature space, SR of discriminatory capabilities of the feature subset gets to the maximal value 1. As the overlapping parts of the two distribution functions increase, SR will decrease. Till the two distribution functions are completely overlapped, SR arrive at the minimal value 0.

Step 7 FS^{2}R function $\varphi(\cdot)$ relates only 3 factors s_d, s_c and s_m. So linear weighted sum function can be chosen as $\varphi(\cdot)$, that is

$$\varphi(s_d, s_c, s_m) = (\omega_d s_d + \omega_c s_c + \omega_m s_m)/(\omega_d + \omega_c + \omega_m) \tag{10}$$

Where $\omega_d, \omega_c, \omega_m$ are weighting coefficients of s_d, s_c, s_m, respectively.

Step 8 FS^{2}R function $\varphi(\cdot)$ in step 7 is chosen as fitness function of QGA to evaluate all individuals in population.

3 Application Example

In our prior work [12-14], 16 features have been extracted from 10 typical radar emitter signals (RESs). The 10 RESs are CW, BPSK, QPSK, MPSK, LFM, NLFM, FD, FSK, IPFE and CSF, respectively. Original feature set is composed of 16 features that are labeled as 1, 2, $\cdots$, 16, respectively. In the beginning, for every RES, 150 feature samples are generated in each SNR point of 5dB, 10dB, 15dB and 20dB. (intrapulse noise) Thus, 600 samples of every RES in total are generated when SNR varies from 5dB to 20dB. The samples are classified into two groups: training group and testing group. The training group, which consists one third of all samples, is applied to make the experiment of feature SR evaluation, feature selection and classifer training. The testing group, represented by other two thirds of samples, is used to test trained classifers. The SRs of 16 features and the ordering results of all SRs are shown in Table 1.

The feature labeled 12 and 6 are respectively the most satisfying and the most unsatisfying features. The parameters are $L=16$, $n=10$, $M=200$, $d_{max}=16$, $d_{min}=1$, $C_{sum}=2163.503$, $C_{min}=6.203$, $P=20$, $m=16$, $\omega_d=0.20$, $\omega_c=0.10$, $\omega_m=0.70$. In each one of 50 experiments using SFSM, the obtained feature subset is composed of feature 5 and 10. Feature SR is 0.9817. To bring into comparison, SFSDC [2,4] and GADC [3] are also used to select the best feature subset from original feature set. Because SFSDC and GADC cannot decide automatically the dimension of the best feature subset, the dimension of feature subset obtained by SFSM is chosen as that of SFSDC and GADC so as to draw a comparison of recognition results of three methods. First of all, the original feature set (OFS) is used to make recognition experiment in which classifiers are neural network classifiers (NNC) whose structure are 16-25-10. 'tansig' and 'logsig' are chosen as transfer functions in the hidden layer and in the output layer respectively. Training algorithm is RPROP [15]. Ideal outputs are "1". Output tolerance is 0.05 and output error is 0.001. Average training generation (ATG) and average accurate recognition rate (A^2R^2) in 50 experiments are shown in Table 2. If FS^2R is

Table 1. SR of each feature in OFS and ordering results (OR)

Feature	1	2	3	4	5	6	7	8
SR	0.798	0.734	0.902	0.922	0.959	0.708	0.887	0.809
OR	13	14	7	6	2	15	8	11
Feature	9	10	11	12	13	14	15	16
SR	0.953	0.941	0.853	0.971	0.922	0.844	0.805	0.950
OR	3	5	9	1	6	10	12	4

Table 2. Comparison results of four methods

Methods	SFSM	OFS	SFSDC	GADC
Feature set	5,10	1~16	4,5	6,7
BPSK	98.58%	100.00%	89.59%	91.93%
QPSK	88.04%	75.25%	58.99%	98.08%
MPSK	100.00%	98.95%	78.23%	85.47%
LFM	100.00%	100.00%	96.03%	100.00%
NLFM	100.00%	100.00%	100.00%	83.53%
CW	100.00%	100.00%	100.00%	100.00%
FD	100.00%	99.97%	100.00%	100.00%
FSK	100.00%	100.00%	100.00%	100.00%
IPFE	100.00%	100.00%	97.12%	94.08%
CSF	100.00%	85.92%	98.88%	95.18%
A^2R^2	98.66%	95.17%	91.88%	94.83%
NNC	2-15-10	16-25-10	2-15-10	2-15-10
ATG	248.10	324.67	4971.80	1146.50

used to compute the SR of OFS, the result is 0.6107. Feature subset obtained by SFSDC is composed of feature 4 and 5 and FS^2R is 0.8907. Feature subset obtained by GADC is composed of feature 6 and 7 and FS^2R is 0.9374. Then, the feature subsets obtained by SFSM, SFSDC and GADC are used to train NNC respectively. The structure of NNC is 2-15-10. ATG and A^2R^2 in 50 experiments of 3 methods are shown in Table 2 when SNR varies from 5dB to 20dB. From Table 2, conclusions can be drawn: (i) In comparison with OFS, SFSM not only lowers significantly the dimension of feature vector and the cost of feature extraction, but also simplifies classifier design and enhances recognition efficiency and accurate recognition rate. (ii) In comparison with SFSDC and GADC, SFSM selects better feature subset and achieves higher A^2R^2 because the ATG of NNC using the feature subset selected by SFSM is much less than that of SFSDC and GADC, and the A^2R^2 of SFSM amounts to 98.66%, which is higher 6.78%, 3.83% and 3.49% than that of SFSDC, GADC and OFS, respectively.

5 Concluding Remarks

This paper introduces feature satisfactory rate and feature set satisfactory rate to consider multiple factors synthetically and designs satisfactory rate functions to evaluate a single feature and feature set. Furthermore, satisfactory feature selection method is proposed to select the most satisfying feature subset with high quality, small dimension and low complexity from original feature set composed of lots of features. SFSM achieves higher A^2R^2 and recognition efficiency than SFSDC, GADC and OFS. Besides, satisfactory rate functions in satisfactory feature evaluation can be changed according to different applications and design purpose, so SFSM has much flexibility.

References

1. Cover, T.M., Vancompenhon, J.M.: On the possible ordering in the measurement selection problem. IEEE Trans. System, Man and Cybernetics. Vol.7, No.9. (1977) 657-667
2. Molina, L.C., Belanche, L., and Nebot, A.: Feature Selection Algorithms: A Survey and Experimental Evaluation. Proc. of Int. Conf. on Data Mining. (2002) 306-313
3. Lü, T.J., Wang, H., and Xiao, X.C.: Recognition of modulation signal based on a new method of feature selection. Journal of Electronics and Information Technology. Vol.24, No.5. (2002) 661-666
4. Jin, F.: Intelligent foundation of neural computing: principles & methods. Chengdu: Southwest Jiaotong University Press. (2000)
5. Jin, F., Hu, F.: Principles of satisfactory-solution for fuzzy neural computing. Journal of the China Railway Society. Vol.18, No.2. (1996) 102-107
6. Xi, Y.G.:. Satisfactory control of complex industrial process. Information and Control. Vol.24, No.1. (1995) 14-20
7. Jin, W.D.: Study on satisfactory optimization and train operation optimization method [Doctoral Dissertation]. Chengdu: Southwest Jiaotong University. (1998)
8. Zhao, D., Jin, W.D.: The application of multi-criterion satisfactory optimization in fuzzy controller design. Proc. of Int. Workshop on Autonomous Decentralized System. (2002) 162-167

9. Zhang, G.X., Jin, W.D., and Hu, L.Z.: Study on satisfactory optimization of parameters of controller in multivariable control system. Control Theory & Apllications. Vol.21, No.3. (2004) 362-366
10. Huang, H.Z., Yao, X.S., and Zhou, Z.R.: Review of the satisfactory degree theory and its application. Control and Decision. Vol.18, No.6. (2003) 641-645
11. Zhang, G.X., Jin, W.D., and Li, N.: An improved quantum genetic algorithm and its application, Lecture Notes in Artificial Intelligence. Vol.2639. (2003) 449-452
12. Zhang, G.X., Jin, W.D., and Hu, L.Z.: Fractal feature extraction of radar emitter signals. Proc. of 3th Asia-Pacific conf. on Environmental Electromagnetics, 2003, 161-164
13. Zhang, G.X., Jin, W.D., and Hu, L.Z.: Complexity feature extraction of radar emitter signals, Proc. of 3th Asia-Pacific conf. on Environmental Electro-magnetics, 2003, 495-498
14. Zhang, G.X., Rong, H.N., Jin, W.D., and Hu, L.Z.: Radar emitter signal recognition based on resemblance coefficient features. Lecture Notes in Artificial Intelligence. Vol.3066. (2004) 665-670
15. Riedmiller, M., Braun, H.: A Direct Adaptive Method for Faster Back Propagation Learning: the RPROP Algorithm. Proc. of IEEE Int. Conf. on Neural Networks. (1993) 586-591

Applying Fuzzy Growing Snake to Segment Cell Nuclei in Color Biopsy Images

Min Hu, Xijian Ping, and Yihong Ding

Information Engineering University, No.837, P.O.Box 1001, Zhengzhou, China 450002
{hhummin, dingyihong}@sina.com

Abstract. This paper proposes a novel cell nucleus segmentation method for color esophageal biopsy image. For each nucleus of cell image, based on color characteristics of cell nucleus, a threshold separating the nucleus can be detected automatically in each *RGB* color component. According to the thresholds, two fuzzy domains are established for each color component with bell-curve and *S*-curve membership functions. Then we propose a novel growing snake to extract cell nucleus boundary. Described in polar coordinates, the proposed snake is driven by the potential energy and the growing energy integrating the fuzzification information of tristimulus components. The proposed model has low computation cost and strong anti-noise ability. The experiments on a number of cell images show encouraging results.

1 Introduction

As one of the essential and critical tasks in automatic detection and quantitative analysis of cytological images, cell segmentation has been extensively studied in the past a few years [1]-[4]. Most of the reported methods are based on grey cell images.

Among various types of cell images, the cell images obtained from biopsy present more difficulties for segmentation, because of the diversity of contained tissues, uneven staining, and overlapped cell clusters. As a global optimization technique, snake (active contour model) seems ideally suited to this problem due to its robustness to noise and boundary gaps. Aiming at the minimization and initialization problems of conventional snake, researchers have made many improvements [5]-[7].

This paper proposes a novel snake-based segmentation method for haematoxylin and eosin (H&E) stained cell images. We choose color images to be processed for they possess more information than monochrome images. Various color image segmentation techniques have been introduced in [8]. As malignant characteristics of cell are mostly contained in nucleus, our work focuses on segmenting cell nucleus. Incorporating with the prior knowledge of the cell images, we choose the thresholds for each nucleus in *RGB* color components in a simple and effective way. To avoid making crisp decisions, we map *RGB* components into two fuzzy domains. A growing snake is built in fuzzy domains for each nucleus to extract its boundary. Described in polar coordinates, the growing snake is more computationally efficient than conventional snake. The tests show that our model has strong boundary attraction ability and the segmentation results on esophageal cell images are very encouraging.

2 Method

2.1 Preprocessing

The color images studied in this paper are H&E stained esophageal cell images. After smoothing the original image with a Gaussian filter, we need to track the localization of each cell nucleus in the image. More than 100 cell images have been observed. It is found that, in all these images, nuclei appear darker than surrounding cytoplasm and extra cellular matters in *RGB* components, and this contrast is much stronger in red component than green and blue components. So Ostu thresholding is applied on red component to segment roughly nucleus regions. Then ultimate erosion is performed on the segmented nucleus regions and simultaneously a distance map is recorded. All the local maximums in the distance map are detected as the localizations of cell nuclei in the image.

To improve the computation efficiency and accuracy, the segmentation procedure of each nucleus is performed on a sub-block including the nucleus. The center of the sub-block is the detected localization point and its size is $2 \times r_{max}+1$, where r_{max} is a predefined invariable and its value should be large enough to include all nuclei of the image in their corresponding sub-blocks.

2.2 Fuzzification

For each nucleus, three thresholds (T_r, T_g, and T_b) are detected to extract the nucleus regions (N_r, N_g, and N_b) in red, green and blue components respectively. N_r (N_g, N_b) is composed of pixels with red (green, blue) component less than T_r (T_g, T_b). Many tests show that Otsu thresholding can extract effectively the nucleus region in red component, but it sometimes fails to act in green and blue components. So we firstly determine T_r using Ostu thresholding in red component and obtain N_r. N_r is then taken as the criterion of choosing T_g and T_b. The value of T_g (T_b) is selected to minimize the size of $(N_r \cup N_g) \cap (\overline{N_r \cap N_g})$ ($(N_r \cup N_b) \cap (\overline{N_r \cap N_b})$), which reflects the difference between N_r and N_g (N_r and N_b).

The nucleus regions segmented with thresholding are crisply defined. This could limit the flexibility of the segmentation system. We use fuzzy set theory to evaluate the degree of each pixel belonging to a region. Thus crisp decisions can be avoided being made in the early processing.

For any point v in the sub-block, we use two membership functions, bell-curve and S-curve functions, to transform its *RGB* color ($I_r(v)$, $I_g(v)$, $I_b(v)$) into two fuzzy spaces. The membership functions are depicted with $\mu'_k(v)$ and $\mu''_k(v)$ ($k=r,g,b$) as Eq.1 and Eq.2, where the value range of $I_k(v)$ is $[a_k, b_k]$ ($k=r,g,b$).

$$\mu'_k(v) = 1 \bigg/ \left(1+ \left| \frac{I_k(v)-a_k}{T_k - a_k} \right|^{2t} \right) \quad (a_k \leq I_k(v) \leq b_k) \tag{1}$$

$$\mu_k''(v) = \begin{cases} \dfrac{(I_k(v)-a_k)^2}{(T_k-a_k)(b_k-a_k)} & (a_k \le I_k(v) \le T_k) \\ 1 - \dfrac{(I_k(v)-b_k)^2}{(b_k-T_k)(b_k-a_k)} & (T_k \le I_k(v) \le b_k) \end{cases} \quad (2)$$

The bell function $\mu_k'(v)$ ($k=r,g,b$), as shown in Fig. 1a, represents the degree of the pixel belonging to the nucleus region in the color component, where t is the ambiguity parameter and the fuzzy set defined with larger t is more approximate to the crisp set.

The S-curve function $\mu_k''(v)$ ($k=r,g,b$) represents the degree of brightness in the component. Compared to linear membership function, the S-curve function has an effect of enhancing the contrast of interior and exterior of the nucleus (Fig. 1b).

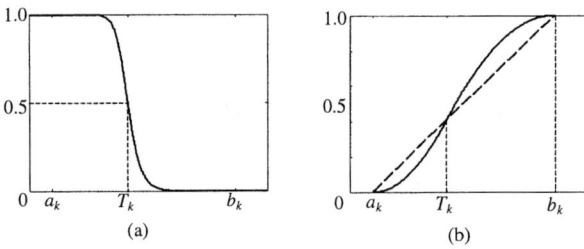

Fig. 1. (a) shows the bell membership function of a component. (b) shows the S-curve membership function (*solid curve*) and the linear membership function (*dotted curve*)

2.3 Growing Snake

Because most cell nuclei have ellipse-like boundaries with low variations of curvature, we describe the contour of the nucleus with n discrete points in polar coordinates (ρ,θ), where the origin point is the localization point detected in Sec. 2.1 and the quantization step sizes of ρ and θ are ρ_s and θ_s respectively. The contour points are denoted as $v_i = (\rho_i,\theta_i)$ ($i=0,1,2,\ldots,n-1$), where $\theta_i(=i\times\theta_s)$ is invariable during the contour deformation.

The basic idea of snake is to search for a curve in the image where the energy function reaches minimum. The energy function of our model is presented as Eq.3, where E_{cont}, E_{curv} and E_{image} denote respectively continuity internal energy, curvature internal energy and potential energy. Besides these conventional energies, we propose a growing energy E_{grow} to overcome the local minimum. α_i, β_i, γ_i and ε_i are positive weighting constants.

$$E = \sum_{i=0}^{n-1}(\alpha_i E_{cont}(v_i) + \beta_i E_{curv}(v_i) + \gamma_i E_{image}(v_i) + \varepsilon_i E_{grow}(v_i)) \quad (3)$$

Greedy algorithm is used to implement the energy minimization. This algorithm iteratively replaces each contour point with the minimum energy point in the neighborhood. In our model, v_i is limited to move along the corresponding radial direction and its neighborhood $\Omega_i = \{(\rho_i - \rho_s, \theta_i), (\rho_i, \theta_i), (\rho_i + \rho_s, \theta_i)\}$. But for snakes in Cartesian coordinates, the size of neighborhood is larger because the point is required to move along two directions (horizontal, vertical). So our model has lower computation cost.

For any $v_j = (\rho_j, \theta_j) \in \Omega_i$, we need to calculate its energy. The calculations of E_{cont} and E_{curv} are in accordance with [6]. The potential energy and the growing energy, as the innovations of our model, are illustrated particularly as the following paragraph.

Potential Energy. Generally potential energy is designed to be in inverse proportion to image gradient. Here, the potential energy is decided by two parts ($C(v_j)$ and $P(v_j)$). $C(v_j)$ presents the fuzzy contrast of the points outside and inside v_j at the radial direction of red component (Eq.4).

$$C(v_j) = \frac{1}{m} \sum_{k=0}^{m-1} \mu_r''(k \times \rho_s, \theta_j) - \frac{1}{R-m+1} \sum_{k=m}^{R} \mu_r''(k \times \rho_s, \theta_j) \quad (v_j \in \Omega_i) \quad (4)$$

In Eq.4, m and R can be obtained respectively by rounding up ρ_j / ρ_s and $r_{max}/\rho_s \cdot m$ and $R-m+1$ denote respectively the numbers of quantization points inside and outside v_j at the direction of θ_i. $\mu_r''(\rho, \theta)$ is the brightness membership mapping to polar coordinates of red component. $C(v_j)$ is then normalized in Ω_i as $C'(v_j)$. According to visual characteristics of nucleus, $C(v_j)$ generally takes the minimum at the nucleus boundary along the radial direction, and its boundary attraction range is broader than conventional potential energy defined by gradient. The other part $P(v_j)$ is inversely proportional to the fuzzy gradient of v_j (Eq.5).

$$P(v_j) = -\nabla \mu_{min}''(v_j) \quad \text{where} \quad \mu_{min}''(v_j) = \min_{k=r,g,b}(\mu_k''(v_j)) \quad (v_j \in \Omega_i) \quad (5)$$

In Eq. 5, ∇ is the gradient operator, and $\mu_{min}''(v_j)$ is the intersection of brightness fuzzy sets in tristimulus components. This operation has an effect of smoothing the variations within the nucleus. $P(v_j)$ is then normalized in Ω_i as $P'(v_j)$. $E_{image}(v_j)$ is calculated by $\omega_1 C'(v_j) + \omega_2 P'(v_j)$, where ω_1 and ω_2 are positive weighting parameters.

Growing Energy. Sometimes E_{image} may get in the local minimum, which leads to v_j converging to spurious border. We adopt the dilation idea of balloon proposed by Cohen [7] to expand the contour from the interior of nucleus. Our superiority is that the growing energy is adaptive. More similarity to nucleus leads to stronger expansibility. The growing energy is defined as Eq. 6.

$$E_{grow}(v_j) = \begin{cases} -\mu_{max}'(v_j) & \text{if } v_j = (\rho_i + \rho_s, \theta_i) \\ 0 & \text{else} \end{cases} \quad (v_j \in \Omega_i) \quad \text{where} \quad (6)$$

$$\mu_{max}'(v_j) = \max_{k=r,g,b}(\mu_k'(v_j))$$

The growing energy gives the energy at the point $(\rho_i + \rho_s, \theta_i)$ a decrement, which gives v_i a priority to move outward. Initially we put v_i inside the nucleus, where generally at least one color component is dark and corresponding $\mu'_{max}(v_j)$ at $(\rho_i + \rho_s, \theta_i)$ is large. Hence there is a strong expansibility to move v_i outward to $(\rho_i + \rho_s, \theta_i)$. When v_i reaches the exterior of the nucleus, the expansibility weakens as $\mu'_{max}(v_j)$ lessens.

3 Results and Conclusion

For each nucleus, the initial position of snake is put on a small circle inside the nucleus. Here we choose the radius of the circle to be 10. The parameters are selected as $n = 100$, $\rho_s = 1$, and $t=8$. After various tests, we decide the weighting parameters as $\alpha_i = \beta_i = 0.5$, $\gamma_i = 0.4$, $\varepsilon_i = 0.7$ and $\omega_1 = \omega_2 = 1$. Fig. 2 shows segmentation results of two cell images, where the localization points of nuclei are marked with white dots, and each nucleus boundary is outlined by connecting all the contour points in turn. From Fig. 2 we can see that the overlapped cell nuclei can be well separated with our method.

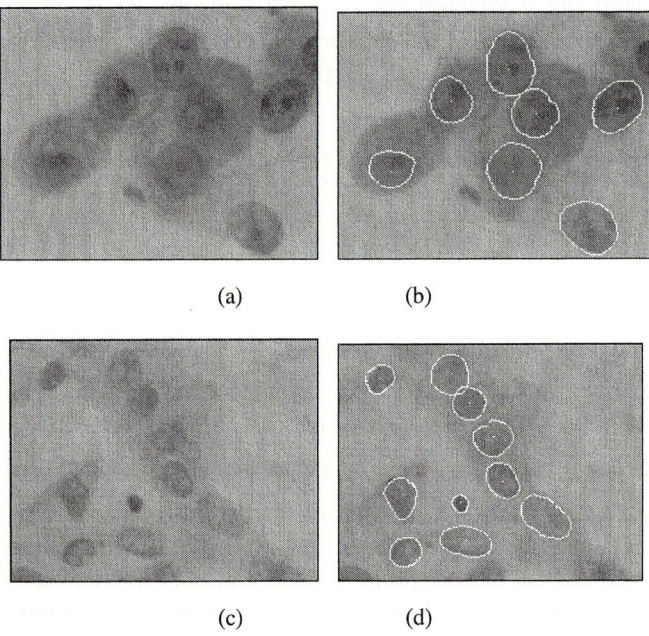

(a)   (b)

(c)   (d)

Fig. 2. (a) and (c) are two color cell images. (b) and (d) are respectively their segmentation results using our method with $r_{max} = 120$

To illuminate the strong boundary attraction ability of our model, we give the results of segmenting a nucleus with different energy definitions (Fig. 3). Fig. 3a is the result using general energy definition as [6]. We can see that most contour points are trapped into the local minima. Fig. 3b shows the result of modifying the potential energy as Eq. 4 and Eq. 5, where the attraction ability of boundary is improved but the contour is still easy to be caught by the strong color variations within the nucleus. Fig. 3c is the segmentation result using the energy definition of our model, where the nucleus boundary is well tracked.

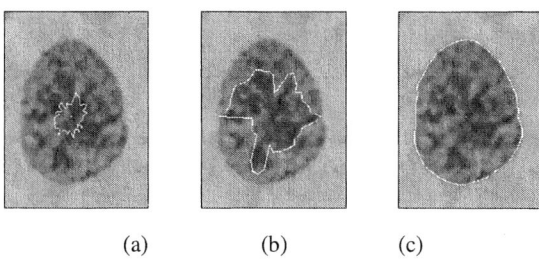

(a)　　　　(b)　　　　(c)

Fig. 3. Segmenting a single cell nucleus with different energy functions. (a) is general energy model. (b) improves (a) by modifying potential energy. (c) is our model both modifying potential energy and imposing growing energy

Conclusively, this paper presents a new cell nucleus segmentation method for H&E stained esophageal cell images. Cooperating with the prior knowledge of the cell images and transforming *RGB* color space to several fuzzy sets with different fuzzy meaning, a growing snake is developed in the fuzzy domain to extract the nucleus boundary. Due to the polar coordinates description, the presented model has higher computation efficiency than general snakes in Cartesian coordinates. Tests on a number of cell images show encouraging results and prove that our model has broadened boundary attraction range.

References

1. Lassouaoui, N., Hamami, L., Zerguerras, A.: Segmentation and Classification of Biological Cell Images by a Multifractal Approach. Int. J. Intell. Syst. 18 (2003) 657-678
2. Jiang, T., Yang, F.: An Evolutionary Tabu Search for Cell Image Segmentation. IEEE Trans. on Systems, Man and Cybernetics. 32 (2002) 675-678
3. Bamford, P., Lovell, B.: Unsupervised Cell Nucleus Segmentation with Active Contours. Signal Processing. 71 (1998) 203-213
4. Mouroutis, T., Roberts, S. J., Bharath, A. A.: Robust Cell Nuclei Segmentation Using Statistical Modeling. Bioimaging. 6 (1998) 79-91
5. Xu, C., Prince, J.: Snakes, Shapes and Gradient Vector Flow. IEEE Trans. on Image Processing. 7 (1998) 359-369
6. Williams, D., Shab, M.: A Fast Algorithm for Active Contours and Curvature Estimation. Computer Vision, Graphics and Image Processing: Image Understanding. 55 (1992) 14-26
7. Cohen, L.D.: On Active Contour Models and Balloons. Computer Vision, Graphics, and Image Processing: Image Understanding. 53 (1991) 211-218
8. Cheng, H., Jiang, X., Sun, Y., Wang, J.: Color Image Segmentation: Advances and Prospects. Pattern Recognition. 34 (2001) 2259-2281

Evaluation of Morphological Reconstruction, Fast Marching and a Novel Hybrid Segmentation Method

Jianfeng Xu and Lixu Gu

Computer Science, Shanghai Jiaotong University,
1954 Huashan Road, Shanghai, P.R. China 200030

Abstract. An evaluation of two traditional segmentation algorithms of Morphological Reconstruction and the Fast Marching method along with a novel hybrid segmentation approach is presented. After introducing the Fast Marching and the Morphological Reconstruction segmentation, we propose a novel hybrid segmentation approach in multi-stage, which is derived from both an improved Fast Marching method and the Morphological Reconstruction. To demonstrate the effectiveness and accuracy of the three methods, we employ an MRI brain image in our experiments, in which "gold standard" is known. The evaluation is measured accordingly in accuracy and speed when running a 2.0 GHz based windows XP PC. The accuracy results of average 0.9738, 0.6302 and 0.9734 measured in similarity indexes of the Morphological Reconstruction, the Fast Marching and the hybrid approach are achieved, respectively. The computing performance required 188.6, 22.3 and 43.4 in seconds accordingly.

1 Introduction

Medical image segmentation is a fundamental technique for computer assisted surgery and therapy. There are many medical image segmentation methods described in the literature. They may be basically divided into two groups: Model-based and Region -based methods. The Snake [1], introduced by Kass et al, provides a general model-based solution to the segmentation problem. Level Set [2] is another classical model-based segmentation method. By introducing an additional dimension, the complicated problems of contour breaking and merging could be effectively handled. However, it brings more computational costs in. Many efforts have been addressed to reduce its complexity. Narrow Band level set [4] and Fast Marching [3] methods improved the situation from different respects. However, the model-based methods are usually fast but sometimes not accurate enough.

In contrast to the model-based methods, the segmentation algorithms, which are operated in region base, can achieve more accurate results. Watershed and Morphological Reconstruction [5] are two main representatives and both derived from mathematical morphology. The second algorithm is also known as a morphological region growing approach with powerful reconstruction ability. Although the region-based algorithms improved the accuracy, they are usually more computationally expensive.

In this paper, we present a hybrid algorithm that integrates the Fast Marching method and the morphological reconstruction to take the advantage of both rapid from model-based method and accurate from region-based approach. Meanwhile, an evaluation of three algorithms using a standard MRI brain image is performed to demonstrate their features.

The rest of this paper is organized as follows: in section 2, we review the Level Set, Fast Marching and Morphological Reconstruction algorithms. In section 3, we introduce the novel hybrid segmentation strategy. In the experiment of section 4, we evaluate the three segmentation approaches based on a standard criterion.

2 Fast Marching and Morphological Reconstruction

2.1 Level set and Fast Marching

The Level Set method is essentially a moving interface problem. It embeds the interface as the level set of a higher dimensional function, so called level set function.

We may locate the front by finding the zero level set of the level set function as:

$$\Gamma(t) = \{ \Phi(x,t) = 0 \} \qquad (1)$$

and differentiate with respect to t, we can get

$$\frac{\partial \Phi}{\partial t} + v |\nabla \Phi(x,t)| = 0 \qquad (2)$$

where the speed function v is related to image features and front characteristic.

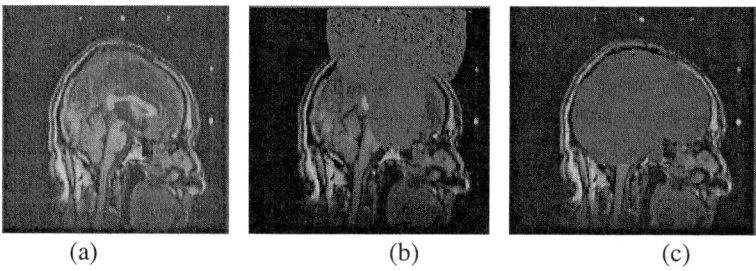

Fig. 1. Examples of the Fast Matching results. (a) Source Image; (b) Traditional Fast Matching; (c) Improved Fast Matching

Fast Marching is a special case of the Level Set, where the sign of the speed Function is always unchanged (positive or negative). Therefore the front is always moving forward or backward. This restriction makes the fast marching approach much more rapid than the more general level set method.

2.2 Improvement of the Fast Marching

The traditional Fast Matching method has a major drawback that it is very hard to control the front during the propagation. In the case of the existing of connection between the object and its neighboring regions, the front can easily lead to an overflow. An example is shown in Fig.1(b).

To prevent the front propagation from overflow, we introduce global information of the front into the speed function. Firstly, we define an average energy of the front as:

$$E_{front}(t) = \frac{1}{N_{front}} \sum_{(x,y,z)\in \Gamma(t)} E(x,y,z) \qquad (3)$$

where $E(x,y,z)$ stands for the image energy at (x,y,z) in the image $I(x,y,z)$, defined as:

$$E(x,y,z) = -|\nabla G_\delta * I(x,y,z)| \qquad (4)$$

Where G_δ is a 3D Guassian function with a standard deviation δ. ∇ represents a gradient operation. And $E_{front}(t)$ is associated with the energy of all the points in the front. We introduce it into the speed function F and redefine it as below:

$$F(x,y,z,t) = F(x,y,z) \cdot e^{\beta \cdot E_{front}(t)} = F(x,y,z) \cdot e^{-\beta \frac{1}{N_{front}} \sum_{(x,y,z)\in \Gamma(t)} |\nabla G_\delta * I(x,y,z)|}, \beta > 0 \qquad (5)$$

When most points along the front approach to the object edge, $E_{front}(t)$ becomes much smaller than 0, which leads $F(x,y,z,t)$ close to 0 to halt the front. By changing the speed function from $F(x,y,z)$ to $F(x,y,z,t)$, the front can be efficiently prevented from overflow. An example of the improved fast marching is shown in Fig.1(c).

2.3 Morphological Reconstruction

The morphological reconstruction is a typical approach to extract seeded regions, which is defined as:

$$S_{i+1} = (S_i \oplus k) \cap |m| \qquad (i = 0,1,2\cdots) \qquad (6)$$

Where, S and k denotes the seed and a small disk shaped structuring element, respectively. $\oplus$ and |m| stand for a dilation operation and the mask accordingly. The mask is achieved via a threshold operation using a histogram analysis.

3 Hybrid Segmentation Strategy

The morphological operation during the reconstruction is very computational expensive. To speed up the propagation, we need to define a good seed (marker), which is close enough to the object edge. In that case, the iteration of the morphological reconstruction could be possibly reduced to a reasonable small value.

The Fast Marching method can help us get a good initial marker. We firstly employ the Fast Marching to quickly propagate the front from a user-defined seed. When the front stopped at a position close to the edge, we switch it from the Fast

Marching procedure to the Morphological Reconstruction as the marker to refine the front fitting into the edge accurately. As the result, the proposed hybrid approach can achieve a result much quicker than the morphological reconstruction with similar accuracy. The procedure of the hybrid segmentation approach is described in Fig 2.

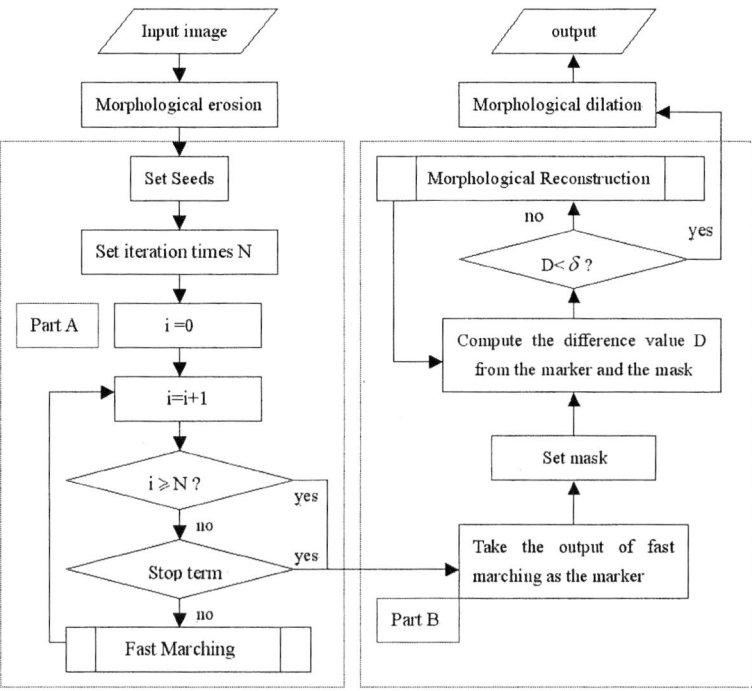

Fig. 2. Flow chart of the hybrid segmentation approach, where part A and B belong to the Fast Marching and the Morphological Reconstruction, respectively. In Part B, δ represent a user defined error tolerance between the marker and the mask

4 Evaluation Experiment

A "TkSegmentation" software was developed to perform the evaluation, which is based on Visualization Toolkit (VTK), Insight Toolkit (ITK) and Python programming environment, and running on a P-IV 2.0 GHz Windows-XP PC.

The source data employed in our experiments includes CT or MRI datasets of brain, heart and kidney studies. A standard CJH27 image volume derived from an average of 27 T1 weighted images of a normal brain was employed in the evaluation study, which is shown in Fig.3(a). CJH27 is a 181 217 181 voxel volume, with isotropic 1 mm^3 voxels. This standard brain was generated from a set of 3D "fuzzy" anatomical models. Each model represents a typical tissue class (white matter, gray matter, CSF, etc.). This model was then used as input for an MR simulator (MNI Brainweb[1]) that produces a realistic MR volume image for which "ground truth" is known with respect to its components.

We evaluated the three segmentation methods in two aspects: the accuracy of the segmented result and the efficiency of the segmentation procedure.

4.1 Accuracy

We first evaluated the accuracy of the segmented results. The example results are shown in Fig.3.(b)-(d). The normal Fast Marching result shown in (b) indicates that most of the surface could not be accurately reconstructed. The Morphological Reconstruction brings us a more accurate result as shown in (c). The result of the proposed approach shown in (d) is quite similar to the one of the Morphological Reconstruction.

In order to quantify the segmentation accuracy, we employed the idea of "similarity index" in our experiment, which is introduced by Zijdenbos [6]. We evaluated the three methods and concluded into the Table.1. The initial seeds were repeatedly defined at various locations three times. The result of our proposed approach is quite similar to the morphological reconstruction, and much better than the one from the Fast Matching method.

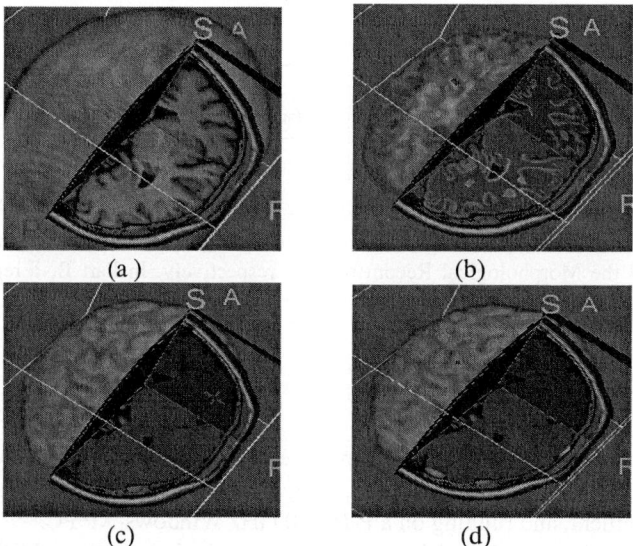

Fig. 3. Segmentation results, where the 2D results are indicated in high-lighted regions. (a) source image; (b) the Fast Matching result; (c) the Morphological Reconstruction result and (d) the proposed hybrid approach

---

[1] http://www.bic.mni.mcgill.ca/brainweb/

4.2 Efficiency

The computing times of the three segmentation algorithms were measured during the experiment. Table 2 describes the results of each method when three initial seeds were repeatedly defined. We can find that even the proposed hybrid method is still more costly than the Fast Matching method, but significantly improved from the Morphological Reconstruction method. It is considered as a near real time algorithm.

The reason why the hybrid approach is much faster than the traditional morphological Reconstruction is that the numbers of the iteration during the reconstruction were significantly reduced. In the example, the total iterations of Morphological reconstruction and Hybrid Approach were 30 and 3 respectively.

Table 1. Comparison of Similarity Index value of different approaches

	1	2	3	average
Morph. Reconstruction	0.9737	0.9731	0.9747	0.9738
Fast Marching	0.6278	0.6298	0.6334	0.6302
Hybrid Approach	0.9737	0.9734	0.9721	0.9734

Table 2. The computing time of the three segmentation method

	1	2	3	average
Morph. Reconstruction	197.5"	184.4"	184.1"	188.6"
Fast Marching	22.3"	19.6"	25.1"	22.3"
Hybrid Approach	41.7"	45.8"	42.6"	43.4"

5 Conclusion

In this paper, we presented an essential evaluation study using two typical segmentation methods and a novel hybrid segmentation approach based on a standard criterion. The evaluation results in both accuracy and efficiency revealed that the proposed hybrid approach takes the advantage of both accurate from region-based algorithm and rapid from the model-based approach, which achieved a near real time segmentation performance and high accurate results.

As future works, we are scheduling to choose more algorithms from both model-based and region-based algorithms into our evaluation study. We are going to design more sophisticated analysis methods other than the similarity index to inspect experimental results more precious, and the evaluation results will be virtually presented.

References

1. Kass, M., Witkin, A., and Terzopolous, D.: Snake: Active Contour Models. Int. J. Compt. Vision, Vol.1 (1988) 321-331
2. Osher, S., and Sethian, J.A.: Fronts Propagating with Curvature Dependent Speed: Algorithms Based on Hamilton-Jacobi Formulation. J. of Comput. Phys., vol. 79 (1988) 12-49
3. Sethian, J.A.: Fast marching method. SIAM Review, 41(2) (1999) 199-235
4. Chop, D.: Computing Minimal Surfaces via Level Set Curvature Flow. Journal of Computational Physics, vol.106 (1993) 77-91
5. Gu, L., and Peters, T.: An Accurate and Efficient Hybrid Approach for Near Real Time 3D Brain Segmentation. Proc. Of 17th International Congress and Exhibition On Computer Assisted Radiology and Surgery, London UK (2003) 1309
6. Zijdenbos, A.P., Dawant, B.M., Margolin, R.A., and Palmer, A. C.: Morphometric Analysis of White Matter Lesions in MR Images: Method and Validation. IEEE Trans. Med. Imag., vol.13 (1994) 716-724

Utilizing Staging Tables in Data Integration to Load Data into Materialized Views

Ahmed Ejaz[1] and Revett Kenneth[2]

[1]Information and Computer Science Department,
King Fahd University of Petroleum and Minerals,
Dhahran, Saudi Arabia, 31261
eahmed@ccse.kfupm.edu.sa
http://www.ccse.kfupm.edu.sa/~eahmed/
[2]Department of Computing and Information System,
University of Luton, Park Square,
LU1 3JU, England
Ken.Revett@luton.ac.uk

Abstract. This paper proposes an approach to data integration and migration from a collection of heterogeneous and independent data sources into a data warehouse schema. Current methodology assumes that the data is loaded into data warehouse using queries in order to extract data from multiple data sources. Extracting data from various data sources requires the establishment of generic data integration methodologies. Sometimes, various data anomalies arise when using complex queries across multiple data sources. A data warehouse can be abstractly seen as a set of materialized views. Selecting views for materialization in a data warehouse is one of the most important decisions making tasks in its design. However, there are few facilities in data integration systems that consider these important issues. In this paper, we propose the approach of introducing staging table's schema that can be utilized to load required data from source tables into corresponding staging tables. Staging tables will be assumed to be temporary tables and each staging table will be empty before loading new data and using simple non-join queries to load data. We then focus on an approach to load data into data warehouse materialized views through staging tables including maintenance of materialized views.

Keywords: Data Integration, Data Migration, Staging Tables, Global Schema, and Relational Schema.

1 Introduction

As the size and complexity of data stored in data warehouses increases, the design and strategy used in staging tables becomes increasingly important. Many organizations are dependent on their data stores and survive on their ability to acquire, analyse and store data in an efficient manner as part of an Enterprise Resource Planning system (ERP). In addition, data integration and data migration is becoming a central issue in World Wide Web based data repositories.

Until recently, most research regarding data integration has focused on managing and highlighting query general issues for migrating data in data warehouse as mentioned in [VM1]. In this paper, strategies are proposed for managing data migration correctly and efficiently managing maintenance of materialized views.

A database relational schema is defined for the implementation of staging tables followed by rules and strategies for migrating data into data warehouse correctly and efficiently. Also, our approach identifies data that is flagged for rejection during the migration process, i.e. before it becomes incorporated into the data warehouse.

2 Framework for Staging Table's Schema

In this section we illustrate formalization of a data integration system, which is based on the relational model. We will follow the nomenclature utilized in the work by Cali in paper [CD1] "Data integration under integrity constraint". We assume to have a fixed (infinite) alphabet Γ of constants, and, if not specified otherwise, we will consider only databases over such alphabet. We adopt the so-called unique assumption, i.e., we assume that different constants denote different objects.

Consider a symbol C for relational schema that consists of

- An alphabet A of predicate (or relation) symbols, each one with the associated arity, i.e., the number of arguments of the predicate (or, attributes of the relation)
- A relational database denoted by DB for a schema C is a set of relations with constants as atomic values, and with one relation γ^{DB} of arity n for each predicate symbol γ of arity n in the A. γ^{DB} is the interpretation of the predicate symbol γ in database DB such that it contains
  1. The set of tuples t that satisfying the predicate γ in DB
  2. The set of attributes as denoted by symbol A satisfying the predicate γ in DB

In this paper we will consider the following two constraints; later on we can validate them with the help of a proposed approach.

- Key Constraints: A relation γ in a schema C, consists of set of tuples such that t belongs to γ. A represents set of attributes in relation γ. Then key constraint over γ is satisfied in a database DB if for each $t_1, t_2 \in \gamma^{DB}$ such that $t_1[A] \neq t_2[A]$, where each $t_i[A]$ is the projection of the tuple t over A.
- Foreign key constraints: Let γ_1 and γ_2 be two relations with primary keys K_1 and K_2 respectively. A subset α of γ_2 is a foreign key referencing K_1 in relation γ_1 such that $\gamma_1[K_1] \subseteq \gamma_2[K_2]$ means for every tuple t_2 in γ_2 there must be a tuple t_1 in γ_1 or $t_1[K_1] = t_2[\alpha_2]$. Also, condition $\Pi_\alpha(\gamma_2) \subseteq \Pi_{K1}(\gamma_1)$ will be satisfied and either $\alpha = K_1$ or α and K_1 must be compatible sets of attributes.

Consider three types of schemas such as source schemas, staging schemas and global schemas for data warehousing. These three schema's layouts are shown in figure 1. A Source schema consists of multiple databases, legacy or operational databases. In this paper the staging schema is introduced to manage the data migration tasks. In the staging schema, data is prepared for conversion into the global schema which is then validated and tested using above constraints. In staging schema, no constraints are applied but algorithm is used to validate all possible checks for constraints. For example, a record or a tuple from a certain table requires foreign key validity on some attribute (s); it will be tested before inserting into target table. Also, we consider staging schema as temporary schema, completely separate or it can be managed within global schema. So as defined in [CD1], G represents global schema, we mean it for staging schema as well. A data integration system Γ is a triple $\Gamma = (G, S, M_{G,S})$, where G is the global schema, S is the source schema, and $M_{G,S}$ is the mapping.

Example 1. Consider a simple example of data integration with single source option containing $\Gamma^1 = (G^1, S^1, M^1_{G,S})$ where G^1 is constituted by the relation symbols

```
customer (Cid, Cname, Caddress, CCity)
orders   (Oid, Otype, Odate, Cid)
sold     (Oid, Pid, Sprice, Qty)
product  (Pid, Pname, Pprice, TaxStatus) and the constraints are
defined as
     key(customer)  = {Cid}
     key(orders)    = {Oid}
     key(sold)      = {Oid,Pid}
     key(product)   = {Pid}
such that
sold[Oid] ⊆ order[Oid], sold[Pid] ⊆ product[Pid],
orders[Cid] ⊆ customer[Cid], sold[Sprice] may not be a subset of
product[Pprice]
```

S^1 consists of four sources. Source s1, of arity n (where n can be different), contains information about customers with their customer id, name, address and city. Source s2, of arity 4, contains order id (invoice id), type, date and customer id (which is a foreign key). Source s3, of arity 3, contains information about product's id, name and price. Finally, Source s4, or arity 3, contains information about products that are sold. The mapping $M^1_{G,S}$ is defined by

```
ρ(customer) = cus(X,Y,Z) <-- s1(X,Y,Z,W,P)
ρ(orders)   = ord(X,Y,Z,W) <-- s2(X,Y,Z,W)
ρ(product)  = prd(X,Y,Z,W) <-- s3(X,Y,Z,W)
ρ(sold)     = sol(X,Y,Z) <-- s4(X,Y,Z)
```

In [CD1] $\rho(\gamma)$ is defined as the associated query. We extend our concept $\rho(\gamma)$ not with associated query but using a well-defined algorithm. In this example, assuming same data integration system $\Gamma = (G, S, M_{G,S})$, we start data at the sources and specify which are the data satisfying the global schema. Also, a source database D for Γ is constituted by one relation γ^{DB} for each source γ in S. A database DB for Γ is said to be legal with respect to D if:

- B satisfies the integrity constraints of G.
- B satisfies $M_{G, S}$ with respect to D i.e., for each γ in G (global database), the set of tuples r^B that assigns to γ is a subset of the set $\rho(\gamma)^D$ computed by the associated query $\rho(\gamma)$ over D. i.e., $\rho(\gamma)^D \subseteq \gamma^B$
- Defining an algorithm containing associated query $\rho(\gamma)$ and programming standards (which indicate the violation of constraints).

3 Comparative Study

In [CD1] and [Li1], data is mapped or loaded using queries. It is usually the case that queries are not designed to measure the validity of the data prior to inclusion into the data warehouse. In addition, most queries fail to provide a complete analysis of the inclusion process in order to determine if the entire data store was loaded successfully or not. These critical issues were not covered in any significant detail in both papers [CD1], [Li1].

A major difficulty in data integration is the frequent presence of data inconsistencies and data incompleteness, requiring a number of complex reasoning tasks and/or algorithms to guarantee the correctness of the final data. Although some of the automated issues are discussed in a paper [KL1]. To resolve this problem, we introduce the concept of *staging schema* as discussed above.

In above definition, the main concern is related with constraints (unique and referential integrity) that can be handled with defined algorithm such that with any constraint violation or inconsistent data (from sources) for any tuple **t** can be recorded in error relation. Such that decision can be made to resolve issue for tuples in error relation. Then cleaned data can be loaded or migrated to data warehouse's materialized views using associated query $\rho(\gamma)$ in database D. In this paper, we restrict our attention how to handle and how to highlight inconsistent data before finally loading of data into materialized view.

4 Managing Staging Schema's Predicates

As defined earlier, a data integration system $\Gamma = (G, S, M_{G, S})$ where G is a staging schema or global schema for data warehouse. Managing mapping $M_{G, S}$ for data migration approach that includes the following predicate standards:

1. Defining predicate γ with arity n, in staging schema as
   $\gamma(R) = \gamma(x_k, y_f, z_n)$, where x_k, y_f, z_n represent keys, foreign key and non-key attributes respectively.
2. Defining a working predicate γ_{wrk} by introducing four standard attributes as
   $\gamma_{wrk} = \gamma(R, w_1, w_2, w_3, w_4)$, where w_1, w_2, w_3, w_4 represent tuple's active validity flag, user update flag, tuple's update date/ time and user name respectively.
3. Similarly, an error predicate can be defined as
   $\gamma_{err} = \gamma(R, w_1, w_2, w_3, w_4, e)$, where e is an attribute that store error message for each tuple t. This predicate contains those tuples those were unable to store in working predicate γ_{wrk}.

4. Finally, defining or creating temporary source predicate s_i of arity n. Temporary source predicate s_i contains data extracted using simple queries through application software gateway as shown in figure 1. In this section, we will discuss further how to migrate data from single or multiple source predicate s_i into working predicate γ_{wrk} using algorithm as give in example-2. During data migration tuples contain inconsistent data. This inconsistent data finally inserted into an error predicate γ_{err}. Tuples in error predicate can be subsequently inserted into working predicate γ_{wrk} by modifying algorithm based upon business rules or error message.

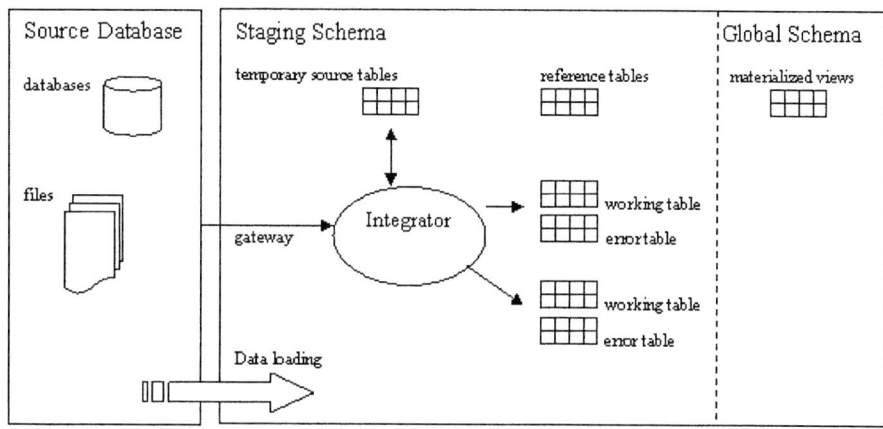

Fig. 1. Architecture of a staging schema in data integration

Example 2. Consider the following algorithm

```
PROCEDURE ProcSetActiveTuple
    --This procedure sets the value for the ACTIVE column 'N'
    --for all the tuples which have
    --USER_UPDATE set to 'N' before performing an update of
    --existing records when this extraction is not the first
    -- one for this SP1 working table. This is the first
    --procedure being called from main procedure
BEGIN
    Set attribute Active = 'Y'
    Modify TargetTable attribute Active = 'N'
END
PROCEDURE ProcExtractSource
BEGIN
    Set value for NoOfRecordsProcessed = 0
```

```
    Set attribute Active = 'Y'
    REPEAT
     Check foreign key exists for referencing attributes
     IF foreign key value does not exist THEN
        Raise exception to insert record in error table
           with error message
     ELSE
        Assign source attribute value to target attribute
     END IF
     Call ProcInsertUpdateTuple
    UNTIL NOT EXISTS (select * from SourceTable)
    Display NoOfRecordsProcessed
    Display number of rejected records from error table
    Display number of records loaded successfully
END
PROCEDURE ProcInsertUpdateTuple
BEGIN
    Set attribute Active = 'Y'
    Check if corresponding record is already exist in target
    table
    IF record exists and user update = 'Y' THEN
        Raise exception to insert record in error table with
           error message
    ELSIF record exists and user update = 'Y' THEN
       Modify record with corresponding source attribute(s)
    ELSE
       Add a record in target table as a new record
    END IF
END
```

5 Conclusions

It is a common opinion that data migration is not an easy task and its complexity varies from system to system. This paper introduced the concept of temporary schema for data warehousing to manage data migration tasks. We emphasized the need to focus on efficient data integration concepts (integrity checking and completeness) when using a set of heterogeneous and autonomous data sources. Data is loaded on the basis of business rules using staging tables and finally made ready data for inclusion into the data warehouse. The concept of using error table helps us to discover and analyze the reason(s) why records were loaded into the error table rather than into the actual working tables. Also, error messages provide brief descriptions about the source(s) of the error both in terms of attribute identity and value that generated the error. Lastly, once the errors have been autonomously resolved, the system can go about the business at hand. Once the data has been successfully loaded into working tables or on materialized views, key constraints and integrity constraints can be applied in data warehouse.

In future, we aim to consider more in-depth issues regarding staging table's architecture. Introducing standards and methods for designing a complete schema including data migration and data integration for data warehouse.

Acknowledgments

I would like to thank King Fahd University of Petroleum & Minerals for their excellent research facilities. Useful discussions with Professor Paul and Professor Salahadin are appreciated.

References

[CD1] Cali, A., Calvanese D., Giacomo G., and Lenzerini M.: Data Integration under Integrity Constraints. CAISE, Springer LNCS **2348** (2002) 262-279.
[KL1] Kim H., Lee T., Lee S., Chun J.: Automated Data Warehousing for Rule-based CRM Systems, School of Computer Science and Engineering, Seoul National University, Korea. Australian Computer Society, Inc. (2003) Volume 17
[VM1] Velegrakis Y., Miller J. R., Popa L., Mapping Adaptation under Evolving Schemas. VLDB Conference **29** (2003)
[Li1] Li C.: Describing and Utilization Constraints to Answer Queries in Data-Integration Systems, Information and Computer Science, University of California, Irvine, (2001)
[TT1] Thomas, G., Thompson, G. R., Chung, C.W., Barkmeyer, E., Carter, F., Templeton, M., Fox, S., Hartman, B.: Heterogeneous Database systems for production use. ACM Computing Survey **22** (1990) 237-266

HMMs for Anomaly Intrusion Detection

Ye Du, Huiqiang Wang, and Yonggang Pang

College of Computer Science and Technology, Harbin Engineering University,
Harbin 150001, China
{duye, hqwang, ygpang}@hrbeu.edu.cn

Abstract. Anomaly intrusion detection focuses on modeling normal behaviors and identifying significant deviations, which could be novel attacks. The existing techniques in that domain were analyzed, and then an effective anomaly detection method based on HMMs (Hidden Markov Models) was proposed to learn patterns of Unix processes. Fixed-length sequences of system calls were extracted from traces of programs to train and test models. Both temporal orderings and parameters of system calls were taken into considered in this method. The *RP* (Relative Probability) value, which used short sequences as inputs, was computed to classify normal and abnormal behaviors. The algorithm is simple and can be directly applied. Experiments on *sendmail* and *lpr* traces demonstrate that the method can construct accurate and concise discriminator to detect intrusive actions.

1 Introduction

Intrusion detection has been an active topic for about two decades, starting in 1980 with the publication of John Anderson's "Computer Security Threat Monitoring and Surveillance"[1]. His paper classifies six categories of intrusive activities and how these activities might be detected. These recommendations led to the development of two approaches, anomaly detection and misuse detection.

Anomaly detection is based on the premise that intrusions are a subset of anomalous activity[2]. Any events that significantly deviate from normal activity are considered to be suspicious. Misuse detection models inaccurate behaviors and constructs abnormal profiles firstly. By comparing actual activities with preformed modes, any matched occurrence indicates system abuse.

In the work reported here, a machine learning method using HMMs (Hidden Markov Models) is brought forward to detect abnormal events. The HMMs are employed to build a behavior model from the training data, which include only normal activities, and then we use it to classify regular and intrusive activities form actual data. In operating systems such as Unix, the privileged processes that only the superuser is authorized to execute make use of system calls to access privileged system resources or services. Such privileged processes are often a major target for intruders[3]. So, system calls corresponding to these processes are adopted as data sets.

The rest of the paper is organized as follows: Section 2 briefly reviews related work. Section 3 describes the way we build up normal behavior profiles. Section 4 presents the HMMs and how to use HMMs in our approach. Section 5 describes the details of our experiments, and analysis of results. Section 6 makes some conclusions and outlines some open issues for future work.

2 Related Work

The techniques used in anomaly detection are statistical approach, immune system, and HMMs. NIDES[4] is a typical system that adopts a statistical approach to compare the conformity of long-term and recent event patterns in the profile. But it is scarcely practical for the limitation of the number of attack descriptions. An immune system could not detect illegal access acquired from legal transactions[5]. People have used Markov models to successfully model computer system[6]-[8], and presented detecting algorithm on the basis of the models. Because of the strict requirement that next state is dependent only upon the current state is hard to hold, HMMs appear. Reference [9] has established HMMs to describe normal states of system, but these models are too complex to be trained.

We have chosen to define normal in terms of short sequences of system calls. Former studies that use system calls as data sets for HMMs training ignore the parameters passed to system calls, and look only at their temporal orderings, which leads to the result that some important characteristics, such as timing information, are lost to be considered.

Therefore, we present a feasible detection method based on HMMs with simple algorithm and high accuracy, which considers both temporal orderings and parameters of system calls.

3 Building Normal Behavior Databases

In a given interval, we record system calls and parameters generated by a particular process. To simplify the next step of extraction, these parameters were pre-processed. This idea was first brought out in Ref.[12]. For one thing, a threshold T is set for parameter p_i ($i=1,2, \ldots ,n$), and then use the following formula to compare.

$$f(p_i) = \begin{cases} 0, & p_i \leq T \\ 1, & p_i > T \end{cases} \quad (1)$$

We replace p_i with $f(p_i)$. Then we get the sequence of system calls and parameters is:

$$G = c_1 f(p_1) c_2 f(p_2) \hbar \ c_n f(p_n)$$

Where $c_1 c_2 \hbar \ c_n$ denotes the sequence of system calls.

Use a sliding window of length r (r is an even number), and a sliding step of 2, to scan sequence G. Then, we build up a normal database of all unique short sequences of length r. Each program of interest has a different database.

The construction of the normal database is best illustrated with an example. Suppose we observe the following trace that has been pre-processed:

$$G_1 = open, 0, lseek, 0, read, 1, lseek, 0, read, 1, lseek, 0, write, 0, close, 1$$

Initialize sliding window length r equals 4, sliding step equals 2, and take out the iterant sequences, we get 5 sequences:

j_1 = open, 0, lseek, 0
j_4 = lseek, 0, write, 0
j_2 = lseek, 0, read, 1
j_5 = write, 0, close, 1
j_3 = read, 1, lseek, 0

To save storage space, these sequences are stored as trees, the structure of which was introduced in Ref.[13]. Root is the first system call function in each short sequence, and leaf is the last one. The corresponding forest of this example includes 4 trees, with each tree rooted at open(j_1), lseek(j_2, j_4), read(j_3), write(j_5) individually.

The scale of normal behavior database is correlated with the number of unique sequences N. As the sequences are stored as trees, the upper bound on the storage requirements is $O(Nr)$, however in practice, they are much lower. In this example, $N=5$, $r=4$, the number of nodes in the forest is 18. In the following experiments, sendmail process contains 767 unique short sequences, sliding window length $r = 6$, has 3896 nodes in the forest. While not adopting tree structure, we would need 4602 nodes.

4 Anomaly Detection Approach Based on Hidden Markov Models

In the monitored system, both normal and abnormal behaviors all can be described by a doubly embedded dynamic process, which corresponds to a HMM. We use the notation $\lambda = (A, B, \pi)$ to indicate the complete parameter set of a HMM, where A denotes the state transition probability distribution, B denotes the observation symbol probability distribution, and π denotes the initial state distribution.

There are three basic problems of HMM to solve[9], while we are only interested in the first: Given the observation sequence $O = O_1 O_2 \cdots O_T$, and a model $\lambda = (A, B, \pi)$, how do we efficiently compute $P(O|\lambda)$, the probability of the observation sequence, given the model?

We use the forward part of the forward-backward procedure to solve the problem. Consider the forward variable define as

$$\alpha_t(i) = P(O_1 O_2 \cdots O_t, q_t = S_i | \lambda) \qquad (2)$$

i.e., the probability of the partial observation sequence, $O_1 O_2 \cdots O_t$, (until time t) and state S_i at time t, given the model λ.

After initialization, induction and termination, we get

$$p(O|\lambda) = \sum_{i=1}^{N} \alpha_T(i) \qquad (3)$$

The process of reasoning is omitted here, and detailed instruction of this part can be found in Ref.[9].

In the condition that all system behaviors are regular, the initial state of system is normal, $q_1=0$, so the state distribution $\pi = \{1, 0\}$. When operating, no matter what state current system is, the next state must be normal. Thus $a_{00}=1$, $a_{10}=1$. For there are only two states here, 0 indicates normal and 1 indicates abnormal,

$$p(O|\lambda) = \sum_{i=1}^{N} \alpha_T(i) = \alpha_T(0) + \alpha_T(1) \tag{4}$$

$\alpha_T(1) = 0$, $\alpha_T(0) = b_0(O_1)b_0(O_2)\hbar\ b_0(O_t)$
Then

$$p\{O|\lambda\} = \prod_{i=1}^{T} b_0(O_i) \tag{5}$$

Suppose $\quad U=\{O_1, O_2, \cdots, O_H\}$
Define

$$V(O_i) = P\{O_i|\lambda\} \tag{6}$$

$$V_{max} = \max\{P\{O_i|\lambda\}, i=1,2,\cdots,H\} \tag{7}$$

$$V_{min} = \min\{P\{O_i|\lambda\}, i=1,2,\cdots,H\} \tag{8}$$

In this paper, we use a sliding window of length n ($n<N$), and a sliding step 1 to pass training sequences. Suppose there are N sequences at first, now we have $N-n+1$ new observation symbols, which include n sequences separately. Then, $H=N-n+1$, $O_1=\{o_1,o_2,\cdots,o_n\}$, $O_2=\{o_2, o_3,\cdots,o_{n+1}\}$, $\cdots$, $O_{N-n+1}=\{o_{N-n+1},o_{N-n+2},\cdots,o_N\}$. We compute RP (Relative probability) as follows:

$$RP = \frac{\sum_{i=1}^{N-n+1} V(O_i) - V_{max} - V_{min}}{(N-n-1)\cdot n}$$

$$= \frac{\sum_{i=1}^{N-n+1} \prod_{j=i}^{i+n-1} b_0(o_j) - V_{max} - V_{min}}{(N-n-1)\cdot n} \tag{9}$$

By comparing RP, we can determine whether it is anomalous. The outcome is normalized over the sliding window length n, in order to minimize the influence made by it, and enable us to compare RP values for different values of n.

5 Experiments and Analysis

Standard HMMs have a fixed number of states, so one must decide on the size of the model before training. Preliminary experiments[11] showed us that a good choice for our application was to choose a number of states roughly corresponding to the number of unique system calls used by the program. We have monitored *sendmail* and *lpr* programs under UNIX operating system with standard C language for 5 weeks, and extracted short sequences as we described in section 3. Training data was all composed of normal sequences, and the testing data was the normal sequences that were not included in training plus traces of the *syslog-remote* and *lprcp* attack.

Firstly introduces two variables r and n. r is the sequence length of system calls. We scanned traces of system calls generated by *sendmail* and *lpr* programs, then built up a database of all unique sequences of length r that occurred during the trace. These sequences were stored in tree structure. n is the length of an observation symbol, which is described in section 4. In test, we set $r = 2, 4, 6, 8$ and $n = 3, 5, 7$, the results are shown in table 1 and table 2.

Table 1. Experiment results of sendmail normal and syslog attack traces

n	r (in normal sequences)				r (in abnormal sequences)			
	2	4	6	8	2	4	6	8
3	0.05362	0.00262	0.00103	0.00056	0.04906	0.00193	0.00081	0.00027
5	0.00830	0.00325	0.00027	2.307E-5	0.00576	0.00105	0.00005	0.267E-5
7	0.00009	6.721E-5	1.020E-5	3.796E-6	8.576E-6	3.012E-6	9.032E-7	3.271E-7

Table 2. Experiment results of lpr normal and lprcp attack traces

n	r (in normal sequences)				r (in abnormal sequences)			
	2	4	6	8	2	4	6	8
3	0.02751	0.00317	0.00156	0.00011	0.01903	0.00133	0.00071	3.170E-5
5	0.00925	1.089E-5	7.909E-6	3.156E-6	0.00477	8.236E-6	2.615E-6	3.805E-7
7	6.270E-5	0.193E-6	8.796E-8	2.176E-8	2.603E-6	3.229E-7	1.095E-8	8.163E-9

The *RP* value of normal is bigger than anomaly greatly, and the difference between them is distinct. Thresholds are set for "normal" output probabilities. Then, if we encounter a trace that produces a below-threshold output, it is flagged as a mismatch. With the increasing of n and r, the outcome is smaller, but the difference is becoming more significant, such as 10 times in *sendmail* traces. Thus, we can get more accurate and precious result to improve the discrimination power of the model, however that will lead to largen the system expense. Computation complexity of this method is $O(2n^2)$, so the choice of n has to be a tradeoff between performance and efficiency.

6 Conclusion

In this paper, we presented a method for constructing HMMs to learn normal and abnormal behaviors and calculate *RP* probability to discriminate normal and anomaly. The principle of it works were described. Temporal orderings and parameters of system calls are both included in data sets, which helps to accurately describe process behaviors. Experimental results clearly demonstrate the effectiveness of our method. The algorithm is simple, general and can be easily incorporated into an intrusion detection system.

In future, we want to investigate more events related to system activity, such as audit trail and network traffic. Our ultimate goal is to implement and deploy our method in the real world.

References

1. Anderson, J.P.: Computer Security Threat Monitoring and Surveillance. James P. Anderson Co., Fort Washington (1980)
2. Carver, C.A.: Adaptive Agent-based Intrusion Response. Texas A&M University (2001) 8-10
3. Yeung, D.Y., Ding, Y.: Host-based Intrusion Detection Using Dynamic and Static Behavioral Models. Pattern Recognition, Vol. 36 (2003) 229-243
4. Anderson, D., Frivold, T., Valdes, A.: Next-generation Intrusion Detection Expert System (NIDES). Computer Science Laboratory, SRI International Menlo Park, CA (1995)
5. Hofmeyr, S.A., Forrest, S.: Architecture for An Artificial Immune System. Evolutionary Computation, Vol. 8 (2000) 443-473
6. Ye, N.: A Markov Chain Model of Temporal Behavior for Anomaly Detection. In: The 2000 IEEE Systems, Man, and Cybernetics Information Assurance and Security Workshop, West Point, NY (2000)
7. Jha, S., Tany, K., Maxion, R.A.: Markov Chains, Classifiers, and Intrusion Detection. In: The 14th IEEE Computer Security Foundations Workshop, Cape Breton, Novia Scotia, Canada (2001)
8. Eskin, E., Wenke, L., Stolfo, S.J.: Modeling System Calls for Intrusion Detection with Dynamic Window Sizes. In: DARPA Information Survivability Conference & Exposition, Anaheim, California (2001)
9. Rabiner, L.R.: A Tutorial on Hidden Markov Models and Selected Applications in Speech Recognition. Proceedings of the IEEE, Vol. 77 (1989) 257-286
10. Tan, X.B., Wang, W.P., Xi, H.S., Yin, B.Q.: A Hidden Markov Model Used in Intrusion Detection. Chinese Journal of Computer Research and Development, vol. 40 (2003) 245-250
11. Warrender, C., Forrest, S., Pearlmutter, B.: Detecting Intrusions Using System Calls: Alternative Data Models. In: The 1999 IEEE Symposium on Security and Privacy, IEEE Computer Society (1999) 133-145
12. Zhang, K., Xu, M.W., Zhang, H., Liu, F.Y.: An Intrusion Detection Method (RHDID) Based on Relative Hamming Distance. Chinese journal of computers, Vol. 26 (2003) 65-70
13. Forrest, S., Hofmeyr, S.A., Somayaji, A., Longstal, T.A.: A Sense of Self for Unix Processes. In: The IEEE Symposium on Security and Privacy, Oakland, CA, USA (1996) 120–129.

String Matching with Swaps in a Weighted Sequence

Hui Zhang[1], Qing Guo[2], and Costas S. Iliopoulos[1]

[1] Department of Computer Science, King's College London Strand,
London WC2R 2LS, England
{hui, csi}@dcs.kcl.ac.uk
[2] Department of Computer Science and Engineering,
Zhejiang University, Hangzhou, Zhejiang 310027, China
gq@sunyard.com

Abstract. A weighted sequence is a string where a set of characters may appear in certain positions with respectively known probabilities of occurrence. We concentrate on the problem of pattern matching with swaps (swapped matching) in a weighted sequence, that is to locate all the positions in the sequence where there exists a swapped match of the pattern that has probability of appearance greater than a predefined constant. In this case, the number of swaps is not limited. We present a method for reducing the problem of swapped matching in a weighted sequence to the problem in a set of maximal factors of the sequence. We then give a almost optimal solution that takes $O(nlogmlog\sigma)$ time.

1 Introduction

As one of the most widely studied problems in computer science, *string matching* shows its direct applicability to various areas which involve word manipulation, such as computer vision, speech recognition and computational biology. In reality, one seldom expects to find an exact match of the pattern, but an approximate match that allows errors to some extent. The most frequent operations that may produce errors mainly consist of mismatches, insertions and deletions. We are more interested in another typical typing error called *swap*, that is the order of two consecutive characters is reversed.

The last decade has seen some progress in string matching problem with swaps. Amir, Aumann, Landau etc.[1] presented the first algorithm whose time complexity beats the naive $O(nm)$ bound to find all text positions where the pattern matches allowing swaps, regardless of the number of swaps. By reducing this problem to the less-than matching with "don't cares" over a two letter alphabet, they obtained $O(nm^{1/3}logmlog\sigma)$ time complexity for a general alphabet Σ, where $\sigma = min(m, |\Sigma|)$. A recent investigation for this problem in [2] introduced a new model named *structural matching*, then defined the problem of *overlapping matching*. A direct application was to reduce the swapped matching problem to overlapping matching, and this reduction gave a solution in $O(nlogmlog\sigma)$ time for a general alphabet Σ.

We devote our attention to this problem, which arises in DNA sequence analysis. In sequencing a large string of *DNA*, researchers are always limited to sequence only a small amount of *DNA* in a single read. Hence, to sequence long strings or an entire genome, *DNA* must be divided into many pieces of short strings that are individually sequenced and then used to assemble the full string sequence. Reassembling *DNA* substrings introduces a degree of uncertainty for various positions in a biological sequence. In some cases scientists determine the appearance of a symbol in a position of a sequence by assigning a probability of appearance for every symbol from a set of characters. Such a sequence is called a *weighted sequence*. In this paper we concentrate our efforts on solving the string matching problem with swaps in a weighted sequence. The motivation comes from the occurrences of swaps in gene mutations and duplications, which might correspond to physiological variance of certain species.

2 Definitions

Let Σ be a finite alphabet consisting of a finite set of characters (or symbols). A *string* or *word* over the given alphabet is a sequence of zero or more symbols of the alphabet. A string x of length n is represented by an array $x[1,n] = x_1 x_2 \cdots x_n$, where x_i is the i-th symbol of x ($x_i \in \Sigma$ for $1 \le i \le n$) and $n = |x|$. The set of all strings over the alphabet Σ(including the empty string) is denoted by Σ^*. A string w is a *substring* or a *factor* of x if $x = uwv$ for $u, v \in \Sigma^*$. For a nonempty substring $w = x[i,j]$, we say that w occurs at position i in string x.

Definition 1. *Let $S = s_1 s_2 \cdots s_n$ be a string over Σ. A swap permutation for S is a permutation $\Pi : \{1,\ldots,n\} \to \{1,\ldots,n\}$ such that:*
(1) if $\Pi(i) = j$ then $\Pi(j) = i$ (characters are swapped).
(2) for all i, $\Pi(i) \in \{i-1, i, i+1\}$ (only adjacent characters are swapped).
(3) if $\Pi(i) \ne i$ then $s_{\Pi(i)} \ne s_i$ (identical characters are not swapped).

We say that $\Pi(S)$ is a *swapped version* of S, denoted by $\Pi(S) = s_{\Pi(1)} s_{\Pi(2)} \cdots s_{\Pi(n)}$, and, a pattern P has a *swapped match* at position i if there exists a swapped version $\Pi(P)$ of P that exactly matches text T starting at position i.

This paper deals with weighted sequences that may have uncertainty in certain positions, which is formulated as below:

Definition 2. *A weighted sequence $X = x_1 x_2 \cdots x_n$ is a continuous set of couples $(s, \pi_i(s))$, where $\pi_i(s)$ is the probability of having the character s at position i. For every position $1 \le i \le n$, $\Sigma \pi_i(s) = 1$.*

For example, considering the *DNA* alphabet $\Sigma = \{A,C,G,T\}$, $X = [(A,0.5), (C,0),(G,0.25),(T,0.25)][(A,0),(C,1),(G,0),(T,0)]$ $[(A,0),(C,0),(G,0),(T,1)]$ represents a biological sequence having three characters: the first one is either A,G,T with respective probabilities 0.5, 0.25 and 0.25, the second one is always a C, while the third symbol is trivially a T, since its probability of presence is 1. This means that in this sequence, one of the following strings: ACT, GCT, TCT might appear with probability 0.5, 0.25 and 0.25 each. We observe that the probability of presence of a string is the cumulative probability, which is calculated by multiplying the relative probabilities of appearance of each character in every position. For instance, the probability of the string ACT to appear at position 1 can be analyzed as follows: $\pi(ACT) = \pi_1(A) * \pi_2(C) * \pi_3(T) = 0.5 * 1 * 1 = 0.5$. Similarly, the definition of a weighted substring can be easily extended.

In biological problems, scientists pay more attention to the pieces with high probabilities in DNA sequences. Hence, we aim to detect all possible positions in a weighted sequence where a weighted substring having a probability of appearance not smaller than a predefined constant starts, which is a swapped match of the pattern. Now we give the formal definition of this problem :

Definition 3. *The problem of swapped matching in weighted sequences is: Given a pattern P of length m, a weighted sequence X of length n, and an integer constant k, find all positions in X where there exists a swapped match of P that has the probability of appearance larger than or equal to $1/k$, where $k > 1$.*

3 Swapped Matching in Weighted Sequences

As a first step we ignore the pattern P, and perform preprocessing for the given weighted sequence X, to find all maximal weighted factors of X. Consider $X = x_1 x_2 \cdots x_n$, where each x_n is represented by a couple $(s, \pi_i(s))$ as defined above, a maximal weighted factor of X, called for short a maximal factor is formally defined as below:

Definition 4. *A maximal factor of a weighted sequence X is a weighted substring $W = X[i,j]$ such that:*
(1) $\pi_i(W) \geq 1/k$.
(2) if $i > 1$, then $\pi_{i-1}(X[i-1,j]) < 1/k$; and if $j < n$, then $\pi_i(X[i,j+1]) < 1/k$.

In other words, a weighted substring W of X is maximal if any substring of X that contains W as a proper substring has the probability of appearance less than $1/k$. Thus a maximal factor cannot be extended either to left or right.

In methodology, we first classify all the positions of X into three categories:

- *solid positions*: if the characters occurring at position i has probability of appearance equal to 1.
- *branching positions*: if none of the possible characters listed at position i has probability of appearance greater than $1 - 1/k$.
- *leading positions*: if one of the possible characters listed at position i has probability of appearance greater than $1 - 1/k$.

The above criteria for labelling positions indicates the following facts: only one possible character appears at solid positions as it does in normal strings, more than one characters might occur at branching positions with respective possibility of appearance, while there exists only one eligible character at leading positions, since all possible characters except one have probability of appearance less than $1/k$ such that they cannot produce a qualified substring, as the cumulative probability would be less than $1/k$.

By definition, we can infer that any maximal factor must start from the leftmost position of a continuous set of solid positions, and end at the rightmost one of another continuous set of solid positions. The reason comes from the fact that a substring might split at branching positions, and its probability of appearance might dwindle away when it meets some branching positions or leading positions. However, it can be extended as far as continuous solid positions reached. This directly leads to our method for detecting maximal factors.

At the first position of each consecutive solid positions, in other words, at each solid position right behind a branching or a leading position unless it is the first position of X, scan X from left to right. Then a list of possible maximal factors starting from this position is produced as follows: whenever we meet a solid or a leading position, we extend the current factors by adding the same one character that occurs here, and the current factors might split when we meet a branching position where there are more than one character. A maximal factor stops when it touches a branching position or a leading position and the cumulative probability has reached the $1/k$ threshold. Note that if the path of a factor starting from a certain solid position is completely included in a determined maximal factor, it should be deleted from the set of maximal factors as it violates the maximality.

As we have analyzed above, the following lemmas can be easily derived:

Lemma 1. *Given a position $i (1 \leq i \leq n)$ of X, there is only a constant number of maximal factors that can occur at position i.*

Proof. Consider a solid position i where a maximal factor W starts. We assume that the number of branching positions W passes is l, recall that at each branching position none of the possible characters has probability of appearance larger than $f = 1 - 1/k$, and without loss of generality, there are no leading positions within W. In order for W to be a qualified maximal factor, its cumulative probability of appearance must be greater than $1/k$, and is mathematically formulated as follows:

$$f^l \geq 1/k \longrightarrow l \leq \frac{log(1/k)}{log(1 - 1/k)}.$$

So taking into account or not the leading positions, the number of branching positions inside a maximal factor is bounded by a constant. Thus the number of different maximal factors that can occur at position i is at most $|\Sigma|^l$, which is also a constant number.

Trivially, it follows that the total number of maximal factors in a weighted sequence of length n is $O(n)$. Further, we come to the following conclusion:

Lemma 2. *The overall lengths of all the maximal factors in a weighted sequence of length n is $O(n)$.*

Proof. W.l.o.g, we assume that the first position of the weighted sequence is a solid position, with all the leading positions removed, treated as solid positions or branching positions. The basis under this assumption is that the presence of the leading positions simply reduces the cumulative probability of a weighted substring, which makes the length of a actual maximal factor even shorter. Now we divide the weighted sequence into several consecutive windows D_j, such that each window starts from a solid position and contains l branching positions as calculated above. That is, except the first window, each window starts from the solid position right behind a branching position, and ends at a branching position, as shown in Fig.1. Notice that $\sum |D_j| = n$ holds.

Consider window D_i, for any branching position inside D_i, at most a constant number of maximal factors includes this position(by Lemma 1) and none of them will exceed window D_{i+1} since it cannot touch more than l branching positions as we mentioned above. So each maximal factor passing through this position will have length at most equal to $|D_i| + |D_{i+1}|$, therefore, all the maximal factors starting from window D_i have the length summed to at most $O(l|\Sigma|^l(|D_i| + |D_{i+1}|)) = O((|D_i| + |D_{i+1}|))$. Summing up for all possible windows, we will have the total length of all the maximal factors equal to $O(\sum(|D_i| + |D_{i+1}|)) = O(n)$.

Fig. 1. Division for the weighted sequence into windows

Based on the combination of previous lemmas, we conclude:

Theorem 1. *Finding all maximal factors in a weighted sequence of length n takes linear time.*

Proof. The time complexity for finding all maximal factors is proportional to the sum of length of those factors, which is $O(n)$ as we discussed above.

We then show how to reduce the swapped matching problem in a weighted sequence to the problem in all maximal factors of X.

Lemma 3. *Any swapped match of P in a weighted sequence X must be a substring of a maximal factor of X.*

Proof. Suppose that Π is the swap permutation providing a swapped match for P in X, and P aligns with position i of X. We just consider the starting position of the swapped match due to the symmetry of the ending position. In the case that i is an internal position of one maximal factor W, whatever p_1 swaps or not and to whichever direction, $\Pi(P)$ is always within W. In the case that i is the starting position of W, p_1 must either swap to right or not swap, because if it swaps to left, the starting position of $\Pi(P)$ is then to the left of W, where the cumulative probability of $\Pi(P)$ will be less than $1/k$ due to the maximality of W, thus not a eligible match by definition 3. Therefore, it is easily followed that, any swapped match of P in X lies within one of maximal factors.

This lemma inspires us to simplify the swapped matching problem in a weighted sequence. That is, we would find all swapped matching of P in the set of maximal factors of X rather than in X itself, since each maximal factor is a normal string without any weighted characters inside. From this point, we can apply the idea of overlapping matching that proposed in [2] directly into our problem, with the major distinction that the match is detected in a set of strings instead of a single string. The easiest way to deal with them is to concatenate all the maximal factors together into one long string, then create an array to store the starting positions of each factor. Following lemma 6 in [2], we can find all swapped matches of P in the concatenated string, thus determine the positions in corresponding maximal factors. Finally, we can state the following:

Theorem 2. *Given the pattern P of length m and a weighted sequence X of length n, the problem of string matching with swaps in a weighted sequence can be solved in $O(nlogmlog\sigma)$ time for a general alphabet Σ, where $\sigma = min(m, |\Sigma|)$.*

Proof. As we have discussed above, the swapped matching problem in a weighted sequence can be reduced to the problem in the set of maximal factors of X. Finding all maximal factors of X takes $O(n)$ time. Using convolution, swapped matching of P in the concatenated string of all maximal factors runs in $O(nlogmlog\sigma)$ time by means of overlapping matching (For more information on the overlapping matching the readers can refer to [2]), since the length of the concatenated string is $O(n)$ as proven by lemma 2. Therefore, the overall time complexity is $O(nlogmlog\sigma)$.

4 Conclusions

Our future direction is focused on the further improvement of the algorithm over a fixed alphabet. We are also interested in the similar problem in the case that k is not a constant, but input by users.

References

1. Amir, A., Aumann, Y., Landau, G.M., Lewenstein, M., Lewenstein, N.: Pattern matching with swaps, Journal of Algorithms, 37(2), (2000) 247-266.
2. Amir, A., Cole, R., Hariharan, R., Lewenstein, M., Porat, E: Overlapping matching, Information and Computation, 181, (2003) 57-74.
3. Amir, A., Farach, M.: Efficient 2-dimensional approximate matching of half-rectangular figures, Information and Computation, 118(1), (1995) 1-11.
4. Iliopoulos, C., Makris, Ch., Panagis, I., Perdikuri, K., Theodoridis, E., Tsakalidis, A.,: Computing the string regularities in biological weighted sequences using the weighted suffix tree, European Conference On Computational Biology (ECCB 2003), (accepted).

Knowledge Maintenance on Data Streams with Concept Drifting

Juggapong Natwichai and Xue Li

School of Information Technology and Electrical Engineering,
The University of Queensland, Brisbane, Australia
{jpn, xueli}@itee.uq.edu.au

Abstract. Concept drifting in data streams often occurs unpredictably at any time. Currently many classification mining algorithms deal with this problem by using an incremental learning approach or ensemble classifiers approach. However, both of them can not make a prediction at any time exactly. In this paper, we propose a novel strategy for the maintenance of knowledge. Our approach stores and maintains knowledge in ambiguous decision table with current statistical indicators. With our disambiguation algorithm, a decision tree without any time problem can be synthesized on the fly efficiently. Our experiment results have shown that the accuracy rate of our approach is higher and smoother than other approaches. So, our algorithm is demonstrated to be a real anytime approach.

1 Introduction

There are many classification applications that require mining on the data streams, such as network sensor monitoring, stock market analysis or server performance tuning. Concept drifting [1] always occurs in streaming data environment because data is generated continuously. In general, concept drifting happens when the knowledge discovered in the past, is not applicable to the current incoming data/events any more, because of the inherent domain changes. More general, any previous truth is no longer valid in the current time.

In dealing with the concept drifting problem, there are many available algorithms so far. They can be categorized into two groups: increment learning approach [2–4], and ensemble classifiers algorithms [5–7]. Most algorithms provide knowledge for the users in form of decision trees [8] representation. However, these algorithms also have the same problem which is any time prediction. More specifically, we do not know what time period the users might have interests to predict from a decision tree. Both groups of algorithms only reflect to new coming knowledge and manipulate it to existing knowledge. Intuitively, this procedure should avoid noise, so it will take some time to justify the new knowledge whether it is noise or not. And, within the time, a decision tree becomes unstable and produces very high error rate. Apparently, this problem can be seen very clearly in incremental classifier. As shown in Figure 1 from CVFDT [2] approach, which we shaded the areas that represent the periods of unstable decision tree. There is not much chance for the users to get a right tree, particularly when justifying period is larger than the size of concept. On the other hand, if an approach of ensemble classifiers is used, it would take time to adjust weight between each classifier to keep track of concept drifting too.

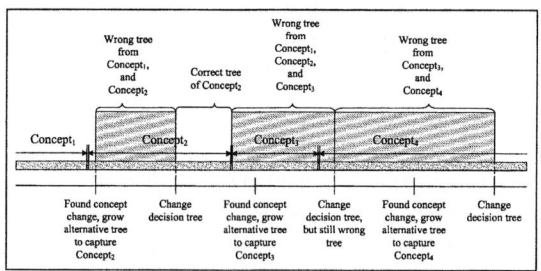

Fig. 1. Not Any Time Problem of Incremental Classifier

In this paper, we propose an approach for knowledge maintenance. Our approach has two phases. Firstly, we maintain knowledge by using an ambiguous Decision Table, (aDT) [9] as an intermediate form of knowledge representation. For reflecting changes, it is essential for counting accuracy of each rule in the decision table and also updating it constantly. So, we also add a set of indicators that is recorded for solving any time problem. This includes current error rate and time stamp for each rule. With ambiguous Decision Table (aDT), ambiguous rules will be kept for handling concept drifting problem. And, for making knowledge as recently as possible, we also implemented sliding window. Secondly, when the users want to predict based on acquired knowledge, we are able to generate a decision tree at any time using simple disambiguation algorithm from aDT.

The remaining sections in this paper are organized as following. In the next section, we introduce basic background related to our work. Section 3, gives the detail about our disambiguation algorithm. The experiment results are presented in section 4. Finally, section 5 concludes this papers and gives an overview of future work.

2 Ambiguous Decision Table with Time Stamp

In this section, aDT [9] and our extension work are reviewed. The aDT consists of two-dimensional array of cells. Each row records a decision rule, with each column within it is an attribute. And, the last column of each rule is an assigned class. Ambiguous decision table will contain two or more rules that conflict with each other e.g. in Table 1 rule 1 and 9, or rule 2 and 10. Our idea is to keep all happened knowledge, and add statistical indicators for each rule to make it ready for decision trees induction. So, we choose CVFDT algorithm [2] that can reflect concept drifting by generating alternative sub-tree to fill in this table.

When a new set of sub-tree is discovered by knowledge feeder, we will transform it into decision rules.

Furthermore, we also add two more columns into the aDT as shown in Table 1.

Table 1. Extended ambiguous Decision Table

Rule No.	Att_0	Att_1	Att_2	Att_3	Att_4	Class	% Error	Time Stamp
1	True	True	False	False	True	True	6	1
2	True	False	False	False	False	False	5	1
...	...	...	...	...	...	...	...	...
9	True	True	False	False	True	False	0	5
10	True	False	True	False	False	False	0	5

In the first, error of corresponding rule is recorded. And, start time stamp is also added here to support our disambiguation algorithm.

While incoming data keeps changing, new knowledge/rule will be derived. We keep making them always up-to-date by sliding window. Consequently, our extended aDT is capable to reflect any discovered knowledge as well as the history.

3 Decision Tree Synthesizing

In this section, we review the second phase of our approach. In the first phase, we collected knowledge along with its history and statistical indicators. When the users want to get the most recently knowledge so far, with additional algorithm we can synthesize decision tree for the users even we do not keep any decision tree. Because of simplicity of our algorithm, synthesizing process can be done on the fly.

So, we presented disambiguation algorithm as Table 2. At the beginning, we have to build the root of a decision tree. From the root of a decision tree to the lowest internal node, we will select all possible candidate attribute at currently state of aDT. All candidate attribute comes from each alternative sub-tree of our feeder. After any selection, we will split the tree by the number of attribute values. Each attribute value will have to be used to expand as a next level node of decision tree. And, this process will be repeated recursively. Because we maintain our decision table as an ambiguous table, we will also face ambiguousness in the synthesizing process. In this paper, we propose to induce all possible options firstly. And then, we eliminate those options with larger error when we induce until we reach a leaf node. This comes from the fact that the accuracy of decision trees come from where the decisions take place. At that point, we sum its error rate up for counting accuracy. The selected attribute at any level of a decision tree will be better than other candidates statistically. No matter whether it is newer than others or it is the oldest one, if it has less error rate than the other attributes at the current time, it would be selected. Although we keep all knowledge within the table, space complexity of our approach is $O(avc)$ where a

Table 2. Disambiguation Algorithm

Inputs: aDT is an extended ambiguous Decision Table,
         i is a level number.
Outputs: DT is a decision tree,
          $DT.error$ is a error rate of decision tree.

Procedure Disambiguation(aDT, i)
Get candidate attributes for level ith from aDT.
While The number of candidates > 0 **do**
   split DT on each candidate attribute.
   For each child of DT on current split **do**
      refine aDT with each attribute value.
      If aDT has only one rule, **then**
         assign Class to current node of DT.
         Return DT.
      End if.
      Get $DT.error$ from **Disambiguation(aDT, $i+1$)**.$error$
   End for.
End while.
Choose DT from candidate with the least $DT.error$ and current
   level node error.
Return DT.

is the number of attributes, v is the maximum number of possible attribute values for any attribute, and c is the number of classes. This can be seen apparently that there is no term related to the number of examples. Moreover, we can assume that real-world streaming data is as large as multi-million examples and is produced continuously. So this asymptotic bound $O(avc)$ is not comparable with size of data streams and our approach can be used without causing efficiency degrading.

4 Experiment Results

We performed experiment using syntactic data set to demonstrate aspects of our algorithm when encountered with concept drifting. We selected hyperplane rotation in a d-dimension problem. A hyperplane in a d-dimensional space is denoted as Equation 1:

$$\sum_{i=1}^{n} a_i x_i = a_0 \qquad (1)$$

where x_i is the coordinate of the ith dimension, and a_i is the weight corresponding to each ith dimension. For example, if $\sum_{i-1}^{n} a_i x_i \geq a0$, we will label

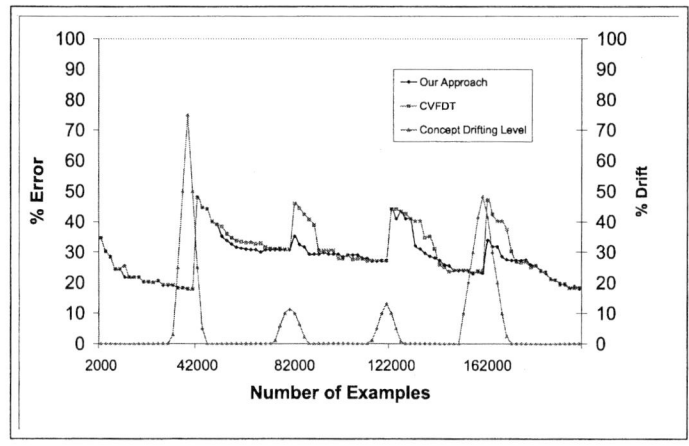

Fig. 2. Error Rate of Syntactic Data Set with Five Concepts

its class as positive. Otherwise it will be labelled as negative. We will also control the values of class distribution, so that both positive class, and negative class will be in the similar amount. The hyperplane rotation problem is an ideal case to demonstrate the handling of concept drifting, because it can adjust the weight of each attribute more smoothly than other ways (see [2] for more explanations).

We performed the experiment by generating 200,000 examples with 50 attributes syntactically. Four concept drifts were generated. We did this through adjustment of weight (a_i) of each attribute. In this way, we obtained five equal sets of syntactic data in this experiment.

Firstly, we made a change radically between Concept One and Concept Two. However, we adjusted weight between Concept Two and Three as smooth as possible. In addition, we wanted to investigate the behavior of our approach when encountered with a concept which was learned in the past. So, we could see efficiency of our approach to exploit previous knowledge. Concept drifting level along data set is shown as a minor vertical axis in Figure 2 .

The result of our approach is shown in Figure 2. Obviously, it is able to see that our approach is better than compared approach. The error-rate of our algorithm dropped sharply when it had learnt a new concept. If there is radical concept drifting, our approach would give dropping of error rate quickly as seen from error rate of Concept Two. And, in the case of slightly changed, concept catching period between Concept Two and Concept Three showed our better performance very clearly. In case of learning a concept which happened in the past, our algorithm can use it to synthesize a decision tree very efficiently too. So, at any time which the users need a decision tree, not much difference accuracy of the decision tree they will get.

5 Conclusion

This paper introduced a novel approach for solving concept drifting problem in mining knowledge from streaming data. With a data structure namely extended ambiguous Decision Table (aDT) [9], we can maintain time-stamped knowledge and handle the problem of concept drifting efficiently. Our work has proved to be capable of capturing emerging concepts in a manner of any time. Currently we are improving our approach in terms of the computational performance, and storage complexity by using a data-mining-ready data structure for a variety of streaming data. And we are also extending our algorithm to synthesize decision trees with the user-defined arbitrary time-ranges validation, compared with the current sliding window approach. For the users, it is essential for them to be able to specify starting time or end time of their decision tree according to the domain knowledge that only user knows. Moreover, our approach can be used for users to make a combination of criteria for different properties of decision tree e.g. the support examples or overall accuracy of the decision trees. In this way our algorithm can also be used as an experimental tool for the human interactive knowledge discovery.

References

1. Schlimmer, J.C., Richard II. Granger, J.: Beyond incremental processing: Tracking concept drift. In: AAAI National Conference on Artificial Intelligence, Philadelphia, PA, USA (1986) 502–507
2. Hulten, G., Spencer, L., Domingos, P.: Mining time-changing data streams. In: ACM SIGKDD International Conference on Knowledge Discovery and Data Mining, San Francisco, CA, USA (2001) 97–106
3. Jin, R., Agrawal, G.: Efficient decision tree construction on streaming data. In: ACM SIGKDD International Conference on Knowledge Discovery and Data Mining. (2003)
4. Kalles, D., Morris, T.: Efficient incremental induction of decision trees. Machine Learning **24** (1996) 231–242
5. Wang, H., Fan, W., Yu, P., Han, J.: Mining concept-drifting data streams using ensemble classifiers. In: ACM SIGKDD International Conference on Knowledge Discovery and Data Mining, Washington, DC, USA (2003)
6. Street, W.N., Kim, Y.: A streaming ensemble algorithm (sea) for large-scale classification. In: ACM SIGKDD International Conference on Knowledge Discovery and Data Mining, San Francisco, CA, USA (2001) 377–382
7. Kolter, J.Z., Maloof, M.A.: Dynamic weighted majority: A new ensemble method for tracking concept drift. In: International Conference on Data Engineering, Bangalore, India (2003)
8. Quinlan, J.R.: C4.5: Programs for Machine Learning. Morgan Kaufmann, San Mateo, CA, USA (1993)
9. Colomb, R.M.: Representation of propositional expert systems as partial functions. Artificial Intelligence **109** (1999) 187–209

A Correlation Analysis on LSA and HAL Semantic Space Models

Xin Yan[1], Xue Li[1], and Dawei Song[2]

[1] School of Information Technology and Electrical Engineering,
The University of Queensland, QLD 4072, Australia
{yanxin, xueli}@itee.uq.edu.au
[2] CRC for Enterprise Distributed Systems Technology (DSTC),
Level 7, General Purpose South,
The University of Queensland, QLD 4072, Australia
dsong@dstc.edu.au

Abstract. In this paper, we compare a well-known semantic space model, Latent Semantic Analysis (LSA) with another model, Hyperspace Analogue to Language (HAL) which is widely used in different area, especially in automatic query refinement. We conduct this comparative analysis to prove our hypothesis that with respect to ability of extracting the lexical information from a corpus of text, LSA is quite similar to HAL. We regard HAL and LSA as black boxes. Through a Pearson's correlation analysis to the outputs of these two black boxes, we conclude that LSA highly co-relates with HAL and thus there is a justification that LSA and HAL can potentially play a similar role in the area of facilitating automatic query refinement. This paper evaluates LSA in a new application area and contributes an effective way to compare different semantic space models.

Keywords: Correlation Analysis, Hyperspace Analogue to Language, Latent Semantic Indexing, Automatic Query Refinement.

1 Introduction

Latent Semantic Analysis (LSA) [1] is a well known model for extracting the relations of words in the context of large text corpus. It is widely used in different area such as human knowledge acquisition and representation. In document indexing and retrieval area, LSA has an competitive performance at high recall. Furthermore, LSA model can economically represent both documents and words as reduced dimensionality vectors in a so called k dimensional space.

With respect of its rationale and many advantages, LSA is also able to measure the similarity between words living in context. However, in literature there is a lack of sufficient contributions that show the ability of LSA to improve retrieval effectiveness. In this paper, we propose a new approach to evaluate potential ability of LSA to do automatic query refinement by comparing it with another model Hyperspace Analogy to Language (HAL) [2], which is the base of an effective query refinement method, HAL-based information flow [3].

1.1 Automatic Query Refinement

Automatic query refinement is a technology in information retrieval area to automatically refine user's query. With respect to query on the web or other large text corpora, user's query terms always imprecisely described their information needs. Without an automatic mechanism, users have to manually delete, alter and add query terms based on their previous query experiences. To an inexperienced user, this manual process usually decreases the retrieval effectiveness.

1.2 HAL-Based Information Flow

In recent years, different automatic query refinement methods ([3], [4], [5]) have been proposed in order to improve user's query. Among those researches, Song and Bruza's HAL-based information flow [3] contributes significantly to the retrieval effectiveness. In their work, they consider query expansion as an information inference process via information flow [6]. A candidate query term can be inferred from user's initial query if the degree of the information flow between the former term and the latter one is above the threshold. In other words, there is a strong degree of inclusion between the information states of the candidate query term and the information states of user's initial query terms. For instance, "grassland" and "swings" may carry the information "Dutton Park", which is a small park in Australia, if and only if the majority of the states of "grassland" (i.e. location, size, owner) and the majority of the states of "swings" (i.e. location, owner) are included in the set of states of "Dutton Park".

The HAL-based information flow is based on HAL semantic space model which automatically constructs a high dimensional semantic space from a given text corpus. In that space, terms in text corpus such as "Dutton Park", "grassland" and "swings" are represented by HAL vectors. A HAL vector is regarded as a representation of the information states of a particular term [3]. An intuitive example is that the term "Dutton Park" can be represented as a string "107 Carmody Road, two swings, one BBQ oven, $400m^2$ grassland, promise of city council".

Through test results of HAL semantic space model in information inference process, the performance of Song and Bruza's HAL-based information flow is better than that of those prominent query refinement methods ([4], [5]).

1.3 Is HAL the Unique Choice?

As a key point in this paper, we believe that HAL model is not the unique choice for HAL-based information flow technology. Instead, LSA model has almost the same effect to extract lexical information from text corpus. It will be another good option for Song and Bruza's query refinement technology. Our hypothesis is firstly based on the observation of the relations and similarities of words extracted from context by LSA and HAL. We find that their measurements of term to term relationship are quite similar. Deerwester [7] used a technical memo example to illustrate LSA model. In his example, a two-dimensional space is used to represent the underlying meaning of 12 terms and 9 documents. Two terms have strong relation if they are "nearby" in that space. The two-dimensional space shows that a query "human computer" is closely relevant to "system", "user", "interface", "EPS", "response" and "time". Through the power of HAL model, we can get a quite similar result that 'human computer" is closely relevant to "system", "user", "interface" and "survey".

1.4 Shall We Construct a Query Model to Test LSA?

It is unnecessary to construct a complete and functional LSA-based query model to compare the precision and recall with that of HAL-based query model (i.e. HAL-based information flow). In this paper we propose a convenient and effective way to testify this presumed similarity between LSA and HAL (Fig. 1). The main idea is using black box to represent a semantic space model. Based on the same inputs (i.e. a common text corpus), we analyze the correlation between the outputs of HAL and LSA (i.e. term by term matrices with each cell entry represents a relevance degree between terms).

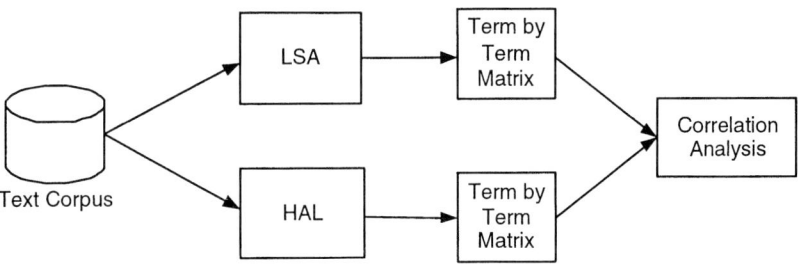

Fig. 1. Black Box Analysis

2 Related Work

2.1 HAL

Given an n-word vocabulary, the HAL space (i.e. a $n \times n$ term-by-term matrix) is constructed by moving a l-length sliding window passing over the beginning to the end of every document in text corpus. Within this window, words are recorded as co-occurring. The strength of the co-occurrence of two words is inversely proportional to the their "distance" which is the number of other words between them. l can be set to a large value to fit all the real structure of data or a small value to reduce the "noise". A HAL vector t_i is generated by making an addition between the ith row and ith column in the HAL space [3]. The strength of association between every pair of HAL vectors derived from HAL space is measured by the cosine similarity function[1] [3].

---

[1] Alternatively, other similarity function,e.g., the Euclidean distance based, could be used.

2.2 LSA

LSA [1] uses Singular Value Decomposition (SVD) algorithm to discover the associations among documents and terms by transforming the standard term by document matrix X to a document matrix D, a singular matrix S and a term matrix T. Each row in T is a term vector and each column is a SVD derived concept. The TS^2T' is a square term by term symmetric matrix in which each cell entry indicates the similarity value between the row and the column. In this paper we use the cosine similarity function to measure the strength of association between every pair of column vectors in the TS^2T' matrix in order to get the results which is comparable to the results of HAL model. Moreover in SVD we can get a matrix approximation X_k of X by keeping the k largest singular values in S and computing the associated singular vectors of X in order to reduce the "noise" in text corpus.

3 Proposed Approach

In our proposed approach, we regard that a semantic space model can be treated as a black box $f\colon R = f(T)$, its input is T, a set of terms in the text corpus. Its output is R, a term by term square matrix where each cell entry r_{ij} corresponds to the relevance degree between t_i and t_j. In this case, we can rewrite the above formula as below:

$$R^{LSA} = f_{LSA}(T)$$
$$R^{HAL} = f_{HAL}(T)$$

In R^{LSA}, r_{ij}^{LSA} is a measure of the relevance degree between term i and term j by using LSA model f_{LSA}; likewise, in R^{HAL}, r_{ij}^{HAL} is another measure made by HAL model f_{HAL}. We construct a scatter diagram to observe the correlation relation between these two models intuitively. In that diagram, a coordinate of a point P_{ij} is r_{ij}^{HAL}, another coordinate is r_{ij}^{LSA}. The second step is using the Pearson's correlation [8] to measure the strength of a correlation relationship between f_{LSI} and f_{HII}. It is beyond the scope of this paper to introduce the technical details of Pearson's correlation.

4 Experimental Results

So far we have tried our correlation analysis approach on two common standard text corpora, CISI and MED. Stemming is not used. We filter the common function words by using a stop word list. In every text corpus we index all the terms by giving them numbers and partition the numbered terms into several intervals. Each interval contains 1000 terms (e.g. interval 1 contains term 1 to 1000).

MED consists of 1033 medical abstracts. After the stop words are removed, a vocabulary of 5825 remains. The average word numbers of MED documents is 72. CISI consists of 1460 information science abstracts. After the stop words

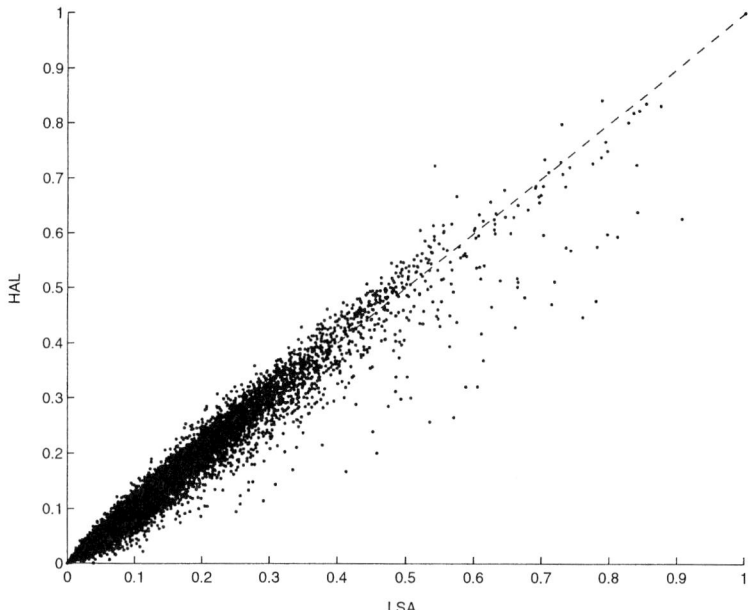

Fig. 2. Scatter Diagrams of MED

are removed, a vocabulary of 5609 remains. The average word numbers of CISI documents is 60. In LSA model, k is set to 1033 for MED and 1460 for CISI in order to fit all the real structure of text corpus. In HAL model, the length of the sliding window is set to a large value, 50, for the same purpose.

To save space, in Fig. 2 and Fig. 3 we only show two scatter diagrams. Each of them is composed by a set of 40,000 points which is the Cartesian product of a small set (i.e. term 201 to 400) in each text corpus vocabulary with itself. The dashed line which begins at the origin point in each diagram is 45 degree to the X axis. HAL and LSA have the same measurement of the relevance between a pair of terms if their respective point is exactly on this dashed line.

In Table 1, we list all the intervals and the respective result of correlation analysis on MED and CISI. The analysis result is named as Pearson's r. A Pearson's r reflects the correlation between the outcomes of HAL and LSA applied on a specific term group or interval.

5 Discussion and Conclusion

Fig. 2 and Fig. 3 intuitively reflects the linear dependence between LSA and HAL on MED and CISI. Table 1 shows that the minimum value of Pearson's r among those samples is 0.9769. The correlation coefficients are all positive and strong. It is an evidence for us to justify the linear dependence between LSA and HAL.

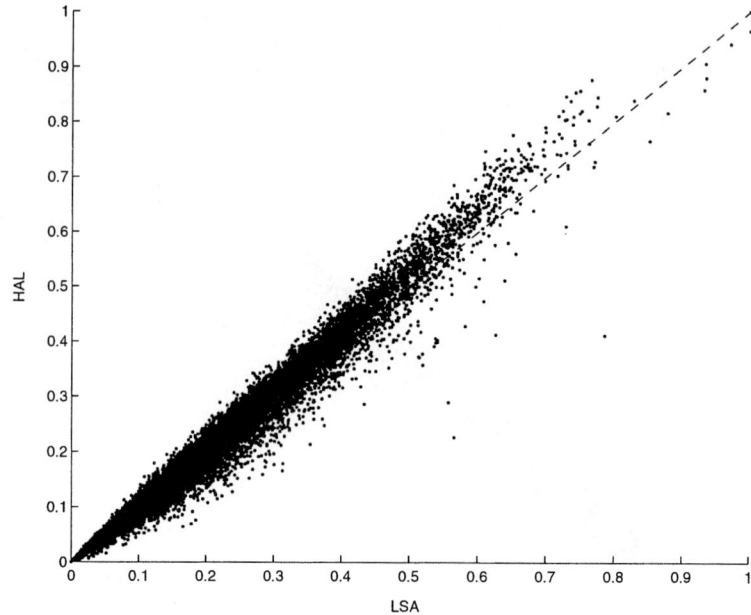

Fig. 3. Scatter Diagrams of CISI

Table 1. MED and CISI Correlation Analysis Result

Text Corpus	Term Intervals	Pearson's r
MED	1 to 1000	0.9783
	1001 to 2000	0.9769
	2001 to 3000	0.9781
	3001 to 4000	0.9795
	4001 to 5000	0.9789
CISI	1 to 1000	0.9842
	1001 to 2000	0.9859
	2001 to 3000	0.9862
	3001 to 4000	0.9867
	4001 to 5000	0.9848

In this paper, we have performed a correlation analysis approach to clarify the quantitative relation between LSA and HAL. The strong correlation prove our hypothesis that with respect to the ability of extracting the lexical information from a corpus of text, LSA is quite similar to HAL. Of course, we cannot neglect the two major differences between these two models. Firstly LSA has a higher computational requirement than HAL. Secondly LSA is easier to deal with static text corpora than dynamic (e.g. WWW). In comparison with LSA, the computation result of HAL is easy to update. Therefore, how to choose these two models largely depends on real application environments.

Acknowledgments

The work reported in this paper has been funded in part by the Co-operative Centre for Enterprise Distributed Systems Technology (DSTC) through the Australian Federal Government's CRC Programme (Department of Education, Science and Training). The authors would like to thank Richard Cole and Peter Bruza for their valuable comments.

References

1. Landauer, T.K., Foltz, P.W., Laham, D.: An introduction to latent semantic analysis. Discourse Processes **25** (1998) 259–284
2. Lund, K., Burgess, C.: Producing high-dimensional semantic spaces from lexical co-occurrence. Behavior Research Methods, Instruments Computers **28** (1996) 203–208
3. Song, D., Bruza, P.: Towards a theory of context sensitive information inference. Journal of the American Society for Information Science and Technology **54** (2003) 326–339
4. Lafferty, C.Z.: Document language models, query models, and risk minimization for information retrieval. In: 24th annual international ACM SIGIR conference on Research and development in information retrieval, ACM Press (2001) 111 – 119
5. Lavrenko, V.W.: Relevance-based language models. In: 24th annual international ACM SIGIR conference on Research and development in information retrieval, ACM Press (2001) 111 – 119
6. Barwise, J.S.: Information Flow : The Logic of Distributed Systems. Cambridge University Press (1997)
7. Deerwester, S., Dumais, S.T., Furnas, G.W., Landauer, T.K., Harshman, R.: Indexing by latent semantic analysis. Journal of the American Society for Information Science **41** (1990) 391–407
8. Wynne, D. J.: Learning Statistics: A Common-Sence Approach. Macmillan Publishing Co., Inc, New York (1982)

Discretization of Multidimensional Web Data for Informative Dense Regions Discovery

Edmond H. Wu[†], Michael K. Ng[†], Andy M. Yip[‡], and Tony F. Chan[‡]

[†]Department of Mathematics, The University of Hong Kong,
Pokfulam Road, Hong Kong
hcwu@hkusua.hku.hk, mng@maths.hku.hk
[‡]Department of Mathematics, University of California,
Los Angeles, CA 90095-1555, USA
{mhyip, chan}@math.ucla.edu

Abstract. Dense regions discovery is an important knowledge discovery process for finding distinct and meaningful patterns from given data. The challenge in dense regions discovery is how to find informative patterns from various types of data stored in structured or unstructured databases, such as mining user patterns from Web data. Therefore, novel approaches are needed to integrate and manage these multi-type data repositories to support new generation information management systems. In this paper, we focus on discussing and purposing several discretization methods for large matrices. The experiments suggest that the discretization methods can be employed in practical Web applications, such as user patterns discovery.

Keywords: Discretization, Dense regions discovery, Web mining, Web information system.

1 Introduction

With the fast growing of Internet in recent years, developing Web-based information systems become an active research area of finding useful information from Web data depending on state-of-art information technologies nowadays. As a key step to ensure data and information quality, discretization is defined as a process that divides continuous numeric values into a set of intervals that can be regarded as discrete categorical values [4]. Although discretization is well studied in literature [1, 6, 5], it is a new challenge to design effective and efficient discretization methods for diverse-type and fast-growing Web data distributed on different Web information systems. In this paper, we focus on purposing practical data discretization solutions to support effective dense regions discovery from large-scale and multidimensional Web databases.

The remaining of this paper is organized as follows. In Section 2, we briefly introduce the concepts and applications of dense region discovery and then address the discretization problems encountered in dense region discovery. In Section 3, we present the details of four solutions purposed for complex Web data discretization. In Section 5, we perform an empirical evaluation on the discretization methods using real Web datasets. Finally, we give the conclusions and suggest our future work.

2 Dense Regions Discovery

Given an n-by-p data matrix X where n is the number of entities and p is the number of attributes of each entity. Let R and C be an index set of a subset of rows and columns of X respectively. Since we do not distinguish between a matrix and its permuted versions, we assume that R and C are sorted in the ascending manner. A submatrix of X formed by its rows R and columns C is denoted by $X(R,C)$. We also identify $X(R,C)$ by the index set $D = R \times C$.

Definition 1 (Dense Regions). *A submatrix $X(R,C)$ is called a maximal dense region with respect to v, or simply a dense region with respect to v, if*

- *$X(R,C)$ is a constant matrix whose entries are v (density), and,*
- *Any proper superset of $X(R,C)$ is a non-constant matrix is non-constant (maximality).*

From the definition of dense regions, we can see that finding dense regions has practical meaning in knowledge discovery. For example, if the rows present customers IDs, the columns present products IDs, then the dense regions are the groups of customers who buy the same products. The information can be used in analyzing customer behavior.

In practice, the matrices for dense regions discovery are large and sparse. Hence, efficient algorithms are needed for mining dense regions. In [9], we present the algorithm for mining dense regions in large data matrices. The essence of the algorithm is that it searches the data matrix for dense regions in both vertical and horizontal directions beginning from a given starting entry (s,t). It returns two dense regions containing (s,t), one is obtained by searching in vertical direction first; the other is by searching in horizontal direction. Due to the limited length of this paper, we just employ the dense regions discovery algorithm in experiment part and omit the detailed introduction of it.

A limitation of existing dense regions discovery algorithms is that they assume the discrete values of entries in matrices. Therefore, in order to deal with data with continuous values, we need to perform data discretization first before we can apply the process of dense regions discovery. Since discretization is quite important to dense regions, we will discuss several considerations for effective and efficient discretization in matrices. The first concern is about how to set the discretization criteria for dense regions discovery. The second concern is related to the computational cost for performing discretization in large data matrices. We should note that for a $n \times n$ matrix, the complexity of sorting all the entries in the matrix is at least $O(n^2 \log(n^2))$. In dense regions discovery applications, the size of matrix n is usually quite large. Hence, we should focus on the feasibility of discretization methods purposed instead of designing complex discretization methods in these cases. Another also quite important issue is the consideration of how many intervals is proper and how to distribute an observed value to an interval, which focuses on the implementation of discretization algorithms.

3   Discretization Models for Dense Regions Discovery

Equal width interval (EWI) method is to discretize data into k equally sized bins, where k is a parameter provided by users. Given a variable v, the upper bound and lower bound of v are v_{max} and v_{min}, respectively. Then, this method uses the formula below to compute the size of bin: $\lambda = (v_{max} - v_{min})/k$. For i_{th} bin, the bin boundary is $[b_{i-1}, b_i]$, where $b_i = b_0 + i\lambda$, $b_0 = v_{min}, i = 1, ..., k-1$. As to the problem of dense regions discovery, given a data matrix X, v_{ij} represents the value of entry x_{ij}, we can directly apply Equal width interval method to discretize X. In such cases, we define $k=max\{1, \lceil \log(mn) \rceil\}$ as the number of intervals of $m \times n$ matrix X. If $m=n$, $k=max\{1, \lceil 2\log(n) \rceil\}$.

There are two main drawbacks of equal width interval method. One is that it may be influenced by outliers, the second one is that it does not use the class information of data. This is the motivation of purposing supervised equal width interval method in discretization. Suppose there are r classes labeled as $C_1, ..., C_r$ in the dataset. As to variable v, the means of each class are $\bar{v}^1, ..., \bar{v}^r$, respectively. The maximal value and minimal value in class C_i are v_{max}^i and v_{min}^i. Then, our supervised adaptive width interval method is to partition $(v_{max}^i - \bar{v}^i)$ and $(\bar{v}^i - v_{min}^i)$ into k parts, individually. Finally, we can divide the dataset into $2k$ bins. The width of the first k bins and the last k bins is calculated by formula 2 and 3 as below: $\lambda_l^i = (\bar{v}^i - v_{min}^i)/k$ and $\lambda_u^i = (v_{max}^i - \bar{v}^i)/k$. For the j_{th} bin of C_i, the boundary b_j^i is:

$$b_j^i = \begin{cases} \bar{v}^i - \lambda_l^i(k-j), & 0 \leq j < k, \\ \bar{v}^i, & j = k, \\ \bar{v}^i + \lambda_u^i(j-k), & k < j \leq 2k. \end{cases}$$

The general form of Supervised Adaptive Width Interval (SAWI) discretization lists as follow: $\lambda_{lj}^i = \rho_{lj}(v^* - \check{v}^i)$ and $\lambda_{uj}^i = \rho_{uj}(\hat{v}^i - v^*)$, where $0 < \rho_{lj} < \rho_{l(j+1)} \leq 1$, $0 < \rho_{uj} < \rho_{u(j+1)} \leq 1$, $\check{v}^i < v^i < \hat{v}^i$, $j = 1, ..., k-1$. In the formula, v^* is a self-selecting value in the data ranging from v_{min} to v_{max}. For instance, besides $\bar{v}$, user can also select median value as v^*. As to ρ_{lj} and ρ_{uj}, the parameters provide the flexibility for user to select suitable length of interval for discretization. The setting of these parameters depends on the datasets and mining tasks. Generally, we can use this formula to do data discretization in particular range of values, e.g., discretize data in the range $[\check{v}^i, \hat{v}^i]$. We can also adopt this approach for iterated discretization. The merit of Supervised Adaptive Width Interval method is that it can overcome the drawbacks of Equal Width Interval method. In some cases, the Supervised Adaptive Width Interval is more useful to identify potential patterns since it considers class label and data distribution information, so the method is more adaptive to diverse large datasets.

Recall that a well known approach in clustering is K-means, the essence of this method is that it attempts to minimize the intra-interval distance and maximize the inter-interval distance of instances. One approach of adopting K-means in discretization is that it begins with an equal width distribution of the observed values over the k intervals, followed by an iterative process where the values near the boundaries change between intervals while this process improves certain criterion until no more improvement can be achieved. However, in some cases, setting the initial partitions of data by equal width intervals may not be a proper partition, especially when we want to control the number of instances in each cluster, e.g., some applications of customer segmentation. Motivated by this, we combine K-means with equal frequency interval for discretization.

The idea is that we first use equal frequency interval to discretize the data into k bins, then we use K-means to adjust the values in each bin. We refine the objective function to minimizes intra-interval variance of values in each bin and maximize the inter-interval variance among bins by adjusting the interval boundaries. The process proceeds until no more improvement can be made or the number of instances of bins exceeds a pre-set limit. The method is especially suitable for dense regions discovery because we can estimate the size of dense regions by controlling the number of entries for each constant value. On the other hand, the underlying clusters information retains, which is quite helpful to find informative dense regions. We summarize the details of the algorithm as follows:

Begin (MEK)
1. Import dataset D with n instances with numerical variable v
2. Sort D into $D' = d_1, ..., d_n$ by v and set number of interval k
3. For each d_i, i=1...,n. Put d_i into bin $b_{\lfloor i/k \rfloor}$
4. Put all n instance into k bins until each bin contains n/k instance
5. Calculate the mean and variance of each bin and the variance among bins
6. Set the number of movable values and maximal difference of instances in bins
7. For each value around the interval boundaries, move it to neighbour intervals
8. Recompute the mean and variance of each bin and the variance among bins
9. If the variance of means among bins is larger and the variance in bins is smaller
10. Save the new partition of intervals, otherwise, go to 7
11. Output the discretization with k adjusted bin-size intervals for D
End

In dense regions discovery, the most important issue is to find subsets from data matrix in which have potentially useful patterns. However, matrices with continuous values are hard to apply dense regions discovery algorithms since most dense regions discovery algorithms assume the values of entries are discrete. Therefore, we need to discretize matrices with continuous values into discrete or categorical entries first. For an entry x_{ij} in matrix X, it belongs to i_{th} row and j_{th} column. If X is a $m \times n$ matrix, then X is a two-dimensional matrix with m and n attributes on each dimension. Specially, if each dimension represents one variable (e.g., rows represent users, columns represent Web pages), then we can regard the value v_{ij} of x_{ij} as a data point in i_{th} row and j_{th} column, individually. In Table 4, v_{ij} denotes the frequency of visiting from Web page i to Web page j. The characteristic of such matrix is that the rows and columns representing the same attribute, e.g., the page visiting frequency. Thus, how to discretize such data matrices become an interesting research problem.

In order to support efficient dense regions discovery from matrices with one entry variable, our idea is to discretize the continuous value v_{ij} into two discrete value by vertical and horizontal searching. It is so called Two-way discretization. The major steps for discretization of X as follows:

Begin
1. Import $m \times n$ matrix X with continuous variable v
2. Sort each row R_i and column C_j by v respectively
3. Set the criteria to discretize every row R_i and column C_j, e.g., equal frequency interval
4. Determine number of intervals k and interval boundaries for R_i and C_j
5. For each x_{ij}, $i = 1..., m, j = 1, ...n$. Discretize x_{ij} by R_i and C_j, store in vr_{ij} and vc_{ij}
6. Output the two-way discretization matrix

End

The virtue of Two-way discretization is that it considers the rows and columns information in matrices, which is also applicable in some particular dense regions discovery applications.

Generally, this discretization method can extend to multidimensional situation. For an entry $x_{d_1,...,dn}$ in N dimensional space $\mathcal{R}^{\mathcal{N}}$, the corresponding discretization result for dimension x_{d_i} is v'_{d_i}. Then, N-way discretization is equivalent to find vector $V' = v'_{d_1}, ..., v'_{dn}$ for all entries. Therefore, the dense regions discovery problem transfers to find subset of entries X_S with respect to common V'_T, where $X_S \subseteq X$ and $V'_T \subseteq V'$.

4 Experiments

We use the Web usage data from ESPNSTAR.com.cn, a sports website in China, to test and validate the performance and effectiveness of our algorithms. Each dataset contains a set of access sessions during some period of time. Table 1 lists the datasets for experiments. ES1, ES2 and ES3 are the log datasets during December, 2002 and ES4 and ES5 are the log datasets from April, 2003.

Table 1. Characteristics of real datasets

Dataset	No.Accesses	No.Sessions	No.Visitors	No.Pages
ES1	583,386	54,300	2,000	790
ES2	2,534,282	198,230	42,473	1,320
ES3	6,260,840	517,360	50,374	1,450
ES4	78,236	5,000	120	236
ES5	7,691,105	669,110	51,158	1,609

4.1 Performance Analysis

In this section, we adopt the Web usage data to evaluate the effectiveness and scalability of the four discretization methods for large matrices.

Experiment 1. In this experiment, we increase the size of dataset for discretization. The continuous variable for discretization is the staying time on each page. After preprocessing of ES1, we got a dataset with 39,500 continuous values. We classify the Web pages into two categories: index pages and content pages. The number of interval k is determine by $\log(n)$, where n is number of values. We adopt EWI, SAWI and MEK for testing. In Fig 1, we can see that there is no significant difference in the running time of three different methods when increasing the number of instances.

Experiment 2. In [7], we suggest a multidimensional model to store and manage information from Web data. In this experiment, we employed the model to extract Web page, user and time profiles from ES4 to do the experiment. ES4 dataset contains the access information of 120 Website members during April, 2003. We increase the number of dimensions(attributes) from the multidimensional model to test the scalability of EWI, SAWI and MEK. Some attributes are generated by other attributes,e.g., AD value. Fig 2 shows the experiment results. We notice that the running time of SAWI obviously longer than the others. It can be explained that each attribute has its own class label, a lot of computation was spent on discretizing data from different classes, individually.

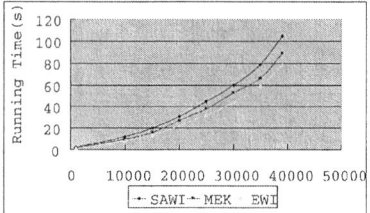
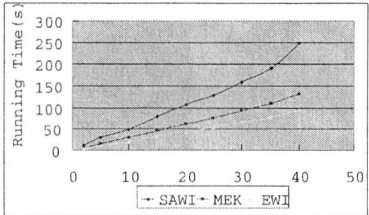

Fig. 1. Increasing Number of Continuous Values

Fig. 2. Increasing Number of Dimensions

Experiment 3. The experiment is to evaluate the influence to performance by increasing the number of intervals k. We used ES2 dataset to generate AD matrix for testing EWI, SAWI, MEK and Two-way. From Fig 3, we notice that MEK method spent longer time to finish when increasing k. The reason is that when the number of bins increases, the cost for performing k-means process will increase obviously. However, in most cases, k is less than 100. Hence, increasing k will not affect the performance of discretization significantly.

Experiment 4. The last experiment is to test the scalability of these discretization methods by increasing matrix size. Using ES5, we want to discretize AI matrix for mining AI dense regions that are labeled significant. The experiment result suggests that due to the time complexity of mining dense regions from high-dimensional data, the running time will increase significantly when increasing the size of matrix. However, if the size of matrix is less than several thousand, the running time is still acceptable. All experiments reported were performed on a PC with a Pentium4 2.0GHz CPU and 256MB memory. Hence, these discretization methods are quite feasible to put into practice.

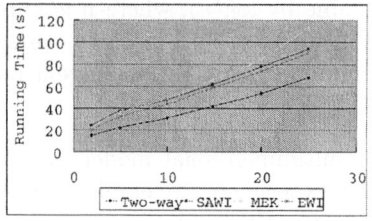 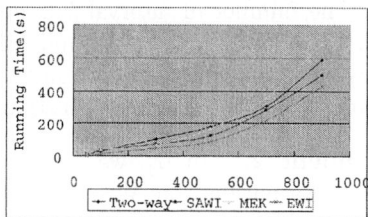

Fig. 3. Increasing Number of Intervals   **Fig. 4.** Increasing Matrix Size

In the summary, The discretization methods for Web data can be used in practical Web information management applications, such as detecting Web usage patterns and analyzing user behavior. In the future, we purpose to integrate the discretization algorithms in an intelligent Web usage mining system.

References

1. Catlett, J. *On changing continuous attributes into ordered discrete attributes.* In Proceedings of the European Working Session on Learning (1991), pp. 164–178.
2. Dougherty, J., Kohavi, R. and Sahami, M. *Supervised and Unsupervised Discretization of Continuous Features.* Proceedings of International Conference on Machine Learning, Tahoe City, CA, 1995, pp. 194-202.
3. Fayyad, U. and Irani, K. *Multi-interval discretization of continuous-valued attributes for classification learning.* Proc. of the 13th Intl. Joint Conf. on Artificial Intelligence, IJCAI-93: Chambery, France, 1993.
4. Gama, J., Torgo, L., and Soares, C. *Dynamic discretization of continuous attributes.* In Proceedings of the Sixth Ibero-American Conference on AI (1998), pp. 160–169.
5. Holte, R. *Very Simple Classification Rules Perform Well on Most Commonly Used Datasets.* Machine Learning, vol. 11, pp. 69-91, 1993.
6. Kerber, R. *Chimerge: Discretization for numeric attributes.* In National Conference on Artificial Intelligence (1992), AAAI Press, pp. 123–128.
7. Wu, E., Ng, M., and Huang, J. *An efficient multidimensional data model for Web usage mining.* Proceeding of the Sixth Asia Pacific Web Conference (APWEB2004), 2004.
8. Yang, C., Fayyad, U., and Bradley, P. *Efficient discovery of error-tolerant frequent itemsets in high dimensions.* Proceedings of the Seventh ACM SIGKDD International Conference on Knowledge Discovery and Data Mining: San Francisco, California, pp. 194–203, 2001.
9. Yip, A., Wu, E., Ng, M., Chan, T. *An efficient algorithm for dense regions discovery from large-scale data stream.* Proc. of the 8th Pacific-Asia Conference on Knowledge Discovery and Data Mining (PAKDD2004), 2004.

A Simple Group Diffie-Hellman Key Agreement Protocol Without Member Serialization[*]

Xukai Zou[1] and Byrav Ramamurthy[2]

[1] Purdue University School of Science at Indianapolis, Indianapolis, IN 46202, USA
xkzou@cs.iupui.edu
[2] University of Nebraska-Lincoln, Lincoln, NE 68588, USA
byrav@cse.unl.edu

Abstract. Many group key agreement protocols (GKA) for secure group communication (SGC) based on the Diffie-Hellman key exchange principle have been proposed in literature. All of these protocols require member serialization and/or existence of a central entity. In this paper, we propose a simple group Diffie-Hellman key agreement protocol which removes these two limitations. Moreover, the new protocol needs minimum (two) rounds of rekeying process and efficiently support high dynamics.

1 Introduction

The *Diffie-Hellman key exchange* [1], proposed by W. Diffie and M. Hellman in 1976, is a novel approach for two parties to obtain a shared secret over an insure channel. Since 1990, the approach has been extended to multi-party group key agreement (GKA) in various ways due to widespread requirement for *secure group communication* (SGC). SGC refers to a setting in which a group of members can send messages to and receive messages from group members, in a way that outsiders are unable to glean any information even when they are able to intercept the messages. The first problem facing SGC is *group key management* (GKM). Moreover, it is required that a GKM protocol provide *forward secrecy* and *backward secrecy* [2] when group member(s) join and/or leave the group (called *dynamics*). The group Diffie-Hellman key agreement is the most typical format of distributed (contributory) GKM protocols for SGC.

Most existing group Diffie-Hellman key agreement protocols [3–9] have an assumption that the group members are serialized, that is, they are numbered as the first member, the second one, ..., until the last one. This assumption is not reasonable in some environments such as ad hoc networks. Moreover, member serialization makes rekeying operations difficult. Recently, a new protocol was proposed in paper [10], which removes this requirement but introduces a central entity (i.e., coordinator). Avoiding a central entity is an initial motivation for distributed GKA. In this paper, we propose a simple group Diffie-Hellman key agreement protocol which removes these two limitations. Moreover, the new

---

[*] This work was partially supported by the U.S. NSF grant CCR-0311577.

protocol has the following additional properties: minimum (two) rounds of rekeying process and efficient support for high dynamics.

In section 2, we present the new proposed protocol. We summarize typical group Diffie-Hellman key agreement protocols and compare the proposed protocol with them in sections 3 and 4 respectively. The paper concludes in section 5.

2 A Simple Group Diffie-Hellman Key Agreement Protocol Without Member Serialization

As in any Diffie-Hellman protocol, let p be a large prime and $g \in \mathbb{Z}_p$ a *primitive element* (i.e., *generator*) of $\mathbb{Z}_p^*$. Both p and g are publicly known. Suppose the group members[1] are $m_0, m_1, ..., m_{n-1}$ and s_i is the private DH share and g^{s_i} is the public DH share of m_i. In the following, all the computations will include a modulo p operation but we omit the operation for simplicity. We call this protocol *NON-SERIAL* and present the protocol next.

2.1 Group Key Formation

Initially the group key can be formed by broadcasting key-related messages in two rounds (six steps) as follows (for every member m_i).

1. Selects its private DH share s_i, computes its public DH share g^{s_i}.
2. Broadcasts (g^{s_i}, m_i) (this is the first round broadcast).
3. Selects another secret number z_i.
4. After receiving the first round broadcast messages, computes $g^{s_i s_j}$ for all $j \neq i$, encrypts z_i with each of $g^{s_i s_j}$ respectively, and packs[2] $((\{z_i\}_{g^{s_i s_0}}, m_0), \cdots, (\{z_i\}_{g^{s_i s_{i-1}}}, m_{i-1}), (\{z_i\}_{g^{s_i s_{i+1}}}, m_{i+1}), \cdots, (\{z_i\}_{g^{s_i s_{n-1}}}, m_{n-1}))$ into one packet.
5. Broadcasts the packet (this is the second round broadcast).
6. After receiving the second round broadcast messages, can compute the group key as $K_G = f(z_0, \cdots, z_{n-1})$ where f is some predefined function. One example for f is XOR, thus $K_G = z_0 \oplus \cdots \oplus z_{n-1}$.

2.2 Rekeying for Joins and Leaves

The rekeying operations with this protocol are very simple and they are all performed in the similar way as the initial key formation. We discuss single join, single leave, multiple joins, multiple leaves, and multiple joins & multiple leaves simultaneously.

– When a member m_j joins, m_j selects its private DH share s_j, computes and broadcasts its public DH share g^{s_j}. At the same time, every member m_i sends its g^{s_i} to m_j. Then every member including m_j executes the steps 3–6

---

[1] The subscriptions denoting members are just for naming purpose. There is no need for their ordering or serialization.
[2] $\{x\}_k$ stands for encrypting x under key k using some secret-key cryptosystem.

as above. Thus the operation for single join involves two rounds. Note: It is feasible that only one representative from the existing group members (and the joining member m_j) executes the steps 3–5. However, the participation of all members in the steps 3–5 will keep the execution of the protocol simple and maintain equality among members.

- When a member m_l leaves, all remaining members perform the steps 3–6. So the operation for single leave involves just one round.
- When multiple members join, the operation is same as single join.
- When multiple members leave, the operation is same as single leave.
- When multiple members joins and leaves simultaneously, each of the joining members selects its private DH share, computes and broadcasts its public DH share. At the same time, every remaining member m_i broadcasts its g^{s_i}. Then all members including the joining members but excluding the leaving members execute the steps 3–6. The operation for multiple joins and leaves also involves two rounds.

3 Summary of Existing Group Diffie-Hellman Key Agreement Protocols

In this section, we summarize typical group Diffie-Hellman key agreement protocols. We use the same notation as before: s_i's are private DH shares and g^{s_i}'s are public DH shares. The group keys established by these protocols can be classified as ((i) through (iv) were proposed by [6,9], [4,5], [8], and [3,7], respectively):

(i) $K_G = g^{s_0 s_1 \cdots s_{n-1}} \mod p$
(ii) $K_G = g^{s_0 s_1 + s_1 s_2 + \cdots + s_{n-1} s_0} \mod p$
(iii) $K_G = g^{s_{n-1} g^{s_{n-2} g^{\cdots s_2 g^{s_1 s_0}}}} \mod p$
(iv) $K_G = g^{g^{g^{s_0 s_1} g^{s_2 s_3}} g^{g^{s_4 s_5} g^{s_6 s_7}}} \mod p$ (suppose $n=8$)

ING Protocol. The earliest attempt for extending the two-party Diffie-Hellman key exchange to a GKA protocol is due to Ingemarsson et al, and is called *ING protocol* [6]. ING consists of $n-1$ rounds and group members perform every round in synchronization. The members are arranged in a cycle. During the 0th round each member m_i computes g^{s_i} and passes it to member $m_{(i+1) \mod n}$. At the end of round $r-1$, member m_i receives $h_{r-1,i-1} = g^{s_{i-r} \cdots s_{i-1}}$ from m_{i-1} and executes round r, i.e., computes $h_{r,i} = (h_{r-1,i-1})^{s_i} = g^{s_{i-r} s_{i-r+1} \cdots s_i}$ and passes $h_{r,i}$ to m_{i+1}. After the final round, every member is in possession of the group key $g^{s_0 s_1 \cdots s_{n-1}}$.

BD Protocol. Burmester and Desmedt proposed an elegant GKA protocol, called *BD protocol*, which requires only two rounds [4,5]:

1. Every m_i computes and broadcasts $b_i = g^{s_i}$. (Round 1).
2. Every m_i computes and broadcasts $X_i = (b_{i+1}/b_{i-1})^{s_i}$. (Round 2, cyclical numbering).
3. Every m_i now computes the key $K_i = b_{i-1}^{n s_i} \cdot X_i^{n-1} \cdot X_{i+1}^{n-2} \cdots X_{i-2}$.

The BD protocol generates key $K_G = g^{s_0 s_1 + s_1 s_2 + \cdots + s_{n-1} s_0} \mod p$.

GDH Protocol. In 1996, Steiner et al. proposed a suite of group Diffie-Hellman protocols called GDH.1, GDH.2, GDH.3 [9]. The key generated by all three protocols is $K_G = g^{s_0 s_1 \cdots s_{n-1}}$.

In GDH.1, there are two stages ($n-1$ rounds each): *upflow* and *downflow*. The upflow stage collects contributions from all group members. In round i, m_i receives an ordered collection of intermediate values from m_{i-1} and raises the last intermediate value, i.e, $g^{s_0 \cdots s_{i-1}}$, to the power of its s_i, appends it to the incoming flow and forwards the new ordered list to m_{i+1}. In the downflow stage, member m_i receives from m_{i+1} an ordered list of intermediate values and raises all of them to the power of its s_i. The last entry of the resulting vector is the group key and m_i forwards all other values to m_{i-1}.

GDH.2 reduces the number of rounds by collecting contributions from all members in stage 1 (upflow) but broadcasting messages by m_{n-1} to all other members in Stage 2 (downflow). In the upflow, every member m_i receives values from m_{i-1}, raises all of the values to the power of its s_i and sends them to m_{i+1}.

GDH.3 reduces the number of exponentiations and involves four stages. The first stage consists of $n-2$ rounds and is used to collect contributions from members $m_0, m_1, \ldots, m_{n-2}$. In the second stage, m_{n-2} broadcasts $g^{s_0 \cdots s_{n-2}}$ to all members, thus m_{n-1} can obtain the key by raising the received value to the power of s_{n-1}. In stage 3, every member m_i extracts its own share s_i from $g^{s_0 \cdots s_{n-2}}$ and sends the result to m_{n-1}. Finally in stage 4, m_{n-1} raises all the values to its s_{n-1} and broadcasts these values to members. After receiving values from m_{n-1}, every member can compute the group key.

Steer Protocol. In 1988, Steer et al. [8] proposed a GKA protocol, called *Steer protocol*, which generates the key best described by the following recursive definition. Let q_m be defined recursively by: (i) $q_0 = s_0$, and (ii) $q_m = g^{s_m q_{m-1}}$ (for $0 < m < n$) Thus, $q_3 = g^{s_3 q_2} = g^{s_3 g^{s_2 q_1}} = g^{s_3 g^{s_2 g^{s_1 q_0}}} = g^{s_3 g^{s_2 g^{s_1 s_0}}}$. The group key generated by Steer's protocol is just $K_G = q_{n-1}$. The protocol processes in two stages with $n-1$ rounds as follows: In stage 1 (broadcast stage), every member m_i computes and broadcasts $b_i = g^{s_i}$ to the group. After receiving all the b_i's, each of m_0 and m_1 is able to compute the key by setting $\ell_1 = g^{s_0 s_1} = b_0^{s_1} = b_1^{s_0}$, and then performing $\ell_j = (b_j)^{\ell_{j-1}}$, $(j = 2, \cdots, n-1)$ recursively. The group key will be $K_G = \ell_{n-1}$. In stage 2, there are $n-2$ rounds. For each round i ($i \in [1, n-2]$), the computation of the group key, by each m_i, $2 \leq i \leq n-1$ is based on two recursive variables γ_j and k_j defined as follows: i) $\gamma_0 = b_0 = g^{s_0}$, ii) $k_j = (\gamma_{j-1})^{s_j}$, and iii) $\gamma_j = g^{k_j}$. After receiving γ_{i-1} from m_{i-1}, m_i computes k_i and γ_i from the above relations, sends γ_i to m_{i+1} and computes the group key by setting $\ell_i = k_i$, and then performing $\ell_j = (b_j)^{\ell_{j-1}}$, $(j = i+1, \cdots, n-1)$ recursively as for stage 1. The group key will be $K_G = \ell_{n-1}$. (Note: m_{n-1} just performs $K_G = \gamma_{n-2}^{s_{n-1}}$.)

TGDH Protocol. The above discussed protocols generate the group key in a *flat way*. The protocol in [3, 7] bases the formation of the group key on a *binary* *tree structure*, called Tree based Group Diffie-Hellman key agreement (TGDH). In TGDH, all members maintain an identical virtual binary key tree and are hosted on leaf nodes respectively. Every node $\langle l, v \rangle$ except for the root node is associated with a secret key $K_{\langle l,v \rangle}$ and a blinded key $BK_{\langle l,v \rangle} = f(K_{\langle l,v \rangle})$ where $f(x) = g^x \mod p$. The key $K_{\langle l,v \rangle}$ of a parent node is the Diffie-Hellman key of its two children, i.e., $K_{\langle l,v \rangle} = g^{K_{\langle l+1,2v \rangle} K_{\langle l+1,2v+1 \rangle}} \mod p = f(K_{\langle l+1,2v \rangle} K_{\langle l+1,2v+1 \rangle})$. Every blinded key $BK_{\langle l,v \rangle}$ is publicly broadcast by some member(s) to the group. Every member m_i selects its own private DH share s_i, computes and publicizes the corresponding public DH share g^{s_i}, and then gradually computes the secret keys and blinded keys from its leaf to the root after receiving the blinded keys (some members also broadcast blinded keys). The root key $K_{\langle 0,0 \rangle}$ is the group key K_G. Suppose there are eight members, $K_G = K_{\langle 0,0 \rangle} = g^{g^{g^{s_0 s_1} g^{s_2 s_3}} g^{g^{s_4 s_5} g^{s_6 s_7}}} \mod p$.

In many GKM protocols, all members stop communication until the new group key is formed (this is called a *block*) during rekeying process. Based on TGDH, the paper [11] proposed a GKM protocol which allows group members to continue their communication seamlessly and without interruption during rekeying and is called Block-free TGDH.

YTCC Protocol. It can be seen that all the above protocols have a requirement of *serialization*. In paper [10], the authors proposed a protocol which removes this limitation. The protocol is called *YTCC* and works as follows.

The group members first select a *coordinator* (e.g., m_c). Then the protocol is executed in two rounds: (1) Each m_i ($i \neq c$) selects its private DH share s_i and broadcasts its public DH share g^{s_i}; (2) The coordinator m_c generates a random number z and its private DH share s_c, computes its public DH share g^{s_c}, computes $g^{s_c s_i}$ for each $i \neq c$, encrypts z with each $g^{s_i s_c}$ respectively, and broadcasts g^{s_c} along with multiple encryptions of z, i.e., $(g^{s_c}, (\{z\}_{g^{s_c s_0}}, m_0), \cdots, (\{z\}_{g^{s_c s_{c-1}}}, m_{c-1}), (\{z\}_{g^{s_c s_{c+1}}}, m_{c+1}), \cdots, (\{z\}_{g^{s_c s_{n-1}}}, m_{n-1}))$. After receiving the broadcast from m_c, each m_i ($i \neq c$) computes $g^{s_i s_c}$, decrypts z, and computes the group key as $K_G = g^{f(g^{s_0}, g^{s_1}, \cdots, g^{s_n}) \circ z}$, where f is some predefined combining function [10] such as a secure *hash function* and $\circ$ is a binary operator such as XOR.

4 Discussions

In Table 1, we provide a comparison of current group Diffie-Hellman key agreement protocols including the new proposed one.

The following primary observations can be concluded from the tables: (1) *NON-SERIAL* and *YTCC* are the only protocols which do not need member serialization; (2) *NON-SERIAL* is one of the four protocols which have the minimum (i.e., 2) rounds of rekeying; (3) from the number of broadcast messages and the number of exponentiations, *NON-SERIAL* is never the worst one among all protocols; (4) in many aspects, *NON-SERIAL* is similar to *BD*. However *BD* requires very strict serialization, thus, not supporting dynamics well. In contrast, *NON-SERIAL* does not require serialization and supports high dynamics;

Table 1. Comparison of group Diffie-Hellman key agreement protocols

Protocol	ING	BD	GDH.2	GDH.3	Steer's	TGDH	YTCC	NON-SERIAL
Rounds	$n-1$	2	n	$n+1$	$n-1$	$log(n)$	2	2
Synchronization	Y	Y	N	N	N	Y	Y	Y
Central entity	N	N	N	N	N	N	Y*	N
Equal duty	Y	Y	N	N	N	N	N	Y
Serialization	Y	Y	Y	Y	Y	Y	N	N
High dynamics	N	N	N	N	N	Y	Y	Y

*: $YTCC$ contains a coordinator which is a central entity.

(5) $YTCC$ is similar to $NON\text{-}SERIAL$ in many aspects. However there are two problems with $YTCC$. It requires that a coordinator be selected prior to the key generation process, which will postpone the group communication. Moreover the *coordinator* will become a single point of failure.

5 Conclusion

We presented a new group Diffie-Hellman key agreement protocol satisfying several highly desired properties for SGC: non member serialization, no central entity, minimum rounds of rekeying, and efficient support for high dynamics. The protocol is very useful in some environments such as ad hoc networks.

References

1. Diffie, W., Hellman, M.E.: Multiuser cryptographic techniques. AFIPS conference proceedings **45** (1976) 109–112
2. Zou, X., Ramamurthy, B., Magliveras, S.S., eds.: Secure Group Communications over Data Networks. Kluwer Academic Publishers, Norwell, MA, USA, ISBN: 0-387-22970-1 (The ebook ISBN: 0-387-22971-X) (2004)
3. Becker, K., Wille, U.: Communication complexity of group key distribution. ACM conference on computer and communication security, CA, USA, 1998 (1998) 1–6
4. Burmester, M., Desmedt, Y.: A secure and efficient conference key distribution system. EUROCRYPT'94, LNCS, Springer, Berlin **950** (1995) 275–286
5. Burmester, M., Desmedt, Y.: Efficient and secure conference-key distribution. Security Protocols Workshop (1996) 119–129
6. Ingemarsson, I., Tang, D., Wong, C.: A conference key distribution system. IEEE Transactions on Information Theory **28** (1982) 714–720
7. Kim, Y., Perrig, A., Tsudik, G.: Simple and fault-tolerant key agreement for dynamic collaborative groups. In Proceedings of the 7th ACM Conference on Computer and Communications Security (ACM CCS 2000) (2000) 235–244
8. Steer, D., Strawczynski, L., Diffie, W., Wiener, M.: A secure audio teleconference system. CRYPTO'88, LNCS, Springer **403** (1990) 520–528
9. Steiner, M., Tsudik, G., Waidner, M.: Diffie-hellman key distribution extended to group communication. ACM Conference on Computer and Communications Security (ACM CCS 1996), New Delhi, India (1996) 31–37

10. Yasinsac, A., Thakur, V., Carter, S., I.Cubukcu: A family of protocols for group key generation in ad hoc networks. IASTED International Conference on Communications and Computer Networks (CCN02) (2002) 183–187
11. Zou, X., Ramamurthy, B.: A block-free tree-based group Diffie-Hellman key agreement for secure group communications. International Conference on Parallel and Distributed Computing and Networks, Innsbruck, Austria (2004) 288–193

Increasing the Efficiency of Support Vector Machine by Simplifying the Shape of Separation Hypersurface

Yiqiang Zhan[1,2,3] and Dinggang Shen[2,3]

[1] Dept. of Computer Science, The Johns Hopkins University, Baltimore, MD
[2] Center for Computer-Integrated Surgical Systems and Technology,
The Johns Hopkins University, Baltimore, MD
[3] Sect. of Biomedical Image Analysis, Dept. of Radiology, University of Pennsylvania,
Philadelphia, PA
{Yiqiang.Zhan, Dinggang.Shen}@uphs.upenn.edu

Abstract. This paper presents a four-step training method for increasing the efficiency of support vector machine (SVM) by simplifying the shape of separation hypersurface. First, a SVM is initially trained by all the training samples, thereby producing a number of support vectors. Second, the support vectors, which make the hypersurface highly convoluted, are excluded from the training set. Third, the SVM is re-trained only by the remaining samples in the training set. Finally, the complexity of the trained SVM is further reduced by approximating the separation hypersurface with a subset of the support vectors. Compared to the initially trained SVM by all samples, the efficiency of the finally-trained SVM is highly improved, without system degradation.

1 Introduction

Support vector machine (SVM) is a statistical classification method proposed by Vapnik in 1995 [1], and it is one of the most interesting developments in classifier design. Given m labeled training samples, i.e. $\{(\vec{x}_i, y_i) | \vec{x}_i \in R^n, y_i \in \{-1,1\}, i=1\cdots m\}$, SVM is able to generate a separation hypersurface that has maximum generalization ability. In application, a testing sample $\vec{x}$ is classified by calculating its distance to the hypersurface:

$$d(\vec{x}) = \sum_{i=1}^{m} \alpha_i y_i K(\vec{x}_i, \vec{x}) + b \qquad (1)$$

where α_i and b are the parameters determined by SVM's learning algorithm, and $K(\vec{x}_i, \vec{x})$ is the kernel function. Those samples $\vec{x}_i$ with nonzero parameters α_i are called "support vectors" (SVs).

SVM is widely used in different application as a powerful classifier. However, SVM usually needs a huge number of SVs to maximize the generalization ability. The huge number of SVs unavoidably increases the computational burden when classifying a new sample by calculating Eq. (1). This disadvantage thus limits the capability of SVM in the applications that require a massive number of classifications or real-time classification. Therefore, it is important to decrease the computational cost of SVM, by reducing the number of SVs.

In this paper, a novel training method is proposed to improve the efficiency of SVM classifier, by selecting appropriate training samples. Since the number of SVs determines the computational cost of SVM and are also highly related to the geometric complexity of the separation hypersurface, the basic idea of our training method is to exclude the samples that incur the separation hypersurface highly convoluted.

2 Methods

Reducing the computational cost of the SVM is equivalent to decreasing the number of the SVs. According to their positions in the feature space, SVs can be categorized into two types. The first type of SVs are the training samples that exactly locate on the margins of the separation hypersurface, i.e., $d(\bar{x}_i) = \pm 1$, as the gray circles/crosses shown in Fig 1. Their number is directly related to the shape of the separation hypersurface, i.e., the more the SVs of this type, the more convoluted the hypersurface. The second type of SVs are the training samples that locate beyond their corresponding margin, i.e., $y_i d(\bar{x}_i) < 1$, as the dashed circles/crosses shown in Fig 1. For SVM, these training samples are regarded as mis-classified samples even though some of them still locate at the correct side of the hypersurface.

SVM usually has a huge number of SVs, when the distributions of the positive and the negative training samples highly overlap with each other. This is because, (1) a large number of the first-type SVs are needed to construct a highly convoluted hypersurface, in order to separate two classes; (2) even the highly convoluted separation hypersurface has been constructed, a lot of confounding samples will be misclassified, and thus selected as the second type of SVs.

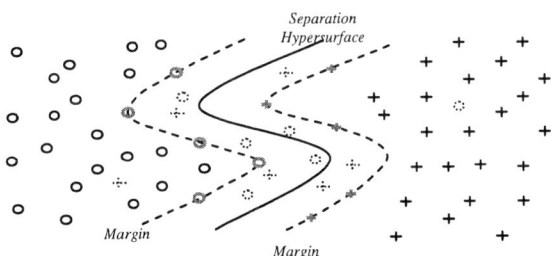

Fig. 1. Schematic explanation of the separation hypersurface, margins and SVs of SVM

Osuna have proposed an effective method to reduce the number of SVs of the trained SVM without system degradation. This method approximates the separation hypersurface with a subset of the SVs using Support Vector Regression Machine (SVRM) [4]. However, in many real applications, while SVM generates a highly convoluted separation hypersurface in the high dimensional feature space, Osuna's method still needs a large number of SVs to approximate the hypersurface. Obviously, an efficient way to further decrease the number of the SVs is to simplify the shape of the separation hypersurface, by sacrificing a very limited classification rate.

An intuitive method to simplify the shape of the hypersurface is to exclude some training samples, thereby the remaining samples are possible to be separated by a less convoluted hypersurface. However, the exclusion of training samples inevitably decreases the variety of the training set and may further influence the classification rate of the finally trained SVM. To minimize the loss of the classification rate, only the training samples that have largest contributions to the convolution of the hyper-surface are preferred to be excluded from the training set. Since the SVs determine the shape of the separation hypersurface, they are the best candidates to be excluded from the training set, in order to simplify the shape of the separation hypersurface.

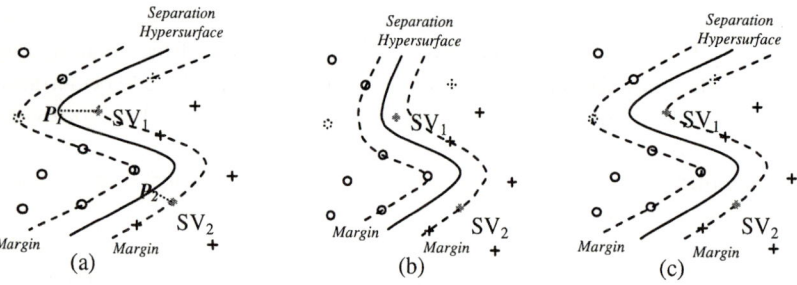

Fig. 2. Schematic explanation of how to selectively exclude the SVs from the training set, in order to effectively simplify the separation hypersurface

Excluding different sets of SVs from the training set will lead to different simplifications of the separation hypersurface. Fig 2 presents a schematic example in the 2-dimensional feature space, where we assume SVs exactly locating on the margins. As shown in Fig 2(a), SVM trained by all the samples has 10 SVs, and the separation hypersurface is convoluted. Respective exclusion of two different SVs, SV_1 and SV_2, denoted as gray crosses in Fig 2(a), will lead to two different separation hypersurfaces as shown in Figs 2(b) and 2(c), respectively. SVM in Fig 2(b) has only 7 SVs, and its hypersurface is less convoluted, after re-training SVM with all samples except SV_1, which was previously selected as a SV in Fig 2(a). Importantly, two additional samples, denoted as dashed circle/cross, were previously selected as SVs in Fig 2(a), but they are no longer selected as SVs in Fig 2(b). In contrast, SVM in Fig 2(c) still has 9 SVs, and the hypersurface is very similar to that in Fig 2(a), even SV_2, which was previously selected as a SV in Fig 2(a), has been excluded from the training set. Obviously, the computational cost of SVM in Fig 2(b) is less than that in Fig 2(c), while the correct classification rates are the same.

It is usually more effective to simplify the shape of the hypersurface by excluding the SVs, like SV_1, which contribute more to the convolution of the hypersurface. For each SV, its contribution to the convolution of hypersurface can be approximately defined as *the generalized curvature* of its projection point on the hypersurface. The projection point on the hypersurface can be located by projecting each SV to the hypersurface along the gradient of the distance function. For example, for SV_1 and SV_2 in Fig 2(a), their projection points on the hypersurface are P_1 and P_2. Obviously, the curvature of the hypersurface at point P_1 is much larger than that at point P_2, which means SV_1 has more contribution to make the hypersurface convoluted. Therefore, it is more effective to "flatten" the separation hypersurface by excluding the SVs, like SV_1, with their projection points having the larger curvatures on the hypersurface.

Based on the above idea and combined with Osuna's method, our training method is designed to have four steps:

Step 1. Use all the training samples to train an initial SVM, resulting in l_1 SVs $\{SV_i^{In}, i=1,2,...,l_1\}$ and the corresponding decision function $d_1(\vec{x})$.

Step 2. Exclude from the training set the SVs, whose projections on the hypersurface have the largest curvatures:

2a. For each SV_i^{In}, find its projection on the hypersurface, $p(SV_i^{In})$, along the gradient of distance function $d_1(\vec{x})$. Then, calculate the generalized curvature of $p(SV_i^{In})$ on the hypersurface, $c(SV_i^{In})$.

2b. Sort SV_i^{In} in the decrease order of $c(SV_i^{In})$, and exclude the top n percentage of SVs from the training set.

Step 3. Use the remaining samples to re-train the SVM, resulting in l_2 SVs $\{SV_i^{Re}, i=1,2,...,l_2\}$ and the corresponding decision function $d_2(\vec{x})$. Notably, l_2 is usually less than l_1.

Step 4. Use the l_2 pairs of data points $\{SV_i^{Re}, d_2(SV_i^{Re})\}$ to finally train the SVRM, resulting in l_3 SVs $\{SV_i^{Fl}, i=1,2,\cdots l_3\}$ and the corresponding decision function $d_3(\vec{x})$. Notably, l_3 is usually less than l_2.

3 Experiments

In our study of 3D prostate segmentation from ultrasound images [2], SVM is used for texture-based tissue classification to differentiate prostate tissues. It is very necessary to speed up the tissue classification algorithm as the real-time segmentation is usually required in clinical applications.

The experimental data are the prostate and non-prostate samples collected from six manually labeled ultrasound images. 3621 samples from one image are used as testing samples, while 18105 samples from other five images are used as training samples. Each sample has 10 texture features, extracted by a Gabor filter bank [2].

In the first experiment, we use our method to train a series of SVMs by excluding different percentages of SVs in Step 2c. As shown in Fig 3(a), after excluding 50% of initially selected SVs, the finally-trained SVM has 1330 SVs, which is only 48% of the SVs (2748) initially selected in the original SVM; but its classification rate still reaches 95.39%. Compared to 96.02% classification rate achieved by original SVM, the loss of classification rate is relatively trivial. If we want to further reduce the computational cost, we can exclude 90% of initially selected SVs from the training set. Our finally-trained SVM has only 825 SVs, which means the speed is triple, and it still has 93.62% classification rate. To further validate the effect of our trained SVM in prostate segmentation, the SVM with 825 SVs (denoted by the white triangle in Fig 3(a)) is applied to a real ultrasound image for tissue classification. As shown in Fig 3(b1-b2), the result of our trained SVM is not inferior to that of the original SVM with 2748 SVs, in terms of differentiating prostate tissues from the surrounding ones.

In the second experiment, we compare the performances of different training methods in reducing the computational cost of the finally-trained SVM and also in correctly classifying the testing samples. The five methods are implemented for comparison; they are (1) a method of slackening the training criterion by decreasing the penalty factor to errors [3]; (2) a heuristic method, which assumes the training samples distributing in a multi-variant Gaussian way, then excludes the "abnormal" training samples distant from the respective distribution centers, and finally trains a SVM only by the remaining samples; (3) a method of excluding the initially-selected SVs from the training set and then training a SVM only by the remaining samples, i.e., our proposed method without using Step 4; (4) Osuna's method [4]; (5) our proposed method. The performances of these five methods are evaluated in Fig 4(a), by the number of SVs used vs the number of correct classifications achieved. By checking the beginning curves of methods 1-5, Osuna's method is the most effective in initially reducing the number of SVs. However, to further reduce the SVs with limited sacrifice of classification rate, our method has better performance than Osuna's method.

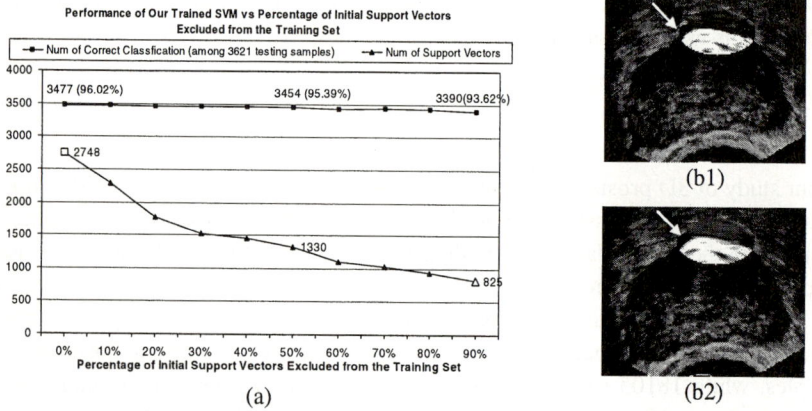

Fig. 3. (a) The performance of the finally-trained SVM changes with the percentages of initial SVs excluded from the training set. (b1-b2) Comparisons of tissue classification results using (b1) the original SVM with 2748 SVs and (b2) our trained SVM with 825 SVs. The tissue classification results are shown only in an ellipsoidal region

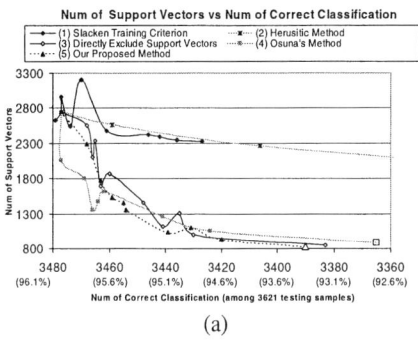 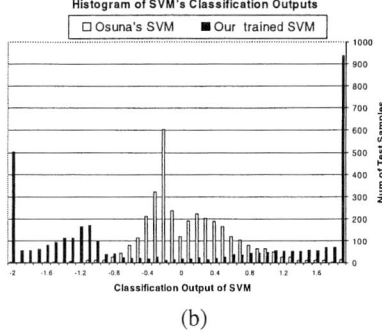

Fig. 4. (a) Comparing the performances of five training methods in increasing the efficiency of SVM. (b) Histograms of classification outputs on a testing dataset respectively from our trained SVM (black bars) and Osuna's SVM (white bars)

The classification abilities of two SVMs, respectively trained by Osuna's method and our method, are further compared. The SVM trained by Osuna's method, as denoted by the white square in Fig 4(a), needs 884 SVs and its classification rate is 92.93%. The SVM trained by our method, as denoted by the white triangle in Fig 4(a), needs only 825 SVs, while its classification rate is 93.62% higher than that produced by Osuna's method. Moreover, our trained SVM actually has much better classification ability than the SVM trained by Osuna's method, once checking the histograms of their classification outputs. As shown in Fig 4(b), the classification outputs of Osuna's SVM concentrate around 0, which means the classification results of the positive and the negative samples are not widely separated. In contrast, most classification outputs of our trained SVM are either larger than 1.0 or smaller than -1.0. This experiment further proves that our training method is better in keeping the classification ability of the finally-trained SVM, after reducing many SVs.

4 Conclusion

We have presented a training method to increase the efficiency of SVM for fast classification, without system degradation. By finding that different SV has different contribution in constructing the separation hypersurface, we proposed a method to exclude the SVs that incur the separation hypersurface highly convoluted from the training set, thereby our finally trained SVM has a less number of SVs and the computational cost is reduced. Combined with Osuna's method, which using SVRM to efficiently approximate the hypersurface, our proposed method can highly increase the classification speed of the SVM, with very limited loss of classification ability. Experiments on real prostate ultrasound images demonstrate the performance of our proposed training method in discriminating the prostate tissues from other tissues. Compared to other four training methods, our proposed training method is able to generate more efficient SVMs, with better classification abilities.

References

1. Vapnik, V.N., *Statistical Learning Theory*, New York: John Wiley & Sons, (1998).
2. Zhan, Y. and Shen, D., "Automated Segmentation of 3D US Prostate Images Using Statistical Texture-Based Matching Method", *MICCAI, 2003*, Nov 16-18, Canada, (2003).
3. Burges, C.J.C., "A Tutorial on Support Vector Machines for Pattern Recognition", *Data Mining and Knowledge Discovery,* Vol. 2, (1998), 121-167.
4. Osuna, E. and Girosi, F., "Reducing the run-time complexity of Support Vector Machines", *ICPR*, Brisbane, Australia, (1998).

Implementation of the Security System for Instant Messengers

Sangkyun Kim[1] and Choon Seong Leem[2]

[1] Somansa, Woolim e-Biz, Yangpyeongdong, Yeongdeungpogu, Seoul, Korea
saviour@somansa.com
[2] Department of Computer and Industrial Engineering, Yonsei University, 134, Shinchondong, Seodaemoongu, Seoul, Korea
leem@yonsei.ac.kr

Abstract. Instant messenger (IM) has grown rapidly to involve billions of users. It can eliminate long trails of voice mails and e-mails, and it is especially valuable to link remote individuals. The previous researches of the bad influences of IM are focused on interruption of desktop computing tasks. We take the security problems of IM into consideration because IM has many security risks that may have severed impacts. In this paper, we provide the requirements analysis of security systems to control IM's risks and our development efforts to implement these security systems in technical aspects.

1 Introduction

Enterprising workers are finding more ways to exploit IM's business value as an effective information processing facility and in the absence of IT supported IM systems, users are signing up for free IM services at an amazing rate. IM provides real-time dialogue and file transfers among multiple users who are connected with the Internet. Everyone connected with the Internet can easily use IM by downloading from the Internet or using OS-embedded IM software. Regardless of whether IM use is sanctioned within the enterprise or not, it exists and is increasing in use. Enterprises should recognize the degree of IM prevalence within their environment and develop guidelines for the uses that are specific to their business [1]. In this paper, we provide the funtional requirements of security system to control the risks of IM with the concept of risk analysis method, and deliver our experience to develop the package software named Messenger-i regarding with these requirements.

2 Requirements Analysis of Security Functions

We analyzed the requirements of IM's security functions with a risk analysis method. Risk analysis is the best method that allows us to determine the protection required for varied assets at the most reasonable cost [2]. Risk is the probability or likelihood of injury, damage, or loss in some specific environment and over some stated period of time. Thus, risk involves two elements: probability and lossing amount. Both risk and risk management are associated with virtually every facet of systems engineering and systems management [3].

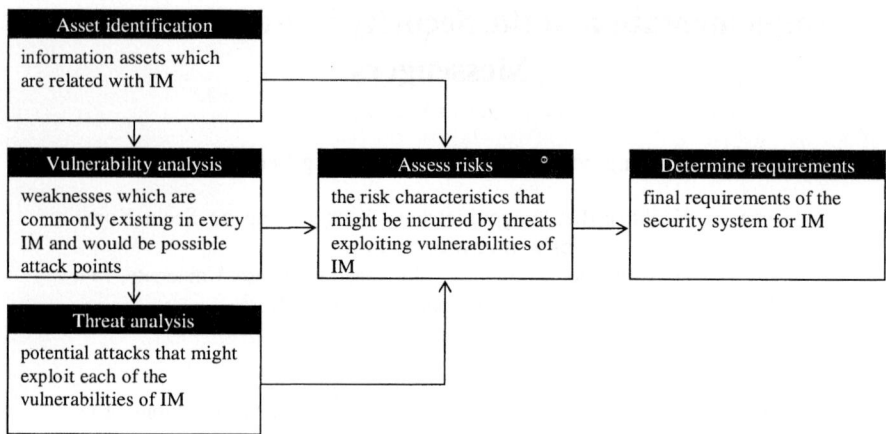

Fig. 1. Process of the requirements analysis

Table 1. Asset identification

Quality	Definition [6]	Asset
Confidentiality	Information which requires protection from unauthorized disclosure	A1: trade secrets deliverable via IM A2: privacy of the end user running IM A3: logged IM messages in the end user PC
Integrity	Information which must be protected from unauthorized, unanticipated, or unintentional modification	A4: computing systems of the end user running IM
Availability	Information or services which must be available on a timely basis to meet mission requirements or to avoid substantial losses	A5: accessability to the private account of IM A6: accessability to the delivered messages that were transferred via IM

Risks are calculated by the significance of the asset value, vulnerability and threat. Mayerfeld defined "Vulnerability is any property of a component that makes it susceptible to compromise [4]." Gary defined a vulnerability as a flaw or weakness in system security procedures, design, implementation, or internal controls that could be exercised (accidentally triggered or intentionally exploited) and result in a security breach or a violation of the system's security policy [5]. Carl defined a threat as any indication, circumstance, or event with the potential to cause the loss of or damage to an asset [2]. We studied several IMs to find risks. The study process which consists of four steps is shown in Fig. 1.

There are three fundamental qualities of information which are vulnerable. Therefore they need to be protected at all times, namely availability, integrity and confidentiality. We studied the characteristics of information asset related with IM with these three quality factors. The results of asset identification are described in table 1. The vulnerabilities of each asset and the threats on each vulnerability are described in table 2.

Table 2. Vulnerability and threat analysis

Asset	Vulnerability	Threat
A1	V1: lack of access controls on outgoing traffic	T1: industrial espionage
A2	V2: unencrypted communication	T2: eavesdropping
A3	V3: absence of controls on logged message	T3: theft of message logging
A4	V4: absence of controls for malicious code	T4: trojans and viruses
A5	V5: lack of password management	T5: theft of identity
A6	V6: absence of backup function of messages	T6: requirement on evidence

We analyzed the risks of IM with table 1 and table 2 according to the process shown in Fig. 1.

R1 (A1*V1*T1): trading files over IM eliminates the file size restrictions of email, and, if the recipient is known, IM is an easier solution to transfer confidential data.

R2 (A2*V2*T2): IM is prone to eavesdropping because it is transmitted in clear text by default. IM services relay messages through a central server. These messages would be prone to interception at the server.

R3 (A3*V3*T3): IM records a messaging session to a text file. Malicious access to this information could be used for social engineering or to gain access to sensitive data.

R4 (A4*V4*T4): the file transfer generally is peer to peer (P2P), so there is no opportunity for enterprise IT security to scan files in transit. When an infected file is opened, the virus spreads to the recipient's machine.

R5 (A5*V5*T5): IM is more prone than e-mail to damage caused by identity theft. Getting access to another's IM identity can be as simple as sitting down at a deserted desktop. Stealing user names and passwords via a keystroke monitor or sniffing is trivial in relation to other cybercrimes.

R6 (A6*V6*T6): IM is not dependant on enterprise information systems. Redundancy of transferred files and messages is required.

3 Implementation of Security Functions

With the risks described above, we determined the functionalities of IM security system, and these are descibed in table 3.

There are many commercial products which provide funtionalities of F2, F3-1, F3-2, F4-1, F4-2, F5-1 and F5-2. These are VPN, desktop firewall, virus vaccine, desktop

Table 3. Functionality of IM security system in technical aspects

Risk	Funtionality
R1	F1-1: central monitoring of incoming and outgoing messages F1-2: filtering of outgining messages
R2	F2: encryption of network traffics
R3	F3-1: screensaver F3-2: desktop encryption
R4	F4-1: virus scanner F4-2: data backup of the end user's desktop
R5	F5-1: strong authentication of log on of the end user's desktop F5-2: screensaver
R6	F6-1: central logging of transferred messages F6-2: reporting and auditing of logged messages

clone software, biometrics and tokens [7]. In this paper, we focus to develop the functionalities of F1-1, F1-2, F6-1 and F6-2 that are not provided by existing commercial products or technologies.

We named the IM management system as Messenger-i. It provides the integrated functionalities: 1) sniffing of network packets of IM; 2) monitoring of real-time connections; 3) generation of duplication (it makes back up of transferred messages via IM); 4) logging; 5) blocking port (it works in tandem with a firewall to block unauthorized connections with a security policy and filtering rule); 6) searching and reporting. The logical diagram of Messenger-i is shown in Fig. 2. It consists of the monitoring agent, database and administration manager.

The development of receiver module in monitoring agent was most difficult to solve: 1) there are no up-to-date international standards of IM; 2) each vendor provides different protocols of IM; 3) communication protocols of IM are closed to the public. We have gone through hardships to analyze communication protocols of each IM. We made a test environment to build various cases of IM usage and to capture communication messages of each case with sniffing technology. For example, one of the analyzed IMs which is dominant in market share of the IM has the five types of communication protocols, and these types are described in table 4. In this case, the types of protocol are determined by port numbers used in transfering files or communicating messages.

We analyzed the communication protocols of six kinds of IM, and made the receiver module to support plug-ins for the continuous addendum of IM protocols.

The monitoring window and sample report of Messenger-i are shown in Fig. 3. With this window, the administrators can monitor the entire connections of real-time IM messages, and execute searching and reporting on logged messages. A sample of the operational statistics in Fig. 3 shows that MSN messenger generates about 95% of IM traffics in the test environment.

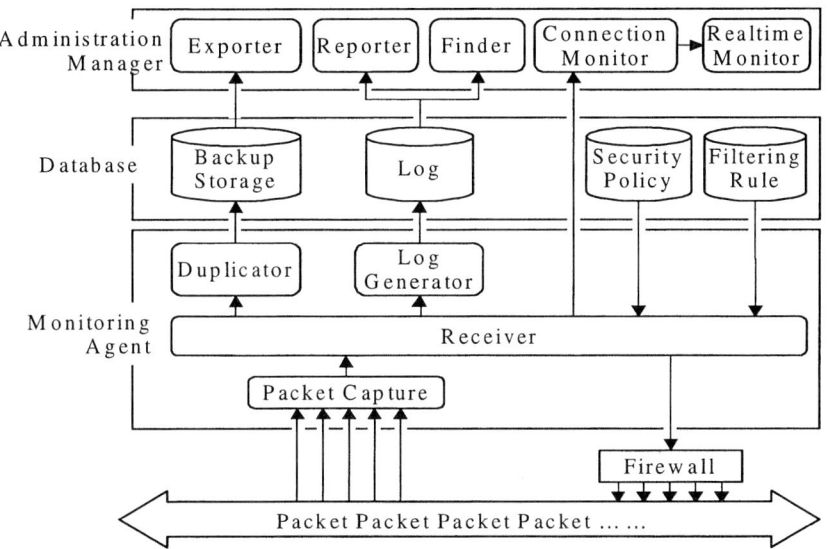

Fig. 2. Logical diagram of the IM management system: Messenger-i

Table 4. Sample of various protocols of market-dominant IM

Message	File	Static port	Proxy	Random port
Static port		type 4	none	type 1
Proxy		none	type 5	type 2
Random port		none	none	type 3

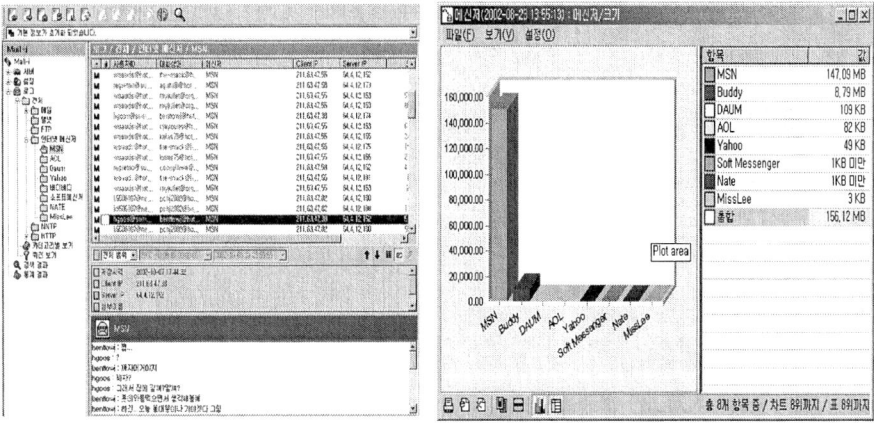

Fig. 3. Monitoring window and sample report of the IM management system: Messenger-i

4 Conclusion

The IM is a carrier of up-and-coming security risks. It provides the ability to transfer both text messages and files. With improper management, IM can put proprietary information in great dangers. However, when properly managed with security controls and integrated into the business workflow, IM has the potential to boost a company's ability to operate as a real-time enterprise.

In this paper, we provide our experience to develop the package software which makes public IM to be secured in the workplace. As mentioned earlier, there are no standardized protocols, and any of IM vendors do not want to make their protocols public. Therefore, as IMs are more broadly spreaded and adopted as a common tool of business communication, the hidden and uncontrollable risks of IMs increase. Up to the present, it's hard to prevent the risks of IMs, but we can manage these risks with detective, recovering and corrective controls. Messenger-i described in this paper is one of the effective contols which provides detective, recovering and corrective controls.

References

1. Grey, M.: Love It or Hate It: Instant Messaging Invades the Enterprise. Gartner (2001)
2. Carl A. Roper: Risk Management for Security Professionals. Butterworth Heinemann (1999)
3. Sage A. P.: Systems Engineering. John Wiley & Sons, New York (1992)
4. Mayerfeld, H.: Definition and Identification of Assets as the Basis for Risk Management. Proceedings 1988 Computer Security Risk Management Model Builders Workshop (1988)
5. Fites et al.: Controls and Security of Computer Information Systems. Computer Science Press (1989)
6. Swanson, M.: Guide for Developing Security Plans for Information Technology Systems. NIST Special Publication 800-18, NIST (1998)
7. Kim, S., Kang, S.: A Study on the Security Vulnerabilities and Defense Mechanism for SET-based Electronic Commerce. The Journal of Korean Institute of CALS/EC, Vol. 4, No. 2 (1999)

Communication in Awareness Reaching Consensus Without Acyclic Condition II

Ken Horie[1],* and Takashi Matsuhisa[2],**

[1] Advanced Course of Electronic and Computer Engineering, Ibaraki National College of Technology, Nakane 866, Hitachinaka-shi, Ibaraki 312-8508, Japan
abc9872000jp@yahoo.co.jp
[2] Department of Liberal Arts and Sciences, Ibaraki National College of Technology
mathisa@ge.ibaraki-ct.ac.jp

Abstract. We present a communication process according to a protocol which is associated with an awareness and belief model. In the model we impose none of the requirements for player's knowledge as those in the standard model with partition information structure. We show that consensus on the posteriors for an event among all players can still be guaranteed in the communication even when the protocol contains a cycle.

Keywords: Awareness, belief, communication process, consensus, protocol, agreeing to disagree.

AMS 2000 Mathematics Subject Classification: Primary 91B50, 91B60; Secondary 03B45.

Journal of Economic Literature Classification: D51, C78, D61.

1 Introduction

This article presents a communication system according to Krasucki [7]. There are more than two players and they interact in pairs with public announcement: Each player has the posterior of an event under his/her private information, and he/she privately announces it to the another player through message. The recipient revises his/her information structure and recalculate the values of posterior under the revised information. The player send the revised posterior to another player according to a communication graph. The recipient revises his/her posterior and send it to another, and so on. The we can show that

Theorem 1. *Suppose that all players have a common prior distribution. Consensus on the limiting values of the posteriors for a publicly aware event among all players can still be guaranteed in the communication even when the protocol contains a cycle.*

---

* Corresponding author and Lecture presenter.
** Partially supported by the Grant-in-Aid for Scientific Research(C)(2)(No.14540145) in the Japan Society for the Promotion of Sciences.

Recently researchers in such fields as economics, AI, and theoretical computer science have become interested in reasoning of belief and knowledge. There are pragmatic concerns about the relationship between knowledge (belief) and actions. Of most interest to us is the emphasis on situations involving the knowledge (belief) of a group of players rather than that of a single player. At the heart of any analysis of such situations as a conversation, a bargaining session or a protocol run by processes is the interaction between players. A player in a group must take into account not only events that have occurred in the world but also the knowledge of the other players in the group.

In some cases we need to consider the situation that the players has common-knowledge of an event; that is, simultaneously everyone knows the event, everyone knows that everyone knows the event, and so on. This notion also turns out to be a prerequisite for achieving agreement: In fact, Aumann [1] showed the famous agreement theorem; that is, if all posteriors of an event are common-knowledge among the players then the posteriors must be the same, even when they have different private information. This is precisely what makes it a crucial notion in the analysis of an interacting group of players.

Because the notion of common-knowledge is defined by the infinite regress of all players' knowledge as above, common-knowledge is actually so infeasible a tool in helping us analyze complicated situations involving groups of players. Thus we would like to remove it from our modelling.

In this regard, Geanakoplos and Polemarchakis [6] investigated a communication process in which two players announce their posteriors to each other. In the process players learn and revise their posteriors and they reach consensus without common-knowledge of an event. Furthermore, Krasucki [7] introduced the revision process mentioned in the above. He showed that in the process, consensus on the posteriors can be guaranteed if the communication graph contains no cycle. The result is an extension of the agreement theorem of Aumann [1].

All of the information structures in the models of Aumann [1], of Geanakoplos and Polemarchakis [6] and that of Krasucki [7] are given by partition on a state space. Bacharach [2] showed that the information partition model is equivalent to his knowledge operator model with the three axioms about the operators: **T** axiom of knowledge (what is known is true), **4** axiom of transparency (that we know what we do know) and **5** axiom of wisdom (that we know what we do not know.) He pointed out that the assumptions for the partition are problematic in decision making, and hence the model of analyzing complicated situations should be also constructed without such strong assumptions.

Matsuhisa and Kamiyama [8] introduced the lattice structure of knowledge for which the requirements such as the three axioms are not imposed, and they succeeded in extending Aumann's theorem to their model. However the knowledge operators satisfy the *monotonicity* property that i knows (believes) all logical consequences of what i knows (believes); i.e., the players have logically omniscient ability.

Matsuhisa and Usami [9] presented an awareness operator model in which the operators are not required to satisfy the monotonicity property. They extended

Aumann's theorem to the awareness model. However the extensions of agreement theorem are established under the common-knowledge (or common-belief) assumption.

The purpose of this article is to introduce a communication process associated with the awareness operator model of Matsuhisa and Usami [9], and to extend the agreement theorem into the model without common-belief assumption in the line of Krasucki [7]. The emphasis is on that we require none of the topological assumptions on the communication graph. Fukuda, Matsuhisa and Sasanuma [5] show Theorem 1 under the assumption that the protocol contains no cycle.

2 The Model

Let N be a set of finitely many *players* and i denote a player. A *state-space* is a finitely non-empty set, whose members are called *states*. An *event* is a subset of the state-space. If Ω is a state-space, denote by 2^Ω the field of all subsets of it. An event F is said to *occur* in a state ω if $\omega \in F$.

2.1 Awareness Structure

A *belief structure* is a tuple $\langle \Omega, (B_i)_{i\in N} \rangle$ in which Ω is a state-space and $(B_i)_{i \in N}$ is a class of operators B_i on 2^Ω, called i's *belief operator*. An *awareness structure* is a tuple $\langle \Omega, (A_i)_{i\in N}, (B_i)_{i\in N} \rangle$ in which $\langle \Omega, (B_i) \rangle$ is a belief structure and (A_i) is a class of i's *awareness* operators A_i on 2^Ω with Axiom **PL** (axiom of plausibility):

PL $B_i F \cup B_i (\Omega \setminus B_i F) \subseteq A_i F$ for every F of 2^Ω.

The *associated* information structure $(P_i)_{i\in N}$ is a class of the mappings P_i of Ω into 2^Ω defined by $P_i(\omega) = \bigcap_{T \in 2^\Omega} \{T \mid \omega \in T \subseteq B_i T\}$. (If there is no event T for which $\omega \in T \subseteq B_i T$ then we take $P_i(\omega)$ to be non-defined.)

We say an event F to be *self-aware* for i if $F \subseteq A_i F$ and say F to be *publicly aware* if $F \subseteq A_E F$. An event T is said to be i's *evident belief* if $T \subseteq B_i T$, and it is said to be *public belief* ω if $\omega \in T \subseteq B_E T$. An event is public belief (or respectively, it is publicly aware) if all players believe it (or all of them are aware of it) whenever it occurs. We can think of public belief as embodying the essence of what is involved in all players making their direct observations. The set $P_i(\omega)$ is the minimal set of all i's evident beliefs containing ω. We will call it i's *evidence set* in ω.

It is worth noting that all of the axioms **K**, **T**, **4**, **5** and **PL** are independent among them. If i's belief operator B_i satisfies **5** then by **PL** the awareness operator A_i becomes the *trivial* operator; i.e. $A_i(E) = \Omega$ for every $E \in 2^\Omega$, and there is no need of awareness.

2.2 Decision Function

Let I be the closed interval $[0, 1]$. By a *set of decisions* we mean a *weak ordered I-additive semigroup* $Z = (Z, +, \cdot, \geq)$, in which $(Z, +, \cdot)$ is an additive semigroup on which the multiplicative group $I^\times = (0, 1]$ acts with the distribution law that $\alpha \cdot (z_1 + z_2) = \alpha \cdot z_1 + \alpha \cdot z_1$ for each $\alpha \in I$, $z_1, z_2 \in Z$, and $(Z, \geq)$ is a non-empty set Z with the weak order $\geq$ called a *preference relation*, which is a binary relation on Z satisfying the two properties:

Completeness. For all $x, y \in Z$, $x \geq y$ or $y \geq x$ (or both);
Transitivity. For all $x, y, z \in Z$, if $x \geq y$ and $y \geq z$ then $x \geq z$.

We say x and y in Z are *indifferent*, denoted by $x = y$, if $x \geq y$ and $y \geq x$. For example: The real line $\mathbf{R}$ equipped with the ordinary order is a set of decisions, and so is a consumption set $\mathbf{R}_+^l$ with l commodities equipped with the ordinary preference order in an economy.

i's *decision function* is a mapping f_i of 2^Ω into Z. It is said to satisfy the *sure thing principle* if it is preserved under disjoint union; that is, for every pair of disjoint events S and T such that if $f_i(S) = f_i(T) = d$ then $f_i(S \cup T) = d$. A decision function f_i is said to be *convex* if for disjoint two events E, F, there are positive numbers $\lambda, \delta \in (0, 1)$ such that $f_i(E \cup F) = \lambda f_i(E) + \delta f_i(F)$ with $\lambda + \delta = 1$. It is *preserved under difference* if for all events S and T such that $S \subseteq T, f_i(S) = f_i(T) = d$ then we have $f_i(T \setminus S) = d$.

If f_i is intended to be a posterior probability, we assume given a probability measure μ which is common for all players and some event X. Then f_i is the mapping of the domain of μ into the closed interval $[0, 1]$ such that $f_i(E) = \mu(X \cap A_i(X)|E)$, where $\mu(E) \neq 0$. We plainly observe that this f satisfies the sure thing principle and is preserved under difference.

2.3 Communication with Awareness Structure

We assume that players communicate by sending *messages*. A *protocol* is the mapping $\mathrm{Pr} : \mathbf{Z}_+ \to N \times N$, $t \mapsto (s(t), r(t))$. Here t stands for *time* and $s(t)$ and $r(t)$ are, respectively, the *sender* and the *recipient* of the communication which takes place at time t. We can consider it as as the directed graph whose vertices are the set of all players N and such that there is an edge (or an arc) from i to j if and only if there are infinitely many t such that $s(t) = i$ and $r(t) = j$. A protocol is said to be *fair* if the graph is strongly-connected; in words, every player in this protocol communicates directly or indirectly with every other player infinitely often. It is said to be *acyclic* if the graph contains no cyclic path.

A *communication process* π of revisions of the values of decision functions $(f_i)_{i \in N}$ is a triple $\langle \mathrm{Pr}, (Q_i^t)_{(i,t) \in N \times \mathbf{Z}_+}, (f_i)_{i \in N} \rangle$, in which $\mathrm{Pr}(t) = (s(t), r(t))$ is a fair protocol such that for every t, $r(t) = s(t+1)$, communications proceed in rounds[1], and Q_i^t is the mapping of Ω into 2^Ω for i at time t that is defined inductively in the following way: We assume given a mapping $Q_i^0 := P_i$. Suppose Q_i^t is defined. Let $d_i^t(\omega)$ denote the decision $f_i(Q_i^t(\omega))$, and $W_i^t(\omega) = \{\xi \in \Omega \mid d_i^t(\xi) = d_i^t(\omega)\}$. The operator B_i^t is the belief operator induced from Q_i^t: i.e., $B_i^t E = \{\omega \in \Omega | Q_i^t(\omega) \subseteq E\}$, and A_i^t is the Awareness operator defined by $A_i^t = B_i^t(E) \cup B_i^t(\Omega \setminus B_i^t(E))$. The recipient $j = r(t)$ at t send the message

---

[1] That is, there exists a natural number m such that for all t, $s(t) = s(t+m)$).

$B_j^t(W_j^t(\omega))$ at $t+1$ when $\omega \in B_j^t(W_j^t(\omega))$, and he does not send the message otherwise. Then $Q_i^{t+1}(\omega)$ is defined as follows:

- If i is not a recipient of a message at time $t+1$, then $Q_i^{t+1}(\omega) = Q_i^t(\omega)$.
- If i is a recipient of a message at time $t+1$, then $Q_i^{t+1}(\omega) = Q_i^t(\omega) \cap B_{s(t)}^t(W_{s(t)}^t(\omega))^2$.

The communication is said to be *cut off in a state* ω if $\omega \notin B_j^t(W_j^t(\omega))$ for some t and the recipient $j = r(t)$ at t.

2.4 Consensus

We note that the limit Q_i^∞ exists in each state where the communication is never cut off[3]. We denote $d_i^\infty(\omega) = f(Q_i^\infty(\omega))$ called the *limiting decision* of f at ω for i. We say that *consensus* on the limiting decisions of a common decision function f can be guaranteed if $d_i^\infty(\omega) = d_j^\infty(\omega)$ for each player i, j and in all the states ω that the communication is never cut off and that Q_i^∞ is defined for all $i \in N$.

3 Proof of Theorem 1

It suffices to show that

Theorem 2. *Let $\pi = \langle \Pr, (Q_i^t), f \rangle$ be a communication process associated with awareness structures where f is the common decision function. Suppose that the common decision function f is preserved under difference, is convex and satisfies sure thing principle. Consensus on the limiting values of the decision function can be guaranteed; i.e., $d_i^\infty(\omega) = d_j^\infty(\omega)$ for every ω and for all i, j.*

In fact: Let X be a publicly aware event. Noting that $X \subseteq A_E(X) \subseteq A_i(X)$, set by f the common decision function defined by $f(E) = \mu(X \cap A_i(X)|E) = \mu(X|E)$ as the revised posterior. We can see that Theorem 1 is a corollary of Theorem 2.

Proof of Theorem 2 (Sketch). Let us consider the case that $(i,j) = (s(\infty), t(\infty))$. For each state ω at which Q_i^∞ is defined, set $H_i = B_i^\infty(W_i^\infty(\omega))$. We can observe the two points: First that $f(H_i) = d_i^\infty(\omega)$, and secondly that $f(H_i) = \sum_{k=1}^m \lambda_k f(\Pi_j^\infty(\xi_k))$ for some $\lambda_k > 0$ with $\sum_{k=1}^m \lambda_k = 1$. By the completeness of the preference relation it follows that for all $\omega \in \Omega$, there is some $\xi_\omega \in \Pi_j^\infty(\omega)$ such that $f(H_i) = d_i^\infty(\omega) \leq d_j^\infty(\xi_\omega)$.

Let us proceed in the general case. Continuing the above process according to the *fair* protocol, we can verify the two facts: For each $\omega \in \Omega$

---

[2] Specifically the sender j sends to i the message that he believes that his decision is $d_j^t(\omega)$.
[3] Because Ω is finite, the descending chain $\{Q_i^t(\omega) \mid t = 0, 1, 2, \ldots\}$ is finite, and so it must be stationary.

1. For every $i \neq j$, $d_i^\infty(\omega) \leq d_j^\infty(\xi)$ for some $\xi \in \Omega$; and
2. $d_i^\infty(\omega) \leq d_i^\infty(\xi) \leq d_j^\infty(\zeta) \leq \cdots$ for some $\xi, \zeta, \cdots \in \Omega$.

By the transitivity of the preference relation it follows that $d_i^\infty(\omega) = d_j^\infty(\omega)$ for every ω and for all i, j. □

4 Concluding Remarks

Our real concern in this article is about relationship between players' beliefs and their decision making, especially when and how the players reach consensus about their decisions. We focus on extending the agreeing theorem of Aumann [1] to an 'awareness and belief' model in the line of Krasucki [7]. We have shown that the nature that consensus can be guaranteed in a communication process is dependent not on common-belief assumption but on the updating with awareness and belief when each player receives information. The emphasis is on that we require none of the topological assumptions on the communication graph. Fukuda, Matsuhisa and Sasanuma [5] showed this theorem under the assumption that the protocol contains no cycle.

It can be seen that the notion of awareness does not play a role in the agreement theorem in the case of decision functions, but it does play an essential role in the probabilistic version of the theorem. In the analysis of decision theoretical situations, it seems to be problematic to keep the tension between allowing for awareness and assuming a common prior distribution that needs to be sorted out at a more fundamental level in revisions of beliefs. Furthermore it is more problematic to remove out the states in which player's evidence set cannot be defined because in such states each player cannot be aware of the evidence set, and so the role of unawareness in updating with awareness has still been unclear in our model. This issue cannot be discussed at all in this article, and there is here a research agenda of potential interest which we hope to pursue further.

References

1. Aumann,R.J.: Agreeing to disagree. Annals of Statistics **4** (1976) 1236–1239.
2. Bacharach,M.: Some extensions of a claim of Aumann in an axiomatic model of knowledge. Journal of Economic Theory **37** (1985) 167–190.
3. Dekel,E., Lipman,B.L. and Rustichini,A.: Standard state-space models preclude unawareness. Econometrica **66** (1998) 159–173.
4. Fagin,R., Halpern,J.Y., Moses,Y. and Vardi, M.Y.: *Reasoning about Knowledge.* The MIT Press, Cambridge, Massachusetts, London, England, 1995.
5. Fukuda, E., Matsuhisa, T. and Sasanuma, H.: Awareness, belief and communication reaching consensus. Journal of Applied Mathematics and Decision Science (In press).
6. Geanakoplos,J.D. and Polemarchakis,H.M.: We can't disagree forever. Journal of Economic Theory **28** (1982) 192–200.
7. Krasucki,P.: Protocol forcing consensus. Journal of Economic Theory **70** (1996) 266–272.

8. Matsuhisa,T. and Kamiyama,K.: Lattice structure of knowledge and agreeing to disagree. Journal of Mathematical Economics **27** (1997) 389–410.
9. Matsuhisa,T. and Usami,S.-S.: Awareness, belief and agreeing to disagree. Far East Journal of Mathematical Sciences **2**(6) (2000) 833–844.
10. Parikh R., and Krasucki, P.: Communication, consensus, and knowledge. Journal of Economic Theory **52** (1990) 178–189.

A High-Availability Webserver Cluster Using Multiple Front-Ends

Jongbae Moon and Yongyoon Cho

Department of Computing, Soongsil University, 156743 Seoul, Korea
comdoct@ss.ssu.ac.kr, yychosslab@hotmail.com

Abstract. A lot of clustering technologies are being applied to websites these days. A webserver cluster can be configured with either a high performance hardware switch or LVS (Linux Virtual Server) software. A high performance hardware switch has good performance but costs a great deal when constructing small and middle-sized websites. LVS, which is free of charge and has good performance, has commonly been used to construct webserver clusters. LVS is hampered by having a single front-end as it can raise a bottleneck with increased requests, and can result in the cluster system being unable to function. In this paper, we suggest new architecture for webserver clusters based on LVS with multiple front-ends which can also act as back-ends. This architecture removes the bottleneck, and is useful in constructing small and middle-sized websites. We also propose a scheduling algorithm to distribute requests equally to servers by considering their load. With this scheduling algorithm, a server will be able to respond directly to a client's request when its load is not too large. Otherwise, the server will redirect the request to the selected back-end with the lowest load. Through our experiments, we show that a webserver cluster with multiple front-ends increases the throughput linearly, while a webserver cluster with a single front-end increases the throughput curvedly. We hope that a webserver cluster with multiple front-ends will be suitable and efficient for constructing small and middle-sized websites in terms of cost and performance.

1 Introduction

With the exponential growth of the Internet, there is an increasing demand for a high performance webserver to provide stable web service to users. To overcome the limitations of a single system power, to supply higher computing power and to assure a stable system, clustering technology has been applied to websites[1]. To configure a webserver cluster system with high performance and high availability, many researches on the DNS (Domain Name Server) and the dispatcher to distribute load have been proposed [2][3][4][5][6][7][8][9]. Switching equipment like L4 has high throughput and reliability, but it also has disadvantages because there are too many additional costs. DNS may be used to provide easy construction and good scalability of cluster systems, but no consideration of server status may cause availability problems.

Webserver clusters based on LVS[10], which is free and guarantees high throughput, have been widely constructed. Because the LVS performs load balancing using a centralized method, it produces a bottleneck at the front-end when user requests increase rapidly. When the front-end does not work, a SPOF (Single Point Of Failure) causes service failure, despite multiple back-ends.

In this paper, we design a webserver cluster system based on the LVS to construct small and middle-sized websites without additional expenses. The system adopts multiple front-ends to resolve LVS's bottleneck problem. It uses a scheduling algorithm that can compute the loads of back-ends that will execute service and assigns client requests to the back-end with the lowest load. The algorithm keeps an equal load balance among back-ends and prevents loads from concentrating on one back-end.

The paper is organized as follows. Section II reviews various configurations of existing webserver clusters. Section III describes the proposed webserver cluster with multiple front-ends and the proposed scheduling algorithm. Section IV presents a comparison study of our proposed webserver cluster system with other webserver cluster systems and presents a comparison study of our proposed scheduling algorithm with other LVS scheduling algorithms. Finally, Section V states our conclusions and presents a summary of our research.

2 Related Works

Existing webserver cluster architectures are classified into four groups depending upon the distribution of user requests[2]: client-based, DNS-based, dispatcher-based, and server-based webserver clusters.

The client-based solution migrates server functionality to the client through a Java applet. An example of this solution is Smart Clients[11]. This solution causes network traffic to increase and delay, because messages are exchanged frequently between applet and server node to monitor node state. And, this solution provides scalability and availability but has a major drawback that the client should modify all applications.

The DNS-based solution is a distributed webserver architecture that uses request routing mechanisms on the cluster side. Thus, this approach is free of the problems of client-based approaches. The Round Robin DNS (RR-DNS) approach[12], implemented by the National Center for Supercomputing Applications (NCSA) to handle increased traffic at its site, is a typical example. The RR-DNS approach is simply constructed and has very good scalability. But, the load distribution under the RR-DNS may become unbalanced because it uses an address-caching mechanism and ignores both server capacity and availability[13][14]. With an overloaded or non-operational server, no mechanism can stop the clients from continuously trying to access the website by its cached address.

The server-based solution uses a two-level dispatching mechanism. The primary DNS of the webserver system initially assigns client request to the webserver nodes. Then, each server may reassign a received request to another server via HTTP redirection. The Scalable server World Wide Web (SWEB) system adopts this solution[15]. In the server-based solution, distributed scheduling provides scalability without introducing a SPOF or potential bottleneck. It also achieves the same fine-grained control on request assignments as dispatcher-based solution does. The server-based solution, however, typically increases users' perceived latency.

To centralize requests scheduling and control client request routing, the dispatcher-based solution has a component that acts as a dispatcher. This solution guarantees transparency of architecture through a single virtual IP (VIP) address unlike a DNS-based solution that deals with addresses at the URL level. The dispatcher-based solution is differentiated by the routing mechanism: packet rewriting and the packet forwarding. The Magicrouter[5] and Cisco Systems' LocalDirector adopt packet rewriting approach. IBM's Network Dispatcher[3] is an example of packet forwarding. The dispatcher can achieve fine-grained load balancing, and the single VIP address prevents the client level caching problems that affect the DNS-based solution. However, a centralized dispatcher can be a bottleneck with increased requests, and a failed dispatcher can disable the cluster system. In the packet rewriting approach, the delay caused by the modifying and rerouting of each IP packet that the dispatcher sends will degrade performance.

The LVS based webserver cluster, which is a kind of dispatcher-based solution and the most common one in clustering, has a single VIP to guarantee transparency in the cluster system. This architecture adopts IP-level load balancing to guarantee a rapid response time. However, a centralized load distribution scheme cannot avoid a bottleneck and a SPOF. The scheduling algorithm in an LVS based webserver cluster does not consider a back-end's load, which produces unbalanced loads between back-ends. The LVS is suitable for small and middle-sized websites because it is free software and provides high throughput.

3 Proposed High-Availability Webserver Cluster System Using Multiple Front-Ends

In this paper, we design a high-availability webserver cluster system with multiple nodes that work as both front-end and back-end. Fig. 1 shows the proposed architecture for a multi LVS webserver cluster.

Each server is composed with the LVS package, heartbeat daemon, load monitor, and load-based scheduling module. Two network cards are used to keep communication between modules independent. RR-DNS is used instead of a hardware switch to reduce additional expenses and to assure transparency of the cluster. In this paper, the LVS is configured via IP Tunneling to increase the scalability of a virtual server. In the LVS via IP Tunneling, the load balancer just schedules requests to the different real servers, and the real servers return replies directly to the users, so the load balancer handles a huge amount of requests. User requests are initially assigned by the DNS to a server. Then, the server checks its load and the threshold value. When the server has a load smaller than the threshold value, it handles user requests to send responses directly to users.

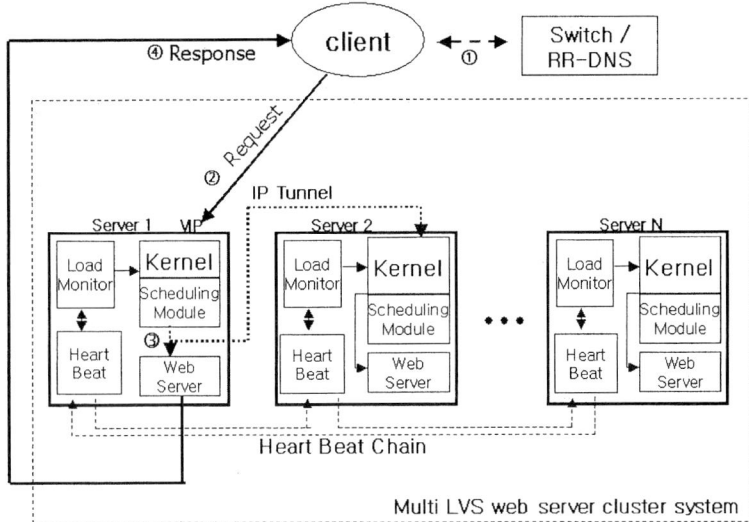

Fig. 1. Proposed architecture of multi LVS webserver cluster

Otherwise, the server with the lowest load is selected by the load-monitor and scheduling module. Then, an incoming request is forwarded to the selected server through an IP tunnel.

LVS via IP Tunneling forwards users' request packets, which are encapsulated within an IP datagram, to the back-end through a tunnel device. Each server must have a kernel with a virtual server patch to support IP encapsulation protocol. To act as a front-end, the services for IP Tunneling need to be configured. To act as a back-end, an IP tunnel needs to be connected with front-end. Tunnel devices for all front-ends are configured so that servers can decapsulate the received encapsulation packets properly. The VIP must be configured on non-arp devices or any alias of non-arp devices. For communication between modules, a private IP address is configured on a network device.

The heartbeat daemon is a component that provides high-availability of a webserver cluster system. This daemon supports two important roles as follows: First, it activates a load monitor when a fault node begins to operate, and it periodically checks whether the load monitor functions. If a load monitor does not run, the heartbeat makes it work. The heartbeat offers stability for a component by checking the load monitor. To do this, a server periodically checks whether the heartbeat works or not, using the Cron linux command. Secondly, the heartbeat daemon has a heartbeat chain to assure stability and high availability of the cluster system. When one server fails, the heartbeat chain detects the node in fault. One of the remaining nodes subsequently runs the work that was assigned to the fault node. To do this, a heartbeat for each node periodically checks the adjoining heartbeat. A heartbeat chain has a ring structure in a single direction. The heartbeat reconstructs the heartbeat chain with information of fault nodes and restored nodes. In this paper, we use an IP fake method to replace a fault node's work. Fig. 2 shows the process structure of the IP fake method. When a cluster is composed with n nodes, the fault rate will be lower than n/2 due to the characteristic of the fake mechanism.

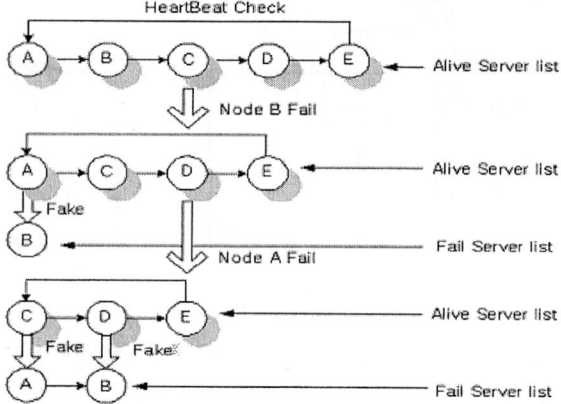

Fig. 2. A fake for a fault node

The load monitor periodically inspects load to transmit measurements to the kernel: When changes of the load are generated, the load monitor transmits load information to other servers as well as its own node. To store load data in a kernel, the LVS data is added with variables such as CPU and DISK. Even with minor changes of load, it may raise network overhead. Therefore, the load monitor is needed to set threshold value. When a load change exceeds threshold value, its load information is sent to other servers. To estimate load, the load monitor makes use of the information on CPU and DISK that is stored in Linux's /proc/stat file. To store the loads and the threshold value in the kernel, we added some variables in Linux kernel's internal data structure.

The load-based scheduling module uses an algorithm based on additional status information such as CPU load, DISK load, and threshold value. The information is collected and is saved in the kernel by the load-monitor. Threshold value is set into the kernel by the system admin. When user requests arrive, the module inspects whether the server's load is smaller than the threshold value. If the load is smaller than the threshold value, the module selects the server and processes it. Otherwise, the module selects the server with the lowest load. The kernel encapsulates the packet within an IP datagram and forwards it to the selected back-end through an IP tunnel. By checking each server's load, the module can specify node overload and prevent a node with too much load from continuously processing user requests.

To estimate throughput of webserver clusters, we categorize webserver clusters into six different cluster systems as follows:

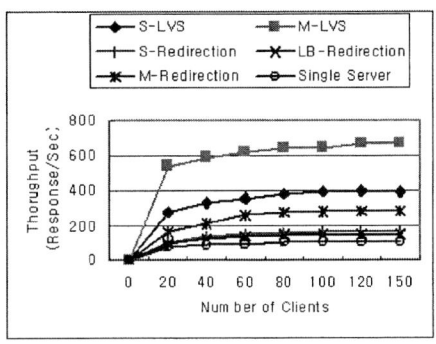 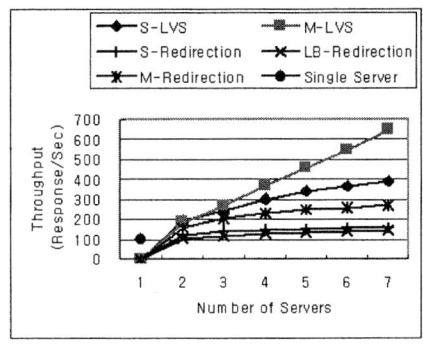

(a) When the number of clients increases

(b) When the number of servers increases

Fig. 3. Throughput for various clustering types

- Single server which is deployed with one node and a commonly used way of supplying web services.
- S-LVS which has one front-end that distributes user requests to many backends.
- S-Redirection involving one server which makes use of HTTP redirections and distributes user requests to an actual service server in Round-Robin.
- LB-Redirection similar to the S-Redirection. Once a front server receives user requests, it redirects them to the server with the lowest load among the cluster servers.
- M-LVS proposed in this paper is a kind of LVS with multiple front-ends.
- M-Redirection similar to the S-Redirection. It has multiple front-ends that distribute users' requests.

In the experiment, the servers consist of 7 nodes of personal computers with P-III 800MHz CPU, 256MB RAM and two 100Mbps Ethernet cards. Linux (Kernel version 2.4.13) was adopted and the ipvs0.9.6 kernel patch was used for LVS. The server is composed of an Apache webserver. We modified the SURGE workload generator[16] to evaluate performance on our server. SURGE produces threads which replace actual user behaviors in order to connect webservers and to produce user requests during a given test time. In this paper, cluster throughput is estimated with log files generated by SURGE for a given time.

3.1 Performance Evaluation of Various Cluster Systems

To estimate the throughput of each webserver cluster, firstly the number of servers was fixed, and the number of clients was increased to force the clients to have maximum performance. Then, the number of clients having maximum performance was fixed and the number of servers was increased.

In Fig. 3(a), the number of servers is fixed at seven, and the number of clients increased. The S-Redirection, the LB-Redirection, and the M-Redirection use two-level scheduling of the HTTP Redirection mechanism. A dispatcher receives all incoming requests and distributes them to the webserver nodes. The dispatcher returns an HTTP redirection status code that indicates to the browser that it should look for the page at another URL. Once a user receives the response, the user transmits requests to the actual server again. The cluster with the application-level scheduling does not increase throughput enough, because it does not assure the server quick response. We can show that S-Redirection generates more hits than a single server because the back-ends of S-Redirection consisted of six servers. The LB-Redirection generates fewer hits than the S-Redirection because it needs the time to select a proper redirection server. The M-Redirection generates almost 1.5 times hits more than the S-Redirection does. This is because multiple retransmission servers for user requests accept more user requests. The LVS webserver cluster responds quickly and handles many more requests because the front-end uses IP-level scheduling. Therefore, it generates many more hits than the cluster with the application-level scheduling. The M-LVS raises throughput 1.5 times more than the S-LVS does. This is because it has multiple front-ends for distribution of user requests.

In Fig. 3(b), the number of clients is fixed while the number of servers is increased from one to seven. We conducted the throughput experiment for S-LVS, S-Redirection, and LB-Redirection, setting only one front-end to distribute user requests and increasing back-ends to serve requests. We conducted the throughput experiment for M-LVS and M-Redirection by increasing front-ends to keep load balancing. Both the S-Redirection and the LB-Redirection raise the throughput a little because the number of back-ends increases. But the throughput does not continue to rise because the servers are overloaded redirecting user requests, though the number of servers is continuously increased. On the other hand, the M-Redirection raises the throughput higher than the S-Redirection according to increase of the front-ends dealing with user requests. It does not raise the throughput greatly and continuously because of network overhead and delay caused by HTTP redirection. The LVS webserver cluster raises the throughput much more than the HTTP redirection when the number of servers increases. The LVS will raise the throughput more than other types of webserver clusters because the front-end uses IP-level scheduling. However, the increasing ratio is not affected because one front-end becomes overloaded and its response is delayed. The M-LVS hits rise almost linearly when the number of servers increases. The M-LVS causes less delay than the HTTP redirection because all of the nodes function as both front-end and back-end at the same time to reduce load. In addition, multiple front-ends can remove the SPOF that is caused by bottleneck.

In this experiment, we found that each webserver cluster raises the throughput when the number of servers handling user requests increases, but a server that balances user requests will not raise the throughput continuously because of limited hits. In addition, multiple servers balancing user requests are established to raise the throughput greatly. The LVS using IP-level scheduling made more hits than application-level scheduling did. Therefore, we know that the suggested webserver cluster delivers better performance.

3.2 Performance Evaluation of Various Types of Scheduling Methods

To evaluate the load-based scheduling method, it is compared with three other types of scheduling methods implemented on LVS: Round-Robin Scheduling (RR), Least-Connection Scheduling (LC), Weighted Least-Connection Scheduling (WLC). We have conducted the simulation with four real servers. In this experiment, server load is generated by a program that is developed in C language. In the WLC scheduling, each server is assigned a performance weight of 2,1,3, and 4 in order of servers.

To evaluate the load-based scheduling method, we implemented two scenarios as follows: First, we conducted simulation experiments with servers that have a small CPU load of 0.1%. Secondly, we conducted simulation experiments with servers whose second server has a large CPU load over 80% and others have a small CPU load of 0.1%.

For each scheduling method under the first scenario, Fig. 4(a) shows the total response. The RR, LC, and Load-based scheduling method have almost the same performance. The WLC scheduling method, however, distributes more requests to the fourth server with the largest weight. The RR scheduling distributes each request sequentially around the pool of real servers. Using this algorithm, all real servers are treated as equal without regard to capacity or load. Thus, all incoming requests are distributed equally among real servers. The LC scheduling distributes more requests to real servers with fewer active connections. In this experiment, all the real servers with the same processing capacity are used and have the same load of CPU. Thus, the LC scheduling distributes almost the same number of requests to each real server. The Load-based scheduling distributes requests almost equally among the real servers. If each server's load is less than the threshold value, the Load-based scheduling does not redirect incoming requests to other server but processes them itself, and responds directly to the users. The WLC scheduling distributes more requests to servers with fewer active connections relative to their capacities. Capacity is indicated by a user-assigned weight, which is then adjusted upward or downward by dynamic load information. In this experiment, we have assigned a weight to each real server of 2,1,3, and 4 respectively. Thus, the WLC scheduling distributes the largest requests to the fourth server of the cluster.

For each scheduling method under the second scenario, Fig. 4(b) shows the total response. The RR scheduling distributes incoming requests almost equally without regard to servers' load. The LC scheduling distributes less requests to a real server with a high load of CPU because the real server with high load of CPU gets less network connections and has poor response time. The WLC scheduling distributes less requests to a real server with low load. In this experiment, however, it is only because the real server with the lowest load has the smallest user-assigned weight.

The Load-based scheduling refers to the threshold value that is assigned in the Kernel. If all the real servers have less loads than the threshold value, the Load-based scheduling works as the RR scheduling does. If the load of a real server

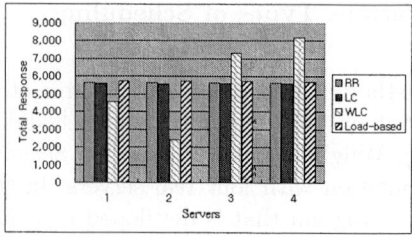 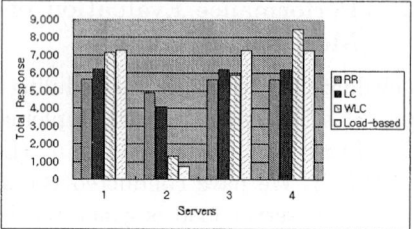

(a) When all servers are assigned a low load

(b) When a specific server (server 2) is assigned a high load

Fig. 4. Throughput for various scheduling methods

is larger than the threshold value, the Load-based scheduling sends requests to the real server with the smallest load of CPU via an IP tunnel. The Load-based scheduling balances the servers' loads by distributing less requests to the real servers with less CPU load. Besides, the Load-based scheduling prevents server faults caused by overload.

4 Conclusions

In this paper, a webserver cluster has been designed for small and middle-sized websites at a low cost. The LVS can be a good solution for small and middle-sized websites because of low prices and high performance IP scheduling. However, the SPOF caused by bottleneck is inevitable in LVS because of a centralized load balancer. In the proposed architecture, both the front-end and the back-end are not separated, which removes the front-end bottleneck, so that scalability is guaranteed. Both the heartbeat and the load monitor check servers' status to provide the cluster system with high-availability. To keep load balance among servers, a dynamic load balancing algorithm assigns less requests to the overloaded servers. As a result, unbalanced loads between servers are effectively solved. In the experiment, while a webserver cluster with a single front-end raises the throughput curvedly, a webserver cluster with multiple front-ends raises the throughput linearly. We also found the IP-level scheduling improves the throughput up to about 1.5 times relative to the application-level scheduling.

References

1. Rajkumar, Buyya: High Performance Cluster Computing Vol. 1. Chap. 36. A Scalable and Highly Available Clustered Web Server, Prentice Hall (1999)
2. Valera, Cardellini, Michele, Colajanni, Philip, S., Yu: Dynamic Load Balancing on Web-Server Systems. Internet Computing, Vol. 3, IEEE (1999) 28–39

3. Guerney, D.H., Hunt, Germán, S., Godlszmidt, Richard, P., King, Rajat, Mukherjee: Network Dispatcher: A Connection Router for Scalable Internet Services. Journal of Computer Networks and ISDN Systems, Vol. 30, Elsevier Science, Amsterdam, Netherlands (1998)
4. Om, P., Damani, P., Emerald, Chung, Yennun, Huang, Chandra, Kintala, Yi-Min, Yang: ONE-IP: Techniques for Hosting a Service on a Cluster of Machines. Journal of Computer Networks and ISDN Systems, Vol. 29, No. 8-13 (1997) 1019–1027
5. Eric, Anderson, Dave, Patterson, Eric, Brewer: The Magicrouter, An Application of Fast Packet Interposing. Class Report, University of California, Berkely (1996)
6. Emiliano, Casalicchio, Michele, Colajanni: A client-aware dispatching algorithm for web clusters providing multiple services Proceedings of the 10th International Conference on World Wide Web (2001)
7. Haesun, Shin, Sook-Heon, Lee, Myong-Soon, Park: Multicast-based Distributed LVS(MD-LVS) for improving scalability and availability. The 8th International Conference on Parallel and Distributed Systems (ICPADS 2001) (2001) 26–29
8. Azer, Bestavros, Mark, Crovella, Jun, Liu, David, Martin: Distributed packet Rewriting and its Application to Scalable Server Architectures. Proceedings of the 1998 International Conference on Network Protocols (INCP '98) (1998)
9. High-availability.com: RSF-1 Technical White paper. Whitepaper (1998)
10. Wensong, Zhang, Shiyao, Jin, Quanyuan, Wu: Creating Linux Virtual Server. Proceedings of the 5th Annual Linux Expo (1999)
11. Chad, Yoshikawa, Brent, Chun, Paul, Eastham, Amin, Vahdat, Thomas, Anderson, David, Culler: Using Smart Clients to Build Scalable Services. Proceedings of the USENIX 1997 Annual Technical Conference (1997)
12. Thomas, T., Kwan, Robert, E., McGrath, Daniel, A., Reed: NCSA's World Wide Web Server: Design and Performance. IEEE Computer, Vol. 28, No. 11. (1995) 68–74
13. Michele, Colajanni, Philip, S., Yu, Daniel, M., Dias: Analysis of Task Assignment Policies in Scalable Distributed Web-Server Systems. IEEE Transaction on Parallel and Distributed Systems, Vol. 9, No. 6. (1998) 585–600
14. Daniel, M., Dias, William, Kish, Rajat, Mukherjee, Renu, Tewari: A Scalable and Highly Available Web-Server. Proceedings of the 41st IEEE International Computer Conference (1996) 85–92
15. Daniel, Andresen, Tao, Yang, Vegard, Holmedahl, Oscar, H., Ibarra: SWEB: Toward a Scalable World Wide Web-server on Multicomputers. Proceedings 10th IEEE International Symposium Parallel Proceeding (1996) 850–856
16. Paul, Barford, Mark, Crovella: Generating Representative Web Workloads for Network and Server Performance Evaluation. Proceeding of the 1998 ACM SIGMETRICS International Conference on Measurement and Modeling of Computer Systems (1998) 151–160

An Intelligent System for Passport Recognition Using Enhanced RBF Network

Kwang-Baek Kim[1], Young-Ju Kim[2], and Am-Suk Oh[3]

[1,2] Dept. of Computer Engineering, University of Silla,
[3] Dept. of Multimedia Engineering, Tongmyong Univ. of Information Technology, S. Korea
{gbkim, yjkim}@silla.ac.kr, asoh@tmic.tit ac.kr

Abstract. The judgment of forged passports plays an important role in the immigration control system and requires the automatic recognition of passports as the pre-phase processing. This paper, for the recognition of passports, proposed a novel method using the enhanced RBF network based on ART2. The proposed method extracts code sequence blocks and individual codes by applying the Sobel masking, the smearing and the contour tracking algorithms in turn to passport images. The enhanced RBF network was proposed and used for the recognition of individual codes, which applies the ART2 algorithm to the learning structure of the middle layer. The experiment results showed that the proposed method has superior in performance in the recognition of passport.

1 Introduction

Due to the globalization and the advance of travel vehicles, the number of passengers of overseas travel is gradually increasing. The current immigration control system carries out manually the passport inspection, and the immigration process takes a long time, putting passengers to inconvenience. The automatic passport inspection has to provide the high precision in executing the critical functions such as the judgment of forged passports, the search for a wanted criminal or a person disqualified for immigration, etc [1]. This paper, for the precise passport inspection, proposed a novel passport recognition method that supports the code extraction using smearing method and contour tracking algorithm and the code recognition using the enhanced RBF network based on ART2.

As the edge extraction methods, the various methods such as Sobel operator, Roberts and Laplacian differential operators etc. are used [2]. Among them, the Sobel operator uses the first-order differential values, so that it is robust to noises and requires small overhead in the processing time [3][4]. So, this paper extracts edges from passport images by using the 3x3 Sobel masking rather than the Sobel operator based on partial differential operation. And this paper applies the smearing method[5][6] and the 4-directional contour tracking algorithm to the edge image for extracting individual codes being recognized: code sequence blocks are extracted by applying the horizontal smearing and the 4-directional contour tracking algorithm to the output of 3x3 Sobel masking, and individual codes are extracted by applying the vertical smearing to code sequence blocks. This paper proposed the enhanced RBF (Radial Basis Function) network that, for the effective learning of new input patterns, carries out two-step learning based on ART2 algorithm: the competitive learning between input layer and middle layer and the supervised learning between middle layer and output layer. And the enhanced RBF network was used to recognize individual codes extracted from passport images.

2 Extraction of Code Blocks and Individual Codes

The passport image consists of the three areas, the picture area in the top-left part, the user information area in the top-right part, and the user code area in the bottom part. This paper, for the recognition of passports, extracts user codes from passport images, and recognizes and digitalizes the extracted codes. The proposed algorithm for passport recognition consists of two phases: the individual code extraction phase extracting individual codes being recognized from the passport image and the code recognition phase recognizing the extracted codes. This section examines the individual code extraction phase. Fig. 1 shows an example of passport image used for experiment in this paper. The user code area has the white background and the two code rows including 44's codes at the bottom part of passport image. For extracting the individual codes from the passport image, first, this paper extracts the code sequence blocks including the individual codes by using the feature that the user codes are arranged sequentially in the horizontal direction.

Fig. 1. An example of passport image

The extraction procedure of code sequence blocks is as follows: First, the 3x3 Sobel masking is applied to the original image generating an edge image. By applying the horizontal smearing to the edge image, the adjacent edge blocks are combined into a large connected block. Successively, by applying the contour tracking to the result of smearing processing, a number of connected edge blocks are generated, and the ratio of width to height of the blocks are calculated. Last, the edge blocks of maximum ratio are selected as code sequence blocks. Fig. 2 shows the edge image generated by applying the Sobel masking to the image in Fig. 1. Fig. 3 shows the results generated by applying the horizontal smearing to the edge image. This paper uses the 4-directional contour tracking to extract code sequence blocks from the results in Fig. 3. The contour tracking extracts outlines of connected edge blocks by scanning and connecting the boundary pixels [7].

The paper uses the 2x2 mask shown in Fig. 4 for the 4-directional contour tracking. The contour tracking scans the smeared image from left to right and from top to

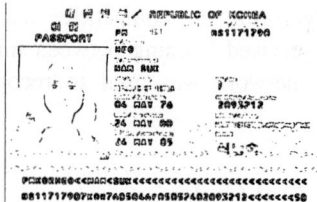 

Fig. 2. Result of 3x3 Sobel masking in Fig 1   **Fig. 3.** Result of horizontal smearing in Fig. 2

bottom to find boundary pixels of edge block. If a boundary pixel is found, the pixel is selected as the start position. The selected pixel is placed at the x_k position of 2x2 mask, and by examining two pixels coming under a and b positions and comparing with the conditions in Table 1, the next scanning direction of the mask is determined and the next boundary pixel being tracked is selected. The selected pixels coming under x_k position are connected into the contour of edge block. By generating outer rectangles including contours of edge blocks and comparing the ratio of width to height of the rectangles, the code sequence blocks with the maximum ratio are extracted.

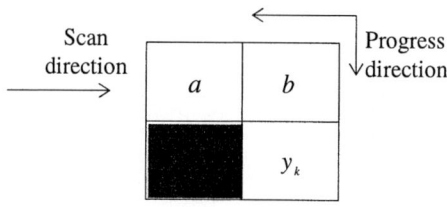

Fig. 4. 2x2 mask for 4-direction contour tracking

Table 1. Progress direction of a and b by 2x2 mask

	a	b	x_k	y_k
Forward	1	0	a	b
Right	0	1	b	y_k
Right	1	1	a	x_k
Left	0	0	x_k	a

For the extraction of individual codes from code sequence blocks, this paper applies the vertical smearing to the blocks and separates each code using the vertical coordinates of vertically smeared edge sub-blocks. And using the horizontal coordinates, the size of each code is determined. The sizes of extracted codes are normalized for the code recognition using the enhanced RBF network.

3 Recognition of Passport Codes Using the Enhanced RBF Network

The RBF (Radial Basis Function) network is the feed-forward neural network that consists of three layers, input layer, middle layer and output layer. The middle layer of the RBF network executes the clustering operation, classifying an input vector set to clusters including only homogeneous vectors. The measurement of homogeneity in clusters is the distance between vectors in clusters. And the classification of an input vector to a cluster means that the distances between the input vector and each vector in the cluster are shorter than or equal to the fixed radius. But, the use of a fixed radius in clustering causes wrong classifications. Therefore the selection of the organization for middle layer determines the overall efficiency of the RBF network [8][9]. In the RBF network, the learning of new patterns not learned previously is apt to classify the patterns to the cluster including similar patterns already learned [10].

So, by using the ART2 algorithm, this paper enhanced the RBF network to classify a new pattern to a new cluster, having no influence on existing clusters. The enhanced RBF network based on ART2 algorithm carries out the two-step learning: the first step of learning is the competitive learning between input layer and middle layer, and the second step is the supervised learning between middle layer and output layer. In the enhanced RBF network, the output vector of middle layer is calculated by Eq.(1), indicating the error between the input pattern and clusters. And, the node with minimum output vector is selected as winner node like Eq.(2).

$$O_j = \frac{1}{N} \sum_{i=0}^{N-1} \left(|x_i - w_{ji}(t)| \right) \tag{1}$$

$$O_j^* = \wedge \{O_j\} \tag{2}$$

Where $\wedge$ is the function calculating the minimum value and w_{ji} is the connection weight between input layer and middle layer. And the similarity test for the selected winner node is like Eq.(3).

$$O_j^* \prec \rho \tag{3}$$

Where ρ is the vigilance parameter given in the proposed network.

If the output vector of winner node is less than the vigilance parameter, the input pattern is classified as the same pattern; otherwise it is classified as the different pattern. In the former case, the connection weight is modified to reflect the similar property of the input pattern to the weight. The adjustment of connection weight is like Eq. (4).

$$w_{j^*i}(t+1) = \frac{w_{j^*i}(t) \times u_n + x_i}{u_n + 1} \tag{4}$$

Where u_n is the number of updated patterns in the cluster corresponding to the winner node.

4 Performance Evaluation

For performance evaluation, this paper implemented the proposed passport recognition system by using C++ Builder tool and experimented on the IBM-compatible PC with Intel Pentium-III 550 MHz CPU and 128MB RAM. In the experiment, 30's passport images with 600x437 pixel size were used. Fig. 5 shows code sequence blocks and individual codes extracted from Fig.1 by the proposed extraction method. And Table 2 shows the total number of code sequence blocks and individual codes extracted from 30's experiment images.

Fig. 5. Extraction result of code sequence block and individual codes

Table 2. Experiment result of the proposed extraction method

Object	Number of objects extracted / Number of failure
Code Sequence Block	60 / 0
Individual Code	2640 / 0

Table 3 shows the result of learning experiment in the enhanced RBF network using 620's individual codes extracted from 15's images. In the learning experiment, when the vigilance parameter was set to 0.2, the enhanced RBF network showed the optimal learning performance and the number of nodes of middle layer is increasing no more.

Table 3. Result of learning experiment in the enhanced RBF network

	Number of nodes created in the middle layer	Number of Epoch
Enhanced RBF network based on ART2	217	4832

This paper divided 30's passport images to two groups: 15's images used in learning and 15's images not used in learning. To evaluate the recognition performance, the enhanced RBF network was applied to each group individually. Table 4 shows the result of recognition experiment on each group. The enhanced RBF network recognized all individual codes extracted from 30's passport images.

Table 4. Result of recognition experiment in the enhanced RBF network

	Image group used in learning	Image group not used in learning
Number of success/ Number of failure	620 / 0	615 / 0

5 Conclusions

The automatic passport inspection has to provide the high precision to execute the critical functions such as the judgment of forged passports, the search for a wanted criminal or a person disqualified for immigration, etc. So, this paper, for the precise passport inspection, proposed a novel passport recognition method that supports the code extraction using smearing method and contour tracking algorithm and the code recognition using the enhanced RBF network based on ART2.

This paper, first of all, extracted the code sequence blocks including only user code strings by applying 3x3 Sobel masking, horizontal smearing and 4-directional contour tracking algorithm sequentially to passport images. Next, by applying the vertical smearing to code sequence blocks, individual codes being recognized were extracted. This paper proposed the enhanced RBF network that, for the effective learning of new input patterns, carries out two-step learning based on ART2 algorithm: the competitive learning between input layer and middle layer and the supervised learning between middle layer and output layer. And the enhanced RBF network was used to recognize individual codes. The experiment for performance evaluation was executed on 30's passport images and the experiment results showed that the proposed code extraction method extracts all identification codes from passport images with no failure and the enhanced RBF network recognizes successfully all individual codes. As the future works, the face authorization method is required for the precise judgment of forged passports, and the research for face authorization is needed.

References

[1] Ryu, J. U., Kim, K. B.: The Passport Recognition by Using Smearing Method and Fuzzy ART Algorithm. Proc. of KKFIS. **12(1)** (2002) 37-42
[2] Sonka, M., Havac, V., Boyle, R.: Image Processing, Analysis and Machine Vision. University Press, Cambridge (1993) 113-121
[3] Gonzalez, R. C., Woods, R. E.: Digital Image Processing. Addison Wesley (1992)
[4] Parker, J. R.: Algorithm for Image Processing and Computer Vision. Wesley Computer Publishing (1996)
[5] Gorman, L.O. and Kasturi, R.: Document Image Analysis Systems. IEEE Computer. **5** (1992) 5-8
[6] Wahl, F. M., Wong, K. Y., Casey, R. G.: Block Segmentation and Text Extraction in Mixed Text/Image Documents. Computer Vision, Graphics and Image Processing. **22** (1982) 375-390
[7] Kim, S. O., Lim, E. K., Kim, M. H.: An Enhancement of Removing Noise Branches by Detecting Noise Blobs. J. of Korea Multimedia Society. **6(3)** (2003) 419-428
[8] Panchapakesan, C., Ralph, D., Palaniswami, M.: Effects of Moving the Centers in an RBF Network. Proc. of IJCNN. **2** (1998) 1256-1260
[9] Watanabe, M., Kuwata, K., Katayma, R.: Adaptable Tree-Structured Self Generating Radial Basis Function and its Application to Nonlinear Identification Problem. Proc. of IIZUKA. (1994) 167-170
[10] Capenter, G. A., Grossberg, S.: ART2: self-organization of stable category recognition code for analog input patterns. OPTICS. **26(23)** (1897) 4919-4930

A Distributed Knowledge Extraction Data Mining Algorithm

Jiang B. Liu[1], Umadevi Thanneru[1], and Daizhan Cheng[2]

[1] Computer Science & Information Systems Department,
Bradley University,
Peoria, IL 61625, U.S.A.
jiangbo@bradley.edu
[2] Institute of Systems Science,
Chinese Academy of Sciences,
Beijing 100080, China

Abstract. We have developed a distributed data mining algorithm based on the progressive knowledge extraction principle. The knowledge factors, the data attributes that are significant statistically or based on a predefined mining function, are extracted progressively from the distributed data sets. The critical data attributes and sample data set are selected iteratively from distributed data sources. The experiments showed that the algorithm is valid and has the potentials for the large distributed data mining practices.

1 Introduction

Online data analysis on distributed large data sets has many applications [1]. With high speed network, organizations will be able to make their business decisions based on the knowledge discovered from dynamically changing data sets that distributed on different data sources. The challenge lies on the distributed data mining algorithms development as how to gain the useful information within the time limit on the network from vast data sets with enormous irrelevant data with respect with the mining goals [2]. In this research, we have developed a practical distributed data mining and selection algorithm based on new progressive knowledge extraction approach. The knowledge factors, the data attributes that are significant statistically or based on a predefined mining function, are extracted progressively from the distributed data sets. The initial knowledge factor set will first be tested, verified, cleaned, and then expanded based on the mining functions. The proposed algorithm is an iterative process. Depending on the applications, it may take many tries for identifying knowledge factor sets and mining function settings to discover the hidden patterns in large distributed data sets. Within each data selection run on the distributed data sets, only the sample data attributes related to the knowledge factors are required. Therefore, this proposed data mining and selection algorithm has the potential to mine the large distributed data sets online with high-speed network connections.

Our distributed selection research has targeted on the remote data analysis and agribusiness fraud detection with the potential data sets connected on the high speed Internet II network [3]. The high performance network connects the remote data centers to our data mining servers for the online data analysis. The data mining system and the algorithms developed are implemented in a client/server computing environment so that the clients can mine the data through distributed data analysis using minimal required data set.

2 New Progressive Knowledge Extraction Data Selection Algorithm

The proposed distributed data mining and selection algorithms are described as follows:

2.1 Data Set Definition

We describe the data set as

$$\{x(t_1, \ldots t_n)\}$$

where $t_1, \ldots t_n$ are data attributes. They are the characteristics of a sample date on a distributed data set (For example, in an agribusiness data set these attributes could be the agent_id, crop_type, farming_practice, and claimed_loss.) x is the concerned value of the data point (x could be 0 or 1. 0 means there is no fraud in the particular agribusiness fraud detection case and 1 means fraud.)

2.2 Expert Knowledge Representation

Expert knowledge is used to describe which factors are significant (For example interesting association rules that are true at a given level of support and confidence.) Suppose we have k different knowledge factors E_i, $i = 1, \ldots k$, initially, which are described by k functions $L_i(t_1^i, \ldots, t_n^i)$ on a sample of data with related partial attributes, then the following table listed their representation.

Table 1. Knowledge Representations

Factor	Function	Significant point
E_1	$L_1(t_1^1, \ldots, t_n^1)$	L_1^0
E_2	$L_2(t_1^2, \ldots, t_n^2)$	L_2^0
...	...	...
E_k	$L_k(t_1^k, \ldots, t_n^k)$	L_k^0

For example, in the agribusiness data analysis, we can let E_1 be the loss ratio and E_2 be the age of the farmer. L_1 then is used to compute the loss ratio against the county average loss and L_2 is used to categorize age of the farmers. Both of them need only related data attributes from the sample data set for their computation.

2.3 Expert Knowledge Based Data Arrangement

To classify the data we can choose several levels of disjoint ranges for each factor. For instance, we may choose $2m_1$ levels for factor 1. That is, we may choose a step δ and form $2m_1$ intervals as

$$I_1^1, I_2^1, \ldots, I_{2m_1}^1 := (-\infty, L_1^0-(m_1-1)\delta], (L_1^0-(m_1-1)\delta, L_1^0-(m_1-2)\delta, \ldots, L_1^0+(m_1-1)\delta, \infty)$$

Thus, for a data $x(t_1, \ldots t_n)$ we can calculate $L_1, \ldots, L_k$. If $L_1 \in I_t^1$, we will simply say that $L_1=t$. The step size δ can be used to handle the granularity problem.

For notational ease, we can construct a tensor product as $L_1 \otimes L_2 \otimes \ldots \otimes L_k$. Now according to the expert knowledge extracted we can classify data into $N = \pi_{i=1}^{k}(2m_i)$ rows. The rows are arranged in alphabetic order. That is,

Table 2. Data Arrangement

$L_1 \otimes L_2 \otimes \ldots \otimes L_k$	Data
1 1 … 1	$x(t_1,\ldots, t_n)_{i_1}, \ldots$
1 1 … 2	$x(t_1,\ldots, t_n)_{i_2}, \ldots$
…	…
$2m_1$ $2m_2$ … $2m_k$	$x(t_1,\ldots, t_n)_{i_N}, \ldots$

For example, in the agribusiness data analysis, let $m_1 = 1$ and $\delta=1$ for E_1 (loss ratio is divided into two ranges) and $m_2 = 2$ and $\delta=20$ for E_2 (farmers are divided into 4 age categories.) Then the tensor product has 8 rows with data sets divided into 8 regions. For instance, row (1, 2) could be the data subset with loss ratio less than county average and farmer age between 25 and 35. Unlike the traditional OLAP MDDB NWAY cube, in which the values of the attributes are used to construct the clusters, here only the discrete factor levels are used to classify the data.

2.4 Significance Verification

To verify the significance of the knowledge factor $E_1, \ldots, E_k$, we first re-organize data in the following way:

For E_i ($i = 1, \ldots, k$):

Step 1: For all sampled data $x(t) := x(t_1,\ldots, t_n)$ from distributed data sets, calculate $L_i(t)$.

Step 2: Construct $2m_i$ data of interval averages as

$$(\tilde{x})_j^i = \frac{\sum_{L_i(t) \in L_j^i} x(t)}{N_j^i}, j = 1, \ldots, 2m_i$$

Step 3: Using $\{(\tilde{x})^i_j \mid j = 1,\ldots, 2m_i\}$ to test the significance of the knowledge factor E_i.

For instance, assume x has multi-normal distribution with respect to $t_1,\ldots, t_n$, and L_i are linear function, then x also has a normal distribution with respect to L_i. Then the standard χ^2 or F test can be used to verify the statistical significance of the knowledge factor E_i.

Finally, we conclude in two things:

(1) Some E_i, say $E_{i1}, \ldots, E_{is}$ are not significant.
(2) For significant E_i the center L^0_i may be shifted.

For example, in the agribusiness data analysis, after the verification, we may decide that E_2 (age of the farmer) is not significant with respect to the fraud detection and the center of E_1 should be shifted from 1 to 1.5 (that is the loss ratio should be 1.5 times large than the county average instead of simply large than the county average.)

In general, one run is not enough to over-throw an expert knowledge factor. We usually need a second phase test.

Second phase test:

(1) Arrange data into s rows by using $L_{i_1}, \ldots, L_{i_s}$.
(2) Choose the sample data as the following:
    (a) Sample same number of data from each of s rows.
    (b) In each row, sample data in a random way.
(3) Add new sampled data into the old data set, repeat the test for $E_{i_1}, \ldots, E_{i_s}$ only.

The second phase test may be repeated several times. In the end, we can conclude that some expert knowledge factors E_t are not significant in this analysis. Then they can be ignored in the next run.

2.5 New Knowledge Revealing

We can now assume $E_1, \ldots, E_s$ are the confirmed expert knowledge factors. A natural question is: are these a complete set? A good data selection process should be able not only to verify the old knowledge factors but also to find the new knowledge factors. The following out of edge searching algorithm can be used to extract these new factors.

Out of Edge Searching:

Step 1: Re-classify the sampled data into 2^s groups as $W = \{U^{\pm}_1 \otimes U^{\pm}_2 \otimes \ldots \otimes U^{\pm}_s\}$. e.g.,

$U^+_1 \otimes U^-_2 \otimes \ldots \otimes U^-_s = \{t \mid L_1(t) > L^0_1, L_2(t) < L^0_2, \ldots, L_s(t) < L^0_s\}$.

Step 2: Using each data in W_i, $i = 1,\ldots, 2^s$ to find a new $\tilde{L}_i$.

Step 3: Using the same testing procedure proposed in the last section to test $\tilde{L}_i$. If for a certain $\tilde{L}_i$ the significance is not lower then the average significance of the expert knowledge level, then it should be considered as a newly discovered expert knowledge factor. (Repeated tests may be necessary to verify the new extract knowledge factor.)

The new data mining and selection algorithm that we have developed above has two major advantages. First, the knowledge factors are verified, cleaned, and expended progressively with the usages of a partial data attributes (a vertical partition) and a sampled data set (a horizontal partition) from the large distributed data set. Therefore it is computational feasibly for our clients to mine the data online. Second, the knowledge functions are determined based on the mining applications and mining methods used. Therefore, we can try different approaches to select the knowledge functions, such as multi-dimensional distribution functions [4-6] or generic learning algorithms/functions [7]. Thus the mining system can be used as a framework for selecting the best mining function for the mining objectives.

3 Sample Distributed Data Analysis Results

To build the distributed data selection infrastructure, we have set up a data mining laboratory with IBM Intelligent Miner [8]. Reducing the mining data set are achieved by using data attributes selections and sample data selection.

3.1 Critical Data Attributes Selection

For a given mining objective, we can identify the critical data attributes from well-sampled data sets. The attributes selected should not change the major mining results. The initial set of the critical data attributes were chosen by the field experts. Our algorithm then applied on the initial set iteratively to identify the final critical data attributes. For this example, we have identified eight critical data attributes, six are the categorical type and two are the numeric type, from a mining mart of fourteen data attributes.

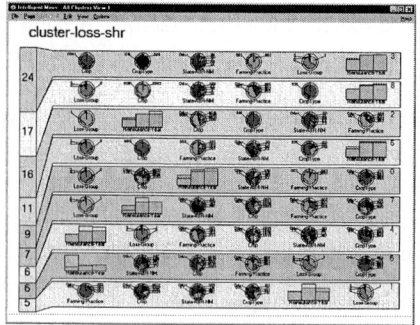
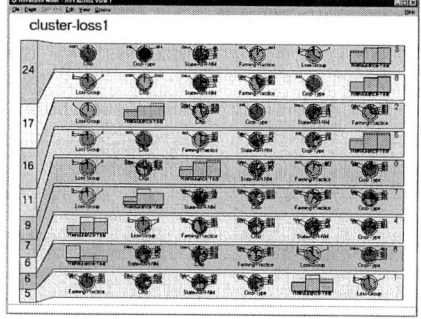

Fig. 1. The output of the clustering object with the reduced data attributes set

Fig. 2. The output of the clustering object with the original data attributes set

The cluster created is based on demographic clustering process. Figure 1 and 2 illustrated the differences between the clustering results from the reduced data attributes set and the original data set. The detailed analysis has shown that there are no major losses of information with respect with the mining objective.

Similarly, we have applied our algorithm with the data association analysis. The detailed analysis also showed that there are no major losses of information in the association data analysis with respect with the mining objective.

To avoid the lost of the critical data attributes from the distributed data analysis practice, our algorithm will constantly check the intermediate results against the mining objective. The reduced data attributes will be restored once the detailed analysis showed the significant loss of information. Figure 3 and 4 illustrated one of such scenarios. Figure 4 shows the detailed clustering data analysis for the original seven data attributes and figure 5 shows such a result for reduced set. The analyses have shown that most crucial information for our clustering data analysis was lost. Therefore the reduced data attribute should be restored back to the critical data attribute set.

3.2 Data Sampling

With large data set, there are inevitably hidden data redundancies. To explore these data redundancies, we applied our algorithm to the data sample selections. The data records are sampled with the random selection using a predefined random function. Figure 5 and 6 illustrated the clustering mining results with the sampled data set and the original data set. The final selected random selected sample data set should reveal the similar mining results as the original data set.

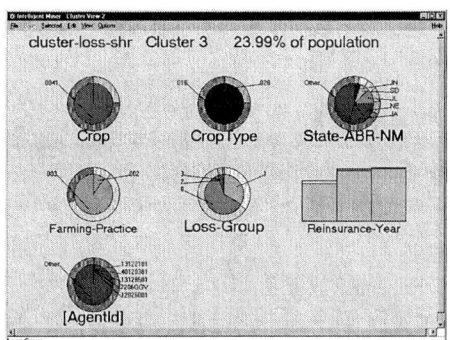
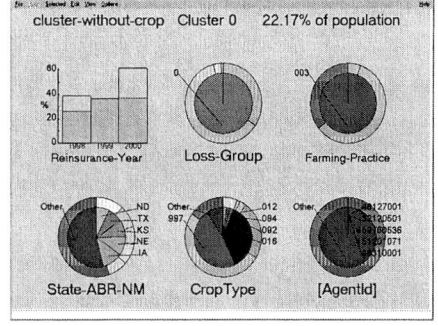

Fig. 3. The detailed clustering analysis with eight data attributes

Fig. 4. The detailed clustering analysis with seven data attributes

4 Conclusion

We have developed a progressive knowledge extraction distributed data mining and selection algorithm. The mining practices demonstrated that the proposed distributed data selection principle and technologies are practical and valid. We believe that many business data mining application can adapt our algorithm in their distributed data analysis practices.

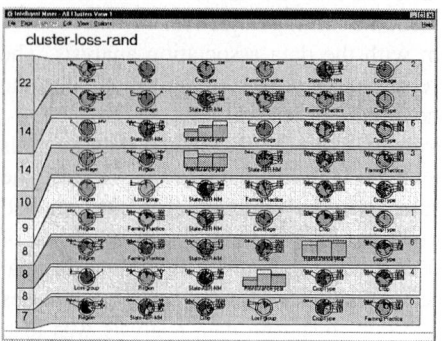 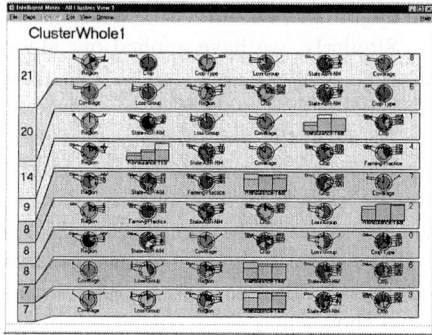

Fig. 5. Clustering results using the sample data set

Fig. 6. Clustering results using the original data set

References

1. Provost, F.: Distributed Data Mining: Scaling Up and Beyond, Advances in Distributed and Parallel Knowledge Discovery, The MIT Press (2000) 3-28.
2. Liu, J. B., Han J.: A Practical Knowledge Discovery Process for Distributed Data Mining, Proceedings of 11th International Conference on Intelligent Systems, Boston (2002) 11-16.
3. Yost, J. K., Liu, J. B., McConnaughay, K. D. M, Winn, W.: HPNC for Bradley University in Science and Engineering Research, NSF Grant 0125067 (2001).
4. Aggarwal, C. C., Yu, P. S.: A New Approach to Online Generation of Association Rules, IEEE Transaction on Knowledge and Data Engineering, Vol. 13 No. 4 (2001) 527-540.
5. Aggarwal, C. C., Yu, P. S.: Mining Associations with the Collective Strength Approach", IEEE Transaction on Knowledge and Data Engineering, Vol. 13 No. 6 (2001) 863-873.
6. Aggarwal, C. C., Yu, P. S.: Redefining Clustering for High-Dimensional Applications, IEEE Transaction on Knowledge and Data Engineering, Vol. 14 No. 2 (2002) 210-225.
7. Aggarwal, C. C., Sun, Z., Yu, P. S.: Fast Algorithm for Online Generation of Profile Associate Rules, IEEE Transaction on Knowledge and Data Engineering, Vol. 14 No. 5 (2002) 1017-1028.
8. IBM Redbooks: Intelligent Miner for Data: Enhance Your Business Intelligence, IBM Corporation (1999).

Image Retrieval Using Dimensionality Reduction

Ke Lu[1], Xiaofei He[2], and Jiazhi Zeng[1]

[1] School of Computer Science and Engineering,
University of Electronic Science & Technology of China, Chengdu, Sichuan 610054, China
[2] Department of Computer Science, University of Chicago,
1100 E 58th Street, Chicago, IL 60637, USA
`xiaofei@cs.uchicago.edu`

Abstract. Image representation has been a fundamental problem for many real world applications, such as image database visualization, browsing, retrieval, etc. In this paper, we investigate the use of Laplacian Eigenmap (LE) for image representation and retrieval. Conventional, Principal Component Analysis (PCA) has been considered effective as to discovering the low dimensional structure of the image space. However, PCA can only discover the linear structure. It fails when the images are sampled from a low dimensional nonlinear manifold which is embedded in the high dimensional Euclidean space. By using Laplacian Eigenmap, we first build a nearest neighbor graph which models the local geometrical structure of the image space. A locality preserving mapping is then obtained to respect the graph structure. We compared the PCA and LE based image representations in the context of image retrieval. Experimental results show the effectiveness of the LE based representation.

1 Introduction

Due to the rapid growth in the volume of digit images, there is an increasing demand for effective image management tools. Consequently, content-based image retrieval (CBIR) is receiving widespread research interest [2][3].

In recent years, much research has been done to deal with the problems caused by the high dimensionality of image feature space [1]. Typically, the dimensions of feature vector range from few tens to several hundreds. For example, a color histogram may contain 256 bins. High dimensionality creates several problems for image retrieval. First, learning from examples is computationally infeasible if it has to rely on high-dimensional representations. The reason for this is known as *curse of dimension*: the number of examples necessary for reliable generalization grows exponentially with the number of dimensions. Learnability thus necessitates dimensionality reduction. Second, in large multimedia databases, high-dimensional representation is computationally intensive and most users are unwilling to wait for results for a long time. Therefore, for storage and efficiency concern, dimensionality reduction is necessary.

Principal Component Analysis (PCA) is a classical technique for dimensionality reduction. It performs dimensionality reduction by projecting the original n-dimensional data onto the $m(<n)$ dimensional linear subspace spanned by the leading

eigenvectors of the data's covariance matrix. Thus PCA builds a global linear model of the data (an m dimensional hyperplane). For linearly embedded manifolds, PCA is guaranteed to discover the dimensionality of the manifold and produces a compact representation in the form of an orthonormal basis. However, PCA fails to discover the underlying structure, if the data lie on a nonlinear sub-manifold.

In this paper, we investigate the use of Laplacian Eigenmap (LE, [1]) for image representation. LE is a recently proposed algorithm for manifold learning. We first build a nearest neighbor graph which models the local geometrical structure of the image space. By using LE, we can obtain a low dimensional representation subspace which best preserves the local information. In many information retrieval tasks, the local information is much more reliable than the global structure. Therefore, the LE based image representation can give optimal performance to image retrieval [2].

The rest of this paper is organized as follows. Section 2 describes Laplacian Eigenmap for dimensionality reduction. In Section 3, we present a theoretical analysis to discuss the connection between Laplacian Eigenmap and Principle Component Analysis. The experimental results are shown in section 4. Finally, we give conclusions in section 5.

2 Laplacian Eigenmap for Image Representation

In this paper, we apply Laplacian Eigenmap [1] to map the images into a low-dimensional space. Consider the problem of mapping the weighted graph G to a line so that connected points stay as close as possible. Here, the vertices of graph G represent the images in high-dimensional Euclidean space, i.e., $x_1, x_2, \ldots, x_n$. If two points are close enough, then there is an edge between them. Let $\mathbf{y} = \{y_1, y_2, \cdots, y_n\}^T$ be such a map. A reasonable criterion for choosing a "good" map is to minimize the following objective function

$$\sum_{ij}(y_i - y_j)^2 W_{ij}$$

W is the weight matrix as follows:

$$W_{ij} = \begin{cases} 1 & \text{if } \mathbf{x}_i \text{ is among } \mathbf{x}_j\text{'s k nearest neighbors} \\ & \text{or } \mathbf{x}_j \text{ is among } \mathbf{x}_i\text{'s k nearest neighbors} \\ 0 & \text{otherwise} \end{cases}$$

The minimization problem reduces to finding

$$\arg\min_{\mathbf{y}, \mathbf{y}^T D \mathbf{y}=1} \mathbf{y}^T L \mathbf{y}$$

where $L = D - W$ is the Laplacian matrix. D is diagonal weight matrix; its entries are column (or row, since W is symmetric) sums of W, $D_{ii} = \Sigma_j W_{ji}$. Laplacian is symmetric, positive semi-definite matrix which can be thought of as an operator on functions defined on vertices of G.

The objective function with our choice of weights W_{ij} incurs a heavy penalty if neighboring points x_i and x_j are mapped far apart. Therefore, minimizing it is an attempt to ensure that if x_i and x_j are "close", then y_i and y_j are close as well.

The algorithmic procedure is formally stated below:

Step 1 [Constructing the adjacency graph] Nodes i and j are connected by an edge if i is among k nearest neighbors of j or j is among k nearest neighbors of i.

Step 2 [Choosing the weights] $W_{ij} = 1$ if and only if vertices i and j are connected by an edge.

Step 3 [Eigenmaps] Assume the graph G, constructed above, is connected; otherwise proceed with Step 3 for each connected component. Compute eigenvalues and eigenvectors for the generalized eigenvector problem:

$$Ly = \lambda Dy \qquad (1)$$

Let $y_0, \ldots, y_{k-1}$ be the solutions of equation 1, ordered according to their eigenvalues, $0 = \lambda_0 < \ldots < \lambda_{k-1}$. We leave out the eigenvector y_0 corresponding to eigenvalue 0 and use the next m eigenvectors for embedding in m-dimensional Euclidean space.

$$x_i \rightarrow (y_1(i), \cdots, y_m(i))$$

where $y_j(i)$ is the ith element of eigenvector y_j.

3 Theoretical Analysis

3.1 Principal Component Analysis

Suppose the sample vectors have a zero mean. Let $X = [x_1, \ldots, x_n]^T$. Thus, the covariance matrix is $C = X^T X$. By eigenvector decomposition, C can be decomposed into the product of three matrices, $C = V\Sigma V^T$, where $\Sigma = diag\{\lambda_1, \lambda_2, \cdots, \lambda_n\}$ are the eigenvalues with descending order and V is a orthogonal matrix, whose column vector is the corresponding eigenvector of C. Thus, $T_{PCA} = V$ is the transformation matrix of PCA.

By applying Singular Value Decomposition (SVD), we can reformulate this problem as follows:

$$\mathbf{X} = U\Sigma V^T$$

where U and V are two orthogonal matrices, whose column vectors are the eigenvector of $\mathbf{XX}^T$ and $\mathbf{X}^T\mathbf{X}$ (the covariance matrix C), respectively, and $\Sigma = diag(\lambda_1, \ldots, \lambda_n)$ are the singular values of $\mathbf{X}$ (also the eigenvalues of $\mathbf{XX}^T$ and $\mathbf{X}^T\mathbf{X}$). Now, we obtain the transformed feature vector in new low-dimensional space:

$$\mathbf{X}' = XT_{PCA} = U\Sigma$$

This indicates that the low-dimensional representation obtained by PCA is just the product of two matrices: (1) the matrix U whose column vectors are the leading *eigenvectors* of weight matrix $W_{PCA} = \mathbf{XX}^T$:

$$W_{PCA}\mathbf{u} = \lambda_{PCA}\mathbf{u},\ U = (\mathbf{u}_1, ..., \mathbf{u}_k) \qquad (2)$$

and (2) the diagonal matrix Σ whose diagonal entries are the leading *eigenvalues* of W_{PCA}, $\Sigma = diag\{\lambda_1, \lambda_2, \cdots, \lambda_k\}$. Note that the weight matrix W_{PCA} is measured by inner product of two image vectors.

3.2 Connection Between Laplacian Eigenmap and PCA

In section 3.1, we have show that the optimal embedding for nonlinear case is obtained by solving a general eigenvector problem below:

$$L\mathbf{y} = \lambda D \Rightarrow (D-W)\mathbf{y} = \lambda D\mathbf{y}$$
$$\Rightarrow W\mathbf{y} = (1-\lambda)D\mathbf{y} \Rightarrow D^{-1}W\mathbf{y} = (1-\lambda)\mathbf{y}$$

Let $W_{Laplacian} = D^{-1}W$, $W_{Laplacian}$ is essentially a normalized weight matrix preserving locality. (Note that the matrix D provides a natural measure on the vertices of the graph. The bigger the value D_{ii} is, the more "important" is that vertex). We rewrite the above formula as follows:

$$W_{Laplacian}\mathbf{y} = (1 - \lambda_{Laplacian})\mathbf{y} \qquad (3)$$

As we can see, the equations (2) and (3) have the same form. The weight matrix used for Laplacian Eigenmap, i.e., $W_{Laplacian}$, is corresponding to the weight matrix used for PCA, i.e., W_{PCA}. The eigenvectors of $W_{Laplacian}$ are corresponding to the eigenvectors of W_{PCA}, which are the low-dimensional representations of data points. This observation shows that, with the same weight matrix, Laplacian Eigenmap and PCA will yield the same result, and $\lambda_{PCA}+\lambda_{Laplacian}=1$. Therefore, the eigenvector of W_{PCA} with the largest eigenvalue is just the eigenvector of $W_{Laplacian}$ with the smallest eigenvalue.

Based on above analysis, we conclude that, the essential difference between Laplacian Eigenmap and PCA is that they choose different weight matrices. PCA uses inner product as a linear similarity measure, while Laplacian Eigenmap uses a nonlinear similarity measure which preserves locality. For PCA, its advantage over Laplacian Eigenmap is that it can produce a transform matrix T_{PCA}. Thus, for a new point, it can be easily map to the new space. The disadvantage is that, it fails to discover the underlying nonlinear structure of data set.

4 Experimental Results

We performed several experiments to evaluate the effectiveness of the proposed approaches over a large image database. The database we use consists of 10,000 images of 79 semantic categories selected from the Corel Image Gallery. It is a large and heterogeneous image set. A retrieved image is considered correct if it belongs to the same category of the query image. Three types of color features and three types of texture features are used in our system. The dimension of the feature space is 435. We designed an automatic feedback scheme to simulate the retrieval process conducted by real users. At each iteration, the system marks the first three incorrect images from the top 100 matches as irrelevant examples, and also selects at most 3 correct images as relevant examples (relevant examples in the previous iterations are excluded from the selection). These automatic generated feedbacks are added into the query example set to refine the retrieval results. To evaluate the performance of our algorithms, we define the retrieval accuracy as follows:

$$Accuray = \frac{\text{relevant images retrieved in top N returns}}{N}$$

4.1 2-D Data Visualization

In this experiment, the dataset contains two image classes. One contains image data of human eyes, and the other contains image data of animal eyes. Each class contains 1500 images in 400-dimensional space, and each image is an 8 bits (256 grey) image of size 20×20. The 2-D embedding results are shown in Figure 1. As can be seen, Laplacian Eigenmap performs much better than PCA.

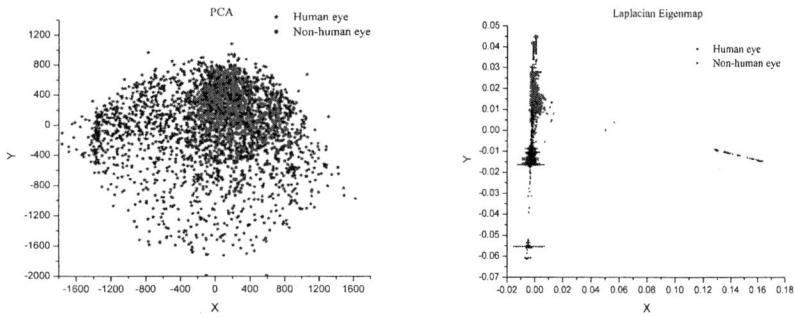

Fig. 1. 2-D embedding results of PCA and Laplacian Eigenmap

4.2 Image Retrieval in Dimensionality-Reduced Space with Different Dimensions

In this subsection, we compared the performance of image retrieval in dimensionality-reduced space with different dimensions, with the results shown in Figure 1 and Figure 2. As the feature space is drastically reduced by PCA, the performance decreases fast. At the 0th iteration, there is no user's relevance feedback introduced. Hence the retrieval result is only based on the feature representation in low-dimensional space. As can be seen, Laplacian Eigenmap performs much better than PCA. In PCA space with 2 dimensions, only about 10% accuracy is achieved after three iterations in the top 10 returns. While in 2-dimensional laplacian space, we can still achieve more than 40% accuracy after *one* iteration. This observation shows that Laplacian Eigenmap algorithm is especially suitable for the case where drastic dimensionality reduction needs to be performed. When the original space is reduced to a 100-dimensional space, we can see that the retrieval accuracy at the 0th iteration is 36% in PCA space and 39% in laplacian space, respectively.

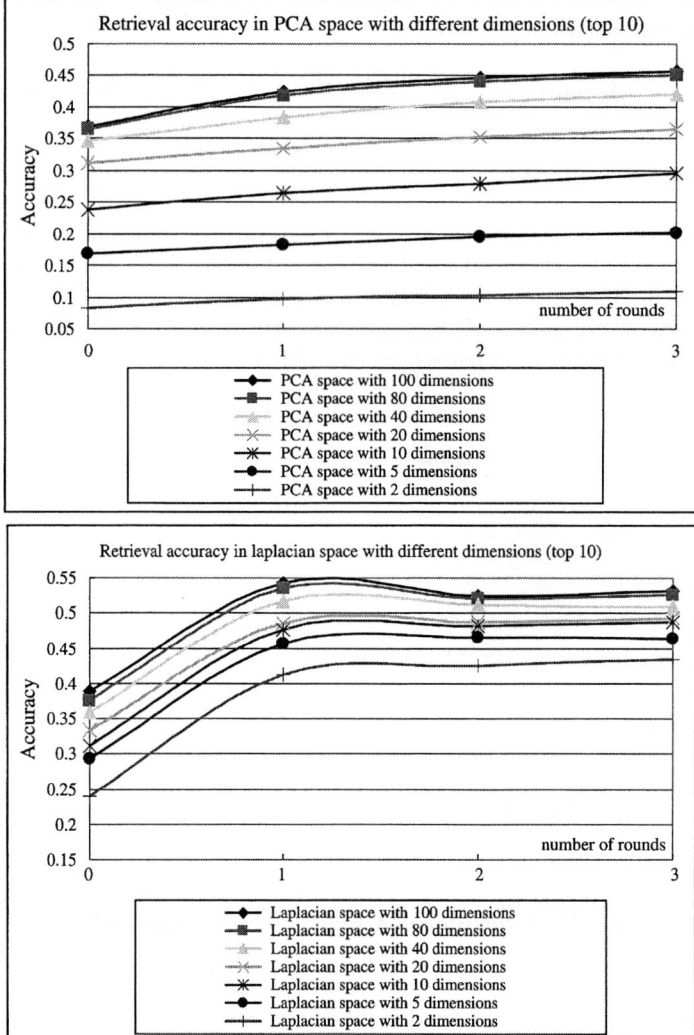

Fig. 2. Retrieval performance in PCA and Laplacian subspaces

5 Conclusions

In this paper, we investigated the use of Laplacian Eigenmap for image retrieval. We presented a theoretical analysis of the connection between Principal Component Analysis and Laplacian Eigenmap. Theoretical analysis shows that PCA discovers the global linear structure of the image space, while Laplacian Eigenmap discovers the nonlinear geometrical structure of the image space. Experimental results show that the Laplacian Eigenmap based image representation can give excellent performance for image retrieval.

References

1. Belkin, M., and Niyogi, P., "Laplacian Eigenmaps and Spectral Techniques for Embedding and Clustering", *Advances in Neural Information Processing Systems* 14, Vancouver, Canada, 2001.
2. He, X., King, O., Ma, W.-Y., Li, M., and Zhang, H.-J., "Learning a Semantic Space from User's Relevance Feedback for Image Retrieval", *IEEE Trans. On Circuit and Systems for Video Technology*, vol 13, No. 1, 2003.
3. Rui, Y., Huang, T. S., and Chang, S.-F., "Image Retrieval: Current Techniques, Promising Directions and Open Issues", *Journal of Visual Communication and Image Representation*, vol. 10, 39-62, 1999.

Three Integration Methods for a Component-Based NetPay Vendor System

Xiaoling Dai[1] and John Grundy[2]

[1] Department of Mathematics and Computing Science,
The University of the South Pacific, Laucala Campus, Suva, Fiji
dai_s@usp.ac.fj
[2] Department of Electrical and Computer Engineering and Department of Computer Science,
University of Auckland, Private Bag 92019, Auckland, New Zealand
john-g@cs.auckland.ac.nz

Abstract. We have developed NetPay, a micro-payment protocol characterized by off-line processing, customer anonymity and relatively high performance and security using one-way hashing functions for encryption. In our NetPay prototypes we have designed and implemented two kinds of NetPay vendor systems which use thin-client user interfaces, a component-based server-side infrastructure and CORBA remote objects for inter-system communications. We describe three alternative ways to integrate NetPay interface facilities into these existing E-Journal web applications. We describe the relative strengths and weaknesses with each approach and our experiences building prototypes of them.

Keywords: Electronic commerce; Micro-payment; System integration.

1 Introduction

World-wide proliferation of the Internet led to the birth of electronic commerce, a business environment that allows the transfer of electronic payments as well as transactional information via the Internet. However, the problem of paying for large-volume, small-value items of information is still to be solved. We have developed a new micro-payment protocol called NetPay [1]. The NetPay protocol allows customers to buy E-coins, worth very small amounts of money, from a broker and spend these E-coins at various vendor sites to pay for large numbers of discrete information or services of small value each. There are a number of other micro-payment systems such as PayWord [5], Millicent [4], and PayFair [6]. However, most existing or proposed micro-payment technologies suffer from problems with communication, security, lack of anonymity and being overly vendor-specific. We have developed the NetPay protocol and supporting architecture to address problems with communication, security, anonymity and vendor-specific. In this paper, we give an overview of existing micro-payment models and point out the problems with these models. We then briefly describe our CORBA-based and component-based NetPay micro-payment system design. We then describe the designs for the three ways to integrate NetPay interface facilities into an existing web application. We conclude with an outline of our further plans for research and development in this area.

2 CORBA-Based NetPay Architecture

We initially developed a software architecture for implementing NetPay-based micro-payment systems for thin-client web applications that used hard-coded vendor facilities for micro-payments [2]. NetPay additions to the web components included pricing for items, pay-per-click for each item, a server-side e-coins spent database, and nightly redemption of spent e-coins with a broker in exchange for real money. The existing web server components' code was modified to interact with the NetPay functions when needed. There are some major disadvantages to this approach:

- It is difficult and time consuming to add NetPay support to existing applications.
- The reusability level is lower. For example, when some NetPay functions and interfaces are fitted into an E-newspaper system developers can't reuse the same NetPay objects into another existing web application without modification.
- Enhancement of the NetPay functions means modifying the web components.

To overcome these disadvantages, a component-based NetPay vendor system was developed. One of the characteristics of such a system is that its components can be plugged into an existing web system. The new system should match the needs of existing systems and have NetPay micro-payment support integrated seamlessly with minimum effort with their existing web application architecture. We need to enhance the existing, domain-specific components of an existing NetPay vendor system and, using plug-and-play, add the NetPay components to the existing web application. In an E-journal example, the journal provider (vendor) would want to charge small amounts on a per-article basis (perhaps varying amounts). The main purpose of the component-based NetPay vendor system was to separate the NetPay EJBs from the particular domain knowledge of the web application, enabling each enterprise bean to be reused in different EJB-based vendor systems via plug-and-play with the existing vendor components. In addition, the E-journal's user interface must be annotated to display E-coin balance, article cost, credit checks and coin debits to the NetPay EJBs.

3 Component-Based NetPay Architecture

We have developed a set of Enterprise Java Beans (EJBs) that capture the functionality of the NetPay micro-payment system [•3]. These include components such as:

- ArticlePrice. This provides pricing for items or services sold by the vendor.
- EwalletController. This provides sever-side e-coin management, either by managing a customer's e-wallet or providing a CORBA interface to an e-wallet hosted on the customer's own PC.
- RedeemController. Provides nightly redemption functionality to broker.

We have used CORBA to enable the NetPay EJB application servers to access the broker application server to obtain touchstone information to verify the e-coins being spent and to redeem spent e-coins. A dispatcher is used to forward web browser requests to JSP pages. Fig. 1. shows some required interactions of existing E-journal system components (grey) with some NetPay components.

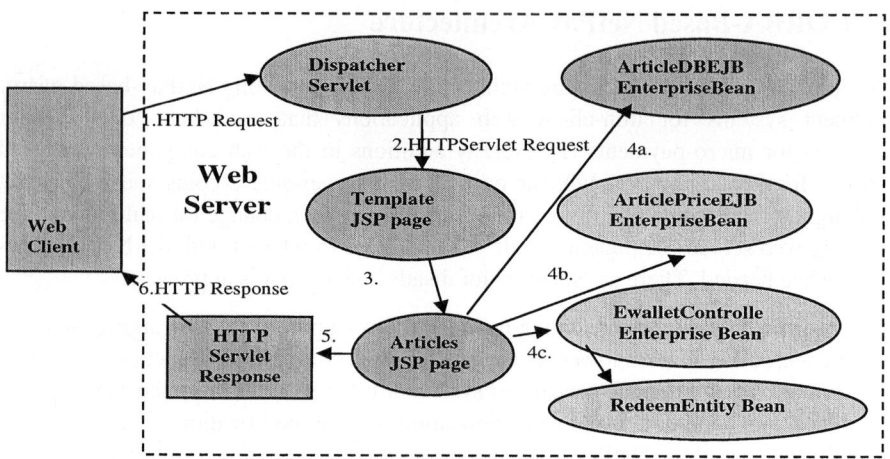

Fig. 1. Web component interaction after modified article.jsp

4 NetPay Integration with E-Journal Web Pages

In order to add our NetPay micro-payment facility to the E-journal, or to other 3rd party J2EE-based applications, we need to be able to add our EJBs to their J2EE server and to detect when pages are being accessed by customers that need to be paid for. We also need to ensure that if the customer attempting to access does not have enough e-coins they are directed to the NetPay broker site to buy some more. There are three main ways to integrate the NetPay user interface facilities: (1) modify the existing system web pages to incorporate NetPay information; (2) generate web pages that display the existing system pages in frames and make appropriate interactions with NetPay EJB components; and (3) generate proxy web pages that interact with NetPay session beans and redirect access to the original web pages.

4.1 Modifying the Existing System Web Pages

In this approach the articles.jsp is modified to retrieve price data from Article-PriceEJB enterprise bean for displaying article price information or retrieve e-wallet data (for server-side NetPay) from e-wallet enterprise bean for displaying e-wallet information. Fig. 1 depicts the interaction between these Web components. A HTTP request (1) is delivered to the dispatcher component which processes and then forwards the HTTPServlet request (2) to the template.jsp. The template.jsp generates the response (3) by including the responses from Articles JSP page. Articles JSP page retrieves article contents from the article enterprise bean (4a) and article price from the article price enterprise bean (4b), and e-wallet data from e-wallet enterprise bean (4c). Articles JSP page transmits responses (5 and 6) to the client for presentation.

The content.jsp is modified to make payment from e-wallet enterprise bean in order to debit e-coins paying for article content and Login.jsp is implemented (for server-side NetPay) to retrieve e-wallet data from e-wallet enterprise bean. Fig. 2 depicts the interaction between these components.

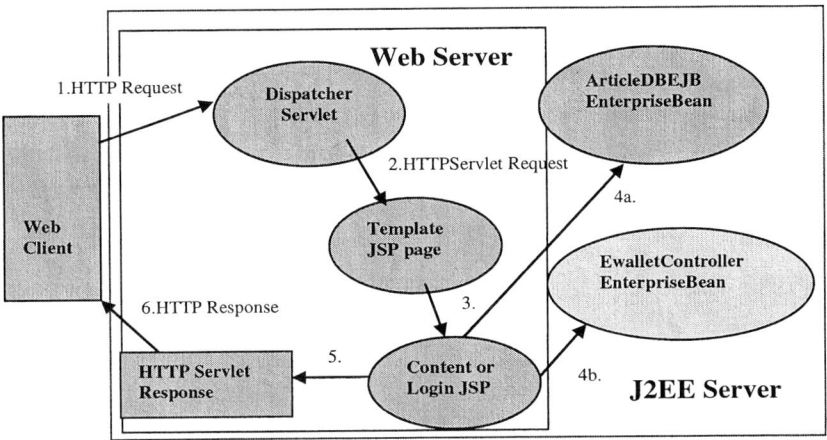

Fig. 2. Web component interaction after modified and implement JSP pages

A HTTP request (1) is delivered to the dispatcher component which processes and then forwards the HTTPServlet request (2) to the template.jsp. The template.jsp generates the response (3) by including the responses from Content or Login JSP page. Content JSP page retrieves article contents from the article enterprise bean (4a) and Login JSP page retrieves e-coin ID and password from the e-wallet enterprise bean (4b). Articles or Content JSP page transmits responses (5 and 6) to the client.

This approach requires updates to the existing system web page implementations. For example, in the journal example system, article.jsp needs to be modified to interact with the price and the e-wallet enterprise beans for displaying the costs of articles and e-wallet balance. content.jsp was modified to debit e-coin from the e-wallet by interacting with the e-wallet enterprise bean before displaying an article content. This can be done easily and possibly by code injection into the existing JSPs by a tool.

4.2 Generating NetPay JSP Pages

In this approach NetPay JSP pages are generated to interface to existing E-journal pages. An HTTP request (1) is delivered to a NetPay JSP page which displays article.jsp in frames, after retrieving article price data from the article price enterprise bean (3) and e-wallet data from the e-wallet enterprise bean (4). The article.jsp retrieves articles' title and author data from the article enterprise bean (2). NetPay JSP pages display the articles and e-wallet information to the client (5). Fig. 3 depicts the interactions between these components. NetPay JSP pages also display content.jsp in frames and interact with the e-wallet enterprise bean in order to debit e-coins.

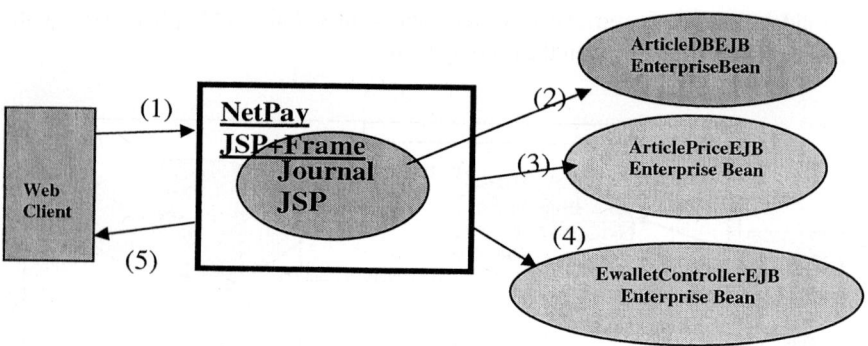

Fig. 3. Generating NetPay JSP pages

This approach has the advantage that no code changes to the original JSPs are needed and the NetPay pricing information can be displayed in a separate frame to existing information. However, this separation of pricing of items from item descriptions may not be ideal when large numbers of items are displayed together e.g. for large search results, table of contents, news headlines etc.

4.3 Generating NetPay Proxy JSP Pages

In this third approach NetPay JSP pages are again generated. These however act as "proxies" to the original web-based system's JSP pages. An HTTP request (1) is delivered to the NetPay proxy pages which obtain article price data from the article price enterprise bean (2) and e-wallet data from the e-wallet enterprise bean (3) for displaying costs of the articles and e-wallet information. NetPay proxy JSP pages then redirect to article.jsp accessing the journal home page (4). When a customer wants to read an article content, NetPay proxy JSP pages interact with the e-wallet enterprise bean to debit e-coins from customer's e-wallet (3) and then redirect to content.jsp which retrieve article content from article enterprise bean (5). Finally NetPay

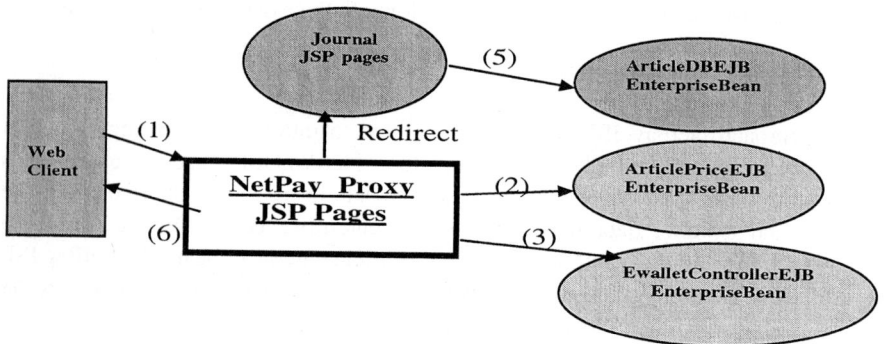

Fig. 4. Generating NetPay proxy JSP pages

proxy JSP pages display the article content to the client (6). Fig. 4 illustrates the interaction between these components.

This approach hides the NetPay functionality from the user without requiring code changes. However, displaying article price information, information about e-coins left in wallet/spent and so on has to be done by the proxy before forwarding to the original pages. For sites with complex multi-JSP page interactions, this can be intrusive.

5 Summary

We have built NetPay vendor Enterprise JavaBeans to provide plug-in vendor micropayment support components and plugged in EJBs into the E-journal's existing application server and developed three techniques to have the E-journal's JSPs to make appropriate function calls to the NetPay EJBs. These allow for minimal code impact to the existing system's infrastructure. The NetPay vendor system components have been designed, implemented, plugged into the E-journal example system, and successfully deployed to a J2EE server running the E-journal web site.

References

1. Dai, X. and Lo, B.: NetPay – An Efficient Protocol for Micropayments on the WWW. Fifth Australian World Wide Web Conference, Australia, 1999 .
2. Dai, X., Grundy, J.: Architecture of a Micro-Payment System for Thin-Client Web Applications. In Proceedings of the 2002 International Conference on Internet Computing, Las Vegas, CSREA Press, June 24-27, 444-450.
3. Dai X. and Grundy J.: Architecture for a Component-based, Plug-in Micro-payment System, In Proceedings of the Fifth Asia Pacific Web Conference, LNCS 2642, Springer, April 2003, pp. 251-262.
4. Manasse, M.: The Millicent Protocols for Electronic Commerce. First USENIX Workshop on Electronic Commerce. New York, 1995.
5. Rivest, R. and Shamir, A.: PayWord and MicroMint: Two Simple Micropayment Schemes. Proceedings of 1996 International Workshop on Security Protocols, LNCS 1189. Springer, 1997, 69—87.
6. Yen, S-M.: PayFair: a prepaid internet ensuring customer fairness micropayment scheme. IEE Procs-E Computers & Digital Techniques, vol.148, no.6, Nov. 2001, pp.207-13.

A Case Study on the Real-Time Click Stream Analysis System

Sangkyun Kim[1] and Choon Seong Leem[2]

[1] Somansa, Woolim e-Biz, Yangpyeongdong, Yeongdeungpogu, Seoul, Korea
`saviour@somansa.com`
[2] Department of Computer and Industrial Engineering, Yonsei University, 134, Shinchondong, Seodaemoongu, Seoul, Korea
`leem@yonsei.ac.kr`

Abstract. The Internet is one of the most significant and efficient ways for communication technology which has the potential to revolutionize business and marketing techniques. To understand efficiently who is visiting our Web site and how they are using it is the primary key for the success of the marketing based on Internet. The traffic analysis systems via Web server log files only provide partial and non-real time information about customer activities on your Web site, it's insufficient to support better business decisions which need detailed and real time information about customers' activities on your Web site. In this paper, we provide our experiences on an implementation of the sophisticated analysis system based on click stream analysis technologies. A case study is also provided to show practical values of this system.

1 Introduction

Business intelligence involves the systematic collection and structuring of data to provide reliable insights into an organization's concerns. Many enterprises operate CRM (Customer Relation Management) for business intelligence in marketing activities. CRM can enable the marketing activities more effectively by creating intelligent opportunities for cross-selling and faster introductions for new products [1, 2, 3]. In e-Business environments, it is the most important in CRM that catch which actions customers take in your Web site. To do this, companies usually spend a lot of time, attention, and money on the Web. However, only few of them really understand the payback on this investment. It's due to the fundamental difference of the Web from other marketing vehicles in two key ways [4]: 1) The Web is anonymous: In traditional marketing, it is relatively easy to construct a demographic and psychographic profile of target audience. With the Web, a company doesn't know much about its visitors because most of them are anonymous; 2) The Web is interactive: Web sites receive information on what pages are being searched by each visitor. More sophisticated sites form a marketing database by collecting demographic and psychographic information as visitors browsing. The ultimate goal of Web traffic analysis is to combine anonymous Web traffic information with traditional demographic and psychographic business information. This information is the foundation of personalized one-to-one marketing techniques, allowing a business to target specific audiences with customized products and services. There are many challenges in performing effective Web traffic analysis [5]. These include: 1) Measuring and modeling data from complex Web sites; 2) Keeping up with ever higher volumes of traffic; 3) Integration of multiple sources of data, including Web data from distributed locations and from other e-Business and enterprise data; 4) Applying appropriate business rules, such as which URLs or types of URLs to be considered, and which traffic to be excluded from the analysis and reports; 5) Interpretation of URLs generated dynamically and difficult to analyze. To solve these problems, we provide our experiences on an implementation of the sophisticated analysis system based on click stream analysis technologies. It includes a design and implementation of the system, and case study which provides practical values of this system.

2 Design and Implementation of the System

The techniques to capture users' profile information are various. These techniques usually include "asking the customer (fill-in-profile, explicit feedback or ratings)" and "watching the customer [6]." Conversational techniques are suggested by Ginsburg [7].

Table 1. Modules' characteristics of the real-time click stream analysis system

Module	Description
Engine: Analyzer	It passively watches network traffic to and from the site's Web servers, providing the benefits of network measurement such as the ability to do network timings, detect cancelled requests, and determine the total amount of information sent by Web servers in fulfillment of requests.
Engine: DB Adaptor	It provides the connectivity to the legacy DB.
Manager: Reporter	It presents statistical information that are gathered and analyzed through Analyzer. It provides various tables and charts.
Manager: OLAP	OLAP (On-Line Analytical Processing) provides additional views of the data to support qualitative analysis.
Manager: Content Manager	It manages configuration information about contents that need to be extracted and analyzed by Analyzer. It is designed to define various information like site-map, contents category, external path to visit, product category, and user's activities.
Manager: Administrator	It configures the various functions of Analyzer in selective and effective manners. Also, it provides administrative functions for Web site manager to manage other modules of the real-time click stream analysis system.
Manager: Monitor	It supports to check Web traffic of a certain site, status of analysis engine (failure of Analyzer), and the status of database.

In this paper, we use "watching the customer" technique. There are essentially three ways to trace and to analyze how users interact with a Web site: 1) Packet sniffing - logging and analyzing the IP traffic between the user (browser) and Web server; 2) Log file analysis - analyzing the log files generated by the Web server; 3) Page tagging - logging traffic to a remote server by tagging each page requested from the site and then analyzing the logged data. The Web server log data is designed for logging and transaction, not for the support of analytics. The log does not contain data that allow customers' activities to be analyzed at a content level [8]. In this paper, we take the packet sniffing mechanism to design the click stream analysis system. Web servers store the records of every visitor action, both clicks and letters typed, in their log files. Although there are other ways to collect click stream data, click stream analysis typically uses the Web server log files to monitor and to measure Web site activity. Click stream analysis reports users' behaviors on a specific Web site: routing, stickiness, the place where users come from. It also can be used for more aggregated measurements, such as the number of hits (visits), page views, and unique or repeating visitors which are valuable to understand how the Web site operates from a technical, user experience and business perspective. Significantly, click stream analysis does not support analysis of an enterprise's total relationship with a visitor or customer. Other information such as users' registration or customer's orders is necessary - obtaining such information requires integration with other data sources. However, once customers or visitors are identified, click stream data can be used to classify them better and to understand the greater relationship deeply [9]. The real-time click stream analysis system consists of two parts: engine and manager. The characteristics of modules of the real-time click stream analysis system are described in Table 1.

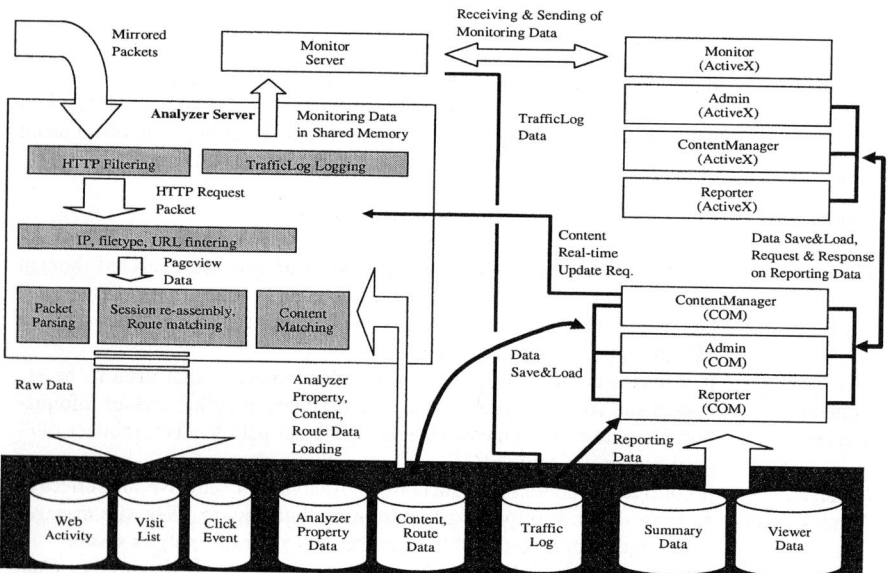

Fig. 1. Data processing diagram between modules of the real-time click stream analysis system

Data processes between these modules are shown in Fig. 1. Here are the overall processes.

Step-1. Contents related to customer activities to be analyzed are configured with Content Manager.
Step-2. All packets are mirrored and gathered by Analyzer.
Step-3. Gathered packets are stored in DB.
Step-4. Stored packets in DB are analyzed with Monitor, Reporter and OLAP.
Step-5. Each module of the real-time click stream analysis system are managed by Administrator.

3 Operation Results

In this case study, we applied the real-time click stream analysis system to a particular project in which XYZ Co., Ltd. wanted to implement Web-based customer activity analysis system. XYZ has operated B2C shopping site for several years.

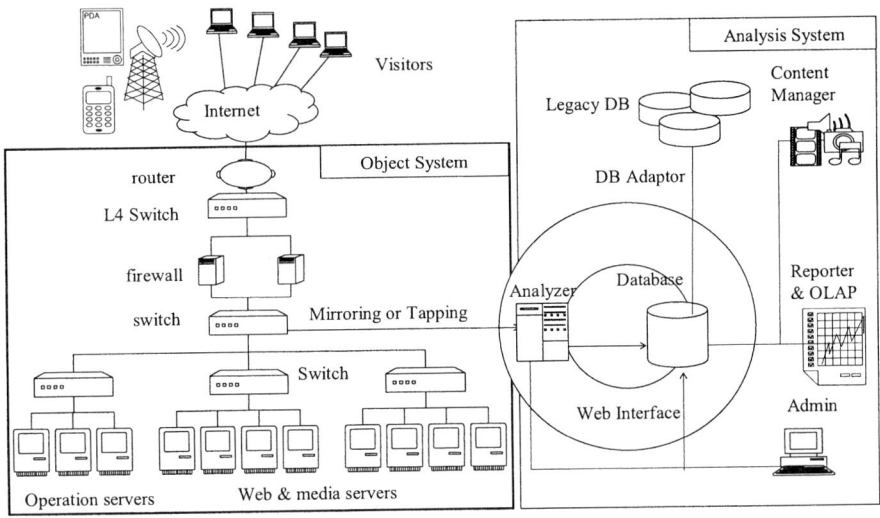

Fig. 2. Operation architecture of the real-time click stream analysis system

The real-time click stream analysis system was installed in XYZ's EIS environments. The operation architecture is shown in Fig. 2. The analyzer was connected to main switching system to gather real-time click stream data. The DB adaptor wasinstalled to operate with customers' information stored in a legacy DB. The contents on B2C site were configured to make customers' activities to be analyzed with the content manager. With the administration manager, reporter and OLAP, administrators of XYZ could manage the real-time click stream analysis system. We have operated the real-time click stream analysis system in XYZ during three months, and acquired various results. The sample reports generated during this period are displayed in Fig. 3. These sample reports were generated by the click reporter module.

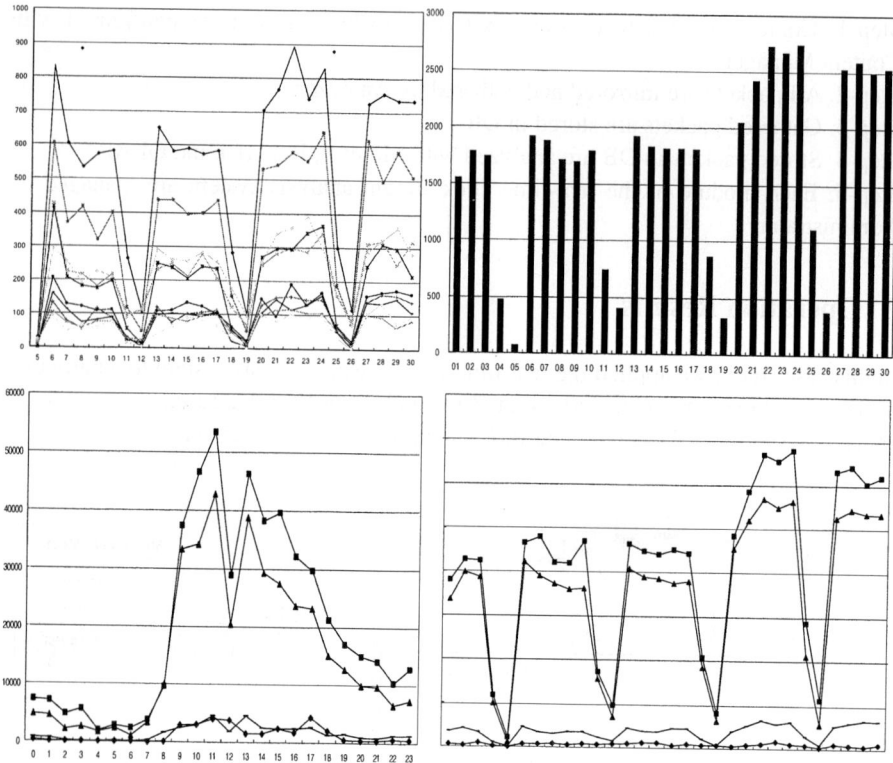

Fig. 3. Operation reports generated by the real-time click stream analysis system installed in XYZ Co., Ltd

The sample reports present significant implications. The implications of each report are as follows: 1) Report-A represents what ISP provides heavier traffics. It's very important because XYZ should invest to increase a network bandwidth of the ISP who delivers heavier traffics than other ISPs. In report-A, we can find that two top-ranked ISPs deliver more than half of total traffics that are generated by ten ISPs; 2) Report-B represents which day of week or month has heavy traffics. In every weekend, number of visitors was decreased significantly, so we suggested XYZ to manage maintenance jobs in every weekend because the most of customers visit their sites on weekdays; 3) Report-C represents daily pageviews of four age groups. The pageviews of all age groups are significantly greater in day time than in night time. This pattern is more remarkable in two age groups (20~29, 30~39). 4) Report-D provides more detailed view of report-B. It classifies daily traffics of report-B into four age groups of report-C.

The effectiveness and efficiency of the real-time click stream analysis system installed in XYZ are summarized in table 2. These results are compared with the results of the benchmarking test of the Web server log-based analysis system which have been written and authorized by the staffs and the managers of XYZ with their past experience of solution evaluation.

Table 2. Effectiveness and efficiency of the system installed in XYZ Co., Ltd

Factor	Real-time click stream analysis	Web server log-based analysis
Analysis time of 1 day streams	30 minutes	6 hours
CPU load on Web server	0% increased	35% increased
Memory load on Web server	0% increased	0.98% increased

4 Conclusion

It has long been said that you cannot manage what you cannot measure. This is truer on the Web - where examining which works and what doesn't directly influences the bottom line [10].

In this paper, we provide our experience on an implementation of the sophisticated analysis system based on click stream analysis technologies. It provides sufficient information on customers' activities to support better business decisions. With this system, the Web site manager can analyze Web traffics easily and effectively to improve the effectiveness of CRM activities. However, to have sufficient information on customers' activities and to utilize these informations for CRM are different issues. The companies which implement the real-time click stream analysis system must try for better utilization of analyzed information. Further researches should be focused on the better utilization of analyzed information generated by the real-time click stream analysis system.

References

1. Grant, A. W. H., Schlesinger, L. A.: Realize Your Customers' Full Profit Potential. Harvard Business Review (1995) 58–72
2. Hill, D.: Love My Brand. Brandweek (1998)
3. Ruediger, A., Grant-Thompson, S., Harrington, W., Singer, M.: What Leading Banks are Learning about Big Databases and Marketing. McKinsey Quarterly (1997)
4. Web Mining Whitepaper: Driving Business Decisions in Web Time. Accrue Software (2000)
5. eBusiness Analysis and Accrue Insight. Accrue Software (1999)
6. Koch, M., Schubert, P.: Personalization and Community Communication for Customer Support. 6th International Conference on Work with Display Units - World Wide Work (WWDU2002), Berchtesgaden, Germany (2002)
7. Ginsburg, M.: Realizing a Framework to Create, Support, and Understand Virtual Communities. Infonomics/Merit Workshop on Digitization of Commerce: e-Intermediation, Maastricht, Netherlands (2001)
8. Knox, M.: Clickstream Dimensions: Information System Requirements. Gartner (2001)
9. Knox, M., Buytendijk, F.: By definition: Web Analytics Need Explaining. Gartner (2001)
10. Buytendijk, F.: Web Analytics: So What? Gartner (2001)

Mining Medline for New Possible Relations of Concepts

Wei Huang[1,2], Yoshiteru Nakamori[1], Shouyang Wang[2], and Tieju Ma[1]

[1] School of Knowledge Science, Japan Advanced Institute of Science and Technology,
Asahidai 1-1, Tatsunokuchi, Ishikawa, 923-1292, Japan
{w-huang, nakamori, tieju}@jaist.ac.jp
[2] Institute of Systems Science, Academy of Mathematics and Systems Sciences,
Chinese Academy of Sciences, Beijing, 100080, China
{whuang, swang}@mail.iss.ac.cn

Abstract. Scientific bibliographies in online databases provide a rich source of information for scientists in support of their research. In this paper, we propose a new method to predict possible relations between a starting, known concept of interest and other concepts by mining scientific literature databases like Medline. The central novel feature of our method is to predict new relations based on finding brothers of middle concepts within the concept hierarchy. The method can help researchers explore new research directions from current scientific literature.

1 Introduction

The availability of scientific bibliographies in online databases is a rich source of information for scientists to support their research. Thus, the process of efficiently discovering knowledge from the huge collection of scientific literature has begun to attract more and more attention. Swanson first proposed that complementary but disjointed non-interactive structures in the literature of science do exist and can lead to novel scientific hypotheses that are worth testing [6-9](see also Figure 1). Various methods were developed to systematically analyze scientific literature in order to generate novel and plausible hypotheses [1-5, 10-14]. Most research on literature-based discoveries is based on Swanson's theory. Two concepts are assumed to be related, if they co-occur in the same scientific literature.

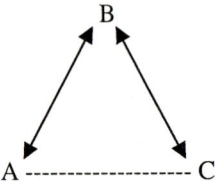

Fig. 1. The relation AB and BC are reported separately in the documents. The implicit relation AC is a putative new discovery

In this paper, we propose a new method to predict potential relations between A and C by mining scientific literature databases like Medline. The remainder of this paper is organized as follows. In Section 2, we describe our method in detail. In Section 3, we apply the method to ten diseases and compare the results with those from the previous research. Some conclusions are given in Section 4.

2 Our Method

As in most of the previous research on literature-based discoveries, we choose biomedicine as the field in which to demonstrate our method.

2.1 Medline and MeSH Tree

Medline is a premier source for bibliographic coverage of biomedical literature produced by the U.S. National Library of Medicine (NLM). An example of a Medline citation is shown as follows:

PMID- 8550811
......
MH - Multiple Sclerosis/*genetics/ immunology
MH - Optic Neuritis/*genetics/immunology
MH - Polymerase Chain Reaction
......

"PMID" and "MH" mean unique identification number, Medical Subject Headings (MeSH) associated to each Medline citation. MeSH terms are the controlled vocabulary created, maintained and provided by NLM. Indexers always choose the most specific MeSH terms available to describe the subject content of a Medline citation. MeSH terms provide a consistent way to retrieve information that may use different terminology for the same concepts. NLM arranges MeSH terms with their associated tree numbers in a hierarchical structure way, called "MeSH Tree". The sample text from MeSH Trees is shown as follows:

> Reptiles;B02.833
>   Alligators and Crocodiles;B02.833.100
>   Dinosaurs;B02.833.150
>   Lizards;B02.833.393
>   Iguanas;B02.833.393.500
>   Snakes;B02.833.672
>   Boidae;B02.833.672.250
>   Colubridae;B02.833.672.280

Brother concepts are concepts that belong to the same direct super-concept in a hierarchical structure of concepts. According to the tree number associated with each MeSH term, we can easily find its brother concepts in MeSH Tree. For example, from the above sample text, we can know that the concept "Alligators and Crocodiles" has at least three brothers, namely "Dinosaurs", "Lizards", "Snakes". They belong to the same direct super-concepts "Reptiles".

2.2 Algorithm

Step 1. Let A be a given starting concept of interest.

Step 2. Find all the middle concepts B that co-occur with A in a given set of scientific literature.

Step 3. Find all targeting concepts C that are brothers of B in the hierarchical structure of concepts.

Step 4. Eliminate those targeting concepts C that already co-occur with A in the given set of scientific literature.

Step 5. The remaining targeting concepts C are candidates for new relations between A and C.

Since relation AB has been already reported in scientific literature, while B and C are similar to each other (brother relation), we suppose that the relation AC will probably hold and be reported in the future.

2.3 Combinatorial Problems and Filtering Functions

Because a starting concept A may co-occur with more than one middle concept B in a body of scientific literature, and a middle concept B may also have more than one brother C, the combination result will produce a considerable number of targeting concepts. In order to deal with this combinatorial problem, it is necessary to add filtering functions, which can be interactively enforced by the users. For example, a filtering function is to set the threshold of the co-occurring frequency of a starting concept A and middle concepts B. The threshold can be set to the average co-occurring frequency of a starting concept A and middle concepts B (AVGS), or 2×AVGS. We may only consider those middle concepts B with a co-occurring frequency exceeding the threshold and search the brother concepts of those B in a hierarchical structure of concepts.

3 Experiment

In order to clearly demonstrate our method, we apply it to ten diseases used as starting concepts A shown in Table 1, and compare the prediction results with those of Hristovski et al [1].

3.1 Experiment Preparation

We divide the Medline database into two segments according to the publication date of literature, namely the old segment (1990-1995) and the new segment (1996-1999). First, we search Medline to collect the citations indexed with the starting concepts in the two segments, respectively. Second, we apply our method on the citations from the old segment to predict targeting concepts that form the new relations with a starting concept. If a targeting concept later co-occurs with the starting concept in at least one citation from the new segment, the new relation between the two concepts is a successful prediction. We use Recall and Precision to measure prediction performance.

$$\text{Recall: } R = \frac{l}{k} \quad (1)$$

$$\text{Precision: } P = \frac{l}{m} \quad (2)$$

where R is the ratio of successfully predicted new relations to all new previously unreported relations; l is the number of successfully predicted new relations; k is the number of relations in the new segment that are not reported in the old one; P is the ratio of successfully predicted new relations to all predicted relations; m is the number of targeting concepts that form new relations with starting concepts, based on the old segment.

3.2 Experiment Results

Table 1 and 2 show the prediction results of new relations between starting concepts and targeting concepts by using our method and Hristovski's method, respectively. Column k is the number of new relations in the new segment that are not present in the old one. Column m is the number of targeting concepts that form new relations with starting concepts, based on the old segment. Column l is the number of successfully predicted new relations. Column R and P are Recall and Precision, respectively. The column headings with subscript 1 and 2 are prediction results when the threshold of the co-occurring frequency of a starting concept A and middle concepts B is set to AVGS and 2×AVGS, respectively.

Here we analyze the results for the starting concept *multiple sclerosis* (MS). MS co-occurs with 807 new concepts in the new segment. When the threshold is AVGS, our method proposes 2424 targeting concepts that form new relations with the starting concept MS. 223 of the 2424 targeting concepts belong to the 807 new concepts. That is to say, our method successfully predicts 223 new relations. Therefore, Recall is 0.28 and Precision is 0.09. When the threshold is 2×AVGS, the number of targeting concepts drops to 1571; 152 of them belong to the 807 new concepts. Therefore, Recall is 0.19 and Precision is 0.10. In other words, increasing the threshold can reduces Recall greatly, but improves Precision a little. It should be noted that we only give examples of setting threshold value to limit the number of new possible relations of concepts. In practical use, researchers can select appropriate threshold values to obtain a suitable number of new relations.

Let's compare the prediction results of the two methods. First, our method's Recall is lower than those of Hristovski's method. However, the values of m_1 and m_2 in Table 1 are both smaller than those in Table 2 for the corresponding diseases. That is to say, our method proposes fewer targeting concepts to form new relations with a starting concept. Second, when the threshold is AVGS, in eight cases our method's Precision is higher; in the other two cases, Precision for the two methods is the same. Third, when the threshold is 2×AVGS, our method's Precision is higher in four cases and lower in four cases; in the last two cases, Precision for the two methods is the same. It indicates that our method performs a little better in term of Precision. In fact, both of the two methods propose a considerable number of targeting concepts. Users need to choose some of the candidates as actual research themes based on their research interest and background knowledge. Therefore, users are more interested in the quality of such candidates instead of the quantity. In other words, Precision is more important than Recall in measuring the prediction performance.

Table 1. The prediction result of our method

Starting Concept A	k	m_1	l_1	R_1	P_1	m_2	l_2	R_2	P_2
multiple sclerosis	807	2424	223	0.28	0.09	1571	152	0.19	0.10
temporal arteritis	210	879	33	0.16	0.04	585	24	0.11	0.04
melanoma	975	2469	272	0.28	0.11	1666	190	0.19	0.11
parkinson disease	761	1741	150	0.20	0.09	1161	109	0.14	0.09
incontinentia pigmenti	50	275	4	0.08	0.01	158	4	0.08	0.03
chondrodysplasia punctata	38	379	6	0.16	0.02	202	4	0.11	0.02
charcot-marie-tooth disease	170	364	28	0.16	0.08	238	16	0.09	0.07
focal dermal hypoplasia	28	142	2	0.07	0.01	98	1	0.04	0.01
noonan syndrome	77	527	28	0.36	0.05	240	12	0.16	0.05
ectodermal dysplasia	198	456	23	0.12	0.05	272	19	0.10	0.07

Table 2. The prediction result of Hristovski et al

Starting Concept A	k	m_1	l_1	R_1	P_1	m_2	l_2	R_2	P_2
multiple sclerosis	635	6848	521	0.82	0.08	3151	366	0.58	0.12
temporal arteritis	187	4735	148	0.79	0.03	1157	72	0.39	0.06
melanoma	692	6272	560	0.81	0.09	2812	392	0.57	0.14
parkinson disease	594	5995	477	0.80	0.08	2322	309	0.52	0.13
incontinentia pigmenti	44	3435	37	0.84	0.01	873	23	0.52	0.03
chondrodysplasia punctata	18	2864	15	0.83	0.01	1046	9	0.50	0.01
charcot-marie-tooth disease	131	3150	105	0.80	0.03	1019	66	0.50	0.06
focal dermal hypoplasia	23	1511	14	0.61	0.01	610	8	0.35	0.01
noonan syndrome	68	3015	59	0.87	0.02	536	23	0.34	0.04
ectodermal dysplasia	124	3301	96	0.77	0.03	967	45	0.36	0.05

4 Conclusions

In this paper, we propose a new method to predict possible relations between a starting, known concept of interest and other concepts by mining scientific literature databases like Medline. The central novel feature of our method is to predict new relations based on finding brothers of middle concepts within the concept hierarchy. The method can help researchers explore new research directions from current scientific literature.

References

1. Hristovski, D., Stare, J., Peterlin, B., Dzeroski, S.: Supporting discovery in medicine by association rule mining in Medline and UMLS. Medinfo. 10(2) (2001) 1344-1348
2. Hristovski, D., Peterlin, B., Mitchell, J.A., Humphrey, S.M.: Improving literature based discovery support by genetic knowledge integration. Stud. Health Technol. Inform. 95 (2003) 68-73
3. Lindsay, R.K., Gordon, M.D.: Literature-based discovery by lexical statistics. Journal of the American Society for Information Science, 50(7) (1999) 574-587
4. Perez-Iratxeta, C., Bork, P., Andrade, M.A.: Association of genes to genetically inherited diseases using data mining. Nature Genetics. 31(3) (2002) 316-9
5. Srinivasan, P.: Text Mining: generating hypotheses from MEDLINE. Journal of the American Society for Information Science and Technology. 55(5) (2004) 396-413
6. Swanson, D.R.: Fish oil, Raynauds syndrome, and undiscovered public knowledge. Perspectives in Biology and Medicine, 30(1) (1986) 7-18
7. Swanson, D.R.: Two medical literatures that are logically but not bibliographically connected. Journal of the American Society for Information Science. 38(4) (1987) 228-233
8. Swanson, D.R., Smalheiser, N.R.: An interactive system for finding complementary literatures: a stimulus to scientific discovery. Artificial Intelligence. 91(2) (1997) 183-203
9. Swanson, D.R., Smalheiser, N.R.: Implicit text linkages between Medline records: using Arrowsmith as an aid to scientific discovery. Library Trends, 48(1) (1999) 48-59
10. Swanson, D.R., Smalheiser, N.R., Bookstein, A.: Information discovery from complementary literatures: categorizing viruses as potential weapons. Journal of the American Society for Information Science and Technology, 52(10) (2001) 797-812
11. Weeber, M., Vos, R., Klein, H., de Jong-van den Berg, L.T.W.: Using concepts in literature-based discovery: simulating Swanson's Raynaud-fish oil and migraine-magnesium discoveries. Journal of the American Society for Information Science and Technology, 52(7) (2001) 548-557
12. Weeber, M., Vos, R., Klein, H., de Jong-van den Berg, L.T.W., Aronson, A.R., Molema, G.: Generating hypotheses by discovering implicit associations in the literature: a case report of a search for new potential therapeutic uses for thalidomide. Journal of the American Medical Informatics Association, 10(3) (2003) 252-259
13. Wren, J.D., Garner, H.R.: Shared relationship analysis: ranking set cohesion and commonalities within a literature-derived relationship network. Bioinformatics. 20(2) (2004) 191-198
14. Wren, J.D., Bekeredjian, R., Stewart, J.A., Shohet, R.V., Garner, H.R: Knowledge discovery by automated identification and ranking of implicit relationships. Bioinformatics. 20(3) (2004) 389-398

Two Phase Approach for Spam-Mail Filtering

Sin-Jae Kang, Sae-Bom Lee, Jong-Wan Kim, and In-Gil Nam

School of Computer and Information Technology, Daegu University,
Gyeonsan., Gyeongbuk., 712-714 South Korea
{sjkang, sblee, jwkim, ignam}@daegu.ac.kr

Abstract. This paper describes a two-phase method for filtering spam mails based on textual information and hyperlinks. Since the body of a spam mail has little text information, it provides insufficient hints to distinguish spam mails from legitimate mails. To resolve this problem, we follows hyperlinks contained in the email body, fetches contents of a remote webpage, and extracts hints (i.e., features) from original email body and fetched webpages. We divided hints into two kinds of information: definite information and less definite textual information. In our experiment, the method of fetching web pages achieved an improvement of F-measure by 9.4% over the method of using an original email header and body only.

1 Introduction

With the popularization of the internet, low cost, and fast delivery of message, email has become an indispensable method for people to communicate each other. Though email brought us such huge convenience, it also caused us trouble of managing the large quantities of spam mails received everyday. Spam mails, which are unsolicited commercial emails or junk mails, flood mailboxes, exposing minors to unsuitable content, and wasting network bandwidth [1]. Most software for email clients provides some automatic spam mail filtering mechanism, typically in the form of blacklists or keyword-based filters. Unfortunately constructing these lists and filters is manual time-consuming process, and is not perfect for a variety of cases in real situation. The spam mail filtering problem can be seen as a particular case of the text categorization problem. Several information retrieval (IR) techniques are well suited for addressing this problem, in addition it is a two-class problem: spam or non-spam. A variety of machine learning algorithms have been used for email categorization task on different metadata [2,4,5,6]. Sahami et al. [2] focuses on the more specific problem of filtering spam mails using a Naïve Bayesian classifier and incorporating domain knowledge using manually constructed domain-specific attributes such as phrasal features and various non-textual features. In most cases, support vector machines (SVM), developed by Vapnik [3], outperforms conventional classifiers and therefore has been used for automatic filtering of spam mails as well as for classifying email text [4,5]. Yang et al. [6] demonstrate that Naïve Bayesian and SVM classifier is by far superior to TFIDF. In particular, the best result was obtained when SVM was applied to the

J. Zhang, J.-H. He, and Y. Fu (Eds.): CIS 2004, LNCS 3314, pp. 800–805, 2004.
© Springer-Verlag Berlin Heidelberg 2004

header with feature subset selection. Accordingly, we can conclude that SVM classifier is slightly better in distinguishing the two-class problem. In this paper, we propose a two-phase filtering system for intercepting spam mails based on textual information and hyperlinks. Our system relies on two basic ideas. First, since the body of a spam mail has little text information on recognizing spam mails, our system follows hyperlinks in the email body, fetches contents of a remote webpage, and regards this webpage as extended email body. Second, a spam mail is classified by using two-phase system. In the first phase, definite information such as sender's URL, email addresses, and spam keyword lists is applied. In the second phase, remaining, that is unclassified, emails are classified using less definite information, extracted from email header and body.

2 Training Phase: Feature Selection and Machine Learning

To extract hints (i.e., features or attributes) about spam mail filtering and use efficiently, we divide hints into two kinds of information: definite information and less definite textual information.

2.1 Definite Information

Definite information for filtering spam mails is sender's information, such as email and URL addresses, and definite spam keyword (including key phrase) lists, such as "porno," "big money" and "advertisement". If an incoming email contains one of sender's information, it has a very high probability of being a spam mail. Therefore, we can regard the email as a spam mail. If an incoming email contains one of definite spam keyword lists in the subject line, or if definite spam keywords are appeared in the body of incoming emails over three times, it is also regarded as a spam mail. We extract sender's information automatically from spam mail corpus, and select definite spam keyword lists manually.

2.2 Less Definite Information

There are many particular features of email, which provide evidence as to whether an email is spam or not. For example, the individual words in the text of an email (not included in definite spam keyword lists), domain type of the sender (e.g. *co*, *com*), receiving time of an email[1], or the percentage of non-alphanumeric characters[2] in the subject of an email are indicative of a spam mail [2]. This phase is based on word vector space representation, in which each feature in the email document mainly corresponds to a single word and then uses the SVM learning algorithm to classify the emails. Since the body of a spam mail has little text information (recently, it often has only image data), our system follows hyperlinks contained in the email, fetches contents of a remote webpage, and regards this webpage

---

[1] Most spam mail is sent at night.
[2] The subject of spam mail is often full of symbol characters such as '!', '@', '#', and '$'.

as extended email body. For extracting textual information (words or phrases in the email), that is hints, each email document is pre-processed to remove symbols, performed morphological analysis, and removed stop-words. Among the extracted features, those with low differential power are not helpful in spam filtering, and thus should be omitted from feature vectors. Feature selection involves searching through all possible combination of features in the candidate feature set to find which subset of features works best for prediction. A few of the mechanisms designed to find the optimum number of features are document frequency thresholding, information gain, mutual information, term strength, and χ^2-test. In comparing two learning algorithm, Yang and Petersen found that, except for mutual information, all these feature selection methods had similar performance and similar characteristics [4,7]. To select features, which have high discriminating power, we experimented with information gain (IG) and χ^2-test (CHI). IG is frequently employed as a term goodness criterion in the field of machine learning. It measures the number of bits of information obtained for category prediction by knowing the presence or absence of a term in a document. CHI measures the lack of independence between a term t and a category c and can be compared to the χ^2 distribution with one degree of freedom to judge extremeness. The feature vector is constituted with the selected features above and other particular features such as domain type of sender, receiving time of an email, etc. Namely, each email is represented by a vector $\vec{x} =< x_1, x_2, ..., x_n >$, where $x_1, x_2, ..., x_n$ are the values of features $X_1, X_2, ..., X_n$. Since SVM performed best when using binary features [4], we used binary representation. In case of the selected features above, the feature values are defined by binary representation which indicates whether a particular word occurs in an email. In other cases, the attribute values are moderately defined by scaling original data according to their own properties. For example, if an email is arrived between 12 pm and 5 am, the feature value of receiving time is 1, otherwise 0. We will finally classify incoming emails into two categories: non-spam and spam mail. However, since each spam category has its own properties and SVM can only compute two-way categorization, we construct three SVM classifiers separately according to the kinds of spam mails. Three SVM classifiers are generated using the feature vectors, where three is the number of spam categories: porn spam, financing spam, and shopping spam. It is more effective than constructing only one SVM classifier for filtering all spam mails.

3 Applying Phase: Two-Phase Spam-Mail Filtering

Incoming emails are processed by using the information and classifiers constructed in the training phase. If an email contains one of the definite information, it is regarded as a spam mail. Otherwise, it is passed to the next SVM applying phase. SVM classifier for porn spam mails is applied first. If an email is classified as a spam mail, the second applying phase is over. If not, it is passed to other SVM classifier in sequence (See Figure 1).

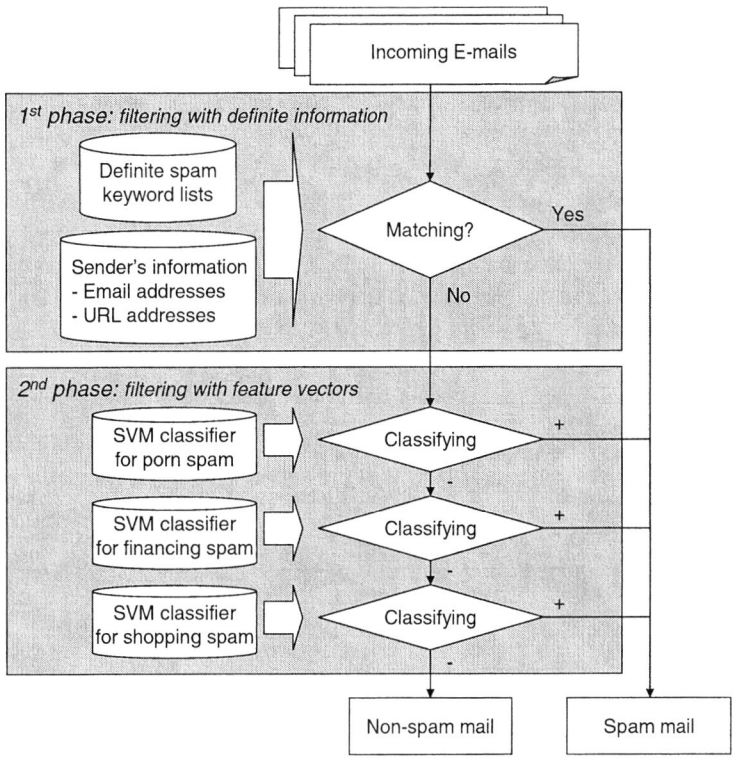

Fig. 1. Applying process for filtering spam mails

4 Experiments

The email corpus used in the experimental evaluation contained a total of 4,792 emails and 4 categories: 2,218 for legitimate mail, 1,100 for porn spam, 1,077 for financing spam, and 397 for shopping spam. To select important features, we used the *weak.attributeSelection* package provided by WEKA [8]. WEKA is a workbench designed to aid in the application of machine learning technology to real world data sets. WEKA contains a number of classification models. The SVM classifier used in this experiment was also provided by WEKA. SVM is tested with its default parameters settings within the WEKA. To evaluate the filtering performance on the email document corpus, we use the recall (R), precision (P), and F-measure (F) commonly employed in the information retrieval. In case of F-measure, we set the equal weight of recall relative to precision. In our experiments, we used ten-fold cross validation to reduce random variation. E-mail corpus was randomly partitioned into ten parts, and each experiment was repeated ten times, each time reserving a different part for testing, and using the remaining nine parts for training. Results were then averaged over the ten runs. Table 1 compared the performance of information gain and χ^2-test in selecting features for filtering porn spam. When compared with the results by Yang [7], it gave almost same results. Table 2 showed the performance of 2nd phase for each SVM classifier. The method of fetching web pages achieved an improvement of F-measure by 9.4% over the method of using an original email header and body only. Compared the two methods, there is little change of precision, but recall was improved a lot when using hyperlinks. We can recognize from these results that fetching web pages plays an important role in collecting more features.

To evaluate the performance of each phase, among 4,792 emails, the 4,335 emails are used for training SVM classifiers, and the remaining 457 emails are used for testing the proposed system's performance. We see from Table 3 that the proposed two-phase method is more effective than the method applying each phase separately, since the 1st phase undertakes some portion of the 2nd phase's workload with very high precision.

Table 1. Experimental results according to the feature selection methods (for porn spam)

Feature selection method	No. of selected features	Recall	Precision	F-measure
Information gain	338	51.5	95.8	67.0
	485	61.3	98.1	75.4
	681	62.7	97.3	76.3
	838	62.7	97.5	76.3
χ^2-test	338	51.4	95.8	66.9
	485	59.0	98.0	73.7
	681	62.7	97.3	76.3
	838	62.7	97.5	76.3

Table 2. Performance of SVM classifier

Object	Category	No. of selected features	R	P	F
Original email only	Porn	269	43.1	96.9	59.7
	Financing	713	63.1	98.3	76.9
	Shopping	92	49.6	90.8	64.2
Original email + Fetched webpages	Porn	1085	64.9	96.2	77.5
	Financing	1224	68.1	94.3	79.1
	Shopping	1233	60.2	90.9	72.4

Table 3. Performance of the proposed system (%)

Applying phase	Recall	Precision	F-measure
1st phase only	43.6	100	60.7
2nd phase only	67.3	99.4	80.3
$1^{st} + 2^{nd}$ phase	73.5	99.5	84.6

5 Conclusion

In this paper, we proposed a two-phase method for filtering spam mails based on textual information and hyperlinks. Since the body of a spam mail has little text information recently, it provides insufficient hints to distinguish spam mails from legitimate mails. To resolve this problem, we utilized hyperlinks contained in the email body. After fetching contents of a remote webpage, we extracted all possible hints from original email body and the fetched webpage. These hints are used to construct SVM classifiers. We divided hints into two kinds of information: definite information and less definite textual information. In case that an email contains one of the definite information, there is no need to perform machine learning algorithms, since it has a very high probability of being spam mails. In other case that the email has no definite information, it is evaluated using the SVM classifiers. We discovered that fetching hyperlinks is very useful in filtering spam mails, and our two-phase method is more effective than the method using machine learning algorithm only, blacklists, or keyword-based filters. This research is very important in that our system can prevent minors from accessing adult material on spam mails by chance, and save valuable time by lightening the email checking work. We will do further research on how to automatically update definite information, find more features having high differential power, and improve the filtering performance. Finally, we will consider that personalized definite information, constructed for each user, is implemented in our system.

Acknowledgement

This work is supported by Brain Korea 21 Information Technology Division, Daegu University.

References

1. Cranor, L. F. and LaMacchia, B. A., "Spam!," Communications of ACM, Vol.41, No.8 (1998) 74-83
2. Sahami, M., Dumais, S., Heckerman, D., and Horvitz, E., "A bayesian approach to filtering junk e-mail," In AAAI-98 Workshop on Learning for Text Categorization (1998) 55-62
3. Vapnik, V., The Nature of Statistical Learning Theory, Springer-Verlag, New York (1995)
4. Drucker, H., Wu, D. and Vapnik, V., "Support Vector Machines for Spam Categorization," IEEE Trans. on Neural Networks, Vol.10(5) (1999) 1048-1054
5. Joachims, T., "Text Categorization with Support Vector Machines: Learning with Many Relevant Features," ECML, Claire Nédellec and Céline Rouveirol (ed.) (1998)
6. Yang, J., Chalasani, V., and Park, S., "Intelligent email categorization based on textual information and metadata," IEICE Transactions on Information and System, Vol.E86-D, No.7 (2003) 1280-1288
7. Yang, Y, and Pedersen, J. P., "A comparative study on feature selection in text categorization," in Fourteenth International Conference on Machine Learning (1997) 412-420
8. Witten, I. H. and Frank, E., Data Mining: Practical machine learning tools and Techniques with java implementations, Morgan Kaufmann (2000)

Dynamic Mining for Web Navigation Patterns Based on Markov Model

Jiu Jun Chen, Ji Gao, Jun Hu, and Bei Shui Liao

College of Computer Science, Zhejiang University, Hangzhou 310027,
Zhejiang, China
rackycjj@163.com

Abstract. Web user patterns can be used to create a more robust web information service in personalization. But the user interests are changeable, that is, they differ from one user to another, and they are constantly changing for a specific user. This paper presents a dynamic mining approach based on Markov model to solve this problem. Markov model is introduced to keep track of the changes of user interest according to his or her navigational behaviors. Some new concepts in the model are defined. An algorithm based on the model is then designed to learn the user's favorite navigation paths. The approach is implemented in an example website, and the experimental results proved the effective of our approach.

1 Introduction

Web user patterns can help in improving the design of website, filtering the information and customizing the web service [1]. The study of user's behavior has become an important issue in web mining technology.

Some works have been done on the discovery of useful user navigational patterns. Most existing approaches focus on finding web association rules, navigational paths or sequential patterns from the log files and web links information. Zaiane et al [2] applied OLAP and data mining technology for pattern mining. Agrawal and Srikant [3] adopted sequential mining techniques to discover web access patterns and trends. In [4], Chen et al. proposed the maximal forward references to break down user sessions into transactions for mining access patterns. And other technologies, such as fuzzy theory, tree construction and support vector machines, are used to improve the performance of the system based on user patterns [5, 6].

Most solutions discover patterns simply according to user's access frequency in web logs. It is inaccurate, (a) pages, which visited frequently, may not show that users have more interest in them, such as page that is only to be utilized the links of a page to another page; (b) web user interests are changeable [7], and it is difficult to track the exact pattern of web users.

In this paper, we will study those two problems. The rest of this paper is organized as follows. In the following section, we first introduce the Markov model to study the navigational characters of web users. Based on the Markov user model, an algorithm is proposed to learn the user's favorite navigational paths, and experimental results are discussed. The conclusions are presented in the final section.

2 Dynamic Mining for Navigation Patterns

2.1 Markov Model

The interest changes of web users can be represented from their navigational behavior, and a Markov model can extract the characters of user's navigational behaviors dynamically. We model the navigational activities as a Markov process for the following reasons: Firstly, the information about the user's navigation pattern is changeable. Secondly, the web user is largely unknown from the start, and may change during the exploration.

Definition 1. State. A state is defined as a collection of one or more pages of the website with similar functions. Besides functional states, the model contains other two special states, Entry and Exit;

Definition 2. Navigational Behavior. Navigational behavior can be viewed as remaining in one state. Two kinds of navigational behaviors are defined: (a) remaining in one state, which can be viewed as reading the contents of web pages; (b) making transitions among states, which represents the requests of pages;

Definition 3. Navigational Paths. In a limited time, the sequence of user's relative web requests in a website is defined as user navigation path. It can be viewed as a sequence set of states;

Definition 4. Transition Probability p_{ij}. It is the probability of transition from state i to state j. A transition occurs with the request for one page that belongs to state j while the user resides in one of the pages belonging to state i. We suppose that if there are n kinds of different transitions to leave one state, the state that has higher transition probability reveals user interest;

Definition 5. Mean Staying Time $\bar{t}_{ij}$. It is the mean time which the process remains in one state before making a transition to another state. The longer staying time, the more interested visiting. We suppose that if there are n kinds of different translations to leave one state, those states that have long staying time reveal user interest. The pages that are only for a user to pass have limited staying time. Although this page have many visited times, it can lower the interest level according to the staying time;

Definition 6. Favorite f_{ij}. It integrates the weight of transition probability and mean staying time while evaluating the interest level of the state visited by the web user. It is defined as formula (1). It can prevent from only mining visited states with high probability and low staying time.

In formula (1), p_{ij} and t_{ij} can be found out by the following methods.

$$\begin{cases} f_{ij} = \dfrac{p_{ij} \times t_{ij}}{(\sum_{j=2}^{n+1} p_{ij})(\sum_{j=2}^{n+1} t_{ij})/n^2} & i,j \in (2, n+1) \\ f_{1j} = F_{threshold} & j \in (2, n+1) \\ f_{i(n+2)} = F_{threshold} & i \in (2, n+2) \\ f = 0 & Others \end{cases} \quad (1)$$

Method to calculate transition probabilities:

(i) For the first request for state s in the session, add a transition from Entry state to the state s, $State_0(Entry) \rightarrow State_s$, and increment $TransitionCount_{0s}$ in a matrix $TransitionCount[i, j]$ by 1, where $TransitionCount[i, j]$ is a matrix to store the transition counts from state i to state j;

(ii) For the rest of user' requests in the session, increment the corresponding transition count of $TransitionCount_{i,j}$ in the matrix, where i is the previous state and j is the current state;

(iii) For the last page request in the session, if the state is not the explicit Exit state then add a transition from the state to Exit state and increment $TransitionCount_{s,(n+2)}$ value by 1;

(iv) Divide the row elements in matrix $TransitionCount[i, j]$ by the row total to generate transition probability matrix P, whose element is $p_{i,j}$:

$$p_{i,j} = \dfrac{TransitionCount_{i,j}}{\sum_k TransitionCount_{i,k}} \quad (2)$$

Method to calculate mean staying time:

(i) To find out the time spent in state i before the transition is made to state j for any transition from state i to state j, except the transition from Entry state, and the transition to Exit state. If this time belongs to the interval k then, increment $StayTimeCount_{i,j,k}$ by 1 in a three-dimensional matrix $StayTimeCount[i, j, m]$, where, $StayTimeCount_{i,j,k}$ is the number of times the staying time is in the interval k at state i before the transition is made to state j;

(ii) Find out the interval total for each transition from state i to state j in $StayTimeCount[i, j, m]$. Divide frequency count in each interval with the interval total to find out the probability of occurrence of the corresponding intervals. Repeat this to generate $StayTime\Pr obability[i, j, m]$, whose element defined as follows:

$$StayTime\Pr obability_{i,j,m} = \dfrac{StayTimeCount_{i,j,m}}{\sum_n StayTimeCount_{i,j,n}} \quad (3)$$

(iii) Multiply each interval with the corresponding probability to generate mean staying times ($\bar{t}_{ij}$), which is the elements of matrix $\bar{T}$.

$$\bar{t}_{ij} = \sum_{m} m \times StayTime \Pr obability_{i,j,m} \quad (4)$$

A matrix of Markov user model, $M_{(n+2)(n+2)}$, as a state to state matrix, is set up from web logs, whose elements include p_{ij}, $\bar{t}_{ij}$, and f_{ij}. In this way, the matrix will capture a compact picture of users' favorite navigational behavior dynamically.

2.2 Algorithm for User Favorite Navigation Path

Based on Markov user model, we design the algorithm, named MFNP (Markov-based favorite navigation paths), which is shown as follows.

```
Algorithm MFNP
Input: M, F_threshold, P_threshold;
Output: Favorite navigation paths;
Variable:
   M: Matrix of Markov user model
   StateVisited[]: Array to store whether the state is
   visited or not;
   FS[]: Array stored the favorite navigation states;
   State[]: Array stated the states of website;
   F_threshold: Threshold of the Transition Probability;
   P_threshold: Threshold of the Favorite;
program FavoritePath(M, P_threshold, F_threshold)
   Initialize StateVisited[] and FS[];
   for k=0 to (n+1) do
      if StateVisited[k]=false then
         FindFavoriteState(M,k);
end
program FindFavoriteState(M,i)
   FS[]=State[i];
   if StateVisited[i]=false then
      StateVisited[i]=true & StateFound=false;
      for j=0 to (n+1) do
         if p_ij >= P_threshold and f_ij >= F_threshold then
            StateFound=true & FindFavoriteState(M,j);
         if StateFound=false then
            Output the elements in array FS[];
   else  Output the elements in array FS[];
end
```

The algorithm is implemented in a used-goods exchanged website, which includes 12 states: (1)Entry; (2)Home; (3)Login; (4)Register; (5)Goods Search; (6)Seller Data; (7)Leave word; (8)Goods List; (9)Goods Register; (10)Customer Support; (11) About Us; (12)Exit. As shown in table 1, no transition can be made to Entry state from any state, and two rows are associated with each state. The upper rows are the transition probability values and the lower rows correspond the mean staying times in minutes.

Table 1. User navigation matrix of the example website

State	2	3	4	5	6	7	8	9	10	11	12
1	0.30	0.13	0.01	0.11			0.35		0.10		
2	0.01	0.21	0.02	0.05			0.23		0.01	0.06	0.43
	0.52	0.86	0.80	2.27			1.20		0.61	1.50	
3		0.36	0.02	0.02	0.10	0.14	0.04	0.22			0.12
		0.54	0.52	0.50	2.36	2.52	1.23	0.55			
4	0.06	0.28	0.48								0.18
	0.5	0.91	0.50								
5				0.42	0.20	0.18	0.10				0.10
				0.53	0.50	2.78	0.68				
6		0.12		0.14	0.05	0.20			0.12		0.37
		2.10		0.50	2.50	0.62			1.85		
7	0.02	0.10		0.45	0.20				0.23		
	3.12	2.78		1.25	0.67				1.32		
8				0.12	0.32		0.34		0.11		0.11
				0.87	0.50		0.50		3.36		
9		0.04			0.33			0.58			0.05
		2.64			0.61			0.55			
10	0.01	0.06	0.01	0.08	0.02	0.06	0.05		0.42	0.05	0.24
	5.60	0.85	0.52	1.37	0.77	1.52	1.56		1.78	2.53	
11	0.21	0.06		0.28			0.23			0.12	0.10
	2.52	7.20		1.21			1.18			2.66	
12											1.00

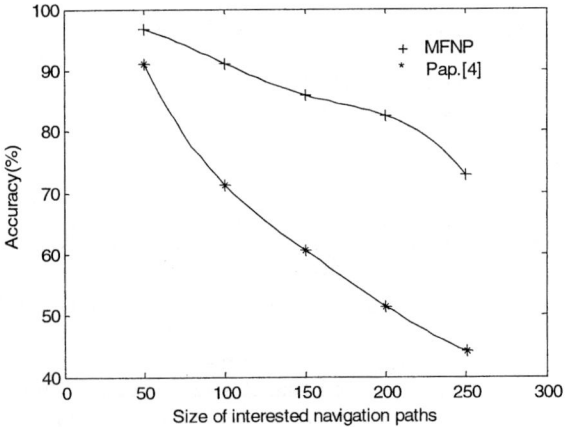

Fig. 1. Accuracy comparison

Probability threshold and favorite threshold are assigned to 0.2 and 3 respectively. According to the algorithm, begin with Entry state. Then search the first Entry row, choose State 8 (Goods List) as favorite state. Because State 8 has not been visited, it begins to search from State 8 row. Then State 8 is chosen. Because State 8 has been visited, the elements in array FP[] (from Entry to Goods List) can be output as the favorite navigational path. Finally, the set of favorite path are included: {(1, 8); (1, 8, 6, 12); (1, 8, 6, 7, 5); (1, 8, 6, 7, 10); (1, 8, 6, 7, 10, 12); (1, 2, 3); (1, 2, 8); (1, 2, 12); (1, 2, 6, 12); (1, 2, 6, 7, 5); (1, 2, 6, 7, 10); (1, 2, 6, 7, 10, 12); (1, 2, 3, 12)}, and the great favorite path is the transition from Entry state to Goods List state.

To evaluate the effectiveness of our method, we compare the accuracy of the algorithm proposed in paper [4] with our model. We implement the algorithms to mine the same number of interested navigation paths. According to those navigating path, we produce predictions as dynamic links. If users click these dynamic links, we think the corresponding algorithm is accurate. Otherwise, it is inaccurate. The experiment result is shown in figure 1. We can find our method is more accurate and effective.

3 Conclusions

In this paper, we proposed a dynamic mining approach to learn the navigation patterns of web user. We constructed a Markov model to track and represent the user's behaviors dynamically. Based on the model, we designed an algorithm to mine user's favorite path. The algorithm was implemented in an example website and the experiment results show that our method is effective.

References

1. Perkowitz, M., Etzioni, O.: Towards adaptive web sites: conceptual framework and case study. Artificial Intelligence, 118(1-2) (2000) 245–275.
2. Zaiane, O., Xin, M., Han, J.: Discovering web access patterns and trends by applying OLAP and data mining technology on web logs. Proceedings on Advances in Digital Libraries Conference, Melbourne, Australia, (1998) 144-158.
3. Agrawal, R., Srikant, R.: Mining sequential patterns. Proceedings of the 11th International Conference on Data Engineering, Taipei, Taiwan, (1995) 3-14.
4. Chen, M. S., Park, J. S., Yu, P. S.: Efficient data mining for path traversal patterns. IEEE Transaction, Knowledge Data Engineering, 10(2) (1998) 209-221.
5. Nasraoui, O., Petenes, C.: An intelligent web recommendation engine based on fuzzy approximate reasoning. IEEE International Conference on Fuzzy Systems, (2003) 1116-1121.
6. Rohwer, J. A.: Least squares support vector machines for direction of arrival estimation. IEEE Antennas and Propagation Society, AP-S International Symposium (Digest), 1 (2003) 57-60.
7. Belkin, N. J., Croft, W. B.: Information filtering and information retrieval: two sides of the same coin. Communication of the ACM, 35(12) (1992) 29-38.

Component-Based Recommendation Agent System for Efficient Email Inbox Management

Ok-Ran Jeong and Dong-Sub Cho

Department of Computer Science and Engineering,
Ewha Womans University,
11-1 Daehyun-dong, Seodaemun-ku, Seoul 120-750, Korea
{orchung, dscho}@ewha.ac.kr

Abstract. This study suggests a recommendation agent system that the user can optimally sort out incoming email messages according to category. The system is an effective way to manage ever-increasing email documents. For more accurate classification, the Bayesian learning algorithm using dynamic threshold has been applied. As a solution to the problem of erroneous classification, we suggest the following two approaches: First is the algorithmic approach that improves the accuracy of the classification by using dynamic threshold of the existing Bayesian algorithm. Second is the methodological approach using recommendation agent that the user, not the auto-sort, can make the final decision. In addition, major modules are based on rule filtering components for scalability and reusability.

Keywords: e-mail classification, dynamic threshold, bayesian algorithm, rule filtering component.

1 Introduction

As information technology develops, the quantity of information the user can have at his disposal has increased exponentially. With it, different recommendation systems have been suggested. These systems enable the user to wade through the information glut. But most of these systems are for personalizing user profiles or for recommending product items the user wants based on his prior search and buying patterns. As part of this type of recommendation system, collaborative filtering can have a major advantage in that it can provide a dynamic link through other users' feedback information [1]. Although the possible applicability of the recommendation system is limitless, the attempt to apply this to email management has not been made as of yet [2, 3]. This study wants to suggest a recommendation agent system for email that can actively manage depending on situation and reflect the user's opinion, rather than a system that can simply auto-sort the messages based on existing text classification. Although such system auto-sorts by category through personalized learning, it is better to have a semi-automatic system that doubles with a recommendation system rather than an automatic sort system to maximize email users' satisfaction.

As a solution to the problem of erroneous classification, we suggest the following two approaches: First is the algorithmic approach that improves the accuracy of the classification by using dynamic threshold of the existing Bayesian algorithm. Second is the methodological approach using recommendation agent that the user, not the auto-sort, can make the final decision. In addition, major modules are based on rule filtering components for scalability and reusability.

2 Related Work

2.1 Learning Algorithm

In the process of classifying email documents, the rules are formed. When classifying the documents by category, learning algorithm can be used. For machine learning method in automatic document classification, there is Naive Bayesian, K-Nearest Neighbor, and TFIDF (Term Frequency Inverse Document Frequency) approach [4]. In this paper, we use the Naive Bayesian approach because it is the most widely in use and appropriate for email document classification. This learning method makes use of a probability model based on the Bayesian Theorem. It types in a vector model for the document to be classified and finds out a class that has the highest possibility for observing the document. The hypothesis here is a Naive Bayesian assumption that all the characteristics of the documents are independent of each other in a given category. The Bayesian algorithm applied in this study improves the accuracy of email documents' classification by varying previously fixed threshold dynamically.

2.2 Feature Extraction for Learning

The Feature Extraction is a way to define again the importance of the learning resources' characteristics by their category. For this, it is necessary to distinguish categories according to their learning resources' characteristics and perform feature extraction based on it. Term weighting method can be done by considering category information to which a set of keywords belongs. In this case, more weight is given to a keyword representing each category [5, 6]. Such machine learning method for feature extraction is applied when there are two different categories, by giving more weight to the keyword for each category. It also makes use of a method, which registers the keyword on the title part as well as the body part. It is the method that takes into duplication account on the title part of email documents.

2.3 The Features of Components

Components are a system element that is independent and divisible into small units that provides service through interface. These components can be replaceable with other elements as required by the system and are part of pre-developed application codes[7] . Based on the definition of the components, it is possible to show their characteristics. Their characteristics include identifiability, replaceability, accessibility through interface, service's fixability provided by interface, concealibility of physical realization, degree of independence, independent reusability of language and development tools, and dynamic reusability. In general, the components can be divided into logical and physical perspectives [4]. The concepts of logical component and physical component based on their features are as follows: The former means the business component which models after real-world business features. Meanwhile, the latter means engineering reconstruction by dividing the business components into independent software. This study thus approaches from logical perspective, improving scalability and reusability by implementing into components the major learning and filtering parts in email management.

3 Email Recommendation Agent System

3.1 Email Classification Method

The reason this study opted for recommendation method as its email classification is that the existing auto-sort method may be problematic in applying the email system. Since email documents are highly personalized, it is difficult to satisfy the user even though they are auto-sorted by category through personalized learning. That is why we suggest in this study a recommendation agent system that can recommend a category according to its rank after classifying the email messages by category. Once the user is recommended a category, he can save the keyword in multiple categories, or manage his email inbox as the category changes with time lapse. That way, it is possible to avoid classifying email messages inadequately. The system is designed so that the user can opt for auto-sort when email messages are overwhelming or when he is satisfied with the reliability of the recommendation system.

3.2 Suggested System

The biggest characteristics of the suggested system are twofold. First, the system is modulized for more efficient interaction among email documents' characteristic extraction, rule generation, and classification by category. In addition, the system is created with rule filtering components. Second, the system adds dynamic threshold function to the existing Bayesian algorithm, thus improving the accuracy of classification that can be the central part of the system function. The overall flow of the system is shown in Figure 1 below. The figure is a system flow chart reflecting all the processes of learning, filtering, rule generation, classification, and recommendation, based on user information.

The recommendation agent system for email management has the following functions: First, once a new message reaches the user's inbox, the system observes how the message is processed and learns from it. This is the process where feature extraction and rule generation are made and the user can set up categories according to his preferences. Second, the system forms rules by applying the Bayesian algorithm to the characteristics derived from the email processing observation module. Third, once an email message based on the new rules

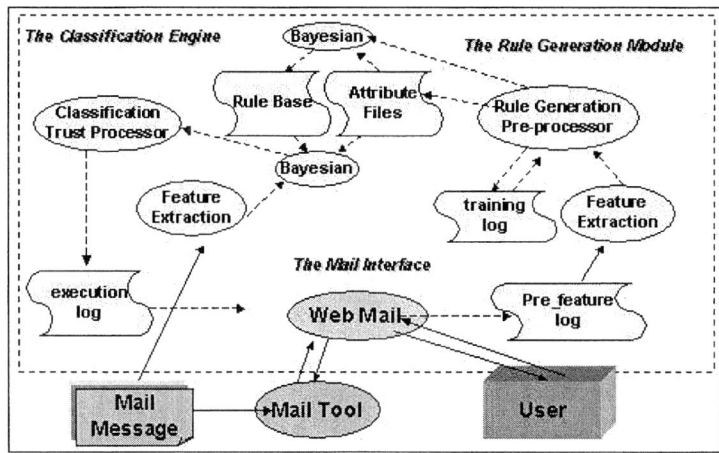

Fig. 1. Modular Design (The Web Mail Interface Module, The Category Rule Generation Module and The Mail Classification Recommendation Module)

reaches the inbox, it is classified by category and then recommends the best category to the user.

These are implemented by COM+, with consideration on scalability with other application systems and on reusability under distributed environment. Major interfaces and methods employed here are illustrated in Figure 2 below. The

```
IRuleFilter
{
HRESULT SetDBOpen(BSTR bstrDBConnect,BSTR bstrDBID,BSTR bstrDBPW);
HRESULT MergeMail(BSTR bstrID,BSTR bstrRuleName,BSTR bstrMailData);
HRESULT CheckFolder(BSTR bstrID, BSTR bstrMailData,[out,retval] BSTR *pRet);
HRESULT CheckFolders(BSTR bstrID, BSTR bstrMailData,[out,retval] BSTR *pRet);
};
SetDBOpen(DBOPEN, DBID, DBPW);
MergeMail(ID, RuleName, MailData);
CheckFolder(ID, MailData);
CheckFolders(ID, MailData);
```

Fig. 2. Rule Filtering Component (COM+)

second feature of the suggested system is that it improves the accuracy of filtering by dynamically varying previously fixed thresholds. This study classifies email documents through the Bayesian learning method, which is the best available learning algorithm for document classification.

Suppose that C is all the category set as in (1) while C_0 is where classification is not possible. Suppose again that D is the set of all the email documents. Then we can define (2) as follows:

$$CategorySet\ C = \{C_1, C_2, ..., C_k\},\ C_0 = unknown\ category \quad (1)$$

$$E-MailDocumentSet\ D = \{d_1, d_2, ..., d_n\} \quad (2)$$

According to the Naive Bayesian classification method, conditional probability of each category c_j of one d_i is given as follows:

$$\Re(d_i) = \{P(d_i|C_1), P(d_i|C_2), P(d_i|C_3), ..., P(d_i|C_k)\} \quad (3)$$

In most systems, each document is classified by the highest probability value as in Equation (4). But this study converts the fixed threshold T, which has been used in the existing Bayesian algorithm, into the dynamic threshold T' through Equation (5). When the dynamic threshold T' is applied, it is shown, the accuracy of the system has been improved.

$$P_{max}(d_i) = max\{P(d_i|C_t)\},\ t = 1, 2, ..., k \quad (4)$$

$$C_{best}(d_i) = \begin{cases} \{C_j | P(d_i|C_j)\} = P_{max}(d_i) & \text{if } P_{max}(d_i) \geq T' \\ & \text{where } T' = 1 - \frac{P_{max}(d_i)}{\sum_{j=1}^{k} P(d_i|C_j)} \\ C_0 & \text{otherwise} \end{cases} \quad (5)$$

4 System Implementation and Result Analysis

4.1 System Implementation

The suggested system is based on Web mail server that does not require a separate mail client program. The Implementation environment is used Windows 2000 Professional as OS, MS-SQL 2000 Server for database control. For rule generation and algorithm implementation, rule filtering components, and other functions, we used MS Visual C++ 6.0, COM+, and ASP and ASP components, respectively. Figure 3 below shows user interface utilized in the system. The user interface in Figure 3 can be used in the user observation process where the user can create and save categories. Besides, the user can create frequently used categories and delete unnecessary ones. The user extracts features, implements internally mail classification based on the established rules, and recommends selected categories with assigned probabilities. The user can save email messages under given categories after reviewing recommended categories.

4.2 Result Analysis

The appropriateness of an information classification system is measured both by recall ratio and precision ratio. The recall ratio refers to the ratio of suitable literature classified by the classification system among suitable literature in the information classification system. The precision ratio implies the ratio of suitable literature among all the classified literature [7]. The performance test for this study is "How accurate does the system recommend categories to the user?"

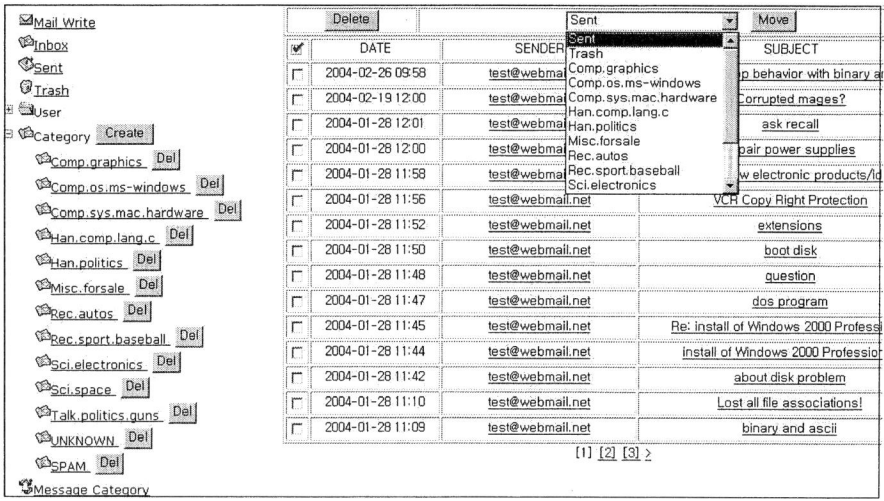

Fig. 3. Mail User Interface

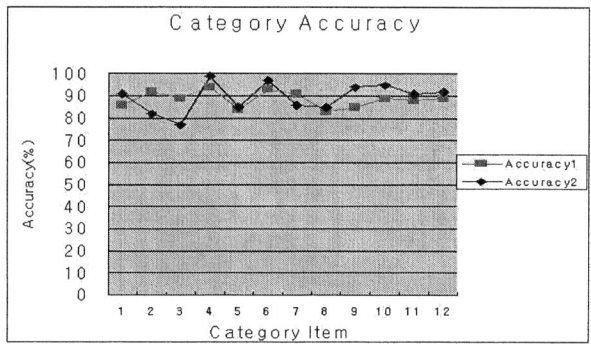

Fig. 4. Category Accuracy

This is measured by the precision ratio and recall ratio, and by testing whether mail contents are classified according to correct categories. For this test, we set up 12 categories, collected sample data for rules, and another set of data for performance evaluation. Since the actual test involved a large amount of data, we attempted to test approximately 10,000 email messages per each category as a single data format. The data format for this test consisted of date, sender, title, number of document lines, empty lines, actual mail contents, and four empty lines. The accuracy test generated the following statistical results: After having the system learns 1,000 emails for each category and carrying out 221,371 incoming emails through the existing algorithm, we obtained the accuracy ratio of 88.6 percent (or 196,137 mails). Meanwhile, the accuracy ratio using the dynamic thresholds was 89.5 percent (or 204,795), which was an improvement of 0.9 percentage points.

5 Conclusions

This paper tries to design and implement a recommendation agent system that may be helpful for the email user. As more and more email messages are being exchanged, the users will demand more convenient customized email interface in their email management. The system makes use of a recommendation method, instead of fully automated sorting method. It attempts to improve scalability and reusability by implementing the all-important filtering part as components. Although the current system is designed for the user to set up the categories himself, the future agent will incorporate both the automatic category setup and recommendation method at the same time.

Acknowledgement

This work was supported by the Brain Korea 21 Project in 2004.

References

1. Pazzani M., Billsus D.: Learning and Revising User Profiles: The Identification of Interesting Web sites. Machin Learning 27. Kluwer Academic Puglishers (1997) 313-331
2. Balabanovic, M., Shoham, Y.: Fab: Content-Based Collaborative Recommendation. CACM 40(3) (1997) 66-72
3. Hill, W., Stead, L., Rosenstein, M., Furnas, G.: Recommending and Evaluating Choices in a Virtual Community of Use. CHI'95 (1995) 194-201
4. Ian, H., Frank, E.: Data Mining. Morgan Kaufmann Publishers. Inc. (2000)
5. Baeza-Yates, R., Ribeiro-Neto, B.: Modern Information Retrieval. Addison-wesley (1999)
6. Cohen, W. W.: Learning Rules that Classify E-Mail, AAAI Spring symposium on Machine Learning in Information Access (1996) 18-25
7. Frye, c.: Understanding Components. Andersen Consulting Knowledge Xchange (1998)

Information Security Based on Fourier Plane Random Phase Coding and Optical Scanning

Kyu B. Doh[1], Kyeongwha Kim[1], Jungho Ohn[1], and Ting-C Poon[2]

[1] Department of Telecommunication Engineering, Hankuk Aviation University,
200-1 Hwajeon-dong, Goyang-city 412-791, Korea
kdoh@hangkong.ac.kr
[2] Bradley Department of Electrical and Computer Engineering, Virginia Polytechnic Institute and State University, Blacksburg, Virginia 24061, USA

Abstract. An information security method is presented by use of electro-optic cryptosystem. We investigate Fourier plane random phase encoding technique, a random phase coding, and holographic coding. The proposed cryptosystem is based on those combined coding technique. The target information is first random phase encoded and then multiplied by a Fourier plane random phase code and further processed into holographic coding. The electro-optic cryptosystem can be implemented in a manner that produced output of the encrypted information is to be directly sent through an antenna to a secure site for digital storage. In addition, the system has a capability of producing the encrypted information in real-time. In secure site, the encrypted information is, together with a decryption key, processed for decryption. If the encryption key and decryption key are matched, the decryption unit will decrypt the information. The proposed electro-optic cryptosystem enables us to store, transmit, and decrypt the encrypted data. We develop the theory of technique and substantiate it with simulation results.

1 Introduction

On the classical front, optical cryptography [1-2], which is based on photons, has a long-standing history. Due to the resent progress in the development of optical component and systems and their increased technical performance, optical cryptography suggests that it has significant potential for security application. In one approach of various kinds of optical data processing technology for information security, the coded image is a phase-amplitude function whose real and imaginary parts can be regarded as realizations of independent random processes [3]. Several studies have been published on the implementation of random phase-encoding and security techniques with optical system [4-8]. While most optical encryption techniques are optically coherent, there is an incoherent optical technique for encryption proposed in recent years [6]. Incoherent optical techniques process many advantages over their counterpart, such as better S/N ratio and insensitivity to misalignment of optical elements. One of the reasons of using holographic encryption is that holographic encryption as opposed to electronic or digital encryption can provide many degrees of freedom for securing information. Another reason is that the encrypted information is difficult to reproduce with the usual reproduction device. When large volumes of information are to be encrypted, such as a 3-D object, the use of holographic encryption methods is probably the most logical choice. The proposed method is based on Fourier plane random phase coding and optical scanning. Advantages of the method include that since it is an scanning method, incoherent objects can be processed without the need of using spatial light modulators to convert incoherent image to coherent image and the system can perform real-time. Another advantage includes that the method is easily extendible to encrypt 3-D information.

2 Theory of Electro-Optic Cryptosystem and Simulation Results

Optical scanning technique, which first suggested by Poon [9], has been used extensively for various fields of optical applications such as incoherent image processing [10], holographic image processing [11,12], 3-D microscopy [13], optical recognition of 3-D object [14], and optical remote sensing [15]. We shall describe the electro-optic cryptosystem based on Fourier plane random phase coding and optical scanning in the encryption stage as well as in the decryption stage. The optical scanning system is based on the two-pupil synthesis processor. Since the mathematical description of the two-pupil optical system has been discussed in [9], we focus on describing the electro-optic cryptosystem for encryption and decryption operation with the results of the two-pupil synthesis system. The cryptosystem includes identical systems in the encryption and the decryption stages. Figure 1 shows the electro-optic cryptosystem for encryption stage. In the encryption system, the target information is scanned by a time-dependent Fresnel plate (TDFP). TDFP is created by the superposition of a plane wave and a spherical wave of different temporal frequency. The data are encrypted optically by Fourier plane random phase mask and input random phase encryption mask. Let $I_0(x,y;z)$ denote the information to be encrypted and $F(k_x, k_y)$ denote the Fourier random phase mask. In lower path, the Fourier random phase

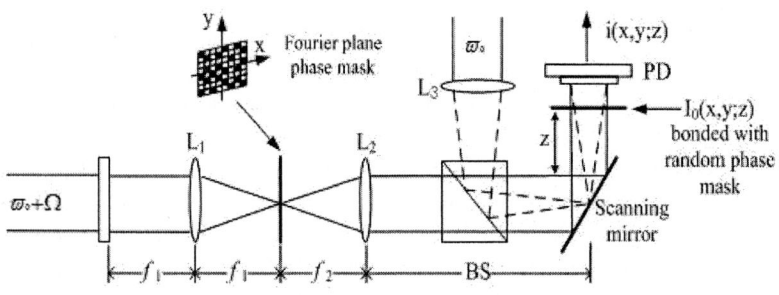

Fig. 1. Electro-optic cryptosystem for encryption

mask, $F(k_x, k_y)$, located in the front focal plane of the lens L_2 is illuminated by laser beam of temporal frequency $w_0 + \Omega$ through lens system. In upper path, the illuminated laser beam of temporal frequency w_0 is focused by lens L_3 on the scanning mirror. The two beam are then combined by the beam-splitter, BS, and used to encrypt the image/document, $I_0(x, y; z)$, located at a distance z away from the scanning mirror. The input random phase mask, $m(x, y)$, is bonded with the target information. The photo-detector, PD, collects all the light information transmitted by the information to be encrypted. The PD is turned at frequency Ω to detect the encrypted information and then produce the current of encrypted information, $i_\Omega(x, y; z)$ expressed as

$$i_\Omega(x, y; z) = Re\{i_{\Omega p}(x, y; z) \exp(j\Omega t)\} \qquad (1)$$

Since the encrypted signal can be made at radio frequency, the signal can be directly sent through an antenna to a secure site for wireless application. To perform encryption on input $I_0(x, y; z)$, we first investigate the Optical Transfer Function (OTF) of the two-pupil synthesis optical system to apply the proposed encryption technique. As we see that the choice of a Fourier plane random phase mask is a good encryption key. For this choice of the mask, we modify the OTF, which have been discussed in [9], as follow:

$$OTF_\Omega(k_x, k_y) = \exp[j\frac{z}{2k_0}(k_x^2 + k_y^2)]F(k_x, k_y) \qquad (2)$$

It can be shown that the spectrum of $i_\Omega(x, y; z)$ is related to the spectrum of $I_0(x, y; z)$ through the following expression:

$$\begin{aligned}\Im\{i_\Omega(x, y; z)\} &= \Im\{I_0(x, y; z)m(x, y)\}OTF_\Omega(k_x, k_y; z) \\ &= [\Im\{I_0(x, y; z)\} \otimes \Im\{m(x, y)\}]OTF_\Omega(k_x, k_y; z)\end{aligned} \qquad (3)$$

where $\otimes$ denotes 2–D convolution, and (3) then becomes

$$\begin{aligned}i_\Omega(x, y; z) &= Re[i_{\Omega p}(x, y; z)\exp(j\Omega t)] \\ &= Re[\Im^{-1}\{\Im\{I_0(x, y; z)m(x, y)\}OTF_\Omega(k_x, k_y; z)\}\exp(j\Omega t)] \\ &= Re[\Im^{-1}\{\Im\{I_0(x, y; z)\} \otimes \Im\{m(x, y)\} \\ &\quad \times OTF_\Omega(k_x, k_y; z)\}\exp(j\Omega t)]\end{aligned} \qquad (4)$$

$i_\Omega(x, y; z)$ is the encrypted image and can be stored by the digital computer. We can interpret equation (4) as the random phase masked information is being encrypted by Fourier plane random phase mask and then the information is being recorded as a digital hologram. After the object has been encrypted, we need to decrypt it. For decryption, since the system is hybrid in nature, it is flexible and either optical decryption or digital decryption could be employed with the system. We let $F_d(k_x, k_y)$ denote as a decryption key. The information is now stored in the digital computer to be used to decrypt the information coming from the encryption site via transmission. To decrypt the information from (4), we provide the use of a electro-optic decryption unit that is basically reversal processing of the encryption unit shown in figure 1. We see that the output of the unit, $I_d(x,y)$, becomes

$$\begin{aligned}
output &= \Im^{-1}[[\Im\{I_0(x,y;z)\} \otimes \Im\{m(x,y)\}] \\
&\quad \times OTF_\Omega(k_x,k_y;z)OTF^*(k_x,k_y;z)F_d(k_x,k_y)\}] \\
&= \Im^{-1}[\Im\{I_0(x,y;z)m(x,y)\}] \\
&= I_0(x,y;z)m(x,y)
\end{aligned} \quad (5)$$

We substantiate it with simulation results. Figure 2 shows the optical transfer function of the electro-optic cryptosystem. The original information to be encrypted is shown in Fig. 3. Figure 4 shows the real part of target information masked by a random phase mask placed immediately in front of the information. The following parameters are chosen that the wavelength of light used is 0.6 μm, the document to be encrypted, $I_0(x,y)$, is located 20 away from the scanning mirror. Figure 5 shows the holographic pattern of the key according to Eq.(2). For decryption, we need to gather information and decryption key according to

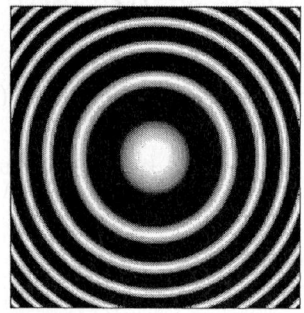

Fig. 2. Optical transfer function of cryptosystem

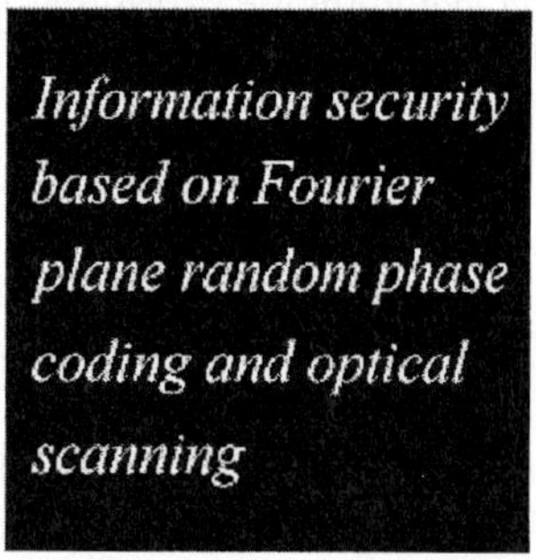

Fig. 3. The original information to be encrypted

the manner as discussed. Figures 6 and 7 show the intensity of decrypted information of document with and without Fourier plane decryption key respectively. Since the decrypted information also has a holographic information as well, the depth information, z, play a role as an additional security key. It is apparent that the choice of phase keys would work.

3   Summary and Concluding Remarks

In summary, an information security method is presented by use of electro-optic cryptosystem. We investigate Fourier plane random phase encoding technique, a random phase coding, and holographic coding. In proposed method, the target information is first random phase encoded and then multiplied by a Fourier plane random phase code and further processed into holographic coding. For the

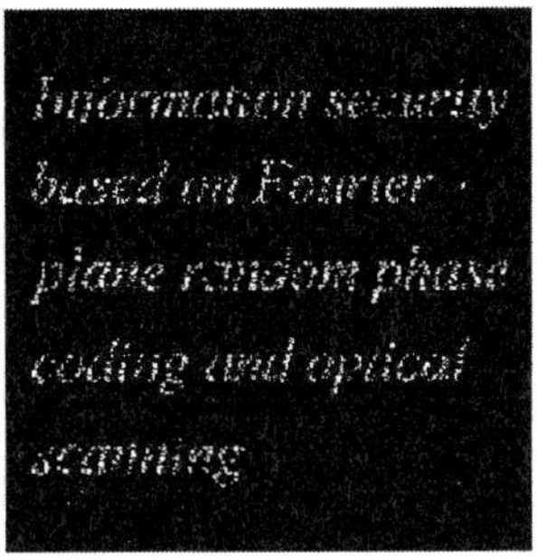

Information Security Based on Fourier Plane Random Phase Coding 823

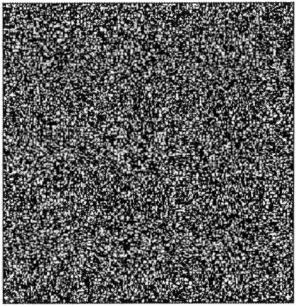

Fig. 4. The real part of target information multiplied by random phase mask

Fig. 5. Holographic pattern of encryption key

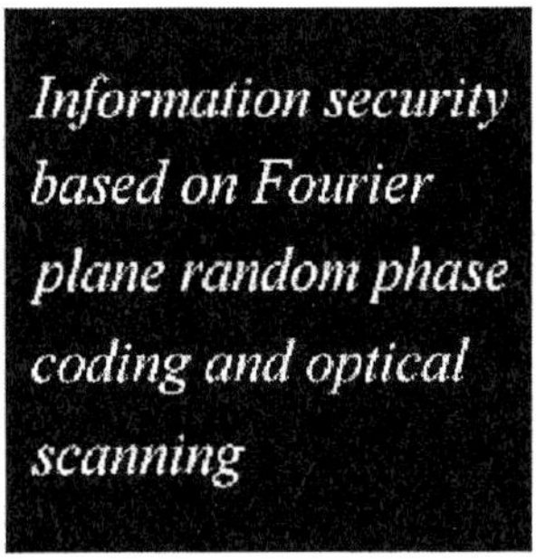

Fig. 6. Decrypted information of document

Fig. 7. Decrypted information of document without decryption key

purpose of holographic coding, the information to be encrypted is coded by a time-dependent Fresnel plate (TDFP). TDFP is created by the superposition of a plane wave and spherical wave of different temporal frequency. The reason of using electro- holographic encryption is that holographic encryption as opposed to electronic or digital encryption can provide many degrees of freedom for securing information. Another reason is that the encrypted information is difficult to reproduce with the usual reproduction device. While most optical encryption techniques are optically coherent, incoherent optical techniques process many advantages over their counterpart, such as better S/N ratio and insensitivity to misalignment of optical elements. Advantages of the method include that since it is an scanning method, incoherent objects can be processed without the need of using spatial light modulators to convert incoherent image to coherent image. In addition, the system have a capability of producing the encrypted information in real-time. Another advantage include that the method is easily extendible to encrypt 3-D information. Since the output of encrypted signal can be made at radio frequency, the signal can be directly sent through an antenna to a secure site for wireless application. It is important to point out that the encryption key and the decryption key are of the same functional form and they are actually the same key. The proposed system enables us to store, transmit, and decrypt the encrypted data digitally.

This work supported in part by Regional Research Center of Korea Science and Engineering Foundation (R12−2001−051−00007−0).

References

1. Diffe, W., Hellman, M.: New directions in cryptography, IEEE Tran. Information Theory, Vol. IT-22., 644-654 (1976)
2. Singh, S.: The code book, Anchor Books, Random House, Inc. New York(1999)
3. Javidi, B., Zhang, G., Li, J.: Experimental demonstration of the random phase encoding technique for image encryption and security verification, Opt. Eng. 35, 2506-2512 (1996)
4. Wang, B., Sun, C.C., Su, W.C., Chiou, A.: Shift-tolerance property of an optical double-random phase-encoding encryption system, Appl. Opt. 39, 4788-4793 (2000)
5. Magensen, P.C., Gluckstad, J.: Phase-only optical decryption of a fixed mask, Appl. Opt. 8, 1226-1235 (2001)
6. Tajahuerce, E., Lancis, J., Javidi, B., Andres, P.: Optical security and encryption with totally incoherent light, Opt. Lett. 26, 678-680 (2001)
7. Tajahuerce, E., Javidi, B.: Encrypting three-dimensional information with digital holography, Appl. Opt. 39, 6595-6601 (2000)
8. Javidi, B., Nomura, T.: Securing information by means of digital holography, Opt. Lett. 25, 28-30 (2000).
9. Poon, T.C.: Scanning holography and two-dimensional image processing by acoutooptic two-pupil synthesis, J. Opt. Soc. Am. A 2, 521-527 (1985)
10. Indeberouw, G., Poon, T.C.: Novel Approaches of Incoherent Image Processing with Emphasis on Scanning Methods, Optical Engineering, 31, 2159-2167 (1992)
11. Doh, K., Poon, T.C., Indebetouw, G.: Twin-image noise in optical scanning holography, Opt. Eng. Vol. 35, no. 6, 1550-1555 (1996)
12. Poon, T.C., Qi, Ying.: Novel real-time joint-transform correlation by use of acoustooptic heterodyning, App. Opt. 42, 4663-4669 (2003)
13. Poon, T.C., Doh, K., Schilling, B., Wu, M., Shinoda, K., Suzuki, Y.: Three-dimensional microscopy by optical scanning holography, Opt. Eng. Vol. 34, No. 5, 1338-1344 (1995)
14. Kim, T., Poon, T.C., Indebetouw, G.: Depth detection and image recovery in remote sensing by optical scanning holography, Opt. Eng. 41(6), 1331-1338 (2002)
15. Schilling, B.W., Templeton, G.C.: Three-dimensional remote sensing by optical scanning holography, App. Opt. 40, 5474-5481 (2001)

Simulation on the Interruptible Load Contract

Jianxue Wang[1], Xifan Wang[1], and Tao Du[2]

[1] School of Electrical Engineering, Xi'an Jiaotong University, Xi'an, 710049, P.R. China
JXWang@mailst.xjtu.edu.cn
[2] School of Electronics and Electric Engineering, Shanghai Jiaotong University,
Shanghai, 200030, P.R. China

Abstract. Interruptible load management (ILM) is an important part of Demand Side Management (DSM), especially in power market. It has deep influences on reliability and economical operation of power system. How to use a rational model to simulate the ILM becomes a hot topic in China. This paper gives the optimal purchase model of interruptible contract and sets a reasonable selected rule, i.e. the interruptible load should be priorly scheduled in the period with maximal shortage capacity. Based on the rule, the heuristic method is employed to solve the problem. Simulation results of real power system demonstrate the usefulness of the proposed model.

1 Introduction

Demand Side Management (DSM) has been promoted worldwide. Considering the customer benefit, DSM provides a pathway for system operation and customer participation. Peak clipping reduces electricity demand during on-peak periods of the day, and it is the most important part of DSM [1]. In many power markets, the peak clipping is implemented through an interruptible load program (ILP) or interruptible load management (ILM). Customer enters into contract with the ISO or the power utility to reduce its demand as and when requested. The ISO/utility benefits by way of reducing its peak load and thereby saving costly reserve, restoring quality of service and ensuring reliability. The customer benefits from reduction in its energy costs and incentives provided by the contract [2].

But how to use a rational model to simulate the ILM is always a difficult problem. The prerequisite in ILM is to determine the contract contents. These have been discussed in papers [3, 4]. According to interruption duration, effective period and minimum curtailment, the interruptible load contract in China could be classified into two main types as shown in Table 1.

Table 1. The types of interruptible load contracts in China

Type	Effective period	Minimum curtailment	Duration
Type A	6 months	500kW	=4hours
Type B	6 months	500kW	=8hours

2 The Optimal Purchase Model of Interruptible Load

2.1 Calculation of Shortage Capacity

Because of deviation in load forecasting and generator outage, the capacities purchased in advance sometimes are not enough. The technology of Capacity Outage Probability Table (COPT) could be used to evaluate the shortage distribution [5]. In traditional method, the probability distribution of load is gotten from the sequential load curve. But in the improved method, the load probability distribution is formed on the basis of load fluctuation. Statistical material shows that the probability distribution of load fluctuation is normal distribution near the predictive load. The standard deviation δ keeps inverse proportion with square root of the load capacity. For example, δ is 3% when the load capacity is 2000MW, and δ is 1% when the load capacity is 20000MW. Using the technology of COPT, we can get the vector of shortage capacity ($g_1, g_2, ..., g_i, ..., g_M$) and the vector of corresponding probability $[p(g_1), p(g_2), ..., p(g_i), ..., p(g_M)]$. Here g_i is the shortage capacity of state i, $p(g_i)$ is the probability of g_i, and M is total number of the states. So the shortage capacity expectation $\Delta C(t)$ of the period t could be expressed as:

$$\Delta C(t) = \sum_{i=1}^{M} g_i \times p(g_i) \tag{1}$$

2.2 The Optimal Purchase Model

Based on the above discussion, the large user could first select the contract type, and then give the bidding information of the interruptible load, including the interruptible capacity, price, and total interruption time. After the system load and the spot price in balance market are forecasted through some method, the electrical utilities could use the optimal model to buy the interruptible load. The objective function of optimal purchase model is described as below:

$$\min f = \sum_{t=1}^{t_n} \sum_{i=1}^{n} S_n(t,i) P_B(t,i) C_B(t,i) \tag{2}$$

Where

t	Index of study intervals	t_n	Number of study intervals
i	Index of large users	n	Number of large users

$S_n(t,i)$ Selected state of the large user i at interval t. 0-1 Variable.
$P_B(t,i)$ Bidding price of the large user i at interval t
$C_B(t,i)$ Bidding capacity of the large user i at interval t

It must be pointed out that the balance market also could reduce the shortage capacity, so the market is regarded as the large-capacity user to participate in the bidding process.

Total Interruptible Capacity Limit. The total capacity of interruptible load is constrained by the total system shortage capacity $\Delta C(t)$:

$$\sum_{i=1}^{n} s_n(t,i) C_B(t,i) \leq \Delta C(t) \quad t = 1,...,t_n \qquad (3)$$

Notes that the $\Delta C(t)$ could be calculated by Eq.1 in the section 2.1.

Duration Limit. The duration time is specified in the contract as:

$$s_n(t+m,i) = 1 \ (m = 1,...,n_d(i)-1) \text{ if } s_n(t-1,i) = 0 \text{ and } s_n(t,i) = 1 \qquad (4)$$

Where $n_d(i)$ is the required duration of every interruption for user i. As shown in Table 1, the value in this model is 4 hours and 8 hours.

Total Interruption Time Limit. The large user always requires the total interruption time is no more than some hours, and this requirement could be expressed as:

$$\sum_{t=1}^{t_n} s_n(t,i) \leq n_s(i) \quad i = 1,...n \qquad (5)$$

Where $n_s(i)$ is total required interruption time of user i.

Time Interval Limit. The large user usually hopes that the curtailment time has an enough interval between two interruptions, and this limit can be expressed as:

$$|r-t| \geq n_i(i) \text{ if } s_n(t-1,i) = 0, \ s_n(t,i) = 1 \text{ and } s_n(r-1,i) = 0, \ s_n(r,i) = 1 \qquad (6)$$

Where $n_i(i)$ is the required time interval of user i between two interruptions.

3 Solution of the Model

Referring to the scheduled method of unit maintenance, the model could be solved. The unit maintenance is to solve how to arrange the units with known repair time to rational period. Some rules such as equivalent reserve or equivalent risk are used to schedule the unit [5].

The optimal schedule of interruptible load is to solve how to arrange the interruptible load with required duration time to rational period. The key point of the solution is to determine the selected strategy. A reasonable selected strategy is presented in the paper that the interruptible load should be priorly scheduled in the period with maximum shortage capacity. After the interruptible load queue is formed according to the bidding price, the interruptible load could be scheduled through merit order method on the basis of this selected strategy. The solution flowchart is given in Fig.1a. And the key process A of arranging the ith interruptible load in Fig.1a is specially described in Fig.1b.

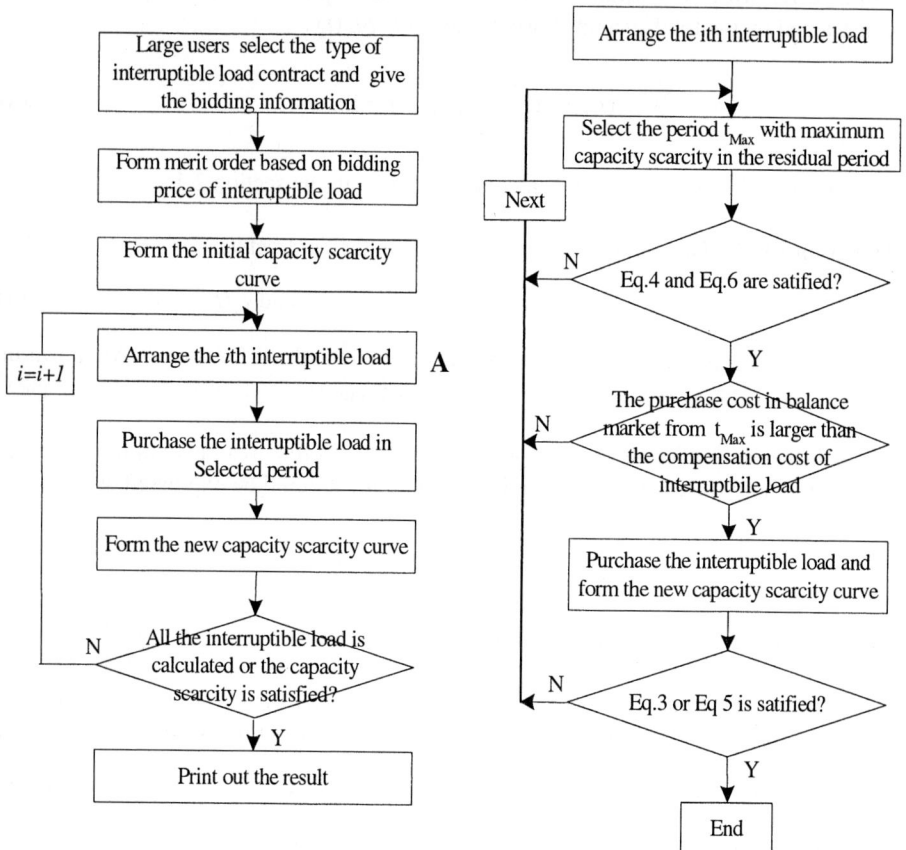

Fig. 1a. The solution flowchart of interruptible load model by merit order method

Fig. 1b. The arrange flowchart of the interruptible load i

4 Case Studies

A real system in China and the related customer surveys in this system are employed to demonstrate the proposed model. The system has 43 large users with more than 500 kW capacity, but only a few users prefer to take part in the ILM if the compensation is no more than 2$/kWh. Table 2 gives the weekly bidding information of large users.

Table 2. The weekly bidding information of large users

User No.	Total interruption time	Duration (hour)	Capacity (kW)	Compensation ($/kWh)
1	2	8	3300	0.66
2	3	4	2200	0.94
3	4	4	4600	0.58
4	2	4	2300	0.75
5	2	4	1800	0.6
6	4	8	800	1.4
7	2	4	2200	1.6
8	2	4	3100	0.88
9	2	4	600	1.2
10	2	8	1600	0.95

Table 3. The utility cost reduction of one week ($)

Mon.	Tues.	Wed.	Thurs.	Fri.	Sat.	Sun.	SUM
9972	7590	7590	6458	6513	630	0	38753

After the implementation of ILM, the utility could reduce the purchase cost as shown in Table 3.

Based on the solution in Fig.1a and Fig.1b, the optimal schedule of interruptible load can be gotten. Users 1,3,4,5,8 have been selected. The first selected large user 3 and the last selected large user 8 are chosen to act as examples in Fig.2a and Fig.2b to illustrate the shortage capacity clipping before and after buying interruptible load.

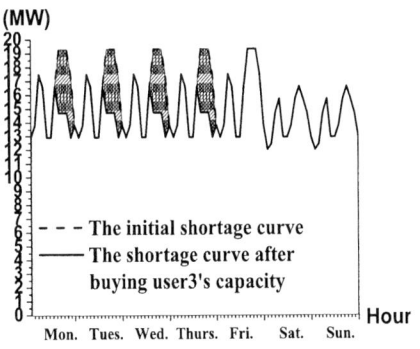

Fig. 2a. The comparison of shortage curve before and after buying user3's capacity

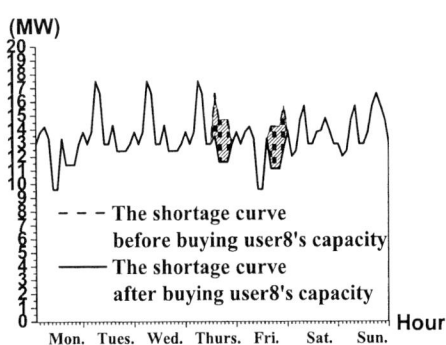

Fig. 2b. The comparison of shortage curve before and after buying user8's capacity

In Fig. 2a and Fig. 2b, the scheduled periods of interruptible load are selected in peak load period and this will effectively reduce the system shortage capacity in load period. Furthermore, the compensation cost is rather high. For the customer with low reliability required, this rational compensation could greatly enhance their enthusiasm to participate in ILM. At the same time, table 3 shows that after the implementation of the ILM, the utility also reduces the purchase cost in high-risk balance market. The cost reduction of utility is $38753 for one week and it reaches $2.01 millions for one year. In summarization, the simulation results show that both participants of the interruptible load management have obvious economic benefits. These will deeply reduce the resistance of the ILM promotion.

5 Conclusions

This paper discusses the content of the interruptible load contract. The contract should be signed per half a year and the duration contains two types: four hours and eight hours. Considering the load stochastic fluctuation, an improved technique of capacity outage probability table is presented to calculate the system shortage capacity. Then the optimal purchase model of interruptible load is given and heuristic method is employed to solve the model. Case study shows that the customers and the utilities both have obvious economic benefits from the ILM implementation. These will deep reduce the resistance of the ILM promotion.

References

1. Malik, A.S.: Simulation of DSM Resources as Generating Units in Probabilistic Production Costing Framework. IEEE Trans. on Power System Vol. 13(4). IEEE(1998) 460-465.
2. Tuan, L.A., Bhattacharya, K.: Interruptible Load Management within Secondary Reserve Ancillary Service Market. IEEE Porto Power Tech Conference. Porto (2001)1-6.
3. Wang, J., Wang, X., Zhang, X.: The Interruptible Load Operation in Power Market and Interim System. Electric Power Automation Equipment, Vol. 24(6). Nanjin(2004) 1-5.
4. Doudna, J.H.: Overview of California ISO Summer 2000 Demand Response Programs. Proceedings of Power Engineering Society Winter Meeting. IEEE(2001) 228-233.
5. Wang, X.: Power System Planning Fundamentals. China Electric Power Press, Beijing (1994) 54-62.

Consistency Conditions of the Expert Rule Set in the Probabilistic Pattern Recognition

Marek W. Kurzynski

Wroclaw University of Technology, Faculty of Electronics, Chair of Systems and Computer Networks, Wyb. Wyspianskiego 27, 50-370 Wroclaw, Poland
Marek.Kurzynski@pwr.wroc.pl

Abstract. The present paper is devoted to the pattern recognition procedure based on the set of expert rules with unprecisely formulated weights understood as appropriate probabilities. Adopting the probabilistic model the different interpretations of rule weight are discussed and the consistency conditions of set of rules are given.

1 Introduction

In statistical pattern recognition, the knowledge about probabilistic characteristics of features and classes must be known so as optimal (Bayes) decision rule could be implemented. If such knowledge is unknown or incompletely defined, a possible approach is to design a system, which will acquire the pertinent information from the actually available data for constructing a decision rule.

This process of knowledge acquisition, called learning, in statistical pattern recognition may be viewed as the problem of estimation (parametric and/or nonparametric) of unknown probability distribution of features and classes. Usually it is assumed that learning data are of numerical nature, ie. they form observations or measurements of features of learning patterns. In this paper we shall focus our attention on another approach, interesting from both theoretical and practical point of view, which supposes that appropriate information is contained in expert knowledge. A typical knowledge representation consists of rules of the form IF A THEN B with the weight (uncertainty measure) α. These rules are obtained from the expert as his/her conditional beliefs: if A is known with certainty then the expert's belief into B is α [1], [2], [3]. Furthermore we suppose, that expert rules are not provided with exact value of α, but only an interval is specified (by its upper and lower bounds), into which this value belongs.

Adopting statistical model we discuss different probabilistic interpretations of weight coefficient for which the conditions of consistency of the set of expert rules are given.

This paper is a sequel to the author's earlier publications [4], [5], [6], [7], [8] and it yields an essential extension of the results included therein.

2 Preliminaries

Let us consider the pattern recognition problem with probabilistic model. This means that vector of features describing recognized pattern $x \in X \subseteq \mathcal{R}^d$ and its class number $j \in \mathcal{M} = \{1, 2, ..., M\}$ are observed values of a couple of random variables X and J, respectively. Its probability distribution is given by *a priori* probabilities of classes

$$p_j = P(J = j), \quad j \in \mathcal{M}, \tag{1}$$

and class-conditional probability density functions (CPDFs) of X

$$f_j(x) = f(x/j), \quad x \in X, \quad j \in \mathcal{M}. \tag{2}$$

Let us now consider the interesting from practical point of view concept of recognition. We assume that *a priori* probabilities (1) and CPDFs (2) are not known, whereas the only information on the probability distribution of J and X is contained in the set of expert rules

$$R = \{R_1, R_2, ..., R_M\}, \tag{3}$$

where

$$R_i = \{r_i^{(1)}, r_i^{(2)}, ..., r_i^{(K_i)}\}, \quad i \in \mathcal{M}, \quad \sum K_i = K \tag{4}$$

denotes the set of rules connected with the i-th class. The rule $r_i^{(k)}$ has the following general form:

IF $w_i^{(k)}(x)$ THEN $J = i$ WITH probability greater than $\underline{p}_i^{(k)}$ and less than $\overline{p}_i^{(k)}$.

$w_i^{(k)}(x)$ denotes a predicate depending on the values of the features x, which determines in the feature space so-called rule-defined region:

$$D_i^{(k)} = \{x \in X : w_i^{(k)}(x) = true\}. \tag{5}$$

Let $X_R \subseteq X$ denote feature subspace covered by the set of rules R. It is clear, that

$$X_R = \bigcup \mathcal{D}, \tag{6}$$

where

$$\mathcal{D} = \{D_i^{(k)}, i \in \mathcal{M}, k = 1, 2, ... K_i\}. \tag{7}$$

The rule $r_i^{(k)}$ denotes some restrictions imposed on probability distribution of X and J, which can be equivalently expressed in the following inequalities:

$$\underline{p}_i^{(k)} \leq p_i^{(k)} \leq \overline{p}_i^{(k)}. \tag{8}$$

Analysis of different relations between decisions and features has led to the threefold interpretation of probability $p_i^{(k)}$ and meaning of bounds $\underline{p}_{-i}^{(k)}$ and $\overline{p}_i^{-(k)}$ and in consequence, to the following kinds of rules:

1. **The 1st-type rule** – now $p_i^{(k)}$ denotes the set of values of *a posteriori* probability of *i*th class for x belonging to the $D_i^{(k)}$:

$$p_i'^{(k)} = \{p_i(x) \text{ for } x \in D_i^{(k)}\} \tag{9}$$

and inequalities (8) refer to all *a posteriori* probability values from the set (9).

2. **The 2nd-type rule** – $p_i^{(k)}$ is *a posteriori* probability of *i*th class on condition that x belongs to the $D_i^{(k)}$, viz.

$$p_i''^{(k)} = P(J = i / x \in D_i^{(k)}) = \int_{D_i^{(k)}} p_i(x) f(x) \, dx \Big/ \int_{D_i^{(k)}} f(x) \, dx. \tag{10}$$

3. **The 3rd-type rule** – $p_i^{(k)}$ denotes mean *a posteriori* probability of *i*th class in the set $D_i^{(k)}$:

$$p_i'''^{(k)} = \int_{D_i^{(k)}} p_i(x) \, dx \Big/ \int_{D_i^{(k)}} dx \tag{11}$$

The sense of weights in expert rules plays the key role in deriving procedures of recognition algorithms and also in formulating of consistency conditions of set of expert rules. This problem will be discussed in the next section.

3 Consistency of the Expert Rule Set

In order to determine consistency conditions let introduce first family of sets $\mathcal{B} = \{B^{(1)}, B^{(2)}, ..., B^{(N)}\}$ where $B^{(n)}$ denote not empty constituents of family $\mathcal{D}$. It is clear, that sets $B^{(n)}$ are disjoint and furthermore $X_R = \bigcup \mathcal{B}$, i.e. family $\mathcal{B}$ forms partition of feature subspace X_R [9].

Let next $I_i^{(n)}$ be the set of indices of rules from R_i fulfilling the conditions $w_i^{(k)}(x)$ for $x \in B^{(n)}$ or equivalently:

$$I_i^{(n)} = \{k : B^{(n)} \subseteq D_i^{(k)} \in \mathcal{D}\} \tag{12}$$

The consistency conditions and the appropriate procedures of consistency checking are different for the rules of the 1st-, 2nd- and 3rd-type.

3.1 Consistency of the 1st-Type Rules

The problem of consistency of the 1st-type rules is quite simple. According to its definition, the 1st-type rule $r_i^{(k)}$ determines the interval to which belongs the *a posteriori* probability $p_i(x)$ at each point of $D_i^{(k)}$. Inconsistency may occur between rules indicating the same class or between rules for different classes. Inconsistency between rules for the same class occurs in the set $B^{(n)}$ if:

$$\overline{p}_i^{(n)} < \underline{p}_i^{(n)}, \text{ where } \overline{p}_i^{(n)} = \min_{k \in I_i^{(n)}} \overline{p}_i^{(k)}, \quad \underline{p}_i^{(n)} = \max_{k \in I_i^{(n)}} \underline{p}_i^{(k)}. \tag{13}$$

It means that among the rules active for $x \in B^{(n)}$ and indicating the ith class there is such a rule for which the upper bound of the probability $p_i(x)$ is less than the lower bound of this probability given in another rule.

Inconsistency between rules for various classes occurs in $B^{(n)}$ if one of the following conditions is satisfied:

- the sum of the lower bounds maxima of the *a posteriori* probabilities for all classes for which there exist active rules in $B^{(n)}$ is greater than 1, i.e.

$$\sum_{i \in \mathcal{M}^{(n)}} \underline{p}_i^{(n)} > 1, \tag{14}$$

where $\mathcal{M}^{(n)} = \{i : I_i^{(n)} \neq \emptyset\}$,

- the set of rules active for $x \in B^{(n)}$ is locally complete ($\mathcal{M}^{(n)} = \mathcal{M}$) and the sum of the minima of the upper bounds of *a posteriori* probabilities is less than 1, i.e.

$$\mathcal{M}^{(n)} = \mathcal{M} \wedge \sum_{i \in \mathcal{M}^{(n)}} \overline{p}_i^{(n)} < 1. \tag{15}$$

Thus, the consistency conditions for the set of the 1st-type rules are as follows:

$$\forall B^{(n)} \in \mathcal{B} \wedge \forall i \in \mathcal{M}: \overline{p}_i^{(n)} \geq \underline{p}_i^{(n)},$$

$$\forall B^{(n)} \in \mathcal{B}: \sum_{i \in \mathcal{M}^{(n)}} \underline{p}_i^{(n)} \leq 1, \tag{16}$$

$$\forall B^{(n)} \in \mathcal{B}: \mathcal{M}^{(n)} \neq \mathcal{M} \vee \sum_{i \in \mathcal{M}^{(n)}} \overline{p}_i^{(n)} \geq 1.$$

3.2 Consistency of the 2nd-Type Rules

In order to define the consistency conditions for the 2nd-type rules set let us introduce the following notation:

$$\tilde{p}^{(n)} = P(x \in B^{(n)}) = \int_{B^{(n)}} f(x)\, dx, \tag{17}$$

$$\tilde{p}_i^{(n)} = P(J = i / x \in B^{(n)}) = \int_{B^{(n)}} p_i(x) f(x)\, dx \Big/ \int_{B^{(n)}} f(x)\, dx, \tag{18}$$

for $n = 1, 2, \ldots, N$, $i \in M$. Then probability (10) can be expressed by (17) and (18) as follows:

$$p_i''^{(k)} = \frac{\sum_{n:\, B^{(n)} \subseteq D_i^{(k)}} \tilde{p}^{(n)} \tilde{p}_i^{(n)}}{\sum_{n:\, B^{(n)} \subseteq D_i^{(k)}} \tilde{p}^{(n)}}. \tag{19}$$

Hence, the set R of the 2nd-type rules is consistent if and only if there exist such $\tilde{p}^{(n)}$ and $\tilde{p}_i^{(n)}$ ($n = 1, 2, \ldots, N$, $i \in M$) for which the following set of inequalities is satisfied:

$$0 \leq \tilde{p}^{(n)} \leq 1, \quad n = 1, 2, \ldots, N,$$

$$0 \leq \tilde{p}_i^{(n)} \leq 1, \quad n = 1, 2, \ldots, N, \; i \in M, \tag{20}$$

$$\underline{p}_i^{(k)} \leq \frac{\sum_{n:\, B^{(n)} \subseteq D_i^{(k)}} \tilde{p}^{(n)} \tilde{p}_i^{(n)}}{\sum_{n:\, B^{(n)} \subseteq D_i^{(k)}} \tilde{p}^{(n)}} \leq \overline{p}_i^{(k)}, \quad i \in M, \; k = 1, 2, \ldots, K_i.$$

3.3 Consistency of the 3rd-Type Rules

Let for $n = 1, 2, \ldots, N$

$$\tilde{\tilde{p}}^{(n)} = \int_{B^{(n)}} p_i(x)\, dx / V^{(n)}, \text{ where } V^{(n)} = \int_{B^{(n)}} dx \text{ - volume of } B^{(n)}, \tag{21}$$

then from (11) we have:

$$p_i'''^{(k)} = \frac{\sum_{n:\, B^{(n)} \subseteq D_i^{(k)}} \tilde{\tilde{p}}^{(n)}}{\sum_{n:\, B^{(n)} \subseteq D_i^{(k)}} V^{(n)}}. \tag{22}$$

Hence, the set R of the 3rd-type rules is consistent if and only if there exist such $\tilde{\tilde{p}}^{(n)}$ ($n = 1,2,...,N$) for which the following set of inequalities is satisfied:

$$0 \leq \tilde{\tilde{p}}^{(n)} \leq 1,$$

$$\underline{p}_i^{(k)} \leq \frac{\sum_{n: B^{(n)} \subseteq D_i^{(k)}} \tilde{\tilde{p}}^{(n)}}{\sum_{n: B^{(n)} \subseteq D_i^{(k)}} v^{(n)}} \leq \overline{p}_i^{(k)}, \quad (23)$$

$n = 1,2,...,N, \; i \in \mathcal{M}, \; k = 1,2,...,K_i$.

4 Final Remarks

In this paper we have focused our attention on the interesting from practical point of view problem of pattern recognition via unprecisely formulated expert rules (knowledge representation). Adopting the probabilistic model of classification, i.e. assuming that feature vector and class number are observed values of random variables, we discuss different probabilistic interpretations of weight coefficient for which the conditions of consistency of the set of expert rules are given.

As far as knowledge representation with uncertainty characteristics is considered, the notion of consistency is based on the assumed properties of uncertainty measure. In the case of the approach being considered the gathered knowledge concerns the probabilistic properties of the population and therefore the consistency conditions were considered in the probabilistic bearing.

References

1. Dubois, D., Lang, J., Possibilistic Logic, In: Handbook of Logic in Artificial Intelligence and Logic Programming, Oxford Univ. Press, (1994) 439-513
2. Mitchell, T., Machine Learning, McGraw-Hill Science (1997)
3. Halpern, J., Reasoning about Uncertainty, MIT Press (2003)
4. Kurzynski, M., Sas, J. and Blinowska, A., Rule-Based Medical Decision-Making with Learning, Proc. 12th World IFAC Congress, Vol. 4, Sydney (1993) 319–322.
5. Kurzynski, M., The Application of Combined Recognition Decision Rules to the Multistage Diagnosis Problem, 20th Int. Conf. of IEEE EMBS, Hong-Kong (1998) 1194-1197.
6. Kurzynski, M., Sas, J., Rule-Based Classification Procedures Related to the Unprecisely Formulated Expert Rules, Proc. SIBIGRAPI Conference, Rio De Janeiro (1998) 241-245.
7. Kurzynski, M., Wozniak, M., Rule-Based Algorithms with Learning for Sequential Recognition Problem, Proc. 3rd Int. Conf. Fusion 2000, Paris (2000) 10-13
8. Kurzynski, M., Puchala, E., Hybrid Pattern Recognition Algorithms Applied to the Computer-Aided Medical Diagnosis, Medical Data Analysis, LNCS 2199, Springer Verlag, (2001) 133-139
9. Kuratowski, K., Mostowski, A., Set Theory, Nort-Holland Publishing Co, Amsterdam (1986)

An Agent Based Supply Chain System with Neural Network Controlled Processes

Murat Ermis[1], Ozgur Koray Sahingoz[2], and Fusun Ulengin[3]

[1] Dept. of Industrial Engineering, Turkish Air Force Academy,
34149 Yesilyurt, Istanbul, Turkey
[2] Dept. of Computer Engineering, Turkish Air Force Academy,
34149 Yesilyurt, İstanbul, Turkey
{ermis, sahingoz}@hho.edu.tr
[3] Dept. of Industrial Engineering, Istanbul Technical University,
80680 Macka, Istanbul, Turkey
ulengin@itu.edu.tr

Abstract. A supply chain refers to any system which consists of multiple entities (companies or business units within an enterprise), that depend on each other in some way in conducting their businesses. In a supply chain, sales forecasting is highly complex due to the influence of internal and external environments. However, a reliable prediction of sales can improve the business strategy. Agent-based technology is considered suitable in providing near-optimal adaptive business and knowledge management strategies to help managers reduce both mental effort and search costs. This paper presents an agent-based approach which supports mobile agents as mediators between system entities. The proposed mobile agent-based system uses the publish/subscribe communication mechanism; therefore, system entities (like customers and suppliers) can dynamically connect and disconnect to the system at any time. The system uses a "Two-Leveled Mobile Agent Structure" and some design details of the sales forecasting process of the system based on a neural network approach are presented.

1 Introduction

A supply chain is a connected series of organizations (e.g., suppliers, equipment manufacturers, distributors, transporters, etc.), resources and activities involved in the criterion and delivery of value, in the form of both finished products and services to end-customers [1,2]. Supply chains exist in both service and manufacturing organizations. Realistic supply chains have multiple end-products with shared components, facilities and capacities. Traditionally, marketing, distribution, planning, manufacturing, and purchasing operate independently along the supply chain. These departments have their own objectives and they are thus often in conflict.

Over the past few decades, we have witnessed an increasing globalization of the economy and thereby also of supply chains. Products are no longer produced and consumed within the same geographical area; even the different parts of a product may, and often do, come from all over the world. This creates longer and more complex supply chains, and therefore it also changes the requirements within supply chain management. Business firms are increasingly embracing integrated supply chains because they promise cost reduction, efficiency, and effective fulfillment of market demand [3]. The rapid development within the information and software engineering sectors has given rise to unprecedented opportunities for integration and coordination, whereby information technology can help overcome the uncertainties of supply chain management. The electronic exchange of information can lead to a reduction of errors and to an increased efficiency of the processes involved. When one company has access to other companies' information in the supply chain, the negative effects of uncertainty (i.e., inaccurate forecasts, higher inventory levels, etc.) can be reduced.

It is not possible to meet extended enterprise requirements under the present configuration of supply chain systems. Therefore, to cope with increasing complexity and organizational issues, intelligent agent architectures are envisaged to build and operate extended manufacturing enterprises. These agents provide a new and important paradigm for developing industrial distributed systems and a number of researchers have attempted to apply agent technology to manufacturing enterprise integration, supply chain management, manufacturing scheduling and control [4–9]. Agents help in the automating of a variety of tasks, including those involved in buying and selling products over the internet. Agent activities in terms of products required and supplied are defined so as to reduce an agent's decision problem as to how to evaluate the trade-offs of acquiring different products. Such agents are designed to reduce a user's information overload or search costs. The difference between such systems and conventional programs is that they are domain-specific, continuously running and autonomous.

This paper presents an agent-based approach for supporting the coordination and control mechanisms of mobile-agent based supply chains for knowledge management with a formal approach by using neural network based analytical tools. The remainder of the paper is organized as follows: in Section 2, we focus on e-supply chain systems in Section 3, we discuss mobile-agent based customer-supplier systems and their subsystems, such as Control/Optimize Services, the sales forecasting process of a supply chain systems based on the neural network approach is given in Section 4; in Section 5, the system performance is evaluated, and Section 6 presents some concluding remarks.

2 E-Supply Chain Systems

Both the functionality and the complexity of the integration level of e-supply chain systems may vary from highly to partially integrated. The highly e-integrated e-supply chain systems generally have fairly complex internal and external operations. They might have many suppliers, spread all over the world, who supply a large variety of parts and components. In order to respond to competitive challenges (e.g., maintaining customer service levels, lowering the inventory, etc.), an e-supply chain system that would efficiently and effectively link-up complex operations can be installed.

Three major players exist in these systems: manufacturers, suppliers, and distributors (Fig.1); logistics and warehousing are also included. The system performs two major tasks geared towards achieving competitive advantage by way of reducing inventory in the chain, avoiding delays, and providing better customer service; the two tasks are: 1) cooperative planning among various members of the supply chain and, 2) responding to customer query [10].

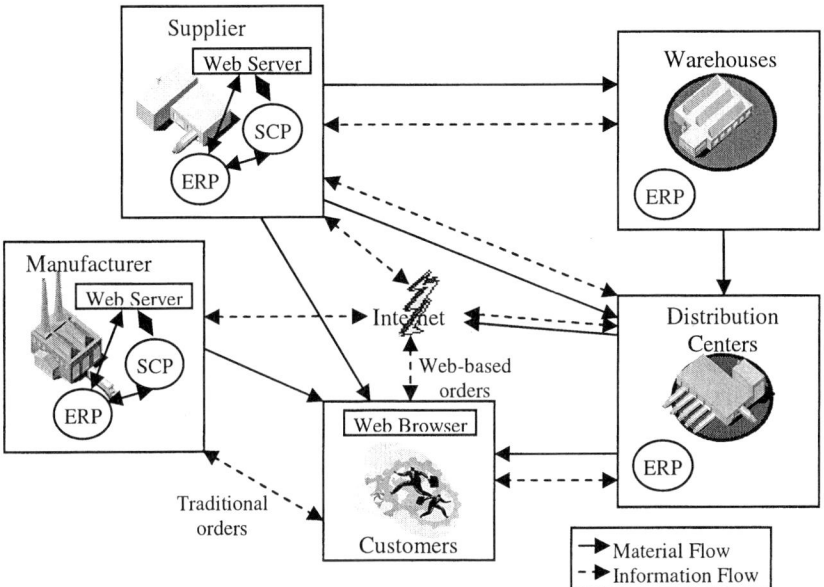

Fig. 1. Order fulfillment process in the integrated supply chain systems

The wide geographic spread of the company's operations, as well as the number of items it sells and different stocking policies for items based on demand, make internal operations exceedingly complex. Due to the involvement of a large number of customers and suppliers, external operations are equally complex. Both internal and external operations are very important for the success of the companies.

To effectively implement the agent-based supply chain systems, two enabling tools need to be developed: first, data and information transformation tools which are used for data gathering from various machinery systems and processes exist at various levels (described in Section 3) and second, sales forecasting tools, which include advanced prediction methods and tools which need to be developed in order to predict the future pattern of sales (described in Section 4).

3 System Design

The procurement and sales forecasting processes of a supply chain system are designed in a large-scale and dynamic environment, in which there can be any number of customers and suppliers at any time. For achieving this goal, our system architecture is based on the publish/subscribe paradigm, which supports the many-to-many interaction of loosely coupled entities. In this model, customers know neither the number of the suppliers nor their addresses. A customer sends its request to the Manager, by means of a mobile agent, and the Manager then dispatches the clones of this mobile agent to the selected suppliers. We define a set of services that need to collaborate with the publish/subscribe infrastructure to address the dynamics of mobile environments. Our work exploits mobile agent technology extensively. Not only does it support the activities of customers and suppliers, it also facilitates the **parallel computation** by running mobile agents on suppliers concurrently.

The system uses mobile agents as mediators between customers and suppliers. Mobile agents belong to two different levels of execution (explained with details in [11]) and responsibilities, as shown in Fig. 2: Manager Level Mobile Agents (MMA) and Supplier Level Mobile Agents (SMA). An MMA is created by a Customer Agent and is sent to the Manager. This MMA creates SMAs and sends them to suppliers in order to search their databases, to select among products and to negotiate with the supplier. We use the *Contract Net Negotiation Protocol* [12], which facilitates distributing subtasks among various agents. The agent wanting to solve the problem broadcasts a call for bids, waits for a reply for some length of time, and then awards a contract to the best offer(s) according to its selection criteria. There is not just one single mobile agent in the system that visits every supplier one by one; instead, clones of the mobile agent are sent to each of the suppliers concurrently. This model of parallel computation is especially important as more suppliers can be searched in a shorter time to provide customers with better choices in their decision-making.

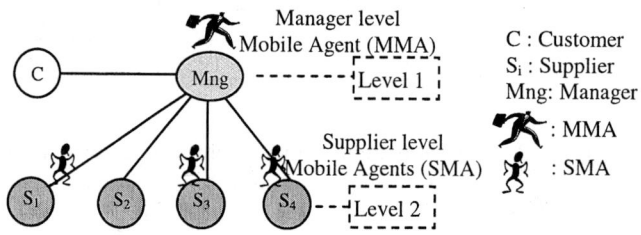

Fig. 2. A Two-Leveled Mobile Agent Model

In many electronic commerce and supply chain systems, a customer (or the system) has a fixed number of suppliers, which are initialized into the system at start up. When a new supplier is to be added, it has to be registered manually by supplying its address and the necessary parameters. In a dynamically changing supply chain environment, a system should have the ability to adapt itself to this dynamic world. To meet this requirement, we have designed an architecture that utilizes the publish/subscribe paradigm for registration and dispatching operations, with the aim of increasing the efficiency and effectiveness of the supply chain system in terms of costs, quality, performance and time, for both customers and suppliers.

In our approach, the supply chain models are composed from software components that represent types of static supply chain agents (like Customer Agent,

Controller Agent and Supplier Agent) and mobile supply chain agents (like Manager Level Mobile Agent and Supplier Level Mobile Agent). Agent descriptions provide an ability to specify both static and dynamic characteristics of various supply chain entities. Each agent is specialized according to its intended role in the supply chain. Since it is vital to reduce both reduce inventory in the chain and avoid delays, as well as to provide better customer service in order to achieve competitive advantage, the sales forecasting process takes on a critical role in the chain. Thus, an accurate and timely sales forecast can be achieved by using non-parametric /semi-parametric approaches, such as artificial neural networks, as illustrated in Section 4.

It is well known that the supply chain is a complex process and even a simple procurement process on a supplier can be complicated by supplying the sub-part of the product from other "tier II suppliers" or from raw material vendors. In this system, we assume that the suppliers can make these products without purchasing components from other suppliers or vendors.

The system has an extendable architecture, as depicted in Fig. 2, and provides all the services which are essential to agent-based commercial activities. Our supply chain sales forecasting process system involves three main actors: 1) customers, who are looking for purchase services from suppliers; 2) suppliers or sellers, who offer the services or products, and; and 3) a Control/Optimize Service (Manager), who facilitates selection of the suppliers and communication between customers and these suppliers.

Customer Subsystem: To request a purchase order from the system, a customer has to initialize a Customer Subsystem on its machine. Customers have to know the address (URL) of the manager agent that they will connect to, just like the URL's of well known web sites (i.e. Yahoo!, Alta Vista).

By allowing users to generate diverse Manager Level Mobile Agents with different behaviors, several transaction scenarios can be realized. A human user interacts with the Customer Agent, who supplies the order's necessary information (such as type and number of products, destination address, delivery date, etc.). The Customer Agent allows the users to control and monitor the progress of transactions and to query past transactions from its database.

The Customer Agent is the main process of the Customer Subsystem. It is a stationary agent created during the initialization step of the subsystem and is responsible for offering an interface to end users for inputting query tasks; it communicates with the Manager Level Mobile Agents, which it creates in order to accomplish his task. When a customer agent receives a purchase request from a user, it creates a Mobile Agent to search for product information and to perform goods or services acquisition in the system. The Customer Agent specifies the criteria for the acquisition of the product and dispatches the MMA to the manager. When this MMA reaches the best deal, it sends a result report to the Customer Agent and this information is added to its database.

Supplier Subsystem: A supplier has to initialize a Supplier Subsystem on its machine to join the system. When a supplier system is created for the first time, it subscribes to the system providing its address and the names of its products. If a supplier starts or stops delivering a product, it again subscribes or unsubscribes its products respectively. Every supplier agent has to know the address of the manager so that it can make a connection. The supplier agent subscribes to the manager by sending its product definitions and waits for customers to make requests for its products.

The Supplier Subsystem includes seven main components, shown in Fig. 3. The *Product Database* contains the services and product details of the supplier. The *Graphical User Interface* is used for interaction (i.e. inquiring about past and current transactions from the agent; a user may also specify selling strategies through this module) with human users. The *Mobile Agent Manager* controls and coordinates the incoming SMAs. The *Priority Manager* checks the waiting requests (SMA) and selects the one with highest priority. The *Class Loader* downloads the necessary classes from the Manager, while the *Supplier Agent* acts on behalf of the user. The *Neural Network* predicts future sales by using historical product data and updates the *Product Database* according to current sales information.

The Supplier Agent is also a stationary agent which is responsible for offering an interface to end users to enter information, control operations and input query tasks. The Supplier Agent processes purchase orders from Customer Agents and decides how to execute transactions according to sales strategies specified by the user. Since organizations differ in the products they sell, a Supplier Agent should be customized before it is placed online.

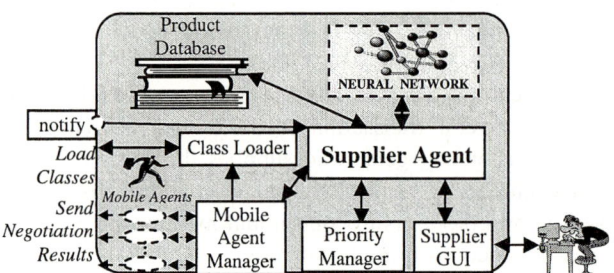

Fig. 3. Supplier Subsystem Architecture

The Control/Optimize Service (Manager): plays the most important role in our proposed Supply Chain system. It mediates between customers and suppliers in an electronic marketplace. The main component of the Control/Optimize Service is the "Manager Agent". The Manager Agent is useful when a supply chain system has many customers and suppliers and the search cost is relatively high, or when trust services are necessary. The inner structure of the Manager is shown in Fig. 4.

There are two main manager modules in the system; Knowledge-Base Manager Modules and Queue Manager Modules. The Knowledge-Base Manager manipulates two important databases, the *Knowledge Base,* which keeps statistical information about suppliers, customers and products in the system, according to message transactions between customer and supplier agents, and the *Subscription Base,* which keeps simple information, such as the names of services or products, which are delivered by suppliers. It also contains supplier *id* and *password* pairs for supplier authorization.

The Queue Manager has access to five major queues in the Manager. These queues are used for execution of manager modules concurrently: The *Class Queue* contains the names and addresses of the classes which need to be downloaded by the MMAs; the *MMA Queue* contains the serialized form of incoming MMAs from Customers; the *SMA Queue* contains the serialized form of SMAs outgoing to Suppliers; the *Result Queue* contains incoming results from the SMAs; the *Decision Queue* contains the decision data of a MMA, which is to be sent to the Customer Agent.

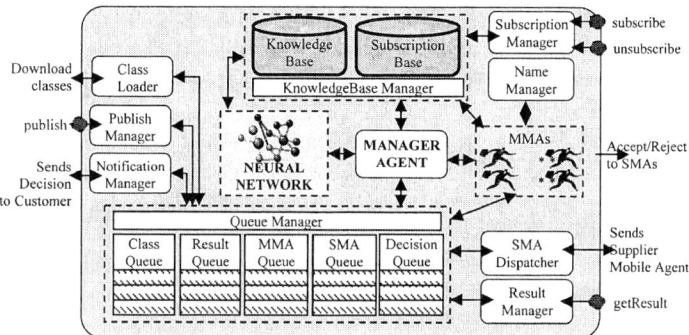

Fig. 4. Inner structure of the Manager

In these queues, each agent selects incoming messages in the system according to a first come first served (FCFS) model. The global list of incoming messages for all queues is sorted in terms of time of activation, with the agent with the earliest message processed next.

The manager consists of nine manager modules, which run in parallel in the system. The tasks of these managers and the execution flow of a procurement process in the Manager is as follows: the *Subscription Manager* gets subscription messages from suppliers via subscribe and unsubscribe methods, verifies their supplier ID and passwords with the Subscription Base, and continues with the necessary actions. The *Publish Manager* receives the serialized forms of incoming MMA's from the Customer Agent via the publish method and adds them to the MMA Queue. It also gets the names and addresses of the necessary class via the same method and adds them to the Class Queue. The *Class Loader* downloads classes in the Class Queue from the Customer Subsystem through a call to the "Download Class" method of the Customer Agent. The *Name Manager* is used for the SMAs' naming operations. The *SMA Dispatcher* sends each SMA in the SMA Queue to its target supplier, each SMA returning the result of its search and negotiation activity to the Manager via the *getResult* method. These results are then received by the *Result Manager* and inserted into the Result Queue. MMAs examine the results in the Result Queue carefully and make a decision after which, they create a decision report and put it into the Decision Queue. The *Notification Manager* sends each decision in the Decision Queue to its target Customer. The key component of the manager (as in supplier subsystem) is supplier sales forecasting, provided by the `Neural Network` module, detailed in Section 4.

The `Manager Agent` is the main component of the Control/Optimize System. It controls all operations in the system, evaluates incoming messages and generates lists of target suppliers by using the information in the Knowledge and Subscription Bases. The Manager Agent provides a platform for incoming mobile agents to run and create SMAs and dispatch them to the necessary suppliers. It also supports registering and dispatching operations in accordance with the publish/subscribe paradigm. The Manager Agent creates and activates a thread for each MMA in the MMA Queue. Each MMA has the authority to read and write both Knowledge and Subscription Base. A MMA searches the suppliers' information in these tables, creates an SMA for each supplier and thereafter puts the serialized form of each SMA into the SMA Queue.

4 The Sales Forecasting Process

An important part of the agent based supply chain system is the sales forecasting process, a process which is highly complex due to the influence of internal and external factors. It is the critical part of the system for a reliable prediction of sales; by calculating this, it can improve the quality of supply chain strategy. The forecasted values can serve as the basis for developing a distribution manufacturing requirements plans.

In view of this, we have selected the Artificial Neural Network (ANN) approach, a model-free approach which has recently been applied in forecasting due to its adequate performance in control and pattern recognition.

In our system, the general pattern of the sales is forecasted by a feed forward neural network with an EBP-learning algorithm. ANN's input layer with some neurons represents the previous sales data. It has been established that a single hidden layer is sufficient for an MLP network to learn all the non-linearities present in input data, provided there is a sufficient number of hidden nodes. Hence, only a single hidden layer architecture is considered. The input neurons use $y = f(x) = x$ while all other neurons generally have the sigmoidal function $y = f(x) = (1+\exp(-x))^{-1}$. Let us assume that the net consists of a p, n_h and n_o number of neurons in the input, hidden and output layers. The weight matrices $\mathbf{W}^{ih}$, for links from input layer i to hidden layer h, and $\mathbf{W}^{ho}$ from hidden layer h to output layer o, respectively, are updated by the following rule:

$$\begin{aligned} \mathbf{W}^{ih} &= \mathbf{W}^{ih} + \Delta\mathbf{W}^{ih}, & \text{where} \quad \Delta\mathbf{W}^{ih}_j &= \alpha\mathbf{ie} + \beta\Delta\mathbf{W}^{ih}_{j-1} \\ \mathbf{W}^{ho} &= \mathbf{W}^{ho} + \Delta\mathbf{W}^{ho}, & \text{where} \quad \Delta\mathbf{W}^{ho}_j &= \alpha\mathbf{hd} + \beta\Delta\mathbf{W}^{ho}_{j-1} \end{aligned} \quad (1)$$

Here, α is the learning rate, β is the momentum factor for weight changes, $\mathbf{i}$ is the input vector, $\mathbf{h}$ is the output vector from hidden layer, $\mathbf{t}$ and $\mathbf{o}$ are the target and the actual outputs, and $\mathbf{d}$ is the vector where $d_k = o_k(1 - o_k)(o_k - t_k)$, for $k=1,2,\ldots,n_o$. The objective is to minimize the cost function E defined as:

$$E = \frac{1}{2}\sum_{k}\left(h_k(1-h_k)\sum_{l=1}^{n_o}W_{lk}^{h_o}d_l\right)^2 \qquad (2)$$

From the above setup and the required data, the ANN can learn the relationship between sales at period t and previous sales from period $t - 1$ to period $t - p$. For training, the weights on the links $\mathbf{W}^{ih}$ and $\mathbf{W}^{ho}$ are initialized randomly. Next, applying an input pattern, the error is computed, and the weights are adjusted accordingly by the above rules. The procedure is repeated for a large number of input samples until the weights are stabilized. After the training is over, the network can be used for prediction.

5 System Performance

We ran a simple benchmark with a single customer, one Manager and a multiple number of suppliers, which were distributed on the system. The computational experiments of the system were performed by the same machines, Intel Pentium IV 1 GHz with 512 MB of SDRAM PCs running the Windows XP Pro operating system. The number of suppliers was varied from 4 to 28 on successive trials. The experiment used a single order source whose order generation rate was slow enough to allow all the listeners to have processed a given order before the next was generated. As we were sure that there was no concurrent order delivery, we could compute the theoretical best average delivery time (Table 1).

Table 1. Order distribution and decision times

# of suppliers	4	8	12	16	20	24	28
Decision times (sec)	0.725	0.871	1.087	1.228	1.349	1.705	1.961

We used the feed forward neural network with a BP-learning algorithm in the sales forecasting subsystem. In the ANN, control parameters which included the number of hidden neurons (HN), number of training epochs (E), learning rate (α) and momentum (β) were set at $HN=4$, $E=1000$, $\alpha=0.1$ and $\beta=0.7$, after conducting a set of experiments. The total number of data points 202. The results obtained are averages of 10 different replications. The mean square error obtained for the above parameter values is 0.000646 after a 5.17 sec execution time. For cross-validation period MSE is obtained as 0.001293. Although, our system has not been tested with a very large number of subscribers and publishers, the results obtained show that the system is scalable.

6 Conclusion

In summary, we have developed a two-level agent architecture that supports the requirements of supply chain applications, such as procurement and sales forecasting, which have continuously evolving needs. Our system's communication infrastructure is based on the publish/subscribe paradigm, which allows participants to join and leave the system dynamically, extending the *Flexibility* and *Adaptability* of the system. By using a two-leveled agent model, we have also made use of parallel computation to enhance performance.

In the proposed system, mobile agents are used as mediators between customers and suppliers. The main advantage of using an agent-based approach is that the agents interact autonomously and power is devolved to the agents, i.e. there is no need for a super-agent to oversee communication and interaction. Our system also addresses some specific requirements like *Reliability*, MMAs and SMAs perform their activity autonomously and through cooperative interaction to accomplish their goals; *Learning*; interaction between agents is stored in a neural network approach with an EBP-learning algorithm; and *Forecasting*, via the use of an Artificial Neural Network approach in sales forecasting process, to obtain more accurate and timely values.

References

1. Copacino, W.C.: Supply Chain Management. St. Lucie Press, Boca Raton, FL (1997)
2. Raman, A., and Singh, J.: i2 Technologies, Inc. Harvard Business School Case, 9-699-042 (1998)
3. Fisher, M.L.: What is the right supply chain for your product? Harvard Business Review, Vol. 75 (1998) 105-116
4. Fox, M.S., Barbuceanau, M., Teigen, R.: Agent-oriented supply chain management. The International Journal of Flexible Manufacturing Systems 12 (2003)165–188
5. Camarinha-Matos, L.M., Afsarmanesh, H.: Enterprise modeling and support infrastructures: applying multi-agent system approaches. Multi-Agent Systems and Applications, Ninth ECCAI Advanced Course and Agent Link's 3rd European Agent Systems Summer School, EASSS 2001, Prague, Czech Republic. Springer and Verlag, (2001) 335–364
6. Bussmann, S.: An agent-oriented architecture for holonic manufacturing control. In: Xirouchakis, P., Kiritsis, D., Valckenaers, P. (Eds.), First International Workshop on Intelligent Manufacturing Systems. Lausanne, Switzerland, (1998) 1–12
7. Lee, J.: E-manufacturing-fundemantals, tools, and transformation. Robotics and Computer Integrated Manufacturing, Vol. 19. (2003) 501-507
8. Karageorgos, A., Mehandjiev, N., Weichhart, G., and Hammerle, A.: Agent-based optimization of logistics and production planning. Engineering Applications Intelligence, Vol.16. (2003) 335-348
9. Maturana, F., Shen, W., and Norrie, D.H.: MetaMorph: an adaptive agent-based architecture for intelligent manufacturing. International Journal of Production Research, Vol.37. (1999) 2159-2173
10. Simchi-Levi, D., Kaminsky, P., and Simchi-Levi, E.: Designing and Managing the Supply Chain. New York: Irwin McGraw-Hill, (2000)
11. Sahingoz, O. K. and N. Erdogan: A Two-Leveled Mobile Agent System for Electronic Commerce, the journal of Aero. and Space Tech. Institute (ASTIN), (2003) 21-32
12. Smith, R.G.: The Contract Net Protocol. High level communication and control in a distributed problem solver. IEEE Transactions on Computers 29 (1980) 1104–1113.

Retrieval Based on Combining Language Models with Clustering

Hua Huo[1,2] and Boqin Feng[1]

[1] Department of Computer Science, Xi'an, Jiaotong University, Xi'an, P.R. China
[2] Institute of Electronics and Information, Henan University of Science & Technology, Luo yang, P.R. China

Abstract. We propose a new retrieval method based on combining language models with clustering. The basic idea of the method is as follows. Firstly, documents in the collection are grouped into clusters by using a clustering algorithm. Secondly, clusters are imported into building language models which are used to estimate how likely a query could be generated from them. Thirdly, language models are smoothed by using a two-stage smoothing method. Our experiments show that the method outperforms both approach "purely" based on clustering and technique "purely" based on language model.

1 Introduction

Information retrieval based on clustering is a process that groups documents into clusters and returns ranked documents from clusters to users [An1]. A main approach is to identify a document set in response to a query using a clustering algorithm. The task for the retrieval system is to match the query against documents and rank documents based on their similarity to the query [TV1]. One language modeling approach to IR is to model the query generation process [CL1]. Researches carried out by a number of groups have confirmed that the language modeling approach is a theoretically attractive and potentially very effective probabilistic framework for studying information retrieval problems [Ka1]. In this paper, we present a new retrieval method based on combining language models with clustering.

2 Retrieval Based on Language Models

A typical retrieval model based on language model is as follows [CL1,SC1]. We call it MBLM (Model Based on Language Model) model. Let Col be the document collection, M_d be the language model of the document d, Q be a given query by users, and $P(Q|M_d)$ be the probability that the query Q can have been generated from the language model M_d. The most common approach to estimating $P(Q|M_d)$ assumes that the query can be treated as a sequence of independent terms, for example $Q = (q_1, q_2, , q_N)$, and thus query probability can be represented as a product of individual term probabilities [Ka1].

$$P(Q|M_d) = \prod_{i=1}^{N} p(q_i|M_d) \tag{1}$$

where q_i is the ith term in the query, and $P(q_i|M_d)$ is specified by the language model. Let q be an arbitrary term in the query Q, $P(q|M_d)$ can be represented as follows.

$$P(q|M_d) = \lambda P_{ML}(q|d) + (1-\lambda)P_{ML}(q|Col)) \tag{2}$$

where $P_{ML}(q|d)$ and $P_{ML}(q|Col)$ are maximum likelihood estimates of term q in the document d and the collection Col respectively. They are represented as follows.

$$P_{ML}(q|d) = \frac{ft(q,d)}{|d|} \tag{3}$$

$$P_{ML}(q|Col) = \frac{ft(q,Col)}{|Col|} \tag{4}$$

where λ is a general symbol for smoothing.

3 Retrieval Based on Combining Language Models with Clustering

In this section, we present a retrieval method based on combining language model with clustering. We call it MBLMC.

3.1 Importing Clustering into Language Models

Let Clu be a cluster of documents in Col. We rewrite equation (2) as follows.

$$P(q|M_d) = \lambda P_{ML}(q|d) + (1-\lambda)P(q|Clu)) \tag{5}$$

where $P(q|Clu)$ is specified by the cluster language model

$$P(q|Clu) = \alpha P_{ML}(q|Clu) + (1-\alpha)P_{ML}(q|Col)) \tag{6}$$

By substituting $P(q|Clu)$ in equation (6) for equation (7), we have

$$P(q|M_d) = \lambda P_{ML}(q|d) + (1-\lambda)(\alpha P_{ML}(q|Clu) + (1-\alpha)P_{ML}(q|Col)) \tag{7}$$

where $P_{ML}(q|d)$, $P_{ML}(q|Clu)$, and $P_{ML}(q|Col)$ are maximum likelihood estimates of the term q in the document d, the cluster Clu, and the collection Col respectively. $P_{ML}(q|d)$ is represented as the equation (3), $P_{ML}(q|Col)$ is represented as the equation (4). $P_{ML}(q|Clu)$ is represented as follows.

$$P_{ML}(q|Clu) = \frac{ft(q,Clu)}{|Clu|} \tag{8}$$

where $ft(q|Clu)$ is the number of times q occurs in the cluster Clu. $|Clu|$ is the size of the cluster Clu, namely, the number of words in the cluster Clu. By combining the equation (1) with the equation (8) we can formulate the new retrieval model based on combining language models with clustering.

3.2 Smoothing of the Language Models

One obstacle in applying language modeling to information retrieval is the problem of zero probability.Many terms are missing in a document or a cluster. If such a term is used in a query, we would always get a zero probability for the entire query [Ka1]. To address the problem, we use a smoothing method in the retrieval based on the MBLMC. There are two symbols for smoothing, namely λ and α, in the equation (8). We can view this as a two-stage smoothing method. But, it is not same as the two-stage smoothing method presented in [ZL1]. The first stage is controlled by α, and the cluster model is smoothed with the collection model. The second stage is controlled by λ, and the document model is smoothed using the smoothed cluster. Bayesian smoothing technique with the Dirichlet prior is used in the two stages, and the technique performs well empirically [ZH1]. λ and α take forms

$$\lambda = \frac{\sum\limits_{w' \in D} ft(w', d)}{\sum\limits_{w' \in D} ft(w', d) + \mu_1} \tag{9}$$

$$\alpha = \frac{\sum\limits_{w' \in Clu} ft(w', Clu)}{\sum\limits_{w' \in Clu} ft(w', Clu) + \mu_2} \tag{10}$$

where μ_1 and μ_2 are Dirichlet smoothing parameters [ZH1].

3.3 Document Clustering

The $Kullback-Liebler(KL)$ divergence is often been used as a distance measure between query and documents, and performs well empirically [XC1]. We modify the KL metric in order to make it fit for determining of closeness of a document d to a cluster Clu.

(1) Modified Kullback-Liebler (KL) Metric

Let d be a document, Clu be a cluster, $KL(d, Clu)$ be the distance metric between d and Clu. $KL(d, Clu)$ is represented as

$$KL(d, Clu) = \sum_{ft(w_i,d) \neq 0} \frac{ft(w_i, d)}{|d|} \log \frac{ft(w_i, d)/|d|}{ft(w_i, Clu) - ft(w_i, d)/(|Clu| - |d|)} \tag{11}$$

where $ft(w_i, d)$ is the number of times word w_i occurs in the document d, $ft(w_i, Clu)$ is the number of word w_i occurs in the cluster Clu. $|d|$ is the size of d and $|Clu|$ is the size of Clu.

(2) K-Means Algorithm

The algorithm consists of four key steps [XC1]. Step 1: The first K documents are taken out from the collection Col, and are treated as the initial clusters. Step 2: Each of the remaining documents of the collection Col is compared with the

K clusters, and then the value of $KL(d, Clu)$ is computed. Then, the document is assigned to the cluster which is closest to the document. Step 3: The results of the step 2 are treated as the initial clusters, and the cluster-membership of documents is re-evaluated based on the new clusters. Each document of the cluster is re-assigned to the new closest cluster unless it is already there. Step 4: If the result of clusters satisfies a given terminating condition, the algorithm terminates, or else the algorithm goes to step 3.

4 Experiments

4.1 Experimental Method

We experiment over five data sets taken from TREC. They are TREC 4, TREC 5, TREC 6, TREC 7, and TREC 8. Three sets of experiments are performed in this paper. The first set of experiments is to select the suitable number of clusters to be generated by the K-means algorithm in the MBLMC on TREC 4. Both parameters, namely μ_1 and μ_2, are set to be 2000, for the retrieval based on the MBLMC with $\mu_1=2000$ and $\mu_2=2000$ performs well empirically. The second set of experiments is to compare the performances of retrieval based on clustering and retrieval based on the MBLM with the performance of retrieval based on the MBLMC. Four different cluster representations are used respectively when we perform the retrieval based on clustering. The third set of experiments examines whether the performance of the MBLMC is sensitive to the smoothing parameters.

4.2 Experimental Results

The 10-points average precision is used as the basis of evaluation throughout our experiments. From table 1, we can observe that K=1500 gives the overall best result on TREC 4, so it is chosen for our experiments on the other data sets. Results of the second set of experiments are shown in Table 2. Among the four cluster representations, the highest ranked representation is the most effective to retrieval based on clustering. We observe that retrieval based on the MBLMC performs better than retrieval based on clustering, and it also does better than retrieval based on the MBLM. Improvements of retrieval based on the MBLMC over retrieval based clustering with the highest ranked cluster representation are observed, respectively with 6.81%, 4.37%, 8.44%, and 6.57% improvement in average precision on the four data sets. The retrieval based on the MBLMC has also improvements of 6.4%, 2.2%, 4.4%, and 4.5% over the retrieval based on the MBLM in average precision on the four data sets, respectively. The plots in Fig.1 show the average precision for different settings of the Dirichlet prior sample size μ_1. It is clear that average precision is much more sensitive to μ_1, especially when it is small. In addition, we also find that the average precision is sensitive to μ_2 by doing a similar experiment for μ_2 while $\mu_1=2000$.

Table 1. Retrieval results of MBLMC ($\mu_1 = \mu_2 = 2000$) on TREC 4

Recall	K=500	K=1000	K=1500	K=2000
0.1	0.4986	0.5012	0.5625	0.5216
0.2	0.4749	0.4806	0.5124	0.4918
0.3	0.4325	0.4451	0.4846	0.4602
0.4	0.3135	0.3325	0.3984	0.3476
0.5	0.2761	0.2911	0.3655	0.3015
0.6	0.1996	0.2356	0.2861	0.2633
0.7	0.1298	0.2036	0.2574	0.2358
0.8	0.0965	0.1102	0.1625	0.1236
0.9	0.0462	0.0671	0.1162	0.0956
1.0	0.0405	0.0412	0.0668	0.0672
Avg.Pre.	0.2508	0.2708	0.3215	0.2908

Table 2. Average precision of Cluster-based retrieval, MBLM and MBLMC

Data set	Cluster-based retrieval K=1500				MBLM (μ=2000)	MBLMC(μ_1 $=\mu_2=2000$)	Chg1%	Chg2%
	Centroid	Highest ranked	Lowest ranked	Medium ranked				
TREC 5	0.2976	0.3096	0.2908	0.2393	0.3108	0.3307	+6.81	+6.40
TREC 6	0.2886	0.2904	0.2804	0.2865	0.2966	0.3031	+4.37	+2.20
TREC 7	0.2945	0.2975	0.2857	0.2916	0.3089	0.3226	+8.44	+4.40
TREC 8	0.2917	0.2936	0.2819	0.2902	0.2995	0.3129	+6.57	+4.50

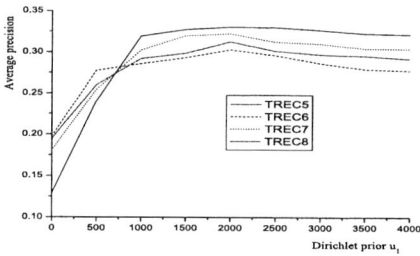

Fig. 1. Plots of average precision of MBLMC for different μ_1 (while μ_2=2000, K=1500)

5 Conclusions

Based on the experimental results, we can make the following conclusions. Firstly, the retrieval based the MBLMC outperforms both retrieval based on clustering and retrieval based on the MBLM. Secondly, using clusters to smooth documents is a more effective approach to solving the problem of zero probability, and to further improving performance of retrieval based on language models. Thirdly, the performance of retrieval based on the MBLMC is sensitive to smoothing parameters.

References

[An1] Anton, L.: Evaluating document clustering for interactive information retrieval. In proceedings of CIKM'01 conference, (2001)33-40.

[CL1] Croft, W.B., Lafferty, J.: Language modeling for information retrieval.In Kluwer International Series on Information Retrieval, 13,(2003)195-197.

[Ka1] Katz, S.M.: Estimation of probabilities from sparse data for the language model component of speech recognizer.IEEE Transactions on Acoustics, speech and Signal processing, (1987)400-401.

[SC1] Song, F., Croft, W.B.: A general language model for information retrieval. In proceedings of ACM-SIGIR'99,(1999)279-280.

[TV1] Tombros, A., Villa, R.: The effectiveness of query-specific hierarchic clustering in information retrieval. Information processing and management,38,(2002)559-582.

[Wi1] Willet, P.: Query specific automatic document classification. International forum on information and documentation, 10,(1995)28-32.

[XC1] Xu, J., Croft, W.B.: Cluster-based language models for distributed retrieval.In proceedings of SIGIR'99,(1999)254-261.

[ZH1] Zaragoza, H., Hiemstra, D.: Bayesian extension to the language model for ad hoc information retrieval, In proceedings of SIGIR'03,(2003)325-327.

[ZL1] Zhai, C., Lafferty, J.: Two-stage language models for information retrieval.In processing of SIGIR '02,(2002)49-56.

Lightweight Mobile Agent Authentication Scheme for Home Network Environments*

Jae-gon Kim, Gu Su Kim, and Young Ik Eom

School of Information and Communication Eng., Sungkyunkwan University,
300 cheoncheon-dong, Jangan-gu, Suwon, 440-746, Korea
{angel77, gusukim, yieom}@ece.skku.ac.kr

Abstract. Recently, interests on the home network, that requires new services and new computing paradigms, have enormously been increased. Applying the mobile agent to the home network is expected to provide a new computing model. The mobile agent authentication is a preceding technology to apply the mobile agent concept to the home network environments. The existing public key based authentication scheme is not suitable to the home network devices which has the limited computation capability. In this paper, we propose a lightweight mobile agent authentication scheme based on the shared key and public key infrastructures.

1 Introduction

The home network can be defined as the interoperable network which interconnects computer systems, home appliances, and mobile devices, and enables remote connections and remote controls[1]. The mobile agent is a computer program which migrates autonomously in the network and works on behalf of the user[2]. By applying the mobile agent concept to the home network environments, we can reduce remote interactions and network traffics among the home network devices[3]. But, for applying the mobile agent to the home network, the mobile agent authentication scheme is necessary. The existing mobile agent systems are using the public key based authentication schemes[4, 5]. But, that is not suitable to the home network devices that have limited computation capabilities.

In this paper, we propose a lightweight mobile agent authentication scheme. Basically, we applied the shared key authentication to reduce the overhead and additionally used the public key authentication to extend this scheme into the multiple domain home networks. This paper is composed of related works in Section 2, requirements and system architecture in Section 3, proposed authentication scheme in Section 4, safety proof in Section 5, and conclusion in Section 6.

2 Related Works

The mobile agent migration requests can be classified into two categories: the request made by the mobile agent itself, and the request made by the platform

---

* This work was supported by National Center of Excellence in Ubiquitous Computing and Networking (CUCN), Korea.

executing the mobile agent. According to who makes the request, the agent itself or the platform, it is determined whether the destination platform should authenticate the agent or the platform. Shimson Berkovits classified these situations into the place hand-off/delegation and the agent hand-off/delegation[6].

The cryptographic algorithms for authentication are classified into the shared key cryptography and the public key cryptography[7]. In the shared key based authentication scheme, the authentication requester encrypts his message with a shared key and the the receiver decrypts the encrypted message with the same shared key. They use the same shared key for both encryption and decryption. The major advantage of the shared key cryptography is that it is much faster than the public key cryptography. But, in order to use the shared key based authentication scheme, the group that shares the key should keep the secrecy of their key, and so, its utilization is very limited. The public key based authentication scheme uses a key pair, public and private. By encrypting with the secret key and decrypting with the corresponding public key, it authenticates that the message is made by the principal who owns the secret key. The advantage of the public key based authentication scheme is that the key management is more simple than the shared key management, because whoever can possess the public key. But, the public key cryptography is much slower than the shared key cryptography.

The home network middlewares, that provide security facilities, include UPnP [8], Jini[9] and HAVi[10]. UPnP, which is proposed by the Microsoft, provides the interoperability between the home network devices and the resource access control facilities. The UPnP security system verifies the signature of the SOAP(Simple Objective Access Protocol)[11] control message with the public key, and if it is valid, it assigns the privilege based on ACL(Access Control List). Jini is proposed by the Sun Microsystems, and it is characterized by the Lookup service and the code mobility. HAVi is proposed by the SONY and has the objective to achieve real time data transmission and interoperability between AV devices through the IEEE1394 technology. Both Jini and HAVi are based on Java, and have the security model provided by Java. These home network middlewares do not provide the mobile agent authentication facilities.

3 Requirements and System Architecture

The requirements for mobile agent authentication in the home network environments are as follows. First of all, both the platform authentication facility and the agent authentication facility are necessary, because the migration request can be made by the platform or the agent itself. Secondly, the authentication scheme in home network devices should be lightweight, because the home network devices generally have limited computation capabilities in comparison with desktop computers. Thirdly, the mobile agent authentication should be performed not only in single home domain environments but also in multi-domain environments.

In our system architecture, a single home network(called single domain) is composed of mobile agents, platforms, platform groups, and DMS(Domain Management Server). Each component is allocated the following IDs: AID(Agent ID), PID(Platform ID), GID(Group ID), and DID(Domain ID).

DMS maintains DPL(Domain Platform List) which is the list used to manage the information about platforms in the single domain. Each entry of DPL is composed of $\{PID, key, groups\}$. PID is the ID of registering platform, key is a shared key between DMS and the platform, and $groups$ are the list of platform group IDs that the registering platform has joined. The shared key is used to make a secure channel between the DMS and the platform.

The platforms in a single domain can be grouped and DMS has DGL(Domain Group List) that maintains information on the platform groups. An entry of DGL represents a platform group, and consists of $\{GID, key, platforms\}$. GID is the platform group ID, key is the group shared key, and $platforms$ are the list of IDs of the platforms privileged to join the group. If a platform joins a platform group, DMS delivers a replication of the DGL entry associated with the platform group to the joining platform.

Each mobile agent has a credential which is composed of $\{AID, HPID, AGID, GA, LAT\}$. $HPID$(Home PID) is the PID of the platform that created the agent, $AGID$(Agent GID) is the GID of the platform group that the agent is currently joining. GA(Global Authenticator) is used for authenticating the agent when the agent migrates across multiple domains and is generated in the following form:

$$GA = E_{K_R}(AID||HPID||D)$$

K_R is the private key of agent's home platform. D is a message digest of the agent execution code, which represents the agent program. As the agent migrates across the platforms not fully trusted, it may not have the secret information such as private key. Therefore, the replay attack prevention technique like the challenge-response protocol[12] can not be applied. Alternately, signing D with K_R ensures the integrity of the action of the agent and prevents GA from being diverted or modified by malicious attackers.

LAT(Local Authentication Ticket) is the authentication ticket authenticating the agent that migrates within a platform group. It represents that the agent has privilege to migrate within the platform group. LAT is generated in the following form, when the agent is created or it joins to the platform group:

$$LAT = E_{K_G}(AID||AGID||D)$$

K_G is the group shared key of the platform group. By decrypting LAT with K_G, platforms can authenticate that the LAT of the platform group is issued to the agent. The role of D is same as that of D in GA.

4 Proposed Authentication Scheme

4.1 Platform Authentication in Multiple Domain Environments

When a platform A tries to send an agent to another platform B located in different domain, the platform B should authenticate the platform A. Figure 1 shows

```
(1) A→B: PID_A || GID || nonce_A
(2) B    : a = find(GID, PID_A)
(3)      : b = E_{K_RB}(H(PID_B || a || GID || nonce_A))
(4) B→A: PID_B || a || GID || nonce_B || b
(5) A    : verify(b, H(PID_B || a || GID || nonce_A))
(7)      : if(a = No)
(8)      :   finish agent migration
(9)      : d = E_{K_RA}(H(PID_A' || nonce_B))
(10) A→B: PID_A' || d
(11) B   : verify(d, PID_A' || nonce_B)
(12)     : if(PID_A = PID_A')
(13)     :   perform agent migration, allocate LAT
(14)     : else
(16)     :   finish agent migration
```

Fig. 1. Platform authentication in multiple domain environments

the procedure of platform authentication in multiple domain environments. This type of platform authentication uses the challenge-response protocol. In step (2), the platform B inspects that the platform A has the privilege to get LAT. A authenticates B with $nonce_A$ from steps (3) to (8) and B authenticates A with $nonce_B$ from steps (9) to (11). If the authentication succeeds, LAT is allocated to the agent for group authentication in single domain environments.

4.2 Agent Authentication in Multiple Domain Environments

Figure 2 shows the procedure of agent authentication in multiple domain environments. When an agent A, generated at platform S, tries to migrate to another platform B located in different domain, platform B authenticates the agent A with GA in step (7) of Figure 2. If the authentication succeeds, LAT is allocated to the agent by the platform B.

```
(1) S    : D_A = H(execution code of A)
(2)      : HPID_A = PID_S
(3)      : GA = E_{K_RS}(AID_A || HPID_A || D_A)
(4) A→B: AID_A || HPID_A || GID_A || execution code of A || G_A
(5) B    : D_A' = H(execution code of A)
(6)      : a = AID_A || HPID_A || GID_A || D_A'
(7)      : verify(GA, a)
(8)      : b = find(GID_A, HPID_A)
(9)      : if(b = No)
(10)     :   finish agent migration
(11)     : else
(12)     :   perform agent migration
(13)     :   allocate LAT
```

Fig. 2. Agent authentication in multiple domain environments

4.3 Group Authentication in Single Domain Environments

When an agent A, to whom LAT was issued by the platform S, tries to migrate to another platform B which is in the same platform group, the platform B authenticates the agent A by checking that the agent A has the LAT of the platform group. Figure 3 shows the procedure of single domain group authentication.

$$
\begin{aligned}
&(1)\ S &&: D_A = H(\text{execution code of } A) \\
&(2) &&: LAT = E_{K_G}(AID_A \,||\, AGID_A \,||\, D_A) \\
&(3)\ A \rightarrow B &&: AID_A \,||\, AGID_A \,||\, \text{execution code of } A \,||\, LAT \\
&(4)\ B &&: D_A' = H(\text{execution code of } A) \\
&(5) &&: verify(LAT, AID_A \,||\, AGID_A \,||\, D_A') \\
&(6) &&: \text{perform agent migration}
\end{aligned}
$$

Fig. 3. Group authentication in single domain environments

5 Safety Proof

5.1 Proof on the Platform Authentication

Because K_{RB}(the private key of B) is known only to the platform B, other platforms can not make b in step (3) of Figure 1. Therefore, it can be verified that the platform B has made b. Also, because K_{RA}(the private key of A) is known only to the platform A, other platforms can not make d in step (9) of Figure 1. Therefore, it can also be verified that the platform A has made d. A malicious attacker C may try to perform a replay attack by eavesdropping the messages transmitted in step (1) and (10). If C transmits the eavesdropped information of step (1) to B, B performs signing to $nonce_A$ and sends it back to C. To masquerade as A, C should generate d in step (9). But C does not know K_{RA} and can not generate d. Also, C may try to masquerade as A by retransmitting the eavesdropped information of step (10) to B. But, because $nonce_B$, that is transmitted by B, always varies, the authentication will fail in step (11). If PID of A is not registered in DGL of step (2), the platform B does not issue LAT. As only DMS manager can add/remove PID in each entry of DGL, LAT can be issued only to the agent trusted by DMS manager.

5.2 Proof on the Agent Authentication

Because K_{RS}(the private key of S) is known only to the platform S, other platforms can not make GA of agent A in step (3) of Figure 2. Therefore, it can be verified that the platform S has created A. A malicious agent C may try to perform a replay attack by eavesdropping the messages transmitted in step (4). C modifies the execution code portion of eavesdropped information of step (4) and requests the agent migration to the platform B. But, because GA containing the message digest of the agent execution code is signed with K_{RS}, the authentication will fail in step (7). Unless the authentication procedure from steps (1) to (7) succeeds, LAT of the platform group can not be issued to A. Also, if PID of S is not registered in DGL of step (8), LAT can not be issued. As only DMS manager can add/remove PID in each entry of DGL, the LAT can be issued only to the agent trusted by DMS manager.

5.3 Proof on the Single Domain Group Authentication

Because K_G(the platform group shared key) is known only to platform group members, non-member platforms can not make the LAT of the platform group in step (3) of Figure 3. Therefore, it can be verified that a platform group member has made LAT. In step (2) of Figure 1 and in step (8) of Figure 2, DMS manager determines whether LAT can be issued to the agent or not. The possession of LAT means that the agent can be authorized to get the privilege to migrate within the platform group. A malicious agent C may try to perform a replay attack by eavesdropping the messages transmitted in step (3). C modifies the execution code portion of the eavesdropped information at step (3) and requests the agent migration to the platform B. But, because LAT that contains the message digest of the agent execution code is signed with K_G, the authentication will fail in step (7).

6 Conclusion

In this paper, we described the requirements for mobile agent authentication in home network environments and proposed a lightweight authentication scheme. We applied shared key authentication scheme to reduce the overhead of authentication and also used public key authentication scheme to extend the scheme into multiple domain home network environments. In the future, based on the proposed authentication scheme, we will develop access control scheme for the resources and services in the home network environments.

References

1. Rose, B.: Home Networks: A Standard Perspective, IEEE Communication Magazine, pp. 78-85, 2001.
2. Karnik, N., M., Tripathi, A., R.: Agent Server Architecture for the Mobile-Agent System, PDPTA '98, pp. 66-73, July 1998.
3. Yoo, J., J., Lee, D., I.: Scalable Home Network Interaction Model Based on Mobile Agents, PerCom '03, pp. 543-546, March 2003.
4. Karnik, N.: Security in Mobile Agent Systems, Ph. D. dissertation, University of Minnesota, 1998.
5. ObjectSpace Inc.: Voyager Security Developer's Guide Table of Contents, http://www.recursionsw.com/products/voyager/voyager.asp.
6. Berkovits, S., Guttman, J., D., Swarup, V.: Authentication for Mobile Agents, Mobile Agents and Security LNCS, pp.114-136, 1998.

7. Lampson, B., Abadi, M., Burrows, M., Wobber, E.: Authentication in Distributed Systems: Theory and Practice, ACM Transactions on Computersystems, pp. 265-310, November 1992.
8. Ellison, C.: UPnP Security Ceremonies Design Document, Intel Corporation, October 2003.
9. Sun Microsystems Inc.: AR - JiniTM Architecture Specification, www.jini.org.
10. HAVi Inc.: HAVi, the A/V digital network revolution, www.havi.org, May 2001.
11. Gudgin, M., Hadley, M., Mendelsohn, N., Moreau, J., Nielsen, H.: SOAP Version 1.2, http://www.w3.org/TR/soap/, June 2003.
12. Abadi, M., Needham, R.: Prudent Engineering Practice for Cryptographic Protocols, IEEE Transactions on Software Engineering, January 1996.

Dimensional Reduction Effects of Feature Vectors by Coefficients of Determination

Jong-Wan Kim, Byung-Kon Hwang, Sin-Jae Kang, Hee-Jae Kim, and Young-Cheol Oh

School of Computer and Information Technology, Daegu University,
Gyeonsan. Gyeongbuk. 712-714 South Korea
{jwkim, bkhwang, sjkang}@daegu.ac.kr

Abstract. This paper presents a method to reduce features less contributing to the classification of user preferred news groups among several news groups by the use of fuzzy inference and coefficient of determination. To this end, we extract a number of representative keywords from example documents through fuzzy inference. From the observation of training patterns, we found that lots of keywords in training patterns are empty. Thus, a new method to train neural network through reduction of unnecessary dimensions by the statistical coefficient of determination is proposed in this paper. Experimental results show that the proposed method is superior to the method using lots of input attributes in terms of within-cluster variance and its standard deviation.

1 Introduction

It is important to retrieve exact news information coinciding with user's need from lots of news documents quickly. To fix user's need for the selecting function exactly desired information from various news documents provided by lots of Usenet news servers, first we have to connect news servers over the Internet and collect news documents. Then we extract a number of terms called representative keywords (RKs) from them through fuzzy inference. Performance of our approach is heavily influenced by the effectiveness of selection method of RKs so that we choose fuzzy inference because it is more effective in handling the uncertainty inherent in selecting RKs within documents [1]. In this paper, Kohonen network is used to train news documents with RKs extracted. Kohonen network, one of unsupervised learning algorithms that do not request user's feedback continuously [2], can classify news groups with only RKs. So we adopted Kohonen network as a training algorithm. However, by observing input patterns used as training vectors of neural network, the sparsity phenomenon that specific keywords chosen in many news groups are empty was found. As you can imagine, if zero values are set in lots of input attributes, then the ones cannot contribute to classify the input patterns. Therefore it is better to exclude the unnecessary attributes. To resolve this sparsity problem, we first select input variables (= chosen RKs) relevant to the target variable (= similar news group) presented by the user and train only these selected input variables. We have a decision that it is more useful than training all of input variables. Resulting from that, we will introduce statistical coefficient of determination that is a method to determine input variables highly relevant to the target variable.

2 Related Works

To extract RKs representing news groups well from example documents and assign them weights are the same problem that the existing linear classifiers such as Rocchio and Widrow-Hoff algorithms [3] find centroid vectors of an example document collection. Both of these algorithms use TF (Term Frequency) and IDF (Inverse Document Frequency) for re-weighting terms but they do not consider term co-occurrence relationship within feedbacked documents. To resolve this problem, the computation of term co-occurrences between these RKs and candidate terms within each example document is required. Three factors – TF, DF (Document Frequency), and IDF – have essentially inexact characteristics, which are used to calculate the importance of a specific term. Since fuzzy logic is more adequate to handle intuitive and uncertain knowledge, we combine the three factors by the use of fuzzy inference. The dimensionality of a training data can be characterized in many ways. A dataset contains a number of patterns, each of which contains several attributes and a class label. The attributes may be considered to be relevant attributes but other attributes may be considered as irrelevant attributes. To determine which of the attributes are relevant to the learning task, nearest neighbor algorithms are presented to calculate an average similarity measure across all of the attributes [4]. These methods are kinds of PCA (Principal Component Analysis) approach, and it follows singular value decomposition (SVD) technique removing those dimensions that have low singular values. PCA involves a mathematical procedure that transforms a number of correlated variables into a smaller number of uncorrelated variables called principal components. Since PCA utilizes these transformed principal components for pattern classification, we could not know which input attributes contribute to classify patterns. However, it is important to determine specific attributes that are contributed to pattern classification in this work. We are going to use statistical coefficient of determination useful to determine the degree of contribution to pattern classification.

3 Dimensional Reduction of Features by FI and CoD

3.1 Feature Extraction by Fuzzy Inference (FI)

For selection of important features or terms representing news groups well, we calculate weights of candidate terms by using the method [1] that it is known to give superior performance to the existing RK extraction methods and assign a priority to select RKs with the weights of candidate terms. Finally RKs are selected according to this priority value for term selection. From now on, we will explain this procedure. Example documents are transformed to the set of candidate terms by eliminating stop words and stemming using Porter's algorithm. The TF (Term Frequency), DF (Document Frequency), and IDF (Inverse Document Frequency) of each term are calculated from this set and are normalized to the NTF, NDF, and NIDF, respectively [1]. They are used as input variables for fuzzy inference. The membership functions of the fuzzy input and output variables should have been fuzzified to the form suitable for fuzzy inference (FI). First, we will define the membership functions (μ) of three fuzzy input variables NTF, NDF, and NIDF as the following expressions:

$\mu_S(x)$ = max(0, 1- x/0.75) and $\mu_L(x)$ = max(0, 1+(x-1)/0.75) for x = NTF variable,
$\mu_S(y)$ = max(0, 1- y/0.35), $\mu_L(y)$ = max(0, 1+(y-1)/0.35), and
$\mu_M(y)$ = min{max(0, 1+(y-0.5)/0.35), max(0, 1-(y-0.5)/0.35)} for y = NDF or NIDF variable.

Similarly the fuzzy output variable TW (Term Weight) that represents the importance of each term has the following membership functions. At this time, the meaning of each fuzzy term {Z, S, M, L, X, XX} is corresponding to Zero, Small, Middle, Large, Xlarge, XXlarge, respectively.

$\mu_Z(TW)$ = max(0, 1-TW/0.2), $\mu_{XX}(TW)$ = max(0, 1+(TW-1)/0.2),
$\mu_S(TW)$ = min{max(0, 1+(TW-0.2)/0.2), max(0, 1-(TW-0.2)/0.2)},
$\mu_M(TW)$ = min{max(0, 1+(TW-0.4)/0.2), max(0, 1-(TW-0.4)/0.2)},
$\mu_L(TW)$ = min{max(0, 1+(TW-0.6)/0.2), max(0, 1-(TW-0.6)/0.2)}, and
$\mu_X(TW)$ = min{max(0, 1+(TW-0.8)/0.2), max(0, 1-(TW-0.8)/0.2)}.

Table 1 gives 18 fuzzy rules to inference the term weight TW, where NTF is considered as primary factor, NDF and NIDF as secondary ones. The first inference rule represents "if NTF=S and NDF=S and NIDF=S then TW=Z" as shown in Table 1. In this case, we assign Z label meaning almost non-relevant to TW, because we think the term is not an important one. The rest of rules were set by the similar way.

Table 1. Fuzzy inference rules are composed of 2 groups according to NTF value

NIDF / NDF	S	M	L		NIDF / NDF	S	M	L
S	Z	Z	S		S	Z	S	M
M	Z	M	L		M	S	L	X
L	S	L	X		L	M	X	XX

NTF = S   NTF = L

Defuzzification is the final step in approximate reasoning, consisting in the replacement of a fuzzy set with a suitable crisp value. The method used most often is the center of gravity method, which defines the output value as a value that divides the area under the curve into two equal sub areas [5]. The center of gravity method is also used to defuzzify the output variable TW in this work. Therefore, the terms with higher TW values are selected as feature vectors to classify news group documents by FI.

3.2 Dimensional Reduction by Coefficient of Determination (CoD)

A statistic that is widely used to determine how well a regression model fits to a given problem is the coefficient of determination [6]. The coefficient of determination (CoD) represents the fraction of variability in target variable y that can be explained by the variability in input variable x. In other words, the CoD explains how much of the variability in the y's can be explained by the fact that they are related to x. The equation for the coefficient of determination, R^2, is defined to the following equation:

$$R^2 = (SST - SSE) / SST \qquad (1)$$

where SST is the total sums of squares of the data and SSE is the sum of squares due to residual errors.

As shown in the equation (1), the bigger the value of R^2 is, the stronger the usability of the regression model is. Thus, it is needed to reduce some variables having low R^2; the terms less contribute to the classification task should be eliminated from RKs. The task is to classify news groups of the target variable well through reduction of less important RKs. By doing that, we can improve pattern classification ratio. We can find the input variables affecting the target variable by using aforementioned regression analysis. Instead of the complete model utilizing all of input variables including unnecessary variables, the reduced model to utilize necessary variables can be a more desirable regression model [6]. To construct this kind of reduced model, we calculate R^2 to select input variables identifying news group with all candidate input variables derived through FI. In order to classify news group documents by using the CoD, a target variable is needed. Thus class labels of news groups based on news group domains are assigned to a target variable in this work. For example, we classified 126 news groups in news.kornet.net of NNTP server based on domain names manually. We classified news groups by considering upper four domain names of a specific news group. Namely han.answers.all has class label 1, han.arts.architecture.all has class label 2, and the rest of news groups have their corresponding class labels. However some news groups are classified into the same one; for example, han.comp.os.windows.apps.all and han.comp.os.windows.nt.admin.all have the same class label 36 because the upper four domain names of these two news groups are equal as "han.comp.os.windows". Finally 114 class labels are assigned to all the experimental data. After we assigned these 114 class labels as values of a target variable temporarily and candidate terms relevant to the target variable as input variables, we calculate the coefficient of determination between every candidate term and the target variable. Backward elimination scheme is chosen to filter input variables in this paper. In the backward elimination scheme, the variable with the lowest coefficient under the predefined threshold value among previously calculated coefficients of determination is eliminated one by one. We can finally get necessary input variables by iterating the backward elimination scheme till all of remaining coefficients of determination is over the predefined threshold value. From the experimental results in section 4, we got an amazing effect that about 80% of all input dimensions are reduced.

4 Experiments

In this paper, we have implemented a Usenet news group classification system in Java. Experiments were performed for 126 news groups. Ten documents for each news group were randomly selected to extract RKs by FI. To evaluate classification performance, Kohonen network is used to cluster news groups reflecting user's intention. The size of Kohonen map is fixed to 5×5 and training is performed for 1000 times. In this work, the parameter values of Kohonen network are not important because we only use the neural network as a classifying tool. Ninety-six terms were extracted as the training data of the base method by the FI only. While much smaller terms in the proposed method were extracted using the FI and the CoD according to the threshold values of the coefficient of determination. That is, 20 terms were extracted in the case of threshold value of 0.005 and 21 terms were extracted at the 0.01. Test vectors were generated with keywords given by user. In order to calculate Euclidean distance between the keywords given by user and the keywords already stored in database, the value of keyword which user did not specify was set "0" and therefore the dimensions of two vectors to be compared were at one. To analyze dimensional reduction effect of the CoD for the case using reduced input dimension and the other case using every input dimension, the within-cluster variance (Vw_j) of the j-th cluster is defined as follows:

$$Vw_j = \frac{1}{|C_j|} \sum_{i \in C_j} \sqrt{[X_i - W_j]^2} \qquad (2)$$

where X_i is the i-th input pattern included in the j-th cluster, W_j is the centroid vector of the j-th cluster, C_j is the set of patterns included in the j-th cluster, and $|C_j|$ is the number of patterns included in the j-th cluster. Therefore Vw_j means the variance between the centroid vector of the j-th cluster and the input patterns included in the j-th cluster. The equation (3) represents the average within-cluster variance (Vw) of all clusters.

$$Vw = \frac{1}{k} \sum_{j=1}^{k} Vw_j \qquad (3)$$

where k is the number of output neurons – the number of clusters.

In also, we calculated Sw, the standard deviation of the average within-cluster variance (Vw) to know the degree of deviation of the proposed method. This Sw can be used to compare the precision of clustering algorithms.

Table 2. Performance evaluation in terms of Vw and Sw

	FI (96)	CoD (21)	Impv. (%)	CoD (20)	Impv. (%)
Vw	0.104	0.049	52.9	0.045	56.7
Sw	0.308	0.207	32.8	0.196	36.4

Since a good pattern classifier reduces the within-cluster variance and its standard deviation, the measures are used to evaluate performance of the proposed dimensionality reduction method. Table 2 shows the experimental results in terms of these two measures. The 96 terms were used to evaluate the performance of the base method using only the FI as shown in Table 2. However, in the case of the threshold value of 0.01 and 0.005, the proposed method used only 21 and 20 terms, respectively; that is about 80 percentage of terms are reduced owing to the CoD. As shown in the Table, the method utilizing both of the FI and the CoD gives the superior performance to the base method using only the FI by over 50 percent in terms of Vw. This means that the FI approach contributes to select feature vectors for example news documents and the CoD approach reducing unnecessary input dimensions affects the news group classification. In also, since the proposed method improved over 30 percent in terms of Sw, we can say that the proposed method gives much stable results.

5 Conclusions

In this paper, we implemented a Usenet news group classification system relevant to the terms may be interesting to user. We first extracted candidate terms from example news documents and chose a number of RKs by the FI. Then we excluded some RKs with low contribution for pattern classification by using the statistical CoD. Finally, Kohonen network was used to train feature vectors of news group documents because it was suitable for clustering keywords chosen. The features of this work can be summarized as follows. First, the proposed method improved precision by automatically extracting RKs from news documents with the help of the FI. Second, pattern classification was improved by using the CoD in statistics to exclude unnecessary attributes for training. Third, we confirmed that the proposed method utilizing the FI and the CoD was superior to the base method using lots of input attributes in terms of the within-cluster variance and its standard deviation. Finally the proposed method brought out an additional effect to reduce training time of neural network owing to the dimensional reduction. In the future, we will extend the proposed method in order to apply other similar fields such as mail filtering. In also, we have to compare our fuzzy inference approach to the other feature selection methods like information gain, mutual information, χ^2-test, and so on.

Acknowledgement

This research was supported in part by the Daegu University Research Grant, 2004.

References

1. Kim, B. M., Li, Q., and Kim, J. W.: Extraction of User Preferences from a Few Positive Documents, Proceedings of The Sixth International Workshop on Information Retrieval with Asian Languages (2003) 124-131
2. Kohonen, T.: Self-Organizing Maps, Springer-Verlag, New York (1995)

3. Lewis, D. D., Schapire, R. E., Callan, J. P., and Papka, R.: "Training algorithms for linear text classifier", Proc. of SIGIR-96, 19th ACM International Conference on Research and Development in Information Retrieval (1996) 298-306
4. Payne, T. R. and Edwards, P.: "Dimensionality Reduction through Sub-Space Mapping for Nearest Neighbor Algorithms," European Conference on Machine Learning (2000) 331-343
5. Lee, C. C.: "Fuzzy logic in control systems: Fuzzy logic controller-part I," IEEE Trans. Syst. Man, Cybern., 20(2) (1990) 408-418
6. Ott, R. L.: An introduction to statistical methods and data analysis, Duxbury Press, Belmont, California (1993)

A Modular k-Nearest Neighbor Classification Method for Massively Parallel Text Categorization

Hai Zhao and Bao-Liang Lu

Department of Computer Science and Engineering, Shanghai Jiao Tong University,
1954 Huashan Rd., Shanghai 200030, China
{zhaohai, blu}@cs.sjtu.edu.cn

Abstract. This paper presents a Min-Max modular k-nearest neighbor (M^3-k-NN) classification method for massively parallel text categorization. The basic idea behind the method is to decompose a large-scale text categorization problem into a number of smaller two-class subproblems and combine all of the individual modular k-NN classifiers trained on the smaller two-class subproblems into an M^3-k-NN classifier. Our experiments in text categorization demonstrate that M^3-k-NN is much faster than conventional k-NN, and meanwhile the classification accuracy of M^3-k-NN is slightly better than that of the conventional k-NN. In practical, M^3-k-NN has intimate relationship with high order k-NN algorithm; therefore, in theoretical sense, the reliability of M^3-k-NN has been supported to some extend.

1 Introduction

In all categorization algorithms that based on vector space model, k-nearest neighbor (k-NN) method is considered as taking on the best performance in some sense. The k-NN method is very concise and easy to use, which simply estimates the test sample's class as the class which most supported in its k nearest neighbor training samples. Because of its advantage, k-NN method has been widely used in text categorization. Many researchers focus their endeavor on the improvement of text categorization ([2], [5], [7]). Commonly, these researches adopt additional information for improving the ultimate classification result.

However, the parallel techniques have been developed for traditional categorization for high-end application. Our research goal is to develop faster and more efficient methods for large-scale text categorization by direct modular classification without reducing the precision of the classification. Therefore we introduce a modular k-NN algorithm, M^3-k-NN.

In fact, many efficient methods have been proposed in the past decade in task decomposition and combination of modular classifiers ([1], [3], [6]), and the effectiveness of these methods have been studied and proved. But, thus far, no divide-and-conquer method has been put forward for k-NN algorithm.

The basic ideal of our method is to adopt the k-NN algorithm in the combination of Min-Max Modular (M^3) Neural Network Model ([6]) and Round Robin Rule (R^3) Learning algorithm ([3]). Our experiments demonstrate that under parallel pattern, R^3-k-NN or R^3-M^3-k-NN has the similar recognition precision to the traditional k-NN and sometime even better. Also, the flexibility of this modular combination method can be found obviously for large-scale parallel processing by using grid or large-scale cluster systems.

2 Algorithm Description

We introduce the modular algorithm, M^3-k-NN and one of its revised versions, R^3-M^3-k-NN. The whole algorithm has been divided into two phases.

In the first phase, we simply use one-against-one task decomposition method. Suppose the number of classes for original classification problem is m, then we will have $m(m-1)/2$ independent base k-NN classifiers. The training samples of each base k-NN classifier come from two different classes. To combine the results, we consider two module combination policies as the following:

P_1 M^3-k-NN voting: In all base k-NN classifiers, if the outputs of $m-1$ base k-NN classifiers support the same class, then the classification result is just this class.

P_2 R^3-M^3-k-NN voting: The result is the class that supported by most base k-NN classifiers.

The policies P_1 and P_2 are exactly the same voting procedures used in [6] and [3], respectively. And policy P_2 is also one of voting strategies for multi-class support vector machine.

In the second phase, we further divide a two-class subproblem from the first phase into a number of smaller two-class subproblems. The decomposition approach is to decompose two training sample sets pair in a two-class classification problem into m_1 and m_2 smaller sample sets, respectively. Thus, we get $m_1 m_2$ smaller two-class classification sub-problems. Let the output coding of class ID is 1 and 0. We define all base classifiers learning from the same training set of class 1 as 'a group'. Then M^3 combination results of all base classifiers includes two stages: Firstly, combination rule ***Min*** is applied in each 'group' to produce a group output. Secondly, all outputs of every groups are applied by combination rule ***Max*** to produce the final combination output.

3 Experiments

We use Yomiuri News Corpus for this study. There are 2,190,512 documents in the full collections from the years 1987 to 2001. Two subsets of the corpus are used in the experiments, which belong to 10 classes. In the first subset, the number of training and test samples is 5,865 and 2,420, respectively. In the second subset, the number of training and test samples is 37,938 and 16,234, respectively.

A χ^2 statistical method ([7]) was used for preprocessing the documents after the morphological analysis was done with ChaSen. The number of features is 500.

The goal of the experiments is to compare recognition accuracy between k-NN and M^3-k-NN under the value of k varying. And we thus do not apply any optimization technology to increase the absolute accuracy. All simulations were performed on a PC with 2.4GHz Pentium 4 CPU and 512MB RAM.

3.1 M^3-k-NN, R^3-M^3-k-NN and k-NN: Accuracy in Training Set 1

We divide training sample sets from each class at the same size 2000 for some modules, if there are still remained samples in a class whose number is less than 2000, we put them into a new module. The value of k varies from 3 to 20. The experiment results are shown in Fig.1.

Obviously, while the value of k increases, the incorrect accuracy of both M^3-k-NN and R^3-M^3-k-NN increases after decreasing, which changes more rapidly than k-NN does. In fact, both M^3-k-NN and R^3-M^3-k-NN get to their peak accuracy earlier than k-NN does. Within the whole interval while the value of k changes, M^3-k-NN, R^3-M^3-k-NN or k-NN performs in a similar manner.

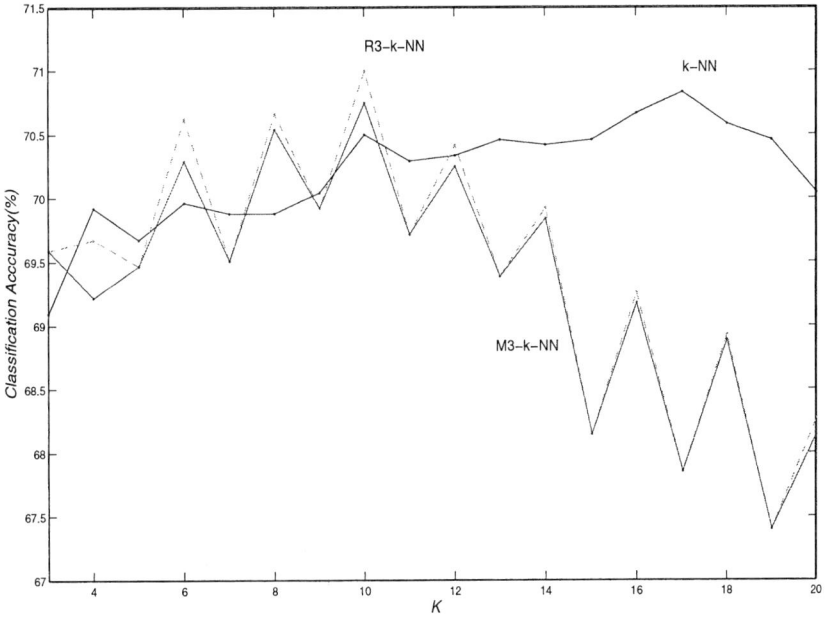

Fig. 1. M^3-k-NN, R^3-M^3-k-NN and k-NN: Accuracy in Training Set 1

3.2 M^3-k-NN and k-NN: Accuracy in Training Set 2

In the following experiment training set 2 is used. Here, we only compare k-NN with M^3-k-NN. The result is shown in Fig.2, from which we can see that M^3-k-NN gets to its peak accuracy earlier than k-NN does, too.

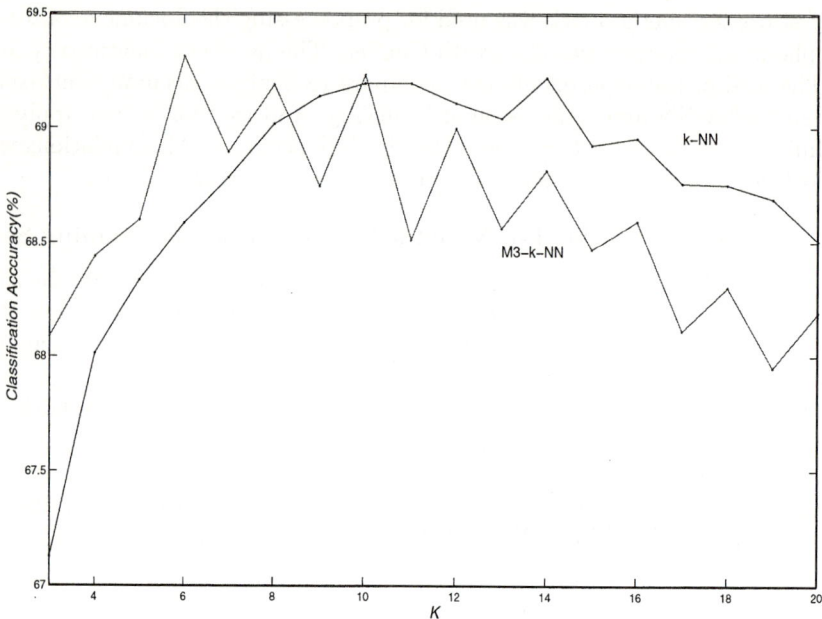

Fig. 2. M³-k-NN and k-NN: Accuracy in Training Set 2

3.3 Comparison of Running Time

Comparison of classification time between two algorithms at each different value of k is shown in Table 1. All experiments run in a PC with Windows XP OS. Thus we estimate M³-k-NN's parallel processing performance by dividing the total running time of M³-k-NN by the number of modules. It is easy to see that the classification time used by M³-k-NN in a parallel model is much less than that of traditional k-NN.

4 Theoretical Analysis of the Algorithm

Parts of our algorithm have been studied in [1] and [3], which shows that the load of a modular algorithm is lighter than the same one without any modularization in the phase 1 of the algorithm described in Section 2. This is an outstanding merit of all one-against-one modularization methods.

For an m-class problem, suppose the number of all training samples is T. If one visiting for a training sample in k-NN algorithm is taken as a basic unit of time complexity, then the time complexity of k-NN algorithm will be about kT.

As to the first phase of M³-k-NN, the time complexity of each k-NN base classifier is near to $2kT/m$. Although the whole time complexity is $(m-1)kT$ under serial processing, the time complexity under parallel processing is only $2kT/m$. This means that the processing time of M³-k-NN will be decreased if only there is a synchronous increase of both the number of classes and the

Table 1. Comparison of classification time between k-NN and M^3-k-NN

k	Training-Set 1 (ms)			Training-Set 2 (ms)		
	k-NN	M^3-k-NN	single M^3-k-NN module	k-NN	M^3-k-NN	single M^3-k-NN module
3	13672	68569	1524	564219	2781059	61801
5	14406	72428	1610	559344	2816960	62599
7	15094	75881	1686	573891	2873521	63856
9	14797	74217	1649	553125	2912828	64730
11	14828	74507	1656	573625	3348766	74417
13	15438	78597	1747	570875	3207375	71275
15	15407	77900	1731	584172	3737203	83049
17	15188	75951	1688	598421	3776235	83916
19	15766	79188	1760	610797	3824140	84981

number of processing units. If the second phase of the algorithm is considered, the parallel processing time complexity will be decreased more. In the extreme case, the value can be $2kT/(mm_1m_2)$.

Now we will give a simple explanation for the similar accuracy between M^3-k-NN and k-NN under the first task decomposition. There will be $m(m-1)/2$ sub-tasks in M^3-k-NN for an original m-class classification problem. A base k-NN classifier learning from samples of class i,j, while $1 \leq i,j \leq m$, or its corresponding output is denoted by X_{ij}.

We consider the composition of k nearest neighbors of a test sample while the original k-NN algorithm that is applied in all m-class training samples. We say, the k nearest neighbors must come from the union set of k nearest neighbors set of each base k-NN classifier's corresponding test sample in M^3-k-NN. Otherwise, suppose there is a sample from class j' outside the set at least, that is to say, it is in k nearest neighbors of m-class training set. Consider there is unique class j' in m classes. If we limit the class upon class i and j' while searching k nearest neighbors, the sample must come from either class i or class j', which contradict our initial suppose. Thus, k nearest neighbors from m-class is the subset of the union set of each base k-NN classifier's corresponding test sample's k nearest neighbors set in M^3-k-NN. This result suggests that there is a fixed ratio between M^3-k-NN's k voting sets and k-NN's. Especially, let k be equal to 1, thus, all samples from M^3-k-NN's voting set will be in the same class. Then, corresponding nearest neighbors in m classes must give the output with the same class number. Namely, M^3-1-NN is equal to 1-NN.

The reason that M^3-k-NN gets to its peak accuracy faster than k-NN does is that less samples are processed in each module for M^3-k-NN, then increasing of the value of k will have more notable effects on M^3-k-NN than k-NN.

However, in [4], another kind of k-NN classifiers combination has been considered, too. The main ideal of that paper is that k-NN classifiers' combinations are made under different feature combinations, and all k-NN classifiers are applied to the whole training text sets, while our method is to cut the whole training sets into different parts and then to perform individual k-NN classification.

5 Conclusions

We present a modular k-NN method, M^3-k-NN and its revised version R^3-M^3-k-NN in this paper, which are applied to large-scale text categorization. Our experiments show that M^3-k-NN admits a flexible k-NN classifiers modularization to realize a complete parallel processing, which will not reduce classification accuracy at the same time. In fact, as the value of k increasing, M^3-k-NN gets to its peak classification accuracy faster than k-NN does. In addition, our analysis suggests that M^3-k-NN algorithm concerns with a high order k-NN algorithm, which makes some theoretical guarantee for M^3-k-NN's reliability.

Acknowledgements

The authors would like to thank Junjie Zhang for his programming. This research is partially supported by the National Natural Science Foundation of China under Grant No. 60375022.

References

1. Allwein, E.L., Schapire, R. E., Singer, Y.: Reducing Multiclass to Binary: A Unifying Approach for Margin Classifiers. Journal of Machine Learning Research, Vol.1 (2000) 113-141
2. Chiang, J.-H., Chen, Y.-C.: Hierarchical Fuzzy-KNN Networks for News Documents Categorization. The 10th IEEE International Conference on Fuzzy Systems, Vol.2 (2001) 720-723
3. Frnkranz,J.: Round Robin Classification. The Journal of Machine Learning Research, Vol.2 (2002) 721-747
4. Han, H., Yand J.Y., Hu, Z.S.: Combination of KNN Classifiers Based on Multiple Correspondence Analysis. Information and Control, Vol.28 (1999) 350-356
5. Lim, H.S.: An Improved KNN Learning Based Korean Text Classifier with Heuristic Information. ICONIP '02, Vol.2 (2002) 731-735
6. Lu, B.L., Ito, M.: Task Decomposition and Module Combination Based on Class Relations: a Modular Neural Network for Pattern Classification. IEEE Transactions on Neural Networks, Vol.10 (1999) 1244-1256
7. Yang, Y., Petersen, J.P.: A Comparative Study on Feature Selection in Text Categorization. ICML'97, (1997) 412-420.

Avatar Behavior Representation and Control Technique: A Hierarchical Scripts Approach[1]

Jae-Kyung Kim[1], Won-Sung Sohn[2], Soon-Bum Lim[3], and Yoon-Chul Choy[1]

[1] Department of Computer Science, Yonsei University,
Shinchon-dong, Seodaemun-gu, 120-749, Seoul, Korea
{ki187cm, ycchoy}@rainbow.yonsei.ac.kr
[2] Computational Design Group,
Carnegie Mellon University, Pittsburgh, PA15213, USA
sohnws@u.washington.edu
[3] Department of Multimedia Science, Sookmyung Women's University,
Chungpa-dong, Yongsan-gu, 140-742, Seoul, Korea
sblim@sookmyung.ac.kr

Abstract. Avatar techniques have rapidly progressed in recent years, and will be widely adopted to various applications. The paper proposes hierarchical approach for representation and control techniques for avatar behavior for simpler avatar control in various domains. We proposed three-layered architecture: task-level behavior, high-level motion, and primitive motion. Thus, the user controls avatar at task-level layer and does not have to concern about low-level animation data. Our goal is to support flexible and extensible representation and control of avatar behavior by hierarchical approach separating application domains and implementation tools.

1 Introduction

As importance of computer usage in our dairy living grows, many of our activities are made in virtual environment. Now, computers are considered to be the crucial intermediary of societal activities of human, not a machinery [1]. Thus, interactive user interface techniques between human and computer, which can induce interests, are becoming more and more significant. The avatar is the most common example of the interface techniques. Avatar application was selected as one of the most important communication to the 21st century by Gartner group[2].

With these avatar applications, research on avatar behavior representation and control is actively going on. As XML being web standard language, avatar behavior script can be based on XML and it is possible to standardize avatar behavior. Many XML-based multi purpose script languages are being developed including AML[3], CML[4], VHML[5], CPSL[6], TVML[7] and STEP[8]. Although, the delicate nature of high-level motion scripts is advantageous, controlling avatar motion is relatively complicated for users.

---

[1] This work was supported by Ministry of Commerce, Industry and Energy.

In this paper, we define the script languages for avatar behavior control by layered approach architecture to provide easy interface to users at various application domain and implementation environments. The proposed script language consists of three-layered architecture: task-level behavior, high-level motion, and primitive motion. We discuss them in the next section in detail.

2 Hierarchical Representation and Control for Avatar Behavior

Our main goal is to gain high extensibility, comparability and reusability for the control of avatar behavior by defining script language for hierarchical representation and control of avatar behavior. The script languages are consisted of task-level behavior which is set of domain-dependent tasks, primitive motion which is supported by character motion engine or motion library, and high-level motion which has independent representation from these domain and motion engine or library. The characteristics of three script languages are shown in Figure 1.

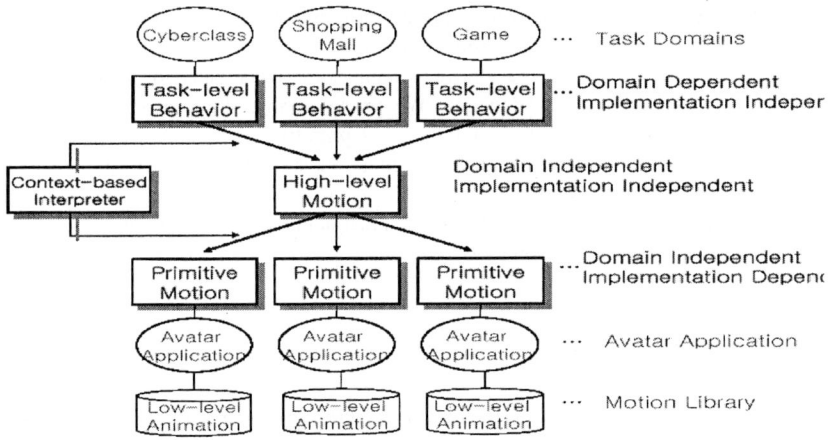

Fig. 1. Overall Procedure of the Proposed System

Overall procedure of the proposed system is shown in Figure 1. Function and attribute of each script languages are discussed in following section.

2.1 Task-Level Behavior Script Language

The definition of avatar task-level behavior in this paper is the necessary behaviors to achieve certain tasks in a specific domain environment. Each domain has different tasks, and the kind of behaviors for performing tasks would be changed. Since tasks are usually given by human scripter, the behavior should be simple and human-readable. This will lead that the design of the scripting language for task-level behavior would be motivated by following requirements: simplicity, abstraction, readability, extensibility, parameterization, and synchronization.

With respect to the design concept, we defined XML DTD for task-level scripting language. Since XML is machine independent and highly extensible, task-level script format is suitable. Task-level behavior DTD can be applied to various domain environments and easily extended when task domain needs additional behaviors. Our language supports required elements for scripting tasks:

Task-level behavior = <behavior name>[<target object>|<target location>] [<purpose>] [<adverb>]* [<synchronization>]+

In task-oriented system[9,10], a task usually divided into sub tasks, and the task is achieved by completing all sub tasks. Similarly, a task-level behavior is separated into sequence of motions[11]. To generate the motion sequences from task-level behavior, we suggested the formal task translation model. Each domain has certain kinds of behaviors and they can be converted to the high-level motion sequences regardless of domain environments. Various task-level behavior scripting languages are translated into a standardized high-level motion scripting language.

Formal Task Translation Model

- Identification of target
  - find location of target object and location of task-level behavior (L_t)
- Locomotive motion (L_m)
  - identification of present avatar location (L_a)
  - spatial distance between L_a and L_t and generate L_m, if $L_a \neq L_t$
- Manipulative motion (M_m)
  - generate M_m for L_t
- Verbal information (V_i)
  - generate verbal speech for avatar behavior if available
- Speed and intensity parameters (S_p and I_p)
  - parameterize L_m and M_m for speed, intensity and duration

In terms of suggested formal task translation model, a task is divided into locomotive and manipulative motions with verbal information and parameters. Finally, the motions are represented in high-level motion script format.

2.2 High-Level Motion

The high-level motion represents and controls the avatar motion by using parameters such as speed, intensity, direction, and so on. We analyzed the parameters which are used in the existing high-level motion scripts, and defined high-level motion script. Also, as stated above, the motion script should be applied to the formal task interpreter. By considering existing scripts and the formal interpreter, we defined the high-level motion scripting language.

First, high-level motion is comprised of two parts: basic avatar characteristic such as name, role, gender, and list of avatar motions. The list is comprised of motion name, and its parameters. Major four elements of the parameters are about spatial, temporal, intensive, and aural information. These four elements contain detailed control information as shown in Figure 2.

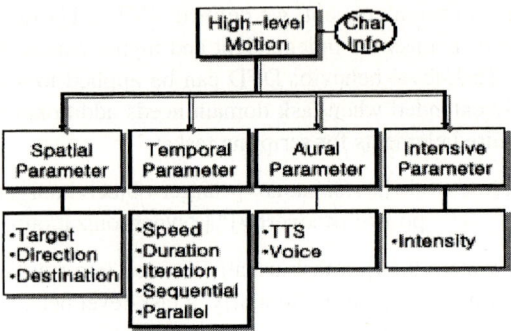

Fig. 2. High-level Motion Parameters

These attributes represent avatar motion independent from both domain and implementation. Also, its parameters are abstracted from physical world such as degree of joint angle or explicit integer values. The script takes a role which is similar to mediator between task-level behavior and primitive motion shown in Figure 1. The high-level scripting language is based on XML DTD for standardized motion representation.

2.3 Primitive Motion

A primitive motion is avatar motions which are supported by avatar engines or motion libraries. Primitive motion is subject to avatar control engines, and it contains physical information for specific implementation tool. For example, a high-level motion script, "<go to="table">" should be expressed like "<walk_to x="100" y="12" z="-2">" at the primitive motion script. The major difference between high-level and primitive motion scripting language is that high-level motion is independent from both domain and implementation while primitive motion is belonged to implementation environment. Although primitive motion controls avatar motion by parameters like high-level motion, the parameters are designed to describe physical values for avatar engine and structure of virtual world.

3 Implementation Results

In real world, the usual settings of school classroom are a lecture, a blackboard, a computer, a table, a door, walls and several lecture-related objects. In task-level scripting language, author command the lecturer by combination of these components and behavior. After the example script is translated to a high-level script language by the suggested formal translator, the high-level script will be loaded to the system and converted to primitive motions and the avatar performs its tasks in Figure 3(a). The primitive motion contains the physical information of the system, so the same task-level script can be properly applied even though physical structure of the virtual world is reorganized as shown in Figure 3(b).

 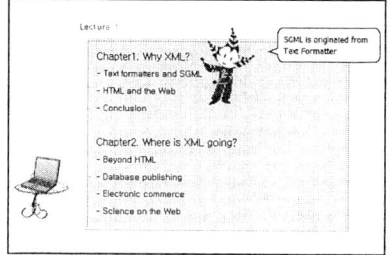

(a) Lecturing in 3D space (b) 2D space and repositioned objects

Fig. 3. The avatar is teaching an algorithm in cyber classroom

The scripting language can be applied to completely different implementation tool or environments. For instance, the same task-level script is loaded to two applications as shown in Figure 3. Because our architecture takes layered approach, application tools and script languages are explicitly decoupled. This makes easier for user to apply script language to various implementations.

4 Evaluation and Discussion

In order to measure the advantages of these scripting languages on creating scenario script and comparability, we applied our system to other implementations. First, users were asked to write down scripts for achieving 10 simple tasks by using each scripting language: AML, CPSL, and proposed script. Second, the scripts user made are applied to avatar implementation. MS Agent engine with basic motion library was selected for the implementation. Then, check the successful task achieved by the each script.

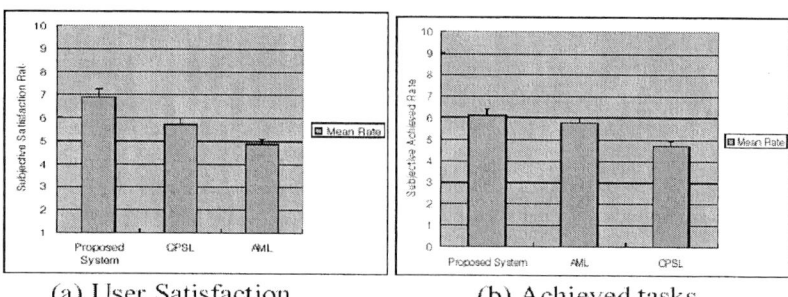

(a) User Satisfaction (b) Achieved tasks

Fig. 4. Subjective evaluation of user satisfaction and achieved tasks

First test uses scales from 1 (the least satisfaction) to 10 (the most satisfaction) to compare the difficulty of creating scenario script and spent time. Second test also uses the same scale to compare the number of correctly accomplished tasks. A group of 20 subjects including 14 male and 6 female graduate students participated in the experiment.

An ANOVA with repeated measures was used to analyze the result. As shown in Figure 4, the users rated that the results of the proposed system were easier and required less time to create the script. Significant effects were found between the time spent for the proposed script and the other script $F(2,57) = 8.98$, $P < 0.05$). Also, the proposed system got more number of achieved tasks. Note that CPSL has the low scores because of its implementation dependency.

5 Conclusion and Future Works

We proposed three-layered architecture including task-level behavior, high-level motion and primitive motion to provide simple interface for avatar control at various application domains to users. Also each layer interacts independently. Task-level behavior is explicitly separated from implementations so that it could be reusable at different tools.

Using this approach, avatar behavior can be controlled more easily in task-level and in high level and primitive motion, it is possible to control and express avatar motion with great reusability and extensibility which does not depend on implementation environment through abstract and physical expression.

In future works, an intuitive graphical user interface for the input of avatar tasks, and avatar motion controls based on avatar-object interaction technique are required for providing more efficient interface to users.

References

1. Prendinger, H.: Life-like Characters. Life-like characters book, Springer-Verlag (2003) 3-17
2. Woo, S.: Virtual Human Trends. Journal of Korea Multimedia Society, Vol. 6. No. 4 (2000)
3. Kshirsagar, S., Thalmann, D., Kamyab, K.: Avatar Markup Language. Proceeding of the workshop on Virtual environments (2002) 169-177
4. Arafa, Y., Mamdani, E.: Scripting embodied agents behaviour with CML. Proceeding of Intelligent User Interfaces (2003) 313-315
5. Marriott, A., Stallo, J.: VHML- Uncertainties and Problems A discussion. Proceeding of Embodied conversational agents for AAMAS2002, Bologna, Italy (2002)
6. Yoshiaki, S., Matsuda, H.: Design and Implementation of Scenario Language for Cyber Teaching Assistant. International conference on Computers in Education (2001)
7. Hayashi, M.: TVML. ACM SIGGRAPH 98 Conference on applications (1998) 292-297
8. Huang, Z., Eliens, A., Visser, C.: Implementation of a scripting language for VRML/X3D-based embodied agents. Proceeding of web technology (2003) 91-100
9. Lester, C., Zettlemoyer, S., Gregoire, P., Bares, H.: Explanatory Lifelike Avatars. Autonomous Agents (1999) 30-45
10. Ricket, J., Johnson, W.: Task-Oriented Collaboration with Embodied Agents in Virtual Worlds. Embodied Conversational Agents, MIT Press (2000) 95-122
11. Thalmann, D.: Autonomy and Task-Level Control for Virtual Actors. Programming and Computer Software, No. 4 (1995)

Analyzing iKP Security in Applied Pi Calculus*

Yonggen Gu, Guoqiang Li, and Yuxi Fu

BASICS**, Department of Computer Science and Engineering,
Shanghai Jiao Tong University, Shanghai 200030, China
{gyg68, liguoqiang, yxfu}@sjtu.edu.cn

Abstract. The security for electronic payments is very important in electronic commerce. Many electronic payment protocols have been proposed recently to meet the security requirements. Since the design of a protocol is a difficult and error-prone task, the use of formal methods that allow for the verification of such protocols has received increasing attention. In this paper we take a look at the iKP from the point of view of the applied pi calculus. The iKP is described in the calculus and some security properties, mainly the authentication and anonymity properties, are verified.

1 Introduction

The security for electronic payments is very important in electronic commerce. Many electronic payment protocols have been proposed recently to meet the security requirements. Such protocols pay considerable attention to the requirements on the security issues like whether customers, sellers or banks can prove their authorizations to each other or whether customers can hold any anonymity etc. IBM Research Lab has developed a family of secure electronic payment protocols [5,6], called iKP (i-Key-Protocol, where i=1,2,3), that are compatible with the existing card-based business models and payment system infrastructures.

The design of an electronic payment protocol is however a difficult and error-prone task. Many popular and widely used electronic payment protocols have been shown to have flaws. For this reason, the use of formal methods for the verification of such protocols has received increasing attention. There are many ways to describe and verify security protocols, such as logic theory, proof theory etc. Using pi calculus [7] and its extensions to describe protocols is attractive because they are simple yet powerful.

Recently, Abadi and Fournet have proposed the applied pi calculus for the analysis of security protocols [3]. The applied pi calculus is one of the extensions of the pi calculus with values passing, primitive functions, and equations among terms. Based on the applied pi calculus, Abadi and his collaborators have

---

* The work is supported by The National Distinguished Young Scientist Fund of NNSFC (60225012) and The National 973 Project (2003CB316905).
** Laboratory for Basic Studies in Computing Science.

analyzed several authentication protocols, mainly treating authenticity and secrecy properties [1, 2, 4]. However they have not considered complex protocols and their properties such as electronic commerce protocols. In this paper, we use the calculus to describe iKP and verify its authentication and anonymity properties. Our contribution is twofold:

- We analyze the security of the iKP in a precise formal framework. We model the authentication and anonymity properties by extending the equational theory of the applied pi calculus. In particular we consider the time stamp, which can extend the analysis of the security protocols.
- We use the applied pi calculus to analyze electronic payment protocols. What we have achieved is to point out a possibility to analyze and verify multi-party protocols. Furthermore iKP protocols are classical electronic payment protocols which have common properties. So our work is beneficial to the formal analysis of future electronic commerce protocols.

In this short paper we assume the reader is familiar with the pi calculus and has some knowledge of the security protocols.

2 Applied Pi Calculus

In this section, we introduce the applied pi calculus briefly for the convenience of the reader. A detailed account can be found in [3]. The grammar for processes is similar to the one in the pi calculus. There is only one difference: Here messages can contain terms (not only names) and names need not be just channel names:

$$P, Q, R ::= \mathbf{0} \mid P|Q \mid !P \mid \nu n.P \mid if\ U = V\ then\ P\ else\ Q \mid u(x).P \mid \overline{u}\langle M\rangle.P$$

Process $\mathbf{0}$ does nothing. $P|Q$ is the parallel composition of P and Q. $!P$ behaves as an infinite number of copies of P running in parallel. The process $\nu n.P$ makes a new name n and then behaves as P. The conditional construct "if $U = V$ then P else Q" is standard, where $U = V$ is an equality test. The input process $u(x).P$ is ready to input from channel u and then evolves as P with the actual message substituting for the formal parameter x. Finally the output process $\overline{u}\langle M\rangle.P$ is ready to output message M on channel u and then evolves as P.

The plain processes are extended with active substitutions:

$$A, B, C ::= P \mid A|B \mid (\nu n).A \mid (\nu x).A \mid \{x = V\}$$

The substitution $\{x = V\}$ replaces the variable x with the term V. $\{x = V\}$ can represent the situation in which a term V has been sent to the environment, but the environment may not have the atomic names that appear in V. A *frame* is an extended process built up from active substitutions by parallel composition and restriction. Informally frames represent the static knowledge gathered by the environment after communications with an extended process. *Evaluation contexts*, denoted by $C[\ ]$, are extended processes with a hole in the place of an extended process.

Structural equivalences, written $A \equiv B$, relate extended processes that are equal by any capture-avoiding rearrangements of parallel compositions, restrictions and active substitutions, and by equational rewriting of any terms in processes. Reductions, written $A \longrightarrow B$, represent silent steps of computation (in particular, internal message transmissions and branching on conditionals). Labelled transitions, written $A \xrightarrow{\alpha} B$, represent interactions with the environment. They consist of message inputs and message outputs, respectively written $A \xrightarrow{a(U)} B$ and $A \xrightarrow{\nu \bar{a}\langle U \rangle} B$. Reductions and labelled transitions are closed by structural equivalence, and consequently by equational rewriting on terms.

We write $A \Downarrow a$ when A can send a message on a, that is, when $A \rightarrow^* C[\bar{a}\langle M \rangle.P]$ for some evaluation context $C[\ ]$ that does not bind a.

Definition 1. *Observational equivalence ($\approx$) is the largest symmetric relation $\mathcal{R}$ on the closed extended processes such that $A \mathrel{\mathcal{R}} B$ implies the following:*
1. *If $A \Downarrow a$ then $B \Downarrow a$;*
2. *If $A \rightarrow^* A'$ then $B \rightarrow^* B' \mathrel{\mathcal{R}} A'$ for some B';*
3. *$C[A] \mathrel{\mathcal{R}} C[B]$ for all closing evaluation contexts $C[\ ]$.*

Definition 2. *Two terms M and N are equal in the frame φ, written $(M = N)\varphi$, if and only if $\varphi \equiv \nu n.\sigma$, $M\sigma = N\sigma$ and $\{n\} \cap (f_n(M) \cup f_n(N)) = \emptyset$ for some name n and substitution σ.*

Definition 3. *Two closed frames φ and ψ are statically equivalent, written $\varphi \approx_s \psi$, if $dom(\varphi) = dom(\psi)$ and, for all terms M and N, it holds that $(M = N)\varphi$ if and only if $(M = N)\psi$.*

We say that two closed extended processes are statically equivalent, written $A \approx_s B$, if their frames are statically equivalent.

3 iKP

IBM Research Lab developed a family of secure electronic payment protocols, iKP (i-Key-Protocol, i=1,2,3), which are compatible with the existing card-based business models and payment system infrastructures. iKP involves three parties: the buyer (who makes the actual payment), the seller (who receives the payment) and the acquirer (who acts as an intermediary between the electronic payment world and the existing payment infrastructure).

1KP requires only the acquirer to possess a public key-pair. Buyers and sellers only need to have authentic copies of the acquirer's public keys. 2KP demands that sellers, in addition to the acquirers, hold public key-pairs and public key certificates. 3KP further assumes that buyers have their own public key-pairs and public key certificates. In this paper, we focus on the security properties of 1KP.

We firstly describe the cryptographic primitives as well as the general protocol structure. SK_X stands for the secret key of Party X (X= Buyer B, Seller S, Acquirer A). We represent $PK(SK_X)$ as its public key. H(.) is a strong collision-resistant one-way hash function which returns strong pseudo-random values; $\{.\}_y$

stands for public-key encryption with $y = Pk(SK_X)$ or stands for signature computed with $y = SK_X$.

To describe 1KP protocol we enumerate a list of quantities used in describing the protocols. The signals and their meanings are defined below:

$Desc$	Description of purchase and delivery address.
$Salt_B$	Random number generated by Buyer, used to salt $Desc$ and consequently to ensure the privacy of the order information on the Seller to Acquirer link.
$Authprice$	Amount and currency.
$Date$	Seller's date/time stamp, used for "coarse-rained" payment replay protection.
$Nonce_S$	Seller's nonce used for more "fine-grained" payment replay protection.
ID_S	Seller's id. This identifies seller to acquirer.
TID_S	Transaction ID, an identifier chosen by the seller that uniquely identifies the context.
Ban_B	Buyer's Account Number.
$Expiration$	Expiration date associated with Buyer's Account Number.
R_B	Random number chosen by buyer to form ID_B.
ID_B	A buyer pseudo-ID which computed as $ID_B = H(R_B, Ban_B)$.
$Respcode$	Response from the clearing network: YES/NO or authorization code.

To describe the protocol flow sententiously, we use signals to represent some composite fields. These are defined below:

$Common$	$Authprice, ID_S, TID_S, Date, Nonce_S, ID_B, H(Salt_B, Desc)$
$Clear$	$ID_S, TID_S, Date, Nonce_S, H(Common)$
$Slip$	$Authprice, H(Common), Ban_B, R_B, Expiration$
$EncSlip$	$E_A(Slip)$
Sig_A	$S_A(Respcode, H(Common))$

There are six interaction flows in 1KP protocol. Based on the above signals, we can explain the flow-by-flow interactions as follows:

- **Initiate** $B \longrightarrow S$: $Salt_B$, ID_B. Buyer forms ID_B by generating a random number R_B and computing $ID_B = H(R_B, Ban_B)$. Then it generates another random number $Salt_B$. Finally it sends Initiate flow.
- **Invoice** $S \longrightarrow B$: $Clear$. Seller retrieves $Salt_B$ and ID_B from Initiate, obtains $Date$, and generates random quantity $Nonce_S$. The combination of $Date$ and $Nonce_S$ is used later by A to uniquely identify this payment. Seller then chooses TID_S and computes $H(Salt_B, Desc)$. Then it forms $Common$ as defined above and computes $H(Common)$. Finally Seller sends Invoice.
- **Payment** $B \longrightarrow S$: $EncSlip$. Buyer retrieves $Clear$ from Invoice and validates $Date$ within a pre-defined time skew. Buyer computes $H(Salt_B, Desc)$ and $H(Common)$ and checks whether it matches the value in $Clear$. Next

Buyer forms *Slip* and encrypt it under the acquirer's public key. Finally Buyer sends it to the seller in the Payment flow.

- **Auth-Request** $S \longrightarrow A$: *Clear*, $H(Salt_B, Desc)$, *EncSlip*. The seller now requests the acquirer to authorize the payment. It then forwards *EncSlip* along with *Clear* and $H(Salt_B, Desc)$.
- **Auth-Response** $A \longrightarrow S$: *Respcode*, Sig_A. The acquirer retrieves *Clear*, *EncSlip* and also $H(Salt_B, Desc)$ from Auth-Request. It then extracts ID_S, TID_S, *Date*, $Nonce_S$ from *Clear* and decrypts *EncSlip*. It obtains *Slip*, from which it extracts *Authprice*, Ban_B, *Expiration*, R_B. Acquirer then re-constructs $H(Common)$ and checks whether it matches $H(common)$ it received. Next acquirer uses the credit card organization's existing clearing and authorization system to obtain on-line authorization of this payment. This entails forwarding: Ban_B, *Expiration*, etc. Finally acquirer sends Auth-Response to Seller.
- **Confirm** $S \longrightarrow B$: *Respcode*, Sig_A. The seller extracts *Respcode* and the acquirer's Signature from Auth-Response. It then verifies the acquirer's signature and forwards both *Respcode* and Signature to the buyer.

4 The Protocol in the Applied Pi Calculus

In this section we express 1KP in the applied pi calculus. We first propose an adequate equational theory, and then define the process that expresses the protocol.

4.1 An Equational Theory

Terms of the applied pi calculus uses not only names, but also variables and function primitives. To define reductions we introduce a named equational theory that consists of algebraic equations on terms.

The following grammar of terms indicates the function symbols and the notation conventions we use:

$T, U, V, V_0, \cdots ::=$	terms
$A, B, S, x_1, x_2, \cdots$	variable
$C_{BS}, C_{SB}, C_{SA}, C_{AS}, \cdots$	name (for channel)
$Nonce_A, K_A, Date, \cdots$	name (for nonces, keys and time)
$f(T_1, \cdots, T_n)$	function application ($f \in \Sigma$)

Here Σ is a set of functions which includes three kinds of functions. Firstly it includes *cryptographic and checker functions*, in which $H(u_1, ...)$ represents Cryptographic hash, $Pk(u)$ represents Public key derivation, $\{T\}_V$ represents Public key encryption or private key signature, $decrypt(w, u)$ represents private key decryption, $checksig(w, u)$ represents public key check signature and $checktime(Time, Expiration)$ represents check time expiration. Secondly it includes *constructing for protocol messages functions*, which in turn includes the followings:

- $Initiate(u_1, u_2)$,
- $Clear(u_1, u_2, u_3, u_4, u_5)$,
- $Common(u_1, u_2, \ldots, u_7)$,
- $Slip(u_1, u_2, u_3, u_4, u_5)$,
- $ARequest(u_1, u_2, u_3)$,
- $AResponse(u_1, u_2)$, and
- $Confirm(u_1, u_2)$.

Lastly the set also includes *projection functions*, which means to select the j-th parameter of the construct for protocol message functions. For example, $Initiate.j(Initiate(u_1, u_2))$ means to select the j-th parameter from $Initiate()$. This kind of functions also includes the followings:

- $Clear.j(Clear(u_1, u_2, u_3, u_4, u_5))$,
- $Common.j(Common(u_1, u_2, \ldots, u_7))$,
- $Slip.j(Slip(u_1, u_2, u_3, u_4, u_5))$,
- $ARequest.j(ARequest(u_1, u_2, u_3))$,
- $AResponse.j(AResponse(u_1, u_2))$, and
- $Confirm.j(Confirm(u_1, u_2))$.

Equational theory is a special feature of the applied pi calculus. An equation represents semantic equality rather than syntactic identity. The equations in the model of this paper are the followings:

$$Decrypt(\{x\}_{Pk(z)}, z) = x$$

$$checksig(\{x\}_z, Pk(z)) = true$$

$$checktime(time, Expiration) = \begin{cases} true & \text{if } time \in Expiration \\ false & \text{otherwise} \end{cases}$$

$$Tuple.j(Tuple(x_1, \cdots x_l)) = x_j \text{ for } j < i, \ Tuple \in \{Initiate, Clear,$$
$$Common, Slip, ARequest, AReponse,$$
$$Confirm\}$$

4.2 The Protocol

Based on the terms and equational theory defined above, we give a formal counterpart to the description of the message flows of 1KP protocol in this subsection.

Messages: We rely on substitutions in order to define the protocol messages. With the construction of the protocol message functions we defined in Section 4.1, the protocol messages can be defined as follows:

$$\sigma_1 \stackrel{def}{=} \{x_1 = Initiate(Salt_B, ID_B)\}$$

$$\sigma_2 \stackrel{def}{=} \{x_2 = Clear(ID_S, TID_S, Date, Nonce_S, H(Authprice, ID_S, TID_S,$$
$$Date, Nonce_S, Initiate.2(x_1), H(Initiate.1(x_1), Desc)))\}$$

$$\sigma_3 \stackrel{def}{=} \{x_3 = \{Slip\}_{Pk(SK_A)}\}$$

$$\sigma_4 \stackrel{def}{=} \{x_4 = ARequest(x_2, H(Initiate.1(x_1), Desc), x_3)\}$$

$$\sigma_5 \stackrel{\text{def}}{=} \{x_5 = AResponse(Respcode, \{Respcode, H(Common)\}_{SK_A})\}$$
$$\sigma_6 \stackrel{\text{def}}{=} \{x_6 = Confirm(AResponse.1(x_5), AResponse.2(x_5))\}$$

Processes: The following is the encoding of the protocol. It includes definitions of the processes for the Buyer, the Seller, and the Acquirer. Moreover all the processes work asynchronously and concurrently.

$$P \stackrel{\text{def}}{=} ((\nu Authprice, Desc)P_B \mid P_S) \mid P_A$$
$$P_B \stackrel{\text{def}}{=} (\nu Salt_B, R_B)(\overline{C_{BS}}\langle x_1\sigma_1\rangle \mid !(C_{SB}(x_2).if\ Clear.5(x_2)$$
$$= H(Authprice, Invoice.1(x_2), Clear.2(x_2), Clear.3(x_2),$$
$$H(R_B, Ban_B), H(Salt_B, Desc))\ then\ (\overline{C_{BS}}\langle x_3\sigma_3\rangle \mid !(C_{SB}(x_6).$$
$$if\ checksig(confirm.2(x_6), Pk(SK_A))\ then\ F_B(x_6)))))$$
$$P_S \stackrel{\text{def}}{=} (\nu ID_S, TID_S, Date, Nonce_S)(C_{BS}(x_1).(\overline{C_{SB}}\langle x_2\sigma_2\rangle \mid C_{BS}(x_3)$$
$$.(\overline{C_{SA}}\langle x_4\sigma_4\rangle \mid C_{AS}(x_5).\overline{C_{SB}}\langle x_6\sigma_6\rangle.F_S(x_5))))$$
$$P_A \stackrel{\text{def}}{=} (\nu SK_A)!(C_{SA}(x_4).if\ Slip.2(Decrypt(ARequest.3(x_4), SK_A)) =$$
$$H(Slip.1(Decrypt(ARequest.3(x_4), SK_A)), Clear.1(ARequest.1(x_4)),$$
$$Clear.2(ARequest.1(x_4)), Clear.3(ARequest.1(x_4)),$$
$$Clear.4(ARequest.1(x_4)), H(Slip.3(Decrypt(ARequest.3(x_4), SK_A)),$$
$$Slip.4(Decrypt(ARequest.3(x_4), SK_A))), ARequest.1(X_4))$$
$$\land\ checktime(datetime,\ Slip(Decrypt(ARequest.3(x_4))))$$
$$then\ ((\nu ARespcode)(\overline{C_{AS}}\langle x_5\sigma_5\rangle.F_A(x_4))))$$

P_B, P_S and P_A represent respectively protocol principals, and P represents the protocol. In addition, functions $F_B(x)$, $F_S(x)$, $F_A(x)$ represent respectively the actions the buyer, the seller, the acquirer is required to perform after finishing the protocol flows.

5 Authentication Properties

In this section we begin our analysis of the protocol with authentication property. We study the executions of the protocol by examining transitions $P \stackrel{\eta}{\longrightarrow} P'$, where η is an arbitrary sequence of labels. When the acquirer debits a certain credit card account by a certain amount, the acquirer must be in possession of an unforgeable proof that the owner of the credit card has authorize this payment. By this security requirement, we should obtain an authentication property stated below.

Theorem 1 (Acquirer Authentication). *Suppose that* $P \stackrel{\eta}{\longrightarrow} P'$. *If the action*

$$\overline{C_{AS}}\langle AResponse(Respcode, \{Respcode, H(Common)\}_{Pk(SK_A)})\rangle \quad (1)$$

occurs in η, *then the following actions*

$$\overline{C_{BS}}\langle\{Slip\}_{Pk(SK_A)}\rangle \text{ and}$$

$$\overline{C_{SA}}\langle ARequest(Clear, H(Salt_B, Desc), \{Slip\}_{Pk(SK_A)})\rangle$$

must occur before (1).

Proof. We can interpret the theorem as a correspondence assertion [8]: Whenever Acquirer debits, we have that the owner of the credit card has authorized the payment, and the seller has asked the payment. We know that if

$$\overline{C_{AS}}\langle AResponse(Respcode, \{Respcode, H(Common)\}_{Pk(SK_A)})\rangle$$

occurs in η, which means that the acquirer has succeeded in checking identities of the buyer and the seller. The acquirer validates the buyer's identity by Ban_B and also validates the seller's by matching of $H(Common)$.

Therefore we certainly have $\overline{C_{BS}}\langle\{Slip\}_{Pk(SK_A)}\rangle$ and $\overline{C_{SA}}\langle ARequest(Clear, H(Salt_B, Desc), \{Slip\}_{Pk(SK_A)})\rangle$. □

The proofs of the seller authentication and buyer authentication are similar.

6 Anonymity Property

In this section, we analyze the anonymity property. In iKP buyers may also want anonymity from eavesdroppers and from sellers. Furthermore buyers may even want anonymity with respect to the payment system provider. iKP does not focus on anonymity and, in particular, offers no anonymity from the payment system provider. This might be desirable for systems that aim to imitate cash, but is not essential for the protocols, like iKP, that follow the credit card-based payment model. However, iKP does try to minimize exposure of buyers' identities with respect to sellers and eavesdroppers. The following theorem states that buyers do not reveal their Ban to sellers and eavesdroppers.

Theorem 2 (Buyers' Anonymity Property). *The buyer has Ban_B anonymity property: For all sellers and eavesdroppers it holds that $P_B(Ban_B) \approx_s P_B(Ban'_B)$.*

Proof. We firstly define the frames. Let $\varphi_1(P_B(Ban_B))$ be

$$\{x_1 = (Salt_B, H(R_B, Ban_B)) \mid$$
$$x_3 = \{Authprice, H(Common), Ban_B, R_B, Expiration\}_{Pk(SK_A)}\}$$

and $\varphi_2(P_B(Ban'_B))$ be

$$\{x_1 = (Salt_B, H(R_B, Ban_B)) \mid$$
$$x_3 = \{Authprice, H(Common), Ban'_B, R_B, Expiration\}_{Pk(SK_A)}\}$$

It is obviously that no sellers and eavesdroppers can know K_A. So it holds that $\varphi_1(P_B(Ban_B)) \approx_s \varphi_2(P_B(Ban'_B))$. Thus $P_B(Ban_B) \approx_s P_B(Ban'_B)$, meaning that the buyers' Ban_B is not exposed to sellers and eavesdroppers.

For the acquirer, $Slip.3(Decrypt(x_3, K_A)) = Ban_B$ is valid in $\varphi_1(P_B(Ban_B))$ but is not valid in $\varphi_2(P_B(Ban'_B))$. So we have $P_B(Ban_B) \not\approx_s P_B(Ban'_B)$, which means that buyers do not have anonymity property to acquirer. □

7 Conclusion

In this paper we use the applied pi calculus of Abadi and Fournet to describe the iKP and verify its authentication and anonymity properties. We believe that the use of the applied pi calculus is particulary appropriate for the analysis of the security protocols. We take the iKP as case study, because they are classical electronic payment protocols compatible with the existing card-based business models and payment system infrastructures. Furthermore, as noted in the introduction, this case study contributes to the development of the ideas and results for the specification and verification of the electronic payment protocols that should be useful beyond the analysis of iKP.

As for future work, we plan to analyze the non-repudiation of iKP using the applied pi calculus and construct an automatic tool to help analyzing the security protocol using the framework advocated in this paper.

References

1. Abadi, M., and Blanchet, B.: Computer-assisted verification of a protocol for certified email. In: Static Anlaysis, 10th International Symposium(SAS'03). Volume 2694 of LNCS., Springer-Verlag (2003) 316–335
2. Abadi, M., Blanchet, B., and Fournet, C.: Just Fast Keying in the Pi Calculus. In: 13th European Symposium on Programming. Volume 2986 of LNCS., Springer-Verlag (2004)
3. Abadi, M., and Fournet, C.: Mobile Values, New Names and Secure Communication. In: 28th ACM Symposium on Principles of Programming Languages(POPL'01). (2001) 104–115
4. Abadi, M., and Fournet, C.: Hiding Names: Private Authentication in the Applied Pi Calculus. In: Proceedings of the International Symposium on Software Security (ISSS'02). Volume 2609 of LNCS., Springer-Verlag (2003) 317–338
5. Bellare, M., Garay, J., Hauser, R., Herzberg, A., Krawczyk, H., Steiner, M., Tsudik, G., and Waidner, M.: iKP - A Family of Secure Electronic Payment Protocols. In: Proceedings First USENIX Workshop on Electronic Commerce, USENIX. (1995)
6. Bellare, M., Garay, J., Hauser, R., Herzberg, A., Krawczyk, H., Steiner, M., Tsudik, G., Herreveghen, E.V., and Waidner, M.: Design, implementation, and deployment of the iKP secure electronic payment system. IEEE Journal of Selected Areas in Communications **18** (2000) 611–627
7. Milner, R., Parrow, J., and Walker, D.: A Calculus of Mobile Processes, parts I and II. In: Information and Computation. (1992) 1–77
8. Woo, T.Y.C., and Lam, S.S.: A semantic model for authentication protocols. In: 14th IEEE Symposium on Research in Security and Privacy, IEEE Computer Society Press (1993) 178–194

General Public Key m-Out-of-n Oblivious Transfer*

Zhide Chen[1] and Hong Zhu[2]

[1] Department of Computer Science, Fudan University, Shanghai 200433, P. R. China
[2] Key Laboratory of Intelligent Information Processing, Fudan University, Shanghai 200433, P. R. China
{02021091, hzhu}@fudan.edu.cn

Abstract. In the m-out-of-n oblivious transfer model, Alice has n messages, Bob has m choices. After the interaction between the two parties, Bob can get m but only m messages from the n messages, Alice cannot know which m messages Bob gets. In this paper, for the first time, we construct a m-out-of-n OT protocol based on the general secure public key system.

Keywords: oblivious transfer, secure protocol, public key system.

1 Introduction

Oblivious transfer (OT) is an important primitive in modern cryptography. It has become the basis for realizing a broad class of cryptographic protocols, such as bit commitment, zero-knowledge proof, and secure multiparty computation [5,6,2]. It has also found its applications for e-commerce [1,8].

OT was first introduced to cryptography by Rabin [10] in 1981. In the protocol, there are two party, one is named Alice, the other one is named Bob. Informally speaking, in an OT, Alice sends a bit to Bob that he receives half the time, Alice does not find out what happened, Bob knows if he gets the bit or nothing.

In 1985, Even, Goldreich and Lempel provided another similar cryptographic tool named 1-out-of-2 oblivious transfer (OT_2^1) [4]. In an OT_2^1, Alice has two messages m_0, m_1 that she sends to Bob in such a way that he can decide to get either of them at his choice but not both. Alice cannot find out which message Bob gets.

m-out-of-n OT (OT_n^m) was introduced by Naor [9] and also by Yu [7] based on the discrete logarithm. However, no one has ever discussed the possibility of OT_n^m based on the general secure public key system (PKS). In this paper, we provide an OT_n^m protocol based on the PKS.

---

* This work is partially supported by a grant from the Ministry of Science and Technology (#2001CCA03000), National Natural Science Fund (#60273045) and Shanghai Science and Technology Development Fund (#03JC14014).

2 Definition

Definition 21. *1-out-of-2 Oblivious Transfer (OT_2^1)*
In an OT_2^1, Alice has input $m_0, m_1 \in \{0,1\}^k$; Bob has input $c \in \{0,1\}$, after the interaction between Alice and Bob, the following requirements should be satisfied:

Correctness: *Bob can get m_c if both Alice and Bob follow the protocol;*
Privacy for Bob: *Alice cannot find out the choice of Bob if Bob follows the protocol;*
Privacy for Alice: *Bob cannot get both m_0 and m_1 if Alice follows the protocol.*

Definition 22. *m-out-of-n Oblivious Transfer (OT_n^m)*
In an OT_n^m, Alice has input $m_1, \cdots, m_n \in \{0,1\}^k$, Bob has m choices $c_1, \cdots, c_m \in \{1, \cdots, n\}$, after the interaction between Alice and Bob, the following requirements should be satisfied:

Correctness: *Bob can get m messages from $m_1, \cdots, m_n$ if both Alice and Bob follow the protocol;*
Privacy for Bob: *Alice cannot find out the choices of Bob if Bob follows the protocol;*
Privacy for Alice: *Bob cannot get more than m messages from $m_1, \cdots, m_n$ if Alice follows the protocol.*

3 OT_2^1 Based on PKS

3.1 PKS

In the public key system (PKS), there are an encrypted algorithm E, a decrypted algorithm D, a public key Key_p and a secret key Key_s. For a secret key Key_s, it is easy to compute the corresponding public key Key_p; However, for a public key Key_p, it is hard to compute the corresponding secret key Key_s, i.e. the probability to get Key_s is very small.

3.2 Intuition of OT_2^1

In an OT_2^1, Alice has input $m_0, m_1 \in \{0,1\}^k$, Bob has input $c \in \{0,1\}$. Firstly, Alice selects a random string $C \in \{0,1\}^k$ and announces it to Bob. Bob constructs a secret key $Key_{s_c} \in \{0,1\}^k$ and two public keys $Key_{p_c}, Key_{p_{\bar{c}}} \in \{0,1\}^k$ ($\bar{c} = 1 - c$), where Key_{p_c}, Key_{s_c} are a pair of public key and secret key in the PKS, Key_{p_0}, Key_{p_1} and C satisfy the following condition:

$$(1\ 1) \begin{pmatrix} Key_{p_0} \\ Key_{p_1} \end{pmatrix} = C \tag{1}$$

Bob announces Key_{p_0}, Key_{p_1} to Alice. Alice checks whether the public keys satisfied Equation (1), if not, reject; else encrypts her two messages m_0 and m_1

with the two public keys, sends $E(m_0, Key_{p_0})$, $E(m_1, Key_{p_1})$ to Bob. Bob gets m_c with the secret key Key_{s_c}.

$$m_c = D(E(m_c, Key_{p_c}), Key_{s_c})$$

3.3 OT_2^1

Protocol 31. $OT_2^1(m_0, m_1)(c)$

1. Alice selects a random string $C \in \{0,1\}^k$, announces it to Bob;
2. Bob constructs a secret key $Key_{s_c} \in \{0,1\}^k$ and two public keys Key_{p_0}, $Key_{p_1} \in \{0,1\}^k$, satisfying $Key_{p_0} + Key_{p_1} = C$, sends Key_{p_0}, Key_{p_1} to Alice;
3. Alice checks whether $Key_{p_0} + Key_{p_1} = C$, if not, reject; else, sends $E(m_0, Key_{p_0})$, $E(m_1, Key_{p_1})$ to Bob;
4. Bob gets m_c with secret key Key_{s_c}.

3.4 Analysis

Correctness: If both Alice and Bob follow the protocol, Bob can always get m_c.

Privacy for Bob: For Alice, the two public keys she receives are just two random strings, so Alice cannot get any information about Bob's choice c.

Privacy for Alice: By the property of PKS, Bob can only construct Key_{p_c}, $Key_{p_{\bar{c}}}$, Key_{s_c} in the following way.

$$Key_{s_c} \longrightarrow Key_{p_c} \longrightarrow Key_{p_{\bar{c}}} \not\longrightarrow Key_{s_{\bar{c}}}$$

It is hard for him to get $Key_{s_{\bar{c}}}$, i.e. the probability that he gets $m_{\bar{c}}$ is very small.

4 OT_n^m

4.1 Intuition of OT_n^m

In this section, we provide an OT_n^m ($m < n$) in a more general way. In the protocol, Alice has input n messages $m_1, m_2, \cdots, m_n \in \{0,1\}^k$, Bob has m choices $c_1, c_2, \cdots, c_m \in \{1, \cdots, n\}$. In OT_2^1, there is a constrain on the public keys, i.e. one of the public key is decided by the other public key. So, in OT_n^m, n-m of the public keys $Key_{p_1}, Key_{p_2}, \cdots, Key_{p_n}$ should be decided by the rest m public keys. The task for us is to find a suitable $(n-m) \times n$ matrix M satisfied.

$$M \begin{pmatrix} Key_{p_1} \\ \vdots \\ Key_{p_n} \end{pmatrix} = \begin{pmatrix} C_1 \\ \vdots \\ C_{n-m} \end{pmatrix} \qquad (2)$$

All computation is in the field $\mathbb{Z}_q$ ($q > n$). The following example will show us the right way to construct a suitable matrix M.

Example 41. In OT_4^2, Let $M = \begin{pmatrix} 1 & 1 & 1 & 0 \\ 0 & 1 & 1 & 1 \end{pmatrix}$, then $\begin{pmatrix} 1 & 1 & 1 & 0 \\ 0 & 1 & 1 & 1 \end{pmatrix} \begin{pmatrix} Key_{p_1} \\ Key_{p_2} \\ Key_{p_3} \\ Key_{p_4} \end{pmatrix} = \begin{pmatrix} C_1 \\ C_2 \end{pmatrix}$.

We get $Key_{p_1} - Key_{p_4} = C_1 - C_2$ and $Key_{p_2} + Key_{p_3} = C_2 - Key_{p_4}$. By the property of OT_2^1, Bob can get only one secret key from Key_{s_0} and Key_{s_3} and only one secret key from Key_{s_1} and Key_{s_2}. This is not allowed by the OT_4^2.

From the above example we can find that, to construct an OT_n^m, we should construct a $(n-m) \times n$ matrix, if remove any m columns from the matrix and we should get a non-singular $(n-m) \times (n-m)$ matrix. While from the above example, we find if we remove the first and the forth column from the 2×4 matrix, we get a singular matrix. The matrix for the OT_n^m is as follows:

$$M = \begin{pmatrix} 1 & 1 & \cdots & 1 \\ 1 & 2 & \cdots & n \\ \vdots & \vdots & \vdots & \vdots \\ 1 & 2^{n-m-1} & \cdots & n^{n-m-1} \end{pmatrix}_{(n-m) \times n}$$

M has the following proposition:

Proposition 41. *Remove any m columns from the matrix M, the remanent matrix M' is non-singular.*

Proof. Remove any m columns from the matrix M, we get a Vandermonde matrix M', M' is non-singular. □

The Vandermonde matrix have the following proposition.

Proposition 42. *If M' is a Vandermonde matrix, then it has a inverse M'' in the field $\mathbb{Z}_q$.*

Proof. Let $M' = \begin{pmatrix} 1 & 1 & 1 & \cdots & 1 \\ x_1 & x_2 & x_3 & \cdots & x_n \\ x_1^2 & x_2^2 & x_3^2 & \cdots & x_n^2 \\ \vdots & \vdots & \vdots & \ddots & \vdots \\ x_1^{n-1} & x_2^{n-1} & x_3^{n-1} & \cdots & x_n^{n-1} \end{pmatrix}$.

Using only elementary row operations, we get:

$$\longrightarrow \begin{pmatrix} 1 & 1 & 1 & \cdots & 1 \\ 0 & x_2 - x_1 & x_3 - x_1 & \cdots & x_n - x_1 \\ 0 & x_2(x_2 - x_1) & x_3(x_3 - x_1) & \cdots & x_n(x_n - x_1) \\ \vdots & \vdots & \vdots & \ddots & \vdots \\ 0 & x_2^{n-2}(x_2 - x_1) & x_3^{n-2}(x_3 - x_1) & \cdots & x_n^{n-2}(x_n - x_1) \end{pmatrix}$$

$$\longrightarrow \begin{pmatrix} 1 & 1 & 1 & \cdots & 1 \\ 0 & x_2 - x_1 & x_3 - x_1 & \cdots & x_n - x_1 \\ 0 & 0 & (x_3 - x_2)(x_3 - x_1) & \cdots & (x_n - x_2)(x_n - x_1) \\ \vdots & \vdots & \vdots & \ddots & \vdots \\ 0 & 0 & x_3^{n-3}(x_3 - x_2)(x_3 - x_1) & \cdots & x_n^{n-3}(x_n - x_2)(x_n - x_1) \end{pmatrix}$$

$$\longrightarrow \begin{pmatrix} 1 & 1 & 1 & \cdots & 1 \\ 0 & x_2 - x_1 & x_3 - x_1 & \cdots & x_n - x_1 \\ 0 & 0 & (x_3 - x_2)(x_3 - x_1) & \cdots & (x_n - x_2)(x_n - x_1) \\ \vdots & \vdots & \vdots & \ddots & \vdots \\ 0 & 0 & 0 & \cdots & (x_n - x_{n-1})(x_n - x_{n-2}) \cdots (x_n - x_1) \end{pmatrix}$$

$$\longrightarrow \begin{pmatrix} 1 & 0 & 0 & \cdots & 0 \\ 0 & x_2 - x_1 & 0 & \cdots & 0 \\ 0 & 0 & (x_3 - x_2)(x_3 - x_1) & \cdots & 0 \\ \vdots & \vdots & \vdots & \ddots & \vdots \\ 0 & 0 & 0 & \cdots & (x_n - x_{n-1})(x_n - x_{n-2}) \cdots (x_n - x_1) \end{pmatrix}$$

As q is a prime and $q > n$, there exist a matrix M'' in $\mathbb{Z}_q$,

$$M''M' = I_{n \times n}$$

So, we have the conclusion of this proposition. □

4.2 OT_n^m

The OT_n^m protocol is described as following:

Protocol 41. $OT_2^1(m_1, \cdots, m_n)(c_1, \cdots, c_m)$

1. Alice selects n-m random strings $C_1, \cdots, C_{n-m} \in \{0,1\}^k$, sends to Bob;
2. Bob constructs $Keys_{c_1}, \cdots, Keys_{c_m}, Key_{p_1}, \cdots, Key_{p_n} \in \{0,1\}^k$ satisfying Equation (2), sends $Key_{p_1}, \cdots, Key_{p_n}$ to Alice;
3. Alice checks whether the public keys satisfy Equation(2), if not, reject; else, sends $E(m_1, Key_{p_1}), \cdots, E(m_n, Key_{p_n})$ to Bob;
4. Bob gets $m_{c_1}, \cdots, m_{c_m}$ with secret keys $Keys_{c_1}, \cdots, Keys_{c_m}$.

4.3 Analysis

Theorem 41. *Protocol 41 is an OT_n^m.*

Proof.
Correctness: $Key_{p_1}, \ldots, Key_{p_n}$ can be looked as the solution of Equation (2). M is an $(n-m) \times n$ matrix so that there are at least m free variables in this equation, then in the protocol, Bob can get at least m secret keys corresponding

to the m public keys, so he can get at least m encrypted messages sent from Alice, i.e. Bob can always get m messages $m_{c_1}, \cdots, m_{c_m}$.

Privacy for Bob: For Alice, the n public keys she receives are just n random strings, so Alice cannot get any information about Bob's choice $c_1, \cdots, c_m$.

Privacy for Alice: Let $Key_{s_{c_1}}, \ldots, Key_{s_{c_m}}$ be the m secret keys that Bob has already known. He can get the corresponding public keys $Key_{p_{c_1}}, \cdots, Key_{p_{c_m}}$. Let $Key_{p_{\bar{c}_1}}, \ldots, Key_{p_{\bar{c}_{n-m}}}$ be the rest public keys, $Key_{s_{\bar{c}_1}}, \ldots, Key_{s_{\bar{c}_{n-m}}}$ be the corresponding secret keys. We will show that Bob can get only m messages from the n messages. We get this by proving that $Key_{p_{\bar{c}_1}}, \ldots, Key_{p_{\bar{c}_{n-m}}}$ are decided by $Key_{p_{c_1}}, \cdots, Key_{p_{c_m}}$. For

$$M \begin{pmatrix} Key_{p_1} \\ \vdots \\ Key_{p_n} \end{pmatrix} = \begin{pmatrix} C_1 \\ \vdots \\ C_{n-m} \end{pmatrix}$$

By the Proposition 41 and Proposition 42, there exists n-m linear functions $f_1, \ldots, f_{n-m}$ and matrix M', M'' such that

$$M' \begin{pmatrix} Key_{p_{\bar{c}_1}} \\ \vdots \\ Key_{p_{\bar{c}_{n-m}}} \end{pmatrix} = \begin{pmatrix} C_1 - f_1(Key_{p_{c_1}}, Key_{p_{c_2}}, \ldots, Key_{p_{c_m}}) \\ \vdots \\ C_{n-m} - f_{n-m}(Key_{p_{c_1}}, Key_{p_{c_2}}, \ldots, Key_{p_{c_m}}) \end{pmatrix}$$

$$\begin{pmatrix} Key_{p_{\bar{c}_1}} \\ \vdots \\ Key_{p_{\bar{c}_{n-m}}} \end{pmatrix} = M'' \begin{pmatrix} C_1 - f_1(Key_{p_{c_1}}, Key_{p_{c_2}}, \ldots, Key_{p_{c_m}}) \\ \vdots \\ C_{n-m} - f_{n-m}(Key_{p_{c_1}}, Key_{p_{c_2}}, \ldots, Key_{p_{c_m}}) \end{pmatrix}$$

So, each of $Key_{p_{\bar{c}_1}}, \ldots, Key_{p_{\bar{c}_{n-m}}}$ are decided by $Key_{p_{c_1}}, \ldots, Key_{p_{c_m}}$, Bob can get only m messages from the n encrypted messages. □

The efficiency is based on the efficiency of the public key system and the integer power in the field $\mathbb{Z}_q$. From [3], using repeating squaring, it only requires $O(\ln q)$ to compute the $(a^b \mod q)$.

5 Conclusion

In this paper, for the first time, we construct a OT_n^m protocol based on the public key system, which satisfying the 3 requirements of OT_n^m.

References

1. Aiello, B., Ishai, Y., and Reingold, O.: Priced Oblivious Transfer: How to Sell Digital Goods. Eurocrypt'01, Lecture Notes in Computer Science, Vol. 2045 (2001) 119-135

2. Crépeau, C., Van de Graaf, J., and Tapp, A.: Committed Oblivious Transfer and Private Multi-party Computations. Adva. in Crypt.: Proc. of Crypto'95 963 (1995) 110-123
3. Cormen, H., Leiserson, E., Rivest, L., Stein, C.: Introduction to Algorithms. Second Edition. MIT Press (2001)
4. Even, S., Goldreich, O., and Lempel, A.: A Randomized Protocol for Signing Contracts. Commu. of the ACM 28 (1995) 637-647
5. Goldreich, O., Micali, S., and Wigderson, A.: How to Play any Mental Game or a Completeness Theorem for Protocols with Honest Majority. Proc. 19th Annu. ACM Symp. on Theo. of Compu. (1987) 218-229
6. Kilian, J.: Founding Cryptography on Oblivious Transfer. Proc. 20th Annu. ACM Symp. on Theo. of Compu. (1988) 20-31
7. Mu, Y., Zhang, J., Varadharajan V.: m out of n Oblivious Transfer. ACISP'02, Lecture Notes in Computer Science, Vol. 2384, (2002) 395-405
8. Naor,M., Pinkas, B.: Oblivious Transfer with Adaptive Queries. Proc. of Crypto'99, Lecture Notes in Computer Science, Vol. 1666 (1999) 573-590
9. Naor, M., Pinkas, B.: Efficient Oblivious Transfer Protocols. 12th Annu. Symp. on Disc. Algo. (2001) 448-457
10. Rabin, M.: How to Exchange Secrets by Oblivious Transfer. Tech. Repo. TR81, Aiken Comp. Lab, Harvard University (1981)

Determining Optimal Decision Model for Support Vector Machine by Genetic Algorithm

Syng-Yup Ohn, Ha-Nam Nguyen, Dong Seong Kim, and Jong Sou Park

Department of Computer Engineering,
Hankuk Aviation University, Seoul, Korea
{nghanam, dskim, syohn, jspark}@hau.ac.kr

Abstract. The problem of determining optimal decision model is a difficult combinatorial task in the fields of pattern classification and machine learning. In this paper, we propose a new method to find the optimal decision model for SVM, which consists of the minimal set of highly discriminative features and the set of parameters for the kernel. To cope with this problem, we adopted genetic algorithm (GA) which provides efficient optimization tool simulating the natural evolution procedures in iterative fashion to select the optimal set of features and set of kernel parameters. In the method, the decision models generated by GA are evaluated by SVM, and GA selects the only good models and gives the selected models the chance to survive and improve by crossover and mutation operation. Combining GA and SVM, we can obtain the optimal decision model which reduces the execution time as well as improves the classification rate of SVM. We also demonstrated the feasibility of our proposed method by several experiments on the sets of clinical data such as KDD Cup 1999 intrusion detection pattern samples and stomach cancer proteome pattern samples.

1 Introduction

Recent pattern classification applications in the areas such as networks, biology, and image recognitions often have high dimensional feature space, and the performance of classification is degraded severely in terms of execution time and classification rate in such cases. Typically, the combinations of a small number of features play an important role in discriminating samples into classes. GA offers a natural way to solve this problem. Much research efforts are concentrated on combining GA and classifiers to improve the performance of classification [1-3]. GA is often applied to select the optimal set of features for the classifiers such as k-nearest neighbors (KNN), probabilistic neural network (PNN), SVM.

Support vector machine (SVM) is a learning method that uses a hypothesis space of linear functions in a high dimensional feature space [4-7]. This learning strategy, introduced by Vapnik [5], is a principled and powerful method and outperformed most of the classification algorithms in many applications. However, the computational power of linear learning machines is limited in the cases of the feature space with nonlinear characteristics. It can be easily recognized that real-world applications require more extensive and flexible hypothesis space than linear functions. By using a proper kernel function, we can overcome the nonlinearity of feature space [4, 7]. Also recent improvements in the area of SVM make it easy to implement and overcome the computation time problem due to a large training set [7].

Genetic algorithm is an optimization algorithm simulating the mechanism of natural evolution procedure [9-11]. Most of genetic algorithms share a common conceptual base of producing improved individual structures through the processes of selection, mutation, and reproduction. GA is generally applied to the problems with a large search space. They are different from random algorithms since they combine the elements of directed and stochastic search. Furthermore, GA is also known to be more robust than directed search methods.

In this paper, we propose a new learning method using GA and SVM. In the new learning method, GA is exploited to derive the optimal *decision model* for the classification of patterns, which consists of the optimal set of features and parameters of a kernel function. SVM is used to evaluate the fitness of the newly generated decision models by measuring the hit ratio of the classification based on the models. Our method is different from the feature selection methods based on GA and SVM recently reported [2, 3] in that they used GA only for feature selection and then manually chose the parameters for a kernel function. We applied GA to obtain the optimal set of features and set of kernel parameters at the same time. In the comparison with other learning methods by extensive experiments on the datasets such as network intrusion pattern samples and proteome pattern samples for cancer identification, proposed learning method achieved faster convergence while searching a decision model and better classification performance for some cases.

This paper is organized as follows. In section 2, our learning method is presented in detail. In section 3, we compare the proposed and other methods by the experiments with the classification of the datasets of network intrusion pattern samples and proteome pattern samples for cancer identification. Finally, section 4 is our conclusion.

2 Learning Method for Optimal Decision Model

The challenging issue of GA is how to map a real problem into a *chromosome*. In our method, the sets of features and parameters for a kernel function should be encoded into a chromosome (Fig. 1). The chromosome consists of binary gene string and multi-valued gene string representing the sets of features and kernel parameters. The binary string is composed of n-bit string and each bit represents an active or inactive state of a feature. The multi-valued gene string represents the index of a kernel function such that 1, 2, and 3 for Inverse Multi-Quadric, Radial, and Neural kernel) and kernel parameters of each kernel (See Fig. 1 and Table 1). The combination of the two gene strings forms a chromosome in GA procedure which in turn serves as a decision model (see Fig. 1).

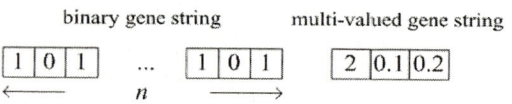

Fig. 1. Structure of a chromosome used in GA

Table 1. The kernel functions are used to experiments

Kernel function	Formula
Inverse Multi-Quadric	$1/\sqrt{\|x-y\|^2 + c^2}$
Radial	$e^{(-\gamma \|x-y\|^2)}$
Neural	$\tanh(s \cdot \langle x,y \rangle - c)$

In our method, GA generates a set of chromosomes, each of which represents a decision model, by evolutionary procedures. The fitness value of each chromosome is evaluated by measuring the hit ratio from the classification of samples with SVM classifier containing the decision model associated with the chromosome. m-fold validation method is used to evaluate the fitness of a chromosome to reduce overfitting [6]. Then only the chromosomes with a good fitness are selected and given the chance to survive and improve into further generations. Roulette wheel rule is used for the selection of chromosome [9]. Some of the selected chromosomes are given the chance to undergo alterations by means of crossover and mutation to form new individuals. One-point crossover is used, and the probabilities for crossover and mutation are 0.8 and 0.015 respectively. This process is repeated for a predefined number of times.

At the end of GA procedure, the decision model with the highest hit ratios is chosen as the *optimal decision model*. The optimal decision model is used to build a SVM for the classification of novel samples and the performance of the model can be evaluated.

3 Experiments

The proposed method is experimented with KDD Cup 1999 and stomach cancer datasets for the evaluation of performance. The experiments are conducted on a Pentium IV 1.8 GHz personal computer. We used SVM module developed by Stefan Rüping [13] for the implementation of the proposed method. The results from the experiments are presented in the following sections.

3.1 KDD Cup 1999 Data

We used the dataset for KDD Cup 1999 contest, in which the competition task was to build a network intrusion detector containing a predictive model capable of distinguishing between "bad" connections, called intrusions or attacks, and "good" normal connections (Available at: http://kdd.ics.uci.edu/databases/kddcup99/kddcup99.html). This database consists of a standard set of data to be audited, which includes a wide variety of intrusions simulated in a military network environment. The set of the features consists of 3 symbolic features and 39 numeral features. Our training set used in

GA procedure consists of 2500 samples randomly selected from the training dataset (kddcup.data.gz). The test dataset (corrected.gz) was applied to the decision model resulted from GA.

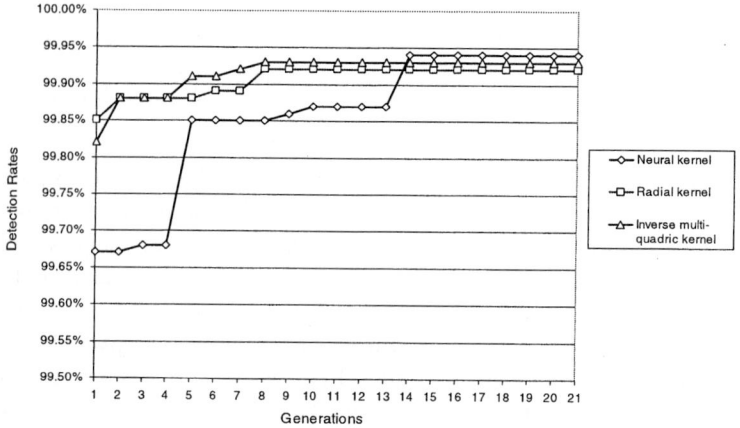

Fig. 2. Detection rates of validation test

We used a simple generalization method: 10-fold cross validation with 2500 samples [6]. In the experiment result, the detection rate designates the accuracy of classification. Three types of kernel functions are used for SVM in the experiment, and the classification rates from during GA procedures are depicted in Fig. 2. After GA was executed for 20 generations, the training with neural kernel achieved the highest detection rate among the kernel functions compared. At this point, the best detection rates obtained are 99.94% in case of neural kernel function, 99.92% in case of radial kernel function, and 99.93% in case of inverse multi-quadric kernel function.

Table 2 shows the indexes of the features in the optimal features set obtained after 20 generations in GA procedures. The sets of selected features are different for each kernel functions used in SVM. In real network environment, GA and SVM together are used to obtain the optimal set of features suitable for the environment, and the SVM with the decision model consisting of the optimal feature set serves as an intrusion detection system.

Table 2. Number of Selected Feature

Kernel function	Selected Feature Number
Inverse multi-quadric	1, 2, 3, 4, 6, 7, 8, 9, 12, 13, 14, 15, 16, 17, 18, 19, 21, 22, 24, 27, 29, 30, 32, 34, 37, 39, 41
Radial	1, 2, 3, 5, 6, 8, 9, 10, 11, 12, 13, 14, 16, 17, 18, 19, 20, 21, 22, 23, 25, 27, 29, 30, 31, 32, 33, 34, 37, 38, 39
Neural	1, 2, 4, 5, 7, 8, 9, 10, 11, 13, 14, 16, 17, 18, 20, 21, 22, 23, 24, 25, 26, 27, 28, 31, 32, 34, 36, 38, 41

The result from the classification of test dataset is depicted in Fig. 3. In case of "GA+SVM", we used the optimal set of features obtained from GA in SVM to classify the test dataset. In case of "SVM only", we used all the features of the dataset for the decision model in the SVM. The result shows the highest detection rates when the system uses GA and SVM with neural kernel function. Furthermore, the detection rate of the proposed system achieved 99.94%, which is higher than the results reported by KDD 1999 contest winner [12].

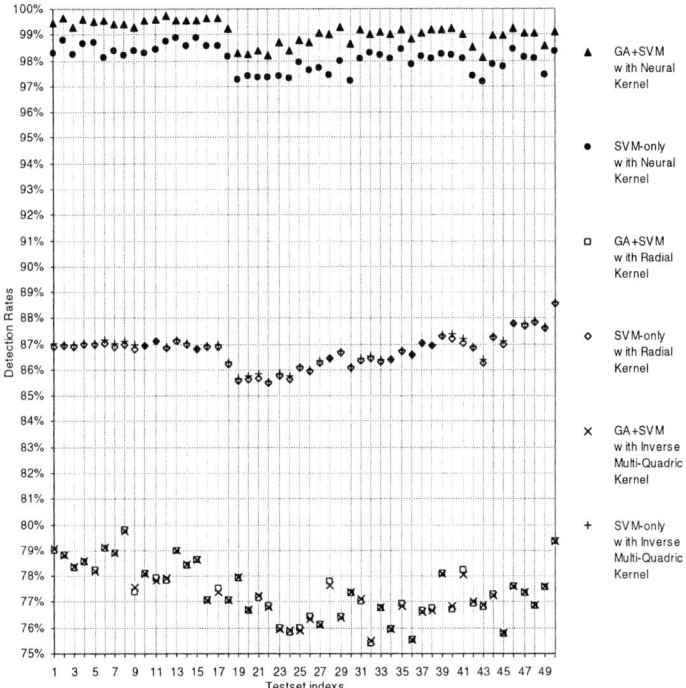

Fig. 3. Summary of the classification of test dataset

In the experiments, proposed method shows the best result when the optimal feature set for neural kernel obtained from GA is used in the decision model. And it also shows that the system can cope with novel attacks very well since it shows detection rates more than 99%. The number of features that SVM should process can be minimized by obtaining the optimal set of features, and the speed and the detection rate of SVM based IDS can be improved. Our experiment results demonstrate the feasibility of exploiting GA and SVM together for IDS in high speed network environment.

3.2 Stomach Cancer Data

The proteome pattern samples used in this case study were provided by Cancer Research Center at Seoul National University in Seoul, Korea. The proteome pattern samples are extracted from the sera from a group of normal persons and a group of the affected persons with stomach cancer. The stomach cancer dataset consists of 67 normal and 70 cancer samples and each sample contains 119 features. The purpose of this experiment is to classify the proteome pattern samples into the normal and the cancer pattern classes.

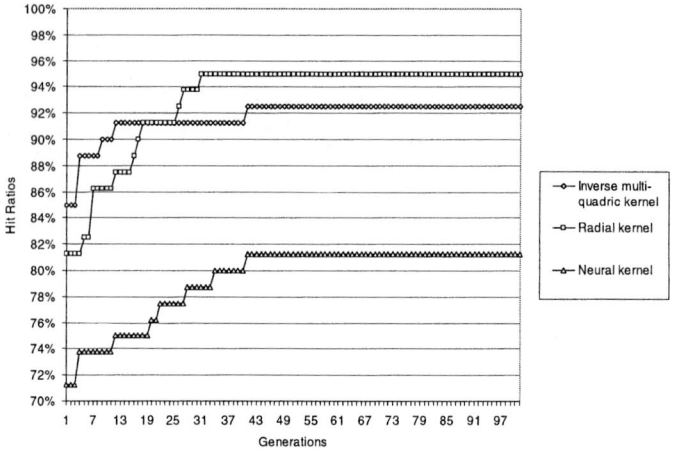

Fig. 4. Hit ratio rates of validation test

We used simple generalization method: 4-fold cross validation with 80 samples [6]. The data was split into a training set of 80 and a test set of 57 samples. The classification rates of validation tests during GA are depicted in Fig. 4. While GA was executed for 100 generations, the learning process using SVM with radial kernel function achieved the highest hit rates among the three kernel functions compared. After 100 generations in GA, the best hit rates are 95% in case of radial kernel function, 92.5% in case of inverse multi-quadric kernel function, and 81.25% in case of neural kernel function.

Table 3. Number of Selected Feature

Kernel function	Selected Feature Indexes
Inverse multi-quadric	1413, 1423, 1523, 2302, 2313, 2317, 3118, 3131, 3306, 3412, 4213, 4314, 4315, 4608, 5103, 5107, 5325, 5336, 5337, 5407, 5513, 5712, 6107, 6303, 6308, 6309, 6415, 7317, 7408
Radial	1413, 1418, 2301, 2419, 2818, 3115, 3412, 3611, 3613, 3618, 4105, 4111, 4213, 4314, 4315, 4319, 4322, 4409, 5308, 5313, 5407, 5712, 6308, 6409, 6420, 6504, 6519, 7313, 7322, 7326
Neural	206, 207, 413, 515, 2301, 2419, 3118, 3309, 3417, 3611, 4213, 4316, 4321, 4429, 4608, 5407, 5409, 5513, 5712, 6115, 6316, 6411, 6415, 6504, 7306, 7309, 7313, 7408, 8301, 8309

The indexes of the features selected by GA for each kernel function are shown in Table 3. The selected features are different for each kernel function used in SVM classifier.

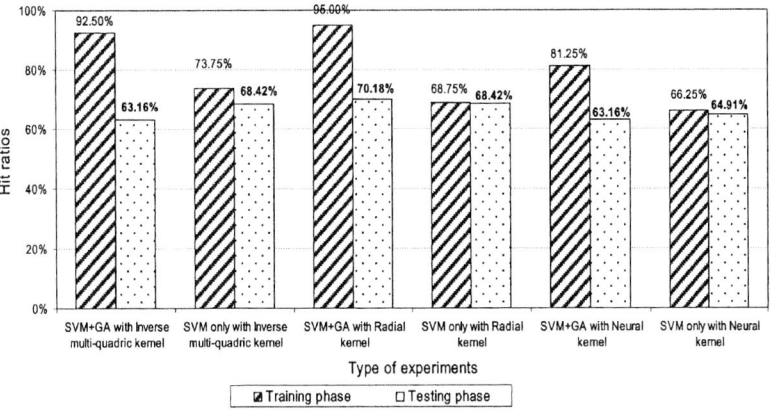

Fig 5. Testing rates are computed by using the optimal decision models obtained in learning phase

The ratios of correct classification are summarized in Fig. 5. The radial kernel function showed the highest hit ratio in case of SVM with GA. In contrast, the test result is not good in case of neural and inverse multi-quadric kernel. According to the result illustrated in Fig. 5, the test results of them even were lower than cases of SVM-only. In this case study, "SVM+GA" with radial kernel showed better classification rate than other cases. The computation time of SVM can be reduced and the classification rates can be improved by using optimal feature set.

4 Conclusion

In this paper, we proposed a novel learning method based on GA to determine the optimal decision model for SVM. In case that the sample spaces of applications are high-dimensional and have complex characteristics, the performance of SVM is often degraded severally. To cope with this problem and improve the performance of classification, we adopted GA technique to select the optimal set of features and the set of kernel parameters at the same time. The proposed learning method showed faster convergence and better classification performance for some datasets than other methods. The experiments indicate that our method is capable of finding the optimal set of features and set of kernel parameters that improve the performance of SVM.

Acknowledgement

This research was supported by IRC (Internet Information Retrieval Research Center) in Hankuk Aviation University. IRC is a Kyounggi-Province Regional Research Center designated by Korea Science and Engineering Foundation and Ministry of Science & Technology. This research also was supported by University IT Research Center (ITRC) project.

References

1. Martin-Bautista, M.J., Vila, M.-A.: A survey of genetic feature selection in mining issues. Evolutionary Computation. Proceedings of the 1999, Vol. 2 (1999) 1321
2. Frohlich, H., Chapelle, O., Scholkopf, B.: Feature selection for support vector machines by means of genetic algorithm. Tools with Artificial Intelligence, Proceedings 15th, IEEE International Conference (2003) 142 – 148
3. Xue-wen Chen: Gene selection for cancer classification using bootstrapped genetic algorithms and support vector machines. The Computational Systems Bioinformatics Conference, Proceedings IEEE International Conference (2003) 504 – 505
4. Cristianini, N. and Shawe-Taylor, J.: An introduction to Support Vector Machines and other kernel-based learning methods. Cambridge (2000)
5. Vapnik, V.N., et. al.: Theory of Support Vector Machines. Technical Report CSD TR-96-17. Univ. of London (1996)
6. HDuda, R. O.H, Hart, HHP. E., Stork, HD. G.: HPattern Classification (2nd Edition). John Wiley & Sons Inc. (2001)
7. Joachims, Thorsten: Making large-Scale SVM Learning Practical. In Advances in Kernel Methods - Support Vector Learning, chapter 11. MIT Press (1999)
8. Minsky M.L. and Papert S.A.: Perceptrons. MIT Press (1969)
9. Michalewicz, Z.: Genetic Algorithms + Data structures = Evolution Programs. 3rd rev. and extended edn, Springer-Verlag (1996)
10. Goldberg, D. E.: Genetic Algorithms in Search, Optimization & Machine Learning. Adison Wesley (1989)
11. Mitchell, M.: Introduction to genetic Algorithms, fifth printing. MIT press (1999)
12. Bernhard P.: Winning the KDD99 Classification Cup. (1999) http://www.ai.univie.ac.at/~bernhard/kddcup99.html
13. Rüping, S.: mySVM-Manual. University of Dortmund, Lehrstuhl Informatik (2000) URL: Hhttp://www-ai.cs.uni-dortmund.de/SOFTWARE/MYSVM/H

A Mobile Application of Client-Side Personalization Based on WIPI Platform

SangJun Lee

Department of Computer Information Communication, Seonam University,
720 GwangchiDong, NamWon, JeonlabukDo, Korea
sjlee@seonam.ac.kr

Abstract. The Internet site is encouraged to take a little time to get the preferred information and bring it to the customer through personalization of customers' information. The mobile shopping users have a strong desire to reduce the shopping hours due to the expensive telecommunication service fees. It requires much time and perseverance to input the personalization information into the server and even to input it repeatedly into a bunch of shopping malls while maintaining contact. In this paper, we develop an application to protect the general personalized information in the mobile client-side based on XML. The mobile shopping using the client-side personalization reduces the connection time and improves the personalization information management through the enhanced data transmission and data input.

1 Introduction

Mobile commerce (M-commerce) implies a purchasing transaction with some monetary value between a wireless network and a mobile device[1]. A full-blown M-commerce future is expected in terms of three reasons. Compared with the personal computer (PC), first, the number of mobile device available today is not only higher but is developed more rapidly than that of the PC. Second, users are much more familiarized with the mobile device than the PC, even with constant connectivity. Third, the spot services with the mobile device can be provided much easier than the PC or TV anytime anywhere.

If the users of the Internet share the preference, interesting goods and purchasing experiences on the website, the website can provide the organized pertinent information by bypassing the need to the corporations.

On the one hand, for corporations, with this personalization they can obtain these following four merits. First, it brings the enhancement of customer's loyalty. Second, it significantly lowers the cost of marketing. Third, it helps to find good consumers. Fourth, it can be possible to offer constantly enhanced and valuable services.

On the other hand, users can acquire three advantages. First, they can save their time to find the wanted information. Second, it is possible for them to make a selection by their preferences. Third, they can enjoy a personalized service.

To be in a win-win situation for the users and the corporations through personalization, the users make their personal information open voluntarily. Since many users are unwilling to reveal their information, a guarantee of protection for their information is a good and necessary consideration [2].

Up to now, as to personalization, even if the research into both the server-centered personalization mobile service[3] and the web host access tool in the digital library[4] have been done, there is no research into mobile client-side personalization before. Considering the comparatively expensive usage cost of communications and limitations of mobile devices, the research for the personalization is momentous.

In this paper, unlike the way to input the personal information into a specific website, keeping the personal information in the mobile client-side through the mobile application is introduced. The download of the application is necessary because of no function of save for personal information in mobile client in the incumbent mobile device.

This paper is organized as follows. In Section 2, we examine the personalization plan and design its information. In Section 3, we implement the client-side personalization on WIPI mobile platform. In Section 4, we evaluate our research. Finally, in Section 5, we make a conclusion.

2 Personalization for Mobile Shopping

2.1 Design of Personalization Information for Mobile Shopping

Like Table 1, referring to Personalization & Privacy Survey[5], we categorize the personalization information items for mobile shopping by its security level.

Table 1. Personalization information for mobile shopping

Category	Items
A Preference	Goods (Clothing,Shoes,Accessories,IT Product), Color, Brand, Specification, Shopping Mall
Body Information	Physical Size(Height,Weight,Waist,Foot)
Contact Information	Name, E-mail, Address, Phone Number, Age, Sex, Job, Income
Doing Secret	ID, Password, Credit Card, Account

2.2 Personalization Strategy

For personalization in mobile shopping, we make a plan in a following way [6].

- Setting the goal: Make the personal information of mobile shopping user easy to save in the client.
- Defining the rule: Only for the websites registered to client-side personalization information, make the basic information uploaded automatically; for the correct password of client-side personalization, input the user automatically; input the ID and Credit card number in case of correct WPKI(Wireless PKI)
- Collecting data: design and input personalization information
- Step-by-step personalization: step-by-step personalization of the information by the user's shopping lists from the mobile commerce site

For privacy protection,

- Apply P3P(Platform for Privacy Preferences Project): Provide the personal information only to the sites registered to client-side personalization. (**A** Preference)
- Apply Anonymous Personalization: Provide the site with the preference or interest without profile or additional information. (**B**ody Information, **C**ontact Information)
- Apply Stepwise Personalization: Ask the information for each step of personalization. (**D**oing Secret)

Since the user is unwilling to reveal the personal information, the personal information is protected in the client using the downloaded mobile application, rather than protected in the server.

3 Application for Client-Side Personalization

3.1 Environment

Mobile platform implies a software system which has a right API(Application Programming Interface) applicable to mobile applications, or other developing environments. The mobile platform consists of two platforms at large: for downloading the application and for developing the device. The former is called wireless Internet platform and its popularized kinds are Sun Java, Qualcomm BREW and WIPI. The latter platform is known as Microsoft Windows Mobile for Smartphone, Symbian OS, PalmOS, Ericsson Mobile Platform, etc.

WIPI(Wireless Internet Platform for Interoperability) is made by KWISF(Korea Wireless Internet Standardization Forum). WIPI is a mobile standard platform specification providing the environment for application program loaded on the wireless devices[7]. WIPI platform has the conceptual structure like Fig.1.

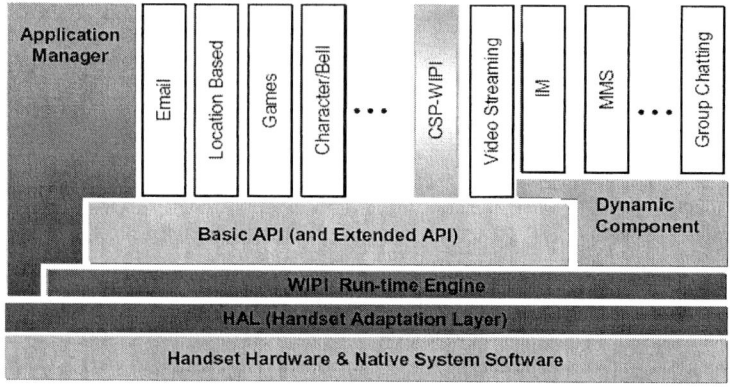

Fig. 1. WIPI Platform Architecture

In this paper, Client-Side Personalization based on WIPI platform(CSP-WIPI) is implemented on specification of WIPI version 2.0 which is published in April 4, 2004. As mobile phone supporting this specification will be appeared in the second half of 2004, we develop mobile application for client-side personalization information on KTF WIPI Emulator[8]. CSP-WIPI is simulated as a mobile application using WIPI Basic API. Since CSP-WIPI is abiding by the standard, the customers can use it regardless of telecommunication companies in Korea.

3.2 Implementation

CSP-WIPI is developed on the WIPI Content Server using the WIPI API. Like Fig.2, the users execute the CSP-WIPI application on the content server after downloading it. Then, the client-side personalization information is created from the user's personal information by its execution.

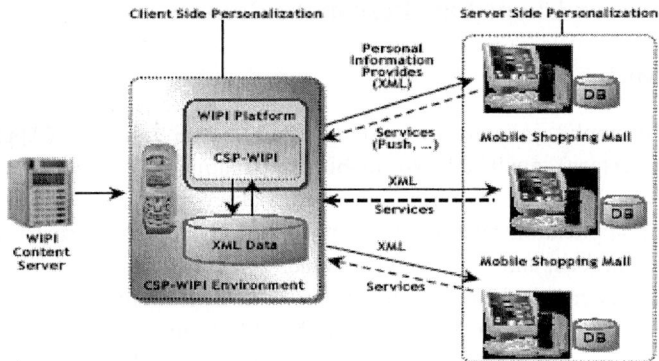

Fig. 2. Architecture of CSP-WIPI System

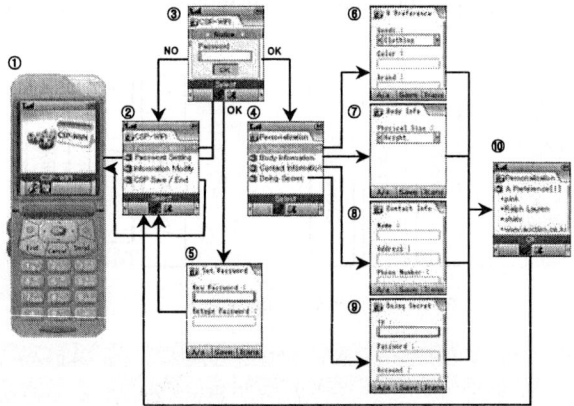

Fig. 3. CPS-WIPI Application Workflow

In mobile shopping, the mobile application for inputting the personalization information is executed in a following way. In Fig. 3, ①Initial Display of CSP-WIPI. After selecting CSP-WIPI among icons in command bar, push Ok.②Initial menu of CSP-WIPI: Personalization Information, Password Setting, Personalization Information Modify, Data Save/ Selecting Application.③Personalization submenu: Setting the value of detailed items ⑥~⑨. ⑩ Confirm the Total Personalization Information. Modifying the existing value: ②Select Information Modify; ⑩,⑥~⑨. Save the client-side personalization information as a XML file into DB.

4 Evaluation

To evaluate the results of research, we make a comparison between a mobile shopping which uses client-side personalization information and the other which does not. In addition, we also compare the protection of personalization information between saved in the client and saved in the server.

4.1 Client-Side Personalization Versus Non-personalization

The time used in communications between shopping mall server and mobile device up to the end of service is compared in case of applying personalization and non-personalization.

The Data Transmission Time means the time taking from the start to the end of service and the Data Input Time means the time for the user to enter the information to get the service from DB.

Table 2. Client personalization versus non personalization

	Personalization scheme	Non personalization scheme
Data Transmission Time	NC * (PTT * 2)	NTII * (PTT *2)
Data Input Time	Number of Category	Number of Total Character

NC : Number of Category   PTT : Packet Transmission Time
NTII : Number of Total Input Item

In this paper, personalization-applied scheme saves the data transmission time by (NTTI-NC) * (PTT * 2) compared with non-personalization scheme. For example, if we use the personalization information consisting of 4 categories and 16 items, and assuming PTT is 1, we can reduce the data transmission time up to 24 seconds.

In case of non personalization, the data input time depends on the number of characters to enter; whereas, in case of personalization scheme, it hardly requires time of data input.

4.2 Client-Side Personalization Versus Server-Side Personalization

There are several ways to keep the personal information safely in the mobile client in communicating with the server. For example, we can protect the data from On-line dictionary attack and shoulder surfing attack by installing a safe password system[9]. The protocol using the inputted password is used to prevent the server from server compromising attack[10]. The Off-line dictionary attack and server compromise attack can be blocked by the password and authenticated key exchange protocol [11].

Table 3. Client-side personalization versus server-side personalization

Classification	Client-side personalization	Server-side personalization
Convenience to manage	Keeping the data in mobile device	Connecting to the server and manage the data
Data Input Times	One-time input	Multiple-time input to servers
Security-related things	Protecting against loss	Same as that of general servers

5 Conclusion

We have grasped the data for personalization of mobile shopping and established the scheme of personalization. In addition, we have developed the mobile application to manage and protect the personalization information. The customers of all communication companies which select WIPI as mobile standard platform are able to download and execute the application; therefore, they can save the personalization information into their mobile devices. Without their personalization information on the server, they can get the right service when connecting with the mobile shopping server with ease.

References

1. May, P.: Mobile commerce, Cambridge University Press. (2001)
2. Personalization Survey, http:://www.zdneton.org
3. Cassel, L., Wolz, U.: Client Side Personalization, Delos Workshop:Personalization and Recommender Systems in Digital Libraries (2001)
4. Wagner, M., Balke, W., Hirschfeld, R., Kellerer, W.:A Roadmap to Advanced Personalization of Mobile Services, Industrial Program of the 10th International Conference on Cooperative Information Systems(CoopIS) (2002)
5. Personalization & Privacy Survey, http://www.personalization.org
6. Personalization Plan, http://www.personalization.co.kr
7. WIPI Specification, http://www.kwisforum.org
8. Ktf WIPI Emulator, http://wipidev.magicn.com/

9. SeungBae, P., MoonSeol K., SangJun L.: New Authentication System, Lecture Notes in Computer Science, Vol. 3032. Springer-Verlag, Berlin Heidelberg New York (2004) 1095-1098
10. SeungBae, P., MoonSeol K., SangJun L.: User Authentication Protocol Based on Human Memorable Password and Using ECC, Lecture Notes in Computer Science, Vol. 3032. Springer-Verlag, Berlin Heidelberg New York (2004) 1091-1094
11. SeungBae, P., MoonSeol K., SangJun L.: Authenticated Key Exchange Protocol Secure against Off-Line Dictionary Attack and Server Compromise, Lecture Notes in Computer Science, Vol. 3032. Springer-Verlag, Berlin Heidelberg New York (2004) 924-931

An Agent Based Privacy Preserving Mining for Distributed Databases

Sung Wook Baik[1], Jerzy Bala[2], and Daewoong Rhee[3]

[1] Sejong University, Seoul 143-747, Korea
sbaik@sejong.ac.kr
[2] Datamat Systems Research, Inc.,
1600 International Drive, McLean, VA 22102, USA
jbala@dsri.com
[3] Sangmyung University, Seoul 110-743, Korea
rhee219@smu.ac.kr

Abstract. This paper introduces a novel paradigm of privacy preserving mining for distributed databases. The paradigm includes an agent-based approach for distributed learning of a decision tree to fully analyze data located at several distributed sites without revealing any information at each site. The distributed decision tree approach has been developed from the well-known decision tree algorithm, for the distributed and privacy preserving data mining process. It is performed on the agent based architecture dealing with distributed databases in a collaborative fashion. This approach is very useful to be applied to a variety of domains which require information security and privacy during data mining process.

1 Introduction

Privacy preserving issues [1-6] became major concerns in inter-enterprise data mining to deal with private databases located at different sites. There have been two broad approaches in the data mining considering privacy preserving issues; 1) the randomization approach to randomize the values in individual records and to reveal only the randomized values for the protection of individual privacy and 2) the secure multi-party computation approach to fully analyze data located at different sites to build a data mining model across multiple databases without revealing the individual records in each database to the other databases. The privacy preserving issues directed towards Distributed Data Mining (DDM) have been applied to process medical, insurance, credit card transaction data and so on.

However, the problem with most of these efforts is that although they allow the databases to be distributed over a network, they assume that the data in all of the databases is defined over the same set of features. In other words, they assume that the data is partitioned horizontally. In order to fully take advantage of all the available data, the data mining tools must have a mechanism for integrating data from a wide variety of data sources and should be able to handle data characterized by the following:

1. Geographic (or logical) distribution.
2. Complexity and multi feature representations.
3. Vertical partitioning/distribution of feature sets.

This paper presents a novel data mining approach for tree induction from vertically partitioned data sets. The approach integrates inductive generalization of a decision tree and agent-based computing, so that decision rules are learned via tree induction from vertically partitioned and distributed data without transferring necessary data to the others for complete analysis. Agents, in a collaborative fashion, generate partial trees and communicate the index information of the data records regardless of privacy among them for synchronization of decision trees being constructed by each agent.

2 Distributed Data Mining Approach for Agents' Collaboration

Agent-based architectures [7-11], have been developed to achieve several distributed data mining techniques. However, they focus on the distributed data mining for analyzing data from homogeneous data sites where the distributed databases are horizontally partitioned and the database schema in every site are same.

In this paper, we present a decision tree algorithm in a distributed database environment and a distributed data mining system, which has an inter-agent communication mechanism on the agent based framework. The decision tree algorithm for the distributed data mining has been extended from a traditional decision tree algorithm [12], so that it is possible to build a decision tree from the distributed databases without transferring any data (or even encrypted data) related to privacy among them. The extended algorithm can naturally resolve the privacy preserving problems. Figure 1 depicts the four layered system architecture for distributed data mining. The distributed learning algorithm of a decision tree in an Agent-Mediator communication mechanism (See Figure 2) is as follows:

1. The Mediator let the distributed and privacy preserving data mining process start by invoking all the Agents.
2. Then, each Agent independently starts the process of mining its own local data. It finds the feature (or attribute) and its value that can best split the data into the various training classes (i.e. the attribute with the highest information gain).
3. The Agent sends the selected attribute, as a candidate attribute, and its associated split value to the Mediator for overall evaluation.
4. Once the Mediator has collected the candidate attributes of all the Agents, it then selects the attribute with the highest information gain as an ultimate attribute to split at the given level of the decision tree. The Agent, whose database includes the attribute with the highest information gain, is called a winner. The other Agents are losers.
5. The Mediator notifies each Agent its status (winner or loser).
6. The winner Agent then continues the mining process by splitting the data using the winning attribute and its associated split value. This split results in the formation of two separate clusters of data (i.e. a group of data satisfying the split criteria and the other group of data not satisfying it).

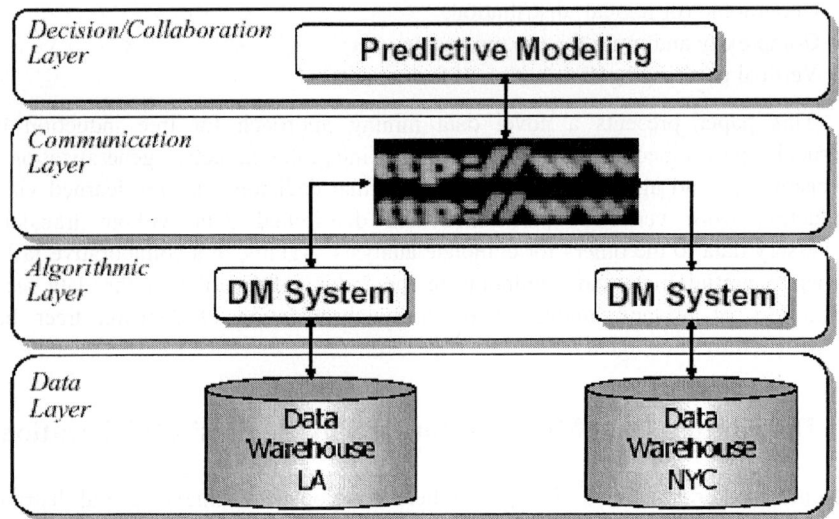

Fig. 1. Four layered system architecture for distributed data mining

7. The Agent passes the associated indices of the data records in each cluster and the winning attribute including its associated split value to the other Agents through the Mediator. Before passing it, the index information, normally represented as a set of integers (i.e. record numbers), is converted to a bit-vector representation. In a bit-vector representation, each individual bit corresponds to the index of a single data record. To further reduce the size of the data being transferred, a bit-vector representation can be compressed.
8. The other (i.e. losers) Agents receive the index information passed to the Mediator by the winner Agent and construct the partial decision trees by splitting their data accordingly.
9. If the decision tree is completely constructed, each Agent generates the classification rules by tracking the attribute/split information coming from the various Agents and terminates the distributed data mining process. Otherwise, the mining process then continues by repeating the process of candidate feature selection by each of the Agents (go to step 2).

3 Experimentation

We have used four kinds of the data sets requiring the protection of individuals' personal privacy for the evaluation of the privacy preserving data mining system dealing with the vertically partitioned and distributed databases. The data sets [13] have been widely used in knowledge discovery and data mining (KDD) fields by many research groups and their descriptions are as follows:

- Health Care Database: The database contains information on medical transactions performed during a three-year period by a US Government health care provider. A

medical transaction in the database is represented as a single billable event, e.g., a drug administration. Each record in the database is represented by a field list describing such information as the patient's age, transaction type, provider's location, cost, etc.
- Congressional Voting Records Database: The database includes votes for each of the U.S. House of Representatives Congressmen on the 16 key votes identified by the congressional quarterly almanac. It lists nine different types of votes; 1) voted for, 2) paired for, 3) announced for, 4) voted against, 5) paired against, 6) announced against, 7) voted present, 8) voted present to avoid conflict of interest, and 9) did not vote or otherwise make a position known.
- Adult Database: The database contains information collected by the US Census Bureau for predicting whether income exceeds $50K/yr based on census data. Each record in the database is represented by a field list describing such information as the adult's age, education, work class, marital status, occupation, race, sex, capital gain, native country, and so on.

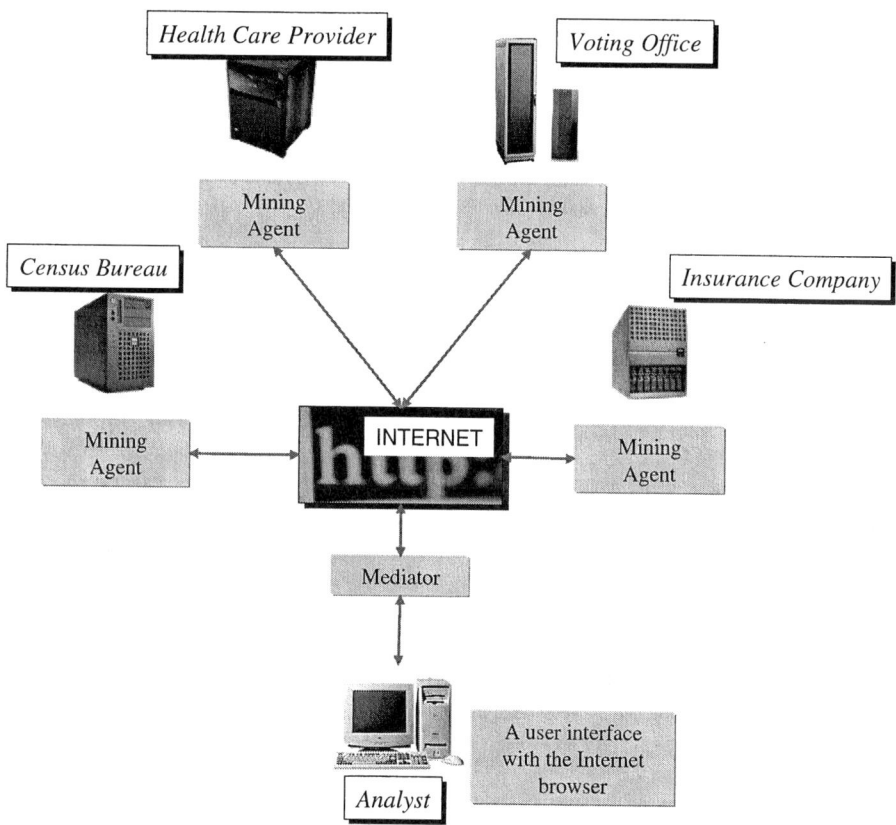

Fig. 2. Agent-Mediator Communication among several sites

- Insurance Database: The database contains information was prepared by the Swiss Life Information Systems Research group. Each record in the database is represented by a field list describing such information as sex, year of birth, marital status, personal state, i.e. nothing special, dead, missed or under revision as disabled person, and disabled person.

Three experiments have been conducted to generate classification rules for two classes (i.e., the above $50K/yr (class A) and the below $50K/yr (class B) for the salary field in the adult database).

Experiment 1: Each database described above was placed on the different site, where each agent resided to access/analyzes the database and to communicate with the others. The distributed data mining process was performed by the collaboration of four agents and a mediator.

Experiment 2: The first two databases and the last two databases were manually combined into two double-sized databases, respectively. Those databases were placed into two different sites. The distributed data mining process was performed by the collaboration of two agents and a mediator.

Experiment 3: All of four databases were manually combined into a database, which has sixty fields. The combined database was processed by non-distributed data mining version.

The results of three experiments have been compared with each other. The comparison has revealed exactly the same rule sets for different versions of the (distributed) data mining.

4 Conclusion and Future Work

This paper focused on the privacy preserving issues and provided a significant approach for the privacy preserving mining for distributed databases through the collaboration of agents and a mediator. The approach is to integrate inductive generalization and agent-based computing so that classification rules can be learned via tree induction from distributed data to be used for prediction. According to experimental results, the classification rules are exactly same for different versions of the (distributed) data mining even though necessary data are not transferred to the others for complete analysis.

As a future work, we need to evaluate the performance [4] of the presented agent-based decision tree algorithm with the comparison of a centralized decision tree algorithm such as C5.0.

References

1. Agrawal, S., Krishnan, V., Haritsa, R. J.: On Addressing Efficiency Concerns in Privacy Preserving Mining. LNCS 2973 (2004) 113-124

2. Malvestuto, M. F., Mezzini, M.: Privacy Preserving and Data Mining in an On-Line Statistical Database of Additive Type. LNCS 3050 (2004) 353-365
3. Lindell, Y., Pinkas, B.: Privacy Preserving Data Mining. LNCS 1880 (2000) 36
4. Krishnaswamy, S., Zaslavsky, A., Loke, S. W.: Techniques for Estimating the Computation and Communication Costs of Distributed Data Mining. LNCS 2329 (2002) 603-612
5. Aggarwal, C. C., Yu, P. S.: A Condensation Approach to Privacy Preserving Data Mining. LNCS 2992 (2004) 183-199
6. Kargupta, H., Datta, S., Wang, Q., Sivakumar, K.: On the privacy preserving properties of random data perturbation techniques. Proceedings of the Third IEEE International Conference on Data Mining (2003) 99-106
7. Kargupta, H., Park, B., Hershberger, D., Johnson, E.: Collective Data Mining: A New Perspective Toward Distributed Data Analysis. Advances in Distributed and Parallel Knowledge Engineering 5 (2000) 131-178
8. Kargupta, H., Hamzaoglu, I., Stafford, B.: Scalable, Distributed Data Mining-An Agent Architecture. Proceedings of the International Conference on Knowledge Discovery and Data Mining. (1997) 211-214
9. Stolfo, S., Prodromidis, A. L., Tselepis, S., Lee, W.: JAM: Java Agents for Meta-Learning over Distributed Databases. Proceedings of the International Conference on Knowledge Discovery and Data Mining. (1997) 74-81
10. Bailey, S., Grossman, R., Sivakumar, H., Turinsky, A.: Papyrus: a system for data mining over local and wide area clusters and super-clusters. Proceedings of the International Conference on Supercomputing. (1999)
11. Klusch, M., Lodi, S., Moro, G.: Agent-Based Distributed Data Mining: The KDEC Scheme. LNAI 2586 (2003) 104–122
12. Quinlan, J. R., Rivest, R. L.: Inferring Decision Trees Using the Minimum Description Length Principle. Information and Computation. 80 (1989)
13. See Web site at http://www.kdnuggets.com

Geometrical Analysis for Assistive Medical Device Design

Taeseung D. Yoo[1], Eunyoung Kim[2], Daniel K. Bogen[3], and JungHyun Han[1,*]

[1] Department of Computer Science and Engineering, Korea University, Korea
[2] School of Information and Communications Engineering,
Sung Kyun Kwan University, Korea
[3] Department of Bioengineering, University of Pennsylvania, USA

Abstract. In the computer graphics field, mesh simplification, multi-resolution analysis and surface parameterization techniques have been widely investigated. This paper presents innovative applications of those techniques to the biomechanical analysis of thin elastic braces. The braces are represented in polygonal meshes, and mesh simplification and multi-resolution analysis enable fast geometric computation and processing. The multi-resolution brace mesh is parameterized for strain energy calculation. The experiment results prove that 3D geometrical analysis works quite well for assistive medical device design.

1 Introduction

An *elastic brace* is a medical device commonly used to restrict the motion of the joints (such as wrist and knee) that suffer from musculoskeletal disorders. While the elastic brace is widely used, its mechanics has been rarely studied. The ultimate goal of the brace research is to design and fabricate *custom-made braces* with the desired *stiffness* such that they limit the joint motions. Stiffness is affected mainly by the brace geometry. Once the relationship between the brace geometry and its stiffness is obtained, it enables us to design a brace of the appropriate geometry that guarantees the stiffness prescribed by a physician.

The relationship is found through sophisticated steps. The joint is scanned to generate a 3D model. When the joint is bent, brace stuck onto it is deformed following the changing surface of the joint. For each bending angle, *strain energy* (SE) is calculated. SE is defined as the mechanical energy stored in the elastic body due to its deformation, and its unit is $N \cdot m$ [1]. From the relationship of SE and deflection, stiffness of the brace is obtained: (1) the derivative of SE with respect to deflection angle is joint *moment*. (2) the derivative of the joint moment with respect to deflection angle is *stiffness*.

The major contribution of our research is to find the relationship between brace geometry and its stiffness. This paper presents how the *mesh simplification* [2] and *surface parameterization* [3] techniques, which have been intensively

---
* Corresponding Author.

investigated in the computer graphics field, can be innovatively applied to SE analysis for efficiency purpose.

2   3D Model and Deformation

The *point cloud* data of a joint surface is generated using Cyberware's 3D scanner. Then, the point cloud is converted into a *triangular mesh* using Paraform's mesh generation packages. Fig. 1-(a) shows the knee mesh model with 25,600 vertices. We have too many vertices to process. As a preliminary step, the simplification algorithms discussed in the next section reduces the vertex number into 5,000 with *geometric fidelity* to the original mesh.

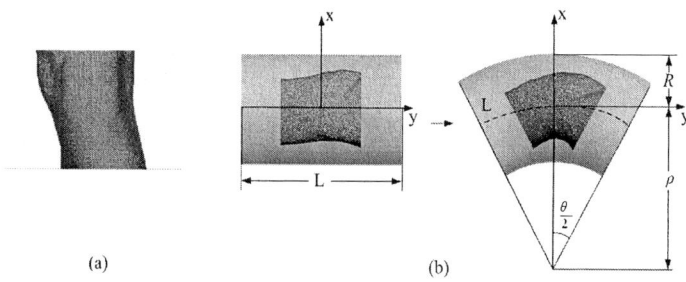

Fig. 1. Knee's mesh model

It is very hard to devise a realistic bending transformation that can be applied to the joints. Therefore, we use *simple beam bending*: the original brace mesh is conceptually embedded into a solid cylindrical beam that undergoes constant curvature bending. The bending beam displaces the vertices of the brace mesh. The formulae in Eq. (1) relate the initial location of an unbent beam's vertex to its transformed location in the bent beam.

$$x' = (\rho + x) \cdot \cos\phi - \rho, \ y' = (\rho + x) \cdot \sin\phi \qquad (1)$$

where (x, y) is the initial location of a vertex, (x', y') is its transformed location, ρ is the radius of curvature, and ϕ is the angular position of the point with respect to the center of curvature of the beam. Note that $\phi = y/\rho$. We use 15 bent models generated with 4° interval out of the range of 0° through 60°.

After the bending transformation, the bent mesh is simplified. The simplified mesh needs to get 'unbent.' For such *inverse bending*, inverses of Eq. (1) are used.

3   Selective Simplification of 3D Mesh

When the joint deforms, some brace regions experience *contraction* while the other regions are *expanded*. Contraction refers to reduction in size. In Fig. 1-(b), the back part of the knee is contracted while the front part is expanded. In the contracted regions, *buckling* usually occurs. Buckling site must be excluded from SE calculation since it stores no energy. The preliminarily reduced number of points is still too huge and significantly degrades the performance in SE calculation. A *selective simplification* strategy is adopted: The higher resolution is retained in the contracted region to detect buckling effectively, and the simplification is focused on the expanded region.

The most popular mesh simplification technique is *edge collapse* or simply *ecol*. See Fig. 2-(a). Garland and Heckbert [4] proposed a simplification algorithm based on *quadric error metrics* (QEM), where the simplified mesh maintains *visual fidelity* to the original mesh. We adopted an *ecol* version of the QEM-based simplification, and modified it: the expanded region must be simplified for fast computation, but the contracted region must maintain the original resolution. A triangle's area can become increased or decreased after the deformation. If two triangles sharing an edge have increased areas, the shared edge is determined to lie in an expanded region. Therefore, the edge is taken as *collapsible*. The *ecol* operations are performed only with the collapsible edge set. Fig. 2-(b) shows the input mesh (5,000 vertices) and the selectively simplified mesh (3,250 vertices).

Most elastic brace design systems are in a stable state if the *strain energy error* (SEE) is kept below 1% [5]. SEE is defined to be $(E_s\text{-}E_o)/E_o$ where E_o is SE calculated with the original mesh and E_s is that with the simplified mesh. Joints have the *generalized cylinder* geometry. Hence, a cylinder model is a good choice for determining how much can be simplified within 1% SEE. A cylinder mesh with 1,050 vertices is bent by the largest bending angle, 60°. Then, *ecol* is repeatedly applied. SEE is calculated for each simplified mesh, and tested if it exceeds 1%. With 1% error limit, we obtain the simplified mesh of 616 vertices, i.e. 41.3% reduction. Since wrist and knee are more complex geometries than cylinder, SEE might be bigger with the same reduction rate. Hence, for general joints, the maximum reduction rate is set to 35%. As the largest bending angle (60°) is used, the reduction rate 35% satisfies 1% error bound for most cases.

4 Strain Energy Calculation

4.1 Surface Parameterization and Strain Energy Calculation

Elastic braces are fabricated using a sheet of elastic membrane, and the traditional choice of the 2D brace shape has been a *rectangle* because every human-

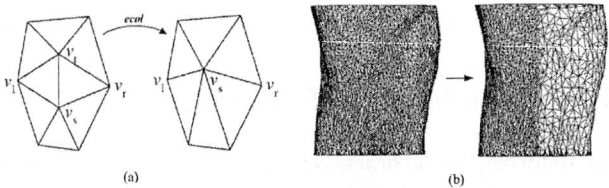

Fig. 2. The *ecol* operation, and knee meshes with different resolutions

body joint has the generalized cylinder geometry and its *flattened* version roughly coincides with a 2D rectangle. We propose to use surface parameterization techniques [3], which map a 3D mesh to a 2D plane.

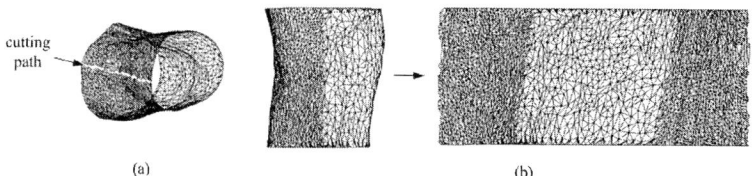

Fig. 3. Brace mesh parameterization

We have adapted Floater's parameterization algorithm [6] to our context. For parameterization, the joint geometry is cut along a vertical path. See Fig. 3-(a). All the *boundary vertices* of the cut surface are *radially projected* to construct the 2D rectangle's boundary such that the inter-vertex distances are preserved (with scaling). Then, the interior vertices are mapped using the Floater's algorithm. Fig. 3-(b) shows the parameterization result.

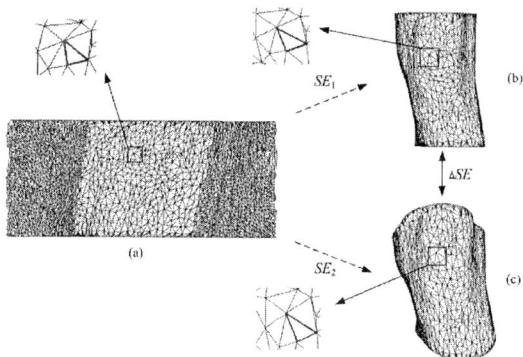

Fig. 4. SE calculation. (a) parameterized plane (b) Unbent joint (c) Bent joint

The parameterization-based SE calculation is illustrated in Fig. 4. Strain energy SE_1 is calculated, which is stored in the 3D brace when the brace is worn in the *un-deformed* joint geometry. Similarly, SE_2 is for the *deformed* joint geometry. The net strain energy ΔSE is the difference between SE_1 and SE_2. SE_1 and SE_2 are calculated as follows: (1) SE density is calculated at each triangle of the joint mesh. (See Section 4.2.) (2) The area of the corresponding triangle on the 2D brace is computed, then the area and the brace thickness are multiplied to produce the triangle volume. (3) The computed volume and density are multiplied to produce SE. (4) The process is repeated for every triangle, and the result is accumulated to obtain the total SE.

4.2 Strain Energy Density Calculation

Our brace is hyperelastic, transversely isotropic, and incompressible. Strain energy density is calculated by *strain energy density function (SEDF)* W_{iso} [7,8] given below:

$$W_{iso} = \frac{1}{2}G_{iso}\left(I_c + II_c^{-1} - 3\right) \quad (2)$$

In Eq. (2), G_{iso} is the rigidity of modulus, $I_c = C_{11} + C_{22}$ and $II_c = C_{11}C_{22}$, where C_{ii}s are the components of the right Cauchy-Green deformation gradient tensor $\mathbf{C}$. Given the deformation gradient tensor $\mathbf{F} = \partial \boldsymbol{x}/\partial \mathbf{X}$ [1] where $\boldsymbol{x}$ and $\mathbf{X}$ are the coordinates of the 3D mesh and of the 2D parameterized plane, respectively, the tensor $\mathbf{C}$ is defined by $\mathbf{F}^T \cdot \mathbf{F}$ [1].

The contracted regions may or may not have the buckling. Whether buckling occurs is determined by the *principal stresses* [9]. Therefore, we need to compute the signs and magnitudes of the principal stresses throughout the mesh. Then, we can partition the contracted regions into buckling and non-buckling sites. The variation of the joint surface geometry is random, and buckling and non-buckling sites may be intermingled. Hence, when such intermingled sites are simplified, we often end up with incorrect partitioning, which eventually leads to incorrect calculation of SE. Thus, we need a high resolution for the contracted regions.

Eq. (2) is good for the expanded regions only. Therefore, we need SEDF for the contracted regions of no buckling. For this purpose, Eq. (2) is extended to Eq. (3) to calculate SE, which is caused only by tensile load [9].

$$W_{iso} = \tfrac{1}{2}G_{iso}\left[\left(C_{11}n_1^2 + 2C_{12}n_1n_2 + C_{22}n_2^2\right) + \frac{2}{\sqrt{(C_{11}n_1^2 + 2C_{12}n_1n_2 + C_{22}n_2^2)}} - 3\right] \quad (3)$$

where n_1 and n_2 are the two components of the principal direction vector [5].

5 Results

We have implemented the proposed algorithms in C++ on a PC with an Intel 2.6GHz Pentium 4 CPU. We have tested two 3D-scanned models: wrist (1,041 vertices) and knee (5,000 vertices). We have three major tasks: (1) mesh simplification, (2) parameterization, and (3) SE calculation. For the wrist model, the three tasks are done in real-time. For the knee model, SE is calculated in real-time but simplification and parameterization take 3 seconds and 12 minutes, respectively. However, the parameterization process with the un-simplified knee takes 31 minutes, i.e. we can save 19 minutes at the cost of 3 seconds. Since we use 15 bent models, total 285 minutes are saved at the cost of 45 seconds.

Fig. 5 shows the normalized SE curves of the original mesh and the simplified mesh for wrist and knee, respectively. The curves (for original and simplified meshes) in the figure are almost identical. The largest error in SE between the original and the simplified is about 0.6% for wrist, and 0.3% for knee.

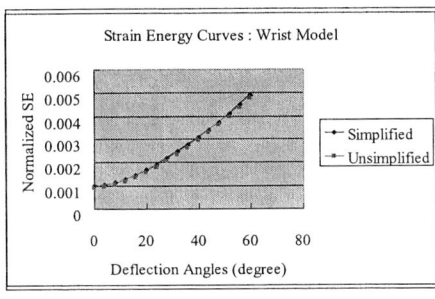 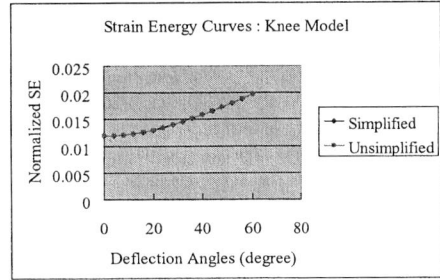

Fig. 5. Strain energy curves

6 Conclusion

This paper presented innovative and successful applications of the mesh simplification and surface parameterization techniques to the analysis of the relationship between the brace geometry and its stiffness. The relationship will enable effective design of braces. Further works may include analysis of *anisotropic* braces with multiple layers of single-direction fibers.

Acknowledgements

This work is supported by the MOST 21st Century Frontier Program: Intelligent Robot Project.

References

1. Mal, A., Singh, S.: Deformation of Elastic Solids. Prentice-Hall (1991)
2. Luebke, D.: A developer's survey of polygonal siplification algorithms. In: IEEE Computer Graphics and Applications. Volume 21(3). (2001) 24–35
3. Floater, M., Hormann, K.: Surface parameterization: a tutorial and survey. In: Advances in multiresolution for geometric modelling. (2004)
4. Garland, M., Heckbert, P.: Surface simplification using quadric error metrics. In: Proc SIGGRAPH '97. (1997) 209–216
5. Yoo, T.: Analytic Mechanics of Hyper-Elastic Wrist Braces. PhD thesis, The University of Pennsylvania, Philadelphia, PA, USA (1999)
6. Floater, M.: Parameterization and smooth approximation of surface triangulations. In: Computer Aided Geometric Design. Volume 14. (1997) 231–250
7. Treloar, L.: The Physics of Rubber Elasticity. The Clarendon Press (1975)
8. Rivlin, R., Topakoglu, C.: A theorem in the theory of finite elastic deformations. In: Journal of Rational Mechanical Analysis. Volume 3. (1954) 581–589
9. Timoshenko, S., Young, D.: Elements of Strength of Materials. Prentice-Hall (1968)

Hybrid Genetic Algorithms and Case-Based Reasoning Systems

Hyunchul Ahn[1], Kyoung-jae Kim[2], and Ingoo Han[1]

[1] Graduate School of Management, Korea Advanced Institute of Science and Technology,
207-43 Cheongrangri-Dong, Dongdaemun-Gu, Seoul 130-722, Korea
{goodguy, ighan}@kgsm.kaist.ac.kr
[2] Department of Information Systems, Dongguk University,
3-26, Pil-Dong, Chung-Gu, Seoul 100-715, Korea
kjkim@dongguk.edu

Abstract. Case-based reasoning (CBR) has been applied to various problem-solving areas for a long time because it is suitable to complex and unstructured problems. However, the design of appropriate case retrieval mechanisms to improve the performance of CBR is still a challenging issue. In this paper, we encode the feature weighting and instance selection within the same genetic algorithm (GA) and suggest simultaneous optimization model of feature weighting and instance selection. This study applies the novel model to corporate bankruptcy prediction. Experimental results show that the proposed model outperforms other CBR models.

1 Introduction

Case-based reasoning (CBR) often shows significant promise for improving the effectiveness of complex and unstructured decision making. Regardless of its many advantages, there are some problems that must be solved in order to design an effective CBR system. Some examples of those problems involve the fact that there are no mechanisms to determine appropriate methods of case retrieval in typical CBR systems. In this aspect, the selection of the appropriate similarity measures, feature subsets and their weights in the case retrieval step has been the most popular research issue.

Recently, simultaneous optimization of several variables in CBR attracts researchers' interest due to its better performance. As a pioneering work, there exists the approach to combine feature selection and instance selection simultaneously [5,7]. Theoretically, feature weighting may improve the effectiveness of CBR systems better than feature selection. Nonetheless, there exist few attempts which try to optimize feature weighting and instance selection simultaneously.

This paper proposes genetic algorithms (GA) to optimize the feature weights and instance selection simultaneously. This study applies the proposed model to the real-world case and presents experimental results from the application.

2 Research Background

2.1 Genetic Algorithms as an Optimization Tool for CBR

CBR is composed of five-step processes; presentation, retrieval, adaptation, validation and update [2]. Among these five steps, step two, case retrieval, is most critical for determining the effectiveness of CBR system. During the retrieval step, similar cases that are potentially useful for solving the current problem are retrieved from the case base. So, how to measure the similarity of the cases and how to combine the similar cases are challenging issues in this step [3]. Especially, feature weighting and instance selection for measuring similarity have been controversial issues in designing CBR system. To determine these uncertain factors of CBR system, there have been many studies that attempt to resolve these problems. Among many methods of instance selection and feature weighting, GA is increasingly being used in the CBR system.

Genetic algorithms are stochastic search techniques that can search large and complicated spaces. It is based on biological backgrounds including natural genetics and evolutionary principle. In particular, GAs are suitable for parameter optimization problems with an objective function subject to various hard and soft constraints [8].

2.2 Feature Weighting

Feature weighting is assigning a weight to each feature according to the relative importance of each one. It is an important factor that determines the performance of artificial intelligence (AI) systems, so it has been the most popular research issue in designing most AI systems including CBR systems.

Regarding feature weighting, Wettschereck et al. presented various feature weighting methods based on distance metrics in the machine learning literature [9]. Kelly and Davis proposed a GA-based feature weighting method for k-nearest neighbor [4]. Similar methods are applied to the prediction of corporate bond rating [8].

2.3 Instance Selection

Instance selection is the technique that selects an appropriate reduced subset of casebase and applying the nearest-neighbor rule to the selected subset. It may increase the performance of CBR systems dramatically if the systems contain many noises. So, it has been another popular research issue in CBR systems for long time. There exist many different approaches to select appropriate instances. For example, Lipowezky suggested linear programming methods for instance selection [6]. Yan suggested ANN-based instance selection method [10] and GA approach was also proposed by Babu and Murty [1].

2.4 Simultaneous Optimization of Feature Weighting and Instance Selection

In general, feature weighting includes feature selection since selection is a special case of weighting with binary weights. Consequently, simultaneous optimization model of feature weighting and instance selection model may improve the performance of the model of feature selection and instance selection. In this manner, Yu et al. proposed simultaneous optimization model of feature weighting and instance selection for collaborative filtering, an algorithm that is very similar to CBR [11]. However, they applied not AI techniques, but an information-theoretic approach to the optimization model. So, in the strict sense of the word, their model is not simultaneous optimization model but sequential combining model of two approaches.

3 GA for Simultaneous Feature Weighting and Instance Selection

To mitigate the limitations of prior studies, this paper proposes GA as a simultaneous optimization tool of feature weighting and instance selection. To test the effectiveness of the proposed model, we compare the results of four different models.

The first model, labeled COCBR (COnventional CBR), uses a conventional approach for reasoning process of CBR. This model considers all initially available instances as an instance subset. Thus, there is no special process of instance subset selection. In addition, relative importance of each feature is also not considered because many conventional CBR models do not have general feature selection or weighting algorithm.

The second model assigns relevant feature weights via genetic search. This study names this model FWCBR (CBR with Feature Weighting). Similar models to it were previously suggested by Kelly and Davis [4] and Shin and Han [8].

The third model uses the GA to select a relevant instance subset. This study names this model ISCBR (CBR with Instance Selection). Babu and Murty proposed similar model to it [1].

The fourth model, the proposed model in this study, employs the GA to select a relevant instance subset and to optimize the weights of each feature simultaneously. This model is named as SOCBR (Simultaneous Optimization of CBR) in this study. The model consists of the following three stages:

Stage 1. For the first stage, we search the search space to find optimal or near-optimal parameters (feature weights and selection variables for each instance). The population (seed points for finding optimal parameters) is initiated into random values before the search process. The parameter to be found must be encoded on a chromosome. The encoded chromosome is searched to maximize the specific fitness function. The objective function of the model is to classify bankrupt or non-bankrupt corporations accurately, and it can be represented by the average prediction accuracy of the test data. Thus, this study applies it to the fitness function for GA. In this stage, the GA operates the process of crossover and mutation on the initial chromosome and iterates it until the stopping conditions are satisfied.

Stage 2. The second stage is the process of case retrieval and matching for a new problem in the CBR system using the parameters that are set in Stage 1. In this stage, 1-NN(one-nearest neighbor) matching is used as a method of case retrieval. And, we use the weighted average of Euclidean distance for the each feature as a similarity measure. This stage is repeated after the process of evolution (crossover, mutation) and the value of the fitness function is updated.

Stage 3. The third stage applies the finally selected parameters - the optimal weights of features and selection of instances - to the hold-out data. This stage is required because GA optimizes the parameters to maximize the average predictive accuracy of the test data, but sometimes the optimized parameters are not generalized to deal with the unknown data.

4 The Research Design and Experiments

4.1 Application Data

The application data used in this study consists of financial ratios and the status of bankrupt or non-bankrupt for corresponding corporation. The data was collected from one of largest commercial banks in Korea. The sample of bankrupt companies was 1335 companies in heavy industry which filed for bankruptcy between 1996 and 2000. The non-bankrupt companies were 1335 ones in heavy industry which filed between 1999 and 2000. Thus, the total number of samples is 2670 companies.

The financial status for each company is categorized as "0" or "1" and it is used as a dependent variable. "0" means that the corporate is bankrupt, and "1" means that the corporate is solvent. For independent variables, we first generate 164 financial ratios from the financial statement from each company. Finally, we get 15 financial ratios as independent variables through the two independent sample t-test and the forward selection procedure based on logistic regression.

4.2 Research Design and System Development

For the controlling parameters of GA search for SOCBR, the population size was set at 100 organisms and the crossover and mutation rate were set at 0.7 and 0.1. And, as the stopping condition, only 1500 trials (15 generations) are permitted.

To compare the result of SOCBR, we also applied other models to the same data set. The compared models include COCBR(Conventional CBR), FWCBR(CBR with feature weighting), and ISCBR(CBR with instance selection). COCBR is 1-NN algorithm whose feature weights are set to 1. FWCBR is 1-NN algorithm whose feature weights are optimized by GA. In the case of COCBR and FWCBR, all the instances are used for reference case-base. However, ISCBR uses only subset of total reference case-base which is selected by GA. As the controlling parameters of GA search for FWCBR and ISCBR, the population size was set at 50 organisms and the crossover and mutation rate were set at 0.7 and 0.1. And, as the stopping condition, about 500 trials (10 generations) are permitted.

5 Experimental Results

In this section, the prediction performances of SOCBR and other alternative models are compared. Table 1 describes the average prediction accuracy of each model.

In Table 1, SOCBR achieves higher prediction accuracy than COCBR, FWCBR, and ISCBR by 5.42%, 2.99%, and 3.55% for the hold-out data.

Table 1. Average prediction accuracy of the models

Model	COCBR	FWCBR	ISCBR	SOCBR
Test data set	-	84.83%	83.71%	85.96%
Hold-out data set	80.75%	83.18%	82.62%	86.17%

The McNemar tests are used to examine whether the predictive performance of the SOCBR is significantly higher than that of other models. This test is used with nominal data and is particularly useful with before-after measurement of the same subjects. Table 2 shows the results of the McNemar test to compare the performances of four models for the hold-out data.

Table 2. McNemar values for the hold-out data

	FWCBR	ISCBR	SOCBR
COCBR	2.361	3.115**	10.453***
FWCBR		0.062	3.309*
ISCBR			4.208**

* significant at the 10% level, ** significant at the 5% level, *** significant at the 1% level.

As shown in Table 2, SOCBR is better than COCBR at the 1% and better than ISCBR at the 5% statistical significance level. But, SOCBR outperforms FWCBR at only 10% statistical significance level.

6 Conclusions

We have suggested a new kind of hybrid system of GA and CBR to improve the performance of the typical CBR system. This paper used GA as a tool to optimize the feature weights and instance selection simultaneously. From the results of the experiment, we show that SOCBR, our proposed model, outperforms other comparative models such as COCBR and FWCBR as well as ISCBR.

However, this study has some limitations. First of all, the number of generations (trial events) in our GA experiments is too small. In fact, the search space for simultaneous optimization of feature weights and feature selection is very large, so we need to increase the number of populations and generations. Secondly, it takes too much computational time for SOCBR. As mentioned, SOCBR iterates case retrieval process whenever genetic evolution occurs. And, in general, case retrieval process in CBR takes much computational time because it should search whole case-base to make just one solution. Consequently, the efforts to make SOCBR more efficient should be followed in future. Moreover, the generalizability of SOCBR should be tested further by applying it to other problem domains.

References

1. Babu, T. R., Murty, M. N.: Comparison of genetic algorithm based prototype selection schemes, Pattern Recognition 34 (2001) 523-525
2. Bradley, P.: Case-based reasoning: Business applications. Communication of the ACM 37 (1994) 40-43
3. Chiu, C.: A case-based customer classification approach for direct marketing. Expert Systems with Applications 22 (2002) 163-168
4. Kelly, J. D. J., Davis, L.: Hybridizing the genetic algorithm and the k nearest neighbors classification algorithm, Proceedings of the Fourth International Conference on Genetic Algorithms (1991) 377-383
5. Kuncheva, L. I., Jain, L. C.: Nearest neighbor classifier: Simultaneous editing and feature selection. Pattern Recognition Letters 20 (1999) 1149-1156
6. Lipowezky, U.: Selection of the optimal prototype subset for 1-NN classification, Pattern Recognition Letters 19 (1998) 907-918
7. Rozsypal, A., Kubat, M.: Selecting representative examples and attributes by a genetic algorithm. Intelligent Data Analysis 7 (2003) 291-304
8. Shin, K. S., Han, I.: Case-based reasoning supported by genetic algorithms for corporate bond rating. Expert Systems with Applications 16 (1999) 85-95
9. Wettschereck, D., Aha, D. W., Mohri, T.: A review and empirical evaluation of feature weighting methods for a class of lazy learning algorithms, Artificial Intelligence Review 11 (1997) 273-314
10. Yan, H.: Prototype optimization for nearest neighbor classifier using a two-layer perceptron, Pattern Recognition 26 (1993) 317-324
11. Yu, K., Xu, X., Ester, M., Kriegel, H-P.: Feature weighting and instance selection for collaborative filtering: an information-theoretic approach, Knowledge and Information Systems 5 (2003) 201-224

Papílio Cryptography Algorithm

Frederiko Stenio de Araújo, Karla Darlene Nempomuceno Ramos,
Benjamín René Callejas Bedregal, and Ivan Saraiva Silva

Universidade Federal do Rio Grande do Norte,
Departamento de Informática e Matemática Aplicada,
59072-970, Natal-RN, Brazil
steniorn@uol.com.br, karla@ppgsc.ufrn.br, {bedregal, ivan}@dimap.ufrn.br

Abstract. Papílio is a Feistel cipher encryption algorithm where the coder process (function F) is based in the Viterbi algorithm. The Viterbi algorithm was proposed as a solution to decode convolutional codes. There are several parameters that define the convolution code and Viterbi algorithm; one of them is the generator polynomial. To use Viterbi algorithm in cryptography, it is necessary to make some modifications. The proposed one does not depend on the parameters of Viterbi nor on the parameters of convolution. In this work we will analyze the cryptographic indices (avalanche, diffusion and confusion) of Papílio considering all possible different polynomials and fix the other parameters.

1 Introduction

A.J. Viterbi in [11] developed the Viterbi Algorithm (VA) in 1967 as a solution for decoding convolutional codes. Convolutional encoder (CE) with Viterbi decoder is a FEC (Forward Error Correction) technique that is particularly suitable to channels where the transmitted signal is corrupted mainly by additive white Gaussian noise [2]. Since then, other researchers have applied VA and CE for other areas of applications such as recognition of handwritten word [8], target tracking [1], image edge detection [7]. Since the CE, increases the length of the input bitstream (it is an injective and not surjective one) and the VA only decode the bitstream generated by CE and some others few which can be recovered (is a partial not injective and not surjective function), this process can not be considered as cryptographic method.

This work apply a modification in the VA, considering specific parameters for the VA and CE, in order to get a bijective function. This bijective function was inserted as the function F in a Feistel cipher with 16 rounds, blocks of 64 bits and keys of 128 bits [9, 10]. To generate the 16 sub-keys, it was used the modified Viterbi (MV). This feistel cipher will be called of Papílio[1]. We will show an study that chooses eight polynomials which provide to Papílio better indices of cryptography and also will provide some evidences that Papílio can be improved in the aspect of the complexity of cryptanalysis and the execution time.

---
[1] The name Papílio was given because the trellis of VA form a butterfly, and *Papílio Thoas Brasiliensis* is the name of a very common sort of butterflies in Brazil.

2 Modified Viterbi

VA attempts to find the closest "valid" sequence to the received bitstream, that is, a sequence which when applied to the CE results in the received bitstream. Notice that two different sequences used as inputs for the convolution encoder result, necessarily, in two different sequences, that compute an injective function. So, the VA can be seen as a decoder. But, VA only decode bitstream generated by CE. Therefore, for VA to be used in the cryptography it is necessary that it processes any input sequence in a bijective way. The VA will be modified to deal with all possible bitstreams, in such a way that can be seen as a bijective function and therefore appropriated for cryptography.

The MV algorithm proposed increases the code space matching VA with CE. For MV deal with any input sequence, independently of the current state, it was created besides output sequence S_0, an output sequence S_1. S_0 presents the result of VA. S_1 exhibits if each output symbol of S_0 was obtained in agreement with the VA, or if it was obtained in an special shape (MV). When an output symbol of S_0 is obtained in agreement with the VA, S_1 generates the bit 0, otherwise S_1 generates the bit 1. The MV algorithm is initialized to zero state and works as VA until an input symbol of bitstream is not appropriate in the current state, i.e. is "invalid". When this occurs, the symbols of not appropriated label are treated separately for the CE with initial state being the current state of MV. It is observed that CE will generate $\lceil \frac{s}{n} \rceil$ additional labels, being $\frac{n}{s}$ the rate of CE. With this procedure the generated labels can be treated by the VA. The application of the VA would generate a size label n for each one of the generated additional labels. However what interests in the code is the generation of an only size label n. The adopted solution consists of considering, to compose the flow S_0, just the first of the $\lceil \frac{s}{n} \rceil$ size labels n (the bitstream S_1 receives the value 1). The continuation of the code process using the VA is adopted as current state the last state of the process of convolution, until a new "invalid" label is found or the code is finished. The bitstreams S_0 and S_1 are independent. At the end of the code the bitstream are concatenated, in way to generate a bitstream of same length of the original. Through MV it's possible create tables that help the code process. For example, the table 1 exhibits the MV taking into account the CE and VA where $n = 1$, $s = 2$, $Q = 3$, $m = 2$ and generator polynomial $G = 111101$.

3 The Encryption Algorithm Papílio

Papílio is a Feistel cipher encryption algorithm where the function F is the function computed by the MV algorithm whose parameters (codification rate $\frac{n}{s}$, Q, m and the polynomial generator) are opens in a first moment. The main characteristics:

Block Length: Actually are considered block of 64 bits. Nevertheless, because the MV does not depends on length of block, its size can be changed in a further implementation turning fix as 128 bit or variable in function of the key;

Table 1. MV of CE with $n = 1$, $s = 2$, $Q = 3$, $m = 2$ and generator polynomial $G = 111101$

Current State	Input	S_0	S_1	Next State	Current State	Input	S_0	S_1	Next State
0	00	0	0	0	2	00	0	1	0
	01	0	1	2		01	1	0	3
	10	1	1	1		10	0	0	1
	11	1	0	2		11	1	1	3
1	00	1	0	2	3	00	0	1	0
	01	0	1	2		01	0	0	1
	10	1	1	1		10	1	0	3
	11	0	0	0		11	1	1	3

Size of the Key: 128 bits. But, its size could be variable or greater;

Number of Rounds: 16. But, this quantity can be reduced to 6 or turned into variable (between 6 to 16) without losing the good cryptographic indices;

Sub-key Generations: Papílio uses 16 sub-keys that are generated from the 128-bit encryption key. The sub-keys are stored temporarily in an array. The scheme for generation is as follows. The first four sub-keys, labelled SC_1, SC_2, SC_3 and SC_4, are generated by applying the MV in the 128-bit initial key, which generates two 64-bit bitstream. Applying MV to the two 64-bit bitstreams it generates four bitstreams of 32-bits that corresponds to the first four sub-keys. To generate the four following sub-keys, the four bitstream are concatenated generating an alone of 128-bits and the procedure to generate first four sub-keys is repeated until all 16 sub-keys are generated.

Decryption: as with most block ciphers, the process of Papílio decryption is essentially the same as the encryption process, except the sub-keys that are employed in reverse order. So, use SC_{16} in the first round, SC_{15} in the second round, and so on until SC_1 is used in the last round. This feature avoids implementing two different algorithms, one for encryption and one for decryption;

Operation Modes: Papílio was implemented in the four usual modes (ECB, CBC, CFB and OFB).

Programming Language: Papílio was implemented in C.

4 The Choice of Better Polynomials

By simplicity and implementation's performance, was considered for MV a CE and VA with the following parameters: codification rate $\frac{n}{s} = \frac{1}{2}$, $Q = 3$ and $m = 2$. With this values we have 64 (2^{sQ}) possible polynomials.

The idea is to analyze considering the behavior of each polynomials regarding to the avalanche effect (in the key and in the block), the diffusion and the confusion properties and to select the eight polynomials with better results.

First Tests and Measure Used: First was made to each polynomials and operation mode a test for confusion and diffusion based on a book of project Gutemberg [4]. We extract from the book the first 3536 characters (including the spaces), despising the 400 first characters to erase the heading. The keys used in this test was (pseudo)randomly generated (all polynomials used the same keys). The test of avalanche effect (in the block and key) was realized on 50 blocks of plaintext and 50 keys randomly generated.

To measure the avalanche effect in the block was used the arithmetic average of Hamming distances between the encryption of a plaintext block and the encryption (with the same key) of the same plaintext block changing a bit in all possible ways. Analogously, to measure the avalanche effect in the key was used the arithmetic average of Hamming distances between the encryption of a text block and the encryption of the same text block changing a bit on the key in all possible ways. The measure of diffusion was calculated using the standard deviation of frequencies of characters in the cyphertext. The confusion was measured using the average of Euclidean distances between the encryptions of plaintext with the original key and the plaintext with the original key changing only an unique bit. This result is divided by the greatest Euclidean distance possible, which allows us to normalize this value obtained a value between 0 and 1.

The avalanche effect in the block for the modes ECB and CBC is, for the most of polynomials, between 0.45 (45% of bits, or more, are changed) and 0.51 which is a very good index, considering that the ideal value is 0.5. For the modes OFB and CFB, the avalanche effect is constant (0.0156), nevertheless it is not a problem of Papílio, but of the modes, because we are measuring only the avalanche in an unique block, and therefore a change of a bit only affect an unique bit. The avalanche effect in the key still is better, because in the modes ECB and CBC 92% of polynomials matched between 0.48 and 0.51 and in the modes CFB and OFB 79% of polynomials matched between 0.48 and 0.51.

The confusion in the modes ECB and CBC, the half of polynomials (50%) are between 0.38 (38%) and 0.41 which is not ideal (the ideal is similar to avalanche, i.e. 0.5 or 50%) but it is reasonable, more over if we consider that the Rijndael algorithm, using the implementation of Rijndael founded in [6] and in the same conditions of test, obtained confusion index of 40.5%. In the modes CFB and OFB, 47% of polynomials are between 0.38 and 0.41. In all operator modes we have more of 8 polynomials with confusion index greater than 40.

The greatest diffusion index for the mode ECB was 0.0285 and 56% of polynomials have lesser than 0.02. In the mode CBC, the greatest diffusion index was 0.0255 and 81% of polynomials have an index lesser than 0.02. In the mode CFB, the greatest diffusion index was 0.0252 and 80% of polynomials have an index lesser than 0.02. Finally, in the mode OFB, the greatest diffusion index was 0.0257 and 80% of polynomials have an index lesser than 0.02. Thus, in any operator mode the symbols in the ciphertext have, practically, the same distribution which allows us to conclude that the statistical frequencies of symbols in the plaintext were destroyed. Therefore, there is not a statistical relation between the frequencies of symbols in the plaintext and the ciphertext.

Similarity of Indices: In order to check if the indices obtained don't depend strongly on texts and key, but only depend on polynomials used, we will make new tests for avalanche effect on the block and diffusion and then we will measure the degree of similarity between the results using the standard deviation of results. For the tests of avalanche was generated 100 series of 50 plaintext blocks and for the diffusion were made 100 series using only 3536 characters despising the 400 firsts of a book of Gutemberg project [4]. For each 10 series was used a different book.

The avalanche affect on the block 98% of the polynomials have an standard deviation lesser than 0.06%, and the diffusion of all polynomials is lesser than 0.35%. Both results are very good, because indicate that Papílio independently of the polynomials is very stable. Since the confusion and avalanche effect on key are, in some sense, subordinated to the avalanche effect on the block, we can conclude that both effects neither depend on the plaintext nor the key used. This also is true for the other modes.

The Winner Polynomials: With the conviction that the Papílio behavior depends quasi exclusively of polynomials, we will make a championship to determine the polynomial which provides to Papílio the best cryptographic indices. Because the confusion strongly relates to avalanche in the block and the avalanche in the key as well as the diffusion obtained in all tests and in all polynomials well indices, beyond diffusion need more computational effort, we opted to only consider the avalanche effect on the block.

The championship consisted in performing 50 news tests for avalanche effect in the block using keys and block generated randomly, at each test the polynomials that achieve the index more proximate of 50% gain a point. To avoid arrive in local optimum when a polynomial had 20 points this would classify for the next stage and the championship continued without it. In the next stage of championship was performed 100 test considering again random keys and blocks. The selected polynomials was those which obtained 50 points. For simplicity we only make the championship for the ECB mode.

5 Final Remarks

The empirical analysis showed that the proposal cipher has very good performance w.r.t. of avalanche, diffusion and confusion properties. However in spite that these properties are interesting and important, just having these properties does not mean that a cipher is secure. In order to conclude that Papílio is reliable, yet is necessary a treatment of the security of Papílio cipher considering modern cryptanalysis methods, such as linear cryptanalysis and differential cryptanalysis. This study will be made in further works.

Considering that by the similarity degree only a few tests will be necessary to analyze the cryptographic indices of Papílio for each polynomial. But even so, we perform a great number of reliable tests, resulting in the choice of eight polynomials. If we analyze the individual avalanche index round to round of each one of these polynomials we will see than we can already obtain very good indices from the round 6. This allows the thoght to reduce the rounds number, decreasing the execution time of Papílio or yet turning it variable, which would difficult the cryptanalysis and would improve the execution time. Since MV can be applied to any length of block, we also can increase the size of the block which also would improve the execution time or variable in term of the key. Since we have eight good polynomials, we also could apply different polynomials to each round (the choice would be in function of the sub-key and current block) which would not increase considerably the computational effort but would increase considerably the cryptanalysis difficulty, once that for each block in the plaintext (fixing the key) we have 2^{48} possible ways to encoder it (considering 16 rounds). But, considering that each polynomials if we changed the start state we will have four different results, then quantity of possible combinations of functions can arrive to 2^{80} for each block!! which will turn eventually impossible the cryptanalysis without knowing the key, more over considering that the combination of polynomials and start states will change to each block, thus the knowledge of a combination for a block not will help to know the ciphertext. So, Papílio is a very flexible cryptographic algorithm and with very good cryptographic indices.

References

1. Demirbas, K.: Target Tracking in the Presence of Interference, Ph.D. Thesis, University of California, Los Angeles, 1981.
2. Fleming, C.: *A Tutorial on Convolutional Coding with Viterbi Decoding*. Spectrum Applications, july, 1999.
3. Forney, G.D. Jr.: Convolutional codes II: Maximum-Likehood Decoding. *Information and Control*, 25(3)177-179, 1974.
4. Project Gutemberg. http://www.veritel.com/gutenberg/index.html. Access in march 2003.
5. Hopcroft, J.E. and Ullman, I.: *Introduction to automata theory, languages and computation*. Addison-Wesley, 1979.
6. OpenSSL project. http://www.openssl.org/ last modification in april, 17 of 2002. Access in april, 10 of 2003.
7. Pitas, I.: *A Viterbi algorithm for region segmentation and edge detection*. Proc. CAIP89, Leipzig, pp. 129-133, 1989.
8. Ryan, M.S. and Nudd, G.R.: *The Viterbi Algorithm*. Department of Computer Science, University of Warwick, Coventry, CV4 7AL, England, February, 1993.
9. Schneier, B.: Applied Cryptography. 2nd Edition, New York, John Wiley & Sons, inc., 1996.
10. Stalling, W.: *Cryptography and Network Security: Principles and Practice*. 2nd edition. Prentice Hall, 1998.
11. Viterbi, A.J.: Error Bounds for Convolutional Codes and na Asymptotically Optimum Decoding Algorithm. *IEEE Transactions on Information Theory*, April 1967

A Parallel Optical Computer Architecture for Large Database and Knowledge Based Systems

Jong Whoa Na

Computer Engineering Dept., Hansei University,
Kun Po Si, Kyung Gi Do, Korea
jwna@hansei.ac.kr

Abstract. We propose a parallel electro-optical computer architecture that uses optical devices to exploit the data-level parallelism available in the database and knowledge base systems. The massive amount of data of the large database and knowledge base are represented in two-dimensional space. The proposed system performs concurrent pattern matching operations using three-dimensional space. The execution speed is theoretically estimated and is shown to be potentially orders of magnitude faster than current electronic systems.

1 Introduction

The ever-increasing data of the multinational companies such as Citibank and GE now require the performance of the supercomputer[1]. Their business processes typically make use of very large database. What are emerging now are the facilities to make an intelligent decision by using the enormous amount of data. Rule-based Database System (RBMS) is an implementation for the intelligent database[2]. The engine of the rule-based system (RBS) is a resolution engine that requires the pattern matching operations. Since the size of the data is enormous, the pattern matching operation is the bottleneck of Rule-based Systems[3]. Also, CYC project implements the common sense reasoning system by incorporating enormous size of knowledge base[4].

Currently, the computers using the Neumann Architecture are used to solve the most of the symbolic applications as well as the numeric applications. These computers use the silicon hardware that has been the source of the remarkable developments in the computer hardware following the Moore's law. However, these trends may end in the near future as the silicon technology approaches the physical limits. In order to overcome this problem, the researchers around the world have investigated novel computing methodologies such as molecular computing, quantum computing, and optical computing. Among these novel technologies, optical computing is more promising candidate since the optical technologies have been around us as the high performance data communication and networking fields such as optical fiber, WDM optical networking system as well as the imaging applications such as CMOS imaging sensors, projection display devices. In addition, due to its inherent parallelism, large spatial and temporal bandwidth, and low crosstalk, optics can be a potential solution for the communication overhead and synchronization problem exposed by electronic parallel RBS. It has been known that optical interconnection networks offer many advantages over electrical counterparts[5].

In this paper, we propose a high performance Optoelectronic Integrated Circuit (OEIC)-based Expert System (OES) tailored for the high performance RBS implementations. Specifically, the architecture is designed for the hardware for CYC that uses the resolution principle. In addition, we tried to exploit the data-level parallelism available in the resolution operations by using the optical components[6]. The OES uses OEIC for parallel inference engine while electronics for the rest of the system. To take advantage of the OEIC properties, we represent facts and rules in two-dimensional space so that the proposed inference engine performs he three-dimensional processing. The inference engine is logically transformed into an optical interconnection network that can carry out inference optically and hence in a highly parallel passion.

2 Organization of the OEIC Expert System(OES)

The major objective of the OEIC-based Expert System (OES) is the exploitation of maximum parallelism in expert systems. Among many operations in RBS, performing the match operation is the most time-consuming operation. The reason is that the size of interconnection network between every fact and every condition elements grows rapidly as the size of the knowledge base increases. Although the processor technology now reaches 4 GHz clock speed, the speed of the internal bus or interconnection network is relatively slower due to the crosstalk, EMI interference, and power dissipation problems. Optical interconnect can be an alternative to this problem owing to the advantage of optical interconnects such as high speed, EMI immunity, and massively parallel structure. By using optical interconnect instead of electrical interconnect, OES can send as many input facts as possible to the condition elements of the rule.

For OES, we have developed a translation algorithm that translates the knowledge base into the data format suitable for the execution in the OES. OES uses a 2-D array called Condition Table (CT) and an 1-D array called Fact Vector (FV) and one 1-D array called conflict set vector (CSV), as shown in Fig. 1.

A rule consists of a condition part and an action part. In Fig. 1, each row of the condition table (CT) represents a rule and each pixel of the row represents a condition element of a rule. The condition part of a rule consists of conjunction of the condition element. In OES, the *Object-Attribute-Value* tuple of a predicate is evaluated as follows: Each entry of the input fact vector (FV) represents the *Value* of a fact that is to be compared with the *Object* of the condition variable. The *Value*, which was an optical signal, is converted into electrical signal using photodiodes. Then the Value and the *Object* are processed according to the *Attribute* in of the tuple is specified by the predicate. In this way, we can exploit the data-level parallelism.

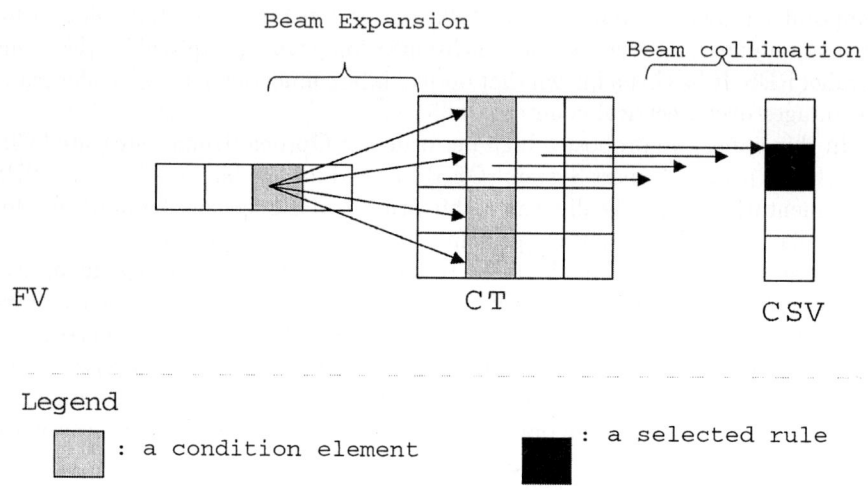

Fig. 1. Knowledge representation and operation principles in OES: FV, CT, and CSV represents fact vector, condition table, and conflict set vector respectively

The evaluation result of each row is converted back into an optical signal via a LASER or LED. These signals form the conflict set vector (CSV) representing the result of the evaluation of the condition elements of a rule in a given microtheory.

In Fig. 1, the *beam expansion* implies sending the source of light representing a fact to multiple destinations representing the condition part of the rule. The light representing the value of a fact is spread (in parallel) into the four condition slots of a rule. Thus, parallel evaluation of the predicate is implemented. Likewise, *beam collimation* implies collecting the results of the evaluation of the condition elements of a rule. By using a collimating optics, the outputs of a rule from the CT are collimated into the CSV.

3 Architecture of OES

3.1 OEIC-Based Match(OM) Module

The FT can be implemented using a two-dimensional VCSEL (Vertical Cavity Surface Emitting Laser) or a LED array. Each pixel of the source array consists of 8-VCSEL or 8-LEDs to represent the 8-bit data representation. The OM module consists of a CT, a beam splitter and two cylindrical lenses. The CT is an OEIC-based smart pixel array where each pixel consists of an 8-photodiodes, an 8-bit register, a comparison logic circuit, and an optical source such as VCSEL or LED. The 8-photodiodes are used to receive the optical input data from the FV. The 8-bit register is used to store the constant part of the condition element of a rule. The contests of the register are compared with the data from the FV at the comparison logic circuit. The comparison logic is designed to perform the magnitude comparison operations such as $=, \neq, >, \geq, <, \leq$. The simulation

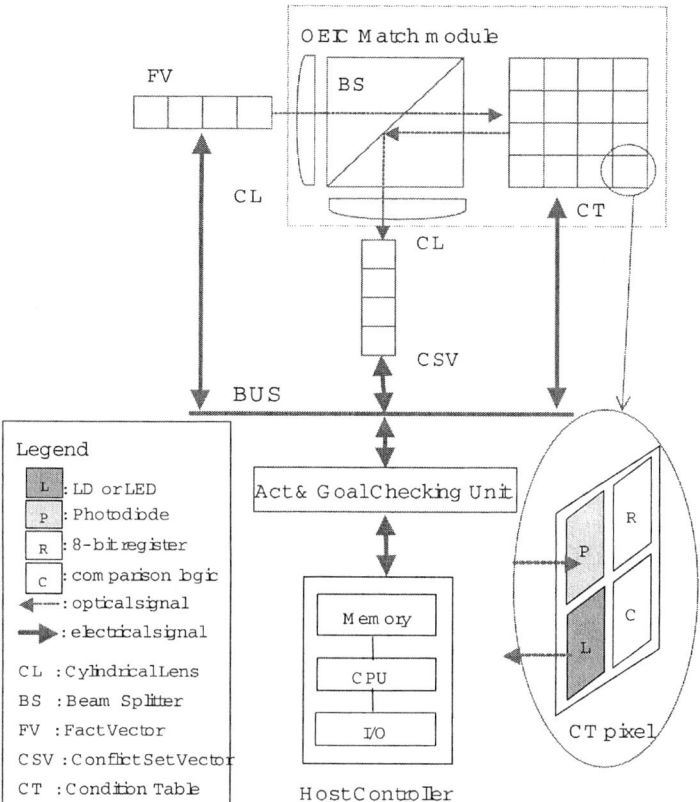

Fig. 2. Architecture of OEIC-based Expert System

model of the comparison logic circuit is shown in Fig. 2. The output of the comparison logic is used to modulate the optical source (VCSEL or LED) of the FV. The optical source represents a condition element of a rule. The optical source is turned on only if the value of the condition element is true. Then the output of the FT is a two-dimensional plane where each pixel represents a Boolean result of the evaluation of a condition element and each row represents a rule.

The cylindrical lens collimates the output of the FT into the one-dimensional CSV (Conflict Set Vector). In CSV, every condition elements of a rule are collimated into a single photodiode. Each pixel of the CSV performs the threshold functions to find out whether the all condition elements are True. If all conditions of a rule are true, the corresponding pixel of the CSV becomes high meaning the rule is selected.

3.2 Action and Goal Checking Module

The electronic AGC module receives the one-dimensional CSV from the OEIC match module to perform a conflict resolution among the selected rules in the conflict set using the conflict resolution strategy. The action part of the chosen rule out of the selected rule is executed and the corresponding variable of the FT are modified. If the fired rule satisfies the goal condition, the AGC stops inference cycle. Otherwise, the OEIC match module initiates next inference cycle.

4 Performance Analysis of OES

4.1 Execution Speed

The execution speed of OES consists of initialization time T_{init}, and cycle time T_{cycle} to complete a single iteration of the main data path. The initialization time includes the setup time for the FT and the CT. Assuming the CT has n^2 elements, the initialization time T_{init} becomes $n^2 \times T_{RegWrite}$, where $T_{RegWrite}$ represents the access time of the registers.

In order to estimate the T_{cycle}, we consider the longest signal path around OES. To complete one cycle, the inference cycle starts with the 8-bit optical sources, 8-bit photodiodes, Comparison Logic, an optical source, a PHOTO-DIODE, and an AGC Logic. Therefore, the overall cycle time T_{cycle} can be calculated as follows:

$$T_{cycle} = 2 \times T_{SRC} + 2 \times T_{PD} + T_{Cmp} + T_{AGC}. \quad (1)$$

Since the LEDs can be modulated at the rate of 800MHz and the SML photodiodes can be modulated at 500MHz, the T_{LED} and the T_{PD} can be modulated at 1.25 ns and 2 ns, respectively. Assuming the current technology allows the Comparison logic and AGC logic circuit to be clocked at $\sim$ 10 ns, the T_{cycle} can be estimated to be 23.25 ns. If we assume that we choose the LED as the optical source, we can easily implement CT of 100×100 smart pixels using the currently available technology. Then, the number of rules that can be evaluated at one second becomes 4.3×10^9 rules/second that exceeds the performance of the rule processing system. Note that the NON-VON is a multiprocessor system that is composed of 32 powerful Large Processing Elements (LPEs) and 16K Small Processing Elements (SPEs), and is estimated to process 850 rules per second[7]. With the processing capability of 6.65×10^5 rules per second, EORBS is expected to achieve a system throughput far better than any electronic RBS can achieve.

5 Conclusion

While rule-based systems are advantageous for solving many AI problems, their slow execution speed has limited their extensive use. As the technology improves, customers expect more intelligence out of the database, which implies the larger size of data/knowledge base. One of the well-known example is the CYC project[8]. Conventional electronic parallel machines contain fast processors, but have limited communication bandwidth, which is critical in parallel processing. Optics can provide the large communication bandwidth as well as fast execution speed required in future RBMS. We have explored the OEIC-based Expert System architecture. The architecture exploits the natural parallelism of database and knowledge base system by using optics and the advantage of optical interconnects. The concurrent rule execution nature of OES can potentially result in a hybrid RBS with significant performance over any existing RBSs.

Acknowledgement

This work was supported by 2004 Hansei University Research Grant.

References

1. Stenstrom, P., et al.: Trends in shard memory multiprocessing. IEEE Computer **Dec.** (1997) 44–50
2. Gottesdiener, E.: Business RULES Show Power, Promise, Application Development Trends. **4** 3 (1997)
3. P. Harmon, P., King, D.: Expert System: Artificial Intelligence in Business. New York: John Wiley and Sons (1985)
4. http://www.cyc.com/cyc/technology/whatiscyc-dir/howdoescycreason
5. Berra, P., Ghafoor, A., Guizani, M., Marcinkowski, S., Mitkas, P.: Optics and supercomputing. Proc. of the IEEE, **77** (1989) 1797–1815
6. Jau, J., Kiamilev, F., Fainman, Y., Lee, S.: Optical expert system based on matrix-algebra formulatio. Applied Optics **27** (1988) 5170–5175
7. Wah, B., Li, G.: Design issues of multiprocessors for artificial intelligence. in Parallel Processing for Supercomputers (K. Hwang eds.), ch. 4, New York: McGraw-Hill (1989) 107–159
8. http://www.cyc.com/cyc/technology/whatiscyc-dir/whatsincyc

Transaction Processing in Partially Replicated Databases

Misook Bae and Buhyun Hwang*

Dept. of Computer Science, Chonnam National University,
300 yongbong dong, kwangju, Republic of Korea
{msbae, bhhwang}@chonnam.chonnam.ac.kr

Abstract. Single master lazy updates propagation methods guarantee weakly consistency for increasing the read availability, but have more chances for update conflicts. We propose an update propagation method based on the balanced tree of replicas in the partially replicated database. The proposed method effectively resolves non-serializabilty resulted from the conflicts of a propagation-transaction and a transaction due to the lazy propagation. To resolve the non-serializable execution, Our method uses the timestamp and the information of the RCTL(the most Recent Committed update-Transaction List) in the status database. We made an experiment on the performance evaluation of our algorithm through the simulation, and proved that it has good performance due to reducing the abort ratio of transactions.

1 Introduction

The replicated system must provide the high availability and preserve the consistency of data. In the single master lazy updates propagation methods that guarantee weakly consistency, the methods can increase the read availability, but have more chances for update conflicts. [1,2,3,4,5] mentioned that the replication consistency can be violated in the lazy master replication database system. The updates propagation methods can increase the workload due to executing the works likewise in all sites that have the replicated data and generate the high abort rate of transactions when the consistency is enforced [3,6].

The weakly consistency scheme allows that a write operation updates the most recent replica, but a read operation can response more stale replica than the most recent replica. It increases the read availability, but has more chances for update conflicts. Because the primary site avoids the update conflicts due to the serializable schedule, but the possibility of conflicts can be increased in other sites except the primary site because the read operations can be executed in all the sites to increase availability. That is, a read only transaction is allowed to read the data in a site even on the way of propagating updates of a committed transaction. In the case that the timestamp of the update transaction is less than the timestamp of the read transaction,

---

* Corresponding author.

the execution in the lazy propagation violates TSO(TimeStamp Ordering) rule in a site when the update transaction arrives after the read transaction has been committed. It is impossible to abort the update transaction because the update transaction already has been committed. We resolve this non-serializability problem and this paper assumes that the updates occur rarely and use a single master technique in which a master copy is updated first. The replications of data are classified into the full replication that replicates all the data and the partial replication that replicates only necessary data according to the degree of replication. The full replication is less efficient and more expensive than the partial replication because it stores all the data at all sites, and includes the data that is not used at some sites by their local user, and uses the system resource to preserve the consistency of all the replicated data. The partial replication saves the replication cost and disk space, and decreases reconciliation time for preserving the consistency of the replication. The partial replication is utilized when a cache is used in mobile environments. The application for the partial replication is expanding now due to the merits as stated above.

2 Transactions for Replicated Databases

2.1 System Model

This depicts the proposed algorithm called CAM-RT (Conflict-Avoidance Method using Replicas Tree). CAM-RT is based on the system architecture like in Fig.1.

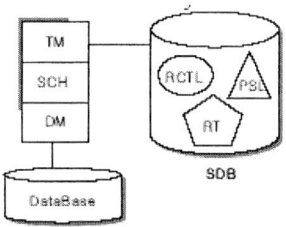

Fig. 1. Architecture of System Model

The architecture is composed of a set of sites connected through a network. Each site consists of a Transaction Manager (TM) and a SCHeduler (SCH). The data are replicated partially. TM assigns a timestamp to each transaction submitted to it. It becomes a coordinator for the transaction. Each status database (SDB) maintains the locations of primary sites (PSL) that execute the update operations and the list of transactions (RCTL) that are most recently committed in the primary site. When the update transaction is committed in the primary site, RCTL is used to inform this update information to its all replica sites before its propagation-transaction arrives at each replica site. After a transaction executes last operation, if there are operations executed in other sites, it executes 2PC protocol to commit it. In case of including read operation in the voting phase of 2PC, if a site received voting result is not primary site, it polls RCTL information to its primary site. RCTL includes the information about the data items that are primary copies, and the identifier and the timestamp of transactions that have been committed most recently in the primary site. Since SDB is frequently updated, each site stores SDB in main memory to decrease the access time. The primary site of each data item maintains replica trees(RT). RT is a tree that consists of sites of the replicas of each data. The primary site of each data is the one being accessed most frequently for the data and becomes the root of RT. The access frequency of a root is higher than that of its subtrees. The sites that have similar access frequency are in the same level in the tree. Each SCH having a primary copy propagates the updates efficiently along the branches of RT. The structure of sites needs to be reconfigured to propagate the updates efficiently using a RT, too [3,7]. Since the updates are propagated along the path of tree, to minimize propagation time, the tree needs to be balanced.

We assume that only a primary site can execute an update operation and propagate the update after the transaction has committed. We also assume that the clock of each site is synchronized to give unique timestamp to each transaction.

2.2 CAM-RT Algorithm

2.2.1 Commit and the Conflict-Avoidance of Transactions

A coordinator of a transaction uses 2PC protocol to commit it. When committing the transaction, if a site that has executed its operations is not a primary site, the site having executed read operations informs its commit decision to its primary site. This produces a serializable history satisfying TSO rule. If there is a transaction that includes write operations, after the transaction has been committed, a trigger to propagate its updates has to be provided.

Our algorithm uses the lazy propagation method. For a data item x, if a propagation transaction PT_i arrives at a site later than a read operation $r_j(x)$ of a transaction T_j and the timestamp of PT_i is less than the timestamp of T_j, it may produce non-serializable execution. To resolve this non-serializability, the transaction that includes read operations must validate its serializability before it commits its execution. To check serializability, our algorithm uses RCTL. RCTL has the list of transactions that have committed most recently in the primary site. It is used to check the conflict of a transaction and a propagation transaction. Though PT_i did not arrive at a site, the TM can check the serializability of them by comparing the timestamp of PT_i with the timestamp of T_j in RCTL. First, it compares $ts(T_j(x))$ with $ts(RCTL(x))$. The $ts(T_j(x))$ means the timestamp of T_j, which reads x, and $ts(RCTL(x))$ means the timestamp of a transaction in RCTL for x. According to the condition of $ts(T_j(x)) > ts(RCTL(x))$ and whether PT_i is arrived or not, it decides to wait for or to commit T_j. For example, for the read operations of T_j, $r_j(x)$, $r_j(y)$, and $r_j(z)$, suppose that $ts(r_j(x)) > ts(RCTL(x))$, $ts(r_j(y)) > ts(RCTL(y))$, and $ts(r_j(z)) > ts(RCTL(z))$. If the timestamp of a transaction T_j is larger than the timestamp of propagation transaction PT_i and PT_i arrived, T_j can commit without delay. If PT_i was not arrived, T_j waits for PT_i until PT_i arrives(arbitrary interval: maximum propagation time from a root to its terminal node in RT), and when PT_i arrives, if $ts(T_j(x)) > ts(RCTL(x))$, PT_i executes and T_j commits. If $ts(T_j(x)) < ts(RCTL(x))$, T_j aborts regardless of the arrival of PT_i according to TSO rule.

2.2.2 Propagation of Updates

When an update transaction commits, the coordinator of the transaction submits its update operations to the corresponding primary sites. The coordinator decides the commit of a transaction using 2PC protocol after the transaction executed its all operations. After completing update operations of the committed transaction at all its primary sites, its propagation transaction is created for each update and is propagated along the its RT. If the coordinator site does not correspond to primary site, it should inform the primary site of the result of decision. The primary site received the commit decision updates its own RCTL information and propagates its update results to its children in the tree after it executes update operations. The timestamp of the committed transaction is assigned to the update operation to be propagated. When the read-only transaction commits, for all data x not in the primary site, it informs the primary site of x of the fact that it has committed.

CAM-RT algorithm is described by pseudo code briefly as follows.

```
begin
When a transaction T is submitted to a site Si;
if (T is a propagation transaction PT)
   perform PT-propagation; /* propagate the updates*/
if (T is a Transaction)
   TM assigns unique timestamp to T;
   perform OP-distribution;/*distribute operations to execute*/
   perform OP-schedule;
 /*schedule operations from scheduling queue according to TSO rule*/
   if (the execution of last operation in T finished)
 /*start 2PC after a coordinator executed the last operation of T */
       perform 2PC-1phase; /*execute 1st-phase of 2PC protocol*/
       perform 2PC-2phase; /*execute 2nd-phase of 2PC protocol*/
end.

/* 2PC-1phase */
begin
if (operations executed in other sites exist)
    a coordinator sends a vote request to all participants;
if (operation executed in the participants == w(x))
    vote YES/NO to coordinator whether executed or not;
if (operation executed in the participants == r(x))
   /*check up of existence of transaction committed while executing*/
       poll RCTL information to the primary site of x ;
       if (ts(T) < ts (RCTL(x))  vote NO to coordinator;
       else
           if ( w_ts(x) == ts(RCTL(x))
                   vote YES to coordinator;   /*PT arrives */
           else {  wait until PT arrives;
                   execute PT as soon as PT arrives;
                   vote YES to coordinator;   }
end;
```

3 Performance Evaluation

To evaluate the performance of the proposed algorithm, we performed simulation experiment using CSIM(C SIMulator) simulator. We use transaction abort ratio and response time as performance measures. The proposed algorithm(CAM-RT) has been compared with existing methods(LHD) while varying three parameters: the number of nodes(sites), the replicas ratio, and the read operation ratio. The unit of response time is milliseconds. The result of simulation is as follows. First, the performance comparison has been done while varying the rate of read operations: 10, 30, 50, 70, and 90%. Read ratio means the rate of read operations to write operations. The result of simulation is shown as in Fig.2. Next, we simulated two algorithms while varying replicas rate; 20%, 40%, 70%, and 100%. The replicas ratio means the number of replicas nodes for the number of total nodes. The result of simulation is shown as in Fig.3. Third, we experimented LHD and CAM-RT while varying the number of nodes;7, 10, and 15. The node means a server or a site. The result of simulation is shown as in Fig.4. The simulation result shows that CAM-RT has better performance than LHD since the transaction abort ratio and the response time are reduced.

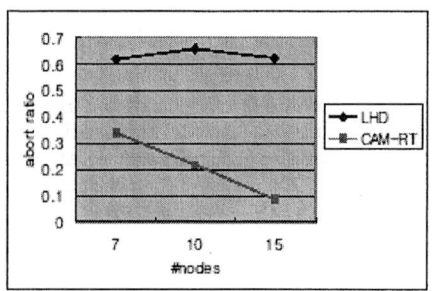 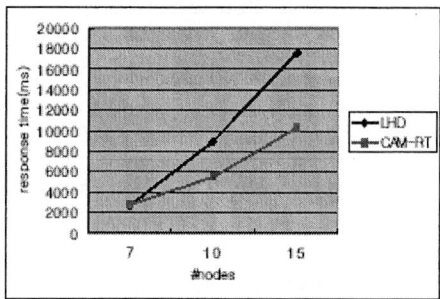

Fig. 2. Abort Ratio and Response Time for the Read Operation Ratio

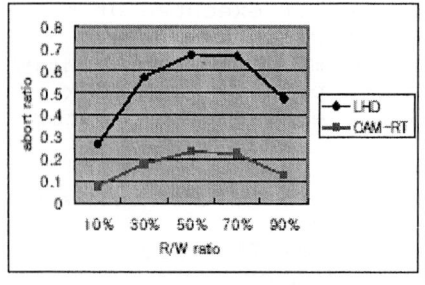 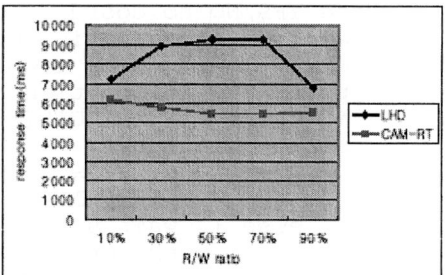

Fig. 3. Abort Ratio and Response Time for the Replicas Ratio(Write-Operation Ratio 0.1)

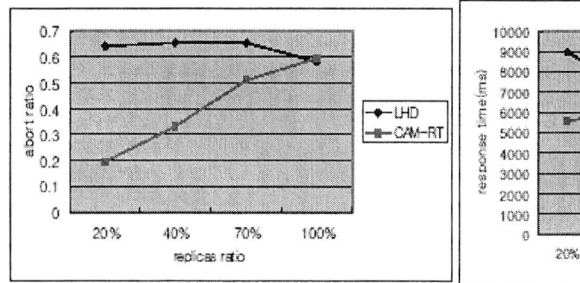 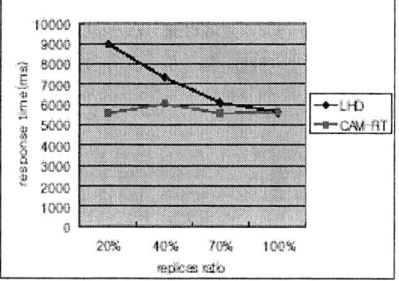

Fig. 4. Abort Ratio and Response Time for the Number of Nodes

4 Conclusion

Recently, there is a tendency that databases are to be scalable and to be partially replicated according to the degree of importance and necessity of data. For preserving consistency in the partially replicated databases, the replication management needs an efficient update propagation method that guarantees serializable execution. This paper proposed the updates propagation method CAM-RT based on the replicas tree of each data. It can be applied to the ubiquitous environment that needs partially replicated databases. The proposed method guarantees the serializable execution of transactions by using timestamp and SDB. It reduces the updates propagation delay since it uses a tree of replicas that needs only to be updated. SDB of each site holds PSL and RCTL.

It is necessary for propagating update operations as fast as possible and for avoiding the conflicts between transactions and lazy propagation transaction. Through the simulation experiment, we compared CAM-RT with a representative method LHD while varying the number of sites, replicas ratio, schedule time, and read operation ratio. We found out performance improvement of our method compared to LHD since our method can reduce the transaction abort ratio and response time.

Acknowledgement

This work was supported by Korea Science and Engineering Foundation (KOSEF #R05-2003-000-10532-0).

References

[1] Jim, G., Pat, H., Patrick, O., Dennis, S.: The Danger of Replication and a Solution. In Procs. of ACM SIGMOD International Conf. on Management of Data, Montreal Canada (1996) 173-182
[2] Todd, A., Yuri, B., Henry, F.K. , Avishai, W.: Replication, consistency and practicality: Are these mutually exclusive?. In Procs. of the ACM SIGMOD International Conf. on Management of Data, Seattle WA (1998) 484-495

[3] Xiangning, L., Abdelsalam, H. and Weimin, D.: Multiview Access Protocols for Large Scale Replication. ACM Transaction on Database Systems Vol.23 No.2. (1998) 158-198
[4] Yuri, B., Henry, F..K.: Replication and Consistency: Being lazy helps sometimes. In Procs. of the ACM SIGACT-SIGMOD-SIGART Symposium on Principles of Database Systems, Tucson Arizona (1997) 173-184
[5] Esther, P., Eric, S.: Update Propagation Strategies to Improve Freshness in Lazy Master Replicated Databases. VLDB Journal (2000) 305-318
[6] Marta, P.M., Ricardo, J.P, B.Kemme, G.Alonso: Scalable Replication in Database Clusters. In Proc. of Distributed Computing Conf. DISC'00 volume LNCS1914, Toledo Spain (2000) 315-329
[7] Theodore, J., Karpjoo, J.: Hierarchical Matrix Timestamps for Scalable Update Propagation. http://w4.lns.cornell.edu/~jeong/index-directory/hmt-paper.ps (1996)

Giving Temporal Order to News Corpus

Hiroshi Uejima[1], Takao Miura[1], and Isamu Shioya[2]

[1] Dept. of Elect. & Elect. Engr., HOSEI University,
3-7-2 KajinoCho, Koganei, Tokyo, 184–8584 Japan
[2] Dept. of Management and Informatics, SANNO University,
1573 Kamikasuya, Isehara, Kanagawa, 259–1197 Japan

Abstract. In this investigation, we propose a new mechanism to give *temporal order* to a news article in a form of timestamps. Here we learn temporal data in advance to extract ordering by means of incremental clustering and then we estimate most likely order to news text. In this work, we examine TDT2 corpus and we show how well our approach works by some experiments.

Keywords: TDT, Stream data, Incremental Clustering, Timestamp Estimation.

1 Motivation and Big Pictures

Recently we have seen many news texts supplied constantly through internet from multiple sources. About some of them we know explicitly when and what kinds of affairs happen, but not about others. This environment drives many researchers to put much attention on grasping their contents easily and quickly. One of the typical approaches is *Topic Detection and Tracking* (TDT)[1,4].

In TDT tasks, data are processed in temporal order and all the data are assumed to carry temporal information explicitly or implicitly. Without any temporal aspects, we can't harmonize any data into consistent states through the tasks. If we can give timestamp to news text with strong confidence, we could capture topics or trends of articles more easily and smoothly so that we could extract closer and deeper relationship among articles.

In this investigation, we discuss how to estimate *timestamp* to a news article n using a collection M of articles where each article of M carries temporal aspects. Intuitively it seems better to look for $m_1 \in M$ which is the most similar to n and then to put the timestamp of m_1 to n. But it is not really nice because the similarity among articles doesn't always mean the similarity among topics. Here we take *event detection and tracking* approach, and estimate timestamp to n based on the event which n belong to. Therefore, our timestamp estimation process is according to following outline. (1)Clustering from (temporal) news texts M into *events*, (2)Assignment the most suitable cluster C(that was created in (1)) to the news n. (3)Looking at C we extract timestamp value and put it to n. From the view point of machine learning, the process above corresponds to *supervised learning* where *training data* means clusters C and the classification problem means timestamp estimation.

In first step of our timestamp estimation, we make clustering M(that are collection of stream data with timestamp) as the materials to predict the timestamp and event(cluster) of n. In TDT activity, it is well-known [8] that temporal clustering works very well with event detection, that is, very often an event corresponds to a temporal cluster[1]. Here we introduce a *Forgetting function*[3] to reflect temporal distance among documents in stream for clustering of M. Moreover, M grows continuously since M is text stream, and we have to think about *incremental clustering*. In this investigation, we discuss a general *batch clustering* algorithm to compare incremental ones. Here we take *Single Pass Clustering*[5, 8] as a incremental method and *k-means* as a batch method.

Second step of our timestamp estimation is assignment of news n to the most suitable cluster created in former step. To estimate timestamp of n based on the event which n mentions, we have to predict the event of n by the cluster assignment. Here we propose two cluster assignment methods, the top 10 method (TOP10) and the *Nearest Neighbor* method (NN).

Finally, we estimate a timestamp of a news article n by result of clustering of collection M and event(cluster) assignment of n, which is the main topic of this paper. Here we give the estimation of the timestamp based on "Similarity of n and some articles m ($m \in M$) that are top 10 articles similar to n and belong to the cluster C that is assigned to n" and "Timestamp distribution of the cluster C". We should examine the timestamp distribution of a cluster because a collection of news articles arise in a short period [1]. This timestamp estimation is called TOP10 method. To evaluate the timestamp estimation based on event, we propose the nearest neighbor method (NN).

In our experiments, we have to evaluate not only timestamp estimation but also clustering of M and event assignment of n since we estimate timestamp to n based on the event which n belong to.

2 Incremental Clustering

In this investigation, we represent a document X as follows: (1)Vector of term weight $\boldsymbol{X} = (t_1, ..., t_n)$, (2)$j$-th value t_j in a document X means *term frequency* $TF(j)$ of the j-th word, (3)Utilizing only *noun* and *proper noun* words for the document expression by `BrillTagger`. *Similarity* of a document X and a cluster C ($sim(\boldsymbol{X}, \boldsymbol{C})$ where the center is V_C is defined by *cosine similarity* of X and V_C.

Forgetting function $w_\lambda(t)$ is defined as $w_\lambda(t) = \lambda^t$ where λ means forgetting speed ($0 \le \lambda \le 1.0$). We define a timestamp $time_C$ of a cluster C as the last timestamp of a document in the cluster C. We denote *current* by $time_{now}$. Then we extend a notion of similarity sim' involving the function λ as follows.

---

[1] Thus, we can say that event detection is deeply related with clustering while topic tracking corresponds to dynamic classification of the clusters.

$$sim'(\boldsymbol{X}, \boldsymbol{C}) = w_\lambda(|time_{now} - time_C|) \times sim(\boldsymbol{X}, \boldsymbol{C}) \qquad (1)$$

In the following, we apply a single pass clustering with sim' to text stream.

3 Assigning Clusters

In this section, we discuss how to assign a cluster (event) to a news text n by clustering result of M as the second step of our timestamp estimation.

- TOP10 method: We select the cluster of n by means of *voting* by top 10 articles similar to n. i.e., we obtain the top 10 articles similar to a n from a collection M and select a cluster C that accounts for the most articles. For example, in a figure 1, the readers see a news document n belongs to C_1 by TOP10 method.
- NN method: We take the most similar document m to a n and select the cluster which m belongs to. For example, in a figure 1, a new document n belongs to C_3 by NN method.

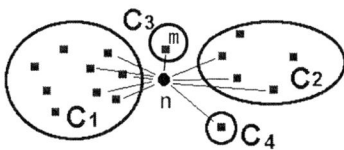

Fig. 1. Selecting a cluster for an article n

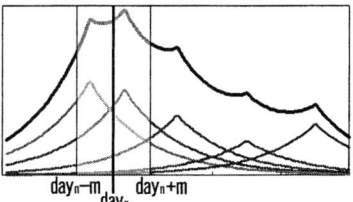

Fig. 2. Timestamp Estimation of 5 articles

4 Estimating Timestamp

In this section, we discuss how to estimate a timestamp of a news article n. By NN method, we put the timestamp of the most similar article to n. For example, in a figure 1, the most similar article m is in C_3, and we estimate the timestamp of n as the one of m. By Top10 method, we look for the top 10 articles similar to n and extract the articles T_C from the 10 articles which belong to the cluster C that is assigned to n. Then we estimate timestamp of n from T_C and n. In a figure 1, among the top 10 articles similar to n, we take 5 articles in a cluster C_1 and estimate the timestamp. Let t_C be an article in T_C. Then we give the estimation of the timestamp by the following rule:

$$TS_{n,t_C,\lambda}(date) = sim(t_C, n) \times distr(C, time_{t_C}) \times w_\lambda(|time_{t_C} - date|) \qquad (2)$$

This equation is illustrated as a curve of timestamp estimation where a function $distr(C, time_{t_C})$ means the number of articles at time $time_{t_C}$ according to a distribution in a cluster C. Here we introduce all the sum to $t_{Ci} \in T_C$ of the timestamp estimation TS obtained by equation 2:

$$TS_{n,T_C,\lambda}(date) = \sum_{t_{ci} \in T_C} TS_{n,t_{Ci},\lambda}(date) \quad (3)$$

For example, in a figure 1, we have 5 articles in T_{C_1} and a figure 2 shows our curve of timestamp estimation.

Here to evaluate the timestamp estimation, we introduce *allowable error* of timestamp which is the maximum difference between true timestamp and estimated. Based on allowable error, we maximize all the sum of difference within the error constraint, and the timestamp value $date_n$ is given by:

$$date_n = \text{MaxArg}_{date} \int_{date-m}^{date+m} TS_{n,T_C,\lambda}(date)$$

For example, in figure 2, given an allowable error m days, we give timestamp ($date_n$) by which we get the maximized sum of timestamp estimation before and after m days.

5 Experimental Results

5.1 Datasets and Protocol

In this experiments, we examine TDT2 corpus. Settings of our experiments are shown below: " Examine news stories in English", " Utilizing transcribed broadcast news stories as *training data* like M, and newswired stories as *test data* like n", "Examining only 4 topics that have are more than 200 articles in both Broadcast and Newswire with confirmation tag YES", "Assuming these news make quick reports and generally we can expect *transaction time* means *valid time*". We process TDT2 data in transaction time order.

Here we assume that k-means results represent situation in reality since we don't have the correct answer of event.

Let $C_1, ..., C_x$ be the results by single pass clustering, $K_1, ..., K_y$[2] be the results by k-means clustering. We want to examine which cluster in $C_1, ..., C_x$ is the most similar to a cluster K_i. To do that, we count the number of articles in both K_i and C_j for $j = 1, .., x$, and select C_j of the maximum. Then we say K_i *corresponds to* C_j. In the same way, given a cluster C, we select a *cluster event* as the most dominant topic in the articles of C.

5.2 Evaluation Criteria

We compare *TOP10* method based on *incremental* and *batch* clustering techniques, and *NN* method with each other (thus 3 cases in total) from the 3 points of views as follows.

---

[2] In our experiment, we give k as the double number of the number of clusters $C_1, ..., C_x$, i.e., $k = 2x$.

- *clustering*(clustering accuracy = β/α): whether incremental approach is really comparable to batch approach and topics or not. where α is the number of the total input articles, and β is the number of articles which are assigned to both K_i and C_j (i,j=1,...) with respect to batch clustering, and it is the number of articles of which topic is same as the one of the cluster assigned from the view point of topics.
- *event assignment*(event accuracy = γ/α): whether clustering results correctly describe events obtained by the refinement of topics of TDT2 or not. where γ is the number of articles of which topic is same as the dominant one of a cluster C assigned with respect to topics, and it is the number of articles of which topic is correctly estimated with respect to batch clustering. When n is assigned to C in incremental clustering and K in batch clustering, we say " topic is correctly estimated".
- *timestamp estimation*(timestamp accuracy = δ/α): whether estimated timestamp falls in the correct date with allowable error or not. where δ is the number of articles of which timestamp is correctly estimated. In this experiments, we give two cases of allowable error (1 week and 1month).

Whenever we apply a single pass clustering algorithm to a collection of broadcasted news articles at a date tm, we give timestamp to all of the newswired articles that have appeared before tm. We examine 3 accuracy in every month.

5.3 Results

First of all let us show clustering accuracy by means of incremental and batch clustering. As the readers see, we have extremely good results here. Note that we say a cluster is *meaningful* if the cluster contains more than 5 articles and we got 21 clusters eventually after the above processes (in 6 months). Thus we give $k = 42$ to $k - means$ clustering.

A table 2 contains the results of event accuracy in incremental and batch approaches. In any aspects, we have excellent results here.

Finally let us show timestamp accuracy in two cases of 1 week allowance and of 1 month allowance (a table 3,4, a figure 3).

Table 1. Clustering Accuracy of TDT2

	Jan.	Feb.	Mar.	Apr.	May	June
k-means	86.16	79.81	77.72	81.98	79.26	88.68
topic	96.89	96.11	96.38	96.55	96.7	97.04

Table 2. Event Accuracy in TDT2

		Jan.	Feb.	Mar.	Apr.	May	June
k-means	TOP10	83.72	85.78	75.95	81.29	81.93	86.03
k-means	NN	80.23	83.74	71.57	75.83	74.62	74.59
topic	TOP10	93.80	96.62	96.53	96.67	97.07	96.81
topic	NN	95.16	96.31	95.90	96.40	96.25	96.69

5.4 Discussion

As shown in a table 1, we got high quality (more than 80%) of clustering accuracy to the comparison with batch approaches. With respect to topics, every test shows more than 96% thus we can say every cluster corresponds to an event.

Table 3. Timestamp accuracy in Allowable Error of 1 Week

		Jan.	Feb.	Mar.	Apr.	May	June
	NN	70.93	57.11	51.91	48.27	44.91	43.59
k-means	TOP10	69.19	59.15	53.96	50.93	49.55	46.6
incremental	TOP10	73.06	60.88	55.30	53.00	49.81	47.38

Table 4. Timestamp accuracy in Allowable Error of 1 Month

		Jan.	Feb.	Mar.	Apr.	May	June
	NN	100	93.72	87.27	80.83	75.13	72.55
k-means	TOP10	100	96.78	90.95	83.69	81.36	77.24
incremental	TOP10	100	98.59	91.09	85.62	82.12	78.33

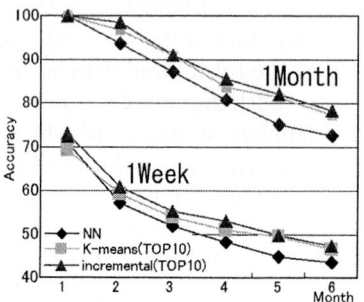

Fig. 3. Timestamp Accuracy

Looking at a table 2, we got good event accuracy, say about 80% by TOP10 and NN methods on incremental clustering compared to the cases of batch clustering. Event accuracy wrt topics also shows almost 95%.

As for timestamp accuracy, we got almost same results by either batch or incremental approach as in figures 3: we got 50% with allowable error 7 days, and more than 70% with allowable error 1 month.

Compared to NN method, TOP10 method is generally superior and suitable for timestamp estimation. In TDT2 corpus, each event depends heavily on temporal aspects, thus the results by TOP10 are not really affected by long allowable error. In fact, in figures 3, the longer error becomes, the bigger difference arises between TOP10 and NN. Let us discuss the correlation of event accuracy and timestamp accuracy. Tables 5 contain the correlation to all the 6 months articles by incremental TOP10 approach. In a case of allowable error of 7 days, we see

Table 5. Incorrect Timestamp

	correct TS	incorrect TS
correct topic	778	830
incorrect topic	9	44

(a) Error 7 days

	correct TS	incorrect TS
correct topic	1262	346
incorrect topic	39	14

(b) Error 30 days

topics are correctly assigned to 1608 (=778+830) articles but timestamps are incorrectly estimated to the half of them. On the other hand, we see topics are incorrectly assigned to 53 (=9+44) articles and timestamps are incorrectly estimated to more than 80% of them. In a case of 30 days we have the similar story, which means timestamp estimation based on event detection works well. In a single pass clustering, we got the best result with a threshold $h = 0.24$ and a forgetfulness function of $\lambda = 0.97$. We have examined several cases and obtained empirically optimal values though we couldn't get any useful characteristics to clustering. This is because TDT2 contains only articles in 6 months and are not enough to analyze them.

6 Conclusion

In this investigation, we have discussed how to estimate timestamp aspects of news articles by clustering news stream in an incremental mode. Our experiments show that we expect about 50% accuracy against 1 week error allowance of TDT2 news corpus.

Acknowledgements

We are very grateful to Prof. Wai Lam in Chinese University of Hong Kong for his sincere help.

References

1. Allan, J., Carbonell, J., Doddington, G., Yamron, J. and Yang, Y.: Topic Detection and Tracking Pilot Study: Final Report, proc. DARPA Broadcast News Transcription and Understanding Workshop (1998)
2. Fukumoto, F. Suzuki, Y. et al.: Detecting Shifts in News Stories for Event Tracking, *IPSJ Journal* 44-7, pp.1766-1777, 2003 (in Japanese)
3. Ishikawa, Y. ,Chen,Y. and Kitagawa, H.: An On-line Document Clustering Method Based on Forgetting Factors, in Proc. 5th European Conference on Research and Advanced Technology for Digital Libraries (ECDL'01),2001
4. National Institute of Standards and Technology (NIST): http://www.nist.gov/speech/tests/tdt/
5. Papka, R. and Allan, J.: On-line new event detection using single-pass clustering, Technical Report UMASS Computer Science Technical Report 98 - 21, Department of Computer Science, University of Massachusetts, 1998
6. Uejima, H., Miura, T. and Shioya, I.: Improving Text Categorization by Synonym and Polysemy, *IEICE Trans.on Info.& Systems* J87-D-I-2, pp.137-144, 2004 (in Japanese)
7. Wayne, C., Doddington,G. et al.: TDT2 Multilanguage Text Version 4.0 LDC2001T57, Philadelphia: Linguistic Data Consortium (LDC), 2001
8. Yang, Y., Pierce, T. and Carbonell,J.: A Study on Retrospective and On-Line Event Detection, Proc. SIGIR-98, ACM Intn'l Conf. on Research and Development in Information Retrieval, 1998

Semantic Role Labeling Using Maximum Entropy*

Kwok Cheung Lan, Kei Shiu Ho, Robert Wing Pong Luk, and Hong Va Leong

Department of Computing, The Hong Kong Polytechnic University, Hong Kong
{cskclan, csksho, csrluk, cshleong}@comp.polyu.edu.hk

Abstract. In this paper, semantic role labeling is addressed. We formulate the problem as a classification task, in which the words of a sentence are assigned to semantic role classes using a classifier. The maximum entropy approach is applied to train the classifier, by using a large real corpus annotated with argument structures.

1 Introduction

The problem of semantic role labeling was first formally studied by Gildea and Jurafsky [1]. They divided the task into two sub-problems: argument identification and role classification. The semantic role classifier was built based on likelihood probabilities, which were maximized by linear interpolation. The inputs to the classifier included syntactic features generated by full parsing. Their work has been extended along various directions. For example, Gildea and Palmer [2] used shallow parsing instead of full parsing for generating the syntactic features. Works have also been done that used other grammar formalisms instead of the traditional phrase-structure grammar, like combinatory categorial grammar [3] and tree adjoining grammar [4], to evaluate how syntactic representation affected classification accuracy. On the other hand, various learning algorithms were applied for training the classifier, such as conditional learning [1], generative learning [5], and discriminative learning [6]. Considerable success has been achieved. Researchers have also attempted to use different types of features, particularly semantic features, to improve the performance of semantic role classification [7].

Existing approaches also differed in the granularity of the elements labeled. In the word-by-word approach [2,6], each individual word is separately classified into one of the semantic role classes. In the constituent-by-constituent approach [1], the words of the input sentence are first grouped into syntactic constituents of varying sizes as identified by full parsing, and a semantic role label is then assigned to each constituent. In the chunk-by-chunk approach [8], syntactic phrases, as generated by shallow parsing, are separately labeled.

---

* The work described in this article was fully supported by a grant from The Hong Kong Polytechnic University (4Z03D).

Previously, Ratnaparkhi [9,10] proposed the Maximum Entropy (ME) approach to natural language processing, viewing the task as a classification problem to assign a class label $a \in A$ to a linguistic context $b \in B$. To tackle the problem, a large data set is employed from which co-occurrences of different combinations of a's and b's are recorded. They are called *observed events*. The classifier is then defined by estimating the probability distribution $p(a,b)$ that is consistent with the observed events. In this way, given a linguistic context $b' \in B$, the probability $p(a|b')$ is estimated for each possible $a \in A$, and b' is assigned to the class a' where $a' = \arg\max_{a \in A} p(a|b')$. In general, more than one probability distribution may be found. Yet, some combinations of a's and b's may appear only a few times in the data set, such that the frequency of occurrence is not statistically significant. In that case, out of the possible p's, the one that maximizes the entropy of p should be selected, based on the Principle of Maximum Entropy [11]. The ME approach has previously been applied to various natural language problems, like part-of-speech tagging [12] and parsing [13].

In this paper, the maximum entropy approach is applied to tackle the semantic role labeling problem, following Ratnaparkhi's formalism closely [9,12,10,13]. A large data set called PropBank [14] is assumed, which contains real sentences annotated with predicate-argument structure. A set of feature functions are designed and co-occurrence statistics are collected from the data set, which are used to approximate a probability distribution for labeling the semantic roles of the words of unseen sentences, based on the word-by-word approach.

2 Modeling Semantic Role Labeling Using ME

2.1 Data Set

Two data sets are commonly employed in semantic role labeling research, namely, FrameNet [15] and PropBank (Proposition Bank) [14]. The strength of FrameNet is that target words are annotated with word senses. Moreover, besides verbs, adjectives and nominal phrases are annotated. The sentences are also shorter, thus avoiding ambiguities. Despite these advantages, we have adopted PropBank in our study for several reasons. First, the sentences in PropBank are longer and more complex. They are thus more suitable for real-world applications. Since PropBank is created from TreeBank [16], it contains richer syntactic information than FrameNet. Moreover, the sentences have a more even coverage of the predicates. So the data is less sparse, which helps classification performance.

However, the formal release of PropBank is not publicly available until recently. As a result, we replicate a subset of PropBank using TreeBank. Since the sentences in TreeBank are represented as phrase structure trees originally, they need to be "flattened" first. Phrase position is then assigned to each word, using a perl script contributed by Tjong and Buchholz [17]. Each word is annotated by two more features: part-of-speech and semantic role. While the part-of-speech label can be directly adopted from TreeBank, the semantic role information has to be derived from the predicate-argument structure of the sentence. The semantic role classes are encoded using the IOB2 format [18]. The total number of semantic role classes is 438. Table 1 shows the top few classes in terms of their occurrence frequencies in the data set. Among the 438 classes, 75 of them (i.e., 17.1%) occur only once in the data set. This causes a sparse data problem to the classifier. Fig. 1 shows an example of an annotated sentence.

We name the resulting data set *Replicated PropBank*, which contains 53,022 annotated sentences. Totally, there are 1,440,594 tokens. So, each sentence consists of about 27 tokens. There are 1,564 unique predicates, with most of them (about 42.8%) occurring fewer than 10 times in the data set. These predicates appear in about 5.03% of the total sentences only. This creates another sparse data problem, making it difficult to generalize on their argument structures.

Table 1. Distribution of the semantic roles in the data set

Semantic role	O	I-ARG1	I-ARG0	B-ARG1	I-ARG1	I-ARGM-ADV	*Others*
Occurrence frequency	743,732	255,041	77,886	44,821	35,516	29,903	253,695
Occurrence percentage	51.63%	17.70%	5.41%	3.11%	2.47%	2.08%	17.61%

Word	An	IBM	spokeswoman	said	the	company	told	customers	Monday	about	the	bugs	and	temporarily	stopped	shipping	the	product	.
POS	DT	NNP	NN	VBD	DT	NN	VBD	NNS	NNP	IN	DT	NNS	CC	RB	VBD	VBG	DT	NN	.
Phrase position	B-NP	I-NP	I-NP	B-VP	B-NP	I-NP	B-VP	B-NP	B-NP	B-PP	B-NP	I-NP	O	B-VP	I-VP	I-VP	B-NP	I-NP	O
Semantic role	O	O	O	O	B-ARG0	I-ARG0	O	O	O	O	O	O	O	B-ARGM-TMP	rel	B-ARG1-PRD	I-ARG1-PRD	I-ARG1-PRD	O

Fig. 1. Example showing how a sentence is annotated

2.2 Encoding Feature Functions

Each sentence in the Replicated PropBank is encoded using the set of features shown below, based on a 5-word sliding window approach [6]:

- Predicate (pred)
- Word (w)
- Part-of-speech (pos)
- Phrase position (p_pos)
- Previous role (r)
- Word position (w_pos)
- Voice (v)
- Path (p)

Consider the sentence shown in Fig. 1. When the sliding window is centered at "customers", the encoded feature vector is as shown in Table 2. The correct semantic role prediction is "O". With reference to the maximum entropy approach, the set of all feature vectors constitute the set of linguistic contexts B while the set of all possible semantic roles form the set of class labels A.

The feature vector, together with the semantic role prediction, form an event. Duplicate events are filtered. The events are then used for generating the binary feature functions f_j's for maximum entropy modeling. In general, for each event, a feature function f_j is defined by pairing one of its 21 features with the semantic

Table 2. Example of encoded feature vector (sliding window centered at "customer")

pred	pred = stopped
w	w_0 = customers w_{-1} = told w_{-2} = company w_{+1} = Monday w_{+2} = about
pos	pos_0 = NNS pos_{-1} = VBD pos_{-2} = NN pos_{+1} = NNP pos_{+2} = IN
p_pos	p_pos_0 = B-NP p_pos_{-1} = B-VP p_pos_{-2} = I-NP p_pos_{+1} = B-NP p_pos_{+2} = B-PP
r	r_{-1} = O r_{-2} = I-ARG0
w_pos	w_pos = before
v	v = active
p	p = NNS → NP → NP → PP → NP → O → VP → VBD

role label (with duplicate feature functions being filtered). For example, given the feature vector in Table 2 and the semantic role label "O", the following feature function is defined:

$$f_j(a,b) = \begin{cases} 1 & : \text{if } a = \text{O \& pos}_0(b) = \text{NNS} \\ 0 & : \text{otherwise} \end{cases} \quad (1)$$

The expectation of f_j is equal to $E_{\tilde{p}}[f_j] = \sum_{i=1}^{N} \tilde{p}(a_i, b_i) f_j(a_i, b_i) = \frac{1}{N} \sum_{i=1}^{N} f_j(a_i, b_i)$, where (a_i, b_i) is the i-th observed event ($\tilde{p}(a_i, b_i)$ being its observed probability of occurrence) and N is the total number of unique events.

2.3 Training the Classifier

Given a set of sentences, feature functions are defined for finding a probability distribution p that satisfies the constraints: $E_p[f_j] = E_{\tilde{p}}[f_j]$ for $j = 1, 2, \ldots, k$ (k being the number of feature functions). In general, more than one p may be found. Based on the Principle of Maximum Entropy [9, 11], the *most appropriate* probability distribution, denoted by p^*, should be the one that maximizes the entropy $H(p)$:

$$H(p) = - \sum_{a \in A, b \in B} p(a,b) \log p(a,b) \quad (2)$$

$$p^* = \arg\max_{p \in P} H(p) \quad (3)$$

Ratnaparkhi [9] showed that p^* should have the following form:

$$p^*(a,b) = \frac{1}{Z(b)} \prod_{j=1}^{k} \alpha_j^{f_j(a,b)} \quad \text{where } 0 < \alpha_j < \infty \quad (4)$$

Here, $Z(b) = \sum_{a \in A} \prod_{j=1}^{k} \alpha_j^{f_j(a,b)}$ is a constant to ensure that $\sum_{a \in A, b \in B} p^*(a,b) = 1$.
The α_j's are called model parameters, with α_j corresponding to the weight for the j-th feature function. They can be found by an iterative algorithm called Generalized Iterative Scaling (GIS) [9].

2.4 Labeling Sentences

After finding p^*, the classifier can be used to assign semantic roles to an unseen sentence. During operation, the words in the sentence are labeled one by one in

```
Let the input sentence be ⟨w₁, w₂, ..., wₙ⟩, where wᵢ are the words;
U ← ∅;
for i = 1 to n do
    if (U = ∅) then
        Encode w₁ to give feature vector b₁;
        U' ← {(⟨a⟩, p*(a|b₁)) | a ∈ A};
    else
        repeat
            Remove an element (S, p*(S)) from U;
            Let aᵢ₋₂ and aᵢ₋₁ be the last two semantic roles of S;
            Encode wᵢ to give feature vector bᵢ (wᵢ₋₂ and wᵢ₋₁ being labeled by aᵢ₋₂ and aᵢ₋₁);
            X ← {(a, p*(a|bᵢ)) | a ∈ A};
            U' ← ∅;
            for j = 1 to N do
                Remove (a', p*(a'|bᵢ)) from X where ∀(a, p*(a|bᵢ)) ∈ X, p*(a'|bᵢ) ≥ p*(a|bᵢ);
                Append a' to S to give S';
                Compute p*(S') ← p*(S) × p*(a'|bᵢ);
                Add (S', p*(S')) to U';
        until (U = ∅);
    for j = 1 to N do
        Remove (S', p*(S')) from U' where ∀(S, p*(S)) ∈ U', p*(S') ≥ p*(S);
        U ← U ∪ {(S', p*(S'))};
Remove (S', p*(S*)) from U where ∀(S, p*(S)) ∈ U, p*(S*) ≥ p*(S);
Output S* as the sequence of semantic roles labeling the words of the sentence;
```

Fig. 2. Beam search algorithm. For a sequence $S = \langle a_1, a_2, \ldots, a_k \rangle$, $p^*(S)$ denotes the joint probability that the word w_i is labeled by a_i, where $i = 1, 2, \ldots, k$

a left-to-right manner. Given a word w, a feature vector b_w is generated, The *most appropriate* semantic role a_w of w can thus be found:

$$a_w = \arg\max_{a \in A} p^*(a|b_w) \qquad (5)$$

However, the feature vector of a word depends on the semantic roles assigned to its preceding words in the sentence. As a result, one is faced with a sequence labeling problem. Beam search is thus employed, following [12], for finding an appropriate sequence of semantic roles for a sentence. The algorithm is shown in Fig. 2, which is characterized by a parameter called beam size, denoted by N. In the experiments (see Section 3), a beam size of 3 was used. Other values were tried also, but no significant impact on performance was noted.

3 Experimental Evaluation

Out of the 53,022 sentences in the replicated PropBank, 50,000 sentences were randomly selected. The data set was shuffled and divided into two sets equally: 25,000 sentences for training and 25,000 sentences for testing. The 25,000 training sentences were first used for finding the target probability distribution p^*, using the GIS algorithm. The number of iterations was set to 100. The resulting classifier was then evaluated, by measuring its performance in assigning semantic roles to unseen sentences in the testing set. With 10-fold cross validation, our system achieved 63.3% precision, 51.4% recall and 56.37% F1-measure, as shown in Table 3. The classification performances for different semantic roles varied. Table 4 shows the top 10 classes in terms of classification accuracy.

Table 3. Evaluation results of our approach

	1	2	3	4	5	6	7	8	9	10	Average
Precision	65.42%	49.74%	64.78%	64.90%	64.75%	64.50%	64.86%	64.61%	64.38%	64.78%	63.27%
Recall	50.54%	64.80%	50.05%	50.00%	49.80%	49.67%	49.84%	49.73%	49.68%	49.91%	51.40%
F1-measure	57.03%	56.28%	56.47%	56.48%	56.30%	56.12%	56.37%	56.20%	56.08%	56.38%	56.37%

Table 4. Classification performances for some of the semantic role classes

Semantic role	ARGM-MOD	ARG2-EXT	ARGM-NEG	ARG0	ARG4-to	ARG3-from	ARGM-EXT	ARG3-on	ARG1	ARG0-by
Precision	87.98%	83.65%	80.28%	75.16%	77.80%	81.85%	82.03%	100.00%	60.32%	59.82%
Recall	73.21%	70.62%	68.81%	64.88%	60.61%	55.92%	49.76%	42.86%	56.15%	54.36%
F1-measure	79.92%	76.58%	74.10%	69.64%	68.14%	66.45%	61.95%	60.00%	58.16%	56.96%

In general, classification performance varies for different number of training sentences used. Thus, several sets of experiments were performed, each using a different number of training sentences. As depicted in Fig. 3, classification performance steadily improves as the number of training sentences is increased.

The effect of the beam size has been studied. A set of 10,000 sentences were randomly selected for training the classifier. Another set of 10,000 sentences were randomly selected from the remaining 40,000 sentences for testing. Classification performance was evaluated using both training and testing sentences. The results in Fig. 4 indicate that the beam size has no significant performance implication.

To further evaluate our system, we compare its performance with other approaches. The results are shown in Table 5. Note that the works by Baldewein et al. [19] and Lim et al. [20], which also applied the maximum entropy approach to the semantic role labeling problem, were only published recently in the CoNLL-2004 shared task (i.e., May 2004). As such, we believe that their works and ours are concurrent works.

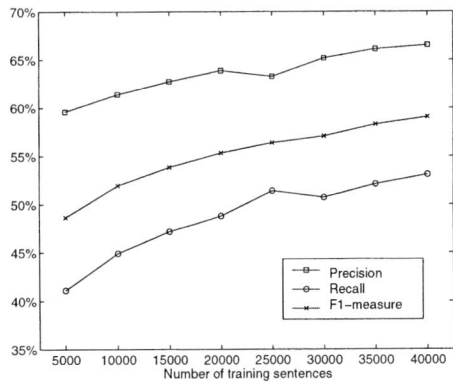

Fig. 3. Performance of our system for different sizes of the training set

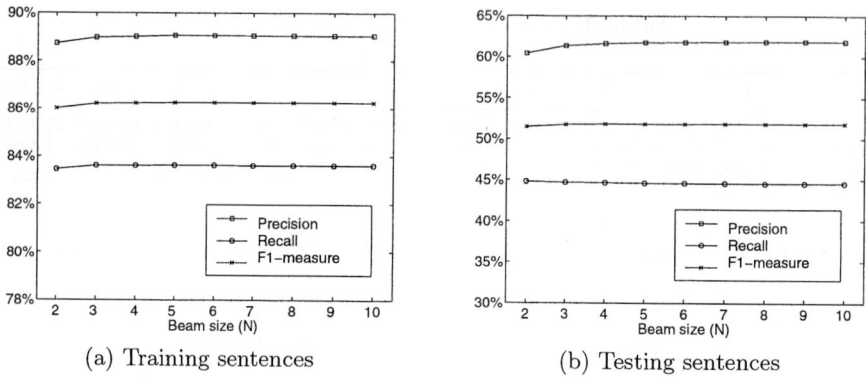

Fig. 4. Effect of the beam size on the performance of the semantic role classifier

Table 5. Performance comparison with other approaches

System	Data set	Method	Granularity	Precision / Recall
Our system	Replicated PropBank	Maximum Entropy	word-by-word	63.3% / 51.4%
Gildea and Palmer's System II [2]	PropBank (Dec. 2001)	Linear Interpolation	constituent-by-constituent	49.5% / 35.5%
Pradhan et al.'s W-by-W Chunker-II [6]	PropBank (Jul. 2002)	Support Vector Machine	word-by-word	66.2% / 54.9%
Baldewein et al.'s system [19]	CoNLL-2004 PropBank (Feb. 2004)	Maximum Entropy	chunk-by-chunk	65.7% / 42.6%
Lim et al.'s system [20]	CoNLL-2004 PropBank (Feb. 2004)	Maximum Entropy	chunk-by-chunk	68.4% / 61.5%

As depicted, the other two works applying the maximum entropy approach outperform ours marginally. We argue that it is mainly due to the larger number of classes involved. In our work, the total number of classes is 438, with a significant proportion of them occurring only once in the data set, causing serious data sparseness problem. In contrast, in the CoNLL-2004 shared task, there were 75 classes only. Hence, the classification task was inherently easier. In addition, their data sets contained richer features, such as clauses and named entities, which are absent from our replicated PropBank. These extra input features can definitely help classification performance. Moreover, they adopted the chunk-by-chunk approach whereby phrases instead of individual words were labeled. But in our approach, each word is separately labeled and a phrase is correctly labeled *only if* all the words it contains are assigned to the correct semantic role classes. Definitely, this makes our task more difficult.

4 Conclusion

In this paper, we have successfully applied the maximum entropy model to build a classifier for semantic role labeling. Preliminary evaluation reveals that it has satisfactory performance. For future work, we will study various ways to improve the system performance, such as incorporating other sources of information to the classifier (e.g., co-reference). Other types of feature functions will be explored as well, to see whether they can boost classification accuracy. We are also working on the application of the semantic role classifier in information extraction.

References

1. Gildea, D., Jurafsky, D.: Automatic labeling of semantic roles. Computat. Linguistics **28** (2002) 245–288
2. Gildea, D., Palmer, M.: The necessity of parsing for predicate argument recognition. In: Proc. ACL-2002, Philadelphia, USA (2002)
3. Gildea, D., Hockenmaier, J.: Identifying semantic roles using combinatory categorial grammar. In: Proc. EMNLP-2003, Sapporo, Japan (2003)
4. Chen, J., Rambow, O.: Use of deep linguistic features for the recognition and labeling of semantic arguments. In: Proc. EMNLP-2003, Sapporo, Japan (2003)
5. Cynthia, A., Levy, R., Christopher, D.: A generative model for semantic role labeling. In: Proc. ECML-2003, Dubrovnik, Croatia (2003)
6. Pradhan, S., Hacioglu, K., Krugler, V., Ward, W., Martin, J., Jurafsky, D.: Support vector learning for semantic argument classification. Technical Report TR-CSLR-2003-03, Center for Spoken Language Research, University of Colorado (2003)
7. Surdeanu, M., Harabagiu, S., Williams, J., Aarseth, P.: Using predicate-argument structures for information extraction. In: Proc. ACL-2003, Sapporo, Japan (2003)
8. Carreras, X., Marquez, L.: Introduction to the CoNLL-2004 shared task: Semantic role labeling. In: Proc. CoNLL-2004, Boston, USA (2004)
9. Ratnaparkhi, A.: Maximum Entropy Models for Natural Language Ambiguity Resolution. PhD thesis, University of Pennsylvania (1996)
10. Ratnaparkhi, A.: A simple introduction to maximum entropy models for natural language processing. Technical Report 97-08, Institute for Research in Cognitive Science, University of Pennsylvania (1997)
11. Jaynes, E.: Information theory and statistical mechanics. Physical Review **106** (1957) 620–630
12. Ratnaparkhi, A.: A maximum entropy model for part-of-speech tagging. In: Proc. EMNLP-1996, Philadelphia, USA (1996)
13. Ratnaparkhi, A.: Learning to parse natural language with maximum entropy models. Mach. Learn. **34** (1999) 157–175
14. Kingsbury, P., Palmer, M.: From TreeBank to PropBank. In: Proc. LREC-2002, Las Palmas, Spain (2002)
15. Baker, C., Fillmore, C., Lowe, J.: The Berkeley FrameNet project. In: Proc. COLING-ACL-1998, Montreal, Canada (1998)
16. Marcus, M., Santorini, B., Marcinkiewicz, M.: Building a large annotated corpus of English: the Penn TreeBank. Computat. Linguistics **19** (1993) 313–330
17. Tjong, K.S.E., Buchholz, S.: Introduction to the CoNLL-2000 shared task: Chunking. In: Proc. CoNLL-2000 and LLL-2000, Lisbon, Portugal (2000)
18. Tjong, K.S.E., Veenstra, J.: Representing text chunks. In: Proc. EACL-1999. (1999)
19. Baldewein, U., Erk, K., Pado, S., Prescher, D.: Semantic role labeling with chunk sequences. In: Proc. CoNLL-2004. (2004)
20. Lim, J.H., Hwang, Y.S., Park, S.Y., Rim, H.C.: Semantic role labeling using maximum entropy model. In: Proc. CoNLL-2004. (2004)

An Instance Learning Approach for Automatic Semantic Annotation

Wang Shu and Chen Enhong

Department of Computer Science and Technology,
University of Science and Technology of China,
Hefei, Anhui, 230027, P.R.China
wangshu@mail.ustc.edu.cn, cheneh@ustc.edu.cn

Abstract. Currently there appear only few practical semantic web applications. The reason is mainly in that a large number of existed web documents contain only machine-unreadable information on which software agent can do nothing. There have been some works devoting to web document annotation manually or semi-automatically to solve this problem. This paper presents an automatic approach for web document annotation based on specific domain ontology. Because complete semantic annotation of web document is still a tough task, we simplify the problem by annotating ontology concept instances on web documents and propose an Ontology Instance Learning (OIL) method to extract instances from structure and free text of web documents. These instances of the ontology concept will be used to annotate web pages in the related domain. Our OIL method exhibits quite good performance in real life web documents as shown in our experiment.

1 Introduction

Semantic Web has shown its usefulness in intelligent information integration by providing a technical means to share and exchange knowledge and information between humans and machines [1]. But the prospect of semantic web applications is still unclear for now. The bottleneck lies in that there exist a large number of web pages containing only machine-unreadable information on which software agents can do nothing. The acquisition of semantic knowledge from web documents is a crucial task. One approach is adding semantic annotation on existed web document. Some works have been done in [2, 3]. S-CREAM is a semi-automatic annotation framework based on knowledge extraction rules, and these rules are learned from a set of manually annotated web documents as training set. This paper will focus on the automatic annotation. Our approach is maintaining an ontology instance base through ontology instance learning (OIL) and adding semantic annotations to web page by matching the concept instances occurred in the pages with those in the instance base. We will focus on the learning method -- OIL (ontology instance learning) considering both the non-grammatical tabular HTML/XML structure and the data-rich free text in web document.

In the following sections we will illustrate our OIL method in detail. Section 2 and 3 will focus on the problem of instance learning from structure and free text. In section 4 we will present experiments to show the instance extraction capability of OIL. Section 5 contains the concluding remarks.

2 Instance Learning from Structure

In our case study, we find that web documents use relational structure patterns to enumerate information related to some specific domain knowledge concepts (see Figure 1). In these structures, we can find instances of related concepts. For example, in the table structure on Figure 2, the concept {TA} has instances as "Jon Bodner" We presuppose that some terms in such list are known as concept instances of our domain specific ontology, we can enrich the instance base of this concept by adding other terms in this list to the instance base. In the following part of this section we will use a frequent tree pattern (FREQT) [6] extraction method to discover instances under such context.

An integral part of lab is learning the Macintosh operating system (System 7.5.3) as well.

In addition, there are some special tools (CD-ROM and scanners) available. There are 10 TAs that teach the lab sections Both the TAs and I have the goal of providing you with high quality instruction and a rich educational experience.

TAs:

Name	Section	Time	Days
Jon Bodner	358	6:10	MW
Nick Leavy	338	3:30	MW
	340	11:00	TR

Fig. 1. A segment of a HTML page

We use the Hepple tagger [7] to transform the name entities to their linguistic taggers, the meanings of these taggers are illustrated in [7]. For example in Figure 2, {Jon Bodner, 358, 6:10, MW} is changed into linguistic tags as {NP, CD, CD, NP} for the frequent tree pattern discovery algorithm.

We have introduced a tree comparing technique based on Hash in [4]. Here it is adopted to simplify the tree comparing process in FREQT, and add linguistic information for every node into the Tree Pattern Finding process.

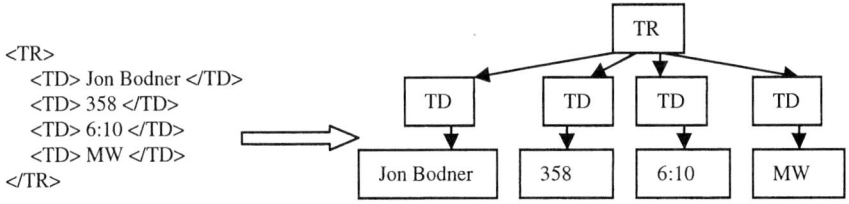

Fig. 2. Finding Frequent Tree Pattern in web document

For every maximum pattern (patterns that do not contain any other patterns), we can find an instance list by which we can enrich our instance base for the corresponding concept.

3 Instance Learning from Free Text

In this section, we will consider how to extract instances from free text in web document. Our approach is based on lexicon-syntactic patterns in form of regular expressions. Generally pattern based approaches are heuristic methods using regular expressions that originally have been successfully applied in the area of information extraction [5]. We observe that the instances are mainly organized in three kinds of patterns; we will design different instance learning methods for each scenario respectively.

3.1 Three Scenarios for Instance Learning in Free Text

Scenario 1: Concept Instance Pattern
One or more instances and their related concept are connected by some specific words. These specific words include "*or other, such as, especially, for example...*" Under such context, if we have known the word representing some certain concept, we can easily find the instances of that concept using predefined lexicon-pattern mapping function. For example:

Fiddle, cello, viola or other stringed instruments ...
Stock market especially NASDAQ, NYSE ...

We can learn that {*fiddle, cello, viola*} are instances of the concept {*stringed instrument*}, {*NASDAQ, NYSE*} are instances of the concept {*stock market*}.
Table 1 defines Concept Instance Pattern to learn instances in such scenario.

Table 1. Concept Instance Pattern

Pattern Name	Lexicon_pattern
Noun Pattern (NP)	{((DT)?(JJ)?(NN(S)?))}
Proper_Noun Pattern1 (PNP1)	{((DT)? (JJ)? (NP(S)?))}
Proper_Noun Pattern2 (PNP2)	{ PNP1 (and|or) PNP1 }
Proper_Noun Pattern3 (PNP3)	{PNP1(,PNP1)? (and|or) PNP1}
Proper_Noun Pattern4	{PNP1(,PNP1)}
Concept_Instance1 (CI1)	{NP such as (PNP1 | PNP2 | PNP3)}
Concept_Instance2 (CI2)	{PNP4 (or | and) other NP1}
Concept_Instance3 (CI3)	{NP (especially | including) (PNP1 | PNP2 | PNP3)}
Concept_Instance4 (CI4)	{such NP as (PNP1 | PNP2 | PNP3)}
Concept_Instance5 (CI5)	{PNP1 (VB|VBD|VBP) NP}

Scenario 2: Juxtaposition Instance Pattern
  Some instances are arranged side by side separated by ",", "*or*" or "*and*". Based on the syntax assumption, these words are instances of the same concept. If one of these

words is known as an instance of a concept *C*, all the other words in the Juxtaposition Instance Pattern might also be instances of concept *C*. For example:

1. ...*Jim, Michael, Susan and Jane* ...
2. ...*French, English, Germany, Chinese*...

If we have known {*Susan*} is an instance of concept {*Student*}, then, we got three more instances of concept {*student*} as {*Jim, Michael, Jane*}. If we have known {*Chinese*} is an instance of concept {*Language*}, then, we learn three more instances of this concept as {*French, English, Germany*}.

We define the following lexicon-syntactic patterns as Juxtaposition Instance Pattern to learn instances in such scenario in Table 2.

Table 2. Juxtaposition Instance Pattern

Pattern Name	Lexicon_pattern
Juxtaposition Instance 1 (JI1)	{NP((,| and) NP)+}
Juxtaposition Instance 2 (JI2)	{PNP1((,| and) PNP1)+}

Scenario 3: *Relation*-word pattern

Definition: relation-word is a verb or verb_proposition phrase which has the most important relation to the instance noun according to its syntax role in a sentence. For example, if the instance is the subject or object of a sentence, the *relation*-word would be the predication verb; if the instance is the object of verb_proposition phrase, the *relation*-word would be the verb_proposition phrase before the instance noun.

According to the different syntax roles of instance noun, we can divide the *relation*-word patterns into following patterns.

Subject_Predication Pattern: instance noun is the subject of a sentence and the *relation*-word is the predication verb in such sentence. Its lexicons-syntactic pattern is {PNP1 VB (D|G)}

Predication_Object Pattern: instance noun is the object of a sentence and the *relation*-word is the predication verb in such sentence. Its lexicons-syntactic pattern is {VB (D|G) PNP1}

Preposition_Object Pattern: instance noun is the object of verb_proposition phrase in a sentence and the *relation*-word is the verb_proposition phrase in such sentence. Its lexicons-syntactic pattern is {VB (D|G)? IN PNP1}

In our approach, first, we will extract the *relation*-word for each existed concept instance; second, according to the *relation*-word pattern, the algorithm use respective pattern matching strategy to learn more instance nouns. A similar method is adopted using ASIUM [5]. In the further step of our approach we will use the verb clustering method to extract instances occurred with the same cluster of verbs, the verb cluster we use is given by ASIUM.

In the following example, we will use Subject_Predication pattern to extract the subjects as concept instances:

1. *Jim travels by car*
2. *Michel travels by train*
3. *Tina drives a saloon car.*

Using this pattern, we can extract the subject {*Jim, Michel, Tina*} and the verb {*travel, drive*}. Supposed that we have known {*Jim*} is an instance of concept {*Student*}, our algorithm will find all the representative predication verbs in all sentences whose subject is {*Jim*}. In this example the predication verb set is {*travel*}. After that, we will find all the subjects in all sentences whose predication verb is {*travel*} and the subject set {*Michel, Jim*}. Now we can extend the instance set of concept {*student*} by adding an instance {*Michel*}. Furthermore, after using verb cluster method in ASIUM, we know that {*travel*} and {*drive*} belong to the same verb cluster set, and get all the subjects in all sentences whose predication verb is {*drive*}. In this example, the subject obtained is {*Tina*}. Thus {*Tina*} can be added as an instance into the instance base of concept {*student*}.

4 Experiments

To evaluate the effectiveness of our algorithm OIL, we designed two experiments on real-life datasets.

The datasets for the first experiment are a collection of HTML pages from the training data in Web_KB project in CMU. We chose the page set which is classified into concept {course} (686 pages totally) to learn the instances on concepts {faculty, person} which include four sub-concepts {Instructor (faculty), Associate Instructor (faculty), Professor (faculty), TA (person)}. We define a simple ontology to describe the structure of faculty in university, some related concept instances are given. In Table 3, we show the results of OIL with given instances, the enriched instances and accuracy. Clearly we can see that considerable instances are learned by HTML/XML structure analysis and free text analysis.

Table 3. Instances of concepts {Instructor, Associate Instructor, TA, Professor}

	Instructor	Associate Instructor	TA	Professor
given instances	20	5	10	10
learned from structure	60	25	0	30
learned from free text	84	8	60	65
accuracy	75.7%	78.8%	61.7%	100%

The datasets chosen for the first experiment are really domain specific. In our second experiment, we will use documents from diverse domains. We adopt Reuters-21578 as our datasets. Reuters-21578 consists of 12344 news articles from all possible domains. The ontology used in this experiment is built from SUMO [9]. But SUMO is really huge so we extract a mini-ontology containing 70 concepts as the ontology used in our experiment. Because documents in Reuters-21578 are data rich free texts without any structure patterns, we only use our Ontology Instance Learning methods from free text.

Among 70 concepts in this ontology, 8 concepts in Table 4 are most frequently occurred. Through this experiment we find that the reasons for the low accuracy of the Relation-word pattern methods are mainly due to the following two factors:

Much complicated grammar structures in real dataset.

The inability to deal with multiword expressions. For instance, "Gulf War" can not be interpreted separately as "Gulf" and "War".

The first problem can be handled by adopting complicated customized lexicon patterns, while the second can be solved by using WordNet [8] which offers synonymous lexicon set and the power to interpret these multiword expressions.

Table 4. Instances of concepts {8 concepts} with accuracy

	Country	Region	Benefit	Institution	Crop	Product	Person	Infrastructure
given instances	3	3	3	3	3	3	3	3
Scenario1	4(100%)	4 (100%)	6 (100%)	4 (100%)	10(100%)	16(100%)	13 (100%)	4(100%)
Scenario2	0	2 (100%)	1 (100%)	2 (100%)	0	3 (100%)	10 (100%)	0
Scenario3	5(45.5%)	4 (30.2%)	9 (55.0%)	4 (83.2%)	6(51.4%)	3(53.7%)	17 (33.3%)	3 (84.5%)

5 Conclusion

Our paper presents an automatic annotation approach for adding semantic annotation on web documents. We describe a method called OIL to extract concept instances from both the non-grammatical tabular HTML/XML structure and the data-rich free text in web document to maintain an instance base for all the concepts in the domain specific ontology.

In our experiment, we find that the instance extraction process on the non-grammatical tabular HTML/XML structure shows quite high performance, because web documents contain a lot of tabular HTML/XML structures to present information. Though the result of instance extraction from data-rich free text is satisfactory, there still exist low accuracy problems in the instances extracting process, partially due to our assumption on the organization of instances in free text and partially due to the obstacle for parsing complicated grammar structures and multiword expressions. We would solve these problems using more powerful NLP techniques with WorldNet in the future.

Acknowledgements

This work was supported by National Natural Science Foundation of China (No.60005004) and the Natural Science Foundation of Anhui Province (No.01042302).

References

1. Tim, B L. James, H. and Ora, L.: The Semantic Web-A new form of Web content that is meaningful to computers will unleash a revolution of new possibilities. Scientific American May 17, 2001.

2. Siegfried, H. Steffen, S. and Alexander, M.: CREAM - creating relationalmetadata with a component-based, ontology-driven annotation framework. In First International Conference on Knowledge Capture
3. Siegfried, H. Steffen, S. and Fabio, C.: S-CREAM Semi-automatic CREAtion of Metadata (2002) 13th International Conference on Knowledge Engineering and Knowledge Management (EKAW02)
4. Wang, Q. Chen, E H. and Wang, S.: Efficient incremental pattern mining from semi-structured dataset. Lecture Notes in Computer Science 3007, Springer 2004.
5. Maedche, A. and Staab, S.: Learning ontologies for the semantic web. In Semantic Web Worshop 2001.
6. Asai, T. Abe, K. Kawasoe, S. and Arimura, H.: Efficient substructure discovery from large semistructured data. In SIAM SDM'02, April 2002.
7. Diana M.: Using a text engineering framework to build an extendable and portable IE-based summarisation system. In Proc. of the A CL Workshop on Text Summarisation, 2002.
8. Miller, G.: WordNet: A lexical database for English. CACM, 38(11):39–41, 1995.
9. Farrar, S. Lewis, W.: A Common Ontology for Linguistic Concepts. In Proceedings of the Knowledge Technologies Conference, 2002.

Interpretable Query Projection Learning*

Yiqiu Han and Wai Lam

Department of Systems Engineering and Engineering Management,
The Chinese University of Hong Kong, Shatin, Hong Kong

Abstract. We propose a novel lazy learning method, called QPL, which attempts to discover useful patterns from training data. The discovered patterns (Query Projections) are customized to the query instance and easily interpretable. As a pattern discovering method, QPL does not require a batch training process and still achieves excellent classification quality. We use some benchmark data sets to evaluate QPL and demonstrate that QPL has a prominent performance and high reliability.

1 Introduction

Suppose we need to predict the class label of an unseen instance, called the *query*. A straightforward learning method is to collect a number of similar known instances to predict the query. This kind of method is described as "instance-based" or "memory-based". For example, the family of k-nearest neighbor (kNN) learning algorithms [5, 2, 3, 6] is one of the most classical and widely adopted learning methods.

In this paper, we propose a novel learning method which can infer useful patterns called *query projections (QPs)* for classification. This learning method is called *Query Projection Learning (QPL)*. A query projection is represented by a set of attribute values, which are shared by the query and some training instances. Given a query and a set of labeled training data, QPL first calculates the maximal QPs shared by the query and different training instances. Then QPL analyzes those QPs and attempts to obtain an appropriate set of QPs as premises for the final prediction, which is made by combining some statistics of the selected QPs.

QPL has several distinct characteristics. First, different from common instance-based learning, QPL has a good interpretability for the class label learning. It can output discovered QPs as the explanation for prediction. Second, unlike many instance-based learning methods such as kNN, QPL does not employ ordinary Euclidean distance metric. The analysis of QPs helps achieve a balance between precision and robustness with a richer hypothesis space. Third,

---

* The work described in this paper was substantially supported by grants from the Research Grant Council of the Hong Kong Special Administrative Region, China (Project Nos: CUHK 4187/01E and CUHK 4179/03E) and CUHK Strategic Grant (No: 4410001).

unlike eager learning algorithms such as decision trees [9] or association rule discovery [7, 8], our method focuses on the local QPs rather than learning a set of rules which form a general classifier. The inferring process is tailored to the query rather than partitioning the whole attribute hyperspace to obtain a global classifier such as a decision tree. Particularly, for real-world problems where the training data need to be frequently updated such as spam email filtering problem, QPL has an advantage of reducing the cost of maintenance and operation.

In Section 2, we introduce the concept of interpretable QPs and how to obtain useful QPs from the query and the training data. We also show how to infer more useful QPs from existing QPs. The detailed algorithm is presented in Section 2.3. The experimental results and discussions on some benchmark data sets are given in Section 3.

2 QPL Framework

2.1 Extracting QPs from Training Data

The learning problem is defined on a class variable C and a finite set $\mathbf{F} = (F_1, \ldots, F_n)$ of discrete random variables, i.e., attributes. Each attribute F_i can take on values from respective domains, denoted by $V(F_i)$. A query instance $\mathbf{t}$ whose class label is to be predicted is represented by the full set of attributes as $\mathbf{t} = (t_1, \ldots, t_n)$ where $t_i \in V(F_i)$. We mainly address discrete-valued attributes in our discussion. To handle continuous-valued attributes, discretization methods [4] can be employed as a preprocessing step. To simplify the discussion without loss of generality, we assume that class variable C is a Boolean variable since a multi-class variable can be broken into a set of binary variables.

Each training instance is transformed into a binary string where each bit is associated with an attribute value. If a training instance shares the same value as the query on a particular attribute, then the corresponding bit is set to "1", otherwise it is set to "0". Thus the query can be represented by a binary string of all "1". Given the query $\mathbf{t}$, a training instance $\mathbf{d} = (d_1, \ldots, d_n)$ becomes a bit string $b_1 b_2 \ldots b_n$ where each bit is defined as an indicator of $d_i = t_i$. After such a transformation, all training instances are mapped into binary strings. Since two or more training instances may have the same binary string, such a binary string can be regarded as a projection of the query. Hence the task of discovering QPs can be accomplished by a series of Boolean operations which can be computed efficiently.

We count the frequency for each binary string and investigate the class distribution among associated training instances. Generally, we prefer the QPs that have a uniform class label and sufficient associated training instances as support. Moreover, useful QPs should have sufficient "1" in its string which means that its associated training instances are very close to the query. Such kind of QPs have a higher utility to help predict the query. The purpose of our learning method is to discover a set of such useful QPs to conduct the prediction reliably.

2.2 Rules of Discovering QPs

Consider two QPs represented by two binary strings, denote by $\mathbf{A} = A_1 A_2 \ldots A_m$ and $\mathbf{B} = B_1 B_2 \ldots B_m$. We define $\mathbf{B}$ is "more-general-than" $\mathbf{A}$ as:

$$\forall i, \quad A_i \wedge B_i = B_i$$

Reversely, we can also say $\mathbf{A}$ "includes" $\mathbf{B}$.

In QPL, two QPs can be compared only when they have this relationship. Given the "more-general-than" or inclusion relationship, QPL attempts to discover an appropriate $\mathbf{u}$ containing set of QPs for learning. We develop the following rules to facilitate the discovery of QPs.

- *Rule of Exclusiveness:* Any two discovered QPs in $\mathbf{u}$ should not have the "more-general-than" relationship. It is obvious that if one QP includes another, then its associated training instances share more characteristics with the query and are more useful in learning. Thus once a particular QP is selected, all QPs included by this QP need not to be considered.
- *Rule of Completeness:* The final set $\mathbf{u}$ of QPs should cover every $t_i \in \mathbf{t}$ to utilize the full attribute information of the query:

$$\forall t_i \in \mathbf{t}, \; (\exists \mathbf{u_i} \in \mathbf{u} \text{ such that } t_i \in \mathbf{u_i}) \qquad (1)$$

In other words, the discovered QPs should offer a complete view for $\mathbf{t}$. Note that in general, the discovered QPs can be overlapping or non-overlapping. This also differentiates QPL from many existing learning methods that are restricted to non-overlapping partitioning over all attributes.
- *Rule of Pruning:* Since any two discovered QPs will be mutually exclusive, whether to select or discard a particular QP triggers a family of QPs to be pruned. Once a QP is discovered, all its children QPs in terms of the subset relationship will be pruned from further search. On the other hand, when a QP is discarded after consideration, all its children QPs will also be pruned from further consideration. When the QP being processed has a large cardinality, the remaining QPs can be reduced significantly, hence the exploration is accelerated.

If a QP is selected according to the above rules, its binary string can be regarded as a premise for an interpretable pattern. Given such a binary string denoted by $b_1 b_2 \ldots b_n$, the corresponding QP can be interpreted as:

$$\text{IF } \wedge_{b_i=1} (F_i = t_i) \quad \text{THEN } C = c$$

where the predictor c is the class label of the training instances satisfying the Boolean expression. Those training instances are supposed to have a uniform class label otherwise their associated QP will not be selected. After a set of such interpretable QPs are discovered for the query, they can be output together with the prediction. These discovered QPs are more interpretable than a list of training instances nearest to the query. They can also be saved as discovered knowledge for future query. When users get the output of prediction from QPL, they can also obtain detailed explanation of the prediction.

2.3 Discovering QPs for Prediction

After the training instances are used to produce a set of QPs denoted by P. QPL attempts to produce a compact set, denoted by M, of useful QPs from the original set P. For small data sets with small attribute dimensions, this process can be done by exhaustive searching. If the data size is large, QPL can discover useful QPs in an efficient way as discussed below.

Since any two discovered QPs cannot include each other, QPL attempts to obtain a compact set of maximal QPs among which no inclusion relationship exists. We denote this intermediate set Q. The processing starts from the QPs in P with the largest cardinalities and moves them into Q. Once a QP with the maximal cardinality is selected into Q, all QPs included by it are also removed from P. This operation continues until no elements were left in P. Then QPL considers the QPs in Q one by one. If any of those QPs have sufficient training instances which have a uniform class label, it will be selected for the final prediction. On the contrary, if a QP with sufficient supporting instances is not capable of producing a reliable prediction, then we remove this QP from Q. Thus we can also ignore all QPs included by it in the next step.

However, in practice, the QPs with the largest cardinality generally have few associated training instances. Thus we may not utilize them directly as useful QPs. If Q has no elements that have both a uniform label and sufficient support, QPL employs an inferring process to discover useful patterns from Q.

The elements of Q are then treated as the seeds for inference and are divided into groups according to the class label. QPL will process these seeds iteratively. At each step, the common subsets of the original seeds are calculated and investigated. If they are not selected into M for final prediction, they will serve as the seeds for the next step. After every group is processed, if M is still empty, then the set Q is directly utilized for final prediction. This may occur when the training data is relatively sparse considering the attribute dimensions.

Finally, we combine the training instances associated with those QPs in M to generate the final prediction for the query. Majority voting is applied among the frequencies summed up from each selected QP. Note that a training instance may contribute to the final prediction for more than once via different QPs. Generally, the more similar a training instance is to the query, the more contribution the training instance can do for the final prediction. The details of QPL is depicted in Figure 1. As shown in this algorithm, the learning process is customized to local QPs for the query. Most steps only involve bitwise Boolean operations on binary strings. Moreover, this algorithm can even infer some new QPs that do not appear in the initial set.

3 Experiments

To evaluate the learning performance of QPL algorithm, we have conducted experiments on 12 benchmark data sets from the UCI repository of machine

```
1    Generate the initial set P of QPs, as discussed in Section 2.1.
2    Initialize three empty QP sets Q, Q', and M.
3    LOOP
4        Move the elements with the maximal cardinality in P to a set Q'.
5        Remove QPs in P that are included by any element of Q'.
6    UNTIL |P| = 0
7    LOOP
8        Q = Q'
9        FOR each element Q_i in Q
10           IF Q_i meets the properties as an interpretable QP,
11               Move Q_i from Q to M.
12       IF |M| = 0
13           FOR every pair (Q_i, Q_j) in Q with the same associated class label
14               IF their common subset C_ij has a uniform class label,
15                   Insert C_ij into Q' and remove Q_i and Q_j from Q' if exist.
16   UNTIL (|M| > 0 or |Q'| = 0)
17   IF |M| = 0
18       Move the elements with the maximal cardinality in Q to M.
19   FOR every class c_i,
20       F_i = 0
21       FOR every elements M_j of M,
22           F_i+ = f_ij where f_ij is the associated frequency of M_j with class c_i.
23   Predict the query t with the class with the maximum F_i.
```

Fig. 1. Pseudo-code of QPL algorithm

learning database [1]. These data sets are collected from different real-world problems in various domains. We partitioned each data set into 10 even portions and then conducted 10-fold cross-validation. In these experiments, we have also investigated the performance of Naive Bayesian, kNN, and Decision Tree (J48) provided by Weka-3-2-6 machine learning software package. All these models used default settings during the entire evaluation process. For kNN, we set $k = 10$.

The results of experiments are depicted in Table 1, showing both the average and the standard deviation of classification accuracy (in percentage) on different data sets. Compared with existing approaches, in most of the data sets, QPL achieves prominent performance. For some data sets such as Labor, Sonar, and Zoo, QPL significantly outperforms all other classical classifiers. The results on the Labor data set also show that QPL excels at handling data sets with poor data quality or incomplete data values. On average, the classification accuracy of QPL on these 12 benchmark data sets is 91.2%, which explicitly shows improvement over other existing models.

Table 1. Classification performance of QPL and other classifiers, measured by classification accuracy (in percentage) and standard deviation of 10-fold cross-validation

Data Set	QPL	kNN	Naive Bayesian	J48
Annealing	97.1±2.53	96.1±1.50	86.5±3.45	98.4±0.78
Breast(W)	96.9±2.57	96.4±3.10	96.0±2.31	95.3±3.69
Credit(A)	84.9±3.09	85.8±3.26	77.7±5.43	86.0±3.81
Glass	72.0±13.43	63.5±7.88	48.5±6.34	67.2±10.08
Heart(C)	82.6±5.40	82.2±8.75	84.5±6.69	79.2±6.75
Iris	95.3±4.12	96.0±5.62	96.0±4.66	95.3±4.50
Labor	96.1±11.67	89.3±12.15	90.0±11.67	78.7±11.46
Letter	95.5±0.94	94.8±0.36	64.2±11.05	87.8±0.75
Mushroom	99.8±0.33	99.9±0.09	95.8±0.79	100±0.00
Sonar	86.3±5.01	73.0±10.19	65.9±12.55	74.1±5.43
Vowel	91.5±2.86	57.2±6.06	61.4±4.32	78.3±4.98
Zoo	96.0±6.09	88.2±6.10	95.2±6.65	92.1±4.18
Average	91.2	85.2	80.1	86

References

1. Blake, C., Keogh, E., and Merz, C.: UCI repository of machine learning databases. http://www.ics.uci.edu/~mlearn/MLRepository.html.
2. Dasarathy, B.: *Nearest neighbor (NN) norms: NN pattern classification techniques.* IEEE Computer Society Press, 1991.
3. Dasarathy, B.: Minimal consistent set (MCS) identification for optimal nearest neighbor decision systems design. *IEEE Transactions on Systems, Man, and Cybernetics*, 24(3):511–517, March 1994.
4. Fayyad, U., and Irani, K.: Multi-interval discretization of continuous-valued attributes as preprocessing for machine learning. In *Proceedings of the 13th International Joint Conference on Artificial Intelligence*, pages 1022–1027, 1993.
5. Friedman, J.: Flexible metric nearest neighbor classification. Technical report, Stanford University, November 1994.
6. Lam, W., and Han, Y.: Automatice textual document categorization based on generalized instance sets and a metamodel. *IEEE Transactions on Pattern Analysis and Machine Intelligence*, 25(5):628–633, 2003.
7. Li, W., Han, J., and Pei, J.: CMAR: Accurate and efficient classification based on multiple class-association rules. In *Proceedings of the IEEE International Conference on Data Mining (ICDM)*, pages 369–376, 2001.
8. Liu, B., Hsu, W., and Ma, Y.: Integrating classification and association rule mining. In *Proceedings of the Fourth International Conference on Knowledge Discovery and Data Mining (KDD)*, pages 80–86, 1998.
9. Mehta, M., Rissanen, J., and Agrawal, R.: MDL-based decision tree pruning. In *Proceedings of the First International Conference on Knowledge Discovery and Data Mining (KDD'95)*, pages 216–221, 1995.

Improvements to Collaborative Filtering Systems

Fu Lee Wang

Department of Computer Science, City University of Hong Kong,
Kowloon Tong, Hong Kong SAR, China
flwang@cityu.edu.hk

Abstract. Recommender systems make suggestions to users. Collaborative filtering techniques make the predictions by using the ratings on items of other users. In this paper, we have studied item-based and user-based collaborative filtering techniques. We identify the shortcomings of current filtering techniques. The performance of recommender systems was deeply affected by user's rating behavior. We propose some improvements to overcome this limitation. User evaluation has been conducted. Experiment results show that the new algorithms improve the performance of recommender systems significantly.

1 Introduction

As the rapid development of internet, the information-overloading problem has become significant. We are overwhelmed by information. As a result, we need some technologies to help us to explore what information is valuable to us. Collaborative filtering is a promising technique to make the recommendations to users [2, 3, 5, 8].

Recommendations are part of our daily life. We usually make decisions based on other users' ratings on the items. Collaborative filtering stores the users' ratings on items in a database. When a user asks the system for recommendation, the system will match the user against the database to discover user's neighbors, which are the people who have similar taste as the user. As the user will probably like the items that his neighbors like, the system ranks the items based on his neighbors' ratings and make recommendations accordingly.

One well-known problem of collaborative filtering is the scalability problem [5, 7]. There may exist a large number of users or a large number of items. As a result, two collaborative filtering techniques have been proposed, namely, user-based collaborative filtering and item-based collaborative filtering. The first filtering technique searches for neighbors among a user population, and then make recommendation based on neighbors' ratings. User-based filtering can avoid the problem of huge number of items. The second filtering technique searches for relationships between items first, and then the system recommends items which are similar to items with a high rating given by the user [7]. Item-based filtering can solve the problem of huge number of users. Both filtering techniques are proved to be useful and practical; therefore, they are widely adopted by recommender systems.

In this paper, we applied two filtering techniques in restaurant recommender system. Hong Kong is known as "Gourmet Paradise", because there are a large number of restaurants offering a wide variety of culinary delights. Some of them retain their exotic flavors, some of them are localized, and some of them even mix different flavors together. It is quite difficult to classify the restaurants into different groups. On the other hand, Hone Kong people go out for dinning frequently. Therefore, restaurant recommender systems are very useful in Hong Kong.

After experiments of the restaurant recommender system have been conducted, we identify the main shortcomings of these techniques. The performance of collaborative filter was deeply affected by the users rating behavior. Some users give their ratings within a narrow range and some users' ratings shift to one side of the scale. We propose new techniques together with standard score that do not suffer from the limitations. Evaluation of the system has been conducted. The experimental results show that new algorithms significantly outperform previously proposed algorithms.

2 User-Based Collaborative Filtering Technique

Collaborative filters help people make choice based on preference of other people. The user-based collaborative filtering technique is based on a simple assumption: predictions for a user should be based the ratings of other users and the similarity between their user profiles [6].

In user-based collaborative filtering technique, a user is matched against the database to find their neighbors by comparing their ratings of items. The correlation r between two users u and v are measured by Pearson correlation coefficient [3], i.e.,

$$r_{u,v} = \frac{\sum_{i \in item(u,v)}(R_{u,i} - \overline{R_u})(R_{v,i} - \overline{R_v})}{\sqrt{\sum_{i \in item(u,v)}(R_{u,i} - \overline{R_u})^2} \sqrt{\sum_{i \in item(u,v)}(R_{v,i} - \overline{R_v})^2}} \quad (1)$$

Where, $R_{u,i}$ is the user u's rating on item i; $R_{v,i}$ is the user v's rating on item i; $\overline{R_u}$ is the mean of ratings on all items given by user u; $\overline{R_v}$ is the mean of ratings on all items given by user v; and $item(u,v)$ is the set of items which are rated by both users u and v.

After the similarities between the users' profiles are obtained, they can be used to compute the prediction on items for user u. A threshold value is chosen filter out dissimilar users, only the users with correction coefficient to user u higher than threshold t are choose as user u's neighbors. The prediction p of user u's rating on item i can be found by the following weighted average of the ratings of those neighbors on the item i [6].

$$p_{u,i} = \overline{R_u} + \frac{\sum_{v \in rater(i) \cap neighbor(u)} r_{u,v}(R_{v,i} - \overline{R_v})}{\sum_{v \in rater(i) \cap neighbor(u)} |r_{u,v}|} \quad (2)$$

Where, the $rater(i)$ is the set of users who have given their rating on item i and $neighbor(u)$ is the set of neighbor of user u.

In the user-based collaborative filtering, the default value of rating on item i is given as the mean of all ratings given by user u. This value is adjusted by the ratings of raters of item i among all the neighbors of user u. However, experimental results show that the user's rating behavior can highly affect the performance of a recommender system. For example, some users give their ratings within a very narrow range, and some users shift their ratings to one side of the scale. The previous filtering technique has addressed the latter problem, because it uses the deviation from the mean instead of the actual value of the ratings to adjust the prediction, and this can solve the problem of dispersion of data.

In the experiment, we found that the predictions of items are close to the mean. In order to solve the problem of dispersion of ratings, we propose to use standard score to replace the deviation in the formula, because standard score is a promising technique in statistics to solve the problem of dispersion of data. The standard score z of data x is measured as the deviation from the mean μ in standard deviations σ:

$$z_x = \frac{x-\mu}{\sigma} \quad (3)$$

The standard score is used to measure how x is compared against the mean of all data, after taking into consideration of the dispersion of data. The standard score allows a comparison of scores drawn from different distributions, and it can eliminate the problem of dispersion of ratings. The improved user-based collaborative filtering together with standard score is given as below:

$$p'_{u,i} = \overline{R_u} + \frac{\sum_{v \in rater(i) \cap neighbor(u)} r_{u,v} \frac{(R_{v,i}-\overline{R_v})}{\sigma_v}}{\sum_{v \in rater(i) \cap neighbor(u)} |r_{u,v}|} \cdot \sigma_u \quad (4)$$

The σ_u and σ_v are the standard deviation of ratings given by users u and v respectively. If no ratings for item i are given by neighbors of user u, the prediction is equivalent to the mean of all the ratings given by user u. Otherwise, this prediction is adjusted by the weighted standard score of all the raters among user u's neighbors.

3 Item-Based Collaborative Filtering

The bottleneck in user-based collaborative filtering is the search for neighbors among a large user population of potential neighbors. Item-based collaborative filtering is proposed to eliminate this problem [7].

In item-based collaborative filtering, the system will first compute the similarity between two items i and j by Pearson correlation coefficient [3], i.e.,

$$r_{i,j} = \frac{\sum_{u \in rater(i,j)} (R_{u,i}-\overline{R_i})(R_{u,j}-\overline{R_j})}{\sqrt{\sum_{u \in rater(i,j)} (R_{u,i}-\overline{R_i})^2} \sqrt{\sum_{u \in rater(i,j)} (R_{u,j}-\overline{R_j})^2}} \quad (5)$$

Where the $rater(i, j)$ is the set of users who have rated both items i and j. The similarities between items are static; therefore it can be preprocessed to save computing time. Similar to filtering of dissimilar users in the user-based collaborative filtering, a threshold value is chosen to filter out dissimilar item pairs. After the similar items are obtained, the system computes the prediction on item i for a user u by computing the weighted sum of ratings given by user u on the items which are similar to item i [7].

$$P_{u,i} = \frac{\sum_{j \in all\ similar\ items} r_{i,j} \times R_{u,j}}{\sum_{j \in all\ similar\ items} |r_{i,j}|} \quad (6)$$

After detail analysis of the item-based collaborative filtering, it is found that the previous formula (Equation 6) is actually equivalent to the following expression.

$$p'_{u,i} = \overline{R_u} + \frac{\sum_{j \in all\ similar\ items} r_{i,j}(R_{u,j} - \overline{R_u})}{\sum_{j \in all\ similar\ items} |r_{i,j}|} \quad (7)$$

However, the above filtering technique will be deeply affected by problem of sparse rating [5]. If there are only a small number of similar items, the prediction of item i will be very close to the mean of ratings given by user u. As a result of sparse rating, the system cannot make recommendations very well because all the items are indifferent, because all of them have a prediction close to the user's mean of ratings. Our improvement was motivated by the idea that if there are only a few similar items, the prediction of item i should be approximated as the mean of ratings on the item i by different users. Therefore, we formulate the item-based collaborative filtering in symmetrical way as the user-based collaborative filtering, i.e.,

$$p'_{u,i} = \overline{R_i} + \frac{\sum_{j \in all\ similar\ items} r_{i,j}(R_{u,j} - \overline{R_j})}{\sum_{j \in all\ similar\ items} |r_{i,j}|} \quad (8)$$

As it might be expected, the item-based collaborative filtering will be affected dispersion of ratings given on items. Taking in to consideration of the dispersion of data, we use the standard score of ratings instead of deviation from the mean to adjust the prediction. The following improved item-based collaborative filter together with standard score is proposed:

$$p'_{u,i} = \overline{R_i} + \frac{\sum_{j \in all\ similar\ items} r_{i,j} \frac{(R_{u,j} - \overline{R_j})}{\sigma_j}}{\sum_{j \in all\ similar\ items} |r_{i,j}|} \cdot \sigma_i \quad (9)$$

Where the σ_i and σ_j are the standard deviation of ratings of all the raters on items i and j respectively. If no similar items are available, the prediction is equivalent to the mean of all the ratings from all raters on item i. Otherwise, this prediction is adjusted by the weighted standard score of all the ratings on similar items by the user u.

4 Experimental Result

The performance of a recommender system can be measured by Mean Absolute Error [1] and the Pearson correlation coefficient [3] between the actual value and the predicted value of item ratings. Both measurements in the experiment have shown that the improved algorithms outperform the pervious algorithms.

In the experiment, 50 subjects are asked to rate 100 selected restaurants with different flavors. If the subject has visited the restaurant before, he need to rate the restaurant on a 1 to 10 scale. For each subject, one quarter of his ratings will be randomly selected and made invisible to the system. The system is then asked to make predictions on those restaurants with invisible rating. The performance of the system can be measured by comparing the pairs of actual value and the predicted value of the ratings for the restaurants with invisible rating.

The Mean Absolute Error (MAE) measures the prediction accuracy of a recommender system [1]. The MAE calculates the mean deviation of the predictions from the actual values by the following formula.

$$MAE = \frac{\sum_{all\ predictions} |actual - predicted|}{number\ of\ predictions} \quad (10)$$

The MAE measurements of four algorithms are shown as Figure 1. The MAE of improved user-based algorithm is less than the previous user-based algorithm by 0.144, which corresponds to 15.1% improvement. The MAE of improved item-based algorithm is less than the previous item-based algorithm by 0.160, which corresponds to 13.5% improvement. To further analyze the results, we have performed the one-tailed paired t-test of the absolute error in improved algorithm against the previous algorithm. The improved user-based algorithm outperforms the previous algorithm at 90% confidence level and the improved item-based algorithm outperforms the previous algorithm at 95% confidence level.

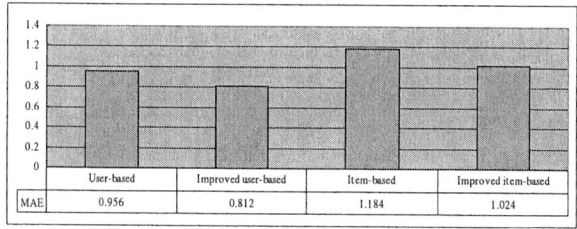

Fig. 1. The MAE between Actual Value and Predicted Value

The performance of a recommender system can be also measured by the Pearson correlation coefficient between the actual value and the predicted value [1]. Table 1 shows the results.

Table 1. The Pearson Correlation Coeffcient between Actual Value and Predicted Value

User-based	Improved user-based	Item-based	Improved item-based
0.8372	0.8564	0.7279	0.7597

In the analysis of Pearson correlation coefficient, it is found that the predicted value is more positively correlated with the actual value in the improved user-based algorithm than the previous user-based algorithm. The similar observation is also made in the item-based algorithm. Combining the result of MAE analysis and the correlation analysis, it is concluded that the improved algorithms outperform the pervious algorithms.

However, if we compare the user-based and item-based algorithm, it is found that previous user-based algorithm outperforms the previous item-based algorithm and the improved user-based algorithm outperforms the improved item-based algorithm. Therefore, the user-based algorithms are more promising techniques in collaborative filtering. Therefore, most efforts are currently devoted into researches of user-based collaborative filtering techniques. As the item-based algorithms can also produce very good results, it is also important to do more researches in this area.

5 Conclusion

As the amount of information available increases exponentially, techniques are required to assist users in finding relevant information. We have reviewed the user-based and item-based collaborative filtering techniques, some improvements have been proposed. We introduce collaborative filtering techniques together with standard score to eliminate the impact of different user's rating behavior. User evaluation has been conducted. The experimental results show that the new filtering techniques outperform the pervious techniques significantly.

References

1. Cayzer S., and Aickelin U. A Recommender System based on the Immune Network, Proc. of the 2002 Congress on Evolutionary Computation, 2002.
2. Hill W., Stead L., Rosenstein M., and Furnas G. Recommending and evaluating choices in a virtual community of use. Proc. on Human Factors in Computing Systems, pp. 194-201, 1995.
3. Kendall M., and Gibbons J., Rank Correlation Methods, fifth ed. New York: Edward Arnold, 1990.
4. Konstan J., Miller B, Maltz D., Herlocker J., Gordon L., and Riedl J. GroupLens: Applying Collaborative Filtering to Usenet News. Communication of ACM 40(3) pp. 77-87, 1997.
5. Lee W. S. Collaborative learning for recommender systems. Proc. Of 18th International Conf. on Machine Learning, pp 314-321. Morgan Kaufmann, San Francisco, CA, 2001.
6. Resnick P., Iacovou N., Suchak M., Bergstorm P., and Riedl J. GroupLens: An Open Architecture for Collaborative Filtering of Netnews. Proc. of ACM 1994.

7. Sarwar B., Karypis G., Konstan J., and Riedl J. Item-based Collaborative Filtering Recommendation Algorithms. Proc. of the 10 International World Wide Web Conference. Hong Kong, 2001.
8. Shardanand U., and Maes P. Social information filtering: algorithms for automating word of mouth. Proc. on Human factors in computing systems, pp. 210-217, 1995.

Looking Up Files in Peer-to-Peer Using Hierarchical Bloom Filters

Kohei Mitsuhashi[1], Takao Miura[1], and Isamu Shioya[2]

[1] Dept. of Elect. & Elect. Engr., HOSEI University,
3-7-2 KajinoCho, Koganei, Tokyo, 184-8584, Japan
[2] Dept. of Management and Informatics, SANNO University,
1573 Kamikamisuya, Isehara, Kanagawa 259-1197, Japan

Abstract. In this investigation, we propose a new mechanism for looking up files in Peer-to-Peer over internet environment so that we can look up files in a location-transparent manner. To do that, we introduce *hierarchical Bloom filters* to local file structures which allow us to query full/part of the structures. We show some experimental system with major features of JXTA where we put attentions on serverless and multi-casting communication, and we examine empirical results on the system.

Keywords: File Look-up, P2P Distributed System, JXTA, Bloom Filter.

1 Managing Data Under P2P Environment

The P2P environment has been targeted for three major application domains, i.e., *data sharing*, *distributed computing*, and *autonomous systems*. By *data sharing* we mean database aspects such as query evaluation techniques, scalability issues and efficient data structuring. For *distributed computing*, we are interested in parallel processing for load-balancing to many computer resources. Non-centralized, non-hierarchical and not pre-determined connections are of our attention. *Autonomous systems* mean that we can join and leave the network any time without any notice of other sites. Here main concern is *durability* of the system.

For data inquery under P2P, we should put our attention on *granularity* for the purpose of data management. For instance, when we keep *record* as a unit, we have to map every *key* in a record to one location. Typical approach is found in *Distributed Hash Table* (DHT) technique [1] where each key item is hashed into IP-based location-space and the record be found there. This is not suitable for our autonomous requirement because we want to keep our own records and migrate them according to our own reasons. The deficiency comes from the granularity and we'd better manage *file* as a unit. Since file is managed by its *name*, we should discuss *file look-up* mechanisms under new environments.

Traditionally several types of file look-up mechanism have been proposed in several P2P systems. The most naive one is the *server* approach where central computers play dominant roles on the jobs and the users come to the servers via ⟨*FileName, Server*⟩ parameters. The second approach is *server hierarchy*. The inqueries are automatically routed to appropriate servers and the users don't care which server they visit. The third approach is found in *peer-to-peer servers* where every server is connected without any governing sites.

Here we take P2P approach paying our attention on efficiency of directory/file look-up and reduction of communication amount. Virtually each query is broadcasted to all the servers thus we might get huge amounts of communication which reduce total throughput of the mechanism. Also we introduce *super peer* sites for controlling purpose, which contradict a notion of *pure* P2P. In an approach based on *DHT to file names*, every file is connected to some site statically. This seems unsuitable for migration purpose where we want to move in and out any files from/to any sites, although many inqueries assume reading files.

We assume each *peer* (local site) works with file look-up using identical principle. That is, every peer may ask other peers of any file look-up requests and should work with the requests for other peers.

In this experimental implementation we discuss two basic ideas, *Bloom filters* for file look-up that is the main issue of this paper, and *packet management based on JXTA* for P2P communication. Here we assume communication activities over wide-spread network but not so heavy such as sensor-net thus JXTA is powerful enough to satisfy the activities.

2 Looking Up Files by Bloom Filters

To look up files in P2P environment, we propose *Bloom filter* which has been devised for distributed database processing[3]. Several investigation with some empirical studies has been made for looking up files so far[4]. In naive queries, all the files are examined thoroughly so that we get mountains of non-relevant information. Usually we assume some techniques to *filter out* desired files. But naive bitmap filters (called *Bloom filter* as described later) are too simple to capture directory structures, although general query processing is quite efficient. Compressed filters are any filters compressed for the purpose of distributed queries among sites, but not useful directly within local sites. Aggregated filters are targeted for summarizing all the files in each local sites but not for (partial) structures.

In this investigation we assume there exists a file management mechanism where all the files are managed hierarchically and we can get to every file by specifying *path* from the top.

Generally we see hierarchical structures to manage files in each site in Windows, Unix, MacOS, IBM Mainframe OS and so on. As described below, we examine several files in our company case. The reader might imagine similar notation as XML/XPath.

(1) `MaterialDept/PartTimeEmp/Address`
describes all the addresses of part time employees in material department.

(2) /Tokyo//Address/
means all the addresses of employees in Tokyo Office. The top /Tokyo describes "Tokyo" appears as the top entry.

(3) /Tokyo/*/Address
means all the addresses in one level lower from the top entry "Tokyo". The character * means a wild card to any file name in one level while // means a wild card to any levels of the structure.

To tackle hierarchical directory structures, basically we take *Bloom filter* approach with both one level and aggregate level filters. Bloom filter is *bit vector* (a sequence of bits), which is obtained by *signature coding*. That is, given a directory and collection of file names in the directory, we calculate each name into a sequence of bits by means of *hash* functions, and we take *logical-or* operation to them. The signature value is called *Local Bloom Filter* (LBF) at the site. Note every subdirectory file has its name and is coded as a part of signature. On the other hand, given a directory, *Global Bloom Filter* (GBF) is generated by taking logical-or to all the descendant LBFs. GBF is recursively defined where a leaf file contains an empty GBF.

Each peer node keeps a pair of LBF and GBF and each peer (site) manages very top level of LBF and GBF. To evaluate any query at a site, the pair is examined efficiently to decide whether the site contains the desired file or not. The mechanism is called *hierarchical Bloom Filter* that we have implement.

Let us exemplify a query C//xyz against hierarchical structure in a figure 1. Let us denote LBF values of C and xyz by BF(C) and BF(xyz) respectively. Also we denote a GBF value and an LBF value of A by GBF(A) and LBF(A) respectively. If GBF(A) contains BF(C) and BF(xyz), this site may contain the partial structure. We ask whether LBF(A) contains BF(C) or not. Since this is the case, we proceed to the next step. GBF(C) contains BF(xyz) but LBF(C) doesn't contain BF(xyz). Thus we examine whether GBF(D) contains BF(xyz) and whether GBF(E) contains BF(xyz). In the former case we have the answer NO, but YES in the latter case. Note there can be no xyz file even if GBF(E) contains BF(xyz), which is called *false positive*. But we can't have the answer YES when a Bloom filter doesn't contain a signature.

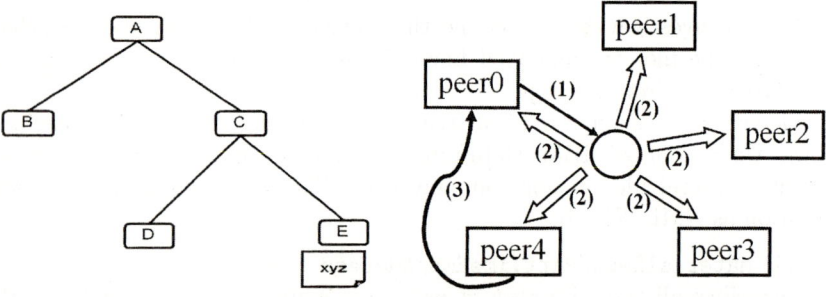

Fig. 1. Hierarchical Bloom Filter   **Fig. 2.** System Architecture

3 System Architecture

To obtain fully P2P system, i.e., serverless, loop-free and multicast communication with low overhead, we utilize the JXTA mechanism[2, 5] as our basic host communication mechanism and discuss how to implement our experimental system. JXTA technology provides a fundamental mechanism called *pipe* by which services and applications are built in a uniform manner. This approach helps us to maintain interoperability among various P2P systems.

In a figure 2, we illustrate architectural framework of communication mechanism in JXTA and the control flows. Rectangles mean Peers, they belong to a PeerGroup that says five peers share one service. A circle means a propagation pipe for host communication. Messages come in from peer0 and they are propagated to all the peers in the same peer-group. On the way back to peer0, peer4 assumes a pointo-to-point pipe to peer0. Thus peer has two ways to hear messages from other sites.

Our basic architecture consists of a P2P communication mechanism using JXTA, GUI-based user interface and distributed file management using hierarchical Bloom filters. As said above, JXTA provides us with multicast communication mechanism (called *propagation pipe*) in P2P mode independent of platforms (operating systems). We design and implement our file management mechanism from scratch on the JXTA communication mechanism.

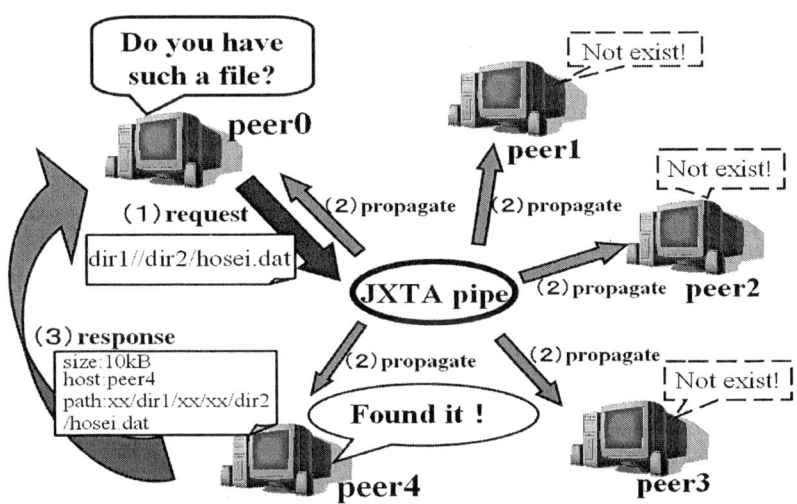

Fig. 3. Whole Flow

In a figure 3, we illustrate architectural framework and the control flows. First one peer called peer0 sends a request message to a common pipe by (1), e.g.,peer0 requests a file which is form of "dir1//dir2/hosei.dat". then the request is propagated as (2). The whole process is managed by means of JXTA multicast communication. Once each peer receives the request message, it works by using hierarchical Bloom filtering technique. When some peer, say, peer4, finds any files to be desired, the peer sends the reply directly to the peer0 as in (3), e.g.,peer4 sends the reply which includes the size of file(10kB), the host name(peer4) and the path from a published top directory (xx/dir1/xx/xx/dir2/hosei.dat). Finally the originating peer peer0 asks peer4 of file transfer. In our experimental system we utilize FTP mechanism by clicking the lines desired.

As said previously, we introduce LBF/GBF mechanism to obtain efficient file look-up. In our implementation, each LBF has 1024 bits length represented as an array of integer. We examine each directory, and we decompose each file name within the directory into four parts, one of which is hashed and *logical-or*ed into 128 bits signature. We decide which part should be used by means of a hash function to the file name. We repeat this operation 8 times ($128 \times 8 = 1024$) and eventually we obtain an LBF. Note that very often we see *similar* names such as XXXX_name.dat and the algorithm can prevent from biased signature values. When there are subdirectories, we generate GBF from their (sub) GBF.

4 Experiments

We have designed and implemented experimental system on three equivalent computers under the environment of Pentium IV 1.8GHz, 256MB memory, 100MB FastEther LAN with FreeBSD 4.6.2 in JAVA SDK 1.3.1. We have examined 77845 files in 800 MB in total without any replica. Here is the summary information (*Files* contain directory).

	Level 1	Level 2	Level 3	Level 4	Level 5	Level 6	Level 7	Level 8	Total
CPU1.Files	31	385	4257	6050	1505	8	0	0	12236
(directory)	24	211	496	147	1	0	0	0	879
CPU2.Files	6	136	1755	14568	17356	5538	148	0	39507
(directory)	6	98	630	1404	432	7	0	0	2577
CPU3.Files	8	130	1063	2276	9559	9953	2750	363	26102
(directory)	8	66	219	722	801	207	14	0	2037
Total Files	45	651	7075	22894	28420	15499	2898	363	77845
(total dir)	38	375	1345	2273	1234	214	14	0	5493

In our system we have randomly selected five files in level 2 and 5 respectively and we give query them as test cases. The files in level 2 and 5 are examined five times in a form of /top/file and dir1/file, /dir2//file denoted by "Level 5 (1)" and "Level 5 (2)" respectively. A table 1 contains all the average results of our queries where BF means Elapsed Time in second for Bloom Filter process, $JXTA$ means Elapsed Time for JXTA Communication, $BF(stdev)$, $JXTA(stdev)$ mean the standard deviation of BF and $JXTA$ respectively. Graphs 4,5 and 6 illustrate the situation.

Looking at the tables and the graphs, we can say that BF values vary but JXTA values stay unchanged for file look-up while the standard deviation on

Table 1. Files in Level 5 and Level 2

Level 5(1)	BF	JXTA	BF(stdev)	JXTA (stdev)
File1	1104.4	214	17.81291666	75.62076434
File2	204.8	193.2	2.588435821	7.395944835
File3	311.6	202.2	1.140175425	16.51363073
File4	200.2	223.4	2.167948339	67.98014416
File5	301	214.8	3	46.55856527
Level 5(2)	BF	JXTA	BF(stdev)	JXTA (stdev)
File1	624.4	260.6	2.792848009	76.34985265
File2	771	196.2	5.099019514	3.346640106
File3	744.2	194.4	3.1144823	5.504543578
File4	770.6	269.4	4.159326869	71.34633838
File5	832	262.2	46.00543446	109.369557
Level 2	BF	JXTA	BF(stdev)	JXTA (stdev)
File1	308.6	313.2	2.701851217	48.03852621
File2	168.4	291.8	1.341640787	43.39585234
File3	304.6	340.4	2.073644135	94.43145662
File4	126.2	352.2	0.447213596	118.9735265
File5	232.6	303.4	11.78134118	38.97178467

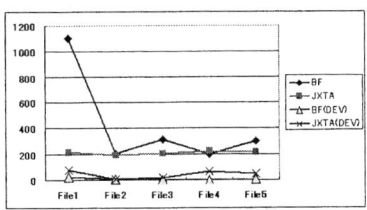

Fig. 4. Files in Level 5(1)

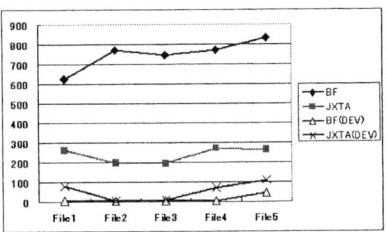

Fig. 5. Files in Level 5(2)

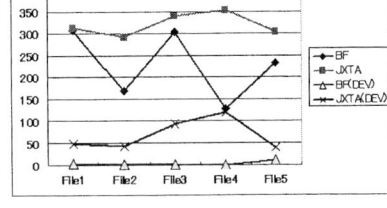

Fig. 6. Files in Level 2

JXTA varies. BF should vary since we examine several files and the results depend on file location, cache buffering or any other situation but the standard deviation values are very small. On the other hand, JXTA may take small amount of time but it varies heavily compared to BF. This may come from some delay of discovery service of pipes.

Finally, to see the feasibility of our system, we have examined whether any peer can join and leave our network or not in our laboratory, and we didn't find any problem.

5 Conclusion

In this investigation, we have proposed P2P file look-up mechanism and implemented experimental system based on JXTA communication package. We have examined file inquery and evaluated the feasibility of P2P availability as well as performance efficiency. By this experiment, we have shown hierarchical Bloom filters play important roles on P2P file look-up. Unfortunately, with current JXTA implementation, very often it takes time for discovery service of pipes thus is not suitable for real-time processing.

References

1. Balakrishnan,H. et al.: Looking Up Data in P2P System, *C.ACM* 46-2, pp.43-48, 2003
2. Gong, Li: Industry Report – JXTA: A Network Programming Environment, IEEE *INTERNET COMPUTING*, 2001 June
3. Kossman, D.: The State of the art in Distributed Query Processing, ACM Comp.Survey 32-4, pp.422-469, 2000
4. Ledlie, J., Serban, L. and Toncheva, D.: Scaling Filename Queries in a Large-Scale Distributed File System, Harvard Univ. TR03-02, 2002
5. Wilson, B.J. et al: JXTA (Voices(New Riders)), Macmillan Computer Pub., 2002

Application of Web Service in Web Mining

Beibei Li and Jiajin Le

Computer Science Department, Donghua University,
200051 Shanghai, China
bb@mail.dhu.edu.cn
lejiajin@dhu.edu.cn

Abstract. To solve the problems we now encounter in web mining, We first propose a new distributed computing strategy—web service. It suggests building a web mining system based on web service, which can share and manage semi-structured data from heterogeneous platforms. Moreover, the system can integrate the mining services and algorithms, improve the efficiency of web mining, and make the mining results easier to access. We also conduct an experiment on selecting useful words to simulate the realization of the web mining system on the Microsoft.NET platform, which demonstrates the importance of Web service in Web mining.

1 Introduction

With the rapid development of online application, the quantities of data stored on the web have increased exponentially since 1990s. As a result of this increase in data quantity, how to discover and obtain useful information faster and more accurate from this vast data resource has become a focus of researchers' interest. Web mining was studied just under this situation, which refers to the procedure of extracting interesting patterns and knowledge from web resources, such as web contents, web structure and web access. It contains three different data mining tasks: Web Contents Mining, Web Structure Mining, and Web Usage Mining [1].

Now the prospects of Web mining technology have attracted the attention of researchers and commercial organizations. They have developed many mining systems and proposed lots of mining algorithms. However, Web mining is a complex procedure and it still faces many problems.

First, as to the Web Contents and Web Structure Mining, most conventional mining algorithms could not get used to the hetero-structured web resources. Because the web resources contain not only structured data in conventional databases, but also non-structured data such as multimedia, sounds, and images. Web mining has extended the objects of data mining from simple and structured data to complicated and semi-structured web data, which hinders those conventional mining algorithms from working effectively.

Secondly, as to the Web Usage Mining, usually the usage records of users are not static. Since some clients use local cache and proxies, the web server could only take the information of those proxies, while missing the users' real information behind them, which prevents catching the behavior of single web user.

Thirdly, Web mining is mostly used by web servers only. So no matter which kind of mining, Web Contents Mining, Web Structure Mining or Web Usage Mining, they all have to be performed by the servers. Therefore, the lack of clear and better design for clients has caused low mining efficiency.

Fourthly, now there is contradiction between the large variety of Web mining algorithms and the specialty of their application fields [2]. The users could use those algorithms provided by certain system only, and never could add, delete or organize any of them, which makes the mining systems quite closed and inflexible.

Lastly, it is difficult for other systems to process the mining results further, because the formats of most mining results are quite closed. And this closure on the output formats is really bad for the improvements and use of those mining results.

So we can think of a new technology nowadays——Web Service. It is a new distributed computing technology and helps solve the problems above.

2 Advantages of Web Service in Web Mining

2.1 The Characters of Web Service

Web service is a new distributed computing model on the Internet [3]. It can be used for coupled network environments and provides an interface, which can be accessed by XML message through the network. This interface defines a group of accessible operations, in which functions can be easily realized just by using this standard interface. Figure 1 shows the related standards and architecture [4].

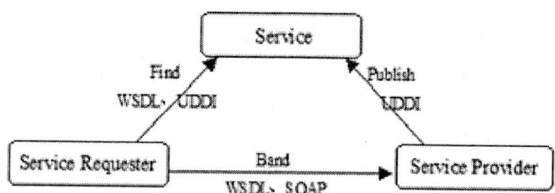

Fig. 1. Web Service Architecture and related standards

Therefore, the key point of Web Service lies in making remote process calling easier and faster by providing a set of Internet standard protocols, such as WSDL, UDDI, SOAP. Meanwhile, it has the compatibility with hetero-structured system platforms, which can eliminate the isolation of information, thus sharing and managing resources over different platforms.

2.2 How to Solve the Problems in Web Mining

From the above discussions and explanation, we can design a new Web mining system. This new system follows all the standard protocols already defined, and integrates various Web mining services based on different kinds of algorithms, operating platforms, and data resources. Figure 2 shows its architecture.

The entire system consists of three layers: the Clients Layer, the Server Layer and the Web Mining Service Providing Layer. The Server Layer is its kernel, which is deployed on the network as the form of web service, finding useful mining services from the Web mining Service Providing Layer according to the clients' particular requests, integrating them with local service, and then providing it to the clients as a whole web service. Obviously, using the web service technology, this new system can easily overcome those problems we encountered in the conventional Web mining process.

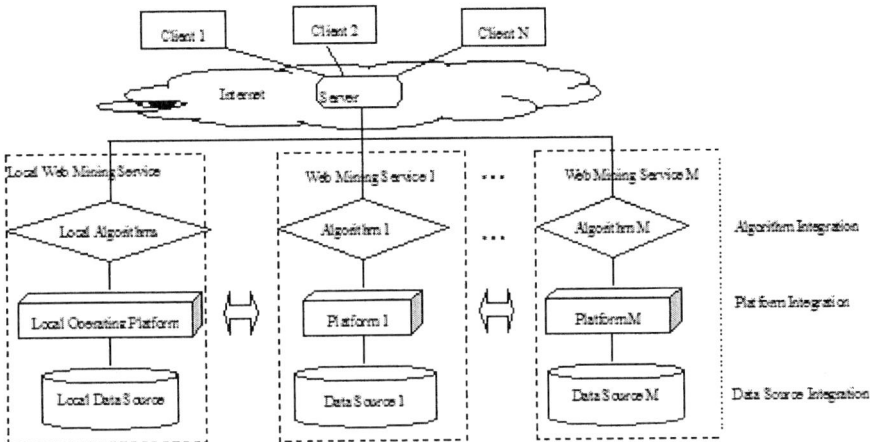

Fig. 2. Architecture of a new Web mining System Based on Web Service

1. Web service technology can solve the problem about sharing and managing the hetero-structured data in Web Contents Mining. Because web service is based on the standard XML language, it can integrate hetero-structured data sources and provide structured descriptions to them, thus making those data from different sources easily bonded together. Therefore, the service requesters could obtain the data directly from the web with the original data formats, rather than allowing the original data formats to get lost in the semi-structured description information with various formats.

2. Web service technology can help clients improve efficiency in the Web Usage Mining. In the proposed new system, the clients can call the mining service interface provided by the server conveniently and directly just like calling its local application procedures, which makes the clients not subordinate and passive any more, but as active as the server. So the Web mining is not confined to the server, and changes from the conventional "One-to-Multeity Mining" to the parallel "Multeity-to- Multeity Mining", which improves the mining efficiency. Meanwhile, by calling the Web mining services, the clients could directly give the feedback of

the local usage pattern analyses to the server, rather than being screened by those proxies or local cache, which makes tracking the behavior of individual users more easily and plays an important role in building personalized recommendation systems.
3. Web Service technology can help strengthen the specialty of the Web mining algorithms. For example, in the Web Structure Mining, the new system can integrate all those Web Structure Mining algorithms that exist in the form of web services, such as HIT, PageRank, improved HIT and etc, and make them form a unified mining service interface. Then the users could select, add, delete or optimize the algorithms according to their own mining tasks, which forms a custom-made mining developing platform and strengthen the specialty of Web mining.
4. Web Service technology can also deal with the results of Web mining effectively. Since the mining results are described and output by the XML language at last, the contents and formats of them are totally separated. The users can figure out the results easily and correctly, use various analyzing tools to visualize them, or make further use of them. Thus this system realizes the sharing of Web mining results over different kinds of systems and platforms.

3 Experiments

3.1 The Implementation of a Simulation System

Web service develops very fast nowadays, and many companies have offered various sorts of platforms to design, develop and deploy this new distributed web application. Such as Sun One, IBM's WebSphere, and Microsoft's .NET. Now we select the .Net platform from Microsoft to develop the Web mining simulation system.

First, we deploy a local web service named "StartWithA" on the local server, which is used to choose the words starting with "A" in certain text. At the same time, we deploy a remote web service named "EndWithD" on another server, which is used to choose the words ending with "D" in certain text. Finally we add its web reference to the local web service.

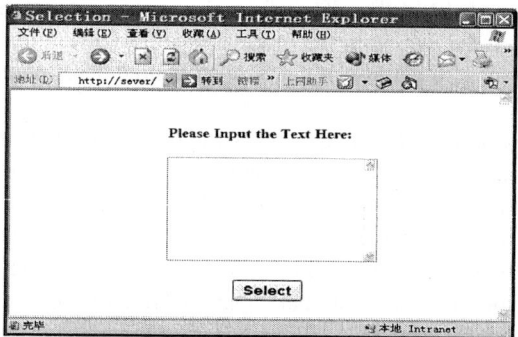

Fig. 3. The Interface of the simulation system

Then we can use those two integrated services through the local server at any client machine, only by adding the web reference of the local web service. The working interface of this simulation system is shown in Figure 3.

We input all the text into the textbox and click the "Select" button. Then the system can automatically find those two web services we already deployed and call them rapidly. Finally it returns the result we want——all the three words, "and", "aloud" and "around", in the text, which both start with "A" and end with "D". Figure 4 shows the result page.

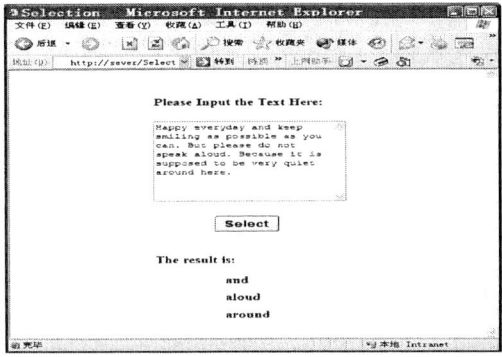

Fig. 4. The result of selection

3.2 The results

Through the process of developing the simulation system we could easily find that when dealing with the real Web mining work, the Web Service technology can not only help the clients by local service, but also integrate a large variety of remote mining services by local server just as fast and convenient as calling local procedure. Meanwhile, by using Web Service as the interface between server and clients, the users could easily write the clients software by themselves, or even integrate all the function provided by this entire system into another system as its subsystem, as long as they conform to the rules when calling the Web Service of this system.

Besides, thanks to the B/S (Browser/Server) structure based on the Web Service, there is no need for the clients to worry about the realization of the mining algorithms, the working platforms, the program languages, or the formats of data description. All the service details are transparent to them and they only need to know the address of the local server. The clients get the requests of the users, submit them to the server, and receive the resulting documents returned from the server. Then the users can choose to see the results in a visualized form.

4 Related Work

Now there are already some Web mining systems based on Web Service. Such as the Web Service systems based on database—Amazon and Google. Both of them are better in the Recall and Precision than some other systems at the same time like Yahoo. Besides, some other network companies have also begun to create certain new services for developers, which can allow the developers access the databases of their companies directly without going through certain application procedures. However, all these services are just based on the elementary data sharing and have not gone deep enough into the data processing and data mining.

The TETRIS system, developed by the Fudan University [5], has made great efforts in the open character based on the service integration, from the whole architecture to mining language. It has provided a flexible platform for developing and applying Web mining. But at present this system is not mature enough in practice, and still needs some improvements.

5 Conclusions

Web data is so complex and enormous, there are many problems in the Web mining process. We propose the Web Service technology and suggest building a new Web mining system based on it, which can share and manage those semi-structured data from different kinds of platforms, strengthen the specialty of mining algorithms, improve the mining efficiency, and eliminate the closed character of mining results.

With the optimization and maturity of Web Service technology, it will surely be used in the Web mining much better. And the various web resources will finally become our treasure, thus making our lives fresher and more vigorous.

References

1. Yun, S., Han, L., Dong, J., Ceh, D.: KDW Survey: Web-based Data Mining. Computer Engineering. 29 (2003) 284-286
2. Zhang, S., Wang, C., Zhou, M., Zou, Y., Wang, W., Shi, B.: TETRIS: An Open Data Mining Application Development Platform Based on Service Integration. Computer Science. 30 (Spp.) (2003) 264-266
3. Zhang, Y., Tang, Y.: The Analysis of Distributed Computing Techniques and Web Service. Modern Computer, 1 (2004) 42-45
4. Kreger, H.: Web Services Conceptual Architecture. http://www-900.ibm.com/developerWorks/cn/webservices/ws-wsca/index.shtml, May (2001)

A Collaborative Work Framework for Joined-Up E-Government Web Services*

Liuming Lu, Guojin Zhu, and Jiaxun Chen

Network & Database Laboratory, Dept. Computer Science,
Donghua University, Shanghai 200051, P.R. China
lmlu@mail.dhu.edu.cn
{gjzhu, jxchen}@dhu.edu.cn

Abstract. One of the main characteristics of e-government is the collaborative work. By the collaborative work we mean that differently functional departments in the government should cooperate with one another in order to accomplish an integrated service required by the citizens. Each functional department works as a part of the integrated service. A web service-based architecture together with a data model of e-documents is proposed to meet the requirements for the collaborative work in e-government. A scenario is given to demonstrate how the suggested architecture works.

Keywords: e-government, collaborative work, web service.

1 Introduction

Today's official business process system lacks the concept of collaboration. By collaboration we mean that differently functional departments in the government should cooperate with one another in order to accomplish an integrated service required by the citizens. Each functional department works as a part of the integrated service. Though a unit of operation in a functional department can be implemented in a service terminal, it sometimes requires cooperation among the departments if a business process involves several departments.

In order to meet the requirements for the e-government collaborative work, we propose a collaborative work framework for join-up e-government web services. We firstly wrap the applications in the departments in modular web services. Adopting web services in e-government enables value-added services by defining services to standardize the description, discovery, and invocation of the applications in the departments. Then, we develop the techniques of integrating web services according to required rules to implement the governmental business process. Finally a data model of e-documents for the collaborative work in

---

* This work is partially supported by a grant from Chinese National Natural Science Fund (#60273051) and Shanghai Science and Technology Development Fund (#03DZ05015).

e-government is designed. Data standards according to the data model about the definition and description of governmental documents, the business process and application security should be provided for the realization of the interoperability.

2 Organizing E-Government Web Services

As Figure 1 shows, basic interactions among e-government web services involve three types of participants: provider, registry, and consumer [1, 2]. Providers include different departments or agencies in the government. The providers publish descriptions of their services, such as operations and network locations in a registry. Consumers, the entities wanting to acquire web services, access the registry to locate services of interest, and the registry returns relevant descriptions, which consumers use to invoke corresponding Web services.

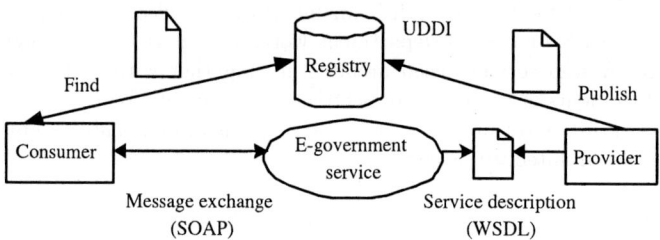

Fig. 1. Interactions among e-government web services

Providers describe the operational features of Web services in the Web Services Description Language (WSDL). Each operation has one of four possible modes:

- One-way, in which the service receives a message.
- Notification, in which the service sends a message.
- Request-response, in which the service receives a message and sends a correlated message.
- Solicit-response, in which the service sends a message and receives a correlated message.

For example, in the collaborative work system of the electronic document transaction a web service offers a request-response operation called gernerateRule. This operation receives a message that includes the document to be processed and returns a message that contains the rule to process the document, such as the involved offices and their executing sequence.

WSDL descriptions are stored in a registry based on Universal Description, Discovery, and Integration (UDDI). The registration of web service includes the URL for communicating with this service and a pointer to its WSDL description.

3 A Collaborative Work Framework in E-Government

Dynamic service composition is the process of creating a new integrated service at runtime from a set of web services. This process includes activities that must take place before the actual composition such as locating and selecting service components that will take part in the composition, and activities that must take place after the composition such as registering the new service with a service registry. The main responsibility of the collaborative work system is to facilitate dynamic service composition. The execution of dynamic service composition is centralized in some systems [3]. At the same time the novel techniques involving peer-to-peer execution of services is employed in several systems such as SELF-SERV [4] etc. Although peer-to-peer can lower cost of ownership and cost sharing by eliminating and distributing the maintenance costs and provide anonymity/privacy, it also raises some concerns about security and management. In order to facilitate the management in e-government, we employ the centralized execution mode. Figure 2 shows the architecture of the collaborative work system, which is described as follows in details.

- The center server for the e-government collaborative work.
  The center controller is to manage the running of the collaborative work system. Requester handler is used to receive the request and involved data from the citizens. And the e-document generator is responsible for converting the data from the citizens into the standard data and generating the corre-

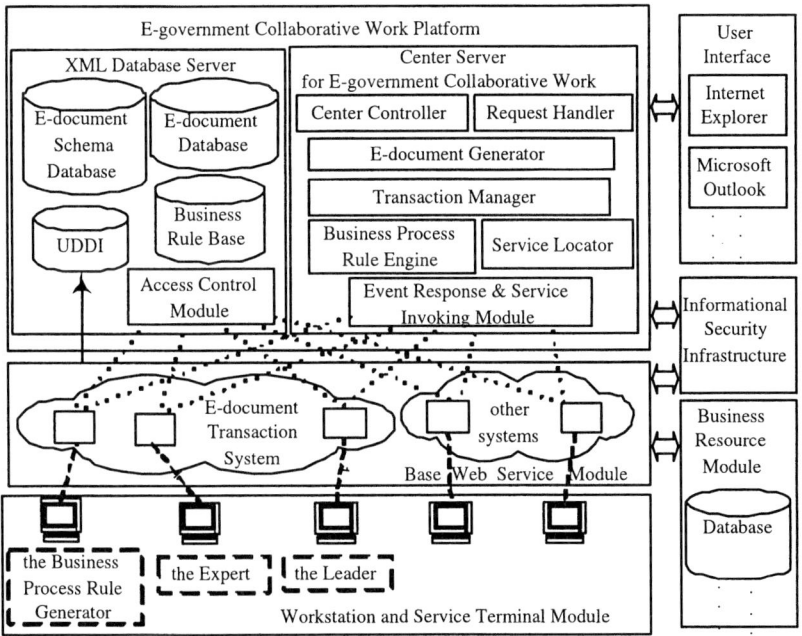

Fig. 2. The architecture of the collaborative work system

sponding e-document according to the schema from the e-document schema database. The service locator can be used to look up WSDL descriptions in the UDDI registry for the rule engine. The main responsibility of the rule engine is to execute the rule. The execution contains the receipt of the event to activate the involved rule and the execution of actions that are defined in the rule. The event response module is responsible for abstracting the significant event, encapsulating the event and submitting it to the rule engine. The service invoking module is used to invoke the corresponding web service based on the action from the rule engine. The transaction manager is to guarantee the ACID (atomic, consistent, durable, isolated) features in the execution of the transactions.
- XML database server.

The XML database server contains an e-document database, an e-document schema database, a UDDI registry, a business rule base, and an access control module. They are described as follows.

1. The UDDI Registry is mainly used to register the collaborated resource such as web services provided by different departments so as to facilitate the service locator to query automatically the required services and the definition of their access interfaces.
2. The e-document schema database is used to store the e-document schemas that are defined in the data standard for the e-government collaborative work. In order to meet the requirements for the collaborative work, the data model in e-documents contains data for business unit requirements, data for controlling the collaborative work, data for ensuring the application security and other data. Data for business unit requirement is provided for business units to implement their corresponding functions. Data for controlling the collaborative work is used in the collaborative work module that contains the components used to integrate the business units, such as the business process rule engine etc. Data for ensuring the application security is used in the application security module whose responsibility is to provide the informational security infrastructure. Other data can include data for index and storage, which facilitates the storage and query of involved data.
3. The e-document database is provided to store e-documents that are instances of the e-document schemas.
4. The access control module is to control the access to the e-document database according to the policy specified in the e-document.
5. The business rule base can store every kind of collaborative rules. Data for controlling the collaborative work in e-documents can be acquired from the business rule base or through the business rule generator, which can also be web service to generate rules.

- The informational security infrastructure.
The informational security stated here includes identity authentication, data encryption, digital signature, access control etc. The certificate register server,

the certificate lookup server and authorization center constitute the infrastructure of the informational security. The main responsibility of the three servers is to manage the key for data encryption & decryption and to authenticate the identity of users.
- The basic web services module and business resource module.
  The basic functions provided by the departments are wrapped with web services. And the web services should be registered in UDDI.The module can take Microsoft .Net as its supporting platform and interact with the service invoking module through the message exchange protocol of SOAP. According to the application requirements of the government business, several application systems are built, such as e-document transaction system, conference management system, information report system etc. Every application system is composed by several web services provided by involved departments. How to execute the web services in every system should be directed by the related rule. Business resource module is composed of several databases that store business data for the application systems.
- The workstation and service terminal module and user interface.
  The workstation and service terminal module provides the operation platform for the office clerks. The workstations can be thin workstations, which means that the business logic is put in the service module. User interface can make use of the universal browsing tools such as Internet Explorer, or the office software such as Microsoft Outlook.

4 A Scenario: The Application for Projects

Now we take the application for projects as an example to demonstrate how the suggested architecture works. Suppose the application for projects is executed through the collaborative work of several office clerks in related departments. The clerks include the business rule generator, the expert and leader. And the execution sequence is the business process rule generator, the expert and leader. Every function provided by the corresponding department is encapsulated in a web service. And the services are registered in the UDDI registry.

We firstly design a special XML Schema used in the scenario according to the data model of e-documents and store the schema in the e-document schema database. When a citizen requests the service of the application for projects, the request handler firstly receives the request and submits it to the center controller, which looks up the corresponding schema in the e-document schema database according to the request. If the center controller finds the schema, it transfers it to the e-document generator, which can generate an e-document based on the XML schema and append some data from the user in the data elements as their values if necessary. Then the e-document is transferred to the center controller, which looks up the rule corresponding to the request in the business rule base. If there is no rule in the rule database, the center controller wants to request the rule generator to generate the rule. The following detailed executing process can be described as follows:

1. As is showed in Figure 2, the business process rule generator receives the XML document from the center server. Then he selects the related data elements and fills in them with detailed data. In the case, the content about the next involved clerks and their execution sequence is appended. If necessary, he should encrypt the sensitive content of data elements, and sign with digital signature. Finally he submits the XML document to the center server.
2. The center server firstly validates the validity of the XML document, then disposes of the content of data elements after making sure that the XML document is valid. It should generate the rule according to the related content of the XML document. After making sure that the web services stated in the rule can actually work together, the center controller submits the rule to the rule engine. At the same time it should find the web services in UDDI registry to acquire the relevant descriptions of web services, which can facilitate the following execution for the rule engine. Finally the center server transfers the XML document to the next clerk according to the rule.
3. The expert receives the XML document and fills in the data elements about the expert's attitude. Then if necessary he should encrypt the appended content and sign. Finally he submits the XML document to the center server. The center server processes it and transfers it to the next clerk, which is the leader in the case.
4. The leader receives the XML document, then decrypts the encrypted data and validates the content if necessary. And he fills in the data elements about the application result. Finally he submits the XML document to the center server.
5. The center server responses to the citizen and stores the e-document in the e-document database.

5 Conclusion

As is illustrated above, the architecture of the collaborative work system for join-up e-government web services together with a data model of e-documents is proposed to meet the requirements for the e-government collaborative work.

References

1. Casati, F., Georgakopoulos, D., and Shan M.: Special Issue on E-Services. VLDB Jour. 24 (2001)
2. Weikum, G.: Special issue on infrastructure for advanced e-services. IEEE Data Engi. Bull. 24 (2001)

3. Casati, F., Ilnicki, S., Jin, L.J., Krishnamoorthy, V., and Shan, M.C.: Adaptive and dynamic service composition in eFlow. Proc. of the Int. Confe. on Adva. Info. Syst. Engi. (2000)
4. Benatallah, B., DUMAS, M., etc: Declarative Composition and Peer-to-Peer Provisioning of Dynamic Web Services. Proc. of the 18th Inte. Confe. on Data Engi. (2002)

A Novel Method for Eye Features Extraction

Zhonglong Zheng, Jie Yang, Meng Wang, and Yonggang Wang

Institute of image processing and pattern recognition,
Shanghai Jiao Tong University, Shanghai, China 200030
{zhonglong, jieyang}@sjtu.edu.cn

Abstract. Extraction of the eye features, like the pupil center and radius, eye corners and eyelid contours from the frontal face images often provides us useful cues in many important applications, such as face recognition, facial expression recognition and 3D face modeling from 2D images. Assuming the rough eye window is known, the color information, Gabor features are extracted from the eye window and then these two features and their mutual relationship have been incorporated into the design of our algorithm. Firstly, after the pupil center is detected in H channel of HSV color space, the pupil radius is estimated and refined. Secondly, eye corners are localized using eye-corner filter based on Gabor feature space. Finally, eyelid curves are fitted by spline function. The new algorithm has been tested on our SJTU database and some other pictures taken from the web. The experimental results show that our algorithm has high accuracy and is robust under different conditions.

1 Introduction

It is essential to detect precisely facial features and their contours when designing automatic face analysis and processing systems. The detection results could be used for the model-based image coding [1], facial expression recognition [2], 3D face recognition [3] and so on. Generally speaking, the detection of eye features is the first step in a recognition system. The eye features consist of pupil center and its radius, eye corners and eyelid contours. Among these features, pupil center and radius are the features easier to be detected and estimated. As we know, deformable contour model is a well-known method to extract object contour in computer vision. [5] and [6] made an attempt to utilize such algorithm to eye contour extraction. As pointed out in [4], deformable contour model is not optimal and stable in eye contour extraction. To overcome the limitations of deformable models, researchers pay much more attention to several landmark points of eyes other than extracting the complete continuous eye contour. Then the eye contour can be fitted by some kind of functions [8][4]. This paper also adopts such strategy and the results are encouraging.

The paper is organized as follows. Based on the assumption that the rough eye window has been detected, Section 2 describes our method in detail: pupil center and radius, eye corners and eyelid contours are precisely detected and estimated step by step. The experimental results are listed in Section 3. The last section draws the conclusion of the paper.

2 Extraction of Eye Features

Most of the existing detection algorithms for eye features extraction are based on the gray level of the images. However, color information could provide extra cues for detection and recognition. Therefore, the input image for the proposed method is color image and each of them contains a single eye. It is assumed that the approximate location of eye is known from previous rough eye detection step.

2.1 Pupil Center Detection and Radius Estimation

Pupil center detection is performed in H channel of HSV color space. Fig 1 shows the original images and the corresponding H channel images. It is interesting that in H channel the pupil is the brightest region comparing with its neighborhoods.

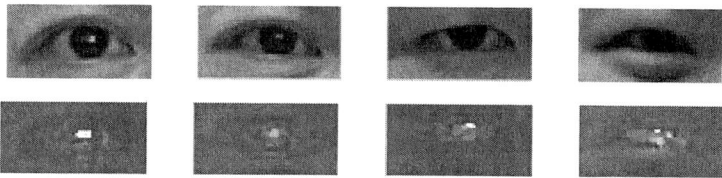

Fig. 1. Original image and corresponding H channel image

The location of the pupil center can be derived from the vertical integral projection and horizontal integral projection in Fig 2. The peak points of the projection curves correspond to the coordinates of the pupil center. This method has at least two advantages: one is its simpleness; the other is its robustness.

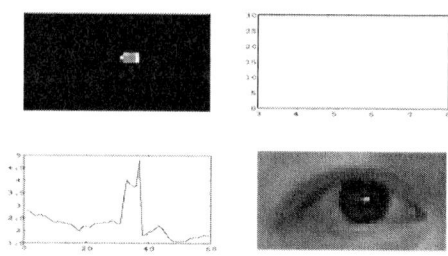

Fig. 2. Location of pupil center by integral projection

Once the approximate location of pupil center is given, a circle with fixed radius is established. The first searching step is circle shift. The operation of this step is to shift the circle in a given neighborhood to obtain a location on which the mean gray level of pixels in the circle is the lowest. The second searching step is expansion or shrinking of the circle. The rule of expansion or shrinking is also to minimize the mean gray value of pixels in the circle. In Fig 3, the first row and second row are the location of eyeball before and after the search strategy.

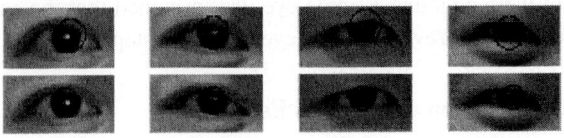

Fig. 3. Location of eyeball before and after the search strategy

2.2 Detection of Eye Corners

Here, we propose an eye corner filter based on Gabor feature space for eye corner detection. The Gabor wavelet can be defined as follows

$$\psi_{\mu,v}(z) = \frac{\|k_{\mu,v}\|^2}{\sigma^2} e^{(-\|k_{\mu,v}\|^2 \|z\|^2 / 2\sigma^2)} [e^{ik_{\mu,v}z} - e^{-\sigma^2/2}], \quad k_{\mu,v} = k_v e^{i\phi_\mu} \quad (1)$$

The Gabor feature representation of an image $I(z)$ is

$$G_{\mu,v}(z) = I(z) * \psi_{\mu,v}(z) \quad (2)$$

where $z = (x, y)$ and $*$ is the convolution operator.

It is obvious that eye corner points are the intersection of the two eyelid curves and the end points of eyelid curves at the same time. Furthermore, the structure is highlighted in Gabor feature space. Motivated by these properties, the paper describes an eye corner filter based on Gabor feature space. The filter is constructed by coefficients at some scales and orientations in Gabor feature space. For example, the filter representing the eye corner near the bridge of the nose is denoted by a 5×5 mask. The center of the mask corresponds to the eye corner manually located while other elements of the mask to the neighbors of the eye corner. The values of the elements in the mask are determined by its Gabor representation. Let $I(x, y)$ be an eye image, $C(x, y)$ be the 5×5 patch image centered at eye corner and $G_{\mu,v}(z)$ be the Gabor representation of $C(x, y)$

$$G = \begin{pmatrix} g_{1,1} & g_{1,2} & \cdots & g_{1,8} \\ g_{2,1} & g_{2,2} & \cdots & g_{2,8} \\ \cdots & \cdots & \cdots & \cdots \\ g_{5,1} & g_{5,2} & \cdots & g_{5,8} \end{pmatrix} \quad (3)$$

at five scales and eight orientations. For illumination invariance, G can be normalized as

$$G' = \frac{G}{\sqrt{\sum_{i,j}|g_{i,j}|^2}} = \begin{pmatrix} g'_{1,1} & g'_{1,2} & \cdots & g'_{1,8} \\ g'_{2,1} & g'_{2,2} & \cdots & g'_{2,8} \\ \cdots & \cdots & \cdots & \cdots \\ g'_{5,1} & g'_{5,2} & \cdots & g'_{5,8} \end{pmatrix} \quad (4)$$

Let f be the mean value of the first two rows coefficients of G'

$$f = \frac{1}{n}\sum_{i,j} g'_{i,j} \quad (5)$$

where $n = 16$, $i = 1, 2$ and $j = 1, 2, \ldots, 8$. Some Gabor representations, f of $C(x, y)$, are shown in Fig 4.

To construct an eye corner (near the bridge of nose) filter, we select 80 corner images described by $C(x, y)$ above. The final eye corner (near the bridge of nose) filter is constructed by calculating the average of f

$$F_{5\times 5} = \frac{1}{N}\sum_{i=1}^{N} f_i \quad (6)$$

where $N = 80$. $F_{5\times 5}$ is shown in Fig 5(a).

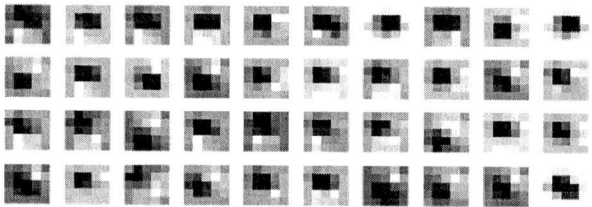

Fig. 4. Gabor representation of eye corner images

The other eye-corner filter is 7×7 in order to contain more information of the corner structure because of its complexity. $F_{7\times 7}$ is shown in Fig 5(b).

To detect eye corners in an eye image, we need to convolve the Gabor representation of the eye image (the selection of scale and orientation is the same as Gabor representation of eye corner images, and normalized by equation (4) and (5)) with certain filter. In order to reduce the computational cost, the convolution operation is calculated on part of the image because the center and radius of the eyeball are given by previous steps. The corner detection results are shown in Fig 6.

Fig. 5. Eye corner filters

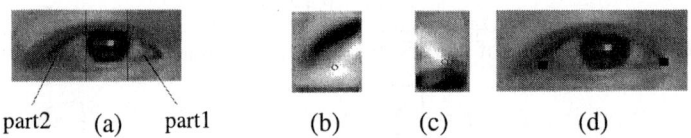

Fig. 6. Corner detection (a) an eye image on gray level (b) convolution of Gabor representation of part2 in (a) with $F_{7\times7}$ (c) convolution of Gabor representation of part1 in (a) with $F_{5\times5}$ (d) final corner detection result

2.3 Eyelid Curve Fitting

Actually, eye corners are the intersection of the two eyelid curves, and the end points at the same time. Now we only need some medial points to fit the curve. It is advisable to detect the points along the column through the center of pupil in the eye image, denoted by red circle in Fig 7 (a). We use a 7×7 window to filter the eye image on gray level: the minimum value in the window is subtracted in order to make the edge more salient. By searching the peak points of column line in Fig 7 (a), we find the medial points. Fitting the eyelid curve with corner points and medial points by spline function, we depict the eyelid curve in Fig 7 (b).

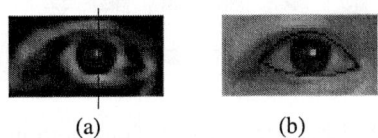

Fig. 7. Medial point detection and eyelid curve fitting

3 Experimental Results

After obtaining the eye corner filter $F_{5\times5}$ and $F_{7\times7}$, we apply them to our test dataset that contains more than 700 images, and to some pictures taken from the web. The images are taken under different lighting conditions and some have a slight rotation. Fig 8 shows part of the experimental results. The detected eyeball and eyelid curves are described in red, and eye corners in green.

The algorithm process is carried out step by step, i.e. if the first step is not accurate, the following detection may be failed. Fortunately, pupil center detection and radius estimation achieve high accuracy over 98%. Thus, the proposed method reached desired performance.

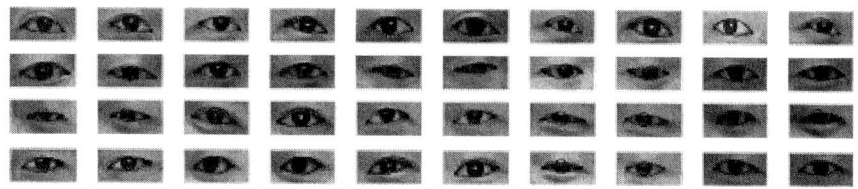

Fig. 8. Part of experimental results

5 Conclusion

A robust and accurate algorithm for eye features extraction is proposed in the paper. Firstly, the center of pupil is detected in H channel of HSV space, and then the radius of eyeball is estimated and refined. Secondly, eye corners are detected by the proposed eye-corner filter constructed in Gabor feature space. Finally, based on the results obtained from the previous steps, eyelid curve is fitted by spline function. The proposed methods have shown its good performance in terms of accuracy and robustness, while being straightforward in implementation.

References

1. Zheng N., Song W., Li W.: Image Coding Based on Flexible Contour Model. Machine Graphics Vision 1 (1999) 83-94
2. Brunelli R., Poggio T.: Face Recognition Through Geometrical Features. In: 2nd European Conf. on Computer Vision (1992) 792-800
3. Blanz V., Vetter T.: Face Recognition Based on Fitting a 3D Morphable Model. IEEE trans. on PAMI. 25 (2003) 1063-1075
4. Vezhnevets V., Degtiareva A.: Robust and Accurate Eye Contour Extraction. Proc. Graphicon (2003) 81-84
5. Lam K.M., Yan H.: Locating and Extracting The Coverd Eye in Human Face Image. Pattern Recogntion 29 (1996) 771-779
6. Yin L., Basu A.: Realistic Animation Using Extended Adaptive Mesh for Model Based Coding. In: Proceedings of Energy Minimization Methods in Computer Vision and Pattern Recognition (1999) 315-318
7. Feng G.C., Yuen P.C.: Multi-cues Eye Detection on Gray Intensity Image. Pattern Recognition 34 (2001) 1033-1046
8. Yulle A., Hallinan P., Cohen D.: Feature Extraction from Faces Using Deformable Templates. International Journal of Computer Vision 8 (1992) 99-111

A *Q*-Based Framework for Demand Bus Simulation

Zhiqiang Liu, Cheng Zhu, Huanye Sheng, and Peng Ding

Department of Computer Science, Shanghai Jiao Tong University,
200030, Shanghai, China
{zqliu, microzc}@sjtu.edu.cn
hysheng@mail.sjtu.edu.cn, dingpeng@cs.sjtu.edu.cn

Abstract. With the development of telecom, it becomes possible to provide people with Location Based Service (LBS) according to their spatial positions by using mobile equipments. Before using such services, people should evaluate them and make suitable strategies. Demand Bus System is attracting attention as a new public transportation system that provides convenient transportation for special services while solving traffic jams in urban areas. Previous researches focused on the algorithms to the dial-a-ride problem, and few works did on evaluating the usability of the demand bus system. In this paper we proposed a Q-based multi-agent simulation framework for this purpose and implemented a prototype system.

1 Introduction

With the development of telecom, it becomes possible to provide people with Location Based Service (LBS) according to their spatial positions by using mobile equipments. Before using such services, people should evaluate them and make suitable strategies. Demand bus system (DBS), also called as demand response system, is thought to be a new way to provide convenient transportation for special services. The user calls a bus center, states the destination and the wanted time period, and then the bus center arranges a suitable bus to pick up the user in some time. It's faster and more convenient than traditional bus, and cheaper than taxi. Different demand bus services had been provided for people all over the world. But people found it's difficult to build perfect demand bus systems to different situations. A rich variety of dial-a-ride problems emerges from the characteristics of the servers, the rides, the metric space, and the objective function. Dial-a-ride problems have been studied extensively in the literature, see for example, Hunsaker and Savelsbergh[1], Diana and Dessouky[2], and Healy and Moll[3].These researches mainly concentrated on the algorithms to solve the problem, and few did on the framework to investigate the effectiveness and the strategy. This has induced the need for suitable methodologies and tools to help in investigating the effectiveness and the strategy of DBS. A useful evaluation tool for achieving these objectives, in particular when lacking real applications is multi-agent based simulation (MABS). In this context, MABS is used as the process of designing and creating a model of a likely or expected public transportation system for conducting numerical experiments. The purpose is to obtain a better understanding of the behaviors of such a system in a given set of conditions, even with uncertain treatment of events. In order to describe the interaction between human and agents, we use the interaction design language-Q, which has been proved to be a good language to build such simulation.

In section 2, we propose a Q-based simulation framework. A prototype system using this framework is implemented in section 3. Conclusion and future work are discussed in the last part.

2 DBS Simulation Framework

2.1 Simulation Framework

The framework is constructed by three parts (Fig 1): Model designers, Multi-agent system and information resources.

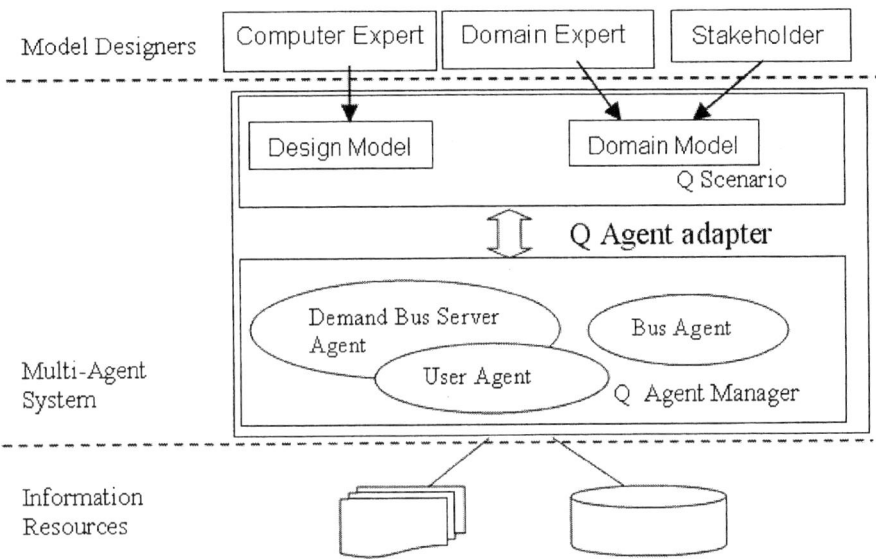

Fig. 1. *Q*-based Framework for Demand Bus Simulation

To understand complex system, researchers, together with domain experts and stakeholders, would build a model to simulate the phenomenon by using participatory methods. Q language provides a bridge for domain experts and computer experts to share their knowledge on the problem. The detail of Q language will be discussed later.

Multi-agent system includes three parts: Q scenario, Q agent adapter and Q agent manager. Q scenario describes agents' states and actions, which are designed according to the simulation model. Q agent manager is in charge of the management of agents and is also the implementation of actions described in scenario. Q agent adapter creates a link between above two parts and also serves as a message layer for the communication between agents.

2.2 Interaction Design Language: Q

Some inter-agent protocol description languages, such as KQML and AgenTalk, often regulate an agent's various actions on the basis of computational model of agent internal mechanisms. Obviously, these agents based on strict computational models have to be designed and developed by computer experts. It is necessary to design a new agent interaction description language, which makes it possible that those non-computer-professional application designers, such as sales managers and consumers, might write scenarios to describe and model the behaviors of agents, so that a practical multi-agent system can be easily established. Under the background, we start working on Q language [4] - a scenario description language for designing interaction among agents and humans. This kind of scenario-based language is to design interaction from the viewpoint of scenario description, but not for describing agent's internal mechanisms.

Q is highly practical and realistic, which can support many totally different agent systems, such as FreeWalk [5] and Crisis Management [6] for its powerful ability of describing human requests to autonomous agents. So it's a good language in demand bus system simulation.

3 Demand Bus Shanghai

3.1 The Problem

Shanghai is one of the biggest cities in China, which has a population of about 13 Millions. People have to spend lots of time back and forth for work because of the long distance and the bad traffic. Traditional buses have too many stops. In rush-hour, many crowded buses still move along fixed route and stop at every stop even no passenger alights in many stops. With the extension of Shanghai, the city is divided into many small parts with different functions. Some of them are living centers, some are working centers, and others are shopping or entertainment centers. Therefore, it's possible to use demand bus system as community transportation in Shanghai.

3.2 Simulation Design

We suppose that there is only one Bus Center that is in charge of the demand bus system. When it receives user's request for demand bus service, it will immediately try to find a suitable bus and inform the task. If there is no bus available, the user's request would be rejected. All buses are treated as independent individuals and can compete for the task to take users. In order to improve the response effect, we designed the multi-agent system according to the Post Price Model [7]. The design process is shown as Fig 2. State transition diagram shows the rules of bus agent will abide by. At the beginning, when a bus comes into system, it would post its price that means how far it would not like to go, how long its working time is or other conditions. The price is used to find a suitable bus to serve a new request. When a user calls bus center, bus center will find one bus according to buses' prices and the objective function.

The objective function in the simulation is to minimize the total trip distance of all users.

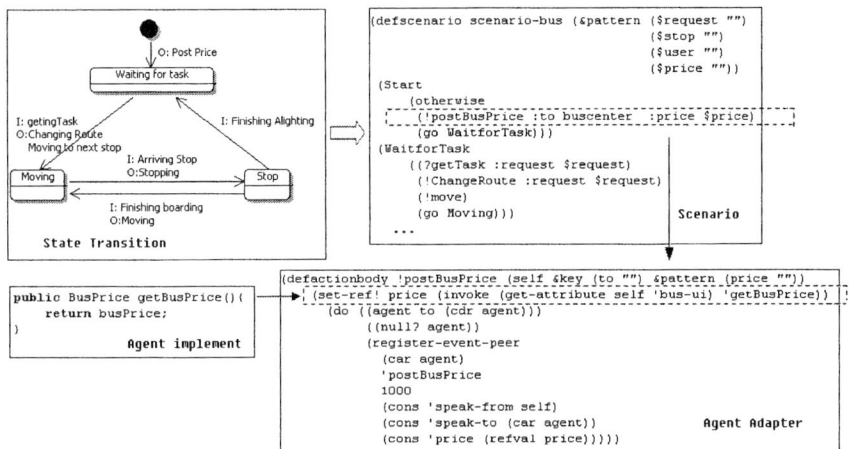

Fig. 2. Illustration of the design process

In the simulation, we supposed that there are no traffic jams; requests for demand bus occur with random frequency in fixed time; departure and destination points are decided randomly and a passenger would use a demand bus and do not transfer between buses.

We designed three situations:

1. Multi-Vehicle, Random user requests and the bus acts as a taxi, each time services only for one user. We do such simulation for the purpose of observing the relationship between the service area and the service usability. It should be pointed out that in this case, the numbers of demand buses are equal to those of the user's requests because we want to decrease the influence of users' waiting-for-bus time (WBT). We did this simulation in four different scale areas. In each area three different situations are tested, one is 10 requests happen randomly in fixed time period, one is 30 and another is 50.
2. Multi-Vehicle, Random user requests and the bus has unlimited capacity. This simulation aims to find the relationship between the service bus numbers and the service usability. In each area, different amount of bus are used. We observed the average service time, the average WBT and the average transition time.
3. In the last case, conditions are the same as the second simulation except that there are two areas represented as the living center and shopping center separately. User's origin and destination would be located in the two centers respectively. We

call this kind of situation as Centralized Position simulation. Three kinds of experiments would be done: in the first case, the distance of two centers is long. In the second, the distance is middle, and in the third, we still use random position. We want to use these experiments to find out the effect of using the demand bus for such situation.

3.3 Experiment Result and Analysis

When the amount of demand requests increased (>30) in a large area (400*400), the Average service time increased obviously. Except the influence of the program, it can also be thought that dividing the whole area into small parts can improve the performance of demand bus system.

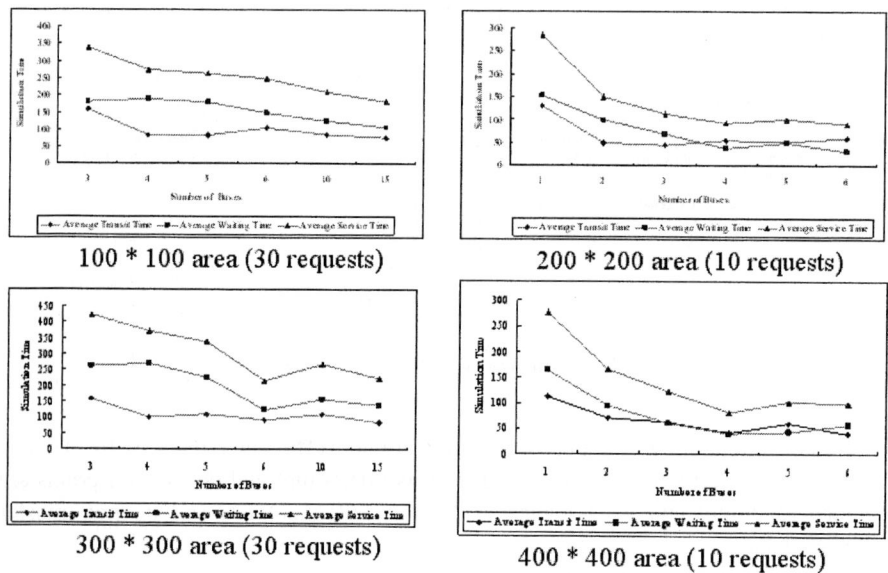

Fig. 3. The simulation result of experiment 2

In the experiment of second situation, the average service time decreased when the number of demand buses increase. But when the amount of demand buses are enough (in the 200*200 and 400* 400 areas cases, to provide service for 10 users, 4 buses is enough.), the average service time remains unchanged. Among the four results, the average service time is mainly influenced by the WBT. The results are shown in Fig3.

In the last experiment, the total distance of all buses with centralized position in the first case is the smallest comparing to the other two. In this case, we can use a small number of buses to serve for the area with similar destinations. It's suitable for Shanghai which has many communities and many residents having the same situations.

4 Conclusion and Future Work

In this paper, we proposed a Q-based multi-agent simulation framework to evaluate the usability of demand bus system. Initial result shows that the system based on such framework can be designed easily and used to evaluate the demand bus system using in Shanghai.

Current work is based on the centralized architecture. It's difficult to re-plan the route and to involve all information of the bus drivers and the demand bus system users in the real system. In our future work, distributed organization would be considered in order that buses can communicate with each other to re-plan some users' route according to drivers' and users' situations.

Acknowledgement. This work was supported by Association of International Education, Japan (AIEJ). The authors would thank Prof. Ishida, Dr. Torii and Dr. Murakami in Kyoto University for their help during the conceiving of this paper.

References

1. Hunsaker B., Savelsbergh M.: Efficient feasibility testing for dial-a-ride problems. Operations research letters. 30 (2002) 169-173
2. Diana M., Dessouky M. M.: A new regret insertion heuristic for solving large-scale dial-a-ride problems with time windows. Transportation Research Part B. 38 (2004) 539–557
3. Healy P., Moll R.: A new extension of local search applied to the Dial-A-Ride Problem. European Journal of Operational Research. 83 (1995) 83-104
4. Ishida T.: Q: A Scenario Description Language for Interactive Agents. IEEE Computer, Vol.35. (2002)
5. Ishida T.: Digital City Kyoto: Social Information Infrastructure for Everyday Life. Communications of the ACM (CACM), Vol. 45, No. 7. (2002) 76-81
6. Ishida T., Nakanishi H., Takata S.: Crisis Management Simulation in Digital City. Journal of the Institute of Systems,Control and Information Engineers. (2002)
7. Buyya R., et al.: Economic Models for Resource Management and Scheduling in Grid Computing. The Journal of Concurrency and Computation: Practice and Experience (CCPE). (2002)

A Revision for Gaussian Mixture Density Decomposition Algorithm

Xiaobing Yang, Fansheng Kong, and Bihong Liu

Artificial Intelligence Institute, Zhejiang University, Hangzhou, China, 310027
konglab@zju.edu.cn

Abstract. Gaussian mixture density decomposition (GMDD) algorithm is an approach to the modeling and decomposition of Gaussian mixtures, and it performs well with the least prior knowledge in most case. However, there are still some special cases in which the GMDD algorithm is difficult to converge or can not gain a valid Gaussian component. In this article, a k-means method for Gaussian mixture density modeling and decomposition is studied. Then based on the GMDD algorithm and k-means method, a new algorithm, called k-GMDD algorithm is proposed. It solves the problems of GMDD caused by the symmetry excellent, and consequently makes the applications of GMDD algorithm more extensive.

1 Introduction

Finite mixture density modeling and decomposition has been widely applied in variety of important practical situations, such as industrial control [1], machine monitoring [3] and economics [2]. A typical example of application of finite mixture density is the analysis of fisheries data [4]. In this example, for different sex, the mean length of a halibut is approximately different linear function of its age. The age and length of each fish can be measured, but the sex is indistinguishable. Estimating the parameters of the two linear growth curves and determining the sex can be regarded as a problem of finite mixture density modeling and decomposition.

A highly robust estimator called model fitting (MF) estimator has been presented by Xinhua Zhuang, Tao Wang and Peng Zhang in 1992 [7]. The MF estimator obtains high robustness through partially but completely modeling the unknown log likelihood function. Using MF estimator, Xinhua Zhuang, Yan Huang, K. Palaniappan and Yunxin Zhao proposed a recursive algorithm called Gaussian mixture density decomposition (GMDD) algorithm for successively identifying each Gaussian component in the mixture. The GMDD algorithm performs well with the least prior knowledge in most case [6].

However, because of some drawbacks, the GMDD algorithm needs still to be improved in order to be more applicable. In this article, a k-means method for Gaussian mixture density modeling and decomposition is studied. Then based on the GMDD algorithm and k-means method, we propose a new algorithm, called k-GMDD algorithm. In the k-GMDD algorithm, the k-means method is used to void the divergence by destroying the symmetry of the data to be decomposed.

2 Fundamental Conceptions and GMDD Algorithm

Let $x_1, x_2, ..., x_N$ be a random sample of completely unclassified observations from n-dimensional Gaussian mixture distribution with q components. If assuming that each observation x_j is generated by an unknown Gaussian distribution $m(x_j; \mu, \Sigma)$ with probability $(1-\varepsilon)$ plus an unknown *outlier* distribution $h(\cdot)$ with probability ε, the probability density function of x_j is given by

$$m(x_j; \mu, \Sigma) = \frac{(1-\varepsilon)\exp(-\frac{1}{2}d^2(x_j))}{(\sqrt{2\pi})^n \sqrt{|\Sigma|}} + \varepsilon h(x_j) \tag{1}$$

where Σ is the covariance matrix, and $d^2(x_j)$ represents the squared Mahalanobis distance of x_j from the unknown mean vector μ, i.e.

$$d^2(x_j) = (x_j - \mu)'\Sigma^{-1}(x_j - \mu) \tag{2}$$

If $\varepsilon = 0$, the density $m(\cdot; \mu, \Sigma)$ becomes a pure Gaussian density. Otherwise, if $\varepsilon > 0$, $m(\cdot; \mu, \Sigma)$ is called a contaminated Gaussian density [5].

The likelihood estimating equations can be written as:

$$\sum_{j=1}^{N} \nabla_\mu \ln m(x_j; \mu, \Sigma) = 0 \tag{3a}$$

$$\sum_{j=1}^{N} \nabla_\Sigma \ln m(x_j; \mu, \Sigma) = 0 \tag{3b}$$

where ∇_μ and ∇_Σ denote respectively differentiations with respect to μ and Σ.

Let g_j stands for $(\exp(-d^2(x_j)/2))/((\sqrt{2\pi})^n \sqrt{|\Sigma|})$, m_j stands for $m(x_j; \mu, \Sigma)$. Ideally, a sample x_j is classified as an *inlier* if it is realized from g_j or as an *outlier* otherwise (i.e., it comes from $h(x_j)$). The probability of a sample x_j being an *inlier* is given by

$$\lambda_j = \frac{(1-\varepsilon)g_j}{m_j} \tag{4}$$

According to the Bayesian classification rule, if we assume the unknown density function h as follows:

$$h(x_1) = h(x_2) = \cdots = h(x_n) = h_0 \tag{5}$$

we would get the same output results. Then, (1) can be rewritten as

$$m_j = (1-\varepsilon)g_j + \varepsilon h_0 \tag{6}$$

then (3) can become

$$\sum_{j=1}^{N} \nabla_\mu \ln((1-\varepsilon)g_j + \varepsilon h_0) = 0 \tag{7a}$$

$$\sum_{j=1}^{N} \nabla_\Sigma \ln((1-\varepsilon)g_j + \varepsilon h_0) = 0 \tag{7b}$$

The gradients $\nabla_\mu \ln((1-\varepsilon)g_j + \varepsilon h_0)$ and $\nabla_\Sigma \ln((1-\varepsilon)g_j + \varepsilon h_0)$ can be derived as $\lambda_j \Sigma^{-1}(x_j - \mu)$ and $-\frac{1}{2}\lambda_j \Sigma^{-1}\{1 - \Sigma^{-1}(x_j - \mu)(x_j - \mu)'\}$, where $\lambda_j = \frac{(1-\varepsilon)g_j}{m_j} = \frac{g_j}{g_j + t}$, $t = \frac{\varepsilon h_0}{1-\varepsilon}$.

Accordingly, the likelihood estimating equations (7) can be written as

$$\sum_{j=1}^{N} \lambda_j (x_j - \mu) = 0 \tag{8a}$$

$$\sum_{j=1}^{N} \lambda_j - \Sigma^{-1} \sum_{j=1}^{N} \lambda_j (x_j - \mu)(x_j - \mu)' = 0 \tag{8b}$$

At each selected partial model "t_s", $s = 0, 1, \cdots, L$, where t_L denotes an upper bound for all potentially desirable partial models, the iterative procedure of GMDD algorithm can be implemented based on (8). In GMDD algorithm, the Kolmogorov-Smirnov (K-S) normality test is used to determine the validity of extracted Gaussian components [6].

3 K-GMDD Algorithm

In most case, the GMDD algorithm performs well with the least prior knowledge [6]. However, there are still some special case in which the GMDD algorithm may be hard to converge or can not gain a valid Gaussian component. For example, when there are two Gaussian components with different mean vector μ. But like covariance matrix Σ and density in n-dimensional space, GMDD may regard them as a whole and be unable to converge or it will get a single Gaussian component which fails to the K-S test. Accordingly, it is still necessary to improve the GMDD algorithm in order to make it be more applicable.

K-means method is a well-known and commonly partitioning method. The k-means method takes the input parameter, k, and partitions a set of objects into k clusters so that the resulting intra-cluster similarity is high whereas the inter-cluster similarity is low. Cluster similarity is measured in regard to the mean value of the objects in the cluster, which can be viewed as the cluster's center.

K-means works well when the clusters are compact clouds that are rather well separated from each other. It is relatively scalable and efficient in processing large data sets because of its low computational complexity. However, the *k*-means method can be applied only in few specific cases for some faults of it. For example, it is sensitive to noise and outlier data points since a small number of such data can substantially influence the mean value [3].

In this article, Gaussian mixture density modeling and decomposition with assist of *k*-means method is studied. Then based on the GMDD algorithm and *k*-means method, a new algorithm, called *k*-GMDD algorithm is proposed. In the *k*-GMDD algorithm, the *k*-means method is used to void the divergence by destroying the symmetry of the data to be clustered. The *k*-GMDD algorithm can be described as the following steps:

Step 1. Let X be the data set to be clustered, set a threshold N_{min} to decided when end the procedure.

Step 2. Execute GMDD algorithm on X until GMDD be unable to converge or can not get a proper Gaussian component, if $S \geq N_{min}$, where S is the size of newest data set X_{new}, go to Step 3. Otherwise, return.

Step 3. Use *k*-means method to split X into k sub-dataset $X_1, X_2, \cdots, X_k$. Assume

$$size(X_1) \leq size(X_2) \leq \cdots \leq size(X_k),$$

then perform GMDD algorithm on $X - X_1$, $X - X_2$, $\cdots$, $X - X_k$ in turn until get a valid Gaussian component G.

Step 4. Subtract G from X, if $S \geq N_{min}$, go to Step 2. Otherwise, return.

If the given data set can be decomposed well by GMDD in Step 2 at first time, the algorithm will not execute Step 3, and become pure GMDD algorithm. Step 3 gives a solution to those cases in which GMDD algorithm can not perform well. The input parameter k in Step 3 can be selected randomly in a certain scope. If in some case, given k selected, valid Gaussian component can not be gained, we should have to change the value of k and try to run Step 3 again. The validity of extracted Gaussian components is still determined by the result of K-S test. If the test succeeds, a valid Gaussian component will be determined, and subtracted from the current data set. Then next valid Gaussian component will be identified in the new size-reduced data set. Individual Gaussian components continue to be estimated recursively until the size of a new data set gets too small.

The *k*-GMDD algorithm is effective and robust. It uses GMDD algorithm clustering, and uses *k*-means method to solve excellently the problems of GMDD caused by the symmetry. Consequently it makes the applications of GMDD algorithm more extensive.

4 Simulation Experiment

In order to illustrate the robustness and efficiency, we compare the *k*-GMDD algorithm with the GMDD algorithm by performing cluster analysis using simulated two-dimensional (2-D) mixture datasets. The generated data consist of three clean Gaussian distribution components under a noisy background consisting of 780 data points uniformly distributed. Fig. 1 and Fig. 2 show the performance of the GMDD algorithm and the k-GMDD algorithm.

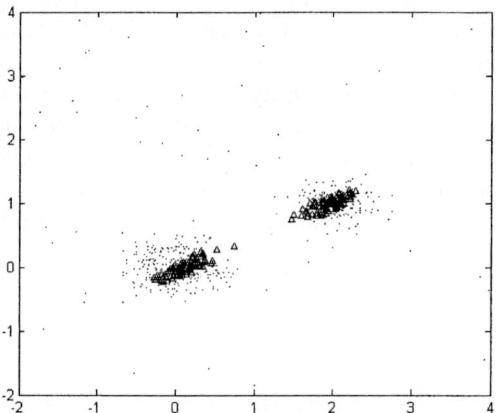

Fig. 1. The GMDD algorithm can not detect a valid component

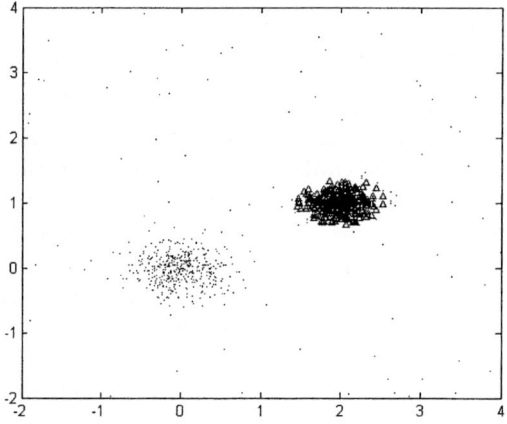

Fig. 2. The k-GMDD algorithm detects a valid component

The data values that are assigned to a cluster are marked by "Δ" in the results, and that unassigned to any cluster are marked by "." symbol. From Figure 1, we can see that it is hard, for the given data set, to converge in GMDD algorithm. On the other hand, the k-GMDD algorithm detected successfully a valid Gaussian component as showed in Figure 2. It is clear that the k-GMDD algorithm will detect easily an other valid Gaussian component after the first gained component having been subtracted from the data set.

5 Conclusions

In this paper a new algorithm called k-GMDD has been proposed for Gaussian mixture density decomposition. This novel algorithm can be robust and effective, which has been illustrated by the simulation experiment. Though the k-GMDD appears to be rather successful, at least by the simulation experiment, there are problems should be further investigated. The value of k may affect badly the performance of the k-GMDD algorithm. Thus, the selection of value of k will be very importable. Application of k-GMDD to high-dimensional data sets is a problem to be researched. Another important topic of k-GMDD is guessing good initializations or reducing the range of random initializations in order to maintain the efficiency of the k-GMDD algorithm. We hope to make some of the above issues the basis of a future investigation.

References

1. Arabie, P., Hubert, L.J., Soete, G. De (eds.): Clustering and Classification, Singapore: World Scientific (1996)
2. Hamermesh, D.S.: Wage bargains, threshold effects, and the phillips curve. In: Quarterly Journal of Economics, vol. 84 (1970) 501-517
3. Han, J., Kamber, M. :Data Mining: Concepts and Techniques, Morgan Kaufmann Publishers, San Francisco (2001)
4. Hosmer, D.W.: Maximum likelihood estimates of the parameters of a mixture of two regression lines. In: Communications in Statistics, vol.3, no. 10 (1974) 995-1005
5. Huber, P.J.: Robust Statistics. New York: Wiley (1981)
6. Zhuang, X., Huang, Y., Palaniappan, K., Zhao, Y.: Gaussian Mixture Density Modeling, Decomposition, and Applications. In: IEEE Trans. Image Processing, vol. 5, no. 9 (1996) 1293-1302
7. Zhuang, X., Wang, T., Zhang, P.: A Highly Robust Estimator through Partial-Likelihood Function Modeling and Its Application in Computer Vision. In: IEEE Trans. Pattern Analysis and Machine Intelligence, vol. 14, no. 1 (1992) 19-35

Discretization of Continuous Attributes in Rough Set Theory and Its Application*

Gexiang Zhang[1,2], Laizhao Hu[1], and Weidong Jin[2]

[1] National EW Laboratory, Chengdu 610036 Sichuan, China
dylan7237@sina.com
[2] School of Electrical Engineering, Southwest Jiaotong University,
Chengdu 610031, Sichuan, China

Abstract. Existing discretization methods cannot process continuous interval-valued attributes in rough set theory. This paper extended the existing definition of discretization based on cut-splitting and gave the definition of generalized discretization using class-separability criterion function firstly. Then, a new approach was proposed to discretize continuous interval-valued attributes. The introduced approach emphasized on the class-separability in the process of discretization of continuous attributes, so the approach helped to simplify the classifier design and to enhance accurate recognition rate in pattern recognition and machine learning. In the simulation experiment, the decision table was composed of 8 features and 10 radar emitter signals, and the results obtained from discretization of continuous interval-valued attributes, reduction of attributes and automatic recognition of 10 radar emitter signals show that the reduced attribute set achieves higher accurate recognition rate than the original attribute set, which verifies that the introduced approach is valid and feasible.

1 Introduction

Rough set theory (RST), proposed by Pawlak [1], is a new fundamental theory of soft computing. [2] RST can mine useful information from a large number of data and generates decision rules without prior knowledge [3,4], so it is used generally in many domains [2,3,5-9]. Because RST can only deal with discrete attributes, a lot of continuous attributes existing in engineering applications can be processed only after the attributes are discretized. Thus, discretization becomes a very important extended research issue in rough set theory. [2,6,7,8] Although many discretization methods, including hard discretization [2,5,6] and soft discretization [7,8], have been presented in rough set theory, the methods are only used to discretize point attribute values (fixed values). However, in engineering applications, especially in pattern recognition and machine learning, the features obtained using some feature extraction approaches usually vary in a certain range (interval values) instead of fixed values because of

---

* This work was supported by the National Defence Foundation (No.51435030101ZS0502).

several reasons, such as plenty of noise. So a new discretization approach is proposed to process the decision table in which the attributes vary continuously in some ranges.

2   Definition of Generalized Discretization

The main point of the existing discretization definition [2] is that condition attribute space is split using selected cuts. The existing discretization methods [2,5-8] always try to find the best cutting set. When the attributes are interval values, it is difficult to find the cutting set. So it is necessary to generalize the existing definition of discretization. The following description gives the generalized discretization definition.

Decision system $S = \langle U, R, V, f \rangle$, where $R = A \cup \{d\}$ is attribute set, and the subsets $A = \{a_1, a_2, \cdots, a_m\}$ and $\{d\}$ are called as condition attribute set and decision attribute set respectively. $U = \{x_1, x_2, \cdots, x_n\}$ is a finite object set, i.e. universe. For any $a_i (i = 1, 2, \cdots, m) \in A$, there is information mapping $U \to V_{a_i}$, where V_{a_i} is the value domain, i.e.

$$V_{a_i} = \{[v_{a_i}^{x_1 \min}, v_{a_i}^{x_1 \max}], [v_{a_i}^{x_2 \min}, v_{a_i}^{x_2 \max}], \cdots, [v_{a_i}^{x_n \min}, v_{a_i}^{x_n \max}]\} \tag{1}$$

Where $v_{a_i}^{x_j \min}, v_{a_i}^{x_j \max} \in R, (j = 1, 2, \cdots, n)$. For attribute $a_i (a_i \in A)$, all objects in universe U are partitioned using class-separability criterion function $J(V_{a_i})$ and an equivalence relation R_{a_i} is obtained, that is, a kind of categorization of universe U is got. Thus, in attribute set A, we can achieve an equivalence relation family P, ($P = \{R_{a_1}, R_{a_2}, \cdots, R_{a_m}\}$) composed of m equivalence relations $R_{a_1}, R_{a_2}, \cdots, R_{a_m}$. So the equivalence relation family P defines a new decision system $S^P = \langle U, R, V^P, f^P \rangle$, where $f^P(x) = k, x \in U, k = \{0, 1, \cdots\}$. After discretization, the original decision system is replaced with the new one. Different class-separability criterion functions generate different equivalence relation families that construct different discrete decision systems.

The core of the definition is that the discretization of continuous attributes is regarded as a function that transforms continuous attributes into discrete attributes. The new definition is an extended version of the old one and emphasizes on the separability of different classes in discretization. The definition provides a good way to discretize continuous interval-valued attributes. Additionally, considering the separability of classes in the process of discretization can simplify the structure of classifier and enhance accurate recognition rate. If $v_{a_i}^{x_j \min} = v_{a_i}^{x_j \max}, (j = 1, 2, \cdots, n)$, the discretization of interval-valued attributes becomes the discretization of fixed-point attributes. If the class-separability criterion $J(\cdot)$ is a special function that can partition the value domain V_{a_i} of attribute $a_i (a_i \in A)$ into several subintervals, that is, the special function decides a cutting-point set in the value domain V_{a_i}, the definition of generalized discretization becomes the common definition of discretization.

3 Discretization Algorithm

The key problem of discretizing interval-valued attributes is to choose a good class-sepability criterion function. So a class-sepability criterion function is given firstly in the following description.

When an attribute value varies in a certain range, in general, the attribute value always orders a certain law. This paper only discusses the decision system in which the attributes have a certain law. To the extracted features, the law is considered approximately as a kind of probability distribution. Suppose that functions $f(x)$ and $g(x)$ that are one-dimensional, continuous and non-negative real, are respectively the probability distribution functions of attribute values of two objects in universe U in decision system. The below class-separability criterion function is introduced.

$$J = 1 - \frac{\int f(x)g(x)dx}{\sqrt{\int f^2(x)dx} \cdot \sqrt{\int g^2(x)dx}} \qquad (2)$$

Function J in (2) satisfies the three conditions of class-separability criterion functions [10]: (i) the criterion function value is non-negative; (ii) the criterion function value gets to the maximum when the distribution functions of two classes have non-overlapping; (iii) the criterion function value equals to zero when the distribution functions of two classes are identical.

Because $f(x)$ and $g(x)$ are non-negative real functions, according to the famous Cauchy Schwartz inequation, we can get

$$0 \leq \int f(x)g(x)dx \leq \sqrt{\int f^2(x)dx} \cdot \sqrt{\int g^2(x)dx} \qquad (3)$$

So the value domain of the criterion function J in (2) is [0,1]. The following description gives the interpretation that function J can be used to justify whether the two classes are separable or not. When the two functions $f(x)$ and $g(x)$ in (2) are regarded respectively as probability distribution functions of attribute values of two objects A and B in universe U, several separability cases of A and B are shown in figure 1. For all x, if one of $f(x)$ and $g(x)$ is zero at least, which is shown in Fig. 1(a), A and B are completely separable and the criterion function J arrives at the maximal value 1. If there are some points of x that make $f(x)$ and $g(x)$ not equal to 0 simultaneously, which is shown in Fig.1(b), A and B are partly separable and the criterion function J lies in the range between 0 and 1. For all x, if $f(x) = k \cdot g(x)$, $k \in R^+$, which is shown in Fig.1(c), $k = 2$, A and B are unseparable completely and the criterion function J arrives at the minimal value 0. Therefore, it is reasonable that the criterion function J is used to evaluate separability of two classes.

According to the above class-separability criterion function, the discretization algorithm of interval-valued attributes in decision system is given as follows.

Step 1. Initialization: deciding the number n of objects in universe U and the number m of attributes.

Step 2. Constructing decision table: all attribute values are arrayed into a two-dimensional table in which all attribute values are represented with an interval values.

Step 3. Choosing a little positive number as the threshold T_h of class separability.

Step 4. For the attribute a_i (in the beginning, $i=1$), all attributes are sorted by the central values from the smallest to the biggest and sorted results are $v_1, v_2, \cdots, v_n$.

Step 5. The position, where the smallest attribute value v_1 in attribute a_i is, is encoded to zero (Code=0) to be the initial value of discretization process.

Step 6. Beginning with v_1 in the attribute a_i, the class-separability criterion function value J_k of v_k and v_{k+1} ($k=1,2,\cdots,n-1$) is computed by the sorted order $v_1, v_2, \cdots, v_n$ in turn. If $J_k \geq T_h$, which indicates the two objects are separable completely, the discrete value of the corresponding position of attribute v_{k+1} adds 1, i.e. Code=Code+1. Otherwise, $J_k < T_h$, which indicates the two objects are unseparable, the discrete value of the corresponding position of attribute v_{k+1} keeps unchanging.

Step 7. Repeating step 6 till all attribute values in attribute a_i are discretized.

Step 8. If $i \leq m$, which indicates there are some attribute values to be discretized, $i = i+1$, the algorithm goes to step 4 and continues until $i > m$, implying all continuous attribute values are discretized.

Step 9. The original decision system is replaced with the discretized one.

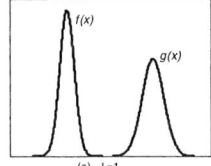 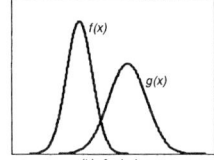 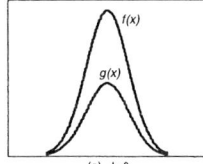

Fig. 1. Three separability cases of functions $f(x)$ and $g(x)$

4 Application Example

10 radar emitter signals (RESs), represented with $x_1, x_2, \cdots, x_{10}$ respectively, are chosen to make the simulation experiment. Attribute set is made up of 8 features [11-13] that are represented with $a_1, a_2, \cdots, a_8$. When signal-to-noise (SNR) varies from 5 dB to 20 dB, 8 features extracted from 10 RESs construct the attribute table shown in Table 1. Table 2 is the discretized result of Table 1 using the proposed approach. In the process of discretization, the parameters $n=10, m=8, T_h=0.99$, and all attribute values are regarded as a *Gaussian* distribution functions with the parameters *expectation* and *variance*. After discretization, the reduction method using discernibility matrix and logic operation [2,3] is used to reduce Table 2. The final result is

$$L = a_7(a_6 + a_3 a_4 + a_5(a_4 + a_1(a_2 + a_3) + a_2 a_8)) \tag{4}$$

Table 1. The attribute table before discretization

Attributes	A_1	a_2	a_3	a_4
x_1	[0.6903,0.7071]	[1.4018,1.4036]	[0.1217,0.1249]	[0.2707,0.2724]
x_2	[0.6711,0.6821]	[1.4296,1.5758]	[0.1151,0.1615]	[0.2136,0.2774]
x_3	[0.5460,0.5976]	[1.5823,1.5857]	[0.2924,0.3040]	[0.2251,0.2261]
x_4	[0.8927,0.8959]	[1.4370,1.4900]	[0.3735,0.6523]	[0.9896,1.000]
x_5	[0.8966,0.9089]	[1.5610,1.7682]	[0.2002,0.2230]	[0.7253,0.7270]
x_6	[0.4194,0.4528]	[1.2190,1.2238]	[0.5616,0.5856]	[0.1410,0.1424]
x_7	[0.3959,0.4129]	[1.2805,1.2841]	[0.0802,0.0818]	[0.0395,0.0404]
x_8	[0.6569,0.6841]	[1.4788,1.4832]	[0.0657,0.0671]	[0.1821,0.1914]
x_9	[0.5211,0.5619]	[1.3938,1.4004]	[0.0780,0.0806]	[0.7066,0.7294]
x_{10}	[0.9048,0.9076]	[1.3841,1.4585]	[0.3317,0.3481]	[0.6250,0.6301]
Attributes	a_5	a_6	a_7	a_8
x_1	[0.4346,0.4365]	[1.9615,2.3013]	[0.9516,1.0209]	[0.4201,0.4334]
x_2	[0.3619,0.4278]	[1.0884,2.9913]	[1.1154,1.2605]	[0.4055,0.4549]
x_3	[0.3741,0.3755]	[0.7260,1.1110]	[0.6909,0.7549]	[0.5343,0.5495]
x_4	[0.8562,0.8672]	[11.738,20.222]	[2.1570,2.9600]	[0.1985,0.2009]
x_5	[0.3304,0.3369]	[6.8174,7.7484]	[2.8992,3.1304]	[0.1986,0.2015]
x_6	[0.2434,0.2439]	[2.6183,2.9668]	[0.1946,0.3095]	[0.7721,0.7866]
x_7	[0.0000,0.0007]	[6.4044,6.7258]	[0.2569,0.3264]	[0.6758,0.6873]
x_8	[0.1149,0.1289]	[2.1474,3.3136]	[1.4230,1.6642]	[0.3932,0.4133]
x_9	[0.5843,0.6121]	[17.351,19.069]	[0.9215,0.9771]	[0.5043,0.5241]
x_{10}	[0.5561,0.5606]	[0.3448,4.668]	[3.1142,3.7371]	[0.2022,0.2108]

In (4), there are 6 reducts corresponding to 6 feature subsets. The complexity of feature extraction is introduced to select the lowest complexity reducts. Finally, the feature subset composed of a_4, a_5, a_7 is regarded as the final result after computing. To test the performances of the reduced result, three-layer BP neural network is used to design classifiers. For every RES, 150 feature samples are generated in each of 5dB, 10dB, 15dB and 20dB. Thus, 600 samples of each RES in total are generated when SNR varies from 5dB to 20dB. The samples are classified into two groups: training group and testing group. The training group, which consists of one third of all samples, is applied to train neural network classifiers (NNC). The testing group, represented by other two thirds of samples, is used to test trained NNC. The structure of NNC is 3-15-10. We choose RPROP algorithm [14] as the training algorithm. Ideal outputs are "1". Output tolerance is 0.05 and output error is 0.001. The average accurate recognition rates of 50 experiments are shown in Table 3. To bring into comparison, the original feature set composed of 8 features is also used to recognize the 10 signals and the results are also shown in Table 3. Attribute reduction method not only simplify the structure of classifiers greatly, but also the average accurate recognition rate amounts to 97.23%, which is 1.18% higher than the original feature set made up of 8 features.

Table 2. The attribute table after discretization

Attributes	a_1	a_2	a_3	a_4	a_5	a_6	a_7	a_8
x_1	2	2	0	0	1	0	2	1
x_2	2	2	0	0	1	0	3	1
x_3	1	2	1	0	1	0	1	2
x_4	3	2	1	1	2	3	5	0
x_5	3	2	0	1	1	2	5	0
x_6	0	0	1	0	1	0	0	4
x_7	0	1	0	0	0	1	0	3
x_8	2	2	0	0	1	0	4	1
x_9	1	2	0	1	1	3	2	2
x_{10}	3	2	1	0	1	0	5	0

Table 3. Accurate recognition rates (%) before reduction (BR) and after reduction (AR)

Sig	x_1	x_2	x_3	x_4	x_5	x_6	x_7	x_8	x_9	x_{10}
BR	96.61	90.85	95.32	96.67	98.35	98.56	100.0	98.76	95.82	89.62
AR	97.05	87.10	98.71	100.0	99.50	97.65	100.0	100.0	99.59	92.71

5 Concluding Remarks

This paper generalizes firstly the definition of discretization based on cut-splitting and gives a generalized definition of discretization using class-separability function. Also, a new discretization approach is proposed to discretize continuous interval-valued attributes. Experimental results show that the introduced approach can discretize continuous interval-valued attributes effectively. The reduced attribute set simplifies the structure of classifiers and achieves higher accurate recognition rate than the original attribute set.

References

1. Pawlak, Z.: Rough sets. Informational Journal of Information and Computer Science. Vol.11, No.5. (1982) 341-356
2. Lin, T.Y.: Introduction to the special issue on rough sets. International Journal of Approximate Reasoning. Vo.15. (1996) 287-289
3. Wang, G.Y.: Rough set theory and knowledge acquisition. Xi'an: Xi'an Jiaotong University Press, (2001)
4. Walczak, B., Massart, D.L.: Rough sets theory. Chemometrics and Intelligent Laboratory Systems. Vol.47. (1999) 1-16
5. Shen, L.X., Tay, F.E.H, Qu, L.S., and Shen, Y.D.: Fault diagnosis using rough set theory. Computers in Industry. Vol.43. (2000) 61-72
6. Dai, J.H., Li Y.X.: Study on discretization based on rough set theory. Proc. of the first Int. Conf. on Machine Learning and Cybernetics. (2002) 1371-1373

7. Roy, A., Pal, S.K.: Fuzzy discretization of feature space for a rough set classifier. Pattern Recognition Letter. Vol.24. (2003) 895-902
8. Susmaga, R.: Analyzing discretizations of continuous attributes given a monotonic discrimination function. Intelligent Data Analysis. Vol.1. (1997) 157-179
9. Kusiak, A.: Rough set theory: a data mining tool for semiconductor manufacturing. IEEE Trans. on Electronics Packaging Manufacturing. Vol.24, No.1. (2001) 44-50
10. Bian, Z.Q., Zhang, X.G.: Pattern recognition. Beijing: Tsinghua University Press, (2000)
11. Zhang, G.X., Jin, W.D., and Hu, L.Z.: Fractal feature extraction of radar emitter signals. Proc. of 3rd Asia-Pacific conf. on Environmental Electromagnetics. (2003) 161-164
12. Zhang, G.X., Hu, L.Z., and Jin, W.D.: Complexity feature extraction of radar emitter signals. Proc. of 3rd Asia-Pacific Conf. on Environmental Electromagnetics. (2003) 495-498
13. Zhang, G.X., Rong, H.N., Jin, W.D., and Hu, L.Z.: Radar emitter signal recognition based on resemblance coefficient features. LNCS. Vol.3066. (2004) 665-670
14. Riedmiller M., Braun H.: A direct adaptive method for faster back propagation learning: the RPROP algorithm. Proc. of IEEE Int. Conf. on Neural Networks. (1993) 586-591

Fast Query Over Encrypted Character Data in Database*

Zheng-Fei Wang, Jing Dai, Wei Wang, and Bai-Le Shi

Fudan University, China
zhengfwang@sohu.com, {Daijing, weiwang1, bshi}@fudan.edu.cn

Abstract. Cryptographic support is an important dimension of securing sensitive data in databases. But this usually implies that one have to sacrifice performance for security. In this paper, we describe an novel cryptographic schemes for the problem of searching on encrypted character data and provide analysis of security for the resulting cryptographic systems.When the character data are stored in the form of cipher, we store not only the encrypted character data, but also characteristic values of character data as an additional field in database.When querying the encrypted character data,we apply the principle of two-phase query. Firstly, it implements a coarse query over encrypted data in order to filter part records not related to querying conditions. Secondly, it decrypts the rest records and implements a refined query over them. Results of a set of experiments validate the functionality and usability of this approach.

1 Introduction

Traditionally, database security has been provided by physical security and operating system security. To the best of our knowledge, however, neither of the methods provides a sufficient support to store and process data in a secure way. Cryptographic support is another important dimension of database security and should be used to guide the storage and access of confidential data in a database system [1–4]. But, how to efficiently query them becomes a challenge after data are encrypted. This usually implies that one has to sacrifice performance for security [5, 6].

It is very interesting to develop a method that directly deals with the encrypted data without decrypting them [7, 8]. But as far as we know, there is still no a secure method to ensure the security of encrypted data. Recently, Song et al[9] presents a new encryption scheme which allows searching the encrypted data without decryption. However, the encryption algorithm used in [9] is not appropriate for database. Hankan et al[10] proposes a way of executing query over the encrypted data in the database-service-provider model. But this approach is only valid for numerical data.

---

* This research is partially supported by National Natural Science Foundation of China under Grant No. 69933010.

In this paper, we propose a framework that can efficiently query over encrypted character data in database. When the character data are stored in the form of cipher, we store not only the encrypted character data, but also their characteristic values as an additional field in database. When querying the encrypted character data, We apply the principle of two-phase query. Firstly, it implements a coarse query over the encrypted character data in order to filter part records no relative with the query conditions. Secondly, it decrypts the rest of the records and implements a refined query over them.

Conventionally, we denote E the encryption function, D the decryption function. The remainder of the paper is organized as following: Section 2 presents the architecture of storage and query. section 3 discusses how encrypted data is stored and queried in database. Section 4 analyzes the security. Section 5 gives our experiment results of querying over the encrypted table from TPC-H benchmark. Section 6 concludes the paper.

2 Architecture of Storage and Query

Our proposed system, whose basic architecture is shown in figure 1, is to add an encryption/decryption layer between the application and DBMS. The purpose of such design is to implement storage and query over the encrypted data without changing the internal architecture of the present DBMS and applications.

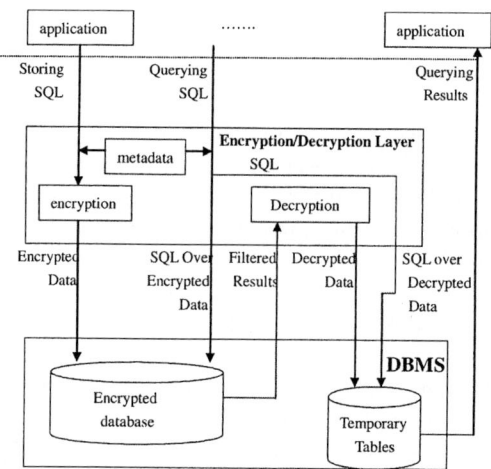

Fig. 1. Architecture of encrypted storage and query over character data

In the encryption/decryption layer of figure 1, the metadata module contains some mapping functions and transforming rules. While storing data, the metadata is used to transform the queries in order to store the characteristic value of the character data ; while querying the encrypted character data, metadata is used to transform user queries into appropriate queries on the encrypted data. The module of encryption and decryption contains encryption functions and decryption functions, which encrypt and decrypt the encrypted data, respectively.

3 Storage and Query Over Encrypted Character Data

In this section, we first introduce a characteristic function of character data, then present the encrypted storage scheme and query algorithm over the encrypted character data.

3.1 Characteristic Function and Encrypted Storage Scheme

Definition 1: if there is a function $PC: s_1 \rightarrow s_2$, where s_1 denotes a string of characters $c_1c_2\cdots c_n$, s_2 is a string of bits $b_0b_1\cdots b_{m-1}$, $b_i = 0$, $0 \leq i \leq m-1$, $n < m$. H denotes a hash function which encodes each connected character pairs $c_1c_2, c_2c_3, \cdots, c_{n-1}c_n$ of s_1 into a number between 0 and $m-1$. then the "signature" of the line $c_1c_2\cdots c_n$ is the string of m bits $b_0b_1\cdots b_{m-1}$, where $b_i = 1$ if and only if $H(c_jc_{j+1}) = i$ for some j, we call PC Pairs Coding Function.

Definition 2: for each relation scheme $R(X_1, \cdots, X_r, \cdots, X_n)$, where X_r field need to be encrypted, the corresponding encrypted relation scheme is $RE(X_1, \cdots, X_{rE}, \cdots X_n, X_{rS})$, where, X_{rE} is the encrypted field, $X_{rS} = PC(X_r)$ and PC is the pairs coding function. We call X_{rS} Pairs Coding Field of X_r, which is also called Index Field.

3.2 Query Over Encrypted Character Data

According to the encrypted storage scheme above, we apply the principle of two-phase to fast query the encrypted character data. Note that, the essential issue of query is how to translate normal query conditions in *where* clause into corresponding conditions over the index field of the encrypted table. This translation function is denoted as Tran( ). We now analyze how to translate the query conditions based on different query categories. In general, we consider query conditions as the following three categories.

(1) Simple Query. It gives a specific string value of a specific attribute. viz. *attribute=value*.

Definition 3: $Tran(a_i = v) \Rightarrow a_i^s = PC(v)$

Where field a_i has been encrypted, string v is the value of the query condition, a_i^s is the corresponding index field of a_i, PC is the pairs coding function.

(2) Contain Query. It gives a specific attribute which contains (or does not contain) a specific string value. viz. *attribute* like *value*, or *attribute* not like *value*.

Definition 4: $Tran(a_i$ like $c_1c_2\cdots c_k) \Rightarrow ((a_i^s)_{H(c_1c_2)} = 1)$ AND $((a_i^s)_{H(c_2c_3)} = 1)$ AND $\cdots ((a_i^s)_{H(c_{k-1}c_k)} = 1)$

Definition 5: $Tran(a_i \text{ not like } c_1c_2 \cdots c_k) \Rightarrow ((a_i^s)_{H(c_1c_2)} = 0) \text{ AND } ((a_i^s)_{H(c_2c_3)} = 0) \text{ AND } \cdots ((a_i^s)_{H(c_{k-1}c_k)} = 0)$

Where H is a hash function of the pairs coding function, $c_1c_2 \cdots c_k$ is the query value, $(a_i^s)_{H(c_{i-1}c_i)}$ denotes the $H_{(c_{i-1}c_i)}$ bit of (a_i^s).

(3) Boolean Query. It consists of the previous two types of queries combined with operation AND, OR, NOT. Typically, there are as following Tran().More complex translating functions result from the combination of these operations.

Definition 6: $Tran((a_i = v_1) \text{ OR } (a_i = v_2)) \Rightarrow Tran(a_i = (PC(v_1))) \text{ OR } Tran(a_i = PC(v_2)))$

Definition 7: $Tran((a_i = v_1) \text{ AND } (a_i = v_2)) \Rightarrow Tran(a_i = PC(v_1)) \text{ AND } Tran(a_i = PC(v_2)))$

Definition 8: $Tran((a_i \text{ like } v_1) \text{ OR } (a_i \text{ like } v_2)) \Rightarrow Tran(a_i \text{ like } (PC(v_1))) \text{ OR } Tran(a_i \text{ like } PC(v_2))$

Query Algorithm: Fquery
First Phase: Coarse Query Phase

(1) Translating the query conditions of SQL using the rules of metadata.
(2) Executing the translated SQL query, returning the records satisfying the translated query conditions and discarding the index field.

Second Phase: Refined Query Phase

(1) Decrypting the records returned in the first phase and storing them in a temporary table.
(2) Modifying the query SQL by replacing the original table with the temporary table.
(3) Executing the modified query SQL and obtaining actual results.

4 Security Analysis

Usually, it is very hard for attackers to directly infer the values of the sensitive field from the values of the index field due to the using of hash function in the process of mapping. However, in the environment of database, it is likely to suffer from the following two attacks.

(1) Statistical Attack. Assuming that the hash value is evenly distributed. When the number of bits m of the index field is increasing, the probability that different character pairs correspond to the same binary values in the corresponding index field is decreasing. That means, it is possible for different character pairs to be mapped to the different values in the index field. In this case, attackers can infer the values of the sensitive field using statistical methods.

(2) Known-plaintext Attack. Similarly, when the number of bits m of the index field is increasing, different character pairs probably correspond to the

different values in the index field, so that the attackers can infer some sensitive values using known-plaintext attack.

The two kinds of attacks are based on the hypothesis that the number of bits m of the index field is great enough. So, only if m is small, it is very difficult for the attackers to figure out the values of the sensitive index in this way.

5 Experiments and Performance Study

In this section, we will evaluate the filtering efficiency and the performance of query in our approach. According to TPC-H benchmark [11], the 10MB database is automatically created, of which lineitem table is used in our experiments. All experiments are conducted on a Pentium IV 2.5GHz PC machine with 512MB RAM. Relevant software platform used are Windows NT as the operating system and SQL Server as the database server. The SQL used in experiments is as following:

*select * from lineitem where comment like "query value"*

5.1 Experiment 1: Filtering Efficiency of Index Field

First, we test the filtering efficiency(FE) of the first phase in algorithm Fquery. We conduct these tests by varying the length of bits m in the index field, the length of the query value and the length of the encrypted character string.

In figure 2(a), the length of the query value is 3 bytes, and the different curves denote the query over the different length of the encrypted character string. In figure 2(b), the length of the encrypted character string is 16 bytes, and the different curves denote the different length of query value. We can find out from figure 2(a) and 2(b): (1) With m increasing, filtering efficiency accordingly improves. The reason is that it is more possible for different character strings to correspond to different values in the index field when m increases. (2) With the length of the encrypted character strings decreasing, filtering efficiency turns better. The reason is that it is more possible for different character strings to correspond to different values in the index field when length of the encrypted character strings decreases. (3) Filtering efficiency increases slowly when m is large than 32. In figure 2(c), m is 32 bits, and the different curves denote the

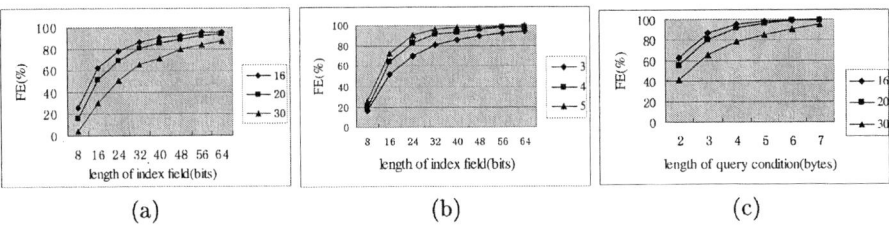

Fig. 2. Effect of filtering efficiency

query over different length of the encrypted character string. We can find out from figure 2(c): (1) With the length of query value increasing, the filtering efficiency improves. (2) Filtering efficiency can reach about 80% when m is 32 bits.

5.2 Experiment 2: Query Performance

Second, we test the query-execution time cost of the algorithm Fquery, and compare the result to the traditional way that is to decrypt all encrypted data before querying them. We conduct these tests by varying the length of bits m of the index field and the length of the query value.

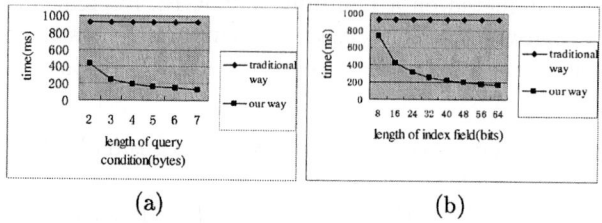

Fig. 3. Effect of query-execution time

In figure 3(a), the length of the query value is 3 bytes. We can find out: (1) With m increasing, the time cost is decreasing. It is due to the decreasing number of records need to be decrypted. (2)If m is larger than 32, the performance will improve slowly. In figure 3(b), the length of the encrypted character strings is 16 bytes. We can find out: (1)The time cost is decreasing with the length of the query value increasing. Obviously, the filtering efficiency will be better if the length of the query value is greater, accordingly, the cost of query time will decrease.(2) The time cost of query in the algorithm Fquery decreases about 75% compared with that in the traditional way.

6 Conclusions

In this paper, we present the architecture of storage and query over the encrypted character data, as well as the query algorithm Fquery. In the future, we would like to explore how to join multi-table through the encrypted field.

References

1. Henry, B.: Considerations in implementing a Database Management System Encryption Security solution. *A Research Report presented to The Department of Computer Science at the University of Cape Town*(2003)
2. George, L.D. et al.: A Database Encryption System with Subkeys. *ACM Transactions on Database Systems*, Vol. 6, No.2(1981) 312–328

3. Hakan, H.: Providing Database as a Service. In *Proceedings of ICDE* (2002) 29–38
4. He, J and Wang, M.: Cryptography and Relational Database Management System. In *Proceedings of IDEAS*(2001) 273–284
5. Oracle.: Oracle9i Database Security for eBusiness. *An Oracle White Paper* (2001)
6. Thomas, F.: Using Encryption for Secure Data Storage in Mobile Database Systems *[Ph D dissertation]* (2002)
7. Ahitub, N. et al.: Processing Encrypted Data. *Communications of the ACM*, September (1987) 777–780
8. Rivest, R. L. et al.: On Data Banks and Privacy Homomorphisms. In *Fundations of Secure Computation* (1978) 169–178
9. Song, D. et al.: Practical Techniques for Searches on Encrypted Data. *IEEE Symposium on Security and Privacy* (2000) 44–55
10. Hakan, H. et al.: Executing SQL over Encrypted Data in the Database-Server-Provider Model. In *Proceedings of ACM SIGMOD* (2002) 216–227
11. TPC-H.: Benchmark Specification. http://www.tpc.org

Factoring-Based Proxy Signature Schemes with Forward-Security*

Zhenchuan Chai and Zhenfu Cao

Department of Computer Science and Engineering, Shanghai Jiaotong University,
1954 Huashan Road, Shanghai 200030, P.R. China
{zcchai, zfcao}@cs.sjtu.edu.cn

Abstract. The property of forward-security can be used to reduce the potential damages caused by key exposure. However, none of the existing proxy signature schemes has taken forward-security into consideration. And most proxy signature schemes are based on discrete logarithm problem. In this paper, we first present a forward-secure proxy signature scheme based on factoring problem. Next, we present a forward-secure threshold proxy signature scheme.

1 Introduction

In 1996, Mambo, Usuda and Okamoto first introduced the concept of proxy signature[1]. In their scheme, an original signer delegates his signing right to a proxy signer in such a way that the proxy signer can sign any message on behalf of the original signer and the verifier can verify and distinguish proxy signature from original signature. Proxy signature is of great use in such a case that a manager needs to authorize his secretary to sign documents on behalf of himself before his leaving for a vacation. Due to its usefulness, proxy signature has drawn great attention, and a lot of researches have been done on it [2–4].

The security of a digital signature could be obtained only under the hypothesis that the secret key is not revealed. So how to solve the key exposure problem[5] is crucial in practical systems. Many solutions have been put forward to minimize the chances of key leakage, such as threshold signatures[7] and proactive signatures[8]. But those solutions that minimize the risk of key exposure do not guarantee that the secret key will never be stolen. Once a key is revealed, security is often compromised for subsequent use of the key. Moreover, the real disaster is that any signatures signed by that key, even the real ones before the key exposure, become unauthentic. To address such problem, forward-security[5, 9] was put forward to reduce the potential damage caused by key exposure. A forward-secure signature scheme requires that it is infeasible for

---

* The paper is supported by the National Natural Science Foundation of China under Grant No. 60072018, the National Science Fund for Distinguished Young Scholars under Grant No. 60225007 and the National Research Fund for the Doctoral Program of Higher Education of China under Grant No.20020248024.

an adversary to forge signatures of past, even if he discovers the secret key for the current time. Forward-security is often obtained by key evolution. The idea of key evolution[6] is illustrated in Fig.1. In a forward-secure signature scheme, the time during which the public key PK is supposed to be valid is divided into T periods. And during each period, the signer has a unique secret key SK_i ($i = 0, \cdots, T$) with his public key fixed. Each secret key of current period is computed from the secret key of last period, using a one-way function $f()$, so the leakage of current key does not lead to exposure of keys of previous periods.

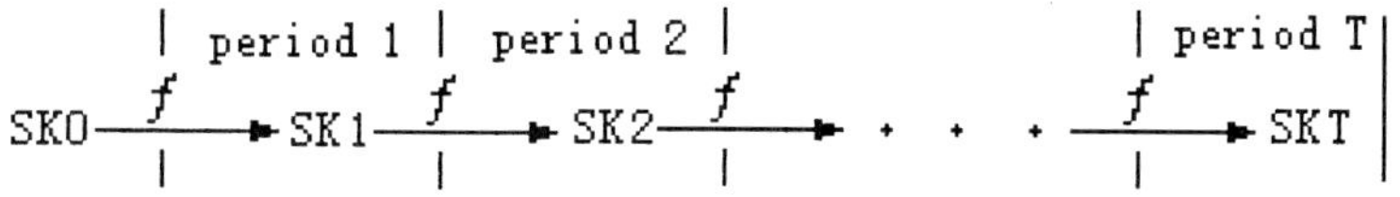

$$SK0 \xrightarrow{f} | \text{period 1} | SK1 \xrightarrow{f} | \text{period 2} | SK2 \xrightarrow{f} \cdots \xrightarrow{f} | \text{period T} | SKT$$

Fig. 1. Key-evolving scheme

In the existing proxy signature schemes, most are based on the discrete logarithm problem, so it is worthwhile to design proxy signature based on factoring[4]. We will present a proxy signature scheme based on factoring. Our proxy signature scheme takes forward-security into account, which is not involved in the existing proxy signature schemes. In our scheme, the original signer and proxy signer need not to update their secrets synchronously. We will also put forward a threshold proxy signature scheme that inherits [9]'merit of short keys.

2 A Forward-Secure Proxy Signature Scheme

In our scheme, a modulus $N = p_1 \times p_2$, where p_1 and p_2 are large primes, is generated by a trusted center, and shared by original singer and proxy signer. But neither the original singer nor the proxy signer knows p_1 or p_2.

It is assumed that the original signer divides the public-key-valid time into T_o periods, and proxy signer divides the time into T_p periods. The original signer is supposed to give his authority to the proxy signer at period 0. And at period 0, the proxy signer accepts the proxy authority. Additionally, there is a secure parameter l and a public hash function $H: \{0,1\}^* \to \{0,1\}^l$.

Suppose the original signer has private key $SK_o = S_{o,0}$ and public key $PK_o = U_o$ at period 0, where $U_o = 1/S_{o,0}^{2^{l(T_o+1)}} (\bmod N)$, and $S_{o,j} (j = 0, \cdots, T_o)$ denotes original signer's secret at period j. Similarly, the proxy signer has his private key $SK_p = S_{p,0}$ and public key $PK_p = U_p$, where $U_p = 1/S_{p,0}^{2^{l(T_p+1)}} (\bmod N)$, and $S_{p,j} (j = 0, \cdots, T_p)$ denotes proxy signer's secret at period j.

Let $T = \max(T_o, T_p)$, and A denotes original signer, P denotes proxy signer.

Proxy Key Generation:

(1) A chooses $r \in_R Z_N^*$, computes $Y = 1/r^{2^{l(T+1)}} (\bmod N)$ and $e = S_{o,0}^{H(m_w, ID)} r (\bmod N)$, where m_w is a warrant, ID is the identity of the proxy signer. Then A sends (e, Y) to P.

(2) P checks the validity of (e, Y) with equation $YU_o^{2^{l(T-T_o)}H(m_w,ID)}e^{2^{l(T+1)}} = 1 (\mod N)$.
(3) If (e, Y) is valid, then P computes his proxy signing key: $S_0 = S_{p,0} \times e = S_{p,0} S_{o,0}^{H(m_w,ID)} r (\mod N)$.

Key Evolution:
At the end of period $j (j = 0, \cdots, \min(T_o, T_p))$, P updates the key as follows: If $j = \min(T_o, T_p)$, then he sets proxy signing key as null string, otherwise he computes the proxy signing key of period $j+1$ as $S_{j+1} = S_j^{2^l} (\mod N)$, then discards S_j. It is worth pointing out that:

$$S_j = S_0^{2^l \times j} (\mod N) = S_{p,0}^{2^l \times j} S_{o,0}^{2^l \times j H(m_w,ID)} r^{2^l \times j} (\mod N) \qquad (1)$$

Proxy Signing Phase:
Signature on M during period j is signed by P via the following steps:

(1) P chooses k from Z_N^* randomly, and computes $K = k^{2^{l(T+1-j)}} (\mod N)$.
(2) P then computes $\sigma = H(j, K, M, U_o)$ and $Z = kS_j^\sigma (\mod N)$.

P outputs the proxy signature as $(j, (Z, \sigma), Y)$, where Y is obtained from A during proxy key generation phase.

Verification Phase:

(1) Receiving the signature $(j, (Z, \sigma), Y)$ on M, the verifier computes $K' = Z^{2^{l(T+1-j)}} Y^\sigma U_o^{\sigma \cdot 2^{l(T-T_o)} H(m_w,ID)} U_p^{\sigma \cdot 2^{l(T-T_p)}} (\mod N)$.
(2) The verifier then checks the validity of the signature with equation $\sigma = H(j, K', M, U_o)$. If the equation holds, then the signature is valid.

Because: by equation (1)

$$S_j^{2^{l(T+1-j)}} = S_{p,0}^{2^{l(T+1)}} S_{o,0}^{2^{l(T+1)} H(m_w,ID)} r^{2^{l(T+1)}}$$
$$= U_p^{-2^{l(T-T_p)}} U_o^{-2^{l(T-T_o)} H(m_w,ID)} Y^{-1} (\mod N)$$

Then $K' = Z^{2^{l(T+1-j)}} Y^\sigma U_o^{\sigma \cdot 2^{l(T-T_o)} H(m_w,ID)} U_p^{\sigma \cdot 2^{l(T-T_p)}}$
$= k^{2^{l(T+1-j)}} S_j^{\sigma 2^{l(T+1-j)}} Y^\sigma U_o^{\sigma \cdot 2^{l(T-T_o)} H(m_w,ID)} U_p^{\sigma \cdot 2^{l(T-T_p)}}$
$= KU_p^{-2^{l(T-T_p)}\sigma} U_o^{-2^{l(T-T_o)}\sigma H(m_w,ID)} Y^{-\sigma} Y^\sigma U_o^{\sigma \cdot 2^{l(T-T_o)} H(m_w,ID)} U_p^{\sigma \cdot 2^{l(T-T_p)}} = K$
Therefore $\sigma = H(j, K, M, U_o) = H(j, K', M, U_o)$.

3 A (t,n) Threshold Proxy Signature Scheme with Forward-Security

Now we modify the above proxy signature scheme into a (t,n) threshold proxy signature scheme with forward-security. In our setting, the original signer A

is supposed to delegate his authority to the proxy group that consists of n members $P_i(i = 1, \cdots, n)$ at period 0. Original signer and proxy group are set to have the same number of periods T just for clarity of notation. Also, there are shared modulus N, a secure parameter l and a public hash function $H : \{0,1\}^* \to \{0,1\}^l$.

At period 0, the original signer A chooses $S_{o,0}$ and d from Z_N^* to generate his private key $SK_o = S_{o,0}^d (\bmod N)$ and public key $PK_o = U_o$, where $U_o = 1/(S_{o,0}^d)^{2^{l(T+1)}}(\bmod N)$, and $S_{o,j}^d$ ($j = 0, \cdots T$) denotes original signer's secret key at period j. The proxy group also has a key pair $\langle PK_p = U_p, S_{p,0}^e \rangle$ where $U_p = 1/(S_{p,0}^e)^{2^{l(T+1)}}(\bmod N)$, but the secret key is distributed among n members. Each member P_i has a shadow of secret key as $S_{p,0}^{g(i)}(i = 1, \cdots, n)$, where $g(x) = e + x \cdot b_1 + x^2 \cdot b_2 + \cdots + x^{t-1} \cdot b_{t-1}(\bmod N)$ is a polynomial over Z_N^* of degree $t-1$. The distribution of the secret key could be done by a trusted dealer.

Let $\lambda_{i_u} = \prod_{j=1, j \neq u}^{t} \frac{-i_j}{i_u - i_j}$ ($i_u \in \{1, \cdots, n\}, u = 1, \cdots, t$).

Proxy Key Generation:

(1) A picks a random r in Z_N^*, computes $Y = 1/(r^d)^{2^{l(T+1)}}(\bmod N)$.
(2) A chooses a polynomial $f(x) = d + x \cdot a_1 + x^2 \cdot a_2 + \cdots + x^{t-1} \cdot a_{t-1}(\bmod N)$ over Z_N^*, and calculates $e_i = (S_{o,0}^{H(m_w, ID)} r)^{f(i)}(\bmod N)(i = 1, \cdots, n)$, then sends e_i to P_i secretly.

Note $d = f(0) = \sum_{u=1}^{t} f(i_u) \lambda_{i_u} (\bmod N)$.
(3) A calculates $E_i = H(e_i)(i = 1, ..., n)$ then broadcasts E_i.
(4) After receiving e_i, $P_i(i = 1, ..., n)$ can verify e_i by checking the equation $E_i = H(e_i)$. If e_i is not valid, then the process is stopped.
(5) Each P_i computes his secret singing key $x_{i,0} = e_i S_{p,0}^{g(i)}(\bmod N)(i = 1, ..., n)$, and then discards $S_{p,0}^{g(i)}$, where $x_{i,j}(i = 1, \cdots, n; j = 0, \cdots, T)$ denotes the secret key P_i holds for period j.

Key Evolution:
At the end of period $j(j = 0, \cdots, T)$, each P_i updates his secret key as follows: If $j = T$, then P_i sets secret key as null string, otherwise he computes the secret key for period $j + 1$ as $x_{i,j+1} = x_{i,j}^{2^l}$, then discards $x_{i,j}$. Note:

$$x_{i,j} = x_{i,0}^{2^{l \times j}} = (e_i S_{p,0}^{g(i)})^{2^{l \times j}} = (S_{o,0}^{H(m_w, ID) f(i)} r^{f(i)} S_{p,0}^{g(i)})^{2^{l \times j}}(\bmod N) \qquad (2)$$

Proxy Signing Phase:
Signature generation for a given M during period j is done by t out of n members of the group, $P_{i_u}(i_u \in \{1, \cdots, n\}, u = 1, \cdots, t)$, as follows:

(1) Each P_{i_u} picks random k_{i_u} in Z_N^*, and computes $K_{i_u} = k_{i_u}^{2^{l(T+1-j)}} (\mod N)$, then sends K_{i_u} to a designated member called dealer.

(2) After collecting all the K_{i_u} ($i_u \in \{1, \cdots, n\}, u = 1, \cdots, t$), the dealer computes $K = \prod_{u=1}^{t} K_{i_u} (\mod N)$ and $\sigma = H(j, K, M, U_o)$, then returns σ to all the participants P_{i_u}.

(3) After receiving σ, each P_{i_u} computes his partial signature on M as $Z_{i_u} = k_{i_u}(x_{i_u,j}^\sigma)^{\lambda_{i_u}} (\mod N)$. Then P_{i_u} sends his partial signature Z_{i_u} to the dealer.

(4) The dealer computes $Z = \prod_{u=1}^{t} Z_{i_u} (\mod N)$, and outputs the signature as $(j, (Z, \sigma), Y)$, where Y is obtained from A during proxy key generation phase.

Verification Phase:

(1) Receiving the signature $(j, (Z, \sigma), Y)$ on M, the verifier computes $K' = Z^{2^{l(T+1-j)}} Y^\sigma U_o^{\sigma H(m_w, ID)} U_p^\sigma (\mod N)$.

(2) The verifier then checks the validity of the signature with equation $\sigma = H(j, K', M, U_o)$. If the equation holds, then the signature is valid.

Because: $Z^{2^{l(T+1-j)}} = \prod_{u=1}^{t} Z_{i_u}^{2^{l(T+1-j)}} = \prod_{u=1}^{t} k_{i_u}^{2^{l(T+1-j)}} (x_{i_u,j}^{\lambda_{i_u}})^{2^{l(T+1-j)}\sigma}$

$\stackrel{equation(2)}{=} K \prod_{u=1}^{t} (S_{o,0}^{f(i_u)\lambda_{i_u} H(m_w, ID)} r^{f(i_u)\lambda_{i_u}} S_{p,0}^{g(i_u)\lambda_{i_u}})^{2^{l(T+1)}\sigma}$

$= K(S_{o,0}^{H(m_w, ID)\sum_{u=1}^{t} f(i_u)\lambda_{i_u}} r^{\sum_{u=1}^{t} f(i_u)\lambda_{i_u}} S_{p,0}^{\sum_{u=1}^{t} g(i_u)\lambda_{i_u}})^{2^{l(T+1)}\sigma}$

$= K(S_{o,0}^{dH(m_w, ID)} r^d S_{p,0}^e)^{2^{l(T+1)}\sigma} \equiv KU_o^{-\sigma H(m_w, ID)} Y^{-\sigma} U_p^{-\sigma} (\mod N)$

Then $K' = Z^{2^{l(T+1-j)}} Y^\sigma U_o^{\sigma H(m_w, ID)} U_p^\sigma \equiv K (\mod N)$

Therefore $\sigma = H(j, K, M, U_o) = H(j, K', M, U_o)$.

4 Security Analysis

In our proxy signature scheme, the proxy signer uses his proxy-signing key S_j to generate a signature on M the same way as Abdalla's scheme [9]. So outsiders could not forge a valid proxy signature because of the security of Abdalla's scheme.

In the verification phase, both the original signer's and proxy signer's public keys are involved, so it is easy to verify that the signature is signed by proxy signer but not by original signer, hence our scheme is distinguishable. Now we examine the possible attacks that the original signer may launch to forge a proxy signature. The original signer may obtain a valid proxy signature on M as $(j, (Z, \sigma), Y)$. Knowing $Z = kS_j^\sigma = kS_{o,j}^\sigma S_{p,j}^\sigma r^{2^{l \times j}\sigma H(m_w, ID)} (\mod N)$, he can extract $kS_{p,j}^\sigma$ from Z with his private key $S_{o,j}$ and r which he generated during proxy key generation phase. The $kS_{p,j}^\sigma$ is somewhat like a signature signed by proxy signer with his own key. But $\sigma = H(j, K, M, U_o)$ contains original signer's public key, so $kS_{p,j}^\sigma$ is not a valid signature signed by proxy signer. And if original signer has the ability to forge a valid signature of proxy singer or a valid proxy signature from $kS_{p,j}^{\sigma}$, then he must have the ability to forge a valid signature under Abdalla's scheme[9], which is proved to be secure under random oracle model. Similarly, although the proxy signer receives (e, Y) during proxy key generation phase, where $e = S_{o,0}^{H(m_w, ID)} r (\bmod N)$, $Y = 1/r^{2^{l(T+1)}} (\bmod N)$, he could not forge a valid original signature.

In our threshold scheme, each member of the proxy group has a shadow of the proxy signing key. To sign a message, t participants compute their own partial signatures on the message with their own shadows, and then multiply their partial signatures to generate a valid proxy signature. During the signing process, the proxy group's signing key is not formed explicitly, so our threshold scheme is a strong threshold scheme. If t members collude at proxy key generation phase, then they may work out $S_{o,0}^{dH(m_w, ID)} r^d$ with $e_i = (S_{o,0}^{H(m_w, ID)} r)^{f(i)} (\bmod N)$ which P_i received from original signer. They may also collude to work out the proxy group's signing key by computing $S_{p,0}^e = \prod_{u=1}^{t} (S_{p,0}^{g(i_u)})^{\lambda_{i_u}}$. Now, if t colluders want to forge some signatures, they are confronted with the same situation of non-threshold scheme we discussed above, so they could not forge a valid original signature. And what they can do is that a single one of the colluders could sign a valid proxy signature without the help of others. But the irony is that a valid signature could be generated by the t colluders without the need to collude. So our scheme is secure against collusion.

5 Conclusion

In this paper, we first propose a forward-secure proxy signature scheme that is based on factoring problem. And then we give a (t, n) threshold proxy signature with forward-security. Both these two schemes do not need the original signer and proxy signer to update their keys synchronously.

References

1. Mambo, M., Usuda, K., Okamoto, E.: Proxy signatures: Delegation of the power to sign messages. IEICE Trans. Fundam., 1996, E79-A, (9), 1338-1354
2. Mambo, M., Usuda, K., Okamoto, E.: Proxy signatures for delegating signing operation. Proc. 3rd ACM Conference on Computer and Communications Security, ACM, 1996, 48-57
3. Zhang, K.: Threshold proxy signature schemes. Information Security Workshop, Japan(1997), 191-197
4. Shao, Z.H.: Proxy signature schemes based on factoring. Information Processing Letters 85(2003) 137-143
5. Abdalla, M., Miner, S., et al.: Forward-secure Threshold Signature Scheme. RSA'01
6. Bellare, M., Miner, S.: A forward-secure digital signature scheme. Advances in Cryptology Crypto'99 Proceedings, Lecture Notes in Computer Science, Vol.1666, Wiener, M. ed., Springer-Verlag 1999

7. Desmedt, Y., Frankel, Y.: Threshold Cryptosystems. Advances in Cryptology-Crypto'89 Proceeding, Lecture Notes in Computer Science, Vol. 435, Brassard, G. ed., Springer-Verlag, 1989
8. Herzberg, A., Jakobsson, M., Jarecki,S.: Proactive public key and signature schemes. Proceeding of the Fourth Annual Conference on Computer and Communications Security, ACM, 1997
9. Abdalla, M., Reyzin, L.: A New Forward-Secure Digital Signature Scheme. Asiacrypt 2000, Lecture Notes in Computer Science

A Method of Acquiring Ontology Information from Web Documents

Lixin Han[1,2,3], Guihai Chen[2], and Li Xie[2]

[1] Department of Mathematics, Nanjing University, China
[2] State Key Laboratory of Novel Software Technology, Nanjing University, China
[3] Department of Computer Science and Engineering, Hohai University, China
lixinhan2002@yahoo.com.cn

Abstract. Ontology plays an important role on the Semantic Web. In this paper, we propose a method, AOIWD, of acquiring ontology information from Web documents. The AOIWD method employs data mining techniques combined with inference engine techniques to develop ontology. One key feature of AOIWD is that it employs association rules and inference mechanism to create association instances of ontology classes. Another key feature of AOIWD is that it employs clustering to create ontology class hierarchy. The method can discover more meaningful relationships and has less human intervention.

1 Introduction

The emergence of the Semantic Web makes it possible for machines to understand the meaning of resources on the web. Thus it provides opportunities for automated information processing. Ontology plays an important role on the Semantic Web by providing a source of shared and well defined terms that can be used in meta-data.

Generally there are two classes of methods for acquiring ontology information from Web documents. Firstly, The domain experts and users build the ontology by themselves [1]. However, The method overly depends on people's experience. It is hard and time-consuming to implement. Secondly, DAML+OIL and other description-logics languages rely on inference engines to create a class hierarchy and to determine class membership of instances, which is based on the properties of classes and instances [2]. This method increases computing workload, and lack of enough knowledge leads to less accuracy.

We propose a method called AOIWD (Acquiring Ontology Information from Web Documents). AOIWD is different from both classes of the above methods in that we use essentially a combination of data mining and inference engine techniques to acquire ontology information. One key feature of AOIWD is that it employs association rules and inference mechanism to create association instances of ontology classes. Another key feature of AOIWD is that it employs clustering to create ontology class hierarchy.

2 AOIWD Method

We propose the AOIWD method of acquiring ontology information from Web documents. The AOIWD method consists mainly of the CAA algorithm for creating association instances, the CCH algorithm for creating class hierarchy and the DACH algorithm for dynamically maintaining class hierarchy.

2.1 CAA Algorithm for Creating Association Instances

Different types of instances determine different expectations for information that might be available. For example, if one person is a professor, we can expect to find pages related to his research interests, publications, research projects etc. Thus the identification of association relationships between concepts is very important.

Data mining techniques can find the existed patterns in the data. However, it is not possible to encode all the relevant relationships as rules, because they are not all usually known. The existed relationships in the knowledge base provide a scope for discovering new relationships. Based on the existed association instances acquired from association rules algorithm, new meaningful relationships can be discovered through an inferential process. Accordingly, based on the idea above, we propose the CAA algorithm to find association instances.

The CAA algorithm is described as the following steps:

Input: a document
Output: association instances
{ S1=nil; // S1 stores a set of association instances
  S={the instances in the document};
  frequent 1-itemsets called L_1={the instances};
  count=2;
  While count$\leq$k   // k is the maximum number of instances in S
  { the frequent itemset called $L_{count-1}$ creates the set of new candidate itemsets
      C_{count};
    order the items in descending order in every candidate itemset;
    according to the predefined support, count the candidate itemsets in C_{count};
    create frequent itemset called L_{count};
    count=count+1;}
  S1=$L_1 \cup L_2 \cdots \cup L_{count}$;
  delete the subset with inclusion relation from S1 in order to acquire association instances;
  define some rules from association instances;
  store these rules into the knowledge base;
  provide the inference mechanism for reasoning about semantic associating the existed rules in the knowledgebase in order to discover more meaningful association instances;}

In contrast to the widely used APRIORI algorithm [3], a sequence number is added to every item in order to order the items in descending order in candidate itemsets in the CAA algorithm. Moreover, some typical documents collected from different Web sites reduce the size of the transaction database. Thus the computing workload is reduced while creating frequent itemsets.

In addition, the data mining algorithms, such as the APRIORI algorithm, do not easily find more meaningful association instances. But the CAA algorithm provides the inference mechanism to find more useful association instances.

2.2 CCH Algorithm for Creating Class Hierarchy

The CCH algorithm creates a class hierarchy based on the similarities between classes. It employs hierarchy cluster method from bottom to top as there are not many classes in the class hierarchy. Compared with the other hierarchy cluster algorithms, the CCH algorithm is not restricted by topology and the algorithm stops once the degree of similarity is below threshold. Therefore, it has a faster convergence rate in building the class hierarchy. Its time complexity is $O(n \cdot \log_2 n)$, where n is the number of classes.

The CCH algorithm is described as the following steps:

```
Input: n classes
Output: a class hierarchy
{ classset = { n classes };
  height=H;
  While similarity>threshold and height>1
  {  according to the Cosine similarity formula, the similarity between the classes
     in classset is calculated;
     according to the similarity, similar classes are regarded as an element in the
     set;
     classset = set;
     pair= first(set) ;
     While set≠nil
     {  if the similarity between the classes in pair> threshold
        then { regard the classes in pair as the subclasses and merge the classes in
               pair in order to get their parent class called fclass;
               classset = classset –{pair};//delete the merged classes from classset
               classset = classset + { fclass };}
        pair =next(set); }
     height= height - 1;}}
```

2.3 DACH Algorithm for Dynamically Maintaining Class Hierarchy

How to dynamically merge the classes into class hierarchy becomes more and more important when ontology changes. We propose a new algorithm, DACH, for dynamically maintaining class hierarchy. Its time complexity is $O(N)$, where N is the layer number.

The DACH algorithm is described as the following steps:

```
Input: class hierarchy, the merged class
Output: merge the class into class hierarchy
{ set = class hierarchy;
```

```
height =1;
using the Cosine similarity formula, calculate the similarity between the merged
class and the node in height layer;
find the node with the maximal similarity called maxnode;
While height<N-1
{ height = height+1;
  find the node connected to maxnode in the height layer;
  using the Cosine similarity formula, calculate the similarity between the
  merged class and the node;
  find the node with the maximal similarity called maxnode;}
if      similarity> threshold
then {  merge the class into maxnode;
        store the class in the leaf node;}
else    store the classes as a leaf node;}
```

3 Prototype Implementation

We adopted the above algorithms in a prototype system PTIDM. Our platform is composed of 2 IBM RS6000 workstations and 2 microcomputers. IBM RS6000 workstations are connected to the ATM network by 155MB/S. The microcomputers are connected to the Ethernet by 10MB/S. All computers are interlinked. We use JAVA program language in order to accomplish platform independence. PTIDM uses DAML+OIL to describe resources and their inter-relations to improve inferential capability. PTIDM can help to acquire ontology information. The CAA algorithm is used to acquire meaningful relationships between different types of instances. Thus, the useful classes may be acquired. According to the similarity between classes, the CCH algorithm is used to create class hierarchy. The DACH algorithm is used to dynamically add the class to a class hierarchy. It is proved that the prototype system has good effect.

4 Related Work

Although RDF [4] employs XML to specify semantic networks of information on web pages, there is not any primitive for creating ontology in RDF. Thus it only has a weak idea of ontology. However, some of the Web ontology languages are based on RDF.

RDF Schema [5] is improved to some extent. There are some primitives for defining ontology in RDF Schema. In RDF Schema, there is a class hierarchy with multiple inheritances. However, RDF Schema lacks inferential capability. In addition, RDF Schema doesn't maintain effective ontology when ontology changes.

In the SHOE language [2], [6], Ontology consists of category and relation definitions. Here classes are called *categories*. Categories form a simple *is-a* hierarchy, and slots are binary relations. It also allows relations among instances or instances and data to have any number of arguments. Horn clauses express intensional definitions. SHOE allows all ontologies publicly available on the Web in order to promote interoperability. It also creates some ontologies in specific domain by ontology extension.

DAML+OIL [2], [7] employs a different way of defining classes and instances. Besides defining classes and instances declaratively, it use Boolean expressions and the restrictions of class membership to create class definitions. It relies on inference engines to create a class hicrarchy and class membership of instances, which are based on the properties of classes and instances.

Protégé-2000 [2] provides an integrated environment for editing ontology. It is a conceptual-level ontology editor and knowledge acquisition tool. Thus, developers only consider concepts and relations in the domain that they are modeling. Developers can also customize Protégé-2000 easily to be an editor for a new Semantic Web language.

5 Conclusion

Ontology plays an important role on the Semantic Web. Ontology provides a source of shared and well defined terms that can be used in meta-data. In this paper, we propose a method, AOIWD, of acquiring ontology information from Web documents. The AOIWD method employs data mining combined with inference engines to acquire ontology information. In the AOIWD method, we propose the CAA algorithm to create association instances, the CCH algorithm to create class hierarchy and the DACH algorithm to dynamically maintain class hierarchy. The method can find more meaningful relationships and has less human intervention.

Acknowledgement

This work is supported by the National Grand Fundamental Research 973 Program of China under No. 2002CB312002, the State Key Laboratory Foundation of Novel Software Technology at Nanjing University under grant A200308 and the Key Natural Science Foundation of Jiangsu Province of China under grant BK2003001.

References

1. Sure, Y. et al. Methodology for Development and Employment of Ontology Based Knowledge Management Applications. SIGMOD Record Vol 31, No 4, December 2002.
2. Noy, N. F. et al. Creating Semantic Web Contents with Protégé--2000. IEEE Intelligent Systems 16(2):60-71, 2001.
3. Agrawal, R. and Srikant, R.. Fast Algorithms for Mining Association Rules. In Proc. of the 20th Int. Conf. on Very Large Databases (VLDB'94), pages 478-499, Santiago, Chile, Sep. 1994. Expanded version available as IBM Research Report RJ9839,June 1994.
4. Lassila, O and Swick, R. R. . Resource Description Framework (RDF) Model and Syntax Specification.W3C Recommendation 22 .www.w3.org/TR/1999/REC-rdf-syntax-19990222/
5. Brickley, D. and Guha,R.V.. Resource Description Framework (RDF) Schema Specification.W3C Candidate Recommendation 27,www.w3.org/TR/2000/CR-rdf-schema-20000327.

6. Heflin, J. and Hendler, J.. Searching the Web with SHOE. In AAAI-2000 Workshop on AI for Web Search. 2000.
7. Hendler, J. and McGuinness, D.L.. "The DARPA Agent Markup Language," IEEE Intelligent Systems, vol. 16, no. 6, Jan./Feb., 2000, pp. 67–73.
8. Han, L. et al. "A Method of Extracting Information From the Web". Journal of The China Society for Scientific and Technical Information. Vol.23, No.1. February, 2004, pp. 45-51.
9. Maedche, A. et al. An Infrastructure for Searching, Reusing and Evolving Distributed Ontologies. WWW2003, May 20-24, 2003, Budapest, Hungary.

Adopting Ontologies and Rules in Web Searching Services*

He Hu and Xiaoyong Du

School of Information, Renmin University of China, Beijing 100872, P.R. China
{hehu, duyong}@ruc.edu.cn

Abstract. Keyword-based search has been popularized by Web search services. However, due to the problems associated with polysemy and synonym, users are often unable to get the exact information they are looking for. In this paper, we propose a web search framework in which ontologies and rules are used to deal with the synonym and polysemy problem to gain a better recall and precision performance than the traditional search services. A demonstration system based on Google's web search APIs is currently under implementation.

1 Introduction

The Web is one of the fastest growing media in the world with millions of pages adding in every day. Buried in this vast, quickly growing collection of documents lies information of interest and use to almost everyone; the challenge is finding it. Currently, search engines are universally employed to find information on the Web. Typically a search engine service works by utilizing keyword (syntactic) matching to find key terms on web sites.

Recall and precision are two of the most widely used measures of information retrieval effectiveness for search services. Recall measures how well a service retrieves all the relevant documents, whereas precision measures how well the system retrieves only the relevant documents. There are two main problems impacting the recall and precision level of search services [1]: Problem of polysemy: search services index words, not semantic units; Problem of synonym: search services may miss relevant pages in the answer, just because they apply a different word for the same meaning.

2 Preliminary

2.1 Ontology Approach and Semantic Web

Current search requests must be specified as key words separated by Boolean operators. Search services can only retrieve data on a purely syntactic basis. It

---

* This research was supported by NSFC of China (project number: 604963205)

is still not possible to embed domain specific knowledge into the search services' queries. Ontology approach can solve this problem by providing a semantic foundation in these systems. Tom Gruber [2] has defined ontology as "a specification of a conceptualization". Ontologies provide a deeper level of meaning by providing equivalence relations between concepts; they can standardize meaning, description, representation of involved concepts, terms and attributes; capture the semantics involved via domain characteristics, resulting in semantic metadata which forms basis for knowledge sharing and reuse.

The Semantic Web has been regarded as the next version of current Web, which aims to add semantics and better structure to the information available on the Web. Underlying this is the goal of making the Web more effective not only for humans but also for automatic software agents. The basic idea is to create an environment for intelligent programs to carry out tasks independently on behalf of the user. Ontologies in fact turn out to be the backbone technology for the Semantic Web; Tim Berners-Lee [3] has portrayed Semantic Web as a layered architecture where ontology layer lies in the middle of the other layers.

2.2 Rule-Based Systems

Research on rule-based systems began in the 1960's. In the early seventies, Newell and Simon proposed a production system model [4], which is the foundation of the modern rule-based systems. The production model is based on the idea that humans solve problems by applying their knowledge (expressed as production rules) to a given problem represented by problem-specific information. The production rules are stored in the long-term memory and the problem-specific information or facts in the short-term memory. A rule-based system usually consists of: a rule set (the rules to evaluate), a working memory (stores state), a matching scheme (decides which rules are applicable) and a conflict resolution scheme (if more than one rule is applicable, decides how to proceed).

3 Ontologies Plus Rules for Search Services

3.1 Ontology Model

The ontology model mainly concentrates on solving synonymy problem. For example, "plane", "aeroplane" and "airplane" all have the meaning of "flying machine"; when a Web user include the term "plane" in his (her) query, our ontology model should automatically take the other two terms into account. The ontology model supported query process is illustrated in Fig. 1:

Nicola Guarino[5] has partitioned ontology into three different kinds according to their level of generality. Various ontology-based system adopted this partition, for example, Teknowledge Ontologies[6] use upper ontology as foundation and mid-level ontology as integration framework to support application ontologies creation and utilization. In our Web search framework, the ontology model is comprised of application and domain ontoloies which are based on SUMO top-level ontologies [7].

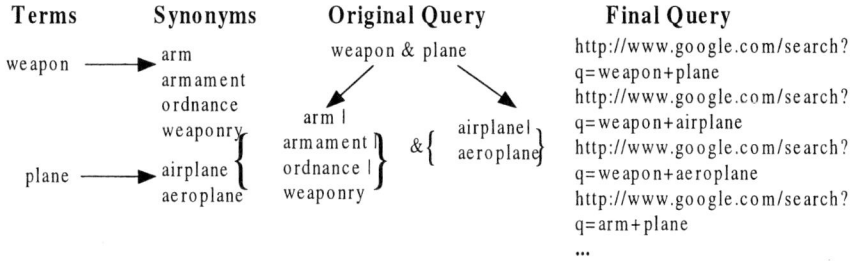

Fig. 1. Query Processing supported by Ontology Model

Unlike the complex ontology structure proposed in [8], the ontology structure of our model is quite simple, ontologies in the framework are represented as hierarchies of terms, and synonymy problem is settled by introducing all the necessary synonymous terms at the same hierarchy level.

3.2   Rule Model

Polysemy problem is the main concern for the design of our rule-based system model. We have built rule-based information filtering systems maintaining user search profiles where the profile consists of a set of filtering rules expressing the users' information filtering policy, each user has his (her) own tailored rules to filter out irrelevant materials. Our model extends the rule approach proposed in [9], which is explained in more detail in section 5. We have designed following predicates in our rule-based system model. The rules are hard coded, and the filtering process is based on simple conventional HTML structures. The bodies of rules consist of the following predicates standing for relations between terms and HTML tags.

ap(region type, word): true if a word appears within a region of region type in a Web page. E.g. ap(title, "plane") returns true if "plane" term appears in title; near(region type, word1, word2, n): This predicate is true if both of words word1 and word2 appear within a sequence of n words somewhere in a region of region type of a Web page. The ordering of the two words is not considered. E.g. near(para, "plane", "ticket", 4). Many other predicates are omitted here.

There are two kinds of filtering rules that users can define in our framework:

Preprocessing rules: For tuning search services' particular behaviors. The actions of preprocessing rules are all supported by search services. They are provided at users' convenience, an experienced user can enter the rules directly in the search query formulas without the help of preprocessing rules. In our framework, different templates are created for each search service, helping user utilizing all the characteristics of each search service. The framework has built a template handling particular Google characteristics; we will continue to build distinct templates for other search services.

Postprocessing rules: Operating on fetched Web pages. They are much more flexible than preprocessing rules, and constitute the kernel part of our rule-based model. These rules will apply directly to the (HTML) pages fetched by search services; further filter the result according to users' special requirements. Postprocessing rules takes web pages as text input and creates output after filtering all the text using all rules in the rule set. The result pages could be greatly shrunk after passing through the users' filtering rules. Postprocessing rules is consisted of predicates defined above. For example, samepara("plane", "ticket", "China", "2004") and near(para, "plane", "ticket", 4); this rule will ignore all the Web pages unless the words "plane", "ticket", "China" and "2004" appear in the same paragraph, and the words "plane" and "ticket" appear within a sequence of 4 words somewhere in a paragraph region of a Web page. The rule-based model is illustrated in the Fig. 2:

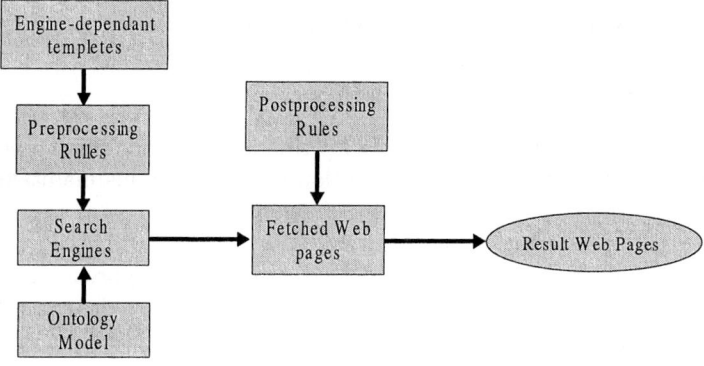

Fig. 2. Query Processing supported by Ontology Model

4 A Web Search Framework

By fusing the ontology model and rule-based system model of the previous section, we propose a Web search framework illustrated below. The framework puts the tasks of ontology selection (creation) and filtering rules compilation on the side of Web users. Some may argue that the architecture brings too much burden to general Web users; however, we believe it is Web users who bear the semantic meaning in their heads and issue the initial query commands; they deserve more involvement than application developers or ontology (knowledge) engineers in the system. This architecture can guarantee maximal flexibility for the whole Web search application. Moreover, carefully designed *User Interfaces* and *Software Wizards* can alleviate the burden of Web users. The structure of our framework and its interactions with Web users and Web search engines is given in Fig. 3:

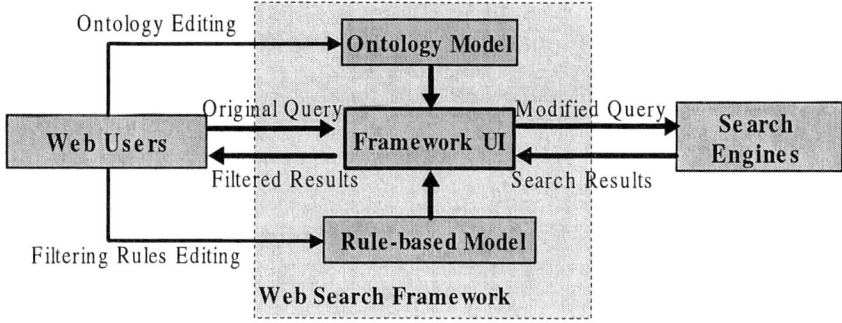

Fig. 3. Query Processing supported by Ontology Model

We are building a demonstrator which tests the intuitions proposed above. In recent years, Google has grown to be one of the most popular search engines that are available on the Web. Google Company has made its index available to other developers through a Web services interface [10]. This allows the developers to programmatically send a request to the Google server and get back a response. Currently the service is still in beta version and is for non-commercial use only. It is provided via SOAP (Simple Object Access Protocol) over HTTP and can be manipulated in any way that the programmer pleases, a developer can program in his or her favorite environment - such as Java, Perl, or Visual Studio.NET.

5 Related Works

Haiping Zhu et al. [11] introduce an approach for semantic search by matching RDF graphs. Their approach relies heavily on RDF files and can not process the large number of HTML files on the Web. Roger et al. propose a smart web query (SWQ) method for the semantic retrieval of web data. It uses domain semantics represented as context ontologies to specify and formulate appropriate web queries to search. Their semantic search filters don't recognize HTML tags and thus can not use the important relations between terms and tags. Masayuki et al. [9] describe a system for collecting Web pages that are relevant to a particular topic through an interactive approach. This system automatically constructs a set of rules to find new relevant pages. Our rule-based model adopts some thoughts proposed in this paper. There are only two rules defined in their paper; our rule-based system extends their work by adding many more rules to provide Web users with more filtering options. Sara Cohen et al. [12] present XSEarch, a semantic search service for XML. XSEarch has a simple query language, suitable for naive users. They also developed advanced indexing techniques to facilitate efficient implementation of XSEarch. However, like [11], their approach limits to XML files only. It is inconvenient for Web users to restrict to only XML files.

Many systems [11, 12] rely on Semantic Web infrastructure to carry out semantic searches; although Semantic Web is developing rapidly, there are still few RDF or OWL (Web Ontology Language) files available on the Web, with many of them for testing purposes only. To convert current available HTML files into semantic language formats like RDF or OWL requires huge human efforts and will not likely be realized in the near future. Our framework is more practical than theirs in this respect.

6 Conclusion and Future Works

We propose a Web search framework using ontologies and rules. The framework uses ontologies to deal with synonym problem, uses filtering rules to deal with polysemy problem, and thus gains a better recall and precision performance than the traditional search services.

References

1. Han, J.W. and Chang, K.C.-C.: Data Mining for Web Intelligence. IEEE Computer bfseries 35 (2002) 64-70
2. Gruber, T.R.: Toward Principles for the Design of Ontologies Used for Knowledge Sharing, International Journal of Human and Computer Studies, bfseries 43 (1995) 907-928
3. Lee, T.B.: James Hendler and Ora Lasilla, The Semantic Web, The Scientific American, May 2001
4. Newell, A., Simon, H.A.: Human Problem Solving, Prentice-Hall, Englewood Cliffs, New Jersey, 1972
5. Guarino, N.: Formal Ontology in Information Systems, in Proceedings of FOIS'98, Trento, Italy, IOS Press, (1998) 3-15
6. Nichols, D. and Terry, A: User's Guide to Teknowledge Ontologies, available on-line at http://ontology.teknowledge.com/Ontology_User_Guide.doc, 2003
7. Niles, I. and Pease, A.: Towards a Standard Upper Ontology, In Proceedings of the 2nd International Conference on Formal Ontology in Information Systems (FOIS-2001), 2001
8. Roger, H.L.C., Cecil, E.H.C. and Veda, C.S.: A smart web query method for semantic retrieval of web data, Data & Knowledge Engineering, bfseries 38 (2001) 63-84
9. Okabe, M. and Yamada, S.: Interactive Web Page Filtering with Relational Learning, Web Intelligence, (2001) 443-447
10. Google Corporation, Google Web APIs (beta), http://www.google.com/apis/, 2004

11. Zhu,H.P., Zhong, J.W., Li J.M. and Yu Y.: An Approach for Semantic Search by Matching RDF Graphs, Proceedings of the Fifteenth International Florida Artificial Intelligence Research Society Conference, AAAI Press, (2002) 450–454
12. Cohen, S., Mamou, J., Kanza, Y. and Sagiv, Y.: XSEarch: A Semantic Search Engine for XML, Proceedings of 29th International Conference on Very Large Data Bases, (2003) 45–56

An Arbitrated Quantum Message Signature Scheme

Xin Lü[1] and Deng-Guo Feng[1,2]

[1]State Key Laboratory of Information Security (Graduate School of Chinese Academy of Sciences), 100039, Beijing, P.R. China
lx@is.ac.cn
[2]State Key Laboratory of Information Security (Institute of Software, Chinese Academy of Sciences), 100080, Beijing, P.R. China

Abstract. Digital signature is an important task in modern cryptography, which concerns about authenticity, integrity and non-repudiation of data on communication channel. In this paper, an arbitrated quantum digital signature scheme is proposed, in which the receiver verifies the signature with the help of an arbitrator. The most important property of the proposed scheme is that it can sign unknown quantum states. The security of the protocol relies on the quantum one-time-pad and the Greenberger-Horne-Zeilinger (GHZ) triplet particles distributed among communication entities. Security analysis shows that the proposed quantum signature is a secure scheme.

1 Introduction

A major future research theme for cryptography is to weaken the assumptions on which security proofs are based, in particular computational intractability assumptions [1]. Quantum cryptography is one candidate for exploring unconditionally secure cryptography protocols. The idea of introducing quantum mechanics to cryptography can be traced back to Wiesner in 1970s and was published in 1983 [2], which proposed that if single-quantum states could be stored for long periods of time they could be used as counterfeit-proof money. Bennett and Brassard [3] gave the first quantum key distribution protocol, known as BB84, which is provable security [4].

Digital signature and message authentication play the important roles in modern cryptography and are widely used in network communication systems. The purpose of a digital signature scheme is to provide a means for an entity to bind its identity to a piece of information. Many signature schemes could be constructed under some unproven computational assumptions, such as the hard-ness of factoring large integers and finding discrete logarithm, in conventional cryptography. However, Shor's celebrated quantum algorithm indicated that the cryptography algorithms based on integer factoring related problems were threatened by quantum computers (if constructed in the future)[5]. Exploring more secure cryptography protocols immune to quantum computers is a new topic for the cryptography researchers.

Gottesman and Chuang proposed a quantum digital signature protocol based on weak quantum one-way functions and claimed that their scheme was secure against quantum attack [7] . The input of that scheme is a classical bits string and the signature keys of the signatory are quantum states. Unfortunately, their signature scheme couldn't sign general quantum superposition states, but only can deal with quantum basis states. Zeng presented an arbitrated quantum signature scheme in the literature [8], the security of which is due to the correlation of the GHZ triplet states and utilization of quantum one-time pad. In an arbitrated signature scheme, all communications involve a so called arbitrator who has access to the contents of the messages. The security of most arbitrated signature schemes depends heavily on the trustworthiness of the arbitrators[9, 10]. Zeng's scheme, however, requires that the signed quantum states must be known to the signatory. It seems impossible to sign general unknown quantum messages [7–9, 11].

In this paper, we propose a quantum signature scheme that can sign unknown quantum states. Due to the main properties of quantum information, signing unknown quantum states must ensure that the states can't be destroyed by the signature process or that the destroyed states can be recovered. In the proposed scheme, a signatory Alice firstly signs some classical bits that are related with the signatory's secret keys and then uses these classical bits to encrypt and encode her quantum states using quantum stabilizer codes. The receiver verifies the signature with the help of a trusted arbitrator. Alice can't disavow her signature, because the signature contains her secret keys' information and this can be confirmed by the arbitrator in the verification phase.

For the rest of this paper we assume that the reader is familiar with the basics of quantum cryptography and quantum computation. Further information about quantum cryptography and quantum error correction codes can be found in Ref. [6].

2 The Proposed Scheme

2.1 Security Requirements

The proposed scheme involves three entities: a signatory Alice, a receiver Bob, and an arbitrator Trent. The security of the signature scheme depends much on the trustworthiness of the arbitrator who has access to the contents of the messages. The existence of the arbitrator ensures that we can sign unknown quantum messages without Alice's deceiving. The general requirements for the quantum digital signature scheme discussed in this article should satisfy:

1. Each user (Alice) can efficiently generate her own signature on messages of her choice;
2. A receiver Bob can efficiently verify whether a given string is a signature of another user's on specific messages with the help of Trent;
3. The signatory can't disavow the messages that she has signed;
4. It is infeasible to produce signatures of other users' messages they haven't signed.

2.2 The Protocol

Initialization

1. Key distribution. Alice, Bob and Trent agree on some random binary strings K_A, K_B as their secret keys. K_A is shared between Alice and Trent, K_B is shared between Trent and Bob. To ensure that the scheme is unconditionally secure, we can generate these keys using quantum key distribution protocols, such as BB84 or EPR protocols [3, 6].
2. Triplet GHZ states distribution. When Trent receives Alice and Bob's request for an arbitrated communication, he creates N triplet GHZ state $|\phi\rangle = |\phi_1, \cdots, \phi_N\rangle$, and

$$|\phi_i\rangle = \frac{1}{\sqrt{2}}(|000\rangle_{atb} + |111\rangle_{atb}) \quad (1)$$

where a, t and b correspond to the particle of Alice's, Trent's and Bob's respectively. Trent distributes each of Alice's and Bob's GHZ particle to Alice and Bob for each GHZ state. After that, Alice, Bob and Trent each has N particles of GHZ pairs.
3. All the entities know a set of quantum stabilizer cods[1] $Q = \{Q_1, Q_2, \cdots, Q_N\}$. Alice will select any one of them, such as Q_y, to encode her quantum states, but others know nothing about the index y.

Signature

1. Alice selects three classical bits strings x, y and t. She constructs a new string as $L = (L_1, L_2, \cdots, L_N) = (x||y||t)$, here "||" means concatenation of two bits strings. Alice performs a controlled unitary operation to her GHZ particles according to L. If $L_i = 1$, she executes a bit flip gate X to ϕ_i. If $L_i = 0$, she does nothing. After that, the ith GHZ state becomes $|\phi'_i\rangle$ when $L_i = 1$

$$|\phi'_i\rangle = \frac{1}{\sqrt{2}}(|100\rangle + |011\rangle) \quad (2)$$

Alice measures her GHZ qubits and gains N classical bits

$$\omega_A = \omega_{A_1}, \omega_{A_2}, \cdots, \omega_{A_N}. \quad (3)$$

2. The aim of this step is to transform N-bit classical string L into quantum states $|R_A\rangle$ using the secrete key K_A distributed in the initial stage. If the ith bit of K_A is zero, namely $K_A^i = 0$, Alice encodes L_i using rectilinear basis $|0\rangle, |1\rangle$. If $K_A^i = 1$, Alice encodes L_i using diagonal basis $|\pm\rangle = \frac{|0\rangle \pm |1\rangle}{\sqrt{2}}$. After transformations, Alice has quantum states

$$|R_A\rangle = M_{K_A}|L\rangle = |l_{A_1}\rangle \otimes |l_{A_2}\rangle \otimes \cdots \otimes |l_{A_N}\rangle \quad (4)$$

---

[1] Introduction about quantum stabilizer cods can be found in the literature [6].

3. Supposing Alice has quantum states $|\psi\rangle$ to sign ($|\psi\rangle$ can be unknown to Alice). She firstly encrypts her classical bits L as $L_A = (x_A||y_A||t_A)$ using K_A and quantumly encrypts (q-encrypts)[2] $|\psi\rangle$ as ρ using classical bits x_A. Then she encodes ρ according to the quantum stabilizer codes Q_{y_A} with syndromes t_A and obtains quantum states π. Alice encodes her measurement results ω_A as quantum states $|\omega_A\rangle$ using rectilinear basis and encrypts $|\omega_A\rangle$ together with $|R_A\rangle$ as quantum states c_1 using K_A[3]. After that, she generates a signature $s = \{\pi, c_1\}$. Here, ρ, τ and c_1 denote the density matrix of the corresponding quantum states. Alice sends the signature s to Bob and publishes ω_A on a public board available only to Bob.

Verification

1. Bob receives the signature s and measures his GHZ particles and obtains $\omega_B = \omega_{B_1}, \omega_{B_2}, \cdots, \omega_{B_N}$. He uses ω_A from the public board to recover L as $L_B = \omega_A \oplus \omega_B = (x_B||y_B||t_B)$. Here, $\oplus$ means bit-by-bit XOR of two bits strings.

2. Bob encrypts L_B as classical ciphertext c_2 using his secrete key K_B and sends $\{s, c_2\}$ to the arbitrator through classical communication channel (for c_2) and quantum communication channel (for s, s is the density matrix of the signature of quantum state $|\psi\rangle$) respectively with the same communication sequence number.

3. Trent receives $\{s, c_2\}$. He decrypts c_2 as classical bits L_B and decrypts s as $|\omega_A\rangle$ and quantum states $|R_A\rangle$ using his secrete keys K_B and K_A respectively. He measures $|\omega_A\rangle$ using rectilinear basis to obtain $\omega_A = \omega_{A_1}, \cdots, \omega_{A_N}$ and measures his GHZ particles to obtain $\omega_T = \omega_{T_1}, \cdots, \omega_{T_N}$. The arbitrator uses ω_A to recover L as $L_{T_1} = \omega_C \oplus \omega_A = (x_{T_1}||y_{T_1}||t_{T_1})$. Trent measures $|R_A\rangle$ according to K_A and obtains $L_{T_2} = (x_{T_2}||y_{T_2}||t_{T_2})$. If $K_A^i = 0$, he measures $|R_A^i\rangle$ using rectilinear basis. If $K_A^i = 1$, he measures $|R_A^i\rangle$ using diagonal basis. The arbitrator compares L_{T_2} to L_B and L_{T_1}. If $L_{T_2} = L_B$, he lets $\lambda_1 = 0$. Otherwise, $\lambda_1 = 1$. Adopting the same method, if $L_{T_2} = L_{T_1}$, he lets $\lambda_2 = 0$. Otherwise, $\lambda_2 = 1$. Trent uses K_A to encrypts L_{T_1} as $L'_A = (x'_A||y'_A||t'_A)$ using the same method as Alice does in step 3 of the signature phase. Trent measures the syndrome t' using the stabilizer code $Q_{y'_A}$ on π and decodes the qubits as ρ. He compares t' with t'_A. If $t'_A = t'$, he sets $\lambda_3 = 0$. Otherwise, $\lambda_3 = 1$. Trent decrypts the quantum states ρ as $|\psi\rangle$ using x'_A.

4. Trent q-encrypts $|\psi\rangle$ as ρ_T using classical bits x_{T_2}. Then he encodes ρ_T according to the quantum stabilizer codes $Q_{y_{T_2}}$ with syndromes t_{T_2} and obtains quantum states π_T. He encrypts $L_{T_2}, \lambda_1, \lambda_2, \lambda_3$ and π_T

---

[2] To obtain unconditionally security, we use quantum one-time-pad method [12] and classical one-time-pad to encrypt quantum message and classical message respectively.

[3] Supposing K_A and K_B are long enough to use.

using K_B as classical bits c_3 and quantum states δ_T. He sends c_3 and δ_T to Bob.

5. Bob decrypts c_3 and δ_T as classical strings $L_{T_2}, \lambda_1, \lambda_2 \lambda_3$ and quntum states π_T using K_B. IF $\lambda_1 = \lambda_2 = \lambda_3 = 0$ and $L_{T_2} = L_B$, he measures the syndrome t'' using the stabilizer code Q_{y_B} on π_T and decodes the qubits as ρ_B. Bob compares t'' with t_B. If $t'' = t_B$, he deciphers the quantum states ρ_B as $|\psi\rangle_B$ and the signature s of quantum states $|\psi\rangle$ is verified. Otherwise, he rejects the signature and stops the protocol.

3 Security Analysis

3.1 Correctness

Theorem 1. *(Correctness) Supposing all the entities involved in the scheme follow the protocol, then Alice's signature passes the verification.*

Proof. The correctness of the scheme can be easily seen by inspection. In the absence of intervention, Alice, Bob and Trent secretly share the GHZ triple states at the end of the initialization phase. Bob recovers Alice's classical bits string $L = (x||y||t)$ and Trent obtains Alice's signature $|R_A\rangle$, L and the encoded quantum states π. Trent will correctly verify Alice's signature and decode Alice's quantum states and secretly send the verification results to Bob. Bob's output will be exactly "yes" at the end of the protocol.

3.2 Security Against Forgery

Theorem 2. *Other entities can forge Alice's signature with only a successful probability at most $1 - \frac{1}{2^{|K_A|}}$.*

Proof. Supposing that an attacker Eve wants to forge Alice's signature. Because Eve doesn't know Alice's private key K_A and doesn't share GHZ particles with Alice and Trent (GHZ particles are supposed distributed securely in the initialization phase), she can't obtain L. If she randomly selects a bits string K'_A and K'_B to execute the protocol, her cheating, will be detected by Trent with an overwhelming probability lager than $1 - \frac{1}{2^{|K_A|}}$. Here, $|K_A|$ means the length of the bits string K_A.

3.3 Security Against Repudiation

Alice can't deny her signature. When dispute between Alice and Bob happens, they will resort to the arbitrator. The signature s contains information about Alice's secrete key K_A, and Trent can confirm this fact in the verification phase.

Supposing Alice intercepts the ciphertext c_3 and δ_T that Trent has sent to Bob in the verification phase, she tries to change the signature of her qubits $|\psi\rangle$ to the signature of a new quantum state $|\varphi\rangle$. Because she knows nothing about Bob and Trent's secret key K_B, she can't prepare a new legal signature of another quantum messages that passes Bob's verification.

Bob can't deny that he has received Alice's signature, because he can't verify the signature without Trent's help.

4 Conclusion

A quantum digital signature scheme is proposed in this paper. One main feature of the protocol is that the signatory can indirectly sign general unknown quantum messages by signing Alice's secret keys and the syndromes of stabilizer codes which are used to encrypt and encode the quantum states. The authenticity of the quantum information is obtained by quantum stabilizer codes. The security of the protocol is studied and results show that the proposed signature scheme is a secure signature scheme. An open problem is that it's still not known whether there exists a general quantum message signature scheme that doesn't need the presence of an arbitrator.

Acknowledgments. This work was supported by the Natural Science Foundation of China under Grant No.60273027; the National Grand Fundamental Research 973 Program of China under Grant No. G1999035802 and the National Science Foundation of China for Distinguished Young Scholars under Grant No.60025205.

References

1. Ueli, M.: Cryptography 2000±10. In: Informatics: 10 Years Back, 10 Years Ahead. Lecture Notes in Computer Science, Vol. 2000, Springer-Verlag, Berlin Heidelberg New York (2001) 63–85.
2. Wiesner, S.: Conjugate coding. SIGACT News. **15** (1983) 78–88.
3. Bennett, C.H., Brassard, G.: Quantum cryptography: public key distribution and coin tossing. In: Proceedings of IEEE International Conference on Computers Systems and Signal Processing, Bangalore, India (1984) 175–179.
4. Mayers, D.: Unconditional security in quantum cryptography. Journal of the ACM. **48** (2001) 351–406.
5. Shor, P.: Polynomial-time algorithms for prime factorization and discrete logarithms on a quantum computer. SIAM Review. **41** (1999) 303–332.
6. Nielson, M., Chuang, I.: Quantum Computation and Quantum Information. Cambridge University Press, 2000.
7. Gottesman, D., Chuang, I.: Quantum digital signatures. Technical Report, http://arxiv.org/abs/quant-ph/0105032, 2001.
8. Zeng, G., Christoph, K.: An arbitrated quantum signature scheme. Physical Review A. **65** (2002) 0423121–0423126.
9. Lee, H., Hong, C., Kim, H. et al.: Arbitrated quantum signature scheme with message recovery. Physics Letters A. **321** (2004) 295–300.
10. Meijer, H., Akl, S.: Digital Signature Scheme for Computer Communication Networks. In: Advances in Cryptography: Crypto 81, Santa Barbara, California (1981) 65–70.

11. Barnum, C., Gottesman, D., Smith, A. et al.: Authentication of Quantum Messages. In: Proceedings of 43rd Annual IEEE Symposium on the Foundations of Computer Science, Vancouver, Canada (2002) 449–458.
12. Boykin, P., Roychowdury, V.: Optimal encryption of quantum bits, Physical Review A. **67** (2003) 0423171–0423175.

Fair Tracing Without Trustees for Multiple Banks[1]

Chen Lin, Xiaoqin Huang, and Jinyuan You

Department of Computer Science and Engineering,
Shanghai Jiao Tong University, 200030 Shanghai, China
{chenlin, huangxq}@sjtu.edu.cn

Abstract. In this paper we present a multiple bank electronic cash system based on group blind signature scheme, which offers a new kind of tracing mechanism. It provides conditional anonymity both for the customers and electronic coins under a judge. The coins can be marked using undeniable signature scheme so that the bank will recognize these coins at deposit. We also use the secret sharing scheme to trace the customer under the permission of a judge. The security of our scheme is analyzed. And compared with other works, our proposed tracing methods offer more privacy and do not need any trusted third parties. Our system is able to prevent from blackmailing, kidnapping, and bank robberies. Also we extend electronic cash system to multiple banks, it is more practical in the real life.

1 Introduction

With the development of E-Commence, People can buy merchandise in home and he can pay electronic coins to the shop by the Internet. How to protect the customer's anonymity and the coin's validity is an important problem. D. Chaum proposed blind signatures for untraceable payments [1]. However, Von Solms and Naccache [2] have shown that unconditional anonymity may be misused for untraceable blackmailing of customers. Also, unconditional anonymity may ease money laundering, illegal purchases, and bank robberies. Several papers [3] proposed the revocable anonymity methods, where one or more trusted third parties are needed. Kugler and Vogt et al proposed a new kind of tracing mechanism [4,7], which guarantees stronger privacy than other approaches. Their coin tracing can be carried out without the help of any trusted third parties. Their payment system allows tracing, but a traced customer will afterwards detect the fact of being legally or illegally traced. If the tracing turns out to be illegal, the customer can prove this violation of his privacy and the bank can be prosecuted. Paper [4] also gives some useful definitions for coin tracing, owner tracing, legal tracing and illegal tracing. According to these definitions in [4], we introduce a new kind of tracing mechanism, which supports the conditional anonymity, both a customer and a bank can reveal the identity of the customer, none of them can

---

[1] This paper is supported by the National Natural Science Foundation of China under Grant No.60173033.

reveal it lonely. The bank can trace the coins and the bank can also sign the coins at different marks.

The rest of the paper is organized as follows: In Section 2 we present some preliminary works. The implementation of a fair electronic cash system is presented in section 3. The security and anonymity aspects are discussed in section 4. Finally we compare our system with other schemes and give the conclusions.

2 Related Works

In this section, we introduce the techniques of group blind signature and secret sharing scheme, which we will use to construct our electronic cash scheme.

2.1 Group Blind Signature

Anna Lysyanskaya [5] presents an electronic cash scheme for distributed banks based on group blind signature technique. A group blind signature scheme consists of the following steps: Setup, Join, Sign, Verify and Open.

Setup: In this phase, the group manager chooses some security parameters and gets the group's public key $Y = (n, e, G, g, a, \lambda, \mu)$.

Join: If Alice wants to join the group, she picks a secret key x and interacts with the group manager. Then she gets her membership certificate $v \equiv (y+1)^{1/e} \pmod{n}$.

Sign: When a user asks a signing request, the signer signs the message m, the signature is as follows:

$$\hat{g} := \tilde{g}^w, \hat{z} := \tilde{z}^w$$
$$V_1 = SKLOGLOG_l[\alpha \mid \hat{z} = \hat{g}^{a^\alpha}](m) \qquad (1)$$
$$V_2 = SKROOTLOG_l[\beta \mid \hat{z}\hat{g} = \hat{g}^{\beta^e}](m)$$

Verify: The signature on the message m consists of $(\hat{g}, \hat{z}, V_1, V_2)$ and can be verified by checking correctness of V_1 and V_2.

Open: Given a signature $(\hat{g}, \hat{z}, V_1, V_2)$ for a message m, the group manager can determine the signer by testing if $\hat{g}^{y_P} = \hat{z}$ for every group member P (where $y_P = \log_g z_P$ and z_P is P's membership key).

2.2 Secret Sharing Scheme

In 1979, Shamir [6] proposed a first (k, n) threshold scheme. We describe it here simply. Suppose p and q are two large primes such that $q \mid (p-1)$ and g is a generator $g \in_R Z_p^*$ of order q. $q > n$.

1. Let S ($S \in Z_p^*$) be a secret value, k be a threshold, and $x_j \in Z_p^*$ ($j = 1, 2, \ldots n$). We can select $x_j = g^j$.

2. Distributor chooses a random polynomial.

$$f(x) = S + a_1 x + a_2 x^2 + \ldots + a_{k-1} x^{k-1} \pmod{q}.$$
$$a_i (i = 1, 2, \ldots, k-1) \in_R Z_p^* \qquad (2)$$

3. Distributor distributes $x_j, D_j = f(x_j)$ to each user j.

4. Any random k sets of (x_j, D_j) of n users can recover secret S from $f(x)$ by using Lagrange interpolation equation $f(x) = \sum_{j=1}^{k} D_j \prod_{l=1, l \neq j}^{k} \frac{x - g^l}{g^j - g^l}$.

3 Our Proposed Scheme

In this section we describe a protocol, which combines secret sharing scheme and group blind signature scheme combined with the Chaum-van Antwerpen undeniable signature [4] in order to make a practical Electronic Cash system.

3.1 Main Idea

First of all, we consider a large of banks composing a group. Also there is a group manager. Each bank participating the group can issue Electronic Cash independently. Each bank can sign the Electronic Cash with undeniable signature, which we call it the marked coins. In our system, we consider three parties, customer, merchant and bank. If the blackmailer blackmails the customer, the customer can communicate with the bank notifying the bank the blackmailing event. Then the bank can mark the Electronic Cash and the victim sends the marked coins to the blackmailer. Later, the customer and the bank can decide if these coins can be accepted. Usually the coins are anonymous, but on the condition of blackmailing, the anonymity of the coins can be abolished.

We also provide owner-tracing scheme in our system. Owner tracing scheme also allows the authorities to identify customers making an illegal purchase, after the illegal seller has been identified because of suspicious purchase. We use the secret sharing scheme to identify the customers. Only the bank and the merchant can collaborate to identify the customer. None of them can complete it independently.

3.2 Protocol

Notations: An RSA public key (n, e), where the length of n is at least $2l$ bits. l-Security parameter, a cyclic subgroup of Z_p^*, g, g_1 is two generators of Z_p^* of order q, where p is a prime and $n|(p-1)$. λ Upper bound on the length of the secret keys and a constant $\mu > 1$.

Initial Step: As in section 2.1 Setup and Join procedures, a bank P joins a group. The bank picks a secret key $x_P \in_R \{0,1,...,2^\lambda -1\}$ and calculates $y_P = a^{x_P} \pmod n$ and $z_P = g^{y_P}$. The bank P's membership certificate is $v \equiv (y+1)^{1/e} \pmod n$.

1. Customer requests coin withdrawal to the bank.
2. Bank selects random number $r \in_R Z_q^*$, makes a new generator $\alpha = g_1^r \pmod p$ and sends it to the Customer.
3. The bank chooses a random number X as a secret mark and calculates $\omega = \alpha^X \pmod p$.
4. Customer selects a random polynomial $f(x) = A_1 + a_1 x \pmod q$, A_1 the secret information (Corresponding to the Customer), $a_1 \in_R Z_p^*$.
5. Customer sends $(x_1, f(x_1))$, g, $c_0 = g^{A_1} \pmod p$, $c_1 = g^{a_1} \pmod p$ to the Bank. Where $x_1 \in Z_P^*$.
6. Customer will send $(x_2, f(x_2))$, g, $c_0 = g^{A_1} \pmod p$, $c_1 = g^{a_1} \pmod p$ to the merchant later, where $x_2 \in Z_P^*$.
7. The secret information A_1 can be recovered by $(x_1, f(x_1)), (x_2, f(x_2))$ using Secret Sharing Scheme. By the A_1, the bank can identify the customer later.

Withdrawal Step: When a customer wants to withdraw a coin from bank P, the bank first asks the user to prove his identity. The customer selects $u_1 \in_R Z_P (g^{u_1} g_1 \neq 1)$ and calculates $I = g^{u_1}$ as his identity. Afterwards, I can be used to verify the identity. Also, the customer and the bank generate secret information A_1, secret key $X_1 \in_R Z_P^*$, and the public key $h = g^{X_1} \mod p$ for the user. These parameters can be preserved in the user's smart card. A_1 Also must be preserved in the user's account in the bank.

1. The customer sends the bank the user's identity I and the coin amount m.
2. For every coin, Customer selects $\delta \in_R Z_q^*$ and calculates $\alpha' = \alpha^\delta \pmod p$, $\omega' = \omega^\delta \pmod p$.
3. For the coin amount m, as in equation (1), the bank signs the message m and the customer gets the group blind signature for m: $(\hat{g}, \hat{z}, V_1, V_2)$.
4. Finally, the customer gets the coin: $(m, \hat{g}, \hat{z}, V_1, V_2, \alpha', \omega')$.

Pay, Deposit and Verification Step. When customer gives coin to merchant, he also has to give $(x_2, f(x_2)), g, c_0, c_1$ to merchant. Then merchant can verify the truth of the shared secret using: $g^{f(x_2)} \stackrel{?}{=} c_0 c_1^{x_2}$. If it is true, he also has to verify the group blind signature as in Section 2.1. If both are passed, then the merchant sends the coin to the bank. The bank verifies group blind signature as the merchant. The bank also verifies $\omega' \stackrel{?}{=} \alpha'^x$. If both are passed, the coin is deposited in the merchant's account and notifies the merchant. The merchant sends the merchandises to the customer.

Later, if the authorities find the customer making an illegal purchase, then we can identify the customer by extracting the secret value A_1 using the banks $(x_1, f(x_1))$ and the customer's $(x_2, f(x_2))$ under the permission of a judger.

$$f(x_1) = A_1 + a_1 x_1, f(x_2) = A_1 + a_1 x_2 \tag{3}$$

4 Discussion of Security and Anonymity

Prevention of Blackmailing. There are three kinds of blackmailing [4]:

- Perfect Crime: When the victim is blackmailed, the victim can send an encryption message with secret key X_1 to tell the bank the blackmailing event. The bank sends the marked coin to the customer and the customer sends the marked coin to the blackmailer.
- Impersonation: The blackmailer gains access to the victim's bank account and withdraws coins by him. The blackmailer communicates directly with the bank but cannot observe the victim's communication with the bank. The customer gives his decryption key to the bank, which can cheat the blackmailer as described in the perfect crime scenario.

Kidnapping the Customer: In this scenario a covert channel is needed to inform the bank about the kidnapping.

Anonymity of Coins. If the coins are not marked at withdrawal phase, the coins remain anonymous. This means that payments with unmarked coins are unconditional anonymous for customer. In the situation of blackmailing, the marked coins are not anonymous.

Anonymity of Customers. In our payment system, we also provide the secret sharing scheme to protect the customer's anonymity. This is a conditional anonymity. The bank and the customer can collaborate to trace the customer under a judge. When we have two sets of $(x_1, f(x_1))$ $(x_2, f(x_2))$, we can use secret sharing scheme to identify the customer.

Multi-bank Properties. In our system, we adopt the group blind signature scheme to compose a group of banks. Each bank can issue the E-cash and everybody can use the group public key to verify the validness of the coins. If there is some problem about the coins, the group manager can identify the identity of the bank. For other people, the coins are anonymous.

5 Comparisons and Conclusions

We have proposed a new anonymous payment system based on group blind signature scheme and secret sharing scheme, offering both the coins' conditional anonymity and the customers' conditional anonymity. In contrast to other systems with revocable anonymity our system doesn't need a trusted third party. Our payment system protects private users against blackmailing attacks, by offering a marking mechanism similar to the well-known marking of banknotes. Also our payment system supports the multiple banks, it is very similar to the real life. So our payment system is a practical electronic cash system.

References

1. Chaum, D.: Blind signature for untraceable payments. In Advances in Cryptology-CRYPTO'82. Plenum (1983) 199-203
2. Von Solms, B. and Naccache, D.: On blind signatures and perfect crimes. Computers and Security. Vol.11, No.6. (1992) 581-583
3. Davida, G., Frankel, Y., Tsiounis, Y. and Yung, M.: Anonymity control in e-cash systems. In Financial Cryptography-FC'97. Lecture Notes in Computer Science, Vol. 1318. Springer-Verlag, Berlin Heidelberg New York (1997) 1–16.
4. Kugler, D. and Vogt, H.: Marking: A Privacy Protecting Approach Against Blackmailing. PKC 2001, LNCS, Vol.1992. Springer-Verlag, Berlin Heidelberg (2001) 137-152.
5. Lysyanskays, A. and Ramzan, Z.: Group blind signatures: A scalable solution to electronic cash, Financial Cryptography' 98, LNCS, Vol.1465. Springer-Verlag (1998) 184-197
6. Shamir, A.: How to share a secret, Comm. ACM, Vol.22 (1979) 612-61.
7. Kim, B.G., Min, S.J. and Kim, K.: Fair tracing based on VSS and blind signature without Trustees. http://www.caislab.icu.ac.kr/paper/ 2003/CSS2003/

SVM Model Selection with the VC Bound*

Huaqing Li, Shaoyu Wang, and Feihu Qi

Department of Computer Science and Engineering,
Shanghai Jiao Tong University, Shanghai 20030, P. R. China
waking_lee@sjtu.edu.cn

Abstract. Model selection plays a key role in the performance of a support vector machine (SVM). In this paper, we propose two algorithms that use the Vapnik Chervonenkis (VC) bound for SVM model selection. The algorithms employ a coarse-to-fine search strategy to obtain the best parameters in some predefined ranges for a given problem. Experimental results on several benchmark datasets show that the proposed hybrid algorithm has very comparative performance with the cross validation algorithm.

1 Introduction

Support vector machine (SVM), due to its powerful learning ability and satisfying generalization ability, is considered to be one of the most effective algorithms for pattern recognition (classification) problems. Generally, it works as follows for binary classification problems: First the training samples are mapped, through a mapping function Φ, into a high (even infinite) dimensional feature space $\mathcal{H}$. Then the optimal separating hyperplane in $\mathcal{H}$ is searched. In implementation, the use of kernel functions avoids the explicit use of mapping functions. However, as different kernel functions lead to different SVMs with probably quite different performance, it turns to be very important, yet very hard, to select appropriate types and parameters of kernel functions for a given problem.

There are mainly two categories of algorithms for SVM model selection. Algorithms from the first category estimate the prediction error by testing error on a data set which has not been used for training, while those from the second category estimate the prediction error by theoretical bounds. At present, the cross validation algorithm, which falls into the first category, is one of the most popular and robust algorithms employed in literatures [3,4]. Though some theoretical bounds have been explored [2,6], the use of the Vapnik Chervonenkis (VC) bound is less reported [5]. However Burges pointed out that, despite its looseness, the VC bound can be very predictive for SVM model selection [1].

In this paper, we propose two algorithms that use the VC bound for SVM model selection. The algorithms employ a coarse-to-fine search strategy to obtain the optimal SVM parameters. Experimental results on several benchmark

---

* This work is supported by the National Natural Science Foundation of China (No. 60072029).

datasets demonstrate that the proposed hybrid algorithm slightly outperforms the cross-validation algorithm. The rest of the paper is organized as follows: Section 2 introduces the VC bound and shows how to make it practical for use. The SVM model selection algorithms are described in Section 3. In Section 4, experimental results are presented and analyzed. Finally, conclusions are given in Section 5.

2 The VC Bound

The VC bound is given by Vapnik [1]:

$$R(\alpha) \leq R_{\text{emp}}(\alpha) + \sqrt{\left(\frac{h(\log(2l/h) + 1) - \log(\eta/4)}{l}\right)} .$$

where $R(\alpha)$ is the generalization error, $R_{\text{emp}}(\alpha)$ is the training error, h is the VC dimension, l is the size of the training set, η is a user-determined parameter, $0 \leq \eta \leq 1$. With probability $1 - \eta$, the above bound holds.

The main difficulty of using the VC bound lies in determining the VC dimension. Burges suggested to ease this difficulty by using the following bound on the VC dimension instead of the dimension itself [1]:

$$h \leq \lceil \frac{D_{\max}^2}{M_{\min}^2} \rceil + 1 .$$

where $D_{\max}$ is the maximal diameter of a set of gap tolerant classifiers, $M_{\min}$ is the minimal margin of the same set of classifiers.

Then the only thing left is to estimate $D_{\max}$. This problem can be described as follows [1]: Given a training set of data points X_i and a function Φ which maps the data points from the original space to a feature space $\mathcal{H}$, we wish to compute the radius of the smallest sphere in $\mathcal{H}$ which encloses the mapped training data. The corresponding formulation is

$$Minimize \quad R^2 , \tag{1}$$

$$subject\ to: \quad R^2 - \|\Phi(X_i) - \mathbf{C}\|^2 \geq 0 \quad \forall\ i .$$

where $\mathbf{C}$ is the (unknown) center of the sphere in $\mathcal{H}$. As the problem resembles that of SVMs training, algorithms for the latter can be modified to solve (1) [5].

3 SVM Model Selection Algorithms

In this paper, we investigate model selection for SVMs with the radius basis function (RBF) kernel. The RBF kernel is defined as:

$$K(X_i, X_j) = \exp(-\sigma(\|X_i - X_j\|^2)) .$$

Hence there are mainly two parameters to be tuned, the penalty parameter C and the kernel parameter σ.

For algorithms that can not employ gradient-descent based method to obtain the optimal parameters, Chung et al. recommended the use of an exhaustive grid-search strategy in some predefined parameter ranges [2]. They also pointed out that trying exponentially growing sequences of C and σ was a practical method to identify good parameters (for example, $C = 2^{-5}, 2^{-3}, \ldots, 2^{15}$, $\sigma = 2^{-15}, 2^{-13}, \ldots, 2^3$). However, a standard grid-search is very computational expensive when dealing with even moderate problems.

In [4], Staelin proposed a coarse-to-fine search strategy based on ideas from design of experiments. Experimental results showed that it was robust and worked effectively and efficiently. The strategy can be briefly described as follows: Start the search with a very coarse grid covering the whole search space and iteratively refine both the grid resolution and search boundaries, keeping the number of samples roughly constant at each iteration. In this paper, a similar search strategy like this is employed for the algorithms proposed.

3.1 Algorithm 1: The Fixed-C Algorithm

Empirically we find that, when employing the VC bound for SVM generalization error estimation, searching simultaneously for the optimal values of C and σ always results in poor SVM models. This is due to the effect the parameter C has on the SVM margin, which is inclined to lead to a very small C. Thereby, we'd better prefix C to some appropriate value and merely search for the optimal value of σ. In [5], Li et al. showed that a relative large C, e.g. 2048, could work for most cases. Note that the fixed-C algorithm has a pleasing byproduct — only one parameter is needed to be tuned.

3.2 Algorithm 2: The Hybrid Algorithm

In SVMs training, the penalty parameter C is very important. It balances the training error and the capacity of the machine. However, the fixed-C algorithm suggests a constant C for all problems. This, to some extent, is unreasonable. Therefore, we propose a hybrid algorithm, which works as follows: First the fixed-C algorithm is performed to choose the optimal value of σ. Then, with σ fixed, the cross validation algorithm is performed to obtain a better C.

4 Experiments

Experiments are carried out on several benchmark datasets[1] to investigate the performance of the proposed algorithms. The search is done in the $\log_2$-space of both parameters. The parameter ranges are $\log_2 C \in \{-5, -4, \ldots, 15\}$ and $\log_2 \sigma \in \{-15, -14, \ldots, 3\}$. Totally five iterations are performed for both the fixed-C algorithm and the fixed-σ cross validation algorithm. At each iteration five points uniformly distributed in the latest range are examined. LIBSVM [3]

---

[1] Dr. Chih-Jen Lin from National Taiwan University kindly shares these datasets with us.

is employed for SVMs training and testing, as well as model selection with the cross validation algorithm. Model selection with the fixed-C algorithm is done with a module developed by us with $C++$ (Ref. [5] for more details).

The experimental results are shown in Table 1. Where *c-ratio* stands for the cross validation ratio, *t-ratio* is the test ratio on the corresponding test datasets, *bound* is the minimal bound obtained by the fixed-C algorithm with a prefixed C of 2048. Datas in the first three columns of the cross validation algorithm are taken from [4]. Algorithm 2 has the same σ with algorithm 1.

Table 1. Performance comparison of algorithms on several datasets

Data sets	Cross Validation [4]				Algorithm 1			Algorithm 2		
	$\log_2 C$	$\log_2 \sigma$	c-ratio	t-ratio	$\log_2 \sigma$	bound	t-ratio	$\log_2 C$	c-ratio	t-ratio
banana	11.25	-1.5	93	87.39	0.75	124.14	85.88	0	92.25	**88.73**
diabetes	5.62	-10.5	79.7	**76.33**	-1.78	0.30	72.67	-1.25	76.28	75.0
image	8.44	-4.6	96.5	**97.92**	0.75	1.03	96.24	1.88	95.38	96.34
ringnorm	7.19	-3.47	98.75	98.20	-3.47	0.77	98.20	-2.5	98.75	**98.43**
splice	12.19	-7.97	87.8	89.52	-5.44	1.15	89.89	0.94	87.6	**89.93**
Titanic	-1.25	-1.5	80.67	77.08	-15.0	0.72	**77.43**	9.06	80.0	**77.43**
twonorm	1.87	-6.00	98	97.07	-3.75	0.92	96.64	-2.5	97.25	**97.61**
waveform	1.56	-5.16	93.5	**89.63**	-2.91	1.07	88.43	0.31	92.0	88.85

From Table 1, we can see that on five datasets the hybrid algorithm behaves best. On the rest three datasets the cross validation algorithm obtains best test ratios. While the fixed-C algorithm only has satisfying performance on datasets *titanic* and *splice*. It is interesting to note that, although the test ratios of the cross validation algorithm and algorithm 2 are very comparative, the obtained C and σ may differ widely, e.g. on datasets *banana* and *titanic*.

Another interesting thing is to compare the computational cost of the three algorithms. To make things simple, we assume that no cache technique like the one used in [4] is employed. Obviously 125 samples need to be examined with the cross validation algorithm to select the best C and σ, while only 25 samples with the fixed-C algorithm and 50 samples with the hybrid algorithm. This makes the hybrid algorithm more appealing.

5 Conclusion

Model selection is very important for a support vector machine (SVM). Several algorithms have been explored in literatures to select appropriate SVM parameters for a given problem. In this paper, we investigate the use of the Vapnik Chervonenkis (VC) bound for SVM model selection. Two algorithms, the fixed-C algorithm and the hybrid algorithm, are proposed. Experimental results on several benchmark datasets show that the hybrid algorithm has very comparative, if not better, performance than the cross validation algorithm. Moreover, the computational cost of the former is much less than the latter, which makes it more appealing.

Future research directions include: (1) Extend current work with other types of kernel functions, such as the polynomial kernel and the sigmoid kernel. (2) Combine the proposed algorithms with other algorithms to obtain even better SVM models. For example, we can use the proposed algorithms to select better initial values for other gradient-descent based algorithms, e.g. those using the radius margin bound.

References

1. Burges, C.J.: A Tutorial on Support Vector Machines for Pattern Recognition. Data Mining and Knowledge Discovery. **2** (1998) 121–267
2. Chung, K.-M., Kao, W.-C., Sun, T., Wang, L.-L., Lin, C.-J.: Radius Margin Bounds for Support Vector Machines with the RBF Kernel. Neural Computation. **11** (2003) 2643–2681
3. Chang, C.-C., Lin, C.-J.: LIBSVM: A Library for Support Vector Machines. (2002) Online at http://www.csie.ntu.edu.tw/~cjlin/papers/libsvm.pdf
4. Staelin, C.: Parameter Selection for Support Vector Machines. (2003) Online at http://www.hpl.hp.com/techreports/2002/HPL-2002-354R1.pdf
5. Li, H.-Q., Wang, S.-Y., Qi, F.-H: Minimal Enclosing Sphere Estimation and Its Application to SVMs Model Selection. IEEE Intl. Symposium on Neural Networks. (2004) to appear
6. Chapelle, O., Vapnik, V., Bousquet, O., Mukherjee, S: Choosing Multiple Parameters for Support Vector Machines. Machine Learning. **46** (2002) 131-159

Unbalanced Hermite Interpolation with Tschirnhausen Cubics

Jun-Hai Yong and Hua Su

School of Software, Tsinghua University, Beijing 100084, P. R. China
yongjh@tsinghua.edu.cn

Abstract. A method for constructing a cubic Pythagorean hodograph (PH) curve (called a Tschirnhausen cubic curve as well) satisfying unbalanced Hermite interpolation conditions is presented. The resultant curve interpolates two given end points, and has a given vector as the tangent vector at the starting point. The generation method is based on complex number calculation. Resultant curves are represented in a Bézier form. Our result shows that there are two Tschirnhausen cubic curves fulfilling the unbalanced Hermite interpolation conditions. An explicit formula for calculating the absolute rotation number is provided to select the better curve from the two Tschirnhausen cubic curves. Examples are given as well to illustrate the method proposed in this paper.

Keywords: Hermite; Pythagorean hodograph; Absolute rotation number.

1 Introduction

Hermite interpolation problem is a fundamental problem in computer aided geometric design (CAGD)[1], and has numerous applications in a lot of areas such as robotics (for path planning), computer graphics, and computer-aided design (CAD) [2]. Yong and Cheng [1] make a good summarization on recent research directions on this problem. A lot of attentions have been paid to build *geometric Hermite curves* [3–7 et al.] with a low degree, high order geometric continuity and high order approximation accuracy. Another research direction is to combine Hermite interpolation with Pythagorean hodograph condition [8–12 et al.] The research in this paper belongs to this direction.

A parametric polynomial curve satisfying the Pythagorean hodograph (PH) condition has at least the following benefits,

(1) the arc length could be expressed in a polynomial form, and
(2) both the curvature and the offset curve are rational.

In this research direction, current research focuses on balanced Hermite interpolation conditions with a PH curve, i.e., constructing a PH curve to interpolate two given end points and take two given vectors as the tangent vectors at the end points.

Up to now, no literature could be found about unbalanced Hermite interpolation with a PH curve. The endpoint position requirement is the same between the balanced Hermite interpolation and the unbalanced one, while the unbalanced Hermite interpolation has the tangent vector requirement only at one end point, i.e., only a vector is given, which should be the tangent vector of the resultant curve at an end point, for example, at the starting point. Unbalanced Hermite interpolation problem is frequently required in a lot of practical applications as well. For example, the curves of this kind could be two end segments of a composition curve, which could be a sketch in the mechanical design or a contour in a GIS (geographic information system). To construct either of the two end segments of the composition curve is usually to build a curve fulfilling unbalanced Hermite interpolation conditions. The resultant curve, which is an end segment of composition curve, should match the tangent condition at one end point to satisfy the C^1 continuity requirement of the composition curve. At the other end point of the resultant curve, the tangent condition is often unnecessary. PH curves with balanced Hermite interpolation conditions could fulfill the unbalanced Hermite interpolation conditions as well. However, Farouki and Neff [9] point out that Quintics are the simplest PH curves with balanced Hermite interpolation conditions. In this paper, we show that a cubic PH curve (called a Tschirnhausen cubic curve as well) is enough to build a PH curve with unbalanced Hermite interpolation conditions. Complex number theory is used to generate the Tschirnhausen cubic curve fulfilling the unbalanced Hermite interpolation conditions. We find that there are totally two resultant Tschirnhausen cubic curves. An absolute rotation number is used to automatically find out a better one from the two resultant curves. An explicit formula is given to calculate the absolute rotation number.

The remaining part of the paper is arranged as follows. According to complex number theory, Section 2 presents the method for generating Tschirnhausen cubics, which are the simplest polynomial curves with the Pythagorean hodograph property, to match the unbalanced Hermite interpolation requirements. We find that there are totally two resultant Tschirnhausen cubic curves. Based on the absolute rotation number, a method for selecting a better curve from the two resultant Tschirnhausen cubic curves is provided in Section 3. The formula for calculating the absolute rotation number is given in this section as well. Some examples and some concluding remarks are provided in the last section.

2 Unbalanced Hermite Interpolation

As shown in Figure 1, this section provides a method for constructing a Tschirnhausen cubic curve $\mathbf{P}(t)$ such that

$$\mathbf{P}(0)=\mathbf{A},\ \mathbf{P}(1)=\mathbf{D}\ \text{and}\ \mathbf{P}'(0)=\mathbf{T},$$

where $\mathbf{A}$ and $\mathbf{D}$ are two given points, and $\mathbf{T}$ is a given vector. The resultant curve $\mathbf{P}(t)$ is presented in a Bézier form. According to the interpolation property of a Bézier curve, $\mathbf{A}$ and $\mathbf{D}$ should be two control points of $\mathbf{P}(t)$. Assume

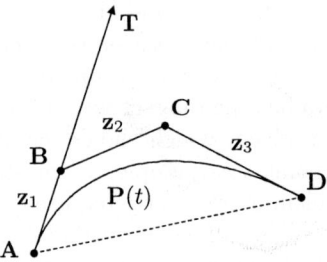

Fig. 1. Unbalanced Hermite interpolation with Tschirnhausen cubic curve $\mathbf{P}(t)$

that the other two control points are $\mathbf{B}$ and $\mathbf{C}$, respectively. Then, the Bézier form of $\mathbf{P}(t)$ is

$$\mathbf{P}(t) = \mathbf{A}B_{0,3}(t) + \mathbf{B}B_{1,3}(t) + \mathbf{C}B_{2,3}(t) + \mathbf{D}B_{3,3}(t),$$

where $B_{i,3}(t)(i = 0, 1, 2, 3)$ are Bernstein basis functions. Our method is based on complex number calculation, i.e., values of $\mathbf{A}$, $\mathbf{B}$, $\mathbf{C}$, $\mathbf{D}$, $\mathbf{T}$ and $\mathbf{P}(t)$ are given by complex numbers. For convenience, for any complex number $\mathbf{z}$, let $arg(\mathbf{z}) \in (-\pi, \pi]$ represent the angle from the positive real axis to $\mathbf{z}$, where counterclockwise and clockwise rotations give positive and negative values, respectively. In this paper, we do not consider the degenerated cases that $\mathbf{A} = \mathbf{D}$, or $\mathbf{T} = \mathbf{D} - \mathbf{A}$, or $\mathbf{T} = 0$. For convenience of presentation, we take these assumptions for granted. With complex analysis, we have the following theorem.

Theorem 1. *There are two Tschirnhausen cubics satisfying the unbalanced Hermite interpolation conditions.*

Proof. Let $\mathbf{z}_1 = \mathbf{B} - \mathbf{A}$, $\mathbf{z}_2 = \mathbf{C} - \mathbf{B}$ and $\mathbf{z}_3 = \mathbf{D} - \mathbf{C}$, as shown in Figure 1, then we have

$$\mathbf{z}_1 + \mathbf{z}_2 + \mathbf{z}_3 = \mathbf{D} - \mathbf{A}, \tag{1}$$

and the tangent vector of $\mathbf{P}(t)$ at the starting point is

$$\mathbf{P}'(0) = 3\mathbf{z}_1 = \mathbf{T}. \tag{2}$$

Reference [8] points out that a Bézier curve $\mathbf{P}(t)$ is a Tschirnhausen cubic curve if and only if

$$\mathbf{z}_1 \mathbf{z}_3 = \mathbf{z}_2^2. \tag{3}$$

Substituting Equations (2) and (3) into Equation (1), we obtain that $\mathbf{z}_2$ is a root of

$$9\mathbf{z}_2^2 + 3\mathbf{T}\mathbf{z}_2 + \mathbf{T}(3\mathbf{A} - 3\mathbf{D} + \mathbf{T}) = 0. \tag{4}$$

Therefore, z_2 has two possible values. Additionally, we have $z_1 = \frac{T}{3}$ and $z_3 = \frac{3z_2^2}{T}$ from the above equations. Thus, the resultant cubic Bézier curve $P(t)$ is determined by the control points A, $B = A + z_1 = A + \frac{T}{3}$, $C = D - z_3 = D - \frac{3z_2^2}{T}$, and D. Because z_2 has two possible values, C has two possible positions. Hence, we obtain the conclusion in Theorem 1. □

The above proof also provides the method for producing the resultant curves. The method for automatically choosing a better one from the two resultant curves is given in the next section.

3 Absolute Rotation Number

Absolute rotation number here is used to select a better curve from two resultant Tschirnhausen cubics. Experience shows that the curve with a better shape is the one with the smaller absolute rotation number in two resultant Tschirnhausen cubics. The value of the absolute rotation number of a given curve is the result of 2π dividing the total absolute rotation angle of the curve tangent vector along the curve. Some more details about the absolute rotation number could be found in Reference [9]. In this paper, an explicit formula for calculating the absolute rotation number of a Tschirnhausen cubic curve is proposed here. It is based on the following lemma.

Lemma 1. *Let a Tschirnhausen cubic curve $P(t)$ be given in a cubic Bézier form with four control points A, B, C and D. And let $z_1 = B - A$, $z_2 = C - B$ and $z_3 = D - C$. Then, the rotation direction of the curve $P(t)$ at the starting point $P(0)$ is the same as the rotation direction from z_1 to z_2, i.e., if $arg\left(\frac{z_2}{z_1}\right) < 0$, the rotation direction of $P(t)$ at $P(0)$ is clockwise; otherwise, it is counterclockwise.*

The above lemma could be proved with the definition of the curvature of a Bézier curve and the difference property of a Bézier curve. Those properties of a Bézier curve could be found in a lot of textbooks about computer graphics or CAGD such as Reference [13]. And then, the explicit formula for calculating the absolute rotation number of a Tschirnhausen cubic curve is given by the following theorem.

Theorem 2. *The absolute rotate number of a Tschirnhausen cubic curve $P(t)$ is given by*

$$\mathcal{R}_{abs} = \frac{1}{\pi}\left|arg\left(\frac{z_2}{z_1}\right)\right|, \tag{5}$$

where the notations here just follow those in Lemma 1.

Proof. According to the derivative property of a Bézier curve, the directions of the tangent vectors of $P(t)$ at the starting point and at the ending point are the same as the directions of z_1 and z_3, respectively. References [10] and [11] have shown that a Tschirnhausen cubic curve does not have any inflection point or cusp. Therefore, $2\pi\mathcal{R}_{abs}$ is the absolute angle from z_1 to z_3 along the curve $\mathbf{P}(t)$. From Equation (3), we have that z_2 is on the line, which bisects the angle from z_1 to z_3. Lemma 1 shows that the rotation direction of $\mathbf{P}(t)$ at the starting point $\mathbf{P}(0)$ is the same as the rotation direction from z_1 to z_2. Therefore, we obtain that z_2 exactly bisects the angle from z_1 to z_3 along the direction of $\mathbf{P}(t)$. Hence, $2\pi\mathcal{R}_{abs}$ is twice of the absolute angle from z_1 to z_2. Thus, we have the conclusion in Theorem 2. □

4 Example and Conclusions

An example is presented here to illustrate the method proposed in this paper. It is shown in Figure 2, where the two given end points are $\mathbf{A} = 0$ and $\mathbf{D} = 1$, and the tangent vector at $\mathbf{A}$ is $\mathbf{T} = 0.25 + 0.25i$. As shown in Figure 2(a), there are two Tschirnhausen cubics matching the above unbalanced Hermite interpolation requirements. They are drawn in solid and in dashed, respectively. The four control points of the solid one are 0, $0.25 + 0.25i$, $0.6469 + 0.2747i$ and 1. And the four control points of the dashed one are 0, $0.25 + 0.25i$, $-0.3969 - 0.0247i$ and 1. Figures 2(b) and 2(c) illustrate the Gauss maps of the solid curve and the dashed curve, respectively. As shown in the curves, z_2 bisects the angle from z_1 to z_3 along the direction of the resultant curve. The solid curve whose absolute rotation number is 0.2302 has a better shape than the dashed curve whose absolute rotation number is 0.8778.

One more example is given in Figures 3. The curves are built with the method provided in the proof of Theorem 1. The degree of the resultant curves fulfilling the unbalanced Hermite interpolation requirement is only 3, while the degree of the PH curves satisfying the balanced Hermite interpolation requirement is at least 5 [9]. The Gauss maps illustrate the bisection property of z_2. Absolute rotation numbers could be calculated according to Theorem 2 in Section 3. The

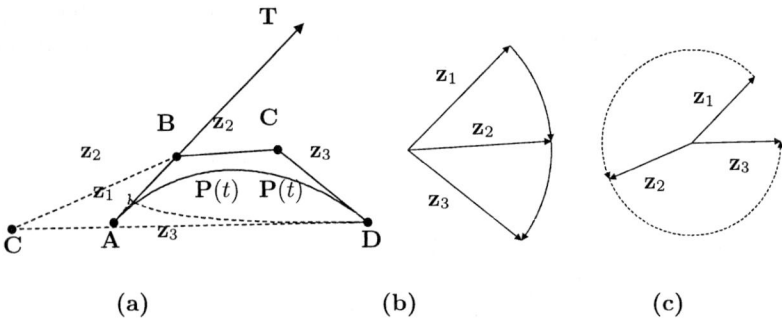

Fig. 2. Example 1: **(a)** two resultant Tschirnhausen cubics; **(b)** Gauss map of the *solid* Tschirnhausen cubic curve; **(c)** Gauss map of the *dashed* Tschirnhausen cubic curve

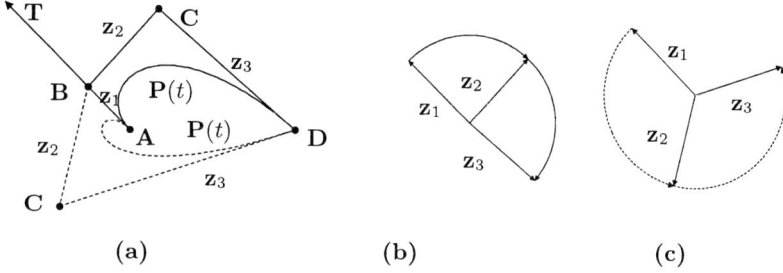

Fig. 3. Example 2: (a) two resultant Tschirnhausen cubics; (b) Gauss map of the *solid* Tschirnhausen cubic curve; (c) Gauss map of the *dashed* Tschirnhausen cubic curve

absolute rotation numbers of the solid curve and the dashed curve in Figure 3(a) are 0.4873 and 0.6744, respectively. Those examples (see Figures 2 and 3) show that the resultant curve (solid in figures) with the smaller absolute rotation number has a better shape than the other resultant curve (dashed in figures).

Acknowledgements

The research was supported by Chinese 863 Program (2003AA4Z3110) and 973 Program (2002CB312106). The first author was supported by a project sponsored by SRF for ROCS, SEM (041501004), and a Foundation for the Author of National Excellent Doctoral Dissertation of PR China (200342).

References

1. Yong, J.H. and Cheng, F.: Geometric Hermite curves with minimum strain energy. Computer Aided Geometric Design **21** (2004) 281–301
2. Yong, J.H. and Zheng, W.: Geometric method for Hermite interpolation by a class of PH quintics. Journal of Computer Aided Design & Computer Graphics (*to appear*)
3. de Boor, C., Höllig, K., and Sabin, M.: High accuracy geometric Hermite interpolation. Computer Aided Geometric Design **4** (1987) 269–278
4. Höllig, K. and Koch, J.: Geometric Hermite interpolation. Computer Aided Geometric Design **12** (1995) 567–580
5. Höllig, K. and Koch, J.: Geometric Hermite interpolation with maximal order and smoothness. Computer Aided Geometric Design **13** (1996) 681–695
6. Reif, U.: On the local existence of the quadratic geometric Hermite interpolant. Computer Aided Geometric Design **16** (1999) 217–221
7. Schaback, R.: Optimal geometric Hermite interpolation of curves. In Dahlen, M., Lyche, T., and Schumaker, L.L., eds.: Mathematical Methods for Curves and Surface II. (1998) 1–12
8. Farouki, R.T.: The conformal map of the hodograph plane. Computer Aided Geometric Design **11** (1994) 363–390
9. Farouki, R.T. and Neff, C.A.: Hermite interpolation by Pythagorean hodograph quintics. Mathematics of Computation **64** (1995) 1589–1609

10. Farouki, R.T. and Sakkalis, T.: Pythagorean hodographs. IBM Journal of Research and Development **34** (1990) 736–752
11. Meek, D.S. and Walton, D.J.: Geometric Hermite interpolation with Tschirnhausen cubics. Journal of Computational and Applied Mathematics **81** (1997) 299–309
12. Meek, D.S. and Walton, D.J.: Hermite interpolation with Tschirnhausen cubic spirals. Computer Aided Geometric Design **14** (1997) 619–635
13. Piegl, L. and Tiller, W.: The NURBS Book. Springer, Berlin (1995)

An Efficient Iterative Optimization Algorithm for Image Thresholding

Liju Dong[1,2] and Ge Yu[1]

[1] School of Information Science and Engineering, Northeastern University,
Shenyang 110004, China
[2] School of Information Science and Engineering, Shenyang University,
Shenyang 110044, China
dong_liju@yahoo.com.cn, yuge@mail.neu.edu.cn

Abstract. Image thresholding is one of the main techniques for image segmentation. It has many applications in pattern recognition, computer vision, and image and video understanding. This paper formulates the thresholding as an optimization problem: finding the best thresholds that minimize a weighted sum-of-squared-error function. A fast iterative optimization algorithm is presented to reach this goal. Our algorithm is compared with a classic, most commonly-used thresholding approach. Both theoretic analysis and experiments show that the two approaches are equivalent. However, our formulation of the problem allows us to develop a much more efficient algorithm, which has more applications, especially in real-time video surveillance and tracking systems.

1 Introduction

Image segmentation plays a very important role in many tasks of pattern recognition, computer vision, and image and video retrieval. Many approaches have been proposed in the literature [1–5]. Image thresholding is one of the most important techniques for image segmentation. Its goal is to automatically find one or more thresholds from the histogram of the image under study. The thresholds divide the image into two or more regions each with similar gray levels. Among many thresholding techniques, the Otsu's method [6] is considered as the most commonly-used one in the survey papers [1–5]. It is also ranked as the best and fastest global thresholding technique in [2] and [3].

In applications such as real-time video surveillance and recognition systems, and image and video retrieval and understanding from large databases, it is desirable to develop as fast algorithms as possible for a task. This paper proposes an efficient approach to image thresholding. We formulate the thresholding as a discrete optimization problem: finding the best thresholds that minimize a weighted sum-of-squared-error objective function. A fast iterative optimization algorithm is proposed to reach this goal based on the histogram. We compare our approach with the Otsu's method. Both theoretic analysis and experiments show that the two methods yield the same segmentation results but our algorithm is much faster.

2 The New Approach

2.1 Formulation of the Problem

The goal of image thresholding is to divide the pixels of an image into two or more regions with similar gray levels, which is similar to data clustering where data are partitioned into clusters with similar properties. Therefore, the widely used sum-of-squared-error criterion in data clustering [7] is modified in this paper to be the objective function of the segmentation.

Suppose that there are L gray levels $\{0,1,...,L-1\}$ in an image. Let n_l denote the number of pixels at level l. If an image contains three different objects each with exactly the same gray level (an ideal case), there will be only three non-zero n_l, $l \in \{0,1,...,L-1\}$, in the histogram of this image (see Fig. 1(a)). However, the practical histogram of a real image with three objects always has much more non-zero n_l on it. These gray levels spread on the histogram in a wide range, as shown in Fig. 1(b). Thus we formulate the image segmentation as finding the clusters on the histogram such that the total deviation of the gray levels from their corresponding cluster centers (centroids) is minimized (see Fig. 1(c)). More formally, we give the following formulation.

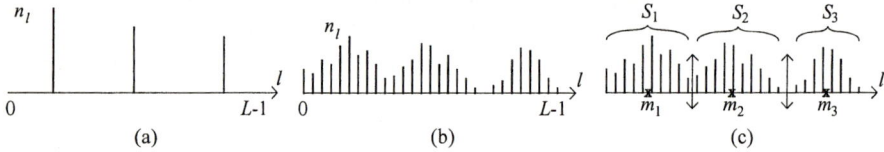

Fig. 1. (a) An ideal histogram with only three non-zero gray levels. (b) A practical histogram with three clusters. (c) One partition of three clusters of the gray levels, S_1, S_2, S_3, where the centroids of the clusters are m_1, m_2 and m_3, and the two thresholds are t_1 and t_2

Definition 1. Suppose that the histogram of an image is divided into c clusters (disjoint subsets) $S_1, S_2, ..., S_c$, as shown in Fig. 1(c). Let $m_1, m_2, ..., m_c$ be the centroids of these clusters. The image thresholding problem is to search for the partition such that the objective function

$$f(m_1, m_2, ..., m_c) = \sum_{i=1}^{c} \sum_{l \in S_i} n_l (l - m_i)^2 \ . \tag{1}$$

is minimized. The $c-1$ thresholds $t_1, t_2, ..., t_{c-1}$ can be obtained from the final partition $S_1, S_2, ..., S_c$.

We call $f(m_1, m_2, ..., m_c)$ a weighted sum-of-squared-error function, where n_l serves as a weighting factor. The centroids are calculated by

$$m_i = \frac{1}{d_i} \sum_{l \in S_i} n_l l \ , \quad d_i = \sum_{l \in S_i} n_l \quad 1 \le i \le c \ . \tag{2}$$

For convenience, let

$$f_i = \sum_{l \in S_i} n_l (l - m_i)^2 .\tag{3}$$

which is the weighted sum of squared errors in cluster S_i.

Assume that a grey level k currently in cluster S_i is tentatively moved to cluster S_j. Then m_j changes to m_j^*, m_i changes to m_i^*, f_j changes to f_j^*, and f_i changes to f_i^*, where m_j^*, m_i^*, f_j^*, and f_i^* can, after some mathematical manipulation, be derived as

$$m_j^* = m_j + \frac{(k - m_j)n_k}{d_j + n_k}, \quad m_i^* = m_i - \frac{(k - m_i)n_k}{d_i - n_k} \tag{4}$$

$$f_j^* = f_j + \frac{d_j n_k (k - m_j)^2}{d_j + n_k}, \quad f_i^* = f_i - \frac{d_i n_k (k - m_i)^2}{d_i - n_k}. \tag{5}$$

From (5), we see that the transfer of gray level k from cluster S_i to S_j can reduce $f(m_1, m_2, ..., m_c)$ if

$$\frac{d_i n_k (k - m_i)^2}{d_i - n_k} > \frac{d_j n_k (k - m_j)^2}{d_j + n_k} \tag{6}$$

If the reassignment of the gray level k is profitable, the greatest decrease in $f(m_1, m_2, ..., m_c)$ is obtained by choosing the cluster for which the value $d_j n_k (k - m_j)^2 / (d_j + n_k)$ is minimal. The above equations and analysis lead to the algorithm presented in the next section.

2.2 The Iterative Algorithm

1. Select an initial partition of c clusters $S_1, S_2, ..., S_c$ of the L gray levels on the histogram and calculate $m_1, m_2, ..., m_c$ and $d_1, d_2, ..., d_c$
2. *changed* ← No
3. for $k = 0, 1, ..., L - 1$ do
4.   if $d_i \neq n_k$ (suppose $k \in S_i$ currently) then
5.   Compute $r_q = \begin{cases} \dfrac{d_i n_k (k - m_i)^2}{d_i - n_k} & q = i \\ \dfrac{d_q n_k (k - m_q)^2}{d_q + n_k} & q \neq i \end{cases}$
6.   if $r_j \leq r_q, j \neq i, \forall q$ then (move k to S_j)
7.   Update m_j and m_i with (4)
8. $d_i \leftarrow d_i - n_k$; $d_j \leftarrow d_j + n_k$; *changed* ← Yes
9. if *changed* = Yes goto Step 2

10. else Find $c-1$ thresholds from the final partition $S_1, S_2, ..., S_c$
11. Return the thresholds

This algorithm reflects the idea of iteratively improvement in minimizing the objective function $f(m_1, m_2, ..., m_c)$ as described in Section 2.1. The optimization procedure repeats until no further improvement is obtained. A good initial partition can reduce the number of iterations. Let the smallest and largest non-zero gray levels on the histogram be l_{min} and l_{max}, respectively. A good initial partition can be obtained by equally dividing $[l_{min}, l_{max}]$ into c clusters.

It is not difficult to find the computational complexity of the algorithm. Step 1 or Step 10 can be computed in $O(L)$ time. The main computation is in Steps 3 to 8, which requires $O(cL)$ time. Therefore, the algorithm runs in $O(cLQ)$ time with Q being the number of iterations, which is the number of times Step 2 is visited. From our experiments, we find that $Q < 15$ in general.

2.3 Comparison with the Otsu's Method

Otsu proposed his method, from the viewpoint of statistical theory, by maximizing a function $\sigma_B^2(t_1, t_2, ..., t_{c-1})$, the detail of which can be found in [6]. In fact, our method and the Otsu's method are equivalent in essence.

Theorem 1. If $\sigma_B^2(t_1, t_2, ..., t_{c-1})$ is maximized by a partition $S_1, S_2, ..., S_c$, then $f(m_1, m_2, ..., m_c)$ in (1) is minimized by the same partition, and vice versa.

The proof can be found in [8]. The optimal thresholds $t_1, t_2, ..., t_{c-1}$ in the Otsu's method are obtained by exhaustive search on the histogram. Its computational complexity is $O(cL^c)$ [8]. Obviously, the complexity $O(cLQ)$ of our new algorithm is much less, where $L = 256$ and $Q < 15$ in general.

3 Experimental Results

Our algorithm is implemented in C and runs on a 1 GHz Pentium III PC. For comparison, we also implement the Otsu's method. More than 80 images have been used to test them. Many of the images are gathered from the web sites on the internet, such as the one in [9] where a number of public image and video databases are available.

For an application of image thresholding, how many clusters are divided from the histogram is application dependent. Most practical applications classify the gray levels on the histogram into two or three clusters. In all the experiments, the two algorithms obtain the same result for each image. Here we give two of them.

Fig. 2(a) is an infrared image. The two-cluster segmentation result by the two algorithms is very good. The thermal infrared image shown in Fig. 3(a) from a surveillance system cannot be handled with one threshold as illustrated in Fig. 3(b). However, the two algorithms can obtain satisfactory result (Fig. 3(c)) by finding the two thresholds 60 and 150.

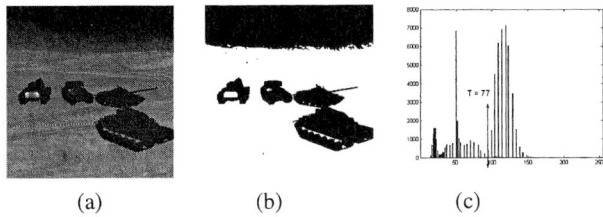

(a)   (b)   (c)

Fig. 2. (a) An infrared image. (b) The two-cluster segmentation result by our or the Otsu's algorithm. (c) Histogram of the image in (a)

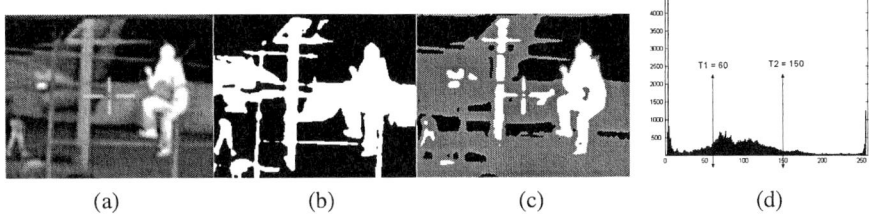

(a)   (b)   (c)   (d)

Fig. 3. (a) A thermal infrared image. (b) Bad two-cluster segmentation. (c) Good three-cluster segmentation where the clusters are denoted by three gray levels. (d) Histogram of image in (a)

Next we compare the computational time taken by the two algorithms. As mentioned in the last section, theoretically, the new algorithm with complexity $O(cLQ)$ is much more efficient than the Otsu's algorithm with complexity $O(cL^c)$. The experiments show that generally $Q<10$ and $Q<15$ when the new algorithm carries out two-cluster and three-cluster segmentation, respectively. Table 1 gives the average running time for each algorithm to handle one image in the cases of segmentation with one threshold and two thresholds.

Table 1. Comparison of the average running time by the two algorithms (in second)

Algorithm	One threshold	Two thresholds
New algorithm	0.0001	0.0008
Otsu's algorithm	0.0011	0.17

From Table 1, we see that both algorithms can find one threshold quickly on an image, and the new algorithm is one order of magnitude faster than the Otsu's. For three-cluster segmentation, the Otsu's algorithm requires 0.17 second to deal with one image, while our algorithm spends only 0.0008 second, which is more than 200 times faster than the Otsu's.

In many real-time automatic video surveillance and tracking systems, there are 30 frames per second usually. Image segmentation is only part of the processing in these systems. Therefore, in order to reach the goal of processing 30 images within 1 second, the time that is allowed to perform segmentation for one image is much less than 0.03 second. In this case, our algorithm is still fast enough for the job, but the Otsu's algorithm is not qualified.

4 Conclusions

Image segmentation by thresholding is the classic technique that is still used widely in many applications of pattern recognition and computer vision. The main advantage is its simplicity and good efficiency, which is a crucial requirement in most real-time systems.

We have presented a new efficient optimization approach to image thresholding. The algorithm iteratively minimizes a weighted sum-of-squared-error objective function, which is expected to finally generate good segmentation of gray levels on the histogram. Our approach is essentially equivalent to the Otsu's method, which is popular and ranked as the best and fastest global thresholding technique in the survey papers [2] and [3]. However, the new formulation of the segmentation in this paper allows us to develop an even far more efficient algorithm. The complexity $O(cLQ)$ of the new algorithm is much lower than the complexity $O(cL^c)$ of the Otsu's, where c is the number of clusters, $L = 256$, and $Q < 15$ in general.

A number of experiments have been conducted to test our algorithm and the Otsu's algorithm. While the two algorithms yield the same segmentation results, our algorithm is more than 10 times and 200 times faster for two-cluster and three-cluster segmentation, respectively. Therefore, our algorithm is more efficient and has more applications, especially in real-time video surveillance and tracking systems.

References

1. Sankur, B., Sezgin, M.: A survey over image thresholding techniques and quantitative performance evaluation. Journal of Electronic Imaging. (to appear).
2. Trier, O.D., Jain A.K.: Goal-directed evaluation of binarization methods. IEEE Trans. Pattern Anal. Machine Intell. 17 (1995) 1191–1201.
3. Trier, O.D., Taxt, T.: Evaluation of binarization methods for document images. IEEE Trans. Pattern Anal. Machine Intell. 17 (1995) 312–315.
4. Pal, N.R., Pal, S.: A review on image segmentation techniques. Pattern Recognition. 26 (1993) 1277–1294.
5. Sahoo, P.K., et al.: A survey of thresholding techniques. Comput. Vis. Graph. Image Process. 41 (1988) 233–260.
6. Otsu, N.: A threshold selection method from grey-level histograms. IEEE Trans. Syst., Man, Cybern. 8 (1979) 62–66.
7. Theodoridis, S., Koutroumbas, K.: Pattern Recognition. Academic Press, London (2003).

8. Dong L.: An iterative algorithm for image thresholding. Technical Report #20031225, Department of Communications Engineering, Shengyang University, China, 2003.
9. Computer vision test images: http://www-2.cs.cmu.edu/~cil/v-images.html.

Computing the Sign of a Dot Product Sum

Yong-Kang Zhu, Jun-Hai Yong, and Guo-Qin Zheng

School of Software, Tsinghua University, Beijing 100084, P. R. China
zhuyk@tsinghua.org.cn
{yongjh, gqzheng}@tsinghua.edu.cn

Abstract. A real number usually cannot be exactly represented by a floating-point number in a computer. Namely, a floating-point number frequently stands for any real number in a specific interval. In this paper, we present a method for computing the sign of a dot product sum. Each initial datum that is a floating-point number is considered as an interval. With interval analysis and floating-point summation methods, an explicit formula for calculating the minimal interval of a dot product sum is presented. Error analysis and some examples are provided as well.

Keywords: Floating-point arithmetic; Rounding error; Interval analysis.

1 Introduction

Sums of floating-point numbers are ubiquitous in scientific computing [1]. Moreover, geometric algorithms are frequently dependent on the sign of a finite sum or a dot product sum [2, 3]. Many people have devoted to studying floating-point summation. In 1999, Anderson [4] presents a floating-point summation method, which iteratively uses the reduction algorithm that makes a cancellation between a positive summand and a negative one, until all the summands have the same sign. This cancellation is exact, which is proved in the paper [4]. And then, using compensated summation method to sum up the remaining summands that have the same sign. Thus, this method could exactly compute the sign of the sum of n floating-point numbers. In the same year, Ratschek and Rokne [3] give a method, named ESSA (Exact Sign of Sum Algorithm), for exactly calculating the sign of a sum of a finite set of floating-point numbers. These two methods are exact when the initial data are just the values of those floating-point summands. However, a real number frequently cannot be exactly represented by a computer floating-point number, due to the length limit of its mantissa. Hence, even if the floating-point computation is exact, the result may be unreliable. Take $A = 1.1$, $B = 9.7$, $C = -10.8$, and $S = A+B+C$ as an example. It seems obvious that the sign of S is zero. If A, B, C are represented by floating-point arithmetic, such as IEEE 754 *double*, then the sign obtained by either of the above two summation methods is negative, which is different from the expected result. In practice, the correct result could be obtained by comparing the sum with a very tiny positive number, ε. If $|s| < \varepsilon$, then the result could be set zero, where s is the exact sum of all the floating-point numbers. But how to choose an appropriate value of ε is difficult and partially depends on the average magnitude of initial data.

A method to calculate the sign of a dot product sum with consideration of the uncertainty of the initial data is presented in this paper. Each floating-point number is represented by an interval, which is the minimal interval that contains the possible real value this floating-point number represents. Then, interval analysis method [5] is used to sum up the intervals to decide the sign. Sections 2 defines the representation error of initial data, and tells how to use an interval to represent a floating-point number. In Section 3, the minimal interval of a product between two floating-point numbers is calculated, and the sign of a dot product sum is computed. Examples and conclusions are given in Section 4.

2 Representation Error

In this paper, only fixed-length floating-point arithmetic is considered. We use β to represent the base (sometimes called the radix), use t to represent the precision, and use the $\text{fl}(a \circ b)$ (where $\circ \in \{+, -, \cdot, /\}$) to represent the floating-point arithmetic. And if x is a real number, then $\text{fl}(x)$ means the floating-point value of x. Additionally, floating-point overflow and underflow are not taken into account. Error of the example in Section 1 is caused by the reason that real numbers 1.1, 9.7, and -10.8 cannot be exactly represented by IEEE 754 *double*. This type of error is defined as representation error: if x is a real number, then the representation error e_r of x satisfies $e_r = x - \text{fl}(x)$.

There are two ways to let the program know whether initial data have representation error. The first one is that all initial data are given with flags, each of which tells whether the corresponding datum is exactly represented. The other way is an estimated method: the representation error is decided by analyzing the mantissa of the floating-point number. Assume x is the initial datum that is a real number. If the last n_m digits of $\text{fl}(x)$'s mantissa are all zero, then we estimate that $x = \text{fl}(x)$. Otherwise, $x \neq \text{fl}(x)$, which means representation error occurs. This method is based on the assumption that if a floating-point number exactly represents a real number, then the precision of it is less than the length of its mantissa. If IEEE 754 *double* ($\beta = 2$, $t = 53$) is used, then $4 \leq n_m \leq 8$ is recommended in practice.

If $x \neq \text{fl}(x)$, then there exists a tiny positive floating-point number ϵ, which satisfies $x \in [\text{fl}(x) - \epsilon, \text{fl}(x) + \epsilon]$. If the floating-point arithmetic is correctly rounded, then the minimal ϵ is $\frac{1}{2}\text{ulp}(x)$. Otherwise, the minimal ϵ is $1\text{ulp}(x)$. Here, ulp means *unit in last place* [6]. In this paper, we assume the floating-point arithmetic is *correctly rounded* [7], so we have $\epsilon = \frac{1}{2}\text{ulp}(x)$. If $x = \text{fl}(x)$, then the floating-point number $\text{fl}(x)$ is represented by the interval, $[\text{fl}(x), \text{fl}(x)]$.

3 Calculating the Sign

In this section, a method that decides the sign of a dot product sum, $S = \sum_{i=1}^{n} A_i B_i$ (A_i and B_i are real numbers), is presented. Because each initial datum, A_i or B_i, is represented by an interval, the dot product sum S is in a certain interval, namely $S = \sum_{i=1}^{n} A_i B_i \in [\underline{S}, \overline{S}]$. The sign of S is determined by analyzing the interval. Firstly, we consider that how to obtain the interval of the product of two real numbers. Given two real numbers A and B, and let $a = \mathrm{fl}(A)$ and $b = \mathrm{fl}(B)$. Without losing generality, we assume that $A > 0$ and $B > 0$. If both a and b have representation error, then we have $A \in [a - \epsilon_a, a + \epsilon_a]$, and $B \in [b - \epsilon_b, b + \epsilon_b]$, where $\epsilon_a = \frac{1}{2}\mathrm{ulp}(a)$ and $\epsilon_b = \frac{1}{2}\mathrm{ulp}(b)$. According to interval analysis [5], we obtain

$$AB \in [a - \epsilon_a, a + \epsilon_a] \cdot [b - \epsilon_b, b + \epsilon_b]$$
$$= [ab - a\epsilon_b - b\epsilon_a + \epsilon_a \epsilon_b, \; ab + a\epsilon_b + b\epsilon_a + \epsilon_a \epsilon_b]. \quad (1)$$

The algorithms are often implemented on a machine using IEEE 754 Standard binary floating-point arithmetic [8]. Real numbers in programs are represented by two types: $float$ (t=24, β=2) or $double$ (t=53, β=2), which are correctly rounded. Here we consider how to calculate the interval (1) with IEEE 754 Standard. The IEEE 754 way of expressing a floating-point number d is

$$d = \pm d_0.d_1 d_2 \ldots d_{t-1} \times \beta^{e_d}, \quad (2)$$

where each d_i is an integer and satisfies $0 \leq d_i \leq \beta - 1$, e_d is the exponent of d, d_0 is the hidden bit, $d_0 \neq 0$ for normalized numbers, and the part $d_1 d_2 \ldots d_{t-1}$ is called the mantissa of d. So ϵ_a and ϵ_b could be calculated by $\epsilon_a = \frac{1}{2}\mathrm{ulp}(a) = 2^{e_a - t}$ and $\epsilon_b = \frac{1}{2}\mathrm{ulp}(b) = 2^{e_b - t}$. Thus, according to Equation (1) we have $AB \in [ab - o - p + q, \; ab + o + p + q]$, where $o = b2^{e_a - t}$, $p = a2^{e_b - t}$, $q = 2^{e_a + e_b - 2t}$. If a does not have representation error, then $\epsilon_a = 0$, whence $o = 0$ and $q = 0$. If b does not have representation error, then $\epsilon_b = 0$, whence $p = 0$ and $q = 0$. Since the exact value of the product of two floating-point numbers can be represented by a sum of two floating-point numbers [9, 10], ab could be exactly computed by $ab = c + d$, where c and d are two floating-point numbers.

Lemma 1. *The minimal intervals that contain the value of AB are:*

$$AB \in \begin{cases} [(c+d+q) - (+o+p), \; (c+d+q) + (+o+p)], & \text{when } a>0, b>0; \\ [(c+d-q) - (-o+p), \; (c+d-q) + (-o+p)], & \text{when } a>0, b<0; \\ [(c+d-q) - (+o-p), \; (c+d-q) + (+o-p)], & \text{when } a<0, b>0; \\ [(c+d+q) - (-o-p), \; (c+d+q) + (-o-p)], & \text{when } a<0, b<0. \end{cases} \quad (3)$$

And these intervals can be exactly obtained with floating-point numbers.

Proof. According to interval analysis [5], the interval of Equation (1) is the minimal interval that contains the product of A and B, where A is any real number that a could represent, and B is any real number that b could represent. Thus, the intervals of Equation (3) are minimal as well. The value of $o = b2^{e_a - t}$ could be obtained by adding $e_a - t$ to the exponent of b without floating-point computation. And the value of $p = a2^{e_b - t}$ could be obtained by adding $e_b - t$ to the exponent of a without floating-point computation. And the value of

$q = 2^{e_a+e_b-2t}$ could be gained by constructing a new floating-point number with mantissa being set zero and exponent being set $e_a + e_b - 2t$. Thus, c, d, o, p and q can be exactly stored in five floating-point numbers, which means the intervals of Equation (3) could be exactly represented with floating-point numbers.

According to the intervals of Equation (3), the interval of $A_i B_i$ could be expressed as the form $[(g_{1i} + g_{2i} + g_{3i}) - (g_{4i} + g_{5i}), (g_{1i} + g_{2i} + g_{3i}) + (g_{4i} + g_{5i})]$.

Theorem 1. *The minimal interval that contains $S = \sum_{i=1}^{n} A_i B_i$ is:*

$$S \in \left[\sum_{i=1}^{n}((g_{1i} + g_{2i} + g_{3i}) - (g_{4i} + g_{5i})), \sum_{i=1}^{n}((g_{1i} + g_{2i} + g_{3i}) + (g_{4i} + g_{5i})) \right]. \quad (4)$$

And the signs of two boundaries of the interval can be calculated exactly.

Proof. According to Lemma 1 and interval analysis, the interval of Equation (4) is the minimal interval that contains the result of $S = \sum_{i=1}^{n} A_i B_i$. Since two boundaries of this interval are both a sum of $5n$ floating-point numbers, the signs of them could be exactly figured out by Anderson's floating-point summation method [4] or ESSA method [3].

Then, the sign of S could be obtained by analyzing the signs of two boundaries of the above interval. The resulting sign s_s could be obtained as follows:

1. Calculate the sign of $\sum_{i=1}^{n}(g_{1i} + g_{2i} + g_{3i})$, say s_1;
2. If $s_1 > 0$, then
   (a) Calculate the sign of $\sum_{i=1}^{n}((g_{1i} + g_{2i} + g_{3i}) - (g_{4i} + g_{5i}))$, say s_2;
   (b) If $s_2 > 0$, then let $s_s = 1$. Otherwise, let $s_s = 0$;
3. If $s_1 < 0$, then
   (a) Calculate the sign of $\sum_{i=1}^{n}((g_{1i} + g_{2i} + g_{3i}) + (g_{4i} + g_{5i}))$, say s_3;
   (b) If $s_3 < 0$, then let $s_s = -1$. Otherwise, let $s_s = 0$;
4. If $s_1 = 0$, then let $s_s = 0$.

Corollary 1. *Assume the minimal interval of S is $[\underline{S}, \overline{S}]$. If $\underline{S}$ and $\overline{S}$ have the same sign, then the sign of S can be obtained exactly by the above method. If $\underline{S}$ and $\overline{S}$ have opposite signs, then the sign obtained by the above method is zero.*

Proof. According to Theorem 1, $[\underline{S}, \overline{S}]$ could be exactly obtained. If $\underline{S}$ and $\overline{S}$ have the same sign, then we have $S > 0$ or $S < 0$. Thus, the above method can produce the exact sign of S. Otherwise, the sign obtained by the method is zero.

With the corollary, if two boundaries of the interval have different signs, the sign of S is set zero. In this case, the actual sign of S is indefinite, because zero falls into the minimal interval of S. There is no error in calculating the sign of the minimal interval. Therefore, this uncertainty is only determined by the limitation of floating-point representation, namely the representation error due to length limit of mantissa. The error analysis of our method is given in the following theorem.

Theorem 2. *When initial data have no representation errors, the error of our method is zero. Assume the sign obtained by our method is s_s.*

1. *If $S = 0$, then $s_s = 0$, i.e. the error of the method is 0.*
2. *If $S \neq 0$ and $s_s \neq 0$, then the error of the method is 0.*
3. *If $S \neq 0$ and $s_s = 0$, then the error of the method is $e = |S|$.*

Let Δ be the length of the interval of Equation(4), we have $S \in [\widehat{S} - \frac{\Delta}{2}, \widehat{S} + \frac{\Delta}{2}]$, where $\widehat{S} \approx S$. If $S \neq 0$ and $s_s = 0$, then we have $|\widehat{S}| \leq \frac{\Delta}{2}$. Therefore, the error only possibly occurs when S is very close to $\frac{\Delta}{2}$ or $|S| < \frac{\Delta}{2}$. Assume δ_i is the interval length of the minimal intervals of Equation(3), we have $\frac{\delta_i}{2} = |o| + |p|$. Let $a_i = \text{fl}(A_i)$ and $b_i = \text{fl}(B_i)$. So, the interval length Δ of Equation(4) satisfies

$$\frac{\Delta}{2} = \sum_{i=1}^{n} \delta_i = \sum_{i=1}^{n} (|a_i 2^{e_{b_i} - t}| + |b_i 2^{e_{a_i} - t}|) = 2^{-t} \sum_{i=1}^{n} (m_{a_i} + m_{b_i}) 2^{e_{a_i} + e_{b_i}}$$

$$< n 2^{e_w - t + 2} = 2n \cdot \text{ulp}(w), \qquad (5)$$

where m_{a_i} and m_{b_i} are mantissas of a_i and b_i, and $e_w = \max(e_{a_i} + e_{b_i})$, $w = \max(|a_i b_i|)$, for all $i = 1, 2, \cdots, n$. Thus, if $|S| < 2n \cdot \text{ulp}(w)$ is satisfied, then the sign of S calculated by our method may be wrong.

Compensated summation [11] produces a relative error of at most nu (u is the unit roundoff) when all the summands have the same sign [1]. It can be used as a tool to accelerate our method when the sign is apparently not zero. Using compensated summation method, all the positive summands are accumulated to s_+, and all the negative summands are accumulated to s_-, where s_+ and s_- are floating-point numbers. If $|s_+ + s_-| > nu$, the sign can be obtained directly. Otherwise, the exact summation method is called to calculate the sign of the boundaries of the interval.

4 Examples and Conclusions

Some examples and some concluding remarks are provided in this section. The floating-point arithmetic used in our examples is IEEE 754 *double*. The first example contains six initial data: 1.1, −0.1, 2.3, 9.8, 9.7, and −22.8. Our new method obtains that the sign of the sum with the six numbers is zero, which matches the sign of the real sum, while Anderson's method [4] and ESSA method [3] produce negative signs against the real value.

The second example is an application of the method proposed in this paper. It determines whether two line segments intersect. The testing problem could be converted into calculating signs of several dot product summations [2]. If the signs could be computed exactly, then the testing result is reliable. In this example, the new method, say M_3, is compared with the other two methods. The first method, say M_1, converts each coordinate into an integer (ex. 0.34 into 34), and then calculates the signs of those dot product summations with the integer arithmetic. It could always produce exact results. The second method, say M_2, uses instances of Java's "class BigDecimal" to represent the floating-point values of coordinates (ex. fl(0.34)), and then calculates the signs with a Java program. It could avoid the error of floating-point computations, but it neglects the representation error of initial data. 1,000,000 random tests are carried out. In each intersection test, eight coordinates of the four end points of the two given line segments are randomly selected from an array, whose elements belong to $\{0.00, 0.01, \cdots, 0.99\}$. We find that M_3 always gives the same results as M_1 does, while M_2 produces different results 197 times. The experiment shows that the two given line segments intersect at an ending point when the differences occur. This example illustrates that the method could well deal with some singular cases.

Thus, the proposed method for calculating the sign of a dot product sum can deal with the representation error of initial data. The method uses an interval, which is as compact as possible, to represent each floating-point number. The minimal interval of the dot product sum is then obtained to determine the the sign of the resulting dot product sum. The accuracy of our method is demonstrated with Theorem 2 in this paper.

Acknowledgements

The authors appreciate the comments and suggestions of the anonymous reviewer. The research was supported by Chinese 863 Program (2003AA4Z3110) and 973 Program (2002CB312106). The second author was supported by a project sponsored by SRF for ROCS, SEM (041501004), and an FANEDD(200342).

References

1. Higham, N.J.: The accuracy of floating point summation. SIAM Journal on Scientific Computing **14** (1993) 783–799
2. Gavrilova, M. and Rokne, J.: Reliable line segment intersection testing. Computer-Aided Design **32** (2000) 737–745
3. Ratschek, H. and Rokne, J.: Exact computation of the sign of a finite sum. Applied Mathematics and Computatoin **99** (1999) 99–127
4. Anderson, I.J.: A distillation algorithm for floating-point summation. SIAM Journal on Scientific Computing **20** (1999) 1797–1806
5. Moore, R.E.: Interval Analysis. Prentice-Hall, Englewood Cliffs, NJ (1966)
6. Higham, N.J.: Accuracy and Stability of Numerical Algorithms. second edn. SIAM. Philadelphia (2002)
7. Jaulin, L., Kieffer, M., Didrit, O., and Walter, E.: Applied Interval Analysis. Springer (2001)
8. ANSI/IEEE New York: IEEE Standard for Binary Floating-Point Arithmetic, Standard 754-1985. (1985)

9. Masotti, G.: Floating-point numbers with error estimates. Computer-Aided Design **25** (1993) 524–538
10. Priest, D.M.: Algorithms for arbitrary precision floating point arithmetic. In Kornerup, P. and Matula, D.W., eds.: Proceedings of the 10th IEEE Symposium on Computer Arithmetic, IEEE Computer Society Press, Los Alamitos, CA (1991) 132–143
11. Kahan, W.: Further remarks on reducing truncation error. Communications of the ACM **8** (1965) 40

Bilateral Filter for Meshes Using New Predictor

Yu-Shen Liu[1], Pi-Qiang Yu[1], Jun-Hai Yong[2], Hui Zhang[2], and Jia-Guang Sun[1,2]

[1] Department of Computer Science and Technology,
Tsinghua University, Beijing 100084, P. R. China
liuyushen00@mails.tsinghua.edu.cn
[2] School of Software, Tsinghua University, Beijing 100084, P. R. China

Abstract. A new predictor of bilateral filter for smoothing meshes is presented. It prevents shrinkage of corners. A major feature of mesh smoothing is to move every vertex along the direction determined by the mean curvature normal with speed defined by the predictor. It prevents unnatural deformation for irregular meshes. In order to remove the normal noise, we use adaptive Gaussian filter to smooth triangle normals.

1 Introduction

Nowadays, mesh smoothing or mesh denoising, whose goal is to adjust vertex positions so that the overall mesh becomes smooth while mesh connectivity is kept, is an important process for many digital geometry applications. Removing noise while preserving important features currently is an active area of research.

Many mesh smoothing algorithms have been developed in the last few years. Taubin [9] pioneered $\lambda|\mu$ algorithm to solve the shrinkage problem caused by Laplacian smoothing. Desbrun et al. [2] extended this approach to irregular meshes using mean curvature flow. However, these techniques are isotropic, and therefore diffuse shape features. Feature-preserving mesh smoothing was recently proposed. Methods presented in [6, 7, 10] achieve this goal by first smoothing the normal field, and then updating vertex positions to match the new normals. The extension from image smoothing to mesh smoothing was explored in [1, 3, 4, 8]. The bilateral filter, which is an alternative edge-preserving image filter [11], has been extended to mesh smoothing in different ways [1, 3, 4]. Since the bilateral filter is simple, fast and well feature-preserving, it is a good choice for smoothing and denoising. However, the bilateral filter is sensitive to the initial normals, and tends to round off corners, which may result in unnatural deformation for irregular meshes.

In this paper, we present a new predictor of bilateral filter which avoids corner shrinkage. This predictor depends on normals of both a vertex and its nearby triangles. We first smooth mesh normals. Then, we move every vertex along the direction determined by the mean curvature normal with speed defined by the new predictor. The major contributions of our work are as follows.

- The new predictor preserves both sharp edges and corners.
- Combination of the new predictor and the method of normal improvement prevents unnatural deformation for highly irregular meshes.

2 Bilateral Filter for Meshes

A bilateral filter is an edge-preserving filter introduced in image processing for smoothing images [11]. It has been extended to mesh smoothing in different ways [1,3,4]. Let $\mathbf{M}$ be a input mesh with some additive noise, and let $\mathbf{s}$ and $\mathbf{p}$ be two points on $\mathbf{M}$. Jones [5] introduces the concept of $predictor(\Pi_\mathbf{p}(\mathbf{s}))$, which defines the denoised position of $\mathbf{s}$ due to $\mathbf{p}$. The bilateral filter for meshes is defined as

$$E(\mathbf{s}) = \frac{\sum_{\mathbf{p}\in N(\mathbf{s})} f(\|\mathbf{p}-\mathbf{s}\|)g(\|\Pi_\mathbf{p}(\mathbf{s})-\mathbf{s}\|)\Pi_\mathbf{p}(\mathbf{s})}{\sum_{\mathbf{p}\in N(\mathbf{s})} f(\|\mathbf{p}-\mathbf{s}\|)g(\|\Pi_\mathbf{p}(\mathbf{s})-\mathbf{s}\|)}, \quad (1)$$

where $N(\mathbf{s})$ is a neighborhood of $\mathbf{s}$, and the weight of $\mathbf{p}$ depends on both the spatial distance $\|\mathbf{p}-\mathbf{s}\|$ and the signal difference $\|\Pi_\mathbf{p}(\mathbf{s})-\mathbf{s}\|$. A *spatial weight* Gaussian f of width σ_f and an *influence weight* Gaussian g of width σ_g are often chosen in practice. Let $\mathbf{n_s}$ be the normal at $\mathbf{s}$, and let $\mathbf{n_p}$ be the normal at $\mathbf{p}$. Here we formally define a displacement signed-distance ds from a current position $\mathbf{s}$ to the predictor $\Pi_\mathbf{p}(\mathbf{s})$.

Fleishman et al. [3] have proposed an extension of bilateral filter to meshes. Their predictor can be written by

$$\Pi_\mathbf{p}(\mathbf{s}) = \mathbf{s} + ((\mathbf{p}-\mathbf{s})\cdot \mathbf{n_s})\mathbf{n_s}, \quad (2)$$

where $\mathbf{p}$ is a vertex in the neighborhood of $\mathbf{s}$. It is illustrated in Fig. 1(a). The predictor does not introduce tangential *vertex drift*. However, it tends to move vertices along the normal direction to round off corners as shown in Fig. 3(b). Considering the case in which the point $\mathbf{s}$ is a corner, Fleishman et al.'s predictor moves great distance from $\mathbf{s}$ to $\Pi_\mathbf{p}(\mathbf{s})$ as shown in Fig. 2(a).

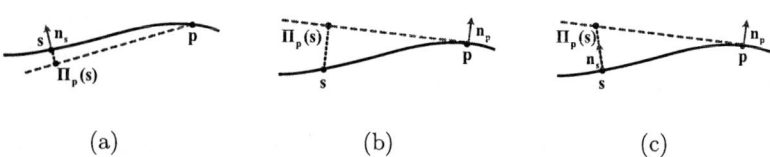

Fig. 1. (a) Fleishman et al.'s predictor. (b) Jones et al.'s predictor. (c) Our predictor

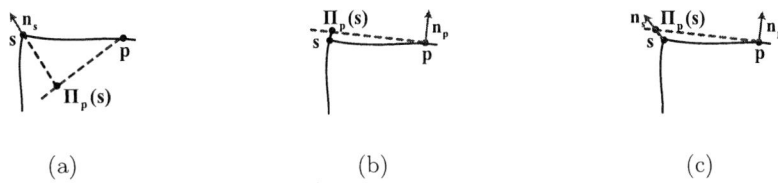

Fig. 2. s is a corner. (a) Fleishman et al.'s predictor. (b) Jones et al.'s predictor. (c) Our predictor

Independently, Jones et al. [4] present a similar algorithm. Their approach projects the central vertex **s** onto the planes of nearby triangles, while that of Fleishman et al. project nearby vertices onto the tangent plane of the central vertex **s**. The predictor of Jones et al. can be written by

$$\Pi_\mathbf{p}(\mathbf{s}) = \mathbf{s} + ((\mathbf{p} - \mathbf{s}) \cdot \mathbf{n_p})\mathbf{n_p}, \tag{3}$$

where **p** is the centroid of a triangle in the neighborhood of **s**. It is illustrated in Fig. 1(b). In Fig. 2(b), Jones et al.'s predictor moves a little distance from **s** to $\Pi_\mathbf{p}(\mathbf{s})$ when **s** is a corner. However, since **s** does not move along the direction of normal $\mathbf{n_s}$, it may introduce tangential vertex drift. This produces unnatural deformation for irregular meshes as shown in Fig. 4(e).

2.1 New Predictor of Bilateral Filter

To avoid unnatural deformation arising from the predictors of Fleishman et al. and Jones et al., we present a new predictor that considers both the vertex normal and its nearby triangle normals. Our approach moves the central vertex **s** to the tangent planes of nearby triangles along the direction of the normal $\mathbf{n_s}$. The new predictor prevents corner shrinkage and tangential vertex drift. Our predictor satisfies

$$\Pi_\mathbf{p}(\mathbf{s}) = \mathbf{s} + d s \mathbf{n_s} \quad \text{and} \quad (\Pi_\mathbf{p}(\mathbf{s}) - \mathbf{p})\mathbf{n_p} = 0,$$

where **p** is the centroid of a triangle in the neighborhood of **s**. By solving the above equations, we obtain

$$\Pi_\mathbf{p}(\mathbf{s}) = \mathbf{s} + \left(\frac{(\mathbf{p} - \mathbf{s}) \cdot \mathbf{n_p}}{\mathbf{n_s} \cdot \mathbf{n_p}}\right)\mathbf{n_s}. \tag{4}$$

It is illustrated in Fig. 1(c). Since our predictor moves vertices along the normal direction, no vertex drift occurs. Due to the combination with nearby triangle normals, corners can be preserved. We consider the case where the point **s** is a corner. Compared with Fleishman et al.'s predictor which tends to round off corners as shown in Fig. 2(a), our predictor is able to preserve corners as

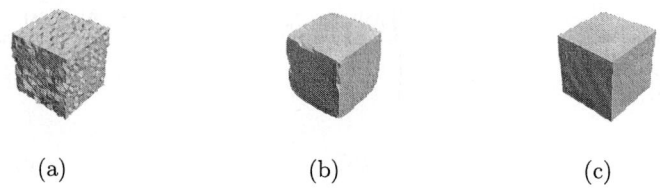

Fig. 3. Smoothing of CAD-like model with large noise. (a) Input noisy model. (b) Fleishman et al.'s result (5 iterations). (c) Our result (5 iterations)

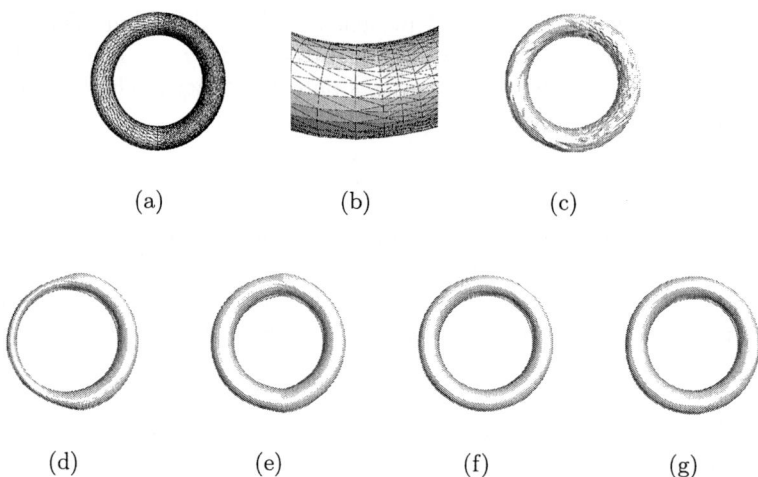

Fig. 4. (a) A torus with different sampling rates. (b) A magnified view of a part of the torus. (c) The torus with additive Gaussian noise in both vertex positions and normals. (d) Fleishman et al.'s method deforms the initial shape. (e) Jones et al.'s method smoothes well but slightly deforms the initial shape. (f) Mean curvature flow smoothes well. (g) Our method smoothes well as (f)

shown in Fig. 2(c). In Fig. 3 we show the smoothing results of a CAD object. In Fig. 3(b), the corners are rounded off by Fleishman et al.'s predictor, while they are preserved by our new predictor as shown in Fig. 3(c). Compared with Jones et al's predictor which introduces tangential vertex drift as shown in Fig. 2(b), our predictor moves vertices along the normal direction as shown in Fig. 2(c). Fig. 4(e) shows the result of vertex drift. Our result achieves better smoothing with respect to the shape as shown in Fig. 4(g).

3 Improving and Smoothing Normals

Fleishman et al. compute the normal at a vertex as the weighted average (by the area of the triangles) of the normals to the triangles in the 1-ring neighborhood of the vertex, where the normal direction depends on the parameterization defined by the areas of the neighborhood triangles. Moving vertex along this direction may result in unnatural deformation for highly irregular meshes (see Fig. 4(d)). To overcome this problem, we use the mean curvature normal. According to [2], a good estimation of the mean curvature normal at vertex **p** is given by

$$\mathbf{Hn}(\mathbf{p}) = \frac{1}{4A} \sum_{i \in V(\mathbf{p})} (\cot \alpha_i + \cot \beta_i)(\mathbf{q}_i - \mathbf{p}), \qquad (5)$$

where A is the sum of the areas of the triangles around **p**, $V(\mathbf{p})$ is the set of adjacent vertex indexes to **p**, $\mathbf{q}_i$ corresponds to the ith adjacent vertex to **p**, and α_i and β_i are the two angles opposite to the edge $\mathbf{pq}_i$. In this paper, we use the unit vector $\mathbf{n}(\mathbf{p}) = \frac{\mathbf{Hn}}{\|\mathbf{Hn}\|}$ as the normal at vertex **p** instead of the normal used by Fleishman et al.. Roughly speaking, our smoothing schemes consist of moving every vertex along the direction determined by the mean curvature normal of Equation (5), with speed defined by the new predictor of Equation (4). This prevents unnatural deformation for irregular meshes (see Fig. 4(g)).

Our predictors are also based on the normals of the nearby triangles. Since the normals are sensitive to noise [4], we smooth normals by adaptive Gaussian filter applied to triangle normals [7].

4 Results and Discussion

We demonstrate our results in Figs. 4-5. The execution time is reported on a Pentium IV 1.70GHz processor with 256M RAM excluding that of loading meshes. All meshes are rendered with flat shading. In Table 1, we indicate model sizes, the number of iterations, running time, and the parameters. The σ_f and

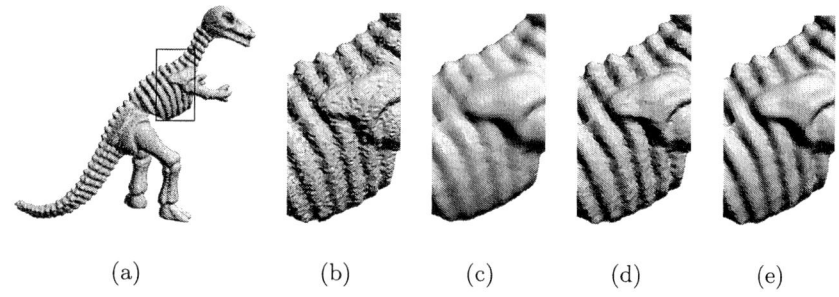

Fig. 5. Results of smoothing the dinosaur model. (a) Input noisy model. (b) A magnified view of (a). (c) Fleishman et al.'s result. (d) Jones et al.'s result. (e) Our result. Notice that details such as the skeletons are better preserved by our method, while flat regions are equivalently smoothed

Table 1. Comparison of smoothing results

Model	Fig.	Verts.	Iterations	Time	σ_f	σ_g
Dinosaur	5(c)	56K	3	1.2s	Interactive	Interactive
	5(d)		non-iterative	22.5s	4.0	0.2
	5(e)		3	7.8s	4.0	0.2

σ_g are expressed as ratios of the mean edge length used by Jones et al.'s [4]. Fig. 4 shows a comparison for smoothing a irregular mesh with other algorithms. In Fig. 5, we compare our result to the results of other bilateral filter algorithm for the dinosaur model.

We have presented a novel predictor of bilateral filter which prevents shrinkage of corners. Based on this predictor and the mean curvature normal, we introduced a new mesh smoothing method which prevents unnatural deformation for irregular meshes. In the future, we wish to find a way to automatically select parameters used in bilateral filter such that smoothing is adaptively achieved.

Acknowledgements. The research was supported by Chinese 973 Program (2002CB312106). The third author was supported by the project sponsored by SRF for ROCS, SEM (041501004) and FANEDD (200342).

References

1. Choudhury, P., Tumblin, J.: The trilateral tilter for high contrast images and meshes. Proc. of the Eurographics Symposium on Rendering (2003) 186-196
2. Desbrun, M., Meyer, M., Schröder, P., and Barr, A.H.: Implicit fairing of irregular meshes using diffusion and curvature flow. In SIGGRAPH'99 Conference Proceedings (1999) 317-324
3. Fleishman, S., Drori, I., and Cohen-Or, D.: Bilateral mesh denoising. In SIGGRAPH' 03 Conference Proceedings (2003) 950-953
4. Jones, T., Durand, F., and Desbrun, M.: Non-iterative, feature-preserving mesh smoothing. In SIGGRAPH' 03 Conference Proceedings (2003) 943-949
5. Jones, T.: Feature preserving smoothing of 3D surface scans. Master's Thesis, Department of Electrical Engineering and Computer Science, MIT (2003)
6. Ohtake, Y., Belyaev, A., and Bogaevski, I.: Mesh regularization and adaptive smoothing. Computer-Aided Design **33** (2001) 789-800
7. Ohtake, Y., Belyaev, A., and Seidel, H.P.: Mesh smoothing by adaptive and anisotropic gaussian filter applied to mesh normal. In Vision, modeling and visualization (2002) 203-210
8. Shen, Y., Barner, K.E.: Fuzzy vector median-based surface smoothing. IEEE Transactions on Visualization and Computer Graphics **10** (2004) 252-265

9. Taubin, G.: A signal processing approach to fair surface design. In SIGGRAPH'95 Conference Proceedings (1995) 351-358
10. Taubin, G.: Linear Anisotropic mesh filtering. Tech. Rep. IBM Research Report RC2213 (2001)
11. Tomasi, C., Manduchi, R.: Bilateral filtering for gray and color images. In Proc. IEEE Int. Conf. on Computer Vision (1998) 836-846

Scientific Computing on Commodity Graphics Hardware

Ruigang Yang

University of Kentucky, Lexington KY 40506, USA
ryang@cs.uky.edu
http://www.cs.uky.edu/~ryang

Abstract. Driven by the need for interactive entertainment, modern PCs are equipped with specialized graphics processors (GPUs) for creation and display of images. These GPUs have become increasingly programmable, to the point that they now are capable of efficiently executing a significant number of computational kernels from non-graphical applications. In this introductory paper we first present a high-level overview of modern graphics hardware's architecture, then introduce several applications in scientific computing that can be efficiently accelerated by GPUs. Finally we list programming tools available for application development on GPUs.

1 Introduction

As the mass-market emphasis in computing has shifted from word processing and spreadsheets to interactive entertainment, computer hardware has evolved to better support these new applications. Most of the performance-limiting processing today involves creation and display of images; thus, a new entity has appeared within most computer systems. Between the system's general-purpose processor and the video frame buffer, there is now a specialized Graphic Processing Unit (GPU).

Early GPUs were not really processors, but hardwired pipelines for each of the most common rendering tasks. As more complex 3D-transformations have become common in a wide range of applications, GPUs have become increasingly programmable, to the point that they now are capable of efficiently executing a significant number of computational kernels from non-graphical applications.

A GPU is simpler and more efficient than a conventional PC processor (CPU) because a GPU only needs to perform a relatively simple set of array processing operations (but at a very high speed). Many problems in scientific computing, such as physically-based simulation, information retrieval, and data mining, can boil down to relatively simple matrix operations. This characteristic makes these problems ideal candidates for GPU acceleration.

In this introductory paper we first present a high-level overview of modern graphics hardware's architecture and its phenomenal development in recent years. Then we introduce a large array of non-graphical computational tasks, in particular, linear algebra operations, that have been successfully implemented on GPUs and obtained significant performance improvements. Finally we list programming tools available for application development on GPUs. Some of them are designed to allow programming GPUs with familiar C-like constructs and syntax, without worrying about the details of the hardware. They hold the promise of bringing the vast computational power in GPUs to the broad scientific computing community.

2 A Brief Overview of GPUs

In this section, we will explain the basic architecture of GPUs and the potential advantages of using GPUs to solve scientific problems.

2.1 The Rendering Pipeline

GPUs are dedicated processors designed specifically to handle the intense computational requirements of display graphics, i.e., rendering texts or images over 30 frames per second. As depicted in Figure 1, a modern GPU can be abstracted as a rendering pipeline for 3D computer graphics (2D graphics is just a special case) [1].

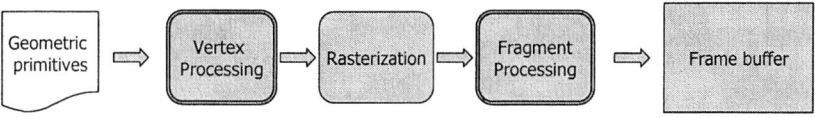

Fig. 1. Rendering Pipeline

The inputs to the pipeline are geometric primitives, i.e., points, lines, polygons; and the output is the *framebuffer*–a two-dimensional array of pixels that will be displayed on screen.

The first stage operates on geometric primitives described by vertices. In this *vertex-processing* stage vertices are transformed and lit, and primitives are clipped to a viewing volume in preparation for the next stage, *rasterization*. The rasterizer produces a series of framebuffer addresses and color values, each is called a *fragment* that represents a portion of a primitive that corresponds to a pixel in the framebuffer.

Each fragment is fed to the next *fragment processing* stage before it finally alters the framebuffer. Operations in this stage include texture mapping, depth test, alpha blending, etc.

2.2 Recent Trend in GPUs

Until a few years ago, commercial GPUs, such as the RealityEngine from SGI [2], implement in hardware a fixed rendering pipeline with configurable parameters.

As a result their applications are restricted to graphical computations. Driven by the market demand for better realism, the current generation of commercial GPUs such as the NVIDIA GeForce FX [3] and the ATI Radeon 9800 [4] added significant programmable functionalities in both the vertex and the fragment processing stage(stages with double-lines in Figure 1). They allow developers to write a sequence of instructions to modify the vertex or fragment output. These programs are directly executed on the GPUs to achieve comparable performance to fixed-function GPUs.

In addition to programable functionalities in modern GPUs, their support for floating point output has been improving. GPUs on the market today support up to 32-bit floating point output. Such a precision is usable for many diverse applications other than computer graphics.

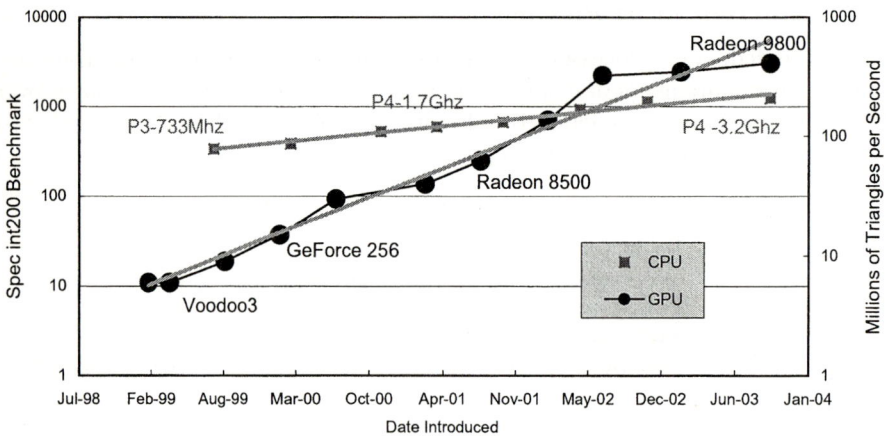

Fig. 2. A graph of performance increase over time for CPUs and GPUs. GPU performance has increased at a faster rate than CPUs. (Data courtesy of Anselmo Lastra)

GPUs have also demonstrated a rapid improvement in performance during the past few years. In Figure 2, we plot the performance increase of both GPUs and commodity Central Processor Units (CPUs). Similar to the number of integer operations per second for CPUs, a typical benchmark to gauge a GPU's performance is the number of triangles it can process every second. We can see that GPUs have maintained a performance improvement rate of approximately 3X/year, which exceeds the performance improvement of CPUs at 1.6X/year. This is because CPUs are designed for low latency computations, while GPUs are optimized for high throughput of vertices and fragments [5]). Low latency on memory-intensive applications typically requires large caches, which use a large silicon area. Additional transistors are used to greater effect in GPU architectures because they are applied to additional functional units that increase throughput [6].

3 Applications of GPUs for General-Purpose Computation

With the wide deployment of inexpensive yet powerful GPUs in the last several years, we have seen a surge of experimental research in using GPUs for tasks other than rendering. For example, Yang et. al. have experimented with using GPUs to solve computer visions problems [7, 8]; Holzschuch and Alonso to speed visibility queries [9]; Hoff et. al. to compute generalized Voronoi Diagrams [10] and proximity information [11]; and Lok to reconstruct an object's visual hull given live video from multiple cameras [12]. Each of these applications obtained significant performance improvements by exploiting the speed and the inherent parallelism in modern graphics hardware.

For the scope of this paper, we introduce several representative approaches to accelerate linear algebra operations on GPUs.

Larsen and McAllister present a technique for large matrix-matrix multiplies using low cost graphics hardware [13]. The method is an adaptation of the technique from parallel computing of distributing the computation over a logically cube-shaped lattice of processors and performing a portion of the computation at each processor. Graphics hardware is specialized in a manner that makes it well suited to this particular problem, giving faster results in some cases than using a general-purpose processor. A more complete and up-to-date implementation of dense matrix algebra is presented by Moravánszky [14].

The paper of Bolz et al. shows two basic, broadly useful, computational kernels implemented on GPUs: a sparse matrix conjugate gradient solver, and a regular-grid multigrid solver [15]. Performance analysis with realistic applications shows that a GPU-based implementation compares favorable over its CPU counterpart. A similar framework for implementation of linear algebra operators on GPUs is by Krüger and Westermann [16], which focuses on sparse and banded matrices.

There are many other algorithms for scientific computing that have been implemented on GPUs, including FFT [17], level set [18, 19], and various types of physically-based simulations [20, 21, 22]. Interested readers are referred to http://www.gpgpu.org for other general-purpose applications on GPUs.

4 GPU Programming Languages

While many non-graphical applications on GPUs have obtained encouraging results by exploiting GPU's fast speed and high bandwidth, the development process is not trivial. Many of the existing applications are written using low level assemble languages that are directly executed on the GPU. Therefore, novice developers are faced with a steep learning curve to master a thorough understanding of the graphics hardware and its programming interfaces, namely *OpenGL* [1] and *DirectX* [23].

Fortunately, this is rapidly changing with several high-level languages available. The first is *Cg* – a system for programming graphics hardware in a C-like language[24]. It is, however, still a programming language geared towards rendering tasks and tightly coupled with graphics hardware.

There are other high-level languages, such as *Brook for GPUs* and *Sh*, which allow programming GPUs with familiar constructs and syntax, without worrying about the details of the hardware. Brook extends C to include simple data-parallel constructs, enabling the use of the GPU as a streaming coprocessor. Sh is a metaprogramming language that offers the convenient syntax of C++ and takes the burden of register allocation and other low-level issues away from the programmer. While these languages are not fully mature yet, they are the most promising ones to allow non-graphics researchers or developers to tap into the vast computational power in GPUs.

5 Conclusion

The versatile programmability and improved floating-point precisions now available in GPUs make them useful coprocessors for scientific computing. Many non-trivial computational kernels have been successfully implemented on GPUs to receive significant acceleration. As graphics hardware continues to evolve at a faster speed than CPUs and more "user-friendly" high-level programming languages are becoming available, we believe communities outside computer graphics can also benefit from the fast processing speed and high bandwidth that GPUs offer. We hope this introductory paper will encourage further thinking along this direction.

Acknowledgments

The author would like to thank Hank Dietz for providing some of the materials in this paper. This work is supported in part by fund from the office of research at the University of Kentucky and Kentucky Science & Engineering Foundation (RDE-005).

References

1. Segal, M., Akeley, K.: The OpenGL Graphics System: A Specification (Version 1.3) (2001) http://www.opengl.org.
2. Akeley, K.: RealityEngine Graphics. In: Proceedings of SIGGRAPH. (1993)
3. NVIDIA Corporation: GeForce FX (2003) http://www.nvidia.com/page/fx desktop.html.
4. ATI Technologies Inc.: ATI Radeon 9800 (2003) http://www.ati.com/products/radeon9800.
5. Lindholm, E., Kilgard, M., Moreton, H.: A User Programmable Vertex Engine. In: Proceedings of SIGGRAPH. (2001) 149–158
6. Harris, M.: Real-Time Cloud Simulation and Rendering. PhD thesis, Department of Computer Science, University of North Carolina at Chapel Hill (2003)

7. Yang, R., Welch, G.: Fast Image Segmentation and Smoothing Using Commodity Graphics Hardware. Journal of Graphics Tools, special issue on Hardware-Accelerated Rendering Techniques **7** (2003) 91–100
8. Yang, R., Pollefeys, M.: Multi-Resolution Real-Time Stereo on Commodity Graphics Hardware. In: Proceedings of Conference on Computer Vision and Pattern Recognition (CVPR). (2003) 211–218
9. Holzschuch, N., Alonso, L.: Using graphics hardware to speed-up visibility queries. Journal of Graphics Tools **5** (2000) 33–47
10. III, K.E.H., Keyser, J., Lin, M.C., Manocha, D., Culver, T.: Fast Computation of Generalized Voronoi Diagrams Using Graphics Hardware. In: Proceeding of SIGGRAPH 99. (1999) 277–286
11. III, K.E.H., Zaferakis, A., Lin, M.C., Manocha, D.: Fast and simple 2d geometric proximity queries using graphics hardware. In: 2001 ACM Symposium on Interactive 3D Graphics. (2001) 145–148 ISBN 1-58113-292-1.
12. Lok, B.: Online Model Reconstruction for Interactive Virtual Environments. In: Proceedings 2001 Symposium on Interactive 3D Graphics, Chapel Hill, North Carolina (2001) 69–72
13. Larsen, E.S., McAllister, D.K.: Fast Matrix Multiplies using Graphics Hardware. In: Proceeding of Super Computer 2001. (2001)
14. Ádám Moravánszky: Dense Matrix Algebra on the GPU. In: Shaderx2: Shader Programming Tips & Tricks With Directx 9. Wordware (2003)
15. Bolz, J., Farmer, I., Grinspun, E., Schrder, P.: Sparse Matrix Solvers on the GPU: Conjugate Gradients and Multigrid. ACM Transactions on Graphics (SIGGRAPH 2003) **22** (2003)
16. Krger, J., Westermann, R.: Linear Algebra Operators for GPU Implementation of Numerical Algorithms. ACM Transactions on Graphics (SIGGRAPH 2003) **22** (2003)
17. Moreland, K., Angel, E.: The FFT on a GPU. In: SIGGRAPH/Eurographics Workshop on Graphics Hardware 2003 Proceedings. (2003) 112–119
18. Strzodka, R., Rumpf, M.: Level set segmentation in graphics hardware. In: Proceedings of the International Conference on Image Processing. (2001)
19. Lefohn, A.E., Kniss, J., Hansen, C., Whitaker, R.T.: Interactive Deformation and Visualization of Level Set Surfaces Using Graphics Hardware. In: Proceedings of IEEE Visualization. (2003)
20. Harris, M., Baxter, W., Scheuermann, T., Lastra, A.: Simulation of Cloud Dynamics on Graphics Hardware. In: Proceedings of Graphics Hardware. (2002) 92–101
21. Kim, T., Lin, M.: Visual Simulation of Ice Crystal Growth. In: Proceedings of ACM SIGGRAPH / Eurographics Symposium on Computer Animation 2003. (2003) 92–101
22. S.Tomov, M.McGuigan, R.Bennett, G.Smith, J.Spiletic: Benchmarking and Implementation of Probability-Based Simulations on Programmable Graphics Cards. Computers & Graphics (2004)
23. Microsoft: DirectX (2003) http://www.microsoft.com/windows/directx.
24. NVIDIA: Cg: C for Graphics (2002) http://www.cgshaders.org/.

FIR Filtering Based Image Stabilization Mechanism for Mobile Video Appliances

Pyung Soo Kim

Mobile Platform Lab, Digital Media R&D Center,
Samsung Electronics Co., Ltd, Suwon City, 442-742, Korea
Phone : +82-31-200-4635, Fax : +82-31-200-3975
kimps@samsung.com

Abstract. This paper proposes a new image stabilization mechanism based on filtering of absolute frame positions for mobile video appliances. The proposed mechanism removes undesired motion effects in real-time, while preserving desired gross camera displacements. The well known finite impulse response (FIR) filter is adopted for the filtering. The proposed mechanism provides the filtered position and velocity that have good inherent properties. It is shown that the filtered position is not affected by the constant velocity. It is also shown that the filtered velocity is separated from the position. Via computer simulations, the performance of the proposed mechanism is shown to be superior to the existing Kalman filtering based mechanism.

1 Introduction

In recent years, video communication and processing play a significant role in mobile video appliances such as mobile phones, handheld PCs, digital consumer camcorders, and so on. Thus, a camera becomes an inherent part to acquire video images. However, image sequences acquired by a camera mounted on a mobile video appliance are usually affected by undesired motions causing unwanted positional fluctuations of the image. To remove undesired motion effects and to produce compensated image sequences that expose requisite gross camera displacements only, various image stabilization mechanisms have been mainly used for the computation of ego-motion [1], [2], the video compression [3], the estimation and tracking of moving mobiles [4]-[7].

In this paper, the image stabilization mechanism for the estimation and tracking of moving mobiles will be considered. Recently, the motion vector integration (MVI) in [4] and the discrete-time Fourier transform (DFT) filtering in [5] are developed. However, in the MVI mechanism, the filtered position trajectory is delayed owing to filter characteristics imposing larger frame shift than actually required for stabilization, in which stabilization performance is degraded. The DFT filtering based mechanism is not suited for real-time application since off-line processing is required. Therefore, in recent, the image stabilization mechanism using the Kalman filtering has been made by posing the optimal filtering problem due to the compact representation and the efficient manner [6], [7]. However, the Kalman filter has an infinite impulse response (IIR) structure that utilizes all past information accomplished by equaling weighting and has a recursive formulation. Thus, the Kalman filter tends to accumulate the filtering error as time goes. In addition, the Kalman filter is known to be sensitive and show even divergence phenomenon for temporary modeling uncertainties and round-off errors [8]-[11].

Therefore, in the current paper, an alternative image stabilization mechanism is proposed. The proposed mechanism gives the filtered absolute frame position in real-time, removing undesired motion effects, while preserving desired gross camera displacements. For the filtering, the proposed mechanism adopts the well known finite impulse response (FIR) filter that utilizes only finite information on the most recent window [10], [11]. The proposed mechanism provides the filtered velocity as well as the filtered position. These filtered position and velocity have good inherent properties such as unbiasedness, efficiency, time-invariance, deadbeat, and robustness due to the FIR structure. It is shown that the filtered position is not affected by the constant velocity. It is also shown that the filtered velocity is separated from the position. These remarkable properties cannot be obtained from the Kalman filtering based mechanism in [6], [7]. Via numerical simulations, the performance of the proposed mechanism using the FIR filtering is shown to be superior to the existing Kalman filtering based mechanism.

The paper is organized as follows. In Section 2, a new image stabilization mechanism is proposed. In Section 3, remarkable properties of the proposed mechanism are shown. In Section 4, computer simulations are performed. Finally, conclusions are made in Section 5.

2   FIR Filtering Based Image Stabilization Scheme

The following fourth order state space model for the image stabilizer is constructed as shown in [6], [7] :

$$x(i+1) = Ax(i) + Gw(i),$$
$$z(i) = Cx(i) + v(i) \tag{1}$$

where

$$x(i) = \begin{bmatrix} x_p(i) \\ x_v(i) \end{bmatrix}, \ w(i) = \begin{bmatrix} w_p(i) \\ w_v(i) \end{bmatrix}, \ A = \begin{bmatrix} I & I \\ 0 & I \end{bmatrix}, \ G = \begin{bmatrix} I & 0 \\ 0 & I \end{bmatrix}, \ C = [I \ \ 0].$$

The state $x_p(i) = [x_{p1} \ x_{p2}]^T$ represents horizontal and vertical *absolute frame position* and the state $x_v(i) = [x_{v1} \ x_{v2}]^T$ represents the corresponding *velocity* in the frame i acquired by a camera mounted on a mobile platform. The process noise $w(i)$ and observation noise $v(i)$ are a zero-mean white noise with covariance Q and R, respectively. Note that noise covariances Q and R can be determined via experiments or left as a design parameter.

The main task of the proposed image stabilization mechanism is the filtering of absolute frame positions in real-time, removing undesired motion effects, while preserving desired gross camera displacements. For the filtering, the well known FIR filter in [10], [11] is adopted. For the state-space model (1), the FIR filter $\hat{x}(i)$ processes linealy the only finite observations on the most recent window $[i - M, i]$ as the following simple form:

$$\hat{x}(i) = \begin{bmatrix} \hat{x}_p(i) \\ \hat{x}_v(i) \end{bmatrix} = HZ_M(i) = \begin{bmatrix} H_p \\ H_v \end{bmatrix} Z_M(i) \qquad (2)$$

where the gain matrix H and the finite observations $Z_M(i)$ are represented by

$$H \triangleq [h(M)\ h(M-1) \cdots h(0)], \qquad (3)$$

$$Z_M(i) \triangleq [z^T(i-M)\ z^T(i-M+1) \cdots z^T(i)]^T. \qquad (4)$$

H_p and H_v are the first 2 rows and the second 2 rows of H, respectively. The algorithm for filter gain coefficients $h(\cdot)$ in (3) is obtained from the following algorithm as shown in [10], [11]:

$$h(j) = \Omega^{-1}(M)\Phi(j)C^T R^{-1}, \qquad 0 \le j \le M,$$

where

$$\Phi(l+1) = \Phi(l)[I + A^{-T}\Omega(M-l-1)A^{-1}GQG^T]^{-1}A^{-T},$$
$$\Omega(l+1) = [I + A^{-T}\Omega(l)A^{-1}GQG^T]^{-1}A^{-T}\Omega(l)A^{-1} + C^T R^{-1}C,$$

with $\Phi(0) = I$, $\Omega(0) = C^T R^{-1} C$, and $0 \le l \le M-1$. Note that gain matrices H_p and H_v require computation only on the interval $[0, M]$ once and is time-invariant for all windows. The finite observations $Z_M(i)$ in (4) can be represented in the following regression form:

$$Z_M(i) = L_M x_p(i-M) + N_M X_v(i) + G_M W_p(i) + V(i) \qquad (5)$$

where $X_v(i)$, $W_p(i)$, $V(i)$ have the same form as (4) for $x_v(i)$, $w_p(i)$, $v(i)$, and matrices L_M, N_M, G_M are defined by

$$L_M \triangleq \begin{bmatrix} I \\ I \\ \vdots \\ I \end{bmatrix},\ N_M \triangleq \begin{bmatrix} 0 & 0 & \cdots & 0 & 0 \\ I & 0 & \cdots & 0 & 0 \\ \vdots & \vdots & \vdots & \vdots & \vdots \\ I & I & \cdots & I & 0 \end{bmatrix},\ G_M \triangleq \begin{bmatrix} 0 & 0 & \cdots & 0 & 0 \\ I & 0 & \cdots & 0 & 0 \\ \vdots & \vdots & \vdots & \vdots & \vdots \\ I & I & \cdots & I & 0 \end{bmatrix}.$$

3 Remarkable Properties

Ultimately, the filtered absolute frame position $\hat{x}_p(i)$ is obtained from (2) as follows:

$$\hat{x}_p(i) = H_p Z_M(i). \qquad (6)$$

The filtered position $\hat{x}_p(i)$ has good inherent properties of unbiasedness, efficiency, time-invariance and deadbeat since the FIR filter used provides these properties. The Kalman filter used in [6], [7] does not have these properties unless the mean and covariance of the initial state is completely known. Among them, the remarkable one is the deadbeat property which the filtered position $\hat{x}_p(i)$ tracks the actual position $x_p(i)$ exactly in the absence of noises. The deadbeat property gives the following matrix equality as shown in [10], [11]:

$$H \begin{bmatrix} C \\ CA \\ \vdots \\ CA^M \end{bmatrix} = A^M$$

and then

$$\begin{bmatrix} H_p \\ H_v \end{bmatrix} [L_M \ \bar{N}_M] = \begin{bmatrix} I^M & MI \\ 0 & I \end{bmatrix}$$

where $\bar{N}_M = [0 \ I \ 2I \cdots MI]^T$. Therefore, the following matrix equalities are obtained:

$$H_p L_M = I, \quad H_p \bar{N}_M = MI, \quad H_v L_M = 0, \quad H_v \bar{N}_M = I, \quad (7)$$

which gives following remarkable properties.

It will be shown in the following theorem that the filtered position $\hat{x}_p(i)$ in (6) is not affected by the velocity when the velocity is constant on the observation horizon $[i - M, i]$

Theorem 1. *When the velocity is constant on the observation window $[i-M, i]$, the filtered position $\hat{x}_p(i)$ in (6) is not affected by the velocity.*

Proof. When the velocity is constant as $\bar{x}_v$ on the observation window $[i - M, i]$, the finite observations $Z_M(i)$ in (5) can be represented in

$$Z_M(i) | \{x_v(i) = \bar{x}_v \text{ for } [i - M, i]\}$$
$$= L_M x_p(i - M) + \bar{N}_M \bar{x}_v + G_M W_p(i) + V(i). \quad (8)$$

Then, the filtered position $\hat{x}_p(i)$ is derived from (6)-(8) as

$$\hat{x}_p(i) = H_p Z_M(i)$$
$$= H_p [L_M x_p(i - M) + \bar{N}_M \bar{x}_v + G_M W_p(i) + V(i)]$$
$$= H_p L_M x(i - M) + H_p \bar{N}_M \bar{x}_v + H_p [G_M W_p(i) + V(i)]$$
$$= x_p(i - M) + MI\bar{x}_v + H_p [G_M W_p(i) + V(i)]. \quad (9)$$

From (1), the actual position $x_p(i)$ can be represented on $[i - M, i]$ as follow:

$$x_p(i) | \{x_v(i) = \bar{x}_v \text{ for } [i - M, i]\}$$
$$= x_p(i - M) + MI\bar{x}_v + \bar{G}_M W_p(i) \quad (10)$$

where $\bar{G}_M = [I \ I \cdots I \ 0]$. Thus, using (9) and (10), the error of the filtered position $\hat{x}_p$ is

$$\hat{x}_p(i) - x_p(i) = H_p[G_M W_p(i) + V(i)] - \bar{G}_M W_p(i),$$

which does not include the velocity term. This completes the proof.

The velocity itself can be treated as variable which should be filtered. In this case, the filtered velocity is shown to be separated from the position term.

Theorem 2. *The filtered velocity $\hat{x}_v(i)$ in (2) is separated from the position term.*

Proof. The filtered velocity $\hat{x}_v(i)$ is derived from (2) and (7) as

$$\begin{aligned}\hat{x}_v(i) &= H_v Z_M(i) \\ &= H_v[L_M x_p(i-M) + N_M X_v(i) + G_M W_p(i) + V(i)] \\ &= H_v[N_M X_v(i) + G_M W_p(i) + V(i)]\end{aligned}$$

which does not include the position term. This completes the proof.

Above remarkable properties of the proposed FIR filtering based mechanism cannot be obtained from the existing Kalman filtering based mechanism in [6], [7]. In addition, as mentioned previously, the proposed mechanism has the deadbeat property, which means the fast tracking ability of the proposed mechanism. Furthermore, due to the FIR structure and the batch formulation, the proposed mechanism might be robust to temporary modeling uncertainties and to roundoff errors, while the Kalman filtering based mechanism might be sensitive for these situations.

The noise suppression of the proposed mechanism might be closely related to the window length M. However, although the proposed mechanism can have greater noise suppression as the window length M increases, too large M may yield the long convergence time of the filtered position and velocity, which degrades the filtering performance of the proposed mechanism. This illustrates the proposed mechanism's compromise between the noise suppression and the tracking speed of the filtered position and velocity. Since M is an integer, fine adjustment of the properties with M is difficult. Moreover, it is difficult to determine the window length is systematic ways. In applications, one way to determine the window length is to take the appropriate value that can provides enough noise suppression.

4 Computer Simulations

The performance of the proposed FIR filtering based image stabilization mechanism is evaluated via a numerical simulation. It was already shown in [7] that the Kalman filtering based mechanism outperforms the MVI mechanism of [4] for the 'bike' sequence acquired by a consumer camcorder mounted on the rear carrier of a moving motorcycle. Therefore, in this paper, the proposed FIR filtering based mechanism will be compared with the Kalman filtering based mechanism for the same 'bike' sequence of [7]. Note that the Kalman filtering with an IIR structure provides a minimum variance state estimate $\hat{x}(i)$, called the one-step predicted estimate, with the estimation error covariance $P(i)$ as shown in [12]:

$$\hat{x}(i+1) = A\hat{x}(i) + [AP(i)C^T(R+CP(i)C^T)^{-1}](z(i) - C\hat{x}(i)),$$
$$P(i+1) = A[P(i)^{-1} + C^T R^{-1} C]^{-1} A^T + GQG^T.$$

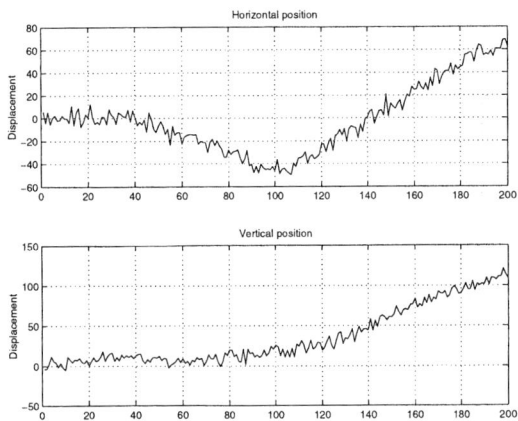

Fig. 1. Actual horizontal and vertical positions

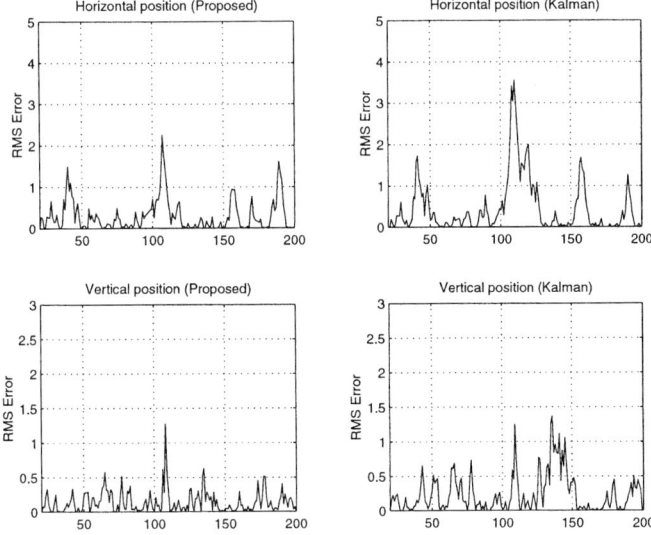

Fig. 2. Filtered position

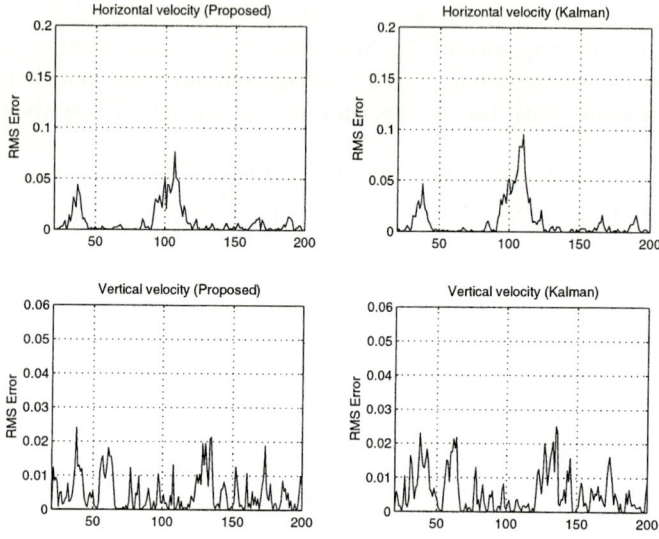

Fig. 3. Filtered velocity

The existing Kalman filtering based mechanism assumes that *a priori* information is exactly known, although this assumption would be unpractical [10]. Thus, the initial state estimate is taken by $\hat{x}(i_0) = [0\ 0\ 0\ 0]^T$ as shown in [7]. On the other hand, the proposed FIR filtering based mechanism doesn't require the initial state estimate as shown in (2). The window length is taken by $M = 20$. For both mechanisms, noise covariances are taken by $Q = 0.1$ and $R = 100$ as [7]. To make a clearer comparison, thirty Monte Carlo runs are performed and each single run lasts for 200 samples. Actual horizontal and vertical positions used in one of thirty runs are plotted in Fig. 1. Fig. 2 and 3 show root-mean-square (RMS) errors of the filtered position and velocity. For both filtered position and velocity, simulation results show that the performance of the proposed FIR filtering based mechanism is superior to the existing Kalman filtering based mechanism.

5 Concluding Remarks

This paper has proposed a new image stabilization mechanism based on filtering of absolute frame positions for mobile video appliances. The proposed mechanism removes undesired motion effects in real-time, while preserving desired gross camera displacements. The well known FIR filter is adopted for the filtering. The proposed mechanism provides the filtered position and velocity that have good inherent properties. It is shown that the filtered position is not affected by the constant velocity. It is also shown that the filtered velocity is separated from the position. Simulation results show that the performance of the proposed FIR filtering based mechanism is superior to the existing Kalman filtering based mechanism.

References

1. Irani, M., Rousso, B., Peleg, S.: Recovery of ego-motion using image stabilization. In: Proc. IEEE Int. Conf. on Computer Vision and Pattern Recognition. (1994) 454–460
2. Rousso, B., Avidan, S., Shashua, A., Peleg, S.: Robust recovery of camera rotation from three frames. In: Proc. IEEE Int. Conf. on Computer Vision and Pattern Recognition. (1996) 796–802
3. Kwon, O., Chellappa, R., Morimoto, C.: Motion compensated subband coding of video acquired from a moving platform In: Proc. IEEE Int. Conf. on Computer Vision and Pattern Recognition. (1995) 2185–2188
4. Engelsberg, A., Schmidt, G.: A comparative review of digital image stabilising algorithms for mobile video communications. IEEE Trans. Consum. Electron. **45** (1999) 591–597
5. Erturk, S., Dennis, T.J.,: Image sequence stabilisation based on DFT filtering. IEE Proc. Vis. Image Signal Process. **147** (2000) 95–102
6. Censi, A., Fusiello, A., Roberto, V.: Image stabilization by features tracking. In: Proc. of Int. Conf. Image Analysis and Processing. (1999) 665–667
7. Erturk, S., Dennis, T.J.,: mage sequence stabilisation based on Kalman filtering of frame positions IEE Proc. Vis. Image Signal Process. **37** (2001) 1217–1219
8. Fitzgerald, R.J.: Divergence of the Kalman filter. IEEE Trans. Automat. Contr. **16** (1971) 736–747
9. Schweppe, F.: Uncertain Dynamic Systems. Englewood Cliffs, NJ:Prentice-Hall (1973)
10. Kwon, W.H., Kim, P.S., Park, P.: A receding horizon Kalman filter for discrete-time invariant systems. IEEE Trans. Automat. Contr. **44** (1999) 1787–1791
11. Kwon, W.H., Kim, P.S., Han, S.H.: A receding horizon unbiased FIR filter for discrete-time state space models. Automatica. **38** (2002) 545–551
12. Kalman, R.E., Bucy, R.S.: New results in linear filtering and prediction theory Trans. ASME J. of Basic Engr. **83** (1961) 95–108

p-Belief Communication Leading to a Nash Equilibrium

Takashi Matsuhisa*

Department of Liberal Arts and Sciences,
Ibaraki National College of Technology, Nakane 866, Hitachinaka-shi,
Ibaraki 312-8508, Japan
mathisa@ge.ibaraki-ct.ac.jp

Abstract. A pre-play communication in the p-belief system is presented which leads to a Nash equilibrium of a strategic form game through messages. In the communication process each player predicts the other players' actions under his/her private information with probability at least p. The players communicate privately their conjectures through message according to the communication graph, where each player receiving the message learns and revises his/her conjecture. The emphasis is on that any topological assumptions on the communication graph are not required in the theorem.

Keywords: p-Belief system, Nash equilibrium, Communication, Protocol, Conjecture, Non-corporative game.

AMS 2000 Mathematics Subject Classification: Primary 91A35, Secondary 03B45.

Journal of Economic Literature Classification: C62, C78.

1 Introduction

This article relates equilibria and distributed knowledge. In game theoretical situations among a group of players, the concept of mixed strategy Nash equilibrium has become central. Yet little is known about the process by which players learn if they do. This article will give a protocol run by the mutual learning leading to a mixed strategy Nash equilibrium of a strategic form game from the epistemic point of belief revision system. We show that

Main Theorem. *Suppose that the players in a strategic form game have the p-belief system with a common prior distribution. In a communication process of the game according to a protocol with revisions of their beliefs about the other*

---

* Partially supported by the Grant-in-Aid for Scientific Research(C)(2)(No.14540145) in the Japan Society for the Promotion of Sciences.

players' actions, the profile of their future predictions induces a mixed strategy Nash equilibrium of the game in the long run.

Let us consider the following protocol: The players start with the same prior distribution on a state-space. In addition they have private information given by a partition of the state space. Beliefs are posterior probabilities: A player p-believes (simply, believes) an event with $0 < p \leq 1$ if the posterior probability of the event given his/her information is at least p. Each player predicts the other players' actions as his/her p-belief of the actions. He/she communicates privately their p-beliefs about the other players' actions through messages, and the receivers update their p-beliefs according to the messages. Precisely, the players are assumed to be rational and maximizing their expected utility according to their p-beliefs at every stage. Each player communicates privately his/her p-belief about the others' actions as messages according to a communication graph as a protocol,[1] and the receivers update their private information and revise their p-beliefs.

The main theorem says that the players' predictions regarding the future p-beliefs converge in the long run, which lead to a mixed strategy Nash equilibrium of a game. The emphasis is on the two points: First that each player's prediction is not required to be common-knowledge among all players, and secondly that the communication graph is not assumed to be acyclic.

2 The Model

Let Ω be a non-empty *finite* set called a *state-space*, N a set of finitely many players $\{1, 2, \ldots n\}$ at least two ($n \geq 2$), and let 2^Ω be the family of all subsets of Ω. Each member of 2^Ω is called an *event* and each element of Ω called a *state*. Let μ be a probability measure on Ω which is common for all players. For simplicity it is assumed that (Ω, μ) is a *finite* probability space with μ *full support*.[2]

2.1 p-Belief System[3]

Let p be a real number with $0 < p \leq 1$. The *p-belief system* associated with the partition information structure $(\Pi_i)_{i \in N}$ is the tuple $\langle N, \Omega, \mu, (\Pi_i)_{i \in N}, (B_i)_{i \in N} \rangle$ consisting of the following structures and interpretations: (Ω, μ) is a finite probability space, and i's *p-belief operator* B_i is the operator on 2^Ω such that $B_i E$ is the set of states of Ω in which i p-believes that E has occurred with probability at least p; that is, $B_i E := \{\omega \in \Omega \mid \mu(E \mid \Pi_i(\omega)) \geq p \}$. We note that when $p = 1$ the 1-belief operator becomes knowledge operator.

---

[1] When a player communicates with another, the other players are not informed about the contents of the message.
[2] That is; $\mu(\omega) \neq 0$ for every $\omega \in \Omega$.
[3] Monderer and Samet [5].

2.2 Game on p-Belief System[4]

By a *game* G we mean a *finite* strategic form game $\langle N, (A_i)_{i \in N}, (g_i)_{i \in N} \rangle$ with the following structure and interpretations: N is a finite set of players $\{1, 2, \ldots, i, \ldots n\}$ with $n \geq 2$, A_i is a finite set of i's *actions* (or i's pure strategies) and g_i is an i's *payoff function* of A into $\mathbb{R}$, where A denotes the product $A_1 \times A_2 \times \cdots \times A_n$, A_{-i} the product $A_1 \times A_2 \times \cdots \times A_{i-1} \times A_{i+1} \times \cdots \times A_n$. We denote by g the n-tuple $(g_1, g_2, \ldots g_n)$ and by a_{-i} the $(n-1)$-tuple $(a_1, \ldots, a_{i-1}, a_{i+1}, \ldots, a_n)$ for a of A. Furthermore we denote $a_{-I} = (a_i)_{i \in N \setminus I}$ for each $I \subset N$.

A probability distribution ϕ_i on A_{-i} is said to be i's *overall conjecture* (or simply i's *conjecture*). For each player j other than i, this induces the marginal distribution on j's actions; we call it i's *individual conjecture* about j (or simply i's conjecture *about* j.) Functions on Ω are viewed like random variables in the probability space (Ω, μ). If $\mathbf{x}$ is a such function and x is a value of it, we denote by $[\mathbf{x} = x]$ (or simply by $[x]$) the set $\{\omega \in \Omega | \mathbf{x}(\omega) = x\}$.

The information structure (Π_i) with a common prior μ yields the distribution on $A \times \Omega$ defined by $\mathbf{q}_i(a, \omega) = \mu([\mathbf{a} = a] | \Pi_i(\omega))$; and the i's overall conjecture defined by the marginal distribution $\mathbf{q}_i(a_{-i}, \omega) = \mu([\mathbf{a}_{-i} = a_{-i}] | \Pi_i(\omega))$ which is viewed as a random variable of ϕ_i. We denote by $[\mathbf{q}_i = \phi_i]$ the intersection $\bigcap_{a_{-i} \in A_{-i}} [\mathbf{q}_i(a_{-i}) = \phi_i(a_{-i})]$ and denote by $[\phi]$ the intersection $\bigcap_{i \in N} [\mathbf{q}_i = \phi_i]$. Let $\mathbf{g}_i$ be a random variable of i's payoff function g_i and $\mathbf{a}_i$ a random variable of an i's action a_i. i's action a_i is said to be *actual* at a state ω if $\omega \in [\mathbf{a}_i = a_i]$; and the profile $a_I = (a_i)_{i \in I}$ is said to be actually played at ω if $\omega \in [\mathbf{a}_I = a_I] := \bigcap_{i \in I} [\mathbf{a}_i = a_i]$ for $I \subset N$. The payoff functions $g = (g_1, g_2, \ldots, g_n)$ are said to be *actually played* at a state ω if $\omega \in [\mathbf{g} = g] := \bigcap_{i \in N} [\mathbf{g}_i = g_i]$. Let $\mathbf{Exp}$ denote the expectation defined by $\mathbf{Exp}(g_i(b_i, \mathbf{a}_{-i}); \omega) := \sum_{a_{-i} \in A_{-i}} g_i(b_i, a_{-i}) \, \mathbf{q}_i(\omega)(a_{-i})$.

A player i is said to be *rational* at ω if each i's actual action a_i maximizes the expectation of his actually played payoff function g_i at ω when the other players actions are distributed according to his conjecture $\mathbf{q}_i(\cdot; \omega)$. Formally, letting $g_i = \mathbf{g}_i(\omega)$ and $a_i = \mathbf{a}_i(\omega)$, $\mathbf{Exp}(g_i(a_i, \mathbf{a}_{-i}); \omega) \geq \mathbf{Exp}(g_i(b_i, \mathbf{a}_{-i}); \omega)$ for every b_i in A_i. Let R_i denote the set of all of the states at which an player i is rational, and R the intersection $\bigcap_{j \in N} R_j$.

2.3 Protocol[5]

We assume that the players communicate by sending *messages*. Let T be the time horizontal line $\{0, 1, 2, \cdots t, \cdots\}$. A *protocol* is a mapping $\Pr : T \to N \times N, t \mapsto (s(t), r(t))$ such that $s(t) \neq r(t)$. Here t stands for *time* and $s(t)$ and $r(t)$ are, respectively, the *sender* and the *receiver* of the communication which takes place at time t. We consider the protocol as the directed graph whose vertices are the set of all players N and such that there is an edge (or an arc) from i to j if and only if there are infinitely many t such that $s(t) = i$ and $r(t) = j$.

---

[4] Aumann and Brandenburger [1].
[5] C.f.: Parikh and Krasucki (1990).

A protocol is said to be *fair* if the graph is strongly-connected; in words, every player in this protocol communicates directly or indirectly with every other player infinitely often. It is said to contain a *cycle* if there are players $i_1, i_2, \ldots, i_k$ with $k \geq 3$ such that for all $m < k$, i_m communicates directly with i_{m+1}, and such that i_k communicates directly with i_1. The communications is said to proceed in *rounds* if there exists a time m such that for all t, $\Pr(t) = \Pr(t+m)$.

2.4 Communication on p-Belief System

A *pre-play communication process* $\pi^p(G)$ with revisions of players' conjectures $(\phi_i^t)_{(i,t) \in N \times T}$ according to a protocol for a game G is a tuple

$$\pi^p(G) = \langle \Pr, (\Pi_i^t)_{i \in N}, (B_i^t)_{i \in N}, (\phi_i^t)_{(i,t) \in N \times T} \rangle$$

with the following structures: the players have a common prior μ on Ω, the protocol Pr among N, $\Pr(t) = (s(t), r(t))$, is fair and it satisfies the conditions that $r(t) = s(t+1)$ for every t and that the communications proceed in rounds. An n-tuple $(\phi_i^t)_{i \in N}$ is a profile of player i's individual conjecture at time t. The information structure Π_i^t at time t is the mapping of Ω into 2^{Ω} for player i that is defined inductively as follows: If $i = s(t)$ is a sender at t, the message sent by i to $j = r(t)$ is $W_i^t(\omega) = [g_i] \cap [\phi_i^t] \cap R_i^t$.

- Set $\Pi_i^0(\omega) = \Pi_i(\omega)$.
- Assume that Π_i^t is defined. It yields the distribution $\mathbf{q}_i^t(a, \omega) = \mu([\mathbf{a} = a] | \Pi_i^t(\omega))$. Whence
  - R_i^t denotes the set of all the state ω at which i is *rational* according to his conjecture $\mathbf{q}_i^t(\cdot; \omega)$; that is, each i's actual action a_i maximizes the expectation of his payoff function g_i being actually played at ω when the other players actions are distributed according to his conjecture $\mathbf{q}_i^t(\cdot; \omega)$ at time t.[6]
  - $\mathcal{Q}_i^t$ denotes the partition of Ω induced by $\mathbf{q}_i^t(\cdot; \omega)$, which is decomposed into the components $\mathcal{Q}_i^t(\omega)$ consisting of all of the states ξ such that $\mathbf{q}_i^t(\cdot; \xi) = \mathbf{q}_i^t(\cdot; \omega)$:
  - $\mathcal{G}_i$ denotes the partition $\{[\mathbf{g}_i = g_i], \Omega \setminus [\mathbf{g}_i = g_i]\}$ of Ω, and $\mathcal{R}_i^t$ the partition $\{R_i^t, \Omega \setminus R_i^t\}$:
  - $\mathcal{W}_i^t$ denotes the join $\mathcal{G}_i \vee \mathcal{Q}_i^t \vee \mathcal{R}_i^t$ that is the refinement of the three partitions $\mathcal{G}_i, \mathcal{Q}_i^t$ and $\mathcal{R}_i^t$. Therefore the message sent by the sender i is $W_i^t(\omega) = [g_i] \cap [\phi_i^t] \cap R_i^t$ if $\omega \in [g_i] \cap [\phi_i^t] \cap R_i^t$. Then:

---

[6] Formally, letting $g_i = \mathbf{g}_i(\omega)$, $a_i = \mathbf{a}_i(\omega)$, the expectation at time t, $\mathbf{Exp}^t$, is defined by $\mathbf{Exp}^t(g_i(a_i, \mathbf{a}_{-i}); \omega) := \sum_{a_{-i} \in A_{-i}} g_i(a_i, a_{-i}) \, \mathbf{q}_i^t(a_{-i}, \omega)$. A player i is said to be *rational* according to his conjecture $\mathbf{q}_i^t(\cdot, \omega)$ at ω if for all b_i in A_i, $\mathbf{Exp}^t(g_i(a_i, \mathbf{a}_{-i}); \omega) \geq \mathbf{Exp}^t(g_i(b_i, \mathbf{a}_{-i}); \omega)$.

- The revised partition Π_i^{t+1} at time $t+1$ is defined as follows:
  - If i is a receiver of a message at time $t+1$ then $\Pi_i^{t+1}(\omega) = \Pi_i^t(\omega) \cap W_{s(t)}^t(\omega)$.
  - If not, $\Pi_i^{t+1}(\omega) = \Pi_i^t(\omega)$.

It is of worth noting that $(\Pi_i^t)_{i \in N}$ is a partition information structure for every $t \in T$. Let B_i^t be the p-belief operator corresponding to Π_i^t defined by $B_i^t E = \{\omega \in \Omega \mid \mu(E \mid \Pi_i^t(\omega)) \geq p\}$. We require the additional condition that: $\bigcap_{i \in N} B_i^t([g_i] \cap [\phi_i^t] \cap R_i^t) \neq \emptyset$ for every $t \in T$. [7] The specification is that each player p-believes his/her payoff, rationality and conjecture at every time t. We denote by ∞ a sufficient large τ such that for all $\omega \in \Omega$, $\mathbf{q}_i^\tau(\cdot\,;\omega) = \mathbf{q}_i^{\tau+1}(\cdot\,;\omega) = \mathbf{q}_i^{\tau+2}(\cdot\,;\omega) = \cdots$. Hence we can write $\mathbf{q}_i^\tau$ by $\mathbf{q}_i^\infty$ and ϕ_i^τ by ϕ_i^∞.

3 The Result

We can now state the main theorem:

Theorem 1. *Suppose that the players in a strategic form game G have the p-belief system with μ a common prior. In the pre-play communication process $\pi^p(G)$ according to a protocol among all players in the game with revisions of their conjectures $(\phi_i^t)_{(i,t) \in N \times T}$ there exists a time ∞ such that for each $t \geq \infty$, the n-tuple $(\phi_i^t)_{i \in N}$ induces a mixed strategy Nash equilibrium of the game.*

Proof. We shall briefly sketch the proof based on the below proposition:

Proposition 1. *If the protocol in $\pi^p(G)$ is not acyclic then for any players $i, j \in N$, both the conjectures $\mathbf{q}_i^\infty$ and $\mathbf{q}_j^\infty$ on $A \times \Omega$ must coincide; that is, $\mathbf{q}_i^\infty(a;\omega) = \mathbf{q}_j^\infty(a;\omega)$ for all $(a;\omega) \in A \times \Omega$.*

Proof of Theorem 1: Set $\Gamma(i) = \{j \in N \mid (i, j) = \Pr(t) \text{ for some } t \in T\}$ and $[\phi^\infty] := \bigcap_{i \in N} \bigcap_{a_{-i} \in A_i} [\mathbf{q}_i^\infty(a_{-i};*) = \phi_i^\infty(a_{-i})]$. For each $i \in N$, denote $F_i := [g_i] \cap [\phi^\infty] \cap R^\infty$. It is noted that $F_i \neq \emptyset$.

We can plainly observe the first point that $\mu([\mathbf{a}_{-j} = a_{-j}] \mid F_i \cap F_j) = \phi_j^\infty(a_{-j})$ for each $i \in N$, $j \in \Gamma(i)$ and for every $a \in A$. Then summing over a_{-i}, we can observe that $\mu([\mathbf{a}_i = a_i] \mid F_i \cap F_j) = \phi_j^\infty(a_i)$ for any $a \in A$. In view of Proposition 1 it can be observed that $\phi_j^\infty(a_i)$ is independent of the choices of every $j \in N$ other than i. We set the probability distribution σ_i on A_i by $\sigma_i(a_i) := \phi_j^\infty(a_i)$, and set the profile $\sigma = (\sigma_i)$. Therefore all the other players j than i agree on the same conjecture $\sigma_j(a_i) = \phi_j^\infty(a_i)$ about i.

We shall observe the second point that for every $a \in \prod_{i \in N} \text{Supp}(\sigma_i)$, $\phi_i^\infty(a_{-i}) = \sigma_1(a_1) \cdots \sigma_{i-1}(a_{i-1}) \sigma_{i+1}(a_{i+1}) \cdots \sigma_n(a_n)$: In fact, viewing the definition of σ_i we shall show that $\phi_i^\infty(a_{-i}) = \prod_{k \in N \setminus \{i\}} \phi_i^\infty(a_k)$. To verify this it suffices to show that for every $k = 1, 2, \cdots, n$, $\phi_i^\infty(a_{-i}) = \phi_i^\infty(a_{-I_k}) \prod_{k \in I_k \setminus \{i\}} \phi_i^\infty(a_k)$: We prove it by induction on k. For $k = 1$ the result is immediate. Suppose it is

---

[7] where we denote $[g_i] := [\mathbf{g}_i = g_i]$, $[\phi_i^t] := \bigcap_{a_{-i} \in A_i} [\mathbf{q}_i^t(a_{-i};*) = \phi_i^t(a_{-i})]$.

true for $k \geq 1$. On noting the protocol is fair, we can take the sequence of sets of players $\{I_k\}_{1 \leq k \leq n}$ with the following properties:

(a) $I_1 = \{i\} \subset I_2 \subset \cdots \subset I_k \subset I_{k+1} \subset \cdots \subset I_m = N$:
(b) For every $k \in N$ there is a player $i_{k+1} \in \bigcup_{j \in I_k} \Gamma(j)$ with $I_{k+1} \setminus I_k = \{i_{k+1}\}$.

Take $j \in I_k$ such that $i_{k+1} \in \Gamma(j)$, and set $H_{i_{k+1}} := [\mathbf{a}_{i_{k+1}} = a_{i_{k+1}}] \cap F_j \cap F_{i_{k+1}}$. It can be verified that $\mu([\mathbf{a}_{-j-i_{k+1}} = a_{-j-i_{k+1}}] \mid H_{i_{k+1}}) = \phi^\infty_{-j-i_{k+1}}(a_{-j})$. Dividing $\mu(F_j \cap F_{i_{k+1}})$ yields that

$$\mu([\mathbf{a}_{-j} = a_{-j}] \mid F_j \cap F_{i_{k+1}}) = \phi^\infty_{i_{k+1}}(a_{-j})\mu([\mathbf{a}_{i_{k+1}} = a_{i_{k+1}}] \mid F_j \cap F_{i_{k+1}}).$$

Thus $\phi^\infty_j(a_{-j}) = \phi^\infty_{i_{k+1}}(a_{-j-i_{k+1}})\phi^t_j(a_{i_{k+1}})$; then summing over $a_{I_k} \in A_{I_k}$ we obtain $\phi^\infty_i(a_{-I_k}) = \phi^\infty_{i_{k+1}}(a_{-I_k-i_{k+1}})\phi^\infty_j(a_{i_{k+1}})$. In view of Proposition 1 it can be plainly observed that $\phi^\infty_i(a_{-I_k}) = \phi^\infty_i(a_{-I_k-i_{k+1}})\phi^\infty_i(a_{i_{k+1}})$, as required.

On noting that all the other players i than j agree on the same conjecture $\sigma_j(a_j) = \phi^\infty_i(a_j)$ about j, we can conclude that each action a_i appearing with positive probability in σ_i maximizes g_i against the product of the distributions σ_l with $l \neq i$. This implies that the profile $\sigma = (\sigma_i)_{i \in N}$ is a mixed strategy Nash equilibrium of G, in completing the proof. □

4 Concluding Remarks

Many authors have studied the learning processes modeled by Bayesian updating. E. Kalai and E. Lehrer [2] (and references therein) indicate increasing interest in the mutual learning processes in games that leads to equilibrium: Each player starts with initial erroneous belief regarding the actions of all the other players. They show that the two strategies converges to an ε-mixed strategy Nash equilibrium of the repeated game.

As for as J.F. Nash [6]'s fundamental notion of strategic equilibrium is concerned, R.J. Aumann and A. Brandenburger [1] give epistemic conditions for mixed strategy Nash equilibrium: They show that the common-knowledge of the predictions of the players having the partition information (that is, equivalently, the **S5**-knowledge model) yields a Nash equilibrium of a game. However it is not clear just what learning process leads to the equilibrium. The present article aims to fill this gap.

Our real concern is with what learning process leads to a mixed strategy Nash equilibrium of a finite strategic form game from epistemic point view. As we have observed, in the pre-play communication process with revisions of players' beliefs about the other actions, their predictions induces a mixed strategy Nash equilibrium of the game in the long run. We have proved this assertion in the p-belief system. Matsuhisa [3] established the same assertion in the **S4**-knowledge model. Furthermore Matsuhisa [4] showed a similar result for ε-mixed strategy Nash equilibrium of a strategic form game in the **S4**-knowledge model, which highlights an epistemic aspect in Theorem of E. Kalai and E. Lehrer [2].

References

1. Aumann, R. J., and Brandenburger, A.: Epistemic conditions for mixed strategy Nash equilibrium, Econometrica **63** (1995) 1161–1180.
2. Kalai E., and Lehrer, E.: Rational learning to mixed strategy Nash equilibrium, Econometrica **61** (1993) 1019–1045.
3. Matsuhisa, T.: Communication leading to mixed strategy Nash equilibrium I, T. Maruyama (eds) *Mathematical Economics*, Suri-Kaiseki-Kenkyusyo Kokyuroku **1165** (2000) 245–256.
4. Matsuhisa, T.: Communication leading to epsilon-mixed strategy Nash equilibrium, Working paper (2001). The extended abstract was presented in the XIV Italian Meeting of Game Theory and Applications (IMGTA XIV), July 11-14, 2001.
5. Monderer, D., and Samet,, D.: Approximating common knowledge with common beliefs, Games and Economic Behaviors **1** (1989) 170–190.
6. Nash J. F.: Equilibrium points in n-person games, Proceedings of the National Academy of Sciences of the United States of America **36** (1950) 48–49.
7. Parikh R., and Krasucki, P.: Communication, consensus, and knowledge, Journal of Economic Theory **52** (1990) 178–189.

Color Image Vector Quantization Using an Enhanced Self-Organizing Neural Network

Kwang Baek Kim[1] and Abhijit S. Pandya[2]

[1] Dept. of Computer Engineering, Silla University, S. Korea
[2] Div. of Information and Computer Engineering, Silla University, S. Korea
Dept. of Computer Science and Engineering, Florida Atlantic University, U.S.A
gbkim@silla.ac.kr

Abstract. In the compression methods widely used today, the image compression by VQ is the most popular and shows a good data compression ratio. Almost all the methods by VQ use the LBG algorithm that reads the entire image several times and moves code vectors into optimal position in each step. This complexity of algorithm requires considerable amount of time to execute. To overcome this time consuming constraint, we propose an enhanced self-organizing neural network for color images. VQ is an image coding technique that shows high data compression ratio. In this study, we improved the competitive learning method by employing three methods for the generation of codebook. The results demonstrated that compression ratio by the proposed method was improved to a greater degree compared to the SOM in neural networks.

1 Introduction

Computer Graphics and Imaging applications have started to make inroads into our everyday lives due to the global spread of Information technology. This has made image compression an essential tool in computing with workstations, personal computers and computer networks [1]. Compression can also be viewed as a form of classification, since it assigns a template or codeword to a set of input vectors of pixels drawn from a large set in such a way as to provide a good approximation of representation. A color Image is composed of three primary components. The most popular choices of color primaries are (R, G, B), (Y, I, Q), and (Y, U, V). The Y component represents the brightness and IQ (IV) components represent the chrominance signals. In this paper, we considered color images in the (R, G, B) domain with a color of one pixel determined by three primary components, the red(R), green (G) and blue (B). Each component is quantized to 8 bits, hence 24 bits are needed to represent one pixel. The number of palette elements to be represented by 24 bits is 2^{24}, but all of colors are not used to represent one image. So it is possible to compress the pixel color of the real image. It is also necessary to compress the pixel color because of the limitation of disk space and the transmission channel bandwidth [2].

In the compression methods introduced until now, the image compression by Vector Quantization (VQ) is most popular and shows a good data compression ratio.

Most methods by VQ use the LBG algorithm developed by Linde, Buzo, and Gray [2]. But this algorithm reads an entire image several times and moves code vectors into optimal position in each step. Due to the complexity of the algorithm, it takes considerable time to execute. To overcome this time consuming constraints, we propose an enhanced self-organizing vector quantization method for color images.

2 Enhanced Self-Organizing Neural Network

2.1 VQ Using a Self-Organizing Feature Map

The Self-Organizing Feature Map has been found to serve as a good algorithm for codebook generation. The SOM algorithm, which is derived from an appropriate stochastic gradient decent scheme, results in a natural clustering process in which the network performs competitive learning to perceive pattern classes based on data similarity. Smoothing of vector elements does not take place in this unsupervised training scheme. At the same time, since it doses not assume an initial codebook, the probability of getting stranded in local minima is also small. The investigations for high quality reconstructed pictures have led us to the edge preserving self-organizing map. This greatly reduces the large computational costs involved in generating the codebook and finding the closest codeword for each image vector. However, from practical experience, it is observed that additional refinements are necessary for the training algorithm to be efficient enough for practical applications [3]. The SOM technique can display, in a single grey level image, the most significant clustering of data in an n-dimensional feature space, without confusing clusters that are distinct in the feature space [4]. This is possible because points which are far apart in the feature space can map to the same grey level, quantized to the same map node, only if the distribution of data near them is very sparse [5]. Thus, the points far apart in grey level belong to different significant clusters. However, this conventional method leaves a large amount of neurons under-utilized after training [6].

2.2 Enhanced SOM Algorithm for Color Image Vector Quantization

In this paper, we improved the SOM algorithm by employing three methods for the efficient generation of codebook. First, the error between the winner node and the input vector and the frequency of the winner node are reflected in the weight adaptation. Second, the weight is adapted in proportion to the present weight change and the previous weight change as well. Third, in the weight adaptation for the generation of the initial codebook, the weight of the adjacent pixel of the winner node is adapted together. In the proposed method, the codebook is generated by presenting the entire image only two times. In the first step, the initial codebook is generated to reflect the distribution of the given training vectors. The second step uses the initial codebook and regenerates the codebook by moving to the center within the decision region. To generate the precise codebook, it needs to select the winner node correctly and we have to consider the real distortion of the code vector and the input vector. For this management, the measure of the frequency to be selected as winner node and the distortion for the selection of the winner node in the competitive learning algorithm are needed. We use the following equation in the weight adaptation.

$$w_{ij}(t+1) = w_{ij}(t) + \alpha(x_i - w_{ij}(t))$$
$$\alpha = f(e_j) + \frac{1}{f_j} \tag{1}$$

α is the learning factor between 0 and 1 and is set between 0.25 and 0.75 in general. $(x_i - w_{ij}(t))$ is an error value and represents the difference between the input vector and the representative code vector. This means weights are adapted as much as the difference and it prefers to adapt the weight in proportion to the size of the difference. Therefore, we use the normalized value for the output error of the winner node that is converted to the value between 0 and 1 as a learning factor. $f(e_j)$ is the normalization function that converts the value of e_j to the value between 0 and 1, e_j is the output error of the jth neuron, and f_j is the frequency for the jth neuron as the winner.

The above method considers only the present weight change and does not consider the previous weight change. So in the weight adaptation, we consider the previous weight change as well as the present one's. This concept corresponds to the momentum parameter of BP. Based on the momentum factor, the equation for the weight adaptation is as follows:

$$w_{ij}(t+1) = w_{ij}(t) + \delta_{ij}(t+1) \tag{2}$$

$$\delta_{ij}(t+1) = \alpha(x_i - w_{ij}(t)) + \alpha\delta_{ij}(t) \tag{3}$$

The algorithm is detailed below:

1. Initialize the network. i.e., initialize weights (w_{ij}) from the n inputs to the output nodes to small random values. Set the initial neighborhood, N_c to be large. Fix the convergence tolerance limit for the vectors to be a small quantity. Settle maximum number of iterations to be a large number. Divide the training set into vectors of size $n \times n$.
2. Compute the mean and variance of each training input vector.
3. Present the inputs $x_i(t)$.
4. Compute the Euclidean distance d_j between the input and each output node j, given by,

$$d_j = f_j \times d(x, w_{ij}(t)) \tag{4}$$

Where f_j is the frequency of the jth neuron being a winner. Select the minimum distance. Designate the output node with minimum d_j to be j^*.

5. Update the weight for node j^* and its neighbors, defined by the neighborhood size N_c. The weights are updated:
   if $i \in N_c(t)$

$$f_{j^*} = f_{j^*} + 1 \tag{5}$$

$$w_{ij*}(t+1) = w_{ij*}(t) + \delta_{ij*}(t+1) \tag{6}$$

$$\delta_{ij*}(t+1) = \alpha(t+1)(x_i - w_{ij*}(t)) + \alpha(t+1)\delta_{ij*}(t) \tag{7}$$

$$\alpha(t+1) = f(e_{j*}) + \frac{1}{f_{j*}} \tag{8}$$

$$e_{j*} = \frac{1}{n}\sum_{i=0}^{n-1}|x_i(t) - w_{ij*}(t)| \tag{9}$$

if $i \notin N_c(t)$

$$w_{ij}(t+1) = w_{ij}(t) \tag{10}$$

The neighborhood $N_c(t)$ decreases in size as time goes on, thus localizing the area of maximum activity. And $f(e_j)$ is normalization function.

6. Repeat by going to step 2 for each input vector presented, till a satisfactory match is obtained between the input and the weight or till the maximum number of iterations are complete.

3 Simulation

Simulation was performed on a personal computer using C++ builder to evaluate the proposed method. Digitized color images of the (R, G, B) domain and a resolution of 128 x 128 were used for the simulation. Fig.1 shows various images used for simulation.

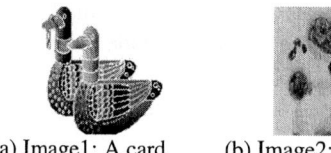

(a) Image1: A card   (b) Image2: A cell image

Fig. 1. Original images used for simulation

Based on simulation results, we can see that the proposed method makes the codebook 5 times faster than the LBG algorithm for the vector quantization. Moreover, the performance of the proposed method is better than the conventional VQ algorithm. That is, the proposed method shows higher data compression ratio than the conventional VQ algorithm. The LBG algorithm reads the entire image data several times and moves code vectors into optimal position in each step. This repetitive process shows the block effect by which the image is recovered. Often in an image, adjacent pixels tend to have similar color and compose a color block. This means that one image is composed of such block. We adapt the weight of adjacent pixels of the winner node together in case of generating the initial codebook. In the proposed method, if an input block and the adjacent pixel have a similarity pixel, the neighboring is adapted. The equation for measuring the similarity is as follows where x_k is the adjacent pixel.

$$|x_i - x_k| < Escrit \qquad (11)$$

That is, if the difference of an input vector from the adjacent pixel is less than *Escrit* which is a criterion for admission, then the neighboring pixel is adapted. Here, the *Escrit* was set to 0.0001. The enhanced self-organizing vector quantization for color images proposed in this paper can decrease the number of times the entire image data is read. Table 1 show the size of the codebook file for the LBG Algorithm, the conventional SOM and the enhanced self-organizing vector quantization for color images.

Table 1. Size of codebook by VQ (byte)

Algorithms / Images	LBG	SOM	Enhanced SOM
Image1	30064	32208	31968
Image2	49376	52080	51648

To Measure a degree of distortion of the reproduction vector, a mean square error (MSE) is generally used. Table 2, which shows the MSE values of images created by using the LBG algorithm, the conventional SOM and the enhanced SOM and from Fig. 2 through 3. Consequently, the transmission time and the memory space reduced than LBG algorithm.

Table 2. Comparison of MSE (Mean Square Error) for compressed images

Algorithms / Images	LBG	SOM	Enhanced SOM
Image1	15.2	13.1	11.2
Image2	14.5	11.3	10.8

Fig. 2 and Fig. 3 show respectively recovered images for original images of Fig. 1. This contribution proposed an improved SOM algorithm. It improves compression and replay rate of image by the codebook dynamic allocation than the conventional SOM algorithm.

(a) SOM (b) Enhanced SOM (c) LBG Algorithm

Fig. 2. The recovered image for Image1

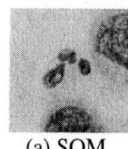 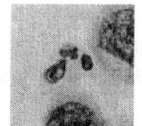 

(a) SOM (b) Enhanced SOM (c) LBG Algorithm

Fig. 3. The recovered image for Image2

4 Conclusion

The proposed method in this paper can be summarized method as follows: Using the enhanced SOM algorithm, the output error concept is introduced into the weight adaptation and the momentum factor is added. The simulation results show that the enhanced SOM algorithm for color image compression produces a major improvement in both subjective and objective quality of the decompressed images. The proposed method is apt to real time application because the codebook is created by reading whole image only twice. Generally, the procreation of the codebook is difficult work in vector quantization of color image. Therefore, we propose a new method that uses enhanced SOM learning algorithm to increase the compression and replay ratio. In the future study, we plan to develop a novel VQ using the fuzzy membership function.

Acknowledgment

This article was supported by the research grant 2004 of Silla University.

References

[1] Rabbani, M., Jones, P. W.: Digital Image Compression Technique. Spie. Optical Engineering Press. (1991) 144-169
[2] Gray, R. M.: Vector Quantization. IEEE ASSP Magazine. (April 1984) 4-29
[3] Liza, J., Kaimal, M. R.: A Self Organizing Neural Network for Second Generation Image Coding. Proceedings of TAINN. (1999) 130-137
[4] Seo, S., Bode, M., Obermayer, K.: Soft Nearest Prototype Classification. IEEE Trans. Neural Networks. **14(2)** (2003) 390-398
[5] Oehler, K. L., Gray, R. M.: Combining Image Compression and Classification using Vector Quantization. IEEE Multimedia. (1997) 36-45
[6] Madeiro, F., Vilar, R. M., Fechine, J. M., Aguiar Neto, B. G.: A Slef-Organizing Algorithm for Vector Quantizer Design Applied to Signal Processing. International Journal of Neural Systems. **9(3)** (1999) 219-226

Alternate Pattern Fill

Xiao-Xin Zhang[1,2], Jun-Hai Yong[1], Lie-Hang Gong[2],
Guo-Qin Zheng[1], and Jia-Guang Sun[1,3]

[1] School of Software, Tsinghua University, Beijing 100084, P. R. China
[2] Department of Mechanical Engineering, PLA University of Science and Technology,
Nanjing 210007, P. R. China
[3] Department of Computer Science and Technology, Tsinghua University,
Beijing 100084, P. R. China

Abstract. An algorithm for alternate pattern fill on arbitrary line segments, circular arcs and elliptical arcs is proposed in this paper. The algorithm creates an external loop which is a minimal loop that contains a given point. And then, all loops that delimit the nested regions for the alternate pattern fill inside the external loop are built. The loop detection is based on a new method for searching the leftmost edges. It requires only values of positions, tangent vectors, curvature radii, and the first derivatives of curvature radii at intersection points and areas of circles and ellipses. After all necessary loops are built, the given pattern is filled in the nested loops inside the external loop. Filling the given pattern in this paper is simplified into filling line segment patterns on some parallel lines.

1 Introduction

Pattern fill is a process of tiling the interior of a closed region with a repetitive pattern. It is a fundamental problem in a lot of fields [1, 2]. In practical applications, a closed region may contain some other closed regions. Thus, alternate pattern fill may be required. Alternate pattern fill is a process of filling nested closed regions alternately.

Up to now, in literature about pattern fill [1] or about loop detection [3, 4], boundaries of closed regions are usually represented by polygons. Most of the methods for pattern fill are scan-line methods or their mutations. Those pattern fill algorithms have some limitations in practical applications. Firstly, boundaries of closed regions often consist of some curves. Those curves have to be converted into line segments. The number of line segments may be gigantic when the precision is critical. Space cost may become too large. Secondly, the pattern fill algorithms assume that the valences of all vertices in boundaries of closed regions should be exactly 2. However, in a geographic map, a river usually has two different end points, from a mountain to another larger river or an ocean.

In this paper, the assumption on valences of vertices is not required any longer. Boundaries of closed regions are composed of line segments, circular arcs and elliptical arcs. Curves of other kinds such as spline curves are converted into arc splines [5] rather than polygonal curves. Experience [5] shows that the number of arcs in the resultant arc splines is much less than the number of line segments in the resultant polygonal curves after conversion. Circular arcs and elliptical arcs are selected here because they are widely used in computer aided design.

The algorithm proposed in this paper is quite different from scan-line method. It consists of three main steps: preprocess, loop detection and filling patterns. These steps will be explained in the following sections.

2 Preprocess

Before building loops, a preprocess is performed on the given edge set **E**. All edges are converted into line segments, circular arcs or elliptical arcs if they are not[5]. Then, all line segments (or circular arcs or elliptical arcs), which are overlapped, are combined into a line segment (or a circular arc or an elliptical arc). Finally, closed circular arcs or elliptical arcs are divided into two equivalent circular arcs or elliptical arcs. Thus, after the preprocess, all edges in **E** have some uniform properties. All edges are line segments, circular arcs or elliptical arcs, the number of intersection points of any two edges in **E** is finite, and every edge in **E** exactly has two distinguished vertices.

3 Searching Leftmost Edge

The loop detection in this paper is based on searching the leftmost edge from **E** with respect to a given directed edge **e**. We define an edge is on the left side of another edge at a given point as follows.

Definition 1. Given a point $\mathbf{p}_s$, and two edges $\mathbf{e}_1$ and $\mathbf{e}_2$, the point $\mathbf{p}_s$ is the ending point of $\mathbf{e}_1$ and on the edge $\mathbf{e}_2$. $\mathbf{e}_1$ and $\mathbf{e}_2$ are line segments, circular arcs or elliptical arcs. Let $\mathbf{T}_1$ and $\mathbf{T}_2$ be the tangent vectors of $\mathbf{e}_1$ and $\mathbf{e}_2$ at $\mathbf{p}_s$, respectively. $\mathbf{e}_1$ is on the left side of $\mathbf{e}_2$ at $\mathbf{p}_s$ if and only if one of the following cases is matched.

Case 1. $\mathbf{T}_1$ is on the left side of $\mathbf{T}_2$.
Case 2. $\mathbf{T}_1$ and $\mathbf{T}_2$ have the same direction. $\mathbf{e}_1$ is a circular arc or an elliptical arc. The center point of $\mathbf{e}_1$ is on the left side of $\mathbf{T}_1$. And $\mathbf{e}_2$ is a line segment, or the center point of $\mathbf{e}_2$ is on the right side of $\mathbf{T}_1$ if $\mathbf{e}_2$ is not a line segment.
Case 3. $\mathbf{T}_1$ and $\mathbf{T}_2$ have the same direction. $\mathbf{e}_1$ is a line segment, and $\mathbf{e}_2$ is a circular arc or an elliptical arc. The center point of $\mathbf{e}_2$ is on the right side of $\mathbf{T}_1$.
Case 4. $\mathbf{T}_1$ and $\mathbf{T}_2$ have the same direction. $\mathbf{e}_1$ and $\mathbf{e}_2$ are circular arcs or elliptical arcs. And both center points of $\mathbf{e}_1$ and $\mathbf{e}_2$ are on the left side of $\mathbf{T}_1$. Let r_i, r'_i and s_i ($i = 1$ and 2) be the curvature radii, the first derivatives of the curvature radii, and the areas of $\mathbf{e}_i$, respectively. If the values of (r_1, r'_1, s_1) and (r_2, r'_2, s_2) are not the same, the first

different value in (r_1, r_1', s_1) is smaller than the first different value in (r_2, r_2', s_2).

Case 5. $\mathbf{T}_1$ and $\mathbf{T}_2$ have the same direction. $\mathbf{e}_1$ and $\mathbf{e}_2$ are circular arcs or elliptical arcs. And both center points of $\mathbf{e}_1$ and $\mathbf{e}_2$ are on the right side of $\mathbf{T}_1$. Let r_i, r_i' and s_i ($i = 1$ and 2) be the curvature radii, the first derivatives of the curvature radii, and the areas of $\mathbf{e}_i$, respectively. If the values of (r_1, r_1', s_1) and (r_2, r_2', s_2) are not the same, the first different value in (r_1, r_1', s_1) is larger than the first different value in (r_2, r_2', s_2).

In Definition 1, we use the following values: positions, tangent vectors, the curvature radii, the first derivatives of the curvature radii, and areas to judge whether an edge is on the left side of another edge at a given point. From above definition, we can search the leftmost edge from the given edge set $\mathbf{E}$ with respect to a given directed edge $\mathbf{e}$. The leftmost edge is just sought in the order mentioned above. Additionally, an edge may be divided into some subedges. A subedge could be considered as an ordinary edge in the algorithm. For convenience and without confusion, a subedge is called an edge as well in this paper.

4 Loop Detection

The loop detection here is to build loops using edges in a given edge set $\mathbf{E}$ for the following main step of filling patterns. At first, an external loop is created. Then, other loops are made subsequently. A point $\mathbf{p}$ is given to identify an external loop, which is the minimal loop among all loops that contain the given point $\mathbf{p}$ and are built by edges in the given edge set $\mathbf{E}$. In order to avoid confusion, it is usually required that the point $\mathbf{p}$ is not on any edge in $\mathbf{E}$. Other loops built for the alternate pattern fill are the nested loops inside the external loop. Here, we do not need to know how the loops are nested. The alternate pattern fill in the next section will identify the nesting hierarchy. The loop detection in this paper is an iterative procedure of building loops. Therefore, the following provides the algorithm for building a loop from a given edge $\mathbf{e}$. It is based on searching the leftmost edge in the previous section.

Algorithm 1. Building a loop from a given edge $\mathbf{e}$ and a given edge set $\mathbf{E}$.
(Note that $\mathbf{e}$ should be in the loop, and that some edges in $\mathbf{E}$, which cannot form loops, will be removed from $\mathbf{E}$.)
Input: an edge set $\mathbf{E}$ after the preprocess, and an edge $\mathbf{e}$.
Output: a loop $\mathbf{L}$ searched from a given edge $\mathbf{e}$ built by $\mathbf{e}$ and edges in $\mathbf{E}$.

1. Let both the edge set $\mathbf{E}_s$ and the loop $\mathbf{L}$ contain exactly an element $\mathbf{e}$. Let $\mathbf{e}_{cur} = \mathbf{e}$ be the current edge.
2. Search the leftmost edge $\mathbf{e}$ with the inputs $\mathbf{E}$ and $\mathbf{e}_{cur}$. Then we obtain the trimmed leftmost edge $\mathbf{e}^*$ and the trimmed current edge $\bar{\mathbf{e}}$ at the intersection point.

3. **If** the output of Algorithm 1 is *null*, then **go to** Step 5. **Otherwise, go to** the next step.
4. Replace the first element in **L** with $\bar{\mathbf{e}}$. Add $\mathbf{e}^*$ into **L** and $\mathbf{E}_s$ respectively. (Note that $\mathbf{e}^*$ becomes the first element of **L** and $\mathbf{E}_s$ after insertion.) **If** the edges in **L** form a loop $\mathbf{L}_1$, then let $\mathbf{L} = \mathbf{L}_1$, and **go to** Step 6 (end of the algorithm). **Otherwise** let $\mathbf{e}_{cur} = \mathbf{e}^*$, and **go to** Step 2.
5. Remove $\mathbf{e}_{cur}$ from $\mathbf{E}_s$ and **E**, and remove the first element of **L** from **L**. If **L** contains no element, then **go to** Step 6 (end of the algorithm). **Otherwise**, let $\mathbf{e}_{cur}$ be the first element of $\mathbf{E}_s$, and **go to** Step 2.
6. **End** of Algorithm 1.

By calling the Algorithm 1, we can build the external loop with respect to the point **p** using edges in **E**. The input edge **e** is constructed as follows. Find a point **q** on an edge in **E** such that **q** is furthest from **p** along the direction of x-axis. Let $\mathbf{E}_1$ be the set of the edges in **E**, which have intersection points with the line segment from **p** to **q**. Sort edges in $\mathbf{E}_1$ in the "leftmost" order. The first element in $\mathbf{E}_1$ is set as input edge **e** of Algorithm 1. If the external loop can not be built from the first element, then the next element is set as input.

Then we can find all loops that delimit the nested regions inside the external loop. The algorithm is as follows. The edge closest to the point **p** is set as input edge **e** of Algorithm 1. If the loop can be built by Algorithm 1, then the loop is appended to the loop set $\mathbf{S_L}$. The edges encountered during search are removed from $\mathbf{E}_2$ which is a set of all edges in **E** and is inside the external loop $\mathbf{L}_{ext}$. The procedure is repeated until $\mathbf{E}_2$ is empty.

5 Filling Patterns

After producing all loops needed by the alternate pattern fill, patterns will alternately be tiled on the nested regions delimited by those loops. Before patterns are filled, it should be known how a pattern is defined and required to be repeated. Usually a given pattern is very small, so we can decompose or approximate a pattern with several line segments. This paper uses the following values to represent a "line segment pattern". α and $\begin{pmatrix} x_0 \\ y_0 \end{pmatrix}$ determine a line passing through $\begin{pmatrix} x_0 \\ y_0 \end{pmatrix}$ and with the direction $\begin{pmatrix} \cos(\alpha) \\ \sin(\alpha) \end{pmatrix}$. $\begin{pmatrix} \Delta x \\ \Delta y \end{pmatrix}$ makes the above line become a cluster of parallel lines passing through $\begin{pmatrix} x_0 + i\Delta x \\ y_0 + i\Delta y \end{pmatrix}$ and with the direction $\begin{pmatrix} \cos(\alpha) \\ \sin(\alpha) \end{pmatrix}$, where i are integers. Each $i \in \mathbb{Z}$ corresponds to a line in the cluster of the parallel lines. The sequence of real numbers determines the lengths of the line segments and the breaks. They are repeated along the parallel lines as shown in the Figure 1.

As shown in Figure 1, pattern fill could be simplified into line segment pattern fill. For each parallel line, let the set $\mathbf{S}_i$ be initiated as a set with all intersection

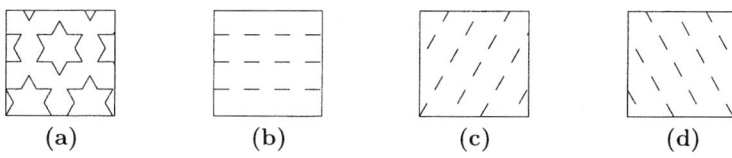

Fig. 1. Pattern fill of stars (a) is simplified into pattern fill of line segments (b, c, d)

points between the parallel line and the edges in the external loop $\mathbf{L}_{ext}$ and in the loop in $\mathbf{S_L}$. If any edge overlaps the parallel line, then $\mathbf{S}_i$ just stores the starting point and the ending point of the edge. If two or more edges have the same intersection points with the parallel line, $\mathbf{S}_i$ records all of them. $\mathbf{S}_i$ is adjusted for some singular cases shown in Figure 2 as follows.

As shown in Figure 2(a), if some edges are tangent with the parallel line, and the tangent points are not vertices of the edges, then remove those tangent points from $\mathbf{S}_i$. In the cases shown in Figure 2(b) and Figure 2(c), two neighboring edges in a loop intersect the parallel line at the join point of the two neighboring edges. If they do not cross the parallel line at the join point(for example, Figure 2(b)), then remove the join point twice from $\mathbf{S}_i$. If they cross the parallel line(for example, Figure 2(c)), then the join point is removed from $\mathbf{S}_i$ only once. The case that any edge overlaps the parallel line (for examples, Figures 2(d) and 2(e)) could be simplified into the cases in Figures 2(b) and 2(c). The method is to ignore the edge that overlaps the parallel line so that the previous edge and the following edge become two "neighboring" edges in the loop.

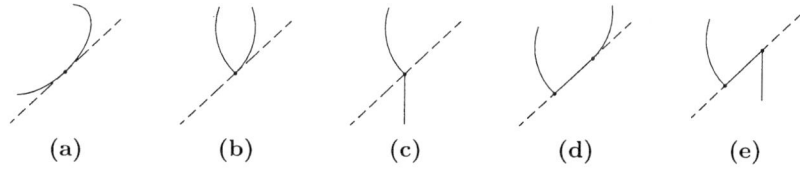

Fig. 2. Some singular cases (edges are in solid, and the parallel line is dashed)

Suppose that $\mathbf{S}_i = \{\mathbf{p}_1, \mathbf{p}_2, \cdots, \mathbf{p}_m\}$ after all the singular cases as shown in Figure 2 are processed. Then, the line segments are drawn from $\mathbf{p}_1$ to $\mathbf{p}_m$ along the parallel line according to the given values for the line segment pattern fill. The line segments between $\mathbf{p}_{2k}$ and $\mathbf{p}_{2k+1}$, where k are an integer, and $2 \leq 2k < 2k+1 \leq m$, are trimmed or removed. The line segments on the edges, which overlap the parallel line, are trimmed or removed as well. In other words, the line segments for filling pattern are drawn on the parallel line from $\mathbf{p}_{2k-1}$ to $\mathbf{p}_{2k}$, and not on the edges of the given edge set.

6 Example and Conclusions

Figure 3 give a practical example. It is a drawing of a toy tractor. The point labelled with $\mathbf{p}$ in the figure is used to identify the regions for the alternate

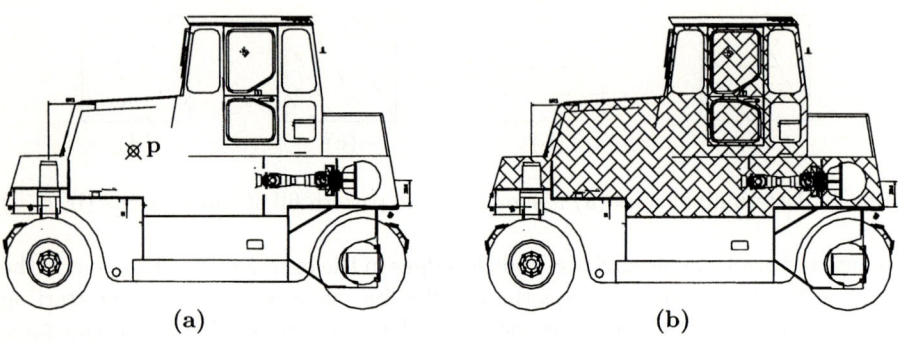

Fig. 3. Example 2(a drawing of a toy tractor): **(a)** before alternate pattern fill; **(b)** after alternate pattern fill

pattern fill. In this example, there are vertices which valences are 1. Some curves are NURBS curves. They are converted into arcs during the preprocess stage. The figure shows that the method in the paper could well deal with it.

The algorithm for alternate pattern fill in this paper has been implemented in our commercial CAD software OpenCAD. The example in the paper is produced with the software OpenCAD as well. The algorithm for alternate pattern fill in the paper could deal with not only line segments but also circular arcs and elliptical arcs. Curves of other kinds will be converted into circular arcs. Edges in the given edge set could be arbitrary. The valences of vertices in the given edge set have no constrain. The loop detection in this paper is based on searching the leftmost edges. While searching the leftmost edges, we only calculate some values at intersection points and areas of ellipses or circles if necessary. Additionally, pattern fill is simplified into line segment pattern fill in this paper.

Acknowledgements

The research was supported by Chinese 863 Program (2003AA4Z3110) and 973 Program (2002CB312106). The second author was supported by a project sponsored by SRF for ROCS, SEM (041501004), and FANEDD(200342).

References

1. Angel, E.: Computer Graphics. Addison-Wesley Publishing Company, New York (1990)
2. Praun, E., Hoppe, H., Webb, M., and Finkelstein, A.: Real-Time Htching. In: Proceedings of SIGGRAPH. (2001) 581–586
3. Gujar, U. and Nagendra, I.: Construction of 3D Solid Objects from Orthographic Views. Computer & Graphics **13** (1989) 505–521

4. Satoh, S., Hiroshi, M., and Sakauchi, M.: An Efficient Extraction Method for Closed Loops Using a Graph Search Technique. IEICE Transactions on Fundamentals of Electronics, Communications and Computer Sciences **78** (1995) 583–586
5. Yong, J.H., Hu, S.M., and Sun, J.G.: A Note on Approximation of Discrete Data by G^1 Arc Splines. Computer-Aided Design **31** (1999) 911–915

A Boundary Surface Based Ray Casting Using 6-Depth Buffers

Ju-Whan Song[1], Ou-Bong Gwun[2],
Seung-Wan Kim[2], and Yong-Guk Kim[3]

[1] School of Liberal Art, Jeonju University,
Jeonju, Jeonbuk, Korea
jwsong@jj.ac.kr
[2] Division of Electronics and Information Engineering,
Chonbuk National University, Jeonju, Jeonbuk, Korea
{obgwun, kswsamson}@chonbuk.ac.kr
[3] School of Computer Engineering, Sejong University, Seoul, Korea
ykim@sejong.ac.kr

Abstract. This paper focuses on boundary surface based ray casting. In general the boundary surface based ray casting is processed in two stages. In the first stage, boundary surfaces are found and stored into buffers. In the second stage, the distance between the viewpoint and the voxels of the area of interest is calculated by projecting boundary surfaces on the view plane, and then the traverse of the volume data space with the distance is started. Our approach differs from the general boundary surface ray casting in its first stage. In contrast to the typical boundary surface based ray casting where all boundary surfaces of volume data are stored into buffers, in our proposal, they are projected on the planes aligned to the axis of volume data coordinates and these projected data are stored into 6-depth buffers. Such a maneuver shortens the time for ray casting, and reduces memory usage because it can be carried out independently from the amount of the volume data.

1 Introduction

Boundary surface based ray casting enables a ray to traverse the voxels of the area of interest directly without traversing the front voxels of this area. It makes voxel traversing fast because the voxels of areas of no interest do not need to be traversed. It is necessary to know the exact distance between a view plane and a boundary surface while traversing voxels in the boundary surface based ray casting. For that purpose, Zuiderveld et al. make a map representing the distance from the viewpoint to the nearest possible voxels of the area of interest area and use the map for voxel traversing[6].

In Polygon Assisted Ray Casting(PARC) algorithm, Sobierajski et al. find the polygons of the boundary surfaces that surround the possible voxels of interested area in the preprocessing and then project the polygons of boundary surfaces on the view plane by Z-buffer algorithm to find the distance in the main processing[3]. The performance of PARC algorithm for the case of a boundary surface with many polygons is degraded as the projection time of the polygons is increased[3]. In boundary cell based rendering, Wan et al. find the boundary cells that surround the region of interest and store these boundary cells in the vector table[4]. The algorithm is only effective for small volume data of less than 128^3.

In this paper, we propose a new boundary surface based ray casting. The algorithm is different from the above algorithms: first, the boundary surface data are once stored in 6-depth buffers by projecting boundary surfaces on the XY, YZ, ZX planes of volume data coordinates; second, projection of boundary surfaces is performed using an incremental algorithm.

2 6-Depth Buffers

In this paper, we propose a novel boundary surface ray casting, in which the distances from a view plane to voxels of area of interest are calculated using depth buffers aligned in 3 directions in the volume data space and volume data are represented by 3 dimensional arrays of cube. In the volume data space, we can define volume data coordinates in which the coordinates axes are principle edges of it as illustrated in Fig. 1. In the coordinates, we build total 6-depth buffers: 2 buffers (near and far) for each direction x, y and z, respectively. Boundary surface data are stored in the 6-depth buffers by projection.

The resolution of the depth buffer for each direction is identical to the resolution of the volume data space for each direction. The far and near depth values for the boundary surface are found by shooting a ray from each plane (XY, YZ, XZ) and calculating the location of the intersection point between a the boundary surface and a ray. They are calculated only once and stored in the head of volume data because the near and far depth values of boundary surface are independent on any viewpoint. The following algorithm finds the near depth values of Z direction on the XY plane. Depth values perpendicular to X, Y directions can be found using a similar method.

```
for all pixels(i, j) on the XY view plane{
  Z(i, j) = inf ;   /* initialize z buffer with maximum value*/
}
for all pixels(i, j) on the XY view plane{
  spawn a ray(i, j);
  traverse cell of the ray(i, j) through the volume data space;
  if( Ray(i, j) meets boundary surface cell for the first time){
    find the distance from XY viewplane to the boundary surface
    cell; Z(i, j) = the distance;
  }
}
```

3   6-Depth Buffers Creation

In Section 2, the depth values are stored in 6-depth buffers if it assumed that the view coordinates are the same as the volume data coordinates. However, changing the viewpoint requires transformation of the near and far depth values of 6-depth buffers. The near depth values from the viewpoint can be calculated by projecting the values of the front 3 depth buffers among 6-depth buffers on a view plane. Fig. 2 illustrates this process. The far depth values of a viewpoint can be calculated using a similar method.

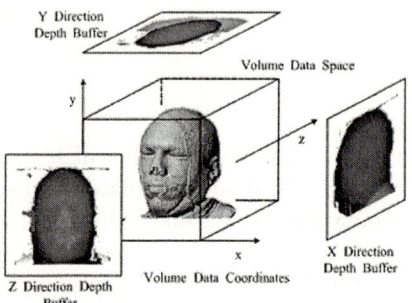

Fig. 1. The concept of 6-depth buffers

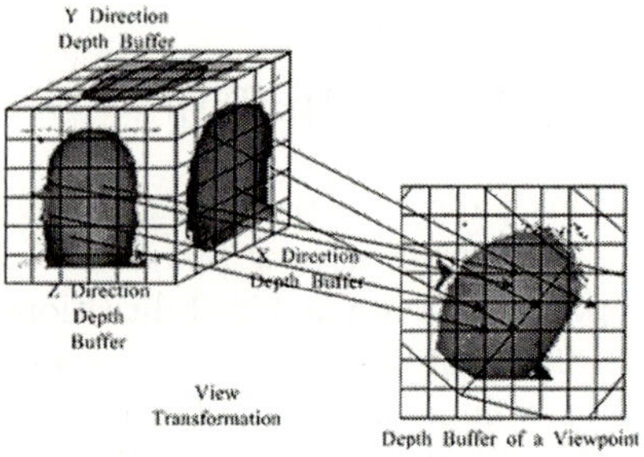

Fig. 2. Finding of the depth value for a viewpoint

The following pseudo code describes the algorithm, in which a location (x, y, z) in the X direction buffer among 6-depth buffers is transformed into a location (x, y, z) of view coordinates,

```
for (yd = 0; yd < ny - 1; yd++)
  for(zd = 0; zd < nz - 1; zd++)
    xv, yv, vv = view_transform(x_buffer(yd, zd), yd, zd);
  }
}
```

Where view transform (x_buffer((yd, zd), yd, zd) is found as follows[5]:

$$\begin{bmatrix} x_v \\ y_v \\ z_v \end{bmatrix} = \begin{bmatrix} cos\alpha cos\beta & cos\alpha sin\beta sin\theta - sin\alpha cos\theta & cos\alpha sin\beta cos\theta + sin\alpha sin\theta \\ sin\alpha cos\beta & sin\alpha sin\beta sin\theta + cos\alpha cos\theta & sin\alpha sin\beta cos\theta - cos\alpha sin\theta \\ -sin\beta & cos\beta sin\theta & cos\beta cos\theta \end{bmatrix} \begin{bmatrix} x_d \\ y_d \\ z_d \end{bmatrix}$$

(1)

Where ny-1, nz-1 is the volume resolution of y, z direction, respectively, and xd, yd, zd is the location of volume data coordinates, respectively, and xv, yv, zv is the location of view coordinates, respectively.

Though the above algorithm represents the transformation from a location of volume data coordinates to a location of view coordinates for x direction, the transformation of a location for y (or z direction) can be obtained in a similar way. The above algorithm is calculated by the incremental algorithm. The first value [xv, yv, zv] is calculated by (1) but the next values [xv, yv, zv+1] or [xv, yv+1, zv+1] are calculated by adding step increment into [xv, yv, zv] or [xv, yv, zv+1].

4 Potential Problems and Their Remedy

We encounter two potential problems with the present approach. The first problem is caused because the footprint of the cells projected on the view plane is smaller than the pixel size as shown in Fig. 3. In this case, some parts of boundary cells not covered by the projected cell are created. The second problem is caused by the insufficient number of depth buffers used to store the boundary surface of the area of interest. When the shape of a boundary surface is concave, then some boundary cells of volume data space cannot be sampled in the 6-depth buffers as shown in Fig. 4. In such a case, holes are generated when view direction changes. Fig. 5 shows the holes generated by the above two problems.

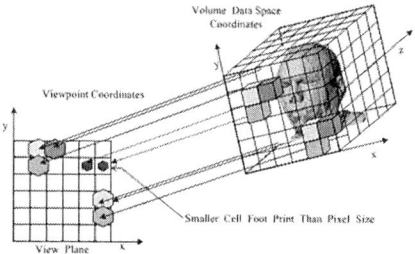

Fig. 3. Holes caused by smaller cell foot print than pixel size

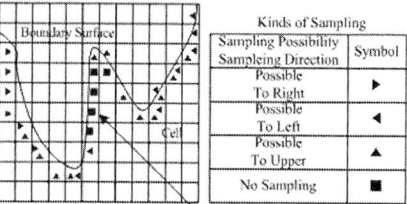

Fig. 4. Holes caused by insufficient number of depth buffer in 2 dimensions

In our case, we remedy this problem as follows. If there is a hole during the process of view transformation of depth buffers, we search the depth values of 4-neighbor pixels of the hole and set the nearest depth value of them as the depth value of the hole. If there are no depth values in the 4-neighbor pixels, we expand the search area. If several depth values from more than one depth buffer are overlapped, the nearest depth value among them is selected and stored into the depth buffer for the view plane. Fig. 6 shows that holes are removed by this method.

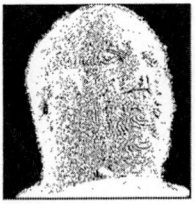 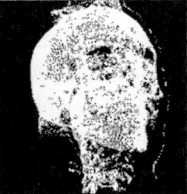  

Fig. 5. Holes caused by smaller cell foot print than pixel size and insufficient number of depth buffers: Black points in the image represent holes

Fig. 6. Boundary surface modified by larger cell foot print than original cell footprint: Holes disappeared

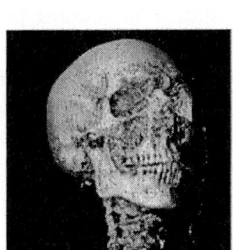 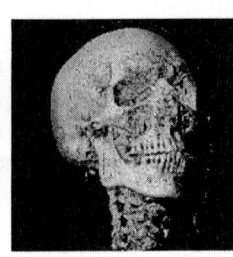 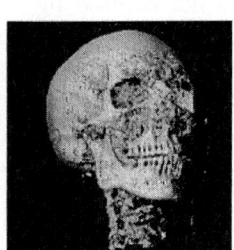

(a) Common ray casting (b)PARC algorithm (c)Proposed algorithm

Fig. 7. Skull images of digital man by common ray casting, PARC algorithm, and proposed algorithm

5 Results and Algorithm Comparison

The proposed algorithm is compared to the common ray casting as well as PARC algorithm on a PC platform. In PARC, x, y, z value of cells (polygon) correspond to the boundary surfaces in the depth buffer. When the number of cells consisting of boundary surfaces is n, and x, y, z component of one cell is represented with 2 bytes; PARC requires nx3x2 bytes memory space. The proposed algorithm always requires the memory space of twice the number of XY, YZ, ZX plane resolution of volume data space regardless of boundary surface complexity, because we know 2 components among x, y, z component from a depth buffer. We have evaluated the result image using the performance measures: MAE (Mean Absolute Error) and RMSE (Root Mean Square error)[2].

Table 1 and Fig. 7 show the experiment results for the CT data of digital man[1]. The size of additional memory of PARC algorithm is 1,493,868 bytes and the size of additional memory for the proposed algorithm is 786,432 bytes. The ray casting time of common ray casting, PARC algorithm and the proposed algorithm is 6.345, 1.221 and 0.582 seconds, respectively. The speed-up ratios of the proposed algorithm to common ray casting and PARC algorithm is 10.902:1

Table 1. Ray casting time and speed-up ratio of skull images of digital man

		†Com.	‡PARC	§Prop.
Additional Memory Capacity(bytes)		0	1,493,868	786,432
Ray Casting Time(seconds)		6.345	1.221	0.582
Speed-up Ratio	Prop.:Com.	10.902 : 1		
	Prop.:PARC	2.098 : 1		
Differences of Result Images	MAE	0.014029(Com. to Prop.), 0.007124(PARC to Prop.)		
	RMSE	0.078967(Com. to Prop.), 0.035247(PARC to Prop.)		

†Com.: Common ray casting algorithm
‡PARC: PARC algorithm
§Prop.: Proposed algorithm

and 2.098:1. The difference of the image generated by common ray casting and the proposed algorithm when measuring MAE and RMSE is 0.014029, 0.078967 respectively. Those by PARC algorithm and the proposed algorithm is 0.007124, 0.035247, respectively. The image generated by the proposed algorithm is very similar to common ray casting. However, the image generated by the proposed algorithm is more similar to the image generated by PARC algorithm than the image generated by common ray casting.

6 Conclusions

In this paper, we proposed a boundary surface based ray casting, implemented it on a PC, and compared it to common ray casting algorithm and PARC algorithm. In contrast to the typical boundary surface based ray casting where all boundary surfaces of volume data are stored into buffers, in our proposal, they are projected on the planes aligned to the axis of volume data coordinates and these projected data are stored into 6-depth buffers. Such a maneuver shortens the time for ray casting, and reduces memory usage because it can be carried out independently from the amount of the volume data. In a future study, we plan to make our algorithm faster by structuring the data in the 6-buffers.

References

1. http://www.nlm.nih.gov/research/visible/visible_human.html. NLM the Visible Human Project
2. Kim, K., Wittenbrink, C. M., Pang, A.: Extended specifications and test data Sets for Data Level Comparisons of Direct Volume Rendering Algorithms. IEEE Transactions on Visualization and Computer Graphics **7** (2001) 299–317
3. Sobierajski, L. M., Avila, R. S.: A Hardware Acceleration Method for Volumetric Ray Tracing. IEEE Visualization 95 (1995) 27–34
4. Wan, M., Bryson, S., Kaufman, A.: Boundary cell-based acceleration for volume ray casting. computers & graphics **22** (1999) 715–721

5. Yang, S., Wu, T.: Improved Ray-casting Algorithm for Volume visualization. SPIE Visualization and Data Analysis **4665** (2002) 319–326
6. Zuiderveld, K. J., Koning, A. H. J., Viergever, M. A.: Acceleration of ray-casting using 3D distance transforms. SPIE Visualization in Biomedical Computing **1808** (1992) 324–334

Adaptive Quantization of DWT-Based Stereo Residual Image Coding*

Han-Suh Koo and Chang-Sung Jeong**

Department of Electronics Engineering, Korea University,
1-5ka, Anam-dong, Sungbuk-ku, Seoul 136-701, Korea
{hskoo, csjeong}@korea.ac.kr

Abstract. General procedure for stereo image coding is to use disparity compensated prediction methods and code residuals separately. Although the characteristics of stereo residuals are different from those of common images, JPEG-like methods which are applied to monocular images are used frequently and less research has been devoted to residual image coding. The focus of this paper is to make stereo image coding more efficient by speculating the characteristics of the residual image. By measuring the edge tendency of residual image, our method modifies the quantization matrix adaptively using discrete wavelet transform.

1 Introduction

A stereo pair is two images which are captured at slightly different positions. Due to the advances in 3D technology, stereo image compression is becoming a crucial research field for various applications that includes teleconferencing, visualization, and robotics. Most stereo image coding algorithms take advantages of disparity estimation and compensation used for compression of monocular image sequences include block matching to predict one image from the other one. Reference image is coded independently, and target image is predicted from reference one. The residual which is the error between actual and predicted target image is coded successively using a suitable transform coder. Although the characteristics of disparity-compensated stereo residuals are different from that of common images, less research has been devoted to residual image coding and JPEG-like methods including Discrete Cosine Transforms (DCT) which are used to compress common monocular images are generally employed.

In this paper, we designed a new stereo residual image coding algorithm which uses wavelet transform. Considering the features of residual images, we proposed a new Discrete Wavelet Transforms (DWT) quantization matrix emphasizing vertical or horizontal structures according to the edge strength of stereo residual image. Our algorithm can adjust entries of quantization matrix in view of the edge tendency.

---

* This work was partially supported by the Brain Korea 21 project in 2004 and KIPA-Information Technology Research Center.
** Corresponding author.

2 Related Researches

DCT-based compressions have been popular so far for monocular image coding, and JPEG-like DCT-based schemes are also frequently used for stereo image residual because they can be easily combined with existing monocular image coding methods. Lately DWT-based compressions attract much attention in image coding research field because they provide many useful features and better performance than DCT-based schemes [1]. Related with this report, stereo image coding algorithms based on wavelet transform have been presented by many researchers [2–4]. As for coding of stereo residuals, mixed image transform coder which uses DCT on the well matched blocks by block matching and Haar filtering on the occluded blocks was proposed [5]. However JPEG coding which uses DCT is a de facto standard and many systems built on JPEG are already used in various applications.

Queiroz et al. proposed the JPEG-like DWT image coder known as DWT-JPEG [6]. They incorporated DWT into the JPEG baseline system for coding general images by designing a new block-scanning order and quantization matrix into JPEG coder block. Nayan et al. showed that DWT-JPEG is superior to DCT in stereo residual coding [7]. By JPEG-like approach, it does not require full-image buffering nor imposes a large complexity increase.

Moellenhoff and Maier presented two special properties for disparity compensated stereo residual images [8,9]. First, in case that disparity compensation is performed well, much of the residual image has intensity values near zero. Second, the remaining contents are narrow vertical edges because of strictly horizontal apparent motion between the two views. They proposed Directional DCT quantization matrix based on anticipated DCT energy distribution to emphasize the linear structure in the image. However optimal quantization matrix for various image condition caused by the baseline of two cameras and a characteristic of subjects cannot be determined easily because DCT quantization matrix is confined within local 8×8 areas. Determining a unified matrix is difficult and finding individual one for each local area is not adequate in rate's aspect within traditional JPEG framework.

3 A DWT Coding of Stereo Residual Image

Most stereo image coding algorithms use block-based disparity estimation and compensation to predict one of stereo pair from the counterpart. One image is coded and transmitted independently using conventional methods for monocular images and the other one is compensated and coded using disparity vectors and residuals. Disparity vectors are coded losslessly by Differential Pulse Code Modulation (DPCM) followed by Variable Length Coding (VLC). Because error-free stereo image pair can hardly be constructed by disparity-compensated prediction, residual errors should be coded to get better results.

A DWT-JPEG has been proposed to provide better coding performance. However, DWT-JPEG quantization matrix does not guarantee the best performance in stereo residual image coding because it was not designed considering the condition of stereo images.

The characteristics of disparity-compensated stereo residuals are different from those of normal images. As mentioned in Sect. 2, Moellenhoff and Maier surveyed two special properties for disparity-compensated stereo residuals. Applying these features, they proposed a quantization matrix for DCT coding of stereo residual images emphasizing the horizontal structures in local 8×8 areas. However we found that the quantization process using this matrix might raise some argument under particular conditions. Because previous method uses a fixed quantization matrix based on anticipated DCT energy distribution, it can not be adapted to various conditions arisen by geometry of cameras and objects. Residual images may have various edge tendency and distribution according to the image type or the performance of disparity compensated prediction.

Based on arguments from DWT-JPEG and Directional DCT, we designed a new stereo residual image coding scheme. Our scheme takes advantage of features of previous works. Figure 1 is an overall flow diagram of our scheme. In ordinary block-based coding scheme, DCT coefficients are grouped within common spatial location and each of them indicates different subbands. While, in DWT, coefficients indicate different spatial locations in common subband. Because overall distribution of intensity in residual image is easily considered by DWT approach, adaptive schemes according to image conditions are devised by DWT rather than DCT. Following DWT to input residuals, the coefficients are reordered by DWT-JPEG scanning [6]. Coefficients in the same location, but different subbands are grouped together in a block like DCT coefficients in JPEG.

For each reordered block, adaptive quantization is processed using the 8×8 matrix defined by

$$Q = A(Q' - 1) + 1, \qquad (1)$$

$$Q' = \begin{bmatrix} c_{k-3} & d^{(1)}_{k-3} & d^{(1)}_{k-2} & d^{(1)}_{k-2} & d^{(1)}_{k-1} & \cdots\cdots & d^{(1)}_{k-1} \\ d^{(2)}_{k-3} & d^{(3)}_{k-3} & d^{(1)}_{k-2} & d^{(1)}_{k-2} & \vdots & & \vdots \\ d^{(2)}_{k-2} & d^{(2)}_{k-2} & d^{(3)}_{k-2} & d^{(3)}_{k-2} & \vdots & & \vdots \\ d^{(2)}_{k-2} & d^{(2)}_{k-2} & d^{(3)}_{k-2} & d^{(3)}_{k-2} & d^{(1)}_{k-1} & \cdots\cdots & d^{(1)}_{k-1} \\ d^{(2)}_{k-1} & \cdots & \cdots & d^{(2)}_{k-1} & d^{(3)}_{k-1} & \cdots\cdots & d^{(3)}_{k-1} \\ \vdots & & & \vdots & \vdots & & \vdots \\ \vdots & & & \vdots & \vdots & & \vdots \\ d^{(2)}_{k-1} & \cdots & \cdots & d^{(2)}_{k-1} & d^{(3)}_{k-1} & \cdots\cdots & d^{(3)}_{k-1} \end{bmatrix}. \qquad (2)$$

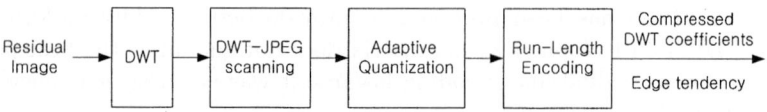

Fig. 1. Flow diagram of proposed stereo residual image coder

In the matrix Q', entry c_{k-3} for DC quantization is set as 2 like Directional DCT for stereo residual. Because intensity of residual image is usually small, DC quantization is done with small value. Entries representing diagonal edges are fixed to the value of that of DWT-JPEG because our scheme is also based on DWT. Accordingly, entries $d^{(3)}_{k-1}$, $d^{(3)}_{k-2}$, and $d^{(3)}_{k-3}$ are set as 55, 12, and 7 respectively. To determine the entries for vertical and horizontal edges, we defined following metric,

$$d^{(l)}_m = d^{(3)}_{m-1} + (d^{(3)}_m - d^{(3)}_{m-1}) \times (1 - \frac{E_l}{E_1 + E_2}), \qquad (3)$$

where m is a level of wavelet tree and E_l is the energy of subband l of the highest level. We called $\frac{E_l}{E_1+E_2}$ as an edge tendency α_l at subband l. By this scheme, smaller factors are assigned to subbands having stronger edge tendency. That is, directional quantization coefficients can be adaptively determined according to the directionality of stereo residual image. In (1), A is a positive value used as a scaling factor to control the bit rate, and bold constants are matrices that act like a constant. Constant matrices are used to avoid quantization failure when A is a decimal fraction. Quantized entries are encoded in Run-Length Encoding (RLE) block.

After encoding process, disparity vector, DWT coefficients, and horizontal edge tendency are transmitted to decoder along with information of first view. Decoder reconstructs stereo image pair using information from encoder. Decoding procedure is virtually reverse order of encoding. Quantization matrix is constructed with edge tendencies. One is transmitted α_l and the other is calculated via $(1 - \alpha_l)$. Wavelet coefficients are restored with the matrix.

4 Experimental Results

We compared our algorithm with DWT-JPEG scheme because the focus of this paper is to evaluate the performance of adaptive quantization matrix in DWT. Figure 2 are stereo images experimented in this paper. The size of test image is 512×512 with 256 gray levels intensity. In all of our experiments, the size of block for disparity compensated prediction and residual coder is chosen to be 8×8 pixels. For DWT, Haar bases were used in our experiments. Search range for finding disparity vectors is confined to 8 pixels for left margin and 4 pixels for right margin. Residuals between original right image and estimated image which is assembled by disparity vectors are coded using residual coder. Using (3), adaptive quantization matrix for each test set is constructed. Entries of quantization matrix Q' is filled with coefficients which are determined from edge tendencies.

We compared Peak Signal-to-Noise Ratio (PSNR) differences of ours named Directional DWT with DWT-JPEG. PSNR is calculated between original right image and the reconstructed image from decoder. Results are compared objectively in Fig. 3. Scaling factor A used in this experiment ranged from 0.6 to

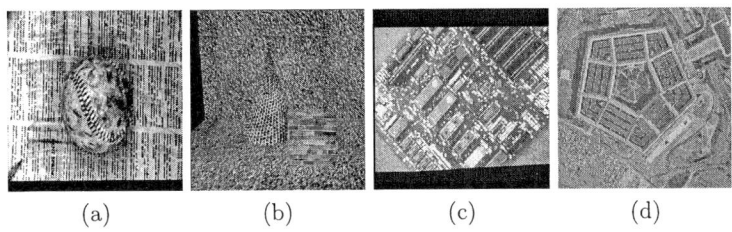

Fig. 2. Experimental image sets (a) Ball image (real) (b) Cone image (synthetic) (c) Apple image (real) (d) Pentagon image (real)

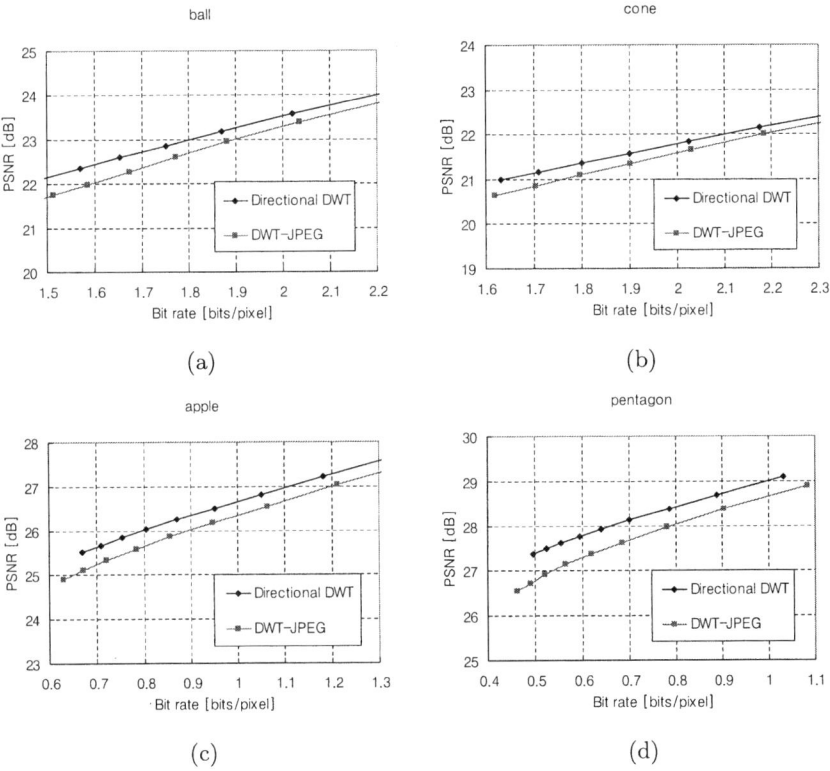

Fig. 3. Performance evaluation of Directional DWT with DWT-JPEG using plots of PSNR (a) Ball image (b) Cone image (c) Apple image (d) Pentagon image

1.4 and discretely varied in 0.1 step. Scaling factor is also multiplied to quantization matrix of DWT-JPEG. Graphs show that PSNR performance varies linearly according to the variation of scaling factor A. Experimental results under various scaling factor demonstrate that Directional DWT performs better than DWT-JPEG through all the observed bit rates though it is not optimized. We can validate that improvements of coding efficiency are achieved by adaptive quantization.

Concerning bit allocation, Directional DWT and DWT-JPEG require bandwidth for disparity vectors and DWT coefficients. In addition to commonly required information, Directional DWT consumes memory to save edge tendency. However Directional DWT shows better performance than DWT-JPEG regardless of additional memory consumption because compression efficiency of Directional DWT is excellent to offset the differences.

5 Conclusion and Future Work

In this paper, we proposed a DWT-based algorithm to code stereo residual image. Our system adopted DWT-JPEG scheme because DWT is known to be superior to DCT in coding efficiency and DWT-JPEG is able to be incorporated into the JPEG framework. Considering edge tendency of residual image, our strategy adaptively modifies quantization matrix to get improved coding efficiency. By experiments with stereo images, we showed that our scheme outperforms compared with previous DWT-based coding. For future work, we will test various wavelet filter banks and optimize our algorithm to get the best performance.

References

1. Xiong, Z., Ramchandram, K., Orchard, M. T., and Zhang, Y. Q.: A Comparative Study of DCT- and Wavelet-Based Image Coding. IEEE T. Circ. Syst. Vid. **9** (1999) 692–695
2. Jiang, Q., Lee, J. J., and Hayse, M. H.: A Wavelet Based Stereo Image Coding Algorithm. Proc. IEEE Int. Conf. Acoust. Speech **6** (1999) 3157–3160
3. Xu, J., Xiong, Z., and Li, S.: High Performance Wavelet-based Stereo Image Coding. IEEE Int. Symp. Circ. Syst. **2** (2002) 273–276
4. Palfner, T., Mali, A., and Muller, E.: Progressive Coding of Stereo Images using Wavelets and Overlapping Blocks. Proc. IEEE Int. Conf. Image Process. **2** (2002) 213–216
5. Frajka, T., Zeger, K.: Residual Image Coding For Stereo Image Compression. Opt. Eng. **42** (2003) 182–189
6. de Queiroz, R., Choi, C. K., Huh, Y., and Rao, K. R.: Wavelet Transforms in a JPEG-Like Image Coder. IEEE T. Circ. Syst. Vid. **7** (1997) 419–424

7. Nayan, M. Y., Edirisinghe, E. A., and Bez, H. E.: Baseline JPEG-Like DWT CODEC for Disparity Compensated Residual Coding of Stereo Images. Proc. Eurographics UK Conf. (2002) 67–74
8. Moellenhoff, M. S., Maier, M. W.: DCT Transform Coding of Stereo Images for Multimedia Applications. IEEE T. Ind. Electron. **45** (1998) 38–43
9. Moellenhoff, M. S., Maier, M. W.: Transform Coding of Stereo Image Residuals. IEEE T. Image Process. **7** (1998) 804–812

Finding the Natural Problem in the Bayer Dispersed Dot Method with Genetic Algorithm

Timo Mantere

Department of Information Technology, Lappeenranta University of Technology,
P.O. Box 20, FIN-53851 Lappeenranta, Finland
tmantere@lut.fi

Abstract. This paper studies how the built-in natural weakness in the image processing algorithm or system can be searched and found with the evolutionary algorithms. In this paper, we show how the genetic algorithm finds the weakness in the Bayer's dispersed dot dithering method. We also take a closer look at the method and identify why this weakness is relatively easy to find with synthetic images. Moreover, we discuss the importance of comprehensive testing before the results with some image processing methods are reliable.

1 Introduction

An image processing system or algorithm may contain some built-in natural weaknesses that normal testing with relatively few standard test images may not reveal. The image-processing researcher may be satisfied with his algorithm if he gets good results with just a few standard test images (Lena, Baboon, *etc.*).

The test image choice has a big influence on the results. This is clearly demonstrated in [1] by selecting an appropriate five-image subset from a 15-image test set, any one of the six Gamut Mapping Algorithms (GMA) tested could appear like the best one. They studied 21 research papers on GMAs and found out that these papers have represented results by using only 1 to 7 test images. The general belief that the system really operates satisfactorily with all possible images requires more extensive testing.

In software testing the software system is usually tested with a much larger test data set. Software faults can be searched and found with automatic testing by *e.g.* random search or evolutionary algorithms. Software test data is given as input and the test results are analyzed from the output. In paper [2] we used image-processing software as a test example. We sent synthetic images to the system and measured how close the processed result images were to the original. In other words, we tested the quality of image processing software. In the case of image processing software the fault we find may not be caused by badly written software, but instead by the image-processing algorithm implemented in it. We also state [2, 3] the importance of extensive testing and explain why it is better to use a large number of simulated test images instead of just standard test images. In study [2] we did not analyze what caused the target software to process the worst test data (hardest images) as badly as it did. The purpose of this is paper is to go back and analyze what made the tested image processing software behave badly with our synthetic test images.

Genetic algorithms (GA) [4] are population-based computerized optimization methods that use the Darwinian evolution [5] of nature as a role model. The solution base of the problem is encoded to individuals. These individuals are tested against our problem, which is represented as a fitness function. The better the fitness value of the individual, the better is its chance to survive and be selected as a parent for new individuals. By using crossover and mutation operations GA creates new individuals. In crossover we select the genes for a new chromosome from each parent using some pre-selected practice, *e.g.* one-point, two-point or uniform crossover. In mutation we change random genes of the chromosome either randomly or using some predefined strategy.

Dithering or digital halftoning [6] is a method where we reduce the number of tones in the image and present the result image with less colors or gray tones. The aim is to do dithering so that the resulting images do not contain any disturbing artifacts or moiré [7]. In this paper, the original images contain 256 possible gray tones and the resulting images are bi-level images, *i.e.* they can only have white and black pixels. The method we use is the thresholding with the Bayer [8] dispersed dot ordered threshold matrix.

Image comparison between original and dithered images is a difficult problem. Much research has been done in this area and several methods of how this should be done have been proposed. Usually the methods try to adapt the human eye visual transformation (HVT) function. The human eye is a kind of lowpass filter that averages the perception from small details. In this paper we use the lowpass filter model based on HVT [9]. The image comparison is done by using Normalized Root Mean Squared Error (NRMSE) [10] difference metric (1), after lowpass filtering dithered images.

$$NRMSE = \sqrt{\frac{\sum_{i=0}^{N-1}\sum_{j=0}^{M-1}[I_o(i,j) - I_r(i,j)]^2}{\sum_{i=0}^{N-1}\sum_{j=0}^{M-1}[I_o(i,j)]^2}} \quad (1)$$

There is still some active research applying Bayer's method, *e.g.* [11] uses the Bayer method as a starting point to develop a dithering algorithm for three-dimensional printing. According to [11], the dispersed dot dither can avoid undesirable low frequency textures in the constant regions and is therefore preferred in applications for 3D printing. Different kinds of pictures may be dithered in the most satisfactory way with different halftoning techniques, *e.g.* hand drawings are totally different than natural images [12], and therefore require different algorithms. We also recognized the fact that different subsets of images are better halftoned with different error-diffusion coefficients when we applied co-evolution to optimize error-diffusion coefficients together with the test images [3]. There are some formerly identified weaknesses in Bayer's method, but despite that it is widely used in digital halftoning, and images that have been halftoned with it are easily recognized from the quasi-period cross-hatch pattern artifacts [12]

that this method produces in the display. This artifact is usually seen as a major
weakness in this technique that is otherwise fast and powerful.

2   The Proposed Method

The proposed method (Fig. 1) consists of a genetic algorithm that generates
parameter arrays. These parameter arrays are used by an image producer application that generates synthetic test images. Images are then sent to the image processing application that halftones the input image using Bayer's ordered
dither method and outputs the result image. The image comparator subprogram
in GA reads the pixels from the original test image and halftoned result image
and compares them. The comparison value is used by GA as a fitness value.

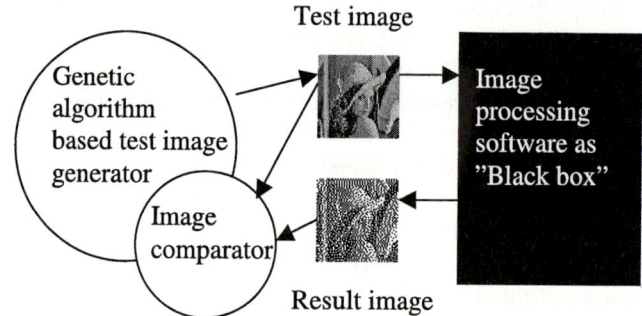

Fig. 1. GA-based image processing software quality testing system

The GA chromosome consists of 217 floating-point numbers, and 210 of them
are divided into 30 groups of 7 parameters. They are used to generate 30 basic
image components; the 7 parameters define their type (lines, rectangles, ovals,
or ASCII characters), position, size, colors, and transparency. One parameter
defines the overall image background tone. The last six are used to generate
chaotic stripes, 3 for vertical and 3 for horizontal stripes. These 3 parameters
for chaotic data are value $a = [2, 4]$, and the initial value $x_0 = [0, 1]$ of Verhulst
equation [13]: $x_{n+1} = a \times x_n \times (1-x_n)$, the third parameter is the offset, *i.e.* how
many x_n values are bypassed in the beginning. Two chaotic data fillings enable
us to generate diagonal stripes and sort of squared artifacts (crossing stripes) to
the test image. We used chaotic data to make images more varied, and because
its appearance can be controlled by parameters, unlike random data.

3   Experimental Results

We run tests with both GA and random search methods in order to verify that
GA really does learn and adapt to the problem parameters, and results are not
due to just random luck. The GA parameters in this case were: population 60,

elitism 50%, mutation percentage 3%, crossover rate 50%, with uniform/single-point crossover ratio 50/50. The test runs with both the GA and the random method were 200 generations long. With random search the "generation" means as many new random test images as new individuals in one GA generation.

Figure 2 shows the development of the highest fitness value with GA and random search, as a function of generations, in the best and worst test runs out of 10 with each method. Fig. 2 shows typical GA development, and it clearly outperforms random search in this problem.

We believe that the reason why it is possible to generate such synthetic images that their darkness changes so dramatically (Fig. 3) during the halftoning process is a natural weakness in the Bayer method. Let us try to identify it.

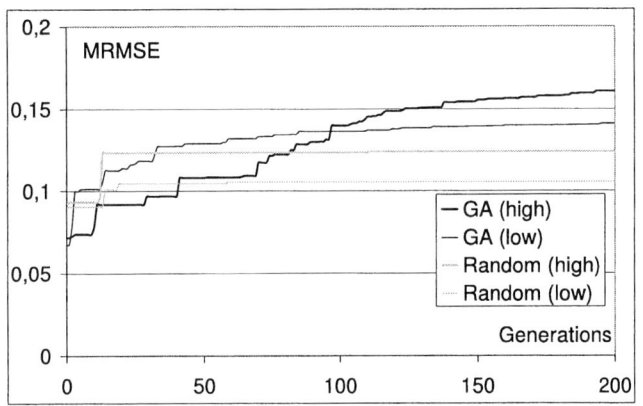

Fig. 2. The development of GA-based optimization and random search while searching harder test images, in the best (high) and the worst test runs (low)

The initial numbers 0, 1, 2, 3 for the Bayer matrix are put as $D^{(2)} = \begin{bmatrix} 0 & 2 \\ 3 & 1 \end{bmatrix}$. The larger matrix is accomplished by (2), e.g. $D^{(4)} = \begin{bmatrix} 0 & 8 & 2 & 10 \\ 12 & 4 & 14 & 6 \\ 3 & 11 & 1 & 9 \\ 15 & 7 & 13 & 5 \end{bmatrix}$, where $\sum(column) = 30$ and $\sum(row) = [\,20\ 36\ 24\ 40\,]$. We can calculate that in $D^{(16)}$ the $\sum(column) = 2040$, but the $\sum(row) = [1360, 2720]$.

At first we suspected that the differences between row sums enable the image darkness to change so dramatically. However, the stripes in our test images (Fig. 3) were clearly vertical, which seems to indicate that there is a different explanation. A closer look reveals that the threshold (or image) values in the vertical columns create this phenomena. When we go through all possible 16×16-pixel one-tone gray images, tones 0 to 255, and threshold them with the Bayer method we saw that the resulting tone values start to change unevenly in the result image rows and columns.

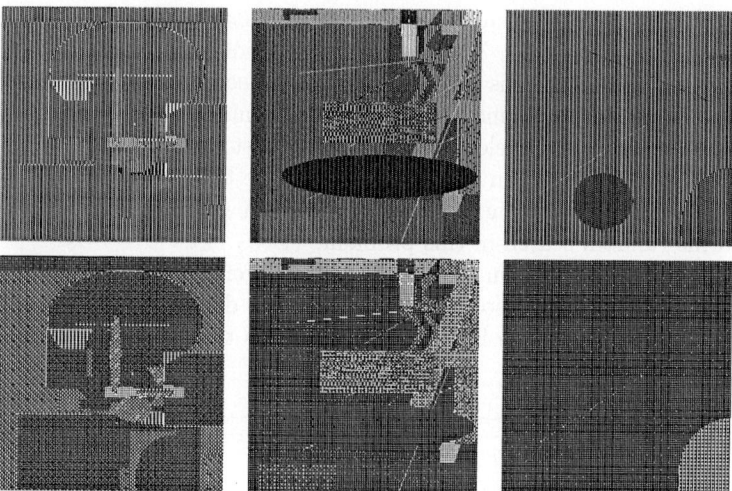

Fig. 3. Examples of GA-generated test images and their halftone result images

$$D^{(n)} = \begin{bmatrix} 4D^{(n/2)} + D^{(2)}_{00}U^{(n/2)} & 4D^{(n/2)} + D^{(2)}_{01}U^{(n/2)} \\ 4D^{(n/2)} + D^{(2)}_{10}U^{(n/2)} & 4D^{(n/2)} + D^{(2)}_{11}U^{(n/2)} \end{bmatrix} \quad (2)$$

If we have a one-tone image where every pixel value is 128, the Bayer method generates an output image where all row and column sums are equal to 2040 (in a 16 × 16 area) and white and black pixels alternate. With a one-tone image, tone=64, the method produces images where every other row and every other column is equal to zero and the rest equal to 2040, this means that every other column is totally black and in every other the black and white pixels alternate. With a one-tone image, tone=192, the method generates an output image where every other row and column is completely white, and in the rest the black and white pixels alternate. These differences mean that if we generate a test image where in every other vertical column all pixels are tone=64 and in every other tone=192, the result image with the Bayer method is exactly the same as with one-tone images, with tone=192 or tone=64, depending on the order of stripes. In other words we have generated a half gray image, but our result image is either 64 tones too dark or too light.

4 Conclusions and Discussion

We demonstrated that synthetic test images could be used to find problematic image features that the tested image processing system is not capable of processing satisfactorily. The genetic algorithm is an optimization method that adapts to the system and "learns" those parameters that generate high fitness function values. In this case the GA learned the natural weakness in the Bayer matrix and was able to generate synthetic test images that cause high tone changes.

In addition, it also learned such image patterns that cause some image components to totally disappear, and also in some cases ghost image artifacts to appear.

We used simulated images and saw that they have the power to find some problems that commonly used standard test images do not discover. As far as we know, no one has reported problems with the Bayer method similar to the ones we found. This can be due to the fact that the method is usually used with natural images. Natural images do not have as large artificial variations, which is probably why it operates well with the subset of natural images. However, we can see that a general method such as this is often used with totally different kinds of images [11, 12]. Testing an image processing system or algorithm with just a few standard test images is not sufficient for assuming that works with all images. We suggest that the system must be tested with an extensive number of test images, both natural and synthetic, before we can fully trust that the system operates sufficiently.

We have also implemented other halftoning methods to the image processing software and found out that other methods seem to have some natural built-in weaknesses as well. The weakness of the commonly used Floyd-Steinberg error diffusion method [14] seems to be its delayed response to the high contrast changes in the borders. Our GA-based test system utilized that problem by generating synthetic images that had many feature borders with large tone changes. In this paper we do not design better halftoning methods. But we believe that testing with synthetic test images is also helpful when designing or improving halftoning methods.

References

1. Morovic, J., Wang, Y.: Influence of test image choice on experimental results. In: Proc. of 11th Color Imaging Conference, Scottsdale, AR, Nov. 3-7 (2003) 143-148.
2. Mantere, T. Alander, J.T.: Testing halftoning methods by images generated by genetic algorithms. Arpakannus 1, FAIS, Espoo, Finland (2001) 39-44.
3. Mantere, T., Alander, J.T.: Testing digital halftoning software by generating test images and filters co-evolutionarily. In: Proc. of SPIE 5267: Intelligent Robots and Computer Vision XXI, Providence, RI, Oct. 27-30 (2003) 257-268
4. Holland, J.: Adaptation in Natural and Artificial Systems. The MIT Press (1992)
5. Darwin, C.: The Origin of Species. Oxford Univ. Press, London (1859)
6. Kang, H.: Digital Color Halftoning. SPIE Press, Bellingham, WA (1999)
7. Lau, D., Ulichney, R., Arce, G.: Blue- and green-noise halftoning models. IEEE Signal Processing Magazine, July (2003) 28-38
8. Bayer, B.: An optimum method for two-level rendition of continuous-tone pictures. In: Proc. of IEEE Int. Conf. on Communications 1, NY, June 11-13 (1973) 11-15
9. Sullivan, J. Ray, L., Miller, R.: Design of minimum visual modulation halftone patterns. IEEE Transactions on Systems, Man, and Cybernetics 21(1), (1991)
10. Fienup J.: Invariant error metrics for image reconstruction. Applied Optics 36(32), (1997) 8352-8357
11. Cho, W., Sachs, E., Patrikalaikis, N., Troxel, D.: A dithering algorithm for local composition control with 3D printing. Computer-Aided Design 35 (2003) 851-867

12. Savazzi, E.: Algorithms for converting photographs of fossils to ink drawings. Computers&Geosciences 22(10), (1996) 1159-1173
13. Addison, P.: Fractals and Chaos. Philadelphia Inst. of Physics Publishing (1997)
14. Floyd, R., Steinberg, L.: An adaptive algorithm for spatial gray-scale. In: Proc. of Social Information Display 17(2), (1976) 75-78

Real-Time Texture Synthesis with Patch Jump Maps

Bin Wang, Jun-Hai Yong, and Jia-Guang Sun

Tsinghua University, Beijing 100084, P. R. China

Abstract. This paper presents a real-time method for synthesizing texture. The algorithm consists of two stages: a preprocess stage and a real-time synthesis stage. In the preprocess stage, jump lists are built for each texture patch in the input sample texture and all the jump lists are stored in patch jump maps. In the synthesis stage, we synthesize the output texture with the aid of stored patch jump maps. Experiments show 200-500 frames of 256×256 high quality textures can be produced within a second.

1 Introduction

Texture synthesis is an important technique in computer graphics and computer vision nowadays. Among lots of texture synthesis algorithms, the methods based on Markov Random Field(MRF) have been intensively investigated recently. Leung and Efros [2] present a nonparametric sampling algorithm which synthesizes the output texture pixel by pixel. A year later, Wei and Levoy [8] significantly accelerate Leung and Efors's algorithm using Gaussian pyramid and tree-structured vector quantization search method. By using the coherence between the neighborhoods of pixels, Ashikhmin [1] presents a method for synthesizing *nature* texture. Unlike Wei and Levoy who search the whole input texture, Ashikhmin only checks several candidate pixels to find the best-fit pixel for current output position. Ashikhmin's algorithm takes only 0.1 seconds to generate a 256×256 texture. It can produce high-quality result for some special textures, such as flower, grass and bark, but is not suitable for other textures, especially textures with regular structures. Zelinka and Garland [10] present the (pixel) jump map method towards real-time texture synthesis. They first find several similar pixels for every pixel in the input texture in the analysis stage, and the algorithm does not calculate the distance of the neighborhood anymore in the synthesis stage. It just randomly chooses a pixel according to the probability in the jump list. The jump map algorithm is very fast, taking only $0.03 - 0.07$ seconds to produce a texture with the size 256×256. However, the behavior of the jump map method is similar to Ashikhmin's algorithm. Although it can generate high quality synthesized results for stochastic sample textures, it meets trouble while synthesizing structured textures.

In this paper, we present a real-time texture synthesis method based on patch jump maps as an extension to the jump map technique. The method has two merits over the old jump map algorithm. First, it can produce much better results for the structured textures than the old one. Second, it takes only 0.002-0.005 seconds to generate a 256 × 256 texture, about 10 times faster than the old one.

2 Patch Jump Maps

Our algorithm fits in the category of patch-based texture synthesis method. Unlike the pixel-based method mentioned in section 1, patch-based methods attempt to copy a selected region from the input texture each time otherwise a pixel. Xu et al. [9] apply cat map chaos transformations to transform a randomly selected patch to a new position and use a simple cross-edge filter to smooth the block boundaries. Efros and Freeman [3] improve Xu's method through an image quilting operation, which tries to find a minimum error boundary cut within the overlap region of two patches. Kwatra el al. [4] present *graphcut* method as an extension to the image quilting algorithm to make the method more flexible. Liang et al. [5] use patch-based sampling strategy to choose the best-fit texture patch from the input sample, paste it into the output texture and perform feather blending in the boundary zones. Wang et al. [7] extend Liang's method to multi-sources 2D directional texture synthesis.

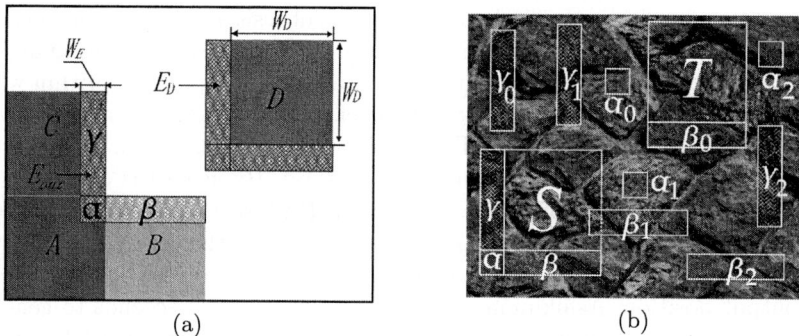

Fig. 1. Illustration of texture analysis

As shown in figure 1(a), suppose A, B and C are three already synthesized patches(size = $W_D \times W_D$) in the output texture. The "L-Shape" zone E_{out}(width = W_E) hatched with white lines is the boundary zone of the already pasted patches and E_D is the boundary zone of candidate patch D(size = $W_D \times W_D$) from sample texture. While pasting D to the target position of the output texture, the two boundary zones would overlap. So a standard patch-based sampling algorithm is to find a patch D in the input texture such that its boundary zone E_D matches E_{out}. That means the distance between E_{out} and E_D should be below a given threshold. L_2 norm [5] is used as the comparison distance metric,

$$d(E_{out}, E_D) = \sqrt{\frac{1}{\Omega} \sum_{i=1}^{A} (p_{out}^i - p_D^i)^2}, \tag{1}$$

where Ω is the number of pixels in the boundary zone, p_{out}^i and p_D^i are the pixel values in the boundary zone.

The boundary comparison for patch-based sampling is a time-consuming operation. Liang et al. use optimized kd-tree, quadtree pyramid and principal components analysis(PCA) to get a good performance. However, we find the algorithm could be much faster if applying the jump map techniques to the patch-based sampling algorithm. Take figure 1 as an example for the constructing of jump map. A strategy for generating the jump map is to use the boundary zone as a whole neighborhood. However, as shown in figure 1(a), E_{out} is composed of part α (the top-right corner of patch A), part β (the top boundary of patch B) and part γ (the right boundary of patch C). So to be more natural, we decompose the boundary zone into three parts and build a jump map for each of them. Suppose patch S is the target patch, and we will build its jumps for the β part of its boundary zone. Let d_{max} denote the distance threshold and T denote another patch in the sample texture. If the distance between the corresponding part of the boundary zones $d_\beta(E_S, E_T) < d_{max}$, then patch T would be considered as a jump of patch S for the boundary part β. Suppose we have found a jump list which contains k jumps for part β of the patch S by searching the whole input texture(Figure 1(b) shows three jumps of them, β_0, β_1 and β_2), then the probability for each jump in the list is calculated as follows:

$$p_i = \frac{\Omega_\beta}{k * \Omega_t}(1 - \frac{d_{\beta_i}}{d_{max}}), \tag{2}$$

where Ω_β is the number of pixels in the β part of the boundary zone, $\Omega_t = \Omega_\alpha + \Omega_\beta + \Omega_\gamma$ and d_{β_i} represents the distance between β_i and β. After all the jump lists are constructed for each patch in the sample texture, we store them as a patch jump map for β part. In a similar way, we build jump maps for α and γ parts, such that three jump maps(called α, β and γ maps) are obtained for an input sample texture.

A brute force search method to construct a jump map usually takes a very long time and is unacceptable. Since the distance metric is actually a vector comparison operation, some existing algorithm could be employed to accelerate the search procedure. We first use PCA method to reduce a high dimension vector[1], and then use an optimized kd-tree to find the matching vectors. With the accelerated algorithm, building a jump map for a 256 × 256 sample texture only takes several minutes.

---

[1] The original dimension is $A \times c$ in our application, where c is the channels of color and it could be reduced to about 25% while retaining 97% original variance.

3 Texture Synthesis with Patch Jump Maps

We have built three patch jump maps for a sample texture and now let's illustrate how to synthesize a texture with the patch jump maps. The algorithm is briefly summarized as follows.

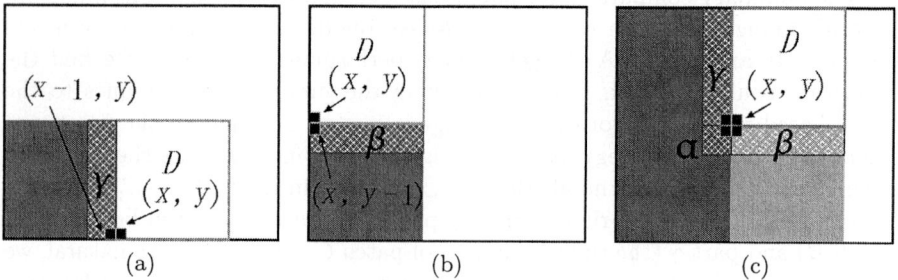

Fig. 2. Illustration of texture synthesis

1. Split the output texture into $n \times n$ patches (patch size = $W_D \times W_D$). Assign every patch a type id (t_{id}). For the patch in the left bottom corner, set its $t_{id} = -1$. Assign 0 to the lowest patches and assign 1 to the left patches. For all other patches, set $t_{id} = 2$.
2. Randomly choose a texture patch from the input sample texture. Paste it to the left bottom corner of the output texture. Record the sample(source) position of every pixel in the output texture.
3. Synthesize the patches with a scan line order, from bottom to top, from left to right.
4. Construct a candidate texture patch set Ψ_D for the current patch to be synthesized. If the patch's id is 0, then only the γ boundary part of already synthesized patch need to be matched(see figure 2(a)). Let (x, y) denote the origin pixel(left bottom corner)'s coordinate of the current patch, the pixel left to it is $(x-1, y)$. Since $(x-1, y)$ has been already synthesized, we suppose its source pixel is (sx, sy). Therefore, the patch with the origin $(sx+1, sy)$ in the input sample texture and its jumps from the γ map form a candidate set Ψ_D for the current patch. Similarly, if the patch's id is 1 and the origin point is (x, y)(see figure 2(b)), then the patch with the origin $(sx, sy+1)$ and its jumps from the β maps form Ψ_D, where (sx, sy) is the source pixel of position $(x, y-1)$. If the patch's id is 2(see figure 2(c)), then Ψ_D are constructed by the jumps from all the three jump maps.
5. Select one texture patch from Ψ_D according to the jump probability. Paste it to the position of current patch and record the sample position of every pixel in the output texture.
6. Repeat steps 3, 4, and 5 until the output texture is fully covered.
7. Perform feather blending (algorithm sees [6]) in the boundary zones.

4 Results and Conclusions

We present a real-time texture synthesis algorithm in this paper by combining the merits of jump map technique and patch-based sampling method. We first generate three different jump maps for a sample texture off-line and then synthesize texture by constructing candidate set Ψ_D from the jump maps dur-

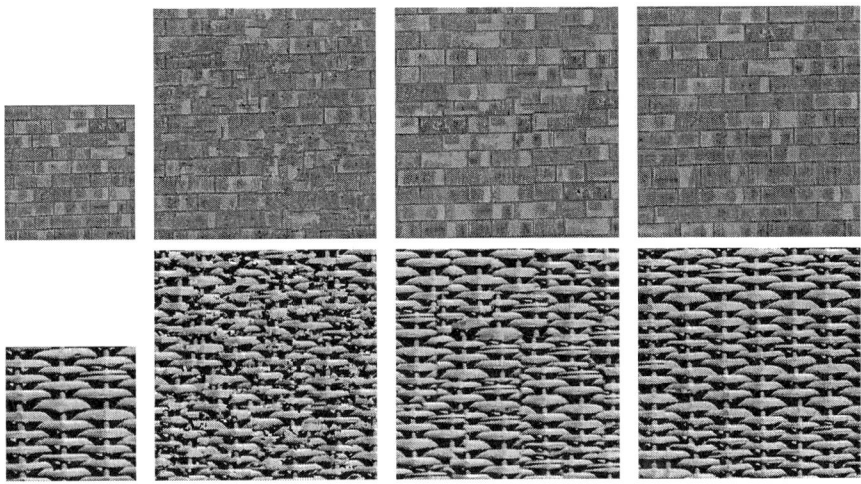

Fig. 3. Comparison among pixel jump map method(left), our method(middle) and patch-based sampling method(right)

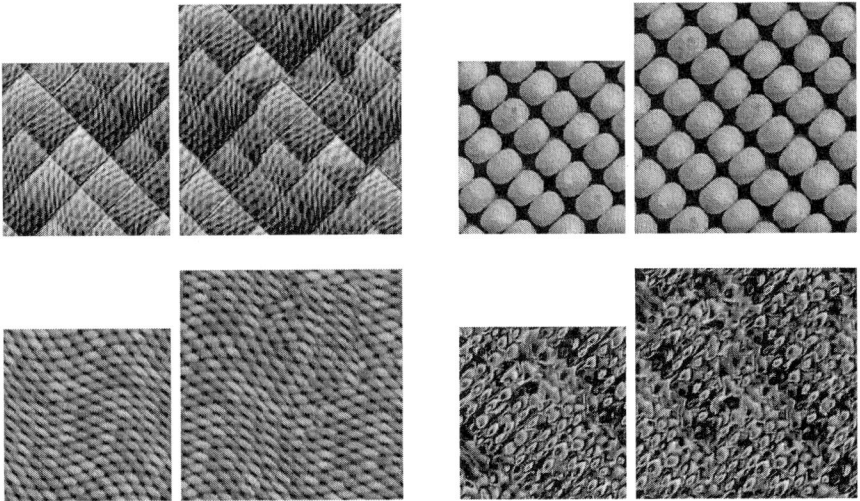

Fig. 4. Some more synthesis examples. In each example, the small(left) image is the input sample texture and the large(right) image is the output texture

ing runtime. The synthesis results of our algorithm are better than the pixel jump map method and are comparable with the patch-based sampling method. Moreover, our algorithm is the fastest one among the three algorithms. It takes 0.002-0.005 seconds[2] to output a 256 × 256 synthesized texture. We also find that if the candidate jumps in each jump list are limited to a fixed number, the performance of the algorithm is non-sensitive to the size of input sample texture.

Figure 3 shows a comparison of our algorithm with pixel jump map method [10] and the patch-based sampling method [5]. We can see that the quality of our results is almost as good as that of the patch-based sampling algorithm and is better than that of the pixel jump map especially for the structured texture. Figure 4 shows some more synthesis results of our algorithm, all the results are 256 × 256, the patch size is 16 × 16 and the width of the boundary zone is set to 4 pixels while building jump maps.

Acknowledgements

The research was supported by Chinese 863 Program (2003AA4Z3110) and 973 Program (2002CB312106). The second author was supported by a project sponsored by SRF for ROCS, SEM (041501004), and FANEDD(200342).

References

1. Ashikhmin M.: Synthesizing natural textures. In Proceedings of the Symposium on Interactive 3D graphics (2001) 217–226
2. Leung T., Efros A.: Texture Syntheis by Non-parametric Sampling. In Proceedings of ICCV'99 (1999) 1033–1038
3. Efros A., Freeman W. T.: Image Quilting for Texture Synthesis and Transfer. In Proceedings of SIGGRAPH'01 (2001) 341–346
4. Kwatra V., Schodl A., Essa I., Turk G., Bobick A.: Graphcut textures:image and video synthesis using graph cuts. ACM Transactions on Graphics **22** (2003) 277–286
5. Liang L., Liu C., Xu Y.-Q., Guo B., Shum H.-Y.: Real-time texture synthesis by patch-based sampling. ACM Transactions on Graphics **20** (2001) 127–150
6. Szeliski R., Shum H.-Y.: Creating Full View Panoramic Mosaics and Environment Maps. In Proceedings of SIGGRAPH'97 (1997) 251–258
7. Wang B., Wang W.-P., Yang H.-P., Sun J.-G.:Efficient example-based painting and synthesis of 2d directional texture. IEEE Transactions on Visualization and Computer Graphics **10** (2004) 266–277
8. Wei L.-Y., Levoy M.: Fast texture synthesis using tree-structured vector quantization. In Proceedings of SIGGRAPH'00 (2000) 479–488
9. Xu Y.-Q., Guo B., Shum H.-Y.: Chaos mosaic: Fast and memory efficient texture synthesis. In *Tech. Rep. 32, Microsoft Research Asia* (2000)
10. Zelinka S., Garland M.: Towards Real-Time Texture Synthesis with the Jump Map. In Proceedings of thirteen eurographics workshop on Rendering (2002)

---

[2] All the experiments data in this paper are obtained on a PC with 2GHz Pentium IV CPU.

Alternation of Levels-of-Detail Construction and Occlusion Culling for Terrain Rendering

Hyung Sik Yoon[1], Moon-Ju Jung[2], and JungHyun Han[3],*

[1] School of Information and Communications Engineering,
Sung Kyun Kwan University, Korea
[2] Samsung SDS, Korea
[3] Department of Computer Science and Engineering, Korea University, Korea

Abstract. Terrain data set is essential for many 3D computer graphics applications. This paper presents a general framework for real-time terrain walk-through application, where the LOD(levels of detail) construction and occlusion culling are alternately selected depending on the terrain geometry. For highly occluded areas, only the occlusion culling is enabled. For less occluded areas, only the LOD rendering is enabled. A small piece of pre-computed information helps the renderer alternate the two methods. The experiment results demonstrate significant savings in rendering time.

1 Introduction

Terrain rendering occupies an important part in many 3D computer graphics applications such as games, and speed-up techniques are crucial because computing resources should also be assigned to other tasks such as 3D character/avatar rendering. The most popular representation for terrain data is uniformly sampled *height field*, a set of height/elevation data sampled in a regular grid. In terrain rendering, the LOD(levels of detail) algorithms are popular that select only a subset of the height field points and produce a coarser mesh from the subset [1] [2] [3] [4] [5] [6]. A well-known speed-up technique in the computer graphics field is *occlusion culling*, which removes the portions of the scene that are occluded and cannot contribute to the final image [7] [8]. Its application to terrain rendering has also been reported [9] [10]. In terrain *walk-through* applications, this paper proposes to alternately select between the LOD construction and occlusion culling depending on the terrain geometry.

2 Hierarchical Structures of Height Field Data

The smallest representable mesh consists of 3 × 3 vertices, called a *block*, as depicted in Figure 1-(a). The *simplification* procedure considers 5 vertices (top,

---

* Corresponding Author.

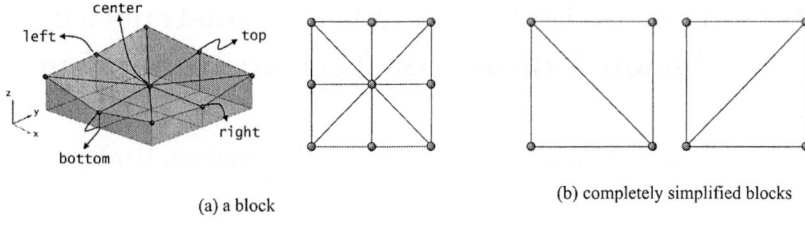

Fig. 1. A quadtree block and its simplification

bottom, left, right and center) for removal. If all of them are removed, we have two possible triangulations in Figure 1-(b).

Figure 2-(a) shows the results of two successive stages of complete simplification. The simplification strategy is compatible with the quadtree structure [11]. Given $(2^k+1) \times (2^k+1)$ vertices, the quadtree will have k levels. A node of the quadtree corresponds to a block.

There can exist 12 triangles in a block, as depicted in Figure 2-(b). A certain simplified block is composed of a subset of the 12 triangles. To describe the status of a block, we use a structure Block, where each triangle in Figure 2-(b) is associated with a 1-bit flag, which indicates its presence/absence.

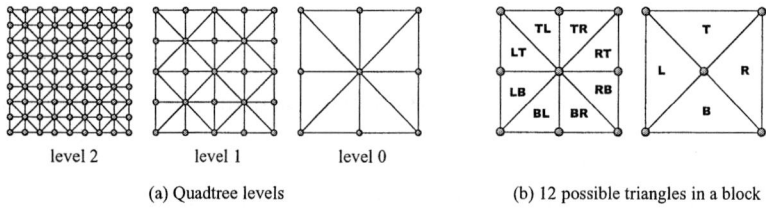

Fig. 2. Quadtree levels and triangle indices

A block is divided into 4 quadrants, and each quadrant is labeled as depicted in Figure 3-(a). Let us see the indexing scheme. The white-vertex block in Figure 3-(a) is positioned inside quadrant 3 ($11_{(2)}$) of the quadtree level 1, and is quadrant 2 ($10_{(2)}$) of level 2. Therefore, its index is $1110_{(2)}=14_{(10)}$. The parent-child relations between blocks are found by shift operations on the indices. A separate array of Block structures is maintained for a quadtree level, and therefore we have k arrays as shown in Figure 3-(b).

3 Occlusion Culling

Occlusion culling requires the terrain data to be rendered in *front-to-back order*. Before a block is rendered, it should be checked if previously drawn blocks occlude the current block. Therefore, those blocks closer to the viewpoint should

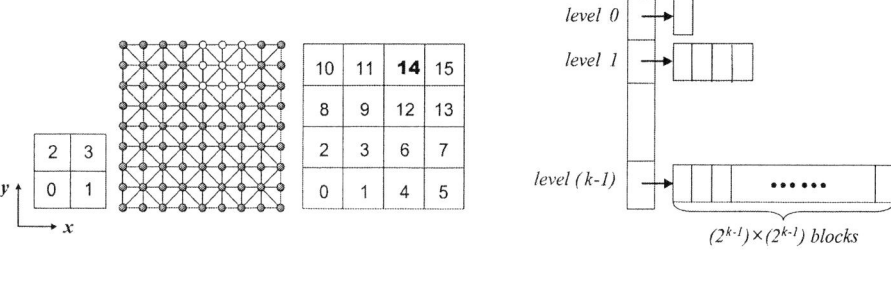

Fig. 3. Block indices and quadtree structure

be rendered first. Figure 4-(a) shows that, given the viewpoint in quadrant 0, the front-to-back ordering is (0,2,1,3). Depending on which quadrant the viewpoint lies in, 4 different orderings are needed. They are stored in a look-up table.

In the quadtree structure, each quadrant is *recursively* traversed and rendered, and the front-to-back ordering should be enforced on all descendants of the quadrant. For example, quadrant 0 in Figure 4-(a) has 4 children as shown in Figure 4-(b), and the correct ordering among the children should be determined. For the purpose, we use the following classifications based on the viewpoint position: the viewpoint's quadrant (quadrant 0 in Figure 4-(a)), the diagonal quadrant (quadrant 3), and the adjacent quadrant (quadrants 1 and 2).

- **The Viewpoint's Quadrant:** The ordering is "retrieved from the look-up table." As shown in Figure 4-(b), the viewpoint is now in quadrant 1, and the look-up table is accessed to retrieve (1,3,0,2).
- **The Diagonal Quadrant:** All descendants of this quadrant simply follow the ordering of their parents, i.e. (0,2,1,3) obtained in Figure 4-(a). Two-level results are illustrated in Figure 4-(d).

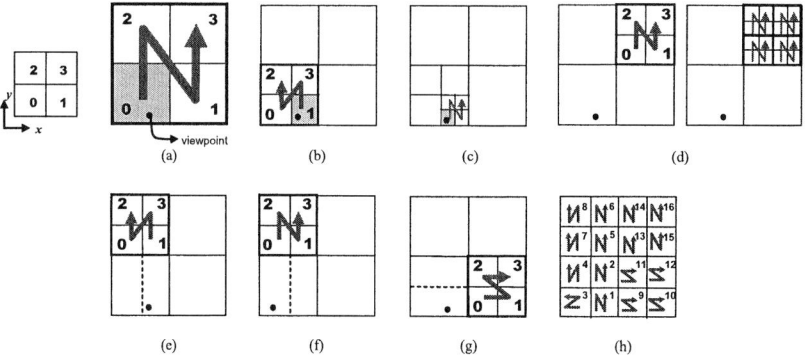

Fig. 4. Front-to-back traversal

– **The Adjacent Quadrant:** Between quadrants 1 and 2 in Figure 4-(a), let us discuss with quadrant 2. The ordering depends on whether the viewpoint is in the right-hand side or in the left-hand side with respect to the quadrant center. As illustrated in Figure 4-(e), the viewpoint is in the right-hand side, and (1,3,0,2) is chosen. If it were in the left-hand side, (0,2,1,3) would be chosen, as shown in Figure 4-(f). (The case of quadrant 1 is shown in Figure 4-(g).) Distinguishing between left- and right-hand sides is done simply by checking the index of the viewpoint's quadrant in the 'same' quadtree level.

The quadtree nodes are rendered in the front-to-back order, and occlusion culling is performed. Occlusion culling is done using *hardware visibility query*. Contemporary graphics hardware supports such visibility query. For a block, its top-most edges of axis-aligned bounding box (AABB) are used for occlusion query. If occlusion query result is '0,' the block is taken as 'occluded.'

4 Rendering Process

One of the most promising LOD algorithms has been reported in [5]. We have implemented their algorithms in the proposed framework. However, suppose a mountain area with high occlusion where only a small subset of (near-side) faces contributes to the final image. Then, the efforts to construct an LOD for the entire area are all in vain. On the other hand, suppose a plain area with few occlusions. In such a case, the cost for occlusion test, which we call *occlusion culling overhead*, significantly degrades the overall rendering performance. Our strategy based on these observations is to choose between LOD construction and occlusion culling such that the chosen method should produce a higher frame rate than the other.

A small piece of pre-computed information, named LOM (**L**OD or **O**occlusion culling guide **M**ap), is generated by pre-rendering the scene twice: one with LOD only, and the other with occlusion culling only. Between the two results, a faster one is selected and recorded into the LOM. The ideal but infeasible method to generate a LOM would be to pre-render the scene at *all* possible positions and directions of the terrain surface. Instead, we have sampled 8 directions (N, E, W, S, NW, NE, SW and SE) 'for every vertex position' of the height field data set. The 8 directions are parallel to the xy-plane, and good for walk-through applications. As we need a 1-bit information for each direction, 1-byte additional information for LOM is needed at each vertex.

LOM-guided rendering is simple. For a walk-through path, the vertex nearest to the current position is retrieved, and the direction (out of 8 sampled directions) is selected which is closest to the viewing direction. If the bit is 0, only LOD construction is enabled. If 1, only occlusion culling is enabled.

5 Test Results and Discussion

We have implemented the proposed algorithms on a PC with an Intel 2.4GHz Pentium 4 CPU, 512MB memory, and two NVIDIA graphics cards: GeforceFX 5900 and GeforceFX 5200. Figure 5-(a) is for a case with an open view. Only LOD is enabled. The top view shows various levels of detail. Figure 5-(b) is for a highly occluded view. Only occlusion culling is enabled. The top view shows only the rendered blocks. The gray area inside the view frustum is taken by the blocks culled out through occlusion culling. Due the *occluding* blocks near the viewpoint (drawn as a white ellipse), most of the blocks which are relatively far from the viewpoint do not have a chance to enter the rendering pipeline.

Figure 6 shows the frame rate graphs for 4 cases of rendering: (1) with the finest-resolution original data (with none of speed-up techniques), (2) with LOD construction only, (3) with occlusion culling only, and (4) with alternation of LOD construction and occlusion culling, guided by the LOM. The results in Figure 6-(a) are obtained using GeforceFX 5900, and those in (b) are using GeforceFX 5200. Note that the LOM-guided alternation takes the higher frame rate between the LOD-only rendering and occlusion-culling-only rendering. Therefore, the stable frame rate of the LOD-only case is almost always guaranteed.

The occlusion-culling-only rendering produces a higher frame rate than the LOD-only case. Note that the relatively higher frame rate is obtained due to

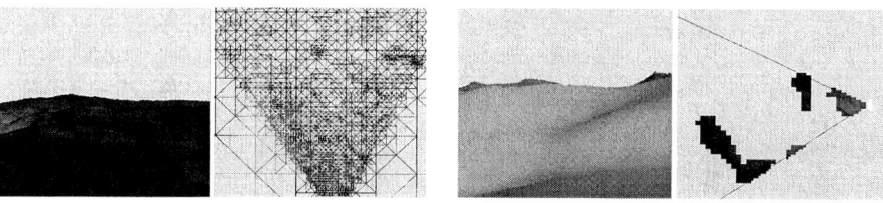

(a) An open view rendered with LOD only   (b) A highly occluded view with occlusion culling only

Fig. 5. Rendering results

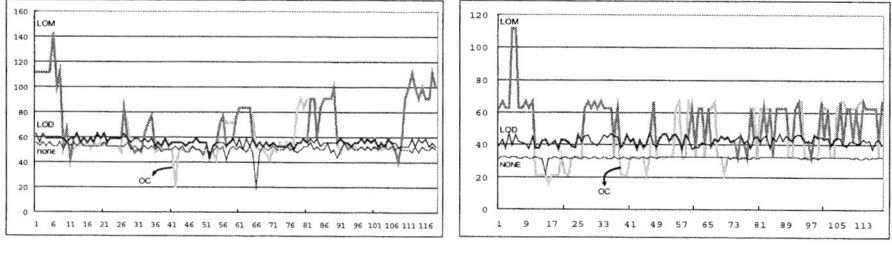

(a) NVIDIA GeforceFX 5900   (b) NVIDIA GeforceFX 5200

Fig. 6. Frame rate graphs

high occlusion, and the occluding triangles are usually at the near-plane side of the view frustum. Then, the *polygonal silhouette* of the terrain is revealed, which is quite annoying. The extra computing power can be used to *smooth* it by using interpolation or subdivision surface techniques. Both of the LOM-guided and LOD-only cases will then show similar frame rates, but the image qualities are quite different.

6 Conclusion

This paper presents a general framework for real-time terrain walk-through application, where the LOD construction and occlusion culling are alternately selected depending on the terrain geometry. The experiment results with current implementations demonstrate significant savings in rendering time. Future research issues include designing more effective LOM structure, incorporating silhouette smoothing techniques into the framework, etc.

Acknowledgements

This work was supported by grant No. 1999-2-515-001-5 from the Basic Research Program of the Korea Science and Engineering Foundation.

References

1. Lindstrom, P., Koller, D., Ribarsky, W., Hodges, L.F., Faust, N., Turner, G.A.: Real-time, continuous level of detail rendering of height fields. In: SIGGRAPH. (1996) 109–118
2. Duchaineau, M.A., Wolinsky, M., Sigeti, D.E., Miller, M.C., Aldrich, C., Mineev-Weinstein, M.B.: ROAMing terrain: real-time optimally adapting meshes. In: IEEE Visualization. (1997) 81–88
3. Rottger, S., Heidrich, W., Slussallek, P.: Real-time generation of continuous levels of detail for height fields. In: Proc. 6th Int. Conf. in Central Europe on Computer Graphics and Visualization. (1998) 315–322
4. Pajarola, R.B.: Large scale terrain visualization using the restricted quadtree triangulation. In: IEEE Visualization. (1998) 19–26
5. Lindstrom, P., Pascucci, V.: Terrain simplification simplified: A general framework for view-dependent out-of-core visualization. IEEE Transactions on Visualization and Computer Graphics **8** (2002) 239–254
6. Pajarola., R.: Overview of quadtree-based terrain triangulation and visualization.(technical report, university of california irvine). (2002)
7. Pantazopoulos, Ioannis; Tzafestas, S.: Occlusion culling algorithms: A comprehensive survey. In: Journal of Intelligent and Robotic Systems: Theory and Applications. Volume 35. (2002) 123–156
8. Daniel Cohen-Or, Yiorgos L. Chrysanthou, C.T.S.F.D.: A survey of visibility for walkthrough applications. In: IEEE Transations on Visualization and Computer Graphics. Volume 9. (2003) 412–431

9. Stewart, A.J.: Hierarchical visibility in terrains. In: Eurographics Rendering Workshop. (1997) 217–228
10. Lloyd, B., Egbert, P.: Horizon occlusion culling for real-time rendering of hierarchical terrains. In: IEEE Visualization. (2002) 403–410
11. Samet, H.: The quadtree and related hierarchical data structures. In: ACM Computing Survey. Volume 16. (1984) 99–108

New Algorithms for Feature Description, Analysis and Recognition of Binary Image Contours

Donggang Yu and Wei Lai

Swinburne University of Technology, PO Box 218, Hawthorn,VIC 3122, Australia

Abstract. In this paper, some new and efficient algorithms are described for feature description, analysis and recognition of contours. One linearization method is introduced . The series of curvature angle, linearity, and bend angle between two neighboring linearized lines are calculated from the starting line to the end line. The series of structural points are described. The useful series of features can be used for shape analysis and recognition of binary contours.

1 Introduction

The description of binary image contour plays an important role for the shape analysis and recognition of image. line segment, curvature angle of lines, the bend angle, and convexity and concavity of the bend angles are useful features to analyze the shape of binary image contour. Many methods and algorithms are developed for the description of contours in the past [1-4]. However, these descriptions cannot form series of sets, or the inter contour of a binary image can not be processed based on these algorithms [3], which make the analysis and understanding of contour shape difficult. Also, no one uses difference code to describe and extract these series of features because there are some spurious contour points. The methods proposed in this paper can make it possible. Some useful structural points are defined to analyze and recognize contour shape. The relevant algorithms are described in Section 2. Finally, a conclusion is given.

2 The Linearization and Description Features

Let the starting point of an binary image be the upper-left corner. Freeman code is used, and the contours are 8-connected.

The chain code set of contour k is represented as:

$$C_k = \{c_0, c_1...c_i, ...c_{n-1}, c_n\} \quad (1)$$

where i is the index of the contour pixels. The difference code, d_i, is defined as:

$$d_i = c_{i+1} - c_i. \quad (2)$$

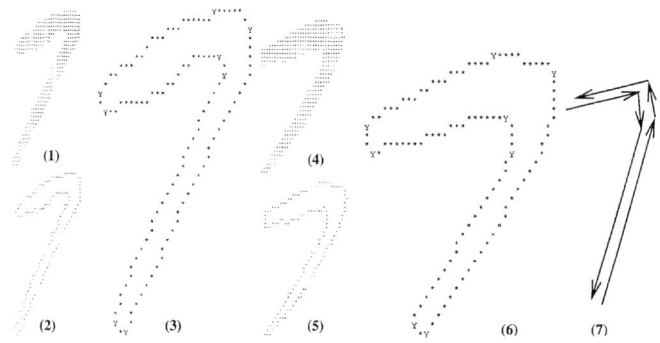

Fig. 1. Original image, contours, smooth following and linearization of two handwritten digits 7

In smoothed contours, $|d_i|$ equals 0 or 1 [4]. Two samples are shown in Figs. 1(3) and 1(6).

2.1 The Curvature Angle of a Linearized Line

The curvature angle is defined as the direction angle between the x coordinate axis and l_{se}, and the angle is formed with starting from the direction of the x coordinate axis to the direction of linearized line, which is determined by the line's element codes, in anti-clock. Let

$$\Delta x = \left| x_k^{ln}[0] - x_k^{ln}[n_k^{ln} - 1] \right| \tag{3}$$

$$\Delta y = \left| y_k^{ln}[0] - y_k^{ln}[n_k^{ln} - 1] \right| \tag{4}$$

in Fig. 2. Let $\angle curve$ be the curvature angle, then it is found as follows (corresponding four quadrants) (see Fig. 2):

Case 1: If $cdir1$ and $cdir2$ are chain codes 0, 1 or 2 (the first quadrant), then $\angle curve = (180°/\pi)\angle se$, where

$$\angle se = tag^{-1}(\frac{\Delta y}{\Delta x}). \tag{5}$$

Case 2: If $cdir1$ and $cdir2$ being chain code 4, 5 or 6 (the third quadrant), then $\angle curve = 180° + (180°/\pi)\angle se$ based on Equation (5) and Fig. 2.

Similarly, the curvature angles can be found in the second and Fourth quadrants.

Case 5: There are eight special cases which are shown in Fig. 3, and $\angle curve$ is found as follows:
(1) If its $cdir1$ is chain code 4, $\angle curve = 180°$ for the case in Fig. 3(1).
(2) If its $cdir1$ is chain code 0, $\angle curve = 360°$ (0°) for the case in Fig. 3(2).
Similarly, the curvature angles of other six cases can be found based on Figs. 3(3-8).

1170   D. Yu and W. Lai

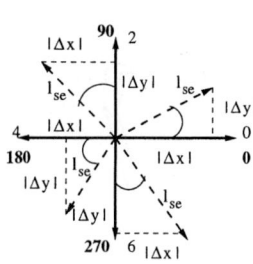

Fig. 2. Finding the curvature angle of linearized lines

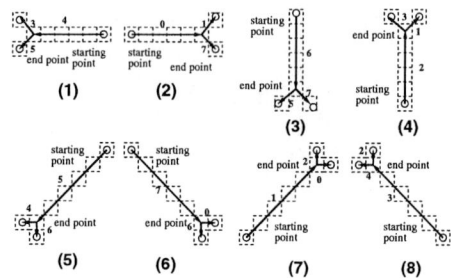

Fig. 3. Eight special cases of the curvature angle

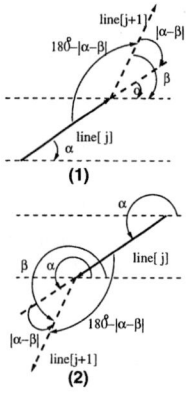

Fig. 4. Finding the bend angle between two neighboring linearized lines

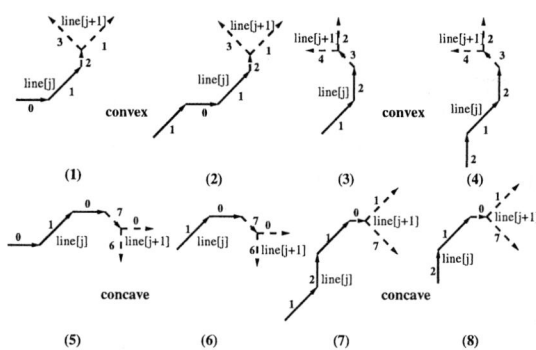

Fig. 5. Detection pattern of the bend angle property

2.2   The Bend Angle and Its Property

The bend angle of linearized lines is defined as the angle between the line j and the line $[j+1]$. It can be calculated based on Fig. 4.

a. The Bend Angle of Linearized Lines

Let $\angle curve[j]$ and $\angle curve[j+1]$ be the curvature angle of lines j and $[j+1]$ respectively, and $\Delta[j, j+1]$ be their curvature angle difference, then

$$\Delta[j, j+1] = \angle curve[j] - \angle curve[j+1], \qquad (6)$$

where $\angle curve[j]$ is α, $\angle curve[j+1]$ is β, and $\Delta[j, j+1]$ is $|\alpha - \beta|$ in Fig. 4 respectively. Let $\angle angle[j]$ be the bend angle between lines j and $[j+1]$, then it can be found based on following equation (see Fig. 4):

$$\angle angle[j] = 180° - |\Delta[j, j+1]|. \qquad (7)$$

b. The Property of Bend Angle of Linearized Lines

Let $l_{cdir1}[j]$ and $l_{cdir2}[j]$ be the first and second element codes of the line j, and $l_{cdir1}[j+1]$ and $l_{cdir2}[j+1]$ be the first and second element codes of the line $[j+1]$ respectively. Thirty two detection patterns of the bend angle property (convex or concave) can be found, here only eight patterns are shown in Fig. 5. One detection rule can be described as follows:

If $l_{cdir1}[j]$ is code 0 and $l_{cdir2}[j]$ is code 1 (see Figs. 5(1) and 5(5)), or $l_{cdir1}[j]$ is chain code 1 and $l_{cdir2}[j]$ is chain code 0 (see Figs. 5(2) and 5(6)), then

- If $l_{cdir1}[j+1]$ is chain code 2 or $l_{cdir2}[j+1]$ is chain code 2, then $\angle angle[j]$ is convex (see Figs. 5(1-2)).
- If $l_{cdir1}[j+1]$ is chain code 7 or $l_{cdir2}[j+1]$ is chain code 7, then $\angle angle[j]$ is concave (see Figs. 5(5-6)).

Similarly, other rules can be found based on other twenty eight patterns.

Based on the above algorithms, series of description features of two sample images (see Figs. 1(3) and 1(6)) can be found from starting point to end point, and shown in Table 1. These series of features can be constructed as a feature vector. For example, the vector of a series of curvature and bend angles of the contour in Fig. 1(3) is:

$200°\_110° \rightarrow 270°\_74° \rightarrow 15°\_119° \rightarrow 315°\_96° \rightarrow 231°\_96° \rightarrow 315°\_81° \rightarrow 53°\_71° \rightarrow 161°\_140°$ (see Table 1(1)).

For Fig. 1(6), the similar vector can be found. These feature vectors can describe the series of directions (concave or convex), and it is shown in Fig. 1(7). Also, the feature vectors of sample images in Figs. 7-9 can be found.

2.3 Structural Points of Smoothed Contours

The structural points are some special points which can be used to represent convex or concave change in the direction of chain codes between two neighboring lines along the contour. Their definition and detection are based on the structure patterns of element codes of two lines. Assume that $line[ln]$ is the current line and that $line[ln-1]$ is the previous line.

Definition 1. *The convex point in the direction of code 4 (represented with the character "∧").*

If the element codes 3, 4 and 5 occur successively as a group of neighborhood linearized lines, then one convex point can be found as follows:
if $cdir1$ of $line[ln]$ is code 4, $cdir2$ is code 5 and the direction chain code of the last pixel of $line[ln-1]$ is code 3, then the first pixel of the current line $line[ln]$ is a convex point which is represented with "∧".

Definition 2. *The concave point in the direction of code 4 (represented with the character "m").*

If the element codes 5, 4 and 3 occur successively as a group of neighborhood linearized lines, then one concave point can be found as follows:

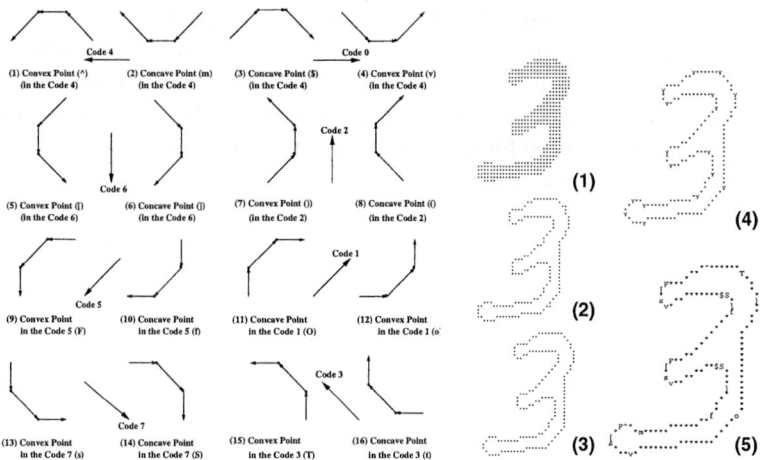

Fig. 6. Structural patterns of structural points

Fig. 7. Example of one handwritten digit 3

if $cdir1$ of $line[ln]$ is code 4, $cdir2$ is code 3 and the direction chain code of the last pixel of $line[ln-1]$ is code 5, then the first pixel of the current line, $line[ln]$, is a concave point which is represented with "m".

Similar to Definitions 1-2, other structural points can be defined and found. These points are convex points "v", "[", ")", "F", "o", "T", "s", and concave points "$", "]", "(", "f", "O", "t" and "S" which are shown in Fig. 6 respectively. These structural points describe the convex or concave change in different chain code directions along the contour, and they can therefore be used to represent the morphological structure of contour regions.

For the outer contour in Fig. 7(5) there is a series of structural points:

"∧" → "F" → "[" → "s" → "v" (convex) → "$" → "S" → "]" → "f" (concave) → "F" → "[" → "s" → "v" (convex) → "$" → "S" → "]" → "f" → "m"(concave) "∧" → "F" → "[" → "s" → "v" → "o" → ")" → "T" → "∧" (convex).

Each convex or concave change consists of a group of convex or concave structural points respectively. For example, the first convex change of the above series consists of convex structural points "∧", "F", "[", "s", and "v".

For the outer contour in Fig. 8(5), the series of structural points is:

"∧" → "F" → "[" → "s" (convex) → "S" → "]" → "f" (concave) → "F" → "[" → "s" (convex) → "S" → "]" → "f" → "m"(concave) "∧" → "F" → "[" → "s" → "v" → "o" → ")" (convex) → "(" (concave) → "T" → "∧" (convex).

There are similar concave changes which contain a group of points "S" and "]" (concave morphological change in chain codes 7 and 6) in both above vectors of structural points. If recognized object images are digit based on prior infor-

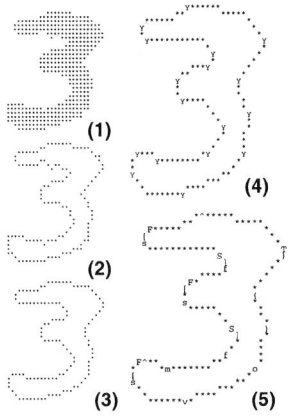
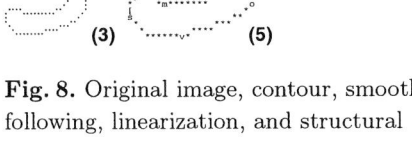

Fig. 8. Original image, contour, smooth following, linearization, and structural points of another handwritten digit 3

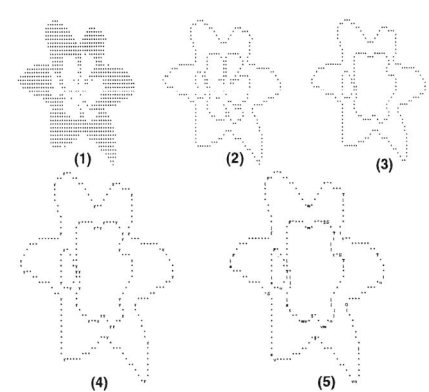

Fig. 9. Example of one lily flower image

mation, then two images are recognized as handwritten digit 3. This is because there is such a morphological structure pattern (two group of points "S" and "]") on the smoothed contours of all type of digits 3 (both print and handwritten digits 3).

For the outer contour in Fig. 9(5), the series of structural points is:

"∧" → "F" → "[" (convex) → "]" → "f" (concave) → "F" → "[" → "s" (convex) → "S" (concave) → "s" → "v" (convex) → "$" (concave) → "v" → "o" → ")" (convex) → "(" → "O" (concave) → "o" → ")" → "T" (convex) → "t" → "(" (concave) → ")" → "T" → "∧" (convex) → "m" (concave) → "∧" (convex).

It is clear, the outer contour is six angles because of six pairs of convex and concave change. If flower images are recognized, it can be recognized as lily flower. For most sorts of lily flower there are six petals which are constructed by sixangles.

Also, the series set of linearized lines, curvature and bend angles give the detail of contour analysis for each convex or concave change.

3 Conclusion

An efficient and new method has been developed to linearize the smoothed contours and to find the series of structural features of linearized lines of contours based on the structure analysis of difference chain code. All features, linearity, curvature and bend angles, and structural points can be use to analyze contours, and to recognize contour shape. Compare our algorithm with other methods [1-4], the best useful contribution is that some series of structural features of linearized lines of contours are found based on our algorithm, but other methods not. Also, these ordered series of structural features (the input of a processing system) make shape analysis and recognition of contours possible. These algorithms have been used in the recognition of document and GIS images.

Acknowledgement. This work is supported by the Australia Research Council SPIRT grant (C00107573).

References

1. Moktarian, F., Mackworth, A. K. A Theory of Multiscale Curvature-Based Shape Representation for Planer Curvature Angles. IEEE Trans. Pattern Analysis Mach. Intell. **14** (8), (1992) 789–805
2. Fu, A. M. N., Yan, H., Huang K. A Curvature Angle Bend Function Based Method to Characterize Contour Shapes. Patt. Recog. **30** (10), (1997) 1661–1671
3. Sonka, M., Hlavac, V. and Boyle R., "Image Processing, Analysis and Machine Vision," Chapman & Hall Computing, Cambridge, 1993.
4. Yu, D., Yan, H. An efficient algorithm for smoothing binary image contours. Pro. of ICPR'96. **2** , (1996) 403–4

A Brushlet-Based Feature Set Applied to Texture Classification

Tan Shan, Xiangrong Zhang, and Licheng Jiao

National Key Lab for Radar Signal Processing and Institute of Intelligent,
Information Processing, Xidian University, Xi'an, 710071, China
tanshan5989@yahoo.com.cn

Abstract. The energy measures of Brushlet coefficients are proposed as features for texture classification, the performance of which to texture classification is investigated through experiments on Brodatz textures. Results indicate that the high classification accuracy can be achieved, which outperforms widely used classification methods based on wavelet.

1 Introduction

The analysis of texture image plays an important role in image processing. Much work has been done to develop proper representations that are effective to texture analysis during the last several decades [1][2][3]. Recently, two spatial-frequency techniques were introduced, namely, Gabor filters [4] and wavelet transforms [5][6], both of which have achieved considerable success in texture classification. Especially, the texture image analysis methods based on wavelet have received more and more attention in that the energy measures of the channels of the wavelet decomposition were found to be very effective as features for texture analysis.

It is well known that the orientation is an important characteristic of texture. Unfortunately, the separable 2-D wavelet transform provides only few orientations other than horizontal, vertical and diagonal ones. Many researchers have been dedicating to the problem, and hope to develop certain new mathematical tool which can provides more orientation information of image than wavelet does [8][9]. In paper [10], the author introduced a new system called brushlet, which is a new kind of analysis tool for directional image. And the ability of brushlet to analyze and describe textural patterns was well demonstrated by compressing richly textured images efficiently in paper [10].

In this paper, we show how the brushlet provide efficient features for texture image classification.

This paper is organized as follows. First, the brushlet transform is discussed briefly. And in section 3, energy measure of brushlet coefficients as texture feature is presented. Then, in the section 4, the effectiveness of brushlet energy measure is investigated experimentally. Finally, the conclusions are drawn in section 5.

2 Brushlet Transform

We will briefly explain the construction of brushlet. A detailed exposition on brushlet may be found in [10].

Brushlet is constructed in the Fourier domain by expanding the Fourier transform of a function into orthonormal basis. Consider a cover $R = \bigcup_{n=-\infty}^{n=+\infty}[a_n, a_{n+1})$, and write $l_n = a_{n+1} - a_n$, $c_n = (a_{n+1} - a_n)/2$. Let r be a ramp function such as

$$r(t) = \begin{cases} 0 & \text{if } t \leq -1 \\ 1 & \text{if } t \geq 1 \end{cases}. \tag{1}$$

And $r^2(t) + r^2(-t) = 1$, $\forall t \in \Re$. Let v is the bump function supported on $[-\varepsilon, \varepsilon]$

$$v(t) = r(\frac{t}{\varepsilon})r(-\frac{t}{\varepsilon}). \tag{2}$$

And b_n is the windowing function supported on $[-\frac{l_n}{2} - \varepsilon, \frac{l_n}{2} + \varepsilon]$,

$$b_n(t) = \begin{cases} r^2(\dfrac{t + \dfrac{l_n}{2}}{2}) & \text{if } t \in [-\dfrac{l_n}{2} - \varepsilon, -\dfrac{l_n}{2} + \varepsilon) \\ 1 & \text{if } t \in [-\dfrac{l_n}{2} + \varepsilon, \dfrac{l_n}{2} - \varepsilon) \\ r^2(\dfrac{-t + \dfrac{l_n}{2}}{2}) & \text{if } t \in [\dfrac{l_n}{2} - \varepsilon, \dfrac{l_n}{2} + \varepsilon) \end{cases}. \tag{3}$$

Write $e_{j,n} = \frac{1}{\sqrt{l_n}} \exp(-2i\pi j \frac{x - a_n}{l_n})$. Then, for $\forall j, n \in Z$, the collection

$$u_{j,n}(t) = b_n(x - c_n)e_{j,n}(x) + v(x - a_n)e_{j,n}(2a_n - x) - v(x - a_{n+1})e_{j,n}(2a_{n+1} - x). \tag{4}$$

is an orthonormal basis for $L^2(R)$.

Let $f \in L^2(R)$. Expanding $\hat{f}$, which is the Fourier transform of f, into the basis $u_{j,n}$, we have $\hat{f} = \sum \hat{f}_{n,j} u_{n,j}$. Then we take the inverse Fourier transform

$$f = \sum \hat{f}_{n,j} w_{n,j}. \tag{5}$$

where $w_{n,j}$ is the inverse Fourier transform of $u_{n,j}$. Obviously, the collection $\{w_{n,j}, n, j \in Z\}$ is an orthonormal basis for $f \in L^2(R)$. We call $w_{n,j}$ brushlet.

Brushlet can be extended to two dimensions case through separable tensor products. Let $\bigcup_{j=-\infty}^{j=+\infty}[x_j, x_{j+1})$ and $\bigcup_{k=-\infty}^{k=+\infty}[y_k, y_{k+1})$ denote two partitions of R, write $h_j = x_{j+1} - x_j$ and $l_k = y_{k+1} - y_k$, then the sequence $w_{m,j} \otimes w_{n,k}$ is an orthonormal basis for $L^2(R^2)$ too, namely, the brushlet in two dimensions.

3 Energy Measures of Brushlet Coefficients

Brushlet transform is of multilevel-type structure as wavelet packet, which can be performed using finer and finer tiling in Fourier plane. The brushlet $w_{m,j} \otimes w_{n,k}$ can extract effectively orientation information of texture. In fact, $w_{m,j} \otimes w_{n,k}$ is an oriented pattern oscillating with the frequency $((x_j + x_{j+1})/2, (y_k + y_{k+1})/2)$ and localized at $(m/h_j, n/l_k)$. The size of the pattern is inversely proportional to the size of the analyzing window, $h_j \times l_k$ in the Fourier space. For the one-level extension, the Fourier plane is divided into four quadrants denoted by Q_{1i} ($i = 1, 2, 3, 4$) first. Then, each one of four quadrants is expanded into four sets of brushlets having the parameter set as follows: $x_0 = -128$, $x_1 = 0$, $x_2 = 127$ and similarly $y_0 = -128$, $y_1 = 0$, $y_2 = 127$ (for the case image is of size 256×256). The four sets of brushlets have the orientation $(\pi/4) + k(\pi/2)$, $k = 0, 1, 2, 3$, as shown in Fig.1. (a). Since the input image is real, the coefficients are antisymmetric with respect to the origin. When expanding function in $L^2(R^2)$ into two-level brushlet, each quadrant of Q_{1i} ($i = 1, 2, 3, 4$), created in the one-level decomposing, is divided into four quadrants again. So, sixteen quadrants are created, denoted by Q_{2i}, $i = 1, 2...16$, then the brushlet coefficients in each Q_{2i} is expanded into orthonormal basis $u_{m,j} \otimes u_{n,k}$. The sixteen sets of brushlet have twelve different orientations as shown in Fig.1. (b). The orientations $(\pi/4) + k(\pi/2)$, $k = 0, 1, 2, 3$ are associated with two different orientations. Note that the sixteen quadrants of brushlet coefficients are antisymmetric with respect to the origin again, as the same as one-level decomposing of brushlet.

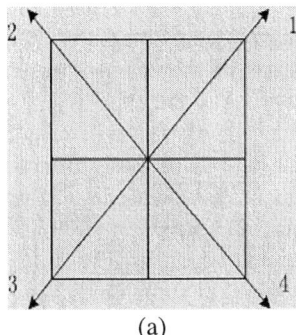 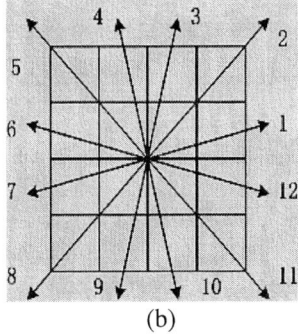

(a) (b)

Fig. 1. (a) Orientation of one-level brushlet extension (b) Orientation of two-level brushlet extension

Since the 2-D brushlet is orthonormal basis in $L^2(R^2)$, obviously, for $\forall f \in L^2(R^2)$ we have the Parseval relation

$$\|f\|_2^2 = \sum_{n,j} \|\hat{f}_{n,j}\|_2^2 . \tag{6}$$

where $\hat{f}_{n,j}$ is the brushlet coefficients being the same as in (5).

We define an energy measure as $E_{l,i} = \sum_{k_1,k_2} |\hat{f}_{n,j}(k_1,k_2)|^2$, which is used to represent the energy of the ith quadrant of the lth level decomposing, where k_1, k_2 is the index of brushlet coefficient in each quadrant. Therefore, we have

$$E = \|f\|_2^2 = \sum_i E_{l,i} \quad for \; \forall l . \tag{7}$$

Our strategy is to compute the energy measures associated with each quadrant first. The energy pattern distributed in brushlet domain should provide unique information and support a representation for texture classification. Thus, a features vector for texture classification consists of a set of energy values. Due to the antisymmetric of the brushlet coefficients, we only compute the energy measure of upper half of Fourier plane. Consequently, in the case of one-level decomposing, the energy feature vector of dimension 2 is obtained; in the case of two-level decomposing, a feature vector of dimension 8 is obtained; and in the case of l level decomposing the dimension of the feature vector is $2 \times 4^{l-1}$. Exactly as the techniques based on wavelet, one can also define different energy measure as the texture feature, for example, the Norm1 Energy Measure, $E = \frac{1}{N}\sum_{k=1}^{N}|C_K|$.

4 Experimental Results

Experiments are carried out to test the performance of energy measures of brushlet for texture classification. Focusing on the effectiveness of the energy measure itself as texture features, the simple classifier, K-NN is used. And the experiments are carried out on test data set from the Brodatz album, which consists of 112 natural textures. Each texture has been digitized, and was stored as a 640×640, 8 bit/pixel digital image. In our experiment, each of the selected texture is divided into 25 nonoverlapping subsamples of size 128×128, 10 for training and 15 for test. Note many of the textures in the Brodatz album are not homogeneous, which usually contaminate the performance of classification algorithm seriously [11]. In some sense, the existing of these textures makes the comparison between different algorithms insignificant. Some examples of such textures are shown in Fig.2.

By removing 34 inhomogeneous texture images from the whole album, a test data set of 78 textures is created. And the whole test data set is of 1950 subsamples. The 34 removed texture is listed as follows: D005, D007 D013, D030, D031), D036, D038, D040, D042, D043, D044, D045, D054, D058, D059, D061, D063, D069, D079, D080, D088, D089, D090, D091, D094, D096, D097, D098, D099, D100, D102, D103, D106, D108, D110.

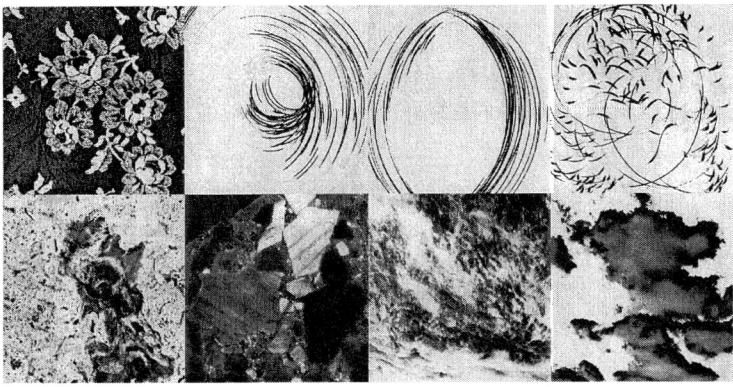

Fig. 2. Examples of inhomogeneous textures in Brodatz album Row 1: D042, D043, D044, D045, Row2: D058, D059, D090, D091

Table 1. 78 textures classification accuracy with Norm1 Energy Measure based on different decomposing method

Brushelt based energy measures	DWT based energy measures	UDWT based energy measures
93.33%	90.85%	92.48%

For comparison, the energy measures based on decimated and undecimated wavelet transform (DWT and UDWT) with three-level decomposing are used for texture classification too. And the wavelet energy measures with Norm1 Energy Measure is extracted from the channels of the wavelet decomposition, as was widely used in the literature for texture analysis [5][6][7]. The Daubechies-6 filter is used for both DWT and UDWT. The results of classification are summarized in Table.1. The experimental results show that the performance of energy measures of brushlet outperforms that of the other two substantially.

5 Conclusion

The performance of brushlet for texture image classification is investigated in this paper. And we have described how the energy measures of brushlet provide efficient features for texture image classification. And the high performance of the new texture feature results from more orientations provided by brushlet coefficients. As has demonstrated by experiments, the new textual features work better than that based on wavelet energy measure. Undoubtedly, Brushlet provides a new mathematic tool for image processing. Other potential application includes texture segment, object identification and image fusion etc.

References

1. Robert, M.H., Shanmugan, K., Dinstein, I.H.: Texture Feature for Image Classification. IEEE Trans. on SMC. 6 (1973) 610–621
2. Chellappa, R., Chatterjee, S.: Classification of Textures Using Gaussian Markov Random Fields. IEEE Trans. on ASSP. 4 (1985) 959–963
3. Mao, J., Jain, A.K.: Texture Classification and Segmentation Using Multiresolution Simultaneous Autoregressive Models. Patt. Recog.. 2 (1992) 173–188
4. Teuner, A., Pichler, O., Hosticka, B.J.: Unsupervised Texture Segmentation of Images Using Tuned Matched Gabor Filters. IEEE Trans. on Image Processing. 4 (1995) 1549–1560
5. Chang, T., Kuo, C.C.J.: Texture Analysis and Classification with Tree-structured Wavelet Transform. IEEE Trans. on Image Processing. 2 (1993) 429–441
6. Unser, M.: Texture Classification and Segmentation Using Wavelet Frames. IEEE Trans. on Image Processing. 4 (1995) 1549–1560
7. Laine, A., Fan, J.: Texture Classification by Wavelet Packet Signatures. IEEE Trans. on Pattern Anal. Machine Intell.. 15 (1993) 1186–1191
8. Donoho, D.L.: Orthonormal Ridgelets and Linear Singularities. SIAM J. Math Anal. 31 (2000) 1062–1099
9. Pennec, E.L., Mallat, S.: Image Compression With Geometrical Wavelets. IEEE Int. Conf. Image Processing. 1 (2000) 661–664
10. Meyer, F.G., Coifman, R.R.: Brushlets : A Tool for Directional Image Analysis and Image Compression. Applied and Computational Harmonic Analysis. 4 (1997) 147–187
11. Haley, G.M., Manjunath, B.S.: Rotation-Invariant Texture Classification Using A Complete Space-Frequency Model. IEEE Trans. on Image Processing. 8 (1999) 255–269

An Image Analysis System for Tongue Diagnosis in Traditional Chinese Medicine

Yonggang Wang, Yue Zhou, Jie Yang, and Qing Xu

Institute of Image Processing & Pattern Recognition,
Shanghai Jiaotong University, Shanghai, China, 200030
{yonggangwang, zhouyue, jieyang, xuqing}@sjtu.edu.cn

Abstract. We introduced a computer-aider tongue examination system, which can reduce the large variation between the diagnosis results of different doctors and quantize the tongue properties automatically in traditional Chinese medicine (TCM) diagnosis. Several key issues and algorithms in this system are discussed, involving some newly proposed image analysis techniques: i) a new tongue color calibration scheme is proposed to overcome the drawbacks of the previous method; ii) a gradient vector flow (GVF) snake based model integrating the chromatic information of the tongue image is used to extract the tongue body; iii) an unsupervised segmentation method of color-texture regions is adopted to quantitatively analyze the distribution of the substance and the coat on the tongue surface. The experimental results show that the presented system has a great deal of potential for the tongue diagnosis.

1 Introduction

In traditional Chinese medicine (TCM), the examination of tongue is one of the most important approaches to retrieving significant physiological information on human body. TCM doctors use information of the tongue colors, the distribution of the coat, the degree of wetness and the shape of the patient's tongue to determine his syndrome and body condition. However, the process of tongue diagnosis, highly relying on the subjective experience and knowledge of the doctors or experts, has impeded the development of TCM to some extent. Moreover, most of the precious experience and cases in traditional tongue diagnosis could not be retained quantitatively. Therefore, it is necessary to build an objective diagnosis system for the examination of tongue.

Recently, several tongue diagnosis systems [1–3] based on image analysis have been developed for the purpose of the diagnosis and treatment of patients, providing a new research approach for tongue diagnosis characterization. However, these systems do not realize completely automatic analysis of the tongue, i.e. need some tiresomely interactive operations. Moreover, their accuracy in recognizing substance and coat colors in the tongue is usually limited. To overcome these drawbacks, we develop a new automatic tongue diagnosis system based on several newly proposed image analysis techniques in this paper.

The paper is organized as follows. Section 2 introduces the architecture of our system. Section 3 discusses the image analysis algorithms involved in this system. Section 4 tests the total system and Section 5 contains the conclusions.

2 System Implementation

The system we designed consists of image acquisition part and software part. The image acquisition part provides a stable and consistent sampling environment, which has several main components as follows: light source, color CCD camera, computer and dark chest with face supporting structure. A Canon-G5 CCD digital camera is used without distinct color distortion and with a high resolution of 1024×768 pixels. The camera is mounted to the side of the dark chest, opposite to and on the horizontal line with the face supporting device. Four standard light sources with 5000K color temperature are fixed at the front of the chest. As an image analysis unit, a Dell Dimension8200 workstation is adopted.

The software part consists of a tongue image analysis module and a database module. The image analysis module includes various algorithms of tongue image processing, such as color calibration, tongue body extraction, quantitative analysis of the substance and the coat of a tongue, and so on. The database module is established to store and manage tongue images and the patient's information.

3 Tongue Image Analysis

3.1 Color Calibration

The purpose of color calibration here is to keep the consistency and the repeatability of the colors transmitted from camera to monitor. In [4], an online color calibration method integrating color evaluation with colorimetric matching was proposed to calibrate the tongue's colors. In the work, several consistent color patches are designed and mounted inside the dark chest. They will be recorded simultaneously when any patient's tongue is captured. To get the standard data of those patches, several subjects' tongue images are captured and rendered on the monitor together with the patches. The experienced doctors adapt the tristimulus values of the tongue image in order to make the rendered tongue look the same as the real tongue, and thus the patches' colors are also adjusted together with the tongue. The adapted color values of the patches are extracted and considered as the standard data. For a tongue image to be calibrated, a polynomial model will be established between the color values of the patches in this image and the standard data. Finally, this model will be used to correct the current tongue image. Our scheme is based on this work.

In the process of retrieving the standard data, however, the method has the following disadvantages: i) The adapted colors of all patches will lean toward the same orientation when the tongue image is adjusted. The isotropic adaption does not agree with the fact that the patches have different color shifts. ii) Putting the tongue out for a long time will make the subject tired and thereby affect the operation. Accordingly, we adapt the rendered color patches directly with the real patches when collecting the standard data. This improvement leads to nonlinear shifts of the patches' color values and a flexible operation. Parts of color patches which we use are shown in Fig. 1. These patches, designed by the experienced doctors, contain most of the usual colors of the substance and the coat. They will be used in the following process of tongue color recognition too.

Fig. 1. Parts of color patches

3.2 Tongue Body Extraction

In the computerized tongue analysis, extracting the tongue body automatically and correctly out of its background is critical to the further processing. Unfortunately, due to the weak edge and the color similarity between the tongue and the lip, the traditional methods, e.g. region growing and gradient based edge detection, fail to segment the whole tongue body. Recently, active contour or snake models, incorporating a global view of edge detection by assessing continuity and curvature, have been utilized to extract the tongue body [3,5]. In [3], a dual-snake model was proposed to deal with the initialization problem existing in the original snakes. The model was tested through an experiment with 400 tongue images and the correct rate came to 88.89%. Pang et al. [5] suggested a bi-elliptical deformable contour (BEDC) scheme for tongue segmentation, which combined a bi-elliptical deformable template and an active contour model. Due to the nature of the snake models, these methods i) still depend on the subjective interference, ii) neglect the color information in tongue image, and iii) have limited accuracy. In our work, we propose a new model based on chromatic partition and gradient vector flow (GVF) snake. At first, chromatic information is used to differentiate tongue from lip and face skin in a coarse scale. Then we adopt the GVF snake to further find the exact edge of the tongue. The GVF snake is a novel snake model proposed recently by Xu and Prince [6], which can effectively solve the two main problems of the original snakes, i.e. sensitivity to initial contour and poor convergence to boundary concavities.

3.3 Quantitative Analysis of Substance and Coat

In the tongue image, the colors of the tongue substance can be classified into light pink, rose pink, red, purple, etc. while the coat colors can be classified into white, yellow, brown, black, etc. Generally, a tongue body image is divided into five parts as shown in Fig. 2. According to the TCM principles, the condition of the different parts indicates the heath information of their corresponding organs of human body. The purpose of quantitative analysis here is to estimate the distribution of substance (and coat) of each color class in each part of the tongue. Most researches have adopted supervised classifiers, e.g. SVM and supervised

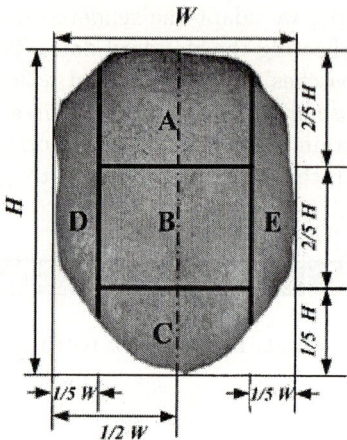

Fig. 2. Five parts of a tongue corresponding to the health condition of different organs

FCM [2, 4], to recognize the colors of the tongue image. These classifiers omit the spatial correlation in the tongue image so as to have poor classification accuracy. Moreover, a great number of samples are required to train the classifiers.

It can be observed that the tongue image is a typical kind of color-texture image and has locally different color distribution. Following this logic, we partition a tongue image into various regions, each of which has homogenous properties such as colors and textures. Then we represent the colors in each region quantitatively so that the distribution of the substance and the coat can be obtained.

JSEG is a new method for unsupervised segmentation of color-texture images presented by Deng and Manjunath [7]. This method involves two steps: color quantization and spatial segmentation. After the first step, the image pixels are represented by their corresponding color class labels. In the second step, a criterion for "good" segmentation is applied to local windows in the class-map, resulting in the so called "J-image". A region growing method is then used to segment the image based on the multiscale J-images.

For each homogenous region obtained through applying the JSEG to the tongue body image, we determine its color class label by virtue of those color patches mentioned in the color calibration section. The minimum distance classifier is used after calculating the statistics between each patch and the region. Finally, we cover the mask given in Fig. 2 on the tongue image and thereby, the color distribution of the substance and the coat will be achieved.

4 Experimental Results

We have developed a prototype implementation of the system and applied it to some samples randomly chosen from our database. The database contains thousands of tongue images. In this section, we will present some useful results.

An Image Analysis System for Tongue Diagnosis   1185

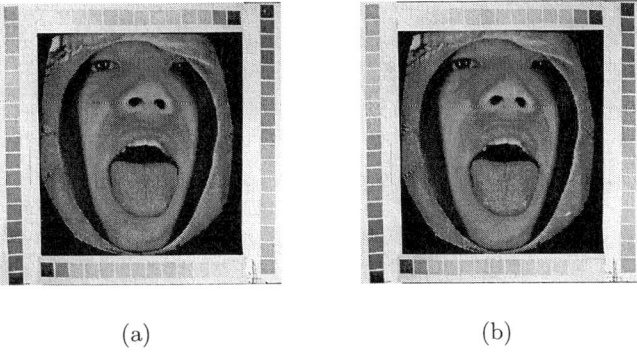

Fig. 3. A color calibration example: (a) an original tongue image and (b) its result

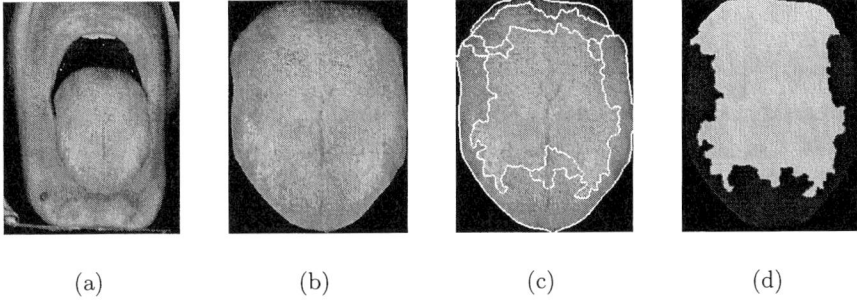

Fig. 4. (a) is an original tongue image; (b), (c) and (d) are the results obtained through tongue body extraction, JSEG-based partition and color recognition of the substance and the coat, respectively

The results through the color calibration method proposed in this paper basically satisfy the TCM doctors who evaluate the system. Fig. 3 is an example of color calibration. Obviously, the original tongue image with a color shift to red has been corrected very well.

In our segmentation experiments, the satisfactory results account for about 93% of the total 1500 tongue images. We also carry out color recognition and quantitative analysis of the substance and the coat with these tongue images. More than 90% recognition results are satisfactory when compared with the results determined by doctors. Fig. 4 shows the results through applying the techniques of tongue extraction, JSEG-based partition and recognition of substance and coat colors, respectively, to a sample image. Note that in Fig. 4, we use pseudocolors to represent different types of substance and coat.

5 Conclusions

We have developed a tongue diagnosis supporting system based on image analysis. The system can provide lots of important information for tongue diagnosis in traditional Chinese medicine. Several key issues in tongue image analysis are described, namely, color calibration, tongue body extraction and quantitative analysis of the substance and the coat. The experimental results show that the system has a great deal of potential for computerized tongue diagnosis. In the next stage, pulse state information will be integrated with the tongue analysis to provide a qualitatively diagnostic result.

References

1. Chiu, C.C.: A novel approach based on computerized image analysis for traditional chinese medical diagnosis of the tongue. Computer Methods and Programs in Biomedicine **61** (2000) 77–89
2. Zhao, Z.X., Wang, A.M., Shen, L.S.: An automatic tongue analyzer of chinese medicine based on color image processing. In: The 4th Int. Conf. on Electronic Measurement and Instruments Conf. Proc., Harbin, China (1999) 830–834
3. Wang, Y.G.: Research on pragmatizing the tongue image analysis instrument of traditional chinese medicine. Master's thesis, Beijing Polytechnic University (2001)
4. Wang, Y.G., Wang, A.M., Shen, L.S.: A study of colour reproduction method of colour image. China Illuminating Engineering Journal **12** (2000) 4–10
5. Pang, B., Wang, K., Zhang, D., et al: On automated tongue image segmentation in chinese medicine. In: IEEE Int. Conf. on Image Processing. Volume I., Rochester NY (2002) 616–619
6. Xu, C., Prince, J.L.: Snakes, shapes, and gradient vector flow. IEEE Trans. on Image Processing **7** (1998) 359–369
7. Deng, Y., Manjunath, B.S.: Unsupervised segmentation of color-texture regions in images and video. IEEE Trans. on Pattern Anal. and Machine Intell. **23** (2001) 800–810

3D Mesh Fairing Based on Lighting and Geometric Conditions for Interactive Smooth Rendering

Seung-Man Kim and Kwan H. Lee

Gwangju Institute of Science and Technology(GIST),
1 Oryong-dong, Puk-gu, Gwangju, 500-712, South Korea
{sman, lee}@kyebek.kjist.ac.kr

Abstract. In this paper, we propose a fairing method of 3D rough meshes based on illuminations and geometric conditions for smooth rendering. In applications of interactive graphics such as virtual reality or augmented reality, rough meshes are widely used for fast optical interactions owing to their simple representation. However, in vertex-based shading, rough meshes may produce non-smooth rendering results; Distinct normal vectors and comparatively longer distance between consecutive vertices increase the difference of radiances. In order to improve the smoothness of the rendering results, the difference of radiances among vertices should be minimized by considering lighting conditions as fairing parameters. We calculated illuminations using diffuse lighting models at each vertex. Then normalized illumination is linearly integrated with the curvedness to prevent the shape distortion. By adapting integrated values to Laplacian weight factors, the difference of radiances is minimized and rendering result is improved, while maintaining the important curved shapes of the rough meshes. The proposed method also improves the compactness of triangles. The comparative study of our method with other existing fairing schemes has been discussed. We also applied our method to arbitrarily simplified meshes for demonstration.

1 Introduction

With the technological advancement of 3D scanners, scanned models are being widely used for several areas of computer graphics and geometric modeling. However the complexity of meshes has increased much faster than the advancement of graphics hardware techniques. Because of hardware limitations, it is necessary to reduce the amount of data by simplifying dense meshes in order to display geometric meshes in real time. Unlike previous studies that consider only geometric conditions, we propose a lighting dependent fairing method for rough meshes, which is constrained by a curvature-based curvedness.

Many fairing algorithms are based on the concept of Laplacian since the storage and computational cost are almost linearly proportional to the number of vertices[1,2,3,4,5]. Also geometric conditions, such as valences, parameters of the transfer function, edge lengths and triangle areas, are used as weight factors to keep the shrinking problems from occurring during mesh fairing. Although the rough meshes, especially simplified ones, are well faired based on geometric conditions, non-smooth rendering may occur, since simplified meshes have different levels of importance determined by lighting conditions in the rendering process. The importance level means the difference of radiances between vertices. The region where radiances have a high variation in the mesh has a higher level of importance. In a vertex-based rendering, the radiance computed at each vertex corresponds to the color of the vertex and inner pixels of the triangle are interpolated from the color of its vertices. High difference of radiances causes undesirable rendering effects to the inside triangles of meshes, especially in highly simplified meshes. Klein[6], Xia[7] proposed illumination dependent refinement method of multiresolution meshes. The method considered only the normal deviation between different levels of detail as refinement metric in the simplification process.

If meshes are faired based on only geometric conditions, triangles in the region of the high importance can be ignored during mesh fairing or vice versa. In addition, simplified meshes have discrete structures and they consist of a small number of triangles, so that they have much higher difference of radiances than dense meshes. Therefore, for smooth rendering with simplified models in real time, we propose a new fairing method that considers lighting conditions as well as geometric conditions.

2 Lighting Dependent Fairing with Geometric Constraints

2.1 Lighting Conditions

Radiances are used to render the objects with two separated components of a specular and a diffuse term. We have chosen the Phong shading model and Lambertian(ideally diffuse) reflection properties due to its simplicity[8]. Equation(1) represents radiances that consider only the diffuse reflection.

$$L_i = \sum_{j=1}^{n} \frac{1}{r_{i,j}^2} \{k_d m_d I_j (\vec{n}_i \bullet \vec{l}_{i,j})\} \quad (1)$$

In equation(1), L_i is the radiance at vertex p_i on the meshes, k_d is the diffuse reflection coefficient, m_d is material properties of the object representing the color value, I_j is the intensity of light sources and n is the number of light sources. And $\vec{n}_i$ represents a normal vector at each vertex, $\vec{l}_{i,j}$ is the unit directional vector from vertex p_i to the light source(j), and $r_{i,j}$ is the distance between a vertex and a light source.

We assume that objects are Lambertian surfaces and a point light source is used for the convenience of visual comparisons, hence the radiance can be defined as below:

$$E_i = \frac{I(\vec{n}_i \bullet \vec{l}_i)}{r_i^2} \quad (2)$$

Here, k_d is set to 1 since surfaces are purely diffuse and m_d is assumed to have a uniform value(here set to 1). From the assumption, radiances(L) are determined by irradiance(E) that is proportional to the light intensity and the inverse of the squared distance(inverse square law). It is also highly affected by the angle of inclined surfaces(cosine's law) in terms of normal vectors and directional lighting vectors.

2.2 Geometric Constraints

While lighting conditions are used as fairing weights, the fairing metric is constrained by the curvedness in order to prevent an undesirable distortion or a loss of important curved shapes. By using the constraint of the curvedness, the movement of vertices in high curvature regions is restricted, since the curvedness represents curved areas in the meshes. In order to use the curvedness $R = \sqrt{(\kappa_1^2 + \kappa_2^2)/2}$ as the constraint of the fairing metric, we should compute the sum of squared principal curvatures κ_1 and κ_2. The curvature might not be defined at vertices properly because meshes are actually comprised of flat triangles, i.e., the mesh is not C^2-differentiable. Therefore we assume that the mesh is a piecewise linear approximation of an unknown continuous surface. The curvatures can be estimated only by using information of the triangular mesh itself[9]. Instead of calculating principal curvatures directly to get the curvedness, we compute the Gaussian curvature $K=\kappa_1\kappa_2$ and the mean curvature $H=(\kappa_1+\kappa_2)/2$ since the curvedness can be expressed by $R = \sqrt{2H^2 - K}$.

The Gaussian curvature of a vertex is related to angles and face areas that are connected to that vertex. The mean curvature H is derived by applying Steiner formula to the triangulated polyhedral surface [10] as shown in Fig. 1.

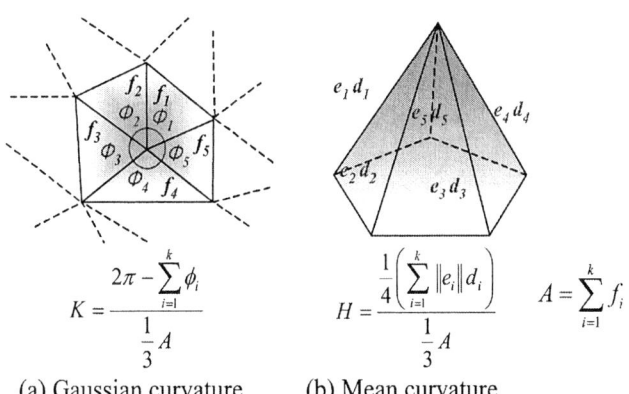

(a) Gaussian curvature   (b) Mean curvature

Fig. 1. Geometric parameters

Here, A is the sum of neighboring triangle's area(f_i) of a vertex, k is the number of neighboring faces, and φ represents the angle at a vertex. $\|e_i\|$ is a length of the edge, and d_i is determined by a dihedral angle which can be calculated at edge e_i using normal vectors of two adjacent triangles.

2.3 Modified Laplacian Fairing Operator

Laplacian fairing techniques are generally based on the membrane energy minimization. For a parametric surface $S:P=P(u,v)$, the membrane energy functional(E_m) is:

$$E_m(S) = \frac{1}{2}\oint_S \left(P_u^2 + P_v^2\right)dS$$

Where, u and v are parameters of parametric surface(S). In minimizing the membrane energy functional, its derivatives correspond to the Laplacian operator $L(P)$: $L(P) = P_{uu} + P_{vv}$. For a discrete mesh, the Laplacian operator $L(P_i)$ at vertex p_i can be linearly approximated[11].

$$L(p_i) = w_{i,j}\sum_{i=0}^{n_i-1}(p_j - p_i)$$

$w_{i,j}$ is general Laplacian weight factors and p_j is neighboring vertices of p_i and n_i is the number of vertices in the mesh.

For each vertex, a lighting dependent weight w_i is defined by the linear combination of the normalized curvedness(R) and irradiance(E) in equation(3).

$$w_i = \mu\frac{E_i}{E_{max}} + (1-\mu)\frac{R_i}{R_{max}}, \quad 0 \leq \mu \leq 1 \qquad (3)$$

E_{max} and R_{max} are the maximum value of E and R respectively. μ is the weight parameter that controls lighting effects and geometric constratins.

Newly defined fairing operator $L(p_i)$ and a faired vertex(p'_i) are in equation(4).

$$L(p_i) = \frac{1}{W_{sum}}\sum_{j=0}^{n-1}w_{i,j}(p_j - p_i), \quad p'_i = p_i + \lambda L(p_i) \qquad (4)$$

where $w_{i,j} = |w_i - w_j|$, $W_{sum} = \sum_{j=0}^{n-1}w_{i,j}$ and $w_{i,j} \geq 0$.

$w_{i,j}$ is the difference of weights at each edge which consists of a vertex and its neighboring vertex, and n is the number of neighboring vertices of p_i. W_{sum} is the summation of all $w_{i,j}$.

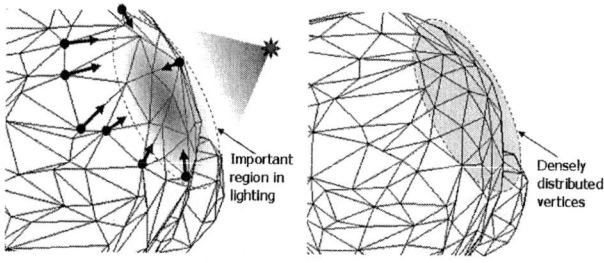

Fig. 2. Schematic diagram of lighting dependent fairing for rough meshes

Fig. 2 shows a schematic diagram of mesh fairing that minimizes the difference of irradiances. The operator $L(p_i)$ gathers the vertices in the direction of the marked arrows shown in the figure into the region where the difference of irradiance is high.

3 Results and Comparisons

We implemented our proposed method using Visual C++ and OpenGL graphics library on Window 2000 platform with Pentium 4 processor(1.5GHz, 512 memory). In order to manage arbitrary mesh data, an efficient data structure is necessary for a fast access to geometric information and connectivities. A half-edge data structure has been widely used for arbitrary meshes [12]. It provides fast, constant-time access to one-ring neighborhood of a vertex. Since the half-edge stores the main connectivity information and always has the same topology; vertices, edges and faces are types of constant size [13]. In the experiment, geometric conditions such as the Gaussian curvature, the mean curvature and the curvedness are computed at compile time in order to reduce a computational load in run time since those parameters are fixed. Whereas, since lighting conditions change dynamically, the computation of lighting conditions and the fairing process are performed at run time as shown in Fig. 3.

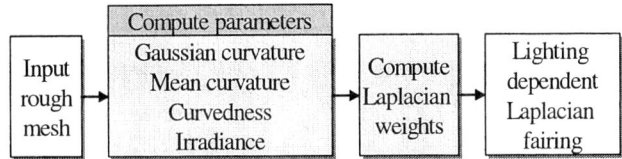

Fig. 3. Flowchart for implementation of lighting dependent mesh fairing

We applied the lighting dependent fairing method to the simplified Stanford bunny model with 800 vertices. Fig. 4 shows fairing results according to the weight parameter(μ). Fig. 4(b) shows fairing weights are determined by the curvedness only, and fig. 4(c) by only radiances. In the experiment, by setting the weight parameter as 0.5 heuristically(fig. 4(d)), we can keep the bunny model from being distorted and appearing small cracks. As a result, the rendering result shows a smoother surface.

In Fig. 5, we set up the weight parameter(μ) and the scaling factor(λ) to 0.5 and 0.7 respectively. The wireframe model of the bunny clearly shows our results with better compactness of triangles, especially in the leg of the bunny. Overall shape is also refined and improved smoothly. In the case of the textured bunny, the mesh is spherically parameterized and mapped by uniform shape of a texture. In the original textured bunny, the textured images are distorted where the triangles of the bad compactness occur. However for the bunny faired by our method, the distortion decreases since the compactness of triangles is refined by minimizing the difference of the radiance.

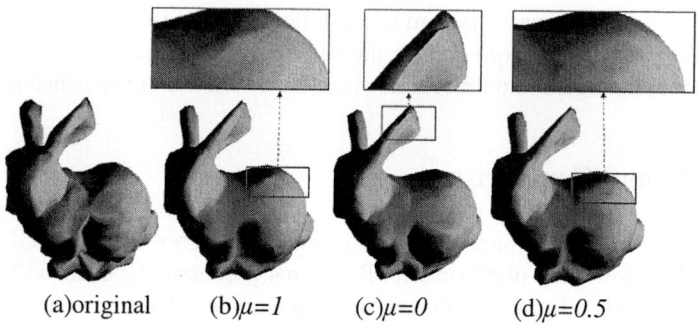

Fig. 4. Fairing results by weight parameters(μ)

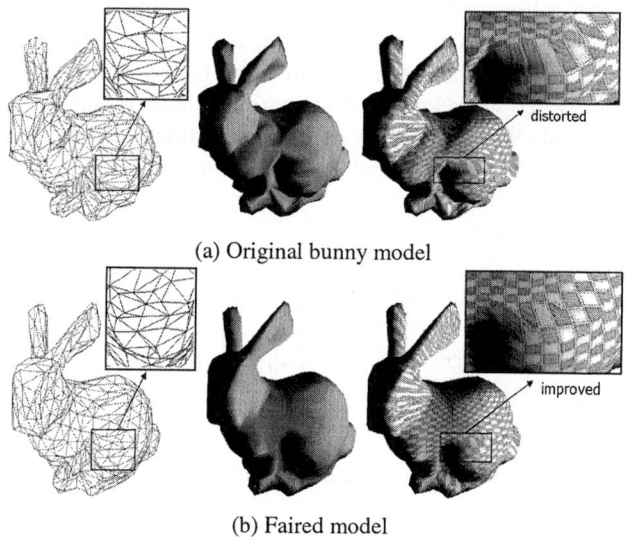

Fig. 5. Fairing results in enhanced rendering quality and parameterization

Fig. 6 shows the computation time for different level of details for the bunny model. The computation time linearly depends on the number of vertices. Based on the graph, the proposed method can be applicable to real time visualization.

We compared our results with different methods(the Laplacian flow, mean curvature flows, and Loop subdivision). Laplacian flow method improves geometric smoothness, but there are problems such as shape shrinkage and loss of details. In order to improve smoothness, our method also prevents from distortion of the shape by using the curvedness as fairing constraints. Fig. 7 shows numerical comparison results by the subdivision sampling scheme [14] that is used to compare the distance between a faired mesh and an original one. The diagonal length of bunny's bounding box is 0.658mm.

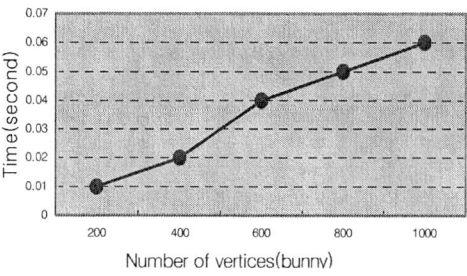
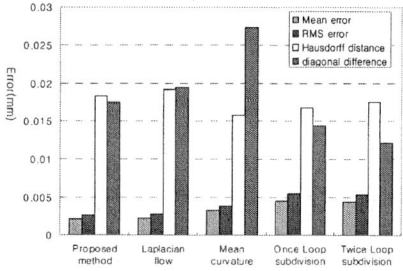

Fig. 6. Computation time   Fig. 7. Numerical comparisons

4 Conclusions

We have proposed a new, lighting dependent fairing method to improve the quality of rendering by enhancing the smoothness of meshes. We use weighted Laplacian to minimize the difference of radiances between consecutive vertices. In addition, the curvedness is used as constraints for retaining curved parts of the meshes. By applying our method to arbitrarily simplified meshes, the rendering result is improved along with triangle's compactness. As a result of advanced compactness, the distortion of textures is reduced. Our method adopts a half-edge data structure and the Laplacian operator that supplies constant time access to one-ring neighborhood vertices. The proposed fairing method can be used for real time applications. The proposed method can be further developed by optimizing the determination of the weight parameter and the scaling. It is also desired to use the estimated lighting conditions from the real world as weight factors for realistic rendering.

References

1. Taubin, G.: A Signal Processing Approach to Fair Surface Design. SIGGRAPH 95 Proceedings (1995) 351-358
2. Kobbelt, L., Campagna, S., Vorsatz, J., Seidel, H.-P.: Interactive Multi-resolution Modeling on Arbitrary Meshes. SIGGRAPH 98 Proceedings (1998) 105-114
3. Fujiwara, K.: Eigenvalues of Laplacians on a Closed Riemannian Manifold and its Nets. Proceedings of the AMS, 123 (1995) 2585-2594
4. Desbrun, M., Meyer, M., Schroder, P., Barr, A.H.: Implicit Fairing or Irregular Meshes Using Diffusion and Curvature flow. SIGGRAPH99 Proceedings (1999) 317-324
5. Guskov, I., Sweldens, W.: Multiresolution Signal Processing for Meshes. SIGGRAPH 99 Proceedings (1999) 324-334
6. Klein, R., Schilling, A., Straßer, W.: Illumination Dependent Refinement of Multiresolution Meshes. In Proceeding of Computer Graphics International, IEEE Computer Society Press (1998) 680-687
7. Xia, J.C., Jihad, E., Varshney, A.: Adaptive Real-Time Level-of-detail-based Rendering for Polygonal Models. IEEE Transactions on Visualization and Computer Graphics, Vol. 3. No. 2 (1997)
8. Moller, T. A., Haines, E.: Real-time Rendering 2nd Edition. A K Peters (2002)

9. Dyn, N., Hormann, K., Kim, S.J., Levin, D.: Optimizing 3D Triangulations using Discrete Curvature Analysis. Mathematical Methods for Curves and Surfaces, Vanderbilt University Press, Nashville (2000) 135-146
10. Lyche, T., Schumaker, L.: Mathematical Methods for Curves and Surfaces. Vanderbilt University Press (2001) 135-146
11. Eck, M., DeRose, T., Duchamp, T., Hoppe, H., Lounsbery, M., Stuetzle, W.: Multiresolution Analysis of Arbitrary Meshes. SIGGRAPH 95 Proceedings (1995) 173-182
12. Lutz, K.: Using Generic Programming for Designing a Data Structure for Polyhedral Surfaces. Proc. 14th Annual ACM Symp. on Computational geometry (1998)
13. Botsch, M., Steinberg, S., Bischoff, S., Kobbelt, L.: OpenMesh-a Generic and Efficient Polygon Mesh Data Structure. OpenSC Symposium (2002)
14. Cignoni, P., Rocchini, C., Scopigno, R.: Metro : Measuring Error on Simplified Surfaces. Computer Graphics Form, Vol. 17. No. 2. (1998) 167-174

Up to Face Extrusion Algorithm for Generating B-Rep Solid

Yu Peng[1], Hui Zhang[2], Jun-Hai Yong[2], and Jia-Guang Sun[1,2]

[1] Department of Computer Science and Technology, Tsinghua University,
Beijing 100084, P. R. China
pengyu00@mails.tsinghua.edu.cn
[2] School of Software, Tsinghua University, Beijing 100084, P. R. China

Abstract. Up to face extrusion (UTFE) is an effective operation to extrude the profile from a sketch plane to a selected face. An up to face extrusion algorithm is presented in this paper. The algorithm first generates the trimmed extrusion through a simple body-surface Boolean operation method, and then generates the resultant solid through a regularized Boolean operation between the original solid and the trimmed extrusion. The algorithm has been implemented in a commercial geometric modeling software system TiGems. Examples are given to illustrate the algorithm.

1 Introduction

Feature-based design emerges as the fundamental design paradigm of geometric model systems. In feature-based geometric modeling system, the user designs with a vocabulary of design elements that are grouped into proto-features based on a planar profile, such as protrusion and cut. When the proto-feature is generated, the resultant solid is built using regularized Boolean operations and the feature operators, i.e., a protrusion is the regularized union, and a cut is the regularized subtraction.

Let S be a closed profile in the sketch plane P. S consists of a set of closed planar curves defining the interior and exterior. The extrusion of S is defined to a solid obtained as follows. The profile S is swept perpendicularly to P, up to a parallel plane P', resulting in one or more ruled surfaces which we called *side face*(s). The interior of the contour in P and in P' defines the planar faces that, together with the side faces, bound a solid. A *blind extrusion* is determined by a dimension d specifying the depth of the extrusion and a direction paralleling the normal of the sketch plane. Up to face extrusion (UTFE) is an operation that extrudes the profile from the sketch plane to the selected face. Chen et al. [3] presented an algorithm to build 'from-to feature', which is a sweep that begins at a face designated as 'from' and ends at a face designated 'to'. In their algorithm, the direction of extrusion must be given by user explicitly, and the method for extending curved surfaces are not given when the 'from face' or the 'to face' does not completely intersect the generated feature. Fig. 1 gives an example of UTFE.

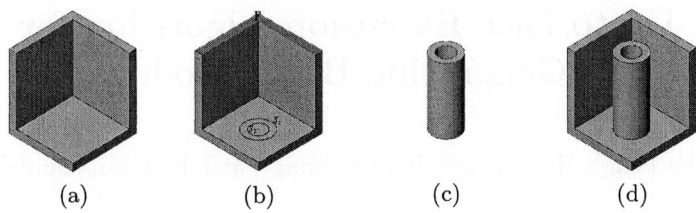

Fig. 1. Up to face extrusion: (a) Original solid. (b) Planar curves S_1, S_2 and selected face **F**. (c) Trimmed extrusion. (d) Resultant solid after UTFE

This paper gives a new UTFE algorithm. The direction of extrusion is determined automatically, and the method of face extension is proposed. A body-surface Boolean operation algorithm is also presented. In this paper, a solid is assumed to be 2-manifold using B-rep, i.e., it is closed topologically and has no isolated faces or edges. Each face of the solid has a surface geometrical data to represent its geometrical shape, and a set of loops (one counterclockwise outer loop and several clockwise inner loops) in 2D parameter domain of the surface to define its boundary. In UTFE, the generated proto-feature is a trimmed extrusion (see Fig. 1(c)), whose top face has the same surface geometrical data as the selected face. The main features of our algorithm are the following.

- Generate the trimmed extrusion.
- Generate the resultant solid using the Boolean operations between the original solid and the trimmed extrusion.

The second step addresses the problem of regularized Boolean operations on two solids, which is outside the scope of this paper but can be found in references [1, 8, 9, 11]. Therefore, we only discuss the first step in the remainder of the paper.

In order to generate the trimmed extrusion, we first generate a blind extrusion with a large depth. The blind extrusion has the same direction as the UTFE, and the depth of the extrusion is large enough to intersect the selected face. The selected face should be extended if it does not completely intersect the blind extrusion. There are three problems in trimmed extrusion generation. The first one shows how to choose the direction and the depth. The second one shows how to extend the selected face. The last one shows how to generate the trimmed extrusion through the trimmed curves and trimmed surface. In the following sections, each problem is explained in detail.

2 Compute Extrusion Direction and Depth

We call the bottom planar face of a blind extrusion a *base face*, which is bounded by the curves in the sketch plane. A *feature line* of a blind extrusion is defined by $P(t) = P_0 + \mathbf{N} \cdot t$, where $t \in (-\infty, +\infty)$, P_0 is an arbitrary point which lies on a curve of the base face, and $\mathbf{N}$ is the normal vector of the sketch plane.

Likewise, we define a *feature line* to represent the selected face. The feature line of a selected face depends on its geometrical surface data. We illustrate the feature lines of some analytic surfaces shown in Fig. 2, respectively. The feature line of plane surface is the projection of the feature line of the selected face (see Fig. 2(a)). The feature line of cylinder surface coincides with the central axis of cylinder (see Fig. 2(b)). The feature line of sphere surface goes through the centre of sphere and is perpendicular to the feature line of the corresponding blind extrusion (see Fig. 2(c)).

We find a point P_1 on the feature line of the blind extrusion and a point P_2 on the feature line of the selected face, where the distance between P_1 and P_2 equals the distance between the two feature lines. Thus, the direction of the blind extrusion is defined by $\mathbf{N}_1 = P_1 - P_0$. The distance between P_0 and P_1 is denoted by d. In this paper, the depth of blind extrusion is given by $2d + f(s)$, where $f(s)$ is a function with respect to the selected face. For example, $f(s)$ equals zero for plane surface, the radius of circle for cylinder surface, and the radius of sphere for sphere surface. Fig. 3 shows an example of a blind extrusion generation with respect to a plane surface, and P_1 coincides with P_2 in this case.

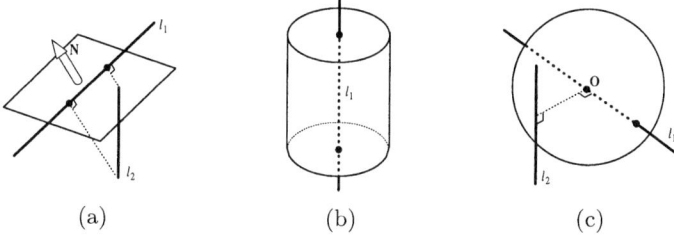

Fig. 2. Feature lines of analytic surfaces are denoted by l_1, and feature lines of blind extrusions are denoted by l_2. The normal vector of plane is denoted by $\mathbf{N}$, and the center of sphere is denoted by $\mathbf{O}$. (a) Feature line of plane surface. (b) Feature line of cylinder surface. (c) Feature line of sphere surface

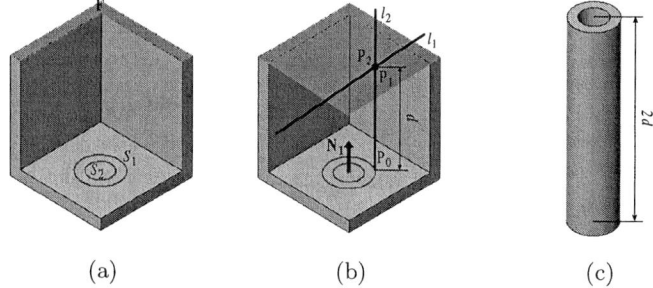

Fig. 3. Blind extrusion generation: Feature line of selected face is denoted by l_1, and feature line of blind extrusion is denoted by l_2. (a) Planar curves S_1, S_2 and selected face **F**. (b) Calculation of direction and depth. (c) Blind extrusion

3 Face Extension

The method of face extension depends on the geometrical surface data of the face. It should be noted that, unlike the faces of analytic surfaces, there is no unique way to extend faces of free-form surfaces. For this reason, the free-form face extension is not handled in this paper. We handle the plane surface extension and some conic surface extensions. The examples of face extension are shown in Fig. 4. The selected face of each example is denoted by **F**, and the planar curves are denoted by S. For a plane surface **F** as shown in Fig. 4(a), we project the 2D bounding box, which includes all the curves of sketch plane, onto the selected plane in the direction of the blind extrusion. The extended face is defined by a face, which is bounded by the projected 2D bounding box (see thick black rectangle in Fig. 4(a)). For a cylinder surface **F** as shown in Fig. 4(b), we first calculate the bounding box of the blind extrusion (see gray cuboid wire-frame in Fig. 4(b)). Next we translate the feature line l_1 of the selected cylinder surface, where the translated line l_2 goes through P_1 defined in Section 2. Then we project l_2 onto top face of the bounding box (see l_3 in Fig. 4(b)) and calculate Q_1 and Q_2 on l_3. Finally we project Q_1 and Q_2 onto l_1 and get Q'_1 and Q'_2. The extended face is bounded by two circles, which uses Q'_1 and Q'_2 as two centers, respectively. Note that a sphere surface bounds a closed region, so we use the boundary of the entire surface in 2D parameter domain as the boundary of the extended face (see Fig. 4(c)).

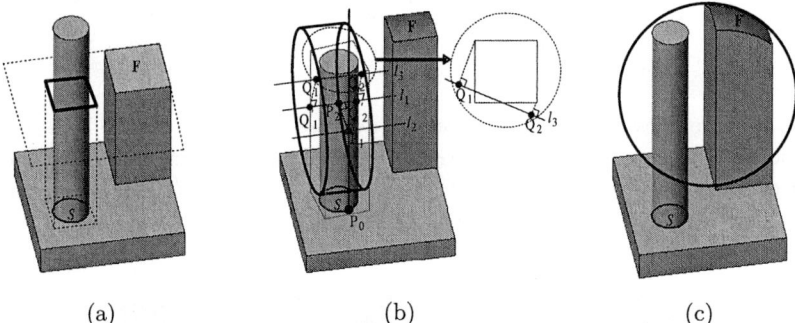

Fig. 4. Face extension for plane surface, cylinder surface and sphere surface: (a) Extended face bounded by thick black rectangle. (b) Extended face bounded by thick black cylinder wire-frame. (c) Extended face bounded by thick black sphere wire-frame

4 Body-Surface Boolean Operations

Before the body-surface Boolean operation, we should determine whether the UTFE is illegal. We project the contour of the extended face onto sketch plane. If the projected contour contains all curves of the sketch plane, the UTFE is legal. Otherwise the UTFE is illegal. The side faces of the blind extrusion are denoted by F_i, where $1 \leq i \leq n$, n is the number of the faces of the blind extrusion solid. In the literature, extensive research have resulted in grand algorithms for Boolean operations on surfaces or non-manifold solid [4, 7, 10]. In this paper, the body-surface Boolean operation consists of three steps.

1. Calculate the trimmed intersection curves between F_i and the extended face and form loops in 2D parameter domain for them.
2. Subdivide F_i and the extended face to some trimmed faces by the loops and classify the trimmed faces of the extended face.
3. Generate the trimmed extrusion.

The first step addresses the problem of curve/surface and surface/surface intersection, which is outside the scope of this paper but can be found in references [2, 5, 6].

In Step 2, we create the trimmed faces by the loops obtained in Step 1. Each trimmed face includes an outer loop and several inner loops and has the same geometrical surface data as the original face. Every trimmed face of the extended face is classified into one of the following three types based on inclusion relation with respect to the closed region of the blind extrusion: Class_In, Class_Out, Class_On.

Now we describe the third step in more detail. We use a face array to store the bounding faces of the trimmed extrusion. First, we add the base face to the face array. Next, for each trimmed face of F_i, we add the trimmed face, which shares one bounding edge of the base face, to the face array. Finally, for each trimmed face of extended face that is classified by Class_In, we search for the one that is nearest to the base face of the blind extrusion. Add the nearest trimmed face to the face array. Thus, the trimmed extrusion is obtained through the face array.

5 Examples and Conclusions

The algorithm has been implemented in a commercial geometric modeling software system TiGems and greatly enhances the capability for creating an extrusion feature. Fig. 5 shows two examples of UTFE using our algorithm.

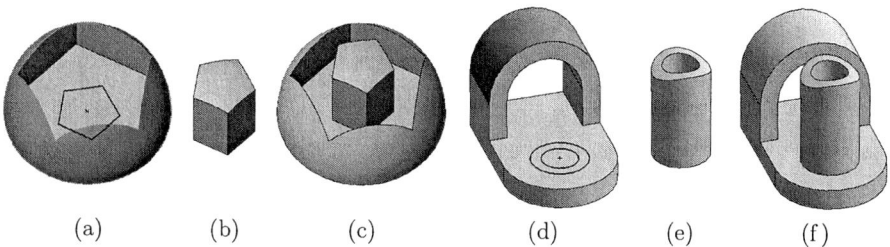

Fig. 5. Examples of UTFE: (a)-(c) is an example of up to sphere surface extrusion, and (d)-(f) is an example of up to cylinder surface extrusion. (a) Original solid, selected face and profile. (b) Trimmed extrusion. (c) Resultant solid. (d) Original solid, selected face and profile. (e) Trimmed extrusion. (f) Resultant solid

We present a UTFE algorithm in this paper. The extrusion direction is determined automatically. The methods of face extension of some analytic surfaces are given, and a simple body-surface Boolean operation algorithm is presented to generate the trimmed extrusion solid.

Acknowledgements

The research was supported by Chinese 863 Program (2003AA4Z3110) and Chinese 973 Program (2002CB312106). The third author was supported by a project sponsored by SRF for ROCS, SEM (041501004), and a Foundation for the Author of National Excellent Doctoral Dissertation of PR China (200342).

References

1. Arbab, F.: Set models and Boolean operations for solids and assemblies. IEEE Computer Graphics & Applications **10** (1990) 76-86
2. Barnhill, R. E., Kersey, S. N.: A marching method for parametric surface/surface intersection. Computer Aided Geometric Design **7** (1990) 257-280
3. Chen, X. P., Hoffmann, C. M.: Towards feature attachment. Computer-Aided Design **27** (1995) 695-702
4. Granados, M., Hachenberger, P., Hert, S., Kettner, L., Mehlhorn, K., Seel, M.: Boolean operations on 3D selective Nef complexes: Data structure, algorithms, and implementation. In Proc. 11th Annu. European Sympos. Algorithms, LNCS 2832 (2003) 654-666
5. Krishnan, S., Manocha, D.: Efficient representations and techniques for computing B-Reps of CSG models with NURBS primitives. In:Woodmark J, eds. Set-theoretic Solid Modeling Techniques and Applications. Winchester, UK: Information Geometers Ltd. (1996) 101-122
6. Manocha, D., Krishnan, S.: Algebraic pruning: a fast technique for curve and surface inter section. Computer Aided Geometric Design **14** (1997) 823-845
7. Marcheix, D., Gueorguieva, S.: Topological operators for non-manifold modeling. Proeedings of the Third International Conference in Central Europe on Computer Graphics and Visualisation **1** (1995) 173-186
8. Mäntylä, M.: Boolean operations of 2-manifolds through vertex neighborhood classification. ACM Transactions on Graphics **5** (1986) 1-29
9. Requicha, A. A. G., Voelcker, H. B.: Boolean operations in solid modeling: boundary evaluation and merging algorithms. Proceedings of the IEEE **70** (1985) 30-44
10. Satoh, T., Chiyokura, H.: Boolean operations on sets using surface data, ACM SIGGRAPH: Symposium on Solid Modeling Foundations and CAD/CAM Applications, Austin, Texas, USA. (1991) 119-127
11. Zhu, X. H., Fang, S. F., Bruderlin, B. D.: Obtaining robust Boolean set operations for manifold solids by avoiding and eliminating redundancy. In: Proceedings of the 2nd ACM Symposium on Solid Modeling, Montreal, Canada (1993)

Adaptive Model-Based Multi-person Tracking

Kyoung-Mi Lee

Department of Computer Science,
Duksung Women's University,
Seoul 132-714, Korea
kmlee@duksung.ac.kr
http://www.duksung.ac.kr/~kmlee

Abstract. This paper proposes a method for tracking and identifying persons from video image frames taken by a fixed camera. The majority of conventional video tracking surveillance systems assumes a likeness to a person's appearance for some time, and existing human tracking systems usually consider short-term situations. To address this situation, we use an adaptive background and human body model updated statistically frame-by-frame to correctly construct a person with body parts. The formed person is labeled and recorded in a person's list, which stores the individual's human body model details. Such recorded information can be used to identify tracked persons. The results of this experiment are demonstrated in several indoor situations.

1 Introduction

With recent advances in computer technology real-time automated visual surveillance has become a popular area for research and development. In a visual surveillance system, computer-based image processing offers means of handling image frames generated by large networks of cameras. Such automated surveillance systems can monitor unauthorized behavior or long-term suspicious behavior and warn an operator when unauthorized activities are detected [2,3]. However, tracking approaches as applied to video surveillance systems must deal adequately with the problems caused by the non-rigid form of the human body and the sensitivities of such systems to dynamic scenes.

Tracking people based on images using a video camera plays an important role in surveillance systems [1,5]. A popular approach of tracking using a fixed camera consists of three steps: background subtraction, blob formation, and blob-based person tracking. To look for regions of change in a scene, a tracking system builds a background model as a reference image and subtracts the monitored scene from the reference frame. However, background subtraction is extremely sensitive to dynamic scene changes due to lighting and extraneous events. Blob formation involves grouping homogeneous pixels based on position and color. However, it is still difficult to form blobs of individual body parts using such spatial and visual information since color and lighting variations causes tremendous problems for automatic blob formation algorithms. The tracking human body parts is performed by relating body parts in consecutive frames [8,9]. Such correspondence-based approaches work well for cases with no occlusion, but are unable to decide upon a person identity if a human body is occluded. In addition, most existing surveillance systems consider only immediate unauthorized behaviors.

In this paper, we propose a framework to track and identify multiple persons in a fixed camera situation with illumination changes and occlusion. Section 2 presents an adaptive background model which is updated with on-line recursive updates in order to cope with illumination changes. In Section 3, blob formation at the pixel-level and a hierarchical human body model at blob-level are proposed. During tracking, the proposed human body model is adaptively updated on-line. Section 4 describes, in terms of the blob and human model, multi-person tracking. Experimental results based on several scenarios and conclusions are presented in Section 5.

2 Adaptive Background Modeling

To detect moving persons in video streams, background subtraction provides the most complete feature data, but it is unfortunately extremely sensitive to dynamic scene changes due to lighting and other extraneous events. To overcome this situation, during tracking, the reference image should be compensated according to the lighting conditions present in the scene. In this paper, we build the adaptive background model, using the mean (μ_B) and standard deviation (σ_B) of the background.

At the pixel level, let I^t represent the intensity value in the t-th frame. At time t, a change in pixel intensity is computed using Mahalanobis distance δ^n:

$$\delta^t = \frac{|I^t - \mu_B^t|}{\sigma_B^t} \qquad (1)$$

where μ_B^t and σ_B^t are the mean and standard deviations of the background at time t, respectively. μ_B^0 is initially set to the first image, $\mu_B^0 = I^0$, and σ_B^0 is initialized by **0**. For each pixel, Eq. (1) is evaluated to separate background as $\delta^t \leq \theta^t$, where θ^t is a difference threshold observed from the sequence of images prior to t. θ^0 is initially set to the largest distance and multiplied at each time step by a decay constant $0 < \gamma < 1$ until it reaches a minimum.

Whenever a new frame I^t arrives, a pixel in the frame is tested to classify background or foreground (moving persons). If a pixel satisfies $\delta^t \leq \theta^t$ at time t, the adaptive background model (μ_B^t and σ_B^t) is updated as follows [6]:

$$\mu_B^t = \alpha^{t-1}\mu_B^{t-1} + (1-\alpha^{t-1})I^t, \text{ and}$$
$$\sigma_B^t = \sqrt{\alpha^{t-1}W + (1-\alpha^{t-1})\{\mu_B^t - I^t\}^2}, \qquad (2)$$

where $W = \{\sigma_B^{t-1}\}^2 + \{\mu_B^t - \mu_B^{t-1}\}^2$. $\alpha^{t-1} = \frac{N_B^{t-1}}{N_B^{t-1}+1}$ where N_B^{t-1} means the number of frames participating in the background model to time t-1.

3 Blob-Based Person Modeling

Before tracking persons, they should be initialized when they start to appear in the video. We construct same colored regions from a foreground image and build a hierarchical human body model. In Sec. 4, this person model will be used for tracking and for identification.

3.1 Blob Formation for Person Tracking

The foreground image derived from adaptive background subtraction (Sec. 2) is applied to find candidate human body parts. To group segmented foreground pixels into a blob and to locate the blob on a body part, we use a connected-component algorithm which calculates differences between intensities between a pixel and its neighborhoods. However, it is difficult to form perfect individual body parts using such a color-based grouping algorithm since color and lighting variations causes tremendous problems for automatic blob formation algorithms. Therefore, small blobs are merged into large blobs and neighboring blobs that share similar colors are further merged together to overcome over-segmentation generated by initial grouping, which is largely due to considerable illumination changes across the surfaces of coherent blobs. Two adjacent blobs will be merged if the intensity difference is smaller than a threshold. Then, some blobs are removed according to criteria, such as, too small, too long, or too heterogeneous incompact blobs. Each blob contains information such as an area, a central position, a bounding box, and a boundary to form a human body.

After completing blob formation, blobs are classified to skin-area and non-skin-area since skin area is important for person detection and tracking. The skin color similarity of a blob B_i is evaluated as follows:

$$P(B_i \mid skin) = \frac{\sum_{x \in B_i} P(x \mid skin)}{Area(B_i)} \qquad (3)$$

where $P(x|skin)$ is the skin color similarity of a pixel x, which is calculated by [7]. $Area(B_i)$ means an area of B_i defined by a pixel number in B_i. If the skin similarity of a blob is larger than a predefined threshold, the blob is considered a skin blob.

3.2 Human Modeling for Multi-person Tracking

As a person can be defined as a subset of blobs, which correspond to body parts, blobs in a frame should be assigned to corresponding individuals to facilitate multiple individual tracking. The person formation algorithm first computes a blob adjacency graph, where each node represents a blob. The criterion of adjacency is based on a minimum distance computed between the associated bounding boxes of each blob. A distance between the vertices of the bounding boxes is computed much faster than adjacency in the pixel domain [4]. The merging criterion between bounding boxes of adjacent blobs is based on a maximum allowed distance. Let P_0 be a subset of blobs B_i. The set of potential person areas is built iteratively, starting from the P_0 set and its adjacent graph. The set P_1 is obtained by merging compatible adjacent blobs of P_0.

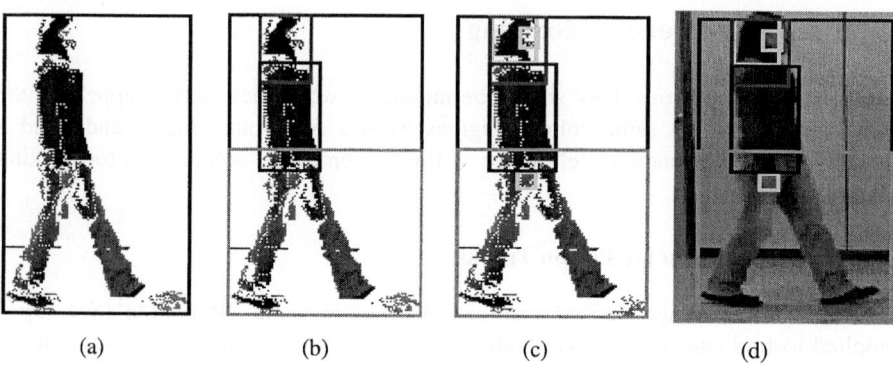

Fig. 1. A hierarchical person model: (a) Bounding region in the high level after background subtraction, (b) three body parts in the middle level, (c) skin blobs in the low level, and (d) a hierarchical human body model applied to a video image frame

Then, each new set of merged blob P_k is obtained by merging the set of merged blob P_{k-1} with the original set P_0. Finally, a candidate person CP_n contains the built sets of merged blob P_k, i.e., $CP_n = \bigcup_{k=0}^{K} P_k$, $n=1...N$ where N is the number of persons. Each candidate person CP_n is formed with a unique label and information on the person, such as, the number of blobs, height, width, centroid, and an adjacent graph of blobs.

To match blobs to body parts, we use a hierarchical human model (Fig. 1). First, we assumed that all individuals in the video are upright or in a slightly skewed standing position. The high level of the model contains a whole human model and its information (Fig. 1(a)). After grouping at the pixel-level, each merged blob is displayed as an average color. Depending on the relative position in a candidate person P_k, each blob is assigned to one of three categories in the middle level: the head, the upper body, and the lower body (Fig. 1(b)). If a category in the middle level contains two more blobs, these blobs are tested for skin similarity to classify the blobs as skin or non-skin area at the low level using Eq. (3) (Fig. 1(c)). The person model is defined by 3 categories and their geometrical relations in the middle level, and as 6 classes and their geometrical and color relations in the low level as follows:

$$CP_n = \left(R_n^0, \{C_n^1, R_n^1\}, \{C_n^2, R_n^2\}, \{C_n^3, R_n^3\}\right) \quad (4)$$

where R^0 means a relation among three parts. C_n^j and R_n^j mean a set of blobs and their relationships of the j-th body part of CP_n, respectively.

3.3 Adaptive Human Modeling

The goal of this paper is to observe existing persons exiting, new persons entering, and previously observed persons re-entering a scene. Since a person is tracked using the model defined by Eq. (4), human body model information is stored to track multiple-persons and to determine whether a specified person has appeared. Even though the total motion of a person is relatively small between frames, large changes in the color-based model can cause simple tracking to fail. To resolve this sensitivity of color-model tracking, we compare the current model to a reference human model. During tracking, the reference model should be compensated according to the scenes lighting conditions. Thus, the reference human model has to adapt to slow change such as illumination changes by model updating.

To maintain a dynamic environment with continually arriving frames, we use online clustering. Let a person CP_n^{t-1} represented by an average (μ_n^{t-1}) and a deviation (σ_n^{t-1}), which are computed up to time t-1 and new blobs B_i^t and their relations Br_i^t are formed in frame t. The minimum difference between the person model CP_n^{t-1} (Eq. (4)) and the new blobs B_i^t is computed as follows:

$$d_n^t = \min_{j=1\cdots 3}\left(\frac{\left\|B_i^t - \mu_n^{t-1,C^j}\right\|_p}{\sigma_n^{C^j}}\right) + \min_{j=0\cdots 4}\left(\frac{\left\|Br_i^t - \mu_n^{t-1,R^j}\right\|_p}{\sigma_n^{R^j}}\right) \quad (5)$$

where μ_n^{t-1,C^j} and μ_n^{t-1,R^j} mean a set of averages of blobs and relations in the j-th body part at time t-1, respectively. σ_n^{t-1,C^j} and σ_n^{t-1,R^j} a set of deviations of blobs and relations, respectively. If the minimum distance is less than a predefined threshold, the proposed modeling algorithm adds blobs B_i^t and relations Br_i^t to corresponding adaptive person model (μ_n^{t-1,C^j} and μ_n^{t-1,R^j}) and updates the adaptive model by recalculating their center and uncertainties [6].

4 Model-Based Multi-person Tracking

Tracking people poses several difficulties, since the human body is a non-rigid form. After forming blobs, a blob-based person tracking maps blobs from the previous frame to the current frame, by computing the distance between blobs in consecutive frames. However, such a blob-based approach for tracking multiple persons may cause problems due to the different number of blobs in each frame: blobs can be split, merged, even disappear or be newly created. To overcome this situation, many-to-many blob mapping can be applied [9]. While these authors avoided situations where blobs at time t-1 are associated to a blob at time n, or vice versa, they adopted a variant of the multi-agent tracking framework to associate multiple blobs simultaneously.

In this paper, we assume that persons CP_n^{t-1} have already been tracked up to frame t-1 and new blobs B_i^t are formed in frame t. Multi-persons are then tracked as follows:

Case 1: If B_i is included in P_k, the corresponding blob in P_k is updated with B_i.
Case 2: If a blob in P_k is separated into several blobs in frame t, the blob in P_k is updated with one blob in frame t and other blobs at time t are appended into P_k.
Case 3: If several blobs in P_k are merged into B_i, one blob in P_k is updated with B_i and other blobs are removed from P_k.

Case 4: If B_i is included in P_k but the corresponding blob does not exist, B_i is added to P_k.

Case 5: If B_i is not included in P_k, the blob is considered as a newly appearing blob and thus a new person is added to the person list (Sec. 3.2).

where including a region into a person with a bounding box means the region overlaps over 90% to the person. Corresponding a blob to the adaptive person model is computed using Eq. (5). In addition to simplify the handling of two lists of persons and blobs, the proposed approach can keep observe existing persons exiting, new persons entering, and previously monitored persons re-entering the scene. One advantage of the proposed approach is to relieve the burden of correctly blobbing. Even though a blob can be missed by an illumination change, person-based tracking can retain individual identity using other existing blobs. After forming persons (Sec. 5), the number of blobs in a person are flexibly changeable. The proposed approach can handle over-blobbing (Case 2) and under-blobbing (Case 3) problems.

5 Results and Conclusions

The proposed multi-person tracking approach was implemented in JAVA (JMF), and tested in Windows 2000 on a Pentium-IV 1.8 GHz CPU with a memory of 512 MB. For 320×240 frames, videos were recorded using a Sony DCR-PC330 camcoder.

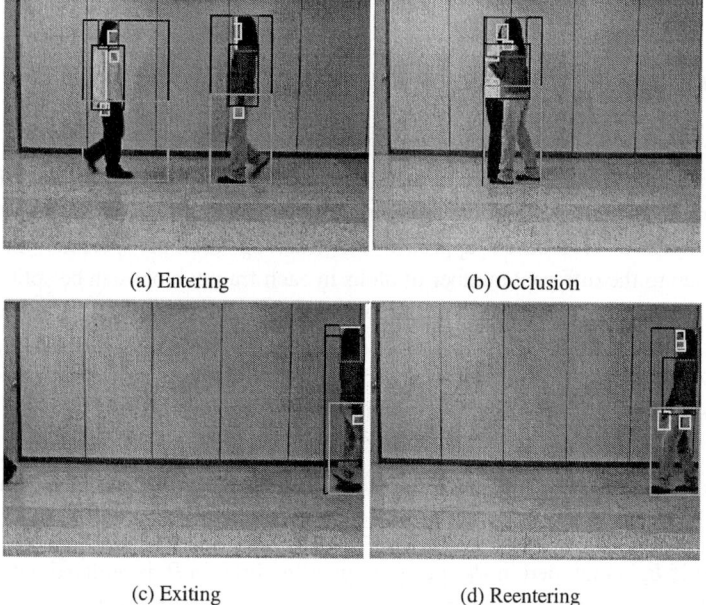

Fig. 2. Experimental results using adaptive person models

The number of persons the algorithm can handle simultaneously and the duration a specific person's information can be held by the algorithm are not limited programmatically. Fig. 2 presents tracking results with scenarios such as entering, leaving, re-entering, and occluding from the top. Whenever a person appears first in the video, a new person model is built and added to the persons' list (Fig. 2(a)). In the event of occlusion, the occluded individual is considered to have exited the scene (Fig. 2(b)). During tracking the person, the proposed adaptive tracking system updates the corresponding person model at each frame, and the system keeps the person's information after the person has exited the scene (Fig. 2(c)). When the person later re-enters the video, the system looks for a model in the previously recorded persons' list (Fig. 2(d)). Then the system tracks and updates the person and the person's model.

The goal of this paper was to track and identify many persons using a video surveillance system which warns of long-term unauthorized behavior and transmits their model-based information between networked cameras. The next phase in our work is to build a networked surveillance system with disjoint cameras and to improve the approach proposed in this paper for wide-area surveillance.

Acknowledgements

This work was supported by Korea Research Foundation Grant (KRF-2004-003-D00376).

References

1. Aggarwal, J.K., and Cai, Q., Human motion analysis: a review, Computer vision and image understanding, Vol. 73, No. 3, (1997) 428–440
2. Foresti, G., Mahonen, P., and Regazzoni, C.S., Multimedia video-based surveillance systems: requirements, issues and solutions, Dordrecht, The Netherlands: Kluwer (2000)
3. Foresti, G., Mahonen, P., and Regazzoni, C.S., Automatic detection and indexing of video-event shots for surveillance applications, IEEE transactions on multimedia, Vol. 4, No. 4, (2002) 459-471
4. Garcia, C., and Tziritas, G., Face detection using quantized skin color regions merging and wavelet packet analysis, IEEE transactions on multimedia, Vol. 1, No. 3, (1999) 264-277
5. Gavrila, D., The visual analysis of human movement: a survey. Computer vision and image understanding, Vol. 73, No. 1, (1999) 82–98
6. Lee, K.-M., and Street, W.N., Model-based detection, segmentation, and classification using on-line shape learning, Machine vision and applications, Vol. 13, No. 4, (2003)222-233
7. Lee, K.-M., Elliptical clustering with incremental growth and its application to skin color region segmentation, Journal of korean information science society, Vol. 31, No. 9 (2004)
8. Niu, W., Jiao, L., Han, D., and Wang, Y.-F., Real-time multi-person tracking in video surveillance, Proceedings of the pacific rim multimedia conference (2003) 1144-1148
9. Park, S. and Aggarwal, J.K., Segmentation and tracking of interacting human body parts under occlusion and shadowing, Proceedings of international workshop on motion and video computing (2002) 105-111

A Novel Noise Modeling for Object Detection Using Uncalibrated Difference Image

Joungwook Park and Kwan H. Lee

Gwangju Institute Science and Technology (GIST),
Intelligent Design and Graphics laboratory, Department of Mechatronics,
1 Oryong-dong, Buk-gu, Gwangju, 500-712, Korea
{vzo, lee}@kyebek.kjist.ac.kr
http://kyebek9.kjist.ac.kr

Abstract. In order to extract the region of an object from an image, the difference image method is attractive since it is computationally inexpensive. However, the difference image is not frequently used due to the noise in the difference image. In this paper, we analyze the noise in an uncalibrated difference image and propose a statistical noise calibration method. In the experiment, the proposed method is verified using a real image.

1 Introduction

In image-based modeling, it is important to separate an object that we are interested in from the background. In background removal methods without considering illumination change, color-based and template matching methods are generally used. In a color-based method, we search the object with the known color information of an object. However, there are two drawbacks in the color-based methods. First drawback is that the color information can be contaminated by environmental changes such as light position and intensity. Second, the color-based methods require additional computational cost due to a large amount of color transformation and matrix calculations [2–4]. In a template matching method, we find the object based on the template set that has information on the object. However, we should design the template set. The use of a template can improve the performance but it also increases the computational cost [5]. A computationally inexpensive background removal method is therefore needed.

One inexpensive background removal method is the difference image method. The difference image gives us information of the changed part in the image. Nevertheless, the difference image has not been frequently used due to the noise occurring from the interaction between the object and the background. Assume that there are two images: one has an object and the other has not. If we compare the backgrounds of both images, the statistical properties of their backgrounds are not the same [1]. The object between camera and background becomes another noise source. As a result, the object degrades the quality of the difference image [6, 7]. Recently, many researchers have been intensively studied the noise modeling considering illumination change [8, 9].

In this paper, however, we focus on the case where an object appears in the image, with a fixed illumination environment and no occurrence of shadows. This affects the change of the background region, but not that of the object region. In section 2, a new noise model is described that verifies the effect introducing an object and improves the quality of difference image generated from uncalibrated images. In section 3, we improve the noise model considering the effect of object appearance. In section 4, we propose the procedure for improving the quality of difference image. In section 5, a numerical example is given for verification of the proposed method. Finally the conclusion is given in section 6.

2 Noise Modeling

Generally the change of noise distribution in the image occurs though images are sequentially taken in the same condition such as no change of foreground, background and illumination. The model of the image taken is defined as follows:

$$I_t(i,j) = I_t^o(i,j) + V_t(i,j), \tag{1}$$

$$V_t(i,j) \sim N(0, \sigma^2), \tag{2}$$

where $I_t^o, I_t, V_t \in \mathbf{R}^{n \times m}$, $i \in [0, n]$, $j \in [0, m]$, and $t \in [0, T]$.

I_t : the t-th image with noise
V_t : the noise in the image
T : the number of image taken

I_t^o : the original image without noise
t ; the t-th image in the sequence
i, j : integer indices

To calculate the noise distribution of sequence images, the model of difference image is defined as follows:

$$DI_{a,b}(i,j) \triangleq I_a(i,j) - I_b(i,j) = V_a(i,j) - V_b(i,j) \; (\because I_a^o = I_b^o), \tag{3}$$

where $DI_{a,b}$ is the difference image of I_a, I_b and a, b are integer indices, which satisfy $a, b \in [0, t]$ and $a \neq b$.

$DI_{a,b}(i,j)$ satisfies the Gaussian distribution because of $V_a(i,j), V_b(i,j)$ as given in eq. (2).

$$DI_{a,b}(i,j) \sim N(0, 2\sigma^2) \tag{4}$$

In the case of detection or recognition of object with difference image, generally a threshold value is used to decide whether a pixel is in the region of an object or not. $DI_{a,b}(i,j)$ is replaced by $|DI_{a,b}(i,j)|$ as below.

$$IsObject(i,j) = \begin{cases} \text{True} & |D_{a,b}(i,j)| \geq \text{Threshold} \\ \text{False} & |D_{a,b}(i,j)| < \text{Threshold} \end{cases} \tag{5}$$

To decide which pixel is noisy, the distribution of $|DI_{a,b}(i,j)|$ is deduced from $DI_{a,b}(i,j)$. The distribution function of $DI_{a,b}(i,j)$ is given by eq.(6) based on eq.(4). The mean of $|DI_{a,b}(i,j)|$ is given by eq.(7).

$$P(z) = \frac{e^{\frac{-z^2}{4\sigma^2}}}{2\sqrt{\pi}\sigma} \tag{6}$$

$$E(|z|) = \int_0^\infty z \frac{e^{\frac{-z^2}{4\sigma^2}}}{\sqrt{\pi}\sigma} dz \tag{7}$$

It should be considered whether all levels of the pixel satisfied the previous assumptions or not. Assume that an image is represented by a gray scale with 8-bit depth. In general, the effect of noise is reduced near the limit level of pixel such as 0 or 255. Therefore the distribution and the noise model in eq.(1) should be changed near the limit level.

$$I_t(i,j) = \begin{cases} 0 & (I_t^o + V_t(i,j) \leq 0) \\ I_t^o(i,j) + V_t(i,j) & (0 < I_t^o(i,j) + V_t(i,j) < 255)) \\ 255 & (I_t^o(i,j) + V_t(i,j) \geq 255) \end{cases} \tag{8}$$

In addition the variance should be defined for the model according to the level of pixel value. Let $p_t(i,j)$ define the pixel value at position (i,j) located in $t-th$ image, and $s_{x,k}$ also defines the set of position where $p_t(i,j)$ in $t-th$ image is equivalent to any pixel level, x. Then the mean of $|DI_{a,b}(i,j)|$ at each pixel level is represented by $E_{DI}(x)$ and the set of means is $Mean_{DI}$.

$$p_t(i,j) = \{x | x \in [0, 255]\} \tag{9}$$

$$s_{x,t} = \{(i,j) | p_t(i,j) = x, \ x \in [0, 255], \text{for all}(i,j)\} \tag{10}$$

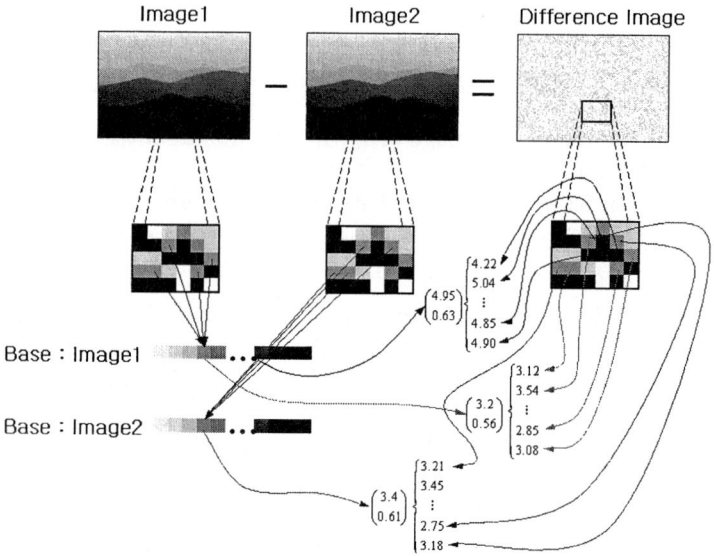

Fig. 1. Concept of the proposed method

$$E_{DI}(x) = \frac{\sum_{k=i}^{T}\sum_{l=0, l \neq k}^{T} \sum_{s_{x,k}} |DI_{k,l}(s_{x,k})|}{\# \text{ of Element}} \quad (11)$$

$$Mean_{DI} = \{E_{DI}(x) | x \in [0, 255], \text{ for all x}\} \quad (12)$$

Fig. 1 shows this concept. Assume that there are two images which are sequentially taken in the same condition. One of the images is temporarily named the base image. Firstly, in the base image, we generate $s_{x,k}$, the set of positions of the pixel for each pixel level and then calculate the mean of the pixel values at the position corresponding to $s_{x,k}$ in difference image. These steps repeat for all levels of pixel. Whole procedure repeats until no more image is considered as the base image.

3 The Effect of Introducing an Object

In the previous section, the noise model is generated in the same condition and is regarded as a kind of random noise. To remove the noise in the difference image, we consider the effect of introducing an object in this section.

Assume that the image with only background is named the background image and the image with the background and an object is the object image. In ideal case the region of background in the object image is not changed when an object enters the image unless the shadow or illumination changes. In real case, however, the image is changed by only introducing an object. In this paper we define the difference generated from the effect as situation-dependent noise (SDN). The SDN is generated due to interaction between the object and the background.

Fig. 2 shows the effect from introducing an object. Fig. 2 (a) shows that the portion of the emitted light from background goes through the lens and it makes the background image. If an object obstructs the way of the emitted light, some of the light from the background is either fully obstructed or partially obstructed. And some of the light is not obstructed as shown in Fig. 2 (b). When some of the light is partially obstructed, the shape of acquired image is not changed, but the value of each pixel is changed in comparison to that of the pixel in the image taken from only the background.

Assume that if the pixel which belongs to the background is able to be seen, the ratio of obstruction is the same for all pixels corresponding to the background. Each pixel value on the region of background in the object image is less than that in the background image. Therefore the effect from introducing an object can be computed from the portion of the difference image, which is regarded as the area corresponding to the portion of background in the object image.

To represent the noise model in the object image we assume that we have two images: one is a background image of which $Mean_{DI}$ is known as given eq.(12). The other is a difference image generated from the background image and object image which is taken in the same background. The portion of background in the difference image is $DI_{bk,obj}^{bk}(i,j)$ and the portion of object is $DI_{bk,obj}^{obj}(i,j)$. When

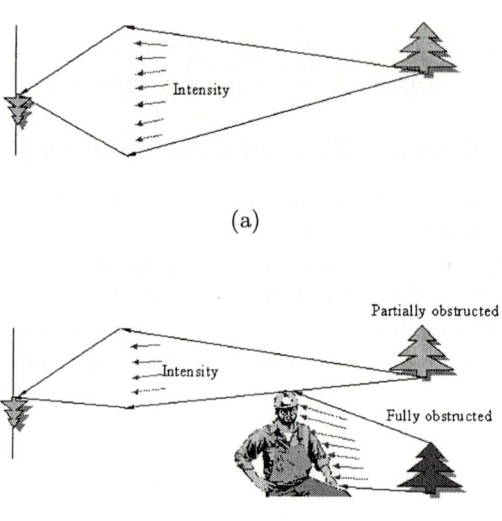

Fig. 2. The effect from introducing an object

the object image is taken, $V_{obj}^{all}(i,j)$ is the variance of noise generated in the same condition as the background image. $V_{obj}^{bk}(i,j)$ is the variance of noise generated from introducing the object. We assume that $I_{obj}(i,j)$ is given as follows:

$$I_{obj}(i,j) = \begin{cases} I_{obj}^{o}(i,j) + V_{obj}^{all}(i,j), & IsObject(i,j) = \text{True} \\ I_{bk}^{o}(i,j) + V_{obj}^{all}(i,j) + V_{obj}^{bk}(i,j), & IsObject(i,j) = \text{False} \end{cases}, \quad (13)$$

where $V_{obj}^{all}(i,j) \sim N(0, \sigma^2)$.

To distinguish the region of object from difference image, the model of difference image is represented as below.

$$DI_{obj,bk}(i,j) = I_{obj}^{o}(i,j) - I_{bk}^{o}(i,j) + (V_{obj}^{all}(i,j) - V_{bk}(i,j)) \quad (14)$$

$$DI_{obj,bk}^{bk}(i,j) = (V_{obj}^{all}(i,j) - V_{bk}(i,j)) + V_{obj}^{bk}(i,j) \ (\because I_{obj}^{o}(i,j) = I_{bk}^{o}(i,j)) \quad (15)$$

4   Procedure of Noise Removal

A simple object detection algorithm is suggested in order to verify the proposed noise model. Once generated the difference image without noise, it is easy to detect the region of the object. Therefore we should first find the noise distribution of difference image with background images. And then difference image is generated from background image and object image. The bounding box is

created by the mean and the variance of the difference image. The noise distribution of the outside of the bounding box($DI^{bk}_{bk,obj}(i,j)$) according to the level of pixel is generated and the noise of the difference image is removed by the noise distribution. The process is performed as follows:
   a) Computation of noise distribution of difference image from background images
   b) Computation of noise distribution resulting from introducing an object
   c) Modification of the difference image using noise distribution
   d) Creation of a bounding edge using the modified difference image

5 Experimental Results

The size of the image used is 640×480 with 8bit-RGB color. Five background images are taken at one image per second. The experiment is carried out inside of a room. In order to capture images the camera for web chatting is used. Fig. 3(b) is one of background images.

Fig. 3(a), (c) show the error maps of the difference image using background images about red and blue color. Fig. 4 illustrates that the noise distribution of the difference image from five background images tends to behave constantly according to each pixel level, except near the limit, 0 or 255.

Fig. 5 shows the bounding box of the object, which is generated from the mean and the variance of the difference image. The difference image is generated from background image and object image. The offset is applied to the bounding box in order to entirely exclude the portion of image corresponding to the object ($DI^{bk}_{bk,obj}(i,j)$) in Fig. 5 (b).

Fig. 6 shows that the error is dependent on the level of the pixel value in the background image. It also illustrates that the distribution of noise resulting from introducing an object tends to behave linearly according to the pixel level, except near the limit, 0 or 255.

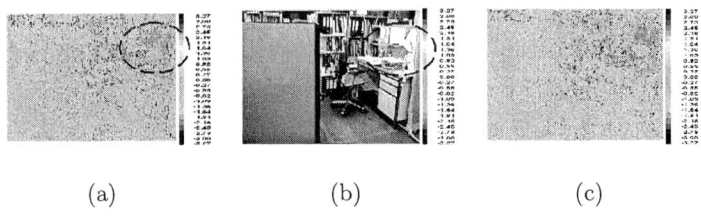

(a)　　　　　　　　　(b)　　　　　　　　　(c)

Fig. 3. The error map of difference image from background images

Fig. 4. The noise distribution of the difference image according to the level of pixel

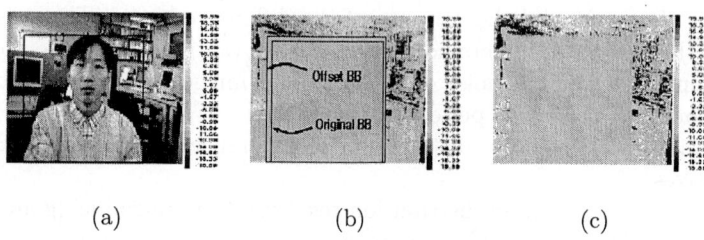

(a) (b) (c)

Fig. 5. The noise distribution of the difference image according to the level of pixel

Fig. 6. The noise distribution resulting from introducing an object according to the level of pixel value

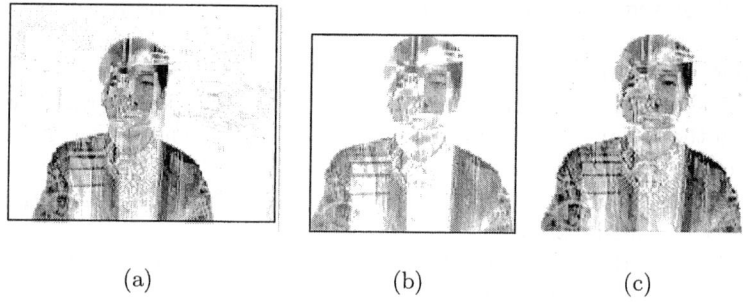

(a) (b) (c)

Fig. 7. The result of using the proposed method

Fig. 7(a) illustrates the difference image without any process. Fig. 7 (b) is the result using the information of the difference image with the threshold of 2σ. Fig. 7 (c) is the result using the proposed method. The result shown in Fig. 7 (c) is more precise and distinguishable than the result shown in Fig. 7 (b).

6 Conclusion

In this paper we proposed a noise reduction method for the difference image. Noise in a difference image is analyzed. Based on noise analysis, we developed the noise estimation process and noise removal algorithm. In the experiment, it is verified by the proposed method that the noise is removed and the loss of the object is minimized in the difference image. The property of noise is used to remove the noise and to find the area of the object. The loss of the detected region of an object is minimized by this procedure.

References

1. Papoulis, A.: Probability, Random Variables, and StochasticProcess. 3rd edn. McGraw-Hill, New York (1991)
2. Gonzalez, R. G., Woods, R. E.: Digital Image Processing. Addison Wesley, Massachusetts (1992)
3. Duda, R. O., Hart, P. E., Stork, D. G.: Pattern Classification. 2nd edn. Wiley-Interscience, New York (2001)
4. Ritter, G. X., Wilson, J. N.: Handbook of Computer Vision Algorithms in Image Algebra. 2nd edn. CRC Press, Boca Raton (2001)
5. Osher, S., Paragios, N.: Geometric Level Set Methods in Imaging, Vision, and Graphics. Springer (2003)
6. Bruzzone, L., Prieto, D. F.: Automatic Analysis of the Difference Image for Unsupervised Change Detection. IEEE Transactions on Geoscience and Remote Sensing, Vol. 38. Issue 3. IEEE(2000) 1171-1182
7. Khashman, A.: Noise-Dependent Optimal Scale in Edge Detection. Proceedings of the IEEE International Symposium on Industrial Electronics 2002, Vol. 2. IEEE(2002) 467-471
8. Stauffer, C.,Grimson, W.E.L.: Adaptive Background Mixture Models for Real-time Tracking. Computer Vision and Pattern Recognition, Vol. 2. IEEE (1999) 246-252
9. Javed, O., Shafique, K., Shah, M.: A Hierarchical Approach to Robust Background Subtraction using Color and Gradient Information. Workshop on Motion and Video Computing 2002, IEEE(2002) 22-27

Fast and Accurate Half Pixel Motion Estimation Using the Property of Motion Vector

MiGyoung Jung and GueeSang Lee*

Department of Computer Science, Chonnam National University,
300 Youngbong-dong, Buk-gu, Kwangju 500-757, Korea
{mgjung, gslee}@chonnam.chonnam.ac.kr

Abstract. To estimate an accuracy motion vector (MV), a two step search is generally used. In the first step, integer pixel points within a search area are examined to find the integer pixel MV. Then, in the second step, 8 half pixel points around the selected integer pixel MV are examined and the best matching point is chosen as the final MV. Many fast integer pixel motion estimation (ME) algorithms can be found by examining less than about 10 search points. However, the half pixel ME requires huge computational complexity. In this paper, We propose a new fast algorithm for half pixel ME that reduces the computational overhead by limiting the number of interpolations of the candidate half pixel points. The proposed method based on the property of MVs and the correlations between integer pixel MVs and half pixel MVs. Experimental results show that the speedup improvement of the proposed algorithm over a full half pixel search (FHPS), horizontal and vertical directions as references (HVDR), chen's half pixel search algorithm (CHPS-1) and a parabolic prediction-based fast half-pixel search (PPHPS) can be up to 1.4 ~ 3.9 times on average. Also the image quality improvement can be better up to 0.05(dB)~ 0.1(dB) compare with CHPS and PPHPS.

1 Introduction

The most popular ME and motion compensation method has been the block-based motion estimation, which uses a block matching algorithm (BMA) to find the best matched block from a reference frame. ME based on the block matching is adopted in many existing video coding standards such as H.261/H.263 and MPEG-1/2/4.

Generally, ME consists of two steps, the integer pixel ME and the half pixel ME. For the first step, the integer pixel ME, many search algorithms such as Diamond Search (DS) [1, 2], New Three Step Search (NTSS) [3], HEXagon-Based Search (HEXBS) [4], Motion Vector Field Adaptive Search Technique (MVFAST) [5] and Predictive Motion Vector Field Adaptive Search Technique (PMVFAST) [6] have been proposed to reduce the computational complexity.

---
* corresponding author.

Most recent fast methods for the integer pixel ME can be found by examining less than about 10 search points. As a consequence, the computation complexity of half pixel ME becomes comparable to that of integer pixel ME and the development of a fast half pixel ME algorithm becomes important. For the second step, the half pixel ME, FHPS that is a typical method, a half pixel search generally needs to check all 8 half pixels to find the final MV. FHPS requires huge computational complexity. Hence, it becomes more meaningful to reduce the computational complexity of half pixel ME. For these reasons, a few methods such as HDVR [7], CHPS-1 [8] and PPHPS [8] have been developed to accelerate the half pixel ME by proposing the prediction models. As a result, the number of half pixels needed for the half pixel MV can be reduced.

In this paper, we propose a new fast algorithm for half pixel ME that reduces the computational overhead by limiting the number of interpolations of the candidate half pixel points. The proposed method based on the property of MVs and the correlations between integer pixel MVs and half pixel MVs. As a result, we reduce the total number of search points used to find the half pixel MV of the current block and improve the ME accuracy.

This paper is organized as follows. Section 2 describes the observation of the property of motion vector. The proposed algorithm is described in Section 3. Section 4 reports the simulation results and conclusions are given in Section 5.

2 Observation of the Property of Motion Vector

This section introduces the observation of the property of MV. We propose a new fast and accurate algorithm for half pixel MV based on the property of MV from several commonly used test image sequences. The property of MV is represented in the following subsections. Table 1 and Table 2 documents the MV distribution probabilities within certain distances from the search window center by exploiting the FS algorithm to three commonly used test image sequences, "Akiyo", "Carphone" and "Table", based on the SAD matching criterion.

2.1 The Ratio of Integer Pixels and Half Pixels in the Final Motion Vector

Table 1 shows that the ratio of integer pixels and half pixels in the final MV is about 24 to 1 in small motion case and about 2 to 1 in large motion case. In this case, the most of integer pixels for the final MV is the search origin (0,0). And the probability of the MVs located at the horizontal and vertical direction of the search center is about 79% (in small motion case) $\sim$ 66% (in large motion case) and the probability of the MVs located at the diagonal direction of the search center is about 21% (in small motion case) $\sim$ 34% (in large motion case). Also, the half pixels for the final MV are concentrated in the horizontal and vertical direction instead of the diagonal direction of the search center. In this paper, The proposed algorithm is based on the ratio of integer pixels and half pixels in the final MV and the cross center-biased distribution property of half pixel MVs.

Table 1. The ratio of integer pixels and half pixels in the final motion vector

		Akiyo		Carphone		Table	
Integer Pixel MV : C(0,0)		95.07%		68.77%		64.01%	
Half Pixel MV		3.93%		30.22%		34.03%	
Diagonal Direction	Cross Direction	21.03%	78.97%	35.02%	64.98%	33.95%	66.05%

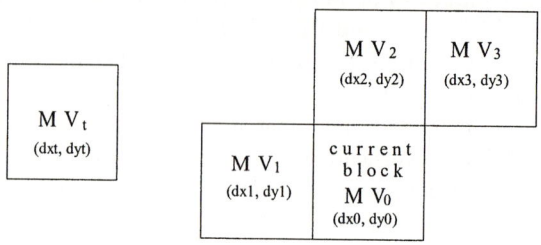

MV$_t$: the MV of the same coordinate block in the reference frame
MV$_0$: the MV of the search origin (0,0) of the current block
MV$_1$: the MV of left block
MV$_2$: the MV of above block
MV$_3$: the MV of above - right block

Fig. 1. Blocks for spatial-temporal correlation information

2.2 Spatial-Temporal Correlation of Motion Vector

Since the time interval between successive frames is very short, there are high temporal correlations between successive frames of a video sequence. In other words, the motion of current block is very similar to that of the same coordinate block in the reference frame in table 2. And also there are high spatial correlations among the blocks in the same frame. The proposed method exploits spatially and temporally correlated MVs depicted in Fig. 1. In Fig. 1, the MV mv_t is the block with the same coordinate in the reference frame, the MV mv_0 is the search origin (0,0) of the current block, the MV mv_c is MV of the current block, and the MVs mv_1, mv_2, and mv_3 are MVs of neighboring blocks in the current frame. If the information of spatially and temporally correlated motion vectors is used to decide one among 8 half pixel search points, the half pixel MV will be found with much smaller number of search points. As indicated in Table 3, about 98.75% (in small motion case) $\sim$ 64.77% (in large motion case) of the MVs are same MV at same coordinates in two successive frames. And about 98.79% (in small motion case) $\sim$ 48.67% (in large motion case) of the MVs are same MV at neighboring blocks in the current frame.

2.3 The Correlations Between Integer Pixel Motion Vectors and Half Pixel Motion Vectors

The integer pixel positions and the half pixel positions are shown in Fig. 2(a). In Fig. 2(a), assume the integer pixel C in the current frame, which is pointed to by the estimated integer MV from the previous frame, has 8 neighbors of

Table 2. Spatial-temporal Correlation of Motion Vector

		Akiyo	Carphone	Table
Spatial Correlation	$mv_1=mv_2=mv_3=mv_c$	88.86%	41.73%	39.97%
	$mv_1=mv_2=mv_3=(0,0)$	(100%)	(99.97%)	(99.93%)
	$mv_1=mv_2=mv_3=mv_t$	88.70%	38.83%	39.60%
	$mv_1=mv_2=mv_3=mv_t=(0,0)$	(100%)	(99.98%)	(99.93%)
Temporal Correlation	$mv_t=mv_c$	93.03%	55.74%	58.93%
	(The ratio of integer pixels)	(96.90%)	(75.98%)	(79.64%)

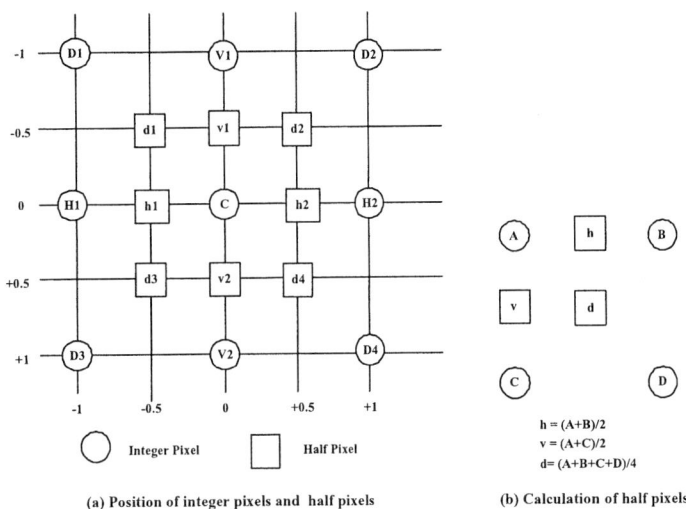

(a) Position of integer pixels and half pixels   (b) Calculation of half pixels

Fig. 2. The position and calculation of half pixel

integer pixels H_1, H_2, V_1, V_2, D_1, D_2, D_3, D_4 and 8 neighbors of half pixels h_1, h_2, v_1, v_2, d_1, d_2, d_3, d_4. If the centered integer pixel point (C) is selected as a minimum distance block at the integer pixel search step, then this point and 8 neighbouring interpolated half pixel candidates will be selected at the half pixel search step. The formulas for the half pixel interpolations are shown in Fig. 2(b). The half pixel MV motion estimation is the most computationally intensive part. In Fig. 2(b), Therefore, we propose a new fast algorithm using the correlations between integer pixel MVs and half pixel MVs.

3 The Proposed Algorithm

We propose a new fast algorithm for half pixel accuracy ME that reduces the computational overhead by limiting the number of interpolations of the candidate half pixel points. The proposed algorithm contain two main steps : In the first step, the final integer pixel MV at the integer pixel search step is examined to check the final integer pixel MV with the search origin (0,0). If the final integer pixel MV is the search origin (0,0), the MVs mv_1, mv_2, mv_3 and mv_t in Fig. 1 are examined. Then, in the second step, if the final integer pixel MV is not the search origin (0,0), 3 candidate half pixel points instead of 8 half pixel points around the selected integer pixel MV are examined using the correlations between integer pixel MVs and half pixel MVs. The proposed method is summarized as follows.

Step 1. Decide Categories to Which the Current Macroblock Belongs.
The final integer pixel MV at the integer pixel search step is examined to check the final integer pixel MV C with the search origin (0,0). If C is the search origin (0,0), the MVs mv_1, mv_2, mv_3 and mv_t are utilized to divide motion flow into two categories as follows.

    I. All motion vectors are equal, which is $mv_1=mv_2=mv_3=mv_t$
    II. Some of motion vectors are not equal

Step 2. Decide the Candidate Search Points.

    Category I: If all motion vectors are equal to the search origin (0,0), which is $mv_1=mv_2=mv_3=mv_t=(0,0)$, the the search origin (0,0) is selected as the final MV without the half pixel search step.

    Category II: Some motion flow of spatial macroblocks are similar to temporal macroblock or not related at all.
    The SAD_0 is the SAD of the search origin (0,0), the SAD_{H1} is the SAD of H_1, the SAD_{H2} is the SAD of H_2, the SAD_{V1} is the SAD of V_1, the SAD_{V2} is the SAD of V_2, the SAD_{D1} is the SAD of D_1, the SAD_{D2} is the SAD of D_2, the SAD_{D3} is the SAD of D_3, and the SAD_{D4} is the SAD of D_4 are calculated. The lowest SAD among SAD_0, SAD_{H1}, SAD_{H2}, SAD_{V1}, SAD_{V2}, SAD_{D1}, SAD_{D2}, SAD_{D3} and SAD_{D4} is calculated. The motion flow is categorized in two cases according to the lowest SAD.

    Case 1 : The lowest SAD is one among SAD_{H1}, SAD_{H2}, SAD_{V1} and SAD_{V2} on horizontal and vertical direction.
    If the SAD of the integer pixels on horizontal and vertical direction is the lowest SAD, the half pixel point correspond to integer pixel point of the lowest SAD and its neighbouring points are selected as the candidate half pixel points. For example, the integer pixel point of the lowest SAD is H_1, the candidate half pixel points are h_1, d_1 and d_3. The lowest SAD among the SAD of C, h_1, d_1 and d_3 is selected as the finial MV.

    Case 2 : The lowest SAD is one among SAD_{D1}, SAD_{D2}, SAD_{D3} and SAD_{D4} on diagonal direction.
    If the SAD of the integer pixels on diagonal direction is the lowest SAD, The half pixel point correspond to integer pixel point of the lowest SAD and its neighbouring points are selected as the candidate half pixel points. For example, the integer pixel point of the lowest SAD is D_1, the candidate half pixel points are d_1, h_1 and v_1. The lowest SAD among the SAD of C, d_1, h_1 and v_1 is selected as the finial MV.

4   Simulation Result

In this section, we show the experiment results for the proposed algorithm. We compared FHPS, HVDR, CHPS-1 and PPHPS with the proposed method in both image quality and search speed. Nine QCIF test sequences are used for the experiment: Akiyo, Carphone, Claire, Foreman, Mother and Daughter, Salesman, Table, Stefan and Suzie. The mean square error (MSE) distortion function is used as the block distortion measure (BDM). The quality of the predicted image is measured by the peak signal to noise ratio (PSNR), which is defined by

$$MSE = \left(\frac{1}{MN}\right) \sum_{m=1}^{M} \sum_{n=1}^{N} [x(m,n) - \hat{x}(m,n)]^2 \tag{1}$$

$$PSNR = 10 \, log_{10} \frac{255^2}{MSE} \tag{2}$$

In Eq. (1), $x(m,n)$ denotes the original image and $\hat{x}(m,n)$ denotes the reconstructed image. From Table 3 and 4, we can see that the proposed method is better than FHPS, HVDR, CHPS-1 and PPHPS in terms of the computational complexity (as measured by the average number of search points per motion vector) and is better than CHPS-1 and PPHPS in terms of PSNR of the predicted image. In terms of PSNR, the proposed method is about 0.06 (dB) better than HVDR in stationary sequences such as Table and about 0.05(dB)~ 0.1(dB) compare with CHPS-1 and PPHPS in Table 3. In terms of the average number of search points per MV, experiments in Table 4 show that the speedup improvement of the proposed algorithm over FHPS, HVDR, CHPS-1 and PPHPS can be up to 1.4 ~ 3.9 times on average. As a result, we can estimate MV fast and accurately.

Table 3. Average PSNR of the test image sequence

Integer-pel ME method	Full Search				
Half-pel Method	FHSM	HVDR	CHPS-1	PPHPS	Proposed
Akiyo	35.374	35.227	35.319	35.046	35.104
Carphone	32.213	32.161	32.181	32.172	32.181
Claire	35.694	35.619	35.655	35.466	35.513
Foreman	31.020	30.940	30.952	30.880	30.942
M&D	32.482	32.398	32.420	32.317	32.320
Salesman	33.706	33.611	33.662	33.514	33.665
Table	32.909	32.843	32.863	32.716	32.902
Stefan	28.095	28.054	28.063	27.934	28.069
Suzie	34.276	34.209	34.262	34.140	34.247
Average	32.863	32.785	32.820	32.687	32.771

Table 4. Average number of search points per half pixel motion vector estimation

Integer-pel ME method	Full Search				
Half-pel Method	FHSM	HVDR	CHPS-1	PPHPS	Proposed
Akiyo	8	5	4	3	1.10
Carphone	8	5	4	3	2.44
Claire	8	5	4	3	1.32
Foreman	8	5	4	3	2.71
M&D	8	5	4	3	1.72
Salesman	8	5	4	3	1.29
Table	8	5	4	3	2.47
Stefan	8	5	4	3	2.71
Suzie	8	5	4	3	2.48
Average	8	5	4	3	2.03

5 Conclusion

In this paper, we propose a new fast algorithm for half pixel accuracy ME that reduces the computational overhead by limiting the number of interpolations of the candidate half pixel points. The proposed method makes an accurate estimate of the half pixel MV using the property of MV and the correlations between integer pixel MVs and half pixel MVs. As a result, we reduce the total number of search points used to find the half pixel MV of the current block and improve the ME accuracy.

Acknowledgement

This work was supported by grant No.R05-2003-000-11345-0 from the basic Research Program of the Korea Science & Engineering Foundation.

References

1. Tham, J.Y., Ranganath, S., Kassim, A.A.: A Novel Unrestricted Center-Biased Diamond Search Algorithm for Block Motion Estimation. IEEE Transactions on Circuits and Systems for Video Technology. **8(4)** (1998) 369–375
2. Shan, Z., Kai-kuang, M.: A New Diamond Search Algorithm for Fast block Matching Motion Estimation.IEEE Transactions on Image Processing. **9(2)** (2000) 287–290
3. Renxiang, L., Bing, Z., Liou, M.L.: A New Three Step Search Algorithm for Block Motion Estimation. IEEE Transactions on Circuits and Systems for Video Technology. **4(4)** (1994) 438–442
4. Zhu, C., Lin, X., Chau, L.P.: Hexagon based Search Pattern for Fast Block Motion Estimation. IEEE Transactions on Circuits and Systems for Video Technology. **12(5)** (2002) 349–355
5. Ma, K.K., Hosur, P.I.: Report on Performance of Fast Motion using Motion Vector Field Adaptive Search Technique. ISO/IEC/JTC1/SC29/WG11.**M5453** (1999)

6. Tourapis, A.M., Au, O.C., Liou, M.L.: Optimization Model Version 1.0, ISO/IEC JTC1/SC29/WG11 **M5866** (2000).
7. K.H., Lee, J.H., Choi, B.K., Lee and D.G., Kim: Fast Two-step Half-Pixel Accuracy Motion Vector Prediction. Electronics Letters, **36**, (2000) 625–627
8. C., Du and Y., He: A Comparative Study of Motion Estimation for Low Bit Rate Video Coding. VCIP2000, **4067(3)**, (2000) 1239–1249

An Efficient Half Pixel Motion Estimation Algorithm Based on Spatial Correlations

HyoSun Yoon[1], GueeSang Lee[1,*], and YoonJeong Shin[2]

[1] Department of Computer Science, Chonnam National University,
300 Youngbong-dong, Buk-gu, Kwangju 500-757, Korea
estheryoon@hotmail.com, gslee@chonnam.ac.kr
[2] Department of Computer Science Engineering, Gwangju University,
592-1 Jinwol-dong Namgu, Gwnagju 503-703, Korea
syj@gwangju.ac.kr

Abstract. Motion estimation is an important part of video encoding systems, because it can significantly affect the output quality and the compression ratio. Motion estimation which consists of integer pixel motion estimation and half pixel motion estimation is very computationally intensive part. To reduce the computational complexity, many methods have been proposed in both integer pixel motion estimation and half pixel motion estimation. For integer pixel motion estimation, some fast methods could reduce their computational complexity significantly. There remains, however, room for improvement in the performance of current methods for half pixel motion estimation. In this paper, an efficient half pixel motion estimation algorithm based on spatial correlations is proposed to reduce the computational complexity. According to spatially correlated information, the proposed method decides whether half pixel motion estimation is performed or not for the current block. Experimental results show that the proposed method outperforms most of current methods in computation complexity by reducing the number of search points with little degradation in image quality. When compared to full half pixel search method, the proposed algorithm achieves the search point reduction up to 96% with only 0.01 ∼ 0.1 (dB) degradation of image quality.

1 Introduction

Recently, great interest has been devoted to the study of different approaches in video compressions. The high correlation between successive frames of a video sequence makes it possible to achieve high coding efficiency by reducing the temporal redundancy. Motion estimation (ME) and motion compensation techniques are an important part of video encoding systems, since it could significantly affect the compression ratio and the output quality. But, ME is very computational intensive part.

---

*corresponding author.

Generally, ME is made of two parts, integer pixel motion estimation and half pixel motion estimation. For the first part, integer pixel motion estimation, many search algorithms such as Diamond Search (DS) [1, 2], Three Step Search (TSS) [3], New Three Step Search (NTSS) [4], Four Step Search (FSS) [5], Two Step Search (2SS) [6], Two-dimensional logarithmic search algorithm [7], HEXagon-Based Search (HEXBS) [8], Motion Vector Field Adaptive Search Technique (MVFAST) [9] and Predictive MVFAST (PMVFAST) [10] have been proposed to reduce the computational complexity. Some fast integer pixel motion estimation algorithms among these algorithms can find an integer pixel Motion Vector (MV) by examining less than 10 search points. For the second part, half pixel motion estimation, Full Half pixel Search Method (FHSM) that is a typical method, examines eight half pixel points around the integer motion vector to determine a half pixel motion vector. This method takes nearly half of the total computations in the ME that uses fast algorithms for integer pixel motion estimation. Therefore, it becomes more important to reduce the computational complexity of half pixel motion estimation. For these reasons, Horizontal and Vertical Direction as Reference (HVDR) [11], the Parabolic Prediction-based, Fast Half Pixel Search algorithm (PPHPS) [12], Chen's Fast Half Pixel Search algorithm (CHPS)[13] and the methods [14–16] have been proposed to reduce the computational complexity of half pixel motion estimation. Since these algorithms do not have any information on the motion of the current block, they always perform half pixel motion estimation to find a half pixel motion vector.

In this paper, we propose an efficient half pixel motion estimation algorithm based on spatial correlations among integer and half pixel motion vectors to reduces more computational complexity. According to spatially correlated information, the proposed method decides whether half pixel motion estimation is performed or not for the current block. Experimental results show that the proposed method reduces the computational complexity significantly when compared to that of FHSM with a little degradation of image quality.

This paper is organized as follows. Section 2 describes the previous works. The proposed method is described in Section 3. Section 4 reports the simulation results and conclusions are given in Section 5.

2 The Previous Works

In Motion Estimation and Compensation, half pixel motion estimation is used to reduce the prediction error between the original image and the predicted image. FHSM that is a typical method, examines eight half pixel points around the integer motion vector 'C' illustrated in Fig. 1. The cost function values of eight half pixel points are calculated to find the best matching point. Finally, the half pixel motion vector is obtained by comparing the cost function value of the best matching point with that of the point 'C'. This method takes nearly half of the total computations in the ME. Therefore, it becomes more important to reduce the computational complexity of half pixel motion estimation. For these reasons, some fast half pixel motion estimation algorithms have been proposed.

In HVDR which is one of fast half pixel motion estimation algorithms, 2 neighboring half pixel points in vertical direction and 2 neighboring half pixel points in horizontal direction around the integer motion vector 'C' illustrated in Fig. 1. are examined to decide the best matching point in each direction. Then, a diagonal point between these two best matching points is also examined. The point having the minimum cost function value among these 5 points and the point 'C' is decided as a half pixel motion vector. In HVDR, only 5 half pixel points are checked to find a half pixel motion vector.

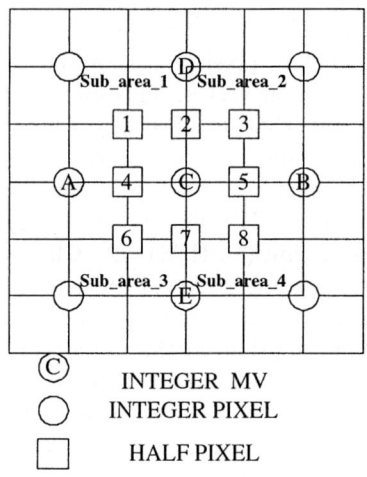

Fig. 1. The Position of integer pixels, half pixels and Subareas

CHPS examines 4 horizontal and vertical half pixel points '2','4','5','7' shown Fig. 1. The best matching point is decided as a half pixel motion vector by comparing the cost function values of these 4 half pixel points and the point 'C'. In CHPH, only 4 half pixel points are checked to find a half pixel motion vector.

PPHPS predicts the possible optimal half pixel point by using the cost function values of 5 integer pixel points 'A','B','C','D','E' shown Fig. 1. The cost function values of the optimal half pixel point and its nearest points are calculated to find the best matching point. The point of the minimum cost function value is decided as a final half pixel MV by comparing the cost function value of this best matching point with that of the point 'C'. In PPHPS, only 3 half pixel points are checked to find a half pixel motion vector.

3 The Proposed Method

In order to reduce the computational complexity of half pixel motion estimation, the proposed method exploits spatial correlations among integer and half pixel

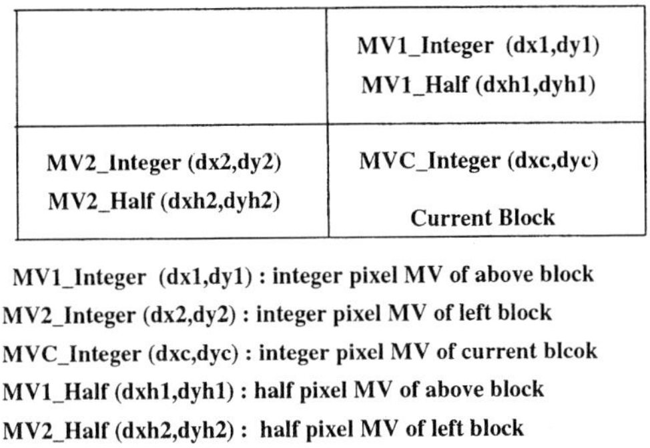

	MV1_Integer (dx1,dy1)
	MV1_Half (dxh1,dyh1)
MV2_Integer (dx2,dy2)	MVC_Integer (dxc,dyc)
MV2_Half (dxh2,dyh2)	Current Block

MV1_Integer (dx1,dy1) : integer pixel MV of above block
MV2_Integer (dx2,dy2) : integer pixel MV of left block
MVC_Integer (dxc,dyc) : integer pixel MV of current blcok
MV1_Half (dxh1,dyh1) : half pixel MV of above block
MV2_Half (dxh2,dyh2) : half pixel MV of left block

Fig. 2. Blocks for Spatial Correlation Information

motion vectors to decide whether half pixel motion estimation is performed or not for the current block.

In other words, the proposed method exploits spatially correlated motion vectors depicted in Fig. 2. to decide whether the half pixel motion estimation is performed or not for the current block. In case half pixel motion estimation is performed, the proposed method predicts the possible subarea by using the cost function values of integer pixel points. According to the position of the possible subarea, three half pixel points in its possible subarea are examined to find a half pixel motion vector. In this paper, we proposed Yoon's Fast Half Pixel Search algorithm (YFHPS) as a search pattern for half pixel motion estimation. At first, YFHPS decides the best horizontal matching point between 2 horizontal integer pixel points 'A', 'B' depicted in Fig. 1. and the best vertical matching point between 2 vertical integer pixel points 'D', 'E' depicted in Fig. 1. And then, the possible subarea is selected by using the best horizontal and vertical matching points. According to the position of the possible subarea, three half pixel points in its possible subarea are examined. Finally, the point having the minimum cost function value among these three half pixel points and the point 'C' that is pointed by the integer MV shown Fig. 1. is decided as a half pixel motion vector. For example, assumes that 'A' and 'D' are the best horizontal and vertical matching points respectively. The Sub_area_1 between these two best matching points is selected as the possible subarea. The half pixel points '1', '2', '4' in Sub_area_1 are examined. The point having the minimum cost function value among these three half pixel points and the point'C' is decided as a half pixel motion vector.

The block diagram of the proposed method appears in Fig. 3. The proposed method is summarized as follows.

Step 1. If MVC_Integer (dxc,dyc), the integer pixel MV of the current block shown in Fig. 2., is equal to (0,0), go to Step 2. Otherwise, go to Step 3.

Step 2. I. If MV1_Integer (dx1, dy1) which is the integer pixel MV of the above block shown in Fig. 2., and MV2_Integer (dx2, dy2) which is the

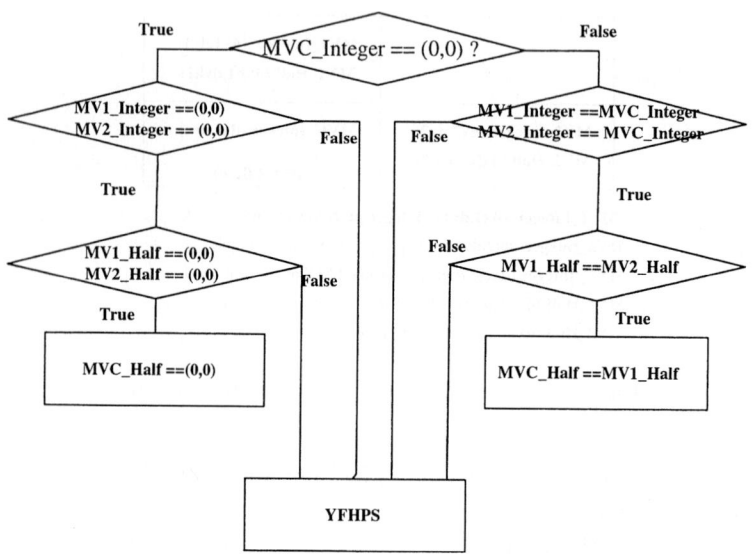

Fig. 3. The Block Diagram of the proposed method

integer pixel MV of the left block shown in Fig. 2., are equal to (0,0), go to II. Otherwise, go to III.

II. If MV1_Half (dxh1, dyh1) which is the half pixel MV of the above block shown in Fig. 2., and MV2_Half (dxh2, dyh2) which is the half pixel MV of the left block shown in Fig. 2., are equal to (0,0), (0,0) is decided as the half pixel MV of the current block. In other words, half pixel motion estimation is not performed for the current block. Otherwise, go to III.

III. YFHPS is performed to find a half pixel motion vector.

Step 3. I. If MV1_Integer (dx1, dy1) and MV2_Integer (dx2, dy2) are equal to MVC_Integer (dxc, dyc), go to II. Otherwise, go to III.

II. If MV1_Half (dxh1, dyh1) is equal to MV2_Half (dxh2, dyh2), (dxh2, dyh2) is decided as the half pixel MV of the current block. In other words, half pixel motion estimation is not performed. Otherwise, go to III.

III. YFHPS is performed to find a half pixel motion vector.

4 Simulation Result

In this section, we show experimental results for the proposed method. The proposed method has been evaluated in the H.263 encoder. Ten QCIF test sequences are used for the experiment: Akiyo, Carphone, Claire, Foreman, Mother and Daughter, News, Salesman, Silent, Stefan and Suzie. The mean square error (MSE) distortion function is used as the block distortion measure (BDM). The quality of the predicted image is measured by the peak signal to noise ratio (PSNR), which is defined by

$$MSE = \left(\frac{1}{MN}\right) \sum_{m=1}^{M} \sum_{n=1}^{N} [x(m,n) - \hat{x}(m,n)]^2 \quad (1)$$

$$PSNR = 10 \, log_{10} \frac{255^2}{MSE} \quad (2)$$

In Eq. (1), $x(m,n)$ denotes the original image and $\hat{x}(m,n)$ denotes the motion compensated prediction image. For integer pixel motion estimation, Full Search algorithm is adopted. For half pixel motion estimation, we compared FHSM, HVDR, CHPS, PPHPS and YFHPS to the proposed method in both of image quality and search speed. The simulation results in Table 1 and 2 show that the search speed of the proposed method is faster than the other methods (FHSM, HVDR, CHPS, PPHPS and YFHPS) while its PSNR is similar to them except

Table 1. Average PSNR for half pixel motion estimation algorithms

Integer-pel ME method	Full search					
Half-pel ME method	FHSM	HVDR	CHPS	PPHPS	YFHPS	Proposed
Akiyo	34.5	34.41	34.46	34.43	34.5	34.40
Carphone	30.88	30.85	30.86	30.88	30.88	30.87
Claire	35.05	35.02	35.03	35.05	35.05	35.04
Foreman	29.54	29.52	29.50	29.51	29.51	29.48
M&D	31.54	31.50	31.54	31.52	31.54	31.46
News	30.59	30.49	30.54	30.57	30.57	30.51
Salesman	32.7	32.64	32.67	32.70	32.70	32.63
Silent	31.81	31.80	31.76	31.79	31.80	31.71
Stefan	23.89	23.85	23.86	23.87	23.87	23.82
Suzie	32.19	32.17	32.15	32.19	32.19	32.18

Table 2. The Number of Search points per half pixel MV

	FHSM	HVDR	CHPS	PPHPS	YFHPS	Proposed
Akiyo	8	5	4	3	3	0.29
Carphone	8	5	4	3	3	2.1
Claire	8	5	4	3	3	1.01
Foreman	8	5	4	3	3	2.5
M &D	8	5	4	3	3	1
News	8	5	4	3	3	0.76
Salesman	8	5	4	3	3	0.48
Silent	8	5	4	3	3	0.97
Stefan	8	5	4	3	3	2.09
Suzie	8	5	4	3	3	2.08
Average	8	5	4	3	3	1.3

for FHSM. In other words, the proposed method can achieves the search point reduction up to 96% with only 0.01 ~ 0.1 (dB) degradation of image quality When compared to FHSM.

5 Conclusion

Based on spatial correlations among integer pixel MVs and half pixel MVs, an efficient method for half pixel motion estimation is proposed in this paper. According to spatially correlated information, the proposed method decides whether half pixel motion estimation is performed or not for the current block. As a result, the proposed method reduce the computational complexity significantly. Experimental results show that the speedup improvement of the proposed method over FHSM can be up to 4 ~ 25 times faster with a little degradation of the image quality.

Acknowledgement

This Study was financially supported by special research fund of Chonnam National University in 2004.

References

1. Tham, J.Y., Ranganath, S., Kassim, A.A.: A Novel Unrestricted Center-Biased Diamond Search Algorithm for Block Motion Estimation. IEEE Transactions on Circuits and Systems for Video Technology. **8(4)** (1998) 369–375
2. Shan, Z., Kai-kuang, M.: A New Diamond Search Algorithm for Fast block Matching Motion Estimation. IEEE Transactions on Image Processing. **9(2)** (2000) 287–290
3. Koga, T., Iinuma, K., Hirano, Y., Iijim, Y., Ishiguro, T.: Motion compensated interframe coding for video conference. In Proc. NTC81. (1981) C9.6.1–9.6.5
4. Renxiang, L., Bing, Z., Liou, M.L.: A New Three Step Search Algorithm for Block Motion Estimation. IEEE Transactions on Circuits and Systems for Video Technology. **4(4)** (1994) 438–442
5. Lai-Man, P., Wing-Chung, M.: A Novel Four-Step Search Algorithm for Fast Block Motion Estimation. IEEE Transactions on Circuits and Systems for Video Technology. **6(3)** (1996) 313–317
6. Yuk-Ying, C., Neil, W.B.: Fast search block-matching motion estimation algorithm using FPGA. Visual Communication and Image Processing 2000. Proc. SPIE. **4067** (2000) 913–922
7. Jain, J., Jain, A.: Displacement measurement and its application in interframe image coding. IEEE Transactions on Communications. **COM-29** (1981) 1799–1808
8. Zhu, C., Lin, X., Chau, L.P.: Hexagon based Search Pattern for Fast Block Motion Estimation. IEEE Transactions on Circuits and Systems for Video Technology. **12(5)** (2002) 349–355

9. Ma, K.K., Hosur, P.I.: Report on Performance of Fast Motion using Motion Vector Field Adaptive Search Technique. ISO/IEC/JTC1/SC29/WG11.**M5453** (1999)
10. Tourapis, A.M., Liou, M.L.: Fast Block Matching Motion Estimation using Predictive Motion Vector Field Adaptive Search Technique. ISO/IEC/JTC1/SC29/WG11.**M5866** (2000)
11. Lee, K.H.,Choi, J.H.,Lee, B.K., Kim. D.G.: Fast two step half pixel accuracy motion vector prediction. Electronics Letters **36(7)**(2000) 625–627
12. Cheng, D., Yun, H., Junli, Z.: A Parabolic Prediction-Based, Fast Half Pixel Search Algorithm for Very Low Bit-Rate Moving Picture Coding. IEEE Transactions on Circuits and Systems for Video Technology. **13(6)** (2003) 514–518
13. Cheng, D., Yun, H.: A Comparative Study of Motion Estimation for Low Bit Rate Video Coding. SPIE **4067(3)**(2000) 1239–1249
14. Sender, Y., Yano, M.: A Simplified Motion Estimation using an approximation for the MPEG-2 real time encoder. ICASSP'95,(1995) 2273–2276
15. Choi, W.I., Jeon, B.W.: Fast Motion Estimation with Modified diamond search for variable motion block sizes. ICIP 2003. (2003) 371–374
16. Li, X., Gonzles, C.: A locally Quadratic Model of the Motion Estimation Error Criterion Function and Its Application to Subpixel Interpolations. IEEE Transactions on Circuits and Systems for Video Technology. **6(1)** (1996) 118–122

Multi-step Subdivision Algorithm for Chaikin Curves

Ling Wu, Jun-Hai Yong, You-Wei Zhang, and Li Zhang

School of Software, Tsinghua University,
Beijing 100084, P. R. China
wu102@mails.tsinghua.edu.cn

Abstract. A Chaikin curve is a subdivision curve. Subdivision begins from an initial control polygonal curve. For each subdivision step, all corners of the polygonal curve are cut off, and a new polygonal curve is thus produced as the input of the next subdivision step. In the limit of subdivision, a Chaikin curve is created. In this paper, a multi-step subdivision algorithm for generating Chaikin curves is proposed. For an arbitrary positive integer k, the algorithm builds the resultant polygonal curve of the kth subdivision step directly from the initial polygonal curve. Examples show that the new algorithm speeds up curve generation in several times.

Keywords: Subdivision scheme; Chaikin curve; Corner cutting.

1 Introduction

Research on subdivision schemes for generating curves and surfaces is popular in graphical modeling [1–3, et al.], animation [4, et al.] and CAD/CAM [5, 6, et al.] because of their stability in numerical computation and simplicity in coding. In 1974, Chaikin [2] proposed the first subdivision scheme for generating subdivision curves. Since then, more and more subdivision schemes come up including the Catmull-Clark subdivision method [1], the 4-point interpolatory subdivision scheme [7], and the $\sqrt{3}$ subdivision algorithm [8].

All the subdivision schemes in the literature to date start from an initial control polygon (i.e., a polygonal curve) or control mesh, denoted by $\mathbf{L}_0$, and refine the control polygon or mesh step by step. At each step, the subdivision scheme is performed on the control polygon or mesh $\mathbf{L}_k$ once, and results in a new control polygon or mesh $\mathbf{L}_{k+1}$, which is the input for the next subdivision step. Here, $\mathbf{L}_k$, where $k = 0, 1, 2, \cdots$, represents the resultant control polygon or mesh of the kth subdivision step. Thus, with any existing subdivision scheme, $\mathbf{L}_k$ must be calculated before $\mathbf{L}_{k+1}$ is generated. In this paper, a multi-step subdivision algorithm is proposed to build $\mathbf{L}_k$ directly from $\mathbf{L}_0$ for the Chaikin subdivision scheme. It offers at least the following two advantages:

(1) with the same initial control polygon $\mathbf{L}_0$, the polygonal curve $\mathbf{L}_k$ built with the multi-step subdivision algorithm in this paper is exactly the same as the result of the Chaikin subdivision scheme after the kth subdivision step, and
(2) it is not necessary to calculate $\mathbf{L}_i$, for all $i = 1, 2, \cdots, (k-1)$, before generating $\mathbf{L}_k$.

Thus, the new algorithm saves the memory space at least for storing the subdivision results of the previous subdivision steps, and experience shows that the new generation method is several times faster than the original Chaikin scheme for producing $\mathbf{L}_k$.

The remaining part of the paper is organized as follows. Section 2 briefly introduces the Chaikin subdivision scheme, which is from Reference [2]. Section 3 provides the multi-step subdivision algorithm for Chaikin curves. Some examples are given in Section 4. The last section addresses some concluding remarks of the paper.

2 Chaikin Subdivision Scheme

This section briefly introduces the subdivision scheme proposed by Chaikin in Reference [2]. Let

$$\mathbf{L}_k = \{\mathbf{P}_{0,k}, \mathbf{P}_{1,k}, \cdots, \mathbf{P}_{n_k,k}\},$$

where $k = 0, 1, 2, \cdots$, be the resultant control polygon of the kth subdivision step with $\mathbf{P}_{0,k}, \mathbf{P}_{1,k}, \cdots, \mathbf{P}_{n_k,k}$ being the control points of $\mathbf{L}_k$. The Chaikin subdivision scheme is a recursive procedure.

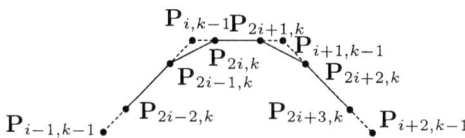

Fig. 1. Subdivision step k: $\mathbf{L}_{k-1}$ (*dashed*) is subdivided, and result in $\mathbf{L}_k$ (*solid*)

When $k = 0$, $\mathbf{L}_0 = \{\mathbf{P}_{0,0}, \mathbf{P}_{1,0}, \cdots, \mathbf{P}_{n_0,0}\}$ is the given initial polygonal curve. Figure 1 illustrates how the kth subdivision step of the Chaikin scheme is carried out. At this step, $\mathbf{L}_{k-1}$ is subdivided, and result in $\mathbf{L}_k$. For each line segment between $\mathbf{P}_{i,k-1}$ and $\mathbf{P}_{i+1,k-1}$, where $i = 0, 1, \cdots, (n_{k-1} - 1)$, of the polygonal curve $\mathbf{L}_{k-1}$, the two points at ratios of $\frac{1}{4}$ and $\frac{3}{4}$ between the endpoints $\mathbf{P}_{i,k-1}$ and $\mathbf{P}_{i+1,k-1}$ are taken as $\mathbf{P}_{2i,k}$ and $\mathbf{P}_{2i+1,k}$ in the polygonal curve $\mathbf{L}_k$. Thus, all points in $\mathbf{L}_k$ are obtained. Connect those points, and $\mathbf{L}_k$ is produced. As shown in Figure 1, a visual relationship between $\mathbf{L}_{k-1}$ and $\mathbf{L}_k$ is that each corner $\mathbf{P}_{i,k-1}$ of the polygonal curve $\mathbf{L}_{k-1}$ is cut by the line segment between $\mathbf{P}_{2i-1,k}$ and $\mathbf{P}_{2i,k}$ of the polygonal curve $\mathbf{L}_k$. Therefore, the Chaikin subdivision scheme is considered as a corner-cutting scheme as well.

3 Multi-step Subdivision Algorithm

This section gives explicit formulae and an algorithm for calculating $\mathbf{L}_k$ from $\mathbf{L}_0$ directly. The idea of the Chaikin subdivision scheme is corner cutting. All corners of the polygonal curve are cut recursively. Thus, as the subdivision steps increase, the resulting polygonal curves become ever smoother. In the limit, a smooth curve is produced. Analyzing the corners of the initial polygonal curve, we find that the points $\mathbf{P}_{2^k(i-1)+j,k}$, where $j = 1, 2, \cdots, 2^k$, of the polygonal curve $\mathbf{L}_k$ depend only on $\mathbf{P}_{i-1,0}$, $\mathbf{P}_{i,0}$ and $\mathbf{P}_{i+1,0}$, as shown in Figure 2. Thus, by mathematical induction, we obtain the theorem below.

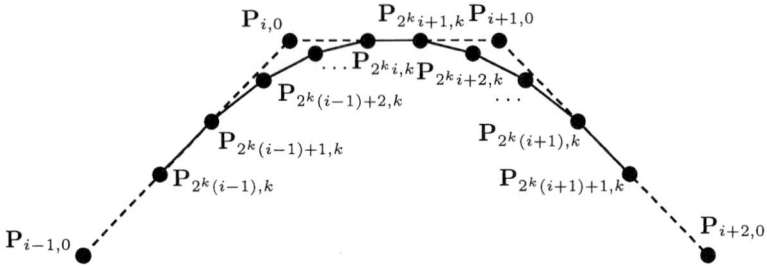

Fig. 2. Multi-step subdivision: $\mathbf{L}_k$ (*solid*) is calculated from $\mathbf{L}_0$ (*dashed*) directly

Theorem 1. *All points in $\mathbf{L}_k$ ($k = 1, 2, \cdots$) can be calculated directly from points in $\mathbf{L}_0$ by*

$$\begin{cases} \mathbf{P}_{0,k} = (2^{-1} + 2^{-(k+1)})\mathbf{P}_{0,0} + (2^{-1} - 2^{-(k+1)})\mathbf{P}_{1,0}, \\ \mathbf{P}_{2^k(i-1)+j,k} = F(j,k)\mathbf{P}_{i-1,0} + G(j,k)\mathbf{P}_{i,0} + H(j,k)\mathbf{P}_{i+1,0}, \\ \mathbf{P}_{2^k n_0 - 2^k + 1, k} = (2^{-1} - 2^{-(k+1)})\mathbf{P}_{n_0 - 1, 0} + (2^{-1} + 2^{-(k+1)})\mathbf{P}_{n_0, 0}, \end{cases} \quad (1)$$

where $i = 1, 2, \cdots, n_0 - 1$, $j = 1, 2, \cdots, 2^k$, and

$$\begin{cases} F(j,k) = 2^{-1} - 2^{-(k+1)} - (j-1)(2^{-k} - j2^{-2k-1}), \\ G(j,k) = 2^{-1} + 2^{-(k+1)} + (j-1)(2^{-k} - j2^{-2k}), \\ H(j,k) = (j-1)j2^{-2k-1}. \end{cases} \quad (2)$$

Proof. Here, the mathematical induction method is used. When $k = 1$, we have

$$\begin{cases} F(1,1) = \frac{1}{4}, \\ G(1,1) = \frac{3}{4}, \\ H(1,1) = 0, \end{cases} \quad \text{and} \quad \begin{cases} F(2,1) = 0, \\ G(2,1) = \frac{3}{4}, \\ H(2,1) = \frac{1}{4}. \end{cases}$$

Thus, from the conclusion in Theorem 1, we obtain

$$\begin{cases} \mathbf{P}_{2i,1} = \frac{3}{4}\mathbf{P}_{i,0} + \frac{1}{4}\mathbf{P}_{i+1,0}, \\ \mathbf{P}_{2i+1,1} = \frac{1}{4}\mathbf{P}_{i,0} + \frac{3}{4}\mathbf{P}_{i+1,0}, \end{cases}$$

for all $i = 0, 1, \cdots, n_0 - 1$. The above equations match the Chaikin subdivision scheme very well. Therefore, Theorem 1 is true when $k = 1$.

Now assume that Theorem 1 is true when $k = m$, where $m = 1, 2, \cdots$. The following will prove that Theorem 1 holds for the case $k = m + 1$. According to the Chaikin subdivision scheme, we have

$$\begin{cases} \mathbf{P}_{0,m+1} = \frac{3}{4}\mathbf{P}_{0,m} + \frac{1}{4}\mathbf{P}_{1,m}, \\ \mathbf{P}_{2^{(m+1)}(i-1)+j,m+1} = \frac{1}{4}\mathbf{P}_{2^m(i-1)+\frac{j-1}{2},m} + \frac{3}{4}\mathbf{P}_{2^m(i-1)+\frac{j+1}{2},m}, & \text{when } j \text{ is odd}, \\ \mathbf{P}_{2^{(m+1)}(i-1)+j,m+1} = \frac{3}{4}\mathbf{P}_{2^m(i-1)+\frac{j}{2},m} + \frac{1}{4}\mathbf{P}_{2^m(i-1)+\frac{j}{2}+1,m}, & \text{when } j \text{ is even}, \\ \mathbf{P}_{2^{(m+1)}n_0-2^{(m+1)}+1,m+1} = \frac{1}{4}\mathbf{P}_{2^m n_0-2^m,m} + \frac{3}{4}\mathbf{P}_{2^m n_0-2^m+1,m}, \end{cases}$$

where $i = 1, 2, \cdots, n_0 - 1$, and $j = 1, 2, \cdots, 2^{m+1}$. In the above equations, substitute all points in $\mathbf{L}_m$ with the expressions which contain only points in $\mathbf{L}_0$ according to the assumption for the case $k = m$, and we demonstrate that Theorem 1 is valid for the case $k = m + 1$. Thus, Theorem 1 is proved. □

From Equations (1) and (2), we have that the coefficients $F(j, k)$, $G(j, k)$ and $H(j, k)$ do not depend on indices i while $\mathbf{P}_{2^k(i-1)+j,k}$ are calculated. Therefore, $F(j, k)$, $G(j, k)$ and $H(j, k)$ can be precomputed and stored in arrays with size 2^k to accelerate the calculation. Thus, we obtain the algorithm as follows for generating $\mathbf{L}_k$ from $\mathbf{L}_0$ directly.

Algorithm 1. Computing $\mathbf{L}_k$ from $\mathbf{L}_0$ directly.
Input. $\mathbf{L}_0 = \{\mathbf{P}_{0,0}, \mathbf{P}_{1,0}, \cdots, \mathbf{P}_{n_0,0}\}$.
Output. $\mathbf{L}_k = \{\mathbf{P}_{0,k}, \mathbf{P}_{1,k}, \cdots, \mathbf{P}_{n_k,k}\}$.

(1) Calculate $F(j, k)$, $G(j, k)$ and $H(j, k)$, for all $j = 1, 2, \cdots, 2^k$, according to Equations (2), and store the results in three arrays;
(2) Compute $\mathbf{P}_{0,k}$ and $\mathbf{P}_{2^k n_0-2^k+1,k}$ with Equations (1);
(3) For $i = 1$ to $n_0 - 1$, do
    For $j = 1$ to 2^k, do
        Calculate $\mathbf{P}_{2^k(i-1)+j,k}$ with Equations (1);
(4) End of the algorithm.

4 Examples

Some examples are provided to illustrate the algorithm proposed in this paper. The first example is shown in Figure 3. The initial polygonal curve as shown in Figure 3(a) is

$$\mathbf{L}_0 = \left\{ \begin{pmatrix} 1 \\ 1 \end{pmatrix}, \begin{pmatrix} 2 \\ 2 \end{pmatrix}, \begin{pmatrix} 3 \\ 2 \end{pmatrix}, \begin{pmatrix} 4 \\ 1 \end{pmatrix} \right\}.$$

Fig. 3. Example 1: (a) $\mathbf{L}_0$; (b) $\mathbf{L}_1$; (c) $\mathbf{L}_2$; (c) $\mathbf{L}_3$

With Theorem 1, we obtain $\mathbf{L}_3$ directly from $\mathbf{L}_0$. The points obtained for $\mathbf{L}_3$ are $\begin{pmatrix}1.4375\\1.4375\end{pmatrix}$, $\begin{pmatrix}1.5625\\1.5625\end{pmatrix}$, $\begin{pmatrix}1.6875\\1.6719\end{pmatrix}$, $\begin{pmatrix}1.8125\\1.7656\end{pmatrix}$, $\begin{pmatrix}1.9375\\1.8438\end{pmatrix}$, $\begin{pmatrix}2.0625\\1.9063\end{pmatrix}$, $\begin{pmatrix}2.1875\\1.9531\end{pmatrix}$, $\begin{pmatrix}2.3125\\1.9844\end{pmatrix}$, $\begin{pmatrix}2.4375\\2.0000\end{pmatrix}$, $\begin{pmatrix}2.5625\\2.0000\end{pmatrix}$, $\begin{pmatrix}2.6875\\1.9844\end{pmatrix}$, $\begin{pmatrix}2.8125\\1.9531\end{pmatrix}$, $\begin{pmatrix}2.9375\\1.9063\end{pmatrix}$, $\begin{pmatrix}3.0625\\1.8438\end{pmatrix}$, $\begin{pmatrix}3.1875\\1.7656\end{pmatrix}$, $\begin{pmatrix}3.3125\\1.6719\end{pmatrix}$, $\begin{pmatrix}3.4375\\1.5625\end{pmatrix}$ and $\begin{pmatrix}3.5625\\1.4375\end{pmatrix}$, respectively. The results are the same as those produced by the method provided in Reference [2]. However, the method in Reference [2] has to calculate $\mathbf{L}_1$ and $\mathbf{L}_2$ before $\mathbf{L}_3$ is computed.

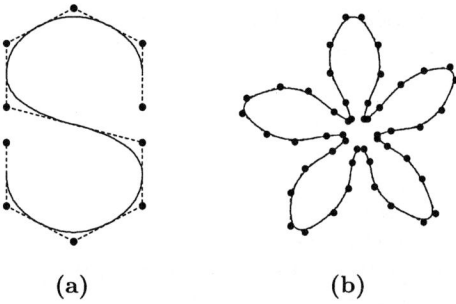

Fig. 4. Chaikin curves: (a) Example 2; (b) Example 3

Table 1. Performance results of the new algorithm and the method in [2]

	Example 2					Example 3			
k	(n_k+1)	$t_C(\text{ms})$	$t_M(\text{ms})$	$\frac{t_C(\text{ms})}{t_m(\text{ms})}$	k	(n_k+1)	$t_C(\text{ms})$	$t_M(\text{ms})$	$\frac{t_C(\text{ms})}{t_m(\text{ms})}$
2	34	111	75	1.48	2	194	128	76	1.68
4	130	379	161	2.35	4	770	398	150	2.65
6	514	780	244	3.20	6	3074	811	228	3.56
8	2050	1320	331	3.99	8	12290	1387	312	4.45
10	8194	2023	433	4.67	10	49154	2324	453	5.13
12	32770	2976	568	5.24	12	196610	4385	822	5.33

Examples 2 and 3 are used to illustrate the efficiency of the new algorithm proposed in this paper. The numbers of control points in the initial polygonal curves of Example 2 (see Figure 4(a)) and Example 3 (see Figure 4(b)) are 10 and 50, respectively. Table 1 gives the time cost of the multi-step subdivision algorithm compared with that of the method in Reference [2] for those two examples. In the table, t_C represents for the time cost used by the method in Reference [2] for generating the resultant polygonal curve $\mathbf{L}_k$, and t_M for the time cost by the new algorithm. All the data are calculated on a laptop personal computer with Celeron 1.7GHz CPU and 256M memory. The programming language is Java with JBuilder environment. As shown in the table, the new algorithm is several times faster than the method in Reference [2].

5 Conclusions

A new algorithm for calculating the resultant polygonal curve of the kth subdivision step for a Chaikin curve is proposed in this paper. It obtains results directly from the input polygonal curve, while the method in Reference [2] has to calculate all resultant polygonal curves from the first subdivision step to the kth subdivision step. Thus, the new algorithm has better space and time complexity than the method in Reference [2]. Examples shows that the new algorithm is several times faster than the method in Reference [2].

Acknowledgements

The research was supported by Chinese 863 Program (2003AA4Z3110) and 973 Program (2002CB312106). The second author was supported by a project sponsored by SRF for ROCS, SEM (041501004), and a Foundation for the Author of National Excellent Doctoral Dissertation of P. R. China (200342).

References

1. Catmull, E. and Clark, J.: Recursively generated B-spline surfaces on arbitrary topological meshes. Computer-Aided Design **10** (1978) 350–355
2. Chaikin, G.: An algorithm for high-speed curve generation. Computer Graphics and Image Processing **3** (1974) 346–349
3. Riesenfeld, R.F.: On Chaikin's algorithm. Computer Graphics and Image Processing **4** (1975) 304–310
4. Zorin, D., Schröder, P., and Sweldens, W.: Interactive multiresolution mesh editing. In: Proceedings of SIGGRAPH. (1997) 259–268
5. Litke, N., Levin, A., and Schröder, P.: Trimming for subdivision surfaces. Computer Aided Geometric Design **18** (2001) 463–481

6. Stam, J.: Exact evaluation of Catmull-Clark subdivision surfaces at arbitrary parameter values. In: Proceedings of SIGGRAPH. (1998) 395–404
7. Dyn, N., Levin, D., and Gregory, J.A.: A 4-point interpolatory subdivision scheme for curve design. Computer Aided Geometric Design **4** (1987) 257–268
8. Kobbelt, L.: $\sqrt{3}$ subdivision. In: Proceedings of SIGGRAPH. (2000) 103–112

Imaging Electromagnetic Field Using SMP Image

Wei Guo[1], Jianyun Chai[2], and Zesheng Tang[1]

[1] Department of Computer Science & Technology, Tsinghua University, Beijing, China
gw99@mails.tsinghua.edu.cn, ztang@must.edu.mo
[2] Department of Electrical Engineering, Tsinghua University, Beijing, China
chaijy@tsinghua.edu.cn

Abstract. This paper proposes a novel texture based method for visualization of electromagnetic fields. According to the characteristics of electric eddy current fields, a scalar potential is constructed to express the field on the surfaces. Then the potential is used in deducing an extension form of the *Sinusoidal function Modulated Potential* (SMP) *image* of general surface eddy current fields. The SMP Image can display the distribution of both the direction and amplitude of the field. The paper also presents a new algorithm based on hardware accelerated texture mapping. The algorithm could render the SMP image in real time and can be further used in dynamic visualization of time-varying electromagnetic field. The rendered dynamic images can reflect the variation of the field with space and time simultaneously. The only requirements for the method are that the vector field is distributed on surface and that its divergent is free, which makes the method suitable for visualization of general divergent-free vector field distributed on surface.

1 Introduction

Visualization of vector fields with computer graphics is very important in interpreting the structure of the field. Texture based visualization methods, such Line Integral Convolution (LIC) [1], produce a representation of the field at much higher resolution compared with the traditional flux line or hedgehog based method. These methods can display much more details of field. But the large computations needed in these methods make them inappropriate in real time applications such as visualization of time-varying field. Fast LIC (FLIC) [2] method greatly improves the rendering speed of LIC images, but it's still not fast enough for real time applications. Another problem with these methods is that amplitude information of the underlying field is missed in the final image [3].

LIC based methods have also been extended in imaging of vector field defined on 3D surfaces and in dynamic imaging of time-varying fields [4, 5, 6]. But they still suffer from the problems of missing of amplitude information and relatively low frame rate.

Sinusoidal function Modulated Potential (SMP) map is a new method for imaging of planar magnetic fields [7]. The method is based on the existence of a scalar magnetic potential. Modulating this potential with a periodic function and mapping the modulated potential onto grayscale pixels, the method renders soft and continuous images of the flux tubes of the magnetic field. Both the amplitude and direction information of the field can be shown.

In this paper, taking eddy current field as an example, we extend the SMP map method for imaging of electromagnetic field distributed on general surfaces. A scalar potential is constructed for eddy current field on conductor surface and the concept of SMP map is extended to the field on the surface. Then a new algorithm based on texture mapping for rendering SMP image on surfaces is presented. The use of texture mapping accelerates the rendering speed of the algorithm greatly. Finally, the application of this method in dynamic rendering of time-vary field is presented with several examples.

2 SMP Map of Eddy Current Field on Surface

2.1 Potential Function of Eddy Current Field on Surface

Eddy currents are induced in a conductor if the conductor is placed in a time-varying magnetic field. Since there is no eddy current flowing into or out of the surface of the conductor, the current density vector consists of only two tangent components at any point P on the conductor surface, i.e., $\vec{J} = J_s \hat{s} + J_\tau \hat{\tau}$ [8]. According to Maxwell equations, the current density vector is always divergent free, namely

$$\nabla \cdot \vec{J} = \frac{\partial J_s}{\partial s} + \frac{\partial J_\tau}{\partial \tau} + \frac{\partial J_n}{\partial n} = \frac{\partial J_s}{\partial s} + \frac{\partial J_\tau}{\partial \tau} = 0 \qquad (1)$$

If we construct a thin uniform shell near the conductor surface, and let all components of current density $\vec{J}$ along any normal line of the surface have the same value with those at the surface point (see Figure 1), then we get the relations $\partial J_s / \partial n = \partial J_\tau / \partial n = \partial J_n / \partial n = 0$ in the shell. With this construction we can define an adjoint vector $\vec{J}'$ in the shell as $\vec{J}' = \hat{n} \times \vec{J} = -J_\tau \hat{s} + J_s \hat{\tau}$. It is obvious that the curl of $\vec{J}'$ is zero in the shell,

$$\nabla \times \vec{J}' = -\frac{\partial J_s}{\partial n}\hat{s} - \frac{\partial J_\tau}{\partial n}\hat{\tau} + (\frac{\partial J_s}{\partial s} + \frac{\partial J_\tau}{\partial \tau})\hat{n} = 0 \qquad (2)$$

Therefore, the adjoint vector $\vec{J}'$ can be expressed by the gradient of a scalar potential Φ as

$$\vec{J}' = \nabla \Phi = \frac{\partial \Phi}{\partial s}\hat{s} + \frac{\partial \Phi}{\partial \tau}\hat{\tau} \qquad (3)$$

And Φ can be calculated by the integral of $\vec{J}'$ as

$$\Phi = \int_L \vec{J}\,' \cdot d\vec{l} = \int_L (\hat{n} \times \vec{J}) \cdot d\vec{l} \qquad (4)$$

where L is an arbitrary path started from a reference point on the conductor surface, and $\partial \Phi / \partial s = -J_\tau$, $\partial \Phi / \partial \tau = J_s$. Since the curl of the adjoint vector $\vec{J}\,'$ is zero, Φ will be only dependent on the distribution of the eddy current and the selection of the reference point. From the construction of the thin shell, we know that the scalar potential can only be defined on the conductor surface for the original eddy current field.

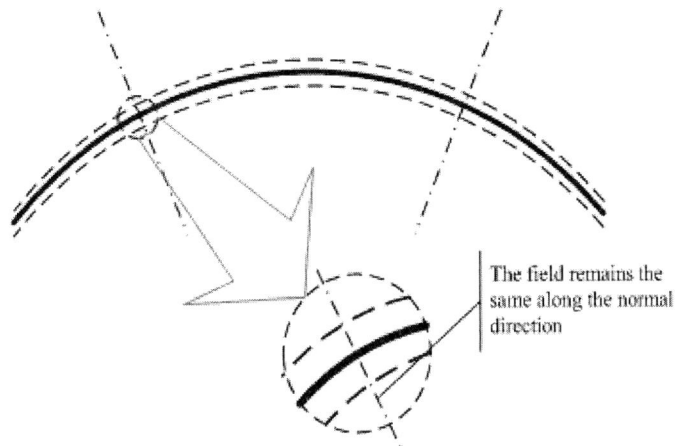

Fig. 1. Construct the thin shell for the definition of the scalar potential

Since the gradient of Φ is the adjoint vector $\vec{J}\,'$, the contour of Φ is perpendicular to $\vec{J}\,'$. Thus from $\vec{J}\,' = \hat{n} \times \vec{J}$ we can say that the direction of the contour of Φ is coincide with the direction of $\vec{J}$.

Another important fact about the potential Φ is that for any two points P and Q on the conductor surface, the difference of their potential represents the flux of eddy current flowing through the path PQ. That is

$$\begin{aligned}\Phi_Q - \Phi_P &= \int_P^Q \vec{J}\,' \cdot d\vec{l} = \int_P^Q (\hat{n} \times \vec{J}) \cdot d\vec{l} \\ &= \int_P^Q \vec{J} \cdot (d\vec{l} \times \hat{n}) = \int_P^Q \vec{J} \cdot (\hat{u}\,dl) = \int_P^Q (\vec{J} \cdot \hat{u})\,dl = \psi\end{aligned} \qquad (5)$$

where $\hat{u} = d\vec{l} \times \hat{n} / \|d\vec{l}\|$, $dl = \|d\vec{l}\|$. $\hat{u}$ is a unit vector along the conductor surface and is perpendicular to $\hat{n}$.

2.2 SMP Map of Eddy Current Field

In the case of planar magnetic field, potential map is generated by mapping potential into grayscale [7]. If the mapping function is periodic function, the generated potential map can display the direction and amplitude information of the filed with an image of flux tubes. With the definition of the potential function on the conductor surface, the concept of potential map can be extended to eddy current fields. The mapping from potential to grayscale is similar.

With different selections for the mapping function, the generated potential maps show different images of the original field. If we choose sinusoidal function as the mapping function $g(\Phi)$, the generated potential map is called Sinusoidal function Modulated Potential (SMP) map. The definition of such mapping function is

$$g(\Phi) = g_{av} + g_m \sin(\omega\Phi + \varphi_0) \qquad (6)$$

where g_{av} and g_m represent the average grayscale and the amplitude of the grayscale respectively; $\omega = 2\pi/\Phi_0$, with Φ_0 being the period of the mapping function; φ_0 represents the initial angle. Similar to the planar magnetic filed potential map in [7], the eddy current potential map can show the image of the eddy current flux tube on the conductor surface intuitively.

Figure 2 shows two SMP images of the eddy current distributed on the surface of a conductor sphere viewed from two different directions. The conductor sphere is placed in a uniform alternating magnetic field. The eddy current field on the conductor surface is also an alternating field. The image in Figure 2 is just the image of the field at a specific moment. From Eq. (5), the flux in each tube is always equal to the period Φ_0 of mapping function. Thus the field is strong if the width of the flux tube is small, and vice versa.

3 Texture Mapping Accelerated Rendering Algorithm

Source data used in visualization is usually obtained from numerical methods such as Finite Element Method (FEM). The surface is represented using polygonal meshes. The representation of the surface with polygonal meshes makes OpenGL a very suitable tool to render it into 2D image.

3.1 Rendering Algorithm

The texture mapping functionality of OpenGL can be utilized to do the potential interpolation and modulation automatically. The algorithm is briefly described as follows:

    i. Compute the potentials of the field on vertices as stated in the previous section;
    ii. Given the periodic modulating function $g(\Phi)$ with period Φ_0, generate a 1D texture with $T(s) = g(s \cdot \Phi_0), s \in [0,1]$ and set the wrapping mode in OpenGL texture mapping as "REPEAT".

iii. Set the 1D texture coordinate for each vertex P as Φ_p/Φ_0, submit the texture coordinates and positions of the vertices to OpenGL to render the image.

The procedure of the potential interpolation and modulation by 1D texture mapping are as follows. First, the interpolation of texture coordinate in each polygon of the meshes is used to automatically interpolate the potentials. Secondly, the "REPEAT" mode of texture mapping means that the texture mapping function $T(s)$ defined on coordinate s in interval [0, 1] is equivalent to a periodical function defined on the whole real axis with period 1.0. Thus, the final texture mapping function becomes

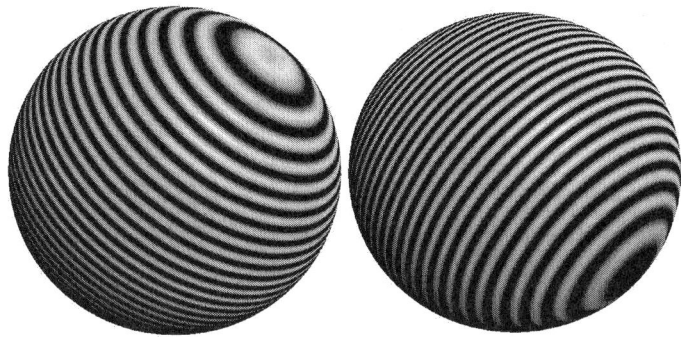

Fig. 2. The eddy current SMP images on the surface of a conductor sphere. A yellow light source with specular highlight effect is used to enhance the visual effect

$$T(s) = g(s \cdot \Phi_0), s \in (-\infty, +\infty) \tag{7}$$

Finally, taking Φ_p/Φ_0 as the texture coordinate for vertex P, the corresponding texel is

$$T(\Phi_p/\Phi_0) = g(\Phi_p/\Phi_0 \cdot \Phi_0) = g(\Phi_p) \tag{8}$$

where $g(\Phi_p)$ is just the modulated potential. Figure 3 shows the process of modulation using texture mapping. The pattern inside the red box in Figure 3(a) is the 1D texture data (2D image is used to show the pattern clearly). The flux tube marked as blue in Figure 3(b) is generated from one period of the 1D texture data which is also marked as blue.

Since texture mapping is supported by almost all graphics hardware, the algorithm is easy to implement and can render the SMP image in real time.

4 Dynamic Images of Time-Varying Field

Using a PC with ATI RADEON™ 7500 graphics card, the images such as in Fig. 2 can be rendered within 20ms. The total number of vertices on the conductor surface is over 16,000 and the total number of facets is over 32,000. Thus the algorithm is believed very suitable for visualization of time-varying eddy current field. The images in Figure 4 show a set of images of the eddy current field at different time. From the images, we can see clearly the variation of the field with time t in addition to the distribution of the field in space in each frame. For example, the amplitudes of the field are decreasing gradually from Figure 4(a) to (c), since the total number of flux tubes becomes smaller and smaller, while the flux in each tube remains unchanged (namely the period Φ_0 of the modulating function). In Figure 4(c), the field almost vanishes. From the animation of these image sequences, one can also observe the variation speed of the field from the changing rate of the number of the flux tubes.

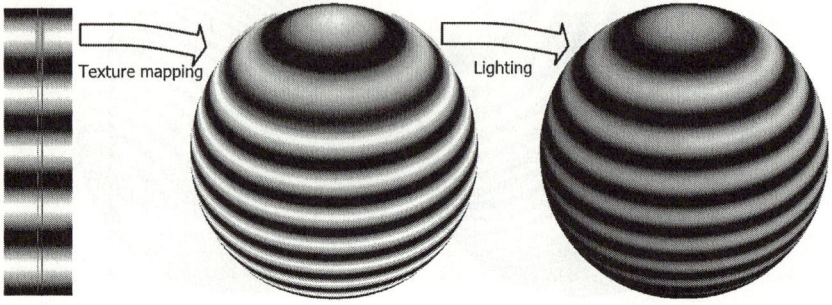

Fig. 3. Utilize texture mapping to automatically modulate the potential

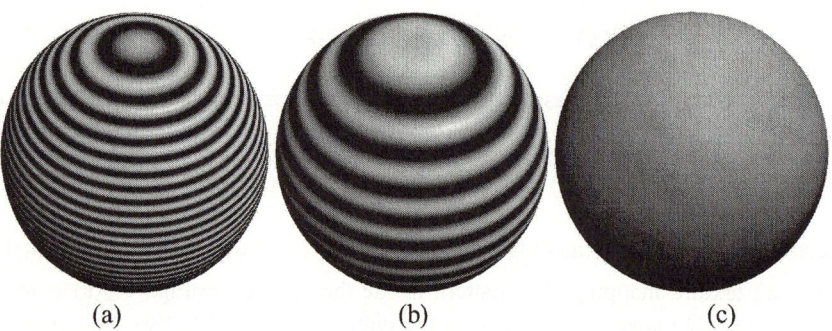

Fig. 4. Dynamic SMP image of eddy current field

The hardware accelerated algorithm can also be applied to planar field visualization. Figure 5 shows a set of images of the time-varying planar magnetic field excited by a pair of current filaments with opposite flow directions. Figure 6 shows a set of images of the time-varying field excited by 3 pairs of current filament with a 120^0 phase difference between each other (three-phase windings).

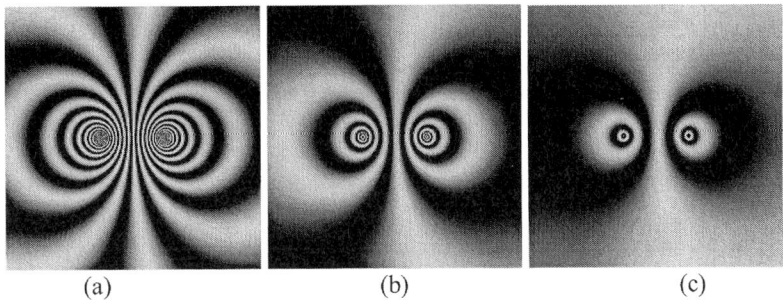

Fig. 5. Dynamic SMP image of planar magnetic field excited by a pair of current filaments

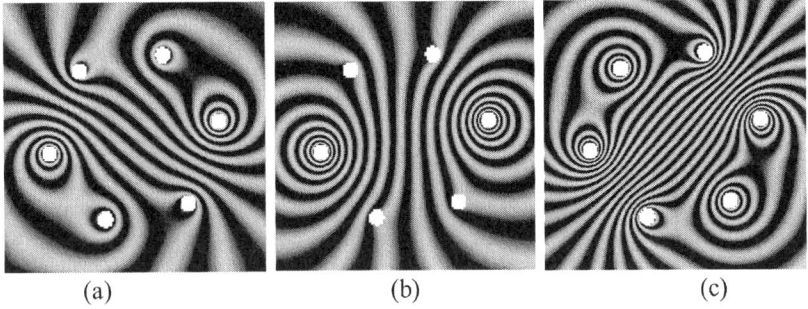

Fig. 6. Dynamic SMP image of planar magnetic field excited by 3-phase windings

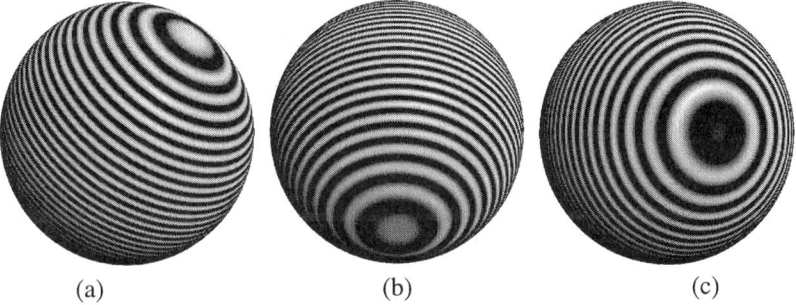

Fig. 7. Dynamic SMP image of eddy current field excited by a traveling magnetic field

The images in Figure 7 show the eddy current field excited by a traveling magnetic field. Two perpendicular uniform alternating magnetic fields in space with a phase difference of 90° in time are used as the source fields. From the images, we find that the total number of flux tubes doesn't vary a lot with time; but the pattern of the eddy current moves with time following the traveling excited magnetic field.

5 Conclusions

Taking eddy current field as an example, SMP method is extended for imaging of electromagnetic field distributed on general surfaces. A new algorithm is also presented for the rendering of the SMP images on surfaces. 1D texture mapping is utilized to do the modulation of the potential automatically. This makes it possible for the SMP image to be rendered in real time and possible for dynamic imaging of time-varying fields. In the rendered dynamic images, the distribution of the flux tubes in each frame reflects the distribution of the direction and amplitude of the field in space, while the changing of the numbers and the patterns of the flux tubes between successive frames gives an intuitive image of the variation of the field with time.

In the deducing of the scalar potential of eddy current field, the only requirements are that the eddy current field is distributed on the surface and its divergent is free. So SMP method can be generalized for imaging of other divergent-free vector fields distributed on surfaces.

References

1. Cabral, B., Leedom, L.: Imaging vector fields using line integral convolution". Proceedings of the ACM SIGGRAPH '93 Conference on Computer Graphics. (1993) 263-270
2. Stalling, D., Hege, H.C.: Fast and Resolution Independent Line Integral Convolution. Proceedings of the ACM SIGGRAPH '95 Conference on Computer Graphics. (1995) 249-256
3. Tang, Z.: Visualization of Three Dimensional Data Field. Tsinghua University Press, Beijing (1999, in Chinese).
4. Battke, H., Stalling, D., Hege, H.C.: Fast Line Integral Convolution for Arbitrary Surfaces in 3D. Visualization and Mathematics, Springer-Verlag, Heidelberg New York (1997) 181-195
5. Shen, H., Kao, D.: A New Line Integral Convolution Algorithm for Visualizing Time-Varying Flow Fields. IEEE Transaction on Visualization and Computer Graphics. (1998) 98-108
6. Sundquist, A.: Dynamic Line Integral Convolution for Visualizing Streamline Evolution. IEEE Transaction on Visualization and Computer Graphics. (2003) 273-282
7. Zhao, Y., Chai, J.: Imaging planar magnetic vector field via SMP. Journal of Tsinghua Univ, 2002, 42(9): 1200-1203. (in Chinese).
8. Smythe, W.R.: Static and Dynamic Electricity. 2nd edn. McGraw-Hill Book Company, Inc., New York (1950)

Support Vector Machine Approach for Partner Selection of Virtual Enterprises*

Jie Wang[1,2], Weijun Zhong[1], and Jun Zhang[2]

[1] School of Economics & Management, Southeast University, Nanjing,
Jiangsu 210096, China
[2] Laboratory for High Performance Scientific Computing and Computer Simulation,
Department of Computer Science, University of Kentucky,
Lexington, KY 40506, USA
jiewang@uky.edu, jzhang@cs.uky.edu

Abstract. With the rapidly increasing competitiveness in global market, dynamic alliances and virtual enterprises are becoming essential components of the economy in order to meet the market requirements for quality, responsiveness, and customer satisfaction. Partner selection is a key stage in the formation of a successful virtual enterprise. The process can be considered as a multi-class classification problem. In this paper, The Support Vector Machine (SVM) technique is proposed to perform automated ranking of potential partners. Experimental results indicate that desirable outcome can be obtained by using the SVM method in partner selections. In comparison with other methods in the literatures, the SVM-based method is advantageous in terms of generalization performance and the fitness accuracy with a limited number of training datasets.

1 Introduction

An important activity in the formation of a virtual enterprise (VE) is the selection of partners. How to select appropriate partners to form a team is a key problem for successful operation and management of VEs. This problem has attracted much attention recently [1, 2].

Partner selection is an unstructured and multi-criterion decision problem. Qualitative analysis methods are commonly used in many research works [3]. However, quantitative analysis methods for partner selection are still a challenge. Existing quantitative methods in the related literatures can be classified into several categories: mathematical programming models, weighting models, genetic algorithms, dynamic clustering, neural network and fuzzy sets. Talluri and Baker [4] proposed a two-phase mathematical programming approach for partner selection by designing a VE, in which the factors of cost, time and distance were considered. The weighting model includes the linear scoring model and analytic hierarchy process (AHP). The linear scoring model assigns weights and scores arbitrarily. In the AHP model, the

---

* This work was supported by grant No. 70171025 of National Science Foundation of China and grant No. 02KJB630001 of Research Project Grant of JiangSu, China.

priorities are converted into the ratings with regard to each criterion using pairwise comparisons and the consistency ratio [5]. Clustering technology can also be used in the partner selection process, which is based on the rule that multiple potential partners can be classified into one class if no significant difference on evaluation criteria exists in these partners. While artificial neural network (ANN) approach seems to be the best one available that efficiently combines qualitative and quantitative analysis to ensure the objective of selection processes, the prerequisite for the ANN approach is a large number of training data and the method may easily lead to local optimum.

This paper proposes a new approach to partner selection process by utilizing the Support Vector Machine technique. Based on well developed machine learning theory, Support Vector Machine (SVM) is a supervised learning technique that has received much attention for superior performances in various applications, such as pattern recognition, regression estimation, time series predication and text classfication. To employ SVM for distinguishing more than two classes, several approaches have been introduced [7]. In this paper, we focus on how to use binary SVM technique in the multiclass problem of partner selection.

2 Design of SVM-Based Partner Selection System

Basically partner selection is a process that produces partner rank in the order of overall scores of their performance according to a certain criterion system. Let us consider a pool of potential partners containing k independent organizations, and a criterion system containing d sub-criteria. Define x_i to be a feature vector of length d for the i-th potential partner.

$$x_i = (x_{i1}, x_{i2}, ..., x_{ij}, ..., x_{id}), (i=1,..., k; j=1,..., d)$$

where x_{ij} is the value of the j-th criterion for the i-th potential partner.

2.1 Selection Criterion System

Partners are selected based on their skills and resources to fulfill the requirements of the VE [4]. The selection process is based on multiple variables such as organizational fit, technological capabilities, relationship development, quality, price, and speed [6]. A three-layer selection criterion is developed as shown in Fig.1. The hierarchical structure includes goal, criteria and sub-criteria. The hierarchy can easily be extended to more detailed levels by breaking down the criteria into sub-criteria.

Some sub-criteria are determined by five rating levels: outstanding, above average, average, below average and unsatisfactory. Since the SVM method requires that each data be represented as a vector of real numbers, such sub-criteria should be converted into numerical data simply by using 5 to represent the best level and 1 for the lowest level.

Support Vector Machine Approach for Partner Selection of Virtual Enterprises 1249

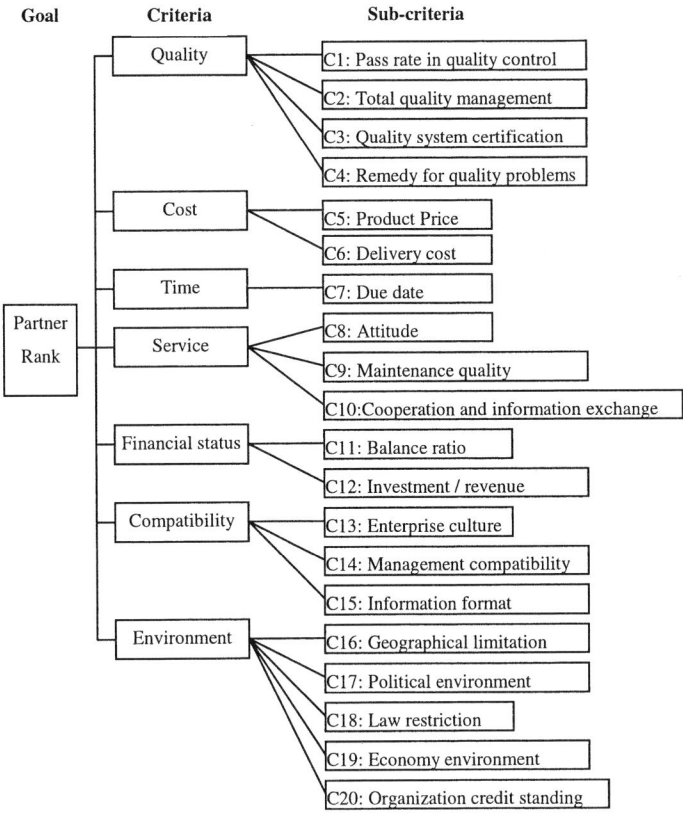

Fig. 1. The three-layer criterion structure for supplier partner selection

2.2 Binary Support Vector Machine

SVM is basically a binary classifier. For a two-class classification problem, given a set of training data (x_i, y_j), for $i=1,2,\ldots,n$, i.e., the input vectors $\vec{x}_i \in \mathbb{R}^d$ with corresponding labels $y_i \in \{1, -1\}$, here +1 and −1 indicate two classes, SVM seeks the solution of the following optimization problem:

$$\min_{w,b,\xi} \quad \frac{1}{2}\omega^T\omega + C\sum_{i=1}^{l}\xi_i$$

$$s.t. \quad y_i(\omega^T\phi(x_i)+b) \geq 1-\xi_i$$

$$\xi_i \geq 0, \quad i=1,\ldots,n$$

where C is a parameter to be chosen by the user. A larger C corresponds to assigning a larger penalty to errors. ξ_i is slack variables when the training data is not linearly separable in the feature space. The decision function is

$$f(x) = (w \cdot x) + b = \sum_{i=1}^{n} \alpha_i y_i (x_i \cdot x) + b$$

2.3 Key Problems of Design

Using the idea of SVM, the class number is equal to the number of potential partners. We need to find a classifier with the decision function, $f(x)$ such that $y = f(x)$, where y is the class label for x. K is a variable determined by the actual number of training data. Obviously the SVM method cannot be applied directly. The following three problems need to be solved when applying the SVM method to the selection process:

1. decompose the ranking problem into classification problems,
2. construct multi-class classification SVM from the binary SVM,
3. transform results of classification into numerical values.

3 System Implementation

3.1 One-to-One Comparison of Potential Partners

Given two partners denoted by i and j, with the feature vectors x_i and x_j of dimension d, their ranks are represented by $F(x_i)$ and $F(x_j)$ respectively. We define the vector of c_{ij} of dimension $2d$ by combining x_i and x_j

$$c_{ij} = (x_{i1}, x_{j1}, ..., x_{im}, x_{jm}, ..., x_{id}, x_{jd}) \ (i,j=1,...,k;\ i \neq j,\ m=1,...,d)$$

Let y_{ij} be the class label, according to the overall score of the partners i and j, we have

$$y_{ij} = \begin{cases} +1, & F(x_i) > F(x_j) \\ 0, & F(x_i) = F(x_j) \\ -1, & F(x_i) < F(x_j) \end{cases} \quad i,j=1,...,k \text{ and } j \neq i$$

Therefore the class number for two partners is 3, which does not change with the total number of training samples. The original training sets represented by (x_i, y_i) is transformed into the format of (c_{ij}, y_{ij}). In this case, the SVM approach can be applied to classify c_{ij} to determine the relative performance of any pair of partners.

3.2 Multi-class Classification SVM

Through the format variation of (c_{ij}, y_{ij}), the selection process can be initiated by a three-class classification process. The one-against-one method is used to construct all the possible classifiers where each one is generated between two classes chosen out of the k classes from the training data [8].

The decision function for the class pair pq is defined by $f_{pq}(\mathbf{x})$. Since $f_{pq}(\mathbf{x})=-f_{qp}(\mathbf{x})$, there exist $k(k-1)/2$ different classifiers for a k-class problem. The "max wins" algorithm is used for class identification of the one-against-one method. In the "max wins" algorithm each classifier casts one vote for its preferred class, and the final result is the class with the most votes.

$$\text{the class of } X = \arg\max_p \sum_{q \neq p, q=1}^{k} \text{sign}(f_{pq}(x))$$

When more than one class have the same number of votes, each point in the unclassifiable region is assigned to the closest class using the real valued decision functions as:

$$\text{the class of } X = \arg\max_p \sum_{q \neq p, q=1}^{k} f_{pq}(x)$$

Based on the above one-against-one algorithm, we developed three binary classifiers between every two classes.

3.3 Transformation of Classification Results

In order to compute the final rank of the partners, we have to decide how to transform the results of classification into the final rank of partners. The idea of Round Robin are utilized here in which each partner is compared with every other partners in the same test dataset. Let n be the total number of the test datasets, x_i be the feature vector of the i-th partner, then c_{ij} is constructed by combining x_i with every other x_j ($j = 1,...,n$ and $j \neq i$). Using 3-class SVM in Section 3.2, the $n-1$ class labels y_{ij} are calculated for each c_{ij}. Define

$$g_i(x_j) = \begin{cases} 2, & y_{ij} = 1, \\ 1, & y_{ij} = 0, \\ 0, & y_{ij} = -1, \end{cases} \quad i, j = 1,...,n \text{ and } j \neq i$$

where $g_i(x_j)$ is the score of the i-th partner with respect to the j-th partner. And define

$$f'(x_i) = \sum_{\substack{j=1 \\ j \neq i}}^{n} g_i(x_j) \qquad i = 1,...,n$$

where $f'(x_i)$ is the final score of the i-th partner. Therefore, the order of potential partners in the test dataset can be determined according to the absolute value of each $f'(x_i)$ in either ascending or descending order.

4 Experimental Analysis

Using the ideas discussed in Section 3, a partner selection system based on the multi-class SVM method was implemented. We point out that in the multiclass SVM method, the kernel function and the parameter adjustment are very important. We used a polynomial function.

$$K(x, x_i) = [(x \cdot x_i) + 1]^q$$

We ran experiments with different kernel parameters and measured accuracy and CPU time for the training time. Table 2 indicates that the values of q and C have direct effect on the accuracy and training time of the system.

Table 1. Accuracies and trainning times for kernel parameter q and penalty coefficient C

Kernel Function	Kernel Parameters	Accuracy(%)	Training time(second)
Polynomial function $K(x, x_i) = [(x \cdot x_i) + 1]^q$	q=2, C=0.1	96.6	31
	q=2, C=0.2	96.6	24
	q=2, C=0.3	97.1	24
	q=3, C=0.1	98.1	22
	q=3, C=0.2	98.1	20
	q=3, C=0.3	98.1	19
	q=4, C=0.1	97.5	24
	q=4, C=0.2	97.6	22
	q=4, C=0.3	97.6	22
	q=5, C=0.1	97.1	31
	q=5, C=0.2	97.1	25
	q=5, C=0.3	97.1	23
	q=8, C=0.1	96.1	56
	q=8, C=0.2	96.1	56
	q=8, C=0.3	95.6	43

5 Conclusion

By transforming the ranking of partners into multiple binary classification problems, we proposed and implemented a new approach to the solution of partner selection of virtual enterprise. Results of our experiments indicate that in comparison with other methods reported in the literatures, the SVM-based method presented in the paper is advantageous in terms of achieving certain fitness accuracy with a limited number of training datasets. More works need to be done in the selection of the kernel function and parameters. Determining the best kernel parameters would be an interesting topic for future research. Furthermore, the comparison of the SVM-based system with those based on other methods such as PCA and Fisher are desirable in order to demonstrate that better accuracy can be obtained by the adoption of SVM in the ranking system.

References

1. Subramai, M., Walden, E.: Economic Returns to Firms from Business-to-Business Electronic Commerce Initiatives: An Empirical Examination. In Proc. 21st Int'l Conf. Information Systems (2000) 229-241

2. Davulcu, H., Kifer, M., et al: Modeling and Analysis of Interactions in Virtual Enterprises. In Proceedings of Ninth International Workshop on Research Issues on Data Engineering (1999)
3. Maloni, M.J., Benton, W.C.: Supply Chain Partnerships: Opportunities for Operations Research. European Journal of Operational Research 101 (1997) 419-429
4. Talluri, S., Baker, R.C.: Quantitative Framework for designing efficient business process alliances. In: Proceedings of 1996 International Conference on Engineering and Technology Management (1996) 656-661
5. Lee, E.K., et al.: Supplier Selection and Management System Considering Relationships in Supply Chain Management. IEEE Transactions on Engineering Management, vol. 48. (2001) 307-318
6. Sarkis, J., Sundarraj, P.: Evolution of Brokering; Paradigms in E-Commerce Enabled Manufacturing. Int. J. Production Economics, vol. 75. (2002) 21-31
7. Platt, J., Cristianini, N., and Shawe-Taylor. J.: Large Margin DAGs for Multiclass Classification. In Advances in Neural Information Processing Systems 12 (NIPS Conference, Denver, CO, 1999) (2000) 547–553
8. David, M.J., Robert, P.W.: Using Two-Class Classifiers for Multiclass Classification. www.ph.tn.tudelft.nl/People/ bob/papers/icpr_02_mclass.pdf

Author Index

Achalakul, Tiranee 38
Ahn, Hyunchul 922
Amodio, Pierluigi 1
Azuaje, Francisco 556

Bae, Misook 940
Bae, Yongeun 219
Baik, Ran 245
Baik, Sung Wook 245, 910
Bala, Jerzy 910
Bedregal, Benjamín René Callejas 928
Black, Norman D. 498, 556
Bogen, Daniel K. 916

Cao, Chunhong 324, 449
Cao, Jiannong 77
Cao, Zhenfu 1034
Chai, Jianyun 1239
Chai, Zhenchuan 1034
Chan, Keith C.C. 77
Chan, Tony F. 718
Chen, Bao-Xing 19
Chen, Guihai 1041
Chen, Guo-liang 213
Chen, Jiaxun 995
Chen, Jiu Jun 806
Chen, Wang 32
Chen, Tiejun 330
Chen, Zhide 888
Cheng, Daizhan 768
Cho, Dong-Sub 812
Cho, Jae-Hyun 613
Cho, June-Sik 589
Cho, Sehyeong 407
Cho, Wanhyun 639
Cho, Yongyoon 752
Choi, Byung-Jae 148, 582
Choi, Jong-Hwa 238
Choi, Soo-Mi 504
Choi, Yoo-Joo 504
Choy, Yoon-Chul 873
Chu, Wanming 51
Chung, Ilyong 219
Chung, Tae-Sun 252
Cui, Xia 44

Dai, Jing 1027
Dai, Weizhong 304
Dai, Xiaoling 782
de Araújo, Frederiko Stenio 928
Deng, Zhong-Shan 437
Ding, Peng 1008
Ding, Yihong 672
Doh, Kyu B. 819
Dong, Liju 1079
Dong, Lingjiao 330
Dong, Shoubin 353
Dong, Xiaoju 336
Du, Tao 825
Du, Xiaoyong 1047
Du, Ye 692

Ejaz, Ahmed 685
Enhong, Chen 962
Eom, Young Ik 853
Ermis, Murat 837

Fei, Minrui 310, 330
Feng, Boqin 847
Feng, Deng-Guo 1054
Feng, Dongqing 330
Fu, Cheng 455
Fu, Heng 443
Fu, Yuxi 336, 477, 879

Gao, Ji 806
Gao, Jing 523
Giannoutakis, Konstantinos M. 111
Gong, Lie-Hang 1127
Gravvanis, George A. 111
Grundy, John 782
Gu, Lixu 678
Gu, Wei 93
Gu, Wen 154
Gu, Yonggen 879
Guo, Qing 698
Guo, Ruiqiang 471
Guo, Wei 1239
Gwun, Ou-Bong 516, 1134

Han, Ingoo 922
Han, JungHyun 916, 1161
Han, Lixin 1041

Han, Yiqiu 969
Han, Zongfen 136
Hasan, M.K. 57
He, Ji-Huan 465, 652, 659
He, Kejing 353
He, Lin 477
He, Wenzhang 379
He, Xiaofei 775
He, Xingshi 425
Heo, Won 238
Hiew, Fu San 200
Ho, Kei Shiu 954
Hong, Chuleui 188, 194
Hong, Sugwon 252
Horie, Ken 745
Hu, He 1047
Hu, Jun 806
Hu, Laizhao 665, 1020
Hu, Min 672
Hu, Yunxia 7
Huang, Wei 794
Huang, Xiaoqin 1061
Hughes, John G. 498
Huo, Hua 847
Huo, Meimei 391
Hwang, Buhyun 940
Hwang, Byung-Kon 860

Ibrahim, Zuwairie 71
Iliopoulos, Costas S. 698

Jang, Hyuk Soo 252
Jang, Tae-won 258
Je, Sung-Kwan 613
Jeong, Chang-Sung 1141
Jeong, Ok-Ran 812
Jiao, Licheng 1175
Jin, Hai 136
Jin, Weidong 665, 1020
Jung, MiGyoung 1216
Jung, Moon-Ju 1161

Kang, Min-goo 258
Kang, Sin-Jae 800, 860
Karaa, Samir 124
Kenneth, Revett 685
Khalid, Marzuki 71
Kim, Dong Seong 895
Kim, Dong-oh 258

Kim, Eui-Jeong 595
Kim, Euntai 538
Kim, Eunyoung 916
Kim, Gu Su 853
Kim, Hee-Jae 860
Kim, HyungJun 207
Kim, Jae-gon 853
Kim, Jae-Kyung 873
Kim, Jeong-Sik 504
Kim, Jong-Wan 800, 860
Kim, Jung-Ja 631
Kim, Kwang-Baek 613, 762, 1121
Kim, Kyeongwha 819
Kim, Kyoung-jae 922
Kim, Kyoungjung 538
Kim, Kyungsoo 431
Kim, Myoung-Hee 504
Kim, Pyung Soo 1106
Kim, Sangkyun 739, 788
Kim, Seung-Man 1187
Kim, Seung-Wan 1134
Kim, Sin-Jae 860
Kim, Sun Kyung 231
Kim, Tae Hee 231
Kim, Tae-Hyung 298
Kim, Wonil 188, 194
Kim, Yeongjoon 188, 194
Kim, Yong-Guk 274, 516, 1134
Kim, Yoon Hyuk 413, 431
Kim, Young-Ho 631
Kim, Young-Ju 762
Koay, Kah Hoe 200
Kong, Fansheng 1014
Koo, Han-Suh 1141
Kukkonen, Saku 399
Kurzynski, Marek W. 831
Kwon, Oh-Kyu 130
Kwon, Taekyoung 274
Kyung, Kyu Min 538

Lai, Wei 1168
Lam, Wai 969
Lampinen, Jouni 399
Lan, Kwok Cheung 954
Lan, Zhiling 280
Lavrenteva, Olga 93
Le, Jiajin 471, 989
Lee, Chung Ki 252
Lee, DongWoo 182
Lee, GueeSang 1216, 1224

Lee, Kwan H. 1187, 1208
Lee, Kyoung-Mi 1201
Lee, Moonkey 266
Lee, Ou-seb 258
Lee, Sae-Bom 800
Lee, SangJun 903
Lee, SeongHoon 182
Lee, SuKyoung 142, 160
Lee, Yongbum 576
Leem, Choon Seong 739, 788
Leong, Hong Va 954
Li, Beibei 989
Li, Guoqiang 477, 879
Li, Huaqing 1067
Li, Lixiong 310
Li, Minglu 172, 510
Li, Qingfeng 419
Li, Wenhui 324, 449
Li, Xue 705, 711
Li, Yaming 51
Li, Yawei 280
Li, Yunfa 136
Li, Zhenquan 118
Liao, Bei Shui 806
Liao, Jun 85
Liao, Lin 77
Lim, Hwa-seop 258
Lim, Soon-Bum 873
Lin, Chen 1061
Liu, Bihong 1014
Liu, Bing 166
Liu, Dezhi 530
Liu, Huafeng 562
Liu, Hui 172
Liu, Jiang B. 768
Liu, Jinde 85
Liu, Jing 437
Liu, Junhong 399
Liu, Xiaoxiao 570
Liu, Xu-Zheng 44
Liu, Yong 346
Liu, Yu 425
Liu, Yu-Shen 1093
Liu, Yuhua 105
Liu, Zhiqiang 1008
Long, Keping 365
Lu, Bao-Liang 867
Lu, Jia 7
Lu, Jian 286
Lu, Ke 775

Lu, Liuming 995
Lü, Xin 1054
Luk, Robert Wing Pong 954
Luo, Jianhua 32, 645
Luo, Limin 550

Ma, Tieju 794
Mantere, Timo 1148
Mao, Jingzhong 105
Matsuhisa, Takashi 745, 1114
McClean, Stephen 498
Mitsuhashi, Kohei 982
Miura, Takao 947, 982
Moon, Jongbae 752

Na, Jong Whoa 934
Nakamori, Yoshiteru 794
Nam, In-Gil 800
Natwichai, Juggapong 705
Ng, Michael K. 718
Nguyen, Ha-Nam 895
Nir, Avinoam 93
No, Jaechun 225
Noh, Seung-Moo 589

Oh, Am-Suk 762
Oh, Young-Cheol 860
Ohn, Jungho 819
Ohn, Syng-Yup 895
Ojha, Piyush C. 498
Ono, Osamu 71
Othman, M. 57

Pan, Yunhe 346
Pandya, Abhijit S. 1121
Pang, Yonggang 692
Park, Byung-In 130
Park, Chang-Woo 538
Park, Dong Sun 492
Park, Hyun Suk 607
Park, Jong Sou 895
Park, Jong-Won 589, 595
Park, Jonghyun 639
Park, Joungwook 1208
Park, Mignon 538
Park, Soonyoung 639
Park, Sung Soon 225
Park, Won Man 413
Peng, Shietung 51

Peng, Yu   292, 1195
Ping, Xijian   672
Poon, Ting-C   819

Qi, Fei-hu   316
Qi, Feihu   359, 601, 1067
Qi, Zhengwei   455
Qin, Zheng   425

Ramamurthy, Byrav   725
Ramos, Karla
  Darlene Nempomuceno   928
Razdan, Anshuman   530
Ren, Qingsheng   359
Rhee, Daewoong   910
Ryu, Yeonseung   252

Sahingoz, Ozgur Koray   837
Sang, Nong   570
Sarochawikasit, Rajchawit   38
Seol, Young Joo   607
Seong, Won   589, 595
Sgura, Ivonne   1
Shan, Tan   1175
Shaw, Chris   498
Shen, Dinggang   732
Shen, Yun-Qiu   99
Sheng, Huanye   1008
Shi, Bai-Le   1027
Shi, Lianshuan   443
Shi, Pengcheng   562
Shi, Yonghong   601
Shin, Byeong-Seok   619, 625
Shin, Dongil   238
Shin, Dongkyoo   238
Shin, Jin Wook   492
Shin, YoonJeong   1224
Shioya, Isamu   947, 982
Shu, Huazhong   550
Shu, Wang   962
Silva, Ivan Saraiva   928
Sohn, Won-Sung   873
Song, Bin   213
Song, Binglin   353
Song, Dawei   711
Song, Guoxiang   379
Song, Hyoung-kyu   258
Song, Ju-Whan   516, 1134
Song, Peiming   645

Sonh, Seung-il   258
Su, Hua   1072
Sulaiman, J.   57
Sun, Jia-Guang   44, 292, 1093, 1127, 1155, 1195
Sun, Zhonghua   544

Tang, Qiling   570
Tang, Zesheng   1239
Thanneru, Umadevi   768
Tian, Guang   316
Tsai, Du-Yih   576
Tsuboi, Yusei   71

Uejima, Hiroshi   947
Ulengin, Fusun   837

Wan, Chunru   166
Wang, Bin   25, 1155
Wang, Fu Lee   975
Wang, Guojun   77
Wang, Haiying   556
Wang, Huiqiang   692
Wang, Jianxue   825
Wang, Jie   1247
Wang, Lipo   154, 166
Wang, Meng   1002
Wang, Paul P.   582
Wang, Shaoyu   1067
Wang, Shouyang   794
Wang, Wei   1027
Wang, Xifan   825
Wang, Yonggang   1002, 1181
Wang, Zhaoan   419
Wang, Zheng-Fei   1027
Wei, Xiaopeng   460
Weng, Wenjie   570
Wiyarat, Thitirat   38
Won, Yonggwan   631
Wu, Aidi   379
Wu, Edmond H.   718
Wu, Ling   1232
Wu, Minna   136
Wu, Yue   465
Wu, Zhaohui   346

Xia, Ling   391
Xiao, Wen-Jun   19
Xie, Chao   136

Xie, Dexuan 64
Xie, Li 1041
Xie, Linsen 645
Xu, Congfu 346
Xu, Jianfeng 678
Xu, Qing 1181
Xu, Zenglin 425

Yan, Dong-Ming 292
Yan, Xin 711
Yang, Gab Seok 492
Yang, Geng 13
Yang, Hoonmo 266
Yang, Jianguo 419
Yang, Jie 1002, 1181
Yang, Ju Cheng 492
Yang, Peng 105
Yang, Ruigang 1100
Yang, Xiaobing 1014
Yang, Xiaolong 365
Ye, Rui-song 286
Yi, Wan 449
Yip, Andy M. 718
Yong, Jun-Hai 44, 292, 1072, 1086
  1093, 1127, 1155, 1195, 1232
Yoo, Seog-Hwan 148, 582
Yoo, Seong-Joon 607
Yoo, Taeseung D. 916
Yoon, HyoSun 1224
Yoon, Hyung Sik 1161
You, Jinyuan 455, 1061
You, Young-hwan 258
Youn, Chunkyun 219
Ypma, Tjalling J. 99
Yu, Donggang 1168
Yu, Ge 1079
Yu, Jian-Yong 465
Yu, Pi-Qiang 1093
Yu, Shengsheng 105
Yu, Shoujian 471

Zeng, Jiazhi 775
Zeng, Jin 359
Zhan, Yiqiang 732
Zhang, Gexiang 665, 1020
Zhang, Hui 292, 698, 1093, 1195
Zhang, Juan 486
Zhang, Jun 523, 1247
Zhang, Lei 510
Zhang, Li 1232
Zhang, Ling 353
Zhang, Min 365, 477
Zhang, Qiang 460
Zhang, Xianggang 85
Zhang, Xiangrong 1175
Zhang, Xiao-Xin 1127
Zhang, You-Wei 1232
Zhang, Zhizhou 477
Zhao, Hai 867
Zhao, Jing 32, 645
Zhao, Yongqiang 510
Zhao, Yu-Mei 465
Zheng, Guo-Qin 1086, 1127
Zheng, Huiru 498, 544
Zheng, Zhonglong 1002
Zhong, Farong 371
Zhong, Weijun 1247
Zhou, Dongsheng 460
Zhou, Feng-feng 213
Zhou, Jian 550
Zhou, Quan 13
Zhou, Xiaobing 310
Zhou, Yue 1181
Zhu, Cheng 1008
Zhu, Guojin 995
Zhu, Hong 888
Zhu, Hongqing 550
Zhu, Teng 304
Zhu, Yong-Kang 1086
Zou, Shengrong 385
Zou, Yu-ru 286
Zou, Xukai 725

Lecture Notes in Computer Science

For information about Vols. 1–3243

please contact your bookseller or Springer

Vol. 3356: G. Das, V.P. Gulati (Eds.), Intelligent Information Technology. XII, 428 pages. 2004.

Vol. 3347: R.K. Ghosh, H. Mohanty (Eds.), Distributed Computing and Internet Technology. XX, 472 pages. 2004.

Vol. 3340: C.S. Calude, E. Calude, M.J. Dinneen (Eds.), Developments in Language Theory. XI, 431 pages. 2004.

Vol. 3339: G.I. Webb, X. Yu (Eds.), AI 2004: Advances in Artificial Intelligence. XXII, 1272 pages. 2004. (Subseries LNAI).

Vol. 3338: S.Z. Li, J. Lai, T. Tan, G. Feng, Y. Wang (Eds.), Advances in Biometric Person Authentication. XVIII, 699 pages. 2004.

Vol. 3337: J.M. Barreiro, F. Martin-Sanchez, V. Maojo, F. Sanz (Eds.), Biological and Medical Data Analysis. XI, 508 pages. 2004.

Vol. 3336: D. Karagiannis, U. Reimer (Eds.), Practical Aspects of Knowledge Management. X, 523 pages. 2004. (Subseries LNAI).

Vol. 3334: Z. Chen, H. Chen, Q. Miao, Y. Fu, E. Fox, E.-p. Lim (Eds.), Digital Libraries: International Collaboration and Cross-Fertilization. XX, 690 pages. 2004.

Vol. 3333: K. Aizawa, Y. Nakamura, S. Satoh (Eds.), Advances in Multimedia Information Processing - PCM 2004, Part III. XXXV, 785 pages. 2004.

Vol. 3332: K. Aizawa, Y. Nakamura, S. Satoh (Eds.), Advances in Multimedia Information Processing - PCM 2004, Part II. XXXVI, 1051 pages. 2004.

Vol. 3331: K. Aizawa, Y. Nakamura, S. Satoh (Eds.), Advances in Multimedia Information Processing - PCM 2004, Part I. XXXVI, 667 pages. 2004.

Vol. 3329: P.J. Lee (Ed.), Advances in Cryptology - ASIACRYPT 2004. XVI, 546 pages. 2004.

Vol. 3328: K. Lodaya, M. Mahajan (Eds.), FSTTCS 2004: Foundations of Software Technology and Theoretical Computer Science. XVI, 532 pages. 2004.

Vol. 3323: G. Antoniou, H. Boley (Eds.), Rules and Rule Markup Languages for the Semantic Web. X, 215 pages. 2004.

Vol. 3322: R. Klette, J. Žunić (Eds.), Combinatorial Image Analysis. XII, 760 pages. 2004.

Vol. 3321: M.J. Maher (Ed.), Advances in Computer Science - ASIAN 2004. XII, 510 pages. 2004.

Vol. 3321: M.J. Maher (Ed.), Advances in Computer Science - ASIAN 2004. XII, 510 pages. 2004.

Vol. 3320: K.-M. Liew, H. Shen, S. See, W. Cai (Eds.), Parallel and Distributed Computing: Applications and Technologies. XXIV, 891 pages. 2004.

Vol. 3316: N.R. Pal, N.K. Kasabov, R.K. Mudi, S. Pal, S.K. Parui (Eds.), Neural Information Processing. XXX, 1368 pages. 2004.

Vol. 3315: C. Lemaître, C.A. Reyes, J.A. González (Eds.), Advances in Artificial Intelligence – IBERAMIA 2004. XX, 987 pages. 2004. (Subseries LNAI).

Vol. 3314: J. Zhang, J.-H. He, Y. Fu (Eds.), Computational and Information Science. XXIV, 1259 pages. 2004.

Vol. 3312: A.J. Hu, A.K. Martin (Eds.), Formal Methods in Computer-Aided Design. XI, 445 pages. 2004.

Vol. 3311: V. Roca, F. Rousseau (Eds.), Interactive Multimedia and Next Generation Networks. XIII, 287 pages. 2004.

Vol. 3309: C.-H. Chi, K.-Y. Lam (Eds.), Content Computing. XII, 510 pages. 2004.

Vol. 3308: J. Davies, W. Schulte, M. Barnett (Eds.), Formal Methods and Software Engineering. XIII, 500 pages. 2004.

Vol. 3307: C. Bussler, S.-k. Hong, W. Jun, R. Kaschek, D.. Kinshuk, S. Krishnaswamy, S.W. Loke, D. Oberle, D. Richards, A. Sharma, Y. Sure, B. Thalheim (Eds.), Web Information Systems – WISE 2004 Workshops. XV, 277 pages. 2004.

Vol. 3306: X. Zhou, S. Su, M.P. Papazoglou, M.E. Orlowska, K.G. Jeffery (Eds.), Web Information Systems – WISE 2004. XVII, 745 pages. 2004.

Vol. 3305: P.M.A. Sloot, B. Chopard, A.G. Hoekstra (Eds.), Cellular Automata. XV, 883 pages. 2004.

Vol. 3303: J.A. López, E. Benfenati, W. Dubitzky (Eds.), Knowledge Exploration in Life Science Informatics. X, 249 pages. 2004. (Subseries LNAI).

Vol. 3302: W.-N. Chin (Ed.), Programming Languages and Systems. XIII, 453 pages. 2004.

Vol. 3299: F. Wang (Ed.), Automated Technology for Verification and Analysis. XII, 506 pages. 2004.

Vol. 3298: S.A. McIlraith, D. Plexousakis, F. van Harmelen (Eds.), The Semantic Web – ISWC 2004. XXI, 841 pages. 2004.

Vol. 3295: P. Markopoulos, B. Eggen, E. Aarts, J.L. Crowley (Eds.), Ambient Intelligence. XIII, 388 pages. 2004.

Vol. 3294: C.N. Dean, R.T. Boute (Eds.), Teaching Formal Methods. X, 249 pages. 2004.

Vol. 3293: C.-H. Chi, M. van Steen, C. Wills (Eds.), Web Content Caching and Distribution. IX, 283 pages. 2004.

Vol. 3292: R. Meersman, Z. Tari, A. Corsaro (Eds.), On the Move to Meaningful Internet Systems 2004: OTM 2004 Workshops. XXIII, 885 pages. 2004.

Vol. 3291: R. Meersman, Z. Tari (Eds.), On the Move to Meaningful Internet Systems 2004: CoopIS, DOA, and ODBASE, Part II. XXV, 824 pages. 2004.

Vol. 3290: R. Meersman, Z. Tari (Eds.), On the Move to Meaningful Internet Systems 2004: CoopIS, DOA, and ODBASE, Part I. XXV, 823 pages. 2004.

Vol. 3289: S. Wang, K. Tanaka, S. Zhou, T.W. Ling, J. Guan, D. Yang, F. Grandi, E. Mangina, I.-Y. Song, H.C. Mayr (Eds.), Conceptual Modeling for Advanced Application Domains. XXII, 692 pages. 2004.

Vol. 3288: P. Atzeni, W. Chu, H. Lu, S. Zhou, T.W. Ling (Eds.), Conceptual Modeling – ER 2004. XXI, 869 pages. 2004.

Vol. 3287: A. Sanfeliu, J.F. Martínez Trinidad, J.A. Carrasco Ochoa (Eds.), Progress in Pattern Recognition, Image Analysis and Applications. XVII, 703 pages. 2004.

Vol. 3286: G. Karsai, E. Visser (Eds.), Generative Programming and Component Engineering. XIII, 491 pages. 2004.

Vol. 3285: S. Manandhar, J. Austin, U.B. Desai, Y. Oyanagi, A. Talukder (Eds.), Applied Computing. XII, 334 pages. 2004.

Vol. 3284: A. Karmouch, L. Korba, E.R.M. Madeira (Eds.), Mobility Aware Technologies and Applications. XII, 382 pages. 2004.

Vol. 3283: F.A. Aagesen, C. Anutariya, V. Wuwongse (Eds.), Intelligence in Communication Systems. XIII, 327 pages. 2004.

Vol. 3282: V. Guruswami, List Decoding of Error-Correcting Codes. XIX, 350 pages. 2004.

Vol. 3281: T. Dingsøyr (Ed.), Software Process Improvement. X, 207 pages. 2004.

Vol. 3280: C. Aykanat, T. Dayar, İ. Körpeoğlu (Eds.), Computer and Information Sciences - ISCIS 2004. XVIII, 1009 pages. 2004.

Vol. 3278: A. Sahai, F. Wu (Eds.), Utility Computing. XI, 272 pages. 2004.

Vol. 3275: P. Perner (Ed.), Advances in Data Mining. VIII, 173 pages. 2004. (Subseries LNAI).

Vol. 3274: R. Guerraoui (Ed.), Distributed Computing. XIII, 465 pages. 2004.

Vol. 3273: T. Baar, A. Strohmeier, A. Moreira, S.J. Mellor (Eds.), <<UML>> 2004 - The Unified Modelling Language. XIII, 454 pages. 2004.

Vol. 3271: J. Vicente, D. Hutchison (Eds.), Management of Multimedia Networks and Services. XIII, 335 pages. 2004.

Vol. 3270: M. Jeckle, R. Kowalczyk, P. Braun (Eds.), Grid Services Engineering and Management. X, 165 pages. 2004.

Vol. 3269: J. Lopez, S. Qing, E. Okamoto (Eds.), Information and Communications Security. XI, 564 pages. 2004.

Vol. 3268: W. Lindner, M. Mesiti, C. Türker, Y. Tzitzikas, A. Vakali (Eds.), Current Trends in Database Technology - EDBT 2004 Workshops. XVIII, 608 pages. 2004.

Vol. 3267: C. Priami, P. Quaglia (Eds.), Global Computing. VIII, 377 pages. 2004.

Vol. 3266: J. Solé-Pareta, M. Smirnov, P.V. Mieghem, J. Domingo-Pascual, E. Monteiro, P. Reichl, B. Stiller, R.J. Gibbens (Eds.), Quality of Service in the Emerging Networking Panorama. XVI, 390 pages. 2004.

Vol. 3265: R.E. Frederking, K.B. Taylor (Eds.), Machine Translation: From Real Users to Research. XI, 392 pages. 2004. (Subseries LNAI).

Vol. 3264: G. Paliouras, Y. Sakakibara (Eds.), Grammatical Inference: Algorithms and Applications. XI, 291 pages. 2004. (Subseries LNAI).

Vol. 3263: M. Weske, P. Liggesmeyer (Eds.), Object-Oriented and Internet-Based Technologies. XII, 239 pages. 2004.

Vol. 3262: M.M. Freire, P. Chemouil, P. Lorenz, A. Gravey (Eds.), Universal Multiservice Networks. XIII, 556 pages. 2004.

Vol. 3261: T. Yakhno (Ed.), Advances in Information Systems. XIV, 617 pages. 2004.

Vol. 3260: I.G.M.M. Niemegeers, S.H. de Groot (Eds.), Personal Wireless Communications. XIV, 478 pages. 2004.

Vol. 3259: J. Dix, J. Leite (Eds.), Computational Logic in Multi-Agent Systems. XII, 251 pages. 2004. (Subseries LNAI).

Vol. 3258: M. Wallace (Ed.), Principles and Practice of Constraint Programming – CP 2004. XVII, 822 pages. 2004.

Vol. 3257: E. Motta, N.R. Shadbolt, A. Stutt, N. Gibbins (Eds.), Engineering Knowledge in the Age of the Semantic Web. XVII, 517 pages. 2004. (Subseries LNAI).

Vol. 3256: H. Ehrig, G. Engels, F. Parisi-Presicce, G. Rozenberg (Eds.), Graph Transformations. XII, 451 pages. 2004.

Vol. 3255: A. Benczúr, J. Demetrovics, G. Gottlob (Eds.), Advances in Databases and Information Systems. XI, 423 pages. 2004.

Vol. 3254: E. Macii, V. Paliouras, O. Koufopavlou (Eds.), Integrated Circuit and System Design. XVI, 910 pages. 2004.

Vol. 3253: Y. Lakhnech, S. Yovine (Eds.), Formal Techniques, Modelling and Analysis of Timed and Fault-Tolerant Systems. X, 397 pages. 2004.

Vol. 3252: H. Jin, Y. Pan, N. Xiao, J. Sun (Eds.), Grid and Cooperative Computing - GCC 2004 Workshops. XVIII, 785 pages. 2004.

Vol. 3251: H. Jin, Y. Pan, N. Xiao, J. Sun (Eds.), Grid and Cooperative Computing - GCC 2004. XXII, 1025 pages. 2004.

Vol. 3250: L.-J. (LJ) Zhang, M. Jeckle (Eds.), Web Services. X, 301 pages. 2004.

Vol. 3249: B. Buchberger, J.A. Campbell (Eds.), Artificial Intelligence and Symbolic Computation. X, 285 pages. 2004. (Subseries LNAI).

Vol. 3247: D. Comaniciu, R. Mester, K. Kanatani, D. Suter (Eds.), Statistical Methods in Video Processing. VIII, 199 pages. 2004.

Vol. 3246: A. Apostolico, M. Melucci (Eds.), String Processing and Information Retrieval. XIV, 332 pages. 2004.

Vol. 3245: E. Suzuki, S. Arikawa (Eds.), Discovery Science. XIV, 430 pages. 2004. (Subseries LNAI).

Vol. 3244: S. Ben-David, J. Case, A. Maruoka (Eds.), Algorithmic Learning Theory. XIV, 505 pages. 2004. (Subseries LNAI).